Handbook of Metal–Ligand Interactions in Biological Fluids

Bioinorganic Chemistry

Volume I

Handbook of Metal–Ligand Interactions in Biological Fluids

Bioinorganic Chemistry

Volume I

edited by

Guy Berthon

Centre National de la Recherche Scientifique (CNRS)
Institut National de la Santé et de la Recherche Médicale (INSERM)
Toulouse, France

Marcel Dekker, Inc. New York • Basel • Hong Kong

ISBN: 0-8247-9296-3

The publisher offers discounts on this book when ordered in bulk quantities. For more information, write to Special Sales/Professional Marketing at the address below.

This book is printed on acid-free paper.

MARCEL DEKKER, INC.
270 Madison Avenue, New York, New York 10016

Current printing (last digit):
10 9 8 7 6 5 4 3 2

PRINTED IN THE UNITED STATES OF AMERICA

PREFACE TO THE HANDBOOK

In recent years, there has been a growing awareness of the potential contribution of chemistry toward progress in medicine, as is evident from the constant development of related interdisciplinary research. In particular, the area called "medicinal chemistry" has seen enormous growth during the past decade. However, this is largely still the sphere of *organic* chemists, in conjunction with their traditional and, more lately, newer partners in biology. In a recent comprehensive attempt to distinguish it from the larger field known as "bioorganic chemistry," for example, medicinal chemistry has been centered mainly on the understanding of structure–activity relationships and drug design from the mechanistic point of view, with physico-chemical implications hardly being mentioned.

Even though presumably to a lesser extent in quantitative terms, *inorganic* chemistry has also seen a huge development with respect to its contribution to medicine, and the new discipline of "bioinorganic chemistry" has greatly helped toward the understanding of fundamental biological processes in which inorganic elements are involved. Among these elements, metal ions are undoubtedly of the greatest importance. The various roles they play in the above biological processes have thus attracted particular attention regarding the health of living organisms, especially human beings.

In this context, the need for a reference book on the interactions between metal ions and endogenous or exogenous chemical substances occurring in the main body fluids has become evident. This has given rise to the present *Handbook of Metal–Ligand Interactions in Biological Fluids*. In practice, this project took shape from a few common-sense observations.

Keeping human beings healthy is, in the first place, a matter for medicine considered as the nonsurgical treatment of disease. Indeed, except for the strictly "mechanical" repair of physiological injuries, recourse to surgery often amounts to an acknowledgment of medicine's inadequacy. Somewhat paradoxically, however, compared with the gigantic advances in surgery due to recent breakthroughs registered in physics, medicine has progressed much more slowly. This can be attributed at least partly to a lack of adaptation of chemistry to its needs. As a matter of fact, there is no equivalent intermediate field between chemistry and medicine in the way biomedical engineering exists between physics and surgery.

First of all, it seems that full advantage has not been taken of all the potential of chemical research. For example, progress in medicine is mainly expected to ensue from the synthesis of new pharmacologically active substances by organic chemists. The tendency to generalize computer-aided applications of principles such as structure–activity relationships may be effective in the design of new analogs of existing drugs; however, research on how these molecules interact with endogenous substances and how they are translocated in vivo is still in its infancy and is mostly based on animal screening. In this respect, pharmacokinetic profiles are undoubtedly of great value in assessing the extent to which a substance is distributed in vivo, but such macroscopic observations do not provide any quantitative information about the absorption, retention and excretion processes at the molecular level. Also, too often, as it logically ensues from the empirical approach used, the exact site and process of action of new drugs remain obscure whereas their elucidation from a molecular point of view would be essential to allow further progress.

On the other hand, the fundamental activities of inorganic chemists tend to be regarded as being too distant from the biological sciences to be applied to the large variety of medical problems they could in fact help solve. For example, the enormous efforts made in recent years to discover the mechanism of action of the (fortuitously discovered) platinum antitumor compounds has undoubtedly proved useful in that they will eventually help in the design of more specific and less toxic drugs of successive generations. However, the fact that the activity of many centers of excellence has been focussed on this almost exclusive target may have been at the expense of the development of other conceptual approaches. In particular, the possible significance of the systematic variation of parameters such as the level of zinc, copper, histamine and polyamines in tumor cells compared to normal ones has meanwhile been largely overlooked.

Given the extreme complexity of the roles of metal ions in vivo, many objectives are still to be met before these elements can be commonly used in the control of health and disease. Combined efforts from both empirical and theoretical sides are required. In short, these should be directed to (i) provide a more comprehensive knowledge of the physiological functions of essential metals as well as a better understanding of their metabolic processes; (ii) help delineate the origin of the toxicity of harmful elements competing with essential ones and find more efficient detoxication treatments; (iii) develop the use of metal ions for therapeutic and diagnostic purposes; (iv) address the still largely unexplored field of the interactions of metal ions with exogenous organic substances, especially drugs.

Undoubtedly inorganic chemists now have at their disposal extremely powerful techniques to solve many problems with medical applications. The amount of specialized information available to them is also constantly increasing. However, chemists are rarely in a position to appreciate the problems encountered in medicine from their own point of view. For example, while "biological relevance" is commonly put forward to advertise articles of fundamental research relative to substances active in life processes, the corresponding results are too often devoid of real relationships with actual medical or biological problems.

On the other hand, medical researchers are becoming more and more conscious of the importance of the interactions between chemical substances whose effects they investigate in vivo or in vitro, but, having no training in fundamental solution chemistry, they are in turn rarely in a position to tackle the problems they are facing in the proper physico-chemical terms.

In many cases, the information necessary to solve medical problems using chemical techniques may even exist, but it remains so poorly accessible to specialists on both sides that it cannot be successfully exploited. Most of the time, it is so scattered, or described in such specific language, that it is a difficult task even for those working in interdisciplinary sciences to make good progress.

Fortunately, there has recently been a progressive trend to publish at the interdisciplinary border, thus bridging the gap between the chemical and medical communities; however, this has too often resulted in collections of separate, highly specialized contributions. The time has now probably come for a systematic effort to "popularize" the appropriate knowledge from and for the benefit of both sides. In this context, *the objective of the present Handbook is to facilitate contact between medicine and chemistry in the field of the metal–ligand interactions occurring in biological fluids.* It would probably be somewhat overenthusiastic to expect a sudden spectacular advance from bringing together such basically different disciplines; therefore, the first aim of this work is to develop a realistic approach based on a progressive and systematic exchange of information.

For this reason, the Handbook is divided into two two-volume sets that condense this information:

The first, which deals with the physico-chemical aspects of metal–ligand interactions in biological fluids, should provide the fundamental knowledge necessary to understand at the molecular level most of the roles played by metal ions in vivo. Mainly intended to raise chemists' interest in medical applications, it should also introduce medical researchers to the techniques of inorganic chemistry so that they can realize how these techniques can help them solve the problems they face.

The second set, which largely reviews the field of the empirical observations relative to metal ions in medicine, should draw special attention to the relationships between particular metal–ligand interactions and their correlative effects, so as to outline directions for inorganic chemists and to facilitate their transposition of medical problems into physico-chemical terms. This should prompt inorganic chemists to play a more active part in medical research, in particular by taking the lead in predicting physiological or pharmacological effects before they have been fortuitously observed.

Some entries will appear in parallel in both parts of the Handbook, *Bioinorganic Chemistry* and *Bioinorganic Medicine*, which may a priori seem redundant. This, on the contrary, proceeds from the declared will to favor the opening of new avenues in interdisciplinary research, which most of the time results from considering the same notions or facts from distinct perspectives.

The above-mentioned intermediate field between chemistry and medicine remains to be created. In regard to the inorganic side, it is hoped that the Handbook will in the future be considered a determining factor of its emergence.

ACKNOWLEDGMENTS

The editor is grateful to the Institut National de la Santé et de la Recherche Médicale (INSERM, Paris, France), the Centre National de la Recherche Scientifique (CNRS, Paris, France) and the Laboratoires Guerbet (GCA, Aulnay-sous-Bois, France) for financial support to this project.

The editor also thanks the temporary part-time assistants who successively collaborated on the copyediting of the book (Mina Aït' Mbark, Zohra Benzina, Mylène Bielsa, Isabelle Garcia, Sandrine Leulier, Peta Rigaill-Visée, and Ghislaine Touches).

Guy Berthon

PREFACE TO *BIOINORGANIC CHEMISTRY*

Volumes 1 and 2

As an introduction to the second set of volumes in this Handbook—*Bioinorganic Medicine*—which is devoted to the macroscopic effects of metal–ligand interactions in human biology and medicine, this pair of volumes—*Bioinorganic Chemistry*—first deals with the fundamental aspects of these interactions at the molecular level, viewed through the current theories and techniques of inorganic chemistry. Examples of applications of the basic principles of this discipline are then given in relation to the different biomedical implications of metal–ligand interactions, with special attention to the potential uses of chemical speciation in medicine.

In biological fluids, no metal ion can really exist in the free state since the so-called "free" dissociated forms commonly evoked by chemists are in fact solvated. Nevertheless, there is a large consensus to formally consider the notion of metal complexation as corresponding to the exchange of the solvent molecules present in the metal coordination sphere with other ligands. Such exchange processes are of primary importance for the behavior of metal ions in vivo, since the nature of the complexes these locally form determines to a large extent their diverse physiological functions and metabolic pathways.

Of course, the nature of the roles metal ions are led to play in vivo derives in the first place from the chemical properties that differentiate them from one another, but specificity in these different roles always stems from the immediate environment in which each metal ion is placed. Thus, the way a given metal ion is handled or used by a living organism depends mainly on the nature of the ligands it can be associated with, and on the kind of interactions in which it is involved.

Part One addresses these fundamental aspects of metal–ligand interactions in biological fluids. After a first chapter introducing the multiple facets of bioinorganic chemistry, the second one outlines the thermodynamic background against which metal ions and ligands can interact in solution. Then, diverse aspects of the nature and character of metal–ligand interactions are developed: semi-empirical rules of affinity between acceptors and donors, fundamental features of the metal–ligand bond according to current theories, and factors of selectivity and specificity in coordination. Finally, the influence of metal–ligand interactions on the distribution of these reactants in biological fluids is described, and the corresponding physiological implications outlined.

Quantitatively, metal ions are predominantly associated with macromolecules in vivo. In addition, most of their physiological functions originate in their specific interactions with such ligands. Because of this recognized importance, Part Two is devoted to the interactions of metal ions with proteins and nucleic acids.

Starting with a general introduction to all types of metal ion–protein interactions and to techniques of their investigation, the first chapter successively considers the different types of proteins known to interact with each of the essential and beneficial metal ions (respectively defined as being necessary to life or to normal health): work proteins, metallo-enzymes with metal-containing active sites, enzymes with metal ions as structural coordinators, and storage and transport proteins.

Following this review of the individual interactions of metal ions with proteins, the next sections focus on the common roles shared by some of the corresponding metal–protein complexes. For example, almost all metal ions that naturally occur in vivo behave as catalysts in activating different sorts of enzymes. A separate section has thus been devoted to metalloenzyme reactivity and its different aspects. In contrast with inert metalloproteins in which metal ions are firmly incorporated, labile metal–protein complexes carry metal ions in exchangeable forms and play important roles in their storage and transport. Therefore they too are the subject of a particular section. Detrimental metal ions (i.e., those that are exclusively toxic at any level) exert most of their deleterious actions by competing with essential elements for the active sites of proteins. Given the important consequences of these interactions for the toxicity of these elements, they have also been treated separately.

Part Two concludes by reviewing the fundamental aspects of the interactions of metal ions with nucleic acids and their constituents, with a special section devoted to zinc proteins involved in genetic information transfer.

Among the different fractions into which metal ions are distributed in vivo, protein-associated pools represent overwhelming proportions. In the case of transition metal ions, for example, the ultrafiltrable part generally accounts for less than 1% and may thus seem negligible. It is, in fact, no less essential than its quantitatively more important counterpart.

If macromolecular fractions are indeed quite indispensable to fulfill the extremely complex and specific tasks developed in Part Two, small species identifiable as low-molar-mass [l.m.m.] complexes (usually $M < 1$ kg mol^{-1})* are no less necessary since they constitute the "dynamic units" without which the whole system could not work. Among other things, these complexes are the only exchangeable forms in which metal ions can move across membranes from one biological compartment to another and from which metalloproteins are synthesized.

Part Three deals with the interactions through which these small species are formed. After a general overview of the basic thermodynamic notions relative to metal complex equilibria, the first chapter treats the kinetic, as well as the structural, aspects of complex formation in solution.

Examples of interactions of metal ions with different kinds of l.m.m. ligands naturally occurring in vivo are then developed, with amino acids, peptides, neurotransmitters, carbohydrates, polyamines, lipids, aminosugars, and aminophosphonates being successively considered.

Following the description of fundamental principles and experimental features of metal–ligand interactions themselves, Part Four deals with the applications of inorganic chemistry techniques to several biomedical problems in which these interactions are involved.

First, metal–ligand interactions play multiple roles in metal ion metabolism. The

---

* In accordance with IUPAC specifications, molar masses are expressed in SI units, i.e., kg mol^{-1} (equivalent to kDa for the molecular masses used mainly by biochemists).

absorption of a metal from the gastrointestinal tract, its accumulation in specific parts of the body, its urinary as well as its biliary excretions, depend largely on the nature of the complexes it may form in corresponding biofluids. The first chapter deals with the physico-chemical interpretation of this dependence, successively addressing general problems of metal ion translocation and particular aspects of metal absorption, accumulation and excretion.

Metal toxicity and its clinical manifestations are extensively treated in the second set of volumes, *Bioinorganic Medicine*. Some of their physico-chemical bases are developed here, special attention being paid to the competitive processes between essential and beneficial metal ions on the one hand and detrimental metal ions on the other hand.

Recently, there has been a growing awareness about the role of oxygen-free radicals in disease and aging. Given their capacity to exist in different oxidation states, transition metal ions are logically involved in the generation and reactivity of these entities in vivo. Physico-chemical considerations relative to this topic are introduced in the following chapter, which also reviews the effects of complexation on the reactivity of iron and copper ions as well as on the protective role of zinc in this context.

During disease treatment, naturally occurring metal ions interact (intentionally or not) with most of organic drugs being administered. Diverse physico-chemical aspects of these interactions with different drug families have thus been successively considered.

Metal complexes themselves may also be administered as pharmaceuticals. Physico-chemical data underlying the pharmacological effects of such drugs are presented in the next chapter. A large section is naturally devoted to platinum anticancer compounds, but copper and gold complexes are also considered.

In addition to their pharmacological applications, metal ions and their complexes may also be used for diagnostic purposes. For example, radiopharmaceuticals have long been in common use for the detection of tumors and other abnormalities. More recently, paramagnetic metal ions and their complexes have been introduced as contrast agents in magnetic resonance imaging. A special chapter reviews the main physico-chemical aspects of these problems.

Whenever one tries to understand the effects of the interactions between a given metal ion and any substance acting as a ligand in terms of metabolism or mechanism of action of either of these reactants, the first objective should always be to know under which chemical form these reactants are present in the main biofluids and at their presumed physiological target. By definition, this objective falls within the field of speciation. Part Five is devoted to this important domain of application of inorganic chemistry to medicine.

In the present context, the experimental approach to speciation begins with the determination of the overall concentrations of metal ions and ligands in biological media. Specific techniques for these determinations are developed in the first chapter.

Since metal ions in the main biofluids are bound to ligands of different size, the first objective of speciation is to discriminate the different fractions into which each of these is distributed, and whenever possible its different complexes within each of these fractions. The next chapter describes the experimental techniques available, and underlines their limits concerning the identification of small labile species.

Since these small labile species are the above-mentioned "dynamic units" through which most metal ions can be translocated in vivo, it is of particular importance to know which of these predominate under specific circumstances. In the absence of experimental access to their distribution, computer simulation must be used. This approach is thus introduced here: diverse applications of computer-aided speciation to bioinorganic medicine are successively given, and future trends are outlined.

CONTENTS OF *BIOINORGANIC CHEMISTRY*

Volume 2

INTRODUCTION TO BIOINORGANIC CHEMISTRY

by

R.J.P. Williams

University of Oxford,
Inorganic Chemistry Laboratory,
South Parks Road
Oxford OX1 3QR
U.K.

I. INTRODUCTION

Bio-inorganic chemistry has two distinct parts. First, it is the study of the chemistry of the compounds of those elements which are found in or are related to compounds found in biology. Second, it is the study of any other element in a compound which can be considered to be able to act as a medicine or a poison in biological systems. In both cases we exclude all those compounds which are clearly organic compounds. Three separate areas are easily delineated: (i) the direct study of the elements in biology which is *biological* inorganic chemistry (Fraústo da Silva and Williams, 1991). It is complicated by many considerations absent from chemistry, e.g. compartmentalization and energization of distribution; (ii) the examination of compounds extracted from biology containing inorganic elements and models of those compounds; (iii) the study of compounds not found in biology which could be said to affect biological systems, i.e. inorganic reagents, drugs, poisons and pollutants. Obviously together these studies include all the elements of the Periodic Table (Fig. 1) but by tradition those based on H, C, N, and O and to some degree with S and P are excluded and their comparable study is called bio-organic chemistry. It is necessay therefore to start from a knowledge of those elements which are important in biological chemistry.

II. BIOLOGICAL INORGANIC CHEMISTRY

The central peculiarities of biological-inorganic chemistry are contained in the following features of living organisms: (i) their elemental content (see Fig. 1) which derives from the availability of elements.

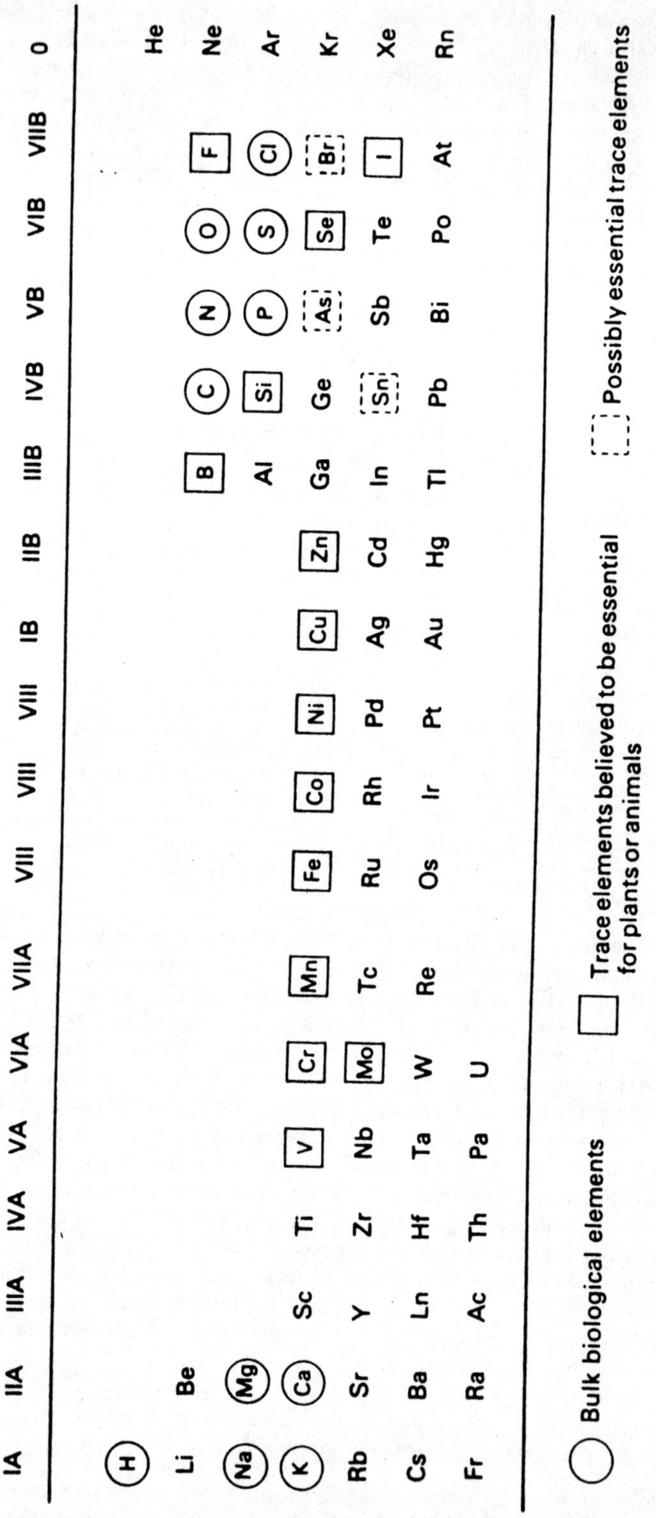

Figure 1.
The elements of greatest significance in biological systems.

in the air and waters of the earth's surface and from the effort life makes to obtain them; (ii) the manner in which chemical elements combine largely with organic matter and in the presence of water. As well as thermodynamic binding constants, such as the stability constants of complex ions and solubility products, there are various routes to kinetically controlled combinations; (iii) the organization of the compounds so formed within the body of the organism; (iv) the functional properties of the compounds in relation to the living state. In this chapter, we shall next tackle these features in turn. Finally, any biological system has to be seen as a flow system which is in a virtual steady state over the short-term but is developmental over the long-term. There is then a continuous, required input and output of material and an input of energy. All this activity is related to the chemical nature of life.

A. The availability of the elements

An approximate idea concerning the availability of the elements today as found by concentrations in the sea is given in Figure 2 (Cox, 1989). Some relatively rare elements are quite available and this applies particularly to the water soluble elements and those in the air. The less available elements than their abundance indicates generally form very insoluble salts. Notice that availability does not relate easily to abundance. In the next figure (Fig. 3), we plot the elemental composition of the human body against availability. We see that biological systems can accumulate elements (Fe) from a low availability pool or reject elements (Al) from a higher availability pool. This selection requires energy input to the uptake or rejection mechanisms in two senses (i) direct pumping into or out of the organism (ii) stabilization in the organism through energized synthesis of selective binding ligands.

Now the above figures concern today's availability and today's composition of a living organism, man. In the period from about 5 to $3x10^9$ years ago, the availability of some elements was quite different and the composition of the then-living unicellular prokaryotes and archaebacteria was correspondingly different. Many of these organisms still exist in niches on the Earth's surface, but away from di-oxygen. Not only does this teach us that

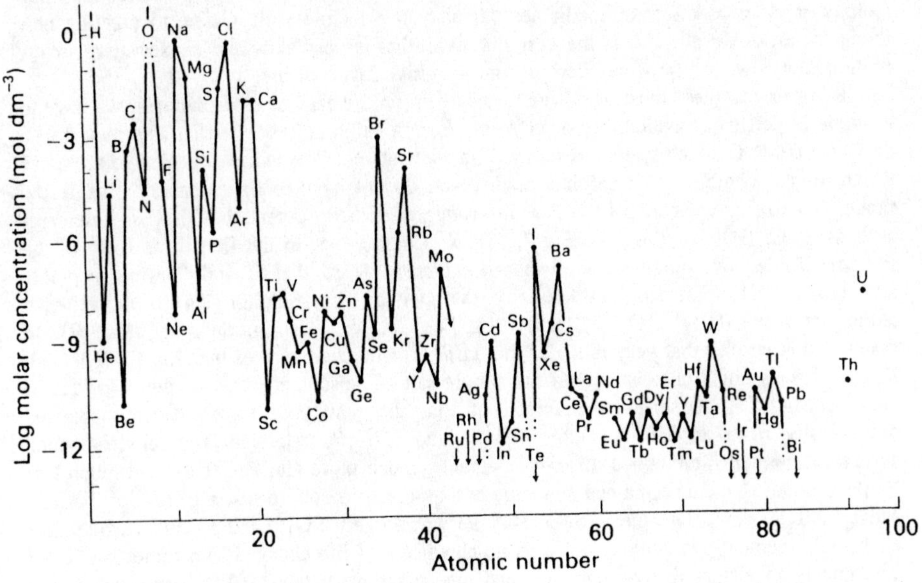

Figure 2.
The concentration of the elements in the sea.

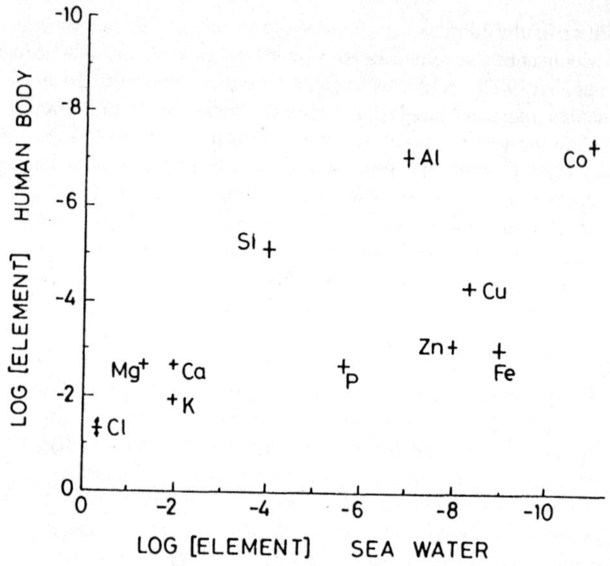

Figure 3.
The concentrations of elements in the sea in comparison with those in the human body.

living forms reflect environmental conditions, which should cause man pause for thought, but also that in principle nature, or today man, could create life forms having very different elements in them than those present today. These two features must open up bio-inorganic chemistry to the study not only of existing biological systems but to (i) the careful examination of man's industry relative to the environment and (ii) the prospect not just of synthesizing models of life's chemical units but of introducing a variety of elements into biology at times as drugs or medicines but also in principle with a view to creating new living forms. After all, if it is the case that evolution is very slow and conservative, many of the features we see now may derive from primitive forms of life.

Returning to these most primitive forms of life in the earliest environments, we have to examine in particular availability of elements from a sulphide (H_2S/S) rich as opposed to a dioxygen (H_2O/O_2) rich system of today. The elements not affected by this change are those which do not combine with sulphur readily and do not have redox chemistry outside the range -0.5 to 0.0 volts at pH=7, i.e. the range of H_2S/S_n potentials. They are elements such as B, C, (H); Na, Mg, Al, Si, P, (S); K, Ca, Sc, Ge, in the first three rows of the Periodic Table. Of course water always dominated life so that H and O were available although not O_2. Elements affected by the switch from sulphur-rich to oxygen-rich atmospheres are (H): N, O; (S), (Cl); Mn, Fe, Co, Ni, Cu, Zn, Se; (Br), Mo, (I). In essence, this implies that very much of the *basic* organic chemistry of life, i.e. C, H, N, O, P, but perhaps not S chemistry, and the simple salt chemistry, i.e. that of the chlorides of Na, K, Mg, Ca, remained unaffected but that the (catalytic) trace elements changed dramatically from an Mn^{2+}, Fe^{2+}, (Ni^{2+}) and MoS_4^{2-}, relatively rich environmental system to one which contained different elements, much more Cu, Zn, (Cd), and much less Fe (Ni). Some elements remained available but appeared in new forms, e.g. MoS_4^{2-} became MoO_4^{2-}, R_2Se could become R_2SeO, H_2S became largely SO_4^{2-} and so on. Through the changes in element availability, the organic chemistry of life changed to a rather small but enormously important degree using O_2 and the new trace catalysts. We next examine the sulphide system in a little more detail, where we first see selectivity of element combination in operation.

B. The solubility of sulphides

The solubility products of the sulphides have been a critical factor in the development of life (Cairns-Smith et al., 1992). It is the very fact that the amount of oxygen formed as an element in the universe was inadequate for the removal of all non-metals and metals as oxides which led to the presence of free metals and metal sulphides in the original reducing atmosphere. We, living today, are used to an oxidizing atmosphere but this planet may be the only such planet in the universe to develop such an atmosphere and certainly it did not start as it is now. (Moreover it is chemically very unstable). The fact that sulphur remained as H_2S meant that the original availability of elements was based in large part on sulphide solubility products (Fig. 4). There is a peculiar feature of these solubilities. To a first approximation they follow the Irving-Williams series as does all complex ion formation with proteins or any inorganic or organic ligand.

Ca Mg	<	Mn	<	Fe Co Ni <	Cu Zn
soluble		increasingly insoluble			most insoluble
not bound		bound			strongly bound

The insolubility of zinc sulphide considerably exceeds that of nickel while copper sulphide is extremely insoluble. Thus the metals Mn, Fe, Co, Ni were available at around 10^{-8}M or greater, but copper and zinc were at 10^{-25} and 10^{-15}M respectively in the original 10^{-3}M H_2S solution. Given the high temperatures and the volcanic activity of the initial surface of the earth, the sea must have been a black colloidal soup of metal, mainly iron and some nickel, sulphides (geologists such as Russell et al. (1992) refer to black smoky chimneys of colloidal sulphides). They would have contained a variety of $(Ni)Fe_nS_n$ particles in all probability. No systematic experiments have been done but it would be interesting to see what organic chemicals form in water at 100°C in the presence of dispersed Fe/S, Ni/S "colloids" through reaction of H_2O, CO, H_2, CO_2, $NH_3$and CH_4. This would be completely disorganized organic synthesis, energized by the particles, since the emerging iron-rich sulphide solutions were and are not stable in sea water, and/or by photo-activation. There can be little doubt that these clusters predated the coming of the iron/sulphur proteins (Holm et al., 1990; Williams, 1965).

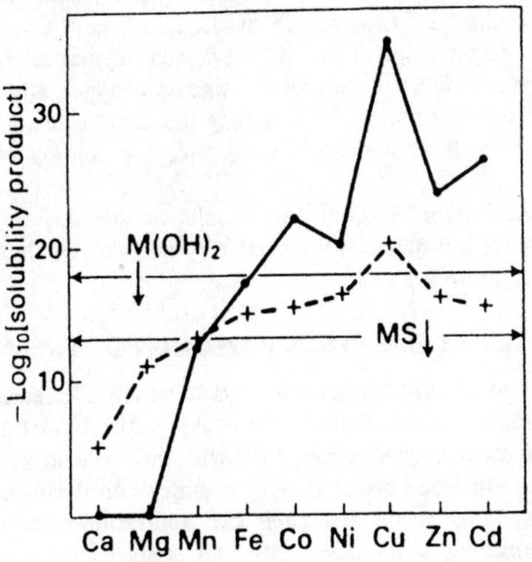

Figure 4.
The solubility products of divalent metal sulphides (full line) and hydroxides (dashed line). The horizontal lines give the precipitation level for both at pH=7 where H_2S is millimolar. Note oxidation to trivalent metals gives a different picture.

It must be understood that any discussion of early life must be speculative, but we can base it on the nature of the archaebacteria which use nickel and cobalt but not copper and little zinc. Their metabolic paths use CO and CH_4 and they have an extensive sulphur chemistry. The archaelogical record is not of great help although clearly iron has been in living systems for as long as any other element including H, C, N and O. Life did not necessarily start from organic progenitors and there are speculative ways of describing development based on iron sulphides (Russell et al., 1992). All life presently known does relate to a DNA (RNA) world, but there is little possibility that the system started by synthesizing them from electric arcs and sparks since this has no limiting volume and no energy supply for reproduction.

C. The coming of dioxygen

Through the switch of inorganic chemicals in the environment due to the coming of dioxygen, and no matter how the change in O_2 content of the air began (and it could have had a purely inorganic origin), the evolution of life changed dramatically. Life became multi-compartmental, newly organized and eventually multi-cellular. Though we cannot trace the origins of life for sure, we may well be able to trace through inorganic chemistry the development of these later stages of life. I shall show that it is the removal of sulphide and consequent loss of iron hydroxide through oxidation together with the increased concentration of copper and zinc, all caused by the advent of O_2, which have allowed a switch from random colony formation to organized cell-cell living forms for example. Bio-inorganic chemistry is central to evolution through catalysis but also via other functions (see below).

Now, as stated, the switch from a sulphur-rich to an oxygen-rich world did also affect the *organic* chemistry of life to a marked, though minor, and extremely important degree. It did not change the major basic DNA, RNA, protein, fat, polysaccharide, relationships with small molecules and many salts. It did change for example the possibility of cross-linking of polymers via -S-S- bridges and other oxidative mechanisms mainly outside cells. It also changed the available oxidation states of nitrogen, carbon, sulphur and halogens in many small molecules, e.g. NO, and produced a variety of organic secondary metabolites, e.g. DMSO and halogenated organic molecules. (It also changed the available chelating centers for metal ions). But these compounds remain only as a small fraction of the whole. Most of them are or were initially poisons and are now signalling devices, hormones etc. Signalling devices were bound to change on the switch to organized multi-cellular devices and in fact such apparatus as that using external calcium advanced beyond all recognition once multi-cellular life evolved while internal phosphorus signals changed slightly. There are in fact regulatory systems of major importance concerning the metal elements, Fe, Cu, Zn, Mo, Se and Ca which could have been developed only after the coming of dioxygen (Williams, 1992).

In conclusion, the internal metabolism of cells to this day is based on a reducing, sulphide-type atmosphere, but much of the chemistry outside cells is more closely related to oxidizing, dioxygen, chemistry.

III. COMBINATION OF ELEMENTS IN BIOLOGICAL SYSTEMS

There is no space here to analyze further the ways in which all elements may combine with one another and fortunately in life there is the heavy restriction of availability. Let us put aside the kinetic controls and energy inputs which allow life to synthesize C/H/N/O/P/S compounds. If we take this huge basic stock of chemicals in the form of a soup of organic molecules of different size, shape and chemical composition we can ask how are the inorganic elements combined with these non-metal elements in compounds. We then get Table 1 (Fraústo da Silva and Williams, 1991).

The result is quite striking. We can divide the metal elements immediately into three groups. Those which bind best to neutral oxygen especially water, those which bind best to oxygen anions, those which bind best to N/S ligands. Biological systems utilize these combinations which relate to the old inorganic analytical group tables which used to be studied and should still be studied by all chemistry and biochemistry students. The procedure allows selective group separation of Na, K: from Mg Ca, Sr, Ba: and from Mn,

Table 1

The Character of Life's Major Elemental Combinations

Element	Combination
Na^+, K^+, Cl^-	H_2O
Mg^{2+}, Ca^{2+}	Combined O in C-O, P-O compounds
	H_2O
(Sr) (Ba)	H_2O, SO_4^{2-}
Mn^{2+}/Mn^{3+}	N, O ligands
Fe^{2+}/Fe^{3+}	RS^-, S^{2-}, N, O ligands
Co^{2+}/Co^{3+}	N (vitamin B_{12})
Ni^{2+}/Ni^{3+}	N, RS^-
Cu^+/Cu^{2+}	N, RS^-
Zn^{2+}	N, RS^-, Combined O
Mo^{6+}/Mo^{5+}	RS^-, S^{2-}, O^{2-}

(Fe), Co, Ni, (Cu), Zn, Mo. The first two form combinations with neutral oxygen donors only, the first only with water and the second with oxygen anions, while the third precipitate with sulphide [N.B. Kinetic control will generate some exceptions, e.g. Mg in chlorophyll]. How do chemical properties separate elements within the three groups?

We start by saying that the three groups of metals are separated due to their electron affinity as ions, i.e. their covalent binding due to the ionization potentials to a given charged state and by the sizes of the ions produced. The two factors come together in figures such as Figure 5. Ions separated on size grounds only are shown in Table 2. Not

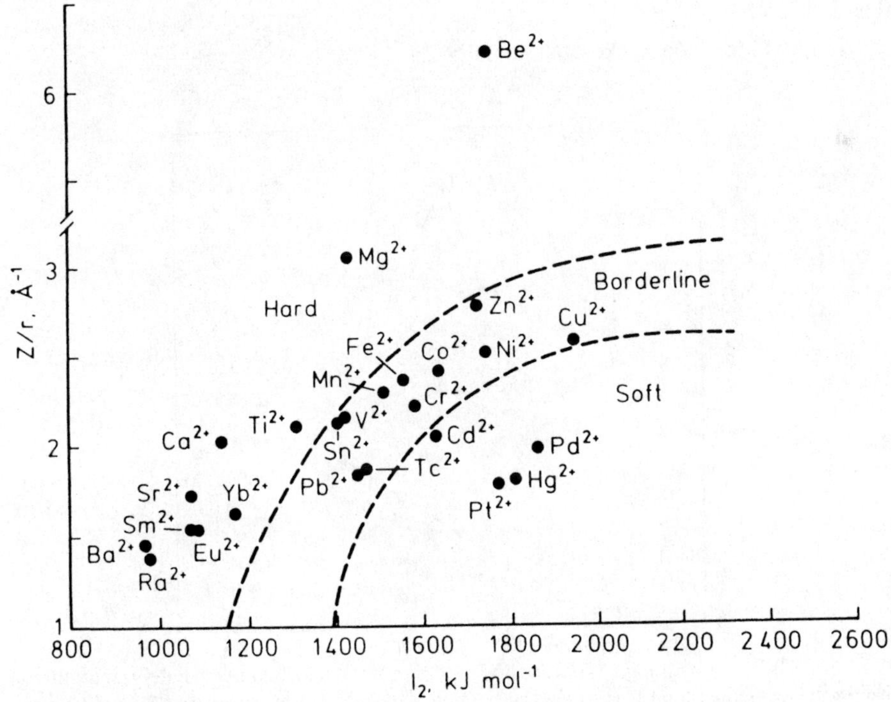

Figure 5.
A plot of ionisation potential, I_2, to the M^{2+} state, i.e. the Lewis acid strength as an acceptor for M^{2+}, against ion size (r), z=2. Similar plots can be made for different ionic states, z = 1 or z = 3.

Table 2

Sizes of Ions

<0.55	0.5 - 0.65	0.6 - 0.755	0.75 - 1.00	>1.00
Mo^{6+}	Mg^{2+} Al^{3+}, Fe^{3+}, Co^{3+} Cu^{2+}, Zn^{2+}	Mn^{2+} Na^{+}, Ca^{2+} Fe^{2+}, Co^{2+}, Ni^{2+} Cu^{+}	K^{+}	Sr^{2+} Ba^{2+}

only can Na^+ be separated from K^+ and Mg^{2+} from Ca^{2+} (and from $Sr^{2+}$and from Ba^{2+}) on size and charge grounds *but* we see that Fe and Co as Fe^{3+} and Co^{3+} can be separated from Mn^{2+}, Ni^{2+} and Zn^{2+} on the one hand and copper (Cu^+) on the other by redox state and size change. Biological systems do just that. When size and charge are similar, biological combination depends largely on electron affinity of ions. The remaining problems are solved using the smaller ligand field stereochemical preferences. We showed the principles using the solubilities of sulphides and can do so again using complex ion formation with ammonia. These principles apply alike to organic reagents or proteins using O-, N-, S- donor centres with or without negative charges, but with larger structures the organic moiety can generate donor systems of fixed hole size and fixed stereochemical disposition of donor atoms. Selectivity is increased (Fraústo da Silva and Williams, 1991).

A. Metal complexes of ammonia

This example of the stability of complex ions is given as an illustration of a general rule which applies to virtually all organic and inorganic ligands in the further fraction of transition metal ions. The Irving-Williams series of stability constants of divalent ions (Fig. 6) is generally

(Ca,Mg) Mn < Fe < Co < Ni < Cu > Zn

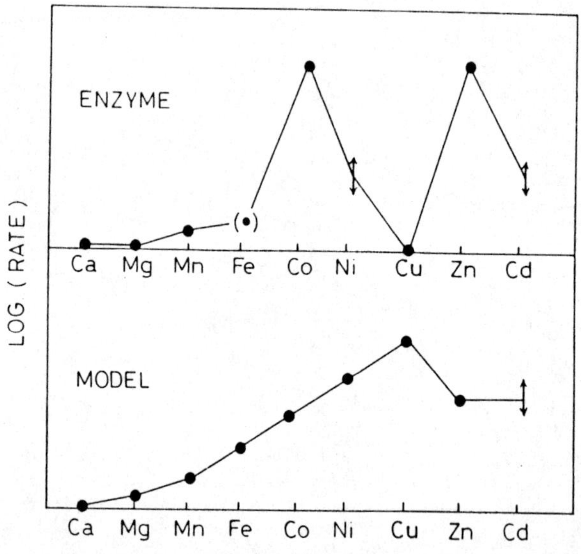

Figure 6.
Under the heading MODEL we show the general relative strength of acid catalysis (RATE) (and here stability constant) for a variety of metal ions. Under the heading ENZYME we show how the MODEL order is not obeyed in enzyme catalyzed reactions in which a metal, Zn, is substituted by various M^{2+} ions, e.g. in carbonic anhydrase.

We shall not be concerned further with how such a series arises (see Fig. 5) but it applies to precipitation and extraction into organic solvents of *all* complex ions as well as to the stability constants in water. It is against this background and the use of ring chelates (see below) that we must see the biological speciation which follows together with the extra energy input (for extraction/pumping of complexes) as follows:

. Convert $Cu^{2+} \rightarrow Cu^{+}$ and remove as a sulphur ligand complex (metallothionein);

. Convert $Fe^{2+} \rightarrow Fe^{3+}$ and remove as an O/N complex (transferrin, ferritin);

. Cobalt can be largely ignored. There is rather little but it can be selected for specifically as low-spin Co^{3+} (vitamin B_{12});

. Select Ni^{2+} from Zn^{2+} on the basis of stereochemical selectivity: Ni^{2+} octahedra; Zn^{2+} tetrahedra. Use sulphur for Zn^{2+} (e.g. zinc fingers) and N/O ligands for Ni^{2+} (the need for this would not be seen in anaerobes where Zn^{2+} is precipitated by sulphide leaving soluble Ni/S systems);

. Manganese is left with weaker N/O ligands (photo-system for O_2 production);

. Calcium and magnesium remain with O ligands and are separated on the basis of size difference as happens with EGTA as opposed to EDTA. The use of different sizes in the form of an organic hole to capture different sizes of cation is a major biological and recent chemical advance which is used to explain the selective use of calcium and the selection of Na^{+} from K^{+}.

Given the concentrations of elements and ligands, the above selection principles based on thermodynamics allow distinctive metal-ligand complexes to be formed and these complexes can now be placed in space by pumps.

B. Metal complexes of ring chelates

A ring chelate as a complexing agent was discussed by Williams (1953a) on the basis of its advantage in being able to select by controlled size. This approach became extremely powerful in the period after 1970 in the hands of organic chemists (Cram, 1987; Lehn, 1987; Pedersen, 1987), but the original idea came from the observations of biological ligands such as porphyrin. Now there is a problem which biological systems also throw up. The same ring or a very similar ring is used to bind Mg (chlorophyll), Fe (several porphyrins), Co (corrin), Ni (F.430) and Cu (Turacin feather pigment). Such combinations show that the metal synthesis is a kinetically controlled path based on *prior* selection of the metal on thermodynamic grounds given above. A second advantage of a rigid ring chelate is slow uptake and release of metal which allows kinetically controlled chemistry. The scheme is:

(i) choose a complex ML which can be any L which is selective for M giving almost specific M_1L_1 in thermodynamic equilibrium as described above, e.g. Mg^{2+} + ATP within the prescribed conditions of cells gives MgATP without interference from other metal ions;

(ii) recognize M_1L_1 very selectively and then *transfer* it to a different compartment or reaction system thereby isolating the reactions of M_1. Ion pumps select ions on the basis of size;

(iii) in the new environment transfer using energy M_1 into L_2 giving M_1L_2, where L_2 is a ligand which is not selective for M_1, and M_1 exchanges slowly from M_1L_2.

(iv) transfer M_1L_2 to where it is needed.

We have now described the ways in which biological systems can recognize M_1, can selectively bind it and then also selectively transfer it so that it functions in a particular compound in a special part of biological space. All the considerations apply to large or small (protein) complexes. This is no more than a highly sophisticated parallel with the operations in a chemical laboratory. The steps are all energized. We can understand much of this chemistry by careful study of extracted species. However, the combinations and spatial separations have been selected for their functional value and it is to the functional value of individual metal elements that we turn next, by way of the examples of copper and zinc, before we ask how the functions are integrated so as to create the flow of activity which is life. In order to gain functional advantage, biological systems transfer elements stepwise from their environment to the required position in a cellular system and in each step selectivity is used (Fig. 7).

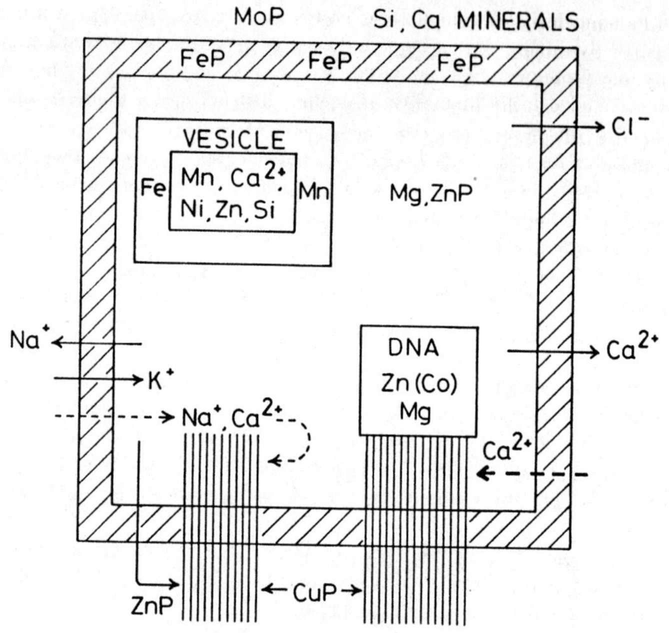

Figure 7.
Some of the divisions of metal ions due to the pumping of elements into compartments, e.g. membranes, extracellular, intracellular and vesicular spaces. Broken arrows show temporary invasion.

C. Compartments for copper and zinc: an example

Given the changes in the redox conditions from inside to outside a cell, there are two separate biological problems of chemistry to be solved. We must consider how elements prefer to combine with one another in the redox potential range 0.0 to -0.5 volts and how they combine when the redox range is extended to between +0.8 volts (after O_2) and -0.5 volts. The situation is made the more complex in that biological control permits the inside of cells, i.e. the cytoplasm, to be kept in the reducing mode 0.0 to -0.5 volts even in the presence of O_2 while the outside of cells in an oxygen atmosphere has a potential in the range 0.0 to +0.8 volts. As a result, the inside of the cell is still a sulphide medium but much less so than it was in the first two billion years of life.

Now, as explained, the coming of dioxygen led to a new competition between the heavy elements, especially for sulphur ligands. In the (primitive) reducing atmosphere where H_2S dominates, iron can still form soluble Fe_nS_n clusters in aqueous solution both inside and outside cells and can even make simple $Fe(RS^-)_4$ protein complexes, rubredoxins, while copper and zinc precipitate. Thus, iron with nickel and molybdenum and sulphur dominated the early transition-metal biochemistry. In the presence of oxygen, external H_2S largely disappears as S_n, or even as SO_4^{2-}, and RS^-goes readily to R-S-S-R. The latter forms a stabilizing protein cross-link. Outside cells, the only simple RS^- complexes which are now stabilized sufficiently so that they are not affected by O_2 are those of copper, e.g. the copper blue proteins. Even zinc cannot bind to RS^- in the presence of oxygen, so that zinc proteins for export, peptidases, phosphatases and so on are largely formed from N/O donor groups. This apparently means that RS^- should be found bound to copper, either as cuprous or cupric, everywhere in biology. The observed system is very different. In fact inside cells, i.e. in the cytoplasm of eukaryotes, copper has not been found bound to RS^- except in metallothionein where cuprous copper is found in clusters. In contrast, in the cytoplasm zinc is found bound to RS^- groups in many proteins and of course the Fe/S clusters are still found. Selectivity for iron is based only on cluster chemistry, and the formation of mononuclear RS^- complexes is now a property of zinc. This can mean but one thing. The strongly binding free copper ions are so reduced in concentration in the cytoplasm by metallothionein and subsequent outward pumping that simple copper-RS^- complexes are excluded from the cytoplasm. By contrast, zinc and iron are homeostatically retained in cells by metallothionein and ferritin. The lowering of the free iron by oxidation allows zinc alone to form mononuclear RS^- complexes while iron can only bind in Fe/S clusters to receptor sites of the correct size and shape. All such systems developed after oxygen entered the atmosphere in quantity. As a consequence, copper evolved in the *extracellular* oxidases generating cross-linked polymers and the possibility of a stable extracellular matrix. The consequence for evolution was multicellular life. To allow growth in such a matrix, degradative catalysts are needed and these have turned out to be Zn-containing enzymes (see Fig. 7).

D. The selection of non-metal anions

The availability of non-metals such as C and O (CO_2), H and O (H_2O and O_2), N(N_2) is through the air and aqueous solution, and cannot be prevented by diffusion restrictions such as membranes. There is a limitation on the availability of dioxygen in a large organism and even in cells with compartments so that circulating systems, e.g. haemoglobin, and storage devices, e.g. myoglobin, are required. The major problem in the handling of these elements is kinetic. The fact that HS^- or RS^- react much more readily than O_2 means that the inside of a cell is made relatively reducing. Elements taken up in high oxidation states can then be incorporated in low oxidation states in cells. Examples are SO_4^{2-}, SeO_4^{2-} and MoO_4^{2-} which are largely found as -SR, -S-S-, -SeR and MoS_nO_{4-n} inside cells. Now these three elements are all taken up, like NO_3^-, Cl^- (halide generally), and HPO_4^{2-} through membrane pumps. There is then selectivity at two levels as for cations - transport in the forms found in the environment and redox reaction in the cell before virtually covalent incorporation except for Cl^-. Leaving Cl^- on one side since it behaves very like Na^+ and K^+ in osmosis and electrolytic charge balances, HPO_4^{2-} is the only other anion which does not enter redox reactions in biology and this gives it a role similar to Ca^{2+} in many respects although it is involved in covalent rather than ionic interactions. In various forms phosphate is a unique anion in biology so that it does not suffer competition. Just in passing, we can note its role in bio-energetics and in biopolymers where there is no parallel with Ca^{2+}. Selection at the early stages of intake against SO_4^{2-} (HPO_4^{2-}) is based presumably on H-bond strengths. By way of contrast, NO_3^- has a different size and shape and differs from HCO_3^- in a parallel manner to the difference between SO_4^{2-} and HPO_4^{2-}.

Finally interest settles on the coordination chemistry of molybdenum. The uptake of MoO_4^{2-} perhaps mimicking SO_4^{2-} is followed by incorporation with thiols or dithiolate. An examination of the chemistry of molybdenum in high oxidation states suggested a strong parallel with Fe^{3+} (Mitchell and Williams, 1962). In oxygen-rich media they are coordinated by oxide, O^{2-}, in MoO_4^{2-} and Fe_2O_3, while in sulphur-rich media both readily associate with sulphur. The form of a joint cluster in nitrogenase is now known. Elsewhere molybdenum is found in a special dithiolate where it still retains some of its oxo chemistry. The fact that this association is unique may well mean that the very small size of Mo(VI) allows it to fit into the chelate ring of a dithiolate better than any other available element.

IV. EXCHANGE RATES

It is not just the thermodynamic properties of metal-ligand interactions that are important in biology. It is required that metal ions are able to exchange from binding sites at rates which are functionally valuable (Table 3). In a survey such as this we can give no more than the barest outline of these fundamental kinetics (Eigen, 1960). The fast exchange is found for Na^+, K^+ and Cl^- in electrolytic circuits which involve exchange from sites in pumps and channels. The second level of relatively fast exchange is found for Mg^{2+} and Ca^{2+} in signalling devices such as muscle triggers, fertilization, and hormone release. There is a parallel exchange of one anion, phosphate, but this is not simple, depending on enzyme phosphorylation and dephosphorylation (kinases and phosphatases) which control in-cell metabolic action. The last two enzymes are linked to Mg^{2+}/Ca^{2+} exchange controls and

Table 3

Matching of the Physical Properties of Metal Ions and Proteins

Function	Metal ion	Protein
Mobile triggers and messengers in cells	Ca^{2+}	Highly charged intracellular proteins
Relatively rigid enzymes	Cu^{2+}, (Fe^{2+}), Zn^{2+}, Mo^{6+}	Lowly charged intracellular proteins
Intermediate mobility: control of calcium kinetics	Mg^{2+} (Mn^{2+})	Kinases, phosphate transferases in cells
Very fast diffusion	Na^+, K^+	Almost none
Hydrolytic enzymes (low mobility)	Ca^{2+}	Extracellular proteins calcium stabilized
Structural units forming mobile *n*-mer	Mg^{2+} (Ca^{2+})	In-cell structures, tubulin

both are then linked to control of expression via DNA and RNA. The rates of exchange of phosphate, of necessity, are somewhat slower than those of Mg^{2+} and Ca^{2+}. There is then a parallel between phosphate control and the straightforward exchange of iron and zinc from regulatory proteins such as the FUR and IRE proteins (Fe) and the zinc fingers. There is a less extensive regulation of copper via the ACE protein and metallothionein. Finally, very slow rates of exchange are established by locking metal ions into cavities either in small ring ligands or in proteins often metalloenzymes. Examples are porphyrin, chlorin, vitamin B_{12} and F-430 amongst ring chelates and almost every zinc or copper metalloenzyme. In so far as there is exchange, it is brought about by the destruction of the ligand or movement of the complex into a new compartment.

These rates are of vast importance and must be seen in the contest of two other rates: (i) the rate of protein conformation change and folding (see below), and (ii) the rate of transfer of either free metal ions or metal ion complexes with small ligands or proteins from compartment to compartment. This second movement is pumping in and out of cells and its rate is dependent on energy supply and can be switched on or off according to need.

V. THE FUNCTIONAL PROPERTIES OF METAL IONS

The Periodic Table was partly founded on the basis of the potential for reaction of the elements. This Table relates then to the basis for functional activity of elements in biology. We expect and find that the simplest cations Na^+ and K^+ which are present in the largest quantities act in osmotic and charge balance and in carrying electrolytic currents. They have an influence on a few structures. The elements, Mg^{2+} and Ca^{2+} as ions, act in *binding* to negatively charged oxygen donors and so control or initiate very many activities and assist in structures. They are not very effective as catalysts. The cations, Zn^{2+} with the proton H^+, are the catalytic Lewis acids of biology largely through binding to N and S donors. The elements Mn, Fe, Co, Ni, Cu and Mo perform redox functions as cations. Anionic chemistry has a reciprocal balance. Chloride carries current and balances osmotic and electrical charge. Sulphate with phosphate and carboxylate give a *bound* charged framework while phosphate has the added function of controlling energy distribution. Thiolate is the centre of much catalytic acid/base and redox activity where selenium is also used now and then. These generalities hold but there are also special important functions of each group of elements outside the general situations.

VI. THE SPECIAL NATURE OF FOLDED PROTEINS AS LIGANDS

Now, in the context of biochemistry, the possible functions of the metal ions are refined by the creation of energized concentration gradients. Inside a selected compartment, a metal has selected partners which heighten its functional potential. It is then essential for the understanding of bio-inorganic chemistry that we note the compartments into which different metal ions go. Again through evolution the protein partner with its particular metal have been refined to achieve effectively the best of all possible functional conditions (Fig. 8). Here we need to be careful since the function is not of necessity independent of time. There are different types of protein which can either be almost rigid, like many a ring chelate, giving slow exchange or relatively flexible, giving fast exchange. The different proteins are used in different ways. We shall find it useful to look on proteins in the following classes (see Table 4):

(i) relatively rigid β-sheets;
(ii) helical proteins (α-proteins);
(iii) very loosely folded and random coil proteins.

A. The entatic state and rack action

The concept that stress could be engendered in proteins, in protein-bound substrates or in a cofactor, metal, bound to protein has arisen in at least three different ways. The concept of induced fit described by Koshland (1973) allows the idea that not only the substrate but the protein, enzyme, uses part of the energy of binding to adjust, i.e. to stress, its internal structures. If this energy is local, then it can be seen that, as far as the substrate of an enzyme is concerned, it is a similar concept to that of the Haldane and Pauling postulate. The latter authors considered that in fair part the enzyme structure forced a bound substrate into a state close to the transition state for reaction. Clearly this is a stressed situation. Note most enzymes are based on β-sheet proteins (Table 4). If the energy stress of the substrate-binding runs into the protein, then the induced fit (stress of the protein) is a model for allosteric change, action at a distance through mechanical change, and these thoughts gave rise to the multi-step cooperativity model of haemoglobin, again due to Koshland, as well as

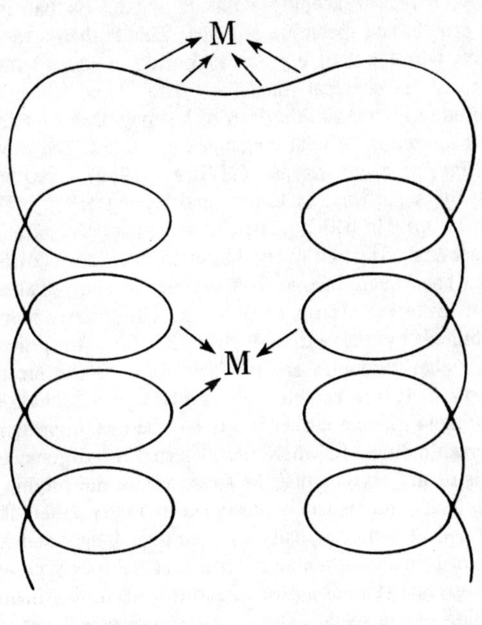

Figure 8.
An illustration of how a metal ion can be exposed and free to exchange or buried in a cavity in an enzyme.

Table 4

Metal-Proteins of Extreme Different Types

Entatic Only	Rack	Random
Copper blue proteins (β)	Haemoglobin (α)	Cytochrome *c* (apo)
Copper oxidases (β)	Calmodulin (α)	Zinc Fingers (apo)
Ferredoxins (β)	Cytochrome P-450 (α)	Osteocalcin (apo)
Rubredoxins (β)	Haemocyanin (α)	Metallothionein (apo)
CuZn Superoxide dismutase(β)	Cytochrome *c*' (α)	Calsiquestrin (apo)
Fe_3S_4 Ferredoxins (β)		

being clearly related to the R→T (relaxed-tense) model. Note that allosteric action is based on helical structures.

In the early nineteen-fifties (Williams, 1953), I described a somewhat different approach and this was treated generally by Vallee and Williams (1968) under the title of the entatic state. In essence, this description of stressed states is that the protein folds in such a stable manner that a metal bound to it is forced into an unusual (entatic) state able to assist catalysis. The statement does not relate to substrates but to the cycle of reaction of the metal in the metalloenzyme. The idea here is that stress is in the catalyst so that the ground state of catalytic groups (metal ions) are close to the transition states for the *enzymes*' catalytic cycles without induction on substrate binding. The protein frame, a β-sheet, is rigid. The copper blue proteins, described by the geometry around the copper, provided one good example. However, it is easy to see that a variety of metalloenzyme ground states could be described as stressed - including the Fe_3S_4 group and iron of haem in haemoglobin in the absence of binding ligands - while at the same time the protein is stressed.

Independently and at roughly the same time as I started to discuss the stressed ground *state* of metalloenzymes in these terms, Lumry and Eyring (1954) described the *rack* hypothesis. The idea here, in keeping with its name, is that in the formation of a metalloprotein both protein and metal are stressed. This is then a rack where stress applied to one object interacts with another, e.g. the rack used to stress human bodies in medieval times. It relates to a mechanical machine. This idea leads directly to allosteric communication on binding oxygen to the iron of haemoglobin protein. In the final oxygen-bound state, the iron is conformationally changed and it and the protein are stressed, in a new way. In my writings on haem enzymes (Williams, 1956) I used very similar mechanical machine language at the same time as Lumry and Eyring (1954). This *rack action* is not equivalent to *the entatic state* in which metal-ligand bonds do not change conformation. The confusion in the recent analysis of Gray and Malmström (1989) is difficult to understand.

Examples which have been uncovered subsequently show that entatic states and rack action even belong to different classes of proteins. The blue copper proteins have entatic ground *states*, generated by the extreme strength given to protein folds by β-sheets especially β-barrels. Other examples are in Table 4. Here the metal ion accepts the site geometry forced upon it by the protein. The protein resists change no matter what the combined state of the metal. If one wishes to say so, the metalloprotein ground state (entatic state) is generated by an induced fit where the distorted metal geometry only is induced. In a rack, however, it is of the essence that the stress enters the protein so the protein cannot be a rigid β-sheet. In fact, proteins can be designed in a very different way so as to respond to metal ion binding and become internally stressed over long distances. The best examples are the totally α-helical proteins of not such great fold stability such as calmodulin. Here in the extreme case, the ground state of a protein and the metal coordination are (considerably) adjusted by the binding of the metal. This is the nature of calcium binding to calmodulin and of oxygen-binding in haemoglobin. There is a running change in the helical structures of the proteins. In fact this rack action is the basic function of the protein and there is bound to be engendered some stress at the metal. The use of the helices to transmit stress is further described in Table 4.

In case the above discussion appears to be somewhat academic, it is of interest to see what can be deduced from it. I therefore make the statements:

(i) purely helical structures are most suitable for mechanical devices. This follows from lack of rigidity. They are composed of a set of "mobile" helices (Table 4);

(ii) purely β-sheet structures are most suitable for recognition and for catalytic activity. This follows from rigidity. The less mobile the protein, the more selective the activity, but the slower are the reaction steps involving exchange;

(iii) combinations of domains from type (i) and type (ii) proteins generate hinges as in kinases (Table 5);

(iv) mixed αβ proteins can be both enzymes and mechanical devices through the use of domains. They can also be allosteric through movement of helices even over β-sheets. This is illustrated in pumps for ion movement or in ATP synthetases;

(v) There are a few helical enzymes and they arise when mechanical coupling is necessary, for example where required order reaction kinetics are essential as in citrate synthetase and cytochrome P-450. In essence all the systems under (v) and many under (iv) demonstrate rack action whether they are enzymes or not. A rack is a mechanical machine involving transmission: entasis refers to a state of a system.

This analysis leads directly to considerations relating to the study by inorganic chemists of metalloproteins. Proteins are often unconventional ligands. As a warning, I now treat electron transfer proteins. However, it is an essential point concerning metal-protein interaction that the protein is thoroughly studied in order to appreciate static local conditions and global dynamics.

B. Electron transfer proteins

The transfer of electrons can be accompanied by relatively very small ligand rearrangement or by large changes. In the copper blue proteins, entatic binding, the change is small and the protein with its metal ion are ideally designed for electron-transfer. These proteins are β -barrels. There is no coupling to protein energetics even at the level of H-bond rearrangements.

Cytochrome *c*, an helical protein (Fig. 9), is often used by inorganic chemists as a protein with which to study the theory of electron transfer reactions (Gray and Malmström, 1989). There is a grave danger in this study in that as shown by X-ray crystal structure determination, NMR structural work and N*H* exchange, cytochrome *c* has different H-bond networks in its two oxidation states. It is a helical protein. It is therefore not possible to use cytochrome *c* safely in the comparison of theories of electron-transfer rates based on one electron-transfer distance alone or on through-bond transmission. In the description of through-bond transmission we have to note whether NH bonds are broken in different ways in one or the other oxidation state.

When reading this volume on bio-inorganic chemistry, it is essential that the reader keeps in his mind the nature of the protein he is examining as well as the nature of the

Table 5

α/β Metalloproteins

Protein	Metal Ions
Nitrogenase	Fe, Mo, Mg, ATP
Kinases	Mg, ATP
Cytochrome oxidase	Cu, Fe, (Zn)
Superoxide dismutases	Mn, Fe
Carboxypeptidase	Zn

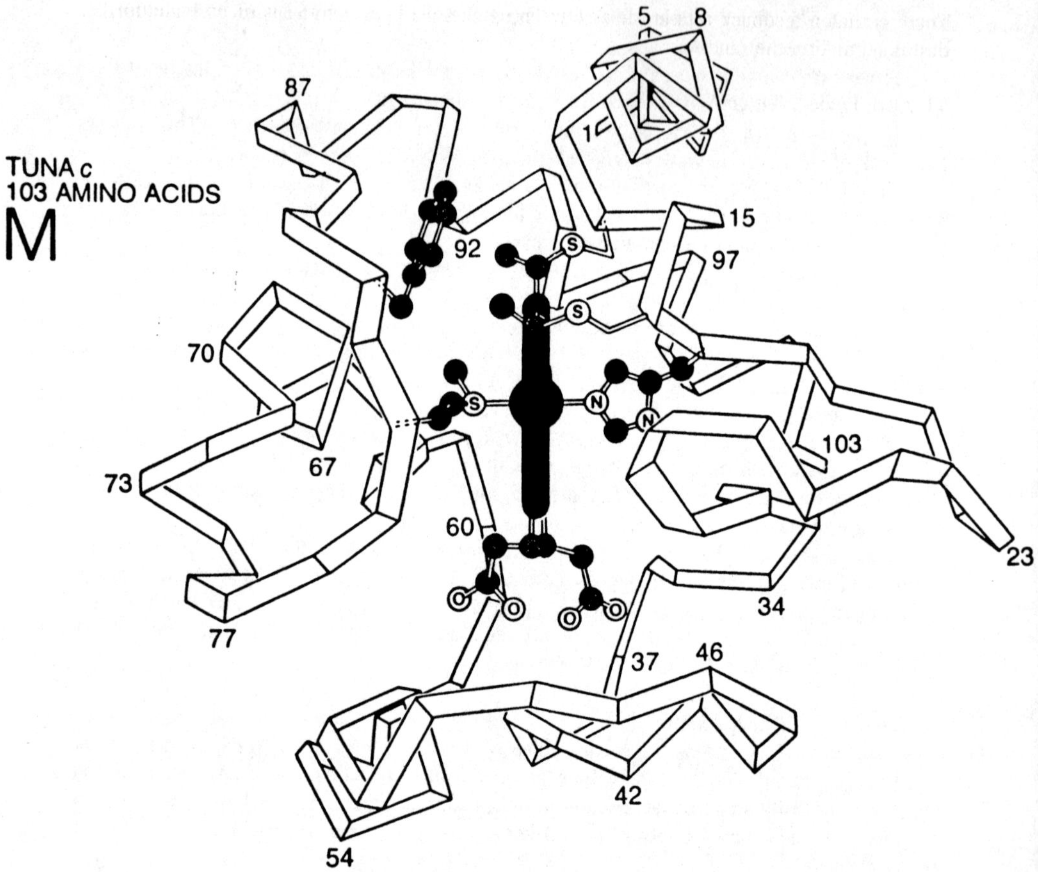

Figure 9.
The fold of cytochrome *c*.

metal. We have illustrated the differences between helical and β-sheet proteins and domains, but there are mixed αβ proteins and there is a third class of proteins (Table 4).

C. The random coil proteins

A protein may be described by two states, R $\rightleftharpoons$ T in haemoglobin, but this is an improbable though useful approximation derived from crystal structure studies. In solution, all proteins are to some degree fluctional and have many states. The extreme limit of the random coil is perhaps never reached, but there are now known to be many proteins which approach this state. Table 4 lists several of them which bind metal ions. They are associated with controls or signalling but never do they serve as enzymes. The reasons for the weakness of the fold can only be guessed at since in the presence of a binding unit, e.g. a metal ion, they fold. The conformation then achieved is recognizable by other biological proteins or pumps or DNA. The protein is then an identity kit for a different metal ion since every metal ion drives a random polymer ligand to a particular conformation. The speculation is that these proteins have very little structure so that (i) the bound form is a reflection of which metal is bound (ii) the on-rate is fast.

Given these features, the proteins become of great value in homeostasis of particular elements such as calcium, zinc and copper. In homeostasis, the free ion concentrations of many different elements are maintained by (i) complex formation by ligands L, (ii) pumping of M, L and ML across compartment boundaries, and (iii) interaction of M, L or ML with DNA so as to regulate synthesis of L where L is often a protein but can be a small ligand.

There is then a feedback relationship between uptake of M and synthesis of its ligand(s), L. In this regulation the random coil protein has its peculiar value.

VII. BINDING TO DNA AND RNA

This discussion brings us to a quite different area of bio-inorganic chemistry: binding to polynucleotides (Williams, 1992). I summarize very briefly here:

(i) direct binding by such ions as K^+ and Mg^{2+} (Fig. 10). The DNA telomer structure depends on K^+ and the ribosome, t-RNA and ribozyme structures depend on Mg^{2+};

(ii) indirect binding via proteins (or complexes). Proteins which control expression by binding to DNA differentially in metal free or metal-bound states are known for (P), Mn, Fe, haem Fe, Fe/S, Cu and Zn. Table 6 shows the basis of the control system. There are links to Ni, Mo and Co.

The situation is summarized in Fraústo da Silva and Williams (1991).

In passing, there are biological complex ions and many synthetic complex ions which bind DNA and which act in a number of modes, e.g. as drugs and finger-print agents. They are described elsewhere in these volumes. Again selectivity is based firstly on the ligand, L, donor atoms, but then on the recognition of ML by DNA (RNA). DNA has also become a target for many researchers using metal ions (or non-metals) not normally found in biology.

This brings us to a brief introduction to the third part of bio-inorganic chemistry.

A. Feedback controls

While discussing controls, we must observe that homeostasis of cellular activity demands that the concentrations of cellular species is held fixed at any stage in a cell cycle. Now, because some 25 elements are involved, the cell must control the uptake of all 25 in concert. In effect, this means that there has to be a feedback mechanism which stops uptake of say iron once it is in balance with twenty-four other elements. We know much of the cross-talk necessary to create this condition (see Fraústo da Silva and Williams, 1991). Here we illustrate a plausible network of communication between many of the elements in Figure 11 where a double-headed arrow implies known feedback relationships.

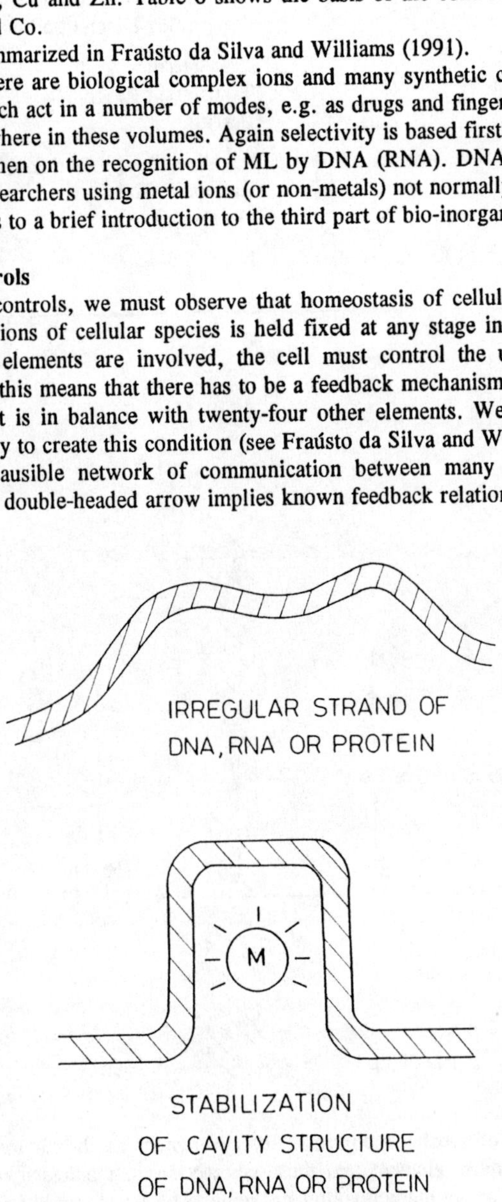

Figure 10.
The K^+ binding of DNA as observed in the telomer.

Table 6

Examples of Regulatory Metalloproteins

DNA-binding Protein	Expressed Enzyme or Protein
ACE 1 Cu(I) Cluster	Metallothionein Cu, Zn Superoxide Dismutase Cu, Zn
FUR Fe(II)	TON B Receptor System (Fe, Co) Enzymes for Siderophore Synthesis Superoxide Dismutase Fe, Mn
IRE (RNA) Fe	Transferrin Transferrin Receptor Ferritin
Vitamin D Receptor Zn	Alkaline Phosphatase (Zn) Calbindin (Ca) Osteocalcin (Ca)
Glucocorticoid Receptor Zn	Na, K Pumps?

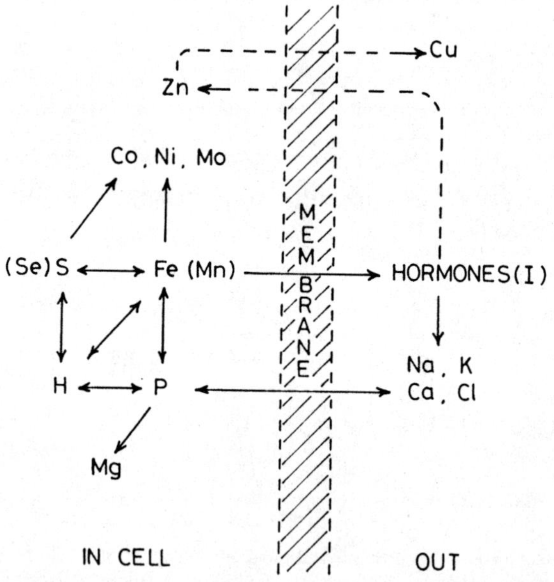

Figure 11.
The interconnection of feedback controls. Every element in the figure has at least one connection to some other element and these connections are adjusted by hormones, e.g. sterols or thyroxine, an iodine-containing hormone. The hormones themselves are part of the feedback. The whole system can drift with time and environment but under constant conditions element levels are all held constant as in a chemostat.

B. Element stores

For a feedback system to be reliable, it needs to be backed up by storage compartments. The storage units will act as buffers so long as exchange is not too slow. In the case of calcium, there are many stores in vesicles and with phosphate in bone. Iron is stored in ferritin, copper and zinc in metallothionein clusters, manganese in vacuoles of plants and so on. There is a large literature on the movement of the elements from stores to functional sites. Of course, simple elements such as C/H/O/N are held in a parallel fashion in polysaccharides, glycogen and starch, and in special proteins.

VIII. BIOMINERALIZATION

A final feature of biological inorganic chemistry is the biominerals (Mann et al., 1989). Just as many of man's mineral materials, clays, cements, mortars and plasters are calcium salts and/or silicates, so are biological minerals mainly based on calcium salts and/or silica. The animals frequently use calcium carbonate (shells) or calcium phosphate (bones) while the plants most frequently use hydrated silica. The great advantage of both Ca^{2+} and $Si(OH)_4$ is the ability to bind to oxygen donors of proteins or of polysaccharides, celluloses. Hence they act as fillers of organic polymer structures, making composites of great strength. This is not their only value since both are adjustable in their growth patterns according to applied stresses. Both are good proton conductors and are pressure sensitive. There is a vast field for exploration here in the use of chemicals. There are other materials which are used to a lesser degree in structures, e.g. $SrSO_4$, $BaSO_4$, FeO(OH), and CaC_2O_4.

Biominerals are extremely valuable in sensors. In plants and animals, gravity is detected using $CaCO_3$ or $BaSO_4$ crystals. Magnetic fields are detected using the permanent magnets made from Fe_3O_4 or Fe_3S_4. It has been mentioned that bone and hydrated silica are sensitive to electric field and mechanical pressure. There may be other surprising features in biology yet to be discovered.

IX. ELEMENTS FOREIGN TO BIOLOGY

Bio-inorganic chemistry has to concern itself with all the 92 naturally occurring elements and even those which man cares to synthesize, e.g. Pu. From the lightest, Li, a drug used in therapy, to the heaviest, U and Pu, there are probably functional uses in medicine, agriculture or industry for all elements since living things can be exposed to them all. Each element has selective potential advantage and will be used by man if not already used by biology. We are concerned here with the effects of the element in different combined forms within biology. The first point is that the selectivity of binding is related to the size of any cation and its electron affinity. We can extend Figure 5 to monovalent and trivalent ions. The selection of binding to non-metals is that $S > N > O$ going toward the top left-hand corner of this figure. Equally we can introduce steric factors. However, these rationalizations are never deeply satisfactory since the transfer and kinetics of insertion of elements can decide their fates. We could not predict the binding of $Pt(NH_3)_2Cl_2$ to guanidine bases of DNA, necessary for inhibiting tumor growth, and in fact much more Pt is bound elsewhere. Like the uses of reagents such as Tc and Gd in diagnostics, the uses of drugs must be based on observation guided by knowledge of inorganic chemistry but not on confident prediction.

X. MODELS

The original attempt to model the functional properties of metal ions in large biological molecules was largely a failure. I know this well enough since for fifteen years I attempted modelling, 1950-1965. However, the period was not wasted since the failures pointed to the sophistication of the biological ligands as described above. It has required a great effort in the analysis of proteins in particular to show the full extent of the problems. There is a temptation then to dismiss "models". I believe that this goes too far. We now know much more, and slowly but surely models become more interesting as this volume will show. Modelling is often the basis of drug action in organic chemistry, and I believe that modelling will greatly assist in the increase of the value of metal therapy in the future. It always increases our knowledge of chemistry.

The volumes of this book contain many examples of all the branches of bio-inorganic chemistry, which is relatively young and much is left to explore.

REFERENCES

The material in this article is largely covered in the first reference Fraústo da Silva, J.J.R., and Williams R.J.P. (1991). *The Biological Chemistry of the Elements*, Oxford University Press, Oxford.

Cairns-Smith, A.G., Hall, A.J., and Russell, M.J. (1992). *Origins Life Evol. Biosphere*, *22*: 161.

Cox, P.A. (1989). *The Elements*, Oxford University Press, Oxford.

Cram, D.J., (1987); Lehn, J.-M. (1987); and Pedersen, C.J. (1987). *Les Prix Nobel 1987*, Almquist and Wiksell, Stockholm, Sweden.

Eigen, M. (1960). *Z. Electrochem.*, *64*: 115.

Gray, H.B., and Malmström, B.G. (1989). *Biochemistry*, *28*: 7499.

Holm, R.H., Cuirlie, S., Weigel, J.A., and Komiya, H., (1990). *Prog. Inorg. Chem.*, *38*: 1.

Koshland, D.E. (1973). *Scientific Amer.*, *52*: 229.

Lumry, R., and Eyring, H. (1954). *J. Phys. Chem.*, *58*: 110.

Mann, S., Webb, J., and Williams, R.J.P. (1989). *Biomineralisation*, VCH Weinheim, Germany.

Mitchell, P.C.H., and Williams, R.J.P. (1962). *J. Chem. Soc.*, 4570.

Russell, M.J., Hall, A.J., and Turner, D. (1992). *Terra Nova*, *1*: 238.

Vallee, B.L., and Williams, R.J.P. (1968). *Proc. Nat. Acad. Sci. USA*, *59*: 498.

Williams, R.J.P. (1953a). *Analyst*, *78*: 586.

Williams, R.J.P. (1953b). *Biol. Rev.*, *28*, 381.

Williams, R.J.P. (1956). *Chem. Rev.*, *56*: 299.

Williams, R.J.P. (1965). *Non-haem Iron Proteins* (A. San Pietro, ed.), Antioch Press, Yellow Springs, Ohio, U.S.A..

Williams, R.J.P. (1992). I.C.C.C. Symposium, Lausanne, Switzerland.

THERMODYNAMIC INTRODUCTION TO SOLUTE-SOLUTE AND SOLUTE-SOLVENT INTERACTIONS IN BIOLOGICAL FLUIDS

by

Terence H. Lilley

Biothermodynamics Laboratory,
Chemistry Department,
The University of Sheffield,
Sheffield S3 7HF
U.K.

I. INTRODUCTION

There is currently a considerable amount of interest, much of it being driven by problems in molecular biology and environmental science, in the 'non-bonding' as well as the 'bonding' interactions which occur in condensed media. Many of these 'non-bonding' problems, or similar ones, have been addressed by solution physical chemists in the past although the language used today is rather different to that used previously; for example, a phrase which is currently popular is 'molecular recognition', but in other language this can be interpreted as the balance which is reached between the tendencies molecular species or chemical groups have to interact with each other, and their propensities to be solvated by the environment in which they are present. In other words, 'molecular recognition' can be considered to be the compromise reached between the intermolecular forces of solvation and those of solute species interaction. It is worthwhile stressing that all chemical events in condensed media are a consequence of such a balance, and to solve the problems of solution chemistry necessitates having information about both solvation and how solvated species interact with each other (Lilley, 1993; 1994).

It is easy to represent the compromise which must be reached between molecular solute solvation and solute-solute interaction, and this is done in Figure 1 for the extreme example in which the solvated species attract each other strongly and so can form a well-characterized associate. Generally, the extent to which association occurs between species depends on how well the two solute species are solvated and on how strongly they interact. The general, qualitative rule is that 'good' solvation inhibits the association and favorable solute species interaction promotes association.

Since both the solvational properties and the 'direct' interactions between solute species are involved in determining the net effects which occur in solution, a knowledge of both is required if predictions or even rationalizations of behavior are to be obtained. It is generally true, of course, that biological fluids are complex multicomponent aqueous solutions, containing a range of different types of solutes such as biopolymers (often but not always bearing charges) which can have very high molar masses and are usually polyfunctional, as well as small molecules and ions of different sorts. This complexity necessarily leads to a wide range of possible and competing interactions, and it is customary

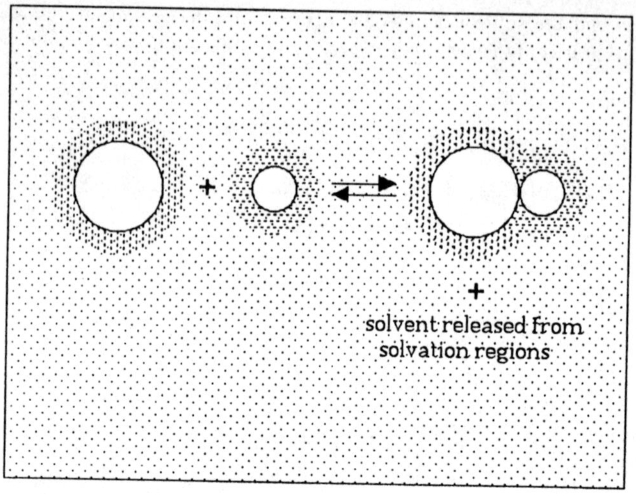

Figure 1.
Diagrammatic representation of the interaction (association) between two species in a solvent. It is assumed that association is relatively intense. The solvation regions (sometimes called co-spheres) around the solutes are indicated by hatching.

to divide these into different categories. As an example and generally speaking, if one considers say ionic equilibria, then these may be subdivided into the following four classes (Prue, 1966):

$NH_4^+ \rightleftharpoons NH_3 + H^+$	Protolytic
$Ag^+ + 2NH_3 \rightleftharpoons Ag(NH_3)_2^+$	Complex ion
$Ca^{2+} + SO_4^{2-} + 2H_2O \rightleftharpoons CaSO_4.2H_2O(s)$	Solubility
$2Ag^+ + C_6H_4(OH)_2 \rightleftharpoons 2Ag + C_6H_4O_2 + 2H^+$	Redox

Usually the solvent is excluded from the written representations of the equilibria, but it is undoubtedly true that there must be some participation of the solvent in any processes occurring in any solvent. In the examples above, for the reactions occurring in water, it is probably self-evident that there will be hydrogen-bonding with the solvent for both the ammonium ion and ammonia and, of course, the proton is known to form the well characterized hydronium ion by reaction with water (there is also evidence for further hydration of the H_3O^+ ion). The silver ion, although forming a well-defined complex with ammonia, is certainly being competed for by other species (e.g. chloride ions) as well as by the solvent. In the solubility equilibrium involving calcium sulfate, since the solid phase is a hydrate, the solvent is specifically included in the equilibrium.

The objective of this chapter is to present a brief summary of some of the thermodynamic approaches to interactions between solutes in solutions. To a large extent the approach adopted is pedagogical in tone and is illustrative rather than comprehensive, and attention will be directed towards presenting some of the bridges between thermodynamics and molecular events.

II. THERMODYNAMICS OF SOLUTIONS OF A SINGLE SOLUTE

In this section a summary of the basic thermodynamic relationships pertinent to interactions in solutions is presented. A more comprehensive description, from a different viewpoint, has been given previously (Lilley, 1988).

If we consider a solution of a single (non-electrolytic) solute, i, in a single component solvent, w, then at constant temperature (T) and pressure (p) the Gibbs energy is given by

$$G = n_w\mu_w + n_i\mu_i \qquad (1)$$

where μ_i and μ_w are the chemical potentials of the solute and solvent, respectively. It is customary to express the latter in terms of the molar Gibbs energy of the pure solvent at the same T and p and the solvent activity (a_w), this being defined most commonly (although necessarily, arbitrarily) in terms of a mole fraction ideal state. Consequently we have

$$\mu_w = \mu_w^o + RT\ln a_w = \mu_w^o + RT\ln x_w + RT\ln f_w \tag{2}$$

where x_w and f_w are, respectively, the mole fraction and the activity coefficient of the solvent. There are alternative ways of representing the solvent activity and a common way is to use osmotic coefficients. In terms of the molal osmotic coefficient (ϕ) the relationship is

$$\ln a_w = -M_w m_i \phi \tag{3}$$

where M_w is the kg molar mass of the solvent and m_i is the molality of the solute in the solution. Consequently combining equations (2) and (3) gives

$$\mu_w = \mu_w^o - RTM_w m_i \phi \tag{4}$$

The chemical potential of the solute is usually expressed in a different way to that for the solvent in that it is customary to use a hypothetical standard state for the standard chemical potential. Usually the standard state chosen is the hypothetical 1 molal state (m^o) and physically this may be considered as a state in which only solute-solvent interactions are occurring at unit molality. (If the molar scale is chosen to represent the solution, then a hypothetical 1 molar standard state may be adopted). The expression for the chemical potential is

$$\mu_i = \mu_i^o + RT\ln\{a_i/m^o\} \tag{5}$$

which may also be written as

$$\mu_i = \mu_i^o + RT\ln\{m_i/m^o\} + RT\ln\gamma_i \tag{6}$$

where now γ_i is the molal activity coefficient which is defined such that as the solution becomes more dilute in the solute then the solute activity coefficient approaches unity, i.e. as $m_i \to 0$, $\gamma_i \to 1$.

Using these definitions for the chemical potentials of the solvent and the solute, then an ideal solution is taken to be one in which the osmotic coefficient and the solute activity coefficient are both unity and so we have

$$\mu_w^{id} = \mu_w^o - RTM_w m_i \tag{7}$$

and

$$\mu_i^{id} = \mu_i^o + RT\ln\{m_i/m^o\} \tag{8}$$

Consequently, if we have an ideal solution containing 1 kg of solvent, the Gibbs energy of this ideal system is

$$G^{id} = M_w^{-1}\mu_w^{id} + m_i\mu_i^{id} = M_w^{-1}\mu_w^o - RTm_i + m_i\mu_i^o + m_iRT\ln\{m_i/m^o\} \tag{9}$$

To all intents, apart from some contributions from the 'concentration' scales used, which in turn are related to cratic entropic contributions, the ideal expression above gives a measure of solute-solvent interactions through the solute standard state term. We will return to this a little later but for the moment suffice it to say that it is not possible to get absolute measures of the Gibbs energy of the system or indeed the partial molar Gibbs energies of the components of the system; all one can determine from experiment are differences in energies.

When one considers a real molecular system then it is useful to define functions, the excess functions, which represent the differences between the properties of the real system and those of an ideal system. As an example, we can define the excess Gibbs energy of a solution containing 1 kg of solvent by

$$G^{ex} = G^{real} - G^{id} \tag{10}$$

and it is sometimes convenient to define partial molar excess quantities such as

$$\mu_i^{ex} = \mu_i^{real} - \mu_i^{id} = RT\ln\gamma_i \tag{11}$$

and

$$\mu_{\mathrm{w}}^{\mathrm{ex}} = \mu_{\mathrm{w}}^{\mathrm{real}} - \mu_{\mathrm{w}}^{\mathrm{id}} = RTM_{\mathrm{w}}m_{\mathrm{i}}(1 - \phi) \tag{12}$$

and so the excess Gibbs energy may be expressed as

$$G^{\mathrm{ex}} = RTm_{\mathrm{i}}\{(1 - \phi) + \ln\gamma_{\mathrm{i}}\} \tag{13}$$

It is, of course, possible to define other thermodynamic functions as appropriate derivatives of the Gibbs energy and similarly other excess properties can be defined through the excess Gibbs energy. As examples we have for the excess enthalpy and excess volume respectively

$$H^{\mathrm{ex}} = \left(\frac{\partial(G^{\mathrm{ex}}/T)}{\partial T^{-1}}\right)_{p,\ composition} = Rm_{\mathrm{i}}\left(\frac{\partial\{(1-\phi)+\ln\gamma_{\mathrm{i}}\}}{\partial T^{-1}}\right)_{p,\ composition} \tag{14}$$

and

$$V^{\mathrm{ex}} = \left(\frac{\partial G^{\mathrm{ex}}}{\partial p}\right)_{T,\ composition} = RTm_{\mathrm{i}}\left(\frac{\partial\{(1-\phi)+\ln\gamma_{i}\}}{\partial p}\right)_{T,\ composition} \tag{15}$$

These excess properties can be related to other properties which come fairly readily from experiment such as the enthalpy of dilution and the density of solutions. As an example, the excess volume is linked into the apparent molar volume of the solute in the following way.

The volume of a solution containing 1 kg of solvent may be written as

$$V = M_{\mathrm{w}}^{-1}V_{\mathrm{w}}^{\mathrm{id}} + m_{\mathrm{i}}V_{\mathrm{i,app}} = M_{\mathrm{w}}^{-1}V_{\mathrm{w}}^{\mathrm{o}} + m_{\mathrm{i}}V_{\mathrm{i,app}}$$

or

$$V_{\mathrm{i,app}} = \{[V - M_{\mathrm{w}}^{-1}V_{\mathrm{w}}^{\mathrm{o}}]/m_{\mathrm{i}}\} \tag{16}$$

where $V_{\mathrm{i,app}}$ is the apparent molar volume of the solute (this is often given the symbol ϕ_{V}). This can be compared with the formal expression for the partial molar volume of the solute,

$$V_{\mathrm{i}} = \left(\frac{\partial \mathrm{V}}{\partial m_{\mathrm{i}}}\right) \tag{17}$$

Using this expression in association with those given in equations (15) and (16) gives after a little manipulation

$$V_{\mathrm{i,app}} = V_{\mathrm{i}}^{\mathrm{o}} + \left(\frac{V^{\mathrm{ex}}}{m_{\mathrm{i}}}\right) \tag{18}$$

and

$$V_{\mathrm{i}} = V_{\mathrm{i}}^{\mathrm{o}} + \left(\frac{\partial V^{\mathrm{ex}}}{\partial m_{\mathrm{i}}}\right) \tag{19}$$

In a similar way we can obtain expressions for enthalpic quantities but in this instance only relative terms can be obtained because of the usual problem of measuring any energy terms absolutely. The enthalpic terms relative to those of the standard state terms are given by the following analogs of equations (18) and (19):

$$H_{\mathrm{i,app}} - H_{\mathrm{i}}^{\mathrm{o}} = \left(\frac{H^{\mathrm{ex}}}{m_{\mathrm{i}}}\right) \tag{20}$$

and

$$H_{\mathrm{i}} - H_{\mathrm{i}}^{\mathrm{o}} = \left(\frac{\partial H^{\mathrm{ex}}}{\partial m_{\mathrm{i}}}\right) \tag{21}$$

The first of these is usually called the relative apparent molar enthalpy and is frequently given the symbol ϕ_L.

All of the expressions given in this section are readily extended to deal with solutions containing ionic species. However, even for solutions containing a single electrolyte, complications and limitations arise because of the demands of overall charge neutrality in solutions. As an illustration of this, if we have a solution of an electrolyte, MX, comprised of the ions M^+ and X^-, then although formally one can write analogs of equation (11) for each of the ionic species, in practice one must deal with the sum of the excess chemical potential contributions.

III. EXPRESSIONS OF NON-IDEALITY

The equations in the above section give formal expressions for some thermodynamic quantities, but links with molecular events are required. In what follows initially we will continue to consider a solution containing only one non-electrolytic solute and then, a little later, will consider rather more complex situations.

In solutions, it is usual to adopt standard states which are directly or indirectly based on Henry's Law and what this means in terms of the molecular situation is illustrated in Figure 2.

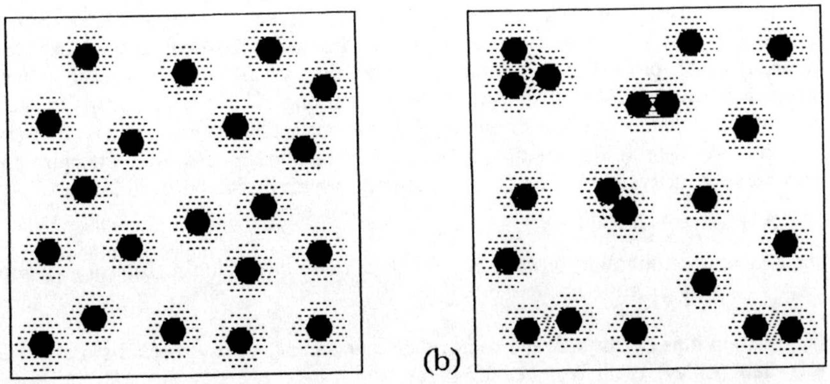

Figure 2.
Schematic representation of (a) an ideal (dilute) solution and (b) a non-ideal solution. The solute is shown as filled circles and the solvation region is shown hatched. The ideal state corresponds to the situation where the solvation regions do not interpenetrate. The non-ideal situation shown in (b) indicates that, in not too concentrated solutions, pairwise interactions between solvated solute molecules are more prevalent than triplet and higher order interactions.

It is apparent from this that a solute behaves ideally when the solvated solute species are sufficiently far apart so that their solvation regions do not inter-penetrate. This clearly will tend to be the situation in very dilute solutions and will be less so as the concentration of solute increases. It is apparent, at least for situations where cooperative interactions between solutes do not occur, that as the concentration of the solution increases there will be an increased tendency for solute molecules to be in proximity to each other. Initially, pairs of molecules will have an increasing propensity to be near each other and as the concentration increases further, triplets, quartets, and higher aggregates will be present. These features can be recognized and formulated by expressing the excess functions as polynomials in solute concentrations. For a solution containing only one solute the expression for the excess Gibbs energy is (Scatchard, 1949; Friedman, 1960)

$$G^{ex} = \sum_{i\geq 2} g_{ni} m_i^n = g_{ii} m_i^2 + g_{iii} m_i^3 + \cdots \qquad (22)$$

where g_{ii}, for example, is a (homotactic, i.e. monitors interactions between like solute species) free energetic coefficient representing the interaction of two solvated species with each other. An analogous expression can be written for solutions containing more than one solute and for example when two solutes, i and j, are present

$$G^{ex} = g_{ii}m_i^2 + g_{jj}m_j^2 + 2g_{ij}m_im_j + g_{iii}m_i^3 + g_{jjj}m_j^3 + 3g_{iij}m_i^2m_j + 3g_{ijj}m_im_j^2 + \text{higher order terms} \quad (23)$$

Now we have as well as homotactic interactions, heterotactic interactions, i.e. interactions between unlike solute species. The activity coefficient of say the i'th species in these two situations is

$$\ln\gamma_i = (2g_{ii}m_i + 3g_{iii}m_i^2 + \cdots)/RT \quad (24)$$

and

$$\ln\gamma_i = (2g_{ii}m_i + 2g_{ij}m_j + 3g_{iii}m_i^2 + 6g_{iij}m_im_j + 3g_{ijj}m_j^2 + \text{higher order terms})/RT \quad (25)$$

When one is dealing with electrolyte solutions then, because of the long-range nature of the interactions, these expressions must be modified. Some comments on this aspect of solutions will be made below.

It is apparent from what has been said that deviations from ideality arise from solute species being in proximity to each other and that such situations can be expressed in thermodynamic properties in terms of interaction coefficients. However, when the interaction between say, two species is relatively strong, then it is customary in chemistry to treat the situation in terms of an equilibrium constant formalism. It should be stressed that the two approaches are simply related to each other and, for example, for the homotactic attractive interaction of two species A, we may write the equilibrium

$$A + A \rightleftharpoons A_2 \quad (26)$$

and relate the equilibrium constant, K_m, to the pairwise interaction coefficient g_{AA} through

$$g_{AA} = -RTK_m \quad (27)$$

Interrelationships of this sort and some other cognate matters have been discussed at length elsewhere (Lilley, 1988; Wood et al., 1978).

IV. EQUILIBRIUM CONSTANTS

The majority of the investigations which address interactions between species in solution do so using the equilibrium constant concept. Although this is not without its difficulties and limitations, some of which have been referred to previously (Lilley, 1988; Wood et al., 1978), it is nevertheless worthwhile spending a little time addressing the concept.

If we consider that the following equilibrium is occurring in solution

$$\nu_A A + \nu_B B + \nu_C C + \cdots\cdots \rightleftharpoons \nu_X X + \nu_Y Y + \nu_Z Z + \cdots\cdots \quad (28)$$

where ν_i is the stoichiometric number of the substance i in the chemical reaction, different equilibrium constants can be defined and these are related to each other but are generally not numerically the same. There are some complexities with regard to what are usually termed 'standard' equilibrium constants which will not be considered here but have been addressed fairly recently by Ewing et al. (1994). The objective here is simply to illustrate the importance of the choice of concentration scale when expressing equilibrium constants. Throughout what follows, T and p are assumed to be constant.

A. Solution composition expressed using mole fractions

Under these circumstances the mole fraction equilibrium constant (K_x) is given by

$$K_x = \frac{(a_X)^{\nu_X}(a_Y)^{\nu_Y}(a_Z)^{\nu_Z}\cdots}{(a_A)^{\nu_A}(a_B)^{\nu_B}(a_C)^{\nu_C}\cdots} \quad (29)$$

where the activities (a_i) of the various reacting solute species are understood to be mole fraction activities and so this equation can be written as

$$K_x = \left(\frac{(x_X)^{\nu_X} (x_Y)^{\nu_Y} (x_Z)^{\nu_Z} \ldots}{(x_A)^{\nu_A} (x_B)^{\nu_B} (x_C)^{\nu_C} \ldots} \right) \times \text{(mole fraction activity coefficient term)} \quad (30)$$

If the solution is sufficiently dilute, then the second parenthetical term approaches unity and we then have what is sometimes termed a semi-ideal solution, i.e. one in which all departures from ideality are taken to come from associative processes.

$$K_x \rightarrow \left(\frac{(x_X)^{\nu_X} (x_Y)^{\nu_Y} (x_Z)^{\nu_Z} \ldots}{(x_A)^{\nu_A} (x_B)^{\nu_B} (x_C)^{\nu_C} \ldots} \right) \quad (31)$$

B. Solution composition expressed using molalities

The analog of equation (31) is

$$K_m \rightarrow \left(\frac{(m_X)^{\nu_X} (m_Y)^{\nu_Y} (m_Z)^{\nu_Z} \ldots}{(m_A)^{\nu_A} (m_B)^{\nu_B} (m_C)^{\nu_C} \ldots} \right) \quad (32)$$

and it is apparent that these two equilibrium constants are generally not the same since molality and mole fraction are related by

$$x_i = \frac{m_i}{M_w^{-1} + \sum_{\text{all j}} m_j} \quad (33)$$

As a consequence, the two equilibrium constants, K_x and K_m, are linked through

$$K_m = K_x \left(M_w^{-1} + \sum_{\text{all j}} m_j \right)^{\nu},$$

$$\text{with } \nu = (\nu_X + \nu_Y + \nu_Z + \ldots) - (\nu_A + \nu_B + \nu_C + \ldots) \quad (34)$$

It is only for the simplest possible situation where $\nu = 0$ that the two equilibrium constants are the same numerically. In all other instances, there are significant differences between them. In very dilute solutions, then the molalities are all small compared to the number of moles of solvent present and so equation (34) becomes

$$K_m = K_x M_w^{-\nu} \quad (35)$$

It is worth noting the numerical differences in two illustrative examples in which the solvent is water. For the complex ion equilibrium

$$Ag^+ + 2NH_3 \rightleftharpoons Ag(NH_3)_2^+ \quad (36)$$

we have $\nu = -2$ and so the mole fraction and molal equilibrium constants are numerically related by $K_m/K_x = M_w^2 = 3.25 \times 10^{-4}$. The van't Hoff isotherm,

$$\Delta G^o = -RT \ln K$$

can be applied to both types of equilibrium constant to give

$$\Delta G_x^o = -RT \ln K_x \quad (37)$$

and

$$\Delta G_m^o = -RT \ln K_m \quad (38)$$

so that generally we have

$$\Delta G_m^o - \Delta G_x^o = \nu RT \ln M_w \quad (39)$$

It is worth stressing that the differences in equilibrium constants which arise from the use of different concentration scales are entropic in origin and these apparent entropic contributions are by no means trivial.

The thermodynamic links between a standard free energy change for a reaction and the corresponding enthalpy and entropy changes are:

$$\Delta H^{o} = \left(\frac{\partial(\Delta G^{o} / T)}{\partial T^{-1}} \right)_p \tag{40}$$

$$\Delta S^{o} = -\left(\frac{\partial \Delta G^{o}}{\partial T} \right)_p \tag{41}$$

The expression for the entropic contribution arising simply from the scale differences is

$$\Delta S^{o}_{m} - \Delta S^{o}_{x} = -\nu R \ln M_{w} \tag{42}$$

For the particular example above, this entropy difference is -67 J K^{-1} mol^{-1}, which is by no means trivial.

As another example we consider the equilibrium

$$MgSO_4 \rightleftharpoons Mg^{2+} + SO_4^{2-} \tag{43}$$

where we have $\nu = +1$ and so the equilibrium constants are numerically related by $K_m/K_x = M_w^{-1} = 55.51$ and the entropic difference arising from the choice of scale is in this instance

$$\Delta S^{o}_{m} - \Delta S^{o}_{x} = +33 \text{ J K}^{-1} \text{ mol}^{-1} \tag{44}$$

C. Solution composition expressed using molarities

For completeness we consider the quite common situation where the solution composition is expressed using molarities.

The expression corresponding to equation (31) is

$$K_c \rightarrow \left(\frac{(c_X)^{\nu_X} (c_Y)^{\nu_Y} (c_Z)^{\nu_Z} \ldots}{(c_A)^{\nu_A} (c_B)^{\nu_B} (c_C)^{\nu_C} \ldots} \right) \tag{45}$$

and since in very dilute solutions the molality of a species is related to its molarity by

$$c_i = m_i d^{o}_{w} \tag{46}$$

d^{o}_{w} being the density of the pure solvent, the relationship

$$K_c/K_m = (d^{o}_{w})^{\nu} \tag{47}$$

is readily established. It is apparent that there are also differences between the standard state changes of thermodynamic properties, and as examples

$$\Delta G^{o}_{c} - \Delta G^{o}_{m} = -\nu RT \ln d^{o}_{w} \tag{48}$$

$$\Delta S^{o}_{c} - \Delta S^{o}_{m} = \nu R \ln d^{o}_{w} \tag{49}$$

If the solvent is water, at ambient temperatures (~ 298K) the difference between the molar and molal changes is very small and when $\nu = 1$ for example, the difference $\Delta S^{o}_{c} - \Delta S^{o}_{m} = -0.02$ J K^{-1} mol^{-1}. For other solvents or for water at other temperatures, more significant differences result and for water at 100°C the difference is -0.35 J K^{-1} mol^{-1} whereas for ethanol as a solvent at 25°C the difference is 2.0 J K^{-1} mol^{-1}.

D. Equilibrium constants and equilibrium quotients

The usual presumption when an equilibrium constant is formulated is that it is indeed constant. Under most circumstances, if the classical expression - using concentrations of the species participating in the equilibrium - is used, then constancy is not observed because of what are usually taken to be non-ideal contributions. To allow for these and irrespective of the concentration scale adopted, 'thermodynamic' equilibrium constants are defined and as an example,

$$K_c = \left(\frac{(a_X)^{\nu_X}(a_Y)^{\nu_Y}(a_Z)^{\nu_Z}\ldots}{(a_A)^{\nu_A}(a_B)^{\nu_B}(a_C)^{\nu_C}\ldots}\right) = \left(\frac{(c_X)^{\nu_X}(c_Y)^{\nu_Y}(c_Z)^{\nu_Z}\ldots}{(c_A)^{\nu_A}(c_B)^{\nu_B}(c_C)^{\nu_C}\ldots}\right)\left(\frac{(f_X)^{\nu_X}(f_Y)^{\nu_Y}(f_Z)^{\nu_Z}\ldots}{(f_A)^{\nu_A}(f_B)^{\nu_B}(f_C)^{\nu_C}\ldots}\right) \quad (50)$$

where the activity coefficient (f_i) terms for the component species are collected together as are the molarities of the components of the assumed equilibrium. The equilibrium quotient (Q) is the first parenthetical term on the right of equation (50) and corresponds to the classical equilibrium constant when the activity coefficient term is unity. This term will also be constant if the activity coefficient term is maintained constant by some means. For equilibria involving ions, it is common for the quotient to be determined experimentally from solutions containing an 'inert' electrolyte, as well as the species of interest. The idea here is that the concentration of the added electrolyte is much greater than that of any of the species participating in the equilibrium and so the activity coefficient term, although not unity, is fixed. A popular medium used in the study of ionic equilibria is 3 molar sodium perchlorate solutions and much of the information we have on ionic equilibria, and particularly where there is a series of equilibria such as those which occur between transition metal ions and ligands, comes from studies in which this medium is used.

It is seldom that one has all of the information necessary to evaluate independently the activity coefficient terms and, almost invariably, thermodynamic equilibrium constants are obtained from experimentally determined equilibrium quotients in conjunction with some extrapolation method. To illustrate the sort of thing which is done, suppose we consider the equilibrium

$$M^{2+} + L^- \rightleftharpoons ML^+ \quad (51)$$

for which the molal equilibrium constant is

$$K_m = \left(\frac{m_{ML^+}}{m_{M^{2+}}m_{L^-}}\right)\left(\frac{\gamma_{ML^+}}{\gamma_{M^{2+}}\gamma_{L^-}}\right) = Q\left(\frac{\gamma_{ML^+}}{\gamma_{M^{2+}}\gamma_{L^-}}\right) \quad (52)$$

Assuming that the quotient is obtainable from experiment, then to determine K_m an expression for the activity coefficient term is needed. In this instance, an approach based on the Debye-Hückel theory of electrolytes is possible although the limiting law expression

$$\ln \gamma_i = -z_i^2 \alpha_{DH} I^{1/2} \quad (53)$$

will only very rarely be applicable. More usually an expression of the form

$$\ln \gamma_i = -z_i^2 \alpha_{DH} I^{1/2}/[1 + A_i I^{1/2}] + B_i I \quad (54)$$

has been used in the past although in recent years it has become more customary to use the approach suggested and popularized by Pitzer, in which an alternative expression for the first term on the right hand side of this equation is used. In equations (53) and (54), α_{DH} is the Debye-Hückel limiting law constant, I is the molar ionic strength and A_i and B_i are adjustable parameters which depend on the species present in the solution. Very often when treating data the term A_i has been put equal to unity (there is some compensation between A_i and B_i) to give the following expression

$$\ln \gamma_i = -z_i^2 \alpha_{DH} I^{1/2}/[1 + I^{1/2}] + B_i I \quad (55)$$

Use of this in the particular equilibrium constant expression [equation (52)] gives

$$\ln K_m = \ln Q + \ln\left(\frac{\gamma_{ML^+}}{\gamma_{M^{2+}}\gamma_{L^-}}\right) =$$

$$\ln Q + 4\alpha_{DH} I^{1/2}/[1 + I^{1/2}] + (B_{ML^+} - B_{M^{2+}} - B_{L^-})I \quad (56)$$

$$= \ln K_m' + (B_{ML^+} - B_{M^{2+}} - B_{L^-})I \quad (57)$$

where K_m' is calculable from the first two terms in equation (56). It is apparent that this expression can be used to obtain the thermodynamic equilibrium constant simply by linear extrapolation. This method relies on the extended Debye-Hückel expression and the B terms in equation (57) being sufficient to represent the activity coefficients. It would seem from perusal of the literature that reliable values for equilibrium constants are obtained in this way only if the ionic strength is less than about 0.1 molar in water at ambient temperatures. The same conceptual approach can be used and the ionic strength range extended if the alternative expression for the Debye-Hückel leading term suggested by Pitzer is adopted and terms other than those linear in ionic strength are incorporated.

A connection can be made between the expressions given for the B terms in equations (54)-(56) and those used for non-electrolyte-non-electrolyte interactions (see section III). If for example we consider a solution of a simple, unassociated electrolyte (NX) of the 1:1 charge type then the contributions to non-ideality from non-Debye-Hückel sources are of the form

$$\ln \gamma_N (\text{nonDH}) = (2g_{NN}m_N + 2g_{NX}m_X)/RT \tag{58}$$

where higher order terms are omitted. This can be written as

$$\ln \gamma_N (\text{nonDH}) = 2I(g_{NN} + g_{NX})/RT \tag{59}$$

and so the similarity with the B term in equation (54) is apparent.

V. CONCLUDING COMMENTS

It is apparent from what has been said so far that when addressing solutions containing solutes which can interact with each other, there are several problems which need to be considered. When one is dealing with solutions which only contain non-electrolytes, then at the present time, it is not possible to make any a priori quantitative predictions about the intensity of solvation of these solutes or indeed about their propensity for interaction with each other. There are qualitative rules which have been established as a result of experience, and there are now tolerably good, experimentally based, predictive schemes for the thermodynamic properties of aqueous solutions of molecules containing a range of molecular functionalities. For example, within a fair degree of accuracy, the molar volumes of many solutes in water can be predicted and the prediction of the energetics of interaction of small peptides with each other is also possible (Lilley, 1988). All too often, however, the phrase "hydrogen-bonding" is invoked to 'explain' any phenomenon occurring in solution.

When it comes to systems containing ions, then the thermodynamic properties of solutions containing both single and mixed strong electrolytes have been rationalized using the semi-empirical approach introduced by Pitzer over the past several years. The interactions between non-electrolytes and electrolytes are in approximately the same situation as those for non-electrolyte-non-electrolyte interactions. It is possible to rationalize the solubility of say a sparingly soluble non-electrolyte, such as a gas, as a function of added electrolyte concentration simply by using an extended form of equation (22). We will not do this here but suffice it to say that a rationalization of the empirical Setschenow expression results. If we consider the interactions of cations with ligands in water, then of course a wide variety of interactions can occur and there must necessarily be a competition between ligands in solutions and the solvent for sites on the metal ion. At the moment, there does not seem to be any completely satisfactory way of predicting in a quantitative way the equilibrium constants for binding of ligands to ions and how these change with solvent or indeed temperature. This is a difficult problem but it seems certain that considerable progress will be made over the next several years as computational power increases.

When one deals with interactions between high-charged density ions in water, then a good many years ago, Bjerrum (1926) suggested that association between ions occurs and he formulated an expression for the association constant using a simple electrostatic approach. The approach he used continues to have some validity and its conceptual basis has been used in recent years (Honig et al., 1993) to address several problems of a biological nature. The physical idea behind the approach is illustrated in Figure 3 and we ask the question "If we consider a system of hard-sphere ions which are in a structureless and non-solvating solvent of fixed permittivity, then what is the probability that the ions will be in proximity and more particularly in this instance what is the probability of ions being between the distance a (the distance of closest approach of the ions) and the distance d (the association distance)?"

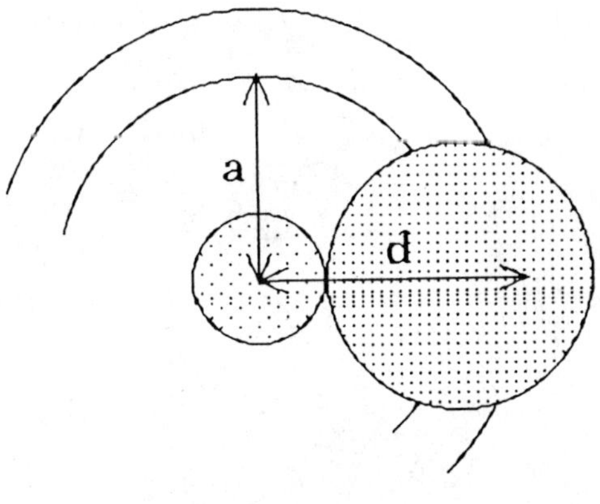

Figure 3.
The means of defining an ion pair. The ions are indicated by the shaded circles. The distance *a* corresponds to the distance of closest approach of the ions and the distance *d* is the (arbitrary) distance after which association is taken to cease.

The answer is, in terms of a molar association constant

$$K_c = 4\pi L_A \int_a^d \exp\left\{\left(z_+ z_- e^2\right) / \varepsilon_0 \varepsilon rkT\right\} r^2 dr \tag{60}$$

Bjerrum chose the association distance as being the distance at which the Coulombic energy is equal to $2kT$ but it should be stressed that this is quite arbitrary and generally, of course, it is never possible to be absolutely sure of the distance when association begins.

The Bjerrum idea has been discussed many times in the past and we will not pursue it here except to say that, notwithstanding its simplicity, it does seem able to predict semi-quantitatively the association constant between ions in a relatively large number of instances. It is straightforward to extend the formalism to deal with other situations in which, for example, non-electrostatic (e.g. Van der Waals, expressed through a Lennard-Jones potential) interactions are included and the expression resulting is

$$K_c = 4\pi L_A \int_a^d \exp\left[\left\{\left(z_+ z_- e^2\right) / \varepsilon_0 \varepsilon rkT\right\} - \left\{Ar^{-12} + Br^{-6}\right\}\right] r^2 dr \tag{61}$$

where *A* and *B* are Lennard-Jones parameters for the species considered. It is clear from this that the presence of the extra attractive interactions will increase the association constant, which is not too surprising.

There have not as yet been any successful attempts to include solvation in quantitative calculations, although there have been attempts (Marcus, 1983) to deal with this indirectly by incorporating dielectric saturation effects near the surface of ions. Again, since the permittivity is reduced, this leads to increases in the association. However, even though quantitative predictions of the effects of solvation on ion association are lacking, for some years now the idea that one can have different types of ion pair has been prevalent and these are illustrated in Figure 4.

It is generally agreed (Eigen and Tamm, 1962) that ions can pass through the whole series of associates and, for example, for magnesium sulfate (Atkinson and Petrucci, 1966; Millero and Masterton, 1974) all of the stages in the associations

$$Mg^{2+}(aq.) + SO_4^{2-}(aq.) \rightleftharpoons Mg^{2+}.H_2O.H_2O.SO_4^{2-}(aq.)$$
$$Mg^{2+}.H_2O.H_2O.SO_4^{2-}(aq.) \rightleftharpoons Mg^{2+}.H_2O.SO_4^{2-}(aq.) + H_2O$$
$$Mg^{2+}.H_2O.SO_4^{2-}(aq.) \rightleftharpoons Mg^{2+}.SO_4^{2-}(aq.) + H_2O$$

have been clearly delineated from a range of experimental techniques.

The general picture which seems to emerge on the associative interactions which can occur between solutes is that although an enormous amount of experimental work has been

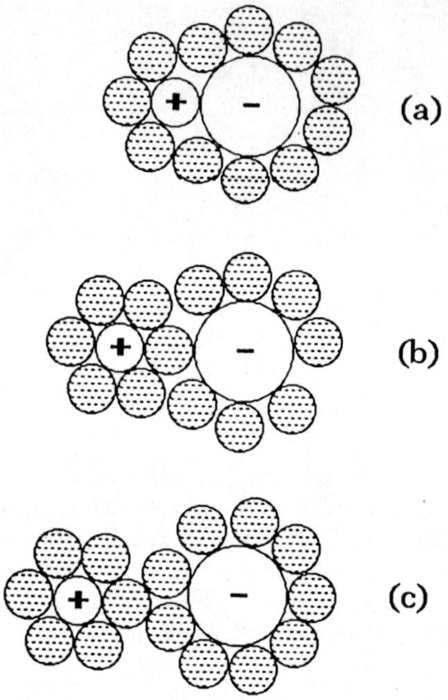

Figure 4.
Diagrammatic representations of different types of ion pairs: (a) contact; (b) solvent-shared; (c) solvent-separated. The ions are shown as being singly charged although generally multiply charged species are more common, at least when water is the solvent. The solvent which is peripheral to the ions is shown as the shaded circles.

done, we are still at the stage where it is difficult to make predictions which have any chance of quantitative success. In particular, it does seem that the influence of the solvent on the interactions between species in solution needs to be studied, since it bears on many areas of chemistry and biological processes.

REFERENCES

Atkinson, G., and Petrucci, S. (1966). *J. Phys. Chem.*, *70*: 3122.

Bjerrum, N. (1926). *Kgl. danske Videnskab. Selskab, Mat.-fys. Medd.*, *7*: No. 9.

Eigen, M., and Tamm, K. (1962). *Z. Elektrochem.*, *66*: 93; 107.

Ewing, M.B., Lilley, T.H., Olofsson, G.M., Rätzsch, M.T., and Somsen, G. (1994). *Pure Appl. Chem.*, *66*: 533.

Friedman, H.L. (1960). *J. Chem. Phys.*, *32*: 1351.

Honig, B., Sharp, K., and Yang, A.-S. (1993). *J. Phys. Chem.*, *97*: 1101.

Lilley, T.H. (1988). *Biochemical Thermodynamics*, 2nd ed. (M.N. Jones, ed.), Elsevier, Amsterdam, Chapter 1.

Lilley, T.H. (1993). *Pure Appl. Chem.*, *65*: 2551.

Lilley, T.H. (1994). *Pure Appl. Chem.*, *66*: 429.

Marcus, Y. (1983). *Z. Naturforsch.*, *38A*: 247.

Millero, F.J., and Masterson, W.L. (1974). *J. Phys. Chem.*, *78*: 1287.

Prue, J.E. (1966). *Ionic Equilibria*, Pergamon Press, Oxford, p. 1.

Scatchard, G. (1949). *Chem. Rev.*, *44*: 7.

Wood, R.H., Lilley, T.H., and Thompson, P.T. (1978). *J. Chem. Soc., Faraday Trans. I*, *74*: 1301.

FACTORS INFLUENCING METAL ION AFFINITIES

by

R. Bruce Martin

Chemistry Department,
University of Virginia,
Charlottesville, Virginia 22903
U.S.A.

I. IONIC RADII

Like size usually plays a greater role than identical charge in promoting metal ion substitutions in biology. Table 1 presents effective ionic radii for many metal ions (Shannon, 1976). For any metal ion the radius increases with coordination number, since the greater number of bonds weakens the strength of any one bond. The radius of the most common coordination number appears in italics in Table 1. The alkali and alkaline earth metal ions exhibit variable coordination numbers without strong directionality in bonding. Because they are of similar size, Ca^{2+} and Na^{+} of differing charges often interchange in biological systems. Ba^{2+} is poisonous because of its antagonism to K^{+} of similar radius and not to Ca^{2+} of identical charge but smaller radius. Ba^{2+} is a muscle poison for which one antidote is intravenous infusion of K^{+}.

II. Ca^{2+} VERSUS Mg^{2+} (Martin, 1984)

To many there are insufficient fundamental differences between Ca^{2+} and Mg^{2+} to explain their widely differing roles in biology. There are, however, ample ways to account for the behavior differences between Ca^{2+} and Mg^{2+} observed in biochemical systems.

(i) Ca^{2+} favors a higher-fold coordination number than Mg^{2+}. The favored coordination number for Mg^{2+} is six-fold octahedral. Ca^{2+} exhibits a greater diversity of coordination numbers, with eight- and six-fold coordination the most common with small ligands. Since coordination in both metal ions is mainly ionic, the greater coordination numbers for Ca^{2+} derive ultimately from its larger size.

(ii) Ca^{2+} is substantially larger than Mg^{2+}. In the same coordination number, the effective ionic radii of Ca^{2+} are at least 23 pm greater than those of Mg^{2+} (Table 1). If the most favored coordination numbers are considered, Ca^{2+} (eight-fold) possesses a 40 pm greater effective ionic radius than Mg^{2+} (six-fold).

(iii) Ca^{2+} to ligand donor atom bond distances vary more than those of Mg^{2+}. Bond distances to oxygen donor atoms typically range from only 20 to 21 pm for Mg^{2+} and a broader 23 to 26 pm for Ca^{2+}. Thus Mg^{2+} prefers to form well-defined six-coordinate octahedral complexes with precise bond lengths and Ca^{2+} forms looser complexes of higher and more variable coordination number, without directionality, and with variable bond lengths.

Table 1

Effective Ionic Radii in Picometers

Ion	Coordination number					
	4	5	6	7	8	9
Be^{2+}	*27*		45			
Al^{3+}	39	48	*54*			
Cr^{3+}			*62*			
Ga^{3+}	47	55	*62*			
Fe^{3+}	49	58	*64*[a]		78	
Ni^{2+}	55	63	*69*			
Mg^{2+}	57	66	*72*		89	
Cu^{2+}	57	65	73			
Co^{2+}	58	67	*74*		90	
Zn^{2+}	*60*	68	74		90	
Li^{+}	59		76		92	
Fe^{2+}	63		*78*[b]		92	
Mn^{2+}	66	75	*83*	90	96	
Lu^{3+}			86		*98*	103
Gd^{3+}			94	100	*105*	111
Cd^{2+}	78	87	*95*	103	110	
Ca^{2+}			100	106	*112*	118
La^{3+}			103	110	116	*122*
Na^{+}	99	100	*102*	112	118	124
Sr^{2+}			118	121	126	131
Pb^{2+}	98		119	123	129	135
Ba^{2+}			135	138	142	147
K^{+}	137		138	146	*151*	155
Tl^{+}			150		159	
Rb^{+}			152	156	161	163
Cs^{+}			167		174	178

[a] High spin. Low spin value is 55. [b] High spin. Low spin value is 61.
From Shannon (1976).

(iv) Though both ions favor coordination at oxygen donor ligands, Mg^{2+} binds more strongly than Ca^{2+} to small ligands with nitrogen donors. To the pair of related tridentate ligands oxydiacetate and iminodiacetate, Ca^{2+} binds more strongly to the former and Mg^{2+} more strongly to the latter N-donor ligand. The stability constant for Mg^{2+} binding is greater to iminodiacetate than to oxydiacetate while for Ca^{2+} the reverse is true. Mg^{2+} binds to four nitrogen donors in chlorophyll, while Ca^{2+} does not.

(v) Ca^{2+} is more comfortable than Mg^{2+} in binding to large multidentate, anionic ligands. Although Mg^{2+} binds more strongly than Ca^{2+} to iminodiacetate, Ca^{2+} binds more strongly to nitrilotriacetate, EGTA, and EDTA. Ca^{2+} binds to EGTA a striking $10^{5.6}$ times more strongly than does Mg^{2+}. In all three ligands there are steric requirements that the larger Ca^{2+} fulfills better than Mg^{2+}. To nucleoside phosphates Mg^{2+} binds slightly more strongly, but within cells there is insufficient free Ca^{2+} to bind. Within cells, Mg^{2+} is associated with phosphates and Ca^{2+} is bound tightly by proteins.

The equilibrium picture shows Mg^{2+} forming more restrictive complexes, usually hexacoordinate with precise bond lengths, favoring smaller chelate groupings, and more readily accepting nitrogen donors. In contrast, Ca^{2+} forms less well-defined complexes with eight, six and other fold coordination with variable bond lengths and favors larger multidentate anionic ligands. These equilibrium stability and structural differences account for differentiation between Ca^{2+} and Mg^{2+} in biological systems.

(vi) Ca^{2+} undergoes substitution at least 10^3 times faster than Mg^{2+} (see Part Three, Chapter 1-B). The slower rate for Mg^{2+} probably results from a more limited capability to aid substitution by reducing or enlarging its coordination sphere. Thus, from both static and dynamic viewpoints, Mg^{2+} forms the more prescribed and tighter complexes.

(vii) Insolubilities of their common phosphate salts provide only a weak differentiation between Mg^{2+} and Ca^{2+}. Inorganic Mg^{2+} phosphates are up to ten times more soluble than Ca^{2+} salts of similar structure. Both metal ions form soluble salts with organic phosphates. The main difference between Mg^{2+} and Ca^{2+} is that, owing to its small size, Mg^{2+} fails to form a structure like the most insoluble of the Ca^{2+} phosphates, hydroxyapatite, $Ca_5(PO_4)_3OH$.

III. STABILITY SEQUENCES

From the many studies on stability constants one finds the order of metal ion stabilities dependent on the ligand. Increasing metal ion stabilities follow the orders (Martell and Smith, 1974-1982):

Glycine, Ca < Mg < Mn < Fe, Cd, Pb < Co, Zn < Ni < < CH_3Hg^+, Cu < < Hg
1,2-Diaminoethane, Mg < < Mn < < Fe < Cd, Co, Zn < < Pb, Ni < CH_3Hg^+ < < Cu < < < < Hg

The ion Ca^{2+} does not form stable amine complexes. Except for methyl mercury, CH_3Hg^+, all the metal ions carry two positive charges. Owing to the strongly chelating bidentate ligands, glycine (gly) and 1,2-diaminoethane (en), CH_3Hg^+ with only a single strong binding site is at a competitive disadvantage in the above series. To a unidentate ligand CH_3Hg^+ binds more strongly than all of the above metal ions except Hg^{2+}.

In the above series each inequality sign stands for an approximate 10-fold increase in stability constant. The two series gly and en are similar, their major difference being a stability constant span from Mg^{2+} to Hg^{2+} of 10^8 for glycine and 10^{14} for en. Generally the increment between metal ions increases on passing from O < N < S donor atoms. The presence of a sulfur donor promotes Cd^{2+} and Pb^{2+} to higher positions than in the above series. The order of increasing sulfur binding strengths is $Zn^{2+} < Cd^{2+} < Pb^{2+} < CH_3Hg^+ < Hg^{2+}$. Sulfhydryl group interactions are the main mode of toxicity of the heavy metal ions (see Part Four, Chapter 2-B). These general bonding features are expressed in the concept of the stability ruler.

IV. STABILITY RULER

For a single ligand the Irving-Williams stability sequence of dipositive metal ions is Mg < Mn < Fe < Co < Ni < Cu > Zn, invariant of ligand. The uniformly progressive part of this sequence from Mg^{2+} to Cu^{2+} defines a *stability ruler* against which more variable metal ions may be compared (Martin, 1986; Martin, 1987). The stability ruler appears across the top of Table 2. For several ligand donor sets Table 2 shows the relative binding strengths of the variable dipositive metal ions Zn^{2+}, Cd^{2+}, Pb^{2+}, and Hg^{2+}. Their placement in 2 corresponds to their stability constants compared with those of the metal ions that define the ruler. The entries for glycine and en agree with the series in the previous section. With most of the metal ions in Table 2, histidine will be tridentate (Martin, 1979b; Sundberg and Martin, 1974). The increment between metal ions and hence the length of the stability ruler increases with the substitutions O < N < S. The ruler length appears in the last column of Table 2 in log K units for binding to each ligand. Thus oxygen donor atoms are least discriminating among metal ions, nitrogen donors intermediate, and sulfur donors most discriminating.

Not only do Co^{2+} and Zn^{2+} display nearly identical radii for the same coordination number in Table 1, they also exhibit similar stabilities for all ligands in Table 2, except for those involving sulfur and hydroxide. The similarities allow the facile and useful substitution of Co^{2+} for Zn^{2+} in many enzymes. When a sulfhydryl group is present, as in 2-mercaptoethylamine, Zn^{2+} binding strengthens, equalling that for Ni^{2+}. Table 2 also shows the relative strengthening of Cd^{2+} and Pb^{2+} binding with the sulfhydryl donor in 2-mercaptoethylamine.

Table 2

Stability Ruler[a]

Ligand	Donors	Mg < Mn < Fe < Co < Ni < Cu	Length	
Oxalate	O,O	Cd Zn Pb	2.1	Hg (2.3)[c]
Glycine	N,O	Cd Pb Zn	6.1	Hg (0.4)
Histidine	N,=N	Cd Pb Zn	8.2	Hg (0.2)
$NH_2(CH_2)_2NH_2$	N,N	Cd Zn Pb	10.1	Hg (0.4)
$NH_2(CH_2)_2S^-$	N,S	Zn Cd Pb	11	Hg (0.3)
Imidazole	=N	Zn Cd	4.0	Hg (1.1)
Ammonia	NH_3	Pb Zn Cd	4.0	Hg (1.2)
Hydroxide	OH^-	Cd Zn Pb	3.7	Hg (1.2)

[a] All dipositive cations in this application. From Martin (1987)
[b] Ruler scale length in difference between log stability constants of Cu^{2+} and Mg^{2+} complexes.
[c] Relative magnitude that the Mg^{2+} to Cu^{2+} log stability constant scale must be extended to reach the Hg value.

To most unidentate ligands Cd^{2+} binds more strongly than Zn^{2+}. Probably because of an unfavorable ring bite size for the relatively large Cd^{2+} (Table 1), Zn^{2+} chelates more strongly to ligands containing O and N donors. Table 2 shows that upon introduction of an S donor atom into a chelate, Cd^{2+} becomes the stronger metal ion binder.

For all donor sets in Table 2, Hg^{2+} binds so strongly that it is off the end of the ruler scale. The number in parentheses after Hg in Table 2 refers to the relative distance by which the length of the whole log stability constant scale from Mg^{2+} to Cu^{2+} must be extended to reach the value for Hg^{2+}. An instructive contrast appears between the length of the extension for bidentate compared to unidentate ligands. The scale extension for Hg^{2+} amounts to 0.2 to 0.4 log units for bidentate ligands and to 1.1 to 1.2 log units for the three unidentate ligands at the end of Table 2. The difference arises because Hg^{2+} prefers linear, two coordination and binds the second donor atom in small chelate rings much more weakly than the first donor atom.

V. METAL ION HYDROLYSIS

The higher the charge density or charge-to-radius ratio, the more likely a metal ion undergoes hydrolysis in aqueous solutions to give hydroxo complexes. Hydroxo complexes may abruptly form polynuclear complexes and precipitates even in solutions more acidic than the pK_a for first hydroxo complex formation. The first five small metal ions in Table 1 hydrolyze even in acidic solutions and form precipitates. In six coordination the charge-to-radius ratio for the first five metal ions is greater than 0.044 while for all the other metal ions in Table 1 the ratio is less than 0.035. The first five metal ions cannot occur to an appreciable extent in the bloodstream (pH 7.4) as the free aqueous ion and must be complexed in some way. Covalency may also promote complex formation and hydrolysis in acidic solutions, as is the case for Hg^{2+}.

Hydroxo complex formation follows the same stability order as other ligands with the strongest binding at the right of the stability ruler in Table 2, where hydroxide ion appears as the last entry. (For consistency all metal ion hydrolysis results are taken from a single authoritative source (Baes and Mesmer, 1976).) The stability ruler Table 2 shows that Zn^{2+} exhibits a relatively strong tendency to form hydroxo complexes and precipitates.

VI. HARD AND SOFT ACIDS AND BASES

Unfortunately for biochemistry, the hard-soft classification of acids and bases has become part of the popular vocabulary, despite serious limitations. According to this precept, "hard" or class a metal ions favor interactions with "hard" bases and "soft" or class b metal ions favor combinations with "soft" bases (Ahrland et al., 1958; Pearson, 1963; 1973). In biochemistry there are only four main types of ligand donor atoms of which O and aliphatic N are classified as hard, aromatic N as borderline, and S as soft bases. All alkali, alkaline earth and lanthanide ions are classified as hard, as are the first five metal ions in Table 1 and Mn^{2+}. Borderline metal ions include the last four of the stability ruler (Table 2) and Zn^{2+} and Pb^{2+}. Soft metal ions include Cd^{2+}, CH_3Hg^+, and Hg^{2+}. The classification scheme has only to three categories each for acids and bases.

Qualitatively we can see some fit between the hard and soft classifications and other results including those in the tables. Hard metal ions interact exclusively with the phosphate groups of nucleotides while borderline and soft metal ions also interact with the nucleic bases (Martin, 1979a; Sigel et al., 1987). The first five metal ions in Table 1, all hard, form insoluble hydroxide (hard) precipitates even in acidic solutions. However, 11 other hard metal ions in Table 1 fail to form hydroxo complexes at pH 7 and many fail to do so even in strongly basic solutions. Some borderline and soft metal ions form hydroxo complexes in neutral solutions, while others do not.

As mentioned in the stability ruler comparisons of Table 2, soft sulfur tends to combine not only with soft metal ions, but also with Pb^{2+} classed as borderline, though not with other borderline metal ions. In fact Pb^{2+} defies classification because it interacts strongly with sulfhydryl groups (soft) and also forms a strong hydroxide (hard) complex. That a single metal ion displays characteristics of all three classes saps the utility of the hard and soft classification scheme (Martin, 1986).

Biology offers metal ions environments inconsistent with and unpredictable under the hard and soft classification scheme. As one example, in the iron-sulfur protein rubredoxin, exclusively sulfur (soft) coordinates to both Fe^{2+} (borderline) and Fe^{3+} (hard).

A quarter of a century after the initial classification of hard and soft acids and bases on a qualitative basis (Pearson, 1963), quantitative hardness values have finally appeared (Pearson, 1988). Hardness is defined as half the difference between the ionization potential and electron affinity, and electronegativity is defined as half the sum of the ionization potential and electron affinity. Softness is defined as the reciprocal of hardness. All four quantities represent gas phase values that are applied to reactions in solution. The quantitative hardness scale along with our expectations may be checked against the properties of metal ions. Extensive results of equilibrium constants exist for aqueous metal ion stabilities with four ligands classified as hard: hydroxide (from pK_a for deprotonation of bound water (Baes and Mesmer, 1976)), fluoride, acetate, and ammonia (Martell and Smith, 1974-1982). A variety of tests failed to reveal any simple quantitative relationship between the solution properties and the hardness parameter. Therefore, Spearman's rank correlation, r_s, was applied. In this non-parametric test just the order of items in two lists are compared. For identical ordering the Spearman rank correlation coefficient is 1.0, for inverse ordering -1.0, and for no ordered correspondence between the two lists 0.0. Table 3 shows the results of applying the test to binding for the four ligands.

The four ligands are among the hardest to exist in aqueous solutions. Therefore, according to the hard and soft precept, they should favor binding to the hardest metal ions. Table 3 shows that the 35 metal ions listed in order of increasing hardness correlate poorly with the order of increasing stability constants for hydroxide binding with $r_s = -0.20$, an almost random value with even the wrong sign! Greater correlations for the order of the hardness parameter with hydroxide binding strengths result when each charge type is considered separately, $r_s = -0.61$ and -0.44 for 15 di- and 15 tri-positive ions, respectively. Again, contrary to expectation, the negative sign shows that interaction with the hard hydroxide is favored by the softest metal ions. The correlation of the hardness parameter works better with the order of fluoride binding strengths to 25 metal ions where $r_s = 0.58$. For 27 acetate cases the correlation is near random ($r_s = -0.13$). For 14 ammonia cases there is a weak correlation with the hardness parameter ($r_s = -0.64$), but again of the wrong sign (Table 3). The softest metal ions interact most strongly with the hard ammonia.

Table 3

Spearman Rank Correlation Coefficients, r_s

Parameter\Stability constant for	Hydroxide	Fluoride	Acetate	Ammonia
Hardness (Hrd)	-0.20	0.58	-0.13	-0.64
Electronegativity (Eln)	0.47	0.82	0.28	-0.26
Eln + Hrd = Ionization potential	0.27	0.77	0.11	-0.40
Eln - Hrd = Electron affinity	0.86	0.77	0.78	0.65

Table 3 also displays the results of three other parameters with the order of binding strengths for the four ligands. There is a better correlation in most cases with electronegativity, ionization potential (= electronegativity + hardness), and electron affinity (= electronegativity - hardness). Indeed, it is only the last parameter, electron affinity, that yields a satisfactory positive correlation for the order of hydroxide, acetate, and ammonia binding strengths. In fact, the pK_a values for deprotonation from 35 aqueous metal ions relate satisfactorily to the electron affinity (EA) by the linear regression $pKa = 15.9 (\pm 0.9) - 0.44 (\pm 0.04) EA$ (units of eV), with a correlation coefficient $r = 0.88$, without the necessity to consider separately different charges.

This analysis shows that the quantitative hardness parameter does not provide any quantitative correlation with the stability constants for a variety of metal ions with four hard ligands. Even qualitative ordering of the hardness parameter with stability constants gives only weak correlations that are the reverse of the expected order for three of the ligands, with only fluoride ion yielding a correlation of the correct sign (Table 3). Since softness is defined as the reciprocal of hardness, the same comments apply. The results of this analysis do not support the common adage that hard metal ions prefer hard ligands while soft metal ions prefer soft ligands. *The hard-soft classification scheme is not only misleading but also leads to wrong conclusions for stability constants in aqueous solutions. Its use should be dropped immediately.*

VII. NON-AQUEOUS SOLVENT MIXTURES

Non-aqueous solvents that lower the dielectric constant increase stability constants between oppositely charged ions, decrease the constants between identically charged ions, and leave relatively unaffected stabilities with one neutral reactant species. For example, on passing from water to a mixed solvent system containing 70% by weight dioxane the dielectric constant drops from 79 to 18. This solvent change increases the pK_a of acetic acid by 3.6 log units and decreases the pK_a of anilinium ion by 1.0 log units. The much larger change in the former case is due to the reaction between oppositely charged proton and acetate becoming much more favored in the lower dielectric medium. In contrast, the latter reaction is merely transfer of a proton from water to amine without the creation or destruction of any charges.

Binding sites in proteins possess lower dielectric constants than the surrounding water. An equivalent solution dielectric constant applicable to binding sites in metalloproteins is calibrated with metal ion stabilities in non-aqueous solvent mixtures. For the zinc ion active sites in bovine carbonic anhydrase and carboxypeptidase A, the equivalent solution dielectric constants are estimated as 35 and 70, respectively. The former enzyme reacts with a smaller substrate and possesses a much less open binding site (Sigel et al., 1985).

VIII. pH-DEPENDENT STABILITIES

If at a given pH a ligand occurs in a protonated form, competition develops between metal ion and proton for a basic ligand site. The concentration of basic ligand available to the metal ion drops in the presence of protonated ligand species. Neutral solutions contain ligands with protonated amine, phenolic, and sulfhydryl groups. In these and other cases the tabulated stability constants overstate the effective binding strength, and its magnitude needs to be reduced to reflect competition from the proton.

The tabulated stability constant K_s refers to the following equilibrium.

$$M + L \rightarrow ML \qquad K_s = [ML]/([M][L])$$

With a protonated ligand, however, the relevant formation reaction becomes $M + HL \rightarrow ML + H^+$. There is the additional acid-base equilibrium of the ligand.

$$HL \rightarrow H^+ + L \qquad K_a = (H^+)[L]/[HL]$$

The fraction of the ligand in the basic form available for metal ion binding is given by

$$\alpha = [L]/([HL] + [L]) = K_a/[(H^+) + K_a] \text{ so that } 0 < \alpha < 1.$$

The conditional, pH-dependent stability constant is given by $K_c = \alpha K_s$, or

$$\log K_c = \log K_s - pK_a - \log [(H^+) + K_a]$$

For pH >> pKa, as for carboxylates in neutral solutions, the basic form of the unbound ligand predominates and $\log K_c = \log K_s$. For pH << pK_a, as for aliphatic amines in neutral solutions, unbound ligand is predominantly protonated and $\log K_c = \log K_s - pK_a + pH$, and the conditional constant displays a pH dependence. When the pH is within 1.5 log units of pK_a the complete equation should be used. By allowing for withdrawal of deprotonated ligand from solution by protonation, the value of the conditional stability constant becomes less than the standard stability constant, $K_c \leq K_s$. (The conditional constant may exceed the tabulated constant for cases where a subsequent deprotonation occurs from a complex (Martin, 1991)).

In Table 4 conditional stability constants at pH 7.0 (represented as K_7) are compared with the tabulated stability constants (K_s) for three metal ions and four unidentate ligands. The four ligands represent carboxylate, imidazole, sulfhydryl, and aliphatic amine donors in biological systems. For acetate, pH = 7 >> pK_a, $\alpha = 1$, virtually all ligand is in its basic, carboxylate form, and the conventional and conditional stability constants are equal. For imidazole 44% of the ligand is in the basic form. For glutathione and ammonia only 1.2 and 0.47%, respectively, of the ligands are in the basic form. Glutathione is the tripeptide γ-L-glutamyl-L-cysteinylglycine, and it serves as a superb model for unhindered sulfhydryl groups in proteins.

The results in Table 4 show that for all three metal ions the ligand stabilities from the conventional constants follow the increasing sequence acetate < ammonia < imidazole < glutathione. This sequence differs from the stability order in neutral solutions because the sequence applies only when all ligands are predominantly in their basic form, which is true

Table 4

Conventional and Conditional (pH 7.0) Stability Constants.[a]

		Ni^{2+}		Zn^{2+}		Cd^{2+}	
Ligand	pK_a	$\log K_s$	$\log K_7$	$\log K_s$	$\log K_7$	$\log K_s$	$\log K_7$
Acetate, O^-	4.7	0.8	0.8	0.7	0.7	1.2	1.2
Imidazole, =N	7.11	3.0	2.6	2.52	2.2	2.8	2.4
Glutathione, RS^-	8.92	4.0	2.1	5.0	3.1	6.16	4.2
Ammonia, NH_3	9.33	2.80	0.5	2.37	0.0	2.65	0.3

From Martin (1986)

only for acetate. For conditional constants at pH 7.0, Table 4 shows that for Ni^{2+} the increasing stability order is ammonia < acetate < glutathione < imidazole while for Zn^{2+} and Cd^{2+} the order is ammonia < acetate < imidazole < glutathione. For all three metal ions in neutral solutions the most basic ammonia has become a weaker ligand than the least basic acetate. The stability order from conditional constants suggests that in neutral solutions Ni^{2+} is apt to be found at imidazole sites and Zn^{2+} and Cd^{2+} at sulfhydryl groups. Of course chelation by adjacent donors may increase stabilities and alter preferred binding sites. Nonetheless the examples in Table 4 serve as models for the analogous groups in proteins and prove that binding strengths and stability orders are pH dependent.

In the nucleosides of the 6-oxopurines, guanosine and inosine, in neutral solutions the much more basic N1 site is protonated while the N7 site is difficult to protonate. Metal ions coordinate at N7 in acid solutions, and as the pH increases, they compete more effectively with the proton and crossover to N1 (Martin, 1979a). A characteristic crossover pH may be defined for each metal ion (Martin, 1985).

The principles developed in this section extend to ligands with several basic sites. For EDTA and catecholates where metal ions displace two protons in neutral solutions, the fraction of unbound, doubly deprotonated ligand becomes

$$\alpha = K_{a1}K_{a2}/[(H^+)^2 + (H^+)K_{a1} + K_{a1}K_{a2}], \text{ where } pK_{a1} < pK_{a2}.$$

Ignored in this section and Table 4 is formation of metal ion hydroxo complexes and precipitates. Formulation of conditional stability constants that incorporate hydroxo complex formation is outside the scope of this section but has been described elsewhere (Martin, 1988).

IX. OTHER FACTORS

Numerous other factors affect metal ion affinities including size and number of chelate rings (Hancock, 1992; Sigel et al., 1981), presence of other ligands (see Part Three, Chapter 1-C3), non-covalent interactions among ligands (Kim and Martin, 1984; Sigel, 1989), ligand field strength (at most 10% of total binding energy), and spin state (important in heme proteins).

As biological systems are not at equilibrium, a more comprehensive discourse would include the ways biology achieves non-equilibrium status. For a recent discussion of speciation, element uptake, and compartmentalization in biology see the book by da Silva and Williams (1991).

REFERENCES

Ahrland, S., Chatt, J., and Davies, N.R. (1958). *Quart. Rev. Chem. Soc.*, *12*: 265.

Baes, C.F., and Mesmer, R.E. (1976). *The Hydrolysis of Cations*, Wiley-Interscience, New York.

da Silva, J.J.R., and Williams, R.J.P. (1991). *The Biological Chemistry of the Elements*, Oxford University Press, Oxford.

Hancock, R.D. (1992). *J. Chem. Educ.*, *69*: 615.

Kim, S.H., and Martin, R.B. (1984). *J. Amer. Chem. Soc.*, *106*: 1707.

Martell, A.E., and Smith, R.M. (1974-1982). *Critical Stability Constants*, Plenum, New York.

Martin, R.B. (1979a). *Metal Ions in Biological Systems*, Vol. 8 (H. Sigel, ed.), Marcel Dekker, New York, p. 57.

Martin, R.B. (1979b). *Metal Ions in Biological Systems*, Vol. 9 (H. Sigel, ed.), Marcel Dekker, New York, p. 1.

Martin, R.B. (1984). *Metal Ions in Biological Systems*, Vol. 17 (H. Sigel, ed.), Marcel Dekker, New York, p. 1.

Martin, R.B. (1985). *Accounts Chem. Res.*, *18*: 32.

Martin, R.B. (1986). *Metal Ions in Biological Systems*, Vol. 20 (H. Sigel, ed.), Marcel Dekker, New York, p. 21.

Martin, R.B. (1987). *J. Chem. Educ.*, *64*: 402.

Martin, R.B. (1988). *Metal Ions in Biological Systems*, Vol. 24 (H. Sigel and A. Sigel, eds.), Marcel Dekker, New York, p. 1.

Martin, R.B. (1991). *Aluminium in Chemistry, Biology, and Medicine* (M. Nicolini, P.F. Zatta and B. Corain, eds.), Cortina International-Verona, Raven Press, New York, p. 3.

Pearson, R.G. (1963). *J. Amer. Chem. Soc.*, *85*: 3533.

Pearson, R.G. (1973). *Hard and Soft Acids and Bases*, Dowden, Hutchinson, and Ross, Stroudsberg, Penn.

Pearson, R.G. (1988). *Inorg. Chem.*, *27*: 734.

Shannon, R.D. (1976). *Acta Crystallogr.*, *A32*: 751.

Sigel, H. (1989). *Pure Appl. Chem.*, *61*: 923.

Sigel, H., Prijs, B., and Martin, R.B. (1981). *Inorg. Chim. Acta*, *56*: 45.

Sigel, H., Martin, R.B., Tribolet, R., Haring, U.K., and Malini-Balakrishnan, R. (1985). *Eur. J. Biochem.*, *152*: 187.

Sigel, H., Tribolet, R., Malini-Balakrishnan, R., and Martin, R.B. (1987). *Inorg. Chem.*, *26*: 2149.

Sundberg, R.J., and Martin, R.B. (1974). *Chem. Rev.*, *74*: 471.

METAL-LIGAND COORDINATION IN BIOLOGICAL FLUIDS

by

P. de Oliveira, H.A.O. Hill and L.-L. Wong

New Chemistry Laboratory
South Parks Road
Oxford OX1 3QT
U.K.

I. INTRODUCTION

The elements known to be essential for the existence of life on Earth are identified in the Periodic Table shown in Figure 1 of Chapter 1 above. Apart from these, a few are on the brink of being either confirmed as essential for particular biological species or dismissed as contaminants or impurities not taking part in any vital process.

Approximately half of the elements essential for life are metals. They are coordinated in a myriad of different biochemical species, ranging from the simplest complex of a magnesium ion with a nucleotide to the most intricate active site of an enzyme. Exceptions are the alkaline metal ions, which move freely in biological fluids where they play their role in electrolytic processes, and some other metal cations that form precipitates with important biological role, such as calcium ion in shells ($CaCO_3$) and bones ($Ca_5(HO)(PO_4)_3$) (Martin, 1984).

Understanding the factors influencing the coordination of metal ions in biological systems might shed some light on how and why nature selected a well-suited species to fulfill a certain task. By and large, these factors apply also to complexation reactions done in the laboratory, often in conditions inappropriate for sustaining life. In the latter cases, an equilibrium is normally reached, which would be a synonym of death in biological systems. In fact, biological systems can be regarded as open systems endowed with the puzzling ability of maintaining and controlling a non-equilibrium state within certain, often quite narrow, parameters. To accomplish this task, they have an absolute requirement for energy. This energy is obtained from redox reactions having as ultimate sources either light or exogenous nutrients.

II. FUNDAMENTAL ASPECTS OF THE METAL-LIGAND BOND IN VIVO

For the sake of simplicity, some of the factors influencing coordination will be approached on a thermodynamic basis, where it is assumed that the metal ion equilibrates between two or more phases. Later on, kinetic phenomena will be taken into account so that a better picture of metal ion complexation in biological systems can be grasped. In fact, kinetic factors are often decisive in the uptake of metal ions, and it is not uncommon that the result predicted on thermodynamic grounds alone turns out to be in direct contrast to that observed in biological systems.

The factors influencing metal ion coordination can be classified in several categories, depending on the chemical process involved. Hence, the selectivity for a certain metal cation (or group thereof) may be due to: metal cation availability; coulombic interaction; matching sizes of the metal cation and the ligand cavity; nature of the donor atoms; preference of the metal cation for a certain geometry; spin state changes; tendency to binding in clusters; and ease of incorporation into proteins.

A. Metal cation availability (Fraústo da Silva and Williams, 1993)
Thermodynamic formation constants (K^{therm}) can differ appreciably in magnitude from the so-called conditional formation constants (K^{cond}). The latter encompass the influence that side reactions have on the main reaction and give a reliable value of the extent of the main reaction under certain, specified conditions.

An uptake factor can be defined as $K^{cond}.[M]$, where [M] is the concentration of free metal ion. This can give a good estimate of the extent of complexation of a certain metal cation M by a specific group of ligands, under a defined set of conditions. It follows that the extent of complexation can be modulated by altering either K^{cond} or [M], or both. It is possible to attain a high degree of selectivity based on thermodynamic (as opposed to kinetic) principles. Nature has endowed living organisms with cunning pathways to carry out this task, as the following example shows.

Consider two metal cations, M and N, and two ligands, X and Y, and their respective complexes MX, MY, NX and NY. The conditional formation constant of MX is given by

$$K_{MX}^{cond} = \frac{[MX]}{[M]' \cdot [X]'}$$

where [M]' and [X]' are the concentrations of M and X, respectively, excluding the amount present as MX. Their values are given by

$$[M]' = [M] \cdot \{1 + K_{MY} \cdot [Y]\}$$

$$[X]' = [X] \cdot \{1 + K_{NX} \cdot [N]\}$$

If the binding of X to M is to be selected, then the formation of NX has to be avoided. This can be accomplished by decreasing [N] *via* coordination of N to Y, preferably to a negligible value. In the end, X will be present in solution as either in the form of its complex with M, or as the free ligand, X.

The concentration of N can be obtained by the expression

$$[N] = \frac{[NY]}{[Y] \cdot K_{NY}}$$

If $K_{NX} \cdot [N]$ is to be negleted when compared to unity, its magnitude should be made smaller than 1%

$$K_{NX} \cdot [N] < 10^{-2}$$

Combining the last two expressions gives

$$\frac{K_{NX} \cdot [NY]}{K_{NY} \cdot [Y]} < 10^{-2}$$

and, provided that Y is in excess relatively to N, it is sufficient that $K_{NY} > 10^2 \cdot K_{NX}$ to decrease [N] to the desired level. The excess of Y over N can be smaller and lead to the same effects if the ratios in the constants are larger. The same reasoning applies to avoid binding of M by Y.

So, if both [M] and [N] are of the same magnitude, it can be expected that just the complexes MX and NY will be found in solution in appreciable amounts provided that

$$\frac{K_{NY}}{K_{NX}} > 10^2 \qquad \frac{K_{MY}}{K_{MX}} < 10^{-2}$$

In conclusion, the presence of Y renders N unavailable to X, and M is not available to Y due to the presence of X. A good example illustrating how different complexing agents can control free metal ion availability is given by blood plasma. Cu(II) strongly coordinates to ceruloplasmin, Fe(II) is picked up by transferrin, which also binds Mn(II), and Zn(II) is encapsulated in an albumin. Ca(II) is partially free and partially coordinated to several calcium-binding sites of some blood serum proteins. These sites would preferentially bind to transition metal ions if these had not been previously removed from the plasma.

B. Coulombic interaction

Considering simple coulombic interaction between charged species, it is expected that metal cations with higher charge will be more strongly coordinated to ligands with highly charged, small donor atoms (Huheey, 1983). In fact, it has been found that ligands with small, singly-charged donor atoms have a strong tendency to complex with trivalent cations, such as Fe^{3+}, Mn^{3+} and Al^{3+}. In biological systems, these ligands would be of the type R_2N^- or RO^-, but since the pH of biological fluids is not far from 7, the former ligand will be mainly protonated. An exception is histidine, where hydrogen bonding between the N-H group and an aspartate or glutamate renders the imidazole more anion-like. As a consequence, just histidine and RO^- type ligands, in the form of hydroxamates, carboxylates, enolates and phenolates, contribute to the coordination of the mentioned metal ions. Organic ligands might compete with hydroxide, and this can explain why Al^{3+} is not taken up by living organisms, but usually exists in the outer medium either coordinated to hydroxide or as a precipitate with this anion (Williams, 1992).

C. Matching sizes of both the metal cation and the ligand cavity

On a very simple basis, it would be expected that a metal complex would be the more stable the higher the charge and the smaller the radius of the metal cation, i.e. the higher its ionic potential, z/r. Exceptions to this rule arise from the participation of other factors in the complexation, such as steric hindrance between the ligands or between different groups of a polydentate ligand.

With inorganic compounds, the most common trend is that the more similar in size the cation and the anion, the more insoluble the respective salt, or the higher the tendency to form ion pairs. This might involve hydrated metal ions in order to match the size of the anion (Phillips and Williams, 1966).

These principles also apply to systems involving ligands of biological nature, such as oligopeptides, proteins, nucleic acids, etc. Usually, they bear a cavity whose dimensions closely match the radius of the metal cation with which they are supposed to complex. The cavity can be part of a cyclic structure, like the antibiotics nonactin and valinomycin that specifically bind K^+, or the chlorin ring in chlorophyll that coordinates Mg^{2+} (see next section), or even a pocket or cleft in a protein, whose tertiary structure folds in such a way as to bring together in a suitable orientation the amino acid residues required to complex a certain metal cation or cluster thereof (Branden and Tooze, 1991). Examples are the calcium-binding protein calmodulin (Chattopadhyaya et al., 1992), a relatively flexible structure that coordinates several calcium ions, folding around them, and the relatively rigid iron/sulfur proteins, which enclose iron/sulfur clusters (Nakamura and Ueyama, 1988).

Ion channels can also be included in this section. These are proteins that span biomembranes, and whose function is to allow the free diffusion of a specific metal cation through them (Keeling and Benham, 1989; Narahashi, 1992). Selectivity can be achieved by matching the channel diameter with the radius of the aquo complex of the metal cation. The selectivity for a metal ion derives from the fact that smaller ions usually have a rather large hydration sphere, and larger ions have already larger dimensions themselves, even in the absence of their hydration shell. As a result, both are larger in diameter than the hole in the channel, so they neither fit nor cross it. Examples are potassium, sodium, calcium and proton channels. Obviously, selectivity is not perfect, and some species do interfere with this transport process; for example cations with no acknowledged biological function, like those of barium (Adams et al., 1982; Brown, 1988) and lanthanides (Tsien et al., 1987) which can block potassium and calcium channels, respectively.

Table 1

Preference of Metal Ions for Certain Donor Atoms in Ligands
(adapted from Fraústo da Silva and Williams, 1993)

Metal ion	Donor atoms
Mg^{2+}	**O** (carboxylate, polyphosphate; negatively charged)
Ca^{2+}	**O** (carboxylate, carbonyl, phosphate)
Mn^{2+}, Fe^{2+}	**O** (carboxylate, phosphate); **N** (His); **S** (Cys, Met, sulphide)
Fe^{2+} (special)	**N** (polypyrroles)
Cr^{3+}, Mn^{3+}	**O** (phenolate, hydroxamate, hydroxide)
Fe^{3+}, Co^{3+}	**O** (carboxylate); **N** (polypyrroles); **S** (Cys, sulphide)
Ni^{2+}	**S** (Cys); **N** (polypyrroles)
Cu^{+}, Cu^{2+}	**N** (His, amines, ionized peptide bond); **S** (Cys)
Zn^{2+}	**N** (His, amines); **S** (Cys); **O** (carboxylate)

D. Nature of the donor atoms

The affinity between metal cations and specific donor atoms from ligands follows the trends "non-polarizable, small cations coordinate to non-polarizable donor atoms", and "polarizable, large cations coordinate to polarizable, large donor atoms" (Huheey, 1983). Table 1 shows several examples of metal cation preferences, such as the contrasting behavior of magnesium and copper ions: the former coordinates to negatively-charged oxygen donor atoms (either in carboxylates or in phosphates), while the latter prefers nitrogen- or sulfur-donor atoms, not necessarily charged.

E. Preference of the metal cation for a certain geometry

Some metal cations show a marked preference for a certain coordination geometry, as can be judged by inspection of Table 2, which compiles results obtained with simple coordination compounds.

Nevertheless, this behavior is very often observed in biochemical species as well. For example, while Co(II) usually appears coordinated in a octahedral geometry (e.g. in cobalamin), Zn(II) prefers a tetrahedral distribution of donor atoms, as found in the enzyme carbonic anhydrase (see next section).

Table 2

Preference of High-Spin Metal Ions for Certain Geometries when Coordinated to Ligands with N and O Donor Atoms
(adapted from Fraústo da Silva and Williams, 1993)

Metal ion	Preferred geometries
Cu^{2+}, Mn^{3+}	tetragonal > 5-coordination > tetrahedral
Ni^{2+}	octahedral
Co^{2+}	octahedral > tetrahedral
Zn^{2+}	tetrahedral > octahedral > 5-coordination
Mn^{2+}	octahedral
Fe^{3+}, Co^{3+}, Cr^{3+}	octahedral

F. Spin state changes

Ions of main group elements such as Li^+ and Ca^{2+} have closed shell, i.e. s^2 or s^2p^6 electronic configurations. For transition metal ions, the d-shell is often not full. Since the orbitals in which the d-electrons reside point to different regions in space, they will have different bonding interactions with ligands and hence different energies. The way in which the d-electrons are distributed amongst these orbitals will affect the metal-ligand interaction, the size of the ion, and hence the stability of the complex. Electron repulsion will tend to keep electrons in different orbitals with their spins parallel to one onother, resulting in the high-spin state. If the metal-ligand interactions are sufficiently strong, different d-orbitals will have such large differences in energy that it becomes energetically more favorable to have two electrons in the same orbital with their spins paired, resulting in the low-spin state (Huheey, 1983).

It follows that a change in the spin state is accompanied by changes in metal-ligand bonding and hence stability of the complex (Phillips and Williams, 1966). It turns out that for low-spin complexes formed with metal ions of high charge and strong donor groups such as N/S and, in some cases, RO^- ligands, the stability follows the trend Co(III) > Fe(III) > Mn(III). For low-spin divalent ions, the order is Fe(II) > Mn(II) > Co(II), but the high-spin order is Co(II) > Fe(II) > Mn(II).

Spin state changes are rarely the only factor responsible for metal selection. Other factors, such as ligand cavity size, nature of donor atoms and kinetic phenomena like ion pumping can be decisive.

G. Tendency to binding in clusters

Some metal cations, such as Fe^{3+}, Mn^{3+} and Cu^{2+}, suffer hydrolysis at pH values around 7, resulting in species like Fe_2O^{4+}, Mn_2O^{4+} and $Cu_2(HO)_2^{2+}$ (Greenwood and Earnshaw, 1984). Clusters like these are found in a few proteins, and their functional value can be as different as the copper dimer in oxidases and the manganese cluster in the dioxygen-producing enzyme in photosynthesis.

Sulphide ions form a series of clusters with iron ions, of the type $[Fe_nS_m]$, where n and m can vary from 1 to 13 in several combinations (Yu et al., 1993). Apart from sulphide, the iron ions are also coordinated to thiolate-S from cysteine residues of the protein backbone where they are encapsulated (Armstrong, 1988).

At first glance, it seems unlikely that clusters will form, since the change in entropy for the process must be rather unfavorable. This is, however, overcome by two main factors. First, there are attractive interactions between highly-charged cations and anions usually present in these clusters. Second, the sizes of the protein cavity and the cluster itself are matched, with their coordinating groups properly oriented.

H. Ease of incorporation into proteins (Fraústo da Silva and Williams, 1993)

All complexes considered so far were assumed to be made from small, monodentate ligands. This is obviously a simplification, especially in the case of proteins, since the energetics of protein folding can markedly influence the coordination process. If the first coordination sphere is represented as ML, L being the set of ligands in a model complex, another situation arises that cannot be overlooked: the transfer of ML from an aqueous medium into a protein (whether embedded in a membrane or not), where the dielectric constant has a lower magnitude, can involve a rather large energy. This fact can be preponderant in the selection for a divalent cation against a trivalent one. As a consequence, the overall stability of a metal-protein complex has contributions both from a formation constant of the complex ML (with the ligands assumed to be free amino acids) and from a transfer coefficient, p, of ML from water into the actual tridimensional protein structure.

I. Kinetic phenomena (Fraústo da Silva and Williams, 1993)

As was stated in the beginning of this chapter, kinetic phenomena have to be taken into account in order to understand and explain complexation processes fully.

A metal cation can be taken up by a cell either in its free state (M) or coordinated to some ligand (ML_1). The first case may simply involve thermodynamically controlled ion

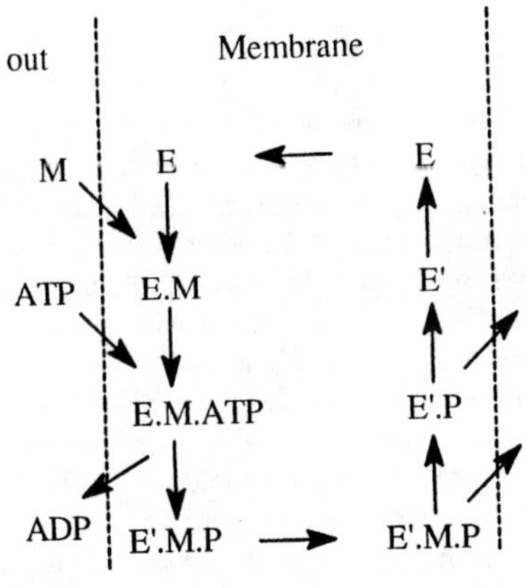

Figure 1.
Pumping of a metal ion through a membrane with the input of energy (adapted from Fraústo da Silva and Williams, 1993).

channels or require a net energy input into the membrane structure so that a suitable transport mechanism may operate. This is usually the case for the uptake of metal complexes. The source of energy is usually ATP and involves the participation of a membrane component capable of changing between at least two different states in a cyclic fashion. These components are often referred to as gates or pumps.

The common mode of operation is the following (Fig. 1): the membrane species E binds M (or ML_1) from the outer medium, the affinity being dependent on K^{cond}, p and D, the partition coefficient between the membrane and an aqueous solution for M (or ML_1). Then, an energy input modifies E to a different state, E', that is in close vicinity of the inner phase. Furthermore, E' interacts with M (or ML_1) with a different affinity, $K^{cond'} \cdot p' \cdot D'$, smaller than that of E and not large enough to retain M (or ML_1), that diffuses away. Finally, E' reverts to E, and the heat loss implies that the transport process is thermodynamically irreversible, although the pathway can be re-used again provided a suitable amount of energy is applied into the system.

Once inside the cell, M can be trapped in a kinetically inert complex ML_2, from which the release of M is highly improbable. The selective incorporation of M into the complex ML_2, even when it is expected that other metal cations will compete for the coordination site, might be due to the preferential complexation of the latter by other ligands that decreases markedly their concentration as free ions. The process might even require the participation of other species that act in a way similar to molecular chaperones (Vandervies et al., 1993; Wiech and Zimmermann, 1993), eventually involving stereochemical interactions or requiring the input of energy. A classic example is the insertion of Mg^{2+} in the chlorin ring of chlorophyll, briefly discussed in the next section.

III. SPECIFIC ASPECTS OF METAL COORDINATION TO MACROMOLECULAR VS LOW-MOLAR-MASS LIGANDS

Metal ions in biological systems are incorporated in coordination compounds that range from simple low-molar-mass (l.m.m.) species to the active sites of enzymes, often including several different metal ions. With the exception of Na^+ and K^+, which are mainly involved in processes of osmotic control and electrolytic phenomena, the ligands in the coordination sphere of all the other metal ions are organic moieties.

In the light of the factors presented in the previous section, some concrete examples of metal ion coordination in vivo will be given here, stressing the peculiarities of each case where appropriate. The biochemical roles of metal ion coordination compounds will be dealt with in the following section.

A. Magnesium and calcium

The elements from Group IIA present in all living organisms are magnesium and calcium. Both are non-polarizable, non-redox active divalent cations showing a marked tendency to bind to small, non-polarizable groups, usually of the O-donor type. However, the fact that Mg(II) has a smaller ionic radius and, consequently, a higher ionic potential (about 50% higher than for Ca(II)) may partially account for some behavioral differences between them: Mg(II) is usually coordinated to negatively charged ligands, almost invariably in an octahedral environment, while Ca(II) interacts easily with neutral O-donor species, such as carbonyl and hydroxyl groups (Fraústo da Silva and Williams, 1993). Ca(II) is also far more flexible as far as coordination numbers and geometry of its complexes are concerned: it is very common to find Ca(II) complexes with coordination numbers higher than 6, namely 7, 8 and 9 (Martin, 1984). Attention should be drawn to the relative lability of Ca(II) complexes (Langford and Gray, 1966; Basolo and Pearson, 1967), a feature brilliantly exploited by Nature in the use of Ca(II) as a messenger, as will be shown in more detail in the next section.

An outstanding characteristic of Mg(II) bioinorganic chemistry is its close association with certain phosphate-derived species. It is interesting to note that this metal cation shows little tendency to bind to phosphate ions of the types HPO_4^{2-} and RPO_4^{2-}, but is easily coordinated to polyphosphates such as pyrophosphate, nucleotides triphosphate, NTP, or to polyphosphate sugars. As a result, the biochemistry of NTP is in reality that of the complexes $[Mg(NTP)]^{m-}$ (Martin and Mariam, 1979; Heaton, 1990; Williams, 1993).

Perhaps the most remarkable Mg(II) complex in biological systems is chlorophyll (Beale and Weinstein, 1991). First, the metal cation is bound to four N atoms of the chlorin ring, when its usual preference is for O-donor groups; the incorporation of chlorophyll into proteins involves a fifth coordination position usually via an N atom of a His residue (Tronrud et al., 1986). Second, the chlorin would preferentially coordinate to most of the divalent metal cations of the first transition series (Fraústo da Silva and Williams, 1993). Nevertheless, the complexation of Mg(II) that takes place in vivo proceeds with a negligible degree of error. This is one of the most striking examples of a complexation process controlled by kinetics. It is believed that the synthesis involves other molecules that might specifically recognize an Mg(II) species, probably $[Mg(NTP)]^{m-}$, and promote the encapsulation of Mg(II) in the chlorin ring:

$$[Mg(NTP)]^{m-} + \text{chlorin} \xrightarrow{\text{molecular chaperones (?)}} \text{chlorophyll } \{= [Mg(\text{chlorin})]\} + NTP^{(m-2)-}$$

an essentially irreversible reaction.

B. Zinc

This element exists in vivo in the non-redox active divalent state. Zn(II) ions have a high electron affinity, evidenced by the magnitude of the second ionization potential for Zn of 17.96 eV compared to, for example, 11.87 eV for Ca. This means that Zn(II) is a good

Table 3

Zn(II) Ligands in the Active Sites of Several Enzymes
(adapted from Vallee and Auld, 1990)

Enzyme	L_1	L_2	L_3	L_4
Carbonic anhydrase	His-N	His-N	His-N	H_2O
Thermolysin	His-N	His-N	Glu-O	H_2O
Carboxypeptidase A	His-N	Glu-O	His-N	H_2O
Phospholipase C	His-N	Glu-O	His-N	H_2O
Alkaline phosphatase	Asp-O	His-N	His-N	H_2O
Alcohol dehydrogenase	Cys-S	His-N	Cys-S	H_2O

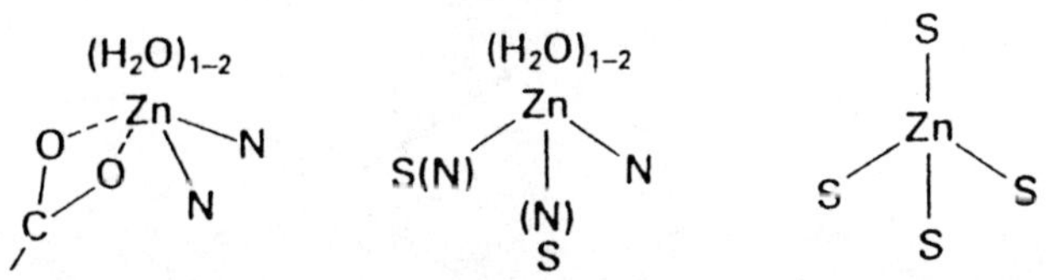

Figure 2.
Spatial arrangement of donor atoms in Zn(II)-binding sites of enzymes (from Fraústo da Silva and Williams, 1993).

Lewis acid and must exhibit a certain tendency to bind to atoms more polarizable and less electronegative than oxygen. In fact, Zn(II) in active sites of enzymes is commonly coordinated to N- and S-donor groups, even if some of the positions are occupied by O-donor ligands, usually carboxylates (Vallee and Auld, 1990). Table 3 gives some examples of coordination spheres of Zn(II) active sites in enzymes.

The donor atoms are positioned in a tetrahedral geometry around the metal cation in most of the cases, but other arrangements are known (Fig. 2). It is interesting to note the presence of a water molecule in the coordination sphere of some of the enzymes. In these cases, and Zn(II) being a Lewis acid, it is highly likely that the water molecule acts, in fact, as a hydroxide ion, capable of attacking other molecules (Vallee and Auld, 1990; Kimura, 1993).

On kinetic grounds, Zn(II) enzymes often have high reaction rates since the metal cation allows rapid binding and releasing of exogenous groups, as well as rapid re-organization of the ligands in its coordination sphere. On the other hand, the metal ion does not dissociate easily from the active site, and there are many cases where its half-life can be as high as several days (Fraústo da Silva and Williams, 1993). However, in other situations the binding to Zn(II) can be more labile, as is the case of the complexes formed with metallothioneins (Vašák and Kägi, 1983), insulin (Kirchgessner and Weigand, 1983) and the so-called zinc fingers (Rhodes and Klug, 1993).

C. Iron

Free iron cations exist as Fe(II) and Fe(III), and these are easily interconvertible. This feature immediately suggests that iron compounds might be useful in mediating redox processes in vivo, which is indeed the case, but several other activities make up the bioinorganic chemistry of this element. Here, an attempt to rationalize the different environments where iron can be found will be made. Several cases of iron complexes exhibiting coordination spheres with O-, N- and/or S-donor groups in different spatial arrangements are known. In addition, iron can transiently attain higher oxidation states, such as Fe(IV)O, during the catalytic cycle of some enzymes (Ortiz de Montellano, 1986; Lippard, 1988; Que Jr., 1993). Therefore, iron species are grouped in the following four different categories based on the features of the active site of proteins or enzymes (Fraústo da Silva and Williams, 1993):

1. *Iron/sulfur proteins*

Iron/sulfur proteins have clusters composed of one to four (Armstrong, 1988; Beinert and Kennedy, 1989), perhaps six (Voordouw et al., 1987), iron cations coordinated to sulfur donors in a tetrahedral geometry. There are always thiolate-S groups from a Cys residue of the protein backbone involved in the cluster, but sulphide ions, S^{2-}, are also present in polynuclear centers. The most common stoichiometries containing both types of ligands are 2Fe/2S, 3Fe/4S and 4Fe/4S (Fig. 3).

In these proteins, high-spin Fe(II) switches to high-spin Fe(III) (and *vice versa*) upon electron transfer. This happens with negligible changes in bond lengths in the cluster, in part due to the special characteristics of sulfur (Cotton and Wilkinson, 1988), and also to the high β-sheet content in these proteins (Nakamura and Ueyama, 1988; Fraústo da Silva and Williams, 1993). This element of protein secondary structure renders proteins relatively

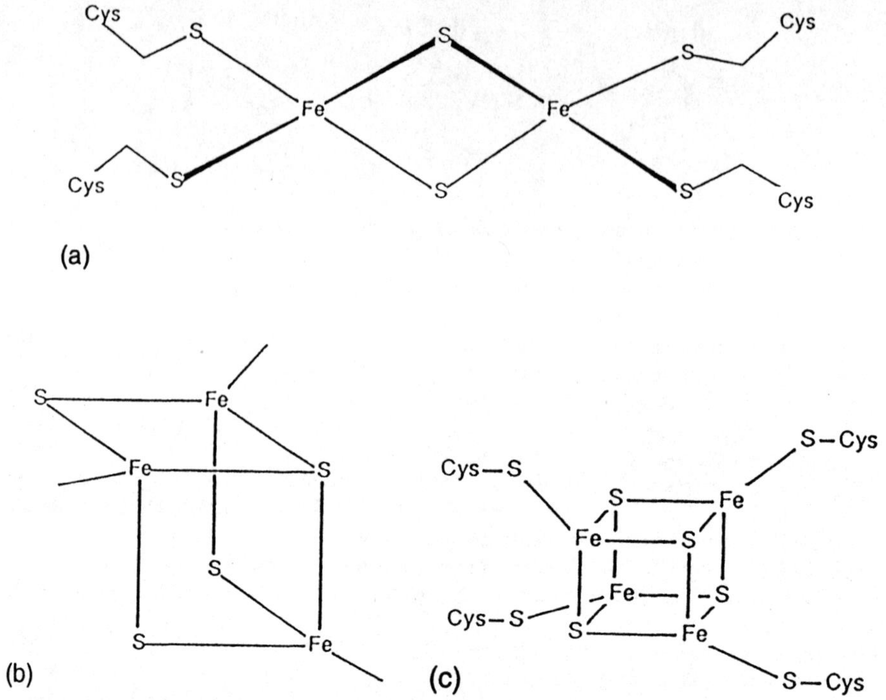

Figure 3.
Three-dimensional structures of iron-sulfur clusters in proteins and enzymes (from Armstrong, 1988).

rigid, and also accounts for the low reorganization energy involved in the redox process (Fraústo da Silva and Williams, 1993). What is more, the rigidity might be essential to stabilize the clusters, in some way decreasing the lability associated with Fe(II).

2. *Proteins containing the oxo-bridged unit Fe-O-Fe*
The oxo-bridged unit Fe-O-Fe is the general feature of this group of proteins. The iron ions coordinate to O- and N-donor ligands of the protein backbone (carboxylate-O or His-N) in an octahedral geometry. The Fe-O-Fe bridge is usually supported by bridging carboxylates. They cycle between high-spin Fe(II), high-spin Fe(III) and in some cases Fe(IV)O during the catalytic process (Lippard, 1988). However, some of the proteins in this group show some structural flexibility, in contrast to the ones in the previous category. This fact might be essential for their biological role, and it does not come as a surprise that they have a higher content of α-helical domains than iron/sulfur proteins.

3. *Mononuclear proteins with O- and N-donor ligands*
The mononuclear iron sites in these proteins have the same type of ligands found in the previous group: carboxylate-O and His-N (Stallings et al., 1983). In the class of non-heme monoiron oxygenases, the iron center can go from Fe(II) to higher oxidation states as Fe(IV)O during the catalytic cycle (Que Jr., 1993).

4. *Heme proteins*
Heme proteins constitute a vast group of biomolecules that carry out various functions in biological systems. Obviously, the heme moiety, a tetragonal iron complex, is the distinctive feature of these proteins. The ligand in the complex is a porphyrin, a dianionic chelating agent with four N as donor atoms, which has a relatively rigid cavity slightly smaller than the ionic radii of Fe(II) or Fe(III) (Fraústo da Silva and Williams, 1993) (Fig.

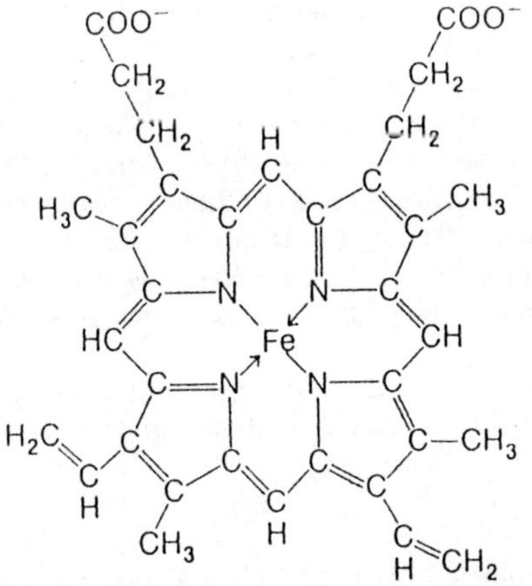

Figure 4.
The structure of heme.

4). The iron atom is inserted in the chelate in a kinetically controlled process similar to chlorophyll biosynthesis (Akhtar, 1991).

The heme moiety is a very stable entity in the sense that the iron does not fall out from the complex. In addition, its reasonably high hydrophobicity renders its incorporation into the hydrophobic core of the protein tertiary structure relatively easy. The heme may be covalently bound to the protein backbone via addition reactions of its vinyl groups, as in *c*-type cytochromes. The axial positions may be occupied by combinations of His-N, Met-S or Lys-N. In many heme proteins which bind small molecules such as H_2O, O_2, CO or NO, the axial ligand may be His-N, Cys-S or Tyr-O (Fraústo da Silva and Williams, 1993).

The environment around the heme in a protein, namely the nature of the donor atoms, the presence or absence of covalent binding to the protein backbone, the protein structure itself, or even different substituents on the porphyrin ring, modulate the properties of the heme. As a result, the oxidation state can be either Fe(II) or Fe(III), and the spin state either high-spin or low-spin (Fraústo da Silva and Williams, 1993). When the protein is performing its biological role, these states interconvert in a cyclic way. In heme enzymes Fe(IV)O is a common intermediate of the catalytic processes (Ortiz de Montellano, 1986).

To finalize this account on iron coordination compounds in biological systems, transport and storage species will be mentioned briefly. Iron ions form complexes with ascorbate, citrate, malate, lactate and tartarate, and it comes as no surprise that all of them are found in vivo (Crichton, 1991). In fact, it is believed that the incorporation of iron into several micro-organisms occurs via a process dependent on a complex of Fe(III) and citrate (Harding and Royt, 1990). Other prokaryotes make use of chelating agents (e.g. ferri-rhodotorulic acid and ferrioxamine B) that bind iron ions exogenously and are incorporated into the organism in a process mediated by receptor proteins (Crichton, 1991). Within higher organisms, iron is transported in serum and in milk bound to transferrins, proteins that coordinate iron via Tyr-O, Asp-O and His-N residues in an octahedral geometry. Curiously, a carbonate anion is coordinated to the iron cation as well (Anderson et al., 1989). Iron can be stored as a precipitate of ferrihydrite encapsulated in the protein shell of ferritin (Harrison et al., 1989), from which it is mobilized when the need arises.

D. Manganese

This metal exists in biological systems in one of the following oxidation states: Mn(II), Mn(III) or Mn(IV) (Brudvig and Crabtree, 1989). Mn(II) usually occurs in six-coordinate, octahedral complexes, both with O- and N/S-donor groups. They show rapid exchange kinetics of ligands and a certain lability towards Mn(II) loss, but trigonal bypyramidal environments are also known (Stallings et al., 1985).

Mn(III) has a $3d^4$ electron configuration and is therefore susceptible to Jahn-Teller distortion in six-fold coordination. It forms complexes with tetragonal geometry, with four short and two long bonds (Cotton and Wilkinson, 1988). As far as the nature of the donor atoms is concerned, Mn(III) tends to bind to small anions of the RO^- type (Chiswell et al., 1987). On kinetic grounds, Mn(III) is not as labile as Mn(II), and so it usually does not fall out from the active sites of proteins (Fraústo da Silva and Williams, 1993).

Mn^{4+} has such a high polarizing power that it suffers hydrolysis, leading to MnO^{2+}. This species coordinates to small anions of the RO^- type in a distorted octahedral geometry. Mn(IV) can eventually form bridged hydroxy/oxo complexes (Greenwood and Earnshaw, 1984).

A mixed-valence (Mn(III) and Mn(IV))) cluster of the type Mn_nO_m is believed to exist in the oxygen-producing photosystem II, but its structure is still elusive (Vincent and Christou, 1987; Rutherford, 1989).

E. Copper

Copper is present in biological systems in the redox states Cu(I) and Cu(II), easily interconvertible and exhibiting high tendency to bind to organic molecules (Phillips and Williams, 1966). The coordination sphere usually includes N-donor ligands, but O-donor groups are also known (Fraústo da Silva and Williams, 1993). The donor atoms are often arranged in a tetragonal geometry, but 5 coordination (Norris et al., 1986) and tetrahedral (Guss and Freeman, 1983) dispositions are also found.

Cu(II) is a paramagnetic ion, and therefore gives rise to an EPR signal. This fact has led to the classification of Cu(II)-containing proteins in three main groups, depending on their spectroscopic features: type 1, type 2 and type 3 (Boas, 1984). Figure 5 shows examples of EPR spectra originated by the first two types. These are mononuclear centers that can still be distinguished by their visible spectra: type-1 centers exhibit an intense blue color (ε = 1,000 - 10,000 $M^{-1}cm^{-1}$), unlike type-2 clusters, which are weakly colored. Type-3 centers are binuclear copper clusters with the two metal ions close together in such a way that exchange coupling takes place, rendering them EPR-silent.

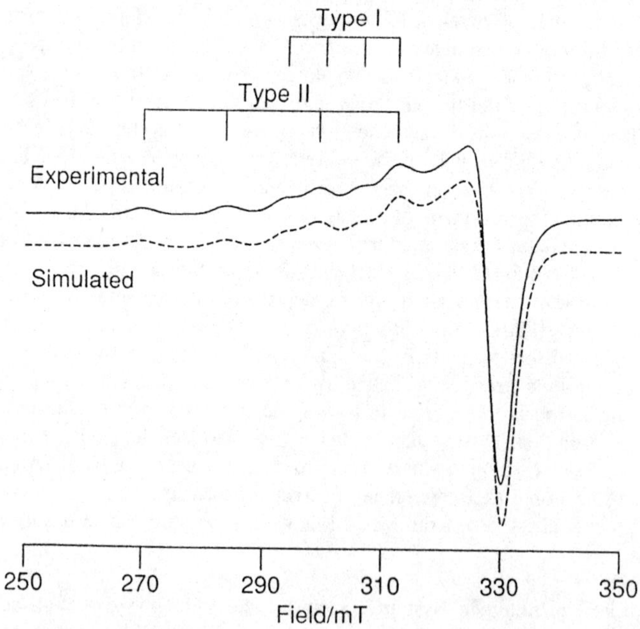

Figure 5.
Simulated and experimental EPR spectra of the type-1 and type-2 copper center-containing enzyme nitrite reductase from *A. xylosoxidans* (from Abraham et al., 1993).

Copper cations are transported in blood bound to ceruloplasmin (Sarkar, 1981), in a four-N environment. When present in excessive amounts, it is sequestered by metallothioneins (Vašák and Kägi, 1983) (which bind seven Cu(I) via thiolate-S ligands) in animals, and by phytochelatins in plants (Grill et al., 1987).

F. Nickel and cobalt

These two elements are grouped under the same entry because their most widespread species in the biological world are two complexes with rings related to porphyrin: Ni(II) in coenzyme F-430 and Co(III) associated with a corrin ligand in vitamin B_{12} (Friedmann et al., 1991).

Both metals show a variety of redox (M(I), M(II) and M(III)) as well as spin (high and low) states, interconvertible while they are engaged in the biochemical reactions where they intervene. Co(II) and Ni(II) complexes are characterized by the small energy differences between 4, 5 and 6-fold coordination, depending on the ligand. For N/S donor groups found in biological systems (F-430 for Ni(II)) the tetragonal geometry is favored, resulting in weak bonds to axial ligands. The facile bond cleavage and change in coordination number, especially those involving carbon-based ligands such as a methyl group, make them of utmost importance in the metabolic pathways where single-carbon groups are transferred.

G. Molybdenum

The bioinorganic chemistry of molybdenum is mainly concerned with redox processes, often with the concomitant transfer of oxo groups (Stiefel, 1993). This comes as no surprise, since this element can easily change its oxidation state from Mo(III) up to Mo(VI), having Mo(IV) and Mo(V) as intermediate states, at relatively low potentials. In its high (Mo(IV), Mo(V), and Mo(VI)) oxidation states, molybdenum forms many compounds with oxo groups (Mo=O), which may be significant in its selection by biological systems.

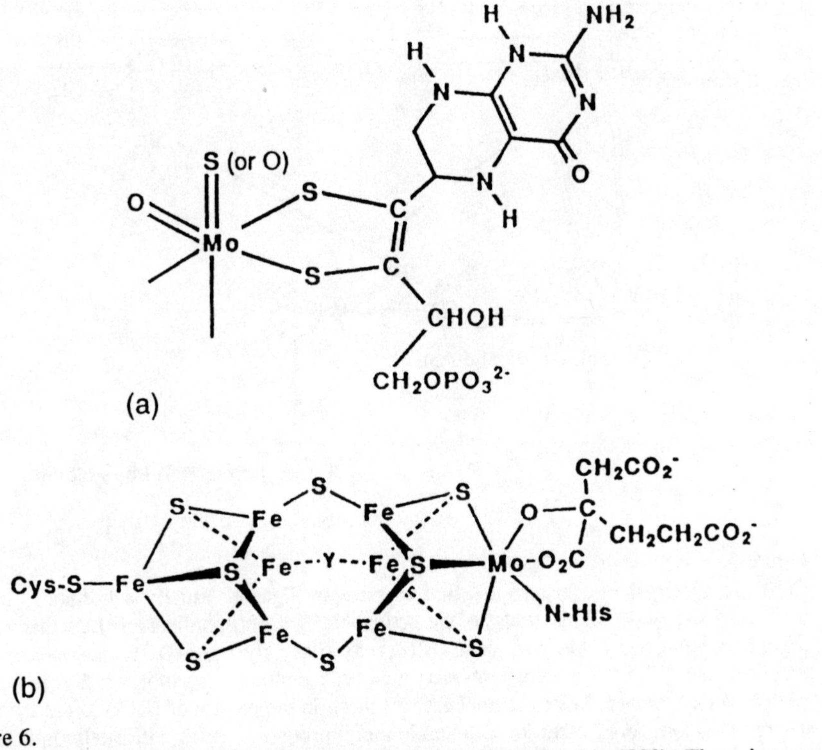

Figure 6.
The structures of molybdopterin (Moco) and FeMoco (from Stiefel, 1993). There is some controversy as to whether the bridging ligand Y is a sulphur or a lighter atom.

In proteins, molybdenum is invariably octahedrally coordinated mainly with sulfur-based donors, and except in nitrogenases, found in close association with molybdopterin (Fig. 6). Molybdenum and molybdopterin constitute the so-called molybdenum co-factor (Moco) site. During catalysis, Moco cycles between Mo(IV) and Mo(VI). The iron-molybdenum co-factor of nitrogenase (FeMoco) consists of a $Fe_7Mo_8S_{8\text{-}9}$ cluster (Fig. 6) in which the octahedral molybdenum is also coordinated to a histidine and a homocitrate.

IV. RELATIONSHIP BETWEEN PHYSICO-CHEMICAL PROPERTIES OF METAL-LIGAND COMPLEXES AND THEIR BIOLOGICAL ROLES

This section briefly presents some of the most outstanding roles of metal ion complexes in vivo. If many of these are now well understood, it should not be forgotten that a great deal still needs clarification or is simply elusive. As a consequence, much has to be explored in the coming years and the outcome is awaited with interest.

A. Calcium

Undoubtedly, the main function of Ca(II) in vivo, as far as its coordination chemistry is concerned, has to do with regulating the metabolic state of cells. Ca(II) is a second messenger, i.e., a chemical species which is released into the cytoplasm as a result of a stimulus from another messenger (Devlin, 1992). It acts by binding to special sites in calcium-binding proteins, known as EF hands or calcium-binding hands (Kretsinger and Nochholds, 1973; Seamon and Kretsinger, 1983). These are sequences of residues relatively rich in O-donor groups, whose spatial arrangement can adopt a geometry suitable to coordinate to Ca(II). On coordination, there is a change in the conformation of the protein

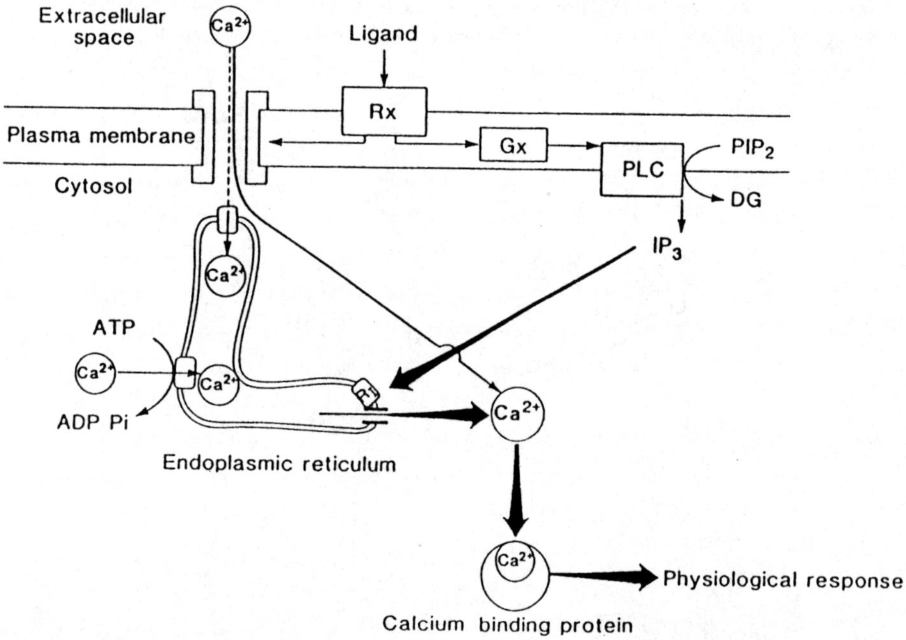

Figure 7.
Ca(II) as a second messenger: the first messenger (ligand), usually a hormone, interacts with the receptor R_x. This leads to the activation of a phospholipase (PLC) that converts phosphatidylinositol-4,5-bisphosphate (PIP_2) to diacylglycerol (DG) and inositol-1,4,5-trisphosphate (IP_3). The latter interacts with a membrane receptor of the endoplasmic reticulum (R_I), where Ca(II) accumulates, resulting in the release of Ca(II) to the cytoplasm. There, this ion coordinates to calcium-binding proteins, which ultimately interact with several biological partners, giving rise to a physiological response (adapted from Khanna et al., 1988).

which endows it with the capacity to trigger a biochemical process by interacting with other proteins (Fig. 7). Examples of such calcium-binding proteins are calmodulins (Chattopadhyaya et al., 1992), parvalbumins (Moews and Kretsinger, 1975), S-100 proteins (Stcharles and Kumar, 1994), annexins (Huber et al., 1992) and troponin C (Satyshur et al., 1988).

The lability of Ca(II) binding to these sites allows the stimulus to come to an end by withdrawing Ca(II) from the cytoplasm, for example by the action of calcium pumps (Harold, 1986).

Other proteins, such as prothrombin (Arni et al., 1993) and osteocalcin (Hauschka, 1980), have γ-carboxyglutamate residues that have high tendency to coordinate to Ca(II), promoting α-helix formation on binding.

On the other hand, Ca(II) forms far less labile complexes with some enzymes. These have active sites made up of residues far apart in the primary structure, which are brought together during protein folding. Some proteins are thermolabile in the absence of Ca(II), e.g. the plant peroxidases (Smith et al., 1990). Others are relatively rigid, keeping the spatial distribution of donor groups even in the absence of Ca(II). Phospholipase A_2 (Dijkstra et al., 1981) and trypsin (Kretsinger, 1976; Martin, 1984) possess such binding sites for Ca(II).

Ca(II) is also believed to play an important role in stabilizing membrane structures. These are rich in components that have O-donor groups, usually glycidic and lipidic moieties, which have some capacity to coordinate to Ca(II). Examples are anionic carbohydrates such as uronates, and lipidic compounds like phosphatidyl serine and several gangliosides (Martin, 1984).

B. Zinc

The properties of Zn(II) seem particularly suited to promote rapid, non-selective hydrolysis (Vallee and Auld, 1990; Kimura, 1993). In fact, several digestive enzymes have Zn(II) in their active sites, and they catalyze the removal of the terminal residue of peptides. It is believed that the Lewis acid properties of Zn(II) and the stereochemical constraint imposed by the protein structure on the substrate backbone come together in such a way as to optimize the catalytic activity (Brown et al., 1984; Fraústo da Silva and Williams, 1993).

Apart from breaking down high-molar-mass substrates, other Zn(II) enzymes deal with small molecules, such as CO_2 in carbonic anhydrase (Erikson and Liljas, 1991) and ethanol in alcohol dehydrogenase (Formicka et al., 1992).

There is some controversy on whether Zn(II) is a hormone on its own, or could trigger in some way a hormone-like response. The reason for this debate lies in the well-established

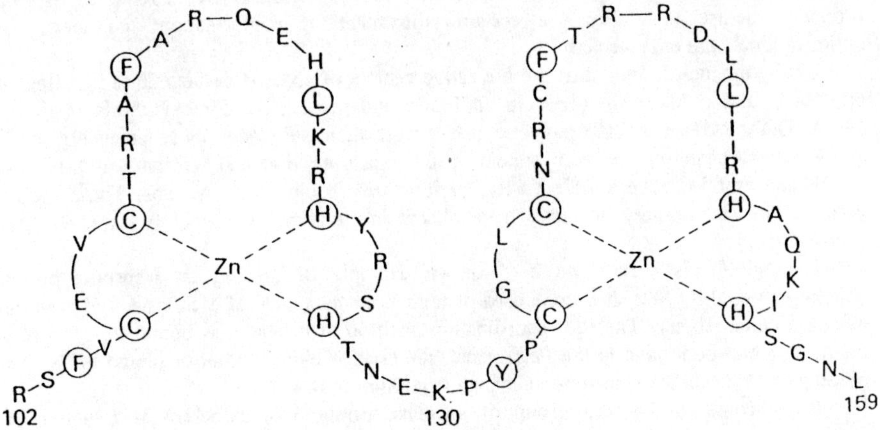

Figure 8.
Schematic representation of zinc fingers. The circled residues are conserved (from Fraústo da Silva and Williams, 1993).

fact that Zn(II), when bound to a certain class of proteins that possess structures known as zinc fingers, interacts with DNA, eventually modulating the expression of genes (Rhodes and Klug, 1993). Each of these proteins has about 30 zinc fingers in a sequence, and each finger is composed of closely-positioned residues that coordinate to Zn(II) (Fig. 8). It is believed that only in the presence of Zn(II) is the protein in the correct conformation to bind to DNA. Furthermore, the lability of the Zn(II)-binding sites is relatively high, and therefore gene expression may be controlled by the concentration of free Zn(II) ions in the nucleus.

In other cases, Zn(II) works as a rather inert cross-linking element, whose function is simply to impart a certain rigidity to a conformation in a protein, as is the case of Cu/Zn superoxide dismutase (Richardson et al., 1976; Fee, 1977).

C. Iron

The different functions iron complexes perform in vivo range from the relatively simple process of electron transfer and oxygen transport to the more complex phenomena of peroxide removal, hormone biosynthesis and xenobiotic degradation.

Iron/sulfur clusters are mostly involved in electron transfer (Cammack, 1979). They exist in proteins whose natural role is simply to shuttle electrons between two other biological partners, as in the ferredoxins (Ikemizu et al., 1994). However, they are also present in multi-prosthetic group enzymes, where the centers work as routes for electron transfer or eventually as electron storage sites, as is the case of several electron transport chain members.

In recent years, there has been mounting evidence supporting the modulation of enzyme activity by binding or dissociating Fe(II) from one of the iron sites in a 4Fe/4S center. Aconitase is a remarkable example, since it is active in the 4Fe/4S form, but inactive when the cluster changes to the 3Fe/4S stoichiometry due to Fe(II) dissociation (Lauble et al., 1992; Beinert and Kennedy, 1989). Obviously, such systems raised doubts concerning the possible role of free iron cations as modulators of enzyme activity. They are also believed to exert their regulatory role at the level of DNA transcription. In fact, several biomolecules that bind to DNA in the presence of an excess of free iron ions have been found. They are known as iron uptake regulatory factors, and control the synthesis of scavenger chelates such as siderophores and ferrioxamines (Fraústo da Silva and Williams, 1993; Ankenbauer and Quan; 1994).

The proteins possessing the group Fe-O-Fe perform various roles in vivo. Examples are the oxygen transport protein of marine organisms, hemerythrin (Holmes and Stenkamp, 1991), some oxidases such as methane monoxygenase (Dalton, 1992; Rosenzweig et al., 1993), that reaches the oxidation state Fe(IV)O during the catalytic cycle, and ribonucleotide reductase (Nordlund and Eklund, 1993). All these have a relatively flexible protein structure that allows a certain adjustment of conformation and thence an optimization of the enzyme activity.

The mononuclear iron sites are the active centers of superoxide dismutase (Stallings et al., 1983) and oxidases involved in antibiotic and prostaglandin biosynthesis (Que Jr., 1993). Other classes include enzymes that render their substrates more amenable to O_2 attack without changing the redox state of iron (Fraústo da Silva and Williams, 1993).

Heme proteins have a wide variety of functions in biological systems. These include transport of metabolites, hormone metabolism, inactivation of xenobiotics and electron transfer.

Hemoglobin is surely the best known example of an oxygen transport protein (Liddington et al., 1992). It is composed of four monomers, two of each type, and each one having a heme moiety. The fifth coordination position of the latter is bound to a His-N of the protein backbone, and in the Fe(II) state, the sixth is either vacant or bound to O_2. The binding of O_2 leads the spin state to change from high to low.

Cytochromes are a vast group of proteins. Some, like cytochromes *b* and *c* are involved in a simple one-electron redox reaction, with the iron ion cycling between Fe(II) and Fe(III) in the low-spin state, the axial coordination sites being occupied by residues of the protein backbone (Moore and Pettigrew, 1990). Others, like *c'*-type cytochromes, scavenge potentially dangerous species as NO (or CO) by irreversibly binding to them

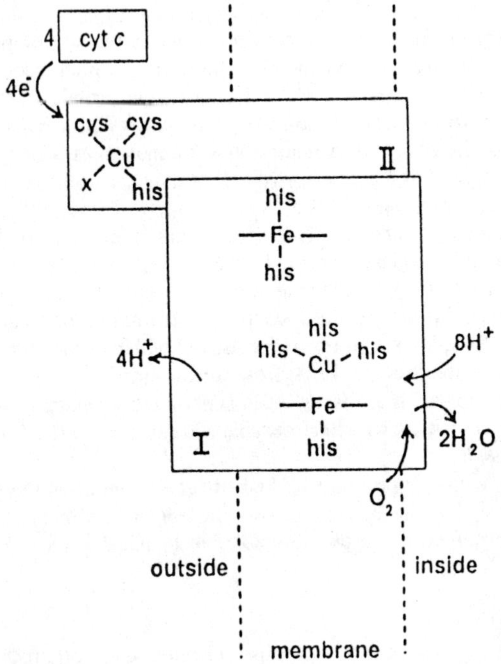

Figure 9.
Schematic model of the role of cytochrome oxidase in O_2 reduction and proton translocation (adapted from Babcock and Wikström, 1992).

(Iwasaki et al., 1991). In fact, xenobiotic inactivation is a field where cytochromes are highly represented. Several examples are mentioned in the book on the P-450 family edited by Ortiz de Montellano (1986). Surprisingly enough, the same groups of enzymes are engaged in hormone metabolism, meaning that the same sort of chemical reactions occurs during both processes.

The important role of nitric oxide, NO, in physiology has recently been recognized. NO acts as a hormone, neurotransmitter, cytotoxic agent, and promotes smooth muscle relaxation (Feldman et al., 1993; Butler and Williams, 1993). In higher organisms, NO biosynthesis from L-arginine is catalyzed by nitric oxide synthetase, a heme enzyme closely related to the P-450 family of mono-oxygenases.

Finally, two more examples are mentioned: catalase (Vainshtein et al., 1986), an enzyme that converts H_2O_2 to O_2 and H_2O, and cytochrome oxidase (Babcock and Wikström, 1992), the last member of the electron transport chain, that reduces O_2 to water with concomitant proton translocation (Fig. 9).

D. Manganese

The best known examples of systems containing manganese are perhaps prokaryotic superoxide dismutase (Stallings et al., 1985) and the oxygen-evolving center of photosystem II (Brudvig and Crabtree, 1989; Rutherford, 1989). The former has a Mn(III) site with trigonal bypyramidal geometry, and the latter is believed to have a cluster with four manganese ions engaged in the production of O_2 from H_2O.

The bacterial Mn-catalases possess two manganese centers which are believed to be bridged by an oxygen atom and carboxylates, and coordinated to N/O donors from the peptide backbone. In these enzymes, the oxidation state of the Mn atoms cycles between Mn(III) and Mn(IV) during the catalytic process (Khangulov et al., 1993).

Also worth mentioning are the enzymes involved in carbohydrate metabolism, usually with Mn(II) centers, such as galactosyl transferase (Lawson et al., 1994), and the Mn-ribonucleotide reductase family (Plönzig and Auling, 1987).

E. Copper

It comes as no surprise that copper is present in the active site of proteins and enzymes involved in redox pathways. A distinction can be made between electron transfer proteins, which possess one type-1 copper center in a rather rigid protein framework (high β-sheet content, e.g. azurin (Norris et al., 1986) and plastocyanin (Guss and Freeman, 1983)), and copper enzymes. The latter are often multisubunit biomolecules with more than one type of copper center, and some having other prosthetic groups as well. Many examples could be cited, but just a few, representative of a larger group, will be mentioned. Copper oxidases keep the oxygen radical bound to them until it has been converted to water. A good example is cytochrome oxidase (Babcock and Wikström, 1992) of the mitochondrial electron transport chain. Copper enzymes are also deeply involved in the nitrogen cycle (Kroneck et al., 1992). All the reactions where nitrogen oxides take part in are catalyzed by a copper enzyme. Examples are nitrite reductases (Godden et al., 1991) and nitrous oxide reductases (von Wachenfeldt et al., 1994) from prokaryotes.

As already mentioned, there is a Cu,Zn-superoxide dismutase, where the Zn(II) site has simply a structural role. The actual enzyme activity lies on the Cu(II) site (Richardson et al., 1976; Fee, 1977).

Hemocyanin is a copper protein well known for its role as an oxygen carrier (Volbeda and Hol, 1989). Its active site is made of two Cu(I) ions coordinated to three His-N each, that on binding oxygen exhibit a purple color due to $[Cu(II)]_2O_2^{2-}$ (Fraústo da Silva and Williams, 1993).

F. Nickel and cobalt

The presence of these two elements is confined to a few metabolic pathways, most of them concerned with the transfer of single-carbon groups (Walsh and Orme-Johnson, 1987; Friedmann et al., 1991). In the metabolic pathways of archaebacteria, Ni-hydrogenases catalyze the reduction of sulphate ions, and of CO_2 to methane. The CO dehydrogenase, an enzyme requiring Ni and Co, transfers a methyl group from Co to Ni, and then onto a metal bound CO molecule to generate ultimately acetyl co-enzyme A (Qiu et al., 1994).

The role of nickel in mammals is limited to urease, which catalyzes the hydrolysis of urea to ammonia and bicarbonate. Two Ni centers are involved in this Lewis acid-catalyzed reaction.

Co-containing vitamin B_{12} enzymes catalyze reactions in which the first step is Co-C bond cleavage to form a carbon-based radical. This radical then generates substrate radicals which rearrange to give the product and regenerate the Co-C bond (Pratt, 1972), as in diol dehydratase. Other vitamin B_{12}-dependent examples are bacterial ribonucleotide reductase, methionine synthetase and methylmalonyl CoA-mutase.

G. Molybdenum

Molybdenum enzymes are found in anaerobic and aerobic bacteria, plants and mammals. They catalyze reactions in energy generation, nutrient assimilation and metabolism of organo-nitrogen compounds, and as such are very important in the global nitrogen, sulphur and carbon cycles (Kroneck et al., 1992; Hille, 1993).

Perhaps the most striking aspect of molybdenum bioinorganic chemistry is its involvement in the nitrogen cycle. Molybdenum nitrogenases contain an Fe-Mo-S cluster of unusual structure, and catalyze the conversion of N_2 to NH_3 under ambient conditions (Thornley and Lowe, 1985; Yates, 1992). The revelation from the recent crystal structures (Kim and Rees, 1992; Bolin et al., 1990) was that the molybdenum center was six-coordinate, with three S, a His-N and a bidentate homocitrate ligands. In this form, the Mo center is coordinatively saturated and will not bind N_2, although reversible loss of one or more ligands from Mo and subsequently N_2 binding and activation cannot yet be ruled out.

The other molybdenum enzymes contain the molybdenum cofactor (Moco), and catalyze the making and breaking of X=O bonds (nitrate reductase, sulphite reductase, arsenite oxidase, formylmethanofuran dehydrogenase) (Stiefel, 1993) and oxidation of C-H bonds (xanthine oxidase). The unusual feature of molybdenum hydroxylases such as xanthine oxidase is that, unlike heme, non-heme, flavin or pterin-containing hydroxylases which use oxygen and consume reducing equivalents from NADH or NADPH, these enzymes use water as the source of oxygen incorporated into the substrates, *and* generate reducing equivalents (Hille, 1993). Nitrate reductase catalyzes the reduction of nitrate to nitrite. The nitrate may be the terminal acceptor in respiration (cf O_2 in aerobic respiration), or the nitrite product may be further reduced via other N_xO_y species by heme and Cu-containing enzymes to ammonia for assimilation.

ACKNOWLEDGEMENT

P. de Oliveira thanks Fundação Calouste Gulbenkian for financial support.

REFERENCES

Abraham, Z.L., Lowe, D.J., and Smith, B.E. (1993). *Biochem. J.*, *295*: 587.

Adams, P.R., Brown, D.A., and Constanti, A. (1982). *J. Physiol. (London)*, *332*: 223.

Akhtar, M. (1991). *Biosynthesis of Tetrapyrroles* (P.M. Jordan, ed.), Elsevier, Amsterdam, p. 67.

Anderson, B.F., Baker, H.M., Dodson, E.J., Norris, G.E., Rumhall, S.V., Waters, J.M., and Baker, E.N. (1989). *J. Mol. Biol.*, *209*: 711.

Ankenbauer, R.G., and Quan, H.N. (1994). *J. Bacteriol.*, *176*: 307.

Armstrong, W.H. (1988). *ACS Symp. Series*, *372*: 1.

Arni, R.K., Padmanabhan, K., Padmanabhan, K.P., Wu, T.P., and Tulinsky, A. (1993). *Biochemistry*, *32*: 4727.

Babcock, G.T., and Wikström, M. (1992). *Nature*, *356*: 301.

Basolo, F., and Pearson, R.G. (1967). *Mechanism of Inorganic Reactions*, John Wiley and Sons, New York.

Beale, S.I., and Weinstein, J.D. (1991). *Biosynthesis of Tetrapyrroles* (P.M. Jordan, ed.), Elsevier, Amsterdam, p. 155.

Beinert, H., and Kennedy, M.C. (1989). *Eur. J. Biochem.*, *186*: 5.

Boas, J.F. (1984). *Copper Proteins and Copper Enzymes*, Vol. I (R. Lontie, ed.), CRC Press, Boca Raton, Florida, p. 5.

Bolin, T.J., Ronco, A.E., Mortenson, L.E., Morgan, T.V., Williamson, W.E., and Xuong, N.-H. (1990). *Nitrogen Fixation: Achievements and Objectives*, (P.M. Ereshoff, L.E. Roth, G. Stacey and W.E. Newton, eds.), Clapman and Hall, New York, p. 117.

Branden, C., and Tooze, J. (1991). *Introduction to Protein Structure*, Garland Publishing, New York.

Brown, D.A. (1988). *Ion Channels* (T. Narahashi, ed.), Plenum Press, New York, p. 55.

Brown, R.S., Huguet, J., and Curtis, N.J. (1984). *Metal Ions in Biological Systems*, Vol. 15 (H. Sigel, ed.), Marcel Dekker, New York, p. 55.

Brudvig, G., and Crabtree, R. (1989). *Prog. Inorg. Chem.* (S. Lippard, ed.), *37*: 99.

Butler, A.R., and Williams, D.L.H. (1993). *Chem. Soc. Rev.*, 22: 233.

Cammack, R. (1979). *Metalloproteins: Structure, Molecular Function and Clinical Aspects* (U. Weser, ed.), Georg Thieme Verlag, Stuttgart, p. 162.

Chattopadhyaya, R., Meador, W.E., Means, A.R., and Quiocho, F.A. (1992). *J. Mol. Biol.*, *228*: 1177.

Chiswell, B., McKenzie, E.D., and Lindoy, L.F. (1987). *Comprehensive Coordination Chemistry*, Vol. 4 (G. Wilkinson, R.D. Gillard and J.A. McCleverty, eds.), Pergamon Press, Oxford, p. 1.

Cotton, F.A., and Wilkinson, G. (1988). *Advanced Inorganic Chemistry*, 5th ed., J. Wiley and Sons, New York.

Crichton, R.R. (1991). *Inorganic Biochemistry of Iron Metabolism*, Ellis Horwood Limited.

Dalton, H. (1992). *Methane and Methanol Utilisers* (J.C. Murrell and H. Dalton, eds.), Plenum Press, New York, p. 85.

Devlin, T.M. (1992). *Textbook of Biochemistry with Clinical Correlations* (T.M. Devlin, ed.), Wiley-Liss, Inc., New York, p. 228.

Dijkstra, B.W., Drenth, J., Hol, W.G.J., and Kalk, K.H. (1981). *J. Mol. Biol.*, *147*: 97.

Erikson, E., and Liljas, A. (1991). *The Carbonic Anhydrases* (S.J. Dodgson, R.E. Tashian, G. Gros and N.D. Carter, eds.), Plenum Press, New York, p. 33.

Fee, J.A. (1977). *Superoxide and Superoxide Dismutases* (A.M. Michelson, J.M. McCord and I. Fridovida, eds.), Academic Press, London, p. 173.

Feldman, P.L., Griffith, O.W., and Stuehr, D.J. (1993). *Chem. Eng. News*, *71*: 26.

Formicka, G., Zeppenzauer, M., Fey, F., and Huttermann, J. (1992). *FEBS Lett.*, *309*: 92.

Fraústo da Silva, J.J.R., and Williams, R.J.P. (1993). *The Biological Chemistry of the Elements*, Clarendon Press, Oxford.

Friedmann, H.C., Klein, A., and Thaner, R.K. (1991). *Biosynthesis of Tetrapyrroles* (P.M. Jordan, ed.), Elsevier, Amsterdam, p. 139.

Godden, J.W., Turley, S., Teller, D.C., Adman, E.T., Liu, M.Y., Payne, W.J., and Legall, J. (1991). *Science*, *253*: 438.

Greenwood, N.N., and Earnshaw, A. (1984). *Chemistry of the Elements*, Pergamon Press, Oxford.

Grill, E., Winnacker, E.L., and Zenk, M.H. (1987). *Proc. Nat. Acad. Sci. USA*, *84*: 439.

Guss, J.M., and Freeman, H.C. (1983). *J. Mol. Biol.*, *169*: 521.

Harding, R.A., and Royt, P.W. (1990). *J. Gen. Microbiol.*, *136*: 1859.

Harold, F.M. (1986). *The Vital Force: a Study of Bioenergetics*, W.H. Freeman and Company, New York.

Harrison, P.M., Artymink, P.J., Ford, G.C., Lawson, D.M., Smith, J.M.A., Treffry, A., and White, J.L. (1989). *Biomineralisation* (S. Mann, J. Webb and R.J.P. Williams, eds.), VCH, Weinheim, p. 257.

Hauschka, P.V. (1980). *Vitamin K Metabolism and Vitamin K-Dependent Proteins* (J.W. Suttie, ed.), University Park Press, Baltimore, p. 227.

Heaton, F.W. (1990). *Metal Ions in Biological Systems*, Vol. 26 (H. Sigel and A. Sigel, eds.), Marcel Dekker, New York, p. 119.

Hille, R. (1993). *ACS Symp. Series*, *535*: 22.

Holmes, M.A., and Stenkamp, R.E. (1991). *J. Mol. Biol.*, *220*: 723.

Huber, R., Berendes, R., Burger, A., Schneider, M., Karshikov, A., Luecke, H., Romisch, J., and Paques, E. (1992). *J. Mol. Biol.*, *223*: 683.

Huheey, J.E. (1983). *Inorganic Chemistry*, 3rd ed., Harper and Row, New York.

Ikemizu, S., Bando, M., Sato, T., Morimoto, Y., Tsukihara, T., and Fukuyama, K. (1994). *Acta Cryst., Sect. D*, *50I:* 167.

Iwasaki, H., Yoshimura, T., Suzuki, S., and Shidara, S. (1991). *Biochim. Biophys. Acta*, *1058*: 79.

Keeling, D., and Benham, C., eds. (1989). *Ion Transport*, Academic Press, London.

Khangulov, S., Sivaraja, M., Barynin, V.V., and Dismukes, G.C. (1993). *Biochemistry*, *32*: 4912.

Khanna, N.C., Tokuda, M., and Waisman, D.M. (1988). *Hormones and their Actions*, Part II (B.A. Cooke, R.J.B. King and H.J. van der Molen, eds.), Elsevier, Amsterdam, p. 63.

Kim, J.S., and Rees, D.C. (1992). *Nature*, *360*: 553.

Kimura, E. (1993). *Pure Appl. Chem.*, *32*: 209.

Kirchgessner, M., and Weigand, E. (1983). *Metal Ions in Biological Systems*, Vol. 15 (H. Sigel, ed.), Marcel Dekker, New York, p. 319.

Kretsinger, R.H. (1976). *Annu. Rev. Biochem.*, *45*: 239.

Kretsinger, R.H., and Nochholds, C.E. (1973). *J. Biol. Chem.*, *248*: 3313.

Kroneck, P.M.H., Beuerle, J., and Schumacher, W. (1992). *Metal Ions in Biological Systems*, Vol. 28 (H. Sigel, ed.), Marcel Dekker, New York, p. 455.

Langford, C.M., and Gray, H.B. (1966). *Ligand Substitution Processes*, W.A. Benjamin, New York.

Lauble, H., Kennedy, M.C., Beinert, H., and Stout, C.D. (1992). *Biochemistry*, *31*: 2735.

Lawson, C.L., Vanmontfort, R., Strokopytov, B., Rozeboom, H.J., Kalk, K.H., Devries, G.E., Penninga, D., Dijkhuizen, L., and Dijkstra, B.W. (1994). *J. Mol. Biol.*, *236*: 590.

Liddington, R., Derewenda, Z., Dodson, E., Hubbard, R., and Dodson, G. (1992). *J. Mol. Biol.*, *228*: 551.

Lippard, S.J. (1988). *Angew. Chem. (Int. Ed. Engl.)*, *27*: 344.

Martin, R.B. (1984). *Metal Ions in Biological Systems*, Vol. 17 (H. Sigel, ed.), Marcel Dekker, New York, p. 1.

Martin, R.B., and Mariam, Y.H. (1979). *Metal Ions in Biological Systems*, Vol. 8 (H. Sigel, ed.), Marcel Dekker, New York, p. 57.

Moews, P.C., and Kretsinger, R.H. (1975). *J. Mol. Biol.*, *91*: 201.

Moore, G.R., and Pettigrew, G.W. (1990). *Cytochromes c*, Springer Verlag, Berlin.

Nakamura, A., and Ueyama, N. (1988). *ACS Symp. Series*, *372*: 292.

Narahashi, T., ed. (1992). *Ion Channels*, Plenum Press, New York.

Nordlund, P., and Eklund, H. (1993). *J. Mol. Biol.*, *232*: 123.

Norris, G.E., Anderson, B.F., and Baker, E.N. (1986). *J. Amer. Chem. Soc.*, *108*: 2784.

Ortiz de Montellano, P.R. (1986). *Cytochrome P-450: Structure, Mechanism and Biochemistry* (P.R. Ortiz de Montellano, ed.), Plenum Press, New York, p. 217.

Phillips, C.S.G., and Williams, R.J.P. (1966). *Inorganic Chemistry*, Oxford University Press, Oxford.

Plönzig, J., and Auling, G. (1987). *Arch. Microbiol.*, *146*: 396.

Pratt, J.M. (1972). *Inorganic Chemistry of Vitamin B_{12}*, Academic Press, New York.

Qiu, D., Kumar, M., Ragsdale, S.W., and Spiro, T.G. (1994). *Science*, *264*: 817.

Que, Jr., L. (1993). *Bioinorganic Catalysis* (J. Reedijk, ed.), Academic Press, New York, p. 347.

Rhodes, D., and Klug, A. (1993). *Scient. Amer.*, *268*: 32.

Richardson, J.S., Richardson, D.C., Thomas, K.A., Silverton, E.W., and Davies, D.R. (1976). *J. Mol. Biol.*, *102*: 221.

Rosenzweig, A.C., Frederick, C.A., Lippard, S.J., and Nordlund, P. (1993). *Nature*, *366*: 537.

Rutherford, A.W. (1989). *Trends Biochem. Sci.*, *14*: 227.

Sarkar, B. (1981). *Metal Ions in Biological Systems*, Vol. 12 (H. Sigel, ed.), Marcel Dekker, New York, p. 233.

Satyshur, K.A., Sambhorao, T.R., Pyzalska, D., Drendall, W., Greaser, M., and Sundaralingam, M. (1988). *J. Biol. Chem.*, *263*: 1628.

Seamon, K.B., and Kretsinger, R.H. (1983). *Met. Ions Biol.*, *6*: 1.

Smith, A.T., Santana, N., Dacey, S., Edwards, M., Bray, R.C., Thornley, R.N.F., and Burke, J.F. (1990). *J. Biol. Chem.*, *265*: 13335.

Stallings, W.C., Powers, T.B., Pattridge, K.A., Fee, J.A., and Ludwig, M.L. (1983). *Proc. Nat. Acad. Sci.*, *80*: 3884.

Stallings, W.C., Pattridge, K.A., Strong, R.K., and Ludwig, M.L. (1985). *J. Biol. Chem.*, *260*: 16424.

Stcharles, R., and Kumar, V.D. (1994). *J. Mol. Biol.*, *236*: 953.

Stiefel, E.I. (1993). *ACS Symp. Series*, *535*: 1.

Thornley, R.N.F., and Lowe, D.J. (1985). *Molybdenum Enzymes* (T.G. Spiro, ed.), Wiley, New York, p. 222.

Tronrud, D.E., Schmid, M.F., and Matthews, B.W. (1986). *J. Mol. Biol.*, *188*: 443.

Tsien, R.W., Hess, P., McCleskey, E.W., and Rosenberg, R.L. (1987). *Annu. Rev. Biophys. Biophys. Chem.*, *16*: 265.

Vainshtein, B.K., Bartels, K.S., Barynin, V.V., Borisov, V.V., Fita, I., Grebenko, A.I., Melikadamyan, W.R., Rossmann, M.G., and Vagin, A.A. (1986). *J. Mol. Biol.*, *188*: 49.

Vallee, B.L., and Auld, D.S. (1990). *Proc. Nat. Acad. Sci. USA*, *87*: 220.

Vandervies, S.M., Gatenby, A.A., Viitanen, P.V., and Lorimer, G.H. (1993). *ACS Symp. Series, 526*: 72.

Vašák, M., and Kägi, J.H.R. (1983). *Metal Ions in Biological Systems*, Vol. 15 (H. Sigel, ed.), Marcel Dekker, New York, p. 213.

Vincent, J.B., and Christou, G. (1987). *Inorg. Chim. Acta, 136*: L42.

Volbeda, A., and Hol, W.G.J. (1989). *J. Mol. Biol., 209*: 249.

von Wachenfeldt, C., de Vries, S., and van der Oost, J. (1994). *FEBS Lett., 340*: 109.

Voordouw, G., Hagen, W.R., Krüse-Wolters, M., van-Berkel-Arts, A., and Veeger, C. (1987). *Eur. J. Biochem., 162*: 31.

Walsh, C.T., and Orme-Johnson, W.H. (1987). *Biochemistry, 26*: 4901.

Wiech, H., and Zimmermann, R. (1993). *ACS Symp. Series, 526*: 84.

Williams, R.J.P. (1992). *Aluminium in Biology and Medicine*, Ciba Foundation Symposium 169 (D.J. Chadwick and J. Whelan, eds.), John Wiley and Sons, Chichester, p. 1.

Williams, R.J.P. (1993). *Magnesium and the Cell* (N.J. Birch, ed.), Academic Press, London, p. 15.

Yates, M.G. (1992). *Biological Nitrogen Fixation* (G. Stacey, R.H. Burris and H.J. Evans, eds.), Chapman and Hall, New York, p. 685.

Yu, Z.D., Zhang, N., Wu, X.J., Gao, Z., Zhu, Q.H., and Kong, F.N. (1993). *J. Chem. Phys., 99*: 1765.

THEORETICAL CONSIDERATIONS FOR THE BINDING OF METALS IN BIOLOGICAL FLUIDS

by

Brian R. Gibney and Vincent L. Pecoraro

Department of Chemistry,
University of Michigan,
Ann Arbor, MI 48109-1055
U.S.A.

I. INTRODUCTION

This chapter describes the theoretical basis that underpins the rational design of metal chelation agents. The ideal molecule will have a high specificity for the the target metal (an absolute thermodynamic criterion), significantly lower affinity for non-target metals (i.e. selectivity, a relative thermodynamic criterion) and rapidly reach equilibrium (a kinetic criterion). Additionally one must consider the following factors for metal sequestration in biological fluids: the drug should be nontoxic, able to reach the target site (e.g. passing the blood brain barrier for Al(III) removal from the brain), remove metals only from target tissues or target proteins (e.g. Fe(III) should be removed from ferritin or transferrin but not from cytochromes or hemoglobin). Finally an ideal drug would be orally active and inexpensive. Because many of the more "biological factors" will be described in other chapters of this treatise, we will focus our discussion on the thermodynamic and kinetic aspects of ligand design.

II. THERMODYNAMICS

The relative stability of a ligand with various metal ions gives rise to ion specificity. When referring to the stability of a metal-ligand complex, one should be cognizant of the difference between thermodynamic and kinetic stability. Thermodynamic stability measures the extent to which the complex is formed when the system has reached equilibrium. Kinetic stability, on the other hand, refers to the rate of the reaction which leads to the attainment of thermodynamic equilibrium. Whereas thermodynamically favored/disfavored products are referred to as stable/unstable, kinetically favored/disfavored products are discussed as inert/labile. A discussion of the factors influencing the thermodynamic stability of metal-ligand complexes will be followed by kinetic considerations.

The stability of a metal-ligand complex is dependent both on enthalpy (ΔH) and entropy (ΔS) terms. These two state functions are related through the free energy expression shown as equation 1. Stronger metal-ligand bonds and charge neutralization associated with an M^{n+} binding to a ligand ($\Delta H < 0$) as well as greater disorder in the system ($\Delta S > 0$) lead to more stable

$$\Delta G = \Delta H - T\Delta S \tag{1}$$

complexes ($\Delta G < 0$ is indicative of a spontaneous process). For metal complexes the increase in entropy upon complexation is not obvious since, at first glance, the binding of a metal ion to a ligand, H_hL appears to be a process leading to greater order. However, metal ions in solution are highly hydrated. This hydration effect is both for the first coordination sphere ligands and second and third coordination solvent molecules that are ordered through hydrogen bonding to first coordination sphere ligands. Thus, the sequestration of the metal leads to a major increase in solution disorder by breaking down the local structure imposed by the Lewis acidity of the metal and by liberating first coordination sphere water molecules as shown in equation 2.

$$M(H_2O)_x^{n+} + L^{m-} \rightleftharpoons ML^{n-m} + x\,H_2O \tag{2}$$

The change in Gibbs Free Energy, ΔG, can be related to the equilibrium constant, K_{eq}, of a reaction by equation 3 (where R is the universal gas constant and T the temperature in Kelvin).

$$\Delta G_{rxn} = -2.303\ RT \log K_{eq} \tag{3}$$

The stability constant, K_n, for a single metal interacting with a series of monodentate ligands, shown generically as equation 4, is given by equation 5. This stability constant can be broken down into a series of stepwise stability constants that reflect the binding affinity

$$p\,M + q\,LH_r \rightleftharpoons M_pL_q + q{*}r\,H^+ \tag{4}$$

$$K_{eq} = [M_pL_q][H^+]^{q*r} / [M]^p[LH_r]^q \tag{5}$$

of the ligand (LH_r) upon sequential addition to the metal to form ML_q. The family of stepwise constants for the formation of ML_3 are given as equations 6-8. Both the stepwise

$$M + LH_r \rightleftharpoons ML + rH^+ \qquad K_1 = ([ML][H^+]^r)/([M][LH_r]) \tag{6}$$

$$ML + LH_r \rightleftharpoons ML_2 + rH^+ \qquad K_2 = ([ML_2][H^+]^r)/([ML][LH_r]) \tag{7}$$

$$ML_2 + LH_r \rightleftharpoons ML_3 + rH^+ \qquad K_3 = ([ML_2][H^+]^h)/([ML_2][LH_r]) \tag{8}$$

and the overall stability constants are proton dependent since the ligand, LH_r, is treated as an acid. An alternative representation of metal ligand stability is provided by the formation constant (β_{MLH}) in which the ligand is treated as a base. The equilibrium in equation 9 can be written as equation 10. In this formalism, the

$$p\,M^{n+} + q\,L^{x-} + r\,H^+ \rightleftharpoons (H_rM_pL_q)^{[(p*n)-(q*x)+r]} \tag{9}$$

$$\beta_{MLH} = [M_pL_qH_r] / [M^{n+}]^p[L^{x-}]^q[H^+]^r \tag{10}$$

protons are included explicitly only if they form part of the final complex.

Each of these three formalisms provide insight into several factors that influence metal complexation. The four examples we will discuss first are the match between metal and ligand donors, chelate effect, preferred metal coordination geometry and the proton dependence of metal complexation.

A. Metal-donor ligands

Metal-ligand complexes can be considered as Lewis acid-base adducts. The ligand (a Lewis base) donates an electron pair to the empty acceptor orbitals of the metal (the Lewis acid). Appropriate ligand donors can be deduced using the theory of hard and soft acids and bases (HSAB) (Pearson, 1963). Hard acids or bases are usually small, highly charged atoms that have electron distributions that are not very polarizable. Soft acids and bases tend to be of lower charge, are larger and have much more diffuse and polarizable electronic distributions. Hard acids favor adduct formation with hard bases and, likewise, soft acids prefer soft bases. Hard acids include alkali metal and alkaline earth cations (these metals have ns^0 configurations so that valence electrons are in the (n-1)p orbitals), lanthanides (these have valence electrons in the 4f orbitals that are deeply buried and unavailable for bonding) and the small, high valent transition metal ions. Soft acids include the larger, lower valent transition metal ions. One way to distinguish hard and soft acids is by the charge to radius ratio (CRR) of the cation. Small highly charged ions such as Fe^{3+} (CRR=4.7) or Mn^{4+} (CRR=6.0) are very hard while larger lower valent ions such as Hg^{2+} (CRR=2.4) or Cu^+ (CCR=1.35) are soft. Hard bases include oxyanions and amines. Soft bases include phosphines, hydrides, alkyls and iodides.

This preference for hard acids binding to hard bases and soft acids binding to soft bases can be seen by a comparison of stability constants for Fe^{3+} with monodentate halides (F^-, K_1=5.4; Cl^-, K_1=1.9 ; Br^-, K_1= 0.33) (Högfeldt, 1982). The progressive decrease, over an order of magnitude, in stability as one goes down group 17 is a result of the greater polarizability of the increasingly larger halides. An opposite trend would be observed for soft metals binding to these monodentate ligands.

B. The chelate effect

A comparison of the stability constants for the equilibrium between Cu^{2+} with either ethylenediamine ($H_2NCH_2CH_2NH_2$, en) or CH_3NH_2 to form the square planar complexes is instructive. With both ligands, substitution of water for the amine is driven by the stronger Cu^{2+}-N bond. The stability constant for the addition of four methylamine molecules to Cu^{2+} to form $Cu(CH_3NH_2)_4{}^{2+}$ is $10^{12.1}$ while complexation of two en molecules to form $Cu(en)_2{}^{2+}$is $10^{19.9}$(Martell and Smith, 1974-1989). This eight order of magnitude increase in stability is referred to as *the chelate effect* (Schwarzenbach, 1952). For most complexes, the chelate effect is almost exclusively entropic in origin and has two components. First, when a monodentate ligand such as methylamine displaces water there is no change in the number of particles in solution. In contrast, ***two*** bidentate ethylenediamine ligands cause the release of ***four*** water molecules. Thus, *complexation of a bidentate ligand to a metal leads to greater solution disorder than for the corresponding monodentate ligands* and, therefore, a more favorable entropic condition. Second, the coordination of the first methylamine ligand slightly disfavors the binding of the subsequent methylamine molecules due to electronic and statistical considerations. Electronically, the effective positive charge on the Cu^{2+} ion decreases as water is displaced by methylamine since methylamine is the stronger base. Binding of additional methylamine is slightly disfavored since the resulting Cu-N interaction is not so strong as for binding of the first methylamine. Statistically, the complexation of the first methylamine can occur at any of four equivalent sites; however, the second molecule has only three equivalent sites for attachment remaining. These two factors (the first enthalpic, the second entropic) explain the decrease in the stepwise stability constants for methylamine binding to Cu^{2+} (K_1= $10^{4.1}$; K_2= $10^{3.4}$; K_3= $10^{2.7}$; K_4= $10^{1.9}$) (Martell and Smith, 1974-1989). Complexation of the second amine of en is adversely affected by these same two terms and, in fact, the entropic factor is even less favored for ethylenediamine (with the amine tethered to the first) since only two of the remaining three coordination sites are accessible. However, *these two terms are far outweighed by the effective increase in concentration of the second amine in the vicinity of the copper after one end of the ethylenediamine molecule has bound to the metal*. The restricted mobility enhances the complexation of the second half of the ethylenediamine molecule leading to displacement of the second water molecule. The optimal chelate ring sizes are 5 and 6 since these lead to the least strained complexes.

The magnitude of the chelate effect is dependent on the chosen standard state. Typically stability constants are reported for one molar or one molal standard states. Under these circumstances, a large chelate effect is often observed and the effective molar concentration argument satisfactorily explains the phenomenon. As Adamson (1954) has pointed out, the chelate effect effectively disappears if the chosen standard state is unit mole fraction rather than one molal. However, the chelate effect is a very useful and important parameter to consider for optimal ligand design since complexation of most cations by ligands occurs far from unit mole fraction conditions.*

The geometrical preorganization of a metal binding cavity also increases the value of the stability constant by decreasing the enthalpy of desolvation of the ligand. Macrocyclic polyamines show stability constants 5-7 orders of magnitude higher than their open chain analogues (Martell and Smith, 1974-1989). Macrobicyclic systems give an additional 3-5 orders of magnitude in the stability constant over macrocyclic systems (Lehn and Sauvage, 1975). This cryptate effect is a function both of the lower desolvation energy of the ligand and the restriction in the orientation of the chelation atoms which increases the entropy of the system. In general, the greater the preorganization of the metal binding cavity and the higher the denticity, up to the coordination number of the metal, the stronger the metal-ligand complex.

In the preorganized cavities of the organic crown ethers, Gokel et al. (1983) have demonstrated for a series of flexible, unsubstituted crown ethers that the hole-size relationship is not obeyed. In these molecules (i.e. 12-C-4, 15-C-5, 18-C-6, 21-C-7 and 24-C-8) for which the hole-size relationship was developed, it was observed that the greatest stability constant in each case was for K^+ (ionic diameter 2.9Å) even though the cavity diameters ranged between 1.2Å - 4.1Å (Hiraoka, 1982). Furthermore, the greatest stability constant in the series was for 18-C-6 regardless of the metal cation studied. These observations demonstrate that cation-crown stability constants are dependent on at least four criteria: (i) the hole-size relationship, (ii) the solvation enthalpies and entropies of the cation and ligand, (iii) the denticity of the ligand, and (iv) the conformations of the free and complexe ligand.

C. Coordination number and geometrical preferences

The coordination number and subsequent geometrical preference of the metal ion of interest strongly influences the stability of the metal-ligand complex. The examples of $Cu(NH_3)_4$ vs $Cu(NH_3)_6$, $Zn(en)_2$ vs $Zn(en)_3$ and $HgCl_2$ vs $HgCl_4$ are particularly enlightening. The d^9 configuration of Cu^{2+} is unstable to a Jahn-Teller distortion that is usually seen as an axial elongation. Thus, Cu^{2+} complexes are either square planar or axially distorted octahedrons. Complexation of the first four ammonia molecules proceed nicely to form $Cu(NH_3)_4^{2+}$ ($K=10^{13}$); however, binding of the final two ammonia molecules is disfavored since they will occupy the axial sites. In fact, the equilibrium constant for $Cu(NH_3)_4^{2+}$ plus NH_3 to form $Cu(NH_3)_5^{2+}$ is very small ($K_5=10^{-0.6}$) leading to a complex with overall lower stability ($K=10^{12.4}$) (Martell and Smith, 1974-1989). Therefore, formation of $Cu(NH_3)_5^{2+}$ and $Cu(NH_3)_6^{2+}$ is achieved only at high ammonia concentrations.

The coordination preference observed for Cu^{2+} results from a strong electronic preference to lower the energy of electrons in the antibonding orbitals (e_g) of an octahedral field; however, coordination geometry preference can also be seen with metals that are not expected to achieve energy stabilization through changes in d electron configuration. The Zn^{2+} ion is d^{10} and bases geometric preference on minimizing ligand repulsion rather than changing the d orbital configuration. The stepwise stability constants for Zn^{2+} + en $\rightleftharpoons$ $Zn(en)^{2+}$, $Zn(en)^{2+}$ + en $\rightleftharpoons$ $Zn(en)_2^{2+}$ and $Zn(en)_2^{2+}$ + en $\rightleftharpoons$ $Zn(en)_3^{2+}$ are $10^{5.7}$, $10^{4.9}$ and $10^{2.0}$, respectively (Martell and Smith, 1974-1989). The precipitous drop on

---

* It is important to note that the difference in the units of the formation constants of $Cu(CH_3NH_2)_4^{2+}$ and $Cu(en)_2^{2+}$, $mol^{-4}\ dm^{12}$ and $mol^{-2}\ dm^{6}$ respectively, reflects the concentration dependence of the chelate effect. To circumvent this difference in units for the stability constants one could compare the stabilities of Fe^{3+} with salicylaldehyde versus benzaldehyde or phenol. In the case of the FeL_3 species formed by $Fe^{3+} + 3L \rightarrow FeL_3$, the salicylaldehyde complex is far more stable than either of the monodentate ligands. Although the units of the stability constants are now equivalent, one must ensure that other factors such as solvation energy, charge, electronic effects, pK_a's and bond strengths between the ligands involved in the comparison are not significantly different for this definition of the chelate effect to be meaningful. For a discussion of this see: May, P.M., and Bulman, R.A. (1983). *Prog. Med. Chem.*, *20*: 225; and Fiabane, A.M., and Williams, D.R. (1977). *The Principles of Bio-Inorganic Chemistry*, The Chemical Society, London, p. 42.

going from $Zn(en)_2^{2+}$ to $Zn(en)_3^{2+}$ is due to a change from tetrahedral to octahedral geometry. Even more dramatic effects are observed with Hg^{2+} which also is a d^{10} ion. The stepwise stability constants for Hg^{2+} reacting with chloride are $10^{6.7}$, $10^{6.5}$, $10^{0.9}$ and $10^{1.0}$ for the formation of $HgCl^+$, $HgCl_2$, $HgCl_3^-$ and $HgCl_4^{2-}$, respectively (Martell and Smith, 1974-1989). Both charge and geometric preference are important in this case. Sequentially adding chlorides decreases the positive charge on mercury and disfavors further negative charge accumulation on the complex. More important, Hg^{2+} can form 5d-6s hybrids that strongly favor linear geometries such that complexation of additional chlorides to generate a tetrahedral compound is energetically costly.

One must be careful to define the oxidation state of the metal to be sequestered since it is common for different oxidation states to have different coordination number and geometrical preferences. The Cu^{2+} is usually square planar while Cu^+ is often tetrahedral. Even more striking, the Pt^{2+} ion has a strong preference for four coordinate, square planar complexes while Pt^{4+} is stabilized greatly in a six coordinate, octahedral environment.

Very recently, Franczyk et al. (1992) have described methods to selectively bind transition metals and actinides using shape selective (stereognostic) recognition of complex ions such as VO^{2+} or UO_2^{2+}. Stereognostic ligands differ from many conventional chelating agents by offering functional groups that recognize molecular shape. For example, molecules that can coordinate cylindrical ions and offer hydrogen bonding groups along the complex ion axis will have added affinity for VO^{2+} or UO_2^{2+}. Similarly, Pb^{2+} has stereochemically active lone pairs that can be exploited to provide an added level for metal recognition. The importance of using stereognostic recognition in the design of a chelating agent will be dependent on the target metal; however, conceptually this approach offers a new viewpoint for the design of selective metal sequestration agents.

D. The competition between protons and metal

The competition between the target metal and protons for the ligand is one of the most important considerations when evaluating a sequestering agent's effectiveness. In theory, β_n provides a measure of the strength of the ligand-metal interaction and the thermodynamic basis for metal ion specificity; however, this constant does not include protons. The hydrogen ion competition impacts the metal binding affinity in three ways. First, the proton release stoichiometry ultimately defines the pH dependence of the reaction. Neutral, non-protonated ligands will show little or no pH dependent binding while molecules that release large numbers of protons will have dramatic pH dependencies. Thus, simple crown ethers may show little decrease in alkali cation affinity due to protons while the tris catecholate ligand enterobactin has a six order of magnitude proton dependence (vide infra). Second, the basicity of the ligand determines the range where proton competition is important. Polyaminocarboxylic acids such as H_4EDTA are relatively acidic and are partially deprotonated at neutral pH. In this case, metal complexation is only modestly affected by protons. In contrast, catecholate or hydroxamate moieties are basic and will express a significant proton dependence at neutral pH values. Third, metal ions, especially hard cations such as Fe^{3+}, can hydrolyze in relatively acidic solution and may form an insoluble oxide by pH 7. It is more difficult for the chelating agent to displace hydroxide than water leading to weaker metal complexation and the slow dissolution of the precipitated metal oxide may effectively limit sequestration from this source.

Care must be exercised when comparing the relative binding affinities of two different ligands for the same metal. A classic example of this is the complexation of ferric ion by the catecholate siderophore enterobactin (Pollack and Neilands, 1970) (ent, $\log \beta_{110}=49$) (Raymond et al., 1984) versus the hydroxamate siderophore desferrioxamine b (DFO, $\log \beta_{110}= 30$) (Schwarzenbach and Schwarzenbach, 1963). Both are formidable binding constants and, on first inspection, one might expect that enterobactin would always be the agent of choice for iron complexation. However, this is not the case since formation constants do not explicitly account for the proton competition for the ligand. Enterobactin encapsulates metals using three catecholate moieties and, at pH 7, releases 6 protons in the process. DFO has three hydroxamate groups and releases only 3 protons upon metal complexation at the same pH. Therefore, enterobactin has an $[H^+]^6$ dependence while DFO has an $[H^+]^3$ proton dependence. Since catecholate ($pK_1 = 8.4$ and $pK_2 = 12$) (Harris et al., 1979a) and hydroxamate (pK=9.3) (Martell and Smith, 1974-1989) ligands are fully protonated at physiologic pH, one can write a competition reaction for the two ligands as given in equation 11 and the corresponding *competition constant*, K_{comp}, as equation 12. Thus, for each unit decrease in pH, enterobactin is 10^3 less effective in sequestering Fe^{3+} relative to

$$Fe^{3+}(DFO)^+ + entH_6 \rightleftharpoons Fe(ent)^{3-} + (DFO)H_3{}^+ + 3H^+ \quad (11)$$

$$K_{comp} = [Fe(ent)^{3-}][(DFO)H_3{}^+][H^+]^3/([Fe^{3+}(DFO)^+][entH_6]) \quad (12)$$

DFO. Remarkably, $K_{comp} = 10^{-14.5}$ even though there is such a great disparity between the formation constants. Thus, the two ligands would have equivalent affinities for Fe^{3+} by pH 5. This equilibrium can be further perturbed by accounting for the existence of protonated metal/ligand species. For example, ferric enterobactin can be protonated in three sequential steps according to equations 13, 15

$$Fe(ent)^{3-} + H^+ \rightleftharpoons Fe(ent)H^{2-} \quad (13)$$

$$K^H{}_1 = [Fe(ent)H^{2-}]/([Fe(ent)^{3-}][H^+]) \quad (14)$$

$$Fe(ent)H^{2-} + H^+ \rightleftharpoons Fe(ent)H_2{}^- \quad (15)$$

$$K^H{}_2 = [Fe(ent)H_2{}^-]/([Fe(ent)H^{2-}][H^+]) \quad (16)$$

$$Fe(ent)H_2{}^- + H^+ \rightleftharpoons Fe(ent)H_3 \quad (17)$$

$$K^H{}_3 = [Fe(ent)H_3]/([Fe(ent)H_2{}^-][H^+]) \quad (18)$$

and 17 to form $Fe(ent)H_3$ (Harris et al., 1979a). The metal chelate protonation constants for each reaction are given as equations 14, 16 and 18, respectively. One must calculate the true $Fe(ent)^{3-}$ concentration through equation 19 for use in equation 12. The $[Fe(ent)^{3-}]_T$

$$[Fe(ent)^{3-}] = [Fe(ent)^{3-}]_T/(1 + K^H{}_1[H^+] + K^H{}_1K^H{}_2[H^+]^2 + K^H{}_1K^H{}_2K^H{}_3[H^+]^3) \quad (19)$$

$$[Fe(ent)^{3-}]_T = Fe(ent)^{3-} + HFe(ent)^{2-} + Fe(ent)H_2{}^- + Fe(ent)H_3 \quad (20)$$

is given by the mass balance equation 20. Similar protonation constants can be defined for protonation of $Fe(DFO)^+$ and would also need to be considered in a thorough comparison of metal complexation by these siderophores.

Because pK_a values and proton stoichiometries vary from ligand to ligand, Raymond has suggested that the most effective way to compare the affinity of different ligands for a metal is by relative pM values. The pM is defined analogously to pH (i.e. pM= -log $[M^{n+}]_F$, where $[M^{n+}]_F$ is defined as all forms of metal that are not complexed to the ligand). The conditions of the calculation must be explicitly stated and usually are pH=7, $[ligand]_T$= 10 mM and $[M^{n+}]_T$= 1 mM (where $[ligand]_T$ and $[M^{n+}]_T$ correspond to the total ligand and metal initially added to the solution). The greater the value of pM, the more thermodynamically effective is the ligand to sequester the metal under these conditions. In the comparison of enterobactin and DFO binding of iron described above, the pM for enterobactin is 35.6 and that of DFO is 26.6 (Harris et al., 1979b).

III. KINETICS

Although far less attention has been focused on the kinetic factors that influence metal complexation, the rate of metal binding to the chelating agent and/or the rate of metal ion release from the targeted biomolecule may be every bit as important for the efficacious removal of metals from biological fluids. We can break down kinetic effects for metal sequestration into four broad classes. First is the mechanism of the complexation reaction which, in the limiting cases, is either associative or dissociative. Second, the inherent rate of ligand exchange leads to complexes that are labile (rapid exchange) or inert (slow exchange). The third factor is the ligand architecture that may trap a normally labile metal within a larger molecular framework making it inaccessible for exchange reactions. Finally, the fourth factor is the conformational changes of macromolecules that may restrict ligand access to the metal.

The kinetic mechanism for ligand complexation can be either associative, when ligand attack into the metal coordination sphere leads to an intermediate with a higher coordination number, or dissociative, whereby a bound ligand is lost to form an intermediate of lower coordination number (Basolo and Pearson, 1967). Metals such as Fe^{3+} usually follow dissociative pathways while Pt^{2+} usually can be described by an associative mechanism. The donor properties of the sequestering agent can be very important in associative processes since bond formation to generate the higher coordination number intermediate is the rate-limiting step. Bulky ligands will slow the observed rate while an increased charge on the initial complex will lead to increased rates. The rate dependence on the entering nucleophile follows the principles of polarizability previously described for hard and soft acid base theory. It should be recognized that the kinetic parameters are affected far more greatly than the equilibrium constants by the polarizability of the nucleophile and the metal substrate. Dissociative processes are far less dependent on the sequestration agent (entering nucleophile) because the rate-limiting step is based on the breaking of a bond in the first coordination sphere of the metal to form an intermediate of lower coordination number. Therefore, ligand design does not directly moderate the rate of the chelation event. Very hard metals will often hydrolyze and then form insoluble polymeric phases. Removal of metals from these crosslinked species is very difficult because it is a heterogeneous reaction (liquid/solid) and because the structure of the polymer (with multiple oxo- and hydroxo-bridged metal species) is difficult to disrupt. This situation is operative in ferritin where complex iron/hydroxy/phosphate polymers form the metal binding core of the protein (Crichton, 1991).

The electronic configuration of the target metal also plays an important role in defining the metal chelation rate. Most aquated metal ions have water exchange rates on the order of 10^3 to 10^9 M sec^{-1} which allows for rapid ligand exchange and achievement of equilibrium (Eigen, 1963). Notable exceptions to this rule are Cr^{3+} and Co^{3+} which have d^3 and low spin d^6 configurations, respectively. The rate of the reaction is dependent on the energy required to achieve the activated complex. For d^3 and low spin d^6 complexes, there is a significant decrease in ligand field stabilization energy as the five coordinate activated complex is generated. This leads to *exchange inert metal* ions that can have half lives of hours to days. Occasionally, the d electron configuration enhances the exchange rate. For example, high spin d^4 (Mn^{3+}) or d^9 (Cu^{2+}) complexes are subject to Jahn-Teller distortions that labilize the axial ligands. Rate enhancements can often be achieved by high acid concentrations that weaken metal-ligand bonds or by redox reagents (usually reductants) that can change the metal from an inert to labile electronic configuration and reduces the bond strength by reducing the electrostatic interaction between metal and donor. This change in metal oxidation number is an obvious determinant of exchange rate since it is related to metal-ligand bond strength. The rate shows a marked decrease when the oxidation number of the metal increases for dissociative processes while the rate increases for associative reactions.

In some cases, the rate of metal exchange can be markedly diminished through the complete encapsulation of a metal ion. Even though one heteroatom from the multidentate ligand may dissociate from the metal, the increased effective concentration of this pendant heteroatom leads to rapid reinsertion into the metal coordination sphere and does not allow an attacking ligand to gain a coordination site to begin sequestering the metal. For example, the rate of removal of Fe^{3+} from ferrioxamine b by catecholate siderophore analogs is very slow, because a hydroxamate group must dissociate in order for a catecholate to bind to the iron (Tufano and Raymond, 1981). Since the reinsertion rate of the hydroxamic acid is so great, there is little chance for the catechol to gain entry into the iron coordination site (Crumbliss, 1990). Other cases where ligands greatly modify exchange rates are with cryptates where metals are completely encapsulated by the cryptand and effectively block attack by competing molecules (Lehn et al., 1970).

Many ligands and metals cause an allosteric effect upon binding proteins (Stryer, 1988). The ΔH of the protein conformational change will effect the stability constant for the binding to the protein and, therefore, the stability constant needed for exogenous ligands to extract the metal ion. For example, transferrin (Bailey et al., 1988) and lactoferrin

(Anderson et al., 1989) show substantial allosteric effects and present the further complication of a synergistic anion binding which increases the complex stability (Schlabach and Bates, 1975). Ferritin encapsulates iron into iron/hydroxy/phosphate cores that are kinetically robust (Crichton, 1991). The combination of channels (Rice et al., 1983) that do not provide easy access of large chelating agents to the interior of ferritin combined with the low reactivity of olated polymers makes iron removal from ferritin difficult. The release of metals from these proteins is slow and is usually the rate determining step (Cowart et al., 1982). Even though Desferal® (desferrioxamine B, DFO) forms a stable ferric ion complex, the kinetics of iron release from transferrin limits its efficacy. The addition of catalysts in vitro such as nitrilotriacetic acid, a tetradentate ligand, enhances the rate of iron removal from proteins into synthetic ligands by enhancing access to the metal-binding region (Bali et al., 1991).

IV. SUMMARY

In this chapter we have described the factors that must be considered when designing an effective metal chelation agent. Foremost is defining what type of metal is targeted. We can consider five metal types. The first class contains hard, high valent metals such as Fe^{3+} or Ga^{3+}. The second class contains hard, low valent metals such as Na^{+} or Li^{+}. The third class contains high coordination number metals such as the actinides (e.g. Pu^{4+}). The fourth class contains soft, low to moderate valent metals such as Hg^{2+} or Pb^{2+}. The fifth class contains ions that have specific shapes such as VO^{2+}, VO_2^{+} or UO_2^{2+}. In order to selectively and effectively sequester metals in any of these classes, both thermodynamic and kinetic properties of the ligand and the metal must be addressed.

Foremost among the thermodynamic considerations is the ligand donor type. Hard acids such as Fe^{3+} or Pu^{4+} require hard donors such as oxyanions for optimal enthalpic stabilization. In contrast, thiolate ligands are better suited for the softer Hg^{2+} or Pb^{2+}. Furthermore, incorporation of anionic donors (to at least neutralize the charge on the metal) leads to stronger electrostatic interactions and decreases the local order of the solvent. Another very important parameter is matching the denticity of the ligand to the preferred coordination number of the metal. Iron(III) is usually six coordinate and octahedral while Pu^{4+} is often eight coordinate with square antiprismatic geometry. The softer metals Hg^{2+} and Pb^{2+} prefer linear and tetrahedral geometries, respectively. In addition to providing a basis for selectivity, multidentate ligands also enhance the metal affinity through the chelate effect. The cavity size of the multidentate ligand should be designed to tightly fit the specified metal, but not be so small as to exclude it due to steric constraints or ligand strain. In certain cases where metal ions have well defined shape, hydrogen bond donors can be incorporated to further enhance metal selectivity. Finally, a balance between the Lewis basicity of a donor and the proton basicity is necessary. While metal binding affinity will often increase as one uses a more basic ligand, competition for the donor by protons may negate this stronger attraction. For example, alkoxides strongly stabilize higher oxidation states; however, an alkoxide rarely binds to a metal in aqueous solution since the proton basicity is so great. Instead, alcohol complexes are isolated.

The mechanism of ligand exchange is probably the most important initial consideration when attempting to sequester a metal ion. Associative processes can be affected by the nucleophilicity and steric properties of the entering agent and so these considerations should be incorporated into ligand design. Dissociative mechanisms are not affected by the entering ligand type and, as such, do not make available as many options for enhancing the metal exchange rate. Of course, if the hope is to stop a metal from exchanging, as would be the case for MRI agents, the dissociative exchange process would be an advantage since one would design a polydentate ligand that rigidly encapsulated the ion so that other ligands could not enter the metal coordination sphere. If one wants to sequester metals from a system, the most important factor often is the kinetic properties of the metallobiomolecule. Conformational changes or the restricted access to environments in the interior of the protein become the rate-limiting step in the metal binding process.

In the practical design of selective ion chelates, several additional properties are desired. The ideal chelating agent should be readily absorbed, either by injection or oral administration, preferentially distributed to the site of projected utility, not be metabolized to an inactive form and upon chelation of the desired metal should be rapidly excreted. Utmost, it must be non-toxic. Approaches for optimizing these pharmacological factors, e.g. log P values, are discussed more thoroughly elsewhere in this treatise.

REFERENCES

Adamson, A.W. (1954). *J. Amer. Chem. Soc.*, *76*: 1578.

Anderson, B.F., Baker, H.M., Dodson, E.J., Norris, G.E., Rumball, S.V., Waters, J.M., and Baker, E.N. (1989). *J. Mol. Biol.*, *209*: 711.

Bailey, S., Evans, R.W., and Garrett, R.C. (1988). *Biochemistry*, *27*: 5804.

Bali, P.K., Harris, W.R., and Nesset-Tollefson, D. (1991). *Inorg. Chem.*, *30*: 502.

Basolo, F., and Pearson, R.G. (1967). *Mechanisms of Inorganic Reactions*, 2nd ed., Wiley, New York.

Cowart, R.E., Kojima, N., and Bates, G.W. (1982). *J. Biol. Chem.*, *257*: 7560.

Crichton, R.R. (1991). *Inorganic Chemistry of Iron Metabolism*, Ellis Horwood, New York, p 263.

Crumbliss, A.L. (1990). *Coord. Chem. Rev.*, *105*: 155.

Eigen, M.(1963). *Pure Appl. Chem.*, *6*: 105.

Franczyk, T.S., Czerwinski, K.R., and Raymond, K.N. (1992). *J. Amer. Chem. Soc.*, *114*: 8138.

Gokel, G.W., Goli, D.M., Minganti, C., and Echegoyen, L. (1983). *J. Amer. Chem. Soc.*, *105*: 6786.

Harris, W.R., Carrano, C.J., and Raymond, K.N. (1979a). *J. Amer. Chem. Soc.*, *101*: 2213.

Harris, W.R., Carrano, C.J., Cooper, S.R., Sofen, S.R., Avdeef, A.E., McArdle, J.V., and Raymond, K.N. (1979b). *J. Amer. Chem. Soc.*, *101*: 6097.

Hiraoka, M.(1982). *Crown Compounds: Their Characteristics and Applications*, Elsevier Scientific, New York.

Högfeldt, E. (1982). *Stability Constants of Metal-Ion Complexes* (E. Högfeldt, ed.), Pergamon Press, New York.

Lehn, J.-M., and Sauvage, J.P. (1975). *J. Amer. Chem. Soc.*, *97*: 6700.

Lehn, J.-M., Sauvage, J.P., and Dietrich, B. (1970). *J. Amer. Chem. Soc.*, *92*: 2916.

Martell, A.E., and Smith, R.M., eds. (1974-1989). *Critical Stability Constants*, Vols. I-VI, Plenum, New York.

Pearson, R.G. (1963). *J. Amer. Chem. Soc.*, *85*: 3533.

Pollack, J.R., and Neilands, J.B. (1970). *Biochem. Biophys. Res. Commun.*, *38*: 989.

Raymond, K.N., Müller, G., and Matzanke, B.F. (1984). *Topics Curr. Chem.*, *123*: 50.

Rice, D.W., Ford, G.C., White, J.L., Smith, J.M.A., and Harrison, P.M. (1983). *Adv. Inorg. Biochem.*, *5*: 39.

Schlabach, M.R., and Bates, G.W. (1975). *J. Biol. Chem.*, *250*: 2182.

Schwarzenbach, G. (1952). *Helv.Chim. Acta*, *35*: 2344.

Schwarzenbach, G., and Schwarzenbach, K. (1963). *Helv. Chim. Acta*, *46*: 1390.

Stryer, L. (1988). *Biochemistry*, 3rd ed., W.H. Freeman, New York.

Tufano, T.P., and Raymond, K.N. (1981). *J. Amer. Chem. Soc.*, *103*: 6617.

THEORETICAL CALCULATIONS OF METAL-LIGAND SELECTIVITY FOR BIOLOGICAL LIGANDS

by

David R. Garmer

Physiology and Biophysics Department,
Mount Sinai School of Medicine,
New York, New York
U.S.A.

I. INTRODUCTION

This section is concerned with the electronic and statistical mechanical factors producing ion binding selectivity in biological systems, and with how theoretical chemistry can be applied to this problem to understand and calculate. Selectivity is logically divided into two problems, first concerned with intrinsic affinities of specific ligand functional groups for specific ions. This is amenable to theoretical understanding and semiquantitative prediction. Secondly, there is the problem of conformation and strain in large chelating ligands coming to dominate selectivity effects. For ligands simpler than globular proteins, these factors can be understood and predicted with molecular mechanics and molecular dynamics methods. Predictions of ion selectivity and preferred binding arrangements in proteins have only been made by analogy to characterized examples. The complicated nature of proteins has then often led to a surprisingly different reality. However, computational simulation, at least for low-molar-mass ligands, is no longer at the point where predictions are indefinite, illegitimate, or excessively difficult. Many experimental studies could be complemented with simulations to learn about how structure is related to behavior. Work in understanding such ligands as macrocycles using computational methods is more extensive than for biosystems and has been discussed in several very accessible reviews (Hancock and Martell, 1989; Hancock, 1989; 1990; Busch and Stephenson, 1990; Brubaker and Johnson, 1984).

II. THEORY OF FUNCTIONAL GROUP SELECTIVITY

The conventional view of the problem of deciding what ligand functional groups make the strongest and most specific complexes with particular cations relies on experience and the hard-and-soft-acids-and-bases concept (HSAB). There are significant numbers of strongly bound complexes involving soft cations combined with hard ligands and a few with hard cations interacting with soft ligands in biological systems. This has led to criticism of the HSAB concept as being very unreliable (Martin, 1986; see also Chapter III-A of this Part One) but exceptions can be understood as the natural result of the physical origins of HSAB behavior.

Recently, qualitatively correct views have been put forward based on quantum chemical theory. Parr and Pearson (1983) partition the general acid-base interaction into an electrostatic interaction plus an energy-lowering charge transfer process. Trends in the free energy gain from electrostatic interaction in a solvated environment may parallel the trends predicted in the gas-phase from the pure Coulombic interaction. More highly charged or highly polar pairs in complexes interact most strongly in solution even though solvation effects may lead to a primary entropic stabilization in many cases. There is also an element of solvation disfavoring ion pairs of drastically different sizes probably best described as a steric effect (see eq. (1) and succeeding paragraphs below).

Charge transfer from a ligand to a cation is an energy lowering process because the unoccupied valence electron orbitals of the dication are lower in energy than the lone pair orbitals of the ligand. Complete transfer of an electron pair to a dication, for example, does not occur because this eliminates the favorable electrostatic interaction the dication has with its shell of ligands. A balance is reached with partial donation of the lone pair charge density, and the degree of transfer thus depends on the properties of the other ligands in addition to the primary dependence on orbital energy properties discussed by Parr and Pearson (1983). They show that this charge transfer is naturally largest in soft-soft interactions. Energy lowering is also maximized because the cation and ligand orbitals are well-matched to mix covalently and form a partial chemical bond of the dative type. The separate solvated cation and ligand engage in comparatively little charge transfer with the solvent molecules and so this is a very dependable source of free energy lowering for soft-soft pairs.

It is possible to quantify the energy lowering in the gas phase due to electrostatic interactions and from the relaxation of electronic structure. An energy decomposition of quantum chemical Hartree-Fock calculations can be utilized (Gresh and Garmer, work in preparation). The relaxation term, labeled as "covalency" in Table 1, includes the charge transfer effect and a component from the internal polarization of the ligand electron density in helping to form a covalent interaction. A comparison of these energy terms for dications interacting with various ligands including water can give some insight into why violations of the primitive HSAB correlation are observed in solution.

The electrostatic energies for dications of similar size are nearly the same for a given ligand. Covalency in each case is greatest for the softer dications. We first consider the process of replacing a hard water ligand with a slightly less hard formate ligand. The electrostatic components of this reaction for the dications Mg^{2+}, Zn^{2+}, Ca^{2+} and Cd^{2+}, respectively, are -235, -235, -210, and -220 kcal/mol and the covalency components are -39, -48, -33, and -43. This leads to the conclusion that there may be a small selectivity for Zn^{2+} over Mg^{2+} or for Cd^{2+} over Ca^{2+} when they are interacting with moderately hard ligands. Therefore, zinc can replace magnesium or cadmium can replace calcium even with hard ligands and the strict primitive HSAB rule is violated. Similar results are obtained for the HSAB-intermediate hydroxyl and ammonia ligands. Progressing to the soft methiolate

Table 1

Energy Terms from a Quantum Chemical Decomposition of Dication-ligand Interactions (expressed in kcal/mol). Species are listed roughly from hardest to softest.

Ion	Interaction	H_2O	$O(CH_3)_2$	OH^-	HCO_2^-	NH_3	SCH_3^-
Ca^{2+}	electrostatic	- 33	- 27	-278	-243	-38	-214
	covalency	- 20	- 34	-53	-53	-23	-63
Mg^{2+}	electrostatic	- 48	- 41	-334	-283	-57	-257
	covalency	- 32	- 52	-52	-71	-38	-88
Zn^{2+}	electrostatic	- 43	- 37	-332	-278	-55	-259
	covalency	- 45	- 68	-75	-93	-61	-132
Cd^{2+}	electrostatic	- 35	- 28	-295	-255	-41	-233
	covalency	- 32	- 51	-68	-75	-48	-118

ligand gives electrostatic components for replacing water of -209, -216, -181, and -198 $^{kcal}/_{mol}$ and covalency components of -56, -87, -43, and -86. There is now a significantly larger covalency component favoring binding the softer dications. These arguments show that soft cation-hard ligand combinations are likely to be of similar stability to hard cation-hard ligand combinations when the cations are of similar size and identical charge. The soft ligands usually involve large second row atoms so that electrostatic stabilization is likely to be weaker than for first row counterparts. In this case, covalency is required to provide stability in solution and soft cations allow for this. Exceptions to the disfavor of hard cations for soft ligands are likely to occur for the more highly charged cations and ligands where the electrostatic stabilization could be exceptional. An example of this is the Fe^{3+}-cysteinate combination in many iron proteins. Similar combinations of a lower charge hard cation such as Mg^{2+} with soft ligands are much less likely according to this rationalization. We note that for similar ion size and charge the softer ions have greater electron affinities. Therefore, electron affinity should correlate with binding strength as found by Martin (see Chapter III-A of this Part One). For different sizes and charges, solvation effects complicate the understanding of this observation.

A semi-empirical equation for solution $\log K_1$ values has been developed by Hancock and Marsicano (1978; 1980)

$$\log K_1 = E_A E_B + C_A C_B - D_A D_B \quad (1)$$

with parameters for various acids and bases derived from a large experimental base. The product of E constants gives a contribution approximately paralleling the expected electrostatic stabilization while the C constant product parallels the covalency. Hancock and Martell (1989) define an absolute hardness as the ratio E/C, so that softer species have a smaller ratio, and report that this ratio correlates best with physical and chemical properties traditionally related to HSAB behavior. Since large E and C values are somewhat mutually exclusive the most favorable complexing occurs for similar E/C ratios. This is in fact in accordance with the quantum chemical view expressed above, because both methods indicate that the amount of covalency generally increases along with the electrostatic interaction for cations of higher charge. Therefore, the ratio of electrostatic and covalency effects is an appropriate measure of relative hard-soft behavior across the range of charges.

D constants were given non-zero values only for about one-half of the cations and ligands surveyed. The product $D_A D_B$ is rationalized as a steric factor from a ligand interacting with the cation's water shell (Hancock and Martell, 1989). Large values of parameter D_A occur for the smallest cations which as such are most susceptible to steric crowding. A sterically large ligand complexed to a large cation is likely to cause a smaller conflict because the interligand distances are larger and the ligand shell is less rigid. Large D_B parameters occur for sterically large ligands, such as iodate, and for ligands with bulky extensions.

The general accuracy of predicted $\log K_1$ values can be seen in the predictions of pK_a values for hydrated cations, listed here with experimental values from Burgess (1988) in parentheses: Fe^{3+}, 2.2(2.0); Zn^{2+}, 9.6(9.5); Mg^{2+}, 11.5(11.4); Ca^{2+}, 12.9(12.6); Cd^{2+}, 9.8(7.9); Hg^{2+}, 2.4(3.7); Ni^{2+}, 9.8(10.0); Pb^{2+}, 8.2(7.9); Ag^{+}, 12.0(11.9). The numerical estimate agrees with the quantum chemical prediction given above that formate ligands have slightly greater affinity for Zn over Mg (0.6 logK units) and for Cd over Ca (1.0 units) which is borne out by experiments. This scheme might succeed in predicting the usefulness of specific ligand substitutions in polyliganded sites. However, as more of the water shell is replaced by sterically complex or anionic ligands there is an increasing likelihood that the ligand shell structure, not cation-specific properties, determines the effects of ligand or metal substitution.

III. OTHER QUANTUM CHEMICAL EFFECTS AND METHODS

The Irving-Williams stability sequence of dication binding, Mg < Mn < Fe < Co < Ni < Cu > Zn, is partly a consequence of many transition metals having a non-spherical d-shell electron density. Binding usually involves coordination to some ligands of a higher ligand field strength than the replaced water. The transition metals undergo geometric and electronic structure distortions to have the largest d-shell electron density directed away from ligands with the higher ligand field strength. These ligands feel a partially deshielded nuclear charge compared with other ligands or with the corresponding spherical cation complexes producing a relatively more stable transition metal complex. It is possible to use ligand field theory to make specific predictions of this effect for particular ligand arrangements (Blake and Cotton, 1964; Berg and Merkle, 1989). Purely quantum chemical methods have also been developed to understand d-shell effects in transition metal complexes. Probably the most popular is the semi-empirical INDO/S method (Anderson et al., 1986) which is fairly accurate and not extremely expensive in computer time.

A few accurate, but expensive, ab initio quantum chemical calculations have examined cation selectivities in biological systems. These used only a first shell of ligands to represent protein-bound and hydrated states, with a Born solvation model accounting for the reaction field response of the environment. One application examined the selectivity of the two calcium binding sites of enzyme Subtilisin BPN' (Krauss and Stevens, 1990). A buried calcium site was found to have a higher binding energy for Mg^{2+} than for Ca^{2+} but the difference did not counterbalance the larger solvation enthalpy of Mg^{2+}. The optimized first shell structures accounted for this contradicting previous opinion that the site was made rigid by the protein to fit calcium specifically. For the other site, comparable affinities were correctly reproduced but the Mg requires a unidentate carboxyl ligand while the Ca prefers bidentate mode. The authors compare the optimized ligand structure around Na^+ to a crystal structure and state that the crystal has non-optimal lengthened coordination distances for the protein ligands. They postulate that this site has a degree of rigidity, possibly due to salt bridges, affecting the coordination structures and selectivity. Since most carboxyl ligands involved in calcium binding do not have other salt bridges, this suggests that a systematic study of the effect of salt bridges on dication binding would be valuable.

Another application considered the substitution of zinc by cobalt, manganese and cadmium in the enzyme carbonic anhydrase (Garmer and Krauss, 1992). In addition to explaining reactivity and pK_a trends for bound water at the active site, the authors calculated the selectivity profile of Zn - Cd > Co > > Mn in agreement with experiment. The observed insensitivity of this order to pH was also obtained because it was unaltered if the hydroxyl involved in the enzymatic reaction was protonated to model the inactive, low-pH form. The selectivity against cobalt as compared to zinc is a weak specific ligand field effect selecting for octahedral coordination. The insight which could only be gained from analysis of quantum chemical calculations was that the selectivity against spherical manganese is a charge transfer effect favoring the hydrated form. The high-spin Mn^{2+} was found to acquire stabilizing charge transfer into its unfilled d^5 subshell from ligands such as water having non-bonding π electrons. This may explain a general preference of manganese for oxygen-donor ligands over nitrogen. Mn^{2+} binding with sulphur-donor ligands should not suffer this source of instability.

IV. MOLECULAR MECHANICS METHODS AND RESULTS

Molecular structure design and analysis can be accomplished by intelligent use of molecular mechanics computer programs. In an application, the user sets up one or several conformations of a molecule and subjects these to energy minimization followed by structural and energy component analyses. It must be assumed that these conformations represent the important ones. Therefore, extensive prior information about the system is a necessity. This might limit the usefulness of molecular mechanics for understanding selectivity problems in large biomolecules because small local changes can easily produce large, intricate conformational shifts.

Molecular mechanics and dynamics methods use potential energy functions (force fields) for energy minimization and energetic comparisons. There are bond stretching forces and bond angle deformation forces which are simple harmonic (spring) functions of interatomic distances and bond angles, respectively. Sinusoidal functions of the torsional angles govern internal rotations of bonded groups. Van der Waals long range interatomic attraction and short range repulsion are represented as Lennard-Jones functions of interatomic distances. Coulombic interaction between atoms computed from atomic partial charges is included in many of the force fields.

There has been extensive work on selectivity effects in chelated ligands and macrocycles using molecular mechanics methods. The force fields employed describe the metal-ligand interaction as a chemical bond using stretching and angle deformation forces. Several extensive reviews of this subject have appeared (Hancock and Martell, 1989; Hancock, 1989; 1990; Brubaker and Johnson, 1984) but we will briefly describe the concepts and general results since they may apply also to low-molar-mass biological ligands.

The molecular mechanics calculations indicated that ligands which are preorganized to bind ions have greater binding affinity. The chelate effect from tying together ligated groups is one type of preorganization giving an entropic gain in stability in complexes which is very predictable in quantitative terms (Hancock and Martell, 1989) while understanding conformational preorganization requires explicit consideration of structures. Selectivity of cations binding with bulky ligands is very much influenced by the magnitude of internal strain energy induced upon binding. Molecular mechanics calculations revealed that the ligand atoms not directly in contact with the metal are usually causing some degree of crowding often not apparent from drawings or models, contributing to strain.

Strain energy within ligands has been quantified by scanning through a range of optimal metal-ligand bond lengths while optimizing the total structure, thus determining a rough energy profile for binding as a function of cation size (Hancock, 1989). If the cation is too large to fit without great strain within a bulky macrocycle it is not usually compressed by the ligand, but rather the ligand distorts to accommodate more optimal coordination distances. This may produce a conformational transition at a specific ion radius and a ligation sphere more open to the environment. Strained coordination also gives somewhat lower ligand field strengths than expected from the set of binding functional groups. An example of a conformational strain effect producing a general trend is seen with five-member chelate rings (including the metal) generally favoring binding large cations while six-member rings favor binding small cations. The least-strain conformations of the chelate rings best accommodate these sizes.

A number of other generalities have emerged from the collaboration of experiments and molecular mechanics (Hancock and Martell, 1989) which may relate to biological ionophore behavior. Addition of groups containing neutral oxygen donors to existing ligands produces a decreased selectivity for smaller cations, particularly if added as five-member chelate rings. These groups may produce excessive crowding around small cations compared with water ligands. If side groups such as methyls can be added to a liganded atom without producing additional strain, as in some macrocycles, there is an increased cation binding stability due to the increased electron donating ability of the ligand. This can be seen in the increased covalency for ether over water in Table 1. However, without some effort of molecular design the increased steric crowding with secondary oxygen and tertiary nitrogen will often overcome the favorable electronic effect. Dominance of this trend was noted above for neutral oxygens and cation size. A good way of eliminating strain is to join the sterically conflicting sites by a bond. The carboxyl functional group is a sterically very efficient ligand because it does not bring bulky side groups close to the coordination site and can bind in unidentate and bidentate modes with roughly the same stability. A mix of neutral oxygen donor sites and carboxylates such as occurring in EF hand proteins may thus favor the moderately large cation Ca^{2+} binding while to some extent excluding small Mg^{2+}.

The force fields applied to biomolecule studies have represented the metal-ligand attachment as a non-bonded interaction. The partial charge interactions give the attachment strength and the Lennard-Jones terms provide repulsion at short range. This approach uses cation non-bonded force field parameters that are added to biomolecular modeling computer codes. For thermodynamic selectivity estimates, the internal energies of complexes are compared with experimental hydration enthalpies and any entropic effects are of course ignored. Only coordinated solvent molecules are considered because including bulk solvent causes the number of local energy minima to explode and adds an unknown uncertainty to analyses.

There have been several studies of this type for biological ion carriers, and workers have also designed and synthesized ligands having properties as predicted by molecular mechanics (Shanzer et al., 1983; Lifson et al., 1983). They have explained cation selectivity trends successfully using the ideas of strain energies for optimal coordination and occurrence of multiple conformations having differing inherent cation size selectivities (Lifson et al., 1984; Masut and Kushick, 1985). A combined theoretical and spectroscopic study (Shanzer et al., 1986) of the biological ionophore enterobactin is particularly interesting in showing that the stable ion binding conformation has internal hydrogen bonds. In aqueous solution these hydrogen bonds are labile so that the molecule may open up to exchange a cation. In a non-polar environment the H-bonds are non-labile and the cation is encapsulated to present a hydrophobic surface for solvation in a membrane.

Gresh and A. Pullman (1986a; 1986b) have carried out similar studies with a more sophisticated force field model called Sibfa developed from quantum chemical theory and calculations. This includes many-body polarization effects and a model for charge transfer from ligands which makes it more physically reasonable than simple pairwise-additive force fields in view of the large amount of electronic relaxation accompanying cation binding (Yamabe et al., 1979). They have studied ionophore ion selectivity in several works briefly reviewed by Gresh and A. Pullman (1985). In one study, two ionophores were proposed to have different Na/K selectivity partly because of increased repulsion of carboxylate groups with Na binding in one (Gresh and A. Pullman, 1986a). There is experimental evidence that the arrangement of carboxylic groups in a coordination sphere can influence cation-binding stability due to variable amounts of ligand repulsion (Marsden et al.; 1988) which could thus be explored theoretically with the Sibfa method.

The Sibfa potential energy function would be even more preferable over pairwise additive representations of the metal-ligand interaction for molecular dynamics applications such as discussed below. Widely used pair additive potentials neglect several important consequences of the ligand polarization and charge transfer on a system's conformational properties. Effects for which there is possible documentation include an underestimation of interligand repulsion giving erroneously high coordination numbers, incorrect representation of relative binding strengths with changing order of donor atoms which could lead to artificial preference for binding sterically small groups such as water, and poor representation of the hydrogen bond strength of ligands due to their extensive polarization by an ion. Pair potentials quite good at simulating hydrogen bonding therefore do not yield good transferability when the non-bonded forces are required to represent interactions with ions (Curtiss et al., 1989; Cieplak et al., 1987).

V. MOLECULAR DYNAMICS AND FREE ENERGY PERTURBATION METHODS

A molecular dynamics calculation consists of the numerical integration of Newton's equations of motion for molecules and solvent at a finite temperature. It is possible to overcome the limited conformational sampling of molecular mechanics and to calculate free energies for physical processes such as ion replacement using the free energy perturbation method (FEP), for a considerable cost in computer time. The basic methods and alternative simulation protocols are discussed by Brooks et al. (1988) Chapters III to V and in the references below. An FEP study looking at ion selectivity involves a series of molecular dynamics calculations in which the ion potential energy function is changed in incremental steps from using parameters representing one ion into parameters suitable for the other. If these steps are small enough, the free energy change in the system due to the parameter change can be obtained as a particular thermodynamic average (Lybrand et al., 1986). Summing the ΔG for all steps, a so-called thermodynamic integration gives the free energy change due to converting one bound ion into another. A parallel thermodynamic integration in water yields the hydration free energy difference for the ions allowing for a prediction of the binding selectivity (and giving a check on the force field quality). The FEP method is still more expensive if used to estimate the actual binding free energy because the conformational changes are larger during a process of binding. Longer simulations are needed to keep the system in equilibrium during the thermodynamic integration steps.

A few estimates of selectivity have been made by simulations for ions interacting with macrocyclic compounds (Lybrand et al., 1986; Mazor et al., 1989) including efforts in which the simulations were run long enough to see conformational transitions due to ion replacement (van Eerden et al., 1988; Auffinger and Wipff, 1991). Estimates of the binding free energy differences of the ions were within a few kcal/mol of the observed values. A study more closely related to biomolecules by Sussman and Weinstein (1989) compared Ca^{2+} and Mg^{2+} binding to one or two molecules of the cyclic peptide cyclo(-L-Pro-Gly-)$_3$ (PG_3). These authors correctly obtained a more favorable 1:1 complex with Mg^{2+} over Ca^{2+} and a more favorable 1:2 complex with Ca^{2+} over Mg^{2+} although the selectivities of 8 $^{kcal}/_{mol}$ and 36 $^{kcal}/_{mol}$, respectively, seem unusually large in comparison with known selectivities for these ions. Both complexes have greater binding energies for Mg^{2+} but the hydration free energy difference overcomes this effect for the 1:2 complex. The selectivity trends were thus explained as a preference for Mg^{2+} to maintain some exposure to water, disfavoring a 1:2 sandwich complex. A detailed structural explanation might be that the calcium ion is favored to sit outside of the peptide ring while magnesium is more enclosed in the 1:1 complex. When the ion potential function developed for this work was applied to simulating an EF hand peptide, binding calcium primarily by carboxylates, the coordination number was too large in comparison with experimental information. A model for ligand polarizability was then added appearing to substantially correct this condition (F. Sussman, unpublished work).

The problem of monocation selectivity for ion channels has been examined using FEP methods (Roux and Karplus, 1991) and with FEP and molecular mechanics (Åqvist and Warshel, 1989) to construct free energy profiles for a comparatively simple and well characterized helical channel, Gramicidin. These studies appear to have achieved the realism and accuracy necessary to explore the dynamical, structural, and selectivity effects. The helix is found to be flexible and accommodates ion flow "like a pig in a python" (Mackay et al., 1984) along with a single file of water molecules. Molecular dynamics has also been used to refine NMR structural data for this channel. The results indicate that the channel peptide carbonyls point inward which was not apparent at the NMR resolution, thus possibly aiding cation conduction (Chiu et al., 1991).

Cation size effects on the reactivity of the metalloenzyme Staphylococcal Nuclease have been studied with an FEP method by Åqvist and Warshel (1990). Their thermodynamic integration is complex and involves changing the potential energy function from representing reactants, to moving to a transition state, and then to products. They find that ions the size of native Ca^{2+} produce the most efficient catalysis which is discussed in terms of the reaction profile.

There have been a few efforts to carry out meaningful molecular dynamics simulations on the selectivity of metalloenzyme active sites toward specific ligands (Shen et al., 1990; Hoops et al., 1991; Merz et al., 1991; Liang and Lipscomb, 1990). Most of the experimentally known structural features were required to be reproduced to calibrate the simulation models. Subsequent FEP simulations using one of them did not give particularly impressive results (Merz et al., 1991 and 1992 correction). The simulation model for Cu,Zn superoxide dismutase (Shen et al., 1990) ignores the very relevant non-spherical d electron distribution of Cu(II). It is not clear from results in these works that significant predictive capability can exist when the strategy is to represent metalloenzyme systems using conventional molecular dynamics force fields such as AMBER, GROMOS and CHARMM without making a more systematic attempt to develop a transferable force field model for the ions than these authors have done. Other work has aimed at predicting the remote electrostatic effect of mutations on processes such as ion binding. A comparison of results from different methods is given by Mehler and Solmajer (1991).

Understanding metal selectivity in protein binding sites certainly involves much more than just first shell behavior. There is experimental evidence from mutation studies showing that ion selectivity can be affected by Asp-Glu exchanges. This implies that there can be structural tension on a binding site causing selectivity (Falke et al., 1991; Snyder et al.,

1990). Since the FEP method can only provide numerical estimates for such problems (probably of limited accuracy) structural and dynamical interpretations of simulations will have to be developed to understand and predict such results if theoretical methods are to make a significant contribution.

ACKNOWLEDGEMENT

The author acknowledges helpful suggestions and manuscript criticism by Dr. Harel Weinstein and Dr. Morris Krauss.

REFERENCES

Anderson, W.P., Edwards, W.D., and Zerner, M.C. (1986). *Inorg. Chem.*, *25*: 2728.

Åqvist, J., and Warshel, A. (1989). *Biophys. J.*, *56*: 171.

Åqvist, J., and Warshel, A. (1990). *J. Amer. Chem. Soc.*, *112*: 2860.

Auffinger, P., and Wipff, G. (1991). *J. Amer. Chem. Soc.*, *113*: 5976.

Berg, J.M., and Merkle, D.L. (1989). *J. Amer. Chem. Soc.*, *111*: 3759.

Blake, A.B., and Cotton, F.A. (1964). *Inorg. Chem.*, *3*: 5.

Brooks III, C.L., Karplus, M., and Pettitt, B.M. (1988). *Proteins: A Theoretical Perspective of Dynamics, Structure, and Thermodynamics*, John Wiley & Sons, New York, Ch. III-V.

Brubaker, G.R., and Johnson, D.W. (1984). *Coord. Chem. Rev.*, *53*: 1.

Burgess, J. (1988). *Ions in Solution: Basic Principles of Chemical Interactions*, Ellis Horwood Limited, Chichester, p. 64.

Busch, D.H., and Stephenson, N.A. (1990). *Coord. Chem. Rev.*, *100*: 119.

Chiu, S.-W., Nicholson, L.K., Brennerman, M.T., Subramaniam, S., Teng, Q., McCammon, J.A., Cross, T.A., and Jakobsson, E. (1991). *Biophys. J.*, *60*: 974.

Cieplak, P., Lybrand, T.P., and Kollman, P.A. (1987). *J. Chem. Phys.*, *86*: 6393.

Curtiss, L.A., Halley, J.W., and Hautman, J. (1989). *Chem. Phys.*, *133*: 89.

Falke, J.J., Snyder, E.E., Thatcher, K.C., and Voertler, C.S. (1991). *Biochemistry*, *30*: 8690.

Garmer, D.R., and Krauss, M. (1992). *J. Amer. Chem. Soc.*, *114*: 6487.

Gresh, N., and Pullman, A. (1985). *Metal Ions in Biological Systems*, Vol. 19 (H. Sigel, ed.), Marcel Dekker, New York, p. 335.

Gresh, N., and Pullman, A. (1986a). *New J. Chem.*, *10*: 57.

Gresh, N., and Pullman, A. (1986b). *New J. Chem.*, *10*: 405.

Hancock, R.D. (1989). *Prog. Inorg. Chem.*, *37*: 187.

Hancock, R.D. (1990). *Accounts Chem. Res.*, *23*: 253.

Hancock, R.D., and Marsicano, F. (1978). *Inorg. Chem.*, *17*: 560.

Hancock, R.D., and Marsicano, F. (1980). *Inorg. Chem.*, *19*: 2709.

Hancock, R.D., and Martell, A.E. (1989). *Chem. Rev.*, *89*: 1875.

Hoops, S.C., Anderson, K.W., and Merz, K.M., Jr. (1991). *J. Amer. Chem. Soc.*, *113*: 8262.

Krauss, M., and Stevens, W.J. (1990). *J. Amer. Chem. Soc.*, *112*: 1460.

Liang, J.-Y., and Lipscomb, W.N. (1990). *Proc. Nat. Acad. Sci. USA*, *87*: 3675.

Lifson, S., Felder, C.E., and Shanzer, A. (1983). *J. Amer. Chem. Soc.*, *105*: 3866.

Lifson, S., Felder, C.E., and Shanzer, A. (1984). *J. Biomol. Struct. Dyn.*, 2: 641.

Lybrand, T.P., McCammon, J.A., and Wipff, G. (1986). *Proc. Nat. Acad. Sci. USA*, *83*: 833.

Mackay, D.H.J., Berens, P.H., Wilson, K.R., and Hagler, A.T. (1984). *Biophys. J.*, *46*: 229.

Marsden, B.J., Hodges, R.S., and Sykes, B.D. (1988). *Biochemistry*, *27*: 4198.

Martin, R.B. (1986). *Metal Ions in Biological Systems*, Vol. 20 (H. Sigel, ed.), Marcel Dekker, New York, p. 21.

Masut, R.A., and Kushick, J.N. (1985). *J. Comp. Chem.*, *6*: 148.

Mazor, M.H., McCammon, J.A., and Lybrand, T.P. (1989). *J. Amer. Chem. Soc.*, *111*: 55.

Mehler, E.L., and Solmajer, T. (1991). *Protein Engineering*, *4*: 903.

Merz, K.M., Jr., Murcko, M.A., and Kollman, P.A. (1991). *J. Amer. Chem. Soc.*, *113*: 4484. See important correction in (1992). *J. Amer. Chem. Soc.*, *114*: 1128.

Parr, R.G., and Pearson, R.G. (1983). *J. Amer. Chem. Soc.*, *105*: 7512.

Roux, B., and Karplus, M. (1991). *Biophys. J.*, *59*: 961.

Shanzer, A., Felder, C.E., and Lifson, S. (1983). *Biopolymers*, *22*: 409.

Shanzer, A., Libman, J., Lifsom, S., and Felder, C.E. (1986). *J. Amer. Chem. Soc.*, *108*: 7609.

Shen, J., Wong, C.F., Subramaniam, S., Albright, T.A., and McCammon, J.A. (1990). *J. Comp. Chem.*, *11*: 346.

Snyder, E.E., Buoscio, B.W., and Falke, J.J. (1990). *Biochemistry*, *29*: 3937.

Sussman, F., and Weinstein, H. (1989). *Proc. Nat. Acad. Sci. USA*, *86*: 7880.

van Eerden, J., Harkema, S., and Feil, D. (1988). *J. Phys. Chem.*, *92*: 5076.

Yamabe, T., Hori, K., Akagi, K., and Fukui, K. (1979). *Tetrahedron*, *35*: 1065.

SPECIFICITY FACTORS IN METAL ION-MACROMOLECULAR LIGAND INTERACTIONS

by

Ivano Bertini, Fabrizio Briganti, and Andrea Scozzafava

Department of Chemistry,
University of Florence,
Via Gino Capponi, 7
50121 Florence
Italy

I. INTRODUCTION

Macromolecular ligands interacting actively with metal ions play essential roles in the metabolism of all the living systems.

The comprehension of the factors governing the formation of these macromolecular complexes appear to be quite necessary in order to understand why some and not other specific metal ions appear to be coordinated to a biological macromolecule and how this selection can occur.

Whereas solution chemistry can be rationalized on the basis of only thermodynamic and/or kinetic considerations, this approach is only partially valid in the case of living organisms. Indeed these systems, as extensively illustrated in Chapter 1, are formed by cells and therefore compartimentalized. In the different compartments we do not generally have a situation of thermodynamic equilibrium (which is incompatible with life and evolution), but rather steady state conditions which are determined by active transport processes, i.e. energy driven processes. Cells can be simplistically viewed as traps for chemical elements: an external membrane stops particle diffusion and interacts with external and internal fluids, whereas an internal concentration of synthesized organic ligands, often divided among different compartments, can specifically bind metal ions (Stryer, 1988). Therefore, the characterization of metal ion-macromolecular ligand interactions is related to the chemical speciation, rejection and uptake processes, as well as internal compartmentalization, all factors which contribute to the ultimate result of a specific and fine control on the concentrations of ligands and metal ions (da Silva and Williams, 1976; 1991).

With the exception of Na^+, K^+, Mg^{2+}, and Ca^{2+} ions, which usually are taken up by ion exchange on the cell surface, the other metal ions are generally very low in concentration as free ions in both soil or sea waters (the most diffuse media). Therefore, the living systems have devised very selective uptake devices for rare and diluted metal ions. Some cells can produce and send outside chelating agents that once combined with specific metal ions (by displacing more loosely bound ligands) enter again the cell. Systems which transport the metal ions into the cell are also present at the membrane surface, making the relative thermodynamic affinity constants very different for different metal ions. Other important trapping mechanisms correspond to the formation of kinetically stable complexes (with covalent bonds) as well as precipitation of combined trapped elements. In all of these cases, the result is twofold because the ways of taking up a particular metal ion also correspond to ways of rejecting the other ions (Williams 1970; 1983).

Inside membranes and cells, a variety of more selective ligands are designed to choose particular metal ions and to protect against damage from other compounds. The pH, the redox potential as well as the relative ion-ligand concentrations can be tremendously different outside the cells with respect to the inside. In particular, metal-to-ligand concentration ratios are usually very high outside the cells whereas inside such ratios are almost 1:1 with a slight excess of ligands, the synthesis of which is under feedback control.

In conclusion, metal ion-macromolecular interactions in living organisms arise through a complex series of molecular events which occur in the different compartments of the organism, according to a sequential mechanism of ligands displacing with the final goal of assuring specific interactions in a given compartment between an evolutionarily selected ligand and another complex, often unknown, containing the metal ion (Harrison and Hoare, 1980).

The aim of this chapter is to review the specificity factors which, with reference to model compounds and metallo-substitution, can lead by themselves to a certain degree of preferred interactions, i.e. selectivity in chemical reactions, stressing again that in biological systems the control on the concentrations and localization of the species is so tremendously rigid that no one of the factors described below, taken alone, could guarantee the required selectivity.

II. FACTORS CONDITIONING SPECIFICITY IN METAL ION - MACROMOLECULAR LIGAND COORDINATION IN VIVO

The selective equilibria of metal ions in biology are finely regulated by the relative *local concentrations (physical selectivity)* of metal ions and specific organic ligands, and by the *effective stability constants (chemical selectivity)* of their complexes. The stability constants are called effective because they take into account the competition of other metal ions and protons for the common ligand as well as the competition of other ligands for the common metal ion. A quick view to the stability constant trends based on the electron acceptor power of metal ions (Irving-Williams series, HSAB theory, see Sillén and Martell, 1964; Jensen, 1980) is useful, but we also need a more detailed analysis of the selectivity factors. The study of model ligands as well as in vitro metallo-substitution of macromolecules of biological relevance has allowed to reveal the importance of each specificity factor (Vallee and Williams, 1968; Bertini and Luchinat, 1984).

For the sake of clarity, we can subdivide the thermodynamic mechanisms of element selection as follows (da Silva and Williams, 1991):

(i) *Ion charges*. Ligands can select metal ions on the basis of the relative charge (Cotton and Wilkinson, 1988). As an example, small, singly charged, first row anions, (hard ligands like F^-, RO^-, R_2N^-), prefer highly charged metal ions (hard metals) like Fe^{3+}, Mn^{3+}, V^{3+}, and Al^{3+}. F^- is uncommon in biological environments and R_2N^- is low in concentration at pH 7; therefore, only RO^- ligands like phenolates, enolates, hydroxamates, carboxylates and phosphates are available for such interactions. Further selection is achieved because inorganic hydroxide competes with RO^- groups removing Al^{3+}.

(ii) *Ion sizes, and preformed holes*. The order of stability of the various metal ion - ligand complexes would correspond to the order of ionic potentials (defined as the ratio between ionic charge and ionic radius) if steric factors, covalent bonds and other components did not operate (Cotton and Wilkinson, 1988). In particular, stereochemical factors generate many exceptions to such roles as in the case of large ligands which cannot fold completely around small metal ions and do not displace their hydration water. In biology, both selections by *complexation* and *macromolecular folding* are observed. A typical example of selective complexation is Mg^{2+} which is choosed by chlorophyll instead of the larger Ca^{2+}. Macromolecular folding, instead, can provide either rigid liganding holes or sites adaptable to many metal ions. In particular, Nature can synthesize ligands such that the constraints on the hole size are powerful selectivity factors. Typical examples are the potassium channels of nerves where selection is based on repulsion of smaller ions like Na^+ which are more strongly hydrated than K^+ and have a high desolvation energy, whereas larger ions cannot enter.

(iii) *Nature of ligands.* The selection based on the affinity of hard metals for oxygen-donor ligands and soft metal ions for sulfur/nitrogen-donor ligands has been very well studied in bioinorganic chemistry (Eichhorn, 1973; Sigel, 1974-; Spiro, 1980-). Generally the selectivity is larger if the polarizability of the donors is larger (S>N>O), i.e. with appropriate ligands we can obtain stability constants different by several orders of magnitude for different metal ions. Ions like Co^{2+} and Ni^{2+} are rare and therefore very selective ligands like vitamin B12 and factor F-430 respectively are necessary to trap them.

(iv) *Stereochemical requirements.* Coordinating sites of particular geometries are also used to select specific metal ions. For transition metal ions in particular, ligand field stereochemical effects can play a key role in specific complex formation. Cu^{2+}, Mn^{3+} prefer tetragonal binding sites whereas Zn^{2+} tends to bind to tetrahedral sites and Fe^{3+} prefers tetrahedral or octahedral geometries (see Table 1). More detailed theoretical treatments are available in the literature (Sigel, 1974-; Spiro, 1980-; see also the above Section C2 in this Chapter 3).

(v) *Spin-state changes.* This point can usually be correlated to ionic sizes and ligand-field stabilization effects because electronic spin changes are known to affect such factors. Anyway it is a further chance to increase selectivity. Typical examples could be the selection of low-spin Co^{2+} and Fe^{2+} in vitamin B12 and cytochromes respectively.

(vi) *Hydrolysis phenomena and formation of clusters.* At pH 7, extensive hydrolysis (and inside cells also sulpholysis) of metal ions is quite diffuse and produces small hydroxy species, polymers and ultimately precipitates. The selective presence of different hydrolysis products depends on the inorganic chemistry of the metal ions but also on the controlled synthesis of specific ligands for such species. Small hydrolyzed units may be easily incorporated into small complexes and proteins. In fact Fe(II,III), Cu(II) and Mn(II) μ-oxo-dimers are found in many proteins like hemerythrin, ribonucleotide reductase, purple acid phosphatase, methane monooxygenase, copper oxidases, and also photosynthetic centers (Sanders-Loehr, 1989). The iron-sulfur proteins are another large group of cluster-containing proteins which are supposed to be formed by trapping polysulphide units produced by sulpholysis inside cells.

As reported in the introduction biological factors also play roles in the achievement of selectivity:

(i) *Regulation of relative metal ion-ligand concentrations.* Particle concentrations are strictly controlled in biological systems by a large variety of mechanisms. We would only

Table 1

Metal Ion Selectivity for Coordination Geometries and Donor Atoms

Donor atoms	Preferred coordination geometries Tetrahedral	 Tetragonal	 Octahedral
Phenolate RO^-	Fe(II,III)	Mn(III)	Fe(II,III)
Thiolate RS^-	Zn(II), Cu(I), Fe(III)	Cu(II)	Mo(V) Ni(II)
Histidine N	Zn(II)	Cu(II) Mg(II)(sq.pl.)	Mn(II), Fe(II), Co(II), Ni(II)
Ionized peptide N^-		Cu(II)	
Carboxylate RCO_2^-	Zn(II)		Fe(II,III), Mn(II), Mg(II), Co(III)

underline the importance of redox potentials and pH control in the distinct biological compartments. The redox states of metal ions are governed by the redox potential of the medium and then they allow selectivity on the basis of radius and charge effects which are strictly correlated to the oxidation states. Buffering is also controlling the metal ions and ligand concentrations by sequestering their excess.

Summarizing, in biological systems, free metal ions and free ligands concentrations are limited by feedback mechanisms controlling uptake and rejection of metal ions and hydrolysis and synthesis of the ligands, respectively. If we know these processes, we can monitor the selectivity by analyzing the effective stability constants.

(ii) *Metal ion transfer from water to macromolecular ligands.* The selectivity of macromolecular ligands for metal ions cannot be treated only as due to first-coordination sphere interactions. In particular we should always consider that the transfer of a ML complex to the inside of a macromolecule with a different dielectric constant requires high energy and is generally very selective. For example, heme proteins, by change of axial ligands or simply by variations of protein folding and heme solvent exposure (i.e. changes of internal dielectric constants), can have redox potentials ranging from about +100 to -500 mV and then select between Fe(II) and Fe(III) (and in principle between M^{2+} and M^{3+}) (Dolphin, 1978). Similar observations can be made in the case of iron sulfur proteins where the composition and the conformation of macromolecular ligands select among Fe-S clusters of different composition and charge (Lovenberg, 1973; 1977; Que, 1987). Actually there is a strong cooperativity between protein conformation and metal ion requirements; therefore, the ligand folding also depends on the properties of metal ions or clusters. The allosteric interactions in dioxygen binding to haemoglobins are a further case where different binding sites can reciprocally interact.

(iii) *Kinetic control.* Metal ion-ligand complex formation depends on all the factors reported above, but the regulation of the input/output steady state rates of metal ions, often assisted by specific ligands, as well as the rates of ligand synthesis and degradation are also very important. Furthermore, kinetic devices as the formation of precipitates, covalent bonding, membrane inglobation etc can select the uptake or rejection of particular elements if the process is thermodynamically unfavored.

Nature utilizes combinations of these mechanisms, which of course are strictly associated, for the selection of the desired metal ion from those available in the specific microenvironment. The uptake of a rare metal ion generally needs a high degree of selectivity. Metal ions such as Na^+ and K^+ which are very abundant in biological fluids form weak complexes with common ligands; they exchange fast and diffuse as simple ions; membranes contain aqueous domains in which such ions are driven using energy pumps. Instead, the less abundant Mg^{2+} and Ca^{2+} ions are more easily retained by thermodynamic traps in biopolymers such as proteins. The rare transition metal ions are generally more strongly trapped by specific ligands. Such macromolecular ligands must thermodynamically compete with the external less selective ligands (like iron-siderophores); they need to be soluble in the membrane phase and pass by simple diffusion.

III. METAL IONS AND MACROMOLECULES IN BIOLOGICAL SYSTEMS

The large utilization of metal ions in natural systems is principally due to their particular and widely different chemical properties like (Williams, 1970; 1983; Harrison and Hoare, 1980; Hughes, 1981; da Silva and Williams, 1991):

(i) generally fast, selective and often strong interactions with organic ligands,
(ii) adjustable bond directions and lengths,
(iii) rapid bond breaking of ionic bonds in low positive oxidation states,
(iv) good π-electron donors in low positive oxidation states,
(v) transport of positive charge by stable ions,
(vi) marked Lewis acidity in many transition metals (polarizing power),
(vii) easy change of oxidation states,
(viii) in high oxidation states properties similar to non-metal elements like slow ligand exchange but fast atom transfer,
(ix) activation of small molecules.

Biological systems are composed by macromolecules belonging to four principal categories: polysaccharides, complex lipids, polynucleotides, and proteins, as well as smaller ligands of great importance like coenzymes, hormones, neurotransmitters and electron transporters.

A. Metal ion - polysaccharide interactions

Polysaccharides have several roles in biology ranging from the formation of animal connective tissues and mucus to vegetal cell walls and skeletons, as well as in storage and recognition processes associated to proteins and lipids (Doi and Edwards, 1986). A large variety of polysaccharides interact with metal ions, and their specificity can be essentially correlated to the presence and the relative number of anionic side chains. In particular, if such side chains include mono-ester phosphate and carboxylate centers, the interaction with metal ions such as Na^+, K^+ and the more specific Mg^{2+} and Ca^{2+} is fundamental for stabilization of connective tissue, cell wall and cell/cell interactions.

B. Metal ion - complex lipid interactions

Among the different lipid molecules present in biological systems, diglycerides forming membranes are the most diffuse and significant (Bergethin and Simons, 1990; Evans and Graham, 1989). In fact, membranes are lipid surfaces consisting of bilayers loosely bound through hydrophobic interactions, their polar heads being exposed to the water environment. Such polar heads are negatively charged with oxygen donors whose scarce selectivity makes difficult to differentiate between specific bonding and general salt effect in the interactions with metal ions. Ca^{2+} has the most selective interactions with membrane lipids, and also interacts with polysaccharides and proteins often bound to membranes (Vance, 1987).

C. Metal ion - polynucleotide interactions

The primary sites of interaction of metal ions with nucleotides are the bases, where heterocyclic and amidate nitrogen donors can form covalent bonds with soft heavy metal centers (Spiro, 1980; Eichhorn and Marzilli, 1981; Berg, 1986). Other charged groups like sugars and diesterphosphates O-donors are possible metal binding sites too.

Ion binding to randomly folded singly stranded polynucleotides is generally very weak, but once bound to proteins (like RNA in ribosomes) such macromolecules can assume local structures and conformations affording high specificity for metal ions. Double helical folded DNA and RNA, instead, already contain selective sites for metal ion interactions and double helix structure conformation and stability are controlled by ions. In fact, metal ion binding to phosphate groups, decreasing the electrostatic repulsion between nucleotides, stabilize the double helix. Also, metal complex intercalation by base-ligand stacking, lowering the free energy of the system, makes the double helix more stable. On the contrary, binding to bases generally disrupts base-pair hydrogen bonding and destabilizes the helix.

Selectivity in metal ion-polynucleotide interactions is also dictated by pH control, the relative reaction rates, the presence of exocyclic base substituents as well as binding to polymeric arrays of base, sugar and phosphate moieties, and by stacking interactions with metal complexes containing planar aromatic ligands (e.g. terpyridine-Pt(II)).

Polynucleotides are very mobile and, therefore, their selectivity towards metal ions is not restricted; the cavities can adjust to very different sizes (e.g. Mg^{2+}/polyamine binding equilibrium).

Bases in folded DNA/RNA molecules react with many heavy metal ions. The binding of cis-dichlorodiamine platinum(II) antitumor drugs to the N7 atom of guanidine has been a fundamental discovery in bioinorganic chemistry and medicine (Spiro, 1980; see also Part Four, Chapter 5-A).

D. Metal ion - protein interactions

As pointed out above, polysaccharides, polynucleotides, and membrane lipids have electrostatic interactions principally with Na^+, K^+, Ca^{2+} and Mg^{2+} whereas proteins bind a large variety of metal ions. Depending on the strength of such interactions, the complexes formed are classified in metalloproteins and metal-protein complexes, the latter being characterized by very low stability constants (Harrison and Hoare, 1980).

Metalloproteins, the largest group of investigated metal ion - macromolecular ligand complexes, are usually divided into four categories on the basis of their functions: transport proteins, storage proteins, redox proteins, and enzymes. Many enzymes, as well as proteins involved in redox processes, need metal ions in order to fulfil their functions and these are usually supplied by metal transport and metal storage proteins and other macromolecular ligands.

For the understanding of the different factors which condition the specificity of metal ion-macromolecular ligand interactions, we need to analyze the types of metal-protein interactions occurring in biological systems. These are usually divided into (Bezrukov et al., 1982):

1. Indirect interaction through the solvent,
2. Indirect interaction through an organic ligand,
3. Indirect interaction through an inorganic ligand,
4. Clathrate complex formation,
5. Direct interactions with aminoacidic residues.
6. Long range interactions

This classification does not pretend to be complete but has to be considered only as a schematic way for better analyzing the types of metal ion-macromolecular ligand interactions.

1. *Indirect interaction through the solvent*

Solvent-mediated metal ion-protein interactions assume a particular importance in aqueous media. In fact, the structural conformation of macromolecular ligands in solution is essentially induced by hydrogen bonds, dipole-dipole interactions, and hydrophobic interactions. Therefore, the lyotropic effect of metal ions includes interactions with charged, polarized, or polarizable groups and also hydrophobic residues. Usually, the addition of metal salts to protein solutions inducing a general weakening of protein-protein electrostatic attractions increases protein solubility. Furthermore, the presence of metal ions weakens the hydrogen bonds and in some cases, depending on the relative affinity of the metal for the peptide side chains, it results in the complete uncoiling of the polypeptide, also reinforced by the weakening of the double bond character of peptide bonds which promotes their free rotation.

Hydrophobic interactions are also important for lyotropic effects. The natural polypeptide organization usually brings about the distribution of hydrophilic residues on the surface and the formation of hydrophobic aggregates in the inside of the macromolecules, resulting in systems with higher entropy and a lower energy state.

2. *Indirect interaction through an organic ligand*

Interactions through organic ligands occur in complexes containing organic cofactors and are characteristic of many proteins like heme-proteins (Dolphin, 1978), the vitamin B12 system in its interactions with the methylmalonyl CoA mutase (Corey et al., 1977), or the Mn(II) binding to creatine kinase occurring through the complexation with adenosine diphosphate (Cohn and Leigh, 1962).

3. *Indirect interaction through an inorganic ligand*

Interactions between polarizable inorganic complexes, like Hg^{2+} and Au^{3+} iodides, and hydrophobic protein regions stabilized by van der Waals forces have been observed (Österberg, 1974). Also, direct interactions between hydrophobic protein pockets and heavy metal ions like Hg^{2+}, Ag^{+} and Cu^{+} are operative and explain the accumulation of such ions in adipose tissues.

4. *Clathrate complex formation*

Storage proteins often utilize interactions involving the formation of supermolecular clathrate complexes (Emery, 1978) like in ferritin where the iron(III) ions interact with the protein as basic hydrogen phosphate gels.

5. *Direct interactions with aminoacidic residues*

The direct coordination of metal ions to the functional groups of the aminoacidic residues of proteins is fundamental for the structure and function of metalloprotein systems (see e.g. Christianson (1991) for zinc enzymes).

Proteins contain a large variety of groups which can bind metals, or which are not directly involved in the ligation processes but favor or prevent the approach of the metal to the possible binding sites.

Soft metal ions like Cu^+, Ag^+, Hg^{2+}, $Pd^{2+}$and Pt^{2+} prefer to bind to imidazole- and sulfur-containing residues whereas hard metal ions like Ca^{2+}, Mg^{2+}, Cr^{3+}, Mn^{2+}, Fe^{3+} and Co^{3+} bind to water molecules as well as hydroxy- and carboxy-containing groups. The behavior of borderline metal ions in the HSAB classification like Fe^{2+}, Co^{2+}, Ni^{2+}, Cu^{2+} and Zn^{2+} depends on the characteristics of the surrounding residues.

Many metal cations can bind any protein with different affinities but certain metal-protein pairs exhibit specific affinities resulting in binding constants several orders of magnitude higher than other non-specific interactions. This is usually the case of the metal-dependent proteins where the metal is indispensable for the formation of an active protein.

A brief analysis of the different metal ion ligands in proteins is useful for the comprehension of the affinity factors in metal-protein interactions (Table 2). Carboxylates (aspartates and glutamates) display syn or anti stereochemistry with metal ions, they can bind in monodentate or bidentate fashions, and can also participate to ring formation with the terminal amino groups. Tyrosine coordination is found in various iron proteins like hemerythrin, acid phosphatases, intradiol dioxygenases, transferrin, etc. Histidines are the most diffuse metal binding centers in proteins because the imidazolic nitrogen is a strong electron donor capable of forming covalent bonds at physiological pH. The highly polarizable and electron donor sulfur-containing ligands (methionine and cysteine) bind preferentially soft metal ions. The peptide carbonyl, which has weak basic properties, principally interacts with alkali or alkaline earth metal ions both in a monodentate fashion

Table 2

Metal Ion Coordination in Proteins

Coordinating groups	Metal ions	Examples	References (Structure)
Carboxylates (Asp, Glu)	Fe(II), Ag(I), Cu(II), Zn(II), Ca(II), Mn(II)	Zn-Ca-Thermolysin Zn-Leucine aminopeptidase	Matthews et al. (1972) Burley et al. (1990)
Sulphydryls (Cys)	Ag(I), Hg(II), Mn(II), Fe(II), Co(II), Ni(II), Zn(II), Pb(II), Cu(II)	Rubredoxin Ferredoxin Metallothionein	Herriot et al. (1970) Sieker et al. (1972) Arseniev et al. (1988)
Imidazoles (His)	Cu(II), Zn(II),Co(II), Fe(II), Ni(II), Cd(II)	Zn-Carbonic anhydrases Cu,Zn-Superoxide dismutase Fe-hemerythrin	Liljas et al. (1972) Tainer et al. (1983) Stenkamp et al. (1976)
Thioesters (Met)	Pd(II), Pt(II), Ag(I), Cu(I), Hg(II)	Cu-Plastocyanin Fe-Cytochrome *c*	Colman et al. (1978) Swanson et al. (1977)
Phenolates (Tyr)	Fe(III), Fe(II)	Fe-Transferrin Fe-Dioxygenases	Bailey et al. (1988) Ohlendorf et al. (1987)
Terminal amino groups	Zn(II), Ag(I), Pt(II) Cu(II)	Pt-α-chymotrypsin Cu-Phosvitin	Sigler et al. (1968) Kozlowski et al. (1988)
Peptide	Ca(II),Co(II),Co(III), Ni(II),Cu(II),Zn(II), Hg(II), Pd(II), Pt(II)	M(II,III)-Polyglycines	Martin (1974)

or bidentate when metal ion chelation with the near amino group is possible. The terminal amino groups are the most usual sites for the non-specific binding of metals despite their high pK values since the amino nitrogen is strongly electron-donating while the oxygen of the proximal carboxyl group can form a chelate ring. Binding to the terminal amino group without chelation is characteristic of Ag(I) and Pt(II). It is also worth mentioning that many amino acid residues are present as bridging ligands in proteins, like a histidine in Cu-Zn-superoxide dismutases, some cysteines in metallothioneins (Kägi and Nordberg, 1979) and zinc fingers.

6. *Long range interactions*
Both electrostatic effects due to simple through space interactions with the bulk protein (Russell and Fersht, 1987; Pantoliano et al., 1988) and hydrogen bond networks involving ligands and the metal ion are supposed to have important roles in the orientation of the metal ligands and in the enhancement of electrostatic effects between the metal ion and the ligands facilitating metal binding (Argos et al., 1978).

IV. USE OF MODEL LIGANDS AND METAL SUBSTITUTION TO IDENTIFY SPECIFICITY FACTORS

A. Model ligands

Model metal complexes with amino acids, peptides, nucleic acids, nucleotides and porphyrins have been useful to obtain information on the reactivity of the coordinated ligands and finally understand into details the metal ion-macromolecular ligand interactions (Sigel, 1974-; Hughes, 1981; Gray and Lever, 1983-).

It is worthwhile to analyze the behavior of simpler model molecules if these display the essential features of the corresponding complex macromolecules without their unwanted complications, but often the protein functions strictly depend on the unusual stereochemistry of the metal ion, on unusual properties of some residue in the protein environment or on long range effects absent in model compounds. The cooperative nature of the interactions of the functional groups independently approaching the metal in the tertiary structure of the macromolecular ligand is also important. Furthermore, this structure can change upon coordination of the metal ion, obscuring the understanding of this process. For these reasons, even though the studies of simpler complexes containing the essential features of original biologically significant macromolecules are always useful, their conclusions should always be applied to real systems with extreme caution. Many of the results discussed above are in fact based on studies of model compounds. A great deal of syntheses and characterizations of porphyrins as models of heme proteins (Dolphin, 1978; Lever and Gray, 1983), of iron-chalconide complexes as analogs of iron-sulfur proteins (Holm and Ibers, 1977; Lovenberg, 1973; 1977; Que, 1987; Matsubara et al., 1987), and many others has been carried out (Sigel, 1974-).

B. Metal substitution

Isomorphous substitution of native metal ions in proteins with other metal ions has been the tool utilized in an enormous number of investigations for the understanding of the specificity factors in metal ion-ligand interactions (Vallee and Williams, 1968). In particular, the substitution of some of the most diffuse metal ions in biology like Zn(II), Mg(II) and Ca(II) with Co(II), Mn(II), and Lanthanides(III) respectively has allowed the use of spectroscopic techniques to reveal the structural and dynamic characteristics of the metal sites (Hughes, 1981). In fact, transition metal ions have specific optical and magnetic properties due to their partially filled d orbitals; electronic absorption, circular dichroism and magnetic circular dichroism spectra, EPR, NMR, magnetic susceptibility, Mössbauer and many other techniques can give specific information on the geometry, the metal-coordinating groups both in the resting and transition states (Darnal and Wilkins, 1980; Bertini and Scozzafava, 1981; Bertini and Luchinat, 1984; 1986; Bertini et al., 1992). The substitute metal ion has to be carefully chosen and the recovery of biological activity is generally a good indication of successful replacement. The catalytic efficiency of transition metal ions can also be correlated to their different chemistry and geometrical requirements, therefore giving insights into kinetic factors. It has been observed that the ionic radius is a significant factor because the metal ion used as a probe has to occupy the same site as the native one, but as already noticed, this is essential only in the case of very rigid preformed holes. It is often much more important for selectivity and maintenance of biological activity to fulfil the requirements in terms of liganding atoms and preferential coordination geometry. In fact, Mg^{2+} (ionic radius 0.065 nm) is functionally replaced by Mn^{2+} which has a similar chemistry, despite its ionic radius of 0.080 nm and its different affinity for N-donors, whereas it is not effectively replaced by Ni^{2+} which, although having an almost identical radius (0.069 nm) and preferential geometries, has a different chemistry. Mn(II) is a good EPR probe and also a good NMR relaxation reagent; other probes for Mg^{2+}, although showing low activity, are Cr(III) and Ln(III) (Mildvan and Grisham, 1974). The lanthanides(III) are also particularly useful as probes for Ca^{2+}. One of the most diffuse metal ions in biological systems which has been subjected to isomorphous substitution is Zn(II) (Bertini et al., 1986). It has a d^{10} diamagnetic electronic configuration; therefore, it has no relevant electronic spectra and magnetic properties. Its substitution with other transition metal ions allows to use spectroscopic methods to recognize coordinating groups and monitor the interactions with substrates or inhibitors. Co(II), which has a very similar chemistry, is the most suitable substitute for Zn(II); therefore, the activity restoration is very high (Vallee, 1981). Co(II) has a d^7 electronic configuration and, with only few exceptions, is always high spin ($S = 3/2$) when substituting Zn(II). This property of Co(II) gives rise to many electronic d-d transitions in the near infrared-visible region which can be analyzed in terms of coordination geometry and relative changes by interactions with physiologically relevant molecules. The relatively large electronic relaxation rates of Co(II) allow the detection of isotropically shifted sharp ^1H NMR spectra of the surrounding ligands, whereas EPR signals are observed only at low temperature. Cu(II) has also been utilized as a probe for Zn(II); generally its derivatives have low biological activity and this can be correlated to the different stereochemical requirements (Table 1) and to a stronger affinity of copper for substrates that does not allow fast turnover. Cu(II) is a good EPR probe whereas the optical spectra do not give significant information. The large nuclear relaxing capacity of Cu(II) also allows to perform multinuclear NMR studies on ligands. Fast electron relaxation rates are also peculiar of Mn(II); thus macromolecules substituted with this ion can be investigated through EPR, electronic absorption spectra as well as multinuclear NMR of the interacting ligands. Cd(II) has also been utilized as a substitute for Zn(II) in proteins; it is generally a good substitute but the pK_a of eventually bound water increases with respect to zinc and, therefore, the activity is shifted to a different non-physiological pH. ^{111}Cd and ^{113}Cd isotopes are good NMR probes, also used for structure determination of interacting macromolecules. Finally, Ni(II)-substituted proteins give informative optical spectra and quite sharp ^1H NMR spectra. Therefore, a large variety of metal ions can be utilized as probes in studying the interactions between metal ions and macromolecular ligands, allowing to obtain detailed information on the specificity factors which regulate such interactions.

V. RELATIONSHIPS BETWEEN METAL-LIGAND INTERACTIONS, CONFORMATION OF MACROMOLECULAR LIGANDS, AND THEIR PHYSIOLOGICAL ROLE

As described in the previous sections, the conformation and configuration of biopolymers is easily modified by any local microchemical change like the addition of low-molar-mass compounds, variations of the metal ion concentration and oxidation state.

The effect of metals on the conformation of various macromolecular ligands has been extensively studied (Sigel, 1974-). The addition of different metal ions can bring to stabilization or either to partial and complete uncoiling and denaturation of the protein, depending on the concentration and properties of the metal ions.

Many enzymes become fully active only in the presence of selected metal ions which participate in the catalytic process (catalytic metal) and/or are essential for the structural conformation of the active enzyme (structural metal) (Eichhorn, 1973; Inorganic Biochemistry Series, 1979-).

Metal-independent enzymes do not exist because every enzyme reacts to the presence of metal ions and generally the optimal activity is reached only over a narrow metal ion concentration range. The principal difference is in the order of magnitude of the metal ion concentration which generates the effects. Li^+, Na^+, and K^+ ions are active at high concentrations because they act indirectly through the solvent, whereas transition metal ions act directly at very low concentrations.

REFERENCES

Argos, P., Garavito, R.M., Eventoff, W., Rossman, M.G., and Brandén, C.I. (1978). *J. Mol. Biol., 126*: 141.

Arseniev, A., Schultze, P., Wörgötter, E., Braun, W., Wagner, G., Vašak, M., Kägi, J.H.R., and Wüthrich, K. (1988). *J. Mol. Biol., 201:* 637.

Bailey, S., Evans, R.W., Garratt, R.C., Gorinsky, B., Hasnain, S., Horsburgh, C., Jhoti, H., Lindley, P.F., Mydin, A., Sarra, R., and Watson, J.L. (1988). *Biochemistry, 27:* 5804.

Berg, J. (1986). *Science, 252:* 485.

Bergethin, P.R., and Simons, E.R. (1990). *Biophysical Chemistry - Molecules to Membranes*, Springer-Verlag, Berlin.

Bertini, I., and Luchinat, C. (1984). *Advances in Inorganic Biochemistry*, Vol. 6 (G.L. Eichhorn and L.G. Marzilli, eds.), Elsevier, New York, p. 71.

Bertini, I., and Luchinat, C. (1986). *NMR of Paramagnetic Molecules in Biological Systems*, Benjamin-Cummings, Menlo Park, CA.

Bertini, I., and Scozzafava, A. (1981). *Metal Ions in Biological Systems*, Vol. 12 (H. Sigel, ed.), Marcel Dekker, New York, p. 31.

Bertini, I., Luchinat, C., Maret, W., and Zeppezauer, M., eds. (1986). *Progress in Inorganic Biochemistry and Biophysics*, Vol. 1, Birkhäuser Verlag, Basel.

Bertini, I., Messori, L., and Viezzoli, M.S., (1992). *Coord. Chem. Rev., 120:* 163.

Bezrukov, M.G., Belousova, A.M., and Sergeev, V.A. (1982). *Russ. Chem. Rev., 51:* 397.

Burley, S.K., David, P.R., Taylor, A., Lipscomb, W.N. (1990). *Proc. Nat. Acad. Sci. USA, 87:* 6878.

Christianson, D.W. (1991). *Adv. Protein Chem., 42:* 281.

Cohn, M., and Leigh, J.S. (1962). *Nature, 193:* 1037.

Colman, P.M., Freeman, H.C., Guss, M.J., Murata, M., Norris, V.A., Ramshaw, J.A.M., and Venkatappa, M.P. (1978). *Nature, 272:* 319.

Corey, E.J., Cooper, N.J., and Green, M.L.H. (1977). *Proc. Nat. Acad. Sci USA, 74:* 811.

Cotton, F.A., and Wilkinson, G., (1988). *Advanced Inorganic Chemistry*, Ellis Horwood, New York.

da Silva, J.J.R.F., and Williams, R.J.P. (1976). *Struct. Bonding, 29:* 67.

da Silva, J.J.R.F., and Williams, R.J.P. (1991). *The Biological Chemistry of the Elements, The Inorganic Chemistry of Life*, Clarendon Press, Oxford.

Darnal, D.W., and Wilkins, R.G., eds. (1980). *Advances in Inorganic Biochemistry*, Vol. 2, Elsevier/ North Holland, New York.

Doi, M., and Edwards, S.F. (1986). *The Theory of Polymer Dynamics*, Oxford University Press, Oxford.

Dolphin, D.H., ed. (1978). *The Porphyrins*, Vols. 1-7, Academic Press, New York.

Eichhorn, G.L., ed. (1973). *Inorganic Biochemistry*, Vols. 1 and 2, Elsevier Publishing Co., Amsterdam.

Eichhorn, G.L., and Marzilli, L.G., eds. (1981). *Advances in Inorganic Biochemistry*, Vol. 3, Elsevier/ North Holland, New York.

Emery, T. (1978). *Metal Ions in Biological Systems*, Vol. 7 (H. Sigel, ed.), Marcel Dekker, New York, p. 114.

Evans, W.H., and Graham, J.M. (1989). *Membrane Structure and Function*, Oxford University Press.

Gray, H.B., and Lever, A.B.P., eds. (1983-). *Physical Bioinorganic Chemistry Series*, VCH Publishers, Inc., New York.

Harrison, P.M., and Hoare, R.J. (1980). *Metals in Biochemistry*, Chapman and Hall Ltd., London.

Herriot, J.R., Sieker, L.C., Jensen, L.H., and Lovenberg, W. (1970), *J. Mol. Biol.*, *50*, 391.

Holm, R.H., and Ibers, J.A. (1977). *Iron-Sulfur Proteins*, Vol. 3 (W. Lovenberg, ed.), Academic Press, New York, Chapter 7.

Hughes, M.N. (1981). *The Inorganic Chemistry of Biological Processes*, John Wiley and Sons, New York.

Inorganic Biochemistry Series (1979-). The Royal Society of Chemistry, London.

Jensen, W.B. (1980). *The Lewis Acid-Base Concepts*, Wiley, New York.

Kägi, J.H.R., and Nordberg, M., eds. (1979). *Metallothionein*, Birkhäuser Verlag, Basel.

Kozlowski, H., Mangani, S., Messori, L., Orioli, P.L., and Scozzafava, A. (1988). *J. Inorg. Biochem.*, *34:* 221.

Lever, A.B.P., and Gray, H.B., eds. (1983). *Iron Porphyrins*, Addison Wesley Publishing Co. Inc., Reading MA.

Liljas, A., Kannan, K.K., Bergsten, P.C., Waara, I., Fridborg, B., Strandberg, B., Carlbom, U., Jarup, L., Lovgren, S., and Petef, M. (1972). *Nature*, *235:* 131.

Lovenberg, W., ed. (1973). *Iron Sulfur Proteins*, Vols. 1-2, Academic Press, New York.

Lovenberg, W., ed. (1977). *Iron Sulfur Proteins*, Vol. 3, Academic Press, New York.

Martin, R.B. (1974). *Metal Ions in Biological Systems*, Vol. 3 (H. Sigel, ed.), Marcel Dekker, New York, p. 129.

Matsubara, H., Katsube, Y., and Wada, K., eds. (1987). *Iron-Sulfur Protein Research*, Japan Scientific Societies Press/Springer Verlag, Berlin.

Matthews, B.W., Jansonius, J.N., Colman, P.M., Shoenborn, B.P., and Duporque, D. (1972). *Nature*, *238:* 37.

Mildvan, A.S., and Grisham, C.M. (1974). *Struct. Bonding*, *20:* 1.

Ohlendorf, D.H., Weber, P.C., and Lipscomb, J.D. (1987). *J. Mol. Biol.*, *195:* 225.

Österberg, R. (1974). *Metal Ions in Biological Systems*, Vol. 3 (H. Sigel, ed.), Marcel Dekker, New York, p. 45.

Pantoliano, M.W., Whitlow, M., Wood, J.F., Rollence, M.L., Finzel, B.C., Gilliland, G.L., Poulos, T.L., and Bryan, B.N. (1988). *Biochemistry*, *27:* 8311.

Que, L., Jr., ed. (1987). *Metal Clusters in Proteins*, American Chemical Society.

Russell, A.J., and Fersht, A.R. (1987). *Nature*, *328*: 496.

Sanders-Loehr, J., (1989). *Iron Carriers and Iron Proteins* (T.M. Loehr, ed.), VCH Publishers, Inc., New York, p. 373.

Sieker, L.C., Adman, E., and Jensen, L.H. (1972). *Nature*, *235:* 40.

Sigel, H., ed. (1974-). *Metal Ions in Biological Systems*, Marcel Dekker, New York.

Sigler, P.B., Blow, D.M., Matthews, B.W., and Henderson, R. (1968) *J. Mol. Biol.*, *35:* 143.

Sillén, L.G., and Martell, A.E. (1964). *Stability Constants of Metal Ion Complexes*, Chemical Society, London.

Spiro, T.G., ed. (1980-). *Metal Ions in Biology*, Wiley Interscience, New York.

Stenkamp, R.E., Sieker, L.C., and Jensen, L.M. (1976). *Proc. Nat. Acad. Sci. USA*, *73:* 349.

Stryer, L. (1988). *Biochemistry*, W.H. Freman and Co., New York.

Swanson, R., Trus, B.L., Mandel, N., Mandel, G., Kallai, O.B., and Dickerson, R.E. (1977). *J. Biol. Chem.*, *252:* 759.

Tainer, J.A., Getzoff, E.G., Richardson, J.S., and Richardson, D.C. (1983). *Nature*, *306:* 284.

Vallee, B.L., and Williams, R.J.P. (1968). *Chem. Brit.*, *4:* 397.

Vallee, B.L. (1981). *Structural and Functional Aspects of Enzyme Catalysis* (H. Eggerer and R. Huber, eds.), Springer Verlag, Berlin Heidelberg, p. 75.

Vance, J.E., ed. (1987). *Biochemistry of Lipids and Membranes*, Benjamin-Cummings, Menlo Park, CA.

Williams, R.J.P. (1970). *Quart. Rev. Chem. Soc. London, 24:* 331.

Williams, R.J.P. (1983). *Pure Appl. Chem., 55:* 1089.

DIFFERENT FRACTIONS OF METAL IONS IN BIOLOGICAL FLUIDS AND THEIR PHYSIOLOGICAL ROLES

by

Mairam Gulumian[1,2], Robert Douglas Hancock[2] and Halina Barbara Röllin[1,3]

[1] National Centre for Occupational Health,
P.O. Box 4788
Johannesburg 2000
South Africa

[2] Centre for Molecular Design,
University of the Witwatersrand,
Wits 2050
Johannesburg
South Africa

[3] Department of Community Health,
University of the Witwatersrand,
Wits 2050
Johannesburg 2001
South Africa

I. INTRODUCTION

For the multiplicity of factors which control the distribution of metal ions, ligands, and their complexes in different biological fluids, it is not possible to predict in which specific fluid compartment they will be found. From a knowledge of metal ions and their distribution into different ligand fractions in each fluid, however, we are able to deduce the reasons for their presence and consequently relate this presence to the functions they perform in biological fluids. The latter knowledge also helps us to understand the movement of metal ions from one fluid compartment to another, as this movement is dependent on the extent of distribution of each metal ion between its corresponding fractions in a particular fluid. Thus, in addition to membranes, the availability of ligands in each fluid is one of the factors which controls metal ion distribution between different fluid compartments. The membranes can allow passage of free and bound metal ions by passive diffusion, active transport, or exocytosis and endocytosis. Ligands on the other hand, through their different donor groups, can selectively combine and affect the absorption, transport, and activity of different metal ions. This combination, as we have seen in the previous chapter, is primarily determined by the affinity of "hard", class a or "soft", class b metal ions for different types of donor groups available on these ligands. The classification of metal ions as "hard" and "soft" is thus an important preliminary guideline to help predict the combination of metal ions with different ligands, and consequently determine their distribution in different biological fluids. The toxicity of the non-essential metal ions, as we shall see later, is also dependent on this combination through their competitive interaction with essential metal ions for the donor groups of different ligands present in these fluids.

Thus, it follows that the control of metal ion distribution into different biological fluids is in most cases very much dependent on the availability of specific ligands in these fluids. Such a control of metal ion distribution is important because the metal ion composition of these fluids must be kept at approximately constant levels to ensure the coordination of different metabolic processes. The importance of the control of metal ion concentrations as well as their distribution in different biological fluids is further emphasized by the fact that essential metal ions are toxic when present in excess and cause adverse effects when deficient. Major metal ions, for example, can disturb fluid balance or interfere with phosphate biochemistry, while minor metal ions can bind to sulfhydryl groups or promote oxygen toxicity.

Different fractions of metal ions present in free and bound forms have different functions to perform. Knowledge of the concentration of these fractions and their distribution in different biological fluids is also of utmost importance in understanding their functions in these fluids. This is inferred from the fact that, in many instances, the total concentration of a metal ion, compared to the concentration of each species in which it exists, might not be a good indicator of its biological significance, function, deficiency, and toxicity.

II. DIFFERENT METAL FRACTIONS IN THE MAIN BIOFLUIDS OF THE NORMAL INDIVIDUAL

Essential metal ions in the normal human body are present either in high concentrations (major metals) such as sodium, potassium, calcium, and magnesium, or in low concentrations (trace or minor metals) such as iron, copper, manganese, zinc, cobalt, and molybdenum. Beneficial metals, chromium and nickel, are present also in small concentrations in the body. These metal ions are selected and distributed on the basis of their potential to perform specific functions in biological fluids. At neutral pH, the redox potential is rather restricted to a range of -0.4 to +0.8V relative to the standard hydrogen electrode. This implies that different metal ions in biological fluids are only stable in certain oxidation states, thus specifying the functions they will perform in these fluids.

Since the chemical properties of metal ions account for their wide ranging utilization, description of some of these properties - as well as the consequent interaction of the major and minor metal ions with various ligands - will emphasize their importance on the distribution of metal ions into different fractions in relation to their functions in biological fluids. Present knowledge about all metal ligands available in biological fluids is far from complete, but those which are known so far will help us form an idea about the distribution of various metal ions into their different fractions in biological fluids of a normal individual.

A. Sodium and potassium

These two group Ia alkali metal ions are only stable at one oxidation state, +1, with ionic radii of 1.02 and 1.38 Å respectively (Shannon, 1976). Both sodium and potassium are classified as very hard ions due to their affinity for hard bases (Pearson, 1963; Hancock and Martell, 1989). Although they can be complexed by neutral oxygen donors, they are weak binders and, therefore, they are rarely functionally significant in the bound state in biological fluids. They both exist mainly as free hydrated cations and thus perform simple electrolytic functions and also are able to neutralize excess negatively charged groups of biological organic molecules such as carboxylate, phosphate, or sulfate groups of proteins, nucleic acids etc.

Sodium is the major *extracellular* (ec) cation. Total body content of sodium in a normal adult is 40-60 mmol/l, of which 60-70% is exchangeable; the remaining 40% is bound to bone matrix and other sites, and therefore is non-exchangeable. Potassium is the main *intracellular* (ic) ion. Therefore, 98% of the total body potassium is present in the ic fluid, of which about 8-15% is exchangeable and the remaining portion is bound (Eastham, 1983).

The distribution of ionic sodium and potassium into the ic and ec biological fluids is dependent on the activity of the membranes separating these fluids. There are slight differences, however, in the ionic distribution of sodium and potassium in the different ec as well as in the ic fluids. Plasma contains sodium as the principal cation with a concentration range of 133-147 mmol/l (Young et al., 1971; Eastham, 1983) but it also contains potassium at a much lower concentration than sodium (3.4-5.6 mmol/l) (Young et al., 1971). The ionic and molecular composition of the cerebrospinal fluid (CSF) differs from plasma for some components and is the same for others. Similarly, the distribution of these ions in the ic fluids of different cells is not identical, and the concentration of these ions can be between 92 to 160 mmol/l for potassium and 4 to 11 mmol/l for sodium (Ericsson, 1982; Lehninger, 1987a).

B. Calcium

In the body, calcium is widely distributed in the ec and ic fluids. The bulk of this element is, however, found in bone (98%), the surface crystals of which are exchangeable with the ec calcium. This exchange is homeostatically controlled by hormones, the concentration of ionic calcium, and the pH of the biological fluids.

The normal concentration of calcium in plasma is about 2.12-2.6 mmol/l and exists in different fractions. About half (53%) is present as ultrafiltrable calcium where 47% is free ionized (electrostatically "bound" and "active" calcium ions) (Siggaard-Andersen et al., 1983), and 6% complexed to phosphate (1.5%), citrate (1.5%), and bicarbonate ions (3%). The remaining calcium (47%) is bound to proteins such as albumin (37%) and globulins (10%) (Kanis et al., 1983). The three fractions, i.e. the ionized calcium, the low-molar-mass (l.m.m.) ligand-bound calcium and protein-bound calcium fractions, are interrelated through several complex equilibria that are influenced by temperature, ionic strength, pH, magnesium ion concentration, total protein concentration, and albumin-to-globulin ratios (Siggaard-Andersen et al., 1981; 1983).

Variations in total as well as in ionized calcium concentrations compared to plasma are seen in different ec fluids, depending on the functional requirements of calcium. Seminal fluid is an example of a compartment where the concentration of ionized calcium is of significance. Although total calcium is much higher than in plasma (5.8 mM), ionized calcium represents only a small portion (0.17 mM). This is very important for the functional properties of the spermatozoa as an increase in free ionized calcium leads to a decrease in their mobility and viability (Arver and Sjöberg, 1983).

There are also variations in calcium concentrations in different ic fluids. Free calcium in the cytosol is normally near 0.1 μmol/l (Siggaard-Andersen et al., 1983) and under very careful control. In the compartments of the ic space, the exact concentration of free calcium is not known but it could be as high as 1 mM. Measurements using electron probe X-ray microanalysis indicate that total calcium can be between 1.8-5.4 mM in vesicles of different cells (Somlyo et al., 1985; Andrews et al., 1988). In general, however, vesicle fluid compared to organelle fluid is high in free calcium, keeping in mind that different vesicles may contain higher or lower concentrations of this element depending on their function.

It is this non-uniformity in the distribution of calcium between different biological fluids, as well as its distribution between the different fractions in a particular fluid, that makes it possible for this metal ion to perform a multitude of functions in different fluids. In all these functions, the chemical properties of calcium are very well utilized. Calcium belongs to the group IIa metals and, as a hard metal ion, prefers hard oxygen donors which include carboxylates, carbonyls and alcohols. This preference minimizes competition from the major cations, magnesium and sodium. Calcium is also a weak binder and therefore exchanges faster from its ligands compared to other divalent cations such as zinc in biological fluids. For these and other properties, calcium, and not any other cation, is involved in numerous metabolic pathways which include, among others, neuromuscular transmission, the secretion of endocrine or exocrine glands, enzymes and plasma clotting activation, and bone modeling, resorption and deposition.

Specific ligands which help calcium achieve these functions are available in the ic and ec fluids. In the cell, for example, there are sets of proteins which can either act as buffers for calcium or have modulatory activity on different cellular functions (activation). These proteins have a common structural feature, the EF-hands (Kretsinger, 1987), which consists of loops containing 12 amino acid residues, and provides the oxygen for the coordination of Ca^{2+} and therefore binds this metal ion selectively and with high affinity. There are four such loops in calmodulin for example. A variant loop of 14 amino acids is also present in the S-100 protein subfamily (Heizmann and Hunziker, 1991; see also Part Two, Chapter 1-B7). However, these proteins are not the only members of this family as many new members are being discovered bringing the total to over 170 proteins (Heizmann and Hunziker, 1991). Other binding sites for calcium are also available in biological fluids. Upon injury, calcium activates a set of proteases by unfolding and binding domains on these proteins which have a specially synthesized amino acid, γ-carboxyglutamic acid (Gla). It therefore follows that not only the distribution of calcium but also the function of calcium in different biological fluids is a consequence of the availability of ligands in these fluids.

C. Magnesium

Magnesium is the fourth cation in terms of abundance in the human body. The body contains about 25 g of magnesium (Lehninger, 1987b), 55-60% of which is present in the bones (Flink, 1976) where it plays an important part in bone structure. Part of this magnesium is readily exchangeable (Eastham, 1983). Extracellular fluid contains 1-2% of the total body magnesium and the remaining 40-50% is present in the ic fluids. Therefore, all cells including red blood cells contain rather high concentrations of magnesium (5-10 mM), and the estimation of its concentration in erythrocytes gives a measure of ic magnesium. Plasma concentration of this metal ion is about 0.75-1.0 mmol/l (Eastham, 1983) of which 90% is ionized. From these figures one can clearly see that magnesium is the metal of choice in the ic fluids. In these fluids, where the concentration of calcium is kept low, magnesium is mainly associated with pyrophosphate and polyphosphate chemistry where it neutralizes their negative anion charges.

Magnesium, being a small ion, can easily perform this function due to its high charge density. The small ionic radius also makes it possible for magnesium to prefer octahedral coordination geometry in contrast to the irregular coordination preferred by the larger metal ion, calcium. Moreover, Mg(II) is classified as a "hard" metal ion and, although it also belongs to group IIa, is rather different from Ca(II). It does not complex with neutral oxygen donors and also has a weak tendency to bind to nitrogen donors. These and other chemical properties, in addition to their concentration differences in the ic fluids, minimize competition between magnesium and calcium and thus let each perform the functions for which it is best suited.

D. Iron

Oxidized iron, Fe(III), is classified as a hard metal ion and can complex very strongly with negative oxygen donors such as carboxylates, and reasonably well with N-donors. Reduced iron, Fe(II), classified as a "borderline" ion, binds rather weakly and has modest affinity for N-donors. Both ions in their complexes show a marked preference for octahedral coordination. These characteristics are fully utilized in all the ligands involved in the uptake, transport, and functions carried out by iron in its two oxidation states present in different biofluids. These functions are multifaceted and include the activation of oxygen by hemoglobin, the synthesis of DNA, and electron transfer through bioenergetic pathways.

Iron is the most abundant of the trace elements in the body where its total content is about 3-5 g (Carl-Gustaf, 1986). Iron in its oxidized form has the tendency to form insoluble hydrolysis and polymerization products. Different ligands which can decrease these two processes can enhance the bioavailability of iron. This in turn implies that the free iron fraction in biological fluids is extremely low. Moreover, free iron is toxic, and although its ic concentration approaches that of magnesium, the major portion of iron is bound to storage proteins such as ferritin and hemosiderin in the trivalent form. Outside the cells, transferrin is the ligand of choice to bind iron again in the trivalent form, where, together with the first two proteins, it binds 20-30% of the total body iron. However, a small fraction of serum iron is also bound to amino acids which have an important physiological role in its biological transport across cellular membranes (Carl-Gustaf, 1986). Serum also contains a low concentration of ferritin, which reflects the total iron body stores, each ng of ferritin/ml serum corresponding to about 8 mg of storage iron (Walters et al., 1973; Underwood, 1977a).

Nucleotides GTP and to a lesser extent ATP present in the ic fluids, including erythrocytes, also bind ferric ions (Bartlett, 1976). Hemoglobin in the latter cells is, however, the ligand which binds the major portion of bodily iron. The remaining 10% is present in myoglobin and in different iron-requiring enzymes found in other cells (Carl-Gustaf, 1986). The availability of these ligands thus makes it possible to control the distribution of iron into the ic and ec fluids homeostatically. This is achieved by the interaction of iron with different donor groups available on these ligands. Several histidines or histidine and carboxylate groups, which can hold iron in an octahedral geometry, are its preferred binding sites. This illustrates well the importance of iron-ligand interactions on the distribution and concentration of iron into different biological fluids.

E. Zinc

Zinc is the second most abundant trace element after iron in the human body, totalling about 2.3 g in adults (Zinc, 1979). Zinc is an essential trace element required, like calcium, in catalysis, control mechanisms, and in the structural cross-linking of biopolymers. With its concentrated electrostatic charge, +2, and its small ionic radius, 0.74 Å (Shannon, 1976), zinc like calcium and magnesium can bind to carboxylate and phosphate groups. It can also bind to thiolates and amines, a property shared with nickel and copper. Unlike the last two metal ions, zinc does not show variable oxidation states. However, zinc assumes multiple coordination geometries and, therefore, can adapt to constraints imposed by biological macromolecules.

Once again, the binding properties of zinc play a great role in its distribution into different biological fluids. There is some experimental evidence that binding of zinc to metallothionein (MT), for example, might control its intestinal absorption (Richards and Cousins, 1976; Menard et al., 1981; Cousins, 1985). In hepatocytes, MT is also utilized for temporary storage or detoxification of zinc, so that in both liver and intestine, this protein is an important ligand in maintaining homeostasis. Zinc in MT is exclusively coordinated by sulfur and its coordination sphere is saturated. Therefore, zinc is not catalytically active. This is possible because MT is a sulfhydryl-rich protein that has cysteine residues as up to one-third of its total 61 amino acid residues (Kissling and Kägi, 1977; see also Part Two, Chapter 1-D5). Metallothionein contains 7 mol zinc/mol protein (Vallee and Auld, 1990).

After absorption, zinc is probably carried in the portal blood bound to transferrin (Evans, 1976). In the liver, it is incorporated into other circulating proteins which are then transferred into blood plasma. In this ec fluid, a concentration of 11.6-19.1 μmol Zn/l (Taylor, 1983) is distributed between protein ligands synthesized by the liver such as α_2-macroglobulin (30-40%) and albumin (60-70%). In human serum, 98% of the total exchangeable zinc is bound to albumin (Giroux and Henkin, 1972), which is believed to play a role in zinc transport (Giroux et al., 1976), but zinc is also complexed to cysteine and histidine (1%) (Burch et al., 1975).

The diffusible serum zinc is composed of the freely exchangeable zinc from albumin, the amino-acid bound zinc, and the low fraction of aquated zinc ions (Henkin et al., 1973; Williams, 1983). The nondiffusible fraction is bound to α_2-macroglobulin (Williams, 1983). Binding of zinc to ligands such as glutathione and its thiol group containing hydrolysis products, cysteine and cysteinylglycine (Ballatori and Clarkson, 1985; Cregus and Varga, 1985) on the other hand can aid the excretion of zinc into the bile (Cregus and Klaassen, 1986). Affinity of zinc to sulfur, nitrogen, and oxygen is also well utilized in its binding to other ligands where it can play important catalytic and structural roles. The latter is especially true for "zinc finger" proteins (Diakun et al., 1986; see Part Two, Chapter 2-C). All these ligands therefore control the transport, distribution, storage, function, and excretion of zinc present in different biological fluids.

F. Copper

The human body contains about 70-80 mg of copper (Cartwright and Wintrobe, 1964; Sass-Kortsak, 1965; Sumino et al., 1975). However, free copper may be toxic and therefore, like iron, copper exists mostly in complexed forms, bound to macromolecules and to l.m.m. ligands in body fluids. As with the other metal ions, the stoichiometry of complex formation and the nature of the binding sites with these ligands are of great importance. As borderline in the HSAB classification, Cu(II) binds to N-donors and also to thiol groups (where it is easily reduced) to form strong complexes. Thus, ic free copper does not exceed 10^{-15} M. Most of ic copper is bound to MTs, which contain seven copper atoms in a thiolate complex (Harris, 1983). Copper in these proteins is, however, relatively more easily exchangeable than in other proteins.

Transport of newly absorbed copper from mucosal cells to blood plasma is also regulated by MT. In this ec fluid, copper is initially bound predominantly to albumin and amino acids, mainly as ternary complexes of histidine and other amino acids. This fraction of copper is then transported to the liver where it is first bound to MT and then incorporated into ceruloplasmin (Cp). Ceruloplasmin (EC 1.16.3.1) is a multifunctional enzyme with a molar mass of 151 kg mol^{-1} which was considered to bind 90-95% of plasma copper (Kasper and Deutsch, 1963) until transcuprein was characterized (Weiss and Linder, 1985). Copper in Cp, coordinated to histidine (and possibly tyrosine) residues, does not, however, exchange with non-Cp copper in vivo (Bush et al., 1956; Scheinberg and Morell, 1957; Sternlieb et al., 1961). Therefore, Cp does not function as a transport vehicle for the ingested copper through the portal blood to the liver. Ceruloplasmin, on the other hand, can be the possible donor of functional ic copper (Hsieh and Frieden, 1975; Udom and Brady, 1980; Li et al., 1980; Geller and Winge, 1980; Harris and DiSilvestro, 1981), a function for which albumin-bound copper apparently cannot substitute (Marceau and Aspin, 1973). For these reasons, namely binding of Cp to the major portion of copper and its possible function as a donor for the ic copper, plasma copper is a more sensitive indicator of the copper status of an animal than whole blood copper (Dreosti and Quicke, 1968; Underwood, 1977b). This is so as erythrocyte copper is not involved in transporting this element to and from the tissues. Moreover, erythrocyte copper content is not affected in copper loading (Gubler et al., 1953) and in copper deficiency (Shields et al., 1961) and, therefore, although red blood cell copper is circulating, it cannot substitute in the functions performed by Cp copper.

G. Manganese

Although manganese can exist in different oxidation states, only Mn(II) and Mn(III) are of any biological consequence. Manganese (II) prefers six-coordination and octahedral geometry and Mn(III) tetragonal symmetry in their complexes. Mn(III) forms relatively more stable complexes than Mn(II). Manganese in these two ionic states shares common properties with other ions present in biological fluids. Mn(III), for example, is very similar to Fe(III) in its binding and kinetics with different ligands. Mn(II), on the other hand, as a rather hard ion (see Part Three, Chapters 1-C1 and 2-A1) with its high affinity for oxygen donor groups, shares many properties with Mg(II) and Ca(II) ions and, therefore, can substitute them in their ligands. This competition is, however, minimized by the low concentration of manganese in biological fluids relative to calcium and magnesium. Total body burden of an adult, for example, is about 10 mg (Šaric, 1986), and its concentration in the ec and in the ic fluids are quite low. Total blood manganese is 0.13-0.22 μmol/l (Šaric, 1986); the major portion is found in the red blood cells (Weissman, 1981) and the rest in plasma at a concentration of 0.01-0.018 μmol/l (Grafflage et al., 1974; Halls and Fell, 1981). In this ec fluid, Mn(II) binds to α_2-macroglobulin and Mn(III) binds to transferrin (Gibbons et al., 1976). A Mn(II)-glycine complex is also known to exist (Brannan et al., 1964; Jokl, 1964). In the bloodstream of animals, Mn(II) can be oxidized on binding to transferrin and then, as Mn(III)-transferrin, is transported across membranes only to revert to Mn(II) on release. In the cytosol, the concentration of manganese is similar to its concentration in the ec fluid, in contrast to its concentration in the vesicles and organelles, especially mitochondria. The latter accumulates manganese needed for the enzymes superoxide dismutase and pyruvate decarboxylase.

H. Cobalt

Trivalent cobalt is a fairly hard ion and it complexes very strongly with nitrogen donors. In biological fluids it is bound to four N-donors in vitamin B_{12}. The other two coordination positions in this vitamin are occupied by a non protein base and a metal-carbon bond. Thus, the activity of cobalt is confined to functions of vitamin B_{12} either in the simple form or as a coenzyme. This association is also important in the uptake of this metal ion in a non-toxic form because the ingestion of cobalt salts produces toxicity (Allen et al., 1978). Transcobalamin II, of molar mass 38 kg mol^{-1}, is involved in vitamin B_{12} transfer in plasma where the concentration of cobalt is about 5.1 nmol/l (Parr and Taylor, 1964).

I. Molybdenum

Molybdenum ions in all oxidation states are classified as soft and, therefore, their affinity for soft donors such as RS^-, and to a lesser extent neutral nitrogen base groups is quite understandable. The strong affinity of molybdenum ions, specially Mo(VI), for S^{2-} is utilized in biological systems where a special ligand, dithiolate in pterin, a l.m.m. extractable ligand, is involved. With this ligand, molybdenum can retain its selective reactions by exchanging and incorporating into proteins without dissociation from the complex. This pterin complex in its reduced form is the active molybdenum cofactor in enzymes such as xanthine oxidase and sulphite oxidase (Berg et al., 1979; Tullius et al., 1979). The need for molybdenum in the biological fluids is therefore quite specific and it is present in rather low concentrations. In plasma, for example, its content is about 41.69 nmol/l (Bowen, 1966; Schroeder and Nason, 1971).

J. Chromium

Chromium has nine oxidation states ranging from +6 to -2 (Mertz, 1969), but only the Cr(III), Cr(V) and Cr(VI) ions are of biological relevance. The oxidation state in which chromium exists plays a very important role in its binding to different ligands to affect its absorption, distribution, and function. Cr(VI), for example, readily passes through cell membranes while Cr(III) does not seem to have this property (Gray and Stirling, 1950; Hopkins and Schwarz, 1964; Mertz, 1969; Sayato et al, 1980; see *Medicine*, Vol. 2, Pt. 4, Chapter 1-B9). Within the cells of different tissues, Cr(III) can complex with a number of ligands. High concentrations of chromium are normally found in nucleic acids (Wacker and Vallee, 1959). Its function is not known, but a stabilizing effect of chromium on RNA has been demonstrated in vitro (Fuwa et al., 1960). The chromium in RNA may be in an organic form.

Chromium exists in a variety of forms in the circulation as well. In plasma at physiological levels, Cr(III) binds to transferrin, a β-globulin fraction (Hopkins and Schwarz, 1964). When excessive amounts are present, other fractions including albumin, α, β, and γ-globulins and fibrinogens, also become important (Jandl and Simmons, 1957). A concentration of 2.69 nmol/l has been reported (Kayne et al., 1978), but plasma chromium is not in equilibrium with the body stores and may not, therefore, adequately reflect tissue loads. In the cytoplasm of red blood cells, Cr(VI) binds to hemoglobin or glutathione (Gray and Stirling, 1950; Mertz, 1969; Sayato et al., 1980).

The toxicity of chromium is also dependent on its oxidation state. In the cytosol of different cells, Cr(VI) is reduced to Cr(III). During this reduction process, the generation of Cr(V) has been demonstrated (Rabenstein et al., 1979; Jennette, 1982; Goodgame et al., 1982). Glutathione can perform this reduction (O'Brien, 1985; Goodgame and Joy, 1986; Kawanishi et al., 1986; Kitagawa et al., 1988). Since this ligand is present in rather high concentrations inside the cell (Knox, 1960), its reduction of Cr(VI) can produce the thiyl radical, resulting in the formation of chromium-glutathione-DNA crosslinks (Borges and Wetterhahn, 1989) and thus producing lesions in DNA.

K. Nickel

Like the other metal ions in the first transition row, nickel occurs commonly in the +2 state. However, unlike the other +2 metal ions, it has a slower rate of ligand exchange in and out of its coordination sphere compared to Zn(II) and Cu(II). Ni(II), as borderline in the

HSAB classification, prefers sulfur and nitrogen donor groups. This affinity is well used in all the ligands in different biological fluids.

Intracellularly, nickel is the component of the enzyme urease present in vesicles of the cell. In this enzyme, nickel is bound to thiolates and catalyzes the hydrolysis of urea. This is a very specialized role and no deficiency state of nickel has as yet been identified in humans. In serum, where nickel is present at a concentration of 51.1 nmol/l (Bowen, 1966; Schroeder and Nason, 1971), different fractions have been identified. Albumin binds nickel at the amino terminal of its molecule: Asp-Ala-His, where histidine is the essential amino acid for binding (Peters, 1975; Glennon and Sarkar, 1982). Nickel is also bound to amino acids, mainly L-histidine (Sarkar, 1984). A major fraction of nickel is bound to nickeloplasmin where it is not readily exchangeable and, therefore, this portion does not appear to play a role in ec transport of nickel (Nomoto et al., 1971; Descy and Sunderman Jr., 1984). In contrast, L-histidine and albumin, which form a ternary complex of albumin-nickel-L-histidine, transport nickel as well as exert a controlling influence on its cellular accumulation (Nieboer et al., 1984). The proposed model of transport of nickel relies on the ability of the amino acid L-histidine to remove it from albumin via this ternary complex and on the ability of the l.m.m. nickel-L-histidine complexes to cross the biological membrane. This is possible as L-histidine has greater affinity for nickel than does albumin (Glennon and Sarkar, 1982).

Intracellular complexes for nickel have also been identified. In kidney, for example, nickel is found to exist as soluble cytoplasmic complexes. A specific nickel-binding glycoprotein has been isolated (Glennon and Sarkar, 1982; Fischer et al., 1984; Cavelier et al., 1985) and its amino acid composition has indicated that it is not MT (Fischer et al., 1984). Similarly, nickel is found not to bind to this protein in the hepatic cytosol (Eaton et al., 1980; Tandon, 1983; Behari et al., 1984; Oskarsson and Tjalve, 1984). In the latter fraction, nickel is reported to bind high-molar-mass sulfhydryl proteins as well as l.m.m. sulfhydryl-containing ligands similar to the glutathione moiety (Van Soestbergern and Sunderman Jr., 1972; Sunderman Jr. et al., 1983; 1984). Nickel is also found to bind with other l.m.m. ligands such as histidine and phosphocreatine (Tandon, 1983; Behari et al., 1984; Oskarsson and Tjalve, 1984).

L. Summary of metal ion fractions

A pattern emerges from the experimental results available so far on individual metal ions. Firstly, metal ions are not present in identical concentrations in all biological fluids and secondly, in a given fluid, they exist in non-exchangeable and exchangeable fractions. The tightly-bound fraction in some metalloproteins, for example, is not exchangeable. The exchangeable fraction, on the other hand, comprises the metal ions loosely-bound to other proteins (in labile equilibrium with ions in solution), the complexes of different l.m.m. ligands and the free (aquated) ions. Although the fraction of the metal bound to metalloproteins is not in equilibrium with the last fraction, a steady state equilibrium may however be achieved through the feedback control of protein synthesis and degradation exerted by the metal ion concentration in the fluids. This in turn will control the uptake and release of the metal bound to the non-exchangeable fraction.

III. SURVEY OF THEIR DIFFERENT PHYSIOLOGICAL IMPLICATIONS

We have seen that biological fluids are not uniform in their metal ion composition. We have also seen that the distribution of metal ions in each fluid into their corresponding fractions is dependent on the chemical properties and on the availability of ligands in each fluid. This lack of uniformity in the total concentration as well as in the pattern of distribution of metal ions between the different fractions, i.e. free and bound metal ions, is essential in determining the physiological roles they play in each biological fluid. These roles include communication, control, regulation and catalysis.

The functions performed by different metal ion fractions cannot, however, be accomplished without an ongoing interaction between these fractions. This interaction is necessary to ensure proper homeostasis on metal ions available in biological fluids. The competitive equilibrium between the l.m.m. and the loosely-bound protein fractions with the free metal ions as well as the feedback control by these fractions on the tightly-bound protein fractions help achieve this homeostasis.

A. Distribution between different biological fluids: physiological implication

1. *Major metal ions*

In the ec fluids, sodium is the major cation, and chloride and bicarbonate are the major anions present. In the ic fluids, potassium and magnesium are the cations of greatest concentration, and proteins and organic phosphates are the principal anions. In the latter fluids, cytosol contains low concentrations of calcium, whereas vesicles and organelles are rich in this metal ion. It follows that the physiological roles that these metals play is a consequence of their energized distribution between different biological fluids. These roles are mostly related to the free ionic fraction of each metal in each fluid and, therefore, are elaborated in relation to this fraction.

2. *Minor metal ions*

Due to their different chemical properties and also to their interaction with different ligands, a pattern of distribution of these metal ions in different biological fluids emerges. Zinc is the most common catalytic metal ion in the cytoplasm of cells, but is less common than Na, K, Ca, and Mg. Iron is more common in membranes, and copper is largely outside the cytoplasm and outside cells. It is, however, essential in the body to reduce free ion concentrations of iron and copper. Therefore, although the cellular content of iron approaches that of magnesium and that of copper exceeds calcium, the minor essential metal ions iron and copper are almost entirely bound - in general, copper is more difficult to exchange than iron. Their functions in the ic or ec fluids are interdependent and are also determined by the ligands to which they bind in these fluids. Manganese is mutagenic and therefore is not allowed in high concentrations in the cytosol of the cell; it is pumped into vesicles where its competitive strength is enhanced.

B. Distribution between different fractions in a biofluid: physiological implications

1. *Free metal ion fraction*

Major metals can exist as free ions in biological fluids and their functions depend greatly on the concentration of this fraction. For sodium and potassium, for example, the differences in their cation concentrations are fundamental for the maintenance of the membrane potential of the cell, vital for nerves impulse conduction and for muscle contraction.

The concentration of calcium ion in the cytosol is several times smaller than its concentration in the ec fluid and the ic organelles and vesicles. These differences in ionic calcium concentrations in different fluids is of critical importance, as the activation of many different cellular processes through hormones, pharmacological agents, and electrical signals depends on the increases in ic free calcium concentration which can be either derived from the ec calcium or from ic calcium stores. These actions, in response to external stimulation of the cell by various means, constitute the trigger role of calcium.

Accumulation of magnesium inside the cell enables it to stabilize or activate ic phosphates. These include nucleic acids and small molecules. The major role of magnesium in the cytoplasm of cells is, however, the neutralization of anion charges such as ATP^{4-}. Other roles include the stabilization of ribosome and microtubule structure. Outside the cell, like calcium, it binds to anionic sites of the plasma membrane. Overall, however, magnesium is deeply involved in every major metabolic path and is also associated with polynucleotide biochemistry.

The minor metal ions might also exist in their free aquated form, but in very much smaller concentrations compared to the bulk metal ions. The importance of this fraction for the trace elements becomes significant during their excess in biological fluids where, in the case of copper and iron, their involvement in the production of oxygen-centered radicals becomes of great relevance to their toxicity.

2. *Low-molar-mass ligand-bound fraction*

In most biological fluids, the size of the labile pool of the metal ion is not the same as the l.m.m. complex fraction. The former also includes the exchangeable metal ions bound to other ligands.

The l.m.m. fraction of iron or copper can illustrate the functions of the l.m.m. complexes in biological fluids. Fructose by complexing iron increases its gastrointestinal absorption (Charley et al., 1963; Davis and Deller, 1967; Bates et al., 1972). Copper-amino acid complexes in serum have a similar property in their ability to enhance the uptake of this metal ion by liver and other cells (Neumann and Silverberg, 1965; Sass-Kortsak et al., 1965; Harris and Sass-Kortsak, 1967). The physicochemical properties imparted by the amino acids determine the recognition by the carrier system on the membranes and thus affect the transport of copper across these membranes (Barnea and Katz, 1990).

These l.m.m. complexes also play a role in transporting the metal ion between biological binding sites. Amino acids, specially L-histidine and L-threonine in serum, can mediate the transfer of copper from albumin which represents the labile-protein bound copper fraction. The transferred copper can then be transported across biological membranes (Sarkar and Wigfield, 1968). Of course this type of transfer is very much dependent on the nature of the l.m.m. complex. Ni(II)-(L-His)$_2$, for example, is thermodynamically and kinetically a very stable complex from which nickel transfer is relatively slow (Tabata and Sarkar, 1992).

In the ic fluids, the l.m.m. fractions can also serve as an intermediate pool to provide metal ions for their incorporation into proteins. Moreover, they can serve as feedback control on the synthesis of these proteins, and in metal excess, can be the cause of metal toxicity. These functions are suggested for the intermediate pool of l.m.m. iron, iron transfer taking place between physiological binding sites via complexes formed by citrate and transferrin (Aisen and Leibman, 1968; Ganzoni et al., 1974). In the transfer of iron from internalized transferrin to ferritin, for example, it was found that the participation of ascorbic acid and ATP is necessary in liver cells and in erythrocytes (Mazur et al., 1960; Mazur and Carleton, 1963). Citrate, sugars and amino acids may also function as intermediate l.m.m. ligands in iron transfer (Miller and Perkins, 1969).

The same l.m.m. pool of iron serves to control protein synthesis. In reticuloendothelial cells, for example, it induces ferritin synthesis (Lipschitz et al., 1971). Moreover, l.m.m. complexes such as ferric ADP and ferric ATP can donate iron to mitochondria (Strickland and Davis, 1965), and ferric citrate can provide iron to reticulocytes for heme synthesis (Morgan, 1971; Barnekow and Winkelman, 1976). On the other hand, Fe(II) l.m.m. ligands are thought to be the source of iron for free radical generation and damage (Floyd, 1983; Ryan and Aust, 1992). It has been suggested that the expansion of the ic l.m.m. transit pool mentioned earlier might be the source of iron for the hepatotoxicity induced during iron overload due to different causes (Bacon and Britton, 1989). Iron bound to these ligands (Bacon and Britton, 1989) or "free" iron generated after the saturation of cellular l.m.m. ligand-binding capacity (Ryan and Aust, 1992) are suggested to be available for peroxidation of lipids as well as generation of hydroxyl radicals.

There are l.m.m. complexes in which the metal ions function as a separate entity, usually in relation to proteins. These metal ions are considered kinetically inert and therefore are not in direct thermodynamic equilibrium with other l.m.m. fractions. Examples include the iron in heme, in hemoglobin and myoglobin. With cobalt and molybdenum, their very presence in special ligands such as vitamin B$_{12}$ and the pterin cofactor where they are retained indefinitely indicates that their exchange is not desirable. This in turn implies that the metal ions bound to such ligands are not involved in the mechanisms of control for the free ion. A similar situation also exists with iron. Although this metal can slowly exchange in most of its complexes, iron in the porphyrin ring does not interact with the iron present outside the ring except in the control of the latter fraction of iron on heme synthesis.

3. *Fraction loosely bound to proteins*

Metal ions loosely bound to proteins represent part of the labile exchangeable pool in biological fluids. Perhaps the best examples are albumin for copper, zinc, nickel, and calcium, transferrin for iron, and most probably MT for copper and zinc. Metallothionein is also under a feedback control mechanism after an increase or a decrease in the levels of these two metal ions in biofluids. These ligands are present in relatively high concentrations and, therefore, act as buffers in addition to their functions in the transport or detoxification of metal ions.

4. *Fraction tightly bound to proteins*
This fraction is largely represented by metal ions found in proteins which carry the metal ion in a thermodynamically irreversible manner. Metal ions are therefore sequestered in them for a specific function and are not available for other biochemical purposes.

C. Conclusion
The physiological implications of the distribution of metal ions into different fractions are therefore far reaching. They determine the distribution of metal ions between biological fluids, help specify the functions performed by the metal ions, and contribute to their homeostasis. It follows that the relative concentration of a metal ion in each fraction is also important, as the fate of the metal ion in a given biofluid is dependent on the fraction in which it is present as well as on the ligand with which it is associated. The major portion of metal ions present in the labile fraction, for example, is associated with proteins. They can therefore act as buffers, and thus control the concentration of the free metal ions in biological solutions. We have seen that the concentration of the latter fraction of transition metal ions is quite low. This low concentration and also the weak binding properties of the major metal ions makes the relative concentration of the l.m.m. fraction very low. This fraction is much lower than the protein fraction, despite the fact that l.m.m. ligands are not in short supply in biological fluids.

As the steady states occurring in vivo operate near the thermodynamic equilibrium, the individual complex species in the labile fractions can be considered in true equilibrium with the free metal ions. Thus, from the concentrations of free ions and the formation constants of the involved metal-ligand complexes, equilibrium distributions of metal ions among their l.m.m. complexes can be estimated. These considerations are the basis of computer simulation studies developed for plasma and other biological fluids (May et al., 1977; see Part Five, Chapters 3 and 4).

ACKNOWLEDGEMENTS

The authors gratefully acknowledge the help of their colleagues, Dr. D.B.K. Rama, Mrs. P. Theodorou, Miss E. Casimiro, and Miss K. Ambaram. They also thank Mrs L. Kotze for her assistance in typing the references.

REFERENCES

Aisen, P., and Leibman, A. (1968). *Biochem. Biophys. Res. Commun.*, *32*: 220.

Allen, R.H., Seetharan, B., Allen, N.C., Podell, E.R., and Alpers, D.H. (1978). *J. Clin. Invest.*, *61*: 1628.

Andrews, S.B., Leapman, R.D., Landis, D.M., and Reese, T.S. (1988). *Proc. Nat. Acad. Sci.*, U.S.A., *85*: 1682.

Arver, S., and Sjöberg, H.E. (1983). *Scand. J. Clin. Lab. Invest.*, *43 (Suppl. 165)*: 123.

Bacon, B.R., and Britton, R.S. (1989). *Chem.-Biol. Interact.*, *70*: 183.

Ballatori, N., and Clarkson, T.W. (1985). *Fundam. Appl. Toxicol.*, *5*: 816.

Barnea, A., and Katz, B.M. (1990). *J. Inorg. Biochem.*, *40*: 81.

Barnekow, A., and Winkelman, G. (1976). *Exp. Hematol.*, *4*: 70.

Bartlett, G.R. (1976). *Biochem. Biophys. Res. Commun.*, *70*: 1063.

Bates, G.W., Boyer, J., Hegenauer, J.C., and Saltman, P. (1972). *Amer. J. Clin. Nutr.*, *25*: 983.

Behari, J.R., Dwivedi, P.P., Misra, M., and Srivastava, R.C. (1984). *Biol. Tr. Elem. Res.*, *6*: 463.

Berg, J.M., Hodgson, K.O., Cramer, S.P. Corbin, J.L., Elsbary, A., Paryadath, N., and Steifel, E.I. (1979). *J. Amer. Chem. Soc.*, *101*: 2774.

Borges, K.M., and Wetterhahn, K.E. (1989). *Carcinogenesis*, *10*: 2165.

Bowen, H.J.M. (1966). *Trace Elements in Biochemistry*, Academic Press, New York.

Brannan, J.R., Dunsmore, H.S., and Nancollas, G.H. (1964). *J. Chem. Soc., Faraday Trans.*, *60*: 304.

Burch, R.E., Hahn, H.K.J., and Sullivan, J.F. (1975). *Clin. Chem.*, *21*: 501.

Bush, J.A., Mahoney, J.P., Gubler, C.J., Cartwright, G.E., and Wintrobe, M.M. (1956). *J. Lab. Clin. Med.*, *47*: 898.

Carl-Gustaf, E. (1986). *Handbook on the Toxicology of Metals*, Vol. II (L. Friberg, G.F. Nordberg, and V.B. Vouk, eds.), Elsevier, Amsterdam, New York, Oxford, p. 276.

Cartwright, G.E., and Wintrobe, M.M. (1964). *Amer. J. Clin. Nutr.*, *14*: 224.

Cavelier, C., Fousseveau, J., and Massin, M. (1985). *Contact Dermatitis*, *12*: 65.

Charley, P.J., Sarkar, B., Stitt, C., and Saltman, P. (1963). *Biochim. Biophys. Acta*, *69*: 313.

Cousins, R.J. (1985). *Physiol. Rev.*, *65*: 238.

Cregus, Z., and Klaassen, C.D. (1986). *Toxicol. Appl. Pharmacol.*, *85*: 24.

Cregus, Z., and Varga, F. (1985). *Acta Pharmacol. Toxicol.*, *56*: 398.

Davis, P.S., and Deller, D.J. (1967). *Austral. Ann. Med.*, *16*: 70.

Descy, M.I., and Sunderman, F.W., Jr. (1984). *Bioinorg. Chem.*, *3*: 87.

Diakun, G.P., Fairall, L., and Klug, A. (1986). *Nature*, *324*: 698.

Dreosti, I.E., and Quicke, G.V. (1968). *Brit. J. Nutr.*, *22*: 1.

Eastham, R.D., (1983). *A Guide to Water, Electrolyte and Acid-base Metabolism*, Wright, P.S.G., Bristol, London, Boston.

Eaton, D.L., Stacey, N.H., Wong, K.L., and Klaassen, C.D. (1980). *Toxicol. Appl. Pharmacol.*, *55*: 393.

Ericsson, F. (1982). *Scand. J. Clin. Lab. Invest.*, *42 (Suppl 163)*: 9.

Evans, G.W. (1976). *Trace Elements in Human Health and Disease*, Vol. I (A.S. Prasad and D. Oberleas, eds.), Academic Press, New York, San Francisco, London, p. 181.

Fischer, T., Fregert, S., Gruvberger, B., and Rystedt, I. (1984) *Contact Dermatitis*, *10*: 23.

Flink, E.B. (1976). *Trace Elements in Human Health and Disease, Essential and Toxic Elements*, Vol. II (A.S. Prasad and D. Oberleas, eds.), Academic Press, New York, San Francisco, London, p. 2.

Floyd, R.A. (1983). *Arch. Biochem. Biophys.*, *225*: 263.

Fuwa, K., Wacker, W.E.C., Druyan, R., Bartholomay, A.F. and Vallee, B.L. (1960). *Proc. Nat. Acad. Sci. USA*, *46*: 1298.

Ganzoni, A.M., Baviera, B., and Hahn, D. (1974). *Experientia*, *30*: 32.

Geller, B.L., and Winge, D.R. (1980). *Fed. Proc.*, *39*: 1768.

Gibbons, R.A., Dixon, S.N., Hallis, K., Russell, A.M., Sansom, B.F., and Symonds, H.W. (1976). *Biochim. Biophys. Acta*, *444*: 1.

Giroux, E.L., and Henkin, R.I. (1972). *Biochim. Biophys. Acta*, *273*: 64.

Giroux, E.L., Duienx, M., and Schechter, P.J. (1976). *Bioinorg. Chem.*, *5*: 211.

Glennon, J.D., and Sarkar, B. (1982). *Biochem. J.*, *203*: 15.

Goodgame, D.M.L., and Joy, A.M. (1986). *J. Inorg. Biochem.*, *26*: 219.

Goodgame, D.M.L., Hayman, P.B., and Hathway, D.E. (1982). *Polyhedron*, *1*: 497.

Grafflage, B., Buttegereit, G., Kübler, W., and Mertens, H.M. (1974). *Z. Klin. Chem. Klin. Biochem.*, *12*: 287.

Gray, S.J., and Stirling, K. (1950). *J. Clin. Invest.*, *29*: 1604.

Gubler, C.J., Lahey, M.E., Cartwright, G.E., and Wintrobe, M.M. (1953). *J. Clin. Invest.*, *32*: 405.

Halls, J.D., and Fell, S.G. (1981). *Anal. Chim. Acta, 129*: 205.

Hancock, R.D., and Martell, A.E. (1989). *Chem. Rev., 89*: 1875.

Harris, D.I.M., and Sass-Kortsak, A. (1967). *J. Clin. Invest., 46*: 659.

Harris, E.D. (1983). *Trace Elements in Health* (J. Rose ed.), Butterworths, London, Boston, Sydney, Wellington, Durban, Toronto, p. 44.

Harris, E.D., and DiSilvestro, R.A. (1981). *Proc. Soc. Exp. Biol. Med., 166*: 528.

Heizmann, C.W., and Hunziker, W. (1991). *TIBS, 16*: 98.

Henkin, R.I., Schulman, J.D., Schulman, C.B., and Bronzert, D.A. (1973). *J. Pediat., 82*: 831.

Hopkins, L.L., Jr., and Schwarz, K. (1964). *Biochim. Biophys. Acta, 90*: 484.

Hsieh, H.S., and Frieden, E. (1975). *Biochem. Biophys. Res. Commun., 67*: 1326.

Jandl, J.H., and Simmons, R.L., (1957). *Br. J. Haematol., 3*: 19.

Jennette, K.W. (1982). *J. Amer. Chem. Soc., 104*: 874.

Jokl, V. (1964). *J. Chromatogr., 14*: 71.

Kanis, J.A., Paterson, A.D., and Russell, R.G.G. (1983). *Biochemistry in Clinical Practice* (D.L. Williams and V. Marks, eds.), William Heinemann Medical Books Limited, London. p. 299.

Kasper, C.B., and Deutsch, H.F. (1963). *J. Biol. Chem., 238*: 2325.

Kawanishi, S., Inoue, S., and Sano, S. (1986). *J. Biol. Chem.*, 261: 5952.

Kayne, F.J., Komar, G., Laboda, H., and Vanderlinde, R.E. (1978). *Clin. Chem., 24*: 2151.

Kissling, M.M., and Kägi, J.H.R. (1977). *FEBS Lett., 82*: 247.

Kitagawa, S., Seki, H., Kametani, F., and Sakurai, H. (1988). *Inorg. Chim. Acta, 152*: 251.

Knox, W.E. (1960). *The Enzymes*, Vol. 2, Academic Press, New York.

Kretsinger, R.H. (1987). *Cold Spring Harbor Symp. Quant. Biol., 70*: 499.

Lehninger, A.L. (1987a). *Principles of Biochemistry*, Worth Publishers, Inc., p. 385.

Lehninger, A.L. (1987b). *Principles of Biochemistry*, Worth Publishers, Inc., p. 779.

Li, T.Y., Kraker, A.J., Shaw, C.F., and Petering, D.H. (1980). *Proc. Nat. Acad. Sci. USA, 77*: 6334.

Lipschitz, D.A., Simon, M.O., Lynch, S.R., Dugard, J., Bothwell, T.H., and Charlton, R.W. (1971). *Brit. J. Haematol., 21*: 289.

Marceau, N., and Aspin, N. (1973). *Biochim. Biophys. Acta, 293*: 338.

May, P.M., Linder, P.W., and Williams, D.R. (1977). *J. Chem. Soc., Dalton Trans.*, 588.

Mazur, A., Green, S., and Carleton, A. (1960). *J. Biol. Chem., 235*: 595.

Mazur, A., and Carleton, A. (1963). *J. Biol. Chem., 238*: 1817.

Menard, P., McCormick, C.C., and Cousins, R.J. (1981). *J. Nutr., 111*: 1358.

Mertz, W. (1969). *Physiol. Rev., 49*: 163.

Miller, J.P.G., and Perkins, D.J. (1969). *Eur. J. Biochem., 10*: 146.

Morgan, E.H. (1971). *Biochim. Biophys, Acta, 244*: 103.

Neumann, P.Z., and Silverberg, M. (1965). *Proc. Can. Fed. Biol. Soc., 8*: 49.

Nieboer, E., Stafford, A.R., Evans, S.L., and Dolovich, J. (1984). *Nickel in the Human Environment* (F.W. Sunderman, Jr., ed.), IARC Sc. Public No. 53, Lyon, France. p. 321.

Nomoto, S., McNeely, M.D., and Sunderman, F.W., Jr. (1971). *Biochemistry, 10*: 1647.

O'Brien, P. (1985). *Inorg. Chim. Acta, 108*: L19.

Oskarsson, A., and Tjalve, H. (1984). *Acta Pharmacol. Toxicol., 45*: 306.

Parr, R.M., and Taylor, D.M. (1964). *Biochem. J., 91*: 424.

Pearson, R.G. (1963). *J. Amer. Chem. Soc., 85*: 3533.

Peters, T., Jr. (1975). *The Plasma Proteins*, Vol. I, 2nd ed. (F.W Putnam, ed.), Academic Press, New York, p. 133.

Rabenstein, D.L., Guevremont, R., and Evans, C.A. (1979). *Metal Ions in Biological Systems*, Vol. 9 (H. Sigel, ed.), Marcel Dekker Inc, New York, p. 103.

Richards, M.P., and Cousins, R.J. (1976). *J. Nutr.*, *106*: 1591.

Ryan, T.P., and Aust, S.D. (1992). *Critical Rev. Toxicol.*, 22: 119.

Šaric, M. (1986). *Handbook on the Toxicology of Metals* (L. Friberg, G.F. Nordberg and V.B. Vouk, eds.), Elsevier, Amsterdam, New York, Oxford, p. 354.

Sarkar, B. (1984). *Nickel in the Human Environment* (F.W. Sunderman, Jr., ed.), IARC Sc. Public No. 53, Lyon, France, p. 367.

Sarkar, B., and Wigfield, Y. (1968). *Can. J. Biochem.*, *46*: 601.

Sass-Kortsak, A. (1965). *Adv. Clin. Chem.*, *8*: 1.

Sass-Kortsak, A., Harris, D.I.M., Goodman, S.J., Hawke, M., and Smuckler, R.H. (1965). *Canad. Med. Ass. J.*, 92: 368.

Sayato, Y., Nakamura, K., Matsui, S., and Ando, M. (1980). *J. Pharm. Dyn.*, *3*: 17.

Scheinberg, I.H., and Morell, A.G. (1957). *J. Clin. Invest.*, *36*: 1193.

Schroeder, H.A., and Nason, A.P. (1971). *Clin. Chem.*, *17*: 461.

Shannon, R.D. (1976). *Acta Cryst.*, *A32*: 751.

Shields, G.S., Markowitz, H., Klassen, W.H., Cartwright, G.E., and Wintrobe, M.M. (1961). *J. Clin. Invest.*, *40*: 2007.

Siggaard-Andersen, O., Thode, J., and Wandrup, J. (1981). "Blood pH, Carbon Dioxide, Oxygen and Calcium Ion", Proceedings of the 1980 Meeting of the IFCC Expert Panel on pH and Blood Gases (O. Siggaard-Andersen, ed.), Private Press, Copenhagen, Denmark, pp. 163-190.

Siggaard-Andersen, O., Thode, J., and Fogh-Andersen, N. (1983). *Scand. J. Clin. Lab. Invest.*, *43 (Suppl. 165)*: 57.

Somlyo, A.P., Bond, M., and Somlyo, A.V. (1985). *Nature*, *314*: 622.

Sternlieb, I., Morell, A.G., Tucker, W.D., Greene, M.W., and Scheinberg, I.H. (1961). *J. Clin. Invest.*, *40*: 1834.

Strickland, E.H., and Davis, D.C. (1965). *Biochim. Biophys. Acta*, *104*: 596.

Sumino, K., Hayakawa, K., Shibata, T., and Kitamura, S. (1975). *Arch. Environ. Health*, *30*: 487.

Sunderman, F.W., Jr., Mangold, B.L.K., Wong, S.H.Y., Shen, S.K., Reid, M.C., and Jansson, I. (1983). *Res. Commun. Chem. Pathol. Pharmacol.*, *39*: 477.

Sunderman, F.W., Jr., Zaharia, O., Reid, M.C., Belliveau, J.F., O'Leary, G.P., and Griffin, H. (1984). *Toxicology*, *32*: 11.

Tabata, M., and Sarkar, B. (1992). *J. Inorg. Biochem.*, *45*: 93.

Tandon, S.K. (1983). *Toxicol. Lett.*, *10*: 71.

Taylor, A. (1983). *Biochemistry in Clinical Practice* (D.L. Williams and V. Marks, eds.), William Heinemann Medical Books Ltd., London. p. 356.

Tullius, T.D., Burtz, D.M., Conradson, S.D., and Hodgson, K.O. (1979). *J. Amer. Chem. Soc.*, *101*: 2776.

Udom, A.O., and Brady, F.O. (1980). *Biochem. J.*, *187*: 329.

Underwood, E.J. (1977a). *Trace Elements in Human and Animal Nutrition*, 4th ed., Academic Press, New York, p. 13.

Underwood, E.J. (1977b). *Trace Elements in Human and Animal Nutrition*, 4th ed., Academic Press, New York. p. 56.

Vallee, B.L., and Auld, D.S. (1990). *Biochemistry*, *29*: 5647.

Van Soestbergern, M., and Sunderman, F.W., Jr. (1972). *Clin. Chem.*, *18*: 1478.

Wacker, W.E.C., and Vallee, B.L. (1959). *J. Biol. Chem.*, *234*: 3257.

Walters, G.O., Miller, F.M., and Worwood, M. (1973). *J. Clin. Pathol.*, *26*: 770.

Weiss, K.C., and Linder, M.C. (1985). *Amer. J. Physiol.*, *249*: E77.

Weissman, N. (1981). *Klin. Wochenschr.*, *59*: 413.

Williams, D.R. (1983). *Chemical Toxicology and Clinical Chemistry of Metals* (S.S. Brown and J. Savory, eds.), Academic Press, London, New York, p. 167.

Young, D.S., Harris, E.K., and Cotlove, E. (1971). *Clin. Chem.*, *17* : 403.

Zinc (1979). Subcommittee on Zinc, Committee on Medical and Biologic Effects of Environmental Pollutants, Division of Medical Sciences Assembly of Life Sciences. National Research Council, University Park Press, Baltimore, p. 123.

DIFFERENT FRACTIONS OF LIGANDS IN BIOLOGICAL FLUIDS AND THEIR PHYSIOLOGICAL ROLES

by

Mairam Gulumian[1,2], Robert Douglas Hancock[2] and Halina Barbara Röllin[1,3]

[1] National Centre for Occupational Health,
P.O. Box 4788
Johannesburg 2000
South Africa

[2] Centre for Molecular Design,
University of the Witwatersrand,
Wits 2050
Johannesburg
South Africa

[3] Department of Community Health,
University of the Witwatersrand,
Wits 2050
Johannesburg 2001
South Africa

I. INTRODUCTION

Nature has mastered ligand design so that physiological effects of metal ions in vivo are very specific. This specificity is due to the different functional groups at the binding sites of ligands that recognize the metal ion and deliver it to its sites of action. Ligands donor atoms participate in metal ion coordination as well as protonation and deprotonation processes. Five important functional groups in biological systems are negative oxygen, neutral oxygen, neutral nitrogen, negative sulfur, and neutral sulfur. Selenium also occurs in ligands available in biological fluids, and is similar to sulfur. One can arrange metal ions in approximate orders of preference for functional groups. These orders will be affected by coordination number and spin state. An example of the effect of coordination number is Zn(II), which, when four-coordinate rather than six-coordinate, shows higher affinities for negative oxygens, sulfurs, and nitrogens. Examples of the effect of spin state are Fe(II) and Fe(III), which when low-spin show greater affinity for neutral nitrogens.

The negative oxygen donor, found for example in carboxylates ($RCOO^-$), phenolates, phosphonates, pyrophosphonates, and other phosphorus-based oxygen donors is a very hard donor group. The strength of binding of metal ions of importance in biological systems is:

Na^+, K^+ << Ca^{2+} < Mg^{2+} < Mn^{2+} < Fe^{2+} < Cd^{2+} < Ni^{2+} < Pb^{2+} < Zn^{2+} < Cu^{2+} < Al^{3+} < Hg^{2+} < Fe^{3+}. In complexes, tetrahedral Zn (II) moves above Cu(II), and Cu(I) is oxidized to Cu(II).

For neutral oxygen, in alcohols ROH, esters R_2O, ethers RCOOR, and amides $RCONH_2$ (Hancock, 1992a; 1992b), classified as very hard donors, the affinity is a function of metal ion size. Only metal ions with an ionic radius above 1 Å have much affinity for neutral oxygens. These metal ions are: Na^+, K^+, Ca^{2+}, Pb^{2+}, and Hg^{2+}.

Neutral nitrogens RNH_2, in pyridines and imidazoles, are hard/borderline donor groups, less hard than oxygen donors. The affinity order is: Na^+, K^+ << Ca^{2+} < Mg^{2+} < Al^{3+} << Mn^{2+} < Fe^{2+} < Pb^{2+} < Cd^{2+} < Zn^{2+} < Ni^{2+} < Fe^{3+} < Cu^{2+} Cu^+ << Hg^{2+}. In complexes with nitrogen donors, Cu(I) is likely to be oxidized to Cu(II), tetrahedral Zn(II) moves up to near Cu(II), low-spin Fe(II) and Fe(III) move to above Cu(II), and the affinity of the ions below Mn^{2+} in the order for nitrogen is negligible.

Negative sulfur RS^-, as in thiols of cysteinate in proteins, is a powerful soft donor with an approximate order: Na^+, K^+ << Al^{3+} < Mg^{2+} < Ca^{2+} << Mn^{2+}, Fe^{2+} < Ni^{2+}, Zn^{2+} << Pb^{2+} < Cd^{2+} < Cu^+ << Hg^{2+}. Cu^{2+} and Fe^{3+} tend to be reduced by negative sulfur donors to Cu^+ and Fe^{2+}, respectively. Affinity for metal ions below Mn^{2+} is negligible.

Neutral sulfurs R_2S, such as thioethers of methionine and also in biotin, a cyclic thioether, are very soft donor groups. Binding to metal ions below Mn^{2+} in the above order is negligible, and very weak to Mn^{2+}, Fe^{2+}, Fe^{3+}, Ni^{2+}, Zn^{2+}, Pb^{2+}, Cd^{2+} and Cu^{2+}. However, interaction with neutral sulfur is important where, with biotin for example, stereospecific biotin-sulfur/metal ion interactions create active enzyme-substrate complexes (Sigel and Scheller, 1982). Strong binding of neutral sulfur occurs to Cu^+ and Hg^{2+}.

Although the R moieties of the above functional groups can vary from a small alkyl group to a large protein, the metal ion affinities of each of these are largely unaffected if steric effects are small. Most ligands, especially macromolecules in biological fluids, have many donor groups. The affinity of ligands with several types of donor groups is approximately the average of those of the groups present. Multidentate ligands bind metal ions much more strongly than the same donor groups in unidentate ligands. This effect, the chelate effect (Schwarzenbach, 1952), explains why many biologically important ligands are multidentate.

Chelate ring size is important in complex stability. The number of atoms in the chelate ring may vary from three upwards. There is a remarkable correlation between chelate ring size, metal ion size, and complex stability. Five-membered chelate rings coordinate best with large metal ions of high coordination number, while six-membered chelate rings coordinate best with small metal ions of low coordination number. Smaller chelate rings of size three and four also coordinate larger metal ions better. When metal ions are the wrong size for the chelate ring, steric strain results, with loss in complex stability.

In biological fluids, the extent to which a ligand binds specific metal ions depends on its donor groups and chelate ring size, as well as the charge, ionic radius, coordination geometry, and concentration of the metal ion. For major ions, concentration is controlled by selective pumping across different membranes. Ligand concentration is also controlled in this way, as well as by rates of synthesis and degradation.

II. LIGAND DISTRIBUTION IN THE MAIN BIOFLUIDS OF THE NORMAL INDIVIDUAL

A. Different types of ligands occurring in biological fluids

Biological fluids are multicomponent systems where interactions of proton-ligand and ligand-metal ions take place. In these multicomponent systems, the concentrations of ligands exceed those of different metal ions, especially the transition metal ions. The latter in turn leads to a competition of ligands for the available metal ions and the outcome depends on the various factors mentioned earlier. At the correct pH and ligand:ligand ratios, however, conditions in these fluids are created such that ternary complexes are also formed. Since fractions of ligands - l.m.m. and macromolecular - available in a fluid will determine the extent of these interactions, it is helpful to know the type of ligands in each fraction. In the l.m.m. ligand fraction, simple organic acids, amino acids and their derivatives, small peptides, carbohydrates, and nucleotides are good examples, in contrast to proteins, nucleic acids, polysaccharides and lipids which constitute the macromolecular fraction.

The donor groups available on each of the components of a given fraction are dependent on the type of ligand in that fraction. Simple acids in the l.m.m. ligand fraction, such as citrate, lactate, oxalate, pyruvate, succinate and ascorbate bind different metal ions in biological fluids. The binding of citrate to Fe(III) is a very good example (Aisen and Leibman, 1968; Ganzoni et al., 1974). Free amino acids, a major fraction of the l.m.m. ligands available, are always present in biological fluids. They all have a carboxyl group and an amino group bonded to the same carbon atom but their side chains are different. The amine and the carboxyl groups of amino acids as well as other additional functional groups interact with different metal ions in order of affinities for functional groups stated earlier.

Simple amino acids can coordinate metal ions through either or both of the amino and carboxylate groups. The amino acids glycine, alanine, and other aliphatic amino acids, are good examples of ligands carrying these two donor groups. These simple aliphatic amino acids form bidentate complexes ML, ML_2, and ML_3 with most of the metal ions.

Donor atoms in the side chains of different amino acids also offer binding sites for metal ions. The increase in the stability constants relative to the bidentate amino acid alanine is a strong indication of the involvement of these side chains. These amino acids can therefore form tridentate complexes. Examples of additional functional groups include -OH in serine and threonine. The alcoholic hydroxy groups in these amino acids are so weakly acidic that they do not undergo dissociation, but their electron attracting affinity lowers the basicity of the amino group. A second carboxyl group in aspartic and glutamic acids also provides the opportunity for these amino acids to form tridentate complexes with different metal ions. Asparagine and glutamine, on the other hand, show a weaker tendency to form tridentate complexes due to the weaker ability of the amide group to participate in the binding. The -SH group in cysteine, a second amine group in arginine and histidine are further examples of side-chain donor groups which can help form tridentate complexes. The ability of cysteine to bind metal ions such as zinc, copper, iron, nickel and molybdenum is well known. There are four sites on histidine that can be involved in protonation equilibrium; carboxylic acid, imidazolium-N(3)-H, side-chain ammonium, and imidazole-N(1)-H. The last group is, however, very weakly acidic and therefore not dissociable in the measurable pH. Histidine is thus a tridentate ligand but also is an ambidentate ligand depending on the participation of the amino group or the carboxylate group in the binding to the metal ion. Amino acid derivatives, which are mostly hormones, also serve as ligands in biological fluids. Examples include histamine, a decarboxylation product of histidine and epinephrine (adrenaline), derived from tyrosine (for details on metal - amino acid interactions, see Part Three, Chapter 2-A).

Dipeptides and oligopeptides formed from different amino acids are also good ligands for different metal ions. Again, the basicity of the donor groups in the peptide molecule is one of the most important factors which influences their binding to these metal ions. In a dipeptide, the functional groups present include the terminal-amino ($-NH_2$) and the carboxylate ($-COO^-$) groups and also the amide group of the peptide linkage (-CONH-). Longer oligopeptides, without side chains, also have these groups in addition to extra amide groups and in consequence make the distance between the terminal groups longer. In general, the longer the peptide, the larger the decrease of the basicity of the amino groups and the acidity of the carboxylic groups. In such a case, only two pK values attributed to the ammonium and carboxylic groups are measured. The amide group, on the other hand, is "neutral" throughout the measurable pH. It can, however, contribute to metal ion binding through oxygen when neutral, or nitrogen after it is deprotonated. The presence of side chains in an oligopeptide significantly modifies its acid-base properties. The effect of these side chains depends on their proximity to the C- or N- terminal of the peptide chain (see Part Three, Chapter 2-B1).

Various functional groups available on the side chains of amino acid residues in peptides also affect the stability of the metal-peptide complexes differently. The alcoholic -OH groups of the amino acids serine and threonine are weakly coordinating side chains.

They do not bind but rather enhance the stability of the metal ion complexes. The phenolic -OH group of L-tyrosine on the other hand, is involved in the coordination depending on the location of the tyrosyl residue. The carboxylic groups of aspartic and glutamic acids enhance the stability of the complexes formed where 5- and 6-membered chelate rings can be formed with the amino acid carboxylate groups. The side-chain amino nitrogen of lysine can serve as a site for polynuclear complex formation. The imidazole-N donor atom in histidine can significantly enhance the stabilities of the corresponding complexes depending on the location of the histidyl residue. The case of L-proline is worth mentioning. This amino acid has no additional donor groups in the side chain. It contains a secondary nitrogen which cannot be protonated when incorporated in a peptide bond. Therefore, the amide nitrogen is not available for metal ion binding and thus serves as a "break point" when it is present in an intermediate or in the C-terminal of the peptide chain. This amino acid is an important residue in many neuropeptides and hormones (Iversen, 1983; Bloom, 1983; Kozlowski et al., 1989; Pettit and Formicka-Kozlowska, 1984; see Part Three, Chapter 2-B2).

Sulfur in peptides occurs in the thioether group in methionine and its derivatives, in the thiol group of L-cysteine, in the disulphide group produced from the oxidation of thiols, and in the thioamides where it appears in the peptide linkage. The thioether functional group is a soft base to which most transition metal ions bind weakly. The thiol group, on the other hand, is a strong base that provides a very effective binding site for common transition metal ions. Glutathione tripeptide is one of the most important thiol-containing ligand. When oxidized, -SH groups in glutathione and other peptides form a disulphide bridge which usually does not take part in coordination, but helps form binuclear complexes (Várnagy et al., 1988). Finally, the thioamides are sulfur-containing strong donor groups which are present in the amide linkage and not on the side chain of the amino acid (Kowalik et al., 1987). (Details on metal ion complexes with sulfur-containing peptides are available in Part Three, Chapter 2-B3). Understanding the rules which govern the binding of metal ions to peptides helps greatly to understand the interactions of metal ions with proteins.

Sugars also act as ligands in biological fluids. Monohydric alcohols and diols do not form very stable complexes with cations and at least three hydroxy groups in a favorable steric arrangement are required for complex formation in such a system. Complex formation with these donor groups in sugars also depend on the type of metal ion where those carrying a +3 or +4 charge show greater affinity. A good example is the ability of Fe(III) to form complexes with simple sugars such as fructose which binds one iron per molecule (Sarkar et al., 1964). A carboxy or an amino group, or both, introduced into a sugar molecule, enhance its ability to form complexes by several orders of magnitude (Sawyer, 1964; Tamura and Miyazaki, 1965; Micera et al., 1985). (For details, see Part Three, Chapters 2-D and 2-G).

Nucleosides and nucleotides represent different groups of l.m.m. ligands in which the pyrimidine bases provide the binding sites at N3. Purine bases, on the other hand, provide such binding sites at N1 and N7. In neutral solutions, the nitrogens at N1 of inosine ($pK_a = 8.7$) and guanosine ($pK_a = 9.2$) are protonated. Therefore, metal ions may coordinate to the weakly basic N7 or compete with the proton for the more basic N1. In the same neutral solutions, adenosine is weakly basic with $pK_a = 3.6$ for N1, both nitrogens at N1 and N7 sites being free to bind metal ions (Martin and Mariam, 1979; Martin, 1985). In nucleotides, phosphate groups offer additional binding sites for metal ions (see Part Two, Chapter 2-A). Binding of magnesium to the latter groups in nucleotides ATP in the cytosol and to UTP in vesicles are excellent examples of such binding where they can act as substrates for many enzymatic reactions.

Proteins are the most abundant and most important macromolecules in biological fluids. The potential ligands available to bind metal ions in proteins include the side chains of amino acid residues, the carbonyl oxygen or amide nitrogen of the peptide bond linking the amino acid residues, the N-terminal amine, and the C-terminal carboxylate group. The tertiary structure of a protein can sterically bring together the side chains of some amino acids and act as exclusive binding sites for metal ions. There may be additional binding to the metal ion by the nonprotein nitrogen ligands of a macrocycle as in heme proteins, or solvents, substrate, or other small molecules. The type of ligand available in a protein determines not only the function of that metal ion in a protein, e.g. catalysis, control, and triggering, but also determines the function performed by a protein in relation to a specific metal ion, e.g. transport, storage, and detoxification.

In nucleic acids, binding of metal ions to phosphate groups has a quite different effect than binding of metal ions to bases. Binding of Mg(II) to phosphate stabilizes the double helix whereas binding of a transition metal ion like Cu(II) to bases leads to the cleavage of the nucleotides.

Polysaccharides are found in the connective tissues of animals, on the external proteins of the outer surfaces where they generate recognizable surfaces on the cell (glycoproteins and proteoglycans, and glycolipids), in mucus, and in storage. The side chains of these macromolecules such as $-CO_2^-$, $-OSO_3^-$, etc provide binding sites for different cations such as Na^+, K^+, Mg^{2+}, and Ca^{2+}.

Membrane lipid two-dimensional surfaces are negatively charged, mostly with O-donor groups. They can therefore interact with cations such as Na^+, K^+, Ca^{2+}, and Mg^+. However, this interaction is not specific but rather electrostatic. The maintenance of membrane structure and function by zinc is also reported through its binding to the acyl chain ester groups of the phospholipid molecules or to the phosphate groups to produce a change in lipid organization (Bettger and O'Dell, 1981; Kaszuba and Hunt, 1990; see Part Three, Chapter 2-F).

B. Relative binding strengths of ligands in biological fluids

The state of functional groups on ligands depends on pH and oxidation potential of the fluid compartment. Functional groups such as imidazole or thiols, for example, tend to deprotonate at typical pH values. Scheme 1 shows typical pH values for protonation/deprotonation of functional groups commonly found in biological systems. These values can be altered by several effects. Firstly, when more than one functional group are close together on a single molecule, there is a decrease in protonation constants for the second and subsequent steps. This is illustrated by the protonation constants of histidine (Scheme 1). The protonation constant of the histidine carboxylate is lowered from pK = 4.9 for typical carboxylates to 1.6. This arises from electrostatic repulsion between the incoming proton and those already bound to the molecule. The state of thiol groups is controlled by the pH. Those with two thiols will have them in $(-SH_2)_2$, $(-S^-)_2$, or $(-SH)(-S^-)$ state depending on pH. Oxidation potential, as well as pH, controls oxidation of thiols to disulfide linkage (-S-S-).

Metal ions compete with the proton for donor groups. Ordinarily, phenols deprotonate at pH 10 (Scheme 2). In presence of a strongly coordinating metal ion such as Fe^{3+}, the proton comes off at pH 2. The ability of metal ions to displace the proton from functional groups is related to the formation constant of the metal-ligand complex - the larger the constant, the lower the pH at which the proton is displaced. At physiological pH, ligands

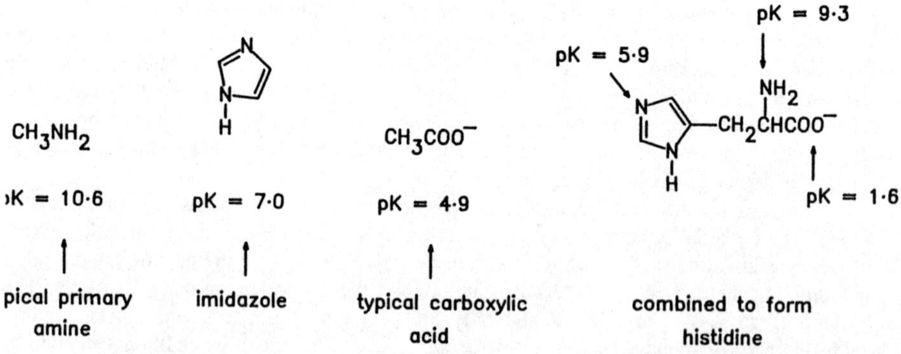

Scheme 1

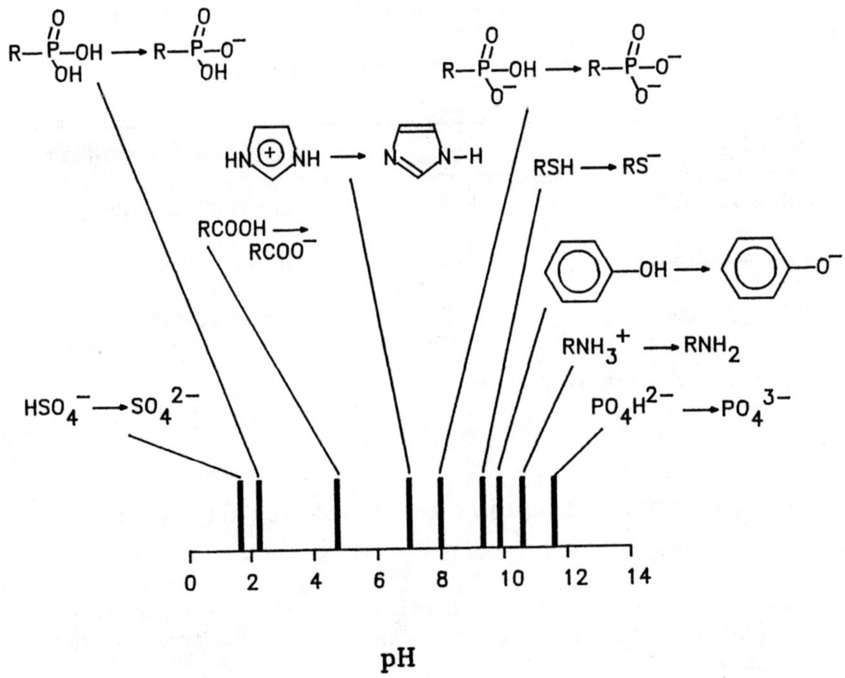

Scheme 2

such as acetic acid or imidazole are already deprotonated, and there is no competition between proton and metal ion. Phosphate groups present on ligands such as ATP, GTP, etc are also weakly protonated at physiological pH.

Membranes separate biological fluids and control which ligands are present. Many ligands are present in a given biological fluid, and multiligand-multimetal ion equilibria occur. Ligands present in a particular fluid compete for a specific metal ion, and also interact to form mixed-ligand complexes with high stability and of great physiological significance.

For example Zn(II) can form a mixed complex containing cysteine and histidine. Statistically there are two ways to form one mixed complex $[Zn(histidine)(cysteine)]^-$, and only one way each for $[Zn(histidine)_2]$ and $[Zn(cysteine)_2]^{2-}$. Statistical arguments suggest that the mixed complex should dominate. One finds that the mixed complex, as for Zn(II) with cysteine and histidine, may be more stable than expected statistically. This extra stabilisation, K_M, of the mixed-ligand complex is the equilibrium constant for the reaction:

$$MA_2 + MB_2 \rightleftharpoons 2MAB$$

For no extra stabilization for MAB, log K_M is zero. Steric factors, π-bonding, and ligand-ligand interactions promote the extra stability of mixed-ligand complexes. Steric effects usually involve ligands that mesh better sterically with each other in the mixed-ligand complex. Formation of π-bonds between π-acceptors such as heteroaromatic nitrogen bases like histamine and donor groups such as catecholates stabilizes mixed-ligand complexes. Finally, hydrogen bonding or hydrophobic interactions between ligands can stabilize mixed-ligand complexes. Mixed-ligand complexes can also give rise to stereoselective effects (Sigel, 1973; 1975; Tribolet et al., 1985; Massoud and Sigel, 1989; Chakraborty and Bhattacharya, 1990; see Part Three, Chapter 1-C3).

Typically, mixed-ligand complexes may have log K_M values up to 4 log units. Some examples are shown in Table 1 (Sóvágó and Gergely, 1976; Sóvágó et al., 1979).

Table 1

Examples of Mixed-ligand Complexes

System	log K_M	Proposed cause of stabilization
Zn(II)/L-Trp/ATP	3.55	ligand-ligand interaction
Cu(II)/histamine/catechol	0.69	π-bond formation
Cu(II)/gly/Me_2en^*	2.76	steric effects (better meshing)

* Me_2en = N,N'-dimethylethylenediamine.

III. SURVEY OF THEIR DIFFERENT PHYSIOLOGICAL IMPLICATIONS

The metal-ion binding of different types of ligands available in a given biofluid and the consequent interaction between ligands through this binding have significant effects not only on the bound metal ions but also on the ligands themselves. Moreover, the physiological implications of such interactions will depend on the metal which facilitates these interactions between ligands.

As far as the effect on bound metal ions is concerned, we have already seen the role of ligands in the facilitated transport of otherwise charged metal ions across membranes. Similarly, the charge on the complex also determines the transport and the fate of the ligand. For example, the fraction of histamine in plasma which is not bound to proteins exists mostly as its mono- or di-protonated forms (Berthon and Kayali, 1982) which are electrically charged. It is unlikely for these species to freely diffuse into tissues (Gudzinowicz et al., 1984). Zinc-histamine-cysteine neutral complex thus represents the diffusible form for histamine in plasma (Berthon and Kayali, 1982; Berthon et al., 1987). The dicarboxylic acids malate, malonate, maleate, and tartrate can also form a ternary complex with zinc and histamine to favor histamine tissue diffusion (Berthon et al., 1987).

The involvement of metal ions in the regulation of the bioactivity of natural peptides is also reported in the literature (Tonoue et al., 1979; Pettit and Formicka-Kozlowska, 1984; Kozlowski et al., 1986). Examples are numerous. Angiotensin is a linear octapeptide which behaves like a hormone and its biological activities are enhanced by certain cations such as Na^+, Ca^{2+}, Mg^{2+}, and Mn^{2+} (Schaechtelin et al., 1974; 1975; Blanc et al., 1978; Gunther et al., 1980). Thymulin, a nonapeptide, requires zinc and its biological activity depends on this ion (Cung et al., 1988). Luteinizing hormone-releasing hormone (LHRH) is a decapeptide amide and its complexes with copper, nickel, and zinc, are more potent than the peptide itself in promoting and releasing the follicle stimulating hormone (FSH) (Kochman et al., 1992).

Possibly the dissociation of metal ions from their ligands, especially proteins, due to the protonation of the donor groups at acidic pH, is a fine example of the effect of metal ions on the ultimate fate of the ligands in biological fluids. It is the acidity that releases the metal ions from their carrying proteins. A good example is MT. Degradation of MT takes place in the lysosomes (pH 5.5). Metal ion release is a prerequisite for this degradation and this release is easily achieved by the low pH of the lysosomes. Saturation of the protein with metal ions protects it from degradation and, therefore, the lower the concentration of the metal ion bound to the protein, the higher is its susceptibility to degradation.

The fate of a ligand can be varied by changing the metal ion bound to it. This is achieved through the ability that metal ions have to bind different functional groups available on ligands and, therefore, mask or unmask various functional groups important for different processes. This is very well illustrated by 3,4-dihydroxyphenylalanine (L-Dopa).

This molecule is biologically important as well as it is used as a therapeutic agent to treat certain diseases. Upon administration, however, it can be metabolized and converted to dopamine after its decarboxylation by the enzyme dopadecarboxylase. This process can be reduced by coordinating, through the amino carboxylate end of the molecule, a metal ion already bound to another ligand so as to produce a mixed-ligand complex. Not all metal ions can achieve this and therefore this type of coordination, (N,O), is determined by the type of the metal ion coordinated and also by the type of the second ligand involved in the mixed-ligand complex (Patel and Bhattacharya, 1984; Kiss and Gergely, 1985). Thus, depending on the metal ion which binds a given ligand, the fate of this ligand can be determined so that it can either be metabolized or made available for activity.

Association of metal ions with ligands also increases the possibility of recognition of these ligands by their respective receptor sites on cellular membranes and thus increase their bioavailability. For example, transferrin attaches itself to special receptor sites on the surface of immature erythrocytes, and molecules which contain iron are more readily taken up than the apo-transferrin molecule which lacks iron.

Recognition between two species facilitated by the formation of mixed-ligand complexes with a metal ion (Sigel et al., 1983) further introduces specificity and selectivity in biological fluids. Activity of many enzymes relies on this type of recognition (Otsuka and Yamanaka, 1988) where the first ligand bound to a metal ion can introduce discriminating qualities toward other ligands to be coordinated (Sigel, 1973).

Nature has thus recognized the potential of mixed-ligand complexes and has utilized it to its advantage. The above example is one of the biological processes that is made possible through the formation of mixed-ligand complexes. The very involvement of these ligand species in this process has far reaching implications. We all know that life would not have been possible without the specificity of enzyme reactions. Despite their significance, not much is known in this field and, therefore, future research efforts directed towards clarification of these mechanisms will not only elucidate further the many aspects of recognition and specificity of enzyme catalysis, but new roles for different mixed-ligand-metal species formed in biological fluids might also be discovered.

ACKNOWLEDGEMENTS

The authors gratefully acknowledge the help of their colleagues, Dr. D.B.K. Rama, Mrs. P. Theodorou, Miss E. Casimiro, and Miss K. Ambaram. They also thank Mrs L. Kotze for her assistance in typing the references.

REFERENCES

Aisen, P., and Leibman, A. (1968). *Biochem. Biophys. Res. Commun.*, *32*: 220.

Berthon, G., and Kayali, A. (1982). *Agents Actions*, *12*: 398.

Berthon, G., Varsamidis, A., Blaquière, C., and Rigal, D. (1987). *Agents Actions*, *22*: 231.

Bettger, W.J., and O'Dell, B.L. (1981). *Life Sci.*, *28*: 1425.

Blanc, E., Sraer, J., Sraer, J.D., Baud, L., and Ardaillou, R. (1978). *Biochem. Pharmacol.*, *27*: 517.

Bloom, F.E. (1983). *Annu. Rev. Pharmacol. Toxicol.*, *23*: 151.

Chakraborty, D., and Bhattacharya, P.K. (1990). *J. Inorg. Biochem.*, *39*: 1.

Cung, M.T., Marraud, M., Lefrancier, P., Dardenne, M., Bach, J.-F., and Laussac, J.-P. (1988). *J. Biol. Chem.*, *263*: 5574.

Ganzoni, A.M., Baviera, B., and Hahn, D. (1974). *Experientia*, *30*: 32.

Gudzinowicz, B.J., Younkin, B.T., Jr., and Gudzinowicz, M.J. (1984). *Drug Dynamics for Analytical, Clinical and Biological Chemists*, M. Dekker, Inc., New York.

Gunther, S., Gimbrone, M.A., and Alexander, R.W. (1980). *Circ. Res.*, *47*: 278.

Hancock, R.D. (1992a). *Perspectives in Inorganic Chemistry* (A.P. Williams, C. Floriani and A.E. Merbach, eds.), VCH Publishers, Weinheim, Basel, p. 129.

Hancock, R.D. (1992b). *J. Chem. Ed.*, *69*: 615.

Iversen, L.L. (1983). *Annu. Rev. Pharmacol. Toxicol.*, *23*: 1.

Kaszuba, M., and Hunt, G.R.A. (1990). *J. Inorg. Biochem.*, *40*: 217.

Kiss, T., and Gergely, A. (1985). *J. Inorg. Biochem.*, *25*: 247.

Kochman, K., Gajewska, A., Kozlowski, H., Masiukiewicz, E., and Rzeszotarska, B. (1992). *J. Inorg. Biochem.*, *48*: 41.

Kowalik, T., Kozlowski, H., Sóvágó, I., Varnagy, K., Kupryszewski, G., and Rolka, K. (1987). *J. Chem. Soc., Dalton Trans.*, 1.

Kozlowski, H., Formicka-Kozlowska, G., Pettit, L.D., and Steel, I. (1986). *Frontiers in Bioinorganic Chemistry* (A.V. Xavier, ed.), Weinheim Verlag GmbH, p. 668.

Kozlowski, H., Radomska, B., Kupryszewski, G., Lammek, B., Livera, C., Pettit, L.D., and Pyburn, S. (1989). *J. Chem. Soc., Dalton Trans.*, 173.

Martin, R.B. (1985). *Accounts Chem. Res.*, *18*: 32.

Martin, R.B., and Mariam, Y.H. (1979). *Metal Ions in Biological Systems*, Vol. 8 (H. Sigel, ed.), Marcel Dekker, Inc., New York and Basel, p. 57.

Massoud, S.S., and Sigel, H. (1989). *Inorg. Chim. Acta*, *159*: 243.

Micera, G., Deiana, S., Dessi, A., Decock, P., Dubois, B., and Kozlowski, H. (1985). *Inorg. Chim. Acta*, *107*: 45.

Otsuka, S., and Yamanaka, T., eds. (1988). *Metalloproteins. Chemical Properties and Biological Effects*, Elsevier, Amsterdam.

Patel, V.K., and Bhattacharya, P.K. (1984). *J. Inorg. Biochem.*, *21*: 169.

Pettit, L.D., and Formicka-Kozlowska, G. (1984). *Neurosci. Lett.*, *50*: 53.

Sarkar, B., Saltman, P., Benson, S., and Adamson, A. (1964). *J. Inorg. Nucl. Chem.*, *26*: 1551.

Sawyer, D.T. (1964). *Chem. Rev.*, *64*: 633.

Schaechtelin, G., Walter, R., Salomon, H., Jelinek, J., Karen, P., and Cort, J.H. (1974). *Mol. Pharmacol.*, *10*: 57.

Schaechtelin, G., Surovec, D., and Walter, R. (1975). *Experientia*, *31*: 346.

Schwarzenbach, G. (1952). *Helv. Chim. Acta*, *35*: 2344.

Sigel, H. (1973). *Metal Ions in Biological Systems*, Vol. 2 (H. Sigel, ed.), Marcel Dekker, Inc., New York, p. 63.

Sigel, H. (1975). *Angew. Chem. Int. Edit.*, *14*: 394.

Sigel, H., and Scheller, K.H. (1982). *J. Inorg. Biochem.*, *16*: 297.

Sigel, H., Fischer, B.E., and Farkas, E. (1983). *Inorg. Chem.*, *22*: 925.

Sóvágó, I., and Gergely, A. (1976). *Inorg. Chim. Acta.*, *20*: 27.

Sóvágó, I., Gergely, A., Harman, B., and Kiss, T. (1979). *J. Inorg. Nucl. Chem.*, *41*: 1629.

Tamura, Z., and Miyazaki, M. (1965). *Chem. Pharm. Bull.*, *13*: 387.

Tonoue, T., Minagawa, S., Kato, N., and Okhi, K. (1979). *Pharmacol. Biochem. Behav.*, *10*: 201.

Tribolet, R., Malini-Balakrishnan, R., and Sigel, H. (1985). *J. Chem. Soc., Dalton Trans.*, 2291.

Várnagy, K., Sóvágó, I., and Kozlowski, H. (1988). *Inorg. Chim. Acta*, *151*: 117.

COMPETITIVE INTERACTIONS AND THEIR PHYSIOLOGICAL CONSEQUENCES

by

Mairam Gulumian[1,2], Robert Douglas Hancock[2] and Halina Barbara Röllin[1,3]

[1] National Centre for Occupational Health,
P.O. Box 4788
Johannesburg 2000
South Africa

[2] Centre for Molecular Design,
University of the Witwatersrand,
Wits 2050
Johannesburg
South Africa

[3] Department of Community Health,
University of the Witwatersrand,
Wits 2050
Johannesburg 2001
South Africa

I. INTRODUCTION

In biological fluids, different metal ions bind various ligands in dynamic steady state equilibria. Competition between metal ions with similar chemical and physical properties for the binding sites of these ligands will lead to changes in these equilibria. Initially, small changes will not much affect the organism and might even help to produce adaptability. Any significant change from these equilibria, produced from interaction between essential or beneficial metal ions as a consequence of their excess or deficiency, will interfere with the functions of different ligands and result in metabolic changes. Similarly, competitive interaction between non-essential metal ions and essential or beneficial metal ions will also interfere with functions of many ligands and produce severe metabolic disturbances. The disturbances produced from the competitive interactions between essential and beneficial metal ions and between these and non-essential metal ions, depend on the type as well as on the dose of the metal ion concerned. Essential metal ions only produce their toxic effects at high doses whereas detrimental metal ions manifest their toxicity at very low doses (for details on the classification used in this Handbook, see *Medicine*, Vol. 1, Part One, Ch.1-C).

Functions performed by low-molar-mass (l.m.m.) and high-molar-mass (h.m.m.) ligands, some of which have been discussed already in the previous section, include absorption, transport, exchange, detoxification, and catalysis.

II. COMPETITIVE METAL-LIGAND INTERACTIONS BETWEEN NATURALLY OCCURRING SUBSTANCES

Interaction of different metal ions starts at the absorption level in the intestine. We have emphasized the importance of ligands on the transport of metal ions across these membranes. Free metal ions present in the gut are hydrophilic and cannot readily diffuse across the lipid bilayer of the cell membrane. To cross these membranes they need to be attached to lipophilic compounds such as heme, or to be associated with ligands which neutralize their charge and also have access to the specific channels in the membranes. Ligands which can perform these functions include sugars, amino acids and small peptides. Competition for these ligands between different essential and beneficial metal ions will therefore determine their absorption. Again, once in plasma, competitive interactions between different essential and beneficial metal ions for l.m.m. ligands will affect their availability, transport across cell membranes, and their insertion into transfer proteins, storage proteins, and metalloenzymes.

Metal ions in biological fluids do not function independently from each other, and an excess or a deficiency of a metal ion will not only affect its own functions, but will also interfere directly or indirectly with the availability and functions of another metal ion. The direct interaction usually results from competitive interaction of chemically similar metal ions for ligand binding sites. On the other hand, the indirect effect is not a consequence of competitive interaction of metal ions for a ligand, but it is a result of impaired functioning of ligands due to the deficiency of a metal ion, which under normal conditions, determines the functions of another metal ion. Experimental evidence for these types of interactions is abundant in the literature.

A. Zinc-copper-iron

Zinc, like copper, has high affinity for donors such as amines and thiolates. In fact, the affinities of these two metal ions for most of the donor groups shown in the previous section are quite similar and, therefore, point to the probability of competitive interaction for these donor groups. Indeed, zinc interferes with copper absorption (van Campen and Scaife, 1967). Interaction of zinc with copper appears to take place in the mucosal cells. Metallothioneins (MTs) present in these cells regulate copper with respect to its absorption as well as its transport into plasma. Zinc can displace copper from sulfhydryl binding sites (van Campen and Scaife, 1967; Evans and Cornatzer, 1970) and thus interferes with these functions of the protein. In copper toxicity, the administration of zinc is therefore recommended to slow the entrance of copper into the system (Harris, 1983). The therapeutic use of zinc in Wilson's disease is based on this property (see *Medicine*, Vol. 2, Pt. Five, Chapter 1-I).

The interference of zinc with copper can also be observed in the hepatic cell (Whanger and Weswig, 1970). Here again cysteine-rich MT is the major copper-binding protein which maintains copper in a harmless form as well as donates copper to some metalloenzymes (Geller and Winge, 1980). Competition between zinc and copper for this ligand will again interfere with the other functions of the protein.

Copper deficiency will in turn have adverse effects on iron. Defective iron absorption in the gastrointestinal tract, restricted flow of iron from reticuloendothelial cells to the plasma, excessive retention of iron in liver, and impaired production of heme will result (Lee et al., 1976). These in turn will have a serious effect on normal haemoglobin production (Hart et al., 1928; Elvehjem, 1935). The involvement of ceruloplasmin, a copper transferring protein, and cytochrome *c* oxidase, a copper enzyme, in some of the above processes are well documented (Osaki and Johnson, 1969; Osaki et al., 1971; Mills, 1980).

B. Magnesium-calcium-potassium

As we have seen, calcium is excluded from the cell and thus magnesium and potassium are the main cations in the cytoplasm. This is so, even though no direct competitive interaction between magnesium and calcium is expected due to differences in their chemical properties; i.e. in their ionic radii, their tendency to bind certain donors, and the geometry of their complexes.

Despite these differences between magnesium and calcium, interaction between these two ions can take place with calmodulin. It is generally accepted that the mechanism by which the calcium modulator protein, calmodulin, regulates Ca^{2+}-dependent functions involves a conformational change in calmodulin upon Ca^{2+} binding. Magnesium is shown to bind to this protein but its binding does not produce a large change in conformation (Sutoo et al., 1988). Thus, this binding might not produce the calcium-calmodulin response in the cell but act as a competitive inhibitor.

Further interactions of these two metals take place with the enzyme calcium adenosine triphosphatase (ATPase) where magnesium acts as a cofactor. It follows that, due to these and other interactions, magnesium deficiency and its consequent depletion in the cytosol will lead to many pathological conditions. Decreased intracellular magnesium can cause a rise in free cytosolic calcium level due to enhanced calcium membrane permeability, probably via the calcium channel and inhibition of calcium efflux (Aikawa, 1981). This mechanism is thought to be the reason for the development of irreversible tissue damage following traumatic brain injury (Vink, 1991/92). Magnesium deficiency is also implicated in ischaemic heart disease (Fehily et al., 1987) and acute myocardial infarction (Speich et al., 1980) where there is also a strong possibility of the above mechanisms operating in these pathological conditions (see Volume 2, Part Four, Chapters 3-B and 3-E). Moreover, the majority of the magnesium in the cell is stored in the inner surface of the cell membrane; hence magnesium deficiency may also influence cell membrane permeability and fluidity (Whang and Whang, 1990) and thus affect cell membrane functions such as the intra- and extracellular balance of other ions, in particular potassium. Magnesium deficiency through its decreased availability to act as a cofactor for ATPase and Na-K-ATPase can interfere in calcium movement and also cause a decrease in electrolyte transport, and thus lead to an increase in vascular tone and blood pressure (Christensson et al., 1977; Bulpitt and Shipley, 1981; Kesteloot and Geboers, 1982; MacGregor et al., 1982; see *Bioinorganic Medicine*, Part Four, Chapter 3-A). It is therefore not surprising that magnesium is an important regulator of cell potassium homeostasis. Cellular potassium levels in hypokalamic patients can be normalized only in the presence of adequate magnesium repletion, despite the production of normal serum potassium levels with potassium treatment (Whang et al., 1985). These observed effects of magnesium deficiency on potassium and calcium are therefore those produced as a consequence of alterations of other processes and are not due to a direct competitive interaction of these metal ions.

C. Nickel-magnesium

Ni^{2+} and Mg^{2+} have almost identical crystal ionic radii (0.69 Å and 0.72 Å) respectively (Shannon, 1976)), prefer octahedral coordination arrangement in biocomplexes, and have affinities for similar donor groups (Nieboer et al., 1984). The carcinogenic activity of nickel is therefore suggested to occur in part through the displacement of magnesium from its cellular functional sites, as the action of nickel appears to involve substitution of magnesium bound to phosphate moieties of DNA and nucleotides. Through this interaction, on the other hand, it is possible to counteract nickel carcinogenicity by magnesium (Carvalho and Ziemer, 1982; Poirier et al., 1984).

With nucleotides, Ni^{2+} is coordinated mainly at the triphosphate group but the involvement of the N7 of the adenine base can be observed (Martin, 1988a). For Mg^{2+} and ATP^{4-}, there is no direct interaction of Mg^{2+} with N7 (Happe and Morales, 1976; Mariam and Martin, 1979; Scheller et al., 1981). Therefore, except in basic solutions, Ni^{2+} cannot promote dephosphorylation of nucleotide triphosphates (Amsler and Sigel, 1976; Sigel and

Amsler, 1976; Sigel et al., 1984). The slow rate of exchange of ligands out of the Ni^{2+} coordination sphere can be the reason for the ineffectiveness of this metal in promoting phosphate hydrolysis (Martin and Mariam, 1979).

D. Nickel-calcium

Nickel is shown to interact with calcium by indirectly interfering with its functions through calmodulin. The calmodulin-dependent phosphoprotein, phosphatase, catalyzes the phosphorylation and dephosphorylation of many proteins. The enzyme consists of two polypeptide chains and has four Ca^{2+}-binding sites. It is activated by divalent metal ions such as Mn^{2+}, Mg^{2+}, Co^{2+}, Ni^{2+} etc where Ni^{2+} is found to have the greatest activating ability (King and Huang, 1983; Pallen and Wang, 1984). The activation of the enzyme by nickel was found to be time dependent and irreversible (King et al., 1985). The irreversible activation of the phosphatase by Ni^{2+} in the presence and absence of calmodulin was thus suggested to be toxic by way of interfering with normal cellular regulation.

These experimental results are some of the interactions between different metal ions and by no means represent all the interactions which take place in biolgical fluids. These examples, however, indicate the importance of these interactions in producing the ill effects observed in metal deficiency or overload. Moreover, the elucidation of the mechanisms of these interactions further emphasizes the importance of monitoring the levels of not only the metal ion under study, but also the metal ions which are affected through these interactions during the deficiency or overload of a particular metal ion.

III. COMPETITIVE METAL-LIGAND INTERACTIONS BETWEEN NATURALLY AND NON-NATURALLY OCCURRING SUBSTANCES

Probably, all of the many ligands with which different metal ions associate in some way or another, determine the essentiality or toxicity of a metal ion. Evidence is mounting which suggests that nonessential metal ions exert their toxic effects through their competitive metal-ligand interaction with essential metal ions and thereby interfere with different ligand functions. The degree of toxicity which will result from ligand-nonessential metal ion interaction will depend on the function of the ligand as well as the percentage of molecules affected, and by their rate of repair or replacement. One noteworthy aspect of non-essential metal ion toxicity is the pattern of distribution which varies from one detrimental metal ion to another. This distribution is mainly controlled, in addition to other factors, by metal-ligand interactions which in turn determine the target tissues and thus the systemic toxicity. Examples include renal toxicity by cadmium, central nervous system toxicity by methyl mercury, haematopoietic system toxicity caused by lead, and central nervous system toxicity due to aluminum.

Many other cells in different organs are also exposed to the non-essential metal ions. These cells, in many instances, tolerate certain doses of detrimental metal ions better than other cells due to the availability or the ability of the tolerant cells to synthesize ligands. This process can therefore act as part of the detoxification process against these metal ions. Those cells which do not have these mechanisms are therefore more susceptible to metal ion toxicity.

Experimental results once again clarify some of these detoxification and toxicity processes brought about by different ligands. The type of ligands involved in these mechanisms varies from one non-essential metal ion to another. This variation is very much dependent on the physicochemical properties of the non-essential metal ions and consequently on their affinity for different functional groups on naturally present ligands in biological fluids. These properties will in turn determine with which essential metal ion the interaction will take place and thus determine the ligands to be involved in such an interaction.

A. Aluminum-iron-magnesium-calcium

Studies with aluminum provide a very good example of how a non-essential metal ion can make use of the endogenous ligands, in its absorption, transport, and availability so as to finally exert its toxic effects on the target organs. Successful competition for the binding sites on different ligands, normally available to carry metal ions which have similar properties, will be the target of such an interaction. Similarity between the ionic radii of Al^{3+}, Fe^{3+}, and Mg^{2+} is one explanation for the interaction between these metal ions. As far as its interaction wlth Ca^{2+} is concerned, Al^{3+}, due to its smaller size, cannot interact and replace this ion in proteins but the interaction can be with l.m.m. ligands (Martin, 1988b).

Experimental and computer simulation studies have indicated that indeed both l.m.m. and macromolecular ligands used by the above essential metal ions play a role in the transport and bioavailability of aluminum. Day et al. (1991) have shown that 95% of the absorbed aluminum in plasma is associated with h.m.m. species of which 80% represented aluminum bound to transferrin, 10% bound to albumin, and 5% bound to other h.m.m. species. The non-protein l.m.m. fraction was shown to contain the remaining 5%. Little if any aluminum was associated with erythrocytes. The binding of Al^{3+} to transferrin and to other ligands was further investigated by Fatemi et al. (1991) who confirmed that aluminum can bind transferrin but with less affinity than Fe^{3+} (Jemil et al., 1992) and, therefore, its release from transferrin was much faster than that of Fe^{3+} (Marques, 1991). This led to the suggestion that this binding may be important if Al^{3+} uptake into cells is largely a result of receptor-mediated endocytosis of $(Al^{3+})_2$-transferrin.

The role of l.m.m. ligands such as citrate and phosphate in aluminum absorption, urinary excretion, and tissue bioavailability were also investigated with animal experiments as well as computer simulation studies (Domingo et al., 1986; Venturini and Berthon, 1989; Daydé et al., 1990; Duffield et al., 1991). Results have shown that the charge carried by the complexes formed between aluminum and the ligands determined the excretion of the metal ion or its transport into tissues.

Once inside the cell, citrate again is the target for competitive metal-ligand interaction between aluminum and other essential metal ions such as iron (May and Williams, 1980). Also, phosphate and adjacent phosphate ligands such as ATP and the phosphate groups of membrane lipophosphates and phosphorylated proteins are expected to be prime candidates. Because of its binding to these groups, aluminum can interact with magnesium and affect calcium indirectly. The binding of aluminum instead of magnesium to ATP, for example, will interfere with numerous functions of this nucleotide (Kiss et al., 1991). Also, the binding of aluminum to 1,4,5-inositol triphosphate at the 4,5-phosphate groups (Berridge and Irvine, 1984; Majerus et al., 1986) is expected to have a profound effect on the function of this intracellular messenger which acts by releasing Ca^{2+} from nonmitochondrial intracellular stores (Sihag et al., 1988). The interaction of aluminum with calcium is further illustrated with studies reported by Provan and Yokel (1992) who suggested that aluminum-induced inhibition of learning may be due to its effect on glutamate release, secondary to changes in calcium channel function.

B. Cadmium-zinc-copper

Cadmium, like zinc, is a member of group II of the periodic Table and, like copper (I), has a similar tendency to form complexes with a tetrahedral geometry. Competitive metal-ligand interaction between cadmium and essential metal ions such as zinc, and most probably also copper, is therefore expected and starts with the absorption of cadmium and continues with its transport, availability, toxicity, and excretion. Metallothionein is the key ligand which has a great role to play in all these metabolic stages for cadmium in the body. Cadmium can completely displace zinc from MT at normal concentrations and on a one to one basis from elevated levels of MT (Day et al., 1984). This protein is found to play a role in the normal metabolism of zinc and copper by supplying these metal ions for macromolecular synthesis in rapidly growing and dividing cells (Brady and Webb, 1981; Mason et al., 1981a; 1981b; Brady, 1983). This binding can thus interfere with the above mentioned function of MT.

Metallothioneins are cellular cytosolic proteins, and do not occur normally in extracellular body fluids. Metallothionein bound to cadmium originating from either intestinal mucosa or hepatic cells which appears in plasma is easily filtered through renal glomeruli and absorbed by tubuli. In the tubular cells of the kidney, the protein is broken down, and cadmium not bound to MT will enter the intracellular (ic) fluid. This cadmium stimulates the production of ic MT which in turn binds cadmium and renders it harmless. The free cadmium in the cell is the most toxic to ic biochemical processes, and the amount of free cadmium will depend on the rate at which Cd-MT enters the cell and the rate at which new MT is synthesized (Nordberg et al., 1985). During chronic exposure to cadmium, MT bound to this metal will eventually lead to damage of the kidney cells, resulting in proteinuria, aminoaciduria, and glucosuria (Kjellström, 1977). Urinary excretion of MT reflects not only cadmium exposure levels but also renal dysfunction caused by long-term cadmium exposure (Tohyama et al., 1988). Exposure to cadmium thus, in the long run, produces damage to liver and kidney and also the intestine due to the ability of these organs to synthesize MT and thus accumulate cadmium.

Understandably, competitive metal-ligand interaction between cadmium and zinc does not end with MT. Cadmium replaces zinc in zinc-dependent enzymes which are crucial for many metabolic processes in the body, e.g. carbonic anhydrase (Barr, 1973), thymidine kinase, DNA-dependent RNA polymerase, and DNA polymerase (Dreosti and Hurley, 1975; Sirover and Loeb, 1976; Hurley, 1977). Also, trans-activating zinc-finger proteins may become distorted when cadmium is bound instead of Zn^{2+} (Evans and Hollenberg, 1988; Sunderman and Barber, 1988). These proteins are thought to play a crucial role in the regulation of genes involved in development and differentiation. The binding of cadmium instead of or in addition to zinc, may affect (onco-)gene expression by stabilizing or destabilizing the DNA binding, thus suggesting a hypothetical molecular mechanism for cadmium-related carcinogenesis (see *Bioinorganic Medicine*, Part Four, Chapter 7-F).

Zinc, magnesium and other essential metals bind DNA on at least two distinct metal-binding sites, phosphate groups, and different electron donor groups of bases in DNA (Eichhorn and Shin, 1968; Williams, 1972; Clement et al., 1973; Furst, 1981). Cadmium can bind DNA in a similar fashion and, therefore, a competitive metal-ligand interaction between cadmium and these essential metal ions is a strong possibility. Indeed, in vitro studies have shown that this binding can be inhibited by zinc and magnesium (Waalkes and Poirier, 1984).

C. Cadmium-calcium

Chao et al. (1984) and Suzuki et al. (1985) speculated that competition of heavy metals such as cadmium with calcium for calmodulin may be a key to their toxicity at the molecular level. Calmodulin is the best known ic calcium-binding protein. It was first reported to be an activator of cyclic nucleotide phosphodiesterase (Cheung, 1970). Indeed, cadmium can interfere with calcium and calmodulin-dependent phosphodiesterase activity, and thus upset catabolic pathways of cellular cyclic nucleotide metabolism (Flik et al., 1987).

Cadmium is also shown to act directly on the cytoskeleton (Kunimoto and Miura, 1985). Interaction of cadmium with calmodulin is thought to be one of the causes of the effect of this metal on F-actin and microtubules. The other mechanism for the action of cadmium on the latter is thought to be due to interference by cadmium in the direct association of calcium with actin-associated proteins (Mills and Ferm, 1989).

D. Lead-zinc-iron

For lead the most stable oxidation state is +2. This oxidation state of lead is intermediate in the HSAB classification. Pb(II) can have coordination number ranging from four to ten, with accompanying changes in ionic radius from fairly small to fairly large. It is thus a very versatile metal ion, bonding reasonably well to all types of donor atoms, and changing size and coordination number to adapt to the requirements of the ligand. It is thus able to compete effectively with a small metal ion such as Zn (II), or a large hard metal ion such as Ca(II).

Lead (organic and inorganic) is a detrimental element but it is present in low concentrations in body fluids. Anemia is a common chronic effect resulting mainly from the effects of lead on heme synthesis. Lead affects five of the seven enzymes involved in heme synthesis but particularly δ-aminolevulinic acid dehydrase (a cytosolic enzyme) and ferrochelatase (a mitochondrial enzyme) (Needleman, 1977). The former is zinc-dependent and therefore by competition with this essential metal ion, lead inhibits this step of porphyrin synthesis. Moreover, by inhibiting ferrochelatase, lead interferes with the incorporation of ferrous iron into the protoporphyrin ring which then incorporates zinc and this results in the accumulation of a zinc protoporphyrin complex (ZPP) in erythrocytes.

E. Lead-calcium

Interaction of lead with calcium occurs at different levels. Lead can replace calcium in the operation of the calcium pump and therefore interferes in the efflux of the latter ion. It can also interfere in the intestinal absorption by competing with calcium for calcium-binding proteins (Mahaffey, 1980; Mahaffey and Rader, 1980). Intracellularly, lead is found to be accumulated inside the mitochondria by an energy-dependent process, with a concomitant inhibition of calcium uptake into these organelles. Respiration is inhibited when a sufficiently high concentration of lead has been taken up (Scott et al., 1971).

Calcium with calmodulin can activate the enzyme phosphodiesterase. This enzyme hydrolyzes 3',5'-cyclic AMP to 5'-AMP. Lead can fully replace calcium in activating phosphodiesterase with calmodulin and is the most effective cation known (Habermann et al., 1983; Chao et al., 1984). It is also more effective than calcium in supporting the calmodulin-dependent phosphorylation of brain proteins (Habermann et al., 1983). Calmodulin has four binding sites for calcium. Lead can displace calcium at these sites and also can bind the protein at sites where calcium is not bound (Mills and Johnson, 1985). Lead can also displace calcium in troponin C (Fullmer et al., 1985).

The interaction of lead with calcium and the consequent disturbance of calcium-dependent pathways are thought to help explain the apparent differences in brain injury from acute compared to low-level chronic inorganic Pb^{2+} intoxication. In acute exposure, it is suggested that the inhibition of mitochondrial uptake of calcium and its subsequent inhibition of respiration can produce cell injury (Pfleger and Wolf, 1975). Also, it is possible that the acute endothelial injury recognized as an early event in acute cerebellar lead encephalopathy is dependent on the abnormal mobilization of ic Ca^{2+} with a resultant increase in free cytosolic calcium. It is shown that elevated ic Ca^{2+} may play an important role in the mechanism of cell death in certain pathologic conditions (Shanne et al., 1979; Shier and DuBourdieu, 1985; Choi, 1987). In contrast, low-level chronic Pb^{2+} exposure is manifested by subtle defects in axodendritic architecture and neurite organization secondary to defective cytoskeletal organization. Also, the abnormal and possibly prolonged activation of calmodulin-dependent processes, especially the activation of protein kinase C, will lead to distorted transmembrane signalling and defective neuronal differentiation (Verity, 1990).

F. Lead-inclusion bodies

In lead toxicity, inclusion bodies are the distinguishing histologic features which occur mainly in the proximal portion of the tubule in the kidney and in the astroglia in the brain. The protein of the inclusion bodies is rich in glutamic and aspartic acids and glycine (Moore and Goyer, 1974). One interesting aspect of these bodies is that cells which have the ability to synthesize them are the ones which are least susceptible to the toxicity of lead. Therefore, inclusion bodies are protective against lead toxicity and they may even have a role to play in its excretion (Goyer et al., 1970).

G. Mercury and sulfhydryl functional groups

Toxicological properties of mercury depend on the chemical form in which this element enters the living organism, e.g. inorganic mercury (elemental mercury and the divalent mercury salts) or organic mercury (methylmercury). The affinity for sulfur and sulfhydryl groups is, however, the major factor underlying the toxicity of mercury compounds. With this affinity, mercury in its different forms may then interact with many essential metal ions.

Despite the high stability of mercury complexes (Stricks and Kolthoff, 1953), it was suggested that mercury in these complexes must be labile to exchange among a multitude of sulfhydryl groups it encounters in biological fluids (Cheesman et al., 1988). Indeed, the labile nature of Hg^{2+}-thiol binding was established from ^{1}H-NMR studies in erythrocytes where glutathione was shown to be a major binding site for Hg^{2+} in these cells (Rabenstein and Isab, 1982). The lifetime of GSH in the GSH-Hg(II)-hemoglobin complex, for example, was less than 30 s. Glutathione is an ubiquitous ligand in biological fluids and its concentration in erythrocytes is 2.2 x 10^3 M (Natelson and Natelson, 1978). Therefore, GSH is a very important ligand in this and other fluids. Glutathione has been implicated in numerous other studies of the toxicology of Hg^{2+} (Ballatori and Clarkson, 1982; 1984a; 1984b; 1985). The study of the kinetics of $Hg(GSH)_2$ ligand exchange is thus of great interest (Cheesman et al., 1988). Results have indicated that at physiological pH and under conditions of thiol-Hg ratio greater than 2:1, in addition to complexes of the stoichiometry $Hg(thiol)_2$, complexes having the stoichiometry $Hg(thiol)_3$ are formed. The rapid exchange of thiol between its free and $Hg(thiol)_2$ forms proceeded via a mechanism involving the $Hg(thiol)_3$ complexes. Considering the ubiquity of glutathione in cellular systems (Meister, 1976), it is likely that these reactions play a major role in the mobility of Hg^{2+} in biological systems (Cheesman et al., 1988).

Methylmercury also has high affinity for thiol groups (Simpson, 1961; Bach and Weibel, 1976). It is found that methylmercury leads to rapid depolymerization of microtubules (Abe et al., 1975). Cells exposed to low concentrations rapidly lost microtubule structure (Miura et al., 1978; Sager et al., 1983). Two protein monomers, α- and β-tubulin, aggregate to form microtubules. Formation of the tubule takes place at one end of the forming tubule and dissociates at the other end (Margolis and Wilson, 1981). Methylmercury reacts with SH groups on tubulin monomers to disrupt the assembly process. The dissociation process continues, thus leading to depolymerization of the tubule. As microtubules are essential for cell division (they are the main component of the mitotic spindle), effects of mercury on cell division can be understood.

Studies have suggested that the toxic effects of methylmercury are a result of its ability to rapidly cross the blood brain barrier (Magos, 1968). This crossing is achieved through a complex formed between methylmercury and cysteine. This complex shows a similarity to methionine (Aschner and Clarkson, 1987) and, therefore, methylmercury toxicity may be due to the structural similarity of this complex to methionine. This, in turn, implies that methionine could inhibit the entry of the cysteine complex into the brain. Once in this tissue, damage produced by methylmercury is specifically in the granule cells of the cerebellum and the neurons in the interstices of the visual cortex (Hunter and Russell, 1954). This specificity of methylmercury toxicity is due to its inhibitory effect on protein synthesis in the brain (Yoshino et al., 1966). This effect is observed well before clinical signs appear. Cheung and Verity (1985) have identified the first stage of synthesis associated with transfer RNA as the most sensitive to methylmercury. The first amino acid to start the polypeptide chain is always methionine. Thus, this indicates that methylmercury not only enters the brain as a complex structurally similar to methionine, but also it may remain in this form in the ic fluids until it enters the ribosomes to arrive at the site of initiation of protein synthesis. (For details on mercury toxicity, see *Medicine*, Vol. 2, Pt. Four, Ch. 1-C3.)

H. Mercury-zinc-copper

The dominating mercury pool in the body is the kidney, where a large portion is soluble and bound to MT (Piotrowski et al., 1974; Cherian and Clarkson, 1976; Komsta-Szumska et al., 1976a). Cherian and Clarkson (1976) reported that mercury vapor increases a MT-like protein in kidney (more than 50 % of mercury being bound to it) but not in liver. Komsta-Szumska et al. (1976a), however, reported that MT is not the only protein which binds mercury. Besides the l.m.m. protein fraction of MT, there were also h.m.m. proteins in the cytoplasm with a molar mass of 60-100 kg mol^{-1}. The remaining mercury was bound to other subcellular fractions, mainly to the nuclei. In the latter fraction, mercury was bound to h.m.m. nonhistone proteins and nucleic acids.

It is found that association of mercury with MT results in altered amounts of zinc and copper (Holt et al., 1980; Brady and Webb, 1981; Webb and Cain, 1982). Furthermore, the addition of Hg^{2+} to loaded kidney mitochondria induced the fast release of accumulated cation and the Ca^{2+} efflux exhibited kinetics that depended on the extent of the binding of

Hg^{2+} to the membrane (Chavez et al., 1989). Regarding its physiological effects, mercury has been shown to alter the metabolism of some of the essential elements including zinc (Sandstead, 1977; Bogden et al., 1980; Barfield and Bevan, 1985; Takenchi et al., 1987).

I. Mercury-cadmium-selenium

Affinity of the two non-essential metal ions for functional groups containing sulfur is substantiated with the experimental results cited earlier. Similar affinity of these metal ions towards selenium made it possible for this essential element to provide a certain degree of protection against the toxicity of these two non-essential metal ions. Selenite (SeO_3^{2-}) administered with these non-essential metal ions forms protein-stabilized, colloidal metal selenides in the blood which seem to be, at least in part, responsible for this protection. This mechanism, however, does not explain all the interactions between cadmium and mercury and the other selenium compounds. Selenomethionine and selenocysteine are more effective for example in the protection against kidney uptake of orally administered mercury (Fang, 1977). In this organ, selenite administration eliminates the binding of mercury to MT but increases its binding to h.m.m. proteins (Chmielnicka and Brzenznicka, 1978, Komsta-Szumska et al., 1976b). A similar effect of selenium is also observed for cadmium. In the kidney, it stimulates the incorporation of cadmium into a h.m.m. proteins and therefore eliminates almost completely the binding of the cation to MT (Chen et al., 1975). Thus, the provision for new high-affinity (Se-containing) binding sites may spare the tissue sulfhydryl binding sites such as MT.

J. Chelation therapy

These experimental results give ample evidence and confirm the involvement of ligands with their functional groups in the toxicity of essential and non-essential metal ions. There is always a delicate "normal" balance on the concentration of essential and beneficial metal ions where ligands have a great part to play. With their deficiency or excess, or with the introduction of non-essential metal ions into biofluids, the balance is certainly disturbed. In these situations, the necessity therefore arises for intervention to help regain this balance through chelation therapy. The latter intervention, in turn, can only be utilized to its fullest potential if knowledge on naturally available ligands and their distribution in biological fluids is exploited. This knowledge becomes specially relevant in the case of acute vs chronic non-essential metal ion intoxication. The targets for therapeutically used chelating agents to treat acute exposure will be different from those used to treat chronic exposure. In the former case, the l.m.m. ligands with the other labile fraction will be of importance to address. In the latter case, however, the challenge is greater for the chelating agent as it has to have the ability to cross the membranes in addition to its ability to compete with the existing ligands and chelate the undesired metal ion and render it available for excretion. Knowledge about ligands and their functional groups will also help in the design of these chelating agents (see Part Four, Chapter 1-B5).

IV. CONCLUSION

In conclusion, we can see how knowledge of metal ion and ligand fractions in relation to their functions in biological fluids can help us predict their distribution in these fluids. In addition, with this knowledge, we are able to elucidate some of the underlying mechanisms of metal ion excess and deficiency as well as toxicity and, therefore, be able to interfere with ligands for therapeutic purposes. Certainly, this knowledge is far from complete, and combined efforts from different disciplines will contribute to the progress in this field.

ACKNOWLEDGEMENTS

The authors gratefully acknowledge the help of their colleagues, Dr. D.B.K. Rama, Mrs. P. Theodorou, Miss E. Casimiro, and Miss K. Ambaram. They also thank Mrs L. Kotze for her assistance in typing the references.

REFERENCES

Abe, T., Haga, T., and Kurokawa, M. (1975). *Brain Res.*, *86*: 504.

Aikawa, J.K. (1981). *Magnesium in Biologic Significance*, CRC Press, Boca Raton, FL.

Amsler, P.E., and Sigel, H. (1976). *Eur. J. Biochem.*, *63*: 569.

Aschner, M., and Clarkson, T.W. (1987). *Toxicologist*, *7*: 158.

Bach, R.D., and Weibel, A.T. (1976). *J. Amer. Chem. Soc.*, *98*: 6241.

Ballatori, N., and Clarkson, T.W. (1982). *Science*, *216*: 61.

Ballatori, N., and Clarkson, T.W. (1984a). *Biochem. Pharmacol.*, *33*: 1087.

Ballatori, N., and Clarkson, T.W. (1984b). *Biochem. Pharmacol.*, *33*: 1093.

Ballatori, N., and Clarkson, T.W. (1985). *Fundam. Appl. Toxicol.*, *5*: 816.

Barfield, K.D., and Bevan, D.R. (1985). *Biochem. Biophys. Res. Commun.*, *128*: 389.

Barr, M. (1973). *Teratology*, *7*: 237.

Berridge, M.J., and Irvine, R.F. (1984). *Nature*, *312*: 315.

Bogden, J.D., Kemp, F.W., Troiano, R.A., Yortner, D.S., Timpone, C., and Giuliani, D. (1980). *Environ. Res.*, *21*: 350.

Brady, F.O. (1983). *Life Sci.*, *32*: 2981.

Brady, F.O., and Webb, M. (1981). *J. Biol. Chem.*, *256*: 3931.

Bulpitt, C.J., and Shipley, M.J. (1981), *Clin. Sci.*, *61*: 855.

Carvalho, S.M.M., and Ziemer, P.L. (1982). *Arch. Environ. Contam. Toxicol.*, *11*: 245.

Chao, S-H., Suzuki, Y., Zysk, J.R., and Cheung, W.Y. (1984). *Mol. Pharmacol.*, *26*: 75.

Chavez, E., Zazneta, C., Diaz, E., and Holquin, J.A. (1989). *Biochim. Biophys. Acta*, *986*: 27.

Cheesman, B.V., Arnold, A.P., and Rabenstein, D.L. (1988). *J. Amer. Chem. Soc.*, *110*: 6359.

Chen, R.W., Whanger, P.D., and Weswig, P.H. (1975). *Bioorg. Chem.*, *4*: 125.

Cherian, M.G., and Clarkson, T.W. (1976). *Chem.-Biol. Interact.*, *12*: 109.

Cheung, M.K., and Verity, M.A. (1985). *J. Neurochem.*, *44*: 1799.

Cheung, W.Y. (1970). *Biochem. Biophys. Res. Commun.*, *38*: 533.

Chmielnicka, J., and Brzenznicka, E.A. (1978). *Bull. Environ. Contam.*, *19*: 183.

Choi, D.W. (1987). *J. Neurosci.*, *7*: 369.

Christensson, T., Hellstrom, K., and Wengle, B. (1977). *Arch. Int. Med.*, *137*: 1138.

Clement, R.M., Sturm, J., and Danne, M.P. (1973). *Biopolymers*, *12*: 405.

Day, F.A., Funk, A.E., and Brady, F.O. (1984). *Chem.-Biol. Interact.*, *50*: 159.

Day, J.P., Barker, J., Evans, L.J.A., Perks, J., Seabright, P.J., Ackrill, P., Lilley, J.S., Drumm, P.V., and Newton, G.W.A. (1991). *Lancet*, *337*: 1345.

Daydé, S., Filella, M., and Berthon, G. (1990). *J. Inorg. Biochem.*, *38*: 241.

Domingo, J.L., Llobet, J.M., Gomez, M., and Corbella, J. (1986). *Res. Commun. Chem. Pathol. Pharmacol.*, *53*: 93.

Dreosti, I.E., and Hurley, L.S. (1975). *Proc. Soc. Exp. Biol. Med.*, *150*: 161.

Duffield, J.R., Edwards, K., Evans, D.A., Morrish, D.M., Vobe, R.A., and Williams, D.R. (1991), *J. Coord. Chem.*, *23*: 277.

Eichhorn, G.L., and Shin, Y.A. (1968). *J. Amer. Chem. Soc.*, *90*: 7323.

Elvehjem, C.A. (1935). *Physiol. Rev.*, *15*: 471.

Evans, G.W., and Cornatzer, W.E. (1970). *Fed. Proc.*, *29*: 695.

Evans, R.M., and Hollenberg, S.M. (1988). *Cell*, *52*: 1.

Fang, S.C. (1977). *Chem.-Biol. Interact.*, *17*: 25.

Fatemi, S.J.A., Kadir, F.H.A., and Moore, G.R. (1991). *Biochem. J.*, *280*: 527.

Fehily, A.M., Yarnell, J.W.G., and Butland, B.K. (1987). *Human Nutr. Appl. Nutr.*, *41A*: 319.

Flik, G., van de Winkel, J.G.J., Pärt, P., Wendelaar Bonga, S.E., and Lock, R.A.C. (1987). *Arch. Toxicol.*, *59*: 353.

Fullmer, C.S., Goldstein, S., and Wasserman, R.H. (1985). *J. Biol. Chem.*, *250*: 6816.

Furst, A. (1981). *Environ. Health Perspect.*, *40*: 83.

Geller, B.L., and Winge, D.R. (1980). *Fed. Proc.*, 39: 1768.

Goyer, R.A., Leonard, D.L., Moore, J.F., Rhyne, B., and Krigman, M.R. (1970). *Arch. Environ. Health*, *20*: 705.

Habermann, E., Crowell, K., and Janicki, P. (1983). *Arch. Toxicol.*, *54*: 61.

Happe, J.A., and Morales, M. (1976). *J. Amer. Chem. Soc.*, *88*: 1077.

Harris, E.D. (1983). *Trace Elements in Health* (J. Rose, ed.), Butterworths, London, Boston, Sydney, Wellington, Durban, Toronto, p. 44.

Hart, E.B., Steenbock, H., Waddell, J., and Elvehjem, C.A. (1928). *J. Biol. Chem.*, *77*: 797.

Holt, D., Magos, L., and Webb, M. (1980). *Chem.-Biol. Interact.*, *32*: 125.

Hunter, O., and Russell, D.S. (1954). *J. Neurol. Neurosurg. Psychiat.*, *17*: 235.

Hurley, L.S. (1977). *Progr. Clin. Biol. Res.*, *14*: 47.

Jemil, S., Fatemi, A., Williamson, D.J., and Moore, G.R. (1992). *J. Inorg. Biochem.*, *46*: 35.

Kesteloot, H., and Geboers, J. (1982). *Lancet*, *i*: 813.

King, M.M., and Huang, C.Y. (1983). *Biochem. Biophys. Res. Commun.*, *114*: 955.

King, M.M., Lynn, K.K., and Huang, C.Y. (1985). *Progress in Nickel Toxicology* (S.S. Brown and F.W. Sunderman, Jr., eds.), IUPAC, Blackwell Sc. Publ., p. 117.

Kiss, T., Sóvágó, I., and Martin, R.B. (1991), *Inorg. Chem.*, *30*: 2130.

Kjellström, T. (1977). *Accumulation and Renal Effects of Cadmium in Man*, Ph.D. Thesis, Dept. of Environ. Hygiene, Karolinska Institute, Stockholm.

Komsta-Szumska, E., Chmielnicka, J., and Piotrowski, J.K. (1976a). *Arch. Toxicol.*, *37*: 57.

Komsta-Szumska, E., Chmielnicka, J., and Piotrowski, J.K. (1976b). *Biochem. Pharmacol.*, *25*: 2539.

Kunimoto, M., and Miura, T. (1985). *Biochim. Biophys. Acta*, *816*: 37.

Lee, G.R., Williams, D.M., and Cartwright, G.E. (1976). *Trace Elements in Human Health and Disease* (A.S. Prasad and D. Oberleas, eds.), Academic Press, New York, p. 373.

MacGregor, G.A., Smith, S.J., Markandu, N.D., Banks, R.A., and Sagnella, G.A. (1982). *Lancet*, *ii*, 567.

Magos, L. (1968). *Brit. J. Ind. Med.*, *25*: 315.

Mahaffey, K.R. (1980). *Nutrient - Lead Interactions* (R.L. Singhal and J.A. Thomas, eds.), Urban and Schwarzenberg, Baltimore, p. 425.

Mahaffey, K.R., and Rader, J.I. (1980). *Ann. N.Y. Acad. Sci.*, *355*: 285.

Majerus, P.W., Connolly, T.M., and Deckmyn, H. (1986). *Science*, *234*: 1519.

Margolis, R.L., and Wilson, L. (1981). *Nature*, *293*: 705.

Mariam, Y.H., and Martin, R.B. (1979). *Inorg. Chim. Acta*, *35*: 23.

Marques, H.M. (1991). *J. Inorg. Biochem.*, *41*: 187.

Martin, R.B. (1988a). *Metal ions in Biological Systems*, Vol. 23 (H. Sigel and A. Sigel, eds.), Marcel Dekker, Inc., New York and Basel, p. 315.

Martin, R.B. (1988b). *Metal ions in Biological Systems*, Vol. 24 (H. Sigel and A. Sigel, eds.), Marcel Dekker, Inc., New York and Basel, p. 1.

Martin, R.B., and Mariam, Y.H. (1979). *Metal Ions in Biological Systems*, Vol. 8 (H. Sigel, ed.), Marcel Dekker, Inc., New York and Basel, p. 57.

Mason, R., Bakka, A., Samarawickrama, G.P., and Webb, M. (1981a). *Brit. J. Nutr.*, *45*: 375.

Mason, R., Brady, F.O., and Webb, M. (1981b). *Br. J. Nutr.*, *45:* 391.

May, P.M., and Williams, D.R. (1980). *Iron in Biochemistry and Medicine*, II (A. Jakobs and M. Worwood, eds.), London, Academic Press, p. 1.

Meister, A. (1976). *Glutathione: Metabolism and Function* (I.M. Arias and W.B. Jakoby, eds.), Raven Press, New York, p. 35.

Mills, C.F. (1980). *Biological Roles of Copper*, Ciba Foundation Symposium 79, Excerpta Medica, Amsterdam, p. 49.

Mills, J.S., and Johnson, J.D. (1985). *J. Biol. Chem.*, *260*: 15100.

Mills, J.W., and Ferm, W.H. (1989). *Toxicol. Appl. Pharmacol.*, *111*: 245.

Miura, K., Suzuki, K., and Imura, N. (1978). *Environ. Res.*, *17*: 453.

Moore, J.F., and Goyer, R.A. (1974). *Environ. Health Perspect.*, *7*: 121.

Natelson, S., and Natelson, E.A. (1978). *Principles of Applied Clinical Chemistry.*, Vol. 2, Plenum Press, New York, p. 36.

Needleman, H.L. (1977). *N. Engl. J. Med.*, *297*: 943.

Nieboer, E., Maxwell, R.L., and Stafford, A.R. (1984). *Nickel in the Human Environment* (F.W. Sunderman, Jr., ed.), IARC Sc. Public No. 53, Lyon, France, p. 439.

Nordberg, G.F., Kjellström, T., and Nordberg, M. (1985). *Cadmium and Health* (L. Friberg, C.-G. Elinder, T. Kjellström and G.F. Nordberg, eds.), CRC Press, Boca Raton, FL.

Osaki, S., and Johnson, D.A. (1969). *J. Biol. Chem.*, *244*: 5757.

Osaki, S., Johnson, D.A., and Frieden, E. (1971). *J. Biol. Chem.*, *246*: 3018.

Pallen, C.J., and Wang, J.H. (1984). *J. Biol. Chem.*, *259*: 6134.

Pfleger, H., and Wolf, H. (1975). *Biochem, J.*, *147*: 359.

Piotrowski, J.K., Trojanowska, B., Wisniewska-Knypl, J.M., and Bolanowska, W. (1974). *Toxicol. Appl. Pharmacol.*, *27*: 11.

Poirier, L.A., Theiss, J.C., Arnold, L.J., and Shimkin, M.B. (1984). *Cancer Res.*, *44*: 1520.

Provan, S.D., and Yokel, R.A. (1992). *Neurotoxicology*, *13*: 413.

Rabenstein, D.L., and Isab, A.A. (1982). *Biochim. Biophys. Acta*, *721*: 374.

Sager, P.R., Doherty, R.A., and Olmsted, J. (1983). *Exp. Cell. Res.*, *146*: 127.

Sandstead, H.M. (1977). *Toxicology of Trace Elements* (R.A. Goyer and M.A. Mehlman, eds.), Hemisphere Publishing Corporation, Washington, p. 241.

Scheller, K.H., Hofstetter, F., Mitchell, P.R., Prijs, B., and Sigel, H. (1981). *J. Amer. Chem. Soc.*, *103*: 247.

Scott, K.M., Hwang, K.M., Yurkowitz, M., and Brierley, G.P. (1971). *Arch. Biochem. Biophys.*, *147*: 557.

Shanne, F.A.X., Kane, A.B., Young, F.E., and Farber, Y.L. (1979). *Science*, *206*: 700.

Shannon, R.D. (1976). *Acta Cryst.*, *A32*: 751.

Shier, W.T., and DuBourdieu, D.J. (1985). *Amer. J. Pathol.*, *120*: 304.

Sigel, H., and Amsler, P.E. (1976). *J. Amer. Chem. Soc.*, *98*: 7390.

Sigel, H., Hofstetter, F., Martin, R.B., Milburn, R.M., Scheller-Krattiger, V., and Scheller, K.H. (1984). *J. Amer. Chem. Soc.*, *106*: 7935.

Sihag, R.K., Jeng, A.Y., and Nixon, R.A. (1988). *FEBS Lett.*, *233*: 181.

Simpson, R.B. (1961). *J. Amer. Chem. Soc.*, *83*: 4711.

Sirover, M.A., and Loeb, L.A. (1976). *Science*, *194*: 1434.

Speich, M., Bousquet, B., and Nicolas, G. (1980). *Clin. Chem.*, *26*: 1662.

Stricks, W., and Kolthoff, I.M. (1953). *J. Amer. Chem. Soc.*, *75*: 5673.

Sunderman, F.W., and Barber, A.M. (1988). *Ann. Clin. Lab. Sci.*, *18*: 267.

Sutoo, D., Akiyama, K., Fujii, N., and Matsushita, K. (1988). *Kitasato Arch. Exp. Med.*, *61*: 149.

Suzuki, Y., Chao, S.H., Zysk, J.R., and Cheung, W.Y. (1985). *Arch. Toxicol.*, *57*: 205.

Takenchi, T., Nakano, Y., Aoki, A., Ohmori, S., and Tsukatani, T. (1987). *J. Radioanal. Nucl. Chem.*, *112*: 259.

Tohyama, C., Mitane, Y., Kobayashi, E., Sugihira, N., Nakano, A., and Saito, H. (1988). *J. Appl. Toxicol.*, *8*: 15.

Van Campen, D.R., and Scaife, P.U. (1967). *J. Nutr.*, *91*: 473.

Venturini, M., and Berthon, G. (1989). *J. Inorg. Biochem.*, *37*: 69.

Verity, M.A. (1990). *Environ. Health Perspect.*, *89*: 43.

Vink, R. (1991 published 1992). *Magnesium Trace Elem.*, *10*: 1.

Waalkes, M.P., and Poirier, L.A. (1984). *Toxicol. Appl. Pharmacol.*, *75*: 539.

Webb, M., and Cain, K. (1982). *Biochem. Pharmacol.*, *31*: 137.

Whang, R., and Whang, D.D. (1990). *J. Amer. Coll. Nutr.*, *1*: 84.

Whang, R., Flink, E., Dyekner, T., Wester. P.O., Aikawa, J., and Ryan, M.P. (1985). *Arch. Intern. Med.*, *145*: 1686.

Whanger, P.D., and Weswig, P.H. (1970). *J. Nutr.*, *100*: 341.

Williams, D.R. (1972). *Chem. Rev.*, *72*: 203.

Yoshino, Y., Mozai, T., and Nakao, K. (1966). *J. Neurochem.*, *13*: 1223.

METAL-PROTEIN INTERACTIONS. GENERALITIES ON METAL ION - PROTEIN INTERACTIONS AND THEIR INVESTIGATION TECHNIQUES

by

Joan Selverstone Valentine

Department of Chemistry and Biochemistry,
University of California. Los Angeles,
Los Angeles, California
U.S.A.

I. OVERVIEW OF METAL-PROTEIN INTERACTIONS

A. Definitions

The term "metalloprotein" refers to a protein that contains a tightly bound metal ion in its normal, functional form. That term is sometimes contrasted with the term "metal-activated protein", which refers to a protein that is activated to carry out a specific function only when a metal ion is bound but which is sometimes present in its metal-free form. The term "metal-binding protein" refers to a protein whose function is, at least in part, to bind and release specific metal ions. In each case, the metal ion alone may be bound directly to the protein or it may be part of a "prosthetic group", a non-amino acid cofactor such as a heme group that binds as a unit to the protein. Proteins that function as biological catalysts are referred to as enzymes, and, likewise, metalloprotein catalysts are termed "metalloenzymes" and metal-activated protein catalysts "metal-activated enzymes".

B. Strong versus weak metal ion binding

Unfolded proteins have the possibility of binding metal ions to any of their basic sites, i.e. to the amino and carboxyl termini and the amino-acid side chains that contain atoms with lone pairs of electrons (see Section D1). For a protein that is folded into its native state, however, binding is likely only where the metal ion has access to the potential binding site. In fact, almost any protein will bind metal ions weakly, the most likely site for weak binding being hydrophilic amino-acid side chains that are on the exterior of the folded protein. The affinities of such sites for particular metal ions can be readily approximated by consideration of small molecule metal ion - ligand interactions. By contrast, strong binding of metal ions to proteins generally occurs when several potentially ligating amino-acid side chains in the interior of the protein are brought into close proximity by the folding of the protein, thus creating what is referred to as a "metal-binding site". The affinities of such metal-binding sites for specific metal ions are frequently very high as a consequence of both the chelate effect (see Part One, Chapter 1-A) and the fact that these highly polar sites are usually imbedded in non-polar, hydrophobic regions of the folded protein and are consequently not very effectively solvated in the metal-free protein. Thus, such sites have

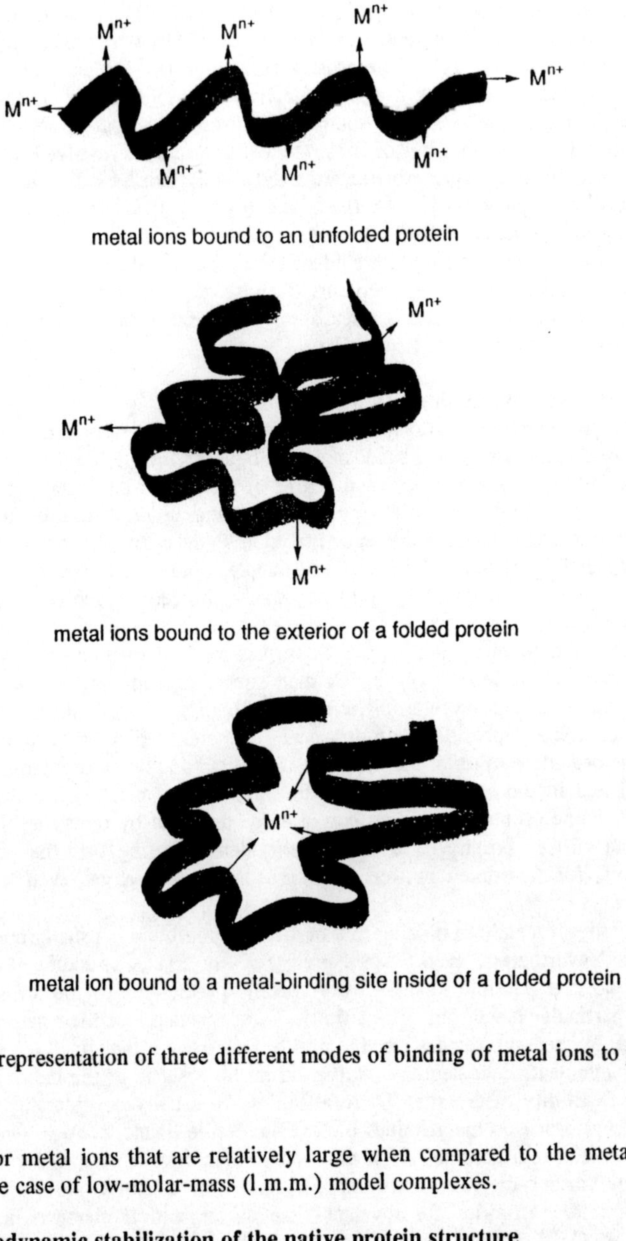

Figure 1.
Schematic representation of three different modes of binding of metal ions to proteins.

affinities for metal ions that are relatively large when compared to the metal ion affinities found in the case of low-molar-mass (l.m.m.) model complexes.

C. Thermodynamic stabilization of the native protein structure

Metal ion - protein interactions play a large number of roles in the functioning of many proteins. In all cases of strong metal ion - protein interactions, the resulting protein-metal complex is thermodynamically stabilized by the interaction relative to the metal-free protein (termed the "apoprotein"). In some cases, stabilization of the protein structure by a tightly bound metal ion seems to be the only role of a particular metal-ion protein interaction, e.g. the structural zinc site in alcohol dehydrogenase in which four cysteine thiolate ligands in an approximately tetrahedral arrangement make up the site (Schneider et al., 1983) and for which no function other than stabilization of the protein structure has been identified. On the other hand, if metal ion binding stabilizes a protein conformation other than that of the apoprotein, binding of the metal ion may be acting as a trigger. For example, the merR protein, a regulatory protein involved in regulation of transcription of the mer operon, which codes for a number of proteins involved in mercury resistance in bacteria, binds to

DNA in the absence or presence of Hg^{2+}. But Hg^{2+} binding to the merR-DNA complex transforms that protein from an inhibitor to an activator of DNA transcription (O'Halloran, 1993). Likewise, binding of Ca^{2+} to the protein calmodulin (McPhalen et al., 1991) causes it to change its properties, enabling it also to function as a regulator of metabolic activities in the cell (see Section B7 below). It is usually wise to reserve judgment on whether a metal ion is functioning only as a stabilizer of the protein structure or is involved in regulation as well. For example, the zinc-finger proteins are stabilized by Zn^{2+} binding in conformations that bind to specific sequences of DNA (Klug and Rhodes, 1987). Evidence has not been forthcoming that the presence of Zn^{2+} actually regulates this binding to DNA, but neither can that possibility be excluded based on evidence available to date.

Metal ion transport and storage proteins likewise are stabilized by strong metal ion binding, and a mechanism to allow release of the metal ions must be part of their evolutionary design (see below and Section B).

D. Specificity of metal ion binding

Metalloproteins are usually remarkably specific in their preferences for particular metal ions, much more so than are typical small-molecule ligands (Fraústo da Silva and Williams, 1991; Glusker, 1991). Size, geometry, and degree of rigidity of the metal-binding site and the relative hardness or softness and charge of the ligating atoms that make up the site are some of the factors that determine this specificity. In some cases, the source of the metal ion - binding specificity is quite obvious. For example, binding sites specific for the hard metal ion Ca^{2+} generally consist largely of negatively charged oxygen-containing ligands, but the coordination number is too high and the site apparently too rigid to accommodate readily the smaller, hard metal ion Mg^{2+}. Sites that evolved to bind the hard metal ion Fe^{3+} likewise tend to be more oxygen-rich than sites designed for the softer metal ion Cu^{2+}, which tend to contain more aromatic nitrogen ligands, i.e. histidines. Sites designed to bind metal-containing prosthetic groups such as heme typically have cavities with affinities for the organic portion of the molecule as well as one or more ligating amino-acid side chains situated in the correct orientation to complex the metal ion in that prosthetic group. Iron-free heme protein derivatives can often be prepared by removing the heme and rebinding the metal-free porphyrin to the protein, demonstrating that the protein has a significant affinity for the organic portion of the prosthetic group alone, even in the absence of the metal ion.

The specificity of metal ion binding to a particular apoprotein is usually understandable in broad terms. Nevertheless, even when a precise X-ray crystal structure of a protein is available, it is not yet possible to make very specific predictions of the affinities of that apoprotein for particular metal ions. The affinities of apoproteins for their native metal ions are determined by several factors, including the nature of the ligating atoms, their geometrical arrangement, their degree of solvation in the absence of the metal ion, and the conformational flexibility of the site. These affinities are usually considerably higher than would be predicted based on the affinities of small-molecule ligand analogs, even when the chelate effect is taken into account. Even the relative affinities of particular protein metal-binding sites for various metal ions are unpredictable. For example, most small-molecule ligands show relative affinities for divalent first-row transition elements in the order $Mn^{2+} < Fe^{2+} < Co^{2+} < Ni^{2+} < Cu^{2+} > Zn^{2+}$ (the Irving-Williams series) (Sigel and McCormick, 1970). The same order is apparently found for metal-substituted derivatives of the zinc enzyme carbonic anhydrase (Lindskog and Nyman, 1964), but in the case of copper, zinc-superoxide dismutase, the affinity of the zinc site for Zn^{2+} is higher than that for Cu^{2+} (Valentine and Pantoliano, 1981), for reasons that are not yet understood.

E. Ligand and substrate binding

Metal ion - protein interactions whose primary purpose is stabilizing a particular protein conformation usually lack open coordination positions on the metal ion, and such sites are frequently imbedded in the protein structure with little or no access to solvent. By contrast, several classes of metalloproteins contain metal ions as an integral part of a ligand binding cavity. The ligand may be dioxygen, O_2, in the case of dioxygen carrier proteins such as hemoglobin, hemocyanin, or hemerythrin, or it may be an enzymatic substrate or cofactor in the case of a metalloenzyme. In each case, one of the functions of the metal ion may be to help in providing a binding site with high affinity and specificity for the ligand in question. In the case of a metalloenzyme, the metal ion frequently plays a further role either in activating the substrate for further reaction or as a template to draw substrates together for reaction with each other.

In the case of dioxygen-binding proteins, several common features are observed, each of which is readily understandable in light of the special nature of the dioxygen ligand. There are few chemical reactions of dioxygen that are readily reversible, other than with metal ion complexes, and the l.m.m. complexes that are known to reversibly bind dioxygen contain low-valent metal ions with open coordination positions. It is therefore not surprising that dioxygen-binding proteins contain the low-valent metal ions Fe^{2+} or Cu^{+} with open coordination positions available to provide a site of interaction for the dioxygen molecule (Karlin, 1993).

The protein also plays an essential role in these dioxygen-binding proteins in ensuring that the reaction of the deoxy form with dioxygen to give the oxy form is reversible. Thus typically they provide the metal ion with a bulky, hydrophobic surrounding that protects the metal ion - dioxygen complexes from further reaction either with water, protons, or other metal ions, thus blocking mechanistic pathways that would lead to irreversible oxidation of the metal ion. Dioxygen-activating enzymes, such as cytochrome P450 or tyrosinase, have metal-binding sites that are very similar to their dioxygen-binding counterparts, e.g. myoglobin or hemocyanin. In those cases, the metalloenzyme contains a substrate-binding site in close proximity to its dioxygen-binding site, and the bound dioxygen ligand is activated for reaction with the substrate by reduction (Valentine, 1994).

Substrate-binding metalloproteins are typically metalloenzymes, and their characteristics vary widely depending on the nature of the reaction catalyzed. Non-redox, hydrolytic enzymes may function by providing a highly Lewis acidic site for substrate binding that polarizes the substrate and makes it susceptible to nucleophilic attack. Alternatively, highly Lewis acidic metal sites are expected to bind hydroxide ligands, even at physiological pH, which may be acting as nucleophiles in attacking substrates bound in close proximity. In each case, the high Lewis acidity of the metal ion when bound to the protein and the proximity of the metal ion to the substrate-binding site are essential features of the enzyme. It is interesting, therefore, to note that the typical metalloenzyme of this type, e.g. carboxypeptidase, contains Zn^{2+} in a tetrahedral site with one bound water molecule (Christianson, 1991). Thus the Zn^{2+} ion has only three protein ligands bound to it. This low coordination number and the hydrophobic nature of the protein surrounding those three ligands are expected to enhance the Lewis acidity relative to typical l.m.m. Zn^{2+} complexes.

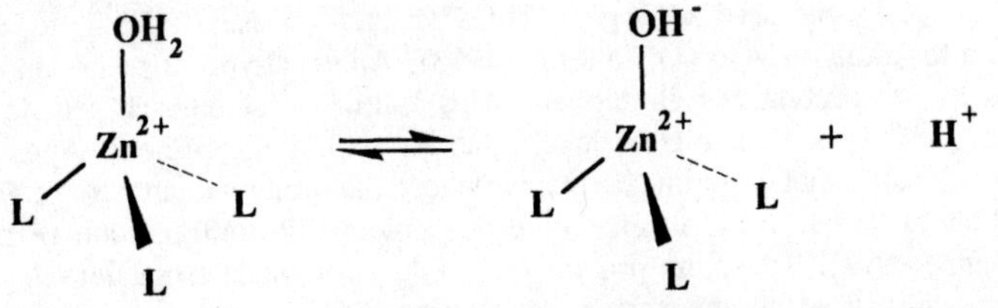

A large number of metalloenzymes exist that catalyze diverse sets of reactions, many of which are more complex than just simple hydrolysis. In those cases, we frequently can only guess how evolution has optimized the function of an individual metalloenzyme by means of metal-protein interactions. This issue is being addressed by studies of synthetic analogs, mutant proteins prepared by site-directed mutagenesis, and *de novo* design. These topics are described in section II.E.

F. Rates of metal ion - ligand exchange

Another important aspect of metalloenzyme active sites is the kinetics of metal ion - ligand exchange. If a substrate must bind to the metal ion as part of the enzymatic mechanism, then the kinetics of ligand addition and ligand exchange at that metal ion may be of enormous importance in determining the overall catalytic rate of that metalloenzyme. In the case of hydrolytic enzymes containing Zn^{2+}, for example, the rate of exchange of water bound to the metal ion with bulk solvent water is very fast, i.e. the metal ion is "labile", indicating that substrates and products may substitute for the water molecule bound to Zn^{2+} in the resting enzyme as well as for each other at a rapid rate. Thus, rapid rates of ligand exchange reactions are typical for metalloenzymes whose catalytic mechanisms require coordination of substrates to and release of products from the metal ions in their active sites. So-called "inert" metal ions, such as low-spin Co^{3+}, low-spin Fe^{2+}, or Cr^{3+}, whose ligand exchange reactions are particularly convenient to study in l.m.m. coordination complexes because they are so slow, are typically not found in such metalloenzymes.

Metal-ligand exchange rates are also particularly important in determining the mode of action of metal transport and storage proteins. In this case, the issue is not how rapidly ligand exchange occurs on a metal ion in a metalloprotein, but how rapidly metal ions are bound to and released from a metal-binding protein.

G. Metal-binding proteins

It is clearly desirable for metal ion storage and transport proteins to have high affinities for the individual metal ions in question, particularly in the case of potentially toxic free metal ions such as $Fe^{2+/3+}$ and $Cu^{1+/2+}$. But a mechanism must exist for relatively rapid release of those metal ions upon delivery to the desired location. Several potential solutions to this dilemma have been proposed in various systems. A metal ion may be labile enough to transfer rapidly and spontaneously from one high affinity site to another, higher affinity site. Alternatively, a small molecule carrier of intermediate affinity may catalyze such a reaction, again provided that the metal ion has sufficient lability to transfer relatively rapidly. Alternatively, transfer of the metal ion may require that its oxidation state be changed in order to reduce the affinity of the protein for the metal ion and/or enhance the rates of exchange. Finally, metal ion release might be achieved by chemical alterations of a portion of the metal-binding site, resulting in a lowered affinity for the metal ions, or by total degradation of the protein (Hausinger, 1990).

Some examples illustrate the enormous diversity of mechanisms that have evolved to ensure that specific metal-binding proteins are able to bind and release their metal ions rapidly in the appropriate locations and at the appropriate ambient metal ion concentrations. The core of the iron-storage protein ferritin, for example, contains an oxy-hydroxy Fe^{3+} polymer that apparently must be reduced to the Fe^{2+} state in order for the metal ions to be released from the protein (O'Halloran, 1993). The iron-transport protein transferrin has a novel mechanism for regulating the binding and release of Fe^{3+}: the binding site has a high affinity for that ion only in the presence of carbonate or bicarbonate anion, which is incorporated as a ligand. The tendency of the carbonate ligand to be converted to CO_2 at low pH is apparently used to regulate binding and release of Fe^{3+} to the protein at appropriate locations in vivo (O'Halloran, 1993). Another type of metal-binding protein is illustrated by the protein metallothionein. This l.m.m., cysteine-rich protein binds Zn^{2+}, Cu^{2+}, or Cd^{2+} ions in thiolate-bridged clusters within the protein. Apometallothionein differs from most metal-binding proteins whose apoproteins contain preformed metal-binding sites in that it exists in an unfolded, random-coil configuration (Fraústo da Silva and Williams, 1991). Thus, the protein must fold as it binds metal ions in a cooperative fashion into clusters within the protein (see Section D5).

H. Electron-transfer proteins

Electron-transfer proteins comprise a special class of metalloproteins whose function is to accept electrons from electron donors and to give them up to electron acceptors. It is not surprising that such proteins are usually transition metal - containing metalloproteins since transition metal ions have the almost unique chemical characteristic of existing in multiple stable forms whose formulations differ from one another only in the number of electrons they contain, e.g. Fe^{2+} and Fe^{3+}. Some electron-transfer metalloproteins have only one apparent purpose, i.e. electron transfer. Examples are cytochrome *c* (see Sections B1 and C2), plastocyanin (see Section B2), and ferredoxins (see Section B1). Alternatively, electron-transfer sites may be found within metalloenzymes, the function of which is to accept and then deliver electrons to the enzymatic active sites, e.g. cytochrome *c* oxidase (see Sections B1 and B2). An interesting example of both classes is found for cytochrome P450. Most members of this class of enzymes require separate electron-transfer proteins, i.e. cytochrome P450 reductases, to deliver electrons to the metalloenzyme that catalyzes hydroxylation of substrates. But in the case of cytochrome P450BM-3, the cytochrome P450 reductase center is contained within the same protein as the hydroxylase center (Fulco, 1991).

Electron-transfer centers in metalloproteins generally share a number of distinctive characteristics. The metal-binding sites contain redox-active transition metal ions, and the metal ions are tightly bound and coordinatively saturated, i.e. ligands from the protein completely surround the site and occupy all of their coordination positions. This characteristic is quite apparent when one compares the structure of the dioxygen-binding protein myoglobin with that of the electron-transfer protein cytochrome *c*. In the former case, the deoxy form contains a coordinatively unsaturated five-coordinate Fe^{2+} center, which is poised to accept O_2 as a sixth ligand. In the latter case, the oxidized, Fe^{3+}, and reduced, Fe^{2+}, states are both six-coordinate.

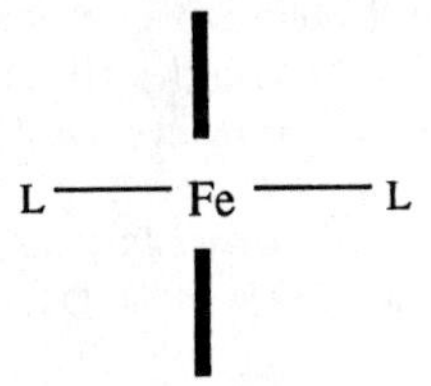

cytochrome *c*

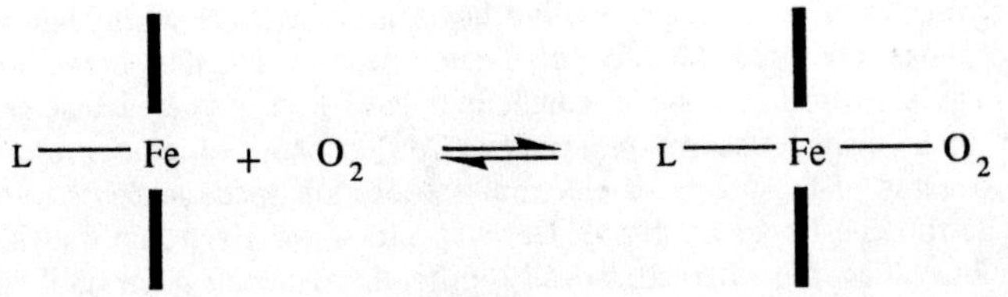

myoglobin

Another striking characteristic of electron transfer metal-binding sites is the relative rigidity of the protein surrounding the sites, as demonstrated by the relatively small differences in the structures of the metal-binding sites between the oxidized and reduced forms of the proteins. The likely explanation for this rigidity is that it enhances the rate of electron tranfer in these systems. The rates of electron transfer tend to be significantly diminished in metal complexes where large changes of the geometry of the first coordination sphere occur upon oxidation or reduction. Electron-transfer metal-binding sites in metalloproteins appear to have evolved to keep such changes to a minimum and thus to enhance the rate of electron transfer.

II. INVESTIGATIVE TECHNIQUES FOR STUDIES OF METAL-PROTEIN INTERACTIONS

A. Metal ions as spectroscopic probes of metal-containing proteins

1. *General comments*

Most of the techniques used for study of metal-free proteins may be profitably applied to metal-containing proteins as well (Campbell and Dwek, 1984). However, a large number of additional techniques are also available specifically for metal-containing proteins, particularly if they contain metal ions or metal ion - containing prosthetic groups with useful spectroscopic and/or magnetic properties. The approach that is used in those cases is to determine the properties of the naturally occurring metal ion within the protein and to compare it with those of other proteins containing the same metal ion and with those of small-molecule complexes with appropriate ligands that enforce various geometries on the metal ion. Such studies are not restricted to the metal ion that naturally occurs within a particular protein, since native metal ions may frequently be removed from a protein and substituted by non-native metal ions. In one case, such a substitution appears to be nearly perfect, i.e. in the case of Co^{2+} substituted for Zn^{2+} in many zinc-containing enzymes, in that the enzyme retains nearly full activity and the structure of the protein in the active site and elsewhere appears to be unchanged (see Section B5). In most cases, however, substitution of one metal ion for another causes at least slight changes in the geometry of the active site and, usually, profound changes in the enzymatic activity. Therefore, information gathered from studies of metal-substituted proteins must be used cautiously, since the geometry of the metal-binding site may have been changed by the substitution.

2. *Electronic absorption spectroscopy*

One of the very useful properties of transition metal-containing proteins is that they are often colored. This property has facilitated their isolation and purification, since their location can frequently be determined by eye during a fractionation. Heme proteins are some of the most intensely colored proteins, their color originating from π-π^* electronic transitions of the porphyrin moiety whose energies fall in the visible range of the spectrum. Other colored metalloproteins generally contain transition metal ions with incompletely filled d orbitals, and their colors are due either to d-d or charge transfer electronic transitions (or both) that frequently fall in the visible range. The energies and intensities of these optical transitions are measured by electronic absorption spectroscopy (Campbell and Dwek, 1984).

Much information can be obtained by comparison of the electronic absorption spectrum of a transition metal - containing protein with those of other proteins and of small-molecule transition metal complexes. For example, characteristic visible absorption spectra are observed for five-coordinate dioxygen-binding heme proteins such as myoglobin that are different from those observed for six-coordinate electron transfer heme proteins. In addition, five- and six-coordinate model compounds have been prepared that show similar differences in their visible absorption spectra. Co^{2+}-substituted zinc proteins provide another good example of the utility of electronic absorption spectroscopy to probe metal-binding sites (Bertini and Luchinat, 1984). Co^{2+} spectra show electronic transitions whose energies and intensities are characteristically different depending on the coordination number and geometry of the ligand environment surrounding the metal ions. Thus, the visible absorption spectra of Co^{2+}-substituted zinc proteins provide information about the nature of the metal-binding site that generally holds true for the colorless native Zn^{2+}-containing protein as well.

3. *Circular dichroism (CD) and magnetic circular dichroism (MCD)*

Chiral species absorb left and right circularly polarized light to different degrees, and CD spectroscopy measures the absorption difference versus the wavelength of the light (Campbell and Dwek, 1984). CD transitions are observed only at wavelengths corresponding to electronic absorption bands. The CD spectra of proteins in the ultraviolet region provide useful information concerning the nature of the secondary structure of the protein, since α-helix and β-pleated sheet structures give different characteristic CD absorptions. For metal-containing proteins that have electronic transitions in the visible range of the spectrum, CD spectroscopy frequently provides information concerning the locations of electronic transitions and aids in their assignment.

MCD spectroscopy measures the CD spectrum of a sample in a magnetic field (see Fig. 2 for an example). The species being measured need not be chiral in order to have a MCD spectrum since the effect of the magnetic field may induce the chirality. MCD spectroscopy has also been invaluable in identifying and assigning electronic absorption transitions. In certain cases, a characteristic MCD spectrum may also provide a fingerprint that permits other, similar chromophores to be identified in metalloproteins.

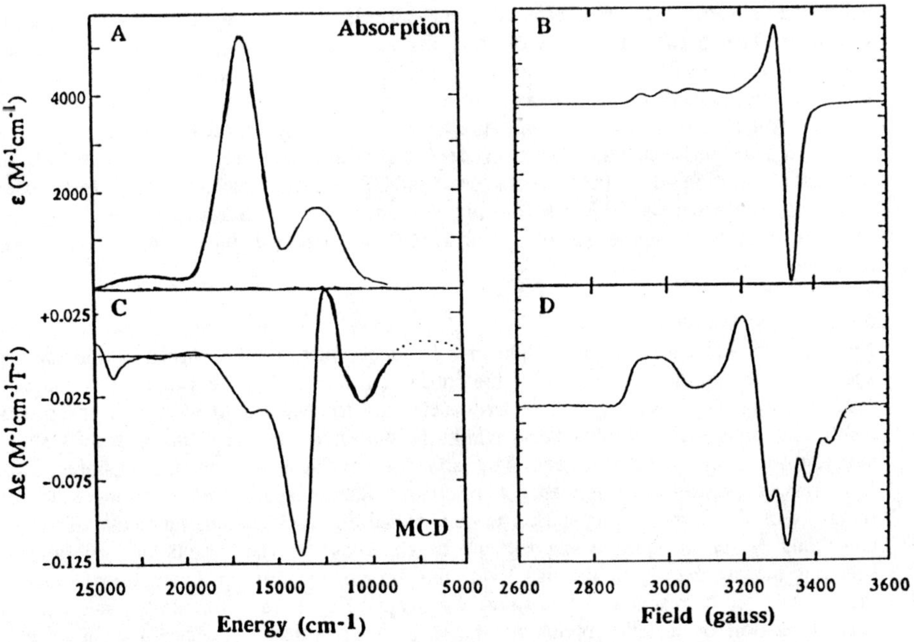

Figure 2.
Examples of electronic absorption, MCD and EPR spectra:
A, electronic absorption spectrum of the blue copper protein spinach plastocyanin; B, EPR spectrum of spinach plastocyanin; C, MCD spectrum of spinach plastocyanin; D. EPR spectrum of the blue copper protein stellacyanin. Adapted from Solomon et al. (1992).

4. *Nuclear magnetic resonance (NMR)*
NMR spectroscopy is a powerful tool for determination of structure in proteins in general, including many metalloproteins, and discussion of the wide range of applications may be found in many excellent books and reviews. It is important to point out here, however, that many proteins containing paramagnetic metal ions are unsuitable for NMR studies other than paramagnetic relaxation studies, because their NMR spectra are excessively broadened by the effect of those metal ions. Alternatively, a protein containing a paramagnetic metal ion may give excellent, paramagnetically shifted NMR spectra. Such spectra frequently provide detailed information concerning the nature of the metal-binding site because the paramagnetic metal ion has shifted only the resonances in close proximity to it into another region of the spectrum. Examples of the application of paramagnetic NMR to biological systems may be found in Bertini and Luchinat (Bertini and Luchinat, 1986).

5. *Electron paramagnetic resonance (EPR)*
EPR is another magnetic resonance technique which has proven invaluable in characterizing proteins containing metal centers with one or more unpaired electrons (Campbell and Dwek, 1984). Different oxidation states of a transition metal ion ordinarily give dramatically different EPR spectra, and thus the presence or absence of an EPR spectrum can be of great use. The characteristic appearance of an EPR spectrum may provide some of the most fundamental information concerning the presence and identity of a transition metal ion in a newly isolated protein. This is particularly true because EPR is a very sensitive technique and usually does not require large samples of the protein to be measured. For certain paramagnetic transition metal ions, the EPR spectrum is particularly sensitive to the coordination number and geometry of the ligand environment, e.g. Cu^{2+}, and detailed information concerning the nature of the site may be deduced from the spectrum (Fig. 2 provides examples of EPR spectra of two different types of "blue" copper proteins). Certain

metal ion clusters, e.g. iron-sulfur clusters, also have very useful characteristic EPR signatures. The absence of EPR signals may also provide a valuable clue as to the nature of particular metal-binding sites. For example, the absence of an EPR spectrum for the two Cu^{2+} ions per subunit in met hemocyanin or met tyrosinase provided very strong evidence that those metal ions were close enough to each other to be magnetically coupled.

Two additional techniques that are related to EPR and have been of particular utility in the characterization of metalloproteins are ENDOR (electron-nuclear double resonance) (Hoffman, 1991) and ESEEM (electron-spin echo envelope modulation) (Jiang et al., 1993). Reviews of these subjects should be consulted for detailed information about these techniques.

6. *Resonance Raman*

Normal Raman spectroscopy uses light of a frequency that is not absorbed by the sample and gives the vibrational spectrum of the entire molecule within the sample. By contrast, resonance Raman spectroscopy uses frequencies that correspond to electronic absorption bands due to particular chromophores within the molecule, and the resulting spectra show selective and dramatic enhancement of vibrations associated with those chromophores. This enhancement permits measurement of resonance Raman spectra of solutions that are relatively dilute, thus enabling spectra to be obtained on relatively small amounts of metal-containing protein samples (Campbell and Dwek, 1984). It also results in a vibrational spectrum of only the region associated with the chromophoric group. Examples of some of the cases where resonance Raman spectroscopy has been particularly valuable in characterization of metalloproteins are heme proteins, where resonance Raman spectra frequently can be used to ascertain the deformations of the heme group within the protein, "blue" copper proteins, which show enhancement of characteristic copper-cysteinate vibrations in their spectra, and iron-tyrosinate proteins, which show enhancement of characteristic tyrosine vibrations.

7. *X-ray absorption fine structure (EXAFS)*

X-ray absorption spectroscopy is a technique in which synchrotron radiation is used to ionize core electrons (Campbell and Dwek, 1984). The fine structure on the absorption peak is caused by interaction of the electron leaving the metal ion with the atoms surrounding it. Thus, this technique provides information about the structure of the region surrounding the metal in metalloproteins. EXAFS gives particularly good information concerning metal-ligand bond distances, but it can also be used to provide other information about the nature of the metal-binding site, particularly when comparisons can be with the EXAFS spectra of model compounds of known structure. EXAFS has proved to be particularly valuable in providing structural information for complex systems before X-ray crystallographic information was available, for example the Fe- and Mo-containing protein nitrogenase (see Section B1) and the Cu- and heme-containing protein cytochrome *c* oxidase (see Sections B1 and B2).

8. *Mössbauer spectroscopy*

Mössbauer spectroscopy is a valuable tool that has been applied to the study of many different types of iron-containing proteins (Campbell and Dwek, 1984). This technique involves absorption of γ-rays by the ^{57}Fe nucleus, whose natural abundance is 2%. The parameters obtained provide considerable information about the oxidation and spin states of the iron atom and the nature of its environment within the protein.

B. X-ray crystallographic determinations of structure

X-ray crystallography is by far the most useful technique to determine structures of all proteins, whether or not they contain metal ions. Much of our understanding of protein chemistry and enzymology is based on information acquired by use of this technique. The principal drawback is the requirement for crystalline samples, which are frequently difficult to obtain, particularly in the case of proteins that are normally localized to membranes within the cell. Even in the case of proteins that crystallize readily, the determination of a protein structure by X-ray crystallography is considerably more involved than the determination of the structure of a small molecule, which has become essentially routine in recent years. Nevertheless, modern advances in techniques for X-ray crystallographic determination of the structures of biological macromolecules continue to improve and to shorten the time required to obtain structures.

C. Electrochemical studies

Transition metal - containing proteins are frequently redox active, and determination of the electrochemical properties of such proteins frequently provides information that is valuable in assessing their functional properties in vivo. Aside from classical redox titrations, the most useful electrochemical technique in the case of metalloprotein is cyclic voltammetry. Several problems arise in studies of metalloproteins using this technique, particularly the tendency of proteins to adsorb to electrode surfaces and the requirement for redox mediators in some cases to effect rapid electron transfer between the metalloprotein and the electrode. Special techniques have been developed to overcome many of these and other problems, and electrochemical techniques are proving to be increasingly useful for studies of metalloproteins (Hill, 1990; see also Section C3).

D. Determination of metal-binding affinities

It is frequently useful and interesting to have information concerning the relative affinities of proteins and various metal ions under various conditions. Such information is particularly important in the case of metal-binding and transport proteins that must bind and release metal ions at specific locations as a part of their function in vivo. It is also important in formulating possible mechanisms for metallation of metalloproteins after they are synthesized in vivo (Hausinger, 1990).

A wide variety of techniques may be employed for the determination of equilibrium constants for metal ion - protein binding (Martell and Motekaitis, 1992). A variety of spectroscopic and potentiometric techniques are available for determination of the concentrations of the species present in solution. These concentrations and the known composition of the solution may be used to calculate equilibrium constants. If the binding constants are too high to determine directly, competition methods may be used, i.e. allowing two different metal ions to compete for the same sites on the protein or allowing a small-molecule chelating ligand to compete with the protein for the same metal ions.

E. Studies of synthetic analogs and modified or synthetic proteins

Studies of synthetic analogs of metalloenzymes are frequently carried out for the purpose of removing the influence of the remainder of the protein on the reactivity and other properties of the metal-binding site alone. Such studies have been enormously important in providing simple models of enzymes and potential catalytic intermediates for comparison of spectroscopic and electrochemical properties, but they rarely, if ever, duplicate the catalytic reactivity of the metalloenzyme. Some of the synthetic analogs that have been particularly valuable in metalloprotein characterization are the sterically blocked "picket fence porphyrins", whose Fe^{2+} complexes react with dioxygen to form stable O_2 complexes that can be structurally characterized, and the iron-sulfur cluster complexes, whose structures and other properties provided explanations for many puzzling characteristics of iron-sulfur proteins (Ibers and Holm, 1980; Karlin, 1993).

Another approach to investigating the role of metal ion - protein interactions in determining the properties of metalloproteins is to study derivatives of such proteins constructed by site-directed mutagenesis. Thus, it is possible to examine the effect of removing or substituting ligands in the metal-binding site of metalloproteins and, if the resulting protein is stable, to examine the consequences. For example, amino acids that act as ligands to the heme in certain heme proteins have been substituted by other amino acids to give metal-binding sites resembling those of other classes of heme proteins, and ligating amino acids in a number of copper proteins have been substituted by other ligating and non-ligating amino acids in order to examine the effects of such substitutions on structure and reactivity.

Another approach that shows promise is de novo design of metalloproteins (Betz et al., 1993). The goal of de novo metalloprotein design is to design and synthesize new metalloproteins, either by synthesizing an entirely new protein that is predicted to fold in such a way that a metal-binding site is created, or by adding a new metal-binding site to a naturally occurring protein. Recent attempts using both approaches have provided encouraging results in that the new metal-binding sites were formed and were found to bind metals. However, these new proteins tended to have disappointingly low metal-binding constants, and the sites appear to be considerably more flexible than is normally found for naturally occurring metalloproteins. It would appear, therefore, that the structures making up the metal-binding sites of many of the naturally occurring metalloproteins have an extra degree of stabilization built into the structures of their metal-binding sites, the basis for which is not yet clear. De novo design of metalloproteins with high binding constants for metal ions and with rigid geometries around the metal-binding sites will become possible when the sources of this extra degree of stabilization are better understood.

REFERENCES

Bertini, I., and Luchinat, C. (1984). *Adv. Inorg. Biochem.*, *6*: 71.

Bertini, I., and Luchinat, C. (1986). *NMR of Paramagnetic Molecules in Biological Systems*, Benjamin/Cummings Publishing Company, Menlo Park, CA.

Bertini, I., Gray, H.B., Lippard, S., and Valentine, J.S., eds. (1994). *Bioinorganic Chemistry*, University Science Books, Mill Valley, CA.

Betz, S.F., Raleigh, D.P., and DeGrado, W.F. (1993). *Curr. Opin. Struct. Biol.*, *3*: 601.

Campbell, I.D., and Dwek, R.A. (1984). *Biological Spectroscopy*, Benjamin/Cummings Publishing Company, Menlo Park, CA.

Christianson, D.W. (1991). *Adv. Protein Chem.*, *42*: 281.

Fraústo da Silva, J.J.R., and Williams, R.J.P. (1991). *The Biological Chemistry of the Elements*, Clarendon Press, Oxford.

Fulco, A.J. (1991). *Annu. Rev. Pharmacol. Toxicol.*, *31*: 177.

Glusker, J.P. (1991). *Adv. Protein Chem.*, *42*: 1.

Hausinger, R.P. (1990). *BioFactors*, *2*: 179.

Hill, H.A.O. (1990). *Pure Appl. Chem.*, *62*: 1047.

Hofffman, B.M. (1991). *Accounts Chem. Res.*, *24*: 164.

Ibers, J.A., and Holm, R.H. (1980). *Science*, *209*: 223.

Jiang, F., Karlin, K.D., and Peisach, J. (1993). *Inorg. Chem.*, *32*: 2576.

Karlin, K.D. (1993). *Science*, *261*: 701.

Klug, A., and Rhodes, D. (1987). *Trends Biochem. Sci.*, *12*: 464.

Lindskog, S., and Nyman, P.O. (1964). *Biochim. Biophys. Acta*, *85*: 462.

Martell, A.E., and Motekaitis, R.J. (1992). *Determination and Use of Stability Constants*, VCH Publishers, New York.

McPhalen, C.A., Strynadka, N.C.J., and James, M.N.G. (1991). *Adv. Protein Chem.*, *42*: 77.

O'Halloran, T.V. (1993). *Science*, *261*: 715.

Schneider, G., Eklund, H., and Cedergren-Zeppezaur, E., and Zeppezaur, M. (1983). *Proc. Nat. Acad. Sci. USA*, *80*: 5289.

Sigel, H., and McCormick, D.B. (1970). *Accounts Chem. Res.*, *3*: 201.

Solomon, E.I., Baldwin, M.J., and Lowery, M.D. (1992). *Chem. Rev.*, *92*: 521.

Valentine, J.S., and Pantoliano, M.W.P. (1981). *Copper Proteins* (T.G. Spiro, ed.), John Wiley & Sons, New York, p. 292.

IRON PROTEINS

by

R.R. Crichton

Unité de Biochimie,
Université Catholique de Louvain,
Place Louis Pasteur 1
B-1348 Louvain-la-Neuve
Belgium

I. INTRODUCTION

Since this is the first of the fourteen chapters devoted to protein interactions with essential and beneficial metal ions, we define the remit which the authors of this section of the Handbook have received: for each metal ion considered, a catalogue should be made of all kinds of protein-interactions presently known, i.e. those involved in specific metalloproteins as well as other more or less specific protein interactions, in the order:

- work proteins
- metalloenzymes with metal-containing active sites
- enzymes with metal ions as structural coordinators
- storage and transport proteins

with a short description of corresponding roles in vivo. Priority should be given to metal ion-protein associations occurring in humans, then in animals, then in other species in order of decreasing relevance to human biology. For the particular case of iron proteins, the task is a daunting, if not impossible one, in view of the multitude of roles that iron-containing proteins play in living organisms (reviewed in Crichton, 1991). Reviews of metalloprotein crystallography and of structural aspects of metal liganding to functional groups in proteins including many iron proteins can be found in Armstrong (1988), Glusker (1991), and of non-heme iron proteins in Howard and Rees (1991).

II. IRON PROTEINS - WORK PROTEINS

A. Introduction

In this category we have included the heme-containing oxygen sequestering and carrier proteins (such as myoglobins, hemoglobins, insect erythrocruorins and plant leghemoglobins), the marine invertebrate oxygen carrier hemerythrin (which contains non-heme iron), electron transporters; these have been separated into hexa-coordinate cytochromes involved in electron-transfer and electron-transferring iron-sulphur proteins, classed either as simple (rubredoxins and ferredoxins) or as complex (containing an additional cofactor or prosthetic group as well as one or more iron sulphur cluster). It might be argued that our classification of electron transferring iron proteins as work proteins rather than metalloenzymes is an arbitrary one, but we feel strongly that where their role is simply that of electron transfer they fit better into the category of work proteins rather than enzymes.

B. Oxygen sequestering and transport proteins

Oxygen can both be sequestered and transported by a number of hemoproteins, known respectively as hemoglobins and myoglobins in mammals, as erythrocruorins in invertebrates and as leghemoglobins in plants. Whatever their state of aggregation (monomer, dimer or tetramer), all hemoglobins - including the leghemoglobin of the nodules of nitrogen-fixing leguminous plants and insect erythrocruorins - have the common 'globin' tertiary fold. The heme group is tightly bound to the protein through a large number of hydrophobic interactions, and there is always a single coordinate bond between the imidazole of the proximal histidine residue of the protein and the iron. All hemoglobins and myoglobins possess an unoccupied sixth coordination site, situated within a hydrophobic pocket, and can reversibly bind molecular oxygen. Monomeric and oligomeric hemoglobins differ in that oxygen binding to the latter often exhibits cooperativity. In the deoxy-form, the high-spin Fe(II) atom is above the heme plane by about 0.08 nm. Upon oxygen binding it is proposed (Perutz, 1970) that the electrons become spin-paired and the smaller low-spin Fe(II) can then fit into the heme plane. This pulls the proximal histidine, transmits oxygenation to other parts of the protein and converts it from the deoxy-T to the

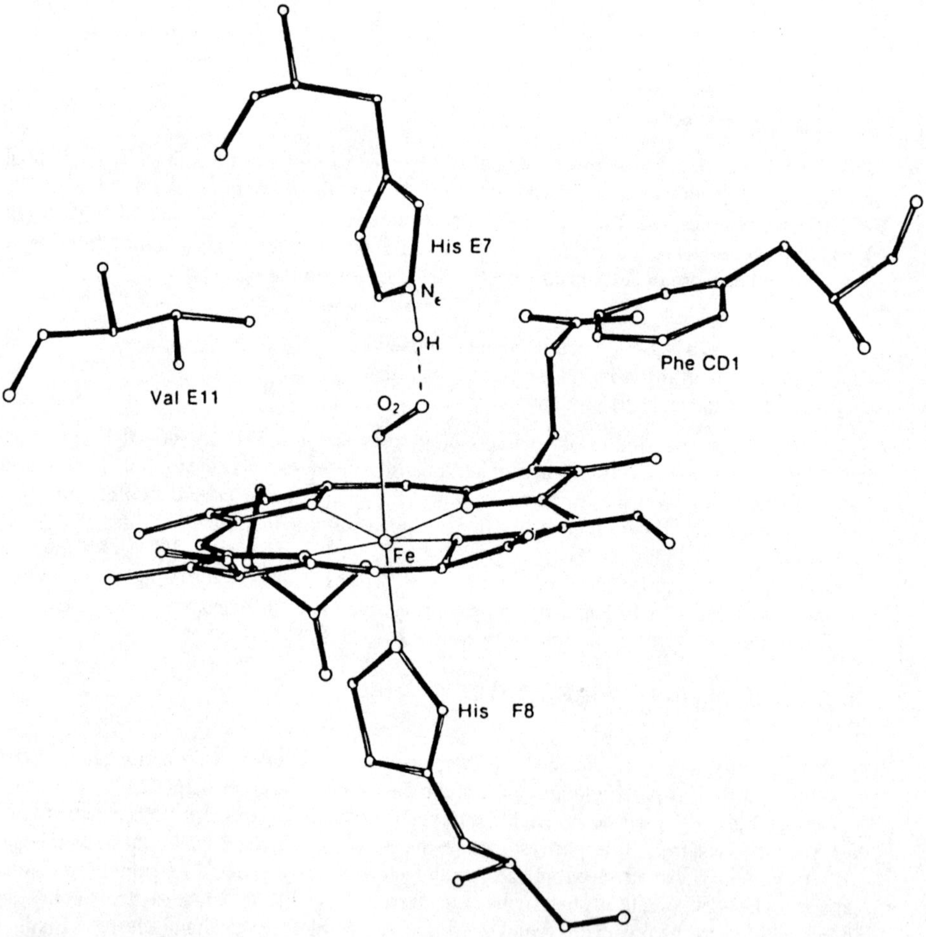

Figure 1.
Arrangement of proximal and distal residues in the iron center of oxymyoglobin (from Perutz, 1989).

oxy-R state. The distal uncoordinated histidine imposes binding of both oxygen and carbon monoxide in a bent configuration: this considerably weakens the affinity of heme for the latter and optimizes that for oxygen. The arrangement of proximal and distal histidines in oxymyoglobin is presented in Figure 1. For a recent account of the role of distal residues in reactions with heme ligands, see Perutz (1989).

Another class of non-heme iron proteins, the hemerythrins, is involved in oxygen storage in marine invertebrates. They contain two iron atoms linked by one oxygen atom as a bridging ligand in a (μ-oxo)bis(μ-carboxylato)-di-iron core, as do a number of other iron metalloenzymes (see later). Figure 2 presents the configuration of the metal-binding site of metazido-myohemerythrin. The two iron atoms are bridged by an oxygen atom and by two

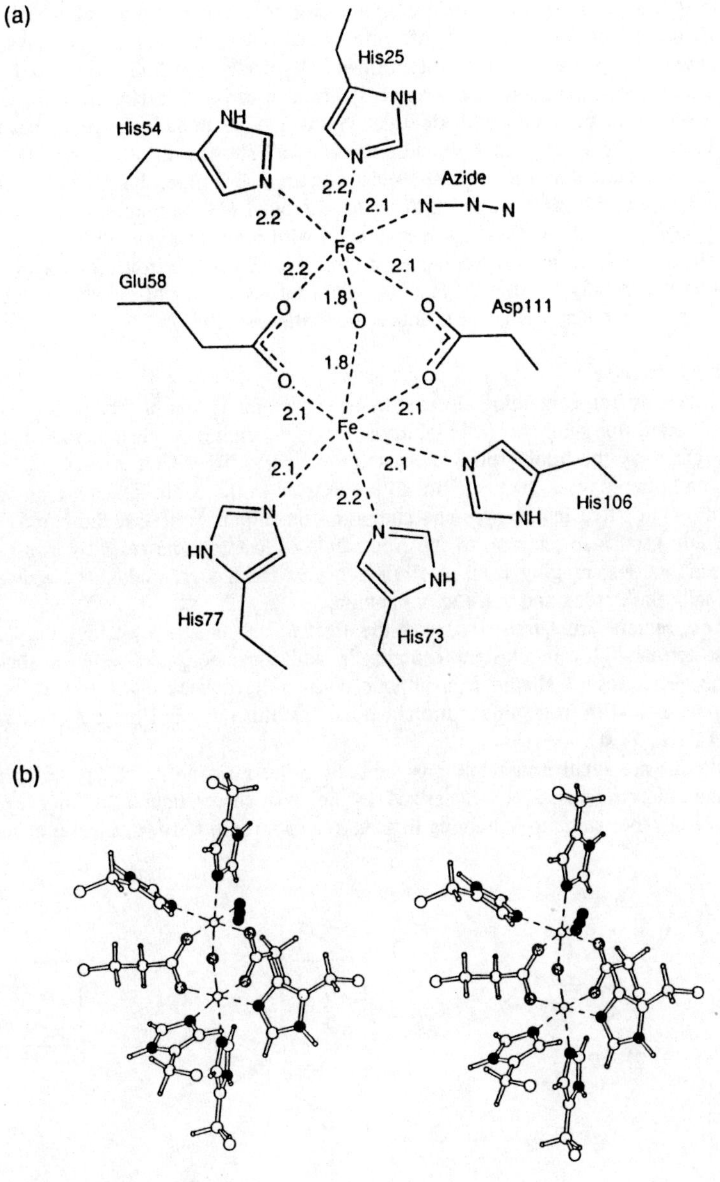

Figure 2.
(a) Atomic designation and (b) stereo view of the dinuclear iron center in myohemerythrin (from Glusker, 1991).

bidentate carboxylate groups from glutamate and aspartate residues. The iron core is further anchored in the protein by five histidine residues, three of which occupy an octahedral face of one iron atom and two of which are coordinated to the other iron atom. The exogenous ligand in the sixth coordination site is azide, but it is assumed that in the oxygenated state a dioxygen molecule occupies the place of azide.

C. Electron transfer proteins
We can distinguish two main classes of electron transfer proteins that contain iron, the cytochromes and the iron-sulphur proteins.

1. *Cytochromes*
Cytochromes are hemoproteins which are widely distributed in nature. Some cytochromes (cytochrome P-450 and cytochrome oxidase) have five coordinate hemes and are discussed in Section III below. In the electron-transferring cytochromes, the iron is bound in a hexa-coordinate low-spin system with two protein ligands (typically histidine and/or methionine) above and below the heme plane. They serve as electron-carrier proteins in mitochondria, endoplasmic reticulum and photosynthetic organelles as well as in bacterial redox chains. At least three classes of cytochromes *a*, *b* and *c* are known. They all have a distinctive three-banded absorption spectrum in the reduced state with an α-absorption band around 550-664 nm, a β-band around 520-546 nm and a Soret band around 400-450 nm. Cytochromes can alternate between an oxidized Fe(III) low-spin state with a single unpaired electron, and a reduced Fe(II) low-spin form with no unpaired electrons. Since the iron remains low spin, electron transfer is greatly facilitated. The best characterized family of cytochromes are the cytochrome *c*'s (for a recent review see Moore and Pettigrew, 1991).

2. *Iron-sulphur proteins*
The second class of iron-containing electron-transfer proteins is that of those which contain iron atoms bound to sulphide (acid-labile sulphur), forming a cluster linked to the polypeptide chain by the thiol groups of cysteine residues. They have a very widespread distribution and often have a characteristic EPR spectrum in the oxidized state. Iron-sulphur proteins can be classified into simple and complex iron-sulphur proteins; the latter contain other prosthetic groups in addition to the iron-sulphur clusters. The roles of iron-sulphur proteins are numerous, ranging from electron transport (such as rubredoxins, ferredoxins) to enzymes with both redox and non-redox functions.

The iron centers are tetrahedral, and the ligands are mostly cysteinyl thiols. The multinuclear tetrahedral iron clusters contain, in addition, inorganic sulfur as bridging ligands. The structures of simple iron-sulphur clusters have been characterized both in model compounds and in iron-sulphur proteins, and are illustrated in Figure 3 (reviewed in Howard and Rees, 1991).

Rubredoxins are small iron-sulphur proteins of molar mass around 6 kg mol^{-1}, found in plants and microorganisms, characterized by an iron center bound by four cysteine sulphur atoms or other sulphur-containing ligands in an approximately tetrahedral geometry.

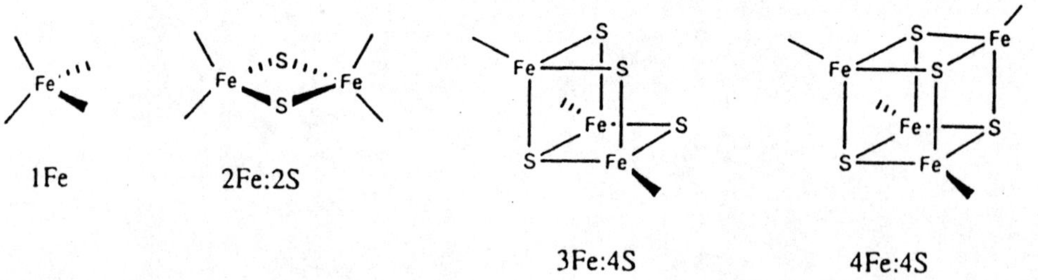

Figure 3.
Tetrahedral iron center types found in non-heme iron-sulfur proteins (from Howard and Rees, 1991).

The reduced protein is colorless and EPR-silent, while oxidized rubredoxin is red with a characteristic EPR signal. Both oxidized and reduced forms are high spin.

The ferredoxins are iron-sulphur proteins which contain 'acid-labile' sulphur, giving H_2S on treatment with mineral acid. They usually contain an equal number of iron and labile-sulphur atoms and have exclusively electron-carrier activity. The binuclear iron clusters can be considered as the fusion of two mononuclear $Fe(SR)_4$ centers with two thiolate ligands replaced by bridging inorganic sulfur atoms, and the tetranuclear as two fused binuclear clusters. The 3Fe/4S clusters appear to be formed from 4Fe/4S clusters by removal of one iron atom. Ferredoxins have a wide range of redox potentials (+350 to -600 mV) and are widely distributed in electron-transfer chains. Although the core structures of the di- and tetranuclear complexes are well established, we do not yet fully understand how the protein fine-tunes the redox potential of an iron-sulfur site, nor how the high-potential iron proteins of purple photosynthetic bacteria (HiPIP) have access to a stable oxidized $(4Fe\text{-}4S)^{3+}$ cluster whereas ferredoxins do not. Extensive studies on model compounds and proteins suggest that the redox potential of the iron-sulfur center is determined by polarity of the pocket surrounding the center, hydrogen bonding from the protein to both the terminal and the bridging ligands of the center, ligand strength (amount of covalency between ligand and iron) and basicity of the terminal ligands (Berg and Holm, 1982; Rees and Farrelly, 1990; Que and True, 1991). In the mitochondrial respiratory chain, the Rieske protein has non-cysteinyl ligands binding the core in addition to thiolate ligands. From multiple spectroscopic techniques (reviewed in Howard and Rees, 1991) it appears that the cluster contains two irons, two cysteinyl sulfurs and two histidinyl nitrogens in a common plane; the nitrogen ligands do not appear to alter the core structure significantly although they have an important effect on the redox potential.

III. IRON PROTEINS - METALLOENZYMES WITH IRON-CONTAINING ACTIVE SITES

A. Heme-containing enzymes

Many heme enzymes are characterized by a penta-coordinate geometry (the four pyrrole nitrogens and a fifth ligand contributed by the protein, which can be histidine, tyrosine or, in cytochrome P-450, cysteine). The iron center can bind and activate a sixth substrate ligand in the sixth metal coordination site. Such enzymes include cytochrome oxidase, peroxidases, catalases, and cytochrome P-450. Their substrates include molecular oxygen, hydrogen peroxide and, in the case of cytochrome P-450's, a multiplicity of organic substrates which can form iron-carbon bonds with the enzyme.

Cytochrome oxidase reduces molecular oxygen to water in a four-electron reduction, and is the terminal component of the widely distributed mitochondrial respiratory chain. The primary oxygen-binding site involves a bimetallic heme iron-copper center.

In the peroxidases, catalases and cytochrome P 450's, the high-spin iron-porphyrin system can go to the radical cation state, at a redox potential close enough to that of the couple Fe(IV)/Fe(III) to allow a ferryl type of iron to participate in chemical reactions such as the activation of oxygen or oxidation of molecules at the expense of hydrogen peroxide.

Catalases and peroxidases both promote H_2O_2 reduction by mechanisms that involve ferryl intermediates (Dawson, 1988). Catalases differ from peroxidases in their ability to use H_2O_2 both as an electron acceptor and a donor, thus catalyzing the disproportionation reaction. Like peroxidases, they can also oxidize a variety of organic substrates. Whereas the proximal heme ligand in peroxidases such as cytochrome *c* peroxidase is a histidine (Poulos and Kraut, 1980), that in catalase is a tyrosine (Fig. 4) (Fita and Rossmann, 1985). Both enzymes have a wide distribution, and catalase is found in copious amounts within the peroxisomes of all animal and many plant cells.

Cytochrome P-450's are a class of monooxygenases catalyzing the reaction below:

$$RH + O_2 + 2e^- + 2H^+ \rightarrow ROH + H_2O$$

in which dioxygen is activated by two electrons and two protons, with the insertion of one oxygen atom into the substrate and formation of water. P-450-dependent monooxygenases are of capital importance in the biosynthesis and catabolism of endogenous compounds such

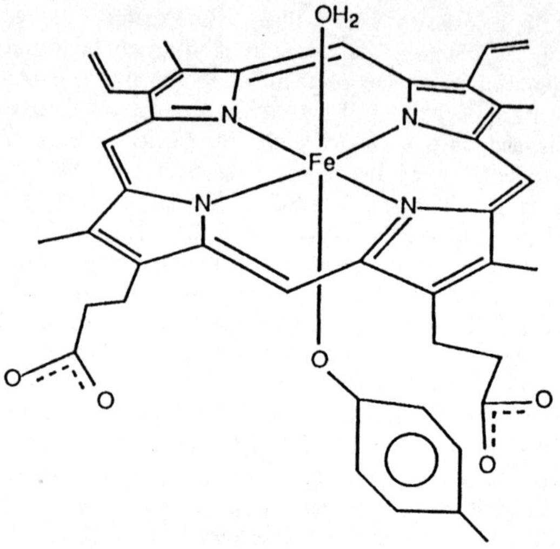

Figure 4.
The iron coordination environment for catalase in the resting state (from Armstrong, 1988).

as fatty acids, steroids, prostaglandins and leukotrienes; they are also involved in the oxidative metabolism of exogenous compounds such as drugs and other environmental products. There are many isoenzymes, each of which can be induced specifically at very high levels in liver (and to a lesser extent in other tissues) by distinct classes of xenobiotics.

The cytochrome P-450's all appear to be b-type heme enzymes with the iron linked to the apoprotein by a cysteine residue. The X-ray structure of the cytochrome P-450 from *Pseudomonas putida* grown on camphor has been determined (Poulos et al., 1985) and a representation of the active site structure of the camphor-bound form is given in Figure 5. Progress in understanding the mechanism of P-450 has been greatly aided by the synthesis of model compounds (reviewed by Mansuy et al., 1989).

B. Iron-sulfur enzymes

Iron-sulphur clusters are also found in association with other cofactors (reviewed in Howard and Rees, 1991). They are present in oxidoreductases such as NADH-coenzyme Q reductase in Complex I and the flavin-containing succinate dehydrogenase in Complex II of the mitochondrial electron transport chain. Other examples of flavoprotein enzymes with Fe-S clusters include trimethylamine dehydrogenase and xanthine oxidases (which also contains Mo). There are also the Mo-Fe, V-Fe and the exclusively Fe-containing nitrogenases in nitrogen-fixing organisms: the X-ray structure of the Mo-Fe nitrogenase has been recently determined (Kim and Rees, 1992); hemoproteins such as nitrite and sulfite reductases; CO dehydrogenase and hydrogenases in bacteria (containing Ni and Fe, although examples of the latter are known which contain only Fe) and the unusual microbial association of W with Fe-S cofactors in formate dehydrogenase. The structure of the active site of sulfite reductase (which catalyzes the reduction of sulfite to sulfide) confirms that both a siroheme group and a 4Fe-4S cluster function as prosthetic groups: they are bridged by a cysteine sulfur atom (Fig. 6). In all of these cases the iron-sulfur cluster essentially plays the role of electron transport. However, a growing class of enzymes employ iron-sulphur clusters to catalyze non-redox functions, such as hydratases and isomerases and hydrolases (endonuclease III). The best studied example of these is aconitase, which transforms citrate to isocitrate in the Krebs' cycle, catalyzing successive reactions of dehydration and rehydration (for a recent review of the mechanism see Lauble et al., 1992). As purified, aconitase is obtained in an inactive 3Fe-4S form, which can be transformed with Fe^{2+}under reducing conditions to give a catalytically active aconitase with a 4Fe-4S cluster. The fourth unique iron atom of the cluster is directly involved in binding the substrate via a hydroxyl

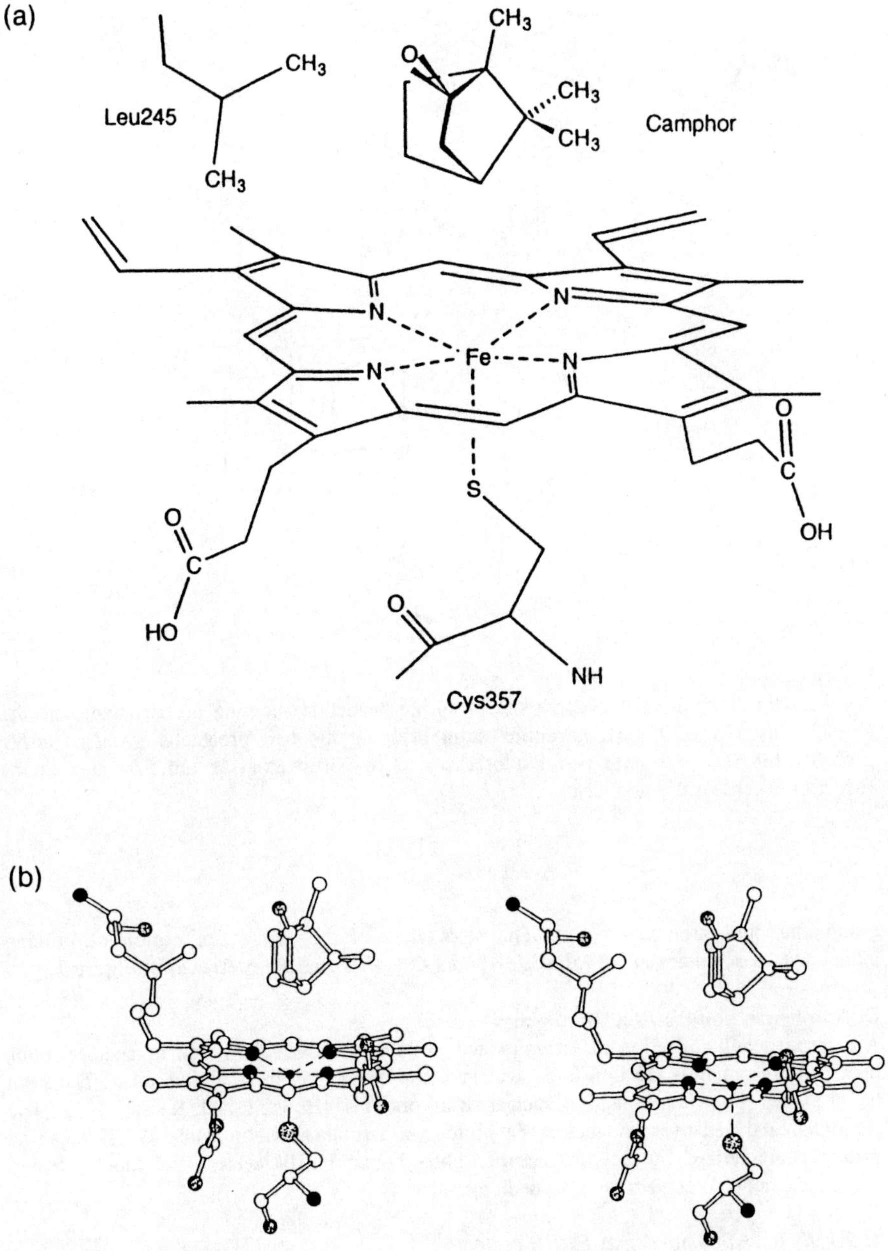

Figure 5.
(a) Atomic designation and (b) stereo view of the iron center of cytochrome P450 with camphor bound (from Glusker, 1991).

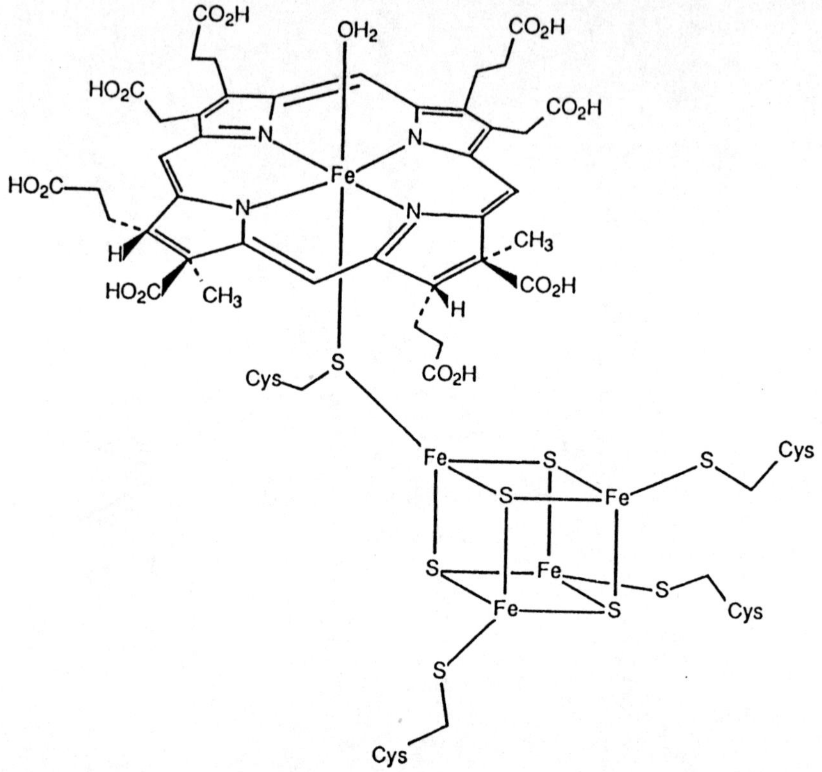

Figure 6.
A schematic diagram of the active site of sulfite reductase showing the siroheme and the 4Fe-4S cluster with a cysteine sulfur atom bridging the two prosthetic groups. Sulfite probably binds to the sixth position of the siroheme iron atom as the first step in the reduction mechanism (from Armstrong, 1988).

group and one carboxyl oxygen. The stereoview of the active site region of porcine mitochondrial aconitase in both the 3Fe-4S and 4Fe-4S forms are presented in Figure 7.

C. Non-heme, non-iron-sulfur enzymes
A heterogeneous collection of enzymes and proteins which contain iron or require it for their activity remain - the iron is present in a non-heme, non iron-sulphur form. They can be classified into mono- and dinuclear iron proteins (Howard and Rees, 1991). The reactions catalyzed by mononuclear Fe proteins are summarized in Table 1. They can be further characterized by the presence of tightly bound Fe(III) centers and loosely bound mononuclear Fe(II) centers (as is done here).

1. *Tightly bound mononuclear Fe(III) centers*
Dioxygenases catalyze the cleavage of molecular dioxygen with subsequent incorporation of both iron atoms into organic substrates. Among them are the bacterial dioxygenases which catalyze the critical ring-opening step in the biodegradation of aromatic compounds and the mammalian homogentisate 1,2-dioxygenase which operates the ring-opening of homogentisate in the oxidative degradation of phenylalanine and tyrosine. They generally contain tightly bound non-heme ferric iron as sole cofactor.

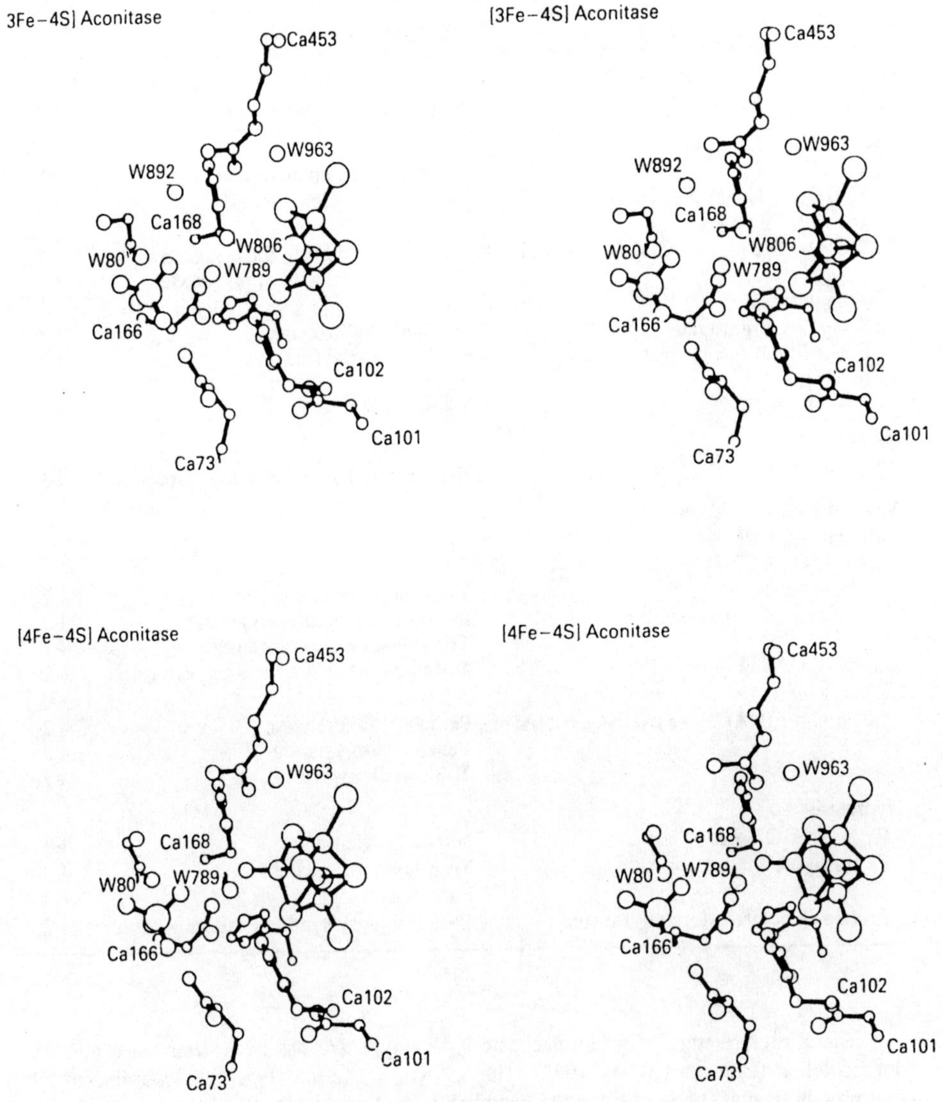

Figure 7.
The Fe-S cluster of aconitase, showing stereo pairs of eight active site residues in the immediate vicinity of the 3Fe-4S and 4Fe-4S clusters together with the bound sulfate ion and five water molecules at a resolution of 0.21nm (from Robbins and Stout, 1989). Ca indicates α-carbon positions and W positions of bound water molecules.

Table 1

Reactions Catalyzed by Mononuclear Fe Proteins

Reaction	Example	Fe state
Oxidase		
$SH_4 + O_2 \rightarrow S + 2H_2O$	Isopenicillin *N*-synthetase	+2
Dioxygenase		
Intradiol	3,4-Protocatechuate dioxygenase	+3
$S(OH)_2 + O_2 \rightarrow S'(COOH)_2$	1,2-Catechol dioxygenase	+3
Extradiol		
$S(OH)_2 + O_2 \rightarrow S'(COOH)(OH)(CHO)$	4,5-Protocatechuate dioxygenase	+2
	2,3-Protocatechuate dioxygenase	+2
2-Oxoglutarate		
SH + 2-oxoglutarate +	Proline 4-hydroxylase	+2
$O_2 \rightarrow SOH + CO_2$ + succinate	Lysine hydroxylase	+2
Peroxidase		
$S + O_2 \rightarrow SOOH$	Lipoxidase	+3
Dismutase		
$2O_2^- + 2H^+ \rightarrow H_2O_2 + O_2$	*Escherichia coli* superoxide dismutase	+3
Mixed-function oxidase		
Monooxygenase		
$SH + O_2 + DH_2 \rightarrow SOH + H_2O + D$		
D = pteridine	Tyrosine monooxygenase	+2
	Phenylalanine monooxygenase	+2
	Tryptophan monooxygenase	+2
D = NADH	4-Methoxybenzoate monooxygenase	+2
Dioxygenase		
$S + O_2 + NADH \rightarrow S(OH)_2 + NAD$	Phthalate dioxygenase	+2
	Benzene dioxygenase	+2
	Toluene dioxygenase	+2
Hydratase		
$RCN + H_2O \rightarrow RCONH_2$	Nitrile hydratase	+3
Fe transport	Transferrin	+3
	Lactoferrin	+3
Protein structure, electron transfer	Bacterial photosynthetic reaction center	+2

The X-ray structure of protocatechuate 3,4-dioxygenase has been determined at 0.28 nm resolution (Ohlendorf et al., 1988). The catalytic ferric ion (Fig. 8) is coordinated by four protein ligands, two tyrosines and two histidines: the iron coordination geometry forms an approximate trigonal bipyramid with a bound water molecule completing the trigonal iron coordination in the equatorial plane. The coordination geometry is similar to the iron superoxide dismutase discussed below.

There are many nonheme-iron oxygenases including other microbial catechol dioxygenases, mammalian homogentisate oxygenase and cysteamine oxygenase, the important plant oxygenase, lipoxygenase, which catalyzes peroxidation of polyunsaturated fatty acids and some of the enzymes involved in prostaglandin biosynthesis.

Many microorganisms have superoxide dismutases which utilize iron(III) in their catalytic site. The X-ray structure (Fig. 9) reveals a coordination geometry reminiscent of that described above for protocatechuate 3,4-dioxygenase, including a water molecule bound to the tetracoordinate iron. However, the ligands are not the same as for the protocatechuate 3,4-dioxygenase.

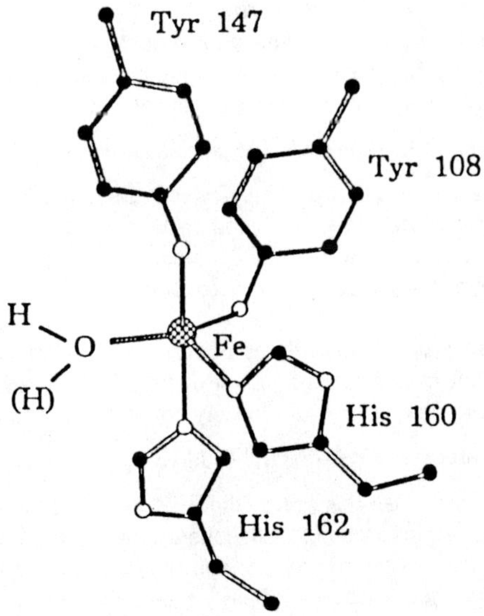

Figure 8.
Stereo-pair of the active center of the Fe superoxide dismutase from *E. coli* (from Carlioz et al., 1988).

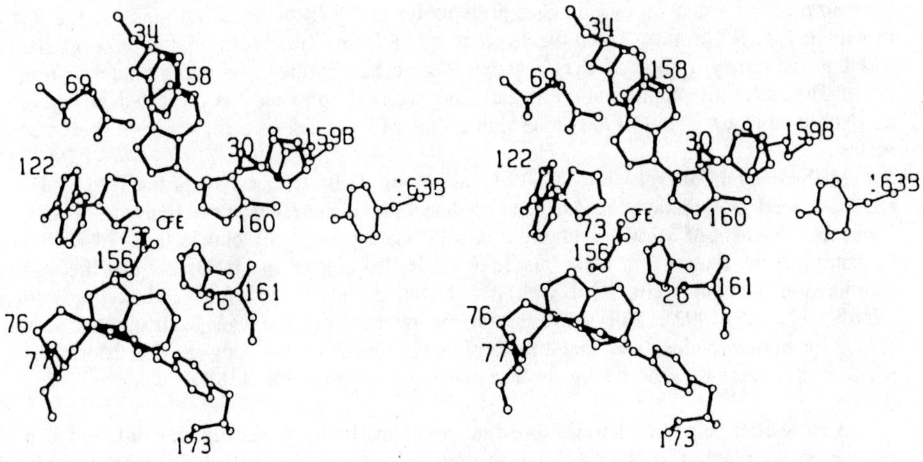

Figure 9.
Active-site region of 3,4-protocatechuate dioxygenase (from Howard and Rees, 1991).

2. *Loosely bound mononuclear Fe(II) centers*

Hydroxylation of proline and lysine residues on nascent chains of procollagen is catalyzed by ferrous iron-containing dioxygenases, which require in addition α-ketoglutarate and have an absolute requirement for ascorbate. The reaction catalyzed can be represented as:

$$XH + \alpha\text{-ketoglutarate} + O_2 \rightarrow XOH + \text{succinate} + CO_2$$

Prolyl 4-hydroxylase and lysyl hydroxylase are involved in the maturation of procollagen. Both enzymes catalyze a reaction in which α-ketoglutarate is decarboxylated with the incorporation of one atom of the O_2 molecule into succinate and the other into the hydroxyl group of either 4-hydroxyproline or lysine residues at specific sites of the procollagen molecule.

The aromatic amino acid hydroxylases of mammals are mono-oxygenases (or mixed-function oxidases) which incorporate one atom of the dioxygen molecule into the substrate and the other into water. They all contain loosely bound Fe(II) and catalyze the reaction :

$$XH + \text{tetrahydropterin} + O_2 \rightarrow XOH + \text{dihydropterin} + H_2O$$

Phenylalanine hydroxylase catalyzes the hydroxylation of L-phenylalanine to L-tyrosine, the first step in phenylalanine catabolism, which also provides the body with an endogenous source of tyrosine. Tyrosine hydroxylase catalyzes the hydroxylation of tyrosine to dihydroxy-phenylalanine (dopa), the rate-limiting step in catecholamine neurotransmitter production (Kaufman and Kaufman, 1985): like phenylalanine hydroxylase, it requires Fe(II) in its catalytic cycle. Tryptophan hydroxylase, involved in the production of serotonin, is also part of the same family (Stoll and Goldman, 1991). Kinetic experiments indicate that tetrahydrobiopterin binds first to the three hydroxylases, followed by binding of dioxygen and an amino acid. The rate-limiting step in catalysis is formation of the hydroxylating intermediate, a peroxytetrahydro-pterin (Haavik et al., 1991) subsequent to dioxygen binding to the catalytic Fe(II). This intermediate is assumed to provoke cleavage of the oxygen-oxygen bond and hydroxylation of the substrate aromatic ring (Fitzpatrick, 1991).

3. *Enzymes with binuclear octahedral iron centers*

A class of proteins which contain a binuclear octahedral iron center (two iron atoms linked by one oxygen atom as a bridging oxo ligand) have been well established recently (for review see Howard and Rees, 1991). They include hemerythrin (see above) and the enzymes ribonucleotide reductase, methane monooxygenase and purple acid phosphatase.

Ribonucleotide reductases (RNR) catalyze the transformation of ribonucleoside phosphates into the corresponding 2-deoxy derivative, a reaction essential for DNA synthesis, and therefore for cell division in almost all living organisms (Crichton, 1991). The enzyme is formed by two dimeric proteins R1 and R2: the substrate binds to the R1 protein and the R2 protein, which functions as the initiator of the radical-based reaction, has a stable free tyrosyl radical at Tyr^{122} which is generated by dioxygen and a binuclear iron center.The active site configuration of the oxo-bridged diiron center is presented in Figure 10 (Nordlund et al., 1990). It has some similarities with the hemerythrin structure presented earlier.

Methane mono-oxygenase (MMO) of methanotropic bacteria catalyzes the NAD(P)H- and O_2-dependent oxidation of methane to methanol at a binuclear iron site (Parkyns, 1990). Sequence alignment of MMO's with the iron-coordinating four helix bundle in R2 of RNR's indicates that the central four helix bundle of R2 is also present in MMO, and that the iron coordination is similar (Fig. 10) with two histidine ligands and four carboxyl groups (Nordlund et al., 1992). This suggests that the mechanism of oxygen activation is very similar in both proteins, and that the local environment of the oxygen site determines whether the reaction catalyzed by the enzyme is an oxygenation (MMO) or an oxidation (RNR).

A number of acid phosphatases found in spleen and in the uterus (uteroferrin), and also in microbial, plant and animal sources, exist in two interconvertible forms: pink, catalytically active, or reduced, and purple, inactive or oxidized, with an Fe(III)-Fe(III) pair (Antanaitis and Aisen, 1986). Although their true biological function is unknown, they

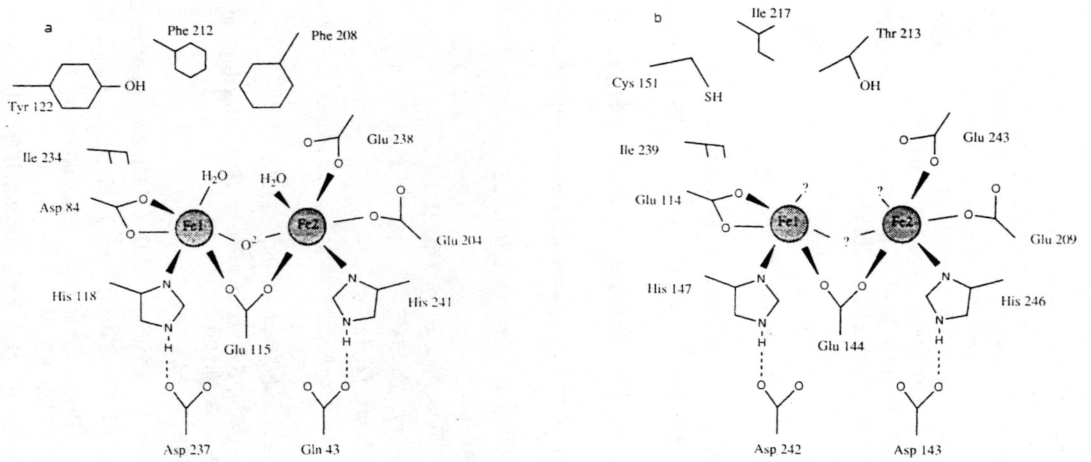

Figure 10.
Schematic view of the iron surroundings in protein R2 of ribonucleotide reductase and the hydroxylase subunit of MMO. (a) The diferric center of ribonucleotide reductase consists of Two His, three Glu one Asp, a bridging μ-oxo group and two water molecules. (b) The corresponding center predicted from amino acid sequence homology in MMO's (from Nordlund et al., 1992).

catalyze the dephosphorylation of phosphoproteins and nucleotides. The magnetic and ESR properties of the mixed-valence forms of these phosphatases from bovine spleen and porcine uterine fluid resemble those of semimet-hemerythrin.

IV. IRON STORAGE AND TRANSPORT PROTEINS

A number of proteins are involved in iron storage and transport, both in mammals and in other multicellular and unicellular species. The storage function is assumed by ferritins in mammals, in invertebrates and in plants, and by hemosiderins in mammals and plants. The role of ferritins (24-subunit proteins enclosing a more or less ordered ferric oxyhydroxide nucleus) is discussed in greater detail in Part Two, Chapter 1-D4 of this volume.

In mammals, iron is transported by transferrins, which usually have a bilobal structure, bind two iron atoms together with a carbonate anion, and are discussed in greater detail in Part Two, Chapter 1-D3 of this volume. There are also receptors for the internalisation of extracellular iron transport molecules (transferrin receptors and siderophore receptors, respectively in mammals and a few other higher organisms, and in bacteria). Transferrin receptors are reviewed in Part Two, Chapter 1-D4 of this volume, and a recent review of siderophore receptors can be found in Crichton and Ward (1992).

REFERENCES

Antanaitis, B.C., and Aisen, P. (1986). *Frontiers in Bioinorganic Chemistry* (A.O. Xavier, ed.), VCH, Weinheim, p. 481.

Armstrong, W.H. (1988). *Metal Clusters in Proteins* (L. Que, ed.), *ACS Symp. 372*: 1.

Berg, J., and Holm, R. (1982). *Iron Sulfur Proteins* (T. Spiro, ed.), Wiley Interscience, New York, p. 3.

Carlioz, A., Ludwig, M.L., Stallings, W.C., Fee, J.A., Steinman, H.M., and Touati, D. (1988). *J. Biol. Chem.*, *263*: 1555.

Crichton, R.R. (1991). *Inorganic Biochemistry of Iron Metabolism*, Ellis Horwood, Chichester, p. 263.

Crichton, R.R., and Ward, R.J. (1992). *Biochemistry*, *31*: 11255.

Dawson, J.H. (1988). *Science*, *240*: 433.

Fita, I., and Rossmann, M.G. (1985). *J. Mol. Biol.*, *185*: 21.

Fitzpatrick, P.F. (1991). *Biochemistry*, *30*: 6386.

Glusker, J.P. (1991). *Adv. Prot. Chem.*, *42*: 3.

Haavik, J., Le Bourdelelles, B., Martinez, A., Flatmark, T., and Mallet, J. (1991). *Eur. J. Biochem.*, *199*: 371.

Howard, J.B., and Rees, D.C. (1991). *Adv. Prot. Chem.*, *42*: 199.

Kaufman, S., and Kaufman, E.E. (1985). *Folates and Pterins* (R.L. Blakeley and S.J. Benkovic, eds.), John Wiley, New York, p. 251.

Kim, J., and Rees, D.C. (1992). *Nature*, *360*: 563.

Lauble, H., Kennedy, M.C., Beinert, H., and Stout, C.D. (1992). *Biochemistry*, *31:* 2735.

Mansuy, D., Battioni, P., and Battioni, J.-P. (1989). *Eur. J. Biochem.*, *184*: 267.

Moore, G.R., and Pettigrew, G.W. (1991). *Cytochromes c Evolutionary, Structural and Physicochemical Aspects*, Springer Verlag, Berlin.

Nordlund, P., Sjöberg, B.-M., and Eklund, H. (1990). *Nature*, *345*: 593.

Nordlund, P., Dalton, H., and Eklund, H. (1992). *FEBS Lett.*, *307*: 257.

Ohlendorf, D.H., Lipscomb, J.D., and Weber, P.C. (1988). *Nature*, *336*: 403.

Parkyns, N.D. (1990). *Chem. Brit.*: 841.

Perutz, M.F. (1970). *Nature*, *228*: 726.

Perutz, M.F. (1989). *Trends Biochem. Sci.*, *14*: 42.

Poulos, T.L., and Kraut, J. (1980). *J. Biol. Chem.*, *255*: 8199.

Poulos, T.L., Finzel, B.C., Gunsalus, I.C., Wagner, G.C., and Kraut, T. (1985). *J. Biol. Chem.*, *260*: 16122.

Que, L., and True, A. (1991). *Prog. Inorg. Chem.*, *38*: 97.

Rees, D.C., and Farrelly, D. (1990). *The Enzymes*, Vol. 19 (D.S. Sigman and P.D. Boyer, eds.), 3rd ed., Academic Press, San Diego, p. 37.

Robbins, A.H., and Stout, C.D. (1989). *Proc. Nat. Acad. Sci. USA*, *86:* 3639.

Stoll, J., and Goldman, D. (1991). *J. Neurosci. Res.*, *28*: 457.

COPPER PROTEINS

by

Ivano Bertini, Luigi Messori and Maria Silvia Viezzoli

Department of Chemistry,
University of Florence,
Florence
Italy

I. INTRODUCTION

Copper is one of the essential trace elements in living organisms. The essentiality of copper is due to its presence in several enzymes and proteins widely distributed in animals and plants where it performs central functions in various fundamental biological processes (Sigel, 1981; Spiro, 1981). Copper proteins play two main functions, namely electron transport and dioxygen transport and activation. Typical are the cases of cytochrome *c* oxidase, the protein which represents the final step in the respiratory chain, and hemocyanins, that act in several organisms as dioxygen transport proteins. Several other fundamental functions of copper proteins have been described: copper oxidases help to produce the extracellular matrices of living eukaryotic organisms (e.g. lysyl oxidase); in plants the copper oxidases, free radical oxidases, perform defensive roles such as the formation of protective films when the tissues are damaged (phenol oxidases). Protection by copper against the toxic effect of free radicals is further found in the cell cytoplasm and outside cell (superoxide dismutase, ascorbate oxidase). Other copper oxidases remove excess of amine neurotransmitters (copper amine oxidases) but are also used for the synthesis of adrenalin, one of the main transmitters of the nervous system (dopamine β-hydroxilase) (da Silva and Williams, 1991).

According to a suggestive hypothesis proposed by da Silva and Williams (1991) copper became an essential element only late in the course of evolution. This development was probably a consequence of the introduction into the environment of an extremely poisonous substance such as dioxygen. Adaptation of living organisms to the presence of dioxygen led to a striking evolution of metabolic pathways: copper(II) became available as copper sulphate resulting from the oxidation of copper sulphides and was integrated into several biomolecules.

Since some copper proteins bear an intense color, they were among the first metalloproteins to be isolated. Then, when isolation and purification techniques were sufficiently developed to permit the obtainment of appropriate samples, spectroscopy played a major role in the investigation and comprehension of the copper active sites (Solomon et al., 1992). It was soon evident that copper sites in proteins exhibit spectral features several of which are unique compared to small copper complexes; these features originate from the unusual geometric and hence electronic structures that can be imposed on the metal ion through its interaction with the protein biopolymer. So the primary goal of researchers in the field was to understand the geometric and electronic structural origin of these spectral features and to evaluate the contribution of electronic structure to the high reactivity and selectivity of these sites in catalysis.

In the initial phase of this research, a basic understanding of the origin of the spectral features was achieved and used to propose models of copper active sites. Then, as results of high resolution protein crystallography became available, it was possible to compare the proposed spectroscopic models with the effective crystal structures and to develop a detailed understanding of the active site electronic structure and its contribution to biological function.

Copper sites are classified into four main types, three involving copper(II) and one involving copper(I) according to structural and spectroscopic properties.

Type 1 copper sites contain tetrahedral copper and are characterized by a very intense absorption band at about 610 nm (ε_{max} = 3000-5000 mol^{-1} dm^3 cm^{-1}) which led to the name of "blue copper proteins". The origin of the band is linked to the existence of one or more copper-sulfur bonds. The EPR parameters for copper(II) in these proteins are also unusual, especially with reference to the hyperfine splitting constant which is approximately one-third ($A_{\parallel}$ = 40-60 x 10^{-4} cm^{-1}) of the value for 'normal' copper(II) complexes. The redox potentials are relatively high (270-370 mV).

Type 2 copper sites are often referred to as 'normal' but this again is an oversimplification. Molar absorbances in the visible region are in the range 100-200 mol^{-1} dm^3 cm^{-1}. Where it has been determined, the coordination geometry appears to be practically square planar. The EPR spectra are characterized by $A_{\parallel}$ values in the range of 110-190 x 10^{-4} cm^{-1}. The redox behavior exhibits a considerable variability.

Type 3 copper is EPR silent although it is made up of two copper(II) ions. These are physically close to each other and the mode of spin pairing that obviously results is still open to discussion even though many aspects have been successfully mimicked in low-molar-mass complexes. An important feature is the virtually synchronous two-electron reduction step associated with this center with which an optical absorption band at about 330 nm is usually associated.

Finally, type 4 copper is copper in the copper(I) state; it is therefore colorless and EPR inactive. Tetrahedral coordination is most common and the flexibility of the site almost certainly has a role to play in the redox stability of the copper. This class comprises copper(II) enzymes in the reduced form and the newly discovered copper(I) thiolate clusters.

On the basis of the knowledge of a certain number of crystal structures it may be hypothesized that type 1 copper is generally used for electron transfer and type 3 copper for binding oxygen. The coordination of type 2 copper is appropriate for ligand exchange and therefore useful for chemical reactivity.

The above classification has dominated the field of copper proteins since the late sixties; it is valid in practice although it often neglects structural differences which are important in specific cases. We believe that the success of this classification reflects the fact that indeed copper centers in biology are reduced to a few structural motifs which appear to be well conserved during the evolutionary history. In other words, copper is the constituent of a limited number of building blocks of specialized function which have been integrated to different extents and under different combinations into biological macromolecules. Several copper proteins bear more than a single copper center of the same or of different types; in other cases, copper domains are combined with iron domains (cytochrome *c* oxidase).

Large part of the above classification is conserved in a recent classification by Solomon et al. (1992), that we will follow in the present review. This classification, with slight modifications, is reported in Table 1, where an updated list of the main copper proteins is presented.

II. OVERVIEW OF COPPER PROTEINS

A survey of the fundamental features of the groups outlined above and of the most representative individual copper proteins is given here. A presentation of the main aspects of copper metabolism, of the proteins involved, and of a newly identified copper type (the copper clusters) is reported in the last part of the article.

Table 1

Copper Proteins

Copper type	Proteins
Blue copper proteins	plastocyanins azurins stellacyanin amicyanin auracyanin cucumber basic blue mavicyanin rusticyanin umecyanin
Normal copper enzymes	copper,zinc-superoxide dismutase amine oxidases, diamine oxidases dopamine β monoxygenase galactose oxidase
Normal+blue copper enzyme	nitrite reductase
Binuclear copper proteins	molluscan hemocyanins arthropodan hemocyanins tyrosinase
Multicopper oxidases	laccase ascorbate oxidase ceruloplasmin
Cu_A-containing enzymes	cytochrome *c* oxidase nitrous oxide reductase
Copper clusters	ACE1 copper metallothionein

A. Blue copper proteins

Blue copper proteins, also known as cupredoxins are small proteins whose function is electron transfer (Adman and Jensen, 1981; Adman, 1985; Sykes, 1991). They comprise plastocyanin, azurin, cucumber basic protein, pseudoazurin, stellacyanins, rusticyanins and other proteins. The copper sites of these proteins have a number of characteristics in common, including an intense absorption band (ε = 2200-5100 $mol^{-1}\ dm^{-3}\ cm^{-1}$) at about 600 nm arising from cysteinate S-Cu(II) charge transfer, a small hyperfine splitting constant in the $g_{\parallel}$ region of their EPR spectra and a relatively high midpoint potential (Solomon et al., 1992). The EPR properties have been shown to be due to tetracoordinated complexes (Gray and Solomon, 1981; Bertini et al., 1980) and the electronic structure has been elucidated on the basis of ligand field calculations (Gray and Solomon, 1981).

The cupredoxins for which high resolution X-ray structures are available comprise poplar plastocyanin (Guus and Freeman, 1983), *Oleander nerium* plastocyanin (Tong, 1991), algal plastocyanin (Collyer et al., 1990) *Alcaligenes denitrificans* azurin (Baker, 1988), cucumber basic blue protein (Guss et al., 1988) and *Alcaligenes faecalis* pseudoazurin (Adman et al., 1989). Despite their different plant and bacterial origins, all six proteins have similar β-sandwich structures and blue copper sites. A sketch of the overall structural organization of cupredoxins is shown in Figure 1. In all cases, the copper is coordinated to one cysteine and two histidines in a distorted trigonal geometry with a weakly ligated axial methionine at a longer distance (Fig. 2). In the azurins, there is also a weakly coordinated peptide carbonyl group serving as fifth ligand.

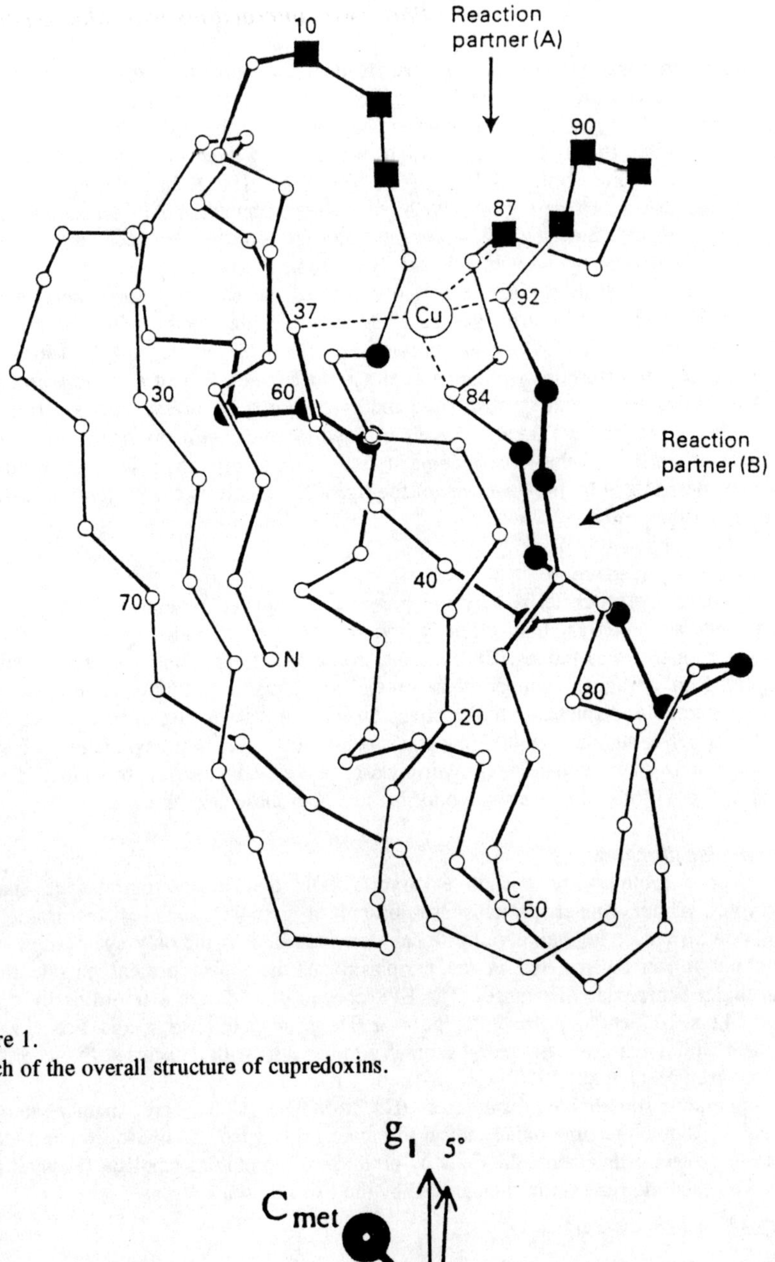

Figure 1.
Sketch of the overall structure of cupredoxins.

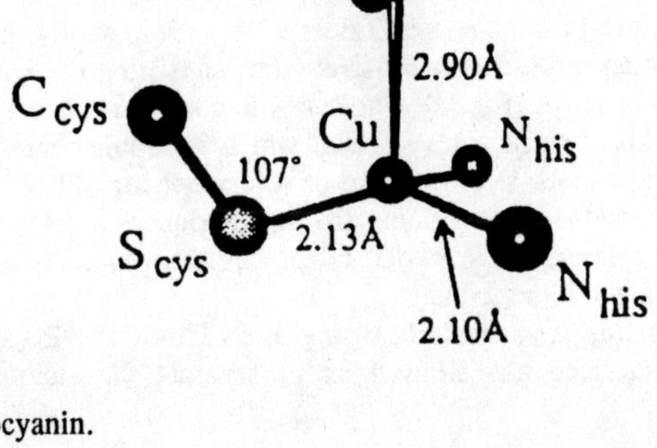

Figure 2.
Active site of plastocyanin.

The electron transfer properties of azurins have been investigated for a long time as a model to understand the molecular basis of biological electron transfer; detailed pathways for the electron through the protein from the copper site to the surface have been proposed (Farver and Pecht, 1989). Now, new insight has been achieved through studies on mutant azurins (van de Kamp et al., 1990a; 1990b; Nar et al., 1991). In particular, it has been demonstrated that a surface region with high content of apolar amino acids, the so called hydrophobic patch, mediates the electron transfer process between azurin and its physiological partners, nitrite reductase and cytochrome *c*551.

Extensive 2D NMR studies have also permitted the complete assignment of the ^1H NMR spectra of reduced azurin (van de Kamp et al., 1992), amicyanin (Kalverda et al., 1991; Lommen et al., 1991) and plastocyanin (Driscoll et al., 1987) which is the prerequisite step to determine with good accuracy the solution structure of these proteins.

Altogether, the several spectroscopic and crystallographic investigations performed on cupredoxins in the last two decades have permitted to elucidate many of the structural and functional properties of the once enigmatic 'blue copper proteins'. The redox potentials seem to be determined by the geometry of the ligands imposed by the protein rather than by the charge of the protein.

B. Normal copper enzymes

These proteins are characterized by the presence of isolated mononuclear type-2 copper centers with a non-tetrahedral, virtually square-planar coordination geometry. Preferred ligands to copper are histidines. EPR and electronic spectra are similar to those commonly encountered in tetragonal copper complexes. This class of proteins comprises, among others, superoxide dismutase, the protein capable of scavenging the toxic superoxide radical, copper-containing amine oxidases, responsible for the degradation of several biogenic amines, and dopamine β-hydroxylase, a key enzyme for the biosynthesis of adrenalin. We will present here some information about these enzymes.

1. *Superoxide dismutase*

Many copper proteins are outside eukaryotic cells and in the periplasmic space of prokaryotes. Other copper proteins are located in granules and vesicles inside cells. Superoxide dismutase is a quite versatile enzyme: indeed, it is the only cytoplasmic copper protein but it can be secreted in the periplasm and it is also present outside the cell (extracellular superoxide dismutase). The EPR spectra of SOD are extraordinarily rhombic ($A_{\|} = 142 \times 10^{-4}$ cm^{-1}, $g_1 = 2.26$, $g_2 = 2.07$, $g_3 = 2.00$ (Brigg and Fee, 1978)) in agreement with a distorted tetrahedral geometry rather than with a distorted five-coordinated chromophore (Banci et al., 1990).

Copper,zinc-superoxide dismutases (Cu,Zn-SODs) which are mainly found in eukaryotes dismute the superoxide radical to molecular oxygen and hydrogen peroxide thus helping to protect cells against the toxic by-products of aerobic metabolism (Roberts et al., 1991). The catalytic reaction is summarized by the two following steps:

$$Cu^{2+} + O_2^- \rightarrow Cu^+ + O_2$$

$$Cu^+ + O_2^- + 2H^+ \rightarrow Cu^{2+} + H_2O_2$$

Each Cu,Zn-SOD molecule is a dimer of identical subunits and each subunit contains about 153 amino acids (depending on the species), one copper ion and one zinc ion. The crystallographic structure of the enzyme (Tainer et al., 1982) shows that the structural core of the subunit is a flattened Greek key β-barrel motif consisting of eight antiparallel strands joined by seven turns or loops (Fig. 3). The catalytic copper ion is bonded to the surface of the β-barrel by four histidine residues - one of which is simultaneously coordinated to the zinc ion - (Fig. 4) and occurs at the bottom of a channel formed by two loops extending from the β-barrel. In addition to forming the superoxide channel, the two loops encode specific functional subdomains involved in precollision electrostatic guidance of the substrate, zinc binding and dimer contact.

The dimer interaction is particularly strong in Cu,Zn-SODs. Recently, the substitution of residues at the interface has allowed us to separate the subunits and to obtain a

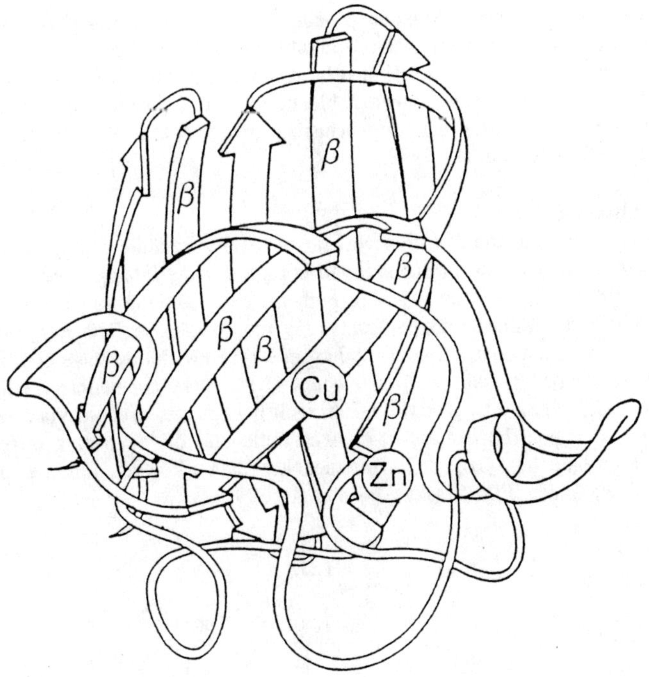

Figure 3.
Overall structure of superoxide dismutase.

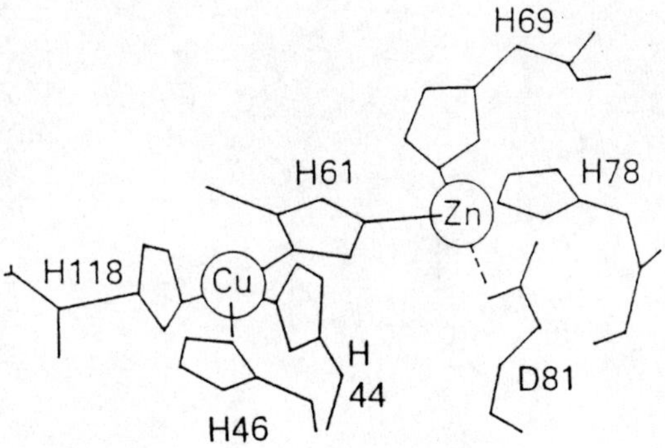

Figure 4.
The active site of superoxide dismutase.

monomeric enzyme which is partially active (Bertini et al., submitted). The Greek key β-barrel found in Cu,Zn-SOD is one of the most versatile structural motifs found in biological systems. Protein domains with this fold are used for recognition and binding, long-lived and stable structural assemblies, electron transfer and catalysis.

Superoxide dismutase has been the subject of intense investigations. 1D and 2D NMR studies on the reduced copper(I) enzyme and the cobalt(II) and nickel(II) substituted derivatives have provided a wealth of information on the solution structure of the enzyme, on its mode of interaction with pseudosubstrate and inhibitors, and on the catalytic mechanism (Bertini et al., 1985; 1992; Banci et al., 1989; 1991).

The expression of human SOD in prokaryotic cells has recently allowed to obtain, through site-directed mutagenesis, a large number of SOD analogs in which some amino acids in the active channel were substituted by different residues. This result has been very important in order to understand, through biochemical and spectroscopic studies, the role played by these residues in the catalytic mechanism (Beyer et al., 1987; Bertini et al., 1990; Getzoff et al., 1992; Banci et al., 1993).

2. *Dopamine β-hydroxylase*

Dopamine β-hydroxylase, found within vesicles of adrenal medulla and noradrenergic nerve cells, catalyzes the conversion of dopamine to norepinephrine (Stewart and Klinman, 1988; Scott et al., 1988). The enzyme exists in both membrane-bound and soluble forms. The soluble enzyme is a tetrameric copper containing monoxygenase. The fully active enzyme requires eight copper ions per tetramer (290 kg mol^{-1}). EPR studies have established that D βH is a type-2 copper protein with two copper(II) centers magnetically non interacting (McCracken et al., 1988). The catalytic mechanism is known to proceed via a redox process in which the two copper(II) centers in each catalytic unit of the resting enzyme are first reduced by ascorbate to a copper(I) intermediate; then dioxygen binding and substrate hydroxylation take place (Fig. 5) (Blackburn et al., 1991).

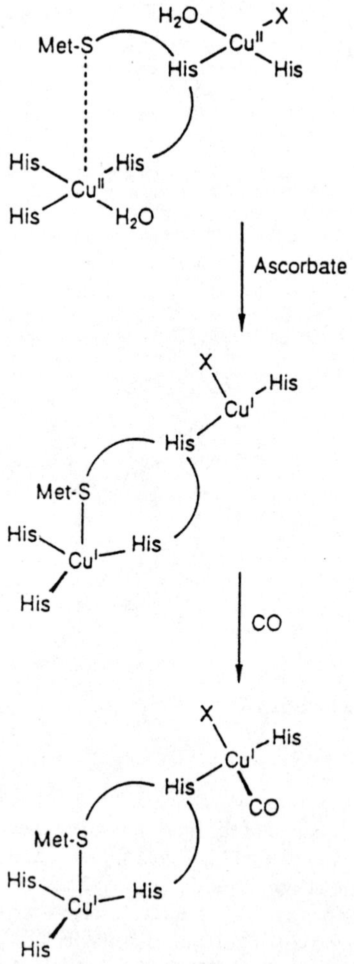

Figure 5.
Dopamine β-hydroxylase reaction.

3. *Copper amine oxidases*

Copper amine oxidases are widely distributed soluble enzymes, which catalyze the oxidative deamination of primary amino groups of several biogenic amines in the presence of molecular oxygen by accepting two electrons from the substrate and transferring them to an oxygen molecule (Dooley, 1987; Mondovi', 1985). The reaction produces the corresponding aldehyde, hydrogen peroxide and ammonia according to the following equation:

$$R\text{-}CH_2\text{-}NH_2 + O_2 + H_2O \rightarrow R\text{-}CHO + H_2O_2 + NH_3$$

The main function performed by copper amine oxidases is the control of polyamine degradation and therefore the regulation of their intracellular concentration. In addition, copper amine oxidases have an important role in the control of the breakdown of histamine and in the oxidative deamination of lysine and hydroxylysine, the primary step for the formation of desmosin and isodesmosin cross links in the connective tissue.

Copper amine oxidases have now been isolated from a variety of eukaryotic sources including yeasts, plants, and mammalian plasma and tissue, all of them showing similar structural features. These enzymes are generally composed of two subunits that appear to be similar if not identical, to a total molar mass of 170-190 kg mol^{-1}. Copper amine oxidases were the first enzymes from eukaryotic organisms that were suggested to contain the novel coenzyme pyrroloquinoline quinone (PQQ). However, recent studies have demonstrated that 6-hydroxydopa (TOPA) instead of PQQ is the true cofactor (Fig. 6) (Janes et al., 1990). All copper-containing amine oxidases isolated to date contain two copper ions per protein molecule. The copper:quinone stoichiometry should be 1:1; hence, the active site is likely to be composed of one copper ion and one quinone, close in space.

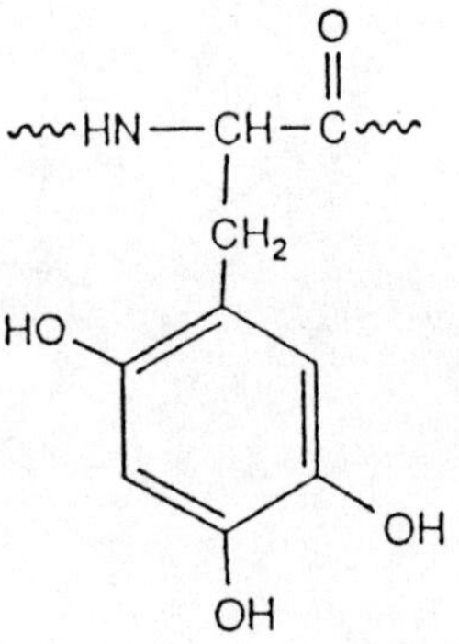

Figure 6.
Active site of copper amine oxidase.

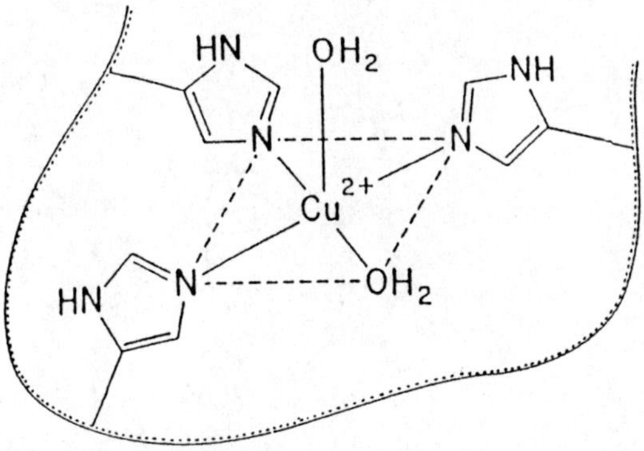

Figure 7.
The cofactor of copper amine oxidases.

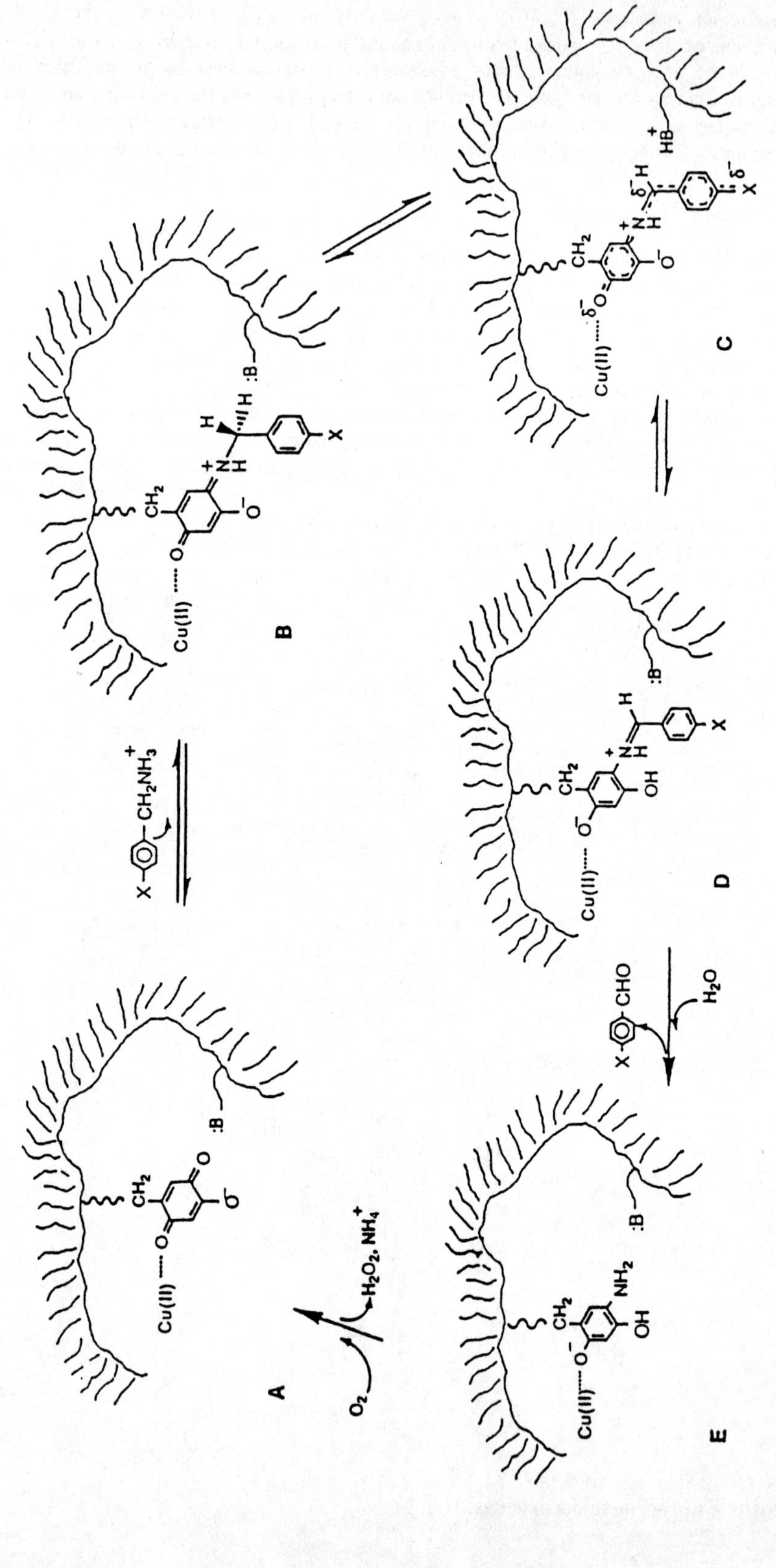

Figure 8.
Catalytic mechanism of amine oxidases.

The proposed structural model for the copper(II) site in amine oxidases is shown in Figure 7 (Dooley et al., 1991). Figure 8 reports the tentative scheme for the overall catalytic mechanism of the enzyme, as proposed by J. Klinman. This mechanism implies that there is no direct binding of the substrate to copper; it also implies that the copper and the quinone interact, as expected from their proximity in the active site (Hartmann and Klinman, 1991).

4. *Galactose oxidase*
Galactose oxidase (Whittaker and Whittaker, 1990) is another type-2 copper enzyme for which the X-ray structure is known (Ito et al., 1991). The enzyme, a monomeric 68 kg mol^{-1} protein that contains a single copper center, catalyzes the two-electron oxidation of primary alcohols to aldehydes. The primary sequence has been determined from the cloned gene and the X-ray structure determined at 1.7 Å resolution (Ito et al., 1991). There are three domains: an amino terminal domain of 150 residues with β-sheet topology; a middle domain; and the carboxy terminal domain. The peculiar structural organization of the second domain is shown in Figure 9, whereas Figure 10 reports the detail of the active site.

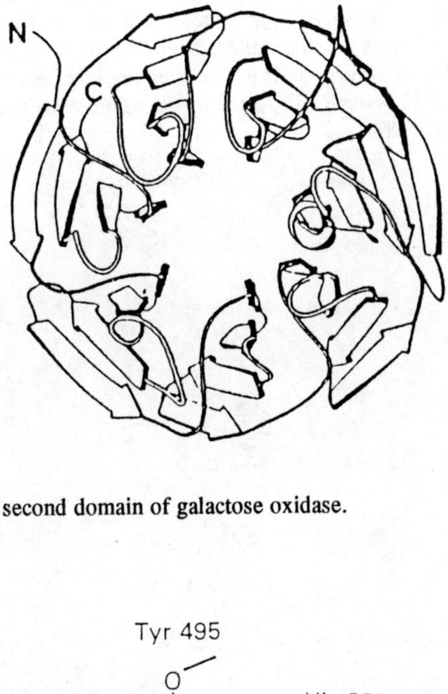

Figure 9.
X-ray structure of the second domain of galactose oxidase.

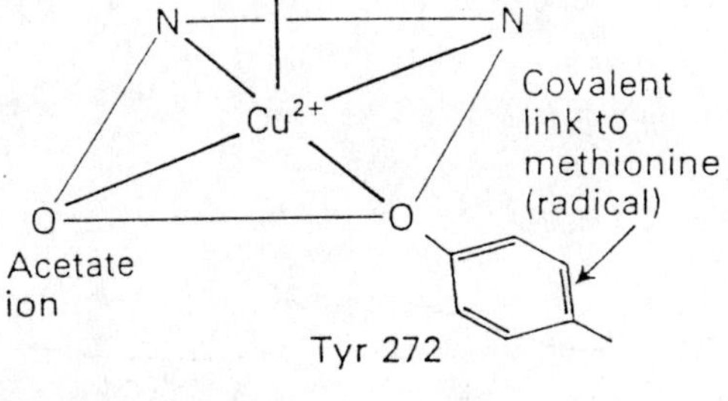

Figure 10.
Detail of the active site of galactose oxidase.

Inspection of the structure shows that the cupric ion is surrounded by several aromatic side chains. Copper has five ligands in square-pyramidal coordination: Tyr 272, His 496, His 581 and an acetate ion form an almost perfect square with Tyr 495 being the fifth axial ligand. In the absence of acetate, this ligand is presumably replaced by a water molecule. Ito has proposed that in galactose oxidase a free radical associated with Tyr 272 plays the role of cofactor. Given the presence of an unusual bond between Cys 228 and Tyr 272, it has been suggested that sulfur has a role in assisting the delocalization of the free radical.

C. Normal + blue copper proteins

1. *Nitrite reductase*

Nitrite reductase is a copper protein containing two copper centers, one type 1 plus one type 2. Nitrite reductase catalyzes the reduction of nitrite to nitric oxide as part of the reductive pathway in which denitrifying bacteria convert nitrate to nitrogen (Shearer and Kohl, 1988; Ferguson, 1987):

$$NO_3^- \rightarrow NO_2^- \rightarrow NO \rightarrow N_2O \rightarrow N_2$$

Elinor Adman has reported the 2.3 Å X-ray crystal structure of nitrite reductase from *Achromobacter cycloclastes* (Godden et al., 1991). The structure clearly shows two copper atoms per monomer of 36 kg mol^{-1} or six per trimer. One copper center is type 1 copper with one cysteine, two histidines and one methionine whereas the other is a type 2 copper with three histidines and one water. Intriguingly two of the histidines come from one monomer of the trimer whereas the third comes from another so that this copper is bound at the interface of two subunits (Fig. 11). The type 2 copper is accessible to solvent via a deep channel. The copper-bound water is displaced by nitrite in large excess suggesting that the type 2 copper is the active site.

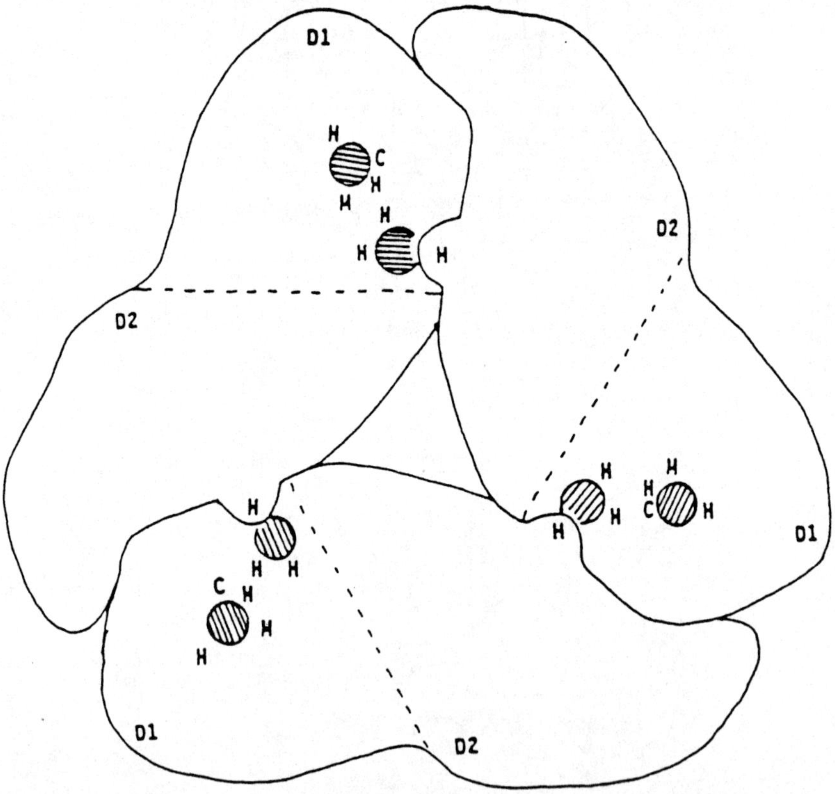

Figure 11.
X-ray structure of nitrite reductase (after Adman).

D. Binuclear copper proteins

The binuclear copper site is found in both hemocyanins (Gaykema et al., 1986; Volbeda and Hol, 1989) and tyrosinases (Himmelwright et al., 1980). Hemocyanins are large allosteric multisubunit proteins constituting one of the three classes of oxygen carriers in animals. They occur in the hemolymph of many arthropods and molluscs in a freely dissolved form and make up the major part of hemolymph proteins. Deoxyhemocyanin contains a binuclear cuprous site which binds dioxygen as peroxide, producing oxyhemocyanins. There is a strong antiferromagnetic coupling between the two cupric centers of oxyhemocyanin so that no EPR spectrum is observed. Results from crystallography, available at 3.2 Å resolution for the deoxy site, indicate that each copper has three histidines but no bridging ligand (Fig. 12).

The overall molecular structure of molluscan and arthropod hemocyanins differ greatly. Molluscan hemocyanins are cylindrically shaped giant molecules consisting of 10 or 20 huge polypeptide chains (400 kg mol^{-1}) each of which comprises seven or eight oxygen binding functional units (50 kg mol^{-1}). Arthropod hemocyanins are composed of hexamers showing a species specific level of aggregation. Hexamers are built up of individual subunits with molar mass about 75 kg mol^{-1}. Each subunit is folded into three domains providing one binuclear copper site for binding one oxygen molecule.

However, in spite of the above large differences in the overall organization, their copper centers appear to be spectroscopically very similar.

Tyrosinase (Himmelwright et al., 1980), having a binuclear site like that of hemocyanin, demonstrates how the same structural motif has evolved from a dioxygen binding function to the enzymatic function of hydroxylation of tyrosine to 3,4 dihydroxyphenylalanine.

E. Multicopper oxidases

The multicopper oxidases (laccase - one type 1, one type 2, one type 3; ascorbate oxidase - two type 1, two type 2, two type 3, and ceruloplasmin - two type 1, one type 2, one type 3) couple four one-electron oxidations of substrate at the type 1 site to the four-electron reduction of dioxygen to water at the type-2/type-3 trinuclear copper cluster site (Reinhammar and Malmström, 1981; Messerschmidt and Huber, 1990).

1. *Ceruloplasmin*

Ceruloplasmin is the only blue copper oxidase of animal species; it is present in the globulin fraction of α_2 vertebrate plasma where it binds the main fraction of plasma copper. The physiological role of the protein in vivo is still unclear. The amino acid sequence of the human protein displays three homologous units, each 350 amino acid long, suggesting a multidomain structure. A model for the overall structure has been recently proposed (Fig. 13) (Fenderson et al., 1991).

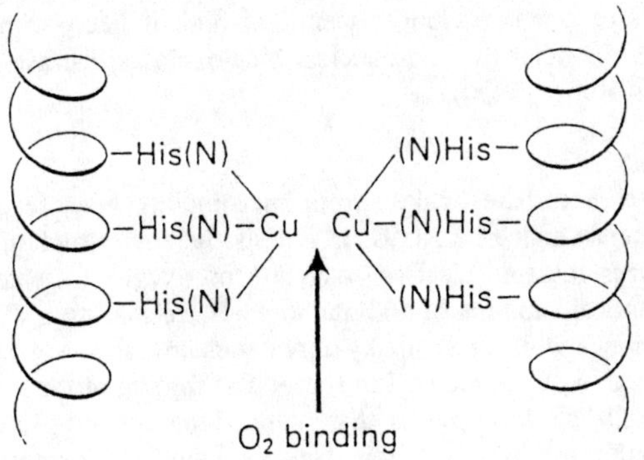

Figure 12.
Scheme of the active site of hemocyanin.

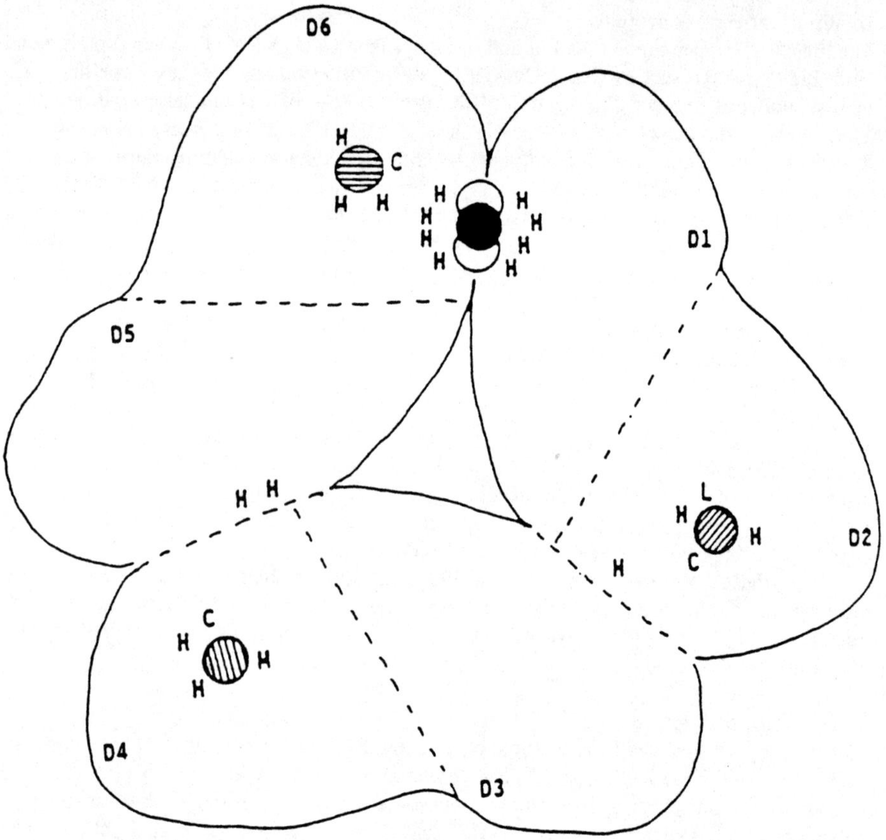

Figure 13.
Proposed arrangement of six domains of ceruloplasmin.

The protein architecture should be such as to accomodate at least five different copper binding sites. Two of them are mononuclear, EPR detectable, blue copper sites which are functionally apt to accept one electron from the substrate. The other three copper atoms were formerly described as an EPR-detectable type-2 mononuclear site plus a pair of EPR silent antiferromagnetically coupled type-3 copper atoms. It has been recently proposed that the latter three copper ions form a magnetically interacting trinuclear cluster where the EPR signal of the type 2 copper is no longer seen. This unit is likely to be the oxygen binding site in analogy with the other two multinuclear blue oxidases, laccase and ascorbate oxidase (Messerschmidt and Huber, 1990).

2. *Ascorbate oxidase*
The 3D structure of ascorbate oxidase from zucchini has been recently solved at 1.9 Å resolution by Messerschmidt et al. (1992). This is the first structure ever reported for an oxidase that catalyzes the four-electron reduction of oxygen to water. Ascorbate oxidase consists of two identical subunits of 552 amino acids, related by a diad axis. Each subunit has a globular shape and is built up by three domains arranged sequentially along the polypeptide chain and highly associated in space. The folding of the three domains is of a β-barrel type and is distantly related to that of the blue copper proteins plastocyanin and azurins. Each subunit has four catalytic coppers bound as mononuclear and trinuclear species. The mononuclear copper is a type 1 copper with two histidines, one cysteine and one methionine as ligands. It is located in the third domain. The trinuclear cluster has eight

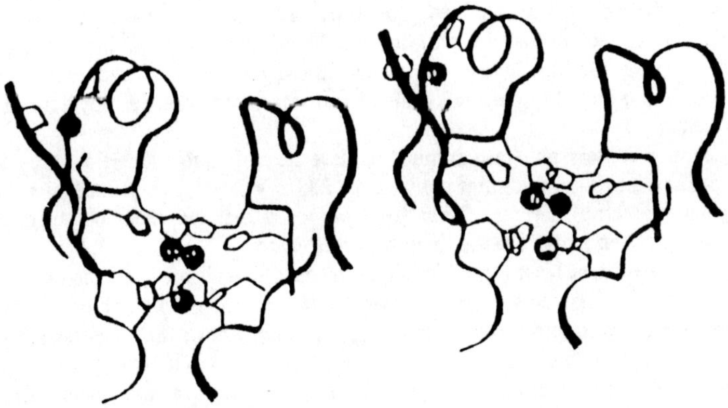

Figure 14.
Overall structure of ascorbate oxidase.

histidine ligands and may be subdivided into a pair of copper atoms with six histidine ligands arranged as a trigonal prisma and a single type 2 copper with two histidine residues. The copper atoms of the type 3 pair are bridged by a hydroxo group; a water or OH moiety binds the type 2 copper. The protein ligands are supplied symmetrically by domains 1 and 3 (Fig. 14).

It is argued that the type 1 copper is the binding site of the substrate and the trinuclear copper center is the binding site for dioxygen. So, type 1 copper would represent the entry for electrons, that are then transferred to the trinuclear cluster, which is able to store three electrons simultaneously. In the model for the mechanism proposed by Messerschmidt, dioxygen binds to the type-3 copper pair as 1,2 μ-peroxo species in the fully reduced enzyme after release of the hydroxo moiety.

3. *Laccase*
Laccase is a multicopper oxidase which is widely distributed in higher plants and fungi (Reinhammar, 1984). The enzyme contains one type-1, one type-2 and one type-3 copper centers. It catalyzes the oxidation of aromatic diamines and diphenols yielding water as the final product of dioxygen reduction. The most extensive studies concerning laccase are of the enzyme from the latex of *Rhus vernicifera* (lacquer tree) which was first discovered as early as 1883 by Yoshida. A host of spectroscopic studies allowed characterization of the main features of the copper sites (Lu et al., 1992).

F. Cu_A-containing copper enzymes
1. *Cytochrome c oxidase*
It has been estimated that nearly 90% of the oxygen consumed by aerobic organisms is reduced to water in the final step of respiration, catalyzed by cytochrome oxidase. Cytochrome *c* oxidases of the aa_3 type (having two a-type cytochromes) are found in a wide variety of aerobic organisms including bacteria, fungi, plants and animals (Chan, Li, 1990).

All aa_3 oxidases contain four redox active metal centers (two iron hemes and two copper ions) and catalyze the four-electron reduction of molecular oxygen to water with reducing equivalents derived from cytochrome *c*:

$$4\ \text{Cyt}\ c^{2+} + O_2 + 4H^+ \rightarrow 4\ \text{Cyt}\ c^{3+} + 2\ H_2O$$

Electrons enter the protein from the cytosol side of the mytochondrial inner membrane whereas the protons are extracted from the matrix; this generates a physiologically relevant proton electrochemical gradient across the inner mytochondrial membrane.

The structure of cytochrome *c* oxidase has been the subject of intense investigations in the last decades (Chan and Li, 1990). There is still uncertainty regarding the polypeptide composition and the minimal molar mass which has been reported to range between 70 and

120 kg mol^{-1}. It has been ascertained, however, that cytochrome *c* oxidase contains four redox active metal centers that are essential for the catalytic activity. One pair, cytochrome a_3 and Cu_B, forms the binuclear cluster where dioxygen is bound and reduced; the other pair, cytochrome *a* and Cu_A, mediates the flow of electrons from ferrocytochrome *c* to the binuclear center.

The copper centers of cytochrome *c* oxidase bear unique properties. Cu_A is the low potential copper, having a midpoint potential of 285 mV. It is a very unusual copper center with enigmatic spectral features; it has been proposed that two cysteine sulfur atoms and two histidine nitrogen atoms are the ligands to this center.

Cu_B is ligated by at least three histidine nitrogens in a fashion similar to that of type-3 copper centers. It lies close to the cytochrome a_3 heme to which it is strongly antiferromagnetically coupled; the bridging ligand is not known but it is possibly a μ-oxo, μ-hydroxyl or μ-chloro species.

A pictorial representation of cytochrome *c* oxidase in the inner mitochondrial membrane, showing the approximate relative positions of the redox active metal centers, is given in Figure 15.

2. *Nitrous oxide reductase*

Nitrous oxide reductase (N2OR) is involved in the conversion of N_2O to N_2, a key reaction within the microbial nitrogen cycle. The N-O bond is cleaved upon transfer of $2e^-/2H^+$ forming one water molecule in analogy to the reduction of O_2 to H_2O, catalyzed by cytochrome *c* oxidase. Two kinds of nitrous oxide reductases both copper dependent have been described. The one contains four copper atoms per subunit of 70 kg mol^{-1}, is acidic and is found in denitrifying and photodenitrifying bacteria (Zumft et al., 1990; Kroneck et al., 1990). According to several spectroscopic results, this enzyme contains a Cu_A type center, as in the case of cytochrome *c* oxidase. The second enzyme has been reported to contain three copper atoms and one heme *c* per subunit of 88 kg mol^{-1} and is basic

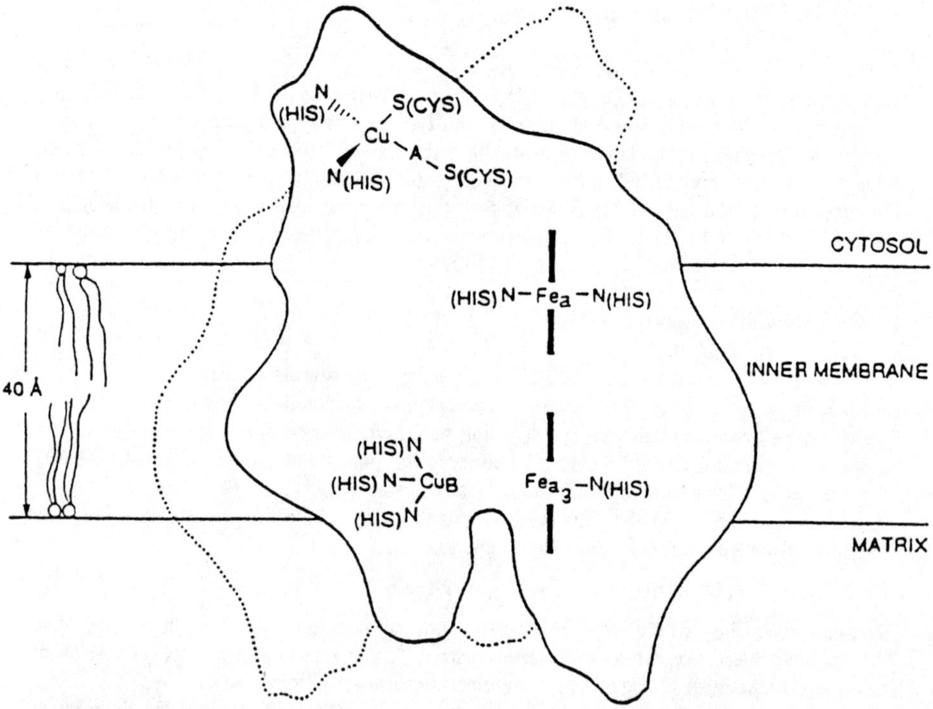

Figure 15.
Model of cytochrome oxidase.

(McEwan et al., 1985). The Cu/Fe enzyme is not sensitive to dioxygen; yet, no X-ray crystal results are available.

III. COPPER METABOLISM AND PROTEINS INVOLVED

Now, we outline the processes through which copper is absorbed, distributed and excreted by living organisms; proteins responsible for copper homeostasis and its regulatory mechanisms will be briefly described.

High levels of copper in living systems do not occur frequently since copper minerals are rather insoluble. Deficiency is more frequent than excess since copper may be removed from dietary sources by precipitation or by complexation. Deficiency leads to connective tissue weakness, probably ascribed to low levels of lysyl oxidase. Copper excess is associated with rare genetic disorders of copper metabolism like Menke's and Wilson diseases (Danks, 1977; Ettinger, 1987; Owen, 1981).

Once copper is ingested, its further uptake is controlled by serum albumin, circulating in the blood, that delivers it to the liver. The copper-binding site of serum albumin utilizes four nitrogens, two of which are peptide nitrogens (see Chapter 1-D2 of this Part Two). If excess copper is present, then there are various powerful chelating peptides and proteins for its removal. It is known that serum transferrin is able to bind copper at its binding sites but it is not clear whether it plays any significant role in copper metabolism (Chasteen, 1983).

Copper is maintained at very low levels in the cytoplasm of cells. The excess of copper is handled by special ligands, the metallothioneins (Kägi and Schäffer, 1988; see also Chapter 1-D5 of this Part Two), which bind seven copper(I) ions in characteristic thiolate clusters. The structure of copper metallothionen is not known in detail but apparently it differs from that of cadmium and zinc derivatives. This would allow bound copper to be recognized as such and exocytosed through vesicles while zinc metallothionein would remain in the cytoplasm. Models of the copper cluster in metallothionein have been proposed on the basis of spectroscopic results, with a tentative Cu_8S_{12} stoichiometry (Fig. 16) (George et al., 1988). More recent studies, however, seem to suggest the occurrence of copper(I) thiolate clusters with Cu_6S_9 stoichiometry in rat metallothionein and Cu_7S_{10} in yeast metallothionein (Winge et al., 1992). In any case, trigonal coordination is typical of each cage cluster with mean Cu-S distances of 2.26 Å. Cu-Cu distances are about 2.7 Å .

Metallothionein synthesis is regulated at a genetic level so that protein concentration is always in excess with respect to zinc, copper and cadmium concentrations. The binding constant of metallothionein for copper is very high; free copper concentration never exceeds 10^{-15} M in the cytoplasm.

In recent years, new insight has been achieved into the mechanisms of regulation of copper in primitive organisms like yeasts. The protein which stimulates the synthesis of metallothionein in yeast is a sulfhydryl-containing protein called ACE1 (Thiele, 1988; Hu et al., 1990; Furst and Hamer, 1988; Culotta et al., 1989). This protein binds several copper(I) equivalents in clusters. In *Saccaromyces cerevisiae* the ACE1 protein (also known

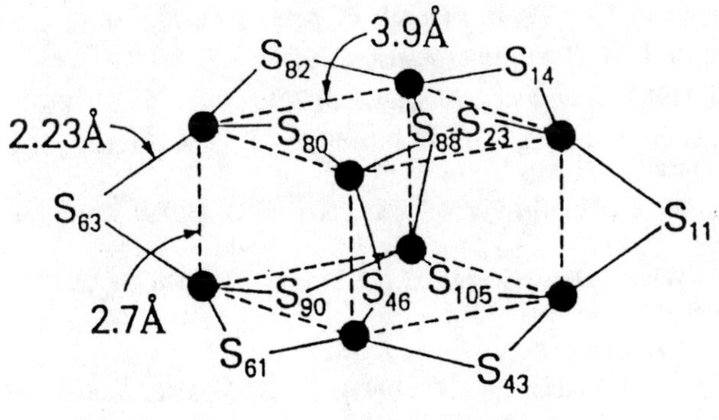

Figure 16.
A preliminary model for the copper cluster in metallothionein.

as CUP2) is involved in the control of the transcription of CUP1, the yeast metallothionein gene, in the presence of copper and silver ions. Among its 225 amino acid residues, ACE1 contains 12 cysteines and an excess of basic residues located in the amino terminal region, the portion of the molecule involved in the metal-dependent DNA binding. The active form of ACE1 has been proposed to contain a copper(I) cysteinyl thiolate polynuclear cluster similar to that of copper metallothionein, with a Cu_6S_{11} stoichiometry; formation of the copper cluster should induce the conformational alteration which is required for DNA binding.

A recent study by Dameron has shown that copper(I) ions in the copper cluster of ACE1, the so-called "Cu fist", are triply coordinated to sulfur in a distorted trigonal coordination environment with average Cu-S distances of 2.26 Å and Cu-Cu distances of 2.68 Å, as found in copper metallothionein (Dameron et al., 1991).

REFERENCES

Adman, E.T. (1985). *Metalloproteins* (P. Harrison, ed.), Verlag Chemie, Weinheim, p. 1.

Adman, E.T., and Jensen, L. (1981). *Isr. J. Chem.*, *21*: 8.

Adman, E.T., Turley, S., Bramson, R., Petratos, K., Banner, D., Tsernoglou, D., Beppu, T., and Watanabe, H. (1989). *J. Biol. Chem.*, *264*: 87.

Baker, E.N. (1988). *J. Mol. Biol.*, *203*: 1071.

Banci, L., Bencini, A., Bertini, I., Luchinat, C., and Piccioli, M. (1991). *Inorg. Chem.*, *29*: 4867.

Banci, L., Bencini, A., Bertini, I., Luchinat, C., and Viezzoli, M.S. (1990). *Gazz. Chim. Ital.*, *120*: 179.

Banci, L., Bertini, I., Bauer, D., Hallewell, R.A., and Viezzoli, M.S. (1993). *Biochemistry*, *32*: 4384.

Banci, L., Bertini, I., Luchinat, C., Piccioli, M., Scozzafava, A., and Turano, P. (1989). *Inorg. Chem.*, *28*: 4650.

Bertini, I., Canti, G., Grassi, R., and Scozzafava, A. (1980). *Inorg. Chem.*, *19*: 2198.

Bertini, I., Lanini, G., Luchinat, C., Messori, L., Monnanni, R., and Scozzafava, A. (1985). *J. Amer. Chem. Soc.*, *107*: 4391.

Bertini, I., Banci, L., Cabelli, D.E., Hallewell, R.A., Luchinat, C., and Viezzoli, M.S. (1990). *Inorg. Chem.*, *29*: 2398.

Bertini, I., Luchinat, C., Ming, L.-J., Piccioli, M., Sola, M., and Valentine, J.S. (1992). *Inorg. Chem.*, *31*: 4433.

Bertini, I., Piccioli, M., Viezzoli, M.S., Chiu, C.-Y., and Mullenbach, G.T., submitted for publication.

Beyer, W.F., Fridovich, I., Mullenbach, G.T., and Hallewell, R. (1987). *J. Biol. Chem.*, 23: 11182.

Blackburn, N.J., Hasnain, S.S., Pettingill, T.M., and Strange, R.W. (1991). *J. Biol. Chem.*, *266*: 23120.

Brigg, R.G., and Fee, J.A. (1978). *Biochim. Biophys. Acta*, *537*: 86.

Chan, S.I., and Li, P.M. (1990). *Biochemistry*, 29: 1.

Chasteen, N.D. (1983). *Adv. Inorg. Biochem.*, *5*: 201.

Collyer, C.A., Guss, J.M., Sugimura, Y., Yoshizaki, F., and Freeman, H.C. (1990). *J. Mol. Biol.*, *211*: 617.

Culotta, V.C., Hsu, T., Hu, S., Furst, P., and Hamer, D. (1989). *Proc. Nat. Acad. Sci. USA*, *86*: 3877.

Dameron, C.T., Winge, D.R., George, G.N., Sansone, M., Hu, S., and Hamer, D. (1991). *Proc. Nat. Acad. Sci. USA*, *88*: 6127.

Danks, D.M. (1977). *Inorg. Persp. Biol.*, *1*: 73.

da Silva, F.J.J.R., and Williams, R.J.P. (1991). *The Biological Chemistry of the Elements*, Clarendon Press, Oxford.

Dooley, D.M. (1987). *Life Chem. Rep.*, *5*: 91.

Dooley, M., McGuirl, M.A., Cote, C.E., Knowles, F., Singh, I., Spiller, M., Brown R.D., III, and Koenig, S.H. (1991). *J. Amer. Chem. Soc.*, *113*: 754.

Driscoll, P.C., Hill, H.A.O., and Redfield, C. (1987). *Eur. J. Biochem.*, *170*: 279.

Ettinger, M.J. (1987). *Life Chem. Rep.*, *5*: 169.

Farver, O., and Pecht, I. (1989). *Proc. Nat. Acad. Sci. USA*, *86*: 6968.

Fenderson, F., Kumar, S., Adman, E.T., Liu, M.-Y., Payne, W.J., and LeGall, J. (1991). *Biochemistry*, *30*: 7180.

Ferguson, S.J. (1987). *Trends Biochem. Sci.*, *12*: 354.

Furst, P., and Hamer, D. (1988). *Proc. Nat. Acad. Sci. USA*, *86*: 5267.

Gaykema, W.P.J., Volbeda, A., and Hol, W.G.J. (1986). *J. Mol. Biol.*, *187*: 255.

George, G.N., Byrd, J., and Winge, D.R. (1988). *J. Biol. Chem.*, *263*: 8199.

Getzoff, E.D., Cabelli, D.E., Fisher, C.L., Parge, H.E., Viezzoli, M.S., and Hallewell, R.A. (1992). *Nature*, *358*: 347.

Godden, J.W., Turley, S., Teller, D.C., Adman, E.T., Liu, M.-Y., Payne, W.J., and Le Gall, J. (1991). *Science*, *253*: 438.

Gray, H.B., and Solomon, E.I. (1981). *Copper Proteins* (T.G. Spiro, ed.), Wiley Interscience, New York, p. 1.

Guus, J.M., and Freeman, H. (1983). *J. Mol. Biol.*, *169*: 521.

Guus, J.M., Merritt, E.A., Phizackerley, R.P., Hedman, B., Murata, M., Hodgson, K.O., and Freeman, H.C. (1988). *Science*, *241*: 806.

Hartmann, C., and Klinman, J.P. (1991). *Biochemistry*, *30*: 4605.

Himmelwright, R.S., Eickman, N., Lu Bien, C.D., Lerch, K., and Solomon, E.I. (1980). *J. Amer. Chem. Soc.*, *102*: 7339.

Hu, S., Furst, P., and Hamer, D. (1990). *New Biol.*, *2*: 544.

Ito, N., Phillips, S.E.V., Stevens, C., Ogel, Z.B., McPherson, M.J., Keen, J.N., Yadev, D.S., and Knowles, P.F. (1991). *Nature*, *350*: 87.

Janes, S.M., Mu, D., Wemmer, D., Smith, A.J., Kain, S., Maltby, D., Burlingame, A.L., and Klinman, J.P. (1990). *Science*, *248*: 981.

Kägi, J.H.R., and Schäffer, A. (1988). *Biochemistry*, *27*: 8509.

Kalverda, A.P., Lommen, A., Wymenga, S., Hilbers, C.W., and Canters, G.W. (1991). *J. Inorg. Biochem.*, *43*: 171.

Kroneck, P.M.H., Riester, J., Zumft, W.G., and Antholine, W.E. (1990). *Biol. Metals*, *3*: 103.

Lommen, A., Wijmenga, S., Hilbers, C.W., and Canters, G.W. (1991). *Eur. J. Biochem.*, *201*: 695.

Lu, J., Bender, C.J., McCracken, J., Peisach, J., Severns, and J.C., McMillin, D.R. (1992). *Biochemistry*, *31*: 6265, and references therein.

McCracken, J., Desai, P.R., Papadopoulos, N.J., Villafranca, J.J., and Peisach, J. (1988). *Biochemistry*, *27*: 4133.

McEwan, A.G., Greenfield, A.J., Wetzstein, H.G., Jackson, J.B., and Ferguson, S.J. (1985). *J. Bacteriol.*, *164*: 823.

Messerschmidt, A., and Huber, R. (1990). *Eur. J. Biochem.*, *187*: 341.

Messerschmidt, A., Ladenstein, R., Huber, R., Bolognesi, M., Avigliano, L., Petruzzelli, R., Rossi, A., and Finazzi Agro', A. (1992). *J. Mol. Biol.*, *224*: 179.

Mondovi', B. (1985). *Structure and Functions of Amine Oxidases*, CRC Press, Boca Raton, FL.

Nar, H., Messerschmidt, A., Huber, R., van de Kamp, M., and Canters, G.W. (1991). *J. Mol. Biol.*, *218*: 427.

Owen, C.A., Jr. (1981). *Wilson's Disease*, Noyes, Park Ridge, NJ.

Reinhammar, B. (1984). *Copper Proteins and Enzymes*, Vol. III (R. Lontie, ed.), CRC Press, Boca Raton, FL, p. 1.

Reinhammar, B., and Malmström, B.G. (1981). *Copper Proteins* (T.G. Spiro, ed.), Wiley, New York, p. 109.

Roberts, V.A., Fisher, C.L., Redford, S.M., McRee, D.E., Parge, H.E., Getzoff, E.D., and Tainer, J.A. (1991). *Free Rad. Res. Commun.*, *12*: 269.

Scott, R.A., Sullivan, R.J., De Wolfe, W.E., Dolle, R.E., and Kruse, L.I. (1988). *Biochemistry*, *27*: 5411.

Shearer, G., and Kohl, D.H. (1988). *J. Biol. Chem.*, *263*: 13231.

Sigel, H. ed. (1981). *Metal Ions in Biological Systems*, Vols. 12 and 13, Marcel Dekker, New York.

Solomon, E.I., Baldwin, M.J., and Lowery, M.D. (1992). *Chem. Rev.*, *92*: 521.

Spiro, T.G., ed. (1981). *Copper Proteins*, Wiley Interscience, New York.

Stewart, L.C., and Klinman, J.P. (1988). *Annu. Rev. Biochem.*, *57*: 551.

Sykes, A.G. (1991). *Struct. Bonding*, *75*: 175.

Tainer, J.A., Getzoff, E.D., Beem, K.M., Richardson, J.S., and Richardson, D.C. (1982). *J. Mol. Biol.*, *160*: 181.

Thiele, D.J. (1988). *Mol. Cell Biol.*, *8*: 2745.

Tong, H. (1991). Ph. D. Dissertation, University of Sydney, Sydney, Australia.

van de Kamp, M., Floris, R., Hali, F.C., and Canters, G.W. (1990a). *J. Amer. Chem. Soc.*, *112*: 907.

van de Kamp, M., Silvestrini, M.C., Brunori, M., van Beeumen, J., Hali, F.C., and Canters, G.W. (1990b). *Eur. J. Biochem.*, *194*: 109.

van de Kamp, M., Canters, G.W., Wijmenga, S.S., Lommen, A., Hilbers, C.W., Nar, H., Messerschmidt, A., and Huber, R. (1992). *Biochemistry*, *31*: 10194.

Volbeda, A., and Hol, W.G.J. (1989). *J. Mol. Biol.*, *209*: 249.

Winge, D.R., Dameron, C., Kurz, B., George, G., Pickering, I., and Dance, I. (1992). *J. Inorg. Biochem.*, *47*: L35.

Whittaker, M.M., and Whittaker, J.W. (1990). *J. Biol. Chem.*, *265*: 9610.

Zumft, W.G., Viebrock-Sambale, A., and Braun, C. (1990). *Eur. J. Biochem.*, *192*: 591.

ZINC PROTEINS

by

Ivano Bertini, Fabrizio Briganti and Andrea Scozzafava

Department of Chemistry,
University of Florence,
Via Gino Capponi, 7
50121 Florence
Italy

I. INTRODUCTION

Although it has been known for more than fifty years that a 70 kg human body contains about 3 g of zinc, this high content was, for a long time, considered as a passive accumulation of an environmental contaminant and not to be of biological relevance (Mc Cance and Widdowson, 1942). In fact, in contrast to copper- and iron-containing proteins, which have been extensively studied in biochemistry and physiology (primarily because they are strongly colored and readily accessible in large quantities), zinc proteins are not easily detectable and not usually present in large amounts.

Actually the role of zinc in biological systems has been studied since 1869 when Raulin discovered that zinc is necessary for the growth of *Aspergillus niger*, a filamentous fungus (Raulin, 1869), but about one hundred years were to pass until zinc was demonstrated to be essential for human growth.

The biochemical function of zinc was discovered around 1939 when Keilin and Mann, isolating the first zinc enzyme, *carbonic anhydrase*, realized that zinc is an essential part of enzymes in terms of catalytic function and structural stability (Keilin and Mann, 1939).

II. ZINC IN BIOLOGICAL SYSTEMS

Zinc is the most diffuse metal ion with catalytic functions in the cytoplasm; Zn(II) ions are known to be essential for nearly 300 enzymes involved in every aspect of metabolism of different species and all phyla (Sigel, 1983; Spiro, 1983; Bertini et al., 1986; Mills, 1988). Zinc is the only metal present in each of the six classes of enzymes and it has been shown to be involved in replication, transcription, and translation of the genetic material of all forms of life (Vallee and Auld, 1991; Coleman, 1992). It also assumes important roles in immunology and virology, and is involved in transport processes and the stabilization of nucleic acids. Also, many digestive enzymes, active outside the cells or in vesicles, contain zinc ions with catalytic and structural functions. Furthermore, zinc has essential roles in filament stabilization (e.g. keratins), in the maintenance of the integrity of subcellular organelles, and in the organization of chromosomes and gametes. The importance of zinc in all these processes recently led da Silva and Williams to propose the hypothesis that this metal ion could be considered as one of the dominant cytoplasmic hormones (da Silva and Williams, 1991) finely regulating the metabolism of all the biological systems.

In summary, the continuously growing knowledge in the fields of biology and bioinorganic chemistry has made it clear that zinc is deeply involved in many biological processes (Williams, 1984; Vallee, 1986; Vallee and Auld, 1990). This raises the question: Why zinc? Why do biological systems chose zinc rather than other metal ions such as copper which is a stronger Lewis acid or iron which is surely more common and readily available? We will try to answer this question in the next paragraphs.

III. ZINC-PROTEIN INTERACTIONS

Proteins can select a metal by distinguishing its ***ionic charge and size, stereochemical requirements*** and, generally, its ***chemical properties***. In the case of zinc(II), its small radius (0.074 nm) produces a high charge density which then allows strong electrostatic interactions with anions although such interactions are reduced in aqueous media because of the competition with hydration water.

Zinc(II) is a metal with ***intermediate hard-soft*** acidic properties (Pearson, 1963), therefore it easily coordinates nitrogen- as well as oxygen- and sulphur-containing ligands. ***Electrostatic effects***, fundamental for protein-metal interactions, also play a role in zinc selection and contribute to the formation of metal-ligand complexes. Zinc(II) can have ***charge-charge*** interactions with anionic groups (thiolates and carboxylates) as well as ***charge-dipole*** interactions with carboxyl oxygens, neutral imidazoles, and water molecules. It has to be pointed out that both interactions are sensitive to separation but charge-dipole interactions although weaker than charge-charge interactions are also sensitive to orientation (Burley and Petsko, 1988). Obviously, these descriptions of the metal-donor bonds are part of the whole theory on coordination chemistry.

It has also been demonstrated that ***molecular orbitals*** interactions favor Zn(II) in the common process of changing the octahedral coordination geometry of the aquo-ion to the tetrahedral binding sites present in many proteins because d^{10} metal ions like zinc(II) have no ligand field stabilization effects (Berg and Merkle, 1989).

Furthermore, ***entropy*** effects can give significant contributions to the thermodynamics of the recognition mechanism during the desolvation of metal ion and protein active site (Berg, 1989).

Even amino acid residues far from the metal binding site can contribute to metal selection. Such ***long range*** interactions are usually classified as ***bulk electrostatic effects*** due to simple through space interactions with the entire protein (Russell and Fersht, 1987; Pantoliano et al., 1988), and ***hydrogen bond networks*** involving ligands to the metal ion. In particular, such networks are supposed to have important roles in the orientation of the metal ligands and in the enhancement of electrostatic effects between the metal ion and the ligands which facilitate zinc binding and modulate the pK_a of the metal-bound water (Argos et al., 1978; Nakagawa et al., 1981). The selection of zinc by protein ligands, in analogy to other metals, is therefore a product of thermodynamic binding and kinetic selection of a pathway to the binding site. For the reasons reported above, zinc is favored with respect to calcium or magnesium (harder metals) if two or more nitrogen or sulphur bases form the metal-binding site. Furthermore, the Irving-Williams order for cations in the divalent state:

$$\mathrm{Cu > Zn \geq Ni > Co > Fe > Mn > Mg > Ca}$$

shows the large affinity of Zn(II) for such ligands with the exception of Cu(II). The selection against copper occurs because cells facilitate the access of Zn(II) and reject Cu(II). Zinc and copper ions enter cells by controlled processes, and inside the cells such metal ions are principally bound to metallothioneins. Due to their different binding strengths zinc is released easily but other kinetic and thermodynamic mechanisms, which are not completely clarified, are present to resolve the competition. All the concentrations of the metal ions are finely regulated in the various compartments of the cell by different mechanisms in such a way as to provide selective uptake of zinc(II) by zinc(II)-requiring proteins (Vallee, 1986).

Zinc enzymes are preferentially formed in the cytoplasm, but zinc(II) is also present to form zinc-proteins in particular vesicles such as in the pancreas which contains a variety of digestive zinc(II)-enzymes (amylases, lipases and peptidases) and in the islets of Langerhans (insulin). To enter these vesicles the proteins have to be unfolded and without zinc. Therefore zinc has to be pumped into these vesicles (Zn(II) > 10^{-8} M) once Cu(II) has been removed by metallothioneins in the cytoplasm. The pumps which play this role are currently unknown.

IV. ZINC COORDINATION CHEMISTRY

For the correct understanding of the role of zinc in proteins we need to analyze in detail the chemistry of this metal and its effects on the macromolecules which interact with it in biological fluids. The identification of the zinc ligands and of their surroundings has implications for the elucidation of the mechanisms of related enzymes and for the design of model systems.

Zinc(II) has special chemical features which can be briefly summarized as follows:
(i) Due to the d^{10} electronic configuration its redox chemistry is completely absent. Zinc(II) is stable and inert to oxidoreduction in biological media where the potential is in continuous flux. Many processes like RNA and DNA polymerization and the easy hydrolysis of oxidizable amino acids are catalyzed by zinc(II)-containing enzymes where zinc(II) is used only as a Lewis acid. Other metals with redox chemistry are not used in such systems because secondary redox reactions could create damaging products, like free radicals, or lead to undesired preferential pathways.
(ii) Another property of zinc(II) is its amphoterism. At neutral pH, for example, such a metal ion can exist in both ***hydrate and hydroxide forms***. Ab initio calculations on models of zinc enzymes have shown that the Zn-OH moiety has a nucleophilicity comparable to that of the free hydroxide group and much larger than that of Zn-OH_2 complexes, independently of the nature and number of the other donor atoms (Bertini et al., 1990). Such studies also show that the energy necessary to deprotonate the water molecule bound to zinc is strongly dependent on the charge of the donor groups and parallels the pattern of the pK_a values. Therefore, Zn-OH_2 can function as a conjugate acid and produce the strong nucleophile Zn-OH^-. This mechanism is clearly elucidated in hydrolytic enzymes like carboxypeptidase A where the carbonyl group of the substrate, hydrogen bonded to Arg 127, is subjected to nucleophilic attack by the zinc-water moiety which in turn is activated by the nearby carboxylate group of Glu 270 (Christianson and Lipscomb, 1989; Bertini and Luchinat, 1993), and it is also observed in model complexes where tetracoordinate zinc ions can lower the pK_a of the bound water to 7 in comparison to ~10 in hexaaquo zinc (Groves and Olson, 1985).

It is interesting to observe that zinc enzymes catalyze reactions which are also carried out by similar enzymes which do not contain any metal ion. Metal ions are generally used in biology when small molecules like CO_2 and H_2O need to be activated, as in the reaction catalyzed by carbonic anhydrases:

$$CO_2 + H_2O \rightarrow HCO_3^- + H^+$$

In fact, amino acid side chains are not able to attack effectively such small molecules whereas in this case, zinc(II) acting as Zn^{2+}-OH^- in a concerted reaction, is able to catalyze such a process.

Zinc(II) is also used in many unspecific terminal carboxy- and amino-peptidases, phosphatases, and DNA-, RNA-polymerases. If we compare such enzymes to the non metal-containing analogues we can observe that zinc is a much stronger functional reagent than amino acid side chains, but on the other hand it shows a much lower selectivity towards larger substrates.
(iii) As suggested above, another very important point is that the ***zinc coordination sphere*** is extremely flexible. The multiplicity of coordination numbers and geometries indicates its adaptability to the characteristics of the ligands. This property is due to the absence of ligand-field stabilization energy characteristic of its d^{10} electronic configuration. As a consequence, zinc readily binds to low coordination number sites (tetra- or penta-coordination) which are more acidic than high coordination sites. Furthermore, the geometry of the ligands can be easily deformed and changed from tetra- to penta- and hexa-coordination favoring the formation of intermediate states during catalysis (Bertini and Luchinat, 1993).

(iv) The flexibility of the zinc(II) coordination sphere is also connected with another clear advantage in zinc(II)-catalyzed reactions: the metal ligands in Zn(II) complexes always show relatively rapid exchange. Proteins and other macromolecules modulate the reactivity of zinc and this feature is important for the translation of chemical structure into multiple biological functions due to a broad range of geometries, stability constants, and reactivities.

Zinc(II) having a d^{10} electronic configuration, is diamagnetic and consequently lacks all the physico-chemical properties characteristic of many paramagnetic metal ions which allow investigation through spectroscopic techniques. This disadvantage led to a search for metal substitution (Vallee and Williams, 1968). Nowadays, Zn(II) can be substituted in vitro by a series of other divalent metal ions such as Co, Cd, and Ni (Rosenberg et al., 1975; Maret et al., 1979; Bertini et al., 1982; 1984; 1992; Bertini and Luchinat, 1993). In particular, cobalt-substituted zinc enzymes have been useful for the investigation of the active site environment and the monitoring of its changes during catalysis through spectroscopic and magnetic studies. Electronic absorption, CD, MCD, EPR, NMR techniques combined with kinetic studies of Co(II)-substituted as well as chemically and genetically modified Zn(II)-enzymes have extended the comprehension of the metal ion-protein interactions and of their mechanisms of action (Auld and Vallee, 1987; Bertini et al., 1982; 1992; Bertini and Luchinat, 1984).

Regarding the known three-dimensional structures of the zinc-enzymes, their analysis provides a reference for the zinc sites of enzyme classes I, II, III, and IV. Unfortunately, no structures of zinc-enzymes of class V (Isomerases) are currently available, whereas for class VI (Ligases) the main chain structure of a tryptic fragment of Methionyl-tRNA synthetase containing a zinc ion with putative structural roles has only recently been refined, but the resolution was not sufficient to identify the zinc ligands (Brunie et al., 1990).

As observed in the reference structures (Table 1) a *tridentate* metal binding site is formed by a combination of aspartate, cysteine, glutamate, or histidine residues (Fig. 1), the

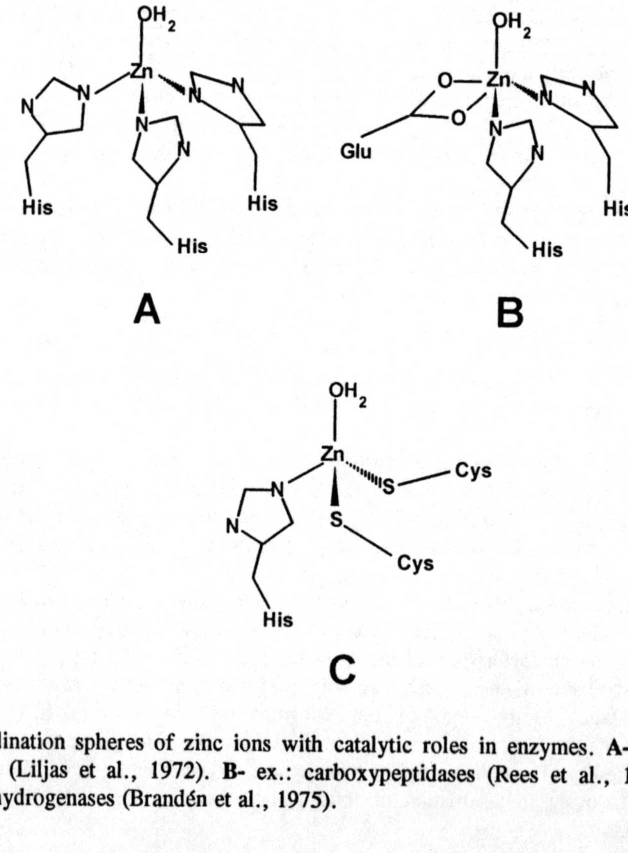

Figure 1.
First coordination spheres of zinc ions with catalytic roles in enzymes. **A-** ex.: carbonic anhydrases (Liljas et al., 1972). **B-** ex.: carboxypeptidases (Rees et al., 1983). **C-** ex.: alcohol dehydrogenases (Brandén et al., 1975).

Table 1

Three-dimensional Structures of Zinc Enzymes

Name	EC #	Zinc ions/mol, Zinc ligands (Role) Molar mass / kg mol^{-1}	References
Class I, Oxidoreductases	*1.*		
Alcohol dehydrogenase (Horse liver)	1.1.1.1	4 Zn/mol 4Cys(Struct); 2Cys-His Water (Cat), 2 x 40	Brandén et al. (1975)
Superoxide dismutase (Bovine erythrocytes)	1.15.1.1	2 Zn/mol 3His-Asp (NonCatal) 2 x 15.6	Tainer et al. (1983)
Class II, Transferases	2.		
Aspartate transcarbamoylase (*E. coli*)	2.1.3.2	6 Zn/mol 4Cys (Struct) 6 x 17+ 4 x 50	Honzatko et al (1982)
DNA Polymerase I Klenow Fragment (*E. coli*)	2.7.7.7	2Zn/mol (or 1Zn+1Mg)/mol 67	Beese and Steitz (1991)
Class III, Hydrolases	*3.*		
Alkaline phosphatase (*E. coli*)	3.1.3.1	4 Zn/mol 1Asp-2His-Wat (Cat), 2Asp-1His-Wat (Cat), 2 x 47	Kim and Wyckoff (1989) Kim and Wyckoff (1991)
Phospholipase C (*Bacillus cereus*)	3.1.4.3	3Zn/mol 2His-Glu-2Wat (Cat) 2His-2Asp-Wat 1His-1Asp-N termTrp NH_2 and -C=O 26	Hough et al. (1989)
Leucine aminopeptidase (Bovine eye lens)	3.4.11.1	12 Zn/mol 2Asp-1Glu (Cat) 2Asp-1Glu (Reg) 6 x 54	Burley et al. (1990)
DD-carboxypeptidase (*Streptomyces albus*)	3.4.15.1	1Zn/mol 3N(His)-Wat (Cat) 22	Dideberg et al. (1982)
Procarboxypeptidase B (Porcine pancreas)		1 Zn/mol (Cat) 2 His-Glu-Wat (Cat) 57	Coll et al. (1991)
Carboxypeptidase A (Bovine pancreas)	3.4.17.1	1 Zn/mol, 2 His-Glu-Wat (Cat) 34.6	Rees et al. (1983)

cont'd

Table 1 (continued)

Name	EC #	Zinc ions/mol, Zinc ligands (Role) Molar mass / kg mol^{-1}	References
Carboxypeptidase B (Bovine pancreas)	3.4.17.2	1 Zn/mol, 2 His-Glu-Wat (Cat) 34	Schmid and Herriot (1976)
Carboxypeptidase A2 (Rat pancreas)	3.4.17.	1 Zn/mol, 2 His-Glu-Wat (Cat) 34	Zhang et al. (1991)
Carboxypeptidase T (*T. vulgaris*)	3.4.17.	1 Zn/mol, 2 His-Glu-Wat (Cat) 36	Teplyakov et al.
Astacin (Crayfish *A. astacus L.*)	3.4.17.	1 Zn/mol 3His-Wat-Tyr (Cat) 22	Bode et al. (1992)
Thermolysin (*B. thermoproteolyticus*)	3.4.24.4	1Zn(4Ca)/mol 2His-Glu-Wat (Cat) 34	Matthews et al. (1972)
Elastase (*Ps. aeruginosa*)	3.4.21.11	1Zn(1Ca)/mol 2His-Glu-Wat (Cat) 32	Thayer et al.(1991)
Endoglucanase CelD (*Clostr. thermocellum*)	3.2.1.	1Zn(3Ca)/mol 2His-2Cys (Struct)	Juy et al. (1992)
Neutral protease (*Bacillus cereus*)	3.4.24.4	1Zn(4Ca)/mol 2His-Glu-Wat (Cat) 34	Pauptit et al. (1988) Stark et al. (1992)
Neutral protease (*Strept. caespitosus*)	3.4.24.4	1Zn/mol 3AA-Wat (Cat) 15	Harada et al. (1991)
β-lactamase II (*Bacillus cereus*)	3.5.2.6	2Zn/mol 3His-Cys-Wat (Cat) 22	Sutton et al. (1987)
Class IV, Lyases	*4.*		
Carbonic anhydrase I (Human erythrocytes)	4.2.1.1	1 Zn/mol 3 His-Wat (Cat) 28.9	Kannan et al. (1975)
Carbonic anhydrase II (Human erythrocytes)	4.2.1.1	1 Zn/mol 3 His-Wat (Cat) 29.2	Liljas et al. (1972)
Carbonic anhydrase III (Bovine muscle cells)	4.2.1.1	1 Zn/mol 3 His-Wat (Cat) 29.6	Eriksson and Liljas (1986)
Class VI, Ligases	*6.*		
Methionyl-tRNA Synthetase (*E. coli*)	6.1.1.10	4Zn/mol unknown lig. (Struct) 4 x 45	Brunie et al. (1990)

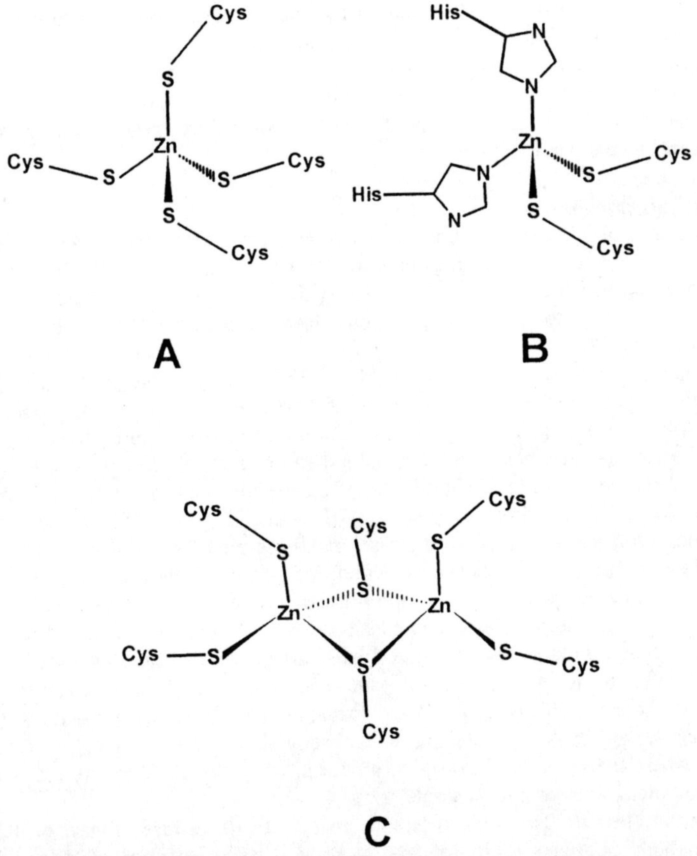

Figure 2.
Scheme of structural/regulatory zinc sites. **A**- ex.: alcohol dehydrogenases. **B**- zinc site from endoglucanase CelD (Juy et al., 1992) or zinc finger from Transcription Factor III A (Miller et al., 1985). **C**- dinuclear zinc finger from GAL4 protein (Pan and Coleman, 1989).

most common being histidine (Vallee and Auld, 1990). Such a type of coordination is characteristic of the catalytic function of zinc(II) and it provides coordination sites open to water and/or substrates which usually bring the metal coordination number to four whereas five-coordination has been recognized in various enzyme-substrate, -product, and -inhibitor complexes during enzymatic turnover (Christianson, 1991).

Tetradentate structural zinc sites with four cysteines generally have high stability constants and allow local conformations and overall structures to be conserved (Fig. 2); the structural zinc ion is generally inaccessible to the solvent. More complex geometries are rarely observed and a few of them will be described later.

V. ROLES OF ZINC IN ENZYMES

The functions of Zinc ions in enzymes are usually classified as follows (Vallee, 1986):
(i) Catalytic: when the metal ion is directly involved in catalysis. Its essentiality for the catalytic process is demonstrated by loss of enzymatic activity caused by the removal of the metal ion (even if the tertiary structure is often maintained).
(ii) Structural: when it is only necessary for the maintenance of the tertiary and often also quaternary structures and it affects activity only in terms of overall conformational stability.

(iii) Regulatory-Modulatory: when the metal ion regulates (as an inhibitor or as an activator) but it is not essential to other protein processes.
(iv) Non-catalytic: in the absence of a clear knowledge of its function.

A combination of such roles is often found in natural systems where it is easy to find proteins and enzymes containing several zinc ions with different functions or single zinc ions with more than one function.

A. Zinc(II) in catalysis

As noted above, Zn(II) can behave in inorganic and organic reactions as a ***Lewis acid*** easily accepting electrons and polarizing groups with the final effect of facilitating the attack on the bound group acting as an acid or increasing the power of a bound base. Let us briefly analyze the catalytic role of zinc in the various classes of enzymes in the light of their three-dimensional structures.

In Class I (Oxidoreductases) the standard of reference of enzymes containing zinc ions with catalytic roles is the crystal structure of horse liver, alcohol dehydrogenase (E.C. 1.1.1.1), which also contains a structural zinc ion (see below). The catalytic zinc is bound to Cys 46 positioned between an α-helix and a β-sheet, to His 67 located at the beginning of another β-sheet, to Cys 174 in the middle of an α-helix, and to a water molecule which completes the metal coordination (Fig. 1C) (Brandén et al., 1975). These residues are strictly conserved in alcohol dehydrogenases requiring zinc for activity (Vallee and Auld, 1990). The role of zinc in the catalyzed alcohol oxidation to aldehyde is well established: it binds the substrate in order to facilitate the formation of an alcoholate ion and orients it in a way so that the hydride transfer to the coenzyme NAD^+ is favored (Eklund and Brändén, 1987); the oxygen atom of the substrate binds directly to the zinc ion by displacing the water molecule. In the reverse reaction, zinc functions as an electron attractor facilitating hydride transfer to the aldehyde. Zinc participates in many other redox reactions in which it usually acts as a Lewis acid helping in the activation and orientation of substrates in processes where the absence of damaging products, often generated by the redox chemistry of transition metal ions, is indispensable.

For what class III (Hydrolases) is concerned, a series of three-dimensional structures relative to both exo- and endo-proteases is known. The structures of A and B bovine carboxypeptidases (EC 3.4.17.1-2) are well characterized and their catalytic mechanisms are discussed in depth. In all the known sequences of this protein family the zinc ion ligands His 69 (bovine carboxypeptidase A numbering), Glu 72, which are positioned at the end of a reverse turn, and His 196 (component of a β-sheet) are strictly conserved (Fig. 1B) as are Glu 270 and Arg 145, which represent surrounding residues essential for activity (Rees et al., 1983). Such enzymes catalyze the hydrolysis of amide bonds of polypeptides and are specific for large hydrophobic side chains. A wealth of kinetic, spectroscopic, and structural data suggests that, contrary to what is expected, the binding of the substrate to the protease cannot implicate a zinc-carbonyl interaction, but the carbonyl group can bind to a second site (Auld and Vallee, 1987; Christianson, 1991). The following hydrolysis of the peptide bond is then favored by the zinc(II)-water nucleophile reinforced by Glu 270.

In endo-proteases like thermolysin (EC 3.4.24.4), which contain one catalytic zinc ion and four Ca(II) ions with structural functions, it has been established that zinc is coordinated to His 142 and His 146, which are present on a α-helix, and to Glu 166 component of a second internal helix, reciprocally disposed in order to allow the binding of a water molecule to the metal ion. The secondary structure supplying the zinc ligands in thermolysin is similar to many other neutral proteases but is markedly different from carboxypeptidases.

From the known three-dimensional structures, it appears that in zinc enzymes which do not require coenzymes for activity the metal binding site is always generated by the first two ligands which are separated by a spacer of a few amino-acids whereas the third zinc ligand which is 20 to 120 amino-acids away has the function of arranging the protein residues in order to bind zinc(II) (Vallee and Auld, 1990). A water molecule is the critical component that completes the first zinc coordination sphere. The catalytic properties of zinc enzymes are generated and finely modulated by the characteristics of the two spacers which make up the active site and its environment.

Among zinc-endopeptidases, the structure of astacin from the crayfish *A. astacus L.* has recently been resolved and shows that the zinc site is different from the other proteases. In fact it is made up by three histidines, a water molecule, and a more remote tyrosine in a trigonal-bipyramidal geometry (Bode et al., 1992). In particular, the third histidine (His 102) is part of a consensus sequence shared by other proteinases such as vertebrate collagenases (Stöcker et al., 1990). During the last few years, this family of metalloproteinases has attracted the interest of the scientific community for their involvement in the mechanism of cancer development. Although the zinc ligands of metalloproteinases like human fibroblast collagenase (hereafter HFC) are not known, the catalytic site seems to be similar in most respects to thermolysin and other zinc-proteases (as seen from primary structures homologies); the fundamental difference is the presence of a removable peptide with a cysteine group bound to the zinc ion in the latent form of HFC which, by displacing the essential zinc-bound water, makes the active site inaccessible to substrates (Windsor et al., 1991). This new zinc coordination which resembles a non-catalytic zinc with four protein ligands has been called a "latent catalytic zinc atom" (Van Wart and Birkendal-Hansen, 1990).

In Class IV (Lyases) the crystal structures of carbonic anhydrases (hereafter CA) I, II from human erythrocytes and CA III from bovine muscle cells are known (EC 4.2.1.1). Two of the zinc ligands (His 94 and His 96, CA I numbering) are on a β-strand whereas a third histidine ligand (His 119) is present on a second β-strand, and a water molecule completes the zinc coordination (Fig. 1A) (Liljas et al., 1972; Kannan et al., 1975; Eriksson and Liljas, 1986). More than a dozen sequences of carbonic anhydrases from different sources are known, and a large invariance in the peptide region containing the zinc ligands is observed (Vallee and Auld, 1990). Such enzymes hydrate carbon dioxide to produce bicarbonate ion and a proton and are among the most efficient catalysts known. The process occurs in two steps: the first involves the attack of the substrate CO_2 whereas the second step corresponds to product dissociation and regeneration of the zinc-hydroxide nucleophile (Coleman, 1967; Silverman and Lindskog, 1988). Carbon dioxide is attracted into the cavity by hydrophobic interactions (Banci et al., 1990) and is oriented in such a way as to be ready for the attack by $Zn\text{-}OH^-$. Then Zn-bicarbonate is formed. The coordinated oxygen bears a proton which is involved in hydrogen bond with a nearby threonine (Thr 199) (Håkansson et al., 1992). It is suggested that an incoming water molecule gives rise to an equilibrium between a four and five zinc coordination (Bertini et al., 1978). The latter coordination allows the product to be released.

B. Structural and protecting roles of zinc

Zinc(II) is found in a series of intracellular proteins in which it assumes particular structural roles affecting local structure and conformation. It can take the place of disulphide bridges with two principal advantages: its stereochemical demand is very small and the absence of redox chemistry allows high stability inside the reducing atmosphere of cells. X-ray structures of zinc enzymes of different classes have allowed to analyze the particular geometry of structural zinc atoms. The classical Zn(II) ions with structural roles are tetrahedrally coordinated to four cysteines (Fig. 2A). For example, one of the two zinc ions present in each subunit of horse liver alcohol dehydrogenase (E.C. 1.1.1.1) is thought to serve this role. In fact this zinc atom, although near to the surface, is not accessible to the solvent and therefore essentially affects protein structure and conformation (Brandén et al., 1975). Also, in Aspartate Transcarbamoylase (E.C. 2.1.3.2) the only zinc ion present is in the regulatory units and is bound to four cysteines; neither substrate nor solvent has access to the metal. These subunits which are in contact with the catalytic ones seem to be responsible for the stabilization of the quaternary structure of the enzyme (Honzatko et al., 1982).

The recently resolved structure of a bacterial cellulase, endoglucanase CelD from *C. thermocellum*, has shown the presence of a tetrahedral zinc binding site composed by two histidines and two cysteines in such hydrolase. This site, which resembles a zinc finger domain (see below) does not seem to be directly involved in catalysis, the role of zinc probably being structural and/or regulatory (Juy et al., 1992).

Zinc in copper-zinc superoxide dismutases provides thermodynamic stability to the protein and may have a regulatory role as the activity profile is different in the absence of zinc (Banci et al., 1991).

Structural and protective functions are also accomplished by zinc in extracellular environments such as in hair and nails where this metal is bound to sulphydryls groups of keratin protecting it from oxidative damage. A similar role is also carried out by zinc(II) in sperm where it appears in two different fractions: in the tail, where it cross-links keratin (extracellular) and in the head, associated to DNA (intracellular)(Roomans et al., 1982).

C. Zinc as a controller and trigger

Many investigations, nowadays, indicate that zinc(II) is probably a major regulatory ion in cell metabolism. The protective and triggering role of zinc in vesicles and special cell systems seems to be one of its major functions. The Zn(II) ion works as a trigger in nucleic acid and protein reactions and in pumped storage systems associated with small chelates or protein containing thiols. Furthermore, it can stabilize and/or inhibit proteins, as well as hormones (like insulin) and zymogens (nerve growth factors). Insulin, which has the function of stimulating the sugar transport from blood to cells, assumes an hexameric quaternary structure in the presence of zinc(II) ions, three dimers being held together by two zinc ions (Baker et al., 1988; Derewenda et al., 1989).

D. Enzymes with multiple zinc sites

Zinc polynuclear assemblies are known to be present in metallothioneins and some zinc fingers (which will be analyzed later) and in a few enzymes like alkaline phosphatase, phospholipase C, the Klenow fragment of DNA polymerase I, and leucine aminopeptidases. Their structures have been recently resolved or refined to a higher resolution and deserve to be described in detail.

The three-dimensional structure of the metal centers of alkaline phosphatase from *E. coli*, a non-specific phosphomonoesterase that contains two zinc and one magnesium ion, is depicted in Figure 3A; zinc 1 is tetrahedrally coordinated to two histidines, one aspartate, and the substrate/inhibitor phosphate (or a water molecule), and zinc 2, more deeply buried into the protein, is bound to one histidine, two aspartates, and the same phosphate molecule (or another water molecule). The two metal ions are only 0.394 nm apart (Kim and Wyckoff, 1991). An extensive hydrogen bonding network formed by the phosphate and several water molecules is present in the active site of this enzyme where both zinc ions are required for fast reactivity whereas magnesium is thought to have a structural role (Chlebowski and Coleman, 1976).

The detailed structure of phospholipase C from *B. cereus* at 0.15 nm resolution also shows the presence of a poly-zinc site. This "all-helix" enzyme belongs to a novel structural class, and the crystalline enzyme contains three zinc ions although atomic absorption and solution state EXAFS studies have indicated the presence of only two tightly bound zinc ions (zinc 1 and 2 in Fig. 3B). This is probably due to the fact that the crystals for this study where grown in the presence of an excess of zinc(II) (Hough et al., 1989). As observed in Figure 4A, zinc 1 and 3, 0.33 nm apart from each other, are bridged by the carboxylate of Asp 122 and by a water/hydroxide molecule whereas zinc 2 has a coordination matching that in carboxypeptidases except for Glu 146 which is monodentate.

The similarities in the active sites of alkaline phosphatase and phospholipase C are probably not accidental because, for example, in *B. cereus* these two enzymes compose a phosphate retrieval system where the product of phospholipase C is hydrolized by alkaline phosphatase.

The presence of multiple metal sites seems to be a common feature of enzymes involved in phosphoryl transfer catalyzed processes. In fact, the crystal structure of the Klenow proteolytic fragment of DNA polymerase I from *E. coli* complexed with a deoxynucleoside monophosphate product shows two bound metal ions 0.39 nm apart (Fig. 4A). The first metal ion is bound to the carboxylate groups of Asp 355, Glu 357, Asp 501, an oxygen atom of the phosphate moiety of the product, and a water molecule in a distorted tetrahedral geometry; the second metal ion is octahedrally coordinated by the carboxylate of

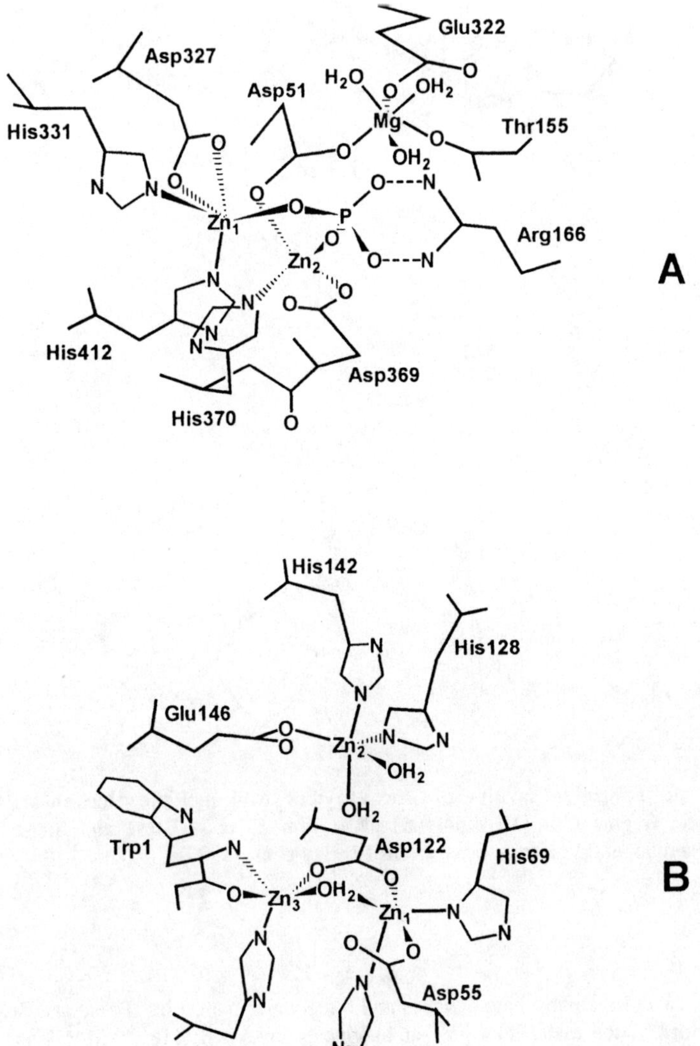

Figure 3.
Scheme of the active cavities of zinc enzymes with multiple zinc sites: **A-** Alkaline phosphatase from *E. Coli* (Kim and Wyckoff, 1991); **B-** phospholipase C from *B. cereus* (Hough et al., 1989).

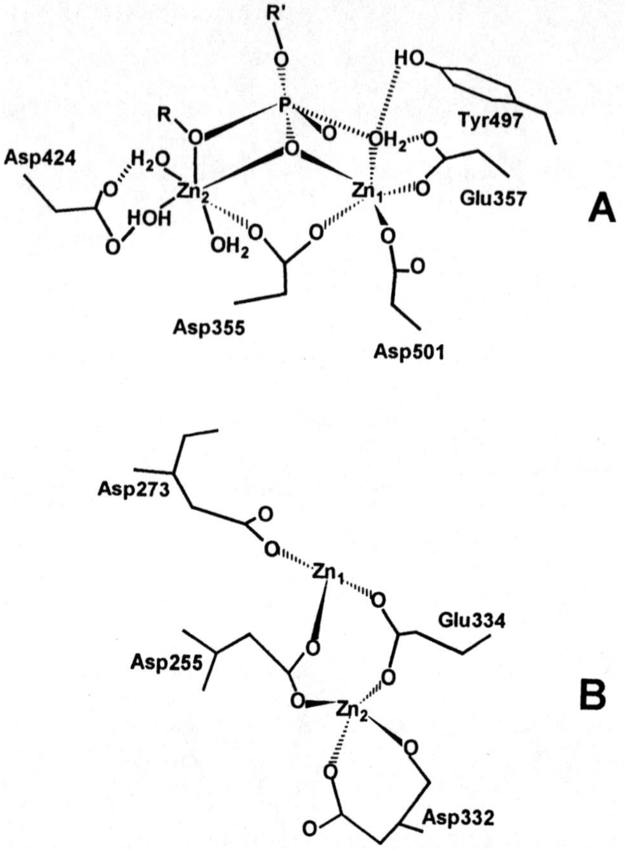

Figure 4.
Scheme of the active cavities of zinc enzymes with multiple zinc sites: **A-** Klenow proteolytic fragment of DNA polymerase I from *E. coli* (Beese and Steitz, 1991). **B-** Leucine aminopeptidase from bovine lens (Burley et al., 1990).

Asp 355, two of the phosphate oxygens and three water molecules (Beese and Steitz, 1991). The identity of the metal ions present in vivo is unknown: Mg^{2+}, Mn^{2+} and Zn^{2+} are essential for activity and interchangeable. However, in the presence of both Mg^{2+} and Zn^{2+} the zinc ions bind preferentially to the first pseudo tetrahedral site whereas magnesium ions bind to the second octahedral site as expected on the basis of their respective coordination requirements.

Among the peptidases, leucine aminopeptidase from bovine eye lens is one of the few known proteins to contain two metal sites per subunit. The crystal structure of the Zn-Zn form has been resolved at 0.27 nm resolution (Burley et al., 1990), although the Zn-Mg form is more active (Thompson and Carpenter, 1976). Therefore, it is not yet clear if the role of zinc ions in this enzyme is catalytic or regulatory (or both). The arrangement of the two zinc ions in leucine aminopeptidase, only 0.288 nm apart, is unique among the known zinc proteases and the implications on the catalytic mechanism are still unknown. Zinc 1 is coordinated to Asp 255, Asp 273 and Glu 334, no fourth ligand being observed in the density map, whereas zinc 2 is bound in a pseudo-tetrahedral arrangement to the carboxylates of Asp 255, Asp 332, Glu 334 and the carbonyl oxygen of Asp 332; note that both Asp 255 and Glu 334 carboxylates are bridging the two metal ions (see Fig. 4B).

VI. ZINC PROTEINS INVOLVED IN GENE EXPRESSION REGULATION

The possible involvement of Zn(II) ions in processes related to nucleic acid replication, transcription and cell division was formulated in the 1970s when zinc deficiency states were seen to be associated to anomalies in the cell cycle and in their DNA and RNA content (Vallee, 1977; Vallee and Falchuk, 1981). At the beginning of the 1980s it was proved that the transcription factor IIIA (hereafter TFIIIA), isolated from *Xenopus laevis*, is a zinc protein (Hanas et al., 1983). This protein binds to the ca 50 nucleotides of the control region of 5S ribosomal RNA genes, controlling their transcription. Later, TFIIIA was shown to consist of 9 repeat sequences of about 30 amino acids, each sequence having two histidines and two cysteines in conserved positions (Ginsberg et al., 1984; Brown et al., 1985). It was therefore proposed that the two histidines and the two cysteines determine a pseudo tetrahedral binding site for the zinc, thus generating peptide domains able to interact with DNA (Miller et al., 1985). These local structures were named "zinc fingers". Recent results indicate that the TFIIIA protein binds only two zinc(II) ions which are essential for the activation of the transcription process. Addition of further zinc to the protein containing two zinc atoms does not enhance binding to DNA or transcription activity (Shang et al., 1989).

The hypothesis that the generation of zinc fingers is the key mechanism for binding to nucleic acids induced a systematic search for primary amino-acidic sequences of several proteins in order to find analogous TFIIIA sequences. A large number of DNA-binding proteins were screened and many were found to contain domains that could form zinc fingers (Berg, 1986; 1988; Sunderman and Barber, 1988). At present a fast growing number of proteins containing zinc finger domains has been found, sequenced, cloned, and the structure determined utilizing X-ray crystallography and in particular multidimensional-multinuclear NMR techniques (South and Summers, 1990; Omichinski et al., 1992 and references therein). Besides the 2 Cys - 2 His containing sequence, it was found that zinc could also be bound to 3 Cys and 1 His, as in the gene 32 protein of bacteriophage T4 (Giedroc et al., 1986) whereas, in the case of the GAL4 protein, the two zinc ions present are bound to 6 cysteines (Hollenberg et al., 1987; Pan and Coleman, 1989), two of which bridging-zinc ligands as observed through ^{1}H-^{113}Cd 2D NMR experiments by Pan and Coleman (1990) (Fig. 2B). The zinc fingers domains are analyzed in detail in Chapter 2-C.

VII. TRANSPORT AND STORAGE ZINC PROTEINS

It seemed well established that the principal zinc binding and transport proteins in serum are albumin, which approximately binds two thirds of the total zinc present, and α_2-macroglobulin, which binds the remaining one-third (Parisi and Vallee, 1970; Smith et al., 1978). At the same time Boyett and Sullivan (1970) reported that transferrin is also involved in serum transport of zinc. Whereas α_2-macroglobulin binds zinc strongly, loosely bound zinc is found in albumin as well as in transferrin, in ceruloplasmin, and bound to amino-acids (Giroux and Henkins, 1972; Charlwood, 1979). Recently, the determination of zinc-transferrin binding constants in the presence of physiological bicarbonate concentrations has indicated that transferrin can compete with albumin for zinc binding therefore playing a primary role in zinc transport (Harris, 1983). Bound zinc complexes are then stored and the metal release is triggered by a decrease in zinc concentration in the particular compartments of cells.

A. Metallothioneins

Zinc(II) appears to be the major metal ion constituent of metallothioneins, whose precise biological function is still unknown (Kägi and Nordberg, 1979). Possible roles as metal storage, as a metal-supplying protein for the biosynthesis of the various metalloenzymes and for maintaining metal homeostasis in living organisms, and also as metal detoxification have been proposed (Kägi and Kojima, 1988).

Seven zinc (or cadmium, or copper) ions are bound per mol by these low-molar-mass metalloproteins forming two separate clusters whose composition is Zn_4Cys_{11} and Zn_3Cys_9 with five and three cysteine residues acting as bridging donors between two zinc ions in

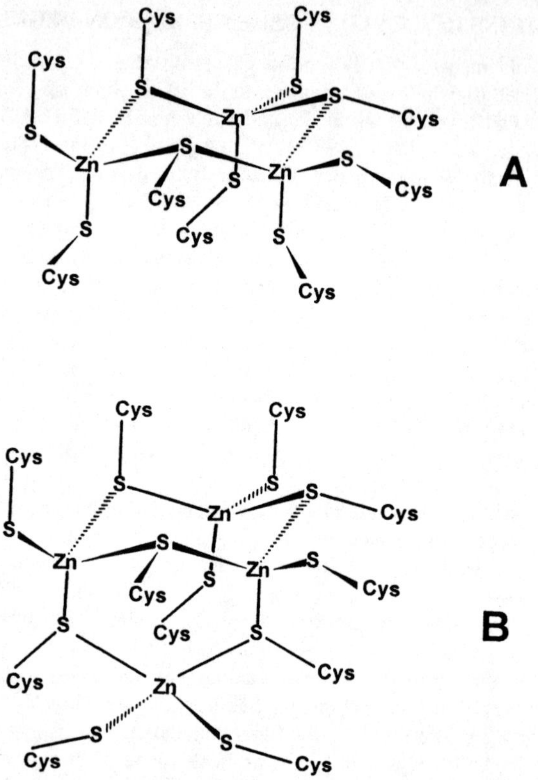

Figure 5.
Scheme of the multiple zinc sites in metallothioneins (Arseniev et al., 1988). **A**- Zn_3Cys_9 site. **B**- Zn_4Cys_{11} site.

each cluster respectively (Fig. 5A and B) (Arseniev et al., 1988). This strange coordination chemistry is reminiscent of other well characterized metal sulphur clusters like the iron sulphur ones, which are very diffused in living organisms. As reported above, DNA-binding proteins also contain similar zinc clusters (Fig. 2B) (Pan and Coleman, 1990), so a possible specific biological role could be associated to these chemical structures. (For more information on metallothioneins, see Section D5 of the present Chapter 1.)

VIII. CONCLUSIONS

The coordination chemistry properties of Zn(II) have been reviewed with the aim of understanding its biological functions. Zinc prefers low coordination numbers (4 and 5), is kinetically labile, and has intermediate hard-soft properties. When bound to water, it lowers the water pK_a to give rise to the strong $Zn\text{-}OH^-$ nucleophile. When bound to cysteines it only has structural roles. It may have a regulatory role besides a structural one when it is not directly involved in catalysis. In several proteins there are more than one zinc ion with different roles and in some cases they are arranged in polymetallic centers. Each of these properties is addressed for those proteins for which the molecular structure is available, and different biological functions are specified.

REFERENCES

Argos, P., Garavito, R.M., Eventoff, W., Rossman, M.G. and Brandén, C.I. (1978). *J. Mol. Biol.*, *126*: 141.

Arseniev, A., Schultze, P., Wörgötter, E., Braun, W., Wagner, G., Va ak, M., Kägi, J.H.R., and Wüthrich, K. (1988). *J. Mol. Biol.*, *201*: 637.

Auld, D.S. and Vallee, B.L. (1987). *Hydrolytic Enzymes* (A. Neuberger and K. Brocklehurst, eds.) Elsevier, New York, p. 201.

Baker, E.N., Blundell, T.L., Cutfield, S.M., Dodson, E.J., Hodgkin, D.C., Hubbard, R.E., Isaacs, N.W.I., Reynolds, C.D., Sabake, K., Sabake, N., and Vijayan, N.M.. (1988). *Phil. Trans. Roy. Soc. London, B 319*: 369.

Banci, L., Bertini, I., Luchinat, C., and Moratal Mascarell, J.M. (1990). *Enzymatic and Model Carboxylation and Reduction Reactions for Carbon Dioxide Utilization* (M. Aresta and J.V. Schloss, eds.), Kluwer, Dordrecht, p. 181.

Banci, L., Bertini, I., Cabelli, D.E., Hallewell, R.A., Tung, J.W., and Viezzoli, M.S. (1991). *Eur. J. Biochem., 196*: 123.

Beese, L.S., and Steitz, T.A. (1991). *EMBO J., 10*: 25.

Berg, J.M. (1986). *Science, 232*: 485.

Berg, J.M. (1988). *Proc. Nat. Acad. Sci. USA, 85*: 99.

Berg, J.M. (1989). *Prog. Inorg. Chem., 37*: 143.

Berg, J.M., and Merkle, D.L. (1989). *J. Amer. Chem. Soc., 111*: 3759.

Bertini, I.; and Luchinat, C. (1984). *Advances in Inorganic Biochemistry*, Vol. 4 (G.L. Eichhorn and L.G. Marzilli, eds.), Elsevier, New York, Chapter 2.

Bertini, I., and Luchinat, C. (1993). *Bioinorganic Chemistry* (I. Bertini, H.B. Gray, S.J. Lippard and J.S. Valentine, eds.), University Science Books, Mill Valley, California, p. 37.

Bertini, I., Canti, G., Luchinat, C., and Scozzafava, A. (1978). *J. Amer. Chem. Soc., 100*: 4873.

Bertini, I., Luchinat, C., and Scozzafava, A. (1982). *Struct. Bonding, 48*: 45.

Bertini, I., Luchinat, C., Maret, W., and Zeppezauer, M., eds. (1986). *Progress in Inorganic Biochemistry and Biophysics*, Vol. 1, Birkhäuser Verlag, Basel.

Bertini, I., Luchinat, C., Rosi, M., Sgamellotti, A., and Tarantelli, F. (1990). *Inorg. Chem., 29*: 1460.

Bertini, I., Messori, L., and Viezzoli, M.S. (1992). *Coord. Chem. Rev., 120*: 163.

Bode, W., Gomis-Rüth, F.X., Huber, R., Zwilling, R., and Stöcker, W.. (1992). *Nature, 358*: 164.

Boyett, J.D., and Sullivan, J.F. (1970). *Metab. Clin. Exp., 19*: 148.

Brown, R.S., Sander, C., and Argos, P. (1985). *FEBS Lett., 186*: 271.

Brandén, C.I., Jörnval, H., Eklund, M., and Furugren, B. (1975). *Enzymes*, Vol. 11 (P.D. Boyer, ed.), Academic Press, New York, p. 103.

Brunie, S., Zelwer, C., and Risler, J.-L. (1990). *J. Mol. Biol., 216*: 411.

Burley, S.K., and Petsko, G.A. (1988). *Adv. Prot. Chem., 39*: 125.

Burley, S.K., David, P.R., Taylor, A., and Lipscomb, W.N. (1990). *Proc. Nat. Acad. Sci. USA, 87*: 6878.

Charlwood, P.A. (1979). *Biochim. Biophys. Acta, 581*: 260.

Chlebowski, J.F., and Coleman, J.E. (1976). *J. Biol. Chem., 251*: 1202.

Christianson, D.W. (1991). *Adv. Prot. Chem., 42*: 281.

Christianson, D.W., and Lipscomb, W.N. (1989). *Accounts Chem. Res.*, 22: 62.

Coleman, J.E. (1967). *J. Biol. Chem., 242*: 5212.

Coleman, J.E. (1992). *Annu. Rev. Biochem., 61*: 897.

Coll, M., Guasch, A., Avilés, F.X., and Huber, R. (1991). *EMBO J., 10*: 1.

da Silva, F.J.J.R., and Williams, R.J.P. (1991). *The Biological Chemistry of the Elements - The Inorganic Chemistry of Life*, Clarendon Press, Oxford, Chapter 11.

Derewenda, U., Derewenda, Z., Dodson, E.J., Dodson, G.G., Reynolds, C.G., Smith, G.D., Sparks, C., and Swensen, D. (1989). *Nature, 338*: 594.

Dideberg, O., Charlier, P., Dive, G., Joris, B., Frere, J.M., and Ghuysen, J.M. (1982). *Nature, 299*: 469.

Eklund, H., and Brändén, C.-I. (1987). *Biological Macromolecules and Assemblies*, Vol. 3 (F.A Jurnak and A. McPherson, eds.), John Wiley & Sons, Chapter 2.

Eriksson, E.A., and Liljas, A. (1986). *J. Biol. Chem.*, *261*: 16247.

Giedroc, D.P., Keating, K.M., Williams, K.R., Konigsberg, W.H., and Coleman, J.E. (1986). *Proc. Nat. Acad. Sci. USA*, *83*: 8452.

Ginsberg, A.M., King, B.O., and Roeder, R.G. (1984). *Cell*, *39*: 479.

Giroux, E.L., and Henkins, R.I. (1972). *Bioinorg. Chem.*, *2*: 125.

Groves, J.T., and Olson, J.R. (1985). *Inorg. Chem.*, *24*: 2715.

Håkansson, K., Carlsson, M., Svensson, L.A., and Liljas, A. (1992). *J. Mol. Biol.*, *227*: 1192.

Hanas, J.S., Hazuda, D.J., Bogenhagen, D.F., Wu, F.Y.-H., and Wu, C.-W. (1983). *J. Biol. Chem.*, *258*: 14120.

Harada, S., Kidatokoro, K., Kinoshida, T., Kai, Y., and Kasai, N. (1991). *J. Biochem. (Tokyo)*, *110*: 46.

Harris, W.R. (1983). *Biochemistry*, *22*: 3920.

Hollenberg, S.M., Giguere, V., Segui, P., and Evans, R.M. (1987). *Cell*, *49*: 39.

Honzatko, R.B., Crawford, J.L., Monaco, H.L., Ladner, J.E., Edwards, B.F.P., Evans, D.R., Warren, S.G., Wiley, D.C., Ladner, R.C., and Lipscomb, W.N. (1982). *J. Mol. Biol.*, *160*: 219.

Hough, E., Hansen, L.K., Birknes, B., Jynge, K., Hansen, S., Horvik, A., Little, D., Dodson, E., and Derewenda, Z. (1989). *Nature*, *338*: 357.

Juy, M., Amit, A.G., Alzari, P.M., Poljak, R.J., Claeyssens, M., Béguin, P., and Aubert, J.-P. (1992). *Nature*, *357*: 89.

Kägi, J.H.R., and Nordberg, M., eds. (1979). *Metallothionein*, Birkhäuser Verlag, Basel.

Kägi, J.H.R., and Kojima, Y., eds. (1988). *Metallothionein II, Experientia, Suppl. 52.*

Kannan, K.K., Notstrand, B., Fridborg, K., Lovgren, S., Orlsson, A., and Petef, M. (1975). *Proc. Nat. Acad. Sci. USA*, *72*: 51.

Keilin, D., and Mann, T. (1939). *Nature*, *144*: 442.

Kim, E.E., and Wyckoff, H.W. (1989). *Clin. Chim. Acta*, *186*: 175.

Kim, E.E., and Wyckoff, H.W. (1991). *J. Mol. Biol.*, *218*: 449.

Liljas, A., Kannan, K.K., Bergstén, P.C., Waara, I., Fridborg, B., Strandberg, B., Carlbom, U., Järup, L., Lövgren, S., and Petef, M. (1972). *Nature*, *235*: 131.

Maret, W., Andersson, I., Dietrich, H., Shneider-Bernlöhr, H., Einarsson, R., and Zeppezauer, M. (1979). *Eur. J. Biochem.* *98*: 501.

Matthews, B.W., Jansonius, J.N., Colman, P.M., Shoenborn, B.P., and Dupourque, D. (1972). *Nature*, *238*: 37.

Mc Cance, R.A., and Widdowson, E.W. (1942). *Biochem. J.*, *36*: 392.

Miller, J., McLachlan, A.D., and Klug, A. (1985). *EMBO J.*, *4*: 1609.

Mills, C.F., ed. (1988). *Human Nutrition Reviews, Zinc in Human Biology*, Springer Verlag, Berlin.

Nakagawa, S., Umeyama, H., Kitaura, K., and Morokuma, K. (1981). *Chem. Pharm. Bull.*, *29*: 1.

Omichinski, J.G., Clore, G.M., Robien, M., Sakaguchi, K., Appella, E., and Gronenborn, A.M. (1992). *Biochemistry*, *31*: 3907.

Pan, T., and Coleman, J.E. (1989). *Proc. Nat. Acad. Sci. USA*, *86*: 3145.

Pan, T., and Coleman, J.E. (1990). *Proc. Nat. Acad. Sci. USA*, *87*: 2077.

Pantoliano, M.W., Whitlow, M., Wood, J.F., Rollence, M.L., Finzel, B.C., Gilliland, G.L., Poulos, T.L., and Bryan, B.N. (1988). *Biochemistry*, *27*: 8311.

Parisi, A.F., and Vallee, B.L. (1970). *Biochemistry*, *9*: 2421.

Pauptit, R.A., Karlsson, R., Picot, D., Jenkins, J.A., Niklaus-Reimer, A.-S., and Jansonius, J.N. (1988). *J. Mol. Biol.*, *199*: 525.

Pearson, R.G. (1963). *J. Amer. Chem. Soc.*, *85*: 3533.

Raulin, J. (1869). *Ann. Sci. Nat. Botan. et Biol. Végétale*, *11*: 93.

Rees, D.C., Lewis, M., and Lipscomb, W.N. (1983). *J. Mol. Biol., 168*: 367.

Roomans,, G.M., Lundevall, E., Bjorndhal, L, and Krist, U. (1982). *Internat. J. Andrology, 5*: 478.

Rosenberg, R.C., Root, C.A., and Gray, H.B. (1975). *J. Amer. Chem. Soc., 97*: 21.

Russell, A.J., and Fersht, A.R. (1987). *Nature, 328*: 496.

Schmid, M.F., and Herriott, J.R. (1976). *J. Mol. Biol., 103*: 175.

Shang, Z., Liao, Y.-D., Wu, F.Y.-H., and Wu, C.-W. (1989). *Biochemistry, 28*: 9790.

Sigel, H., ed. (1983). *Metal Ions in Biological Systems*, Vol. 15, Marcel Dekker, New York.

Silverman, D.N., and Lindskog, S. (1988). *Accounts Chem. Res., 21*: 30.

Smith, K.T., Failla, M.L., and Cousins, R.J. (1978). *Biochem. J., 184*: 627.

South, T.L., and Summers, M.F. (1990). *Advances in Inorganic Biochemistry - Metal-Ion Induced Regulation of Gene Expression*, Vol. 8 (G.L. Eichhorn and L.G. Marzilli, eds.) Elsevier, New York, p. 199.

Spiro, T.G., ed. (1983). *Metal Ions in Biology*, Vol. 5, Wiley Interscience, New York.

Stark, W., Pauptit, R.A., Wilson, K.S., and Jansonius, J.N. (1992). *Eur. J. Biochem., 207*: 781.

Stöcker, W., Ng, M., and Auld, D.S. (1990). *Biochemistry, 29*: 10418.

Sunderman, F.W., Jr., and Barber, A.M. (1988). *Ann. Clin. Lab. Sci., 18*: 267.

Sutton, B.J., Artymiuk, P.J., Cordero-Borboa, A.E., Little, C., Phillips, D.C., and Waley, S.G. (1987). *Biochem. J., 248*: 181.

Tainer, J.A., Getzoff, E.G., Richardson, J.S., and Richardson, D.C. (1983). *Nature, 306*: 284.

Teplyakov, A., Polyakov, K., Obmolova, G., Strokopytov, B., Kuranova, I., Osterman, A., Grishin, N., Smulevitch, S., Zagnitko, O., Galperina, O., Matz, M., and Stepanov, V. (1992). *Eur. J. Biochem., 208*: 281.

Thayer, M.M., Flaherty, K.M., and McKay, D.B. (1991). *J. Biol. Chem., 266*: 2864.

Thompson, G.A., and Carpenter, F.H. (1976). *J. Biol. Chem., 251*: 1618.

Vallee, B.L. (1977). *Biological Aspects of Inorganic Chemistry* (D. Dolphin, ed.), Wiley, New York, p. 37.

Vallee, B.L. (1986). *Progress in Inorganic Biochemistry and Biophysics, Zinc Enzymes* (I. Bertini, C. Luchinat, W. Maret and M. Zeppezauer, eds.), Birkhauser Verlag, Basel, p. 1.

Vallee, B.L., and Auld, D.S. (1990). *Biochemistry, 29*: 5647.

Vallee, B.L., and Auld, D.S. (1991). *Methods in Protein Sequence Analysis* (H. Jörnvall, J.O. Höög and A.-M Gustavsson, eds.), Birkhäuser Verlag, Basel, p. 363.

Vallee, B.L., and Falchuk, K.F. (1981). *Philos. Trans. R. Soc. London, B 294*: 185.

Vallee, B.L., and Williams, R.J.P. (1968). *Chem. Brit., 4*: 397.

Van Wart, H.E., and Birkendal-Hansen, H. (1990). *Proc. Nat. Acad. Sci. USA, 87*: 5578.

Williams, R.J.P. (1984). *Endeavour, 8*: 65.

Windsor, L.J., Birkendal-Hansen, H., Birkendal-Hansen, B., and Engler, J.A. (1991). *Biochemistry, 30*: 641.

Zhang, F., Kobe, B., Stewart, C.B., Rutter, W.J., and Goldsmith, E.J. (1991). *J. Biol. Chem., 266*: 24606.

MANGANESE-DEPENDENT PROTEINS

by

Murray Korc

Departments of Medicine and Biological Chemistry,
University of California,
Irvine, California 92717
U.S.A.

I. INTRODUCTION

This review focuses on manganese-protein interactions. Manganese is an essential trace element that exists in several oxidative states . Mn^{2+} is the preferred state in metal enzyme complexes and in metalloenzymes, where it often exhibits a high affinity for imidazole. In this regard, it is similar to Fe^{3+}. Mn^{3+} appears to be the form in which manganese is bound to certain proteins such as transferrin. Because of its unique chemical characteristics, its ability to interact with ATP and GTP, and its negligible energy expense, manganese may act as an intracellular integrator of many biological processes in eukaryotic cells (Williams, 1982).

II. MANGANESE PROTEINS - WORK PROTEINS

A. Metalloproteins containing manganese

In contrast to trace elements such as iron, there are relatively few manganese-containing proteins. With the exception of transport proteins that will be discussed later, none of the manganese-containing proteins can be classified as true work proteins. A protein called avimagnin was originally isolated from avian liver and it was thought to be a manganese-containing protein devoid of enzymatic activity (Scrutton, 1971). This protein is now believed to be manganese-dependent superoxide dismutase (Mn-SOD), which will be discussed later. A manganese-containing peanut protein termed manganin has been reported, but its role is not known (Dickert and Kozacky, 1969). Finally, plant lectins such as Concanavalin A (Agrawal and Goldstein, 1968), phytohemagglutinins from *phaseolus lunatus* (Galbraith and Goldstein, 1970), and the *Dolichos biflorus* lectin (Borrebaeck et al., 1981) are also manganese-containing proteins.

III. MANGANESE PROTEINS - METALLOENZYMES

A. Introduction

In general, there are two types of interactions between manganese and enzymes. In a limited number of cases, manganese is an essential component of enzymes. In most instances, however, manganese functions as a cofactor by interacting with a phosphate-containing substrate such as ATP, or by interacting directly with the protein. As a result of their similar configurations, Mg^{2+} and Mn^{2+} can often substitute for each other in their interactions with ATP. In the case of Mg^{2+}, the interaction is with the β and γ phosphate moieties, whereas in the case of Mn^{2+}, the interaction is with all three phosphates (Cohn and Hughes, 1962).

McEuen (1981) proposed that enzyme activation by Mg^{2+} and Mn^{2+} can be classified as Mg^{2+}-preferring or Mn^{2+}-preferring, or as optimal with either divalent cation. Transition elements such as cobalt or chromium, which are located next to manganese on the periodic table, can also sometimes substitute for manganese in enzyme activation reactions. However, as will be discussed, there are some enzymes whose activation is specifically dependent on manganese.

B. Manganese-containing enzymes

Manganese may be an integral component of several enzymes. It is an essential component of pyruvate carboxylase, an enzyme with a molar mass of 500 kg mol^{-1} that catalyzes the carboxylation of pyruvate to oxaloacetate (Scrutton et al., 1973). Each mole of pyruvate carboxylase, which is located in the mitochondria, possesses four atoms of Mn^{2+} (Scrutton et al., 1973) on its surface (Fung and Mildvan, 1973). Manganese allows for the activation of the methyl group of pyruvate, which can then accept carbon dioxide in the presence of ATP, thereby leading to the generation of oxaloacetate, ADP, and inorganic phosphorus. This reaction is important in gluconeogenesis.

Manganese is also an essential component of Mn-SOD, which has a molar mass of 80 kg mol^{-1} and four atoms of Mn^{2+} (Weisiger and Fridovich, 1973). Mn-SOD is a mitochondrial enzyme that catalyzes the conversion of the superoxide anion radical to oxygen and hydrogen peroxide (Hassan, 1988), thereby protecting the cell from damaging oxygen radicals. Mn-SOD activity is increased in the liver and kidney of rats following prolonged ingestion of ethanol (Dreosti et al., 1982), and Mn-SOD mRNA levels are induced by tumor necrosis factor and interleukin-1, which may be the mechanism underlying their protective effect against radiation (Wong and Goeddel, 1988). Mn-SOD may also have the capacity to suppress the malignant phenotype of certain cancer cells (Church et al., 1993).

Arginase is another manganese-containing enzyme (Hirsch-Kolb et al., 1971). This cytosolic enzyme has a molar mass of 120 kg mol^{-1} and participates in the urea-forming cycle. It contains four atoms of Mn^{2+} per mole of enzyme (Hirsch-Kolb et al., 1971). The removal of two of the manganese molecules leads to irreversible loss of enzymatic activity, whereas the removal of the other two manganese moieties results in only partial and reversible loss of activity. The increased activity of hepatic arginase in streptozotocin-diabetic rats may be due to an adaptive response that is mediated by elevated hepatic manganese levels in these animals (Bond et al., 1983).

Calmodulin-dependent protein phosphatase is a heterodimer consisting of subunit A (60 kg mol^{-1}) which possesses the catalytic activity of the enzyme and subunit B (19 kg mol^{-1}) which binds four atoms of calcium but is inactive (Klee et al., 1979). The catalytic activity of subunit A of this phosphatase requires Mn^{2+}, and may be a manganese-containing metalloenzyme (Merat et al., 1984). Interestingly, Co^{2+} and Ni^{2+} can partially substitute for Mn^{2+} in vitro, whereas Mg^{2+}, Ca^{2+}, and Zn^{2+} cannot mimic the effects of Mn^{2+} (Merat et al., 1984).

Manganese is also an essential component of several bacterial and plant enzymes. Pseudocatalase, from *Lactobacillus plantarum*, has an approximate molar mass of 172 kg mol^{-1}, and is believed to contain 7 to 9 atoms of manganese in the form of Mn^{3+} (Kono and Fridovich, 1983). Alpha-isopropyl malate synthetase from *alcaligenes eutrophus H 16* contains Mn^{2+} as the functional metal ion (Wiegel, 1978). Extradiol-cleaving dioxygenase from *Bacillus brevis* also contains Mn^{2+} (Sakurai et al., 1983). In contrast, the oxygen-evolving center of photosynthesis in plants which participates in the generation of oxygen from water contains four manganese atoms per enzyme in the form of both Mn^{3+} and Mn^{4+} (Dismukes and Siderer, 1981).

C. Manganese-dependent enzymes

As discussed above, the activation of a variety of enzymes can often be accomplished by different divalent cations. Therefore, the demonstration of enzyme activation by manganese in vitro does not necessarily imply a physiological role in vivo. Schramm (1982) proposed several criteria to help delineate the in vivo role of manganese in enzyme activation. Thus, at physiological substrate concentrations, the kinetic constants for the interactions with manganese must approximate the concentration of free manganese in the cell which he proposed may be modulated physiologically; the binding sites for manganese must prefer manganese over magnesium; and the amount of exchangeable manganese within the cell must be equal or greater than the molarity of the enzyme in question.

It is often difficult, if not impossible, to establish Schramm's strict criteria. Nonetheless, it is generally accepted that manganese can activate numerous enzymes in vivo. Furthermore, the activation of a number of enzymes has been shown to be specifically dependent on manganese. Thus, a class of enzymes known collectively as glycosyltransferases are specifically dependent on the availability of manganese in vivo, as underscored by the deficient synthesis of glycosaminoglycans in manganese-deficient animals (Leach, 1971). This group of enzymes includes ganglioside galactosyl transferase (Yip and Dain, 1970), xylosyl transferase (McNatt et al., 1976), and galactosyl transferase (Tsopanakis and Herries, 1978). Interestingly, only one of the two metal-binding sites in galactosyl transferase is specific for manganese (Tsopanakis and Herries, 1978).

More recently, kinases that are specifically activated by manganese have been demonstrated in HL-60 cells (Elias and Davis, 1985), stomach (Goldering et al., 1987) and pancreas (Knight et al., 1991). In the pancreas, manganese activates calmodulin-dependent protein kinase III, an enzyme that enhances the phosphorylation of elongation factor 2 (Knight et al., 1991). In vitro, this effect was not mimicked by Co^{2+}, Ni^{2+}, Sr^{2+}, Co^{2+}, Ca^{2+}, Ba^{2+} or Mg^{2+} (Knight et al., 1991). In *Bacillus megaterium*, the enzyme 3-p-glycerate mutase is active only in the presence of manganese (Kuhn et al., 1993).

A number of enzymes that are not specifically dependent on manganese nonetheless preferentially use manganese over magnesium as a cofactor. Thus, guanylate cyclase is preferentially activated by manganese (Ramaro and Garbers, 1988). Phosphoenolpyruvate carboxykinase is an enzyme that decarboxylates oxaloacetate to phosphoenolpyruvate and that is also preferentially activated by manganese (Bentle and Lardy, 1976). However, in the presence of a cytoplasmic protein called ferroactivator, iron can effectively compete with manganese for activation of the enzyme (MacDonald et al., 1978). Manganese is also the preferred divalent cation for activating a number of tyrosine kinases, including the insulin receptor tyrosine kinase (Czech, 1985; Rosen and Lebwohl, 1988) and the fibroblast growth factor receptor kinase (Kuo et al., 1990). This may be due, in part, to the ability of manganese, in contrast to magnesium, to act as a nonenzymatic catalyst that allows the transfer of phosphate from ATP to tyrosine (Schieven and Martin, 1988).

In a less specific manner, manganese has also been implicated in the activation of numerous additional enzymes. These include the catalytic component of adenylate cyclase (Walton and Baldessarini, 1976; Bender and Neer, 1983; Brown and Casillas, 1986; Jackman and Bobik, 1989), catechol-O-methyl transferase (Tagliamonte et al., 1970), farnesyl pyrophosphate synthetase (Benedict et al., 1965), glycosyl synthetase (Nakano and Kimura, 1987), phosphodiesterases (Rennie and Upton, 1981), and manganese-dependent peroxidase (Pease et al., 1989).

Manganese has been shown to modulate the activity of several enzymes that are involved in phosphatidylinositol (PI) synthesis. PI is generated when phosphatidic acid is converted to cytidine diphosphate-diglyceride (CDP-DG), and the CDP is replaced by free inositol through the actions of CDP-DG: inositol transferase, whose activity is enhanced by manganese (Paulus and Kennedy, 1960). Manganese also enhances the incorporation of inositol into a distinct phosphatidylinositol pool (Schoepp, 1985; Labarca et al., 1985) that is not directly coupled to agonist-induced phospholipid hydrolysis. Conversely, manganese inhibits myoinositol-1-phosphatase, an enzyme that hydrolyzes myoinositol-1-phosphate (Hallcher and Sherman, 1980). Manganese also inhibits the enzymatic activity of inositol polyphosphate 1-phosphatase (Inhorn and Majerus, 1987), which hydrolyzes $Ins(1,3,4)P_3$ to $Ins(3,4)P_2$, and $Ins(1,4)P_2$ to Ins(4)P (Inhorn and Majerus, 1987). Cyclic GMP-dependent protein kinase is another example of an enzyme whose activity is inhibited by manganese (Nakazawa and Sano, 1975). Together, these observations suggest that manganese may be involved in regulating ligand-mediated activation of signal transduction pathways (Korc et al., 1991).

At times, manganese-mediated enzyme activation may also appear to lead to opposing metabolic effects. Thus, manganese activates phosphorylase kinase (Hallenbeck and Walsh, 1983), an enzyme that activates glycogen phosphorylase which catalyzes glycogen breakdown. However, manganese also activates glycogen synthetase (Huang and Robinson, 1977), an enzyme that enhances glycogenesis.

IV. MANGANESE STORAGE AND TRANSPORT PROTEINS

Most of the manganese that is absorbed from the gastrointestinal tract is bound to plasma α_2-macroglobulin in the portal circulation, each globulin molecule carrying one atom of manganese (Cotzias and Bertinchamps, 1960). Free and bound manganese in the portal blood flow are then taken up by the liver and excreted into the bile (Cotzias and Bertinchamps, 1960). Within the liver, manganese is especially abundant in nuclei and mitochondria (Sakurai et al., 1985). In the mitochondria, manganese is present as hydrated Mn^{2+} and Mn^{3+}, and as a matrix precipitates in the form of $Mn_3(PO_4)_2$ (Bragadin et al., 1983).

Hepatic Mn^{3+} can bind efficiently to the transport protein transferrin, in spite of the fact that the association constant of iron for transferrin is greater than that of manganese (Worwood, 1974). Mn^{2+} also binds to transferrin, but with a lower affinity (Scheuhammer and Cherian, 1985). Once it is bound by transferrin, manganese is released into the circulation where it can be taken up by other tissues (Gibbons et al., 1976; Scheuhammer and Cherian, 1985).

REFERENCES

Agrawal, B.B.L., and Goldstein, I.J. (1968). *Arch. Biochem. Biophys.*, *124*: 218.

Bender, J.L., and Neer, E.J. (1983). *J. Biol. Chem.*, *258*: 2432.

Benedict, C., Kett, J., and Porter, J. (1965). *Arch. Biochem. Biophys.*, *110*: 611.

Bentle, L.A., and Lardy, H.A. (1976). *J. Biol. Chem.*, *251*: 2916.

Bond, J.S., Failla, M.L., and Unger, D.F. (1983). *J. Biol. Chem.*, *258*: 8004.

Borrebaeck, C.A.K., Lönnerdal, B., and Etzler, M.E. (1981). *Biochemistry*, *20*: 4119.

Bragadin, M., Pozzan, T., and Azzone, G.F. (1983). *Eur. J. Biochem.*, *134*: 385.

Brown, M.A., and Casillas, E.R. (1986). *Arch. Biochem. Biophys.*, *244*: 719.

Church, S.L., Grant, J.W., Rindour, L.A., Oberley, L.W., Swanson, P.E., Meltzer, P.S., and Trent, J.M. (1993). *Proc. Nat. Acad. Sci. USA*, *90*: 3113.

Cohn, M., and Hughes, T.R. (1962). *J. Biol. Chem.*, *237*: 176.

Cotzias, G.C., and Bertinchamps, A.J. (1960). *J. Clin. Invest.*, *39*: 979.

Czech, M.P. (1985). *Annu. Rev. Physiol.*, *47*: 357.

Dickert, J.W., and Kozacky, E. (1969). *Arch. Biochem. Biophys.*, *134*: 473.

Dismukes, C.G., and Siderer, Y. (1981). *Proc. Nat. Acad. Sci. USA*, *78*: 274.

Dreosti, I.E., Manuel, S.J., and Buckley, R.A. (1982). *Brit. J. Nutr.*, *48*: 205.

Elias, L., and Davis, A. (1985). *J. Biol. Chem.*, *260*: 7023.

Fung, C.H., and Mildvan, A.S. (1973). *Biochemistry*, *12*: 620.

Galbraith, W., and Goldstein, I.J. (1970). *FEBS Lett.*, *9*: 197.

Gibbons, R.A., Dixon, S.N., Hallis, K., Russell, A.M., Sanson, B.F., and Symonds, H.W. (1976). *Biochim. Biophys. Acta*, *444*: 1.

Golderning, J.R., Oddsdottir, M., and Modlin, I.M. (1987). *Biochem. Biophys. Res. Commun.*, *142*: 559.

Hallcher, L.M., and Sherman, W.R. (1980). *J. Biol. Chem.*, *255*: 10896.

Hallenbeck, P.C., and Walsh, D.A. (1983). *J. Biol. Chem.*, *258*: 13493.

Hassan, H.M. (1988). *Free Rad. Biol. Med.*, *5*: 377.

Hirsch-Kolb, H., Kolb, H.J., and Greenberg, D.M. (1971). *J. Biol. Chem.*, *246*: 395.

Huang, K., and Robinson, J.C. (1977). *J. Biol. Chem.*, *252*: 3240.

Inhorn, R.C., and Majerus, P.W. (1987). *J. Biol. Chem.*, *262*: 15946.

Jackman, G.P., and Bobik, A. (1989). *Biochem. Pharmacol.*, *38*: 1091.

Klee, C.B., Crouch, T.H., and Krinks, M.H. (1979). *Proc. Nat. Acad. Sci. USA*, *76*: 6270.

Knight, A.B.S., Kohr, W., and Korc, M. (1991). *Biochim. Biophys. Acta*, *1092*: 196.

Kono, Y., and Fridovich, I. (1983). *J. Biol. Chem.*, *258*: 6015.

Korc, M., Chandrasekar, B., and Siwik, S.A. (1991). *Endocrinology*, *129*: 39.

Kuhn, N.J., Setlow, B., and Setlow, P. (1993). *Arch. Biochem. Biophys.*, *306*: 342.

Kuo, M.-D., Huang, S.S., and Huang, J.S. (1990). *J. Biol. Chem.*, *265*: 16455.

Labarca, R., Janowsky, A., and Paul, S.M. (1985). *Biochem. Biophys. Res. Commun.*, *132*: 540.

Leach, R.M., Jr. (1971). *Fed. Proc.*, *30*: 991.

MacDonald, M.J., Bentle, L.A., and Lardy, H.A. (1978). *J. Biol. Chem.*, *253*: 116.

McEuen, A.R. (1981). *Inorganic Biochemistry* (H.A.O. Hill, ed.), Burlington House, London, p. 249.

McNatt, M.L., Fiser, F.M., Elders, M.J., Kilgore, B.S., Smith, W.G., and Hughes, E.R. (1976). *Biochem. J.*, *160*: 211.

Merat, D.L., Hu, Z.Y., Carter, T.E., and Cheung, W.Y. (1984). *Biochem. Biophys. Res. Commun.*, *122*: 1389.

Nakano, Y., and Kimura, K. (1987). *Biochem. Biophys. Res. Commun.*, *142*: 475.

Nakazawa, K., and Sano, M. (1975). *J. Biol. Chem.*, *250*: 7415.

Paulus, H., and Kennedy, E.P. (1960). *J. Biol. Chem.*, *235*: 1303.

Pease, E.A., Andrawis, A., and Tien, M. (1989). *J. Biol. Chem.*, *264*: 13531.

Ramaro, C.S., and Garbers, D.L. (1988). *J. Biol. Chem.*, *263*: 1524.

Rennie, P.I.C., and Upton, J.D. (1981). *Hormone Metab. Res.*, *13*: 145.

Rosen, O.M., and Lebwohl, D.E. (1988). *FEBS Lett.*, *231*: 397.

Sakurai, H., Shimomura, S., and Yoshimura, T. (1983). *Biochem. Biophys. Res. Commun.*, *115*: 618.

Sakurai, H., Nishida, M., Yoshimura, T., Takada, J., and Koyama, M. (1985). *Biochim. Biophys. Acta*, *841*: 208.

Scheuhammer, A.M., and Cherian, M.G. (1985). *Biochim. Biophys. Acta*, *840*: 163.

Schieven, G., and Martin, G.S. (1988). *J. Biol. Chem.*, *263*: 15590.

Schoepp, D.D. (1985). *J. Neurochem.*, *45*: 1481.

Schramm, V.L. (1982). *Trends Biochem. Sci.*, *7*: 369.

Scrutton, M.C. (1971). *Biochemistry*, *21*: 3897.

Scrutton, M.C., Grimminger, D., and Wallace, J.C. (1973). *J. Biol. Chem.*, *247*: 3305.

Tagliamonte, A., Tagliamonte, P., and Gessa, G.L. (1970). *J. Neurochem.*, *17*: 733.

Tsopanakis, A.D., and Herries, D.G. (1978). *Eur. J. Biochem.*, *83*: 179.

Walton, K.G., and Baldessarini, R.J. (1976). *J. Neurochem.*, *27*: 557.

Weisiger, R.A., and Fridovich, I. (1973). *J. Biol. Chem.*, *248*: 3582.

Wiegel, J. (1978). *Biochem. Biophys. Res. Commun.*, *82*: 907.

Williams, R.J.P. (1982). *FEBS Lett.*, *140*: 3.

Wong, G.H.W., and Goeddel, D.V. (1988). *Science*, *242*: 941.

Worwood, M. (1974). *Iron Metabolism in Biochemistry and Medicine* (A. Jacobs and M. Worwood, eds.), Academic Press, New York, p. 336.

Yip, G.B., and Dain, J.A. (1970). *Biochim. Biophys. Acta*, *206*: 252.

COBALT-CONTAINING PROTEINS

by

Sven Lindskog

Department of Biochemistry,
University of Umeå,
S-90187 Umeå
Sweden

I. INTRODUCTION

In comparison with iron, copper and zinc, cobalt plays a minor role in biology. In mammals, only a small number of functions have been described at the molecular level. These do not involve direct cobalt-protein interactions, but cobalt sequestered in vitamin B_{12} and its derivatives. A few reports in the literature concern naturally occurring cobalt in non-B_{12} proteins, but most of them have only been found in bacteria.

The main body of literature on cobalt-protein interactions concerns the use of Co(II) as a spectroscopic probe in proteins normally containing zinc, copper or iron. The slow ligand exchange rates typically displayed by Co(III) complexes have also been utilized.

II. COBAMIDE-BINDING PROTEINS

Cobalt in vitamin B_{12} and its derivatives is bound to four nitrogen atoms of a porphyrin-like, *corrin* ligand (Fig. 1). A 5,6-dimethylbenzimidazole nucleotide is covalently linked to the corrin ring system and contributes one of its imidazole ring nitrogens as a fifth cobalt ligand in an axial position to form the *cobalamin* structure. In methanogenic bacteria, the presence of other *cobamides*, having 5-hydroxybenzimidazole or adenine rather than 5,6-dimethylimidazole, has been demonstrated (Di Marco et al., 1990).

The term *coenzyme* B_{12} usually refers to 5'-deoxyadenosylcobalamin, where cobalt is bound to the 5'-carbon atom of the deoxyadenosine moiety (Fig. 1), but methylcobamides are also known to function as cofactors or intermediates in some enzymatic reactions. The physical properties of these compounds suggest that they are best described as carbanion complexes of Co(III). The functions of these coenzymes involve the dissociation of the cobalt-carbon bond. In principle, this can occur in three different ways, namely heterolytic dissociation into a carbanion and Co(III)-cobamide, homolytic dissociation into a neutral free radical and Co(II)-cobamide, and heterolytic dissociation into a carbonium ion and a highly reactive Co(I)-cobamide.

Ten coenzyme B_{12}-dependent enzymes have been described (Table 1). Most of them are found exclusively in bacteria, but one, methylmalonyl-CoA mutase, occurs also in mammals. While ribonucleotide reductase catalyzes a redox reaction, the other coenzyme B_{12}-dependent enzymes catalyze reactions involving the interchange of a substituent on one carbon atom and a hydrogen atom on an adjacent carbon atom as indicated in equation 1. In

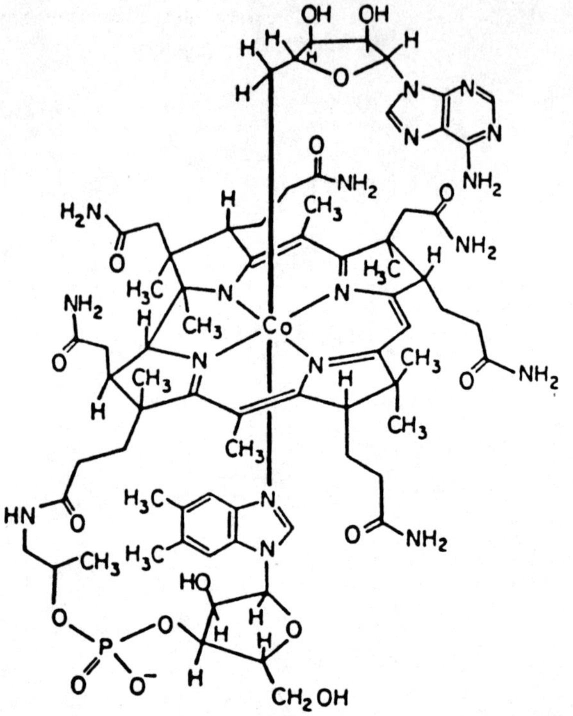

Figure 1.
Coenzyme B_{12}. From Halpern (1985). Courtesy of Science. Copyright 1985 by the AAAS.

Table 1

Coenzyme B_{12}-Dependent Enzymes

Enzyme	EC number
Glutamate mutase	5.4.99.1
Methylmalonyl-CoA mutase	5.4.99.2
α-Methyleneglutarate mutase	5.4.99.4
L-β-Lysine mutase	5.4.3.3
D-α-Lysine mutase	5.4.3.4
Ornithine mutase	5.4.3.5
Ethanolamine deaminase	4.3.1.7
Diol dehydrase	4.2.1.28
Glycerol dehydrase	4.2.1.30
Ribonucleotide reductase*	1.17.4.2

*in eukaryotes and many bacteria (including *E. coli*) ribonucleotide reductase (EC 1.17.4.1) is a non-B_{12} enzyme containing non-heme iron and a tyrosine free radical.

the cases of diol dehydrase, glycerol dehydrase and ethanolamine deaminase, this rearrangement is followed by the release of water or ammonia.

$$\begin{array}{cc} X & H \\ | & | \\ -C_1 - & C_2 - \\ | & | \end{array} \rightarrow \begin{array}{cc} H & X \\ | & | \\ -C_1 - & C_2 - \\ | & | \end{array} \qquad (1)$$

It seems well documented that the first step in these rearrangement reactions is the homolytic cleavage of the cobalt-carbon bond in the coenzyme. The 5'-deoxyadenosyl radical then abstracts a hydrogen atom from the substrate molecule to form 5'-deoxyadenosine and a substrate radical. The mechanism of the subsequent group migration is unclear (Abeles and Dolphin, 1976; Babior and Krouwer, 1979) but might occur on the free radical stage without participation of the coenzyme (Halpern, 1985). Eventually a hydrogen atom is donated back to the product radical from 5'-deoxyadenosine, and the coenzyme can return to its resting state. The lability of the cobalt-carbon bond, thus, is of crucial importance for the function of coenzyme B_{12}. Halpern (1985) has summarized the role of cobalt in the coenzyme as that of a "reversible free radical carrier".

Methylcobalamin appears as an intermediate in the reaction catalyzed by methionine synthetase (EC 2.1.1.13), namely methyl group transfer between N^5-methyltetrahydrofolate and homocysteine to form tetrahydrofolate and methionine (Barker, 1972). The enzyme occurs in *E. coli* as well as in mammals and requires adenosylmethionine and a reducing system (such as a reduced flavin) as additional cofactors. The methyl group is probably transferred as CH_3^+ so that the cobalt ion shuttles between the +3 and +1 oxidation states during substrate turnover (Barker, 1972).

Methylcobamides are also involved in a number of bacterial methyl transfer reactions, for example, in the formation of acetate from CO_2 and H_2 in *Clostridium thermoaceticum* and in the formation of methane from CO_2, methanol or acetate in methanogenic bacteria (Barker, 1972; Di Marco et al., 1990).

It should also be mentioned that methylcobalamin has been implicated in the methylation of various toxic metal ions. Model studies suggest that methylation of Hg(II) occurs by transfer of CH_3^- implying heterolytic dissociation of the cobalt-carbon bond without change of the formal oxidation state of the cobalt ion (Ridley et al., 1977).

Vitamin B_{12}, the physiological form of which probably is aquo- or hydroxycobalamin, cannot be synthesized by mammals but must be provided either by intestinal microorganisms (ruminants) or in the diet. The uptake and transport of cobalamins require a number of specialized proteins. In man, dietary cobalamin binds to the gastric glycoprotein "*intrinsic factor*" which transports the vitamin to the ileum, where a mucosal receptor facilitates the transfer of cobalamin to the blood. In blood plasma a number of B_{12}-binding proteins (*transcobalamins*) carry out further transport to tissue cells (Sennett et al., 1981).

III. OTHER COBALT-CONTAINING PROTEINS

A few non-B_{12}, cobalt-containing proteins have been reported, mainly from bacterial sources. Methylmalonyl-CoA carboxyltransferase from *Propiobacterium shermanii* is a biotin enzyme containing functionally important Co(II) when the bacteria have been grown in cobalt-enriched media (Northrop and Wood, 1969). Electron spin resonance spectra suggest that the enzyme contains high-spin Co(II) in a distorted octahedral environment (Fung et al., 1974).

A cobalt-containing nitrile hydratase, catalyzing the hydration of aliphatic nitriles to amides, has been purified from urea-induced cells of a strain of *Rhodococcus rhodochrous* (Nagasawa et al., 1991). The enzyme might have a $\alpha_{10}\beta_{10}$ structure with one tightly bound cobalt atom per subunit.

Cobalt porphyrin-containing proteins have been isolated from *Desulfovibrio gigas* (Moura et al., 1980) and *D. desulfuricans* (Hatchikian, 1981). It has been suggested that the porphyrin is a Co(III)-isobacteriochlorin (Battersby and Sheng, 1982). An enzyme from the green alga *Botryococcus braunii* catalyzing the formation of hydrocarbons by release of CO from aliphatic aldehydes has recently been reported to contain cobalt, probably in the form of a cobalt porphyrin (Dennis and Kolattukudy, 1992).

IV. COBALT AS A STRUCTURAL PROBE IN METALLOPROTEINS

Metal substitution is a powerful experimental approach in the study of metalloproteins. The spectroscopic and magnetic properties of Co(II) have been utilized extensively in a large variety of such investigations. In particular, the electronic spectra of Co(II)-substituted proteins have been used to extract structural information and to monitor events occurring in the protein molecule in the vicinity of the metal ion, but nuclear magnetic resonance, electron spin resonance, magnetic circular dichroism and resonance Raman spectroscopy have also been employed.

The substitution of Co(II) for Zn(II) has turned out to be especially useful since Zn(II) is a spectroscopically "silent" metal ion (Lindskog, 1970; Bertini and Luchinat, 1983a) and the Co(II) derivatives usually have "native-like" functional properties. Intensive studies have been performed on Co(II)-substituted carbonic anhydrase (Bertini et al., 1982; Bertini and Luchinat, 1983b; Lindskog, 1983; Silverman and Lindskog, 1987), alcohol dehydrogenase (Makinen and Yim, 1981; Dunn et al., 1982; Pocker, 1989), alkaline phosphatase (Banci et al., 1988; Coleman, 1992) and carboxypeptidase (Vallee et al., 1983; Auld et al., 1992), but many additional zinc metalloenzymes and -proteins have been investigated by this approach (Bertini and Luchinat, 1983a). Since a detailed account of these studies is beyond the scope of this chapter, the reader is referred to a few selected references on angiotensin (Bicknell et al., 1987), dihydroorotase (Brown and Collins, 1991), glyoxalase I (Sellin et al., 1983), phospholipase C (Bicknell et al., 1986a), β-lactamase (Bicknell et al., 1986b), *E. coli* RNA polymerase (Giedroc and Coleman, 1986), gene 32 protein from bacteriophage T4 (Giedroc et al., 1986) and zinc finger peptides (Green and Berg, 1989; Michael et al., 1992).

Not only Zn(II) but also Cu(II) in copper-zinc superoxide dismutase has been replaced by Co(II), and extensive spectroscopic studies have been performed on various Co(II)-substituted derivatives (Banci et al., 1987; 1989; Bertini et al., 1991; Desideri et al., 1984; Salvato et al., 1989).

Co(II)-substituted blue copper proteins have also been studied (Solomon et al., 1976; Tennent and McMillin, 1979; Hauenstein and McMillin, 1981) as well as Co(II)-substituted copper sites in hemocyanin (Lorösch and Haase, 1986) and tyrosinase (Rüegg and Lerch, 1981).

The use of Co(II) as a probe of non-heme iron sites can be exemplified by studies on ovotransferrin (Bertini et al., 1986), hemerythrin (Zhang and Kurtz, 1992) and rubredoxin (May and Kuo, 1978; Moura et al., 1991).

While reduced heme proteins are not accessible to electron spin resonance studies, their Co(II) analogues are. This has been utilized in several studies. The reader is referred to the papers by Wagner et al. (1981) and Lee et al. (1992) and references therein.

The versatility of Co(II) as a probe of metal binding sites can also be illustrated by studies on the Mg(II)-dependent enzymes enolase (Rose et al., 1984) and pyruvate kinase (Kwan et al., 1975).

Exchange-inert Co(III) has been utilized to probe the importance of ligands in the first coordination sphere for protein function as well as a spectroscopic probe (Shinar and Navon, 1974; 1979; Urdea and Legg, 1979; Van Wart and Vallee, 1978; Werber and Lanir, 1981).

REFERENCES

Abeles, R.H., and Dolphin, D. (1976). *Accounts Chem. Res.*, *9*: 114.

Auld, D.S., Bertini, I., Donaire, A., Messori, L., and Moratal, J.M. (1992). *Biochemistry*, *31*: 3840.

Babior, B.M., and Krouwer, J.S. (1979). *CRC Crit. Rev. Biochem.*, *6*: 35.

Banci, L., Bertini, I., Luchinat, C., Monnanni, R., and Scozzafava, A. (1987). *Inorg. Chem.*, *26*: 153.

Banci, L., Bertini, I., Luchinat, C., Viezzoli, M.S., and Wang, Y. (1988). *J. Biol. Chem.*, *263*: 11263.

Banci, L., Bertini, I., Luchinat, C., Piccioli, M., Scozzafava, A., and Turano, P. (1989). *Inorg. Chem.*, *28*: 4650.

Barker, H.A. (1972). *Annu. Rev. Biochem.*, *41*: 55.

Battersby, A.R., and Sheng, Z.C. (1982). *J. Chem. Soc. Chem. Commun.*, 1393.

Bertini, I., and Luchinat, C. (1983a). *Metal Ions in Biological Systems*, Vol. 15 (H. Sigel, ed.), Marcel Dekker, Inc., New York and Basel, p. 101.

Bertini, I., and Luchinat, C. (1983b). *Accounts Chem. Res.*, *16*: 272.

Bertini, I., Luchinat, C., and Scozzafava, A. (1982). *Struct. Bonding (Berlin)*, *48*: 45.

Bertini, I., Luchinat, C., Messori, L., Monnanni, R., and Scozzafava, A. (1986). *J. Biol. Chem.*, *261*: 1139.

Bertini, I., Luchinat, C., Piccioli, M., Oliver, M.V., and Viezzoli, M.S. (1991). *Eur. Biophys. J.*, *20*: 269.

Bicknell, R., Hanson, G.R., Holmquist, B., and Little, C. (1986a). *Biochemistry*, *25*: 4219.

Bicknell, R., Schäffer, A., Waley, S.G., and Auld, D.S. (1986b). *Biochemistry*, *25*: 7208.

Bicknell, R., Holmquist, B., Lee, F.S., Martin, M.T., and Riordan, J.F. (1987). *Biochemistry*, *26*: 7291.

Brown, D.C., and Collins, K.D. (1991). *J. Biol. Chem.*, *266*: 1597.

Coleman, J.E. (1992). *Annu. Rev. Biophys. Biomol. Struc.*, *21*:441.

Dennis, M., and Kolattukudy, P.E. (1992). *Proc. Nat. Acad. Sci. USA*, *89*: 5306.

Desideri, A., Cocco, D., Calabrese, L., and Rotilio, G. (1984). *Biochim. Biophys. Acta*, *785*: 111.

Di Marco, A.A., Bobik, T.A., and Wolfe, R.S. (1990). *Annu. Rev. Biochem.*, *59*: 355.

Dunn, M.F., Dietrich, H., MacGibbon, A.K.H., Koerber, S.C., and Zeppezauer, M. (1982). *Biochemistry*, *21*: 354.

Fung, C.-H., Mildvan, A.S., and Leigh, J.S. (1974). *Biochemistry*, *13*: 1160.

Giedroc, D.P., and Coleman, J.E. (1986). *Biochemistry*, *25*: 4969.

Giedroc, D.P., Keating, K.M., Williams, K.R., Konigsberg, W.H., and Coleman, J.E. (1986). *Proc. Nat. Acad. Sci. USA*, *83*: 8452.

Green, L.M., and Berg, J.M. (1989). *Proc. Nat. Acad. Sci. USA*, *86*: 4047.

Halpern, J. (1985). *Science*, *227*: 869.

Hatchikian, E.C. (1981). *Biochem. Biophys. Res. Commun.*, *103*: 521.

Hauenstein, B.L., and McMillin, D.R. (1981). *Metal Ions in Biological Systems*, Vol. 13 (H. Sigel, ed.), Marcel Dekker, Inc., New York and Basel, p. 319.

Kwan, C.-Y., Erhard, K., and Davis, R.C. (1975). *J. Biol. Chem.*, *250*: 5951.

Lee, H.C., Ikeda-Saito, M., Yonetani, T., Magliozzo, R.S., and Peisach, J. (1992). *Biochemistry*, *31*: 7274.

Lindskog, S. (1970). *Struct. Bonding (Berlin)*, *8*: 153.

Lindskog, S. (1983). *Zinc Enzymes* (T.G. Spiro, ed.), Wiley Interscience, New York, p. 77.

Lorösch, J., and Haase, W. (1986). *Biochemistry*, *25*: 5850.

Makinen, M.W., and Yim, M.B. (1981). *Proc. Nat. Acad. Sci. USA*, *78*: 6221.

May, S.W., and Kuo, J.Y. (1978). *Biochemistry*, *17*: 3333.

Michael, S.F., Kilfoil, V.J., Schmidt, M.H., Amann, B.T., and Berg, J.M. (1992). *Proc. Nat. Acad. Sci. USA*, *89*: 4796.

Moura, I., Teixeira, M., Le Gall, J., and Moura, J.J.G. (1991). *J. Inorg. Biochem.*, *44*: 127.

Moura, J.J.G., Moura, I., Bruschi, M., Le Gall, J., and Xavier, A.V. (1980). *Biochem. Biophys. Res. Commun.*, *92*: 962.

Nagasawa, T., Takeuchi, K., and Yamada, H. (1991). *Eur. J. Biochem.*, *196*: 581.

Northrop, D.B., and Wood, H.G. (1969). *J. Biol. Chem.*, *244*: 5801.

Pocker, Y. (1989). *Metal Ions in Biological Systems*, Vol. 25 (H. Sigel, ed.), Marcel Dekker, New York and Basel, p. 335.

Ridley, W., Dizikes, L., and Wood, J. (1977). *Science*, *197*: 329.

Rose, S.L., Dickinson, L.C., and Westhead, E.W. (1984). *J. Biol. Chem.*, *259*: 4405.

Rüegg, Ch., and Lerch, K. (1981). *Biochemistry*, *20*: 1256.

Salvato, B., Beltramini, M., Ricchelli, F., and Tallandini, L. (1989). *Biochim. Biophys. Acta*, *998*: 14.

Sellin, S., Eriksson, L.E., Aronsson, A.E., and Mannervik, B. (1983). *J. Biol. Chem.*, *258*: 2091.

Sennett, C., Rosenberg, L.E., and Mellman, I.S. (1981). *Annu. Rev. Biochem.*, *50*: 1053.

Shinar, H., and Navon, G. (1974). *Biochim. Biophys. Acta*, *334*: 471.

Shinar, H., and Navon, G. (1979). *Eur. J. Biochem.*, *93*: 313.

Silverman, D.N., and Lindskog, S. (1987). *Accounts Chem. Res.*, *21*: 30.

Solomon, E.I., Rawlings, J., McMillin, D.R., Stephens, P.J., and Gray, H.B. (1976). *J. Amer. Chem. Soc.*, *98*: 8046.

Tennent, D.L., and McMillin, D.R. (1979). *J. Amer. Chem. Soc.*, *101*: 2307.

Urdea, M.S., and Legg, J.I. (1979). *Biochemistry*, *18*: 4984.

Vallee, B.L., Galdes, A., Auld, D.S., and Riordan, J.F. (1983). *Zinc Enzymes* (T.G. Spiro, ed.), Wiley Interscience, New York, p. 26.

Van Wart, H.E., and Vallee, B.L. (1978). *Biochemistry*, *17*: 3385.

Wagner, G.C., Gunsalus, I.C., Wang, M.-Y.R., and Hoffman, B.M. (1981). *J. Biol. Chem.*, *256*: 6266.

Werber, M.M., and Lanir, A. (1981). *Isr. J. Chem.*, *21*: 34.

Zhang, J.-H., and Kurtz, D.M. (1992). *Proc. Nat. Acad. Sci. USA*, *89*: 7065.

ESSENTIAL ALKALI, ALKALINE EARTH, AND LANTHANIDE METAL-PROTEIN INTERACTIONS

by

George Eisenman[1], Osvaldo Alvarez[2] and Frederick S. Lee[1]

[1]Department of Physiology,
UCLA Medical School,
Los Angeles, CA 90024
U.S.A.

and

[2]Departamento de Biologia,
Facultad de Ciencias,
Universidad de Chile,
Casilla 653
Santiago
Chile

I. INTRODUCTION

This chapter summarizes the essential interactions that occur between ions, water molecules, and ion-binding sites in proteins. It discusses the energetic and structural factors that contribute to the ion selectivity in protein ion-binding sites whose structure can not only rearrange to accommodate ions of differing size and charge, but whose hydration state can also change as a function of the size and charge of the ion bound. Only cations are discussed explicitly, but similar principles can be expected to apply to anions as well.

The structural and energetic consequences of exchanging one ion for another of different size and charge can be studied by Molecular Dynamics (MD) simulations. We summarize here the results of our recent Free Energy Perturbation (FEP) studies for the interaction of monovalent and divalent cations with valinomycin in order to analyze the individual energy terms that contribute to the experimentally measured selectivity in a depsipeptide prototype system held together with the forces present in proteins and whose structure is allowed to rearrange under the same force fields that exist in proteins. We also use MD simulations to explore the structural and energetic changes that can occur on exchanging ions in a deformable viral Ca-binding site in which ions and water molecules are simultaneously present.

Since the selectivity of this site for trivalent lanthanide cations is known as well as its hydration state (Burroughs et al., 1992), it offers a model system of considerable interest. As an initial step toward simulating this, we have developed a preliminary set of Lennard-

Jones 6-12 parameters for the trivalent lanthanide cations from simulations of experimental hydration free energies and radial distribution functions using the conventional unpolarizable rigid SPC water molecule (Appleby, Alvarez and Eisenman, unpublished results). Though these parameters will probably need to be revised to include polarizability of the water molecule, they provide the necessary first step for any simulations of lanthanide binding to the viral binding site and more generally proteins.

Finally, since polarizability is expected to be increasingly important as modelling of ion binding is extended to ions of higher charge, we extend our simulations to include ion-inducible dipole interactions and show that it is necessary to include inducible dipoles when simulating the experimentally measured mono- and divalent selectivity of valinomycin in methanol.

A. Principles of ionic selectivity

1. *Ion-water vs ion-site energies*

Ionic selectivity depends on a balance between the free energies of interaction of the ions with the binding structure and their free energies of hydration. Specifically, selective binding to an ion-binding site depends upon how the free energy of interaction of the ion with the protein, relative to its free energy of hydration, varies with the size and charge of the ion (Eisenman, 1961; 1962; 1965). Both of these energy terms can be studied by Molecular Dynamics simulations which give insight into their molecular origins.

2. *Energies and forces involved*

In the simplest case where only an ion (but no water molecules) can enter the site, four major non-bond energy terms contribute to the energy of ion-protein interaction. This situation exists for ion-valinomycin complexes in low dielectric solvents; and experimental values for the free energy differences underlying ion selectivity have been measured accurately (see Eisenman et al., 1991). The *ion-protein* energies are all non-bonded and consist of: (i) *Coulomb*: an inverse 1st power electrostatic interaction between the ion and all the partial charges of the protein atoms; (ii) *Polarizability*: an inverse 2nd power induced-dipole interaction between the ion and the polarizable atoms of the protein; (iii) *Van der Waals attraction*: an inverse 6th power attractive interaction between the ion and the protein; (iv) *Van der Waals repulsion*: an inverse 12th power repulsive interaction between the ion and the protein.

There is also a *conformational energy* due to rearrangements of the protein atoms that occur when different ion species are bound. This consists of four bonded-energy terms (*bond stretching, bond angle bending, dihedral angle torsion*, and *inversion*) together with the above listed non-bonded interactions between the protein atoms themselves. If water molecules also are present within the site, corresponding ion-water, water-protein, and water-water energy terms must be calculated in addition.

3. *Structural implications of ion-water-site interactions*

Besides the implications for selectvity of ion-site-water interactions, these interactions also determine the local structure (e.g. the number and arrangement of water molecules that exist in the equilibrium conformation of the site, as will be discussed in III.C.1).

II. A TENTATIVE SET OF LENNARD-JONES PARAMETERS FOR THE TRIVALENT LANTHANIDE CATIONS

The interactions of cations with the usual ion-binding ligands in proteins (generally oxygen atoms) are characterized in terms of Coulomb, polarizability, and van der Waals potentials, the latter described using Lennard-Jones A and B parameters. Aqvist (1990) has published a complete set of Lennard-Jones parameters for the monovalent and divalent cations; but a corresponding set for the trivalent lanthanide cations does not exist. Since such a set is necessary as a first step in simulating lanthanide-protein interactions, we therefore present it in Table 1. This table gives a preliminary set of Lennard-Jones A and B parameters for the lanthanides (Appleby, Alvarez and Eisenman, unpublished results) obtained by MD/FEP simulations of experimental hydration free energies and radial distribution functions using

Table 1

A Tentative Set of Lennar-Jones Parameters for fhe Trivalent Lanthanide Cations fnd the Calculated Ionic Hydration Energies and Radii[(a)]

Ion	Hydration energy	Ionic radius	A	B
La	-759.3 (-758.5)	1.14 (1.14)	1035.00	76.50
Ce	-769.3 (-769.5)	1.12 (1.12)	930.99	75.69
Pr	-779.3 (-779.5)	1.10 (1.10)	851.05	77.37
Nd	-788.7 (-788.3)	1.10 (1.08)	785.00	78.38
Pm	-796.8 (-797.2)	1.06 (1.06)	722.25	78.58
Sm	-804.5 (-805.0)	1.03 (1.04)	671.41	79.64
Eu	-811.4 (-812.3)	1.02 (1.03)	634.38	80.27
Gd	-819.1 (-819.2)	1.02 (1.01)	592.86	81.13
Tb	-827.1 (-825.7)	1.00 (1.00)	554.66	82.13
Dy	-832.1 (-832.0)	1.00 (0.99)	522.53	81.32
Ho	-838.5 (-838.2)	0.98 (0.98)	495.65	82.16
Er	-845.0 (-844.4)	0.95 (0.96)	465.72	83.77
Tm	-850.5 (-850.6)	0.96 (0.95)	450.64	83.76
Yb	-856.1 (-856.9)	0.94 (0.94)	428.35	84.10
Lu	-864.0 (-863.6)	0.92 (0.93)	398.63	84.57

[(a)] Energies and distances are in units of kcal mol^{-1} and Å. A and B (in units of kcal$^{1/2}$ mole$^{-1/2}$ Å^6 and kcal$^{1/2}$ mol$^{-1/2}$ Å^3, respectively) are parameters for the Lennard-Jones potential: $V_{LJ} = \Sigma_{ij} A_i A_j r_{ij}^{-12} - B_i B_j r_{ij}^{-6}$ where A_i, A_j, B_i, B_j are respectively the A and B parameters for the *i*th and *j*th atoms, and r_{ij} is the distance between th two atoms. In parentheses are the experimental hydration energies and target ionic radii.

unpolarizable rigid SPC model water molecules. The agreement between experimental (target) free energies and radii with the calculated values obtained from the simulations can be seen in the table. Though the A and B parameters will probably need to be revised if the model water molecule is refined to include polarizability, they provide the necessary first step for any simulations of lanthanide binding to proteins.

III. MD / FEP SIMULATIONS WITHOUT INDUCED DIPOLES

A. Successful simulation in vacuo of valinomycin 2-phase salt extraction selectivity for monovalent cations

We have shown (Eisenman and Alvarez, 1991; Eisenman et al., 1992a) that it is possible to use MD/FEP procedures in vacuum to simulate the experimentally observed selectivity of valinomycin, a prototype peptide-like molecule, whose 2-phase salt extraction selectivity is independent of dielectric constant over a wide range, indicating that the complexes are "isosteric" (Eisenman et al., 1972; 1973) which validates the use of an in vacuo simulation.

B. Analysis of energies that contribute to valinomycin selectivity

To deepen our understanding of the factors involved in ion selectivity in proteins, we have used MD simulations to analyze the component energy terms that make up the overall valinomycin selectivity. Figure 1 summarizes our findings (Eisenman and Alvarez, 1992).

The free energies are presented relative to Cs; and the similar heights of the Na, K, and Rb energies between the 1st and 2nd set of bar graphs show that the differences in potential energy (dE) between the species closely parallel the free energy differences (dG). These energies underlie the observed non-lyotropic, Rb-preferring selectivity sequence

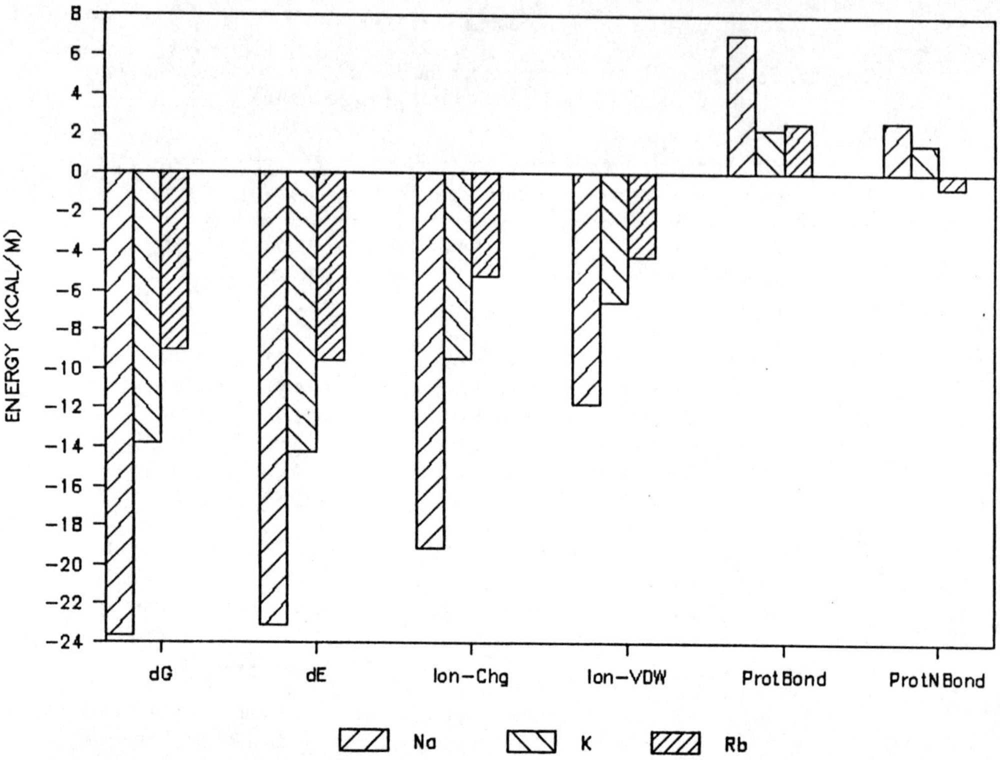

Figure 1.
Analysis of the Individual Components of the Internal Energy of Binding for the Indicated Ions. The values of Free Energy (dG) were obtained by FEP/MD simulations in which ions were reversibly "mutated" from Na to Cs and back to Na. The values of Internal Energy (dE) were calculated for the corresponding structures after steepest descent energy minimization at 1 degree Kelvin. All energies are given in kcal mol^{-1}.

Rb> K> Cs>> Na, which was simulated quantitatively (the selectivity is obtained immediately from dG by subtracting the differences in free energy of hydration).

The rest of the figure analyzes the four components that make up dE. As described above, these are the two non-bonded *ion-protein* interaction energies (ion-charge (Ion-Chg) and ion-van der Waals (Ion-VDW)) and the structural *protein-protein* interaction energies: both bonded (ProtBond) and non-bonded (ProtNBond). It is clear from the magnitudes of the various contributions to the total energy that, although the Ion-Chg term is the largest, there is no energy term that is negligible. However, the last 3 terms come close to cancelling and, reasoning from the different sources of selectivity in frozen vs movable systems, we have suggested that this could be understood as a kind of "stress-strain compensation" (Eisenman and Alvarez, 1992). In such a mechanism the *stress* due to high power van der Waals forces (both attractive and repulsive) of ions that do not exactly fit into the structure is redistributed by rearranging adjacent atoms which are subject to less sharply spatially dependent force fields. This of course constitutes a *strain*. The worse the misfit, the larger the strain, regardless whether the misfit is due to an ion being too large or too small. We suggest that such "stress-strain compensation" is not special to valinomycin and is likely to occur generally in ion binding to proteins.

C. Simulations for water-containing ion binding sites in proteins

1. *The 5-fold Ca and lanthanide binding site in the satellite tobacco necrosis virus (stnv) as an example*

Using the Lennard-Jones paramcters for Mg, Ca, Sr, and Ba determined by Aqvist (1990), Eisenman et al. (1992b) performed Molecular Dynamics simulations on the pentameric Ca binding, channel-like structure that exists on the 5-fold axis of the Satellite Tobacco Necrosis Virus (Jones and Liljas, 1984) using the GROMOS (Berendsen et al., 1981; van Gunsteren and Berendsen 1987) and MOLARIS (Warshel and Creighton 1989) force fields in the presence of flexible SPC water molecules. Although ion-induced dipole terms were not included in these simulations, the structural changes found can still be usefully interpreted. Eisenman et al. examined the role played by the partial charges of the carbonyl ion-binding groups in determining both selectivity and local structure (i.e. ion coordination and hydration) in the region of the "selectivity filter" formed by the ring of 5 ion-binding backbone carbonyl ligands. They found for high values of -C=O partial charge that the crystallographic 7-fold coordination of the Ca ion by a ring of 5 carbonyl oxygens and 2 water molecules is retained for Ca and Mg, but that for the larger Sr and Ba ions the coordination number is increased to 3 by an additional external water molecule. For high values of partial charge (comparable to the partial charge of -0.82 on the SPC water oxygen), both structure and free energy were reversible in "mutating" from Ca to Mg or from Ca to Sr and Ba and returning to Ca. However, for lower values of partial charge, including the default value of -0.383 for the GROMOS carbonyl, the free energy was reversible but the structure was not. This was found to be because the ions bind more strongly to water than to the carbonyl group for oxygen partial charges much lower than that of the SPC water molecule, and this imbalance was sufficient to disrupt the ring of carbonyls to yield an energetically almost equivalent locally hydrated state. For example, for low values of C=O partial charge, most ions hydrate almost completely, dilating the C=O ring like an iris; while for high values of partial charge exceeding 0.5, the coordination by 5 C=O's and 2 H_2O's is retained after equilibration with Ca or Mg, with ion-oxygen distances compatible with Pauling crystal radii. Replacing Ca by larger ions or a model lanthanide adds a 3rd (or more, in the case of the lanthanide) water of coordination. But these changes in hydration state and local structure are reversible on mutating back to Ca from larger or smaller ions.

We conclude that ion - ligand - water competition determines the local hydration state in a viral Ca-binding site and suggest that this conclusion should apply quite generally to ion-binding sites in proteins. Almost certainly, the very high values of partial charge necessary to preserve the reversibility of structure and the correct amount of local hydration reflect the necessity to increase the partial charge to compensate for the missing ion-induced dipole interactions in these simulations, as will be discussed in the following section.

IV. MD / FEP SIMULATIONS WITH INDUCED DIPOLES

A. The importance of ion-induced dipole interactions when dealing with divalent cations and ions of higher charge

Although simulating monovalent cation binding to the binding site typified by Valinomycin can apparently be done without including polarizability, as described above, this does not mean that differences in polarization produced by different monovalent species are really negligible. For more highly charged species, like the divalent alkaline earths and trivalent lanthanides, such an assumption needs to be checked. Indeed, Aqvist et al. (1992; 1993) have suggested that the high partial charge needed to be assigned to the GROMOS carbonyl ligands (0.58 electron units) in simulations of monovalent cation binding to valinomycin in methanol was "indicative of a significant polarization of the carbonyls by the field from the ion." It seems clear that in order to model the unique selectivity of lanthanide binding to the 5-fold Ca-binding site of the Satellite Tobacco Necrosis Virus measured by Burroughs et al. (1992) the molecular dynamics force field will need to include polarizability through ion-induced dipole effects. This section summarizes our recent findings supporting this view (Lee et al., 1993a; 1993b).

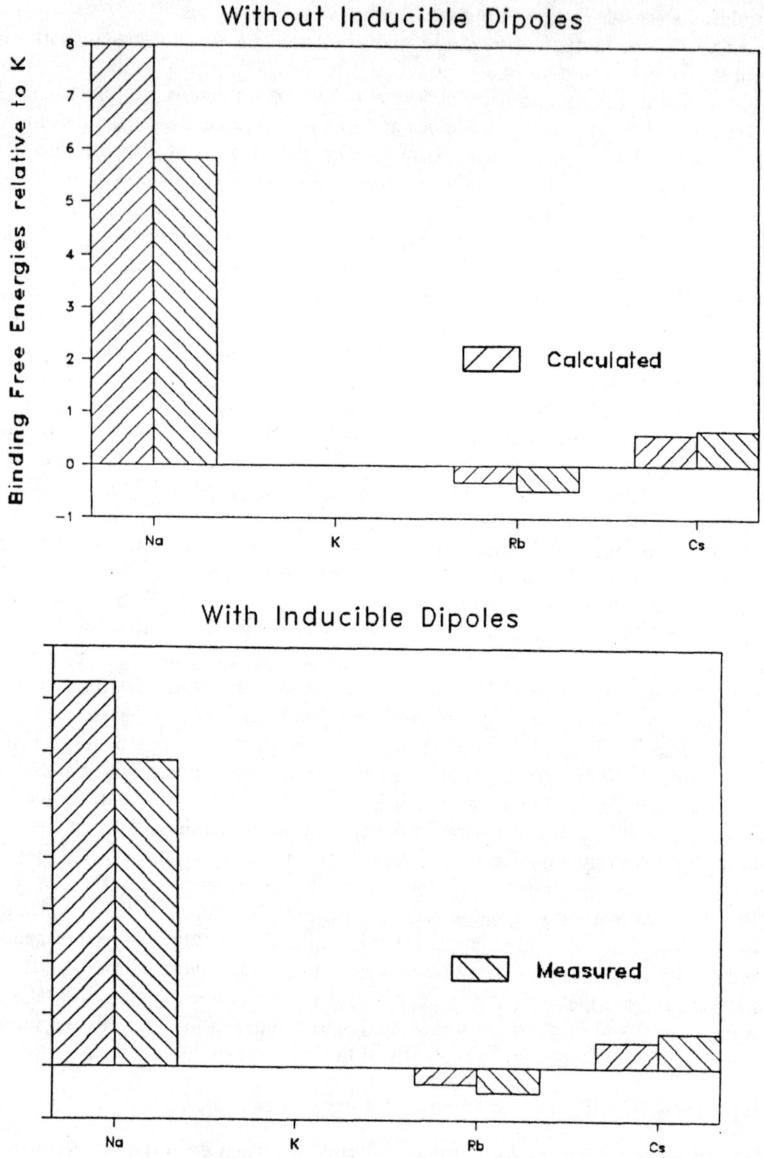

Figure 2.
Comparison of Calculated and Measured Ionic Selectivity Among Monovalent Cations. Experimental free energies of selectivity relative to K in methanol (in kcal mol^{-1}) are compared with values calculated using the MOLARIS force field without inducible dipoles at the *left* and using the MOLARIS force field with inducible dipoles at the *right*. The partial charges of 6 ester carbonyl ligands was adjusted to give the best fits shown. The values for the calculations without inducible dipoles at the left were taken to be +/-0.43 electron units; for the calculations with inducible dipoles at the right were +/-0.35 electron units. The polarizabilities for all atoms in the calculations at the right were 1.1, except for hydrogens and cations, whose values were 0.0. In both parts of the figure the measured selectivity sequence (Rb> K> Cs> Na) and magnitude is well simulated; but it can be seen that including inducible dipoles improves slightly the agreement between measured and calculated values, particularly for Na. In addition, including inducible dipoles decreases the value of partial charge that gives the best fit, as expected intuitively.

1. *Experimental selectivity data for mono- and divalent cations*
Since ionic binding data exist for both mono- and divalent ions in methanol (Grell et al., 1975), valinomycin could be taken as a model system for examining the effects of polarization on mono- and divalent ion-binding to proteins. The ionic selectivity of valinomycin was examined by MD/FEP calculations of the relative binding free energies of the ions using a conventional MD force field MOLARIS (Warshel and Creighton, 1989). As a first step, we assumed that the free energy differences of the ionic complexes in vacuum would be quite close to those deduced from complexation measurements in methanol. This allowed us to perform simulations in vacuum to compare with experimental data in methanol. The justification of this procedure is the finding that the ratios of stability constants in methanol for K, Rb, Cs, NH_4 and Tl agree perfectly with the ratios of the corresponding 2-phase salt extraction equilibrium constants into low dielectric solvents (see Fig. 3 of Eisenman et al., 1991). We will assume that the same holds true among the divalent cations. However, to calculate the energy difference between monovalent and divalent cations, we will have to evaluate an energy contribution from the bulk solvent which will require inclusion of the solvent in the simulations, as we will discuss below. For the present purpose of assessing the importance of including polarizability in the force field for simulations among singly charged ions and among doubly charged ions, the in vacuo simulation are adequate.

2. *Refinements of the MOLARIS force field*
All simulations were based on the FEP/MD procedure described elsewhere (Lee et al., 1993a). Before we could assess the importance of polarization effects in this model system, we needed to establish a version of the MOLARIS force field that could reproduce the monovalent selectivity and K-crystal structure without having to use harmonic constraints to preserve the structures, as had previously been necessary (Eisenman et al., 1991; 1992; Aqvist et al., 1992; 1993), because such constraints were found to have significant effect on the free energy calculations. Several refinements and a new feature of the standard MOLARIS non-polarizable force field were needed to accomplish this. In particular, we needed to include the new hydrogen-bond potential of MOLARIS as well as a more complete set of torsions than in the default set of MOLARIS. For the 6 ester carbonyl ligands we also needed to use smaller C=O ligands (GROMOS instead of MOLARIS Lennard-Jones parameters) and non-standard partial charges of +/-0.43 electron units for these ligands. (It must be emphasized that these carbonyls are different from amide carbonyls; so they need not have identical charges to the standard amide carbonyls used in the rest of the molecule).

With these refinements we were able to preserve the exact structure of the original K-valinomycin complex as well as obtain reasonable H-bond distances and changes in structure for the various mono- and divalent ion replacements (see Lee et al., 1993b for details).

3. *Monovalent cations*
For the monovalent cations, Figure 2 (left) depicts the binding energies of Na, K, Rb, and Cs relative to K without including polarizability; while Figure 2 (right) shows the results when polarizability is included. The agreement between theory and experiment is satisfactory in both cases, yielding the correct selectivity sequence Rb> K> Cs>> Na and reasonably good quantitative agreement which is somewhat improved for Na by the inclusion of ion-inducible dipole effects.

4. *Divalent cations*
In the case of the divalent cations, illustrated in Figure 3, where values relative to Ca were obtained using the identical charge and polarizibilty parameters as for the monovalent cations, the importance of including ion-inducible dipole effects is striking. Without inducible dipoles the results in Figure 3 (left) show a severe discrepancy between theory and experiment. Not only are the magnitudes wrong, but even the sequence is incorrect (the measured sequence is Ba> Ca> Sr> Mg, while the calculated one is Ba> Sr> Ca>

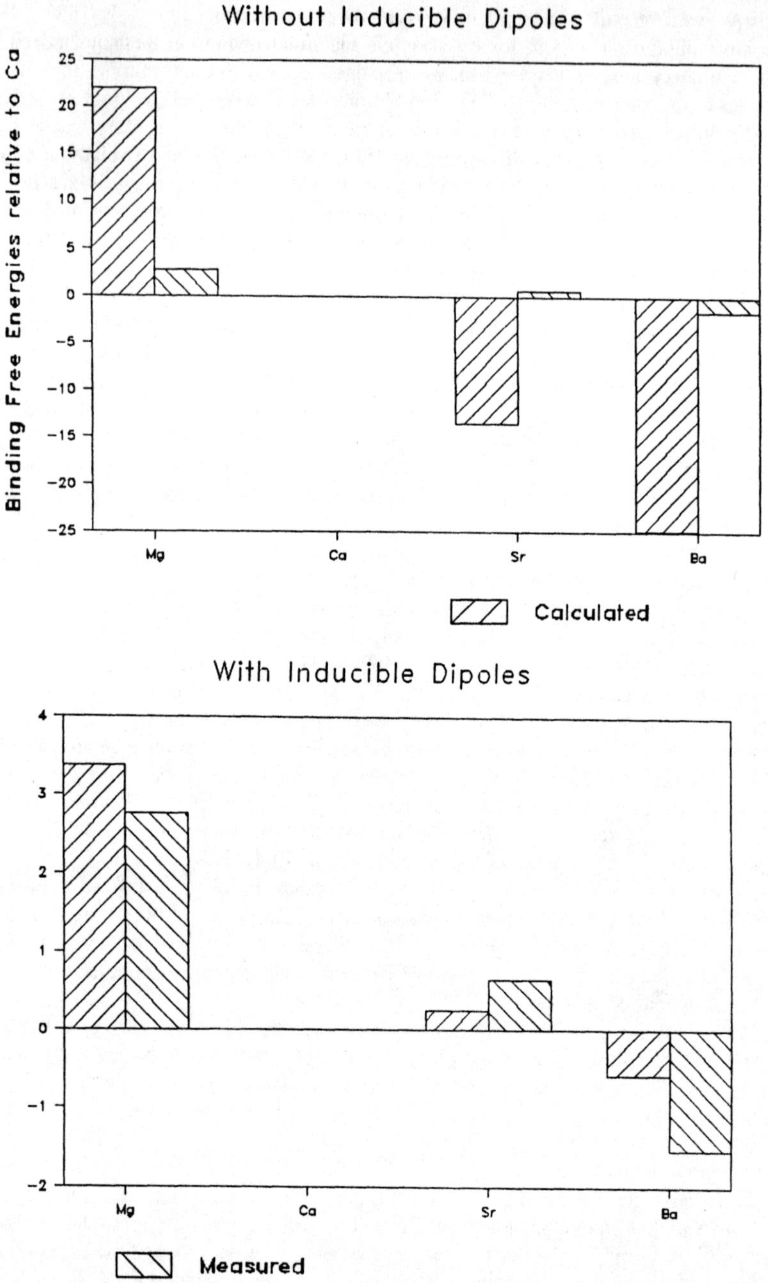

Figure 3.
Comparison of Calculated and Measured Ionic Selectivity Among Divalent Cations. Experimental and calculated free energies of selectivity relative to Ca in methanol are compared in the same manner as Figure 2. The *left* figure shows the results without inducible dipoles. The *right* figure shows the results with inducible dipoles. The same parameters of partial charges and polarizabilities are used for divalents as previously for monovalents. The calculations at the left without inducible dipoles disagree seriously in both sequence and magnitude from the measured values. Those at the right, including inducible dipoles, agree both as to selectivity sequence (Ba> Ca> Sr> Mg) and magnitude. Note the greatly expanded energy scale in the right hand figure.

Mg). By contrast, including inducible dipoles, as shown in Figure 3 (right), leads to satisfactory agreement both as to sequence and magnitude (note the vastly expanded energy scale in the right hand part of Fig. 3).

5. *Divalent vs monovalent cations*

Despite these nice findings, all is not perfect in the in vacuo simulations, as can be seen in Figure 4, which compares the expected "measured" differences between two divalent cations and K with the results of simulations with and without including inducible dipoles.

It is apparent that including inducible dipoles removes more than half the energetic discrepancy between theoretical and measured values. The remaining discrepancy may not be as bad as it seems since the "measured" values could only be estimated imprecisely because they require knowledge of the differing bulk (Born) contributions for mono- vs divalent species, which could only be estimated crudely without including explicit methanol molecules in the simulations. We are hopeful that studies now in progress extending the simulations to include methanol will resolve this discrepancy.

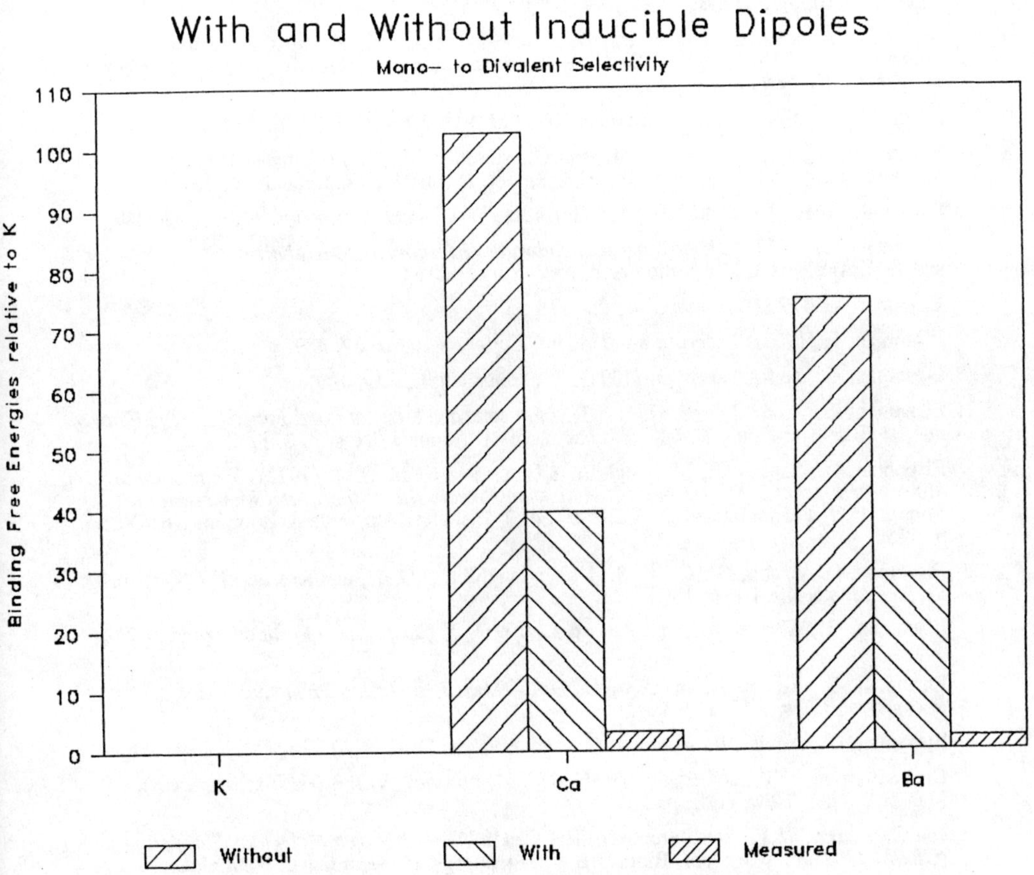

Figure 4.
Effect of Including Inducible Dipoles on the Simulated vs Measured Selectivity Between Divalent and Monovalent Cations. Binding free energies (in kcal mol^{-1}) of Ca and Ba relative to K calculated with, and without, including inducible dipoles are plotted and compared with "measured" values. The "measured" values are imprecise because they are based on rough estimations of Born's energy contributions for mono- and divalent species.

B. Discussion and conclusions

It was possible to simulate monovalent selectivity of valinomycin quite well without including inducible dipoles by using a somewhat higher partial charge for the ester carbonyl oxygen (-0.43 instead of -0.35 electron unit). This finding is in agreement with previous suggestions by Aqvist et al. (1992). However, reversibility seems to be better both as to energy and structure when inducible dipoles are included in the force field and this suggests that inducible dipole effects, though small for monovalent cations, should no longer be neglected even for monovalents. When dealing with divalent ions, the present work shows that inducible-dipole effects are not only too substantial to be accounted for by simply increasing partial charges, but also must be included in the force field in order to reproduce the experimentally measured divalent selectivity of valinomycin. This will be even more important when dealing with the trivalent lanthanides.

ACKNOWLEDGEMENT

The research was supported by USPHS Grant GM 24749 and by FONDECYT GRANT 1134-1992.

REFERENCES

Aqvist, J. (1990). *J. Phys. Chem., 94*: 8021.

Aqvist, J., Alvarez, O., and Eisenman, G. (1992). *Membrane Proteins: Structures, Interactions and Models* (A. Pullman, B. Pullman and J. Jorttner, eds.), Kluwer Academic Publishers, Netherlands, p. 367.

Aqvist, J., Alvarez, O., and Eisenman, G. (1993). *J. Phys. Chem., 96*: 10019.

Berendsen, H.J.C., Postma, J.P.M., van Gunstaren, W.F., and Hermans, J. (1981). *Intermolecular Forces* (B. Pullman, ed.), Reidel, Dordricht, Netherlands, p. 331.

Burroughs, S.E., Eisenman, G., and Horrocks, W.D. (1992). *Biophys. Chem., 42*: 249

Eisenman, G. (1961). *Symposium on Membrane Transport and Metabolism* (A. Kleinzeller and A. Kotyk, eds.), Academic Press, New York, p. 163.

Eisenman, G. (1962). *Biophys. J., 2*: 259.

Eisenman, G.(1965). *Excerpta Medica, Int. Congr. Ser. No. 87*: 489.

Eisenman, G., and Alvarez, O. (1991). *J. Membr. Biol., 119*: 109.

Eisenman, G., and Alvarez, O. (1992). *Biomembrane Structure and Function - The State of the Art* (B.P. Gaber and K.R.K. Easwaran, eds.), Adenine Press, p. 321.

Eisenman, G., Szabo, G., McLaughlin, S.G.A., and Ciani, S.M. (1972). *Symposium on Molecular Mechanisms of Antibiotic Action on Protein Biosynthesis and Membranes* (E. Munoz, F.G. Ferrandiz and D. Vasquez, eds.), Elsevier, Amsterdam, London, New York, p. 545.

Eisenman, G., Szabo, G., Ciani, S.M., McLaughlin, S.G.A., and Krasne, S. (1973). *Prog. Surface Membrane Sci. 6*: 139.

Eisenman, G., Alvarez, O., and Aqvist, J. (1991). *J. Chem. Soc., Faraday Trans., 87*: 2099.

Eisenman, G., Aqvist, J., and Alvarez, O. (1992a). *J. Inclusion Phenomena Mol. Recognition Chem., 12*: 23.

Eisenman, G., Aqvist, J., Appleby, G., and Alvarez, O. (1992b). *Biophys. J., 61*: A144.

Grell, E., Funck, T., and Eggers, F. (1975). *Membranes*, Vol. 3 (G. Eisenman, ed.), Marcel Dekker, New York, p. 1.

van Gunsteren, W.F., and Berendsen, H.J.C. (1987). *Groningen Molecular Simulation (GROMOS) Library Manual*, Biomos B.V. Nijenborgh 16, Groningen, Netherlands.

Jones, T.A., and Liljas, L. (1984). *J. Mol. Biol., 177*: 735.

Lee, F.S., Alvarez, O., and Eisenman, G. (1993a). *Biophys. J. 62: A, 35*: 4.

Lee, F.S., Alvarez, O., and Eisenman, G. (1993b). *J. Phys. Chem.*, submitted.

Warshel, A., and Creighton, S. (1989). *Computer Simulation of Biomolecular Systems* (W.F. van Gunsteren and P.K. Weiner, eds.), Escom, Leiden, p. 120.

CALCIUM-PROTEIN INTERACTIONS

by

Frank L. Siegel and Lynda S. Wright

Departments of Pediatrics and Biomolecular Chemistry,
The Waisman Center,
University of Wisconsin,
Madison, WI 53705
U.S.A.

I. INTRODUCTION

The rapidly growing family of eukaryotic calcium-binding proteins continues to attract the interest of biochemists and cell biologists. These proteins help to underwrite the central role of calcium in biology; they are a diverse group, with calcium affinities ranging from millimolar (calsequestrin) to nanomolar (α-lactalbumin and actin), tailored to biological function and specified by a variety of calcium-binding domains. Members of the prokaryotic subtilisin protease family have higher calcium affinities, with dissociation constants estimated at 10^{-12} M for thermitase (Briedigkeit and Frommel, 1989). Calcium-binding proteins have been classified according to either structure or function. Structurally, seven classes of calcium-binding proteins have been identified: EF-hand proteins, gla proteins, proteins with β-hydroxyasparagine-containing EGF-like domains, proteins containing bound nucleotides, annexins, serine proteases and calsequestrin-like proteins. While the physiological function of many calcium-binding proteins remains uncertain, it is reasonable to recognize several distinct functional categories: trigger or modulator proteins, calcium buffer proteins, calcium storage proteins, calcium-dependent or calcium-stabilized enzymes and extracellular calcium-binding proteins. The trigger proteins (Kretsinger, 1975; Dalgarno et al., 1984) are exemplified by calmodulin, which undergoes a conformational change when it binds calcium; this 'activated calmodulin' is then able to bind to protein targets, altering their biological (enzymatic) activities. Trigger proteins have calcium affinities in the micromolar range and bind calcium when the intracellular calcium concentration increases from its basal level of 0.1 μM to about 10 μM in response to excitatory stimuli. The buffer proteins include parvalbumins and calbindin, which may function to maintain calcium homeostasis in excitable cells. The calcium storage proteins constitute a third group, which includes calsequestrin and calreticulin. These proteins constitute sites of calcium storage in membrane compartments and are an integral part of internal messenger signalling systems. The extracellular proteins participate in blood clotting and bone mineralization. The 'trigger' and 'buffer' proteins tend to have high affinity, low capacity calcium binding, while the storage proteins are low affinity and high capacity. The annexins are calcium and phospholipid-binding proteins which may mediate membrane fusion during exocytosis. A detailed treatment of all aspects of this major topic is beyond the scope of this review, which is designed to provide an overview of classification and descriptions of the major calcium-binding proteins.

II. CALCIUM BINDING SITES

In most calcium-binding proteins, calcium is coordinated to oxygens of amino acid side chains and bound water; thus aspartic acid, glutamic acid, asparagine, glutamine, tyrosine, serine and threonine are conserved residues in calcium-binding regions. Much of what we know about structure-function relations in calcium-binding proteins comes from site-directed mutagenesis (Jost et al., 1992), NMR (Forsen, 1986), other spectroscopic methods, and X-ray crystallography (Strynadka and James, 1989). Most, but not all, calcium-binding sites lie in a continuous domain such as an EF-hand, but discontinuity in calcium-binding sites has been noted, for example, in proteinase K (Bajorath et al., 1988). Calcium coordination numbers range from four to twelve; in relatively few calcium-binding proteins the coordination number is six (octahedral); in most it is seven (pentagonal bipyramid), with a bond distance of about 2.4 Å. A detailed description of calcium-binding structures is found in a recent review by McPhalen et al. (1991).

III. EF-HAND PROTEINS

Members of this major subset of calcium-binding proteins contain one or more EF-hand helix-loop-helix domains (Kretsinger, 1980). EF-hand proteins have been classified into 29 subfamilies by Nakayama et al. (1992) and 25 subfamilies by Kretsinger; an alternative classification proposed by Kligman and Hilt (1988) recognizes two groups, the calmodulin family, with four EF-hands and the S-100 family, with two. Of the 25 subfamilies in the Kretsinger classification, 22 have four EF-hand domains and one each have five, six or eight EF-hands (Kretsinger, 1992). The calcium-binding loop domains are designated by Roman numerals and the flanking helices are described alphabetically. Calmodulin, for example, contains four helix-loop-helix calcium-binding domains, with helices designated A - H in domains I - IV. Six residues in the EF-hand structure and one or more molecules of water can donate oxygen ligands to calcium binding; these residues and their positions within the EF hand are designated: X (1), Y (3), Z (5), -Y (7), -X (9) and -Z (12). The -Z position lies in the F helix; all other residues are in the loop region. Residue X is aspartate and the -Z residue is glutamate. Consensus amino acids at each position of EF hand calcium-binding loops have been determined (Bairoch, 1989) and are shown in Table 1.

Kretsinger (1977) has noted that the presence of an EF-hand structure does not guarantee calcium binding. Conservation of the oxygen-containing amino acid residues X,Y,Z,-X,-Y and -Z is important, and the EF-hand must give a helix-loop-helix secondary structure. Thus, Lundberg et al. (1992) have shown that spectrin contains two conserved

Table 1

Consensus Sequence of Calcium-Binding in EF-Hand Domains

Position	Amino acids	CaM I	TnC I
1 (X)	D	D	D
2	any	K	A
3 (Y)	D,N,S	D	D
4	not I,L,V,F,Y,W	G	G
5 (Z)	D,E,N,S,T,G	D	G
6	D,N,Q,G,H,K,R	G	G
7 (-Y)	not G,P	T	D
8	I,L,V,M,C	I	I
9 (-X)	D,E,N,Q,T, S,G,C,A	T	S
10	any	T	T
11	any	K	K
12 (-Z; in F helix)	D,E	E	

EF-hand sequences, only one of which binds calcium; the EF-hand that does not bind calcium was not predicted to form a helix-loop-helix structure. Myosin light chains, for example, contain four EF-hands and yet bind either one or no calcium ions. Calcium binding in proteins with multiple EF hand domains shows binding cooperativity (Akke et al., 1991).

Several EF-hand proteins exhibit a dumbbell structure on X-ray crystallography, with one pair of EF-hand domains at either end of a long central helix. Calmodulin, troponin C and calbindin are examples of proteins with this structure. In studies of troponin C fragments, Leavis et al. (1978) have shown a positive interaction between adjacent EF-hands which increases the calcium-binding affinity. Williams (1992) has discussed the importance of these paired EF-hands, noting that single calmodulin hands have relatively low affinity for calcium (K_d of 10^{-4} M), while paired hands bind calcium with an affinity two orders of magnitude greater (Minowa and Yagi, 1984). Calcium binding to EF-hand domains increases the alpha-helical content of calcium-binding proteins (Reid, 1983), exposing hydrophobic surfaces which may interact with target proteins. Sekharadu and Sundaralingam (1988) have developed an algorithm which relates calcium affinity of paired EF-hands with beta-sheet hydrophobicity, net charge of binding ligands and hydrophobicity of the helices.

A. Calmodulin

Calmodulin, isolated as an activator of cyclic nucleotide phosphodiesterase by Cheung (1970), was one of the first EF-hand proteins to be characterized and remains one of the few with defined biological functions. The crystal structure of calmodulin at 2.2 Å resolution was reported by Babu et al. (1988); this protein has four EF-hands organized in the dumbbell shape described in the preceding section. Mackall and Klee (1991) have found that the central helix is most susceptible to proteolysis by trypsin when the calcium sites are partially occupied, indicating that a partially saturated calmodulin is the most flexible configuration and that this flexibility may allow facile interaction of calmodulin with target proteins. Of the four calcium-binding domains (Crouch, 1980), domains III and IV are high affinity (K_d of 10^{-6} M) and domains I and II are lower affinity sites (K_d of 10^{-5} M). Studies of calcium binding to tryptic fragments of calmodulin showed that the separated globular domains have calcium binding activities similar to these domains in the intact protein, indicating a lack of interaction between the two EF-hand pairs (Linse et al., 1991). Most calmodulin-activated enzymes (Klee et al., 1980) are either phosphotransferases (adenylate cyclase, protein kinases) or phosphohydrolases (Ca^{2+}-ATPase, calcineurin), but reports from Lopez-Girona et al. (1992) and others suggest that calmodulin may activate DNA polymerase alpha, allowing progression through the G1 phase of the cell cycle. While it is assumed that all of the biological activities of calmodulin are mediated by calcium binding, a recent publication by Geiser et al. (1991) presents an interesting enigma. These workers constructed mutant yeast calmodulins which have amino acid substitutions in each of their calcium-binding domains and are unable to bind calcium. It had previously been shown that deletion of the calmodulin gene is lethal to yeast, yet these mutant calmodulins showed normal growth. The authors of this publication offer several explanations for this unexpected finding, including the possibility that calmodulin function in specialized tissues of higher organisms requires calcium binding, while growth and division of yeast cells requires only apo-calmodulin. Davis (1992), in reviewing this work, also notes the precedent for calmodulin binding to target proteins in the absence of calcium.

B. Troponin C

Troponin C (TnC) is a calcium-binding subunit of the muscle protein troponin (Grabarek et al., 1992). TnC binds four mol/mol of calcium with a K_d of 10^{-7} M (Zot and Potter, 1987); the binding of calcium to troponin triggers muscle contraction by allowing the interaction of myosin and actin. X-ray crystallographic studies have shown that TnC is a dumbbell-shaped molecule, with two EF-hand domains (I and II) near the amino-terminus and the other two EF hands (III and IV) located near the carboxyl-terminus. A central helix separates these two calcium-binding regions of the molecule (Herzberg and James, 1988; Satyshur et al.,

1988). Using NMR spectroscopy, Shaw et al. (1991) have shown that in the absence of calcium, troponin C EF-hands are in a random coil configuration which undergoes a transition to the helix-loop-helix structure upon calcium binding. Calcium binding also promotes the cooperative interaction between members of an EF-hand pair in troponin C.

C. Parvalbumin

The parvalbumins are calcium-binding proteins of 12 k* which contain two EF hand calcium-binding domains (Kretsinger, and Nockholds, 1973). Calcium and magnesium bind to parvalbumin with a K_d of 10^{-9} M; thus, this protein is most likely always in the two magnesium state at the usual intracellular cation concentrations. Parvalbumin, originally found in muscles of fish and amphibians (Pechere et al., 1971), has more recently been found in the nervous system (Baimbridge et al., 1992) and also testis, kidney and adipose tissue. The physiological role of the parvalbumins has remained a matter of conjecture; current opinion contends that they function as calcium buffers to protect against 'calcium overload' during neuronal activity.

D. Calcineurin

Calcineurin is a calcium-dependent phosphoprotein phosphatase which binds both calcium and calmodulin. Structurally, calcineurin is an 80 k protein comprised of two subunits; subunit A (61 k) which binds one mol/mol of calmodulin and subunit B (19 k) which binds four calcium ions in calmodulin-like EF-hand domains (Armstrong, 1989); this subunit has a one-third sequence identity with calmodulin. Calcineurin substrates include DARPP-32, an inhibitor of phosphatase type 1 and the voltage-gated calcium channel.

E. Oncomodulin

Oncomodulin is an 11.7 k protein which shares 50% amino acid sequence homology with parvalbumin and like parvalbumin contains two EF-hand calcium-binding domains. Whereas parvalbumin can bind either calcium or magnesium, the binding sites on oncomodulin are calcium-specific and the two binding sites display cooperativity. NMR studies by Golden et al. (1989) of recombinant oncomodulin in which aspartate-59 was replaced by glutamate (as in parvalbumin) implicate aspartate-59 in the CD domain in conferring calcium specificity to this EF hand. Oncomodulin is found in tumors and placenta (Brewer and MacManus, 1985); its function is not known, but it can activate DNA synthesis, cyclic nucleotide phosphodiesterase and calcineurin (MacManus, 1981), albeit at non-physiological concentrations of oncomodulin.

F. Calbindin

The calbindins were first identified as Vitamin D-inducible calcium-binding proteins of intestinal mucosa; in the chick, calbindin is a 28 k protein (calbindin-D_{28k} and in mammals, calbindin has a molar mass of 9 k (calbindin-D_{9k}). Calbindin-D_{9k} contains two calcium-binding EF hands, while calbindin-D_{28k} contains six EF hands, four of which are calcium-binding domains; the remaining two EF hands have lost calcium-binding activity through non-conservative amino acid replacements. The two calcium-binding sites in calbindin-D_{9k} are non-equivalent. Calcium binding to the carboxyl terminal EF-hand involves three carboxylate anions and one carbonyl oxygen, while at the amino terminal EF-hand, calcium is coordinated to a single carboxylate oxygen and four carbonyl oxygens. Despite these differences, both sites bind calcium with equal affinities; $K_d = 10^{-7}$ M (Shelling and Sykes, 1985). Calbindin-D_{28k} has also been found in mammalian brain and kidney, but in brain this protein is not inducible by vitamin D. Inducible Calbindin-D_{9k} is also present in bone and cartilage, where it is localized in chondrocytes (Balmain, 1991).

G. Calretinin

Calretinin is a 29 k calcium-binding protein found in retina cerebellum (Gabrielides et al., 1991) and cochlear nucleus (Deshesne et al., 1991). Parmentier and Lefort (1991) have

---

* k = kg mol^{-1}, i.e. SI unit of molar mass (or kDa, i.e. unit of molecular mass, as mainly used by biochemists)

cloned calretinin and found an open reading frame of 271 codons, corresponding to a protein of 31.25 k which had 50% sequence identity to calbindin-D_{28k}, another major protein in cerebellum. Calretinin contains six EF-hand domains, five of which are calcium binding sites.

H. S100 proteins

S100 protein was first isolated by Moore (1965) as a brain protein soluble in saturated ammonium sulfate solutions. S-100 is a family of proteins whose biological functions are unknown (Hilt and Kligman, 1991; Donato, 1991) comprised of homo- and heterodimers of 10.5 k subunits α- and β-. The dimeric S100 proteins are designated S-100_{ao}, S-100_a and S-100_b. In contrast to the calmodulin superfamily, S-100 proteins have only two calcium-binding EF-hand domains. The carboxyl-terminal EF hand consists of the canonical 12 amino acid calmodulin-like composition and binds calcium with a K_d of 20-50 μM. The atypical, cationic, EF hand, which is conserved in all members of this family, contains 14 amino acid residues and has reduced affinity for calcium (K_d of 200-500 μM).

S-100b, the β-isoform homodimer, has a single high affinity Zn^2 binding site, is found at high levels in glial cells and is identical to neurite extension factor (NEF) which promotes neurite growth and neuron survival. Binding of zinc increases the affinity of S-100β for calcium (Baudier et al., 1983); the binding sites for zinc and calcium are different as shown in fluorescence studies (Leung et al., 1987). S-$100a_0$, the α-subunit homodimer, a minor form found in neurons has calcium-binding properties similar to S-100b.

I. Calcyclin

Calcyclin (2A9, 5B10), a homodimer homologous to S-100β, which binds two mol/mol of calcium, was first identified from a cDNA clone of an mRNA which is expressed only when fibroblasts progressed from G_0 to G_1 and is implicated in control of the cell cycle (Hirschorn et al., 1984). It is identical to the prolactin receptor-associated protein (Murphy et al., 1991). Numerous cDNAs (18A2, pEL68, p9Ka, 42A, mts1) have been identified as S-100 homologs expressed specifically during proliferation or differentiation (Hilt and Kligman, 1991). Several other tissue-specific S-100 homologs have been described. S-100C has been cloned from porcine cardiac muscle and has 40.9% homology to S-100α (Ohta et al., 1991). S-100L has been purified from bovine lung (Glenney et al., 1989). A placental isoform, S-100P, has 49% homology to S-100β (Emoto et al., 1992).

J. Calgizzarin

Calgizzarin was isolated by Todoroki et al. (1991) from chicken gizzard smooth muscle, bovine aorta and rabbit lung by napthalenesulfonamide affinity chromatography on W-7-Sepharose. Calgizzarin is an EF-hand protein of 13 k, which, on gel filtration migrates as a 30 k protein, indicating that it has a rod-like structure. SDS-PAGE of calgizzarin in the absence of 2-mercaptoethanol gave two bands, of molar masses 13 and 26 k, suggesting that calgizzarin can dimerize by disulfide bond formation. Watanabe et al., have cloned calgizzarin cDNA and have shown the presence of two putative EF-hand structures in this protein, making it a member of the S-100 family. They also report that calgizzarin shows a 43% homology to S-100α, 39% homology to S-100β and 40% homology to annexin II light chain, p10.

K. Grancalcin

Grancalcin is a protein cloned from an HL60 cDNA library and shown to be an abundant protein in neutrophils and monocytes. Grancalcin contains four EF-hand domains, two of which are highly conserved and are presumed to be calcium-binding sites (Boyhan et al., 1992). While studies of calcium binding have not been published, grancalcin is presumed to be a calcium-binding protein by virtue of its conserved EF hand, its calcium-dependent translocation to neutrophil granules and plasma membranes and its sequence homology (57.8%) to sorcin, a known EF-hand calcium-binding protein which is present in heart and several other tissues.

L. Sorcin

Sorcin is a 44 k homodimeric cytoplasmic calcium-binding protein (Meyers, 1991). Heart contains the highest concentration of sorcin; this protein is also found in kidney, brain and muscle and is over-expressed in multidrug-resistant cell lines (Van der Bliek et al., 1986). Sorcin contains four EF-hand domains, two of which are homologous to calmodulin EF-hands. Calcium-binding activity has been shown by gel overlay which indicated that sorcin is the most abundant calcium-binding protein in drug resistant cells.

M. Recoverin, visinin and S-modulin

The activation of retinal guanylate cyclase, a key step in vision, occurs when the calcium concentration is lowered from 300 nM to 10 nM. Retinal calcium-binding proteins underwrite this process by binding guanylate cyclase and stimulating this enzyme (Lagnado and Baylor, 1992). Recoverin is an EF-hand protein of 23 k, purified from bovine retina (Dizhoor et al., 1991; Lambrecht and Koch, 1991; Polans et al., 1991) and recently cloned and sequenced (Ray et al., 1992). Sequence analysis indicates that recoverin contains three EF-hands (Persechini et al., 1989; Strynadka and James, 1989). Recoverin is localized to rod outer segments. A recoverin-like protein has been purified from frog rod outer segments and named S-modulin (Kawamura et al., 1992). This protein is a 26 k EF-hand protein with 83% homology to recoverin and 65% homology to visinin. Since comparisons were made between frog S-modulin, chick visinin and bovine recoverin, the author's claim that S-modulin is distinct from recoverin must await further verification. Visinin is a 24 k protein from retinal cones; it contains three EF-hand domains as determined by sequencing visinin cDNA (Yamagata et al., 1990). Calcium binding to visinin was demonstrated by the calcium-45 overlay method.

N. Neurocalcin

Okazaki et al. (1992) have isolated a protein from brain which binds to an affinity matrix of immobilized calmodulin antagonist W77. This 23 k protein, neurocalcin, contains three EF-hand domains and exists as four homologous isoforms (Terasawa et al., 1992). Neurocalcin was cloned and sequenced and shown to have greater than 90% sequence homology to recoverin, which was postulated to be a member of the neurocalcin family. Unlike recoverin and visinin, neurocalcin is neuron-specific and is not found in the retinal photoreceptor layer.

O. Spectrin

Spectrins are a family of cytoskeleton proteins which form helical fibers to which actin is attached. Spectrin, in turn, is linked to the plasma membrane through binding to integral membrane proteins such as band 3 of the erythrocyte membrane. Spectrins contain EF-hand domains, suggesting that these proteins may bind calcium. Lundberg et al. (1992) have reported that brain spectrin binds two calcium ions, with affinities of 50 μM and 150 μM for each of two binding sites. Calcium binding to erythrocyte spectrin was found to be an order of magnitude weaker.

P. Crystallins

The crystallins are a group of three proteins in mammalian lens, designated α-, β- and γ. Another crystallin, the δ form, is found in avian and reptilian lens. Calcium causes lens to become opaque, and Sharma et al. (1989) have reported that β- and δ-crystallins bind calcium. The δ-isoform binds four ions with a K_d of 4.3×10^{-3} M and contains one EF-hand domain, but no other crystallins contain EF-hands. Although both β- and δ-crystallins bind calcium, only δ-crystallin contains an EF-hand calcium-binding region; binding of calcium to β-crystallin was postulated to occur at a region of the molecule which is rich in anionic

Q. Calpain

Calcium-activated neutral proteases (calpain) are heterodimeric calcium-binding proteins, each consisting of 80 kDa and a 30 kDa subunits. There are two isoenzymes, calpains I (μ) and II (m), which differ in their large (catalytic) subunits. Both the 30 kDa and 80 kDa subunits contain four EF-hand structures in their C-terminal domains (Nishimura and Goll, 1991), but depending upon species, only two or three of the EF hands in each subunit bind calcium (K_d = 35-79 μM).

R. Calvasculin

Calvasculin was isolated from bovine aorta by Watanabe et al. (1992), using affinity chromatography on the immobilized isoquinolinesulfonamide, W-66. Calvasculin is a homodimer of 25,000 Da, comprised of 11,000 Da subunits. This protein does not activate myosin light chain kinase, CaM kinase II or cyclic nucleotide phosphodiesterase and it has no effect on microtubule assembly. Western blots indicate that calvasculin is present in aorta, but not brain, testis, lung or heart. Calvasculin binds two moles of calcium with relatively low affinity (K_d = 5 x 10^{-3} M in the presence of 0.1 M KCl). Isoelectric focusing of calvasculin gave several bands, with pI values of 4.65-5.1. No data on calvasculin structure is available; it is tentatively placed in the EF-hand category since it was isolated using calmodulin-antagonist affinity chromatography.

IV. CALCIUM STORAGE PROTEINS

The endoplasmic reticulum contains at least four calcium-binding proteins which can bind up to 30 calcium ions at relatively low affinity. Sambrook (1990) has calculated that these proteins could account for the storage of 3 mM calcium in the ER.

A. Calsequestrin

Sarcoplasmic reticulum is a major calcium storage site in both skeletal and cardiac muscle, and calsequestrin is thought to be the site of calcium storage and release within the SR (MacLennan and Wong, 1971; Collins et al., 1990). The calsequestrins from these two tissues differ structurally but both bind calcium with high capacity and low affinity. Calsequestrin from skeletal muscle has a molecular mass of 63,000 Da, while cardiac calsequestrin has a molecular mass of 55,000 Da (alkaline SDS-PAGE; 44,000 Da in neutral SDS-PAGE) (Campbell et al., 1991). Cardiac calsequestrin differs from skeletal muscle calsequestrin with respect to its lower molecular mass, slightly lower content of aspartic plus glutamic acid residues (32% vs 37%) and contains a high mannose oligosaccharide content as evidenced by a 1,000 Da reduction in molecular mass when digested with β-N-glycosidase. Skeletal muscle calsequestrin binds twice as many calcium ions as the cardiac protein. Skeletal muscle calsequestrin is a very acidic protein which binds 31 calcium ions with a K_d of 0.79 mM in the absence of magnesium and binds 23 calcium ions in the presence of 3 mM magnesium. Krause et al. (1991) have found that calcium and magnesium bind to different sites on calsequestrin. Calsequestrin contains neither EF-hand domains nor repetitive acidic amino acid sequences, which suggests that the large negative charge density of calsequestrin is responsible for calcium binding.

B. Calreticulin

Calreticulin is a calcium-binding protein in the endoplasmic reticulum of non-muscle cells (Khanna and Waisman, 1986; Michalak et al., 1992) and in the sarcoplasmic reticulum of smooth muscle (Milner et al., 1991) and skeletal muscle (Smith and Koch, 1989). Like calsequestrin, calreticulin is a high-capacity, low affinity calcium-binding protein. A single high affinity site (K_d = 1.6 μM) and 20-50 low affinity sites (K_d = 0.3-2 mM) have been identified. In addition, 14 zinc binding sites (K_d = 0.3 mM) have been found. Calsequestrin contains three structural domains: an amino-terminal half (N-domain), a proline-rich region (P-domain) and the carboxy terminal quarter of the protein (C-domain). The C-domain, which has a high concentration of acidic amino acids, contains the high-capacity, low-affinity calcium-binding sites (Baksh and Michalak, 1991). The high affinity site lies in the P-domain, which does not contain a high concentration of acidic amino acids or an EF-hand structure and may be a novel calcium binding site. The N-domain does not contain calcium-binding sites. Calreticulin may constitute the major site of calcium storage within endoplasmic reticulum, playing a role like that of calsequestrin in sarcoplasmic reticulum.

C. Calnexin (pp90)

Calnexin is a major calcium-binding protein of the endoplasmic reticulum (Wada et al., 1991). This non EF-hand protein has a high degree of sequence homology with calreticulin, and it has been suggested that calnexin, together with another calcium-binding protein, SSR α, regulate Ca^{2+}-dependent retention mechanisms for luminal proteins of the ER. Whereas calreticulin is a luminal protein, calnexin is an integral membrane protein. The cytoplasmic domain of calnexin is phosphorylated, raising the possibility that this protein is involved in signal transduction which senses intraluminal calcium concentration and participates in GTP-mediated calcium release.

D. Chromogranin A

Chromogranins are major protein constituents of adrenal medullary chromaffin vesicles, organelles which sequester catecholamines (0.5 mM), ATP (1 mM) and calcium (40 mM). Chromogranin A, present at 1 mM, is thought to be the site of calcium storage; this protein is structurally similar to calsequestrin and calreticulin. Both chromogranin A and calsequestrin contain about 30% acidic amino acids, clustered near the carboxyl-terminus, have similar molar masses, and both show low affinity, high capacity calcium binding. At pH 5.5 (which is thought to be the pH of the chromaffin vesicle), chromogranin binds 55 mol/mol of calcium with a K_d of 4.4 mM (Yoo and Albanesi, 1991).

V. ANNEXINS

The annexins are a group of proteins which bind calcium and phospholipids; these proteins had previously been known as lipocortins, calelctrins, calcimedins, chromobindins and calpactins. In a recent review, Moss et al., have classified these proteins as ten annexins (I - X), each of which represents a distinct gene product. Alternative nomenclature was suggested by Rojas et al., (1992), who identified eight annexin families. The annexins differ from one another with respect to the number of calcium binding sites, but all contain conserved sequences of 17 amino acid residues, a loop-helix domain called the endonexin fold (Geisow et al., 1986) within either four or eight annexin repeats of 70-80 residues each. There are highly conserved residues within each endonexin fold: gly, thr, (asp,asn,arg) at positions 4-6 and arg at position 17. The crystal structure of human annexin V at 2 Å resolution was recently reported by Huber et al. (1990). Calcium binding to annexins is phospholipid-dependent; annexin II (calpactin), for example, binds two calcium ions with a K_d of 4.5×10^{-6} M in the presence of phosphatidylserine and 0.5 mM in the absence of phospholipid (Glenny, 1986).

VI. GLA PROTEINS

All Gla proteins contain multiple residues of the modified amino acid, γ-carboxyglutamic acid (gla), which bind calcium with low affinity (Suttie, 1989). This posttranslational modification was first described in prothrombin (Stenflo et al., 1974) and later identified in clotting factors VII, IX, X, C, and S (Stenflo and Suttie, 1977). The microsomal enzyme vitamin K-dependent carboxylase catalyzes this reaction, using vitamin K as an oxidative substrate. Warfarin and dicoumarol inhibit the regeneration of the reduced form of vitamin K, providing the basis of their anti-coagulant properties.

The vitamin K-dependent plasma proteins have a highly conserved N-terminus consisting of numerous gla residues which form the low affinity calcium-binding sites, followed by a region consisting of two EGF-like domains (two kringle domains in prothrombin), and the carboxyl terminus made up of a serine protease. A second higher affinity calcium-binding site is associated with the EGF domains and has no gla residues present. Jackson (1988) provides a summary of calcium-binding coefficients for plasma gla proteins. Estimation of calcium-binding is complicated by the large number of low affinity sites per molecule, cooperativity between these sites, and the presence of non-gla calcium-binding sites.

After vascular injury, the intrinsic and extrinsic coagulation pathways are mobilized by a sequence of highly regulated proteolytic activations of the serine protease domains of factors VII, IX, and X (each requiring calcium and phospholipid), and culminating with the conversion of prothrombin to thrombin, which initiates fibrin clot formation. Regulation of anticoagulation pathway involves at least two other homologous gla proteins, C and S, which effect the calcium-dependent proteolytic inactivation of clotting factors Va and VIIIa.

A. Prothrombin

Prothrombin, a 70 k protein, synthesized as the precursor preprothrombin, is converted to the active protein by vitamin K-dependent carboxylation of glutamate residues. Prothrombin binds 10 to 11 calcium ions; six have a K_d of 200 μM and five weaker binding sites have a K_d of 1 mM (Bloom and Mann, 1978). Upon activation, prothrombin loses an amino-terminal 23 k peptide (intermediate 3) which contains the six higher affinity sites. In this amino-terminal peptide, the first ten glutamate residues are present as carboxyglutamate; selective chemical modification of gla indicated that the first calcium-binding site is discontinuous, requiring gla residues 7, 8 and 33 (Zapata et al., 1988). When prothrombin binds calcium, it is able to also bind to membrane phospholipids, where factor Xa-catalyzed activation takes place, releasing thrombin, which in turn converts fibrinogen into fibrin. The amino-terminal fragment of prothrombin is unable to bind phospholipids at normal blood calcium concentrations; the intact protein structure is required.

B. Hydroxyasparagine-containing proteins

Protein S is a gla-containing, vitamin K-dependent plasma protein with unusual calcium-binding properties. Protein S plays a key role in blood coagulation, participating in the metabolism of factors Va and VIIIa (Walker, 1988). There are two classes of calcium-binding sites on protein S; the low affinity gla sites previously described for prothrombin. Protein S, like prothrombin, contains 11 gla residues in an amino terminal domain; these account for calcium binding with a K_d of 0.5 mM. Protein S also contains four EGF (epidermal growth factor) homologous domains which bind calcium with very high affinity; $K_d = 2 \times 10^{-9}$ at low ionic strength (Dahlback et al., 1990). These EGF domains in protein S contain the unusual amino acid hydroxyasparagine, the result of a posttranslational modification catalyzed by a 2-oxoglutarate-dependent oxygenase (Stenflo et al., 1989). While hydroxyasparagine residues are not required for protein S to function in blood coagulation, they may participate in high affinity calcium binding (Nelson et al., 1991). EGF domains containing hydroxyasparagine bind calcium with an affinity two to three orders of magnitude greater than other vitamin K-dependent calcium-binding proteins. Dahlback (1990) also points out that in each of the hydroxyasparagine-containing EGF domains there is a cluster of three or four negatively charged amino acid residues which are likely to form a novel type of calcium binding site. Other proteins which contain hydroxyasparagine EGF domains are the LDL receptor, thrombomodulin and transforming growth factor β_1-binding protein. Calcium binding to these proteins has not yet been characterized.

C. Other blood proteins

Bovine Factor VII has two calcium-binding sites with a K_d of 0.1 mM and at least four sites of K_d 1.7 mM. Human Factor IX has a maximum of 16 calcium-binding sites and K_d of 0.73 mM. Factor X contains 20-39 sites with K_d in the millimolar range, and one site with K_d 0.05 mM. Bovine protein C has 16 sites with K_d 0.9 mM, but bovine protein C, activated by proteolysis, has 5-9 sites with K_d of 0.4 mM. After cleavage by thrombin, bovine protein S binds 2-3 calcium ions with K_d of 0.017 mM.

D. Gla proteins in bone

Gla proteins have also been isolated from bone tissue (as reviewed by Price, 1988). Osteocalcin (bone gla protein, BGP) is produced by the bone-forming osteoblast cells as a precursor of 11 k and is secreted as a 5.5 k peptide. Osteocalcin is localized exclusively to bone and dentine and is the most abundant noncollagenous protein in vertebrate bone. Although its biological functions are unknown, osteocalcin stimulates osteoclast formation and is a potent inhibitor of hydroxyapatite formation, with a K_d for hydroxyapatite of 10^{-7}

M. This action is lost when its three gla residues are decarboxylated to glutamic acid (Price, 1988). MGP mRNA is found in many tissues, but the protein (8.7 k) is expressed only in bone, dentine, and cartilage. The fact that decarboxylation of the gla residues of MGP or the presence of an arginyl-glycyl-aspartyl (RGD)-containing peptide (the amino acid sequence motif for cell adhesion) inhibit the binding of osteosarcoma cells and chondrocytes to an MGP-treated surface indicates that MGP may be implicated in cell adhesion-extracellular matrix interactions (Loeser and Wallin, 1992). The anticoagulant gla protein S also has been recently found to be expressed by bone matrix osteoblasts (Maillard et al., 1992).

VII. α-LACTALBUMINS

The α-lactalbumins are calcium-binding proteins of mammalian milk; all members of this class contain one calcium-binding site. This site, the 'α-lactalbumin elbow' resembles the EF-hand structure in that it is a helix-loop-helix motif (Acharya et al., 1991). The calcium affinities for several lactalbumins and the homologous lysozymes were determined by calcium-43 NMR (Aramini et al., 1992). The lactalbumins were found to bind calcium with higher affinity than any other eucaryotic calcium-binding proteins, with a K_d of 5×10^{-8} for α-lactalbumin. X-ray crystallography (Acharya et al., 1991) indicates that calcium is bound in a pentagonal bipyramid, coordinated within the a-lactalbumin elbow to the side chain carboxyl anions of three conserved aspartic acid residues 82, 87 and 88, the carbonyl oxygens of lysines 79 and 84 and two water molecules.

VIII. ACTIN

Actin is an abundant, highly conserved divalent cation-binding protein which can bind nucleotides (ATP and ADP) and either calcium (K_d = 1.5 - 15 nM) or magnesium to a single high affinity site and 5 - 9 sites of lower affinity (K_d = 0.15 mM). Magnesium binding causes G-actin to polymerize into the F-actin of muscle thin filaments which plays a key role in muscle contraction and cell motility (Mannherz, 1992). While both cations can bind to actin, calcium binds with higher affinity, and the affinity of magnesium is an order of magnitude lower (Carlier et al., 1986). The conformational and functional consequences of binding calcium and magnesium are quite different, however, as calcium-binding inhibits actin polymerization, but polymerization is stimulated by magnesium. Actin binds to a number of proteins, including DNase I, profilin and gelsolin; crystal structure data is available for the actin-DNaseI complex, and the nature of the calcium-binding site comes from this data. At the high affinity site, calcium is bound to the β-phosphate of ADP or the β- and γ-phosphates of ATP and is coordinated to aspartic acid residues 11 and 154, lysine-18 and glutamine-137 (Mannherz, 1992). Actin also binds calcium, magnesium and potassium with low affinity (K_d of 0.15 mM) at 5 to 9 sites.

IX. OTHER CALCIUM-BINDING PROTEINS

A number of enzymes stabilized by calcium are also calcium-binding proteins. Bovine trypsin (Cliffe andf Grant, 1981), for example, is a serine protease with a single calcium-binding site of moderatley high affinity ($K_d = 3 \times 10^{-5}$ M). Trypsinogen contains two calcium-binding sites, one of which is lost upon activation. The common high affinity site was identified by X-ray crystallography of trypsin (Bode and Schwager, 1975) and show to be octahedral, with coordination of calcium to amino acids which lie in a loop formed by residues 60 to 80; glutamine-70, asparagine-72, valine-75, glutamic acid-77, glutamine-80 and two water molecules. Another calcium-requiring enzyme is phospholipase A2; this enzyme binds one calcium with relatively low activity ($K_d = 4 \times 10^{-3}$ M). The crystal structure of phospholipase A2 has been obtained at 1.5 Å resolution and the calcium-binding site was shown to be octahedral, with calcium coordinated to three amino acids, tyrosine-28, glycine-30, glycine 32 and aspartic acid-49 and also to oxygens of two water molecules (Dijkstra et al., 1981).

X. TISSUE DISTRIBUTION - A WINDOW TO FUNCTION?

Several specialized tissues contain specifically localized calcium-binding proteins. In brain, for example, parvalbumin, calbindin or calretinin, either singly or in combination, have been used as anatomical markers for specific neuronal populations where they may function to buffer intracellular calcium levels (Baimbridge et al., 1992). Calcium-modulated proteins which are linked to internal messenger systems bind calcium in one step of a signalling cascade. Calcium-binding proteins linked to internal messenger mechanisms are of several broad types, one type with a calcium affinity of about 1 μM, which would bind calcium when cells are stimulated, and a second type, of much lower affinity, which would represent a calcium storage site in reticular membranes. The membrane systems which comprise the endoplasmic reticulum and sarcoplasmic reticulum contain luminal calcium-binding proteins which, by virtue of their localization and calcium-binding properties, are thought to represent calcium storage sites. These membrane systems also contain membrane-bound calcium-binding proteins which may be involved in protein trafficking. The visual system contains calcium-binding proteins which may control guanylate cyclase and dark currents; thus, one can anticipate building logical correlations between the roles of tissue-localized calcium-binding proteins and their functions.

REFERENCES

Acharya, K.R., Ren, J.S., Stuart, D.I., Phillips, D.C., and Fenna, R.E. (1991). *J. Mol. Biol., 221*: 571.

Akke, M., Forsen, S., and Chazin, W.J. (1991). *J. Mol. Biol., 220*: 173.

Aramini, J.M., Drakenberg, T., Hirakoi, T., Ke, Y., Nitta, K., and Vogel, H.J. (1992). *Biochemistry, 31*: 6761.

Armstrong, D.L. (1989). *Trends Neurosci., 12*: 117.

Babu, Y.S., Bugg, C.E., and Cook, W.J. (1988). *J. Mol. Biol., 204:* 191.

Baimbridge, K.G., Celio, M.R., and Rogers, J.H. (1992). *Trends Neurosci., 15*: 303.

Bairoch, A. (1989). *PROSITE: A Dictionary of Protein Sites and Patterns*, University of Geneva, Geneva.

Bajorath, J., Hinrichs, W., and Sanger, W. (1988). *Eur. J. Biochem., 176*: 441.

Baksh, S., and Michalak, M. (1991). *J. Biol. Chem., 266*: 21458.

Balmain, N. (1991). *Clin. Orthop., 265*: 265.

Baudier, J., Haglid, K., Haiesch, J., and Gerard, D. (1983). *Biochem. Biophys. Res. Commun., 114*: 1138.

Bloom, J.W., and Mann, K.G. (1978). *Biochemistry, 17*: 4430.

Bode, W., and Schwager, P. (1975). *FEBS Lett., 56*: 139.

Boyhan, A., Casimir, C.M., French, J.K., Teahan, C.G., and Segal, A.W. (1992). *J. Biol. Chem., 267*: 2928.

Brewer, L.M., and MacManus, J.P. (1985). *Develop. Biol., 112*: 49.

Briedigkeit, L., and Frommel, C. (1989). *FEBS Lett., 253*: 83.

Campbell, J.A., Biggart, J.D., and Elliott, R.J. (1991). *Biochem. Soc. Trans., 19*: 387S.

Carlier, M.F., Pantaloni, D., and Korn, E.D. (1986). *J. Biol. Chem., 261*: 10778.

Cheung, W.Y. (1970). *Biochem. Biophys. Res. Commun., 38*: 533.

Cliffe, S.G.R., and Grant, D.A.W. (1981). *Biochem. J., 193*: 655.

Collins, J.H., Tarcsafalvi, A., and Ikemoto, N. (1990). *Biochem. Biophys. Res. Commun., 167*: 189.

Crouch, T.H., and Klee, C.B. (1980). *Biochemistry, 19*: 3692.

Dahlback, B., Hildebrand, B., and Linse, S. (1990). *J. Biol. Chem., 265*: 18481.

Dalgarno, D., Klevit, R.E., Levine, A.B., and Williams, R.J.P. (1984). *Trends Pharmacol., 4*: 266.

Davis, T.N. (1992). *Cell Calcium, 13*: 435.

Dechesne, C., Winsky, L., Kim, H.N., Goping, G., Vu, T.D., Wenthold, R.J., and Jacobowitz, D.M. (1991). *Brain Res.*, *560*: 139.

Dijkstra, B.W., Kalk, K.H., Hol, W.G.J., and Drenth, J. (1981). *J. Mol. Biol.*, *147*: 97.

Dizhoor, A.M., Ray, S., Kumar, S., Niemi, G., Spencer, M., Brolley, D., Walsh, K.A., Philipov, P.P., Hurley, J.B., and Stryer, L. (1991). *Science*, *215*: 915.

Donato, R. (1991). *Cell Calcium*, *12*: 713.

Emoto, Y., Kobayashi, R., Akatsuka, H., and Hidaka, H. (1992). *Biochem. Biophys. Res. Commun.*, *182*: 1246.

Forsen, S., Vogel, H.J., and Drakenberg, T. (1986). *Calcium and Cell Function*, *6*: 113.

Gabrielides, C., McCormack, A.L., Hunt, D.F., and Christakos, S. (1991). *Biochemistry*, *30*: 656.

Geiser, J.R., van Tuinen, D., Brockerhoff, S.E., Neff, M.M., and Davis Trisha, N. (1991). *Cell*, *65*: 949.

Geisow, M.J., Fritsche, U., Hexham, J.M., Dash, B., and Johnson, T. (1986). *Nature*, *320*: 636.

Glenny, J. (1986). *J. Biol. Chem.*, *261*: 7247.

Glenney, J.R., Kindy, M.S., and Zokas, L. (1989). *J. Cell. Biol.*, *108*:569.

Golden, L.F., Corson, D.C., Sykes, B.D., Banville, D., and MacManus, J.P. (1989). *J. Biol. Chem.*, *264*: 20314.

Grabarek, Z., Tao, T., and Gergely, J. (1992). *J. Muscle Res. Cell Motil.*, *13*: 383.

Herzberg, O., and James, M.N. (1988). *J. Mol. Biol.*, *203*: 761.

Hilt, D.C., and Kligman, D. (1991). *Novel Calcium-Binding Proteins* (C.W. Heizmann, ed.), Springer-Verlag, New York, p. 65.

Hirshorn, R.R., Aller, P., Yuan, Z.A., Gibson, C.W., and Baserga, R. (1984). *Proc. Nat. Acad. Sci. USA*, *81*: 6004.

Huber, R., Schneider, M., Mayr, I., Romisch, J., and Paques, E.P. (1990). *FEBS Lett.*, *275*: 15.

Jackson, C.M. (1988). *Current Advances in Vitamin K Research* (J.W. Suttie, ed.), Elsevier, New York, p. 305.

Jost, M., Thiel, C., Weber, K., and Gerke, V. (1992). *Eur. J. Biochem.*, *207*: 923.

Kawamura, S., Takamatsu, K., and Kitamura, K. (1992). *Biochem. Biophys. Res. Commun.*, *186*: 411.

Khanna, N.C., and Waisman, D.M. (1986). *Biochemistry*, *25*: 1078.

Klee, C.B., Crouch, T.H., and Richman, P.G. (1980). *Annu. Rev. Biochem.*, *49*: 489.

Kligman, D., and Hilt, D.C. (1988). *Trends Biochem. Sci.*, *13*: 437.

Krause, K.H., Milos, M., Luan-Rilliet, Y., Lew, D.P., and Cox, J.A. (1991). *J. Biol. Chem.*, *266*: 9453.

Kretsinger, R.H., and Nockholds, C.E. (1973). *J. Biol. Chem.*, *248*: 3313.

Kretsinger, R.H. (1975). *Calcium Transport in Contraction and Secretion* (E. Carafoli, W. Clementi, W. Drabakowski and A. Margeth, eds.), North-Holland Publishing, Amsterdam, p.469.

Kretsinger, R.H. (1977). *Calcium-Binding Proteins and Calcium Function* (R.H. Wasserman, R.A. Corradino, E. Carafoli, R.H. Kretsinger, D.H. MacLennan and F.L. Siegel, eds.), North-Holland, New York, p. 63.

Kretsinger, R.H. (1980). *CRC Crit. Rev. Biochem.*, *8*: 119.

Kretsinger, R.H. (1992). *Science*, *258*: 50.

Lagnado, L., and Baylor, D. (1992). *Neuron*, *8*: 995.

Lambrecht, H.G., and Koch, K.W. (1991). *EMBO J.*, *10*: 793.

Leavis, P.C., Rosenfield, S.S., Gergely, J., Grabarek, Z., and Drabikowski, W. (1978). *J. Biol. Chem.*, *253*: 5452.

Leung, I.K.M., Mani, R.S., and Kay, C.M. (1987). *FEBS Lett.*, *214*: 35.

Linse, S., Helmersson, A., and Forsen, S. (1991). *J. Biol. Chem.*, *266*: 8050.

Loeser, R.H., and Wallin, R. (1992). *J. Biol. Chem.*, *267*: 9459.

Lopez-Girona, A., Colomer, J., Pujol, M.J., Bachs, O., and Agell, N. (1992). *Biochem. Biophys. Res. Commun.*, *184*: 1517.

MacLennan, D.H., and Wong, P.T.S. (1971). *Proc. Nat. Acad. Sci. USA*, *68*: 1231.

MacManus, J.P. (1981). *FEBS Lett.*, *126*: 245.

Mackall, J., and Klee, C.B. (1991). *Biochemistry*, *30*: 7242.

Maillard, C., Berruyer, M., Serre, M.C., Dechavanne, M., and Delmas, P.D. (1992). *Endocrinology*, *130*: 1599.

Mannherz, H.G. (1992). *J. Biol. Chem.*, *267*: 11661.

McPhalen, C.A., Strynadka, N.C.J., and James, N.G.M. (1991). *Advances in Protein Chemistry*, Vol. 42 (C.B. Anfinsen, J.T. Edsall, F.M. Richards and D.S. Eisenberg, eds.), Academic Press, New York, p. 77.

Meyers, M.B. (1991). *Novel Calcium-Binding Proteins* (C. W. Heizmann, ed.), Springer-Verlag, New York, p. 385.

Michalak, M., Milner, R.E., Burns, K., and Opas, M. (1992). *Biochem. J.*, *285*: 681.

Milner, R.E., Baksh, S., Shemanko, C., Carpenter, M.R., Smillie, L., Vance, J.E., Opas, M., and Michalak, M. (1991). *J. Biol. Chem.*, *266*: 7155.

Minowa, O., and Yagi, K. (1984). *J. Biochem.*, *96*: 1175.

Moore, B. (1965). *Biochem. Biophys. Res. Commun.*, *19*: 739.

Murphy, L.C., Gong, Y., and Reid, R.E. (1991). *Novel Calcium-Binding Proteins* (C.W. Heizman, ed.), Springer-Verlag, New York, p. 169.

Nakayama, S., Moncrief, N.D., and Kretsinger, R.H. (1992). *J. Mol. Evol.*, *34*: 416.

Nelson, R.M., VanDusen, W.J., Friedman, P.A., and Long, G.L. (1991). *J. Biol. Chem.*, *266*: 20586.

Nishimura, T., and Goll, D.E. (1991). *J. Biol. Chem.* *266*: 11842.

Ohta, H., Sasaki, T., Naka, M., Hiraoka, O., Miyamoto, C., Furuichi, Y., and Tanaka, T. (1991). *FEBS Lett.*, *295*: 93.

Okazaki, K., Watanabe, M., Ando, Y., Hagiwara, M., Terasawa, M., and Hidaka, H. (1992). *Biochem. Biophys. Res. Commun.*, *185*: 147.

Parmentier, M., and Lefort, A. (1991). *Eur. J. Biochem.*, *196*: 79.

Pechere, J.F., Capony, J.P., and Ryden, L. (1971). *Eur. J. Biochem.*, *23*: 421.

Persechini, A., Moncrief, N.D., and Kretsinger, R.H. (1989). *Trends Neurosci.*, *12*: 462.

Polans, A.S., Buczylko, J., Crabb, J., and Palczewski, K. (1991). *J. Cell Biol.*, *112*: 981.

Price, P.A. (1988). *Current Advances in Vitamin K Research* (J.W. Suttie, ed.), Elsevier, New York, p. 259.

Ray, S., Zozulya, S., Niemi, G.A., Flaherty, K.M., Brolley, D., Dizhoor, A.M., McKay, D.B., Hurley, J., and Stryer, L. (1992). *Proc. Nat. Acad. Sci. USA*, *89*: 5705.

Reid, R.E. (1983). *J. Theor. Biol.*, *114*: 353.

Rojas, E., Arispe, N., Haigler, H.T., Burns, A.L., and Pollard, H.B. (1992). *Bone Miner.*, *17*: 214.

Sambrook, J.F. (1990). *Cell*, *61*: 197.

Satyshur, K.A., Rao, S.T., Pyzalska, D., Drendel, W., Greaser, M., and Sundaralingham, M. (1988). *J. Biol. Chem.*, *263*: 1628.

Sekharudu, Y.C., and Sundaralingham, M. (1988). *Protein Eng.*, *2*: 139.

Sharma, Y., Rao, C.M., Narasu, M.L., Rao, S.C., Somasundarum, T., Gopalakrishna, A., and Balasubramanian, D. (1989). *J. Biol. Chem.*, *264*: 12794.

Shaw, G.S., Hodges, R.S., and Sykes, B.D. (1991). *Biochemistry*, *30*: 8339.

Shelling, J.G., and Sykes, B. (1985). *J. Biol. Chem.*, *260*: 8342.

Smith, M.J., and Koch, G.L. (1989). *EMBO J.*, *8*: 3581.

Stenflo, J., Fernlund, P., Egan, W., and Roepstorff, P. (1974). *Proc. Nat. Acad. Sci. USA*, *71*: 2703.

Stenflo, J., and Suttie, J.W. (1977). *Annu. Rev. Biochem.*, *46*: 157.

Stenflo, J., Holme, E., Lindstedt, S., Chandramouli, N., Huang, L.H., Tam, J.P., and Merrifield, R.B. (1989). *Proc. Nat. Acad. Sci. USA*, *86*: 444.

Strynadka, N.C.J., and James, M.N.G. (1989). *Annu. Rev. Biochem.*, *58*: 951.

Suttie, J.W. (1989). *Vitamin K-Dependent Proteins and Their Metabolic Roles* (H. Saito and J.W. Suttie, eds.), Elsevier, New York, p. 1.

Terasawa, M., Nakano, A., Kobayashi, R., and Hidaka, H. (1992). *J. Biol. Chem.*, *267*: 19596.

Todoroki, H., Kobayashi, R., Watanabe, M., Minami, H., and Hidaka, H. (1991). *J. Biol. Chem.*, *266*: 18668.

Van der Bliek, A.M., Meyers, M.B., Biedler, J.L., Hes, E., and Borst, P. (1986). *EMBO J.*, *5*: 3201.

Wada, I., Rindress, D., Cameron, P.H., Ou, W.-J., Doherty, J.J., II, Louvard, D., Bell, A.W., Dignard, D., Thomas, D.Y., and Bergeron, J.J.M. (1991). *J. Biol. Chem.*, *266*: 19599.

Walker, F.J. (1988). *Semin. Thromb. Hemostasis*, *14*: 216.

Wasserman, R.H., Corradino, R.A., and Taylor, A.N. (1968). *J. Biol. Chem.*, *243*: 3978.

Watanabe, Y., Kobayashi, R., Ishikawa, T., and Hidaka, H. (1992). *Arch. Biochem. Biophys.*, *292*: 563.

Williams, R.J.P. (1992). *Cell Calcium*, *13*: 355.

Yamagata, K., Goto, K., Kuo, C.H., Kondo, H., and Miki, N. (1990). *Neuron*, *2*: 469.

Yoo, S.H., and Albanesi, J.P. (1991). *J. Biol. Chem.*, *266*: 7740.

Zapata, C.A., Berkowitz, P., Noyes, C.M., Pollock, J.S., Deerfield, D.W., Pedersen, L.G., and Hiskey, R.G. (1988). *J. Biol. Chem.*, *236*: 8150.

Zot, A.S., and Potter, J.D. (1987). *Annu. Rev. Biophys. Biophys. Chem.*, *16*: 535.

MAGNESIUM-PROTEIN INTERACTIONS

by

Theodor Günther

Institute of Molecular Biology and Biochemistry
Free University of Berlin
Arnimallee 22
14195-Berlin
Germany

I. INTRODUCTION

As a rule, Mg^{2+} can interact with each protein that contains negatively charged groups to which Mg^{2+} has access. Depending on the structural relationships of these negatively charged groups with their environment, the interaction of Mg^{2+} with proteins will express different affinities and specificities. Mg^{2+}- protein interactions may be direct or indirect, e.g. by competition with Ca^{2+} at common binding sites. Obviously there is a great number of proteins interacting with Mg^{2+}, particularly Ca^{2+}-binding proteins and Mg^{2+}-dependent enzymes.

In this contribution, proteins have therefore been classified according to their functions and, because of limited space and systematic information, representative examples of Mg^{2+}-protein interactions have been considered.

II. SERUM PROTEINS

A. Albumin

K_a of Mg^{2+} binding to albumin (69 k) was calculated to be 5 x 10^2 M^{-1} (for literature see Blincoe and Woodin, 1984). Measurement of K_a by Sephadex gel filtration resulted in 3 x 10^2 M^{-1} (Vormann and Günther, 1986). This constant includes enrichment of Mg^{2+} at the surface of the albumin molecule and binding of Mg^{2+}.

The binding groups of albumin are not identified. Since serum albumin can bind free fatty acids and free fatty acids can bind Mg^{2+}, Mg^{2+}-binding to albumin may be increased in states of enhanced lipolysis. Similar relationships hold for serum Ca^{2+}.

B. α-Fetoprotein

α-Fetoprotein (68 k), the major protein in fetal serum, which has about 40% homology to albumin, expresses similar properties as serum albumin, its Mg^{2+} affinity being three times higher (Vormann and Günther, 1986).

C. Prothrombin

Prothrombin (60 k) is an essential vitamin K-dependent protein in the coagulation cascade. It contains 10 γ-carboxyglutamyl residues near the amino terminus which chelate Ca^{2+} and Mg^{2+}.

Equilibrium dialysis and Scatchard plot analysis revealed the binding of 7 Mg^{2+} or 7 Ca^{2+} per molecule of human prothrombin, and 7 Mg^{2+} or 10 Ca^{2+} per bovine prothrombin. Binding constants for all 7 Mg^{2+} binding sites were nearly identical, amounting to 960 M^{-1} for human and 868 M^{-1} for bovine prothrombin. Binding constants for Ca^{2+} ranged from 44 M^{-1} to 8160 M^{-1} for different Ca^{2+}-binding sites (Deerfield et al., 1987). Thus, Mg^{2+} and Ca^{2+} binding takes place in the physiological range of $[Mg^{2+}]_o$ = 0.5 mM and $[Ca^{2+}]_o$ = 1.2 mM.

Mg^{2+}- and Ca^{2+}-binding to prothrombin probably mediates binding of prothrombin to membrane phospholipids.

III. PROTEOHORMONES

A. Gastrin

Human gastrin, a heptadecapeptide produced by duodenal and jejunal mucosal cells, stimulates HCl secretion of the stomach or insulin secretion. The physiological action of the various gastrins depends on the C-terminal sequence Trp-Met-Asp-Phe-NH_2. Gastrin can bind 3 Mg^{2+} cooperatively with K_a = 6 x 10^6, 10^6 and 5 x 10^5 M^{-1}.

Mg^{2+} is probably bound to the carboxyl groups of 5 neighboring glutamyl residues and to the C-terminal Trp and Phe residues.

Binding of Mg^{2+} (or Ca^{2+}) to gastrin induces a conformation change which seems to be necessary for its physiological functions, presumably in the interaction of gastrin with its membrane receptors (Peggion et al., 1984).

B. Avian thymic hormone (ATH)

Thymic hormone (12 k), produced by thymus reticulum cells, is involved in the differentiation of T cells in the thymus. Its structure is similar to parvalbumin. However, polyclonal antibodies to ATH did not cross-react with parvalbumin from chicken muscle. ATH displays two indistinguishable metal binding sites. K_a amounts to 1 x 10^7 M^{-1} for Ca^{2+} and to 1.5 x 10^5 M^{-1} for Mg^{2+} (in 0.15 M NaCl, 25 mM Hepes, pH 7.4) (Serda and Henzl, 1991). Whether ion-binding plays a role in the function of ATH is not defined.

IV. MATRIX PROTEINS

A. Osteocalcin

Osteocalcin (5.2 - 5.9 k, species-dependent) is produced by osteoblasts depending on the presence of vitamin K. Osteocalcin contains 3 γ-carboxyglutamyl residues (Gla) (positions 17, 21, 24) by which it is adsorbed to hydroxyapatite of bone (1-2 molecules of osteocalcin per microcrystal, 1-2% of total bone protein). Serum osteocalcin increases when bone formation or bone turnover rate is elevated. Osteocalcin binds 3 Ca^{2+} (K_a = 3 x 10^2 to 1 x 10^3 M^{-1}). Mg^{2+} competes with Ca^{2+}. Binding of Ca^{2+} or Mg^{2+} induces α-helical conformation in osteocalcin (half-maximal concentration is 0.75 mM for Ca^{2+} and 0.14 mM for Mg^{2+}). However, Mg^{2+} binding to osteocalcin inhibits its binding to hydroxyapatite. Osteocalcin inhibits precipitation of hydroxyapatite from supersaturated solutions and the transition of brushite to hydroxyapatite. Osteocalcin-deficient bone is poorly resorbed (Hauschka, 1985; Wians et al., 1983; Hauschka and Carr, 1982). Thus, Mg^{2+} can interact in bone metabolism via osteocalcin-binding to hydroxyapatite.

B. Matrix Gla protein (MGP)

80% of γ-carboxyglutamate (Gla) in bone is found in osteocalcin, the rest is MGP. MGP (15 k) contains 5-6 Gla and, in contrast to osteocalcin, is associated with the collagenous bone matrix (Price et al., 1983). Binding of Ca^{2+} and Mg^{2+} to MGP may be analogous to osteocalcin.

V. Ca^{2+}-BINDING PROTEINS

Ca^{2+}-binding proteins may be subdivided into two groups with distinct structural features:

(i) The Ca^{2+}-modulated proteins include: calmodulin, troponin C, parvalbumin, oncomodulin, S-100 proteins, calbindins, myosin light chains, calpain, calcineurin, etc.

There are about 170 proteins belonging to this family (Heizmann and Hunziker, 1991; Strynadka and James, 1989).
(ii) Annexins, also known as lipocortins, calcimedins, calpactins, etc, interact with phospholipids and cellular membranes in a Ca^{2+}-dependent manner (Heizmann and Hunziker, 1991; Heizmann, 1992; Villereal and Palfrey, 1989).

Calmodulin, troponin C and the other members of this family exhibit a common structural motif, the EF-hand. The various Ca^{2+}-binding proteins of this family contain 2-8 of these domains. Each of these domains consists of a loop of 12 amino acids (a variant loop with 14 amino acids is present in the S-100 protein subfamily) 5 of which have a carboxyl (or a hydroxyl) group in their side chain and are precisely spaced to coordinate a Ca^{2+} ion or, with less affinity, a Mg^{2+} ion. The 6th ligand is a water molecule. Each loop is flanked by two α-helices.

No sequences homologous to the EF-hand sequence of Ca^{2+}-binding proteins can be found in the primary structure of the annexins, suggesting that they contain a different Ca^{2+}-binding structure.

A. Calmodulin

As measured by equilibrium dialysis, calmodulin can bind 4 Ca^{2+}, 3 with $K_a = 5 \times 10^6\ M^{-1}$ (class A) and one with $K_a = 10^6\ M^{-1}$ (class B). Mg^{2+}-binding occurs at the same loops. Three Mg^{2+} are bound with $K_a = 7 \times 10^3\ M^{-1}$ and one Mg^{2+} with $K_a = 5 \times 10^4\ M^{-1}$ (Wolff et al., 1977). According to Milos et al. (1986) 4 Mg^{2+} are bound with $K_a = 174 M^{-1}$. Binding of metal ions to the apoprotein induces a profound structural change, being less when Mg^{2+} is bound instead of Ca^{2+}. Metal ion binding changes the protein backbone to more hydrophobicity, thus mediating binding of the metal-calmodulin complex to target proteins.

At physiological $[Mg^{2+}]_i$ (1 mM) and $[Ca^{2+}]_i$ (10^{-7} M), calmodulin may exist as tetra-Mg-complex. After stimulation of responding cells, $[Ca^{2+}]_i$ is increased, thus forming the Ca_{3A}-Mg_{1B}-complex of calmodulin and to a smaller extent the Ca_{3A}-Ca_{1B} complex of calmodulin. Since the tetra-Mg-complex of calmodulin revealed unable to activate phosphodiesterase, this transition may represent a regulatory mechanism for calmodulin-dependent enzymes (Wolff et al., 1977; Scharff, 1981; Ohki et al., 1993). However, there is contradiction on the identity of Ca^{2+} and Mg^{2+} binding sites of calmodulin (Milos et al., 1986).

B. Troponin C

During muscle contraction, Ca^{2+} is bound to troponin C, inducing an allosteric conformation change which, by interaction with troponin I and troponin T, is transmitted to tropomyosin inducing muscle contraction.

Troponin C exhibits 6 divalent cation binding sites:
(i) Two high-affinity Ca^{2+} binding sites that also bind Mg^{2+} competitively ($K_a^{Ca} = 2 \times 10^7\ M^{-1}$).
(ii) Two sites with lower affinity for Ca^{2+} that do not bind Mg^{2+} ($K_a^{Ca} = 2 \times 10^5\ M^{-1}$).
(iii) Two sites that bind Mg^{2+} but not Ca^{2+} ($K_a^{Mg} = 5 \times 10^3\ M^{-1}$).

The role of the two specific Mg^{2+} binding sites is unknown. When troponin C is bound to troponin I or in troponin (complexes of troponin C, I and T), the Ca^{2+} and Mg^{2+} binding constants become 10 times higher, suggesting a conformational change in troponin C upon interaction with troponin I. When Mg^{2+} is bound to the high-affinity Ca^{2+}-binding sites, troponin C acquires a different structure. It may be that Mg^{2+} binds to the high-affinity sites during the relaxed state of muscle, when $[Ca^{2+}]_i < 10^{-7}$ M, and serves to stabilize this state (Potter and Gergely, 1975; Levine et al., 1978).

C. Parvalbumin

Parvalbumins (11-12 k) are localized in the cytosol of fast twitch muscles and in a subpopulation of neurons. It contains 3 loops. One of them was degenerated during evaluation, so that parvalbumins contain two Ca^{2+}/Mg^{2+} binding loops (Heizmann and Hunziker, 1991). Binding constants were determined by means of equilibrium dialysis. In the absence of Ca^{2+}, α-parvalbumins bind 2 Mg^{2+} with $K_a = 5 \times 10^4\ M^{-1}$ and β-parvalbumins bind 2 Mg^{2+} with $K_a = 3 \times 10^4\ M^{-1}$ at 2 independent sites. In the absence of Mg^{2+}, α-parvalbumins bind 2 Ca^{2+} with $K_a = 10^8\ M^{-1}$ and β-parvalbumins bind 2 Ca^{2+} with $K_a = 4.5 \times 10^8\ M^{-1}$. Ca^{2+} and Mg^{2+} compete for common binding sites (Haiech et al., 1979; Strynadka and James, 1989). Substitution of Mg^{2+} by Ca^{2+} induces structural changes of the protein (Blancuzzi et al., 1993). The interactions of Mg^{2+} and Ca^{2+} with parvalbumins and troponin C may play a role in muscle contraction and relaxation (Haiech et al., 1979).

D. Oncomodulin

Oncomodulin (11.5 k) is a parvalbumin-like protein which is expressed in the cytotrophoblast of fetal placental tissue and in tumor cells. It is able to stimulate cyclic nucleotide phosphodiesterase in a calmodulin-like manner.

Contrary to parvalbumin, oncomodulin contains a single high-affinity Ca^{2+}/Mg^{2+} site (EF site) and a single low-affinity Ca^{2+}-specific site (CD site) (Hapak et al., 1989).

In the absence of Mg^{2+}, K_a^{Ca} amounted to 2.2×10^7 and $1.7 \times 10^6\ M^{-1}$ and in the absence of Ca^{2+}, K_a^{Mg} amounted to $4 \times 10^3\ M^{-1}$. Mg^{2+} competes with Ca^{2+} for the high-affinity site (Cox et al., 1990).

E. Calbindin

Calbindins are low-molar-mass proteins (9 k - 28 k, species-dependent). Their biosynthesis is induced by 1,25-dihydroxycholecalciferol. They bind Ca^{2+} and prevent toxicity of intracellular Ca^{2+} during transcellular Ca^{2+} transport across mucosa cells in intestinal Ca^{2+} resorption or across osteoclasts in parathormone-induced bone resorption.

It was reported that calbindin expresses 2 (Bryant and Andrews, 1984), 4 (Bredderman and Wasserman, 1974) or 6 (Heizmann and Hunziker, 1991) high-affinity Ca^{2+} binding sites which also can bind Mg^{2+}. Binding constants were estimated by equilibrium dialysis to be $2 \times 10^6\ M^{-1}$ for Ca^{2+} and to be $2.8 \times 10^2\ M^{-1}$ for Mg^{2+} ($[K^+] = 0.150$ M, pH 6.8), whereas Bryant and Andrews (1984) determined $K_a = 2 \times 10^8\ M^{-1}$ for Ca^{2+} and $K_a = 10^4\ M^{-1}$ for Mg^{2+}-binding (20 mM Tris buffer, pH 7.5, no K^+). The difference in binding constants may be caused by the different pH and ionic strength.

Whether Mg^{2+}-binding to calbindin plays a physiological role or functions only as an unspecific competitor remains open.

F. General remarks

Taken together, the reviewed Ca^{2+}-binding proteins of this family have similar properties with respect to Ca^{2+}- and Mg^{2+}-binding. Therefore, other members of this family also expressing the EF-hand motif, which have not yet been investigated for Mg^{2+}-binding and Mg^{2+}-Ca^{2+} competition, may have similar binding properties.

As a common function of these proteins, metal coordination changes their structure into a more hydrophobic surface so that they can bind to target proteins, thus regulating their biological activity.

VI. CYTOSKELETAL PROTEINS

A. Tubulin

The microtubulus system, which is part of the cytoskeleton and involved in processes like exocytosis/endocytosis and mitosis, consists of two globular proteins, α-tubulin (53 k) and β-tubulin (55 k). Both of these form a heterodimer, which is polymerized to flat filaments and assembled to cylindrical microtubuli. In this process 1 GTP pro dimer is hydrolysed to GDP. Each heterodimer has two guanine nucleotide binding sites: the first (exchangeable site or E-site), on the β-subunit, binds GTP or GDP; the second, in the α-subunit, binds GTP in a non-exchangeable manner (N-site). Upon microtubule assembly, GTP bound in the E-site is hydrolyzed to GDP while the N-site GTP is unchanged. The heterodimer also contains one tightly bound Mg^{2+} associated with the N-site GTP whereas a second Mg^{2+} is tightly bound to the E-site GTP. Weaker Mg^{2+}-binding sites may be found at the C-terminal regions.

Binding constants of GTP, GDP and Mg^{2+} were determined by means of the micropartition method and Scatchard plot analysis. K_a for Mg^{2+}-binding to the E-GTP-site amounted to 1.2 x 10^6 M^{-1}, whereas K_a for Mg^{2+}-binding to the E-GDP-site or to tubulin with a vacant E-site amounted to 10^3 M^{-1}. It was suggested that Mg^{2+} is coordinated to one oxygen from each of the β- and γ-phosphates of GTP, two protein side chains and two H_2O ligands, in analogy to p21-protein (see X). In the tubulin GDP-Mg^{2+} complex, two H_2O may replace the γ-phosphate of the nucleotide and one of the protein residues as ligands.

One function of Mg^{2+} in microtubule formation is the binding of GTP to the E-site. K_a for GTP-binding to the E-site amounted to 6 x 10^6 M^{-1} in the presence of 1 mM Mg^{2+} and only to 2 x 10^4 M^{-1} in the absence of Mg^{2+}. Binding of GDP to the E-site (K_a = 2 x 10^6 M^{-1}) was independent of Mg^{2+} (Mejillano and Himes, 1991).

However, increasing $[Mg^{2+}]_i$ in intact cells by microinjection of $MgSO_4$ caused a reversible disassembly of the microtubuli arrays and of the F-actin stress fibres. The effect is not caused by direct actions of Mg^{2+} on tubulin but seems to be caused by loss of microtubule associated proteins, by activation of cytoskeletal-associated kinases or nucleotidases (Prescott et al., 1988).

B. Actin

Actin filaments (F-actin) are found as polymers of α-actin in muscle and as β- and γ-actin in other cell types. G-actin (monomeric actin) has a high-affinity binding site for nucleotides and two classes of cation binding sites which could be separated on the basis of their Ca^{2+} affinity, a single high-affinity site K_a = 10^6 M^{-1} and 3 identical moderate-affinity sites with K_a = 5 x 10^4 M^{-1} (Zimmerle et al., 1987). Binding of Mg^{2+} to the same sites is about 5 - 10 times lower. Absolute values from different authors differ by a factor of 100 (Estes et al., 1987).

When Ca^{2+} is replaced by Mg^{2+} at the moderate-affinity site(s), the conformation of actin is changed, which may play a role in the formation of actin filaments (Shu et al., 1992).

C. Gelsolin

Gelsolin (83 k) is a cytoplasmic protein secreted in plasma and involved in actin filament formation. It expresses two classes of Ca^{2+}-binding sites K_a = 1 x 10^5 M^{-1} and K_a = 10 M^{-1}. Mg^{2+} can be bound to the low-affinity Ca^{2+}-binding site.

Metal binding changes the secondary structure of gelsolin which is different when 2 Ca^{2+} or 1 Ca^{2+} and 1 Mg^{2+} are bound (Doi et al., 1990).

VII. CELL ADHESION PROTEINS

A. Integrins

Integrins are transmembrane-spanning receptor proteins, consisting of two protein chains (α, β). Integrins bind fibronectin, vitronectin, laminin, collagen, etc by means of the sequence Arg, Gly, Asp, and mediate the interaction of matrix proteins with the cytoskeleton. These processes are involved in embryonic differentiation, wound healing and spreading of tumor cells. Receptor binding and subsequent cellular reaction, e.g. spreading (Hynes, 1992) of HeLa cells, was enhanced by Mg^{2+} and Mn^{2+} but not by Ca^{2+} (Beacham and Jacobson, 1990). Also, other integrin receptor / ligand interactions are specifically regulated by divalent cations (Shimizu and Mobley, 1993).

The sequence of the extracellular domain of the α-subunit of each integrin contains several (probably 4) sites that are homologous to Ca^{2+}-binding sites in other proteins such as calmodulin. These sequences represent the Ca^{2+}- and Mg^{2+}-binding sites (Ruoslahti, 1988).

B. Cadherins

Cadherins (E-, N-, P-cadherin) are membrane-associated glycoproteins that mediate Ca^{2+}-dependent cell-cell adhesion among cells with the same type of cadherin. The Ca^{2+}-binding sites are located in the extracellular domain (Takeishi, 1990). E-cadherin binds Ca^{2+} with K_a = 5 x 10^2 M^{-1} (Ozawa et al., 1990). Ca^{2+}-binding induces a conformation change in the extracellular domain which is a prerequisite for cadherin-mediated cell-cell adhesion. Probably, Mg^{2+} and Ca^{2+} compete for common binding sites.

VIII. MEMBRANE RECEPTORS

Membrane receptors are membrane-integrated proteins which coordinate the function of multicellular tissues and organisms by hormones, cytokines, growth factors, etc. A great number of membrane receptors are influenced by Mg^{2+}.

A. Insulin receptor

The insulin receptor has tyrosine kinase activity. After insulin binding to its receptor, Mg^{2+} is operating together with ATP as kinase substrate (see XII). Additionally, Mg^{2+} is bound to a regulatory site and couples the receptor with the kinase (Vicario et al., 1988).

B. G-protein coupled receptors

1. *β-Adrenergic receptor*

Mg^{2+} (and also Ca^{2+} and Mn^{2+}) caused a 5-fold increase in the binding of β-agonist to its receptor in a membrane preparation from frog erythrocytes. Half maximal effect was obtained at 0.4 mM Mg^{2+}. Binding of β-adrenergic antagonist was not affected by Mg^{2+}. The increase in the amount of agonist binding was due to a 20-fold increase in the apparent affinity of the agonist for the receptor (Williams et al., 1978).

Similar results were obtained with a plasma membrane preparation from S49 lymphoma cells. Ca^{2+} was ineffective. Mg^{2+} increased agonist affinity with a half-maximal effect at 2-3 mM (Bird and Maguire, 1978).

In both preparations Mg^{2+} was only effective when the hormone-receptor adenylate cyclase system was complete. These results suggest that Mg^{2+} affects agonist receptor affinity via involvement of G-proteins. (see IX). For detailed discussion see Johansson et al. (1992).

2. *α -Adrenergic receptor*

Binding of α-agonist was doubled by Mg^{2+}, whereas Ca^{2+} increased α-agonist binding by 20% only. The effect of Mg^{2+} was caused by doubling the affinity of the α-receptor.

The increase of α-agonist binding was steapest up to 1 mM Mg^{2+}. There was an optimum at pMg 2, and then binding inhibition at higher Mg^{2+} concentrations (Glossmann and Presek, 1979).

3. *Prostaglandin receptor*

Similar to β-agonist binding, the binding of ^{3}H-prostaglandin E_1 was increased 5-fold by Mg^{2+}, halfmaximal effect being obtained at 0.3 mM Mg^{2+} (Williams et al., 1978).

4. *Muscarinic cholinergic receptor*

Mg^{2+} increased the affinity of acetylcholine binding to chicken heart muscarinic receptors. These receptors exist in 3 agonist affinity states. Mg^{2+} caused an increase in the proportion of high-affinity receptors and a decrease in that of low-affinity ones (McMahon and Hosey, 1985).

5. *Dopaminergic receptor*

Similar to other receptors of this group, Mg^{2+} increased both agonist binding (4-fold) and affinity of agonist binding. High Mg^{2+} increased the percentage and affinity of high-affinity binding sites (Shibley and Creese, 1983).

6. *A1-Adenosine receptor*

Mg^{2+} enhanced binding of adenosine to A_1 receptors in rat striatum (Parkinson and Fredholm, 1992).

7. *5-Hydroxytryptamine 1 A (5-HT1A) receptor*

Increased physiological $[Mg^{2+}]_o$ enhanced the binding affinity of 5-hydroxytryptamine (5-

HT) to the 5-HT_{1A} receptor of Purkinje cells and the inhibitory functions of 5-HT. Reciprocally, decreased $[Mg^{2+}]_o$ reduced 5-HT_{1A} receptor affinity and function, whereas the 5-HT_{1B} receptor subtype was hardly affected by $[Mg^{2+}]_o$ (Darrow et al., 1991) .

8. *Formylpeptide chemotactic receptor*
N-Formylmethionine-containing peptides, generated as by-products or degradation products of bacterial protein synthesis or released from eukaryotic mitochondria upon tissue damage, stimulate cellular functions of neutrophils and macrophages, e.g. by chemotaxis, release of granular enzymes and superoxide generation. Formylpeptides act via specific membrane spanning receptors. Mg^{2+} enhanced binding of fMet-Leu[^{3}H]-Phe to membranes from HL-60 cells. Half-maximal effect occurred at 0.15 mM Mg^{2+}.

The effect of Mg^{2+} was caused by increasing the number of high-affinity receptors up to 9-fold depending on Mg^{2+} concentration, whereas the number of low-affinity receptors remained unaltered (Gierschik et al., 1989).

Since the receptor-mediated uptake of formylpeptides into intact neutrophils was stimulated by Mg^{2+}, it is probable that Mg^{2+} interacts with this receptor at an extracellular domain.

9. *Oxytocin receptor*
Oxytocin is bound to its receptors located on the membrane of myometrium cells with at least 3 different binding constants. Affinity and capacity of oxytocin binding depend on $[Mg^{2+}]_o$. Oxytocin binding at half-saturation of binding sites in the presence of physiological Ca^{2+} is maximal at 1 mM Mg^{2+}. Additionally, the high-affinity oxytocin binding site displays a positive cooperativity which is amplified by increasing $[Mg^{2+}]_0$ (Plisca and Kohlhauf-Albertin, 1991).

10. *Vasopressin receptor*
Arginine vasopressin (AVP) is a nonapeptide of the hypophysis. Its vasoconstrictor and antidiuretic actions are mediated by the activation of V_1 and V_2 receptors. Moreover, AVP activates glycogen phosphorylase via V_1-receptors located on liver plasma membranes. V_2 receptors are operating via cAMP and G_s proteins, and V_1 receptors via inositol trisphosphate (Ca^{2+}) and diacylglycerol (for literature see Gonzalez et al., 1992).

Binding of ^{3}H-AVP to V_1 receptors of isolated plasma membranes was stimulated up to 6.5 fold by Mg^{2+}. $[Mg^{2+}]_o$ was active in the physiological range (0.5 mM). Maximal stimulation of AVP-binding was obtained at 1 mM Mg^{2+}. Higher Mg^{2+} concentrations were inhibitory. $[Ca^{2+}]_o$ in its physiological range of 1.2 mM had no significant effect, whereas unphysiological high $[Ca^{2+}]_o$ inhibited competitively (Wang et al., 1990).

11. *Other receptors*
Other receptors with similar Mg^{2+}-dependencies are the opiate receptors (Wei and Sulakhe, 1980; Rodriguez et al., 1992).

All the above-mentioned receptors are operating via GTP-binding proteins. The actions of GTP are dependent on Mg^{2+}, and by these interactions, Mg^{2+} can retrogradely influence hormone-receptor binding (see IX).

There are about 100-150 G-protein-coupled receptors (Birnbaumer et al., 1990). For most of these, the effect of Mg^{2+} on hormone receptor-binding has not been investigated so far.

One exception seems to be the glucagon receptor. Binding of glucagon to its receptor was independent of Mg^{2+} (Rojas and Birnbaumer, 1985). However, recent experiments with permeabilized hepatocytes have shown that Mg^{2+} (probably from the intracellular side) enhanced glucagon binding to its receptor (Post et al., 1992).

IX. G-PROTEINS

G-proteins, heterotrimers consisting of α-, β- and γ-subunits, are a family of membrane-associated GTP-binding proteins that transduce receptor-mediated signals to membrane-bound enzymes such as adenylate cyclase, phospholipase A_2 and phospholipase C (Gilman,

1987; Birnbaumer et al., 1990), or to ion channel proteins (Brown, 1991). After contact with the hormone-receptor-complex, G-proteins dissociate into α- and βγ-subunits. The stimulatory subunit $G_{s\alpha}$ binds GTP in exchange for GDP and activates the subsequent enzyme reaction, e.g. adenylate cyclase, whereas the inhibitory $G_{i\alpha}$ binds GTP and inhibits adenylate cyclase. After hydrolysing GTP to GDP by the GTPase activity of G_α the GDP form of G_α reassociates with $G_{\beta\gamma}$.

As shown with isolated subunits, Mg^{2+} is bound to G_α-GTP. The apparent dissociation constant of G_α-GTP-Mg^{2+} is 5 nM, similar to the halfmaximal Mg^{2+} concentration for GTPase activity. Thus, the effects of Mg^{2+} are stimulation of GTP binding to G_α and activation of GTPase (Higashijima et al., 1987a; 1987b). Mg^{2+} is also required for dissociation of G_s and for membrane release of G_s (Ransnäs et al., 1992). For further effects of Mg^{2+} in the action of G-proteins, see Gilman (1987). The $G_{i\alpha}$ from porcine atrial membranes, which is stimulated by cholinergic agonists, is operating analogously. K_m for Mg^{2+} of Mg^{2+}-activated GTPase amounts to 5 μM (Hilf and Jakobs, 1989).

Similar effects of Mg^{2+} were found in the action of G-protein in amplification and transduction of photon capture in rod outer segments (Yamazaki et al., 1987).

X. P21-PROTEINS

The products of the ras genes, the p21-proteins, express some sequence homology with G-proteins (and bacterial elongation factor Tu). They are believed to function as switches in signalling events of cell growth and differentiation. They bind Mg^{2+} complexes of guanine nucleotides with high affinity (K_a = 10^{11}-10^{12} M^{-1}), are associated with the plasma membrane, and express weak GTPase activity which triggers the conformational change necessary for signal transduction (Barbacid, 1987, Wittinghofer and Pai, 1991). Although nucleotides bind to $p21^{ras}$ in the absence of Mg^{2+}, the interaction is much stronger in its presence. Mg^{2+} is surrounded by six oxygen atoms in a nearly perfect octahedral arrangement. The ligands are donated by the β- and γ-phosphates of GTP, the side chains of Ser 17 and Thr 35, and two water molecules which occupy the axial positions. Asp 33, Asp 57 and the α-phosphates interact with these two waters. After hydrolysis of GTP, two more water molecules are bound to Mg^{2+} (Wittinghofer and Pai, 1991).

XI. CHANNEL / TRANSPORT PROTEINS

Besides the action of Mg^{2+} on Ca^{2+} and K^+ channels via G-proteins (Brown, 1991; Birnbaumer et al., 1990), Mg^{2+} can interact directly with ion channels and transport proteins.

A. Voltage-dependent blockage of ion channels by Mg^{2+}

After binding of NMDA to its receptor, which is a subspecies of the glutamate receptor of neurons, Na^+ and Ca^{2+} channels are opened, thus mediating influx of Na^+ and Ca^{2+} and efflux of K^+.

These channels are blocked when Mg^{2+} enters from the extracellular side and is bound within them (Asher and Nowak, 1988). This process depends on the membrane potential and occurs at physiological $[Mg^{2+}]_o$. It may, therefore, be changed in states of hypo- and hypermagnesemia.

In analogy to extracellular Mg^{2+}, intracellular Mg^{2+} also can enter ion channels depending on voltage and block various ion channels of neurons such as the NMDA-gated Na^+ and Ca^{2+} channels, the ATP-dependent K^+ channel, the K^+ inward rectifier, the muscarinic K^+ channel and the voltage-dependent Na^+ channel. The voltage dependencies of the blocks and the affinities of Mg^{2+} at 0 mV vary greatly among these channel types (Johnsen and Asher, 1990; Agus and Morad, 1991; Matsuda, 1991). In analogy to NMDA-gated channels, extracellular and intracellular Mg^{2+} also blocked histamine-operated cation channels of arterial endothelial cells (Yamamoto et al., 1992).

B. Mg^{2+} interaction with Ca^{2+} transport proteins

1. *Na^+/Ca^{2+} antiport*

Mg^{2+} can compete with Ca^{2+} in the Na^+/Ca^{2+} antiport without being transported by this system and thus inhibit it (DiPolo and Beauge, 1985; Lyu et al., 1991).

2. *Sarcoplasmic Ca^{2+} transport*

Sarcoplasmic reticulum (SR) forms an intracellular membrane compartment that regulates the contraction-relaxation cycle of muscle by releasing and reabsorbing Ca^{2+}. Release of Ca^{2+} from SR is triggered by an action potential at the neuromuscular junction that is communicated to SR via the T system.

Ca^{2+} release is performed by a channel which is activated by Ca^{2+} and nucleotides, and inhibited by Mg^{2+}. At 5 x 10^{-6} M Ca^{2+}, halfmaximal inhibition occurred at 0.1 mM Mg^{2+} (physiological $[Mg^{2+}]_i$ in muscle cells is 1 mM). Mg^{2+} has been suggested to inhibit Ca^{2+} release by binding competitively to the Ca^{2+} activating sites and by blocking the channel as it binds to the transport sites (Meissner et al., 1986).

C. Mitochondrial K^+/H^+ antiport

Removal of Mg^{2+} from mitochondria increased K^+/H^+ exchange in this organelle (Beavis and Garlid, 1990; Jung and Brierly, 1986) indicating an interaction of Mg^{2+} with this still undefined transport system.

D. Other transport systems

Intracellular Mg^{2+} can inhibit stilbendisulfonate-sensitive Cl^- efflux (Russell and Brodwick, 1988), and can also influence the Na-K-2Cl-cotransporter, K-Cl-cotransporter, and Na^+/H^+ antiporter. For literature, see Parker et al. (1989).

E. Mg^{2+} transport

Since Mg^{2+} is transported across cell membranes, Mg^{2+} must interact with its own transport proteins:

(i) Mg^{2+} is transported out of the cell, depending on $[Na^+]_o$, by electroneutral Na^+/Mg^{2+} antiport which is gated by increased $[Mg^{2+}]_i$, indicating two different modes of Mg^{2+} protein interaction in this transport process (Günther et al., 1990b).

(ii) Mg^{2+} efflux from erythrocytes can occur irrespective of $[Na^+]_o$ in combination with Cl^- efflux for charge compensation (Günther and Vormann, 1990a).

(iii) Mg^{2+} uptake in hepatocytes is performed by Na^+, Mg^{2+}/anion symport (Günther and Höllriegl, 1993).

XII. ENZYMES

The most significant interaction of Mg^{2+} with proteins is the activation of Mg^{2+}-dependent enzymes.

There may be about 300 Mg^{2+}-activated enzymes in a liver cell and about 800 bacterial restriction endonucleases, most of them being dependent on Mg^{2+}. Most Mg^{2+}-dependent enzymes are reacting with negatively charged substrates such as ATP, polyphosphates, citrate, isocitrate, etc (for analogy, see G-proteins and p21-proteins). Thus, Mg^{2+} is involved in all energy-forming and energy-consuming metabolic reactions. For all Mg^{2+}-dependent enzymes, the catalytic activity as a function of pMg expresses a bell-shaped curve composed of two sigmoidal curves: (i) activation, and (ii) inhibition of enzyme activity as a function of pMg. The difference of both curves gives an optimum at pMg = 3 $\pm$ 0.5 for various glycolytic enzymes (Günther, 1981).

The activation of enzymes by Mg^{2+} can be classified according to the sequence of reactions:

(i) Mg^{2+} is bound to the negatively charged substrate (ATP^{4-}), forming the Mg substrate (MgS) complex, which is the real substrate. Thereafter, MgS is bound to the enzyme (E), forming the enzyme-substrate complex EMgS, in which the substrate is transformed to the product. In this reaction mechanism, the main actions of Mg^{2+} is charge reduction of the substrate and, in some cases, additionally binding of the substrate to the active center of the enzyme.

(ii) Mg^{2+} is bound to the enzyme, establishing its active conformation (active center) which can bind the substrate, forming MgES. Thereafter, in MgES, S is transformed to the product. Also, in this type, Mg^{2+} can be involved in binding of the substrate to the active center.

Enzymes which react according to type (i) are: e.g. hexokinase, phosphoglycerate kinase, creatine kinase, adenylate kinase.

Enzymes which react according to type (ii) are: e.g. enolase, pyruvate kinase, pyrophosphatase, pyruvate carboxylase (Ebel and Günther, 1980; O'Sullivan, 1973).

An example for an enzyme of type (i) is adenylate kinase (Fry et al., 1985). The enzyme (194 amino acids) has two binding sites: the MgATP-binding site and the AMP-binding site which binds uncomplexed AMP or ADP. By means of X-ray crystallography and NMR, a model of the enzyme, its MgATP-binding sites and its reaction mechanism was elaborated (Fry et al., 1985; 1986).

Another example for the spatial structure of EMgS (type (i)) is $NADP^+$-dependent isocitrate dehydrogenase (E.C.1.1.1.42), in which isocitrate is transformed to oxalosuccinate by dehydrogenation, and its subsequent decarboxylation to α-ketoglutarate. By means of X-ray reflection data for crystals of the enzyme-Mg-isocitrate complex, it was found that Mg^{2+} is coordinated to 6 oxygen ligands in a roughly octahedral arrangement. Two ligands are provided by the bidendate chelation of Mg^{2+} via the 07 of the α-carboxylate and the α-hydroxyl oxygen of isocitrate, in a manner suitable for stabilization of a negative charge on the hydroxyl oxygen during both the dehydrogenation and decarboxylation steps. Two more ligands are provided by water molecules 458 and 462. The latter 2 ligands are provided by the side chains of aspartate 283 and 307 (Hurley et al., 1991).

In both types of Mg^{2+}-dependent enzymes, Mg^{2+} can support the catalytic reaction by polarization of the reactive groups (Yoshikawa et al., 1987). In some Mg^{2+}-dependent enzymes, both types of activation are represented simultaneously: MgS is the substrate, and the enzyme needs a second Mg^{2+} bound at a different site to induce an allosteric change and thus the active conformation (Malmström and Rosenberg, 1959; Ebel and Günther, 1980). In the mitochondrial F_0F_1 ATPase, Mg^{2+}-nucleotide is the substrate and the enzyme complex simultaneously needs an additional Mg^{2+} (Bulygin and Vinogradov, 1991). The same holds for pyrophosphatase (Sosa et al., 1992).

In other enzyme systems, e.g. Na^+K^+ ATPase, both effects of Mg^{2+} occur consecutively. First, MgATP is the substrate for enzyme phosphorylation; thereafter Mg^{2+} induces the $E_1P \rightarrow E_2P$ allosteric transition of the phosphorylated enzyme (EP) and thus the translocation of bound Na^+ within the reaction cycle (Flatman and Lew, 1981; Glynn and Karlish, 1975).

For the function of Mg^{2+} in bacterial restriction endonucleases (EcoRI and Eco RV) see Jeltsch et al. (1992).

XIII. CONCLUDING REMARKS

In many Mg^{2+}-protein interactions, the effect of Mg^{2+} is not specific. Other divalent cations like Ca^{2+}, Mn^{2+}, Zn^{2+} etc can mimic the effect of Mg^{2+}. There are examples (NMDA-receptor) in which the effect of Mg^{2+} can be mimicked by polyamines (Sacaan and Johnson, 1990). The most specific effect of Mg^{2+} is the activation of enzymes. In some Mg^{2+}-dependent enzyme reactions, Ca^{2+} or Mn^{2+} are also active either with the same or with less efficacy as/ than Mg^{2+}. High Mg^{2+} specificity compared to Ca^{2+} is caused by the 3000 times lower rate of water substitution from the inner hydration shell of Mg^{2+} (Eigen and Hammes, 1963).

In all binding reactions with high affinity, Mg^{2+} may be bound in an octahedral chelate or complex. Some ligands may be bound by outer and some by inner sphere complexes. In most of these binding reactions, the spatial structure of the complexes has not been defined so far. However, there is growing knowledge on the amino acid sequence of the Mg^{2+}-binding region of Mg^{2+}-dependent enzymes (Black and Hruby, 1992). The above-reported Mg^{2+}-protein interactions are dependent on the Mg^{2+} concentration (pMg) in the corresponding cellular compartment according to the law of mass action.

$[Mg^{2+}]_i$ is buffered and regulated by Mg^{2+} efflux and Mg^{2+} influx, and is rather constant (Günther, 1990). Therefore, $[Mg^{2+}]_i$ has a permissive and not a regulatory function. However, $[Mg^{2+}]_i$ may acquire a regulatory function by Mg^{2+}-protein interactions at constant $[Mg^{2+}]_i$ when other factors, e.g. membrane potential, pH_i or $[Ca^{2+}]_i$, are changed.

Since intracellular Mg^{2+} is compartmentalized, resulting in micro-heterogenicity of $[Mg^{2+}]_i$ (Günther, 1990), a quantitative description of Mg^{2+}-protein interactions within the cell by means of Mg^{2+} affinity constants which are often measured under other conditions (pH, ionic strength, $[Ca^{2+}]_i$) will only be approximative.

$[Mg^{2+}]_o$ can be changed to a higher extent than $[Mg^{2+}]_i$, e.g. by Mg deficiency leading to hypomagnesemia, or by therapeutic Mg^{2+} infusion. States of hypo- or hypermagnesemia have significant physiological effects by changing directly Mg^{2+}-protein interactions or indirectly by competition with Ca^{2+} in Ca^{2+}-protein interactions at the extracellular surface of cell membranes.

XIV. ABBREVIATIONS

K_a, apparent association constant; $[Mg^{2+}]_o$, $[Mg^{2+}]_i$, concentration of extracellular or intracellular free Mg^{2+}; $[Ca^{2+}]_o$, $[Ca^{2+}]_i$, concentration of extracellular or intracellular free Ca^{2+}; $[Na^+]_o$, extracellular Na^+ concentration; k, kg mol^{-1}, i.e. SI unit of molar mass (or kDa, i.e. unit of molecular mass, as mainly used by biochemists).

REFERENCES

Agus, Z.S., and Morad, M. (1991). *Annu. Rev. Physiol.*, *53*: 299.

Asher, P., and Nowak, L. (1988). *J. Physiol.*, *399*: 247.

Barbacid, M. (1987). *Annu. Rev. Biochem.*, *56*: 779.

Beacham, D.A., and Jacobson, B.S. (1990). *Exp. Cell Res.*, *189*: 69.

Beavis, A.D., and Garlid, K.D. (1990). *J. Biol. Chem.*, *265*: 2538.

Bird, S.J., and Maguire, M.E. (1978). *J. Biol. Chem.*, *253*: 8826.

Birnbaumer, L., Abramowitz, J., and Brown, A.M. (1990). *Biochim. Biophys. Acta*, *1031*: 163.

Black, M.E., and Hruby, D.E. (1992). *J. Biol. Chem.*, *267*: 6801.

Blancuzzi, Y., Padilla, A., Parello, J., and Cavé, A. (1993). *Biochemistry - USA*, *32*: 1302.

Blincoe, C., and Woodin, T.S. (1984). *J. Inorg. Biochem.*, *21*: 209.

Bredderman, P.J., and Wasserman, R.H. (1974). *Biochemistry - USA*, *13*: 1687.

Brown, A.M. (1991). *FASEB J.*, *5*: 2175.

Bryant, D.T.W., and Andrews, P. (1984). *Biochem. J.*, *219*: 287

Bulygin, V.V., and Vinogradow, D. (1991). *Biochem. J.*, *276*: 149.

Cox, J.A., Milos, M., and McManus, J.P. (1990). *J. Biol. Chem.*, *265*: 6633.

Darrow, E., Strahlendorf, J.C., and Strahlendorf, H.K. (1991). *Eur. J. Pharmacol.*, *209*: 19.

Deerfield, D.W., Olson, D.L., Berkowitz, P., Koehler, K.A., Pedersen, L.G., and Hiskey, R.G. (1987). *Biochem. Biophys. Res. Commun.*, *144*: 520.

DiPolo, R., and Beauge, L. (1985). *J. Gen. Physiol.*, *84*: 895.

Doi, Y., Kim, F., and Kido, S. (1990). *Biochemistry - USA*, *29*: 1392.

Ebel, H., and Günther, T. (1980). *J. Clin. Chem. Clin. Biochem.*, *18*: 257.

Eigen, M., and Hammes, G.G. (1963). *Adv. Enzymol.*, *25*: 1.

Estes, J.E., Selden, L.A., and Gershman, L.C. (1987). *J. Biol. Chem.*, *262*: 4952.

Flatman, P.W., and Lew, V.L. (1981). *J. Physiol.*, *315*: 421.

Fry, D.C., Kuby, S.A., and Mildvan, A.S. (1985). *Biochemistry - USA*, *24*: 4680.

Fry, D.C., Kuby, S.A., and Mildvan, A.S. (1986). *Proc. Nat. Acad. Sci. USA*, *83*: 907.

Gierschik, P., Steisslinger, M., Sidiropoulos, D., Herrmann, E., and Jakobs, K.H. (1989). *Eur. J. Biochem.*, *183*: 97.

Gilman, A.G. (1987). *Annu. Rev. Biochem.*, *56*: 615.

Glossmann, H., and Presek, P. (1979). *Naunyn-Schmied. Arch. Pharmacol.*, *306*: 67.

Glynn, I.M., and Karlish, S.J.D. (1975). *Annu. Rev. Physiol.*, *37*: 13.

Gonzalez, C.B., Reyes, C.E., Caorsi, C.E., and Troncoso, S. (1992). *Biochem. Int.*, *26*: 759.

Günther, T. (1981). *Magnesium-Bull.*, *3*: 91.

Günther, T. (1990). *Metal Ions in Biological Systems*, Vol. 26 (H. Sigel and A. Sigel, eds.), Marcel Dekker, New York and Basel, p. 193.

Günther, T., and Vormann, J. (1990a). *FEBS Lett.*, *271*: 149.

Günther, T., Vormann, J., and Höllriegl, V. (1990b). *Biochim. Biophys. Acta*, *1023*: 455.

Günther, T., and Höllriegl, V. (1993). *Biochim. Biophys. Acta*, *1149*: 49.

Haiech, J., Derancourt, J., Pechère, J.-F., and Demaille, J.G. (1979). *Biochemistry - USA*, *18*: 2752.

Hapak, R.C., Lammers, P.J., Palmisano, W.A., Birnbaum, E.R., and Henzl, M.T. (1989). *J. Biol. Chem.*, *264*: 18751.

Hauschka, P.V. (1985). *Chem. Biol. Mineral Tiss.* (W.T. Butler, ed.), EBSCO Media, Birmingham, p. 149.

Hauschka, P.V., and Carr, S.A. (1982). *Biochemistry - USA*, *21*: 2538.

Heizmann, C.W., and Hunziker, W. (1991). *TIBS*, *16*: 98.

Heizmann, C.W. (1992). *Gen. Physiol. Biophys.*, *11*: 411.

Hilf, G., and Jakobs, K.H. (1989). *Europ. J. Pharmacol.*, *172*: 155.

Higashijima, T., Ferguson, K.M., Smigel, M.D., and Gilman, A.G. (1987a). *J. Biol. Chem.*, *262*: 757.

Higashijima, T., Ferguson, K.M., Sternweis, P.C., Smigel, M.D., and Gilman, A.G. (1987b). *J. Biol. Chem.*, *262*: 762.

Hurley, J.H., Dean, A.M., Koshland, Jr. D.E., and Stroud, R.M. (1991). *Biochemistry - USA*, *30*: 8671.

Hynes, R.O. (1992). *Cell*, *69*: 11.

Jeltsch, A., Alves, J., Maass, G., and Pingoud, A. (1992). *FEBS Lett.*, *304*: 4.

Johansson, L.-H., Persson, H., and Rosengren, E. (1992). *Pharmacol. Toxicol.*, *70*: 192.

Johnsen, J.W., and Asher, P. (1990). *Biophys. J.*, *57*: 1085.

Jung, D.W., and Brierley, G.P. (1986). *J. Biol. Chem.*, *261*: 6408.

Levine, B.A., Thornton, J.M., Fernandes, R., Kelly, C.M., and Mercola, D. (1978). *Biochim. Biophys. Acta*, *535*: 11.

Lyu, R.-M., Smith, L., and Smith, J.B. (1991). *J. Membrane Biol.*, *124*: 73.

Malmström, B.G., and Rosenberg, A. (1959). *Adv. Enzymol.*, *21*: 131.

Matsuda, H. (1991). *Annu. Rev. Physiol.*, *53*: 289.

McMahon, K.K., and Hosey, M.M. (1985). *Mol. Pharmacol.*, *28*: 400.

Meissner, G., Darling, E., and Eveleth, J. (1986). *Biochemistry - USA*, *25*: 236.

Mejillano, M.R., and Himes, R.H. (1991). *Arch. Biochem. Biophys.*, *291*: 356.

Milos, M., Schaer, J.-J., Comte, M., and Cox, J.A. (1986). *Biochemistry - USA*, *25*: 6279.

Ohki, S., Iwamoto, U., Aimoto, S., Yazawa, M., and Hikichi, K. (1993). *J. Biol. Chem.*, *268*: 12388.

O'Sullivan, W.J. (1973). *Inorganic Biochemistry* (G.L. Eichborn, ed.), Elsevier Scientific Publishing Company, Amsterdam, Oxford, New York, p. 582.

Ozawa, M., Engel, J., and Kemler, R. (1990). *Cell*, *63*: 1033.

Parker, J.C., Gitelman, H.J., and McManus, T.J. (1989). *Amer. J. Physiol.*, *257*: C1038.

Parkinson, F.E., and Fredholm, B.B. (1992). *J. Neurochem.*, *58*: 941.

Peggion, E., Mammi, S., and Palumbo, M. (1984). *Biopolymers*, *23*: 1225.

Pliska, V., and Kohlhauf Albertin, H. (1991). *Biochem. J.*, *277*: 97.

Post, S.R., Miyazaki, H., and Tager, S. (1992). *J. Biol. Chem.*, *267*: 25776.

Potter, J.D., and Gergely, J. (1975). *J. Biol. Chem.*, *250*: 4628.

Prescott, A.R., Comerford, J.G., Magrath, R., Lamb, N.J.C., and Warn, R.M. (1988). *J. Cell Sci.*, *89*: 321.

Price, P.A., Urist, M.R., and Otawara, Y. (1983). *Biochem. Biophys. Res. Commun.*, *117*: 765.

Ransnäs, L.A., Jasper, J.R., Leiber, D., and Insel, P.A. (1992). *Biochem. J.*, *283*: 519.

Rodriguez, F.D., Bardaji, E., and Traynor, J.R. (1992). *J. Neurochem.*, *59*: 467.

Rojas, F.J., and Birnbaumer, L. (1985). *J. Biol. Chem.*, *260*: 7829.

Ruoslahti, E. (1988). *Annu. Rev. Biochem.*, *57*: 375.

Russell, J.M., and Brodwick, M.S. (1988). *J. Gen. Physiol.*, *91*: 495.

Sacaan, A., and Johnson, K.M. (1990). *Mol. Pharmacol.*, *38*: 705.

Scharff, O. (1981). *Cell Calcium*, 2: 1.

Serda, R.E., and Henzl, M.T. (1991). *J. Biol. Chem.*, *266*: 7291.

Shibley, D.R., and Creese, I. (1983). *J. Biol. Chem.*, *258*: 4957.

Shimizu, Y., and Mobley, J.L. (1993). *J. Immunol.*, *151*: 4106.

Shu, W.P., Wang, D., and Stracher, A. (1992). *Biochem. J.*, *283*: 567.

Sosa, A., Ordaz, H., Romero, I., and Celis, H. (1992). *Biochem. J.*, *283*: 561.

Strynadka, N.C.J., and James, M.N.G. (1989). *Annu. Rev. Biochem.*, *58*: 951.

Takeishi, M. (1990). *Annu. Rev. Biochem.*, *59*: 237.

Vicario, P.P., Saperstein, R., and Bennun, A. (1988). *Biosystems*, *22:* 55.

Villereal, M.L., and Palfrey, H.C. (1989). *Annu. Rev. Nutr.*, *9*: 347.

Vormann, J., and Günther, T. (1986). *Biol. Tr. Elem. Res.*, *9*: 37.

Wang, H., Gopalakrishnan, V., McNeill, J.R., Sulakhe, P.V., and Triggle, C.R. (1990). *Brit. J. Pharmacol.*, *100*: 5.

Wei, J.W., and Sulakhe, P.V. (1980). *Eur. J. Pharmacol.*, *62*: 382.

Wians, F.H., Krech, K.E., and Hauschka, P.V. (1983). *Magnesium*, 2: 83.

Williams, L.T., Mullikin, D., and Lefkowitz, R.J. (1978). *J. Biol. Chem.*, *253*: 2984.

Wittinghofer, A., and Pai, E.F. (1991). *TIBS*, *16*: 382.

Wolff, D.J., Poirier, P.G., Brostrom, C.O., and Brostrom, M.A. (1977). *J. Biol. Chem.*, *252*: 4108.

Yamamoto, Y., Chen, G., Miwa, K., and Suzuki, H. (1992). *J. Physiol.*, 450: 395.

Yamazaki, A., and Bitensky, M.W., Garcia-Sainz, J.A. (1987). *J. Biol. Chem.*, *262*: 9324.

Yoshikawa, K., Shinohara, Y., Terada, H., and Kato, S. (1987). *Biophys. Chem.*, *27*: 251.

Zimmerle, C.T., Patane, K., and Frieden, C. (1987). *Biochemistry - USA*, *26*: 6545.

NICKEL-PROTEIN INTERACTIONS

by

Robert L. Blakeley

Department of Biochemistry
University of Queensland
Brisbane, Queensland 4072
Australia

I. INTRODUCTION

A specific function for nickel has not yet been found in human beings or any other vertebrate animal, even though an induced deficiency of nickel is detrimental in chickens, cows, goats, miniature pigs, rats and sheep (Nielsen, 1984). Nickel-protein interactions that have been described thus far in vertebrate animals are concerned only with binding and/or transport of nickel(II) (Nieboer et al., 1988).

Nickel has well established roles as an essential active-site metal ion in a few enzymes in many organisms other than vertebrates. A biological function for nickel was first defined by the discovery of stoichiometric amounts of nickel at the active site of the enzyme urease from jack beans (Dixon et al., 1975a; 1975b). The early 1980s saw a modest explosion in nickel enzymology, with the discovery of nickel in several partially purified bacterial enzymes as reviewed in Sigel and Sigel (1988) and Lancaster (1988).

Nickel ions are very versatile, with a number of readily accessible oxidation states, biological ligands and geometries of coordination (Andrews et al., 1988a; 1988b; Cammack, 1992). There are as yet no crystal structures available for naturally occurring nickel enzymes, but a variety of other evidence provides insights into the nature of nickel-protein interactions.

II. HYDROGENASES

Hydrogenases are a group on bacterial enzymes which catalyze the redox interconversion of hydrogen molecule (H_2) and hydrogen ions:

$$H_2 \rightleftharpoons 2H^+ + 2e^-$$

Hydrogenases are used either to produce hydrogen or to utilize it, depending on the organism's metabolic requirements and hence its electron donor/acceptor (Cammack, 1992).

There are at least two major classes of hydrogenases, both of which contain iron-sulfur clusters (4Fe-4S) which serve as electron transfer agents. Members of one of these classes have no other attached cofactor (the [Fe]-hydrogenases), while members of the other additionally contain nickel (the [NiFe]-hydrogenases). A sub-set of the latter class are the [NiFeSe]-hydrogenases in which a selenocysteine residue replaces a particular cysteine residue. For the nickel-containing hydrogenases, the essential catalytic chemistry involves hydrogen (H_2) complexed to the nickel ion (Cammack, 1992; Przybyla et al., 1992).

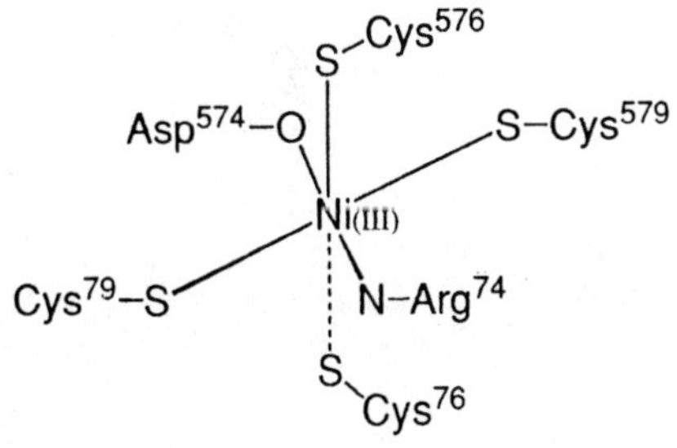

Figure 1.
Model of the active-site nickel environment in [NiFe]-hydrogenase showing proposed side chain ligands (adapted from Przybyla et al. (1992)). The numbering refers to residues in [NiFe]-hydrogenase I of *Escherichia coli*. In [NiFeSe]-hydrogenases, Cys-576 is replaced by selenocysteine. Redox alterations, and replacement of Cys-76 by hydrogen (H_2) take place during catalysis (Przybyla et al., 1992).

[NiFe]- and [NiFeSe]-hydrogenases often have at least two different types of subunits, and, wherever tested, hydrogenase activity resides on the large subunit (45-65 kg mol^{-1}) which contains roughly one nickel ion. On the basis of 15 gene sequences, the large subunit contains a potential nickel-binding motif built around the invariant signature sequences RXCXXC in the N-terminal region and DP<u>C</u>XXC in the C-terminal region. The underlined cysteine residue is selenocysteine in the [NiFeSe]-hydrogenases (Przybyla et al., 1992).

Site-directed mutagenesis experiments on [NiFe]-hydrogenase I of *Escherichia coli* led to a suggestion of six-coordinate nickel at the active site of the [NiFe]- and [NiFeSe]-hydrogenases as shown in Figure 1 (Przybyla et al., 1992). However, the observed diamagnetism of nickel(II) in the [NiFeSe]-hydrogenase in *Desulfovibrio baculatus* supports a five-coordinate structure (Wang et al., 1992).

The coordination of a deprotonated arginine side chain nitrogen to a metal ion in a protein (Fig. 1) will, if confirmed, be the first such example in metalloenzymology. Some properties of guanidine derivatives coordinated to cobalt(III) and platinum(II) have been discussed recently by Fairlie and Jackson (1990).

It has been suggested that the [NiFe]-hydrogenase from *Thiocapsa roseopersicina* may have a "novel Fe,S,Ni cluster" with linkage of the nickel ion to an iron-sulfur cluster via a bridging ligand (Maroney et al., 1991; Baidya et al., 1992).

Formation of active *Rhodospirillum rubrum* hydrogenase requires that Ni(II) be available in the medium during protein synthesis, which suggests that the nickel ion may be involved in the folding of the nascent polypeptide chain (Koch et al., 1992).

III. CARBON MONOXIDE DEHYDROGENASE

Carbon monoxide dehydrogenase catalyzes the interconversion of carbon monoxide and carbon dioxide in anaerobic bacteria (Ragsdale et al., 1990; Cammack, 1992):

$$CO + H_2O \rightleftharpoons CO_2 + 2H^+ + 2e^-$$

Carbon monoxide dehydrogenase from the purple photosynthetic bacterium, *Rhodospirillum rubrum*, is a membrane protein consisting of a single 61.8 kg mol^{-1} polypeptide which contains 1.4 ± 0.5 Ni, 0.8 ± 0.1 Zn, 9.5 Fe and 8.3 S atoms per protein monomer (Bonam and Ludden, 1987). This is the only nickel enzyme thus far in which the isolated nickel-free enzyme can be readily activated by addition of Ni^{2+} ion. The function of this carbon monoxide-inducible enzyme is the oxidation of CO to CO_2, and it interacts reversibly with a 22-kg mol^{-1} membrane-bound iron-sulfur protein which mediates electron transport to a membrane-bound hydrogenase (Ensign and Ludden, 1991).

In acetogenic bacteria (e.g. *Clostridium thermoaceticum*), carbon monoxide dehydrogenase acts as an 'acetyl Coenzyme A synthase' in the Harlan Wood pathway in

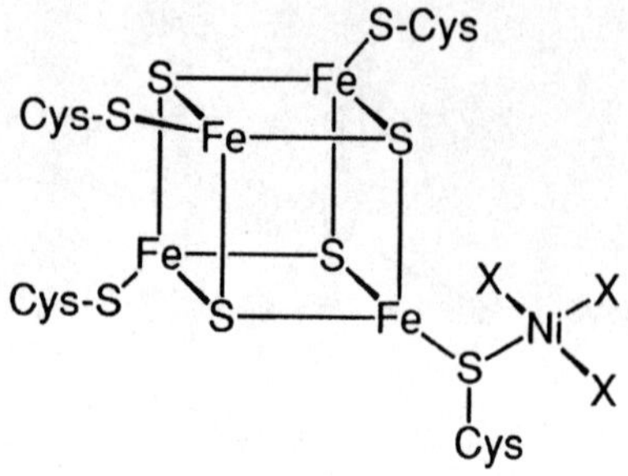

Figure 2.
Model of the active-site nickel environment in carbon monoxide dehydrogenase from *C. thermoaceticum* (Lindahl et al., 1990) and *R. rubrum* (Tan et al., 1992) based on EXAFS and other techniques (adapted from Cammack, 1992). The other ligands (X) of nickel may be S(Cys) or N/O from other side chains.

which CO_2 is reduced to CO which subsequently reacts with a corrinoid methyl donor and acetyl coenzyme A (Ragsdale et al., 1990; Cammack, 1992).

$$H_3C\text{-corrin} + CO + HS\text{-CoA} \rightleftharpoons H_3C\text{-C(O)-S-CoA} + \text{corrin}$$

Methanogens such as *Methanobacterium thermoautotrophicum* also synthesize acetyl coenzyme A via this reaction, while acetoclastic methanogens such as *Methanothrix soehngenii* utilize acetate and produce methane via a reversal of this reaction (Eggen et al., 1991).

Carbon monoxide dehydrogenase from *C. thermoaceticum* is an $(\alpha\beta)_3$ trimer (α, 71 kg mol^{-1}; β, 78 kg mol^{-1}) and contains 2 Ni, 1 Zn and several iron-sulfur clusters per α β pair. The binding sites for nickel and acetyl Coenzyme A may be on different subunits. The substrates CO and CO_2 bind directly to nickel ion (Diekert, 1988; Ragsdale et al., 1990; Cammack, 1992), and a minimum model for the nickel site is shown in Figure 2.

IV. METHYL COENZYME M REDUCTASE

This enzyme catalyzes the final step in methane production in the cell membrane of methanogenic bacteria:

$$H_3C\text{-S-}CH_2CH_2SO_3^- + \text{MRF-SH} \rightarrow CH_4 + \text{MRF-S-S-}CH_2CH_2SO_3^-$$

where $HS\text{-}CH_2CH_2SO_3^-$ is coenzyme M and MRF-SH is a recently characterized 'methyl reducing factor' which is unstable by virtue of a carboxylic phosphoric anhydride linkage (Sauer et al., 1990; Cammack, 1992).

Methyl coenzyme M reductase is an $\alpha_2\beta_2\gamma_2$ hexamer, and the subunit sizes (α, 61 kg mol^{-1}; β, 46 kg mol^{-1}; γ, 20 kg mol^{-1}) are known from the total gene sequences for five methanogenic bacteria (*Methanosarcina barkeri*, *Methanococcus vannielii*, *Methanobacterium thermoautotrophicum* (Marburg), *Methanococcus voltae* and *Methanothermus fervidus*). Associated with the hexamer (probably with each α subunit) are two tightly but apparently non-covalently bound molecules of the nickel(II)-containing 'coenzyme F_{430}' (Friedmann et al., 1990) (Fig. 3).

When coenzyme F_{430} is bound to the *resting* enzyme, the Ni(II) ion appears to be six-coordinate but without sulfur ligands (Shiemke and Scott, 1988; Shiemke et al., 1988). Albracht et al. (1988) proposed on the basis of extensive EPR experiments that the *active* enzyme involves Ni(I) in coenzyme F_{430} complexed by the thiolate anion of HS-HTP (a then known fragment of MRF-SH). Cammack (1992) extended their mechanism to suggest a subsequent methyl-nickel intermediate which accepts a proton to form methane and regenerate the Ni(I) species.

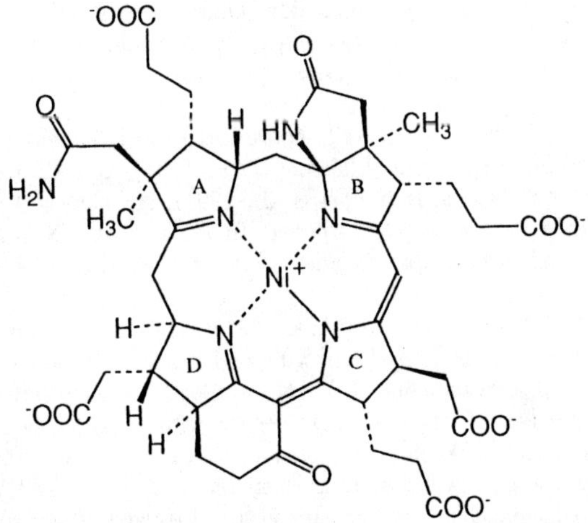

Figure 3.
The structure of coenzyme F_{430} (adapted from Färber et al., 1991). Configurations at ring D were recently revised on the basis of a re-investigation by NMR spectroscopy and X-ray crystallography (Färber et al., 1991).

V. UREASE

Urease catalyzes the hydrolysis of urea to form ammonium carbamate:

$$H_2N\text{-}C(O)\text{-}NH_2 + H_2O \rightarrow H_2N\text{-}COO^- + {}^+NH_4$$

Carbamate decomposes spontaneously to NH_3 and CO_2 and the latter hydrates to form bicarbonate as a final product (Andrews et al., 1984; Andrews et al., 1988b).

Urease is found in bacteria, mycoplasma, fungi, yeast, algae, higher plants, invertebrates and (arising from microorganisms or plants) in soil: no matter what the source, all ureases appear to have Ni(II) at the active site as judged by nutrition studies and/or inhibition by hydroxamic acids or phosphorodiamidates (Blakeley and Zerner, 1984; Andrews et al., 1988b). Many phosphorodiamidate esters ($(H_2N)_2P(O)OR$) and phosphoric triamides are powerful inhibitors of urease by virtue of being enzymatically degraded to form a phosphorodiamidate-nickel complex at the active site. This complex breaks down very slowly to regenerate active enzyme (Andrews et al., 1986).

The sequence of the 840 amino acid residues of jack bean urease (Takishima et al., 1988) has been confirmed by its cDNA sequence (Riddles et al., 1991), the only difference being position 242 for which the codon AAG corresponds to lysine in place of arginine. The native enzyme has six identical subunits, and it contains two nickel ions and one active site per 92-kg mol^{-1} subunit (Andrews et al., 1984; 1988b).

Many enzymatic properties of bacterial ureases match closely those of jack bean urease (Mobley and Hausinger, 1989; Todd and Hausinger, 1989). However, several bacterial ureases have three types of subunits (α, β, and γ), each of which is highly homologous to a segment of the 840-residue jack bean peptide: γ, 1 to ~102; β, ~130 to ~240; α, 271 to ~840 (Jones and Mobley, 1989; Mulrooney and Hausinger, 1990; Mörsdorf and Kaltwasser, 1990; Blanchard, 1990) In *Helicobacter pylori* urease, the γ and β peptides are fused (Clayton et al., 1990; Labigne et al., 1991). All ureases have the signature sequence M-V-C-H-H-L-[DN] (Barioch, 1992) where the underlined residue is the active-site cysteine (Cys-592 in jack bean urease).

The quaternary structures $(\gamma\beta\alpha)_2$ and $(\gamma_2\beta_2\alpha)_2$ have been reported for various bacterial ureases. The former is suggested by homology with jack bean urease. Where tested, these bacterial ureases contain two nickel(II) ions per α subunit.

There is only one invariant cysteine residue among the various sequenced ureases, and it corresponds to Cys-592 in the jack bean enzyme. This thiol is not coordinated to nickel ion, as first shown by the lack of charge-transfer transitions in the absorption spectrum of jack bean urease (Blakeley et al., 1983).

Jack bean urease contains octahedrally-coordinated nickel(II) with N and O ligands, on the basis of its absorption spectrum (Blakeley et al., 1983), EXAFS (Alagna et al., 1984) and magnetic circular dichroism (Finnigan et al., 1991). An increased number of histidine residues is accessable to diethyl pyrocarbonate in the apoenzyme from *K. aerogenes* as compared with the holoenzyme, consistently with possible ligation by histidine residues (Lee et al., 1990).

Dixon et al. (1980) originally postulated that the two nickel ions in each subunit of urease were not greater than 6 Å apart, allowing both to take part in the hydrolysis of urea. When the inhibitor β-mercaptoethanol binds to nickel, strong ferromagnetic coupling develops between the two nickel ions, consistently with a bridge being formed between them (Finnigan et al., 1991). Weak antiferromagnetic coupling between the nickel ions in β-mercaptoethanol-free jack bean enzyme has been reported (Clark and Wilcox, 1989), but was not found in an independent study of jack bean urease and *K. aerogenes* urease (Day et al., 1993). An earlier EXAFS study (Alagna, 1984) did not detect adjacent nickel ions, which suggests that the two nickel ions are separated by at least 4 Å in the free enzyme (Cammack, 1992). Preliminary crystallographic studies with both jack bean and *K. aerogenes* urease are very promising (Jabri et al., 1992).

VI. NICKEL STORAGE AND TRANSPORT PROTEINS

The serum concentration of nickel in human beings is comparable to that of the ultratrace elements chromium, manganese, cobalt and molybdenum (0.1-0.5 μg/l) and much less than that of the trace elements iron, copper and zinc (1000 μg/l) (Andrews et al., 1988a). Much of the circulating nickel(II) in human beings is bound to the N-terminal residues of serum albumin in a four- or five-coordinate complex (Fig. 4).

Nickel(II) has been substituted for the normal metal ion(s) in lectins, azurin, metallothionein, haemoproteins, and conalbumin (Andrews et al., 1988a). In addition, a number of mammalian nickel-binding proteins has been reported, but little is yet known of the binding sites (Drake, 1988; Andrews et al., 1988a).

Three types of putative nickel-binding proteins have been recently described in bacteria. The gene for each one of them is associated with that of a nickelloenzyme, and heterologous expression should allow the nickel-protein interactions to be defined.

A. Genes for bacterial ureases occur in a cluster, and expression of several nearby genes is necessary for the functional incorporation of nickel(II) into urease in *Klebsiella aerogenes* (Lee et al., 1992) and *Helicobacter pylori* (Cussac et al., 1992).

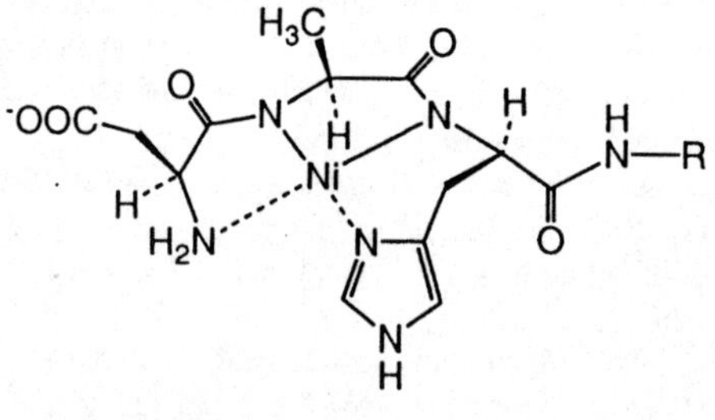

Figure 4.
Nickel(II) coordination in its complex with human serum albumin at pH 7 (adapted from Glennon and Sarkar, 1982). The structure was deduced by comparison of Ni(II)-albumin absorption spectra with those of the Ni(II) complex of the tripeptide N-methylamide L-Asp-L-Ala-L-His-NHMe which mimics the N-terminal sequence of the protein, The side chain of the N-terminal aspartic acid residue may be a fifth ligand (Glennon and Sarkar, 1982).

B. Nickel uptake by *Alcaligenes eutrophus* is mediated by a high-affinity nickel transport protein. On the basis of the cloned gene, the 33-kg mol^{-1} protein has five hydrophobic transmembrane segments (Eitinger and Friedrich, 1991).

C. A 0.6 kbase nickel metabolism locus has been identified in the genome of *Bradyrhizobium japonicum*. The gene product appears to be a protein involved in incorporation of Ni(II) into hydrogenase (Fu and Maier, 1991; 1992).

VII. MODEL NICKEL COMPLEXES AND "UN-NATURAL" NICKELLOENZYMES

Many insights into nickel-protein interactions have been derived from studies with model nickel complexes, and some of these have been summarized in books and reviews, e.g. Sigel and Sigel (1988), Lancaster (1988), Cammack (1992).

"Un-natural" nickelloenzymes have been prepared from metalloenzymes from various sources. Nickel(II) gives essentially normal enzymatic activity when substituted for the native active-site metal ion in at least 12 metalloenzymes (Andrews et al., 1988a). X-ray diffraction data for Ni(II)-carboxypeptidase A reveals that it has the same ligands as the native Zn(II) ion (side-chain nitrogens of His-69 and His-196, both side chain oxygens of Glu-72 and a water molecule) (Hardman and Lipscomb, 1984).

For most of the active "unnatural" nickelloenzymes, the replaceable divalent active-site metal ion has a catalytic role as a Lewis acid. Nickel(II) evidently binds in much the same way and serves the same function as the normal metal ions (Andrews et al., 1988a). Despite the general availability of the nickel ion, it appears to be utilized by enzymes only where a more readily available metal ion is chemically inadequate. Nickel is a versatile element, and it seems likely that many new nickel-protein interactions will be characterized in coming years.

REFERENCES

Alagna, L., Hasnain, S.S., Piggott, B., and Williams, D.J. (1984). *Biochem. J.*, *220*: 591.

Albracht, S.P.J., Ankel-Fuchs, D., Böcher, R., Ellermann, J., Moll, J., van der Zwaan, J.W., and Thauer, R.K. (1988). *Biochim. Biophys. Acta*, *955*: 86.

Andrews, R.K., Blakeley, R.L., and Zerner, B. (1984). *Advances in Inorganic Biochemistry*, Vol. 6 (G.L. Eichhorn and L.G. Marzilli, eds.), Elsevier, New York, p. 225.

Andrews, R.K., Dexter, A., Blakeley, R.L., and Zerner, B. (1986). *J. Amer. Chem. Soc.*, *108*: 7124.

Andrews, R.K., Blakeley, R.L., and Zerner, B. (1988a). *Metal Ions in Biological Systems*, Vol. 23 (H. Sigel and A. Sigel, eds.), Marcel Dekker, New York, p. 165.

Andrews, R.K., Blakeley, R.L., and Zerner, B. (1988b) *Bioinorganic Chemistry of Nickel* (J.R. Lancaster, Jr., ed.), VCH Publishers, Inc., Florida, p. 141.

Baidya, N., Olmstead, M.M., Whitehead, J.P., Bagyinka, C., Maroney, M.J., and Mascharak, P.K. (1992). *Inorg. Chem.*, *31*: 3612.

Barioch, A. (1992). PROSITE: a Dictionary of Protein Sites and Patterns, University of Geneva, Ninth release.

Blakeley, R.L., and Zerner, B. (1984). *J. Mol. Catal.*, *23*: 263.

Blakeley, R.L., Dixon, N.E., and Zerner, B. (1983). *Biochim. Biophys. Acta*, *744*: 219.

Blanchard, A. (1990). *Mol. Microbiol.*, *4*: 669.

Bonam, D., and Ludden, P.W. (1987). *J. Biol. Chem.*, *262*: 2980.

Cammack, R. (1992). *Bioinorganic Catalysis* (J. Reedijk, ed.), Marcel Dekker, New York, p. 189.

Clark, P.A., and Wilcox, D.E. (1989). *Inorg. Chem.*, *28*: 1326.

Clayton, C.L., Pallen, M.J., Kleanthous, H., Wren, B.W., and Tabaqchali, S. (1990). *Nucl. Acid Res.*, *18*: 362.

Cussac, V., Ferrero, R.L., and Labigne, A. (1992). *J. Bacteriol.*, *174*: 2466.

Day, E.P., Peterśon, J., Sendova, M., Todd, M.J., and Hausinger, R.P. (1993). *Inorg. Chem.*, *32*: 634.

Diekert, G. (1988). *Bioinorganic Chemistry of Nickel* (J.R. Lancaster, Jr., ed.), VCH Publishers, Inc., Florida, p. 299.

Dixon, N.E., Gazzola, C., Watters, J.J., Blakeley, R.L., and Zerner, B. (1975a). *J. Amer. Chem. Soc.*, *97*: 4130.

Dixon, N.E., Gazzola, C., Blakeley, R.L., and Zerner, B. (1975b). *J. Amer. Chem. Soc.*, *97*: 4131.

Dixon, N.E., Riddles, P.W., Gazzola, C., Blakeley, R.L., and Zerner, B. (1980). *Can. J. Biochem.*, *58*: 1335; erratum *59*: 564.

Drake, H.L. (1988). *Bioinorganic Chemistry of Nickel* (J.R. Lancaster, Jr., ed.),VCH Publishers, Inc., Florida, p. 111.

Eggen, R.I.L., Geerling, A.C.M., Jetten, M.S.M., and de Vos, W.M. (1991). *J. Biol. Chem.*, *266*: 6883.

Eitinger, T., and Friedrich, B. (1991). *J. Biol. Chem.*, *266*: 3222.

Ensign, S.A., and Ludden, P.W. (1991). *J. Biol. Chem.*, *266*: 18395.

Fairlie, D.P., and Jackson, W.G. (1990). *Inorg. Chem.*, *29*: 140.

Färber, G., Keller, W., Kratky, C., Jaun, B., Pfaltz, A., Spinner, C., Kobelt, A., and Eschenmoser, A. (1991). *Helv. Chim. Acta*, *74*: 697.

Finnegan, M.G., Kowal, A.T., Werth, M.T., Clark, P.A., Wilcox, D.E., and Johnson, M.K. (1991). *J. Amer. Chem. Soc.*, *113*: 4030.

Friedmann, H.C., Klein, A., and Thauer, R.K. (1990). *FEMS Microbiol. Rev.*, *87*: 339.

Fu, C., and Maier, R.J. (1991). *Appl. Environ. Microbiol.*, *57*: 3502.

Fu, C., and Maier, R.J. (1992). *Arch. Microbiol.*, *157*: 493.

Glennon, J.D., and Sarker, B. (1982). *Biochem. J.*, *203*: 15.

Hardman, K.D., and Lipscomb, W.N. (1984). *J. Amer. Chem. Soc.*, *106*: 463.

Jabri, E., Mann, H.L., Hausinger, R.P., and Karplus, P.A. (1992). *J. Mol. Biol.*, *227*: 934.

Jones, B.D., and Mobley, H.L.T. (1989). *J. Bacteriol.*, *171*: 6414.

Koch, H.-G., Kern, M., and Klemme, J.-H. (1992). *FEMS Microbiol. Lett.*, *91*: 193.

Labigne, A., Cussac, V., and Courcoux, P. (1991). *J. Bacteriol.*, *173*: 1920.

Lancaster, J.R., Jr., ed. (1988). Bioinorganic Chemistry of Nickel, VCH Publishers, Inc., Florida.

Lee, M.H., Mulrooney, S.B., and Hausinger, R.P. (1990). *J. Bacteriol.*, *172*: 4427.

Lee, M.H., Mulrooney, S.B., Renner, M.J., Marcowicz, Y., and Hausinger, R.P. (1992). *J. Bacteriol.*, *174*: 4324.

Lindahl, P.A., Ragsdale, S.W., and Münck, E. (1990). *J. Biol. Chem.*, 265: 3880.

Mobley, H.L.T., and Hausinger, R.P. (1989). *Microbiol. Rev.*, *53*: 85.

Mörsdorf, G., and Kaltwasser, G. (1990). *FEMS Microbiol. Lett.*, *66*: 67.

Maroney, M.J., Colpas, G.J., Bagyinka, C., Baidya, N., and Mascharak, P.K. (1991). *J. Amer. Chem. Soc.*, *113*: 3962.

Mulrooney, S.B., and Hausinger, R.P. (1990). *J. Bacteriol.*, *172*: 5837.

Nieboer, E., Rickey, T.T., and Sanford, W.E. (1988). *Metal Ions in Biological Systems*, Vol. 23 (H. Sigel and A. Sigel, eds.), Marcel Dekker, New York, p. 91.

Nielsen, F.H. (1984). *Annu. Rev. Nutr.*, *4*: 21.

Przybyla, A.E., Robbins, J., Menon, N., and Peck, H.D., Jr. (1992). *FEMS Microbiol. Rev.*, *88*: 109.

Ragsdale, S.W., Baur, J.R., Gorst, C.M., Harder, S.R., Lu, W.-P., Roberts, D.L., Runquist, J.A., and Schiau, I. (1990). *FEMS Microbiol. Rev.*, *87*: 397.

Riddles, P.W., Whan, V.A., Blakeley, R.L., and Zerner, B. (1991). *Gene*, *108*: 265 plus corrigendum.

Sauer, F.D., Blackwell, B.A., Kramer, J.K.G., and Marsden, B.J. (1990). *Biochemistry*, *29*: 7593.

Shiemke, A.K., and Scott, R.A. (1988). *J. Amer. Chem. Soc.*, *110*: 1645.

Shiemke, A.K., Kaplan, W.A., Hamilton, C.L., Shelnutt, J.A., and Scott, R.A. (1988). *J. Biol. Chem.*, *264*: 7276.

Sigel, H., and Sigel, A., eds. (1988). *Metal Ions in Biological Systems*, Vol. 23, Marcel Dekker, New York.

Takishima, K., Suga, T., and Mamiya, G. (1988). *Eur. J. Biochem.*, *175*: 151.

Tan, G.O., Ensign, S.A., Ciurli, S., Scott, M.J., Hedman, B., Holm, R.H., Ludden, P.W., Korszun, Z.R., Stephens, P.J., and Hodgson, K.O. (1992). *Proc. Nat. Acad. Sci. USA*, *89*: 4427.

Todd, M.J., and Hausinger, R.P. (1989). *J. Biol. Chem.*, *264*: 15835.

Wang, C.-P., Franco, R., Moura, J.J.G., Moura, I., and Day, E.P. (1992). *J. Biol. Chem.*, *267*: 7378.

CHROMIUM-PROTEIN INTERACTIONS

by

Akiko Yamamoto Ph. D.

Department of Hygiene,
Gunma University School of Medicine,
3-39-22, Showa-machi, Maebashi
Gunma
Japan 371

and

Osamu Wada M. D.

Department of Hygiene and Preventive Medicine,
Faculty of Medicine, University of Tokyo
7-3-1 Hongo, Bunkyo-ku
Tokyo
Japan 113

I. INTRODUCTION

Chromium has been recognized to be a beneficial element in mammals since Schwartz and Mertz (1959) first elucidated the significance of this metal in normal glucose utilization, although the protein(s) which participates in this process has yet to be established. Chromium, like other micronutrients, is toxic when given in excessive amounts.

The interactions of chromium with proteins are highly dependent upon its valence state. The chromate form, Cr(VI), is one of the best documented human and animal carcinogens (*Medicine*, Vol. 2, Part Four, Ch. 7-E). In order to avoid the mutagenic effect of Cr(VI), it must be reduced to physiologically stable form, Cr(III), before it reaches the target cells as Cr(III), unlike Cr(VI), cannot enter the nucleus. There are many low- and high-molar-mass Cr-reducing substances in the body. When the reducing capacity of the extracellular fluid is saturated, Cr(VI) compounds enter cells and react with active enzymes and other substances to produce Cr(III). During this reduction process, Cr passes through the penta (V) and tetra (IV) forms and a great deal of energy is released (Hartford, 1986). These compounds are capable of damaging cellular macromolecules (the relevant proteins are cited in section III).

Chromium-induced DNA lesions, including the formation of Cr-DNA and DNA-protein or inter- and intrastrand cross-links and the DNA strand breaks, disrupt the normal

Table 1

Enzymes Interacting with Chromium

Enzymes	E C	Cofactor	Mode of Cr action	References
Activated by Cr				
Trypsin	3.4.4.4.	Cr		Mertz (1969)
Chymosin (Rennin)	3.4.23.4.	Mn	Replaces Mn	Maze and Maze (1941)
Phosphoglucomutase	5.4.2.2.	Mg	Requirement for a second metal	Stockland (1949)
Phosphoenolpyruvate carboxykinase (GTP)	4.1.1.32.	Mn	Substrate analog (Cr-GTP)	Kramer and Nowak (1988)
Pyruvate kinase (enorization of pyruvate)	2.7.1.40.	Mg	Substrate analog (Cr-ATP)	Dunaway-Mariano et al. (1979)
Inhibited by Cr				
ATPase	3.6.1.3.	Mg or Ca	Exchange for Mg	
Ca-ATPase			or Ca and formation	Vilsen and Andersen (1992)
Na, K-ATPase			of Cr- bridges	Vilsen et al. (1987)
Pyruvate carboxylase	6.4.1.1.	Mg or Mn	(βP, γP of Cr-ATP	Armbruster and Rudolph (1976)
Kinases	2.7. group	Mg	with enzyme)	Cleland and Mildvan (1979)
Glutathion reductase	1.6.4.2.		Oxidation of	Nieboer and Jusys (1988)
Aminolevulinate dehydratase	4.2.1.24.		active site	Nieboer and Jusys (1988)
Urease	3.5.1.5.		(sulfhydryl group)	Henry and Smith (1946)

function of DNA as a template for transcription and replication (Wetterhahn et al., 1989) with mutagenic and carcinogenic consequences (Nieboer and Shaw, 1988) (cross-linking proteins are cited in section V). The mechanism through which DNA-Cr(III) and protein cross-links are formed within cells may involve a sequence of events, e.g. the initial capture of Cr(III) by an acidic protein and a subsequent slower reaction with DNA (Kortenkamp et al., 1992).

Chromium(III) has affinity for histidyl, methionyl and cysteinyl groups of proteins, nitrogens of DNA bases and oxygen of phosphoryl and carboxylic groups (Miller and Costa, 1990). Thus, chromium binds to many types of proteins specifically or non-specifically and forms stable inert coordination complexes, some of which damage the structure and function of normal proteins and DNA. Some (listed in Table 1) are used as inhibitors of various enzymes for which Mg-nucleotide complexes are substrates in kinetic studies (Van Wart, 1988) and others have been used in studies to determine the metabolic fate of proteins such as hemoglobin.

The following is a catalog of chromium-protein interactions (some of which is summarized in Table 1) with some explanatory notes.

II. WORKING PROTEINS

A. Low-molar-mass chromium-binding substance (LMCr)

LMCr is an anionic Cr-containing peptide with a molar mass of 1.5 kg mol^{-1}. It has been found in mammalian liver, cow's milk and human urine. Its stimulation of glucose uptake by the rat epididymal fat pad in vitro is comparable to that of glucose tolerance factor (GTF) with respect to insulin action. It has been suggested that LMCr, which is involved in the detoxification of chromium, may also play an essential nutritional role (Yamamoto et al., 1989; 1988).

B. Cr-P70

The 70 kg mol^{-1} Cr-bound protein Cr-P70 was found to be a nuclear protein that was induced by Cr(III) in partially hepatectomized rats. It bound to the activated nucleolar chromatin and loosened its higher-order structure, resulting in an increase of the b-form fraction of chromatin DNA and an enhancement of hepatic nuclear RNA synthesis during the early stage of liver regeneration. Therefore, Cr has been suggested to play a role in the cell proliferation process (Okada et al., 1989).

III. CHROMIUM- REDUCING PROTEINS

The following are some of the best-characterized classes of electron-transfer proteins involved in Cr(VI) reduction.

A. NADPH-cytochrome P-450 reductase and cytochrome P-450

NADPH-cytochrome P-450 reductases possess Cr(VI)-reducing activity and cytochrome P-450 exhibits catalytic activity in this system. Mikalson et al. (1991) demonstrated that the reactive species of Cr(V) and Cr(IV) formed during enzymatic conversion of chromate (VI) interact with and bind to both cytochrome P-450 and its reductase as does Cr(III).

B. Diaphorase

Diaphorase is a NADPH-dependent enzyme that catalyzes the reduction of Cr(VI) to Cr(III). This reducing reaction has been interpreted as a detoxifying mechanism of carcinogenic Cr(VI), especially in the respiratory tracts (De Flora et al., 1989).

C. Hemoglobin

The reduction of chromate by hemoglobin results in the formation of Cr(III)-hemoglobin complex (Kitagawa et al., 1982).

IV. CHROMIUM-OXIDIZING PROTEINS

A. Plastocyanines

Plastocyanines are blue copper-containing proteins that function as electron carriers in the chloroplasts found in French beans, *Phaseolus vulgaris*, (Eley and Moore, 1983) and cucumber, *Cucumis sativus* (Farver and Pecht, 1988). These proteins are reduced by Cr(II) to produce a substitution-inert Cr(III)-containing complex.

V. PROTEINS THAT CROSS-LINK TO DNA

A. P 97

P 97 is an acidic 97 kg mol^{-1} nuclear protein that is cross-linked to DNA by Cr(III) in Chinese hamster ovary cells (Miller and Costa, 1990). Exposure to Cr(III) induces P97-DNA cross-links only in isolated nuclei, whereas chromate (VI) treatment results in significant cross-link formation only in intact cells. These observations provide evidence that Cr(III) is responsible for some of the DNA-protein cross-links that are formed when chromates are taken up and reduced within the cell.

B. Actin

Actin was found to be the major protein cross-linked to the DNA of intact Chinese hamster ovary cells treated with potassium chromate (VI). In addition to actin, several other major proteins (e.g. 53 kg mol^{-1}, and 50 kg mol^{-1}) are cross-linked to DNA following exposure to chromate. Incubation of Cr(III), but not Cr(VI), with purified actin and DNA results in DNA-actin cross-links (Miller et al., 1991).

C. Albumin

The formation of DNA-bovine serum albumin (BSA) cross-links has been detected by treating DNA with Cr(III) in the presence of BSA. This cross-linking formation was only detected at acidic pH, but not at neutral pH. Physiological meanings of such a lesion are still unclear (Kortenkamp et al., 1992).

D. Ferric-uptake-regulation protein (Fur)

The Fur protein is a negative regulator of the expression of the aerobactin operon and of several other siderophore-mediated, high-affinity iron transport systems in *E. coli*. This protein requires Fe(II) or one of certain other divalent metals as a cofactor for maximum activity. Chromium(III) can replace Fe(II) and has been used in studies to map the anion-binding regions that may be involved in DNA binding and to determine possible modes of action through which Fur controls expression (Saito and Williams, 1991).

VI. ENZYMES

To our knowledge, there are, to date, no reports about Cr-metalloenzymes. Enzymes which are known to interact with chromium are listed in the table.

VII. TRANSPORT AND STORAGE PROTEINS

Transferrin and albumin affect chromium absorption. Transferrin binds newly absorbed chromium, whereas albumin substitutes for transferrin when it is saturated with either chromium or iron (Yamamoto et al., 1984). At very high concentrations, chromium can bind non-specifically to other plasma proteins, such as α-, β-, and γ-globulins and lipoproteins (Hopkins and Schwartz, 1964). Urinary excretion of chromium is shown to be conducted mainly by LMCrs (see section C).

A. Transferrin

Transferrin is responsible for the transportation of iron from its absorption site to the utilization site. This protein possesses two binding sites, A and B, with different pH-dependent affinities for iron. Trivalent chromium has a high binding affinity for plasma transferrin (Saito et al.,1980; Yamamoto et al., 1984), and binds exclusively to site B (Ducros, 1992). Tyrosine appears to be the most likely ligand for the binding of Cr(III) to transferrin, as is the case for iron (Bacley and Evan, 1988). Thus, there is antagonism between Cr(III) and Fe(III) under conditions of high iron concentrations, such as in hemochromatosis (Ani and Moshtaghie, 1992) which could possibly result in Cr deficiency.

B. Albumin

Chromium(III) was detected in the albumin region of the gels during electrophoresis and gel filtration. This indicates that it binds strongly to albumin (Merritt et al., 1984; Yamamoto et al., 1984).

C. Low-molar-mass chromium-binding substance (LMCr)

A LMCr (see Section II.A) with a high affinity for Cr(III) is responsible for a reversible and significant transfer of chromium to transferrin, but not to albumin (Yamamoto et al, 1984). An Apo-LMCr has been found in human, mouse and dog serum and organs, and binds invading Cr(VI) into a stable form of Cr(III)-LMCr. Because of the high urinary excretion and renal clearance rates of Cr(III)-LMCr, it has been suggested that LMCr may play a role in the detoxification of chromium (Wada et al., 1983).

VIII. OTHER PROTEINS

A. Collagen

Skin collagen is tanned by Cr(III), but not Cr(VI), into leather by cross-linking, which involves predominantly the carboxyl groups of the collagen polypeptide chain, although the polymerization of hydrated Cr(III) by hydroxide ions is also required (Mertz, 1969). It should be borne in mind that tanning involves total saturation of protein with Cr and that such a situation does not normally occur in the living organism.

B. Actin

Under physiological conditions, ATP-actin contains Mg(II) as a tightly bound divalent metal ion. Competition between Ca(II) and Mg(II) for binding to a single high-affinity site on actin has been demonstrated. G-actin binds Cr-ATP, a substituted inert analog of metal-ATP complexes, in competition with ATP. Thus, Cr-ATP-actin is polymerized efficiently into filaments with concomitant hydrolysis of Cr-ATP into Cr-ADP-Pi that remain tightly bound to F-actin (Valentin-Ranc and Carlie, 1989).

The Cr-ATP complex has been found to be a useful tool to study the conformation of F-actin frozen in the intermediate state (F-actin-ADP-Pi) of the hydrolysis reaction.

C. cytochrome *b5* and cytochrome c

Results of NMR spectrum analysis have shown that propionate-bearing edge of the haem moiety of microsomal cytochrome *b5* in its interprotein complex with horse cytochrome *c* is accessible to Cr(III). The binding site differs from the protein-protein binding region (Hartshorn et al., 1987) (useful as a probe).

D. Conarachin II

A groundnut protein, conarachin II is dimerized slowly at pH 5.3 with Cr(III) and the product is stable when dialyzed against phosphate buffer (Naismith, 1958).

E. Egg albumin

At physiological pH, egg albumin was found to bind Cr(III) strongly, but not Cr(VI) (Mertz, 1969).

F. Myoglobin

Chromium-myoglobin was synthesized as a Cr-substituted myoglobin and has been used in studies of the stability of this protein (McLendon and Murphy, 1980).

G. Inorganic pyrophosphatase (3.6.1.1.)

This enzyme strongly binds Cr(III). Cr-PPi cannot be a substrate for the yeast enzyme and does not inhibit the Mg-dependent activity of this enzyme (Sperow and Butler, 1976).

REFERENCES

Ani, M., and Moshtaghie, A.A. (1992). *Biol. Tr. Elem. Res.*, *32*: 57.

Armbruster, D.A., and Rudolph, F.B. (1976). *J. Biol. Chem.*, *251*: 320.

Bacley, S., and Evan, R.W. (1988). *Garrat Biochem.*, *27*: 5804.

Bresler, S.E., and Rozentsveig, N.A. (1961). *Biokhimiya*, *16*: 84.

Broughall, J.M., and Reid, R.A. (1974). *Biochem. Soc. Trans.*, *2*: 498.

Cleland, W.W., and Mildvan, A.S. (1979). *Advances in Inorganic Biochemistry* (G.L. Eichhorn and L.G. Marzilli, eds.), Elsevier North Holland Inc., New York, p. 163.

De Flora, S., Morelli, A., Basso, C., Sera, D., Camoirano, A., and Zanacchi, P. (1989). *Biol. Tr. Elem. Res.*, *21*: 179.

Ducros, V. (1992). *Biol. Tr. Elem. Res.*, *32*: 65.

Dunaway-Mariano, D., Benovic, J.L., Cleland, W.W., Gupta, R.K., and Mildvan, A.S. (1979). *Biochemistry*, *18*: 4347.

Eley, C.G.S., and Moore, G.R. (1983). *Biochem. J.*, *215*: 11.

Farver, O., and Pecht, I. (1988). *Prog. Biol. Res.*, *274*: 269.

Hartford, W.H. (1986). An Update 1986, "Proceedings, Chromium Symp. 1986", Industrial Health Foundation, Pittsburg, pp. 9-42.

Hartshorn, R.T., Mauk, A.G., Mauk, M.R., and Moore, G.R. (1987). *FEBS Lett.*, *213*: 391.

Henry, R.J., and Smith, E.C. (1946). *Science*, *104*: 426.

Hopkins, L.L., and Schwarz, K. (1964). *Biochim. Biophys. Acta*, *90*: 484.

Kitagawa, S., Seki, H., Kametani, F., and Sakurai, H. (1982). *Chem.-Biol. Interact.*, *40*: 265.

Kortenkamp, A., Curran, B., and O'Brien, P. (1992). *Carcinogenesis*, *13*: 307.

Kramer, P., and Nowak, T. (1988). *J. Inorg. Biochem.*, *32*: 135.

Langenbeck, W., Augstin, M., and Schafer, C. (1961). *Hoppe-Seyler Z. Physiol. Chem.*, *324*: 54.

Maze, P., and Maze, P.J., Jr. (1941). *C. R. Soc. Biol.*, *135*: 808.

McLendon, G., and Murphy, P. (1980). *J. Biol. Chem.*, *255*: 4035.

Merritt, K., Brown, S.A., and Sharkey, N.A. (1984). *J. Biomed. Mater. Res.*, *18*: 1005.

Mertz, W. (1969). *Physiol. Rev.*, *49*: 163.

Mikalson, A., Alexander, J., Wallin, H., Ingelman-Sundberg, M., and Anderson, R.A. (1991). *Carcinogenesis*, *12*: 825.

Miller, C.A., Cohen, M.D., and Costa, M. (1991). *Carcinogenesis*, *12*: 269.

Miller, C.A., and Costa, M. (1990). *Mutat. Res.*, *234*: 97.

Naismith, W.E.F. (1958). *Arch. Biochem. Biophys.*, *73*: 255.

Nieboer, E., and Jusys, A. (1988). *Advances in Science and Technology*, Vol. 20 (E. Nieboer and J.O. Nriagu, eds.), Wiley, New York, p. 21.

Nieboer, E., and Shaw, S.L. (1988). *Advances in Science and Technology*, Vol. 20 (E. Nieboer and J.O. Nriagu, eds.), Wiley, New York, p. 399.

Okada, S., Tsukada, H., and Tezuka, M. (1989). *Biol. Tr. Elem. Res.*, *21*: 35.

Saito, T., and Williams, R.J.P. (1991). *Eur. J. Biochem.*, *197*: 43.

Saito, T., Levis, A.G., and Paulon, C. (1980). *Caryologia*, *33*: 239.

Schwartz, K., and Mertz, W. (1959). *Arch. Biochem. Biophys.*, *85*: 292.

Sperow, J.W., and Butler, L.G. (1976). *J. Biol. Chem.* *251*: 2611.

Stickland, L.H. (1949). *Biochem. J.*, *44*: 190.

Valentin-Ranc, C., and Carlier, M.-F. (1989). *J. Biol. Chem.*, *264*: 20871.

Van Wart, H.E. (1988). *Meth. Enzymol.*, *158*: 95.

Vilsen, B., and Andersen, J.P. (1992). *J. Biol. Chem.*, *267*: 3539.

Vilsen, B., Petessen, J., and Jorgensen, P.L. (1987). *J. Biol. Chem.*, *262*: 1051.

Wada, O., Wu, G.Y., Yamamoto, A., Manabe, S., and Ono, T. (1983). *Environ. Res.*, *32*: 228.

Watterhahn, K.E., Hamilton, J.W. Aiyer, J., Borges, K.M., and Floyd, R. (1989). *Biol. Tr. Elem. Res.*, *21*: 405.

Yamamoto, A., Wada, O., and Ono, T. (1984). *J. Inorg. Biochem.*, *22*: 91.

Yamamoto, A., Wada, O., and Manabe, S. (1989). *Biochem. Biophys. Res. Commun.*, *163*: 189.

Yamamoto, A., Wada, O., and Suzuki, H. (1988). *J. Nutr.*, *118* : 39.

METAL-PROTEIN INTERACTIONS IN ENZYMES CONTAINING MOLYBDENUM AND TUNGSTEN

by

K.V. Rajagopalan

Department of Biochemistry,
Duke University Medical Center,
Durham, North Carolina 27710
U.S.A.

I. INTRODUCTION

While the presence of molybdenum in several enzymes has long been known, the nature of the forces involved in the tight association of the metal with host proteins has only lately been clarified. Two different Mo-containing, dissociable prosthetic groups each with its own unique features, the Fe-Mo cofactor of nitrogenase and the Mo-pterin cofactor of all the other molybdoenzymes, have been discovered. Other than the tetrapyrrole-based porphyrins, corrins and chlorins, the two Mo-containing structures are the only dissociable metallophores specifically designed for protein-metal associations. Of great interest also is the recent discovery of several tungsten-containing enzymes, mechanistically related to the pterin-containing molybdoenzymes, in hyperthermophilic archaea. This chapter summarizes the state of knowledge on the bonding chemistry of molybdenum and tungsten in biological systems.

II. MOLYBDENUM ENZYMES

In all molybdoenzymes, the metal functions as a component of a larger organometallic prosthetic group. Of the known molybdoproteins, the enzyme system known as nitrogenase contains a unique mixed-metal complex, known as Fe-Mo cofactor. All of the other known enzymes contain one of a family of molybdenum pterin cofactors. A representative list of molybdenum-containing enzymes is shown in Table 1. Most of these enzymes are complex proteins containing two or more tightly-bound prosthetic groups. While a great deal is known about the physical and chemical properties of many molybdoenzymes, no crystal structure has yet been published other than for nitrogenase. All of the information on the ligand fields of Mo in diverse enzymes has been derived through the use of several types of spectroscopic techniques. A vast literature exists on the application of electron paramagnetic resonance (EPR) to the paramagnetic Mo(V) species of molybdoenzymes (Bray, 1985; 1988). More recently the technique of X-ray absorbance spectroscopy (XAS) has yielded crucial insight into the ligand field of the metal (Garner and Bristow, 1985). For Mo complexes, the absorption edge provides definitive information of the presence or absence of terminal oxo groups, i.e. Mo=O bonds, whereas analysis of the extended X-ray absorbance fine structure (EXAFS) region provides spherically-averaged information on the

Table 1

Cofactor Compositions of Molybdoenzymes[1]

Enzyme	Source	M_r	Subunit composition	Cofactors per molecule	Pterin
Nitrogenase	*Azotobacter vinelandii*				
Fe-protein		60	A_2	4Fe:4S	None
Mo-Fe-protein		230	A_2B_2	2Mo:32Fe:30S	None
Xanthine oxidase	Bovine milk	300	A_2	2Mo:2FAD:8Fe:8S	MPT
Xanthine dehydrogenase	Chicken liver	300	A_2	2Mo:2FAD:8Fe:8S	MPT
Xanthine dehydrogenase[a]	*Pseudomonas aeruginosa*	ND	ND	(Mo:FAD:Fe:S)	MPT
Xanthine dehydrogenase[b]	Plants	ND	ND	(Mo)	ND
Purine hydroxylase II	*Aspergillus nidulans*	300	A_2	2Mo:2FAD:8Fe:8S	ND
Aldehyde oxidase	Rabbit liver	340	A_2	2Mo:2FAD:8Fe:8S	MPT
Aldehyde dehydrogenase	*Pseudomonas testosteroni*	186	A_2	1Mo;1FAD:4Fe:4S	ND
Pyridoxal oxidase	*Drosophil melanogaster*	ND	ND	ND	ND
Nicotinate hydroxylase	*Clostridium barker*	ND	ND	(Mo,FAD,Fe,Se)	ND
Carbon monoxide oxidase	*Pseudomonas carboxydoflava*	300	$A_2B_2C_2$	2Mo:2FAD:8Fe:8S	MCD
Sulfite oxidase	Rat liver	120	A_2	2Mo:2HEME	MPT
Sulfite oxidase	*Thiobacillus novellus*	40	A	1Mo:1HEME	ND
Arsenite oxidase	*Alkaligenes faecalis*	92	A	1Mo:2Fe:2S	ND
Formate dehydrogenase	*Escherichia coli*	90	A_2B_2	4Mo:56Fe:52S:4Se:4HEME	MGD
	Clostridium pasteurianum	117	AB	2Mo:24Fe:28S	ND
	Methanobacterium formicicum	177	AB	1Mo:24Fe:29S:2Zn	MGD
Nitrate reductase					
Assimilatory	*Neurospora crassa*	290	A_2	1Mo:2HEME*	ND
	Aspergillus nidulans	200	A_4	1Mo:1FAD:1HEME	ND
	Chlorella vulgaris	360	A_4	4Mo:4FAD:4HEM	MPT
	Spinach	190	A_4	(Mo,HEME,FAD)	ND
Respiratory	*Escherichia coli*	200	AB	1Mo:12Fe:12S	MGD
Trimethlamine N-oxide reductase	*Escherichia coli*	200	A_2	2Mo	ND
Dimethylsulfoxide reductase	*Rhodobacter sphaeroides*	82	A	1Mo	MGD
	Escherichia coli	155	ABC	(Mo,Fe)	ND
Tetrathionate reductase	*Proteus mirabilis*	ND	ND	ND	ND
Chlorate reductase	*Proteus mirabilis*	ND	ND	ND	ND
Biotin sulfoxide reductase	*Escherichia coli*	ND	ND	ND	ND
Quinoline oxidoreductase[c]	*Pseudomonas putida*	ND	ND	(Mo)	MGD
Formylmethanofuran oxidoreductase[d]	*Methanobacterium*	146	ABC	(Mo:Fe:S)	MGD

* Activity dependent on FAD in assay.
1 Except where indicated, see Rajagopalan (1991) for references.
a Johnson et al. (1991).
b Nguyen (1986).
c Hettrich et al. (1991).
d Karrasch et al. (1990).

chemical identities, the number and the distances of atoms in the ligand field of a metal. The EXAFS data for a number of molybdoenzymes are summarized in Table 2, and will be discussed further in later sections.

A. Nitrogenase (Stephens, 1985; Pienkos et al., 1980)

1. *Biochemistry*

The reduction of N_2 to ammonia is carried out by some free living microorganisms and symbiotic bacteroids. The enzyme catalyzing the reaction, nitrogenase, has been purified from several sources. In all cases two specific proteins are required for the activity. The Fe-protein is a homodimer of 30 kg mol^{-1} subunits and contains one Fe_4S_4 cluster. The Fe,Mo protein is an A_2B_2 tetramer of 230 kg mol^{-1} and contains 2 Mo, 32 Fe and 30 S. Electrons derived from oxidized metabolites are transferred singly to the Fe-protein by ferredoxin or flavodoxin, and thence to the Fe,Mo protein. The Fe,Mo protein is capable of accumulating several electrons and catalyzing the concerted 6-electron reduction of N_2 to 2 NH_3.

2. *The Fe-Mo cofactor*

The multiplicity of metal content of the Fe,Mo protein could have made sorting out the metal centers of the protein a potentially difficult problem, had it not been for the crucial finding of Shah and Brill (1977) that a Mo-containing prosthetic group in the protein from *Azotobacter vinelandii* could be extracted into N-methylformamide. The material contained Mo, Fe and S in a ratio of 1:8:6 and was designated the Fe-Mo cofactor. The cofactor was capable of restoring nitrogenase activity to extracts of the UW45 mutant of *A. vinelandii* containing inactive nitrogenase (Nagatani et al., 1974). These findings made the Fe-Mo cofactor a suitable target for extensive physicochemical characterization. In particular, the absorption edge of the XAS of the cofactor as well as the native Mo,Fe-protein showed the absence of Mo=O bonds. EXAFS analysis (Table 2) confirmed this and also indicated the

Table 2

Mo K-edge EXAFS Data for Molybdoenzymes

Enzyme	Source	Mo-O		Mo-S	
		No.	Distance (Å)	No.	Distance (Å)
Nitrogenase	*A. vinelandii*				
Fe-Mo protein				3-4	2.36
Fe-Mo cofactor		2-3	2.10	3-4	2.35
Xanthine dehydrogenase	Chicken liver	1	1.70*	1 2	2.15 2.47
Sulfite oxidase	Chicken liver	2	1.68*	2-3 1	2.41 2.86
Xanthine oxidase	Bovine milk	1	1.75*	1 2	2.25* 2.49
Assimilatory nitrate reductase	*Chlorella vulgaris*	2	1.72*	2-3	2.44
Dissimilatory nitrate reductase	*Escherichia coli*			2	2.37
Formate dehydrogenase	*Clostridium pasteurianum*	3	1.74*		
Aldehyde ferredoxin oxidoreductase[1]	*Pyrococcus furiosus*	2	1.74* (W-O)	3	2.41 (W-S)

*Terminal ligand.

[1]W-L_{III} XAS.

presence of 3-4 Mo-S and 2-3 Mo-O bonds, of 2.35 Å and 2.1 Å length, respectively (Stiefel and Cramer, 1985). Best fit for computer simulation of the EXAFS spectrum were obtained by including 3 Mo to Fe distances of 2.66 Å. Though synthetic metal clusters with metal compositions and Mo-S, Mo-O and Mo-Fe distances similar to those of the natural Fe-Mo cofactor could be made, none of them had the ability to reconstitute the inactive nitrogenase in the *A. vinelandii* UW45 mutant (Holm and Simhon, 1985).

Though the Fe-Mo cofactor can be isolated in large amounts, its structure has remained a mystery. For a long time it was not known whether the cofactor contained any organic component. In 1989 Ludden and coworkers showed that homocitrate is an essential constituent of the active cofactor (Hoover et al., 1989). This finding raised the possibility that the Mo-O bonds detected by EXAFS could reflect bonding of homocitrate to Mo by the hydroxyl and/or carboxyl groups of the latter. The exact role of Mo in the catalytic activity of the enzyme is not clear, since there exist two alternative nitrogen-fixing enzymes in *Azotobacter* and other organisms, one containing vanadium instead of molybdenum, and the other containing neither metal (Bishop and Premakumar, 1992), both containing extractable cofactors akin to Fe-Mo cofactor. EXAFS analysis of the V-Mo cofactor has shown that the coordination sphere of V is quite similar to that of Mo in Fe-Mo cofactor (George et al., 1988).

A more direct visualization of the Fe-Mo cofactor has been afforded by the recent successful X-ray crystallographic studies of the *A. vinelandii* nitrogenase proteins (Georgiadis et al., 1992; Kim and Rees, 1992). The electron density map of the cofactor has been used to construct a model of the cofactor, as shown in Figure 1. The cofactor is seen to consist of one 4Fe:3S cluster and one 1Mo:3Fe:3S cluster bridged by three non-protein ligands, two of which are designated to be S atoms of sulfide ions and the third as yet undefined atom (Y in Fig. 1). Six of the seven Fe sites have trigonal geometry while the other Fe is bound to a cysteine thiol and has tetrahedral geometry. The Mo site has octahedral coordination, with three bridging S atoms, two O atoms donated by the hydroxyl and one of the carboxylates of homocitrate, and a sixth ligand from a histidine residue. The EXAFS data on the Fe-Mo cofactor were used in deriving these assignments since the metal atoms themselves were not distinguished one from another. Only two protein residues, the cysteine bound to one Fe, and a histidine serving as a ligand for Mo, are used for the prehension of the cofactor, explaining the ease with which the cofactor can be extruded into N-methylformamide. A prominent feature of the structure is a cavity of ~4 Å diameter which could conceivably accommodate the substrate dinitrogen molecule. The full coordination of the Mo center suggests that it may not be involved in substrate binding, unless ligand exchange is invoked.

B. Oxo transferases

1. *Biochemistry*

All of the enzymes listed in Table 1 with the exception of nitrogenase mobilize oxo groups, either from substrate to water or from water to substrate. The term oxo transferase is finding increasing acceptance as a descriptor for these enzymes. Though the reactions catalyzed by these enzymes involve transfer of only two electrons, most of them contain two

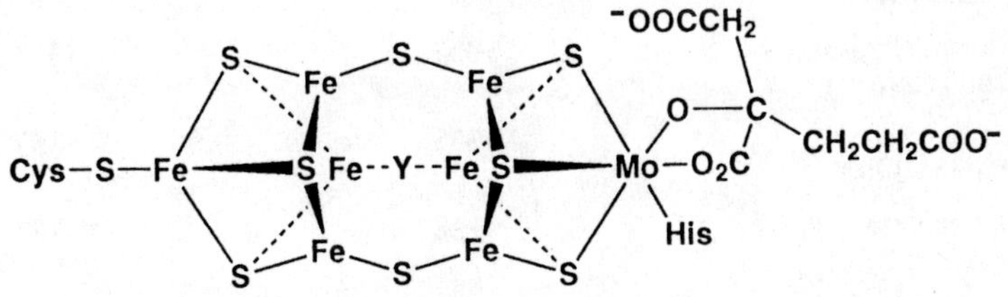

Figure 1.
Schematic model of the Fe-Mo cofactor fitted to electron density map, showing the approximate locations of Fe, Mo and S.

or more electron carriers as components of internal electron transport systems. The additional prosthetic groups include heme, flavin and Fe/S centers, all of which are strong chromophores and tend to mask the electronic absorption spectra of the molybdenum centers. However, they do not interfere with the EPR susceptibility of the Mo(V) species. The extensive literature covering the EPR studies on molybdoenzymes has been reviewed elsewhere (Bray, 1985; 1988). Qualitatively, the shapes of the Mo(V) EPR signals of enzymes such as xanthine dehydrogenase, sulfite oxidase, xanthine oxidase, aldehyde oxidase and nitrate reductase indicated the presence of thiol ligands in the Mo coordination field. Quantitatively, the diamagnetic-paramagnetic diamagnetic transitions going from Mo(VI) to Mo(V) to Mo(IV) have been examined to derive the oxidation-reduction potentials corresponding to the two 1-electron changes. The results of some of these studies are shown in Table 3. It can be seen that the potentials range from about -450 mV to +250 mV in different enzymes. The physicochemical basis for this large span in potentials has not been elucidated.

The Mo(V) EPR signals of sulfite oxidase and xanthine oxidase are extremely sensitive to alterations in the composition of the medium. As shown in Figure 2, the shape of the signal in sulfite oxidase is strongly influenced by anions and by changes in pH (Kessler et al., 1974). In the case of xanthine oxidase, the Mo(V) center displays at least two distinct transient EPR signals during rapid-freeze turnover, termed "Very Rapid" and "Rapid" (Bray, 1988). Other variant forms of the Mo(V) signal in xanthine oxidase have been described as "Slow" and two different "Inhibited" forms (Bray, 1988).

The N-terminal heme domain of rat liver sulfite oxidase can easily be cleaved and separated from the C-terminal molybdenum domain by mild treatment with trypsin (Johnson and Rajagopalan, 1977). This made it possible to observe for the first time the electronic absorption spectrum of a native molybdenum center (Fig. 3). The similarity of the absorption spectrum to those of Fe/S centers, both in quality and in magnitude, suggested that the observed spectrum reflected coordination of the Mo to thiol ligands (Johnson and Rajagopalan, 1977). The absorption spectrum was detectably perturbed by changes in pH and in ionic environment, as seen with the Mo(V) EPR signal. Thus both the Mo(VI) and Mo(V) forms of the Mo center of sulfite oxidase are sensitive to environmental variations.

X-ray absorbance spectroscopy has proved to be a powerful technique for probing the coordination state of Mo in different oxidation states. The results of EXAFS of sulfite oxidase and xanthine dehydrogenase are presented in Figure 4 (Cramer et al., 1981). The oxidized form of sulfite oxidase contains Mo(VI) and displays two Mo=O bonds. Upon

Table 3

Oxidation-reduction Potentials of Molybdenum Enzymes

Enzyme	E_1 (mV) $Mo^{6+} \rightarrow Mo^{5+}$	E_2 (mV) $Mo^{5+} \rightarrow Mo^{4+}$
Aldehyde oxidase (rabbit liver)	-359	-351
Xanthine dehydrogenase (chicken liver)	-357	-337
Xanthine oxidase (bovine milk)	-355	-355
Sulfite oxidase (chicken liver)	+38	-163
Nitrate reductase		
Chlorella vulgaris	-34	-54
Escherichia coli	+180	+220

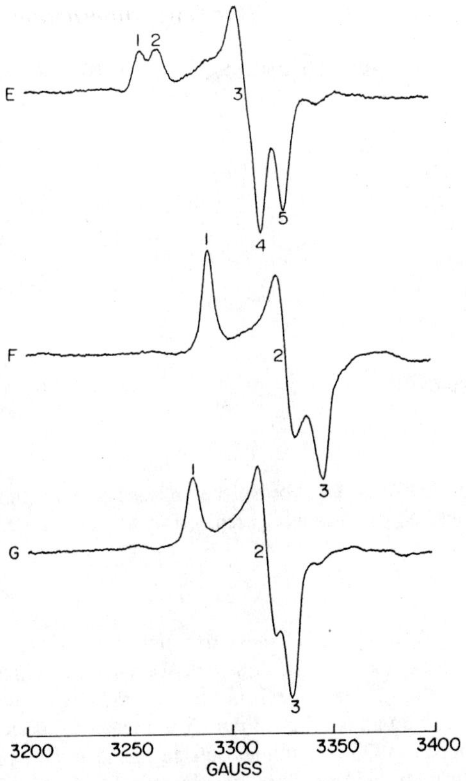

Figure 2.
Effect of pH and anions on the Mo(V) EPR signal in sulfite oxidase. *Top*, in Tris-HCl, pH 7.0; *middle* Tris-HCl, pH 9.0, *bottom*, potassium phosphate, pH 7.0.

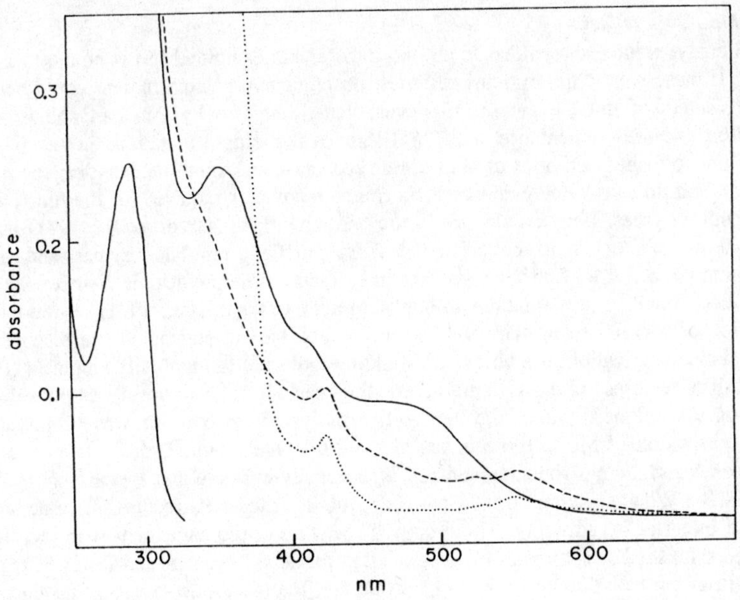

Figure 3.
Absorption spectrum of the Mo-containing domain of rat liver sulfite oxidase. *Solid* line, oxidized enzyme in Tris-HCl, pH 7.0; *dashed line*, reduced with sulfite; *dotted line*, reduced with dithionite.

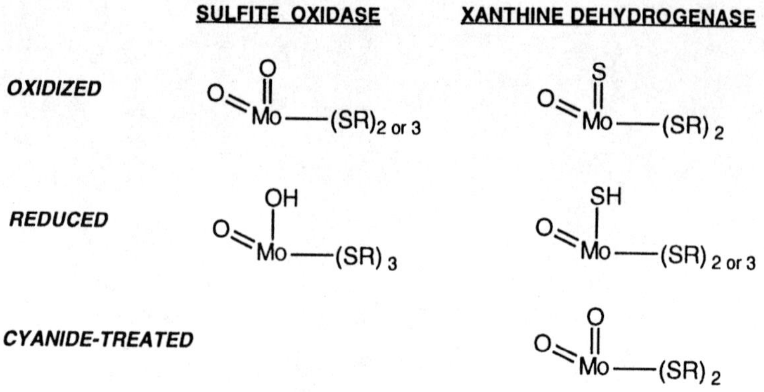

Figure 4.
Molybdenum K-edge EXAFS data for sulfite oxidase and xanthine dehydrogenase from chicken liver. The reduced samples were prepared by addition of dithionite.

reduction to the Mo(IV) state, one of these oxo ligands is replaced by an OH ligand. In contrast to sulfite oxidase, oxidized xanthine dehydrogenase contains one Mo=O bond and one Mo=S bond. This sulfide ligand is in fact the "cyanolysable S", removed as thiocyanate and replaced by an oxo ligand during the inactivation of the enzyme by cyanide (Massey and Edmondson, 1970). A similar sulfide ligand is also present in milk xanthine oxidase and liver aldehyde oxidase. The specific activity of any purified xanthine oxidase preparation is directly related to the content of the intact "sulfo" form of the enzyme (Massey, 1973). The results of EXAFS studies on several oxo transferases are shown in Table 2. Only molybdoenzymes effecting hydroxylation at a C atom appear to have the terminal sulfide ligand.

2. *The Mo-pterin cofactor*

EXAFS analyses have shown that in all oxo transferases examined Mo is bound to 2-3 thiol ligands. If these were from cysteine residues, no other non-protein moiety could be part of the Mo centers of these enzymes. However, genetic studies by Pateman and Cove with *Aspergillus nidulans* (Pateman et al., 1964) showed that single mutations in any of several genetic loci abolished activities of both nitrate reductase and xanthine dehydrogenase in that organism, and thus provided evidence for a common cofactor required for the functioning of Mo in both enzymes. Later studies by Nason and coworkers (Nason et al., 1971) indicated that such a cofactor could be separated from purified molybdoenzymes and used to reconstitute the activity of nitrate reductase in extracts of the pleiotropic *Neurospora crassa* mutant *nit-1*. Finally, investigations in our laboratory (Rajagopalan, 1991) on the nature of association of Mo to the protein of liver sulfite oxidase culminated in the discovery and structural characterization of a unique Mo-binding molecule termed molybdopterin (Fig. 5). Two distinct features of molybdopterin are the pterin ring and the 4-carbon side chain containing the unique dithiolene group. Active molybdenum cofactor was proposed to be formed by ligation of Mo to the S atoms of the dithiolene group. Unlike most of the other known cofactors, the molybdenum cofactor is extremely unstable and appears not to exist in the free state. When separated from a molybdoprotein, it loses its reconstitutive activity and structural integrity very rapidly. The proposed structure of the cofactor was in fact deduced from the established structures of two degradation products, Form A and Form B, that were derived from the active molecule, and of urothione, the metabolic excretion product of the cofactor in animals. Isolation of an inactive dialkylated derivative of molybdopterin, camMPT, allowed verification of the structure of the 4-alkyl side chain of the pterin (Kramer et al., 1987; Kruger and Meyer, 1987). The structures of the various derivatives of the cofactor are shown in Figure 5. The same cofactor was subsequently found in milk

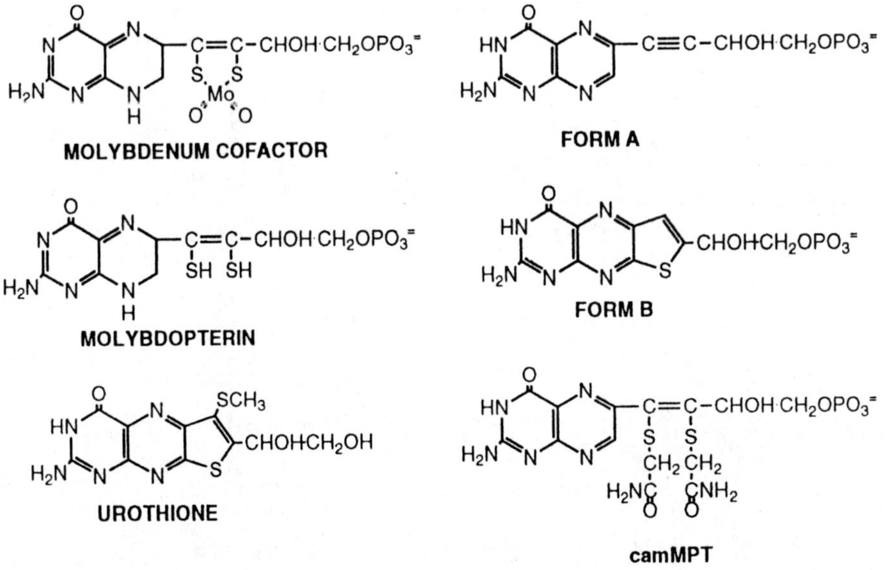

Figure 5.
Structures of the molybdenum cofactor deduced from structural characterization of degradation products generated from sulfite oxidase. The structures of molybdopterin and molybdenum cofactor are shown with the pterin ring in the quinonoid dihydro form. All other structures are shown with fully oxidized pterin.

xanthine oxidase, chicken liver xanthine dehydrogenase and rabbit liver aldehyde oxidase (Rajagopalan, 1991).

In the course of their studies on the enzyme CO dehydrogenase from *Pseudomonas carboxydoflava*, Meyer and coworkers proposed the existence of a larger form of cofactor containing two phosphate groups (Kruger et al., 1987). Failure to obtain camMPT from DMSO reductase of *Rhodobacter sphaeroides* led to the discovery and structural characterization of molybdopterin guanine dinucleotide (MGD) in our laboratory (Johnson et al., 1990a). Shortly thereafter the Mo cofactor of CO dehydrogenase was found to contain molybdopterin cytosine dinucleotide (Johnson et al., 1990b). These dinucleotide structures are shown in Figure 6. Two other variants of molybdopterin containing AMP or IMP as the nucleotide component have since been identified (Borner et al., 1991). The distribution of molybdopterin variants in a number of enzymes is shown in Table 1. While dinucleotides have not yet been reported in higher organisms, it is to be noted that molybdopterin itself is present in some bacterial enzymes.

The proposed *cis* Mo-dithiolene ring is the marquee feature of the molybdenum cofactor. Strong evidence for the *cis* configuration was provided by the finding that alkylation of the dithiolene thiols by iodoacetamide in anaerobically denatured sulfite oxidase is prevented by arsenite (Hageman and Rajagopalan, 1985), a reagent known to form a stable complex with cis dithiols, but not known to react with any protein side-chain residues including individual thiols. The C=C bond and the Mo-S bonds should be detectable by Resonance Raman (RR) spectroscopy. However, this goal has been made difficult to attain since most of the molybdoenzymes contain other strongly interfering cofactors. Recently, the molybdoenzyme DMSO reductase from *Rhodobacter sphaeroides* has been purified and shown to contain only the Mo-pterin cofactor as a prosthetic group (Bastian et al., 1991). The absorption spectrum of DMSO reductase shown in Figure 7 is characterized by absorption bands in the 600nm to 800nm range, ideally suited for RR. The RR spectra of the enzyme in both the Mo(VI) and Mo(IV) states have provided strong evidence for the proposed dithiolene coordination of Mo (Gruber et al., 1990). Magnetic

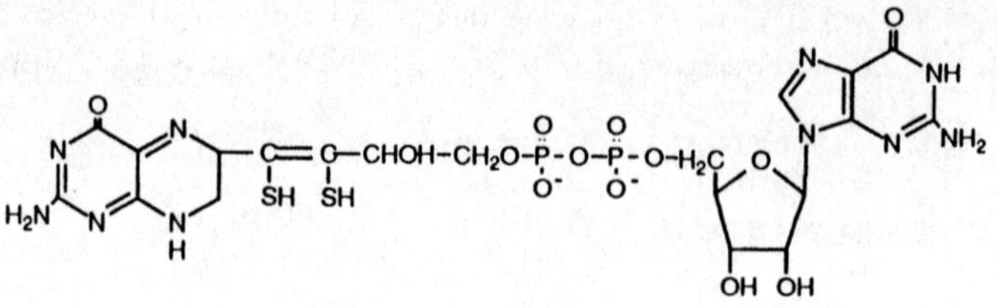

qH$_2$-MOLYBDOPTERIN GUANINE DINUCLEOTIDE

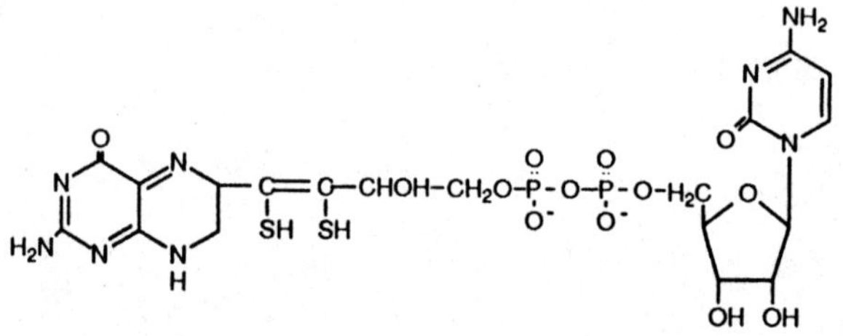

qH$_2$-MOLYBDOPTERIN CYTOSINE DINUCLEOTIDE

Figure 6.
Structures of molybdopterin guanine dinucleotide (MGD) and molybdopterin cytosine dinucleotide (MCD). The pterin ring is shown in the quinonoid dihydro form.

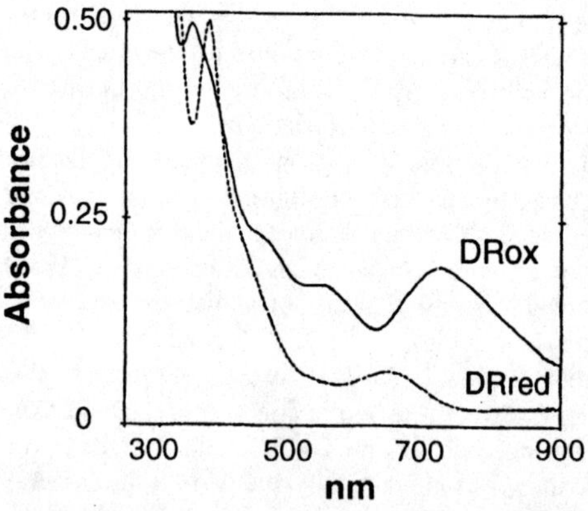

Figure 7.
Absorption spectrum of DMSO reductase from *Rhodobacter sphaeroides*. DRox, oxidized enzyme in 0.05 M phosphate, pH 7.8; DRred, enzyme reduced with dithionite.

circular dichroism (MCD) spectroscopy on the Mo(V) form of DMSO reductase also provides support for the existence of Mo-dithiolene charge transfer in the enzyme (Finnegan et al., 1993).

As described above, the EXAFS data on several molybdoenzymes suggest that the coordination field of Mo contains three thiol ligands. Presumably at least one of these is provided by a cysteine residue of the protein. It is also possible that the coordination field is not static and that the metal undergoes ligand exchange during catalysis. Xanthine oxidase (Fridovich and Handler, 1958), xanthine dehydrogenase (Weaver et al., 1970) and aldehyde oxidase (Rajagopalan and Handler, 1964) are known to be syncatalytically inactivated by the -SH reagents p-chloromercuribenzoate, quinones and F-dinitrobenzene, respectively, suggesting that the reactive thiol becomes free only during the catalytic cycle of each of these enzymes. These findings possibly constitute evidence for alterations in molybdenum ligation dependent on catalysis. Comparison of the peptide sequences of sufite oxidase from chicken and rat and the deduced amino acid sequences of several plant and fungal nitrate reductases shows extensive sequence identities among their molybdenum domains, with two invariant cysteines (Garrett and Rajagopalan, unpublished data). It remains to be seen whether one or both of these residues could serve as ligand(s) for Mo. In the case of sulfite oxidase, examination of the mechanism of the suicide inactivation of the enzyme with arsenite as the reducing substrate has shown that the loss of activity results from removal of Mo from the enzyme (Gardlik and Rajagopalan, 1991). Conceivably the excess arsenite in the medium is able to bind to two Mo thiolate ligands that become accessible during catalysis. Identification of the exchangeable thiol ligands should provide crucial information on the dynamic nature of the Mo ligand field during catalysis.

Fischer and coworkers have examined the reaction of $Mo(VI)O_2Cl_2$ with tetrahydrobiopterin (Fischer et al., 1991). X-ray crystallography of the product, trichloro(1,5-quinoid-7,8-dihydro-6H-L-biopterin)oxo-Mo(IV), revealed the structure shown in Figure 8.

As mentioned earlier, the uniqueness of molybdopterin resides not so much in the fact that it is a pterin, but in the presence of the dithiolene moiety. Since this feature has remained immutable in all oxo transferases containing Mo or W and is not present in any other type of biological molecule, any model lacking the dithiolene should be viewed with caution. Further, the lack of information on the stereochemical aspects of the pterin side chain and the oxidation state of the ring make any model speculative at best. It is conceivable, however, that the activity-dependent changes in the Mo ligand field mentioned above could include bonding of the Mo to ligands provided by the pterin ring, possibly involving *cis-trans* isomerization of the dithiolene group. Perhaps this could explain the essentiality of the pterin ring as a component of the molybdenum cofactor.

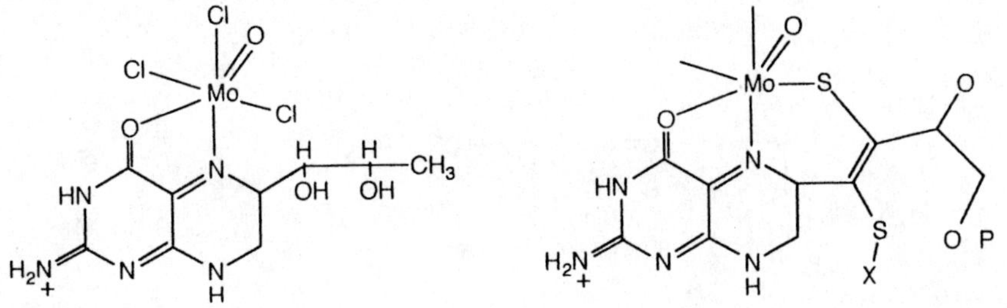

Figure 8.
Proposed model for the molybdenum cofactor based on the crystal structure of q-dihydrobiopterin-Mo complex. *Left*, Mo coordination in the crystal; *Right*, proposed model for molybdenum cofactor.

Table 4

Compositions and Functions of Tungsten Enzymes[1]

Enzyme	Source	M_r	Subunit composition	Cofactors per molecule	Pterin
Formaldehyde-Fd oxidoreductase	*Thermococcus litoralis*	280	A_4	(W:Fe:S)	MPT
Aldehyde-Fd oxidoreductase	*Pyrococcus furiosus*	85	A	(W:Fe:S)	MPT
Carboxylic acid reductase	*Clostridium formicoaceticum*	134	A_2	(W:Fe:S)	ND
Carboxylic acid reductase	*Clostridium thermoaceticum*	300	A_3B_3C	(W:Fe:S)	ND
Aldehyde oxidoreductase	*Clostridium thermoaceticum*	ND	ND	(W)	ND
Formate dehydrogenase	*Clostridium thermoaceticum*	340	A_2B_2	(W:Fe:S)	ND
Formylmethanofuran dehydrogenase	*Methanobacterium wolfei*	130	ABC	(W:Fe:S)	MGD

[1]See Mukund and Adams (1993) for references.

III. TUNGSTEN ENZYMES

A highly interesting recent development has been the discovery of a group of tungsten-containing enzymes in hyperthermophilic and moderately thermophilic organisms (Table 4). These tungsten enzymes catalyze reactions analogous to those catalyzed by molybdoenzymes in mesophilic organisms (Mukund and Adams, 1993). Aldehyde oxidoreductase from *Pyrococcus furiosus* was analyzed for the presence of molybdopterin using established procedures for the preparation of Form A and of camMPT. Form A was indeed obtained from the enzyme, demonstrating the presence of the molybdopterin moiety (Johnson et al., 1993). Alkylation of the pterin with iodoacetamide and analysis of the product by HPLC showed that the enzyme contained molybdopterin, not one of the dinucleotides. Analogous data were obtained from four tungsten enzymes that were examined (Johnson et al., 1993). The structure of the tungsten cofactor is shown in Figure 9. In view of the likelihood that hyperthermophilic bacteria and archaea represent the earliest forms of life, these findings demonstrate that the molybdopterin structure has remained unchanged during the entire evolutionary history. Apropos of this, it is significant

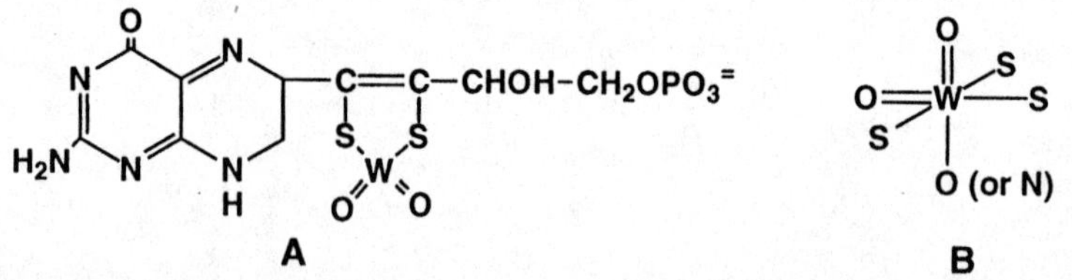

Figure 9.
Structural features of the W cofactor. *A*, the W cofactor of tungsten-containing enzymes from hyperthermophilic archaea. *B*, EXAFS data for the tungsten center of aldehyde ferredoxin oxidoreductase from *Pyrococcus furiosus*.

that *Methanobacterium wolfei*, a moderate thermophile, contains two different formylmethanofuran dehydrogenases, one containing and induced by molybdenum and the other containing and induced by tungsten. The moderate thermophiles could thus represent the transition from a tungsten-dependent high temperature existence to the molybdenum-dependent life forms of today. Both enzymes contain MGD rather than MPT, suggesting that the transition to prokaryotic life forms was associated with the emergence of molybdopterin dinucleotides.

EXAFS studies on aldehyde ferredoxin oxidoreductase have provided evidence for a tungsten ligand field shown in Figure 9. It is to be noted that though the tungsten enzyme hydroxylates a C atom of the substrate, it apparently lacks the Mo=S bond. The coordination field of tungsten is practically indistinguishable from that of molybdenum in sulfite oxidase. In fact, the interchangeability of the two metals was demonstrated several years ago when it was found that the catalytically inactive sulfite oxidase isolated from livers of tungsten-fed rats contained tungsten and that the line shape and g value of the EPR signal of W(V) were in accord with exact substitution of W for Mo (Johnson and Rajagopalan, 1976). The protein-bound tungsten was released upon incubation of the enzyme at 37 °C (Jones et al., 1977). Thus substitution of W for Mo does not per se engender thermostability.

IV. CONCLUSION

Recent developments have generated a great deal of information on the biochemistry of metal protein interactions in enzymes containing Mo and its congener W. However, a great deal more remains to be understood. In the case of nitrogenase, the basis for the presence of Mo in the Fe-Mo cofactor is still not understood. Determination of the crystallographic structures of the two alternative forms of the enzyme may be necessary to gain the necessary insight on the mechanism of conversion of N_2 to ammonia. In the case of oxo transferases, several critical aspects of the structural and functional features of the Mo-pterin cofactors, such as the reduction state of the pterin ring, the stereochemistry of the side chain and the protein ligand(s) to Mo remain to be clarified. X-ray crystallographic studies on a number of enzymes would seem to be absolutely essential for better understanding the biochemistry of these enzymes.

REFERENCES

Bastian, N.R., Kay, C.J., Barber, M.J., and Rajagopalan, K.V. (1991). *J. Biol. Chem.*, *266*: 45.

Bishop, P.E., and Premakumar, R. (1992). *Biological Nitrogen Fixation* (G. Stacey, R.H. Burris, and H.J. Evans, eds.), Routledge, Chapman and Hall, New York, p. 736.

Borner, G., Karrasch, M., and Thauer, R.K. (1991). *FEBS Lett.*, *290*: 31.

Bray, R.C. (1985). *The Enzymes*, Vol. 12 (P.D. Boyer, ed.), Academic Press, New York, p. 299.

Bray, R.C. (1988). *Quart. Rev. Biophys.*, *21*: 299.

Cramer, S.P., Wahl, R., and Rajagopalan, K.V. (1981). *J. Amer. Chem. Soc.*, *103*: 7721.

Finnegan, M.G., Hilton, J., Rajagopalan, K.V., and Johnson, M.K. (1993). *J. Inorg. Chem.*, *32*: 2616.

Fischer, B., Strahle, J., and Viscontini, M. (1991). *Helv. Chim. Acta*, *74*: 1544.

Fridovich, I., and Handler, P. (1958). *J. Biol. Chem.*, *231*: 899.

Gardlik, S., and Rajagopalan, K.V. (1991). *J. Biol. Chem.*, *266*: 16627.

Garner, C.D., and Bristow, S. (1985). *Molybdenum Enzymes* (T.G. Spiro, ed.), Wiley-Interscience, New York, p. 343.

George, G.N., Coyle, C.L., Hales, B.J., and Cramer, S.P. (1988). *J. Amer. Chem. Soc.*, *110*: 4057.

Georgiadis, M.M., Komiya, H., Chakrabarti, P., Woo, D., Kornuc, J.J., and Rees, D.C. (1992). *Science*, *257*: 1653.

Gruber, S., Kilpatrick, L., Bastian, N.R., Rajagopalan, K.V., and Spiro, T.G. (1990). *J. Amer. Chem. Soc.*, *112*: 8179.

Hageman, R.V., and Rajagopalan, K.V. (1985). *Nitrogen Fixation and CO_2 Metabolism* (P.W. Ludden and J.E. Burris, eds.), Elsevier Science Publishing Co., New York, p. 133.

Hettrich, D., Peschke, B., Tshisuaka, B., and Lingens, F. (1991). *Biol. Chem. Hoppe-Seyler*, *372*: 513.

Holm, R.H., and Simhon, E.D. (1985). *Molybdenum Enzymes* (T.G. Spiro, ed.), Wiley-Interscience, New York, p. 1.

Hoover, T.R., Imperial, J., Ludden, P.W., and Shah, V.K. (1989). *Biochemistry*, *28*: 2768.

Johnson, J.L., and Rajagopalan, K.V. (1976). *J. Biol. Chem.*, *251*: 5505.

Johnson, J.L., and Rajagopalan, K.V. (1977). *J. Biol. Chem.*, *252*: 2017.

Johnson, J.L., Bastian, N.R., and Rajagopalan, K.V. (1990a). *Proc. Nat. Acad. Sci. USA*, *87*: 3190.

Johnson, J.L., Rajagopalan, K.V., and Meyer, O. (1990b). *Arch. Biochem. Biophys.*, *283*: 542.

Johnson, J.L., Chaudhury, M., and Rajagopalan, K.V. (1991). *BioFactors*, *3*: 103.

Johnson, J.L., Rajagopalan, K.V., Mukund, S., and Adams, M.W. (1993). *J. Biol. Chem.*, *268*: 4848.

Jones, H.P., Johnson, J.L., and Rajagopalan, K.V. (1977). *J. Biol. Chem.*, *252*: 4988.

Karrasch, M., Borner, G., Enssle, M., and Thauer, R.K. (1990). *Eur. J. Biochem.*, *194*: 367.

Kessler, D.L., Johnson, J.L., Cohen, H.J., Rajagopalan, K.V. (1974). *Biochim. Biophys. Acta*, *334*: 86.

Kim, J., and Rees, D.C. (1992). *Science*, *257*: 1677.

Kramer, S.P., Johnson, J.L., Ribeiro, A.A., Millington, D.S., and Rajagopalan, K.V. (1987). *J. Biol. Chem.*, *262*: 16357.

Kruger, B., and Meyer, O. (1987). *Biochim. Biophys. Acta*, *912*: 357.

Kruger, B., Meyer, O., Nagel, M., Andreesen, J.R., Meincke, M., Bock, E., Bluemle, S., and Zumft, W.G. (1987). *Fems Microbiol. Lett.*, *48*: 225.

Massey, V. (1973). *Iron-Sulfur Proteins* (W. Lovenberg, ed.), Academic Press, New York, p. 301.

Massey, V., and Edmondson, D.E. (1970). *J. Biol. Chem.*, *245*: 6595.

Mukund, S., and Adams, M.W.W. (1993). *J. Biol. Chem.*, *268*: 13592.

Nagatani, H.H., Shah, V.K., and Brill, W.J. (1974). *J. Bacteriol.*, *120*: 697.

Nason, A., Lee, K.Y., Pan, S.S., Ketchum, P.A., Lamberti, A., and DeVries, J. (1971). *Proc. Nat. Acad. Sci. USA*, *68*: 3242.

Nguyen, J. (1986). *Physiol. Veg.*, *24*: 261.

Pateman, J.A., Cove, D.J., Rever, B.M., and Roberts, D.B. (1964). *Nature*, *201*: 58.

Pienkos, P.T., Shah, V.K., and Brill, W.J. (1980). *Molybdenum and Molybdenum-containing Enzymes* (M.P. Coughlan, ed.), Pergamon Press, New York, p. 385.

Rajagopalan, K.V. (1991). *Advances in Enzymology and Related Areas of Molecular Biology*, Vol. 64 (A. Meister, ed.), John Wiley & Sons, New York, p. 215.

Rajagopalan, K.V., and Handler, P. (1964). *J. Biol. Chem.*, *239*: 2027.

Shah, V.K., and Brill, W.J. (1977). *Proc. Nat. Acad. Sci. USA*, *74*: 3243.

Stephens, P.J. (1985). *Molybdenum Enzymes* (T.G. Spiro, ed.), Wiley Interscience, New York, p. 117.

Stiefel, E.I., and Cramer, S.P. (1985). *Molybdenum Enzymes* (T.G. Spiro, ed.), Wiley-Interscience, New York, p. 89.

Weaver, R.F., Rajagopalan, K.V., Handler, P., Jeffs, P., Byrne, W.L., and Rosenthal, D. (1970). *Proc. Nat. Acad. Sci. USA*, *67*: 1050.

VANADATE - PROTEIN INTERACTIONS

by

Debbie C. Crans

Department of Chemistry,
Colorado State University,
Fort Collins, Colorado 80523
U.S.A.

I. INTRODUCTION

In its interactions with proteins, vanadate differs from other metal ions in that it is an anionic form of vanadium in oxidation state five. The structural and electronic analogy between vanadate monomer and phosphate explains why vanadate behaves as a phosphate analog in many biological systems. The interactions of vanadate with enzymes catalyzing reactions centered at phosphorus as well as enzymes catalyzing other condensation, hydrolysis, isomerization, oxidation and reduction reactions will be summarized here. Interactions of vanadate with enzymes accepting phosphate as a substrate or generating phosphate as a product involve phosphatases and ATPases. Additional enzyme-catalyzed reactions centered at phosphorus include phosphotransfer reactions carried out by kinases and mutases, and hydrolysis reactions carried out by ribonucleases. Dehydrogenases, isomerases, and aldolases are examples of enzymes catalyzing reactions involving substrates or cofactors containing phosphate groups, and interactions of vanadate with these enzymes will be described. Other mechanisms by which vanadium interacts with enzymes, including redox chemistry and interactions of vanadium(IV), are briefly summarized. However, the vanadium-dependent haloperoxidases and nitrogenases as well as the photolytic cleavage of proteins by vanadate are reviewed elsewhere (Chasteen, 1990). First, we turn to the chemistry dictating the reactions and interactions of anionic vanadium(V) with proteins.

II. AQUEOUS CHEMISTRY OF VANADATE

Vanadate is very sensitive to pH, vanadium concentration and environments because of the many forms vanadium(V) can take under physiological conditions. At submicromolar concentrations, vanadium(V) exists as the monoanion ($H_2VO_4^-$, pKa about 8.0) or as the dianion (HVO_4^{2-}, pK_a about 12.5) (Pope, 1983; Pettersson et al., 1985). At very high pH the trianion (VO_4^{3-}) is the major species and at low pH a cationic species, VO_2^+, will prevail. Above 10^{-4} M concentrations, vanadate solutions consist of several oligomeric species at neutral and basic pH. These species can include vanadate monomer (V_1 as $H_2VO_4^-$, HVO_4^{2-}, VO_4^{3-}), dimer (V_2), tetramer (V_4) and pentamer (V_5) which all interconvert rapidly in solution (Crans et al., 1990a). The species in the vanadate solutions vary with pH, vanadium concentration and composition. These are determined using ^{51}V NMR spectroscopy since each compound has a different chemical shift (Fig. 1). Stock solution of any combination of these species will immediately convert to the respective equilibrium mixture upon addition to an enzyme assay or administration to a mammal. The rapidly converting colorless vanadate derivatives are in contrast to the orange-yellow vanadate decamer, which forms under acidic conditions or when acid is added to the colorless vanadate solutions (Pope, 1983). Although the decamer is thermodynamically unstable above pH 6, it remains intact over limited time periods.

In addition to the oligomerization reactions, vanadate monomer reacts with compounds containing free hydroxyl (Gresser and Tracey, 1985) or carboxylate groups (Tracey et al., 1990). Reactions with amine groups have also been reported (Crans and Shin, 1988; Rehder, 1988). The reaction of $H_2VO_4^-$ with hydroxyl groups is shown in (1).

$$ROH + H_2VO_4^- \rightleftharpoons ROVO_3H^- + H_2O \quad (1)$$

These types of reactions can cause significant problems in vanadate studies, because complexes form with many commonly used buffers. In Figure 1, the spectra of a solution of 10 mM vanadate at pH 7.8 is shown in 150 mM imidazole, with 35 mM triethanolamine (TEA) and 30 mM tricine, respectively. The spectra of the two latter buffers each show a new signal which is attributed to vanadate-triethanolamine (V-TEA, -483 ppm) and vanadate-Tricine (V-Tricine, -504 ppm) complexes, respectively. The chemical shift of V_1 is -555 ppm, V_2 -569 ppm, V_4 -574 ppm and V_5 -582 ppm. Hepes is also an excellent buffer to use in studies with vanadate; a more detailed description of appropriate buffers has been given elsewhere (Crans et al., 1989a).

Vanadate can interact with enzymes by all the reactions described above: (i) vanadate can interact non-covalently with an enzyme; (ii) vanadate can form a covalent complex with an enzyme; (iii) covalent substrate complexes can form which interact with enzymes; (iv) a substrate (or cofactor) forms a complex with the enzyme which subsequently binds the vanadate, or, conversely, the vanadate binds to the enzyme before the substrate. At this time, examples of two types of mechanisms (non redox) have been described (Chasteen, 1983; 1990; Rehder, 1991; Nechay et al., 1986; Butler and Eckert, 1989; Nour-Eldeen et

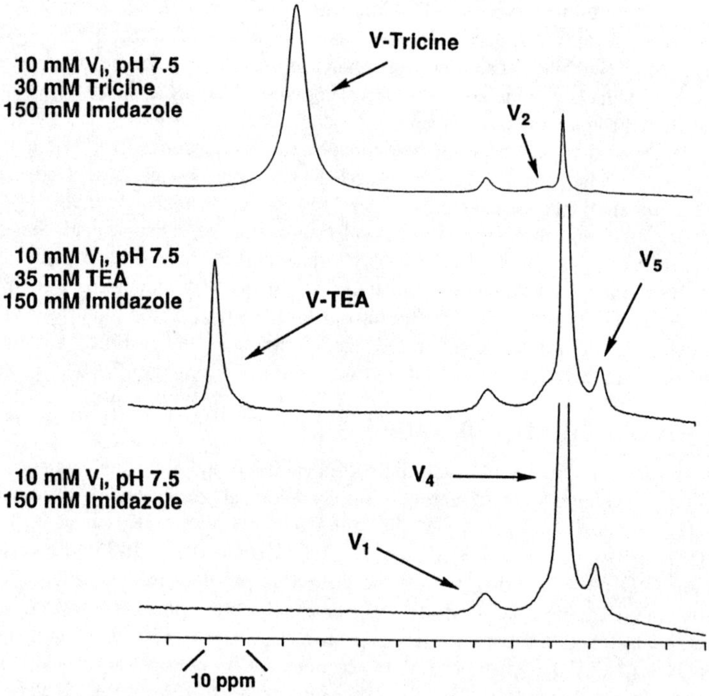

Figure 1.
The ^{51}V NMR spectra of 10 mM vanadate solutions containing 150 mM imidazole, 35 mM triethanolamine (TEA) and 30 mM Tricine.

al., 1985; Ray and Post, 1990, Ray and Puvathingal, 1990; Liu et al., 1992; Drueckhammer et al., 1989; Crans et al., 1992a; 1992b).

In reducing environments, the vanadium atom in vanadate reduces to oxidation state four (Pope, 1983). The presence of free thiol groups at low or neutral pH can generate vanadium(IV) and disulfide (Cohen et al., 1987). Vanadium(IV), often in the form of vanadyl cation, can also interact with enzymes (Chasteen, 1983). The presence of thiol groups introduces redox mechanisms by which the enzymes can interact with vanadate.

III. VANADATE INTERACTIONS WITH PHOSPHATASES AND ATPASES

Over 100 years ago, vanadate was reported to have a digitalis-like action on heart muscles, stimulating many studies in the following years (Gamagee and Larmuth, 1876; reviewed in Ramasarma and Crane, 1981). Recently, vanadate has been reported to induce insulin-like effects on isolated cells, tissue, the purified insulin receptor, and in diabetic rats (reviewed in Gresser et al., 1987). Several hypotheses have been advanced explaining the vanadate-induced effects on systems regulated by phosphorylation of tyrosine or by mechanisms involving inhibition of key enzymes such as the Na^+K^+ATPase (Chasteen, 1990; Gresser et al., 1987). Interest in vanadium biochemistry was reignited in 1975 by the discovery of a potent inhibitor of Na^+K^+ATPase (vanadyl cation) in commercial preparations of Sigma's equine muscle ATP preparations (Charney et al., 1975; Cantley et al., 1977) and in 1980 with the insulin-mimetic activities of vanadate (Schechter and Karlish, 1980).

The Na^+K^+ATPase belongs to a group of intrinsic membrane enzymes that have two principal conformations, E1 and E2 (Nechay et al., 1986; Boyd and Kustin, 1984). In the E1 conformation, the enzyme reversibly accepts a phosphate group from ATP and, in the E2 conformation, it accepts a phosphate group from P_i. Vanadate monomer interacts with the E2 conformation. The inhibition by vanadate is influenced by cations such as Mg^{2+}, K^+ and Na^+ and is reversed by catecholamines. Even if the inhibition constant of 4 nM determined in the laboratory (Cantley et al., 1978) holds for physiological conditions, the estimated intracellular vanadate concentration of 3×10^{-11} M appears to be too low to effect Na^+K^+ pump (Nechay et al., 1986). Other enzymes in this group include several Ca^{2+} and H^+K^+ATPases. These enzymes distinguish themselves from the Na^+K^+ATPase in a lower affinity for vanadate (Nechay et al., 1986).

The hydrolysis of alkyl phosphate groups to free hydroxyl groups and inorganic phosphate is catalyzed by phosphatases. Depending on its pH activity profile, the particular enzyme is classified as either an acid or an alkaline phosphatase. The two types of phosphatases convert similar substrates. However, the mechanisms of reactions are different. Vanadate inhibits both types of phosphatases with micromolar or lower K_i values. The K_i values for the alkaline phosphatases from human liver, intestine and kidney are 0.9, 0.5 and 0.6 mM (Seargeant and Stinson, 1979). With two exceptions, the human prostatic acid phosphatase (Crans et al., 1989b) and the human transmembrane leukocyte antigen related (LAR) protein tyrosine phosphatase (Cho et al., 1991), the inhibiting form of vanadate is the monomer. Vanadate inhibits most acid and some alkaline phosphatases significantly more than phosphate (Chasteen, 1983; Crans et al., 1989b; Cho et al., 1991; Lopez et al., 1976; VanEtten et al., 1974; Stankiewicz and Gresser, 1988). Accordingly, it has been proposed that vanadate binds as a transition state analog mimicking the pentacoordinate phosphorus in the transition state of the alkyl phosphate hydrolysis reaction. Vanadate inhibition of protein phosphatase is of particular interest because some selectivity in inhibiting tyrosine protein phosphatases has been observed (reviewed in Gresser et al., 1987). The effect of vanadate on serine or threonine protein phosphatases is significantly less, although there appear to be some differences depending on the substrate used in the studies (Gresser et al., 1987).

IV. VANADATE INTERACTIONS WITH RIBONUCLEASES, MUTASES AND KINASES

A complex of vanadate and uridine was discovered to potently inhibit ribonuclease A more than two decades ago (Linquist et al., 1973). The complex was termed a transition state analog (TSA) because the pentacoordinate vanadium mimicked the transition state of the hydrolysis of the cyclic phosphate. The complex was used for preparation of RNA until more potent ribonuclease inhibitors were discovered. About ten years later, two groups characterized the vanadate-uridine-ribonuclease A complex crystallographically (Alber et al., 1983; Wlodawer et al., 1983). Since then, the enzymatic and spectroscopic characterizations of ribonuclease T have demonstrated that analogous vanadate-guanosine complex is a potent inhibitor of this enzyme (Rehder et al., 1988).

The complex of 6-vanadioglucose-1-phosphate inside phosphoglucomutase has recently emerged as another TSA (Ray and Post, 1990, Ray and Puvathingal, 1990). Kinetic and spectroscopic evidence suggest that this complex contains an unusually tightly bound inhibitor, which would be consistent with a pentacoordinate vanadium TSA. 6-Phosphoglucose-1-vanadate does not show similar high affinity for the enzyme and this is consistent with the binding of the inhibitor in the active site of the enzyme. A TSA is also proposed for the complex between vanadate and 3-phosphoglycerate (2-vanadio-3-phosphoglycerate) inside phosphoglycerate mutase (Liu et al., 1992).

The interactions of kinases with vanadate have been the focus of studies both in vivo and in vitro. Unfortunately, both protein kinase activation and inhibition have been reported in the presence of vanadate for the same enzyme systems (reviewed by Gresser et al., 1987). Studies of isolated enzymes have proven difficult since the substrates for these enzyme are often ill defined. Furthermore, the assay conditions are not without reducing compounds or other components which potentially could affect the observed enzyme activities. It is very likely that some of the controversy is generated because of assay methods and conditions. Studies with pure enzyme preparations of pyruvate kinase, hexokinase and phosphofructokinase have shown that vanadate monomer has no effect on these enzymes, but that the vanadate decamer is an inhibitor (Boyd et al., 1985). In the case of the protein kinases, studies with pure enzyme preparations must be carried out to resolve these controversies.

V. VANADATE INTERACTIONS WITH OTHER ENZYMES

Using a combination of enzyme kinetics and ^{51}V NMR spectroscopy, oligomeric vanadate anions were found to inhibit several dehydrogenases (Crans and Schelble, 1990; Crans and Simone, 1991, Crans et al., 1990b; 1993), aldolase (Crans et al., 1992b) and bind to superoxide dismutase (Wittenkeller et al., 1991). Activity and binding studies with superoxide dismutase from erythrocytes did not show similar affinity for vanadate (Serra et al., 1992). The vanadate tetramer was first reported to be an inhibitor for 6-phosphogluconate dehydrogenase from bakers yeast, rat liver and human erythrocytes with submillimolar K_i values (Crans et al., 1990b). Since then the tetramer has been found to inhibit glucose-6-phosphate dehydrogenase (Crans and Schelble, 1990), glycerol-3-phosphate dehydrogenase (Crans and Simone, 1991), superoxide dismutase (Wittenkeller et al., 1991) and alcohol dehydrogenase (Crans et al., 1993). The vanadate dimer inhibits 3-phosphoglycerate mutase (Stankiewicz et al., 1987), glycerol-3-phosphate dehydrogenase (Crans and Simone, 1991), glucose-6-phosphate dehydrogenase (Crans and Schelbe, 1990) and aldolase (Crans et al., 1992b). The inhibition by these oligomeric anions is significant only under conditions where the vanadate concentrations are in the millimolar concentration range. These types of interactions are not likely to be important under physiological conditions. Many studies with vanadate are conducted at millimolar concentrations where the oligomeric anions interact with proteins, and the oligomers must be considered in such studies. The interaction of vanadate with proteins has been used, for example, in crystallization of Na^+ K^+ and Ca^{2+} ATPases (Dux and Martonosi, 1983; Zampighi et al., 1984).

Recent studies showed that vanadate could substitute for phosphate in organic phosphate substrates (Nour-Eldeen et al., 1985; Drueckhammer et al., 1989; Crans et al., 1992a; Crans et al., in preparation). These studies have been conducted with human, mammalian and non-mammalian enzyme sources. Glucose-6-phosphate dehydrogenase accepts glucose-6-vanadate as a substrate and converts it to product (Nour-Eldeen et al., 1985; Drueckhammer et al., 1989). Other enzymes that can convert the corresponding vanadate derivative include glycerol-3-phosphate dehydrogenase (Craig, 1986;

Drueckhammer et al., 1989) phosphoribose isomerase (Drueckhammer et al., 1989), ribulose-5-phosphate-3-epimerase (Drueckhammer et al., 1989) and fructose 1,6-bisphosphate aldolase (Drueckhammer et al., 1989; Crans et al., in preparation). In the few cases in which the Michaelis-Menten parameters have been reported, the organic vanadate derivative has a similar K_m to the natural substrate (Crans et al., 1992a; 1993). Perhaps the most physiologically interesting finding is the observation that NAD, in the presence of vanadate, behaves as an NADP (presumably as NADV). Glucose-6-phosphate dehydrogenase, 6-phosphogluconate dehydrogenase and alcohol dehydrogenase specific for NADP can convert substrate to product using NAD in the presence of vanadate (Crans et al., 1992a; 1993).

Vanadate has been reported to interact with glyceraldehyde-3-phosphate dehydrogenase through redox chemistry (Benabe et al., 1987). This interaction was proposed to be responsible for the effects of vanadate on the glycolysis. Other enzymes have also been found to undergo redox chemistry with vanadate; aldolase (Crans et al., 1992b) and purple acid phosphatase (Crans et al., 1992c). However, the presence of thiol groups in the active site is not a guarantee that such interaction will occur, since glycerol-3-phosphate dehydrogenase does not undergo redox chemistry with vanadate even though it is believed to bind in the active site (Crans and Simone, 1991).

VI. VANADYL INTERACTIONS WITH ENZYMES

Vanadyl cation (VO^{2+}) also interact with proteins and in some cases in a manner similar to that of vanadate. It is a potent inhibitor for phosphatases (Chasteen, 1983), ribonucleases (Berger and Birkenmeier, 1979) and reverse transcriptase (Krug and Berger, 1991) although it has significantly lower affinity for Na^+K^+ATPase. It activates NADP-requiring dehydrogenases in the presence of NAD (Crans et al., 1992a). In addition, vanadyl cation can substitute for Mg^{2+} in pyruvate kinase (Lord and Reed, 1987; 1990) and S-adenosyl methionine synthetase (Markham, 1984), and for Zn^{2+} in carboxypeptidase (DeKoch et al., 1974).

VII. SUMMARY

The interactions of vanadate with proteins can occur by several different mechanisms which depend on the enzyme and the form of vanadate. Advantage should be taken of the potent biological effects of vanadate, and recently vanadium(V)-based drugs have emerged which are more potent insulin-mimetics than vanadate alone (Posner et al., 1990; McNeill et al., 1992). However, the complex interactions of vanadate with proteins suggest that much more must be known before the beneficial biomedical interactions of this metal are likely to be fully realized.

ACKNOWLEDGEMENTS

I thank NIH for funding this work, and that described from our laboratory in the references.

REFERENCES

Alber, T., Gilbert, W.A., Ponzi, D.R., and Petsko, G.A. (1983). *Ciba Found. Symp.*, *93*: 4.

Benabe, J.E., Echegoyen, L.A., Pastrana, B., and Martinez-Maldonado, M. (1987). *J. Biol. Chem.*, *262*: 9555.

Berger, S.L., and Birkenmeier, C.S. (1979). *Biochemistry*, *18*: 5143.

Boyd, D.W., and Kustin, K. (1984). *Adv. Inorg. Biochem.*, *6*: 311.

Boyd, D.W., Kustin, K., and Niwa, M. (1985). *Biochim. Biophys. Acta*, *827*: 472.

Butler, A., and Eckert, H. (1989). *J. Amer. Chem. Soc.*, *111*: 2802.

Cantley, L.C., Jr., Josephson, L., Warner, R., Yanagisawa, M., Lechene, C., and Guidotti, G. (1977). *J. Biol. Chem.*, *252*: 7421.

Cantley, L.C., Jr., Cantley, L.G., and Josephson, L. (1978). *J. Biol. Chem.*, *253*: 7361.

Charney, A.N., Silva, P., and Epstein, F.H. (1975). *J. Appl. Physiol.*, *39*: 156.

Chasteen, N.D. (1983). *Struct. Bonding*, *53*: 105.

Chasteen, N.D., ed. (1990). *Vanadium in Biological Systems*, Kluwer Academic Publishers, Boston.

Cho, H., Ramer, S.E., Itoh, M., Winkler, D.G., Kitas, E., Bannwarth, W., Burn, P., Saito, H., and Walsh, C.T. (1991). *Biochemistry*, *30*: 6210.

Cohen, M.D., Sen, A.C., and Wei, C.-I. (1987). *Inorg. Chim. Acta*, *138*: 179.

Craig, M. (1986). M. Sc. Thesis, Simon Fraser University.

Crans, D.C., and Schelble, S.M. (1990). *Biochemistry*, *29*: 6698.

Crans, D.C., and Shin, P.K. (1988). *Inorg. Chem.*, *27*: 1797.

Crans, D.C., and Simone, C.M. (1991). *Biochemistry*, *30*: 6734.

Crans, D.C., Bunch, R.L., and Theisen, L.A. (1989a). *J. Amer. Chem. Soc.*, *111*: 7597.

Crans, D.C., Simone, C.M., Saha, A.K., and Glew, R.H. (1989b). *Biochem. Biophys. Res. Commun.*, *165*: 246.

Crans, D.C., Rithner, C.D., and Theisen, L.A. (1990a). *J. Amer. Chem. Soc.*, *112*: 2901.

Crans, D.C., Willging, E.M., and Butler, S.K. (1990b). *J. Amer. Chem. Soc.*, *112*: 427.

Crans, D.C., Simone, C.M., and Blanchard, J.S. (1992a). *J. Amer. Chem. Soc.*, *114*: 4926.

Crans, D.C., Sudhakar, K., and Zamborelli, T.J. (1992b). *Biochemistry*, *31*: 6812.

Crans, D.C., Simone, C.M., Holz, R.C., and Que, L., Jr. (1992c). *Biochemistry*, *31*: 11731.

Crans, D.C., Marshman, R.W., Nielsen, R., and Felty, I. (1993). *J. Org. Chem.*, *58*: 2244.

Crans, D.C., Sudhakar, K., and Felty, I. in preparation.

DeKoch, R.J., West, D.J., Cannon, J.C., and Chasteen, N.D. (1974). *Biochemistry*, *13*: 4347.

Drueckhammer, D.G., Durrwachter, J.R., Pederson, R.L., Crans, D.C., Daniels, L., and Wong, C.-H. (1989). *J. Org. Chem.*, *54*: 70.

Dux, L., and Martonosi, A. (1983). *J. Biol. Chem.*, *258*: 2599.

Gamagee, A., and Larmuth, L.N. (1876). *J. Anat. Physiol. Norm. Pathol. Homme Anim.*, *11*: 235.

Gresser, M.J., and Tracey, A.S. (1985). *J. Amer. Chem. Soc.*, *107:* 4215.

Gresser, M.J., Tracey, A.S., and Stankiewicz, P.J. (1987). *Adv. Prot. Phosphatases*, *4*: 35.

Krug, M.S., and Berger, S.L. (1991). *Biochemistry*, *30*: 10614.

Lindquist, R.N., Lynn, J.L., Jr., and Lienhard, G.E. (1973). *J. Amer. Chem. Soc.*, *95*: 8762.

Liu, S., Gresser, M.J., and Tracey, A.S. (1992). *Biochemistry*, *31*: 2677.

Lopez, V., Stevens, T., and Lindquist, R.N. (1976). *Arch. Biochem. Biophys.*, *175*: 31.

Lord, K.A., and Reed, G.H. (1987). *Inorg. Chem.*, *26*: 1464.

Lord, K.A., and Reed, G.H. (1990). *Arch. Biochem. Biophys.*, *281*: 124.

Markham, G.D. (1984). *Biochemistry*, *23*: 470.

McNeill, J.H., Yuen, V.G., Hoveyda, H.R., and Orvig, C. (1992). *J. Med. Chem.*, *35*: 1489.

Nechay, B.R., Nanninga, L.B., Nechay, P.S.E., Post, R.L., Grantham, J.J., Macara, I.G., Kubena, L.F., Phillips, T.D., and Nielsen, F.H. (1986). *Fed. Proc.*, *45*: 123.

Nour-Eldeen, A.F., Craig, M.M., and Gresser, M.J. (1985). *J. Biol. Chem.*, *260*: 6836.

Pettersson, L., Andersson, I., and Hedman, B. (1985). *Chem. Scr.*, *25*: 309.

Pope, M.T. (1983). *Heteropoly and Isopoly Oxometalates*, Springer-Verlag, New York.

Posner, B.I., Shaver, A., and Fantus, I.G. (1990). *New Antidiabetic Drugs* (C.J. Bailey and P.R. Flatt, eds.), Smith-Gordon, New York, p. 107.

Ramasarma, T., and Crane, F.L. (1981). *Curr. Top. Cell. Reg.*, *20*: 247.

Ray, W.J., Jr., and Post, C.B. (1990). *Biochemistry*, *29*: 2779.

Ray, W.J., Jr., and Puvathingal, J.M. (1990). *Biochemistry*, *29*: 2790.

Rehder, D. (1988). *Inorg. Chem.*, *27*: 4312.

Rehder, D. (1991). *Angew. Chem. Int. Ed. Engl.*, *30*: 148.

Rehder, D., Holst, H., Quaas, R., Hinrichs, W., Hahn, U., and Saenger, W. (1989). *J. Inorg. Biochem.*, *37*: 141.

Seargeant, L.E., and Stinson, R.A. (1979). *Biochem. J.*, *181*: 247.

Serra, M.A., Pintar, A., Casella, L., and Sabbioni, E. (1992). *J. Inorg. Biochem.*, *46*: 161.

Schechter, Y., and Karlish, S.J. (1980). *Nature (London)*, *284*: 556.

Stankiewicz, P.J., Gresser, M.J., Tracey, A.S., and Hass, L.F. (1987). *Biochemistry*, *26*: 1264.

Stankiewicz, P.J., and Gresser, M.J. (1988). *Biochemistry*, *27*: 206.

Tracey, A.S., Li, H.L., and Gresser, M.J. (1990). *Inorg. Chem.*, *29*: 2267.

VanEtten, R.L., Waymack, P.P., and Rehkop, D.M. (1974). *J. Amer. Chem. Soc.*, *96*: 6782.

Wittenkeller, L., Abraha, A., Ramasamy, R., de Freitas, D.M., Theisen, L.A., and Crans, D.C. (1991). *J. Amer. Chem. Soc.*, *113*: 7872.

Wlodawer, A., Miller, M., and Sjölin, L. (1983). *Proc. Nat. Acad. Sci. USA*, *80*: 3628.

Zampighi, G., Kyte, J., and Freytag, W. (1984). *J. Cell. Biol.*, *98*: 1851.

IRON-CONTAINING METAL CLUSTERS

by

Brian G. Fox and Eckard Münck

Carnegie Mellon University,
Department of Chemistry,
4400 Fifth Avenue
Pittsburgh, PA 15213
U.S.A.

I. INTRODUCTION

A variety of metal clusters are found in proteins, where the ligands to the metal ions of the cluster are provided primarily by amino acid residues. This chapter will focus on clusters which contain iron. Iron-containing clusters function as electron carriers, catalytic, regulatory or storage sites, or as structural determinants. Two distinct types of iron-containing clusters will be described here: iron-sulfur clusters, composed of iron and inorganic sulfide and coordinated to cysteine ligands (and, in a few cases, histidine, aspartate or hydroxyl ligands); and diiron-oxo clusters, having oxygen-containing bridging ligands and a variable ligation sphere of nitrogen and oxygen derived from histidine, glutamate, aspartate, and water. These clusters have a direct relation to human biology since enzymes and proteins containing these structures are involved in a wide variety of metabolic processes.

Detailed insights into the structure of protein-bound metal clusters has arisen from the intensive application of physical techniques such as electron paramagnetic resonance (EPR), Mössbauer, electron nuclear double resonance (ENDOR), paramagnetic NMR, magnetic circular dichroism (MCD), optical, resonance Raman, and X-ray absorption fine structure (EXAFS) spectroscopies as well as magnetic susceptibility studies. The interpretation of this spectroscopic data has been greatly aided by the availability of spectroscopic studies of corresponding synthetic iron clusters. These techniques have provided information about the redox states and ligation environments of the cluster. The elucidation of the spin coupling phenomenon has been a significant triumph in the application of spectroscopic techniques to the study of metal cluster structures in proteins. In many cases, the use of X-ray crystallography has provided further, detailed refinements of this structural understanding. For the most part, spectroscopic and crystallographic studies of the cluster content and structure have been performed with proteins obtained from non-human sources due to the large amounts of highly purifed protein required for the application of these techniques. However, with the advance of molecular biological methods, the direct structural characterization of recombinant human proteins has become feasible.

Throughout this article, extensive reference will be made to volume 38 of Advances in Inorganic Chemistry, dedicated solely to iron-sulfur proteins, and to recent review articles by R. Cammack (Cammack, 1992) and by J. B. Howard and D. C. Rees (Howard and Rees,

1992) on iron-sulfur proteins and by L. Que and A. True on diiron-oxo clusters (Que and True, 1990). These reviews contain extensive citations to the original literature and tables that list many properties of interest. Moreover, the article by Howard and Rees contains the relevant references to crystallographic work as well as stereo views of the protein structures.

II. IRON-SULFUR CLUSTERS

A. Rubredoxins

Although not strictly an iron-sulfur protein (it lacks labile sulfide), rubredoxin contains an iron atom in a tetrahedral coordination environment of sulfur atoms provided by four cysteine residues (see Fig. 1A), thus providing a model for the coordination environment found in the clusters of interest (Howard and Rees, 1992; Eaton and Lovenberg, 1973). The 1.2 Å X-ray structure of the *Clostridium pasteurianum* rubredoxin (M = 6.8 kg mol^{-1}) has revealed a distorted tetrahedral environment of sulfur ligands provided by a relatively conserved binding motif formed from two copies of $Cys(Xaa)_2Cys$ (Watenpaugh et al., 1980). The distortion from tetrahedral symmetry is essentially matched by synthetic models complexes suggesting that the geometry is controlled, in large measure, by coordination chemistry rather than by protein conformational constraints.

While rubredoxins are among the most thoroughly studied of metalloproteins, their proposed function as electron carriers has not been well established. The best evidence for this function has been obtained for the *Pseudomonas oleovorans* rubredoxin, which acts as an intermediate electron carrier in the ω-hydroxylation of fatty acids (Lode and Coon, 1971). As an electron carrier, the iron site of rubredoxin reversibly cycles between the Fe^{3+} and Fe^{2+} states. The protein environment modulates the redox potential of these proteins (E°' = +20 mV to -60 mV relative to NHE). The optical spectrum of ferric

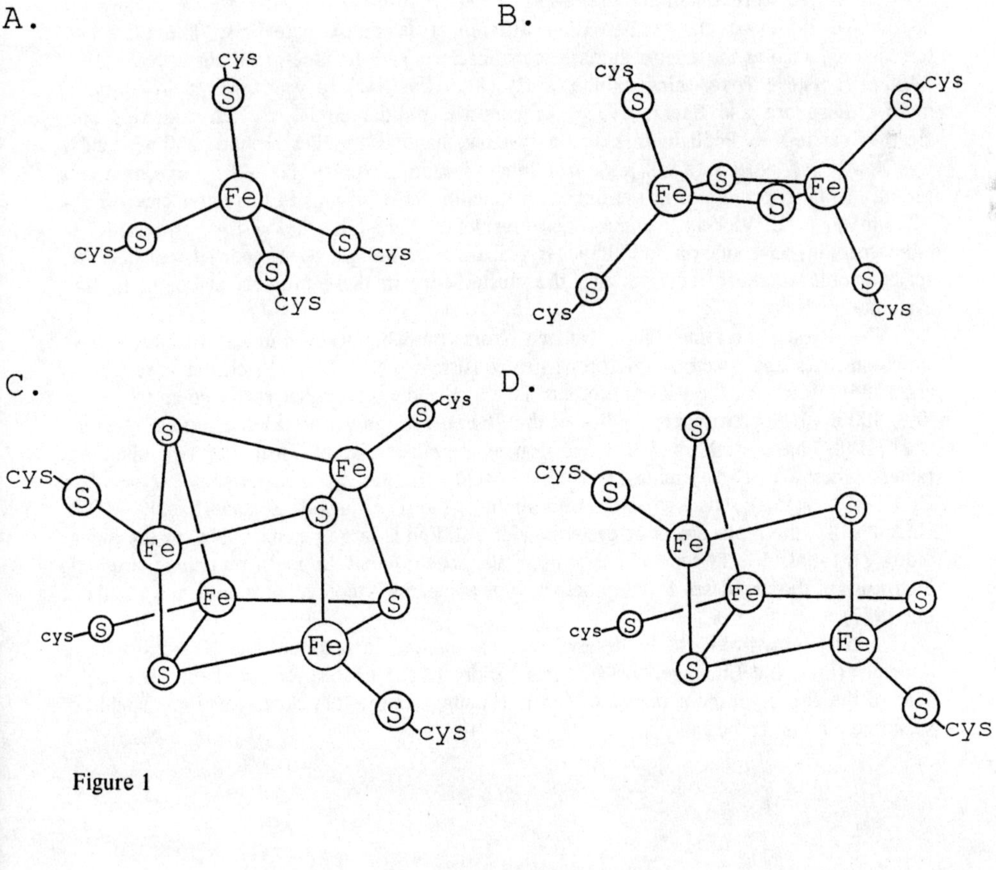

Figure 1

rubredoxin is characterized by strong sulfur to iron charge transfer bands between 400 nm and 500 nm ($\varepsilon \approx 10\ mM^{-1}\ cm^{-1}$). Rubredoxins have been studied extensively with a variety of spectroscopic techniques, and the two stable oxidation states serve as useful spectroscopic benchmarks for the study of iron environments of iron-sulfur clusters.

B. [2Fe-2S] Ferredoxins

Proteins containing a cluster with a [2Fe-2S] core (see Fig. 1B), termed ferredoxins, serve as electron carriers in many systems (Cammack, 1992; Howard and Rees, 1992). The [2Fe-2S] ferredoxins were first identified in the chloroplast of higher plants. About 20 years ago, a variety of spectroscopic techniques including EPR, ENDOR and Mössbauer spectroscopies (Sands and Dunham, 1975; Palmer, 1973) were used to delineate the basic features of the [2Fe-2S] cluster core. Later, the X-ray structures of the ferredoxins from *Spirulina platensis* (Fukuyama et al., 1981) and *Aphonothece sacrum* (Tsukihara et al., 1990) were reported.

The iron atoms in the [2Fe-2S] ferredoxins are linked by two sulfide bridges and separated by ~2.7 Å. The sulfide bridges provide an efficient pathway for superexchange (antiferromagnetic coupling; see (Palmer, 1973)) which dominates the spectroscopic features of the [2Fe-2S] clusters. Thus, although each of the two Fe^{3+} sites of the oxidized $[2Fe\text{-}2S]^{2+}$ cluster are high-spin (d^5, 5 unpaired spins, $S = 5/2$), the cluster has a ground state with $S = 0$. Upon reduction of the cluster, one iron site becomes high-spin Fe^{2+} (d^6, 4 unpaired spins, $S = 2$). Antiferromagnetic coupling between the Fe^{3+} and Fe^{2+} sites yields a ground state with $S = 1/2$. This state yields the prominent "$g = 1.94$" EPR signal (Sands and Dunham, 1975; Bertrand et al., 1991) ($g_x = g_y = 1.94$, $g_z = 2.02$ for putidaredoxin from *Pseudomonas putida*). It is noteworthy that the electron which enters the $[2Fe\text{-}2S]^{2+}$ cluster upon reduction is localized on one iron site (a fact established by Mössbauer spectroscopy) rather than being shared by both sites (Münck et al., 1972). This trapped valence Fe^{2+} site is a characteristic feature of $[2Fe\text{-}2S]^{1+}$ ferredoxins. In contrast, the [4Fe-4S] clusters exhibit extensive electron delocalization (see II.C. below).

The ferredoxins containing a [2Fe-2S] cluster coordinated by four cysteine residues can be divided into two classes based on differences in redox potentials. The plant-type ferredoxins, such as those from spinach, cyanobacteria and the blue-green algae, have E°' ≈ -400 mV. These ferredoxins exhibit a $Cys(Xaa)_2Cys(Xaa)_{7\text{-}9}Cys(Xaa)_3Cys$ iron-binding motif (Matsubara and Saeki, 1992). In contrast, putidaredoxin and adrenodoxin, both electron carriers in P450 hydroxylation systems, have E°' values around -250 mV and a $Cys(Xaa)_4Cys(Xaa)_2Cys(Xaa)_{29}Cys$ iron-binding motif. Human [2Fe-2S] ferredoxin , a recently isolated protein of as yet unknown function, also belongs to this latter class (E°' ≈ -270 mV; L. E. Vickery, unpublished results). Although the X-ray structures of putidaredoxin, adrenodoxin and human ferredoxin have not yet been reported, considerable spectroscopic evidence indicates that the cluster core in these proteins is bound to four cysteines.

The Rieske proteins, first isolated from mitochondral complex III, also have stoichiometries and spectroscopic properties consistent with a [2Fe-2S] cluster core (Fee et al., 1984). However, the Rieske proteins have a considerable higher redox potential (E°' ≈ 0 to 300 mV). Spectroscopic studies of the *Thermus thermophilus* Rieske protein (Gurbiel et al., 1989) have established that one iron of the cluster is coordinated to two histidines rather than two cysteines. Amino acid sequence comparisons reveal a $Cys(Xaa)_2Cys(Xaa)_{16}His(Xaa)_2His$ iron-binding motif (Matsubara and Saeki, 1992). Undoubtedly, the replacement of cysteine with histidine is largely responsible for the higher redox potential. Rieske-type clusters have also been found in multicomponent aromatic hydrocarbon dioxygenases in conjunction with a separate, mononuclear iron site (Kuila et al., 1987).

Spinach dihydroxyacid hydratase contains another variation of the [2Fe-2S] cluster structure (Flint and Emptage, 1988). Perturbations of the EPR spectrum of the $[2Fe\text{-}2S]^{1+}$ state of the cluster in the presence of substrates suggests that the cluster may be capable of substrate binding and catalysis.

C. [4Fe-4S] clusters in electron carriers and enzymes

Clusters with a [4Fe-4S] core (see Fig. 1C) are found in a large variety of proteins. The main function of [4Fe-4S] ferredoxins is intra- and inter-protein electron transport. In virtually all instances (with a few notable exceptions), the [4Fe-4S] cluster has a tetracysteinate coordination. However, changes in the protein environment adjacent to the cluster apparently allow the redox potentials of the [4Fe-4S] core to span a potential range from -650 mV to +450 mV (Armstrong, 1992). It is customary to indicate the oxidation state of the cluster core by a superscript. Thus, the $[4Fe\text{-}4S]^{2+}$ cluster contains formally 2 Fe^{2+}, 2 Fe^{3+} and 4 S^{2-} atoms; a 1-electron reduction yields the $[4Fe\text{-}4S]^{1+}$ state whereas a 1-electron oxidation yields the $[4Fe\text{-}4S]^{3+}$ state. All three states have been observed in biological systems, but a given protein employs only one redox couple. The majority of [4Fe-4S] proteins used as electron carriers in metabolic reactions employ the $1^+/2^+$ couple at potentials from ca. 0 to nearly -650 mV. The High-Potential Iron-Sulfur Proteins (HiPIP) found in purple sulfur bacteria and some other species shuttle between the +2 and +3 states at potentials ranging from ca +50 to +450 mV. Detailed comparisons of resonance Raman and crystallographic data have shown that the lower redox state is stabilized by a greater number (as many as five) of N-H....H hydrogen bonds, allowing dispersal of negative charge in the electron-rich +1 state (Backes et al., 1991).

The local spins of the four iron sites of all [4Fe-4S] clusters are coupled by antiferromagnetic and double-exchange interactions. Thus, the observed electronic spin refers to the cluster as a whole. The ground state of the $[4Fe\text{-}4S]^{2+}$ cluster has spin $S = 0$, resulting in an EPR-inactive state, while the +1 and +3 states are half-integer spin paramagnetic states which generally exhibit EPR signals. The ground state of oxidized HiPIP ($[4Fe\text{-}4S]^{3+}$) has $S = 1/2$ and yields a characteristic EPR signal near $g = 2$ with $g_{av} = (g_x + g_y + g_z)/3$ greater than 2.0. In contrast, most other [4Fe-4S] proteins stabilize the $[4Fe\text{-}4S]^{1+}$ state, which has $S = 1/2$ and EPR g-values near $g = 2$ but with $g$av less than 2.0, for example $g_x = 1.88$, $g_y = 1.94$, $g_z = 2.06$. EPR spectroscopy, which led to the discovery of iron-sulfur proteins, is undoubtedly the most useful spectroscopic tool for studying the $[4Fe\text{-}4S]^{1+}$ state in purified proteins as well as cell extracts and tissues. Although the g-values of the $[4Fe\text{-}4S]^{1+}$ and $[2Fe\text{-}2S]^{1+}$ states are quite similar, the temperature dependence of the EPR signals can give a rough indication of the cluster type. Thus, signals of $[4Fe\text{-}4S]^{1+}$ are generally not detectable above 40 K, whereas most $[2Fe\text{-}2S]^{1+}$ clusters are detectable at temperatures above 70 K (for EPR information, consult Hagen (1992) or Palmer (1973)).

D. Aconitase

Until 1983, it was widely believed that electron transport was the only function of the [4Fe-4S] clusters. However, by combination of biochemical studies with Mössbauer spectroscopy, it has been demonstrated that active bovine heart aconitase contains a site differentiated [4Fe-4S] cluster (Kent et al., 1982) and that substrates and inhibitors bind directly to one iron site of the cluster (designated Fe_a in Fig. 2) (Kent et al., 1985). A proposal for the structure of the active site cluster of the mitochondrial aconitase, as deduced by catalytic studies, Mössbauer, EPR, and ENDOR spectroscopies, and as refined by X-ray crystallography, is shown in Figure 2 (Kennedy and Stout, 1992). The as-isolated, inactive aconitase contains a [3Fe-4S] cluster (see Fig. 1D) that is converted to the catalytically active [4Fe-4S] form in the presence of Fe^{2+} and reducing agents. Under these conditions, Fe_a is incorporated into a vacant cluster site to form a cubane structure. In the absence of substrates, a solvent-derived hydroxyl moiety acts as the fourth ligand to provide a distorted tetrahedral coordination around Fe_a. Upon complexation with substrates, Fe_a is converted into an octahedral site, with three exogenous coordinations to Fe_a provided by the solvent-derived hydroxyl (which becomes protonated to form a coordinated water molecule), the substrate hydroxyl, and the α- or β-carboxylate oxygen of either isocitrate or citrate. Catalytic and site-directed mutagenesis experiments in combination with X-ray crystallographic analysis implicate Ser-642 as the enzymic base utilized for proton abstraction during the catalytic reaction. Recently obtained evidence also indicates that site differentiated iron-sulfur clusters may be involved in the coordinated posttranscriptional regulation of gene expression involved in iron metabolism (Haile et al., 1992; Constable et

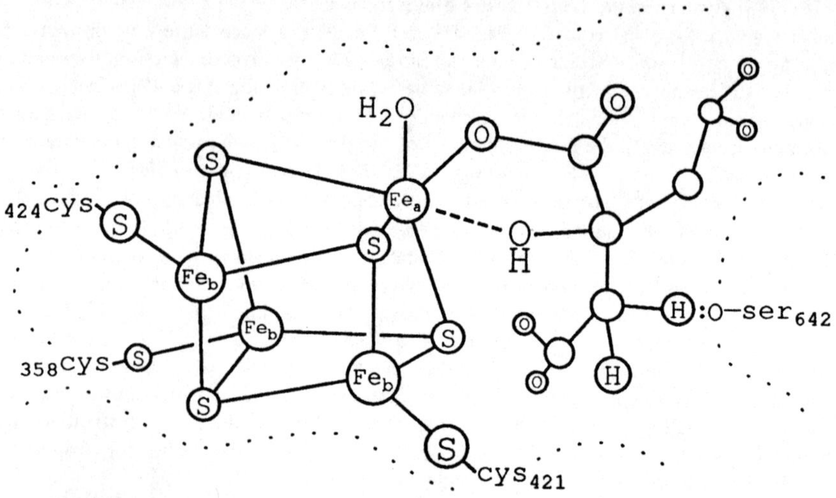

Figure 2

al., 1992). Examination of the deduced primary amino acid sequence for the human iron regulatory element binding protein has revealed a high degree of homology for residues corresponding both to iron cluster ligands and to essential active site catalytic residues of the mitochondrial aconitase (Rouault et al., 1991). This mRNA binding protein also exhibits aconitase activity (Kaptain et al., 1991).

E. [3Fe-4S] Clusters

Clusters with [3Fe-4S] cores were discovered in 1979 by application of Mössbauer and EPR spectroscopy to a ferredoxin from *Azotobacter vinelandii* (Emptage et al., 1980). Subsequent work by Beinert and collaborators showed that the inactive form of aconitase contained a [3Fe-4S] cluster which could be converted into a [4Fe-4S] cluster when the enzyme was activated in the presence of Fe^{2+} (Kent et al., 1982), suggesting that [3Fe-4S] clusters were structurally related to the [4Fe-4S] cubanes. This suggestion was subsequently confirmed by crystallographic studies of aconitase (Robbins and Stout, 1989), *Desulfovibrio gigas* ferredoxin II (Kissinger et al., 1991) and *Azotobacter vinelandii* ferredoxin (Stout, 1988). While the [3Fe-4S] cluster of aconitase is most likely an artifact reflecting a cluster site not protected against oxidation by a protein ligand, many other electron carriers and enzymes contain [3Fe-4S] clusters. Johnson et al. (1988) have shown that fumarate reductase contains one [2Fe-2S], one [3Fe-4S] and one [4Fe-4S] cluster and that the [3Fe-4S] cluster undergoes redox reactions during catalytic turnover. Whether the [3Fe-4S] clusters of other multicluster enzymes such as the Ni-hydrogenases (Adams, 1992; Cammack, 1988) function as electron carriers or participate in substrate binding remains to be established. In some cases, [3Fe-4S] clusters can be converted into [4Fe-4S] clusters by chemical manipulation. Following the work of Moura et al. (1986), who showed that Co^{2+} can be incorporated into the vacant site of the $[3Fe\text{-}4S]^0$ cluster of *Desulfovibrio gigas* Ferredoxin II to yield $[CoFe_3S_4]^{1+}$, other metals such as Zn, Cd, Ga, Ni, and Tl have been incorporated into the [3Fe-4S] cluster core. The review articles by R. H. Holm, F. A. Armstrong, and M. W. W. Adams in volume 38 of Advances in Inorganic Chemistry provide additional information about mixed-metal clusters. Since recent crystallographic evidence indicates that the cofactor of nitrogenase contains a $MoFe_3S_4$ fragment (Kim and Rees, 1992), it is intriguing to speculate whether [3Fe-4S] structures may serve as precursors to as yet unknown clusters.

F. Other iron-sulfur clusters

A variety of complex iron-sulfur clusters are utilized for specialized purposes, generally related to the bacterial metabolism of small molecules, that are not obviously related to human biology (Cammack, 1992). These clusters include the cofactor centers ($MoFe_7S_8$ plus homocitrate) and P-clusters (two bridged [4Fe-4S] clusters) of nitrogenase (Kim and Rees, 1992), the H-clusters of the Fe-hydrogenases (Adams, 1992), the polyferredoxins of methanogenic bacteria (Reeve et al., 1989), and a [6Fe-6S] cluster from *Desulfovibrio vulgaris* (Pierik et al., 1992; Moura et al., 1992). All Ni-containing hydrogenases also contain at least one [4Fe-4S] cluster (Adams, 1992; Cammack, 1988). In addition, the [4Fe-4S] clusters occur in a variety of electron carriers and enzymes in conjunction with other prosthetic groups such as [2Fe-2S] and [3Fe-4S] clusters, FAD, FMN or pterin organic cofactors, mononuclear metal sites, or in combination with iron-sulfur clusters of as yet unknown stoichiometry (such as the Fe-hydrogenases or the CO dehydrogenases). A most remarkable chromophore has been observed in *Escherichia coli* sulfite reductase and spinach nitrite reductase, where a [4Fe-4S] cluster is covalently linked to the Fe of a siroheme (McRee et al., 1986; Christner et al., 1984). Glutamine 5-phosphoribosyl-1-pyrophosphate amidotransferase (Vollmer et al., 1983) and endonuclease III (Cunningham et al., 1989) each contain one [4Fe-4S] cluster with an as yet unspecified function. A recent X-ray study of endonuclease III suggests a structural role for the cluster (Kuo et al., 1992).

III. THE BINUCLEAR CLUSTERS OF IRON-OXO PROTEINS

Two examples of the diiron-oxo cluster structure are shown in Figure 3. This type of cluster is a commonly occurring, catalytically versatile structure found in hemerythrin, ribonucleotide reductase, the methane monooxygenase (MMO) hydroxylase, the purple acid phosphatases including uteroferrin, rubrerythrin (see Que and True (1990) for a review), and the plant stearoyl ACP Δ9 desaturase (Fox et al., 1993b). In contrast to the [2Fe-2S]

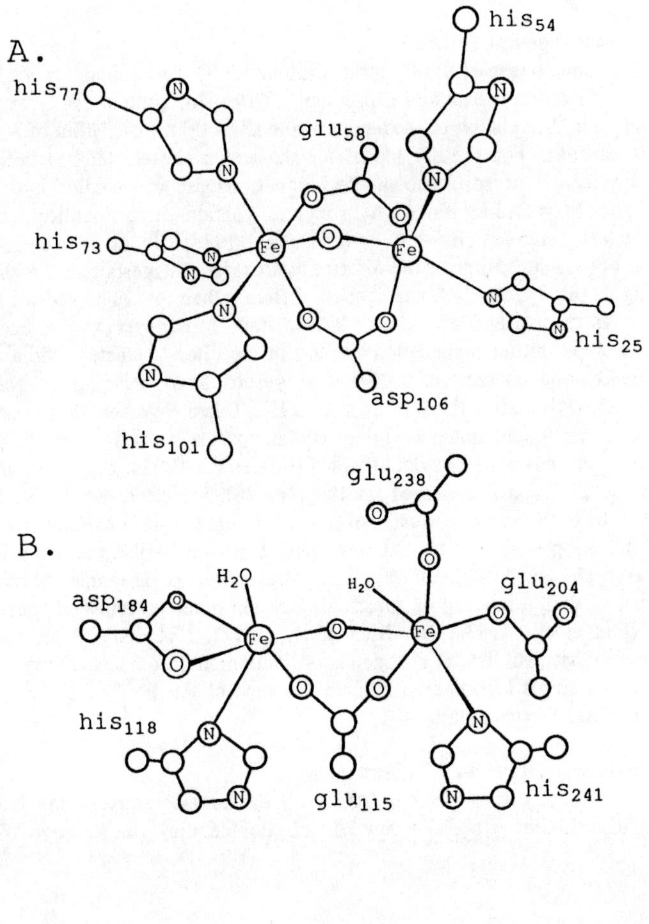

Figure 3

clusters, the diiron-oxo clusters can be observed in three stable redox states commonly designated as the diferric, mixed-valence, and diferrous states. Spectroscopic studies show that the iron atoms are spin-coupled in each of these states. The diiron-oxo clusters exhibit greater variability in ligand composition, coordination geometry, and chemical nature of the bridging atomic species than the iron-sulfur clusters, which likely contributes to the observed catalytic versatility.

A. Hemerythrin and ribonucleotide reductase

Hemerythrin is an O_2-transport protein obtained from marine worms. The structure of the diiron-oxo cluster of hemerythrin is shown in Figure 3A (Stenkamp et al., 1984). The diiron-oxo core is triply-bridged by a single oxygen atom derived from water and by two carboxylate residues. The remainder of the ligands are histidines. One iron site of the diiron cluster is 6-coordinate, while the other site is 5-coordinate. The open coordination site is utilized for reversible binding during O_2-transport.

The stucture of the diiron-oxo cluster of ribonucleotide reductase is shown in Figure 3B (Nordlund et al., 1990). This cluster catalyzes the O_2-dependent formation of an essential tyrosyl radical required for formation of deoxyribonucleotides. In ribonucleotide reductase, the diiron-oxo cluster is doubly-bridged by a single oxygen atom also derived from water and by a single carboxylate residue. The remainder of the protein-derived iron ligands are histidine, aspartate and glutamate. Both iron atoms have an open coordination site occupied by a water molecule.

For the diiron-oxo clusters in hemeryrthrin and ribonucleotide reductase, characteristic optical absorption features from 300 nm to 350 nm ($\varepsilon \approx 5\ mM^{-1}\ cm^{-1}$) arise from charge transfer interactions between the doubly substituted bridging oxygen atom and iron. The doubly substituted oxygen atom bridge also gives rise to a strongly antiferromagnetically coupled cluster in the diferric state.

B. Methane monooxygenase (MMO)

The hydroxylase component of the bacterial protein MMO also contains a variation of the diiron-oxo cluster structure (Fox and Lipscomb, 1990). This protein catalyzes the NADH- and O_2-dependent hydroxylation of methane and the adventitious oxidation of a wide variety of other hydrocarbons. The present model for the active site cluster has been assembled through the application of spectroscopic techniques. Mössbauer studies indicate that the dimeric ($(\alpha\beta\gamma)_2$; $M = 245\ kg\ mol^{-1}$) hydroxylase contains two identical, independent, antiferromagnetically coupled cluster units (Fox et al., 1988; 1993a). However, the absence of a detectable optical spectrum for the diferric hydroxylase suggests that the chemical state of the bridging atomic species is fundamentally different than that observed for hemerythrin or ribonucleotide reductase (Fox et al., 1989). Both Mössbauer and integer-spin EPR studies indicate weak antiferromagnetic coupling in the diferric cluster, while no evidence for a short Fe-O bond characteristic of a doubly substituted bridging oxygen atom was obtained from EXAFS studies (DeWitt et al., 1991). Taken together, these results suggest the presence of a triply substituted bridging oxygen species, such as could be provided by hydroxo, alkoxo, or end-on carboxylato ligands (Fox et al., 1993a; DeWitt et al., 1991). In distinction to the EPR signal observed for the $[2Fe\text{-}2S]^{1+}$ cluster, which has at least one g-value above 2.0, the diiron-oxo cluster of the MMO hydroxylase exhibits a characteristic EPR signal with all g-values below 2.0 (Fox and Lipscomb, 1990) in the mixed-valence state. EPR, ^1H-, ^2H-, and ^{14}N-ENDOR and Mössbauer studies show that the mixed-valence cluster is sensitive to the binding of substrates and the other protein components required for catalysis (Fox et al., 1993a; Hendrich et al., 1992). ^{14}N-ENDOR studies have also provided evidence that two distinct nitrogen coordination environments are present, possibly corresponding to a single histidine coordinated to each of the Fe^{2+} and Fe^{3+} sites in the mixed-valence MMOH hydroxylase.

C. Stearoyl acyl carrier protein Δ9 desaturase

The plant stearoyl acyl carrier protein (ACP) Δ9 desaturase catalyzes the NADPH- and O_2-dependent insertion of a position specific *cis* double bond in stearoyl ACP to form

A. Ribonucleotide Reductase

-SETIHS——DEALHL-

118 241

B. MMO hydroxylase α subunit

-DELRHM——DEIRHT-

147 246

C. Stearoyl ACP Δ9 desaturase

-EENRHG——DEKRHE-

179 265

Figure 4

oleyl ACP. Recently, this proteins has been shown to contain a catalytically essential diiron-oxo cluster (Fox et al., 1993b). Combined biochemical and spectroscopic analysis shows that both subunits of the holoprotein (α_2; M = 83 kg mol^{-1}) contain an identical diiron-oxo cluster. An X-ray crystal structure determination is in progress (Schneider et al., 1992).

D. Iron-binding site motifs for O_2-activating diiron-oxo clusters

As outlined above, three classes of proteins containing diiron-oxo clusters that are capable of catalytic O_2-activation have been identified: ribonucleotide reductase (Elgren et al., 1991); MMO hydroxylase (Andersson et al., 1991); and stearoyl ACP Δ9 desaturase (Fox et al., 1993b). For each of these three catalytically distinct enzyme classes, only the diferrous form of the diiron-oxo cluster is kinetically competent to interact with O_2. Figure 4 shows the amino acid sequence near to the crystallographically defined iron ligands for ribonucleotide reductase (Nordlund et al., 1990) and the comparable homologous sequences observed in the MMO hydroxylase (Cardy et al., 1991) and the stearoyl ACP Δ9 desaturase (Shanklin and Somerville, 1991). Several other non-heme oxygenase proteins also exhibit these iron-binding sequences, suggesting an iron-binding motif for the diiron-oxo clusters.

REFERENCES

Adams, M.W.W. (1992). *Adv. Inorg. Chem.*, *38*: 341.

Andersson, K.K., Froland, W.A., Lee, S.-K., and Lipscomb, J.D. (1991). *New. J. Chem.*, *15*: 411.

Armstrong, F.A. (1992). *Adv. Inorg. Chem.*, *38*: 117.

Backes, G., Mino, Y., Loehr, T.M., Meyer, T.E., Cusanovich, M.A., Sweeney, W.V., Adman, E.T., and Sanders-Loehr, J. (1991). *J. Amer. Chem. Soc.*, *113*: 2055.

Bertrand, P., Guigliarelli, B., and More, C. (1991) *New J. Chem.*, *15*: 445.

Cammack, R. (1988). *Adv. Inorg. Chem.*, *32*: 297.

Cammack, R. (1992). *Adv. Inorg. Chem.*, *38*: 281.

Cardy, D.L.N., Laidler, V., Salmond, G.P.C., and Murrell, J.C. (1991). *Mol. Microbiol.*, *5*: 335.

Christner, J.A., Münck, E., Kent, T.A., Janick, P.A., Salerno, J.C., and Siegel, L.M. (1984). *J. Amer. Chem. Soc.*, *106*: 6786.

Constable, A., Quick, S., Gray, N.K., and Hentze, M. (1992). *Proc. Nat. Acad. Sci. USA*, *89*: 4554.

Cunningham, R.P., Asahara, H., Bank, J.F., Scholes, C.P., Salerno, J.C., Surerus, K.K., Münck, E., McCracken, J., Peisach, J., and Emptage, M.H. (1989). *Biochemistry*, *28*: 4450.

DeWitt, J.G., Bentsen, J.G., Rosenzweig, A.C., Hedman, B., Green, J., Pilkington, S., Papaefthymiou, G.C., Dalton, H., Hodgson, K.O., and Lippard, S.J. (1991). *J. Amer. Chem. Soc.*, *113*: 9219.

Eaton, W.A., and Lovenberg, W. (1973). *Iron-sulfur Proteins*, Vol. II (W. Lovenberg, ed.), Academic Press, New York, p. 131.

Elgren, T.E., Lynch, J.B., Juarez-Garcia, C., Münck, E., Sjöberg, B.-M., and Que, L., Jr. (1991). *J. Biol. Chem.*, *266*: 19265.

Emptage, M.H., Kent, T.A., Huynh, B.H., Rawlings, J., Orme-Johnson, W.H., and Münck, E. (1980). *J. Biol. Chem.*, *255*: 1793.

Fee, J.A., Findling, K.L., Yoshida, T., Hille, R., Tarr, G.E., Hearshen, D.O., Dunham, W.R., Day, E.P., Kent, T.A., and Münck, E. (1984). *J. Biol. Chem.*, *259*: 124.

Flint, D.H., and Emptage, M.H. (1988). *J. Biol. Chem.*, *263*: 3558.

Fox, B.G., and Lipscomb, J.D. (1990). *Biological Oxidation Systems* (C.C. Reddy, G.A. Hamilton and K.M. Madyastha, eds.), Academic Press, New York, p. 367.

Fox, B.G., Surerus, K.K., Münck, E., and Lipscomb, J.D. (1988). *J. Biol. Chem.*, *263*: 10553.

Fox, B.G., Froland, W.A., Dege, J.E., and Lipscomb, J.D. (1989). *J. Biol. Chem.*, *264*: 10023.

Fox, B.G., Hendrich, M.P., Surerus, K.K., Andersson, K.K., Froland, W.A., Lipscomb, J.D., and Münck, E. (1993a). *J. Amer. Chem. Soc.*, *115*: 3688.

Fox, B.G., Shanklin, J., Münck, E., and Somerville, C. (1993b). *Proc. Nat. Acad. Sci. USA*, *90*:2486.

Fukuyama, K., Hase, T., Matsumoto, S., Tsukihara, T., Katsube, Y., Tanaka, N., Kakudo, Y., Wada, K., and Matsubara, H. (1981). *Nature*, *286*: 522.

Gurbiel, R.J., Batie, C.J., Sivaraja, M., True, A.E., Fee, J.A., Hoffman, B.M., and Ballou, D.P. (1989). *Biochemistry*, *28*: 4861.

Hagen, W.R. (1992). *Adv. Inorg. Chem.*, *38*: 165.

Haile, D.J., Rouault, T.A., Tang, C.K., Chin, J., Harford, J.B., and Klausner, R.D. (1992). *Proc. Nat. Acad. Sci. USA*, *89*: 7536.

Hendrich, M.P., Fox, B.G., Andersson, K.K., Debrunner, P.G., and Lipscomb, J.D. (1992). *J. Biol. Chem.*, *267*: 261.

Howard, J.B., and Rees, D.C. (1992). *Adv. Protein Chem.*, *42*: 299.

Johnson, M.K., Kowal, A.T., Morningstar, J.E., Oliver, M.E., Whittaker, K., Gunsalus, R.P., Ackrell, B.A.C., and Cecchini, G. (1988). *J. Biol. Chem.*, *263*: 14732.

Kaptain, S., Downey, W.E., Tang, C., Philpott, C., Haile, D., Orloff, D.G., Harford, J.B., Rouault, T.A., and Klausner, R.D. (1991). *Proc. Nat. Acad. Sci. USA*, *88*: 10109.

Kennedy, M.C., and Stout, C.D. (1992). *Adv. Inorg. Chem.*, *38*: 323.

Kent, T.A., Dreyer, J.-L., Kennedy, M.C., Huynh, B.H., Emptage, M.H., Beinert, H., and Münck, E. (1982). *Proc. Nat. Acad. Sci. USA*, *79*: 1096.

Kent, T.A., Emptage, M.H., Merkle, H., Kennedy, M.C., Beinert, H., and Münck, E. (1985). *J. Biol. Chem.*, *260*: 6871.

Kim, J., and Rees, D.C. (1992). *Science*, *257*: 1677.

Kissinger, C.R., Sieker, L.C., Adman, E.T., and Jensen, L.H. (1991). *J. Mol. Biol.*, *219*: 693.

Kuila, D., Fee, J.A., Schoonover, J.R., Woodruff, W.H., Batie, C.J., and Ballou, D.P. (1987). *J. Amer. Chem. Soc.*, *109*: 1559.

Kuo, C.-F., McRee, D.E., Fisher, C.L., O'Handley, S.F., Cunningham, R.P., and Tainer, J.A. (1992). *Science*, *258*: 434.

Lode, E.T., and Coon, M.J. (1971). *J. Biol. Chem.*, *246*: 791.

Matsubara, H., and Saeki, K. (1992). *Adv. Inorg. Chem.*, *38*: 223.

McRee, D.E., Richardson, D.C., Richardson, J.S., and Siegel, L.M. (1986). *J. Biol. Chem.*, *261*: 10277.

Moura, I., Moura, J.J.G., Münck, E., Papaefthymiou, V., and LeGall, J. (1986). *J. Amer. Chem. Soc.*, *108*: 349.

Moura, I.,Tavares, P., Ravi, N.,Huynh, B.H., Liu, M.-Y., and LeGall, J.(1992) *J.Biol. Chem.*, *269*: 4489.

Münck, E., Debrunner, P.G., Tsibris, J.C.M., and Gunsalus, I.C. (1972). *Biochemistry*, *11*: 855.

Nordlund, P., Sjöberg, B.-M., and Eklund, H. (1990). *Nature*, *345*: 593.

Palmer, G. (1973). *Iron-sulfur Proteins*, Vol. II (W. Lovenberg, ed.), Academic Press, New York, p. 285.

Pierik, A.J., Hagen, W.R., Dunham, W.R., and Sands, R.H. (1992). *Eur. J. Biochem.*, *206*: 705.

Que, L., Jr., and True, A.E. (1990). *Progress in Inorganic Chemistry: Bioinorganic Chemistry* (S.J. Lippard, ed.), John Wiley and Sons, New York, p. 97.

Reeve, J.N., Beckler, G.S., Cram, D.S., Hamilton, P.S., Brown, J.W., Krzycki, J.A., Kolodziej, A.F., Alex, L., Orme-Johnson, W.H., and Walsh, C.T. (1989). *Proc. Nat. Acad. Sci. USA*, *86*: 3031.

Robbins, A.H., and Stout, C.D. (1989). *Proc. Nat. Acad. Sci. USA*, *86*: 3639.

Rouault, T.A., Stout, C.D., Kaptain, S., Harford, J.B., and Klausner, R.D. (1991). *Cell*, *64*: 881.

Sands, R.H., and Dunham, W.R. (1975). *Q. Rev. Biophys.*, *7*: 443.

Schneider, G., Lindqvist, Y., Shanklin, J., and Somerville, C. (1992). *J. Mol. Biol.*, *225*: 561.

Shanklin, J., and Somerville, C. (1991). *Proc. Nat. Acad. Sci. USA*, *88*: 2510.

Stenkamp, R.E., Sieker, L.C., and Jensen, L.H. (1984). *J. Amer. Chem. Soc.*, *106*: 618.

Stout, C.D. (1988). *J. Biol. Chem.*, *263*: 9256.

Tsukihara, T., Fukuyama, K., Mizushima, M., Harioka, T., Kusunoki, M., Katsube, Y., Hase, T., and Matsubara, H. (1990). *J. Mol. Biol.*, *216*: 399.

Vollmer, S.J., Switzer, R.L., and Debrunner, P.G. (1983). *J. Biol. Chem.*, *258*: 14284.

Watenpaugh, K.D., Sieker, L.C., and Jensen, L.H. (1980). *J. Mol. Biol.*, *138*: 615.

METALLOPROTEINS AND DIOXYGEN METABOLISM

by

Indrajit Sanyal and Kenneth D. Karlin

Department of Chemistry,
Johns Hopkins University,
Baltimore, MD 21218
U.S.A.

I. INTRODUCTION

Aerobic organisms utilize dioxygen (molecular oxygen; O_2) as their primary source of energy, by capturing and storing that potential available from the reduction of O_2 to water, this being highly thermodynamically favorable, i.e. E° = 0.82 Volts vs NHE; 1 atm O_2 at pH 7 (Sawyer, 1991). The other use of O_2 is as a source of oxygen atoms, for use in biological substrate oxygenation reactions. Because ground state dioxygen is a triplet, it is rather inert (kinetically) to reaction with normally singlet ground state organic molecules; thus nature utilizes either a flavin or metal ion catalyst (e.g. iron or copper) to 'activate' O_2 (Hayaishi, 1974; Niederhoffer et al., 1984; Ingraham and Meyer, 1985). Here, we summarize basic information concerning the active-site chemistry for protein O_2-carriers, oxidases and oxygenases involved in dioxygen utilization/processing (Table 1). Enzymes processing reduced dioxygen species such as superoxide (O_2^-, e.g. Fe, Mn or Cu,Zn-superoxide dismutases) and hydrogen peroxide (e.g. peroxidases) are covered in other chapters.

II. REVERSIBLE O_2 CARRIERS

A. Hemoglobin and myoglobin

Hemoglobin (Hb) and myoglobin (Mb) act as the reversible dioxygen carrier proteins in blood and muscle tissue of vertebrates, respectively (Ingraham and Meyer, 1985). The dioxygen binding site in hemoglobin contains four Fe(II)-protoporphyrin IX complexes located in two α and two β polypeptide chains (Perutz, 1976). Myoglobin, on the other hand, has one Fe(II)-protoporphyrin IX unit buried inside a single polypeptide chain of 153 amino acids (globin). The crystal structure of deoxy-Hb shows a five-coordinate high-spin Fe(II) complex where the iron is 0.42 Å out of the porphyrin plane with an axially bound 'proximal' histidine group. The O_2 binds in a bent fashion (∠Fe-O-O = 115°) in oxy-Hb and resonance Raman studies (Barlow et al., 1973) are consistent with a Fe(III)-O_2^- ($\nu_{O\text{-}O}$ = 1105 cm^{-1}) formulation.

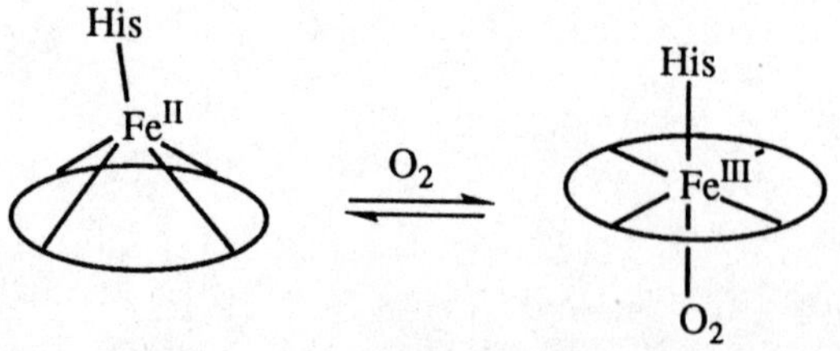

Scheme 1

Table 1

Dioxygen-metabolizing Metalloproteins

Heme iron	Non-heme iron	Copper
Dioxygen transport	$xM^{n+} + O_2 \rightleftharpoons (M)_x(O_2^{2-})$	
Hemoglobin Myoglobin	Hemerythrin Myohemerythrin	Hemocyanin
Monooxygenases	$XH_2 + O_2 \rightarrow XO + H_2O$ or $X + O_2 + DH_2 \rightarrow XO + H_2O + D$ (D = e^- donor)	
Cytochrome P-450	Methane-monooxygenase Fe-phenylalanine-hydroxylase	Tyrosinase Dopamine β-hydroxylase Cu-phenylalanine-hydroxylase Peptidylglycine α-amidating-monooxygenase Methane monooxygenase
Dioxygenases	$S + O_2 \rightarrow SO_2$	
Indoleamine 2,3-dioxygenase	Protocatechuate-3,4-dioxygenase Lipoxygenase	Quercetinase
Oxidases	$O_2 + 4e^- + 4H^+ \rightarrow 2H_2O$	
Cytochrome *c* oxidase		Laccase Ascorbate oxidase
	$XH_2 + O_2 \rightarrow X + H_2O_2$	Galactose oxidase Amine oxidase

A distal histidine present in both Hb or Mb stabilizes the Fe(III)-O_2^- moiety by forming a hydrogen bond. The reduced high-spin Fe(II) ion in deoxy-Hb and Mb has too big an ionic radius to fit inside the porphyrin and the iron atom is out of plane. Upon oxidation to Fe(III) the ionic radius is lessened and the iron atom moves in plane. This movement of iron into the porphyrin ring upon dioxygen binding causes the movement of the distal histidine which results in a change in the conformation of the other chain. This, in turn, increases the affinity of O_2 in the second chain. This cooperative effect gets transcribed to all the four subunit chains in Hb resulting in great enhancement of O_2-binding affinity (for a review, see Brunori et al., 1982).

B. Hemerythrin

The nonheme counterpart of hemoglobin is hemerythrin, a reversible O_2 carrier found in marine invertebrates (Que and True, 1990; Wilkins, 1992; Lippard, 1988). Typically, it is an octamer (M = 108 kg mol^{-1}) having one diiron site per subunit.

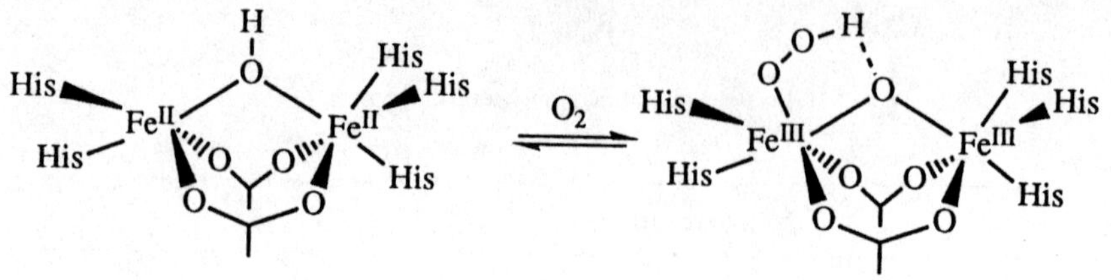

Scheme 2

The crystal structure of deoxy-hemerythrin (deoxy-Hr) shows a triply-bridged high-spin diiron(II) center (Fe...Fe = 3.32 Å) (Holmes et al., 1991). Other than two carboxylato bridges (Asp and Glu) there is also an asymetrically bridging hydroxyl group (Fe1-O = 1.88 Å, Fe2-O = 2.15 Å), ∠ Fe-O-Fe_{av} = 111°). One of the Fe(II) ions is five-coordinate with two histidine ligands, whereas the other is six-coordinate with three histidines. Deoxy-Hr binds O_2 reversibly by a two-electron oxidative addition process, generating a peroxo-diferric moiety. Resonance Raman (Shiemke et al., 1986) and crystallographic studies (Holmes et al., 1991) on oxy-Hr show that the bound dioxygen is present in a hydroperoxide form and is hydrogen bonded to the bridging oxo atom. Magnetic studies show that the two ferric atoms are weakly antiferromagnetically coupled (J = -77 cm^{-1}), presumably through the oxo bridge (Que and True, 1990). In the case of the diferric met-Hr form, the coupling is stronger (J = -130 cm^{-1}), mainly due to the absence of a hydrogen bond to the oxo bridge.

C. Hemocyanin

Hemocyanin (Hc) is a reversible O_2-carrier found in hemolymph of arthropods and molluscs (Solomon et al., 1992; Karlin and Gultneh, 1987). They are large multisubunit proteins. An early structure (crystallized at low pH) of the active site of deoxy-Hc contains a dicopper(I) center (Cu...Cu = 3.54 Å, X-ray crystallography) where each copper ion is bound to three imidazole ligands derived from side-chain histidines (Volbeda and Hol, 1989). One of the copper-His bonds (2.5 Å) is longer than the other two (1.96 Å), imparting a distorted T-shaped geometry. However, a new deoxy-hemocyanin structure displays a trigonal planar tri-coordination which is more typical for copper(I) ion, with Cu(I)...Cu(I) = 4.6 Å (Hazes et al., 1993). Deoxy-Hc reacts reversibly with O_2 by a two-electron oxidative addition process to generate a dicopper(II)-peroxo bridged species ($\nu_{O\text{-}O}$ ~ 750 cm^{-1}) (Freedman et al., 1976).

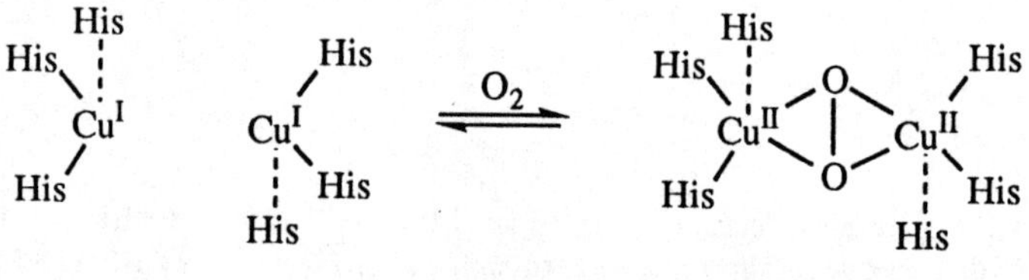

Scheme 3

A recent crystal structure of oxy-Hc by Magnus and Ton-That (1992) shows that the peroxo group is bridging in a side-on μ-η^2:η^2 configuration between the two coppers (Cu(II)...Cu(II) = 3.6 Å), a binding mode previously proposed by Blackburn et al. (1988) and Kitajima et al. (1992) based on model studies. The movement of copper ions accompanying the deoxy to oxy-Hc transformation, and the associated movement of histidyl ligands, are probably responsible for the "cooperative" dioxygen binding of the multisubunit protein (Hazes et al., 1993). Oxy-Hc is blue in color and has strong charge-transfer absorption bands at λ_{max} = 345 (ε, 20,000 $M^{-1}cm^{-1}$), 485 (seen in circular dichroism spectra) and 570 (ε, 1000 $M^{-1}cm^{-1}$) nm. The oxy site is also EPR silent and diamagnetic, reflecting strong magnetic coupling between the two copper(II) centers.

III. MONOOXYGENASES

A. Cytochrome P-450

Cytochrome P-450 enzymes are found in a variety of organisms and have a broad substrate specificity (Ortiz de Montellano, 1986). The bacterial camphor hydroxylase P-450_{CAM} has been studied by X-ray crystallography (Raag and Poulos, 1991) and is found to contain a heme iron with a cysteine sulfur as an axial ligand in the active site. A threonine residue is also in close proximity, and this is thought to stabilize an iron-peroxo complex (vide infra) by forming a hydrogen bond with the bound dioxygen (Gerber and Sligar, 1992; Raag and Poulos, 1991).

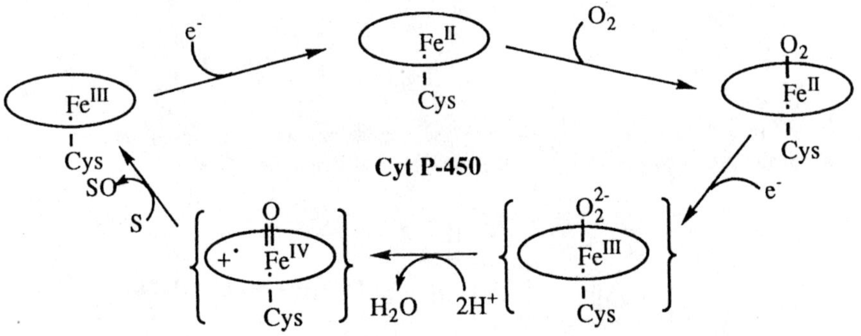

Scheme 4

Initially, the iron is in a low-spin ferric state; it becomes transformed to a high-spin state upon addition of the substrate. An electron is added (from NADPH) with concomitant reduction to a high-spin ferrous state. This binds dioxygen with addition of another electron to form a putative Fe(III)-peroxo species. A probable heterolytic O-O bond cleavage, accompanied by loss of water, generates a higher valent active iron-oxo ('ferryl') species which is responsible for the oxygen atom transfer to the substrate (Mansuy, 1987; Dawson, 1988; Groves and Watanabe, 1988).

B. Nonheme monooxygenases

1. *Methane monooxygenase*

Nonheme-iron containing methane monooxygenase from methanotropic bacteria catalyzes the hydroxylation of methane to methanol. It is a soluble enzyme and consists of three components (Dalton et al., 1990; Fox et al., 1991). The A component is a dimer, each subunit having a dinuclear iron site which contains the monooxygenase activity. The B and C components are the electron donor and transfer sites. EXAFS (Dewitt et al., 1991), EPR and Mössbauer spectroscopic studies (Fox et al., 1988) on the A site suggest an antiferromagnetically coupled high-spin diferric site with Fe...Fe = 3.4 Å. However, the coupling is weak suggesting a hydroxo or alkoxo bridge between two irons, unlike the oxo-bridge found in a related enzyme, ribonucleotide reductase (Stubbe, 1991; Nordlund et al., 1990). Very recently, the first X-ray structure of a methane monooxygenase enzyme (from *Methylococcus capsulatus* (Bath)) was solved, confirming a dinuclear iron coordination with considerable ligation from carboxylato groups (Rosenzweig et al., 1993).

MMO and cytochrome P-450 catalyze many similar reactions. Like cytochrome P-450, a high valent iron-oxo intermediate has been proposed for MMO. This species can abstract a H atom from CH_4 to generate a $CH_3^{\bullet}$ radical and recent spin-trapping experiments support this speculation (Deighton et al., 1991).

2. *Phenylalanine hydroxylase*

Mammalian aromatic amino acid hydroxylases use pteridine-dependent enzymes for activity.

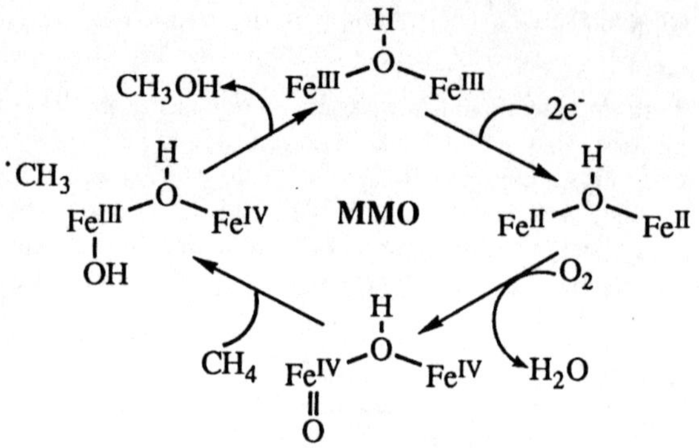

Scheme 5

One of those, phenylalanine hydroxylase, contains an iron atom along with a tetrahydrobiopterin cofactor in its active site (Dix and Benkovic, 1988).

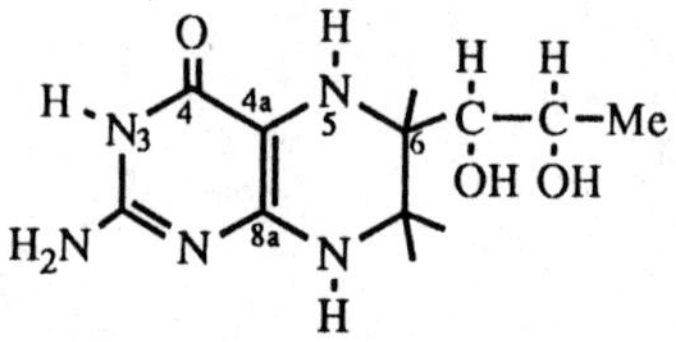

Scheme 6

This protein catalyzes the hydroxylation of phenylalanine to tyrosine. Initially in the catalytic step, pterin-hydroperoxide adduct (4a-OOH, based on flavin chemistry) formation has been proposed by Dix and Benkovic (1988). Substrate oxygenation is accompanied by cofactor oxidation to dihydropterin, which is reduced back to the tetrahydropterin by another enzyme, dihydropterin reductase. The function of the metal may be to bind O_2 (1:1 Fe(II)-O_2), or facilitate tetrahydropterin semiquinone and/or peroxy pterin-metal formation.

C. Copper monooxygenases

1. *Tyrosinase*

These proteins widely occur in bacteria, fungi, plants and mammals and are responsible for melanin pigment formation. Tyrosinase (Tyr) catalyzes the *o*-hydroxylation of phenols to catechols with further oxidation to quinones (Jolley et al., 1972; Wilcox et al., 1985). Tyrosinase has a dicopper(I) site in its deoxy form similar to hemocyanin. The protein binds O_2 to form a dicopper(II)-peroxo moiety which is directly accessible to substrates (Solomon et al., 1992). Spectroscopic similarities between oxy-Tyr and oxy-Hc leads to the proposal of a side-on peroxo bridging in oxy-Tyr. Based on theoretical calculations, Ross and Solomon (1991) have suggested that such a bond leads to a less negative (than usual) peroxide due to a stronger σ donor interaction with the coppers; this contributes to electrophilic activation. Also, there is some σ^* character present in the HOMO resulting in a weaker O-O bond. Both these factors are important for the substrate activation in tyrosinase.

2. *Dopamine β-hydroxylase (DβH) and peptidylglycine α-amidating monooxygenase (PAM)*

Dopamine β-hydroxylase is a Cu-dependent monooxygenase found in the mammalian adrenal medulla and it catalyzes the ascorbate and O_2-dependent stereospecific benzylic hydroxylation of dopamine to the neurotransmitter norepinephrine (Stewart and Klinman, 1988). Two coppers form the functional unit; one is a substrate and O_2 [or CO (Pettingil et al., 1991)] binding site, the other receives reducing equivalents from ascorbate. The primary ligands to the coppers are histidines but in the reduced form a methionine is also bound to one of the copper (Blackburn et al., 1991). It has been recently proposed by Blackburn et al. (1991) that there is a Cu-Met-His-Cu connection which is the primary electron transfer pathway from the electron-accepting copper center to the substrate/O_2-binding copper center.

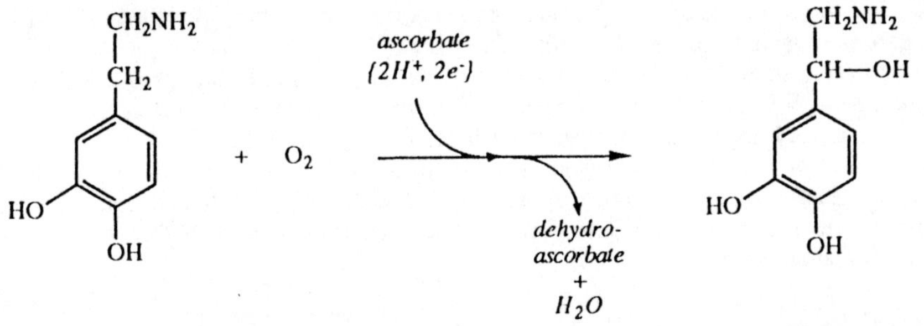

Scheme 7

Initially during a catalytic cycle, the two coppers are in fully reduced Cu(I) form. With O_2 and substrate binding, electron transfer to dioxygen generates a mononuclear "hot" Cu(II)-OOH species which hydroxylates the substrate through a homolytic O-O bond cleavage pathway with the possible involvement of a proximal tyrosine radical (Stewart and Klinman, 1989; Klinman et al., 1992).

Another Cu-ion, O_2- and ascorbate-dependent monooxygenase, peptidylglycine α-amidating monooxygenase (PAM), is isolated from mammalian tissues (Eipper et al., 1991; Merkler et al., 1992).

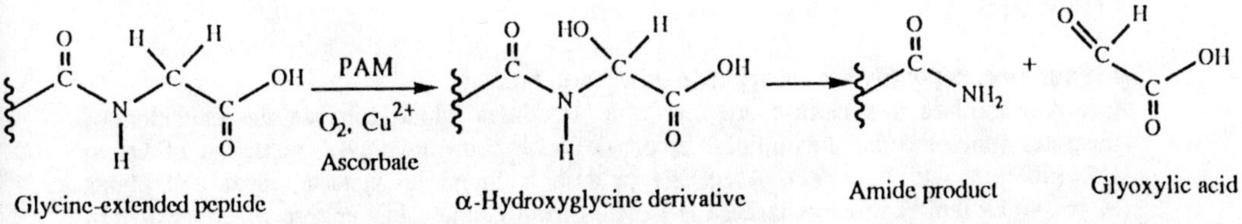

Scheme 8

The primary function of this enzyme is to produce α-amidated peptides from their glycine extended pro-hormone precursors. This enzyme is bifunctional and the first step involves the hydroxylation of C-terminal glycine with further N-dealkylation to give the products.

3. *Other copper monooxygenases*

Copper-phenylalanine hydroxylase (Cu-PAH) is a bacterial enzyme and contains one Cu(II) ion and a tetrahydrobiopterin cofactor in its active site (Pember et al., 1989). Recent EXAFS studies by Blackburn et al. (1992) suggest that the reduced form of Cu-PAH contains a three-coordinate copper with two histidine ligands and a possible pterin ligation through the 5-N position. Here, a similar mechanism as occurs in the iron-dependent enzyme has been proposed.

There is a membrane-bound form of methane monooxygenase (MMO) from methanotrophic bacteria that is a copper enzyme; in depth characterization of this enzyme has just begun. Based on recent spectroscopic and magnetic studies, Chan et al. (1992) have proposed a symmetrical tri-copper active site where the copper(II) ions (i.e. in the oxidized form) are ferromagnetically coupled.

IV. DIOXYGENASES

Dioxygenases are a major class of enzymes which incorporate both atoms of O_2 into a substrate. The most well characterized enzyme in this class is protocatechuate 3,4-

dioxygenase (PCD) which catalyzes the cleavage of catechols to give corresponding dicarboxylic acids. Based on crystallographic (Ohlendorf et al., 1988) and spectroscopic (Que, 1989; Lipscomb and Orville, 1992) studies on PCD and other related catechol dioxygenases, a five-coordinate high-spin iron has been proposed in the active site. The ligands to the ferric ion are two tyrosines, two histidines and one water or hydroxide ion. The role of the iron is to bind the substrate molecule which is then directly attacked by O_2 to give an organic peroxide intermediate (Jang et al., 1991).

Lipoxygenase is a nonheme iron enzyme and catalyzes the production of 13-hydroperoxy-9,11-*cis,trans*-octadecadienoic acid from linoleic acid (Nelson and Cowling, 1988). Quercetinase is a 2,3-dioxygenase which cleaves quercetin to give 2-protocatechuoyl phloroglunicol and CO. This enzyme contains copper which serves to activate the substrate for O_2 attack (Balogh-Hergovich et al., 1991). Indoleamine 2,3-dioxygenase is a heme enzyme and it cleaves the indole portion of L-tryptophan to form N-formyl kynurenine (Hayaishi et al., 1990).

V. OXIDASES

A. Cytochrome *c* oxidase

Cytochrome *c* oxidase is located in the inner mitochondrial membrane in eukaryotes and in plasma membrane in bacteria. Cytochrome *c* oxidase (C*c*O) acts as the redox center in the electron transfer sequence from cytochrome c to O_2, the terminal step in the respiratory chain (Babcock and Wikström, 1992). The active site of C*c*O contains two copper ions (Cu_A and Cu_B) and two heme irons (heme *a* and a_3). The protein catalyzes the four electron reduction of O_2 to H_2O. A peroxo-bridged Fe_{a3}-Cu_B intermediate has been proposed for this enzyme. The other two metal ions are the primary electron acceptors. The O_2 reduction process is coupled to translocation of protons vectorially across the inner membrane. This proton pump maintains the electrochemical potential across the membrane which drives ATP synthesis.

B. Four-electron oxidases: ascorbate oxidase and laccase

Ascorbate oxidase and laccase are the "blue" oxidases which catalyze the oxidation of substrates (polyphenols, polyamines, ascorbate) with concomitant $4e^-$ reduction of O_2 to H_2O (Solomon et al., 1992). Ascorbate oxidase is found in bacteria, algae and plants ($M = 140$ kg mol^{-1}) whereas laccase is isolated from Chinese laquer tree *Rhus vernicifera* ($M = 110$ kg mol^{-1}). The X-ray structure of ascorbate oxidase (Messerschmidt et al., 1992) shows one "blue", one "normal" and two antiferromagnetically coupled copper atoms. The "normal" copper and the dinuclear copper atoms form a trinuclear unit that is ~ 13 Å away from the "blue" copper.

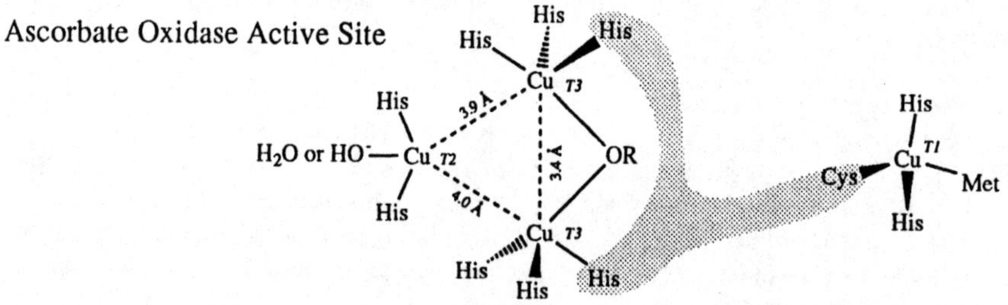

Scheme 9

The "blue" site on the protein is responsible for electron transfer to the trinuclear site where O_2 binds and is reduced. There is a conserved His-Cys-His sequence joining the two discreet sites; this acts as the electron shuttle. A similar site has been proposed for laccase. Cole et al. (1991) have recently suggested the formation of a hydroperoxide intermediate at the trinuclear site in laccase.

C. Two-electron oxidases

1. *Galactose oxidase*

Galactose oxidase is a extracellular enzyme secreted by the fungus *Dactylium dendroides* (M = 68 kg mol^{-1}). This enzyme catalyzes the stereospecific oxidation of primary alcohols (galactose) with 2e$^-$ reduction of O_2 to H_2O_2 (Whittaker and Whittaker, 1990).

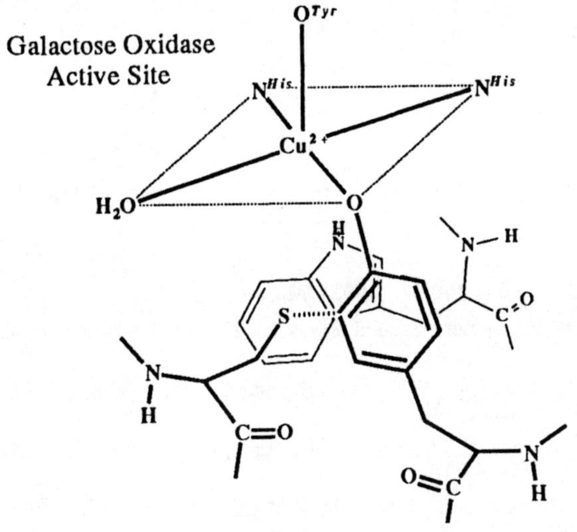

Scheme 10

X-ray crystallographic studies on this protein show a mononuclear Cu(II) site with a unique 'built in' cofactor which is a covalently linked Tyr-Cys moiety stacked with a Trp (Ito et al., 1991). This novel cofactor is probably responsible for stabilizing a ligand-protein radical, i.e. (L)Cu(I) + O_2 + 2H$^+$ $\rightarrow$ (L$^{+\cdot}$)Cu(II) + H_2O_2.

2. *Amine oxidase*

Another class of radical-stabilizing cofactor based copper oxidases are amine oxidases. These can be isolated from a variety of organisms and consist of two subunits (M = 170-190 kg mol^{-1}). Their primary function is the oxidative deamination of amines with concomitant 2e$^-$ reduction of O_2. Based on spectroscopic and mechanistic studies, a quinone cofactor derived from tyrosine (Mu et al., 1992), 6-hydroxy dopa (TOPA) has been proposed along with a "normal" Cu(II) ion in the active site and a proton activation mechanism has been hypothesized by Janes et al. (1990).

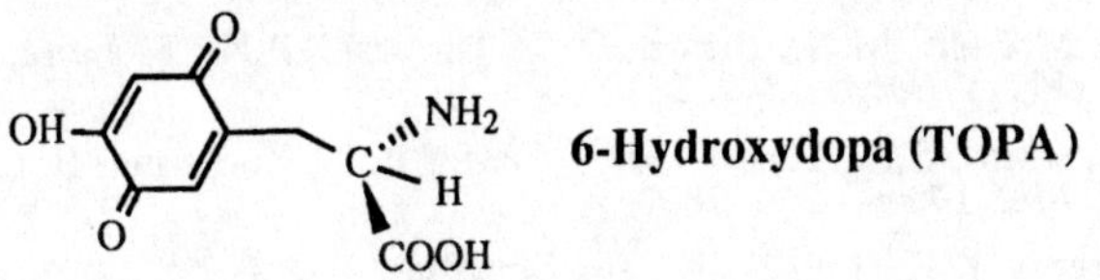

Scheme 11

Detection of a Cu(I)-semiquinone intermediate during the reaction cycle by Dooley et al. (1991) has led to the proposition that the primary function of copper is to facilitate reoxidation of the reduced cofactor after its reaction with amine.

VI. CONCLUSIONS

Nature has evolved a number of ways using iron or copper in which to reversibly bind or activate dioxygen. Yet, with regard to basic function and stoichiometry of reaction (e.g. reducing equivalents required, hydrogen peroxide or water as O_2-reduction product), heme-iron, non-heme iron and copper oxygenases and oxidases effect very similar reactions. The diversity of structures and mechanisms which clearly exist must reflect evolutionary trends, as well as adaptations to specific needs regarding substrate. There is considerable continuing interest in the study of dioxygen metabolism by metalloproteins by biochemists, biophysicists and inorganic chemists.

REFERENCES

Babcock, G.T., and Wikström, M. (1992). *Nature, 356*: 301.

Balogh-Hergovich, É., Speier, G., and Argay, G. (1991). *J. Chem. Soc. Chem. Commun.*, 551.

Barlow, C.H., Maxwell, J.C., Wallace, W.J., and Caughey, W.S. (1973). *Biochem. Biophys. Res. Commun., 55*: 91.

Blackburn, N.J., Strange, R.W., Farooq, A., Haka, M.S. and Karlin, K.D. (1988). *J. Amer. Chem. Soc., 110*: 4263.

Blackburn, N.J., Hasnain, S.S., Pettingill, T.M., and Strange, R.W. (1991). *J. Biol. Chem., 266*: 23120.

Blackburn, N.J., Strange, R.W., Carr, R.T., and Benkovic, S.J. (1992). *Biochemistry, 31*: 5298.

Brunori, M., Giardina, B., and Kuiper, H.A. (1982). *Inorg. Biochem., 3*: 126.

Chan, S.I., Nguyen, H.-H.T., Shiemke, A.K., and Lidstrom, M.E. (1992). *J. Inorg. Biochem., 47*: 10.

Cole, J.L., Avigliano, L., Morpurgo, L., and Solomon, E.I. (1991). *J. Amer. Chem. Soc., 113*: 9080.

Dalton, H., Smith, D.D.S., and Pilkington, S.J. (1990). *FEMS Microbiol. Rev., 87*: 201.

Dawson, J.H. (1988). *Science, 240*: 433.

Deighton, N., Podmore, I.D., Symons, M.C.R., Wilkins, P.C., and Dalton, H. (1991). *J. Chem. Soc. Chem. Commun.*, 1086.

Dewitt, J.G., Bentsen, J.G., Rosenzweig, A.C., Hedman, B., Green J., Pilkington, S., Papaefthymiou, G.C., Dalton, H., Hodgson, K.O., and Lippard, S.J. (1991). *J. Amer. Chem. Soc., 113*: 9219.

Dix, T.A., and Benkovic, S.J. (1988). *Accounts Chem. Res., 21*: 101.

Dooley, D.M., McGuirl, M.A., Brown, D.E., Turowski, P.N., McIntire, W.S., and Knowles, P.F. (1991). *Nature, 349*: 262.

Eipper, B.A., Perkins, S.N., Huste, E.J., Johnson, R.C., Keutmann, H.T., and Mains, R.E. (1991). *J. Biol. Chem., 266*: 7827.

Fox, B.G., Surerus, K.K., Münck, E., and Lipscomb, J.D. (1988). *J. Biol. Chem., 263*: 10553.

Fox, B.G., Liu, Y., Dege, J.E., and Lipscomb, J.D. (1991). *J. Biol. Chem., 266*: 540.

Freedman, T.B., Loehr, J.S., and Loehr, T.M. (1976). *J. Amer. Chem. Soc., 98*: 2809.

Gerber, N.C. and Sligar, S.G. (1992). *J. Amer. Chem. Soc., 114*: 8742.

Groves, J.T. and Watanabe, Y. (1988). *J. Amer. Chem. Soc., 110*: 8443.

Hayaishi, O. (1974). *Molecular Mechanisms of Oxygen Activation*, Academic Press, New York.

Hayaishi, O., Takikawa, O., and Yoshida, R. (1990). *Prog. Inorg. Chem., 38*: 75.

Hazes, B., Magnus, K.A., Bonaventura, C., Bonaventura, J., Dauter, Z., Kalk, K.H., and Hol, W.G.J. (1993). *Protein Sci.*, *2*: 597.

Holmes, M.A., Le Trong, I., Turley, S., Sieker, L.C., and Stenkamp R.E. (1991). *J. Mol. Biol.*, *218*: 583.

Ingraham, L.L., and Meyer, D.L. (1985). *Biochemistry of Dioxygen* (E. Frieden, ed.), Plenum Press, New York.

Ito, N., Phillips, S.E.V., Stevens, C., Ogel, Z.B., McPherson, M.J., Keen, J.N., Yadav, K.D.S., and Knowles, P.F. (1991). *Nature*, *350*: 87.

Janes, S.M., Mu, D., Wemmer, D., Smith, A.J., Kaur, S., Maltby, D., Burlingame, A. L., and Klinman, J.P. (1990). *Science*, *248*: 981.

Jang, H.G., Cox, D.D., and Que, L., Jr. (1991). *J. Amer. Chem. Soc.*, *113*: 9200.

Jolley, R.L., Jr., Evans, L.H., and Mason, H.S. (1972). *Biochem. Biophys. Res. Commun.*, *46*: 878.

Karlin, K.D. and Gultneh, Y. (1987). *Prog. Inorg. Chem.*, *35*: 219.

Kitajima, N., Fujisawa, K., Fujimoto, C., Moro-oka, Y., Hashimoto, S., Kitagawa, T., Toriumi, K., Tatsumi and K., Nakamura, A. (1992). *J. Amer. Chem. Soc.*, *114*: 1277.

Klinman, J.P., Berry, J.A., and Tian, G. (1992). *J. Inorg. Biochem.*, *47*: 18.

Lippard, S.J. (1988). *Angew. Chem. Int. Ed. Engl.*, *27*: 344.

Lipscomb, J.D., and Orville, A.M. (1992). *Metal Ions in Biological Systems*, Vol. 1 (H. Sigel and A. Sigel, eds.), Marcel Dekker, New York, p. 243.

Magnus, K., and Ton-That, H. (1992). *J. Inorg. Biochem.*, *47*: 20.

Mansuy, D. (1987). *Pure Appl. Chem.*, *59*: 759.

Merkler, D.J., Kulathila, R., Consalvo, A.P., Young, S.D., and Ash, D.E. (1992). *Biochemistry*, *31*: 7282.

Messerschmidt, A., Ladenstein, R., Huber, R., Bolognesi, M., Avigliano, L., Petruzzelli, R., Rossi, A., and Finazzi-Agró, A. (1992). *J. Mol. Biol.*, *224*: 179.

Mu, D., Janes, S.M., Smith, A.J., Brown, D.E., Dooley, D.M., and Klinman, J.P. (1992). *J. Biol. Chem.*, *267*: 7979.

Nelson, M.J., and Cowling, R.A. (1988). *J. Amer. Chem. Soc.*, *112*: 2820.

Niederhoffer, E.C., Timmons, J.H., and Martell, A.E. (1984). *Chem. Rev.*, *84*: 137.

Nordlund, P., Eklund, H., and Sjöberg, B.-M. (1990). *Nature*, *345*: 593.

Ohlendorf, D.H., Lipscomb, J.D., and Weber, P.C. (1988). *Nature*, *336*: 403.

Ortiz de Montellano, P.R., ed. (1986). *Cytochrome P-450. Structure, Mechanism and Biochemistry*, Plenum Press, New York.

Pember, S.O., Johnson, K.A., Villafranca, J.J., and Benkovic, S.J. (1989). *Biochemistry*, *28*: 2124.

Perutz, M.F. (1976). *Brit. Med. Bull.*, *32*: 195.

Pettingil, T.M., Strange, R.W., and Blackburn, N.J. (1991). *J. Biol. Chem.*, *266*: 16996.

Que, L., Jr. (1989). *Iron Carriers and Iron Proteins* (T.M. Loehr, ed.), VCH, New York, p. 467.

Que, L., Jr., and True, A.E. (1990). *Prog. Inorg. Chem.*, *38*: 97.

Raag, R. and Poulos, T.L. (1991). *Biochemistry*, *30*: 2674.

Rosenzweig, A.C., Frederick, C.A., Lippard, S.J., and Nordlund, P. (1993). *Nature*, *366*: 537.

Ross, P.K., and Solomon, E.I. (1991). *J. Amer. Chem. Soc.*, *113*: 3246.

Sawyer, D.T. (1991). *Oxygen Chemistry*, Oxford University Press, New York.

Shiemke, A.K., Loehr, T.M. and Sanders-Loehr, J. (1986). *J. Amer. Chem. Soc.*, *108*: 2437.

Solomon, E.I., Baldwin, M.J., and Lowery, M.D. (1992). *Chem. Rev.*, *92*: 521.

Stewart, L.C., and Klinman, J.P. (1988). *Annu. Rev. Biochem.*, *57*: 551.

Stubbe, J. (1991). *Curr. Opin. Struct. Biol.*, *1*: 788.

Volbeda, A., and Hol, W.G.J. (1989). *J. Mol. Biol.*, *209*: 249.

Whittaker, M.M., and Whittaker, J.W. (1990). *J. Biol. Chem.*, *265*: 9610.

Wilcox, D.E., Porras, A.G., Hwang, Y.T., Lerch, K., Winkler, M.E., and Solomon, E.I. (1985). *J. Amer. Chem. Soc.*, *107*: 4015.

Wilkins, R.G. (1992). *Chem. Soc. Rev.*, *21*: 147.

MECHANISMS OF METALLOENZYMES

by

Robert W. Hay

Chemistry Department,
University of St. Andrews,
St. Andrews, KY16 9ST
U.K.

I. INTRODUCTION

It has been estimated that some 30% of enzymes are metalloenzymes or are activated by metal ions. As a result, there has been considerable interest by inorganic chemists in the structure of biometallic sites and in attempts to develop metal complexes as models for the more complex enzyme systems. Because of the very large number of metalloenzymes, the present chapter will concentrate on those metalloenzymes which have been studied by X-ray crystallography. In these cases, detailed information concerning the structure of the active site is available.

The metal ion requirement of an enzyme is frequently determined by removal of the metal ion (for example by dialysis against a chelating agent such as 1,10-phenanthroline). If the resulting apoenzyme (metal-free enzyme) is catalytically inactive and the activity can be restored by the addition of the metal ion, this observation provides very strong evidence for the involvement of the metal ion at the active site of the enzyme. The technique also allows the preparation of so-called "artificial enzymes" by the addition of other metal ions to the apoenzyme. For example, zinc(II) can be removed from carbonic anhydrase and the resulting apoenzyme reacted with cobalt(II) to give an "artificial" cobalt(II) carbonic anhydrase. Such "artificial enzymes" frequently display some catalytic activity but normally less than that of the natural enzyme. This technique has been used to probe the active site of "spectroscopically silent" d^{10} zinc(II) enzymes by replacement with the d^7 cobalt(II) which has useful spectroscopic and magnetic properties.

II. ZINC(II) ENZYMES

Zinc(II) is widely employed as a Lewis acid in many enzymes such as carbonic anhydrase, carboxypeptidase and alkaline phosphatase. Zinc is one of the most abundant of the trace metals and over 100 enzymes require zinc for activity, although only a relatively small number have been studied in detail. Some typical examples of zinc enzymes are summarized in Table 1.

The most common type of catalysis in organic chemistry is acid-base catalysis. Normally enzymes operate within a quite narrow pH range close to neutrality, although some enzymes such as pepsin function at quite acidic pH values while others such as alkaline phosphatase require quite alkaline pH's. Enzymes are thus not able to employ specific acid (H_3O^+ catalysis) or specific base (OH^- catalysis). However, they can use

Table 1

Examples of Zinc-dependent Enzymes

Enzyme	Source
Oxidoreductases	
Alcohol dehydrogenase	Vertebrates, plants microbes
Superoxide dismutase	Vertebrates, plants, fungi, bacteria
D-Glyceraldehyde 3-phosphate dehydrogenase	
Transferases	
Transcarboxylase	*E. coli*
RNA polymerase	*S. cerevisiae*, bacteria
Mercaptopyruvate sulphur transferase	Rat liver, *E. coli*
Hydroxylases	
Alkaline phosphatase	Mammals, bacteria
Phosphodiesterase	Snake venom
Leucine aminopeptidase	Mammals, fungi, bacteria
Carboxypeptidase A	Vertebrates, Crustaceans
Carboxypeptidase B	Mammals, Crustaceans
Dipeptidase	Mammals, bacteria
Elastase	Kidney
Lyases	
Aldolase	
Carbonic anhydrase	Animals, plants
Glyoxalase I	Mammals, *S. cerevisiae*
Isomerase	
Phosphomannose isomerase	*S. cerevisiae*
Ligase	
Pyruvate carboxylase	*S. cerevisiae*, bacteria

Data from: Londesborough, J., and Suoranta, K. (1983). *J. Biol. Chem.*, *258*: 2966.

general acid and general base catalysis, for example, by using -COOH or $-COO^-$ groups respectively from glutamic and aspartic acid side chains.

Although zinc(II) is employed as a Lewis acid, it can also act as a source of hydroxide ion by ionization of a coordinated water molecule

$$Zn-OH_2 \rightleftharpoons Zn-OH + H^+$$

As would be predicted, hydroxo-zinc nucleophiles have a much lower nucleophilicity than hydroxide ion. Normally there is a good linear relationship between the reactivity of a nucleophile (k_N) and the pK_a of the conjugate acid (Fig. 1), provided that the attacking atom remains the same. Zinc(II) hydroxo complexes can, however, be very effective in reactions in which they can act as intramolecular nucleophiles as shown in (1) for the hydrolysis of an ester. The zinc(II) is acting as a source of the nucleophile and as a Lewis acid polarizing the carbonyl function of the ester.

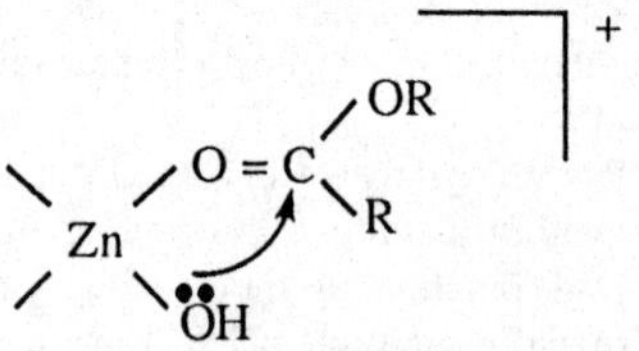

Structure (1)

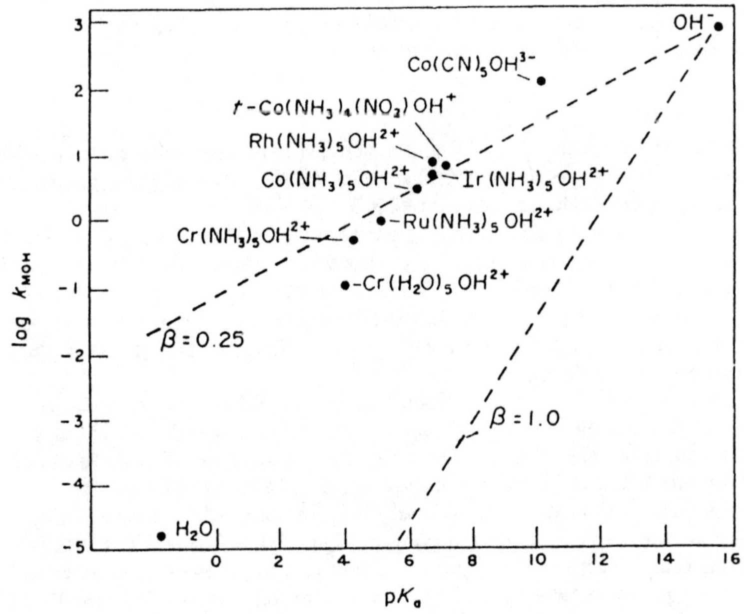

Figure 1.
Plot of log k_{MOH} (i.e. k_N) vs pK_a (of MOH_2^{n+}) for the reaction of various metal hydroxo complexes with propionic anhydride at 25°C and μ = 1.0 M (reproduced with permission from "Biological Aspects of Inorganic Chemistry" Wiley, New York, 1966).

The main reasons for the wide occurrence of zinc(II) as a Lewis acid can be summarized under the headings:

(i) zinc(II) is a d^{10} ion and therefore has no ligand-field effects associated with it to determine a particular coordination number or geometry;

(ii) ligand exchange processes on zinc(II) are rapid, so that substrates and products can be readily introduced and removed;

(iii) zinc(II) is a quite effective Lewis acid and is active in many model systems;

(iv) zinc(II) does not hydrolyze to form hydroxo complexes at low pH. The pK_a of the aqua zinc(II) ion is ca 8.8;

(v) zinc(II) has no redox chemistry under biological conditions. A metal ion such as copper(II) which is very effective as a Lewis acid is involved in many redox reactions, and it appears that Nature prefers a metal ion to adopt a single defined role either as a Lewis acid or as a redox active species.

A. Carbonic anhydrase

Carbonic anhydrase is a zinc metalloenzyme present in animals, plants and certain microorganisms which catalyzes the reversible hydration of carbon dioxide (equation 1). In addition, the enzyme also catalyzes the hydration of many aldehydes (equation 2) and the hydrolysis of various esters (e.g. 4-nitrophenyl acetate) and sultones.

$$CO_2 + H_2O \rightleftharpoons HCO_3^- + H^+ \qquad (1)$$

$$CH_3CHO + H_2O \rightleftharpoons CH_3CH(OH)_2 \qquad (2)$$

A number of slightly different enzymes, carbonic anhydrases A, B and C occur in different organisms. The most well characterized enzymes are the bovine and human carbonic anhydrases B, which are monomeric and contain 1 atom of tightly bound zinc per 30 kg mol^{-1} molar mass.

The turnover number for CO_2 hydration is ca 10^6 s^{-1}, that is one mole of the enzyme can hydrate 10^6 moles of CO_2 per second at 37°C. The hydration of CO_2 in the absence of a catalyst is quite slow with a rate constant of 0.037 s^{-1} at 25°C. The enzymic rate acceleration is at least 10^7-fold for reaction (3)

$$OH^- + CO_2 \xrightarrow{k} HCO_3^- \quad (3)$$

The human isoenzymes B and C and the bovine B isoenzyme have been studied by X-ray diffraction at high resolution (Kannan et al., 1975). The enzymes are ellipsoidal in shape with approximate dimensions 41 x 42 x 55 $Å^3$. The zinc ion lies near the bottom of a cleft 16Å deep, which reaches nearly to the center of the molecule and is bound by the imidazole donors of three histidine residues, His-94, His-96 and His-119. The fourth site is occupied by a water molecule giving a distorted tetrahedral geometry. The coordinated water is linked by hydrogen bonding to Thr-199, which in turn is hydrogen bonded to Glu-106 which is close enough to exert a catalytic role. The general structural features of the molecule are shown in Figure 2.

The structures of the human carbonic anhydrase B-imidazole complex establish the presence of five-coordinate zinc, the aqua group being retained (Kannan et al., 1977). Imidazole acts as a competitive inhibitor for the hydration of CO_2. Sulfonamides of the general formula XSO_2NH_2 displace the aqua ligand and also act as inhibitors.

The K_m value for binding CO_2 is nearly independent of pH, but values of k_{cat} increase with pH and indicate that the group having a pK_a of 7.1 is involved at the active site and controls the catalytic activity. This group is almost certainly a zinc-bound water molecule as originally suggested in 1959 by Davis. Numerous model studies using zinc(II) complexes also support this view (see for example, Woolley, 1975; Kimura and Koike, 1991).

The detailed mechanism of action of carbonic anhydrase has been the subject of much discussion. A possible mechanistic scheme is outlined in Figure 3. There is now increasing support for the view that CO_2 is weakly bound at a fifth coordination site (Stein et al., 1977; Stein, 1985). The competition between imidazole and CO_2 also suggests that CO_2 is bound. Anionic inhibitors function by displacing hydroxide.

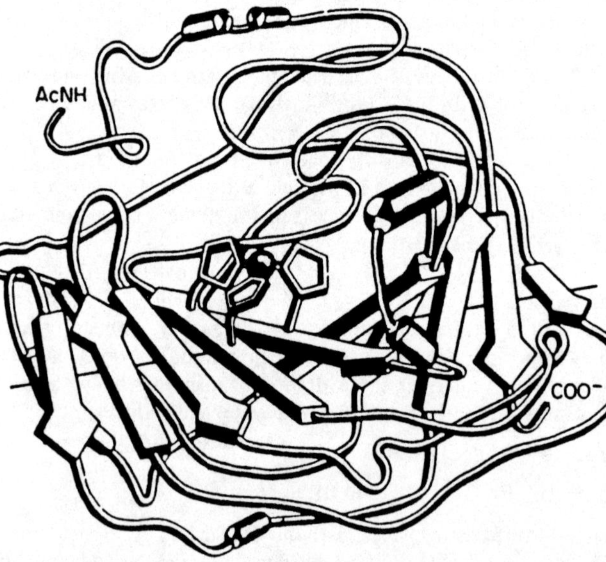

Figure 2.
The structure of carbonic anhydrase. Cylinders represent helices and arrows β-sheet. (Reproduced with permission from "Isozymes I. Molecular Structure", Academic Press, New York, 1975, p. 575).

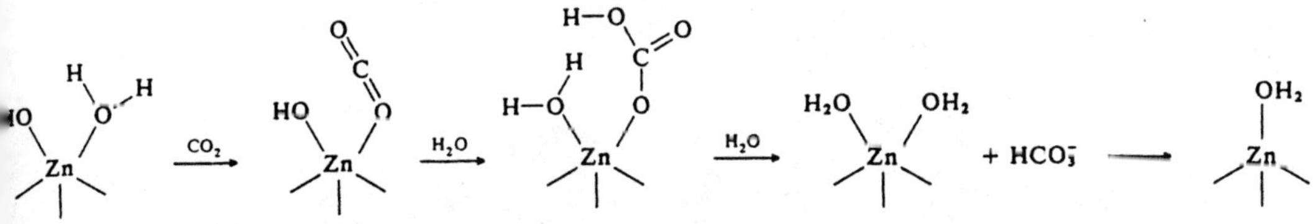

Figure 3.
A mechanistic scheme for carbonic anhydrase.

B. Carboxypeptidase
The carboxypeptidases are released from their inactive precursors in the pancreatic juice of animals. The best studied example is bovine carboxypetidase A, which contains one mole of zinc per molar mass of 34.5 kg mol^{-1}. These enzymes cleave the C-terminal amino acid from peptides and proteins, when the side chain of the C-terminal residue is aromatic or branched aliphatic of the L-configuration.

Bovine CPA is an ellipsoidal molecule, with dimensions 50 x 42 x 38 $Å^3$ (Rees et al., 1983). The zinc is bound in a cleft by three protein residues, two imidazoles (His-69, His-196) and the bidentate carboxyl group of Glu-72. The remaining position is occupied by a water molecule to give five coordinate zinc(II) (Fig. 4).

A number of other amino acid residues appear to be involved in the active site, Glu-270; Tyr-248 and Arg-145. Detailed mechanisms have been suggested on the basis of the crystallographic data on a stable enzyme-substrate complex with the dipeptide glycyl-L-tyrosine. Figure 5 shows the conformational changes which are believed to occur when Gly-L-Tyr is bound to the active site. The guanidinium group in Arg-145 moves about 1Å closer to the carboxylate group of the substrate, while the phenolic group of Tyr-248 moves about 12Å, with a twisting of the C-C chain so that the OH group lies close to the peptide N. These movements allow Tyr-248 to donate a proton to the NH group of the peptide bond undergoing hydrolysis. The carboxylate group of Glu-270 is in a position to provide general base catalysis for attack by water or direct nucleophilic attack on the carbonyl group of the peptide linkage.
In the zinc hydroxide mechanism, the carbonyl group is directed away from zinc, while zinc-bound hydroxide attacks the substrate (Fig 6). It should be appreciated that these mechanistic conclusions are based on a static solid state model and the assumption that the features of the enzyme in solution are conserved on crystallization. It is not certain that the enzyme which crystallizes from solution is the catalytically active species. Indeed there is

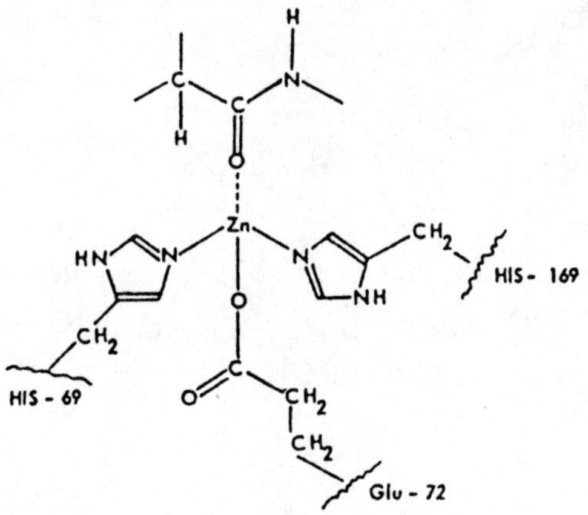

Figure 4.
The zinc(II) site in carboxypeptidase A.

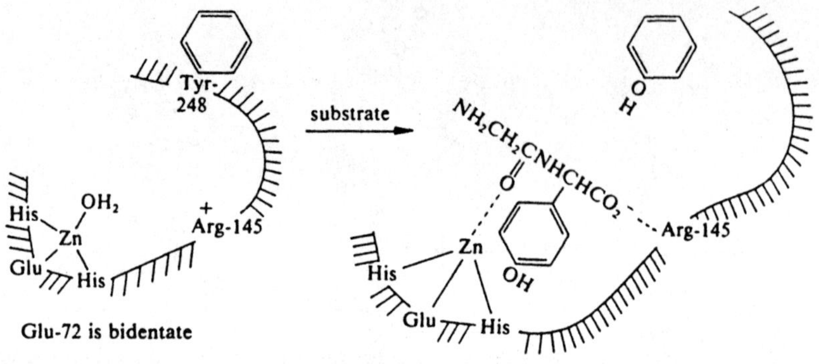

Figure 5.
The conformational changes induced in carboxypeptidase on binding to the pseudosubstrate glycyl-L-tyrosine.

evidence that the kinetics of carboxypeptidase in solution differ from those of enzyme crystals, with the crystalline enzyme being 1000-fold less active with some substrates (Spilburg et al., 1977).

The zinc in carboxypeptidase may be replaced by a variety of metal ions, although most work has been carried out with the cobalt(II) enzyme. The crystal structures of the cobalt(II) and nickel(II) enzymes have been determined to 1.7Å resolution (Hardman and Lipscomb, 1984). The detailed geometry of the cobalt(II) site is identical to that of the zinc(II) site within experimental error, while there are some changes in the nickel enzyme in the position of the nickel(II) and water molecule. The nickel appears to be octahedral with Glu-72 acting as a bidentate ligand and the sixth coordination site vacant.

Kinetic studies on various metallocarboxypeptidase have provided a useful insight into the role of the metal ion in peptidase and esterase activities. Some typical kinetic results are summarized in Table 2. For peptidase activity the K_m value for substrate binding depends only slightly on the nature of the metal ion, but the rate constant is metal dependent. The opposite situation occurs with esterase activity, suggesting that the primary role of the metal ion in peptide hydrolysis is associated with catalysis rather than binding, while the converse is true for ester hydrolysis.

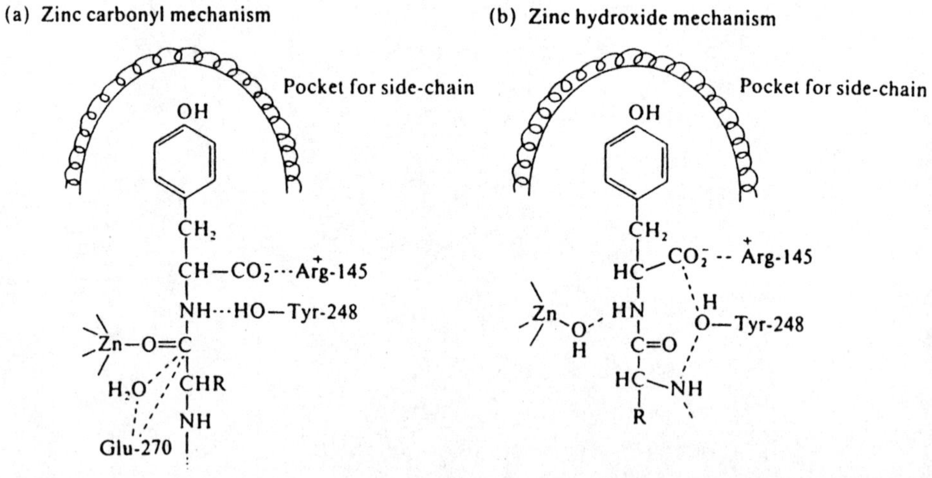

Figure 6.
Possible mechanisms for the action of the enzyme carboxypeptidase on a peptide substrate.

Table 2

Hydrolysis of Bz-$(Gly)_2$-L-Phe and its Phenyl Ester by Metallocarboxypeptidases at pH 7.5 and 25°C

Metal	Peptide		Ester	
	k_{cat}/min^{-1}	10^3 K_m/M	10^{-4} k_{cat}/min^{-1}	10^4 K_m/M
Co(II)	6000	0.66	3.9	3.03
Zn(II)	1200	1.00	3.0	3.33
Mn(II)	230	0.36	3.6	15.2
Cd(II)	41	0.77	3.4	83.3

C. Alkaline phosphatase

Alkaline phosphatase is a dimeric metalloenzyme found in both prokaryotes and eukaroytes (Coleman and Gettings, 1983). It has two zinc(II) and one magnesium in each active site region. The crystal structure of native *E. coli* alkaline phosphatase complexed with inorganic phosphate (the enzyme-product complex) has been determined to 2.0Å resolution and establishes that the three metal atoms are in close proximity (Kim and Wyckoff, 1991) (Fig. 7). The Zn_A-Zn_B separation is 3.94Å. The catalytic site zinc atom (Zn_A) is pentacoordinate with His-331, His-412, the two oxygens of Asp-327 and one of the phosphate oxygens acting as donors. The coordination polyhedron can be described as pseudo-tetrahedral with both oxygens of the chelating Asp-327 occupying one apex. The presence of an additional exchanging water molecule in the coordination sphere has been indicated by NMR studies (Schultz et al., 1992). The zinc atoms are bridged by the inorganic phosphate (at Zn-O distances of 2.12 and 2.23Å).

Coleman and Gettings (1983) have reviewed the hydrolysis of monoesters of phosphoric acid. Alkaline phosphatase is a non-specific phosphomonoesterase which functions via a phospho-seryl intermediate to give inorganic phosphate, or to transfer the phosphoryl group to other alcohols (Scheme 1) (Sowadski et al., 1985).

The left-hand side of the Scheme shows the formation of the phospho-seryl intermediate (E-P) while the right-hand side relates to its hydrolysis to release inorganic phosphate (P_i). If an alcohol R_2OH is added, it is possible to transfer phosphate from R_1 OP to R_2OH. The X-ray results support the view that the mechanism involves a pentacovalent phosphorane intermediate which is stabilized by the two zinc atoms. It is noteworthy that phospholipase

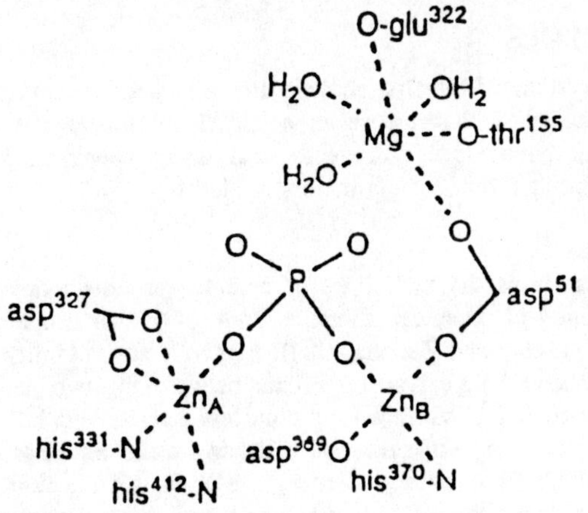

Figure 7.
The metal constellation in *Escherichia coli* alkaline phosphatase.

C (Byberg et al., 1992; Hough et al., 1989) and the exonuclease site of the large fragment of DNA polymerase I (Freemont et al., 1988; Beese and Steitz, 1991) have similar metal constellations and carry out related hydrolytic reactions with phophodiester substrates.

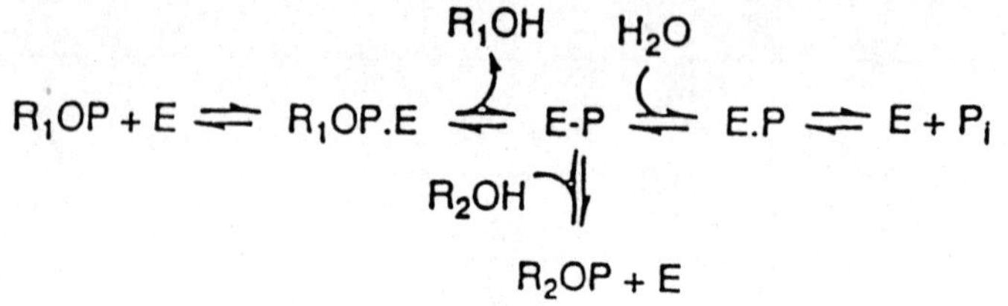

Scheme 1.
Reactions of alkaline phosphatase.

Recently there have been a large number of investigations of the metal-ion and metal-complex promoted and catalyzed reactions of phosphate esters (Hay and Govan, 1991; Morrow and Trogler, 1989; Rawji and Milburn, 1988; Norman and Cornelius, 1982). Koike and Kimura (1991) have studied the hydrolysis of phosphate monoesters using the mononuclear zinc(II) complexes [ZnL(OH)] (L = 1, 5, 9-triazadodecane and 1, 4, 7, 10-tetra-azadodecane). They suggest that in alkaline phosphatase it is possible that one zinc will bind the phosphate ester while the second provides the Zn-OH required for hydrolysis. The structure of a dinuclear zinc center with a phosphate bridge has been reported by Hikichi et al. (1992) and suggested as a possible model for the dinuclear site in alkaline phosphatase. Reaction of tris(3,5-diisopropyl-1-pyrazoyl)boratozinc(II) hydroxide (ZnLOH) with 4-nitrophenyl phosphate leads to ester hydrolysis and formation of the complex (2)

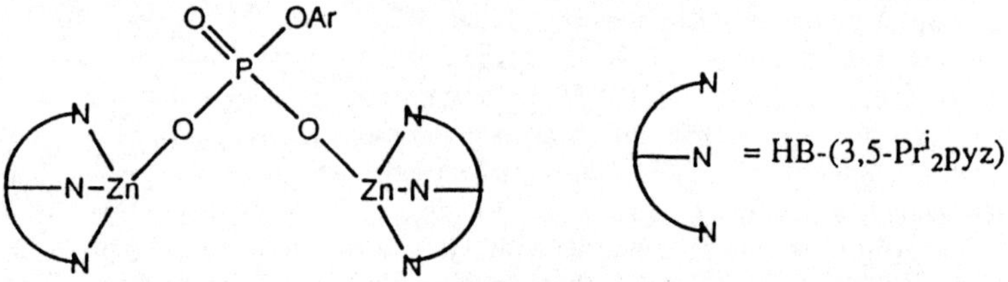

Structure (2)

The active centers of typical zinc(II) enzymes are summarized in Figure 8.

III. NICKEL ENZYMES

In 1926, Sumner crystallized the first enzyme urease and it was not until 1975 that Dixon and coworkers established that urease is a nickel metalloenzyme. Currently ureases, hydrogenase, carbon monoxide dehydrogenase and methyl coenzyme M reductase cofactor F430 are known to contain functionally important nickel.

A. Jack bean urease

Jack bean urease catalyzes the hydrolysis of urea to carbon dioxide and ammonia. The enzyme is a hexamer of identical subunits each of which contains two nickel ions (Cammack, 1988; Blakeley and Zerner, 1984). Magnetic susceptibility measurements have indicated that a weak exchange reaction occurs between the two paramagnetic nickel(II) centers, providing speculative evidence for a dinuclear active site (Clark and Wilcox, 1989). A mechanism involving a concerted reaction at the two nickel sites has been proposed (Fig. 9) (Dixon et al., 1980; Blakeley and Zerner, 1984). In this reaction scheme, one nickel center is thought to polarize the carbonyl group of the urea and the second nickel provides a metal-bound hydroxide which attacks the metal-bound carbonyl group. Such a mechanism

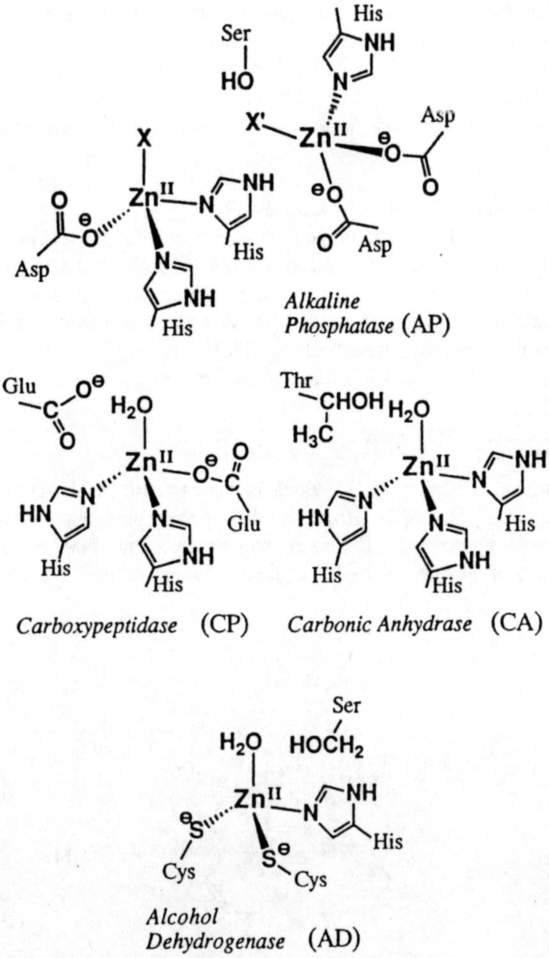

Figure 8.
Active centers of typical zinc(II) enzymes.

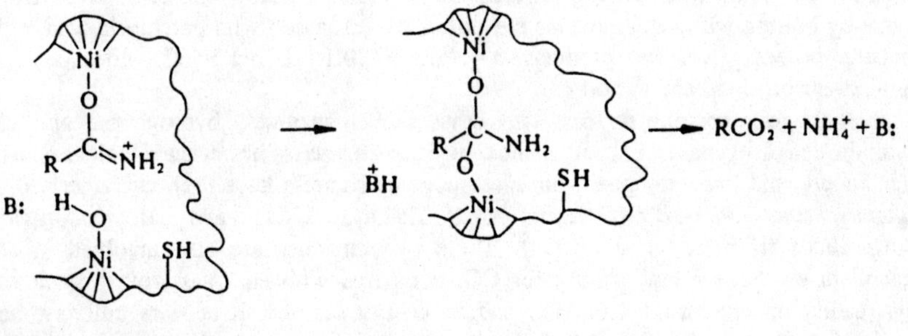

Figure 9.
Possible mechanism for the action of urease.

would require that the two metal centers are within 6.0Å of each other. A preliminary structural investigation has been carried out by X-ray absorption spectroscopy (XAS) and provides support for this proposal (Clark et al., 1990). Crystals of the urease from *Klebsiella aerogenes* has recently been reported as being suitable for X-ray structural analysis (Jabri et al., 1992) and definitive structural information should soon become available.

B. Nickel in methanogenic bacteria. Factor F430

The requirement for nickel by methanogenic bacteria was long overlooked, partly because the stainless steel needles used in the anaerobic culture technique allowed sufficient nickel to dissolve to sustain growth. A yellow, low molar mass (l.m.m.) nickel-containing compound with an absorption maximum at 430 nm and an absorption coefficient of 22,800 M^{-1} cm^{-1} has been isolated from these bacteria. This factor (F430) is the prosthetic group of methyl coenzyme reductase, which catalyzes the conversion of methyl cofactor M into methane

$$MeSCH_2CH_2SO_3^- \xrightarrow{H_2} CH_4 + HSCH_2CH_2SO_3^-$$

Factor 430 is a nickel tetrapyrrole (3) which is diamagnetic in non-coordinating solvents, indicating square planar nickel(II). However, it becomes paramagnetic when dissolved in water and methanol, suggesting addition of solvent in either both axial sites to give an octahedral complex, or in one axial site to give a five-coordinate complex with the nickel out of plane.

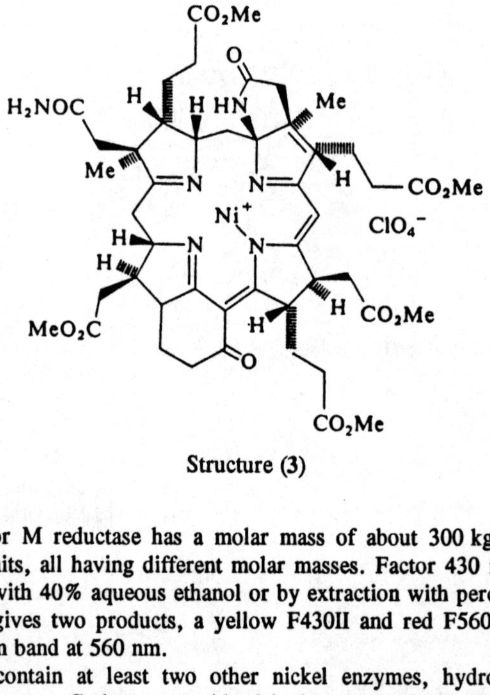

Structure (3)

Methyl cofactor M reductase has a molar mass of about 300 kg mol^{-1} and contains three pairs of subunits, all having different molar masses. Factor 430 is obtained from the enzyme by boiling with 40% aqueous ethanol or by extraction with perchloric acid at pH 2. The latter process gives two products, a yellow F430II and red F560, which contains an additional absorption band at 560 nm.

Methanogens contain at least two other nickel enzymes, hydrogenase and carbon monoxide dehydrogenase. Carbon monoxide dehydrogenases from acetogenic bacteria are better known, but two examples from methanogenic bacteria have been characterized, from *Methanosarcina barkerii* (Kryzycki and Zeikus, 1984) and *Methanobrevibacter arboriphilicus* (Hammel et al., 1984). These two enzymes are also involved in acetate metabolism but have a low affinity for CO, compared with enzymes from organisms that grow readily on the carbon from CO and an energy scource. It appears unlikely that the primary function of the CO dehydrogenase is associated with the reactions of carbon monoxide.

C. Nickel in carbon monoxide dehydrogenase from acetogenic bacteria

Carbon monoxide is a substrate for a number of bacteria and serves as the sole carbon and energy source under aerobic conditions for the carboxydo bacteria, which have a molybdenum-containing carbon monoxide oxidase. Carbon monoxide is used under anaerobic conditions by acetogenic bacteria. These acetogenic bacteria can form acetate from CO_2, in which the methyl group of acetate is formed from CO_2 by reduction via formate and one-carbon derivatives of H_4 folate. They can also be grown on CO, which may be oxidized to CO_2 by carbon monoxide dehydrogenase or converted into acetate.

The enzymes from *C. thermoaceticum* (Ragsdale et al., 1983a) and *A Woodii* (Ragsdale et al., 1983b) have been compared. They both have molar masses of about 450 kg mol^{-1} with six subunits $(\alpha\beta)_3$, but can exist as dimers $\alpha\beta$. The larger subunits appear to be identical in both cases, but the smaller ones differ. Nickel is present in concentrations which suggest two Ni per molecule of dimeric enzyme. There appears to be two or three [4Fe-4S] clusters per dimeric enzyme, while the *A. Woodii* enzyme may have a three-iron center.

It has been suggested that the nickel is bound to a l.m.m. factor, different from F430 of methanogens and which may involve a corrinoid structure. The nickel is thought to cycle between Ni(II) and Ni(III), with the CO bound to the metal. ESR data have been interpreted in terms of a nickel(III) species with a bound radical derived either from CO or CO_2. A Ni-C species is a viable intermediate in the catalytic conversion of CO to CO_2. Electron transfer from Ni(II) could generate a carbon monoxide radical anion. Electrophilic attack by a proton on the oxygen atom would enhance nucleophilic attack of hydroxide on the carbon, with formation of a formate radical species. This species could either decompose or be incorporated into acetate synthesis.

D. Hydrogenase

Nickel is not required for all hydrogenases, but many hydrogenases have a nickel requirement. Hydrogenases also contain iron-sulfur centers. ESR is an important technique for detecting and characterizing nickel centers, while Mössbauer spectra indicate the presence of diamagnetic iron-sulfur clusters.

Nickel in hydrogenase is bound to a protein and cycles between Ni(II) and Ni(III) during the reaction with dihydrogen. Conclusive evidence for Ni(III) comes from studies on hydrogenase isolated from *M. thermoautotrophicum* (Albracht et al., 1982) grown on media enriched with ^{61}Ni(II), an isotope with a nuclear spin of 3/2 and observing the expected hyperfine splitting in the ESR spectrum.

A comparison of the ESR spectra of Ni(III) complexes of the peptides N-mercaptoacetylglycyl-L-histidine and N-mercaptoacetylglycylglycylglycine with the Ni center in hydrogenase has led to the suggestion (Sugiura et al., 1993) that the nickel in hydrogenase contains one cysteine in its coordination sphere. The potential of the Ni(III)/Ni(II) couple in hydrogenase is higher than the $2H^+/H_2$ couple by about 0.2V, so that there are mechanistic problems in proposing a role for Ni(III)/Ni(II) in the activation of dihydrogen. In addition, only some 50% of the nickel can be readily oxidized to Ni(III), suggesting that there is a second ESR undetectable nickel which may have an important role in the activity of the hydrogenase. Some type of nickel dimer cannot be excluded.

IV. COPPER ENZYMES

Copper has an essential role in a number of enzymes, notably those involved in the catalysis of electron transfer, in the transport of dioxygen and the catalysis of its reactions. Hemocyanin is the copper-containing dioxygen carrier, while the important role of copper in oxidases is exemplified in cytochrome oxidase, the terminal member of the mitochondrial electron transfer chain, the multicopper blue oxidases such as laccase, ascorbate oxidase and ceruloplasmin and the non-blue oxidases. Copper is also involved in the copper,zinc-superoxide dismutase and a number of hydroxylases such as tyrosinase and dopamine-β-hydroxylase. Tyrosinase and hemocyanin have similar binuclear copper centers.

A. Copper centers

The copper centers in the multicopper blue oxidase have been classified into three groups, which may be extended to include other copper proteins. Type-1 copper is responsible for the deep blue color of the blue oxidases and the blue electron-transfer proteins. Type-1 copper centers have an intense absorption band near 600 nm, with an absorption coefficient some 100 times larger than those commonly observed in normal coordination compounds of copper(II), together with other absorptions at about 450 and 750 nm. The ESR spectra of the type-1 center has small hyperfine coupling constants.

Type-2 copper is found in the multicopper blue oxidases and is similar to the typical tetragonal copper (II) sites found in normal coordination compounds. It is e.s.r active and has a normal d-d spectrum which is often difficult to detect in the presence of type-1 copper due to its intense bands. Type-3 copper consists of a pair of copper(II), which are diamagnetic due to antiferromagnetic interaction. It is characterised by a strong absorption band near 330 nm with an absorption coefficient in the range 3000 to 5000 M^{-1} cm^{-1}. The type-3 center is associated with redox reactions of dioxygen, as it can transfer two electrons and so bypass the formation of reactive superoxide.

B. Tyrosinase

Tyrosinase is widely found in micro-organisms, plants and animals where it acts as a mono-oxygenase catalyzing the *o*-hydroxylation of monophenols to *o*-diphenols, and as a two-electron oxidase catalyzing the oxidation of *o*-diphenols to *o*-quinones (Wilcox et al., 1985; Solomon 1981). Chemical and spectroscopic evidence indicates that the tyrosinases have a coupled dinuclear active site similar to that found in the hemocyanins. EXAFS studies on tyrosinase from *Neurospora crassa* reveal a Cu ... Cu separation of 3.63Å in oxy-tyrosinase (Fig. 10) with values of 3.39 and 3.43Å in met-aquo- and met-mimosine-tyrosine, respectively (Woolery et al., 1984). The deoxy protein [copper(I)] binds dioxygen to give oxytyrosinase, which is a copper(II) peroxide species with antiferromagnetic coupling between the two copper(II) centers. The oxybinuclear site is diamagnetic to the most sensitive detection techniques.

Photo-oxidation studies have implicated His-188, His-192, His-289 and either His-305 or His-306 as ligands to copper. The pair of copper(II) centers may be replaced by two cobalt(II) centers, but an oxyderivative cannot be prepared (Ruegg and Lerch, 1981). It appears probable that the oxytyrosinase is the hydroxylating species.

C. Blue oxidases

Laccase (M ca 65 kg mol^{-1}) is the simplest member of this family and contains four copper(II) atoms (one type-1, one type-2 and two type-3) (Fee, 1975). Dimeric ascorbate oxidase with identical laccase-like subunits contains eight copper(II) and it was suggested that it was a dimer of two identical laccase-like subunits (Jacobson et al., 1988). Cumulative spectroscopic and azide binding studies on *Rhus* laccase have led to the suggestion that the type-2 and type-3 centers define a trinuclear copper cluster site (Fig. 11) (Cole et al., 1990).

Recent X-ray crystallographic work on oxidized ascorbate oxidase from zuchini (Messerschimdt et al., 1989) has provided much needed information on this type of site. Two crystal forms were analyzed, one a dimer (M 140 kg mol^{-1}) and one a tetramer (M 280 kg mol^{-1}). Each subunit has four copper atoms present bound as mononuclear and trinuclear units. The mononuclear copper, which is bound by two histidines, one cysteine and one methionine, is type-1 resembling plastocyanin. This type-1 site is separated from the trinuclear site by ca 15Å. The trinuclear site (Fig. 12) may be divided into a type-2 copper and a pair of type-3 coppers held in an approximately isosceles triangular array. The type-3 coppers are ligated by three histidine ligands and form a trigonal prism with an

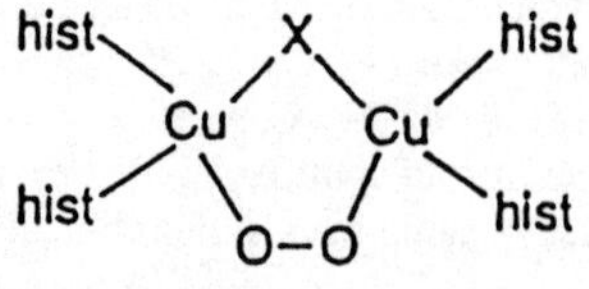

Figure 10.
EXAFS derived site for tyrosinase (Woolery et al., 1984).

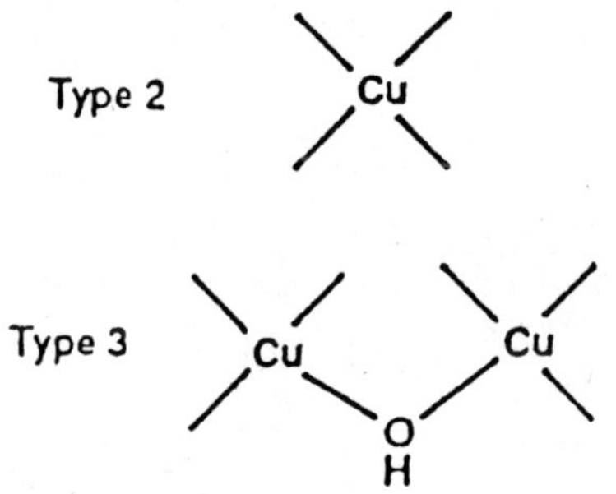

Figure 11.
Suggested trinuclear structure in laccase.

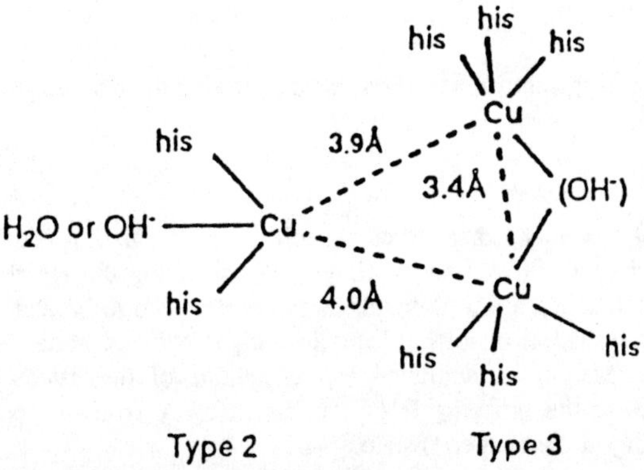

Figure 12.
The trinuclear site in ascorbate oxidase.

intermetallic separation of 3.4Å. The crystallographic data indicates the occurrence of an oxo- or hydroxo-bridging ligand.

D. Copper,zinc-superoxide dismutase

The most studied dismutase is that isolated from bovine erythrocytes. It has a molar mass of 31.4 kg mol^{-1} and is made up of two identical subunits, each containing one copper and one zinc. X-ray diffraction studies (Tainer et al., 1983; Getzoff et al., 1983) have shown the copper to be bound by four imidazole donors (His-44, His-46, His-61 and His-118) and a water ligand, in a distorted five-coordinate geometry (Fig. 13). The zinc(II) has a tetrahedral stereochemistry with three imidazole ligands (His-61, His-69 and His-78) and one oxygen donor (Asp-81). His-61 acts as a bridging imidazolate group.

Each subunit has a deep channel on the outside surface, with the copper slightly exposed on the floor of the channel and close to the completely buried zinc. The amino acid residues making up this channel are highly conserved in enzymes from different species, suggesting that the channel is important. The site for superoxide binding has been identified as a small pit over the copper(II).

During catalysis of superoxide disproportionation, the copper center is reversibly oxidized and reduced by successive encounters with superoxide, giving O_2 and H_2O_2.

$$ECu(II) + O_2^- \rightarrow ECu(I) + O_2$$

$$ECu(I) + O_2^- + 2H^+ \rightarrow ECu(II) + H_2O_2$$

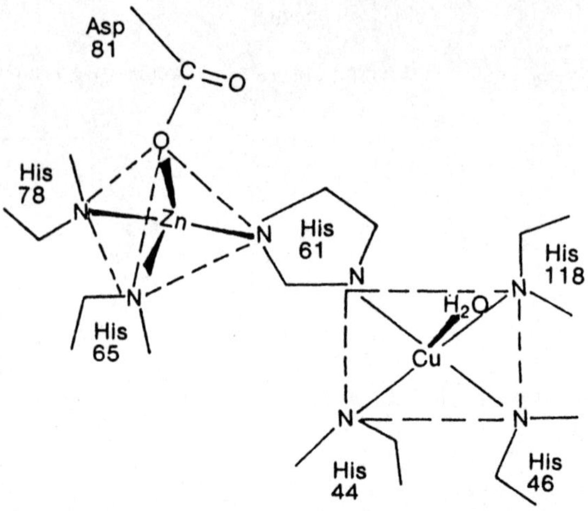

Figure 13.
Schematic drawing of the active site of bovine superoxide dismutase.

The zinc(II) almost certainly has a structural role in the formation and stabilization of the active site (Lippard et al., 1977) and indirectly in enhancing the reactivity of the copper. The detailed X-ray data has allowed details of the mechanism to be elaborated. The protein environment with tetrahedral distortion leads to a high redox potential for copper (0.42V). The first catalytic reaction is facilitated by relaxations of distortions in the zinc-ligand geometry which cause the bridging His-61 to move away from the copper with copper-bridge bond breaking in the copper(I) state.

It is interesting that the rate constant for the superoxide-dismutase catalyzed disproportionation of superoxide is within an order of magnitude of the diffusion-controlled limit. Clearly, the number of non-productive collisions between SOD and O_2^- is low. The large rate enhancement has been discussed in terms of the electrostatic potential in and around the active site channel (Getzoff et al., 1983). The presence of positively charged residues at critical positions allows the incoming negatively charged superoxide ion to be guided towards the active site copper, with minimization of non-productive collisions.

V. MANGANESE ENZYMES

Enzymes with a specific requirement for manganese are summarized in Table 3. Some of these enzymes such as pseudocatalase, manganese superoxide dismutase, ribonucleotide reductase and the oxygen-evolving center in photosystem II are redox active. The resting pseudocatalase and superoxide dismutase have Mn(II) as does purple acid phosphatase. In these three enzyme systems and in ribonucleotide reductase manganese appears to act as a functional catalytic counterpart to iron.

Purple acid phosphatase from the sweet potato contains manganese, the purple color arising from an intense absorption band near 515 nm. The dimeric enzyme may contain one (Sugiura et al., 1981) or two (Fujimoto et al., 1980) manganese(II), depending on the variety of sweet potato. The manganese (and iron) acid phosphatase belong to the group of metal-tyrosinate proteins which are receiving increasing attention. Resonance Raman studies have shown that coordination of the tyrosine phenolate anion to the Mn(III) center occurs (Sugiura et al., 1980). The Mn(III) acid phosphatase has an intense band in the visible spectrum at 515-570 nm assigned to tyrosine-Mn(III) charge transfer. The linear relation between the intensity of the absorption at 515 nm and the phosphatase activity indicates that the metal is essential for catalysis. On acid denaturation, the enzyme displays a spectrum characteristic of high-spin Mn(II)

Table 3

Proteins and Enzymes with a Specific Requirement for Manganese

System	Comments
Pseudocatalase	*Lactobacillus plantarum* Mn(III)
Superoxide dismutase	Mn(III)
Photosystem II	O_2 evolution in photosynthesis
Ribonucleotide reductase	
Acid phosphatase	Mn(III)
Phosphoglycerate phosphomutase	
Phosphoenolopyruvate carboxykinase	
Galactosyl transferase	
Glucosyl transferase	
Arginase	
Phosphoprotein phosphatase	

A. Pseudocatalase

The major function of catalase is the decomposition of peroxides, but it can also act as a peroxidase. The iron catalases usually have molar masses of about 240 kg mol^{-1} with four identical subunits, each containing protoporphyrin IX, with high-spin Fe(II) and an amino acid residue in the axial site. Microorganisms which are unable to synthesize heme produce a catalase which is not a heme catalase as it is not inhibited by cyanide. The pseudocatalase from *Lactobacillus plantarum* contains manganese (Kono and Fridovich, 1983). The enzyme is pink in concentrated solution and has a molar mass of some 175 kg mol^{-1}, with six non-covalently associated subunits, each containing one Mn(III) in the resting enzyme. A possible mechanism could involve oxidation of Mn(III) to Mn(IV) by hydrogen peroxide, then Mn(IV) oxidizing a second molecule of hydrogen peroxide.

B. Manganese superoxide dismutase

Superoxide dismutases catalyze the disproportionation of the superoxide ion to dioxygen and peroxide and are widespread in nature.

$$2O_2^- \rightarrow O_2 + O_2^{2-}$$

Superoxide undergoes spontaneous disproportionation, but the catalytic effect of superoxide dismutase is remarkable, leading to a rate enhancement of some 10^9-fold.

The manganese dismutases contain Mn(III) and occur either as dimers or tetramers with subunits of molar mass about 20 kg mol^{-1}. The Mn/subunit ratio varies from 0.5 through 1.0 to 2.0. It is interesting that the anaerobic bacterium *Propionbacterium shermanii* can synthesize either Fe- or Mn-containing superoxide dismutase with an apparently identical protein dependent on the metal supplied (Meier et al., 1982). Little is known about the mechanism of action of the iron and manganese superoxide dismutases, except that it appears to be different from that of the Cu,Zn enzyme. There appears to be differences between the Mn and Fe enzymes.

VI. IRON ENZYMES

Iron is extremely important in biology and is present in a range of hemoproteins, including the oxygen carriers hemoglobin and myoglobin, the cytochrome electron transfer catalysts, and enzymes such as cytochrome P-450, cytochrome oxidase, peroxidase and catalase.

The emphasis which has been initially placed on the study of the hemoproteins and the iron-sulfur proteins has rather distracted attention until recently from other proteins where the iron is directly bound by the protein. A number of these proteins involve dimeric iron centers in which there is a bridging oxo group. These sites are found in hemoerythrin, the ribonucleotide reductases, uteroferrin and purple acid phosphatase. There are also a number of proteins in which the iron is bound by tyrosine ligands as in catechol dioxygenases, uteroferrin, and purple acid phosphatase. A tyrosine radical is involved in ribonucleotide reductase.

A. Ribonucleotide reductase

Ribonucleotide reductase catalyzes the conversion of ribonucleotides to deoxyribonucleotide di- or triphosphates, the first step in DNA synthesis (Sjoberg and Grassland, 1983; Stubbe, 1989; Lammers and Follman, 1983) (Scheme 2).

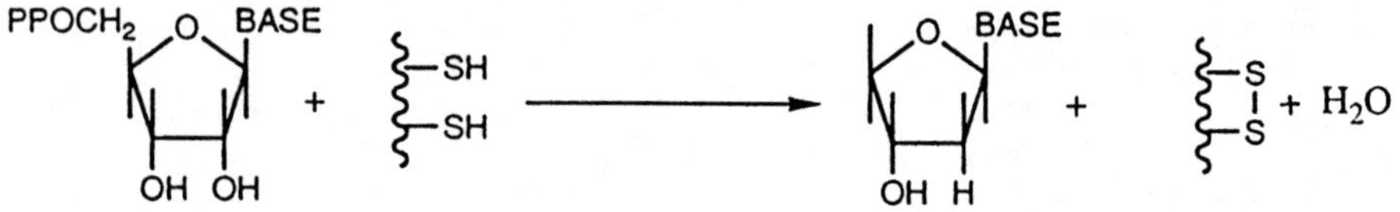

Scheme 2.
Typical reaction catalyzed by ribonucleotide reductase.

The iron protein is found in animals, viruses and some prokaryotic organisms. The enzyme from *Escherichia coli* consists of two subunits B_1 and B_2. The B_1 subunit has a molar mass of 160 kg mol^{-1} and is made up of two polypeptide chains and contains redox active SH groups which play a role in the reaction and has two substrate-binding sites. Subunit B_2 has a molar mass of 78 kg mol^{-1} and also has two peptide chains. This subunit contains a dimeric oxo-bridged iron site associated with a remarkably stable tyrosine radical. The formation of the active enzyme from B_1 and B_2 requires the presence of magnesium ions. The di-iron site in ribonucleotide reductase B_2 from *E. coli* is shown in Figure 14. The reaction catalyzed by ribonucleotide reductase has been suggested to proceed via a free radical mechanism involving the cyclic transfer of the pre-existing radical to the substrate and back. The di-iron center in B_2 displays similar spectral properties to those found in met-hemerythrin. The X-ray structure (Nordlund et al., 1990) of *E. coli* reductase B_2 at 2.2Å resolution (Fig. 14) confirms the presence of an Fe-O-Fe bridge as in met-hemerythrin and establishes that there are two dinuclear iron centers separated by ca 25Å in the dimeric molecule.

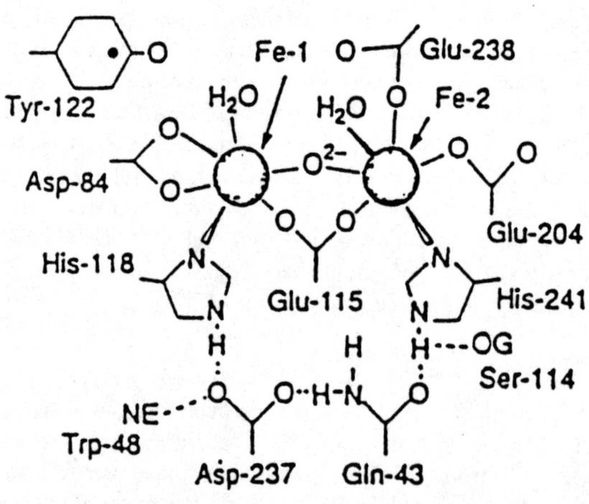

Figure 14.
The di-iron site in ribonucleotide reductase B_2 from *Escherichia coli* (Nordlund et al., 1990).

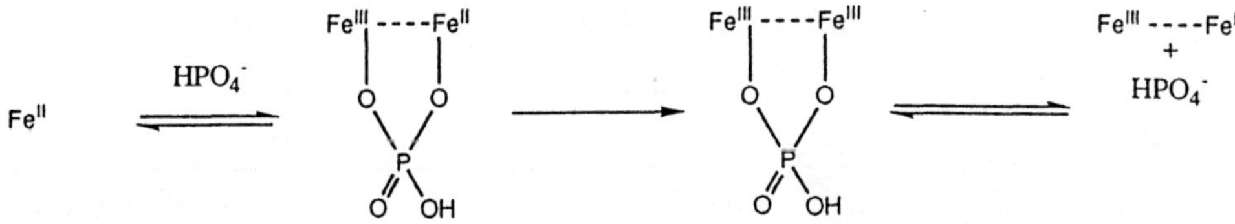

Figure 15.
Suggested interaction of HPO_4^- with pink acid phosphatase. Formation of A takes place via unidentate HPO_4^- binding to Fe(III).

B. Methane mono-oxygenase

The enzyme methane mono-oxygenase catalyzes the conversion of methane to methanol in methanortropic bacteria for which methane serves as the sole source of energy and carbon (Dalton, 1980; Green and Dalton, 1989).

$$CH_4 + O_2 + NADH + H^+ \rightarrow CH_3OH + H_2O + NAD^+$$

There are two types of methane mono-oxygenase, a membrane-bound enzyme containing copper and a water-soluble enzyme containing non-heme iron (Dalton, 1990). ESR studies on the hydroxylase component of methane mono-oxygenase from *Methylooccus capsulatus* (Bath) indicated that it contained a di-iron center which might be similar in type to those found in hemerythrins and ribonucleotide reductase B_2 (Woodland et al., 1986). However, subsequent studies involving EXAFS, Mössbauer spectroscopy and magnetochemistry have suggested that there are marked differences in the sites. A novel type of Fe-Fe bridge is indicated in this enzyme which may contain an alkoxo-, hydroxo- or monodentate carboxylato-bridge, together with one, or two, bridging carboxylates (deWitt et al., 1991).

C. Purple acid phosphatases

The purple acid phosphatases (PAP's) are glycoproteins which catalyze the hydrolysis of phosphate esters at low pH. The enzymes have a distinct purple color and are not inhibited by tartrate. Beef spleen purple acid phosphatase is usually isolated as the oxidized enzyme-phosphate complex, with the two iron atoms present as Fe(III), and there is a single tightly bound phosphate anion. Mild reduction gives a pink form with one Fe(II) and one Fe(III). The ESR spectra of reduced PAP's indicate the presence of a dinuclear iron-oxo unit as the spectra resemble those for the Fe(II)/Fe(III) states of hemoerythrins (Antanaitis and Aisen, 1982; Antanaitis et al., 1983). The chemistry of the enzyme has recently been reviewed (Wilkins, 1992).

The interaction of HPO_4^- with pink acid phosphatase is shown in Figure 15. It is believed that a weak complex with the Fe(II)/Fe(III) form is reversibly formed and that this is converted into the oxidized strong adduct. Similar behavior of a substrate $ArOPO_3H^-$ would facilitate hydrolysis, leading to phosphate and phenolate formation.

VII. MOLYBDENUM ENZYMES

Molybdenum is the only transition element of the second transition series known to have a biological role. The element is present in at least 15 enzymes which are in Table 4, and in several uncharacterized proteins.

A number of the enzymes listed in Table 4 are hydroxylases and in many cases the substrate coordinates to the molybdenum and electron transfer follows. A number of reviews on the topic are available (Coughlan, 1980; Newton and Otsuka, 1980; Steifel, 1977; see also Chapter 1-B11 above). The chemistry of molybdenum of importance to biology can be summarized under the headings (i) a range of oxidation states which can be stabilized in aqueous solution by the common ligands of biology, (ii) the formation of oxo compounds and their sulfur analogues, (iii) the ability to participate in atom transfer reactions, and (iv) the availability of higher coordination numbers.

Table 4

Molybdenum in Enzymes

Enzyme	Source	M_r	Mo	Other cofactors
Xanthine oxidase	Bovine milk	300	2	$4Fe_2S_2$, 4FAD
Xanthine dehydrogenase	Animals, microbes	250	2	$4Fe_2S_2$, 4FAD
Aldehyde oxidase	Hog milk	270	2	$4Fe_2S_2$, 4FAD
Sulfite oxidase	Animals, *Thiobacillus novellus*	110	2	2 heme
Nitrate reductase (dissimilatory	*Escherichia coli*	200	1	$4Fe_4S_4$
Nitrate reductase (assimilatory)	*Chlorella vulgaris*	360	4	4 heme, 4FAD
Formate dehydrogenase	*E. Coli*		2-4	Fe_nS_n, heme, Se
Carbon monoxide oxidase	*Pseudomonas carboxydovorans*	300	2	$4Fe_2S_2$, 2FAD, Se
Nicotinic acid hydroxylase	*Clostridium barkerii*	300	2	Fe/S, FAD, Se
Chlorate reductase	*Proteus mirabilis*			
Biotin sulfoxide reductase				
Pyridoxal oxidase	*Drosophila spp.*		2	($4Fe_2S_2$, 4FAD)
Tryptophan side-chain oxygenase	*Pseudomonas spp.*			Heme
Nitrogenase	*Range of microbes*		2	Mo, Fe, S cluster, iron-sulfur cluster
Trimethylamine oxidase	*E. coli*			

ESR measurements have been of particular value in studying molybdoenzymes. Molybdenum(VI), Mo(V) and Mo(IV) have d^0, d^1 and d^2 configurations. Molybdenum(V) species are therefore ESR detectable and are usually easily recognizable with a slightly anisotropic set of g values in the range of $g = 1.96 - 1.99$. In addition, Mo(V) intermediates produced during an enzymatic reaction display superhyperfine splitting from a proton. Molybdenum(V) model compounds with coordinated N-H groups display superhyperfine coupling of the same order of magnitude as that found in molybdenum enzymes (Pariyadeth et al., 1976).

Molybdenum hydroxylases catalyze the two-electron oxidation of purines, aldehydes and pyrimidines, sulfite and nicotinic acid in the general reaction

$$RH + H_2O \rightarrow ROH + 2H^+ + 2e^-$$

Xanthine oxidase, xanthine dehydrogenase and aldehyde oxidase all have relatively low redox potentials and a unique "cyanolyzable" sulfur atom and so are considered together.

A. Xanthine oxidase and xanthine dehydrogenase

Xanthine oxidase and xanthine dehydrogenase are two closely related enzymes which differ in the nature of their terminal electron acceptors, the oxidase using O_2 and the dehydrogenase NAD^+. Xanthine oxidase is probably the most well studied molybdenoenzyme, and a typical reaction is the xanthine-oxidase catalyzed oxidation of xanthine to uric acid (Scheme 3). There is evidence that the molybdenum is the site for bonding and reduction of the xanthine. The enzyme contains Mo(VI) in the resting form, while Mo(IV) and Mo(V) are implicated during catalysis.

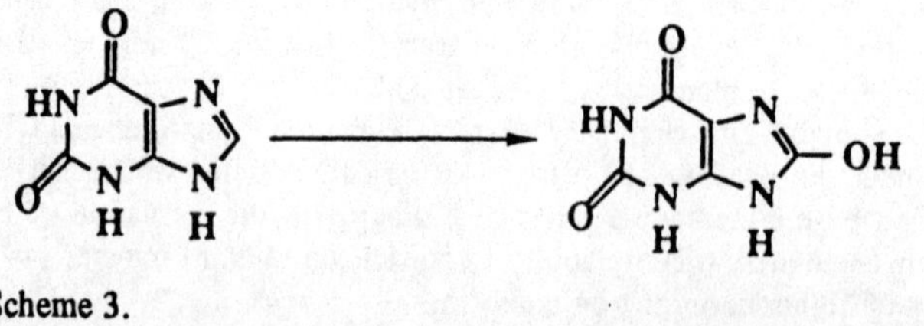

Scheme 3.
The xanthine-oxidase catalyzed oxidation of xanthine to uric acid.

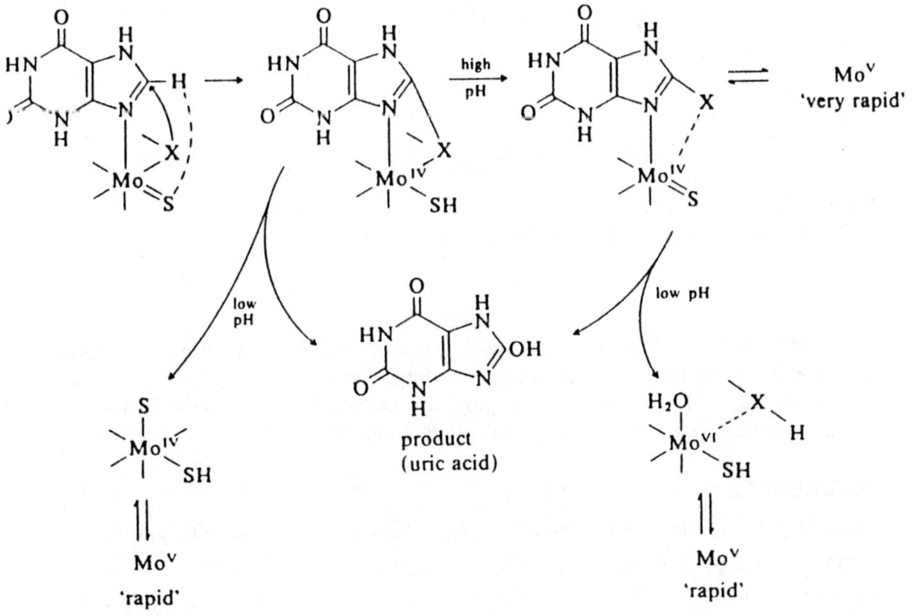

Figure 16.
The mechanism of action of xanthine oxidase.

Xanthine oxidase, as normally isolated, is a mixture of the active enzyme and a catalytically inactive degradation product desulfo xanthine oxidase. Loss of the sulfur atom to give the desulfo enzyme leads to marked changes in the absorption spectrum and redox potential of the molybdenum (Cammack et al., 1976), and it is believed that a Mo = S group is the source of the abstracted sulfur. A possible mechanistic scheme for the enzyme is outlined in Figure 16.

B. Sulfite oxidase

Sulfite oxidase catalyzes the oxidation of sulfite to sulfate, with cytochrome *c* as the final electron acceptor. Sulfite is oxidized at the molybdenum site, which is then reoxidized by the heme. The enzyme is a dimer, with each subunit having a molar mass of 55 kg mol^{-1}.

The redox potential for Mo in sulfite oxidase and nitrate reductase is higher than in other molybdenum enzymes. Both enzymes are inhibited by cyanide in their reduced states, and this process is reversed by oxidation and thiocyanate is not released. As a result, there does not appear to be a terminal sulfur atom to be removed by cyanide. The EXAFS results for sulfite oxidase (Cramer et al., 1981) indicate that there is loss of an oxo group in transferring from the oxidized to the reduced form. This result suggests a simple mechanism in which the molybdenum site acts as an oxo-transfer reagent converting sulfite to sulfate. The molybdenum(IV) generated then accepts an oxygen from water on reoxidation to Mo(VI).

C. Nitrate reductase

Nitrate reductases have been isolated from a variety of sources, plants, bacteria and fungi and always contain molybdenum. Two types of nitrate reductases can be recognized (i) the assimilatory nitrate reductases which catalyze the reduction of nitrate to nitrite (which is ultimately reduced to ammonia), and (ii) the dissimilatory nitrate reductases which use nitrate as a terminal electron acceptor in the absence of O_2.

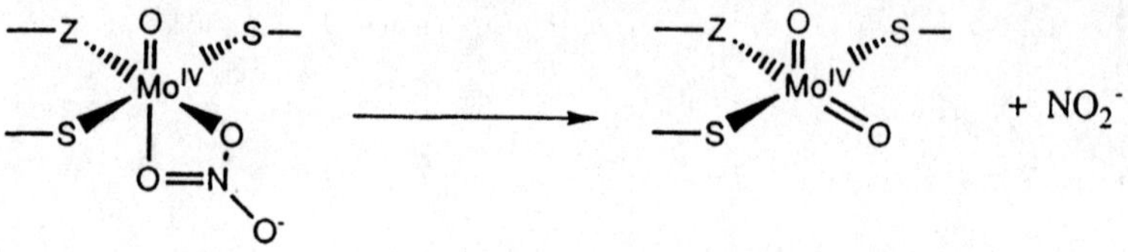

Figure 17.
A possible mechanism for nitrate reductase.

Nitrate reductase from *Chlorella*, an assimilatory enzyme is a tetramer of molar mass 360 kg mol^{-1}. Each subunit contains one Mo, one heme and a FAD. A possible mechanism is shown in Figure 17, in which an oxo group is transferred from nitrate to Mo(IV) with the production of nitrite and Mo (IV) (Hewitt and Notton, 1981).

REFERENCES

Albracht, S.P.J., Graf, E.-G., and Thauer, R.K. (1982). *FEBS Lett.*, *140*: 311.

Antanaitis, B.C., and Aisen, P.C. (1982). *J. Biol. Chem.*, *257*: 5330.

Antanaitis, B.C., Aisen, P.C., and Lilienthal, H.R. (1983). *J. Biol. Chem.*, *258*: 3166.

Beese, L.S., and Steitz, T.A. (1991). *EMBO J.*, *10*: 25.

Blakeley, R.L., and Zerner, B. (1984). *J. Mol. Catal.*, *23*: 263.

Byberg, J.R., Jorgensen, F.S., Hensen, S., and Hough, E. (1992). *Protein-Struct. Funct. Genet.*, *12*: 331.

Cammack, R. (1988). *Adv. Inorg. Chem.*, *32*: 297.

Cammack, R., Barber, M.J., and Bray, R.C. (1976). *Biochem. J.*, *157*: 469.

Clark, P.A., and Wilcox, D.E. (1989). *Inorg. Chem.*, *28*: 1326.

Clark, P.A., Wilcox, D.E., and Scott, D.A. (1990). *Inorg. Chem.*, *29*: 579.

Cole, J.L., Clark, P.A., and Solomon, E.I. (1990). *J. Amer. Chem. Soc.*, *112*: 9534.

Coleman, J.E., and Gettings, P. (1983). *Metal Ions in Biology*, Vol. 5 (T.G. Spiro, ed.), John Wiley, New York, Chapter 5.

Coughlan, M.P., ed. (1980). *Molybdenum and Molybdenum-containing Enzymes*, Pergamon, New York.

Cramer, S.P., Wahl, R., and Rajagopalan, K.V. (1981). *J. Amer. Chem. Soc.*, *103*: 7721.

Dalton, H. (1980). *Adv. Appl. Microbiol.*, *26*: 71.

Dalton, H., Smith, D.D.S., and Pilkington, S.J. (1990). *FEMS Microbiol. Rev.*, *87*: 201.

Davies, R.P. (1959). *J. Amer. Chem. Soc.*, *81*: 5674.

DeWitt, J.G., Bentsen, J.G., Rosenzweig, A.C., Hedman, B., Green, J.,Pilkington, S., Papaefthymiou, G.C., Dalton, H., Hodgson, K.O., and Lippard, S.J. (1991). *J. Amer. Chem. Soc.*, *113*: 9219.

Dixon, N.E., Gazzola, C., Blakeley, R.L., and Zerner, B. (1975). *J. Amer. Chem. Soc.*, *97*: 4131.

Dixon, N.E., Riddles, P.W., Gazzola, C., Blakeley, R.L., and Xerner, B. (1980). *Can. J. Biochem.*, *58*: 1335.

Fee, J.A. (1975). *Struct. Bonding (Berlin)*, *23*: 1.

Freemont, P.S., Friedman, J.M., Beese, L.S., Sanderson, M.R., and Steitz, T.A. (1988). *Proc. Nat. Acad. Sci. USA*, *85*: 8924.

Fujimoto, S., Ohara, A., Uehara, K. (1980). *Agric. Biol. Chem.*, *44*: 1659.

Getzoff, E.D., Trainer, J.A., Weiner, P.K., Kollman, P.A., Richardson, J.S., and Richardson, D.C. (1983). *Nature*, *306*: 287.

Green, J., and Dalton, H. (1989). *Biochem. J.*, *259*: 167.

Hammel, K.E., Cornwell, K.L., Diekert, G.B., and Thauer, R.K. (1984). *J. Bacteriol.*, *157*: 975.

Hardman, K.D., and Lipscomb, W.N. (1984). *J. Amer. Chem. Soc.*, *106*: 463.

Hay, R.W., and Govan, N. (1991). *J. Chem. Soc. Chem. Commun.*, *714*.

Hewitt, E.J., and Notton, B.A. (1981) *Molybdenum and Molybdenum Enzymes* (M.P. Coughlan, ed.), Pergammon Press, New York, p. 273.

Hikichi, S., Tanaka, H., Moro-oka, Y., and Kitajima, N. (1992). *J. Chem. Soc. Chem. Commun.*, 814.

Hough, E., Hansen, L.K., Birkness, B., Jynge, K., Hansen, S., Hordvik, A., Little, C., Dodson, E.J., and Derewenda, Z. (1989). *Nature*, *338*: 357.

Jabri, E., Lee, M.H., Hausinger, R.P., and Karplus, P.A. (1992). *J. Mol. Biol.*, *227*: 934.

Jacobson, R.R., Tykelar, Z., Farooq, A., Karlin, K.D., Liu, S., and Zubieta, J. (1988). *J. Amer. Chem. Soc.*, *110*: 3690.

Kannan, K.K., Notstrand, B., Fridborg, K., Lovgran, S., Ohlsson, A., and Petef, M. (1975). *Proc. Nat. Acad. Sci. USA*, *72*: 51.

Kannan, K.K., Petef, M., Fridborg, H., Cid-Dresner, H., and Lovgren, S. (1977). *FEBS Lett.*, *73*: 115.

Kim, E.K., and Wyckoff, H.W. (1991). *J. Mol. Biol.*, *218*: 449.

Kimura, E., and Koike, T. (1991). *Comments Inorg. Chem.*, *11*: 285.

Koike, T., and Kimura, E. (1991). *J. Amer. Chem. Soc.*, *113*: 8935.

Kono, Y., and Fridovich, I. (1983). *J. Biol. Chem.*, *258*: 6015.

Krzycki, J.A., and Zeikus, J.G. (1984). *J. Bacteriol.*, *158*: 231.

Lammers, H., and Follman, H. (1983). *Struct. Bonding (Berlin)*. *54*: 29.

Lippard, S.J., Burger, A., Ugurbil, K., Pantoliano, M., and Valentine, J.S. (1977). *Biochemistry*, *16*: 1136.

Meier, B., Barra, D., Bossna, F., Calabrese, L., and Rotillio, G. (1982). *J. Biol. Chem.*, *257*: 13977.

Messerschimdt, A., Rossi, A., Ladenstein, R., Huber, R., Bjolgnesi, M., Gatti, G., Marchesini, A., Petrazelli, R., and Finazzo-Agro, A. (1989). *J. Mol. Biol.*, *206*: 513.

Morrow, J., and Trogler, W.C. (1989). *Inorg. Chem.*, 2330.

Newton, W.E., and Otsuka, S., eds. (1980). *Molybdenum Chemistry of Biological Significance*, Plenum, London.

Nordlung, P., Sjoberg, B.-M., and Eklund, H. (1990). *Nature*, *345*: 593.

Norman, P.R., and Cornelius, R.D. (1982). *J. Amer. Chem. Soc.*, *104*: 2356.

Parayadeth, N., Newton, W.E., and Steifel, E.I. (1976). *J. Amer. Chem. Soc.*, *98*:538.

Ragsdale, S.W., Clark, J.E., Ljungdahl, L.G., Lundie, L.L., and Drake, H.L. (1983). *J. Biol. Chem.*, *258*: 2364.

Ragsdale, S.W., Ljungdahl, L.G., and Der Vertanian, D.V. (1983b). *J. Bacteriol.*, *155*: 124.

Rawji, G.H., and Milburn, R.M. (1988). *Inorg. Chim. Acta*, *150*: 227.

Rees, D.C., Lewis, M., and Lipscomb, W.N. (1983). *J. Mol. Biol.*, *168*: 369.

Ruegg, C., and Lerch, K. (1981). *Biochemistry*, *20*: 1256.

Schultz, C., Bertini, I., Viezzoli, M.S., Brown, R.D., Koening, S.H., and Coleman, J.E. (1992). *Inorg. Chem.*, *28*: 1490.

Sjöberg, B.-M., and Grassland, A. (1983). *Adv. Inorg. Biochem.*, *5*: 87.

Solomon, E.I. (1981). *Copper Proteins* (T.G. Spiro, ed.), Wiley-Interscience, New York, Chapter 2.

Sowadski, J.M., Handshumacher, M.D., Murthy, H.M.K., Foster, B.A., and Wyckoff, H.W. (1985). *J. Mol. Biol.*, *186*: 417.

Spilburg, C.A., Bethune, H.L., and Vallee, B.L. (1977). *Biochemistry*, *16*: 1142.

Steifel, E.I. (1977). *Prog. Inorg. Chem.*, 22: 1.

Stein, P.J. (1985). *Biochemistry, 24*: 2459.

Stein, P.J., Merrill, S.P., and Henkins, R.W. (1977). *J. Amer. Chem. Soc., 99*: 3194.

Stubbe, J.A. (1989). *Annu. Rev. Biochem., 58*: 593.

Sugiura, Y., Kawabe, H., and Tanaka, H. (1980). *J. Amer. Chem. Soc., 102*: 6581.

Sugiura, Y., Kawake, H., Tanaka, S., Fujimoto, S., and Ohara, A. (1981). *J. Biol. Chem., 256*: 10664.

Sugiura, Y., Kuwahara, J., and Suzuki, T. (1983). *Biochem. Biophys. Res. Commun., 115*: 878.

Sumner, J.B. (1926). *J. Biol. Chem., 70*: 97.

Tainer, J.A., Getzoff, E.D., Richardson, J.S., and Richardson, D.C. (1983). *Nature, 306*: 284.

Wilcox, D.E., Borras, A.G., Hwang, Y.T., Lerch, K., Winkler, M.E., and Solomon, E.I. (1985). *J. Amer. Chem. Soc., 107*: 4015.

Wilkins, R.G. (1992). *Chem. Soc. Rev.*, 171.

Woodland, M.P., Patel, D.S., Cammack, R., and Dalton, H. (1986). *Biochim. Biophys. Acta, 873*: 237.

Woolery, G.L., Power, L., Winkler, M., Solomon, E.I., Lerch, K., and Spiro, T.G. (1984). *Biochim. Biophys. Acta, 788*: 155.

Woolley, P. (1975). *Nature, 258*: 677.

ELECTRON TRANSFER PROCESSES IN METALLOPROTEINS

by

George McLendon

Department of Chemistry
University of Rochester
Rochester, NY 14627
U.S.A

I. INTRODUCTION

Metal ions are particularly well suited for redox processes: they can possess multiple stable oxidation states which can be differentially stabilized by specific ligand environments. This affords biology the ability to modulate the redox potentials of metalloenzymes over a very wide range (Fig. 1) without necessarily encountering the problematic reactivity associated with odd electron organic radicals. It is therefore no surprise that metal ions are used in a variety of contexts to carry out one (or multi) electron transfer processes (Table 1). The biological function of these electron transfer processes range from an ubiquitous role in energy transduction (as exemplified by the cytochromes of the electron transport system) to intermediary metabolism involving oxidation reduction reactions (Sigel and Sigel, 1991).

In redox metalloenzymes, the metal may directly participate in substrate activation, as in the oxidases and peroxidases which activate oxygen and may use bound oxygen atoms for further transformations of organic substrates. These topics are dealt with in some detail in other contributions (see Sections B14, C6 and C8 in this Chapter 1) and so will not be specifically treated here. In a second class of redox metalloenzymes, electrons are shuttled

Table 1

Examples of Metalloredox Proteins

General class	Example	Physiological reaction partner
Peroxidases	Cytochrome *c* peroxidase	Cytochrome *c*
Oxidases	Cytochrome oxidase	Cytochrome *c*
Oxygenases	Cytochrome P450	Cytochrome b_5 (microsomal)
Electron transport		
Cytochromes	Cytochrome *c*	Cytochrome c_1/Cytochrome oxidase
FeS Proteins	Ferredoxin	Ferredoxin reductase
Flavoproteins	Flavodoxin	Fe Protein nitrogenase
"Blue" proteins	Plastocyanin	Photosynthetic reaction center

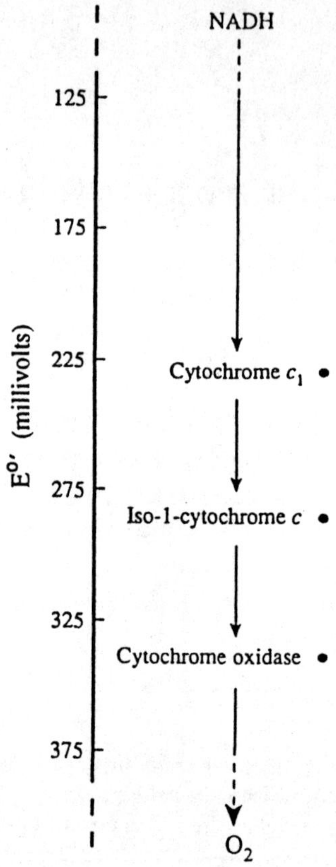

Figure 1.
Electron transport energies of oxidative phosphorylation.

to a reactive active site (e.g. flavin) from a redox active metal, which acts as a redox "mediator" (for a discussion of redox mediators, see Section C3 in this Chapter 1). A third category of redox interaction is germane to this overall monograph: metal ions may be mobilized for storage/release via oxidation reduction as in ferritin. Some discussion of this aspect is contained in other sections in this monograph.

In the short space of this review, it would be unrealistic and inappropriate to attempt to exhaustively survey all of these types of proteins. Instead, in the spirit of other chapters in this volume, I will briefly survey the basics of electron transfer processes which are common to all of the single electron transfer proteins. I will then indicate how these processes may be modulated by specific ligand effects, conformational effects, and by specific protein-protein recognition. These discussions will largely focus on simple electron transfer proteins, particularly cytochrome *c* and related systems, which are not treated elsewhere in this volume.

II. FUNDAMENTALS

The fundamental theory which describes electron transfer was first clearly laid out by Marcus (Marcus, 1956; Marcus and Sutin, 1985) over thirty years ago and subsequently confirmed in qualitative and quantitative detail (Marcus received the Nobel Prize in 1993 for this work). This theory has been reviewed in detail elsewhere. For present purposes, we only focus on a few key aspects.

Theory suggests (and experiments have confirmed) that if electron transfer to or from a compound causes a significant change in geometry, there will be a large activation energy associated with "reorganization energy" required by this geometry change. Taube (1970) pointed out that such effects will generally occur when electrons are added to or removed from metal bonding orbitals. A classic example is provided by redox reactions of high spin chromium:

$$Cr^{III}(t_2g^3) \rightarrow Cr^{II}(t2g^3\ eq^1)$$

A more common biological example is provided by $Cu^{II} \rightarrow Cu^{I}$ ($d^{10} \rightarrow d^9$) couples which invariably involve a metal-ligand antibonding electron.

Proteins might minimize the associated activation energy by providing a coordination geometry which is intermediate between that preferred by the oxidant (e.g. square planar for d^9 Cu(II)) and the reductant (e.g. tetrahedral (or even three coordinate) for d^{10} Cu(I)). Thus, an obvious approach to minimizing redox-associated structural change (and associated slow rates) is to use redox groups which are delocalized so that the geometry change on oxidation-reduction is small. It is thus no accident that many of the proteins involved in electron transfer are just such delocalized redox chromophores. Multicenter Fe clusters provide one such example; heme porphyrins a second example. Note that for the heme group, geometry changes are best minimized if the Fe group remains low spin (coordinate) so that only nonbonding $(t2g)^6 \rightarrow (t2g)^5$ electrons are exchanged. This is indeed the case for the redox mediator proteins like cytochrome *c* and cytochrome b_5.

Other factors also contribute to the "reorganization energy," λ. A key contribution comes from motion of the medium for a small molecule required to stabilize the charge via maximal charge-dipole interactions (Marcus, 1956; Marcus and Sutin, 1985). This medium is the solvent for a small molecule. The surrounding protein matrix may also reorganize for a redox chromophore buried in a protein.

Marcus' second insight was that this reorganization energy can be compensated by the available free energy, ΔG, of the reaction (Marcus, 1956; Marcus and Sutin, 1985). This compensation occurs (in the simplest model) as a parabolic function: the activation energy, $\Delta G^{\neq} = (\Delta G - \lambda)^2/4\lambda$. (In more complex models (Miller et al., 1984; Jortner, 1976), quantization of the vibrational modes which reorganize is considered and this simple parabolic form is altered.)

This insight leads to a surprising prediction: as reaction ΔG increases, $\Delta G^{\neq}$ decreases and the rate increases. (No surprise, yet.) However, when $\Delta G = \lambda$, the rate reaches a maximum. Any further increase in ΔG ($\Delta G > \lambda$) actually leads to a *decrease* in rate! This initially counter intuitive "inverted region" behavior has been amply documented (Miller et al., 1984; Jortner, 1976). In biology, the "inverted region" makes a significant contribution to the efficiency of initial charge separation in the photosynthetic center. The large macrocyclic chlorophyll molecules show little geometry change with oxidation-reduction, and the surrounding (hydrophobic) protein undergoes little reorganization, as well. Thus, λ is small. In fact, ΔG charge separation $\sim \lambda \sim 0.40$. By contrast, the charge recombination reaction occurs with a large ΔG, $\Delta G \sim 1V >> \lambda$. This reaction thus falls in the inverted region and is thereby slowed by orders of magnitude, permitting the high efficiency charge separation required for photosynthetic harvesting of solar energy (Fig. 2).

Finally, given the primacy of reaction free energy, ΔG, in determining electron transfer rates, it is necessary to briefly review those structural features which influence the redox potential which controls ΔG. The first and most obvious effects on potential in metalloproteins come from the metal-ligand interactions. As a first rule, "softer" ligands (e.g. S) will favor the lower oxidation state, while harder ligands (e.g. O) will favor a higher oxidation state. For example, for the basic cytochrome *c* structure, replacement of the native neutral sulfur donor, methionine, by an oxygen donor leads to a large shift in redox potential from ca +270 mV to ca 50 mV (Raphael and Gray, 1991).

A second insight into control of redox potential is based on the free energy for transfer of a charge into a dielectric cavity. Rees pointed out (1985) that transfer of a charged Fe(III) heme from water to a low dielectric cavity would destabilize the charged Fe(III) heme and stabilize the neutral Fe(II) heme. At a more microscopic level, merely changing dipoles associated with specific amino acids can significantly change the Fe(II)/Fe(III) potential. Thus, the replacement in cytochrome *c* of Asn^{52} by Ile^{52} shifts the potential by 50 mV;

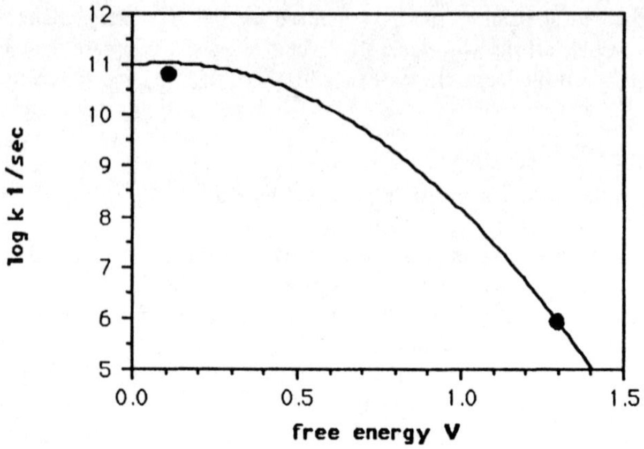

Figure 2.
Energetics rate plot for photosynthetic charge separation.

combining several such replacements can give shifts > 100 mV (Komar-Panicucci et al., 1992).

One final aspect of electron transfer is generally important in redox proteins. As already noted, the redox active group is often buried well within the protein matrix. As a result, electron transfer between one protein and another must occur over a "long distance"; i.e. long with respect to van der Waals contact (McLendon, 1988). For physiological protein-protein couples, these distances may range up to 24 Å between the metal atoms as in the cytochrome *c*/cytochrome *c* peroxidase couple (Fig. 3) (McLendon et al., 1992; Leonard and Yonetani, 1973). Consider the metal heme delocalization previously discussed, a more germane distance may be the heme edge-heme edge distance of ca 15Å. In order for an electron to traverse this significant distance, it must "tunnel" from the reductant to oxidant. (In this context tunneling is simply taken to mean that the electron has no significant resistance time localized on the intervening medium - there are no stable radical ion intermediates in the transfer process.) A (growing) consensus is that this tunneling occurs via "through-bond" delocalization via Hückel-type electron exchange from bond to bond, thus the common term "superexchange" (Miller et al., 1984; Jortner, 1976). Since through bond delocalization falls off rapidly with the number of bonds, it is expected that the probability (i.e. rate) of donor to acceptor electron transfer will also diminish rapidly with the distance (or equivalently, number of bonds) which separate them. In fact, many studies have shown that electron transfer rates decrease (essentially) exponentially with distance: k_{et} $\alpha_{exp}{}^{-\beta R}$ (Sigel and Sigel, 1991; Miller et al., 1984; Jortner, 1976; McLendon, 1988). (The simplest theory of "tunneling through a barrier" makes an equivalent prediction.) In one limiting view, this superexchange may occur via a very specific bonded pathway of covalent bonds and hydrogen bonds which link the oxidant and reductant. Some evidence for such pathways has been found by Gray and coworkers, building on a theoretical formalism developed by Beratan for enumerating such "pathways."

In a long and elegant series of studies, the Caltech group (and others) have attached redox active ruthenium centers to specific histidine residues on cytochrome *c* (Nocera et al.; 1984; Wuttke et al., 1992; Isied, S., 1991), as well as other proteins (Kostic et al., 1988). Reaction can be initiated by rapid reduction of the Fe(III) cytochrome *c* center, by flash photolysis of an external photoactive reductant, or pulse radiolysis, or, most recently, by photolysis of the bound ruthenium derivative. The results from many studies are hard to fit to a "universal distance model" but are better described by invoking a specific "pathway" for electron transfer to each individual derivative (Nocera et al., 1984; Wuttke et al., 1992; Isied, 1991).

In practice, the difference in electronic coupling between one path and another may be small, so that several pathways may contribute equally to the rate. In the limit where *many paths* contribute, the overall probability will scale simply with overall ("line of sight")

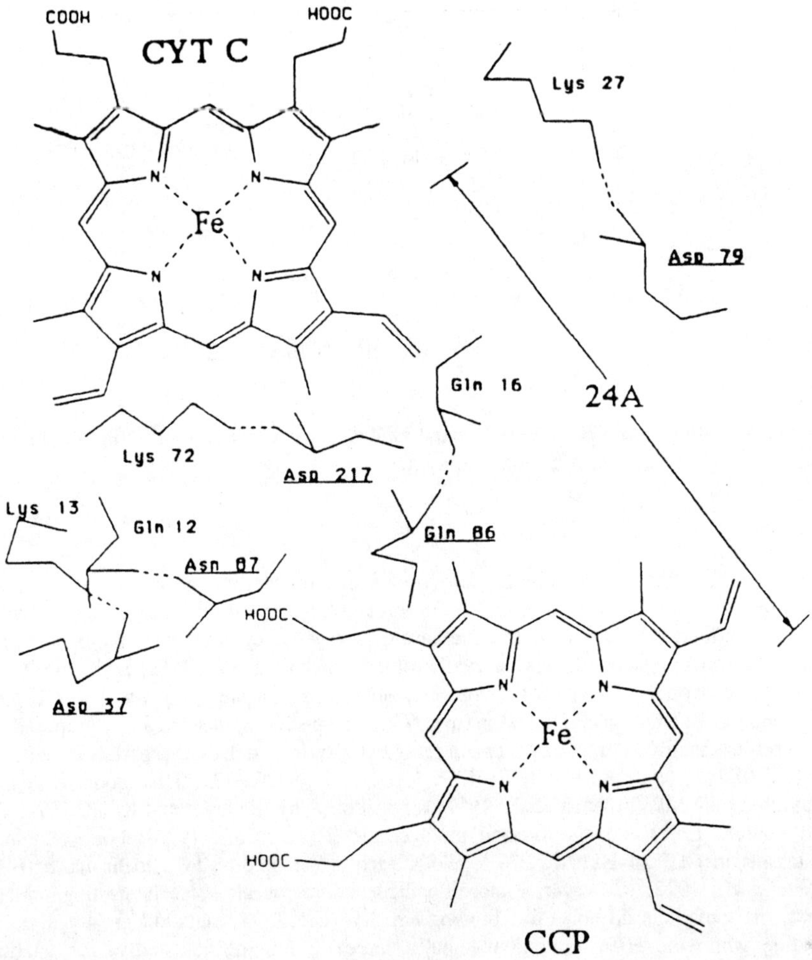

Figure 3.
Model structure for a kinetically active cyt-*c*:CCP complex.

distance. Some evidence for this limit has been presented by Dutton et al. (Moser et al., 1992).

III. RECOGNITION AND SPECIFICITY

In the previous section, we noted that redox metalloproteins seem designed so that the active metal is sequestered from the environment. Since electron transfer rates decrease rapidly (roughly exponentially) with distance, this sequestering ensures that *inter*molecular electron transfer will be relatively "slow" (i.e. less than collision controlled). In biological terms, by "burying" the redox chromophore, a direct "collision" of the oxidant and reductant is precluded. This is surely no accident. The sequestered redox group can only react with molecules which bind long enough to the protein surface to permit reaction over a long distance, thereby minimizing accidental, and biologically disastrous, redox "short circuits." In general, then, only those molecules which can specifically recognize and bind to a redox partner for ca 10^{-3} sec will be able to react (Sigel and Sigel, 1991; McLendon, 1988).* Thus, for protein to protein electron transfer, recognition and binding play a critical

* Obviously the situation is different for the large, separate class of intramolecular electron transfer proteins like cytochrome oxidase which contains multiple redox centers, so that electrons are funneled directly to the substrate (O_2) site; thereby preventing deleterious "short circuits."

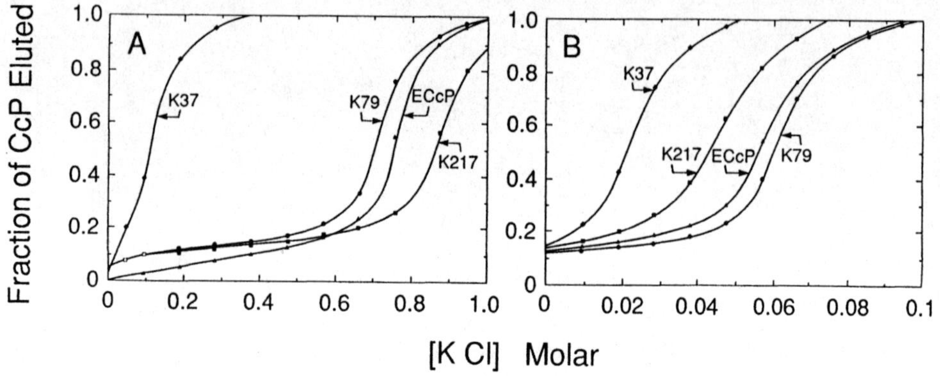

Figure 4.
Panels A (reduced) and B (oxidized) show binding curves to the cyt-*c* affinity column as a function of potassium chloride concentration.

role. As with other macromolecular recognition, electrostatic interactions appear to play an important (but not the only) role in recognition and binding. In this regard, the best studied redox metalloenzyme couple is the cytochrome *c* (cyt-*c*): cytochrome *c* peroxidase (CCP) system. Specific lysine residues on cytochrome *c*, including K72, K13, K81, and K27 are found to be strongly conserved in evolution, suggesting that they play a role in recognition and binding. For the physiological partner, CCP, a specific model has been proposed (Fig. 3) in posits specific interactions between these cyt-*c* lysines and CCP aspartates: (cyt-*c*)K72 → (CCP)E217, (cyt-*c*)K13 → (CCP)E37, (cyt-*c*)K27 → CCPE79. This model is generally supported by NMR (Moench et al., 1992), crystallography (Pelletier and Kraut, 1992; Kraut and Poulos, 1981) and site-directed mutagenesis (Hake et al., 1992). For example, the mutations in CCP of E217K, E37K, E79K each affect binding to cytochrome *c* (Fig. 4) (Hake et al., 1992). However, a more detailed picture reveals some interesting subtleties. First, the mutations do not affect binding equally. The E37K mutation greatly diminishes binding while the E79K mutation actually increases binding! Secondly, the qualitative effects of mutation differ from (substrate) Fe(II)cyt-*c* and (product) Fe(III)cyt-*c*: Fe(II)cyt-*c* and Fe(III)cyt-*c* appear to recognize different subsites within the overall binding domain. Thus, a redox reaction at the heme center of cytochrome *c* appears to be transmitted to the binding interface.

It is clear that we are only beginning to understand these critical aspects of molecular recognition in the context of metalloprotein electron transfer.

REFERENCES

Hake, R., Corin, A., and McLendon, G. (1992). *J. Amer. Chem. Soc.*, *114*: 5442.

Isied, S. (1991). *Metal Ions in Biological Systems*, Vol. 27 (H. Sigel and A. Sigel, eds.), Marcel Dekker, New York, p. 1.

Jortner, J. (1976). *J. Chem. Phys.*, *64*: 4860.

Kostic, N., Magalit, R., Che, C., and Gray, H.B. (1988). *J. Amer. Chem. Soc.*, *105*: 7765.

Komar-Panicucci, S., Bakker, G., Bixler, J., and McLendon, G. (1992). *J. Amer. Chem. Soc.*, *114*: 5443.

Kraut, J., and Poulos, T. (1981). *J. Biol. Chem.*, *255*: 10322.

Leonard, J.J., and Yonetani, T. (1973). *Biochemistry*, *13*: 1465.

Marcus, R. (1956). *J. Chem. Phys.*, *24*: 966.

Marcus, R., and Sutin, N. (1985). *Biochim. Biophys. Acta*, *74*: 765.

McLendon, G. (1988). *Accounts Chem. Res.*, *21*: 160.

McLendon, G., Zhang, Q., Wallin, S., Miller, R., Billstone, V., Spearks, K., and Hoffman, B. (1993). *J. Amer. Chem. Soc.*, *115*: 3665.

Miller, J. R., Calcaterra, L.T., and Closs, G. (1984). *J. Amer. Chem. Soc.*, *106*: 3047.

Moench, S., Chonia, S., Lou, B., Emmons, J., and Satterle, J. (1992). *Biochemistry*, *31*: 3661.

Moser, C., Keske, J., Warnke, K., Farid, R., and Dutton, L. (1992). *Nature*, *305*: 796.

Nocera, D., Winkler, J., Yocum, X., Bordignon, E., and Gray, H.B. (1984). *J. Amer. Chem. Soc.*, *106*: 5145.

Pelletier, H., and Kraut, J. (1992). *Science*, *258*: 1748.

Raphael, A., and Gray, H.B. (1991). *J. Amer. Chem. Soc.*, *113*: 1038.

Rees, D. (1985). *Proc. Nat. Acad. Sci. USA*, *82*: 3082.

Sigel, H., and Sigel, A., eds. (1991). *Metal Ions in Biological Systems*, Vol. 27, Marcel Dekker, New York.

Taube, H. (1970). *Electron Transfer Reactions of Complex Ions*, Academic Press, New York.

Wuttke, D., Bjerrum, M., Winkler, J., and Gray, H.B. (1992). *Science*, *256*: 1007.

THE ELECTROCHEMISTRY OF METALLOPROTEINS

by

Pedro de Oliveira and H. Allen O. Hill

New Chemistry Laboratory
South Parks Road
Oxford OX1 3QT
U.K.

I. INTRODUCTION

All living organisms satisfy their energy needs by employing redox reactions that they efficiently undertake and control. The electrons in these reactions travel sequentially along a series of chemical species that together constitute an electron transfer chain. At some points of this chain, the energy derived from the electron transfer is coupled to the synthesis of energy-rich compounds, such as ATP, that will be subsequentially consumed in metabolic pathways. Several other redox processes take place in living organisms, related to toxin disposal, parasite killing or biosynthesis.

Intimately involved in these events are proteins and enzymes having at least one metal atom in their native structure. These metalloproteins, which behave as electron shuttles, and metalloenzymes which participate in redox processes, have been the subject of a number of several electrochemical investigations (Betso et al., 1972, and references therein; Hawkridge and Kuwana, 1973). However, all the early methods required the use of mediators, i.e. electrochemically active, low-molar-mass (l.m.m.) species that convey electrons between a working electrode and the biomolecules. Although these indirect methods can provide reliable thermodynamic information, a number of problems may arise when kinetic parameters of multicomponent systems are desired, since the mediator will not discriminate between the various redox centers. As a consequence, the quest for *direct* metalloprotein electrochemistry, which allows these problems to be circumvented, started in the middle of the century.

The first attempts of direct (un-mediated) metalloprotein electrochemistry made use of bare metal electrodes. This straightforward way of trying to apply the experimental procedures employed with l.m.m. chemical species to the study of proteins turned out to be rather unsuccessful. In fact, the polarographic techniques gave poor results with cytochrome *c* (cyt *c*) (Betso et al., 1972, and references therein), leading to irreversible protein adsorption on the mercury electrode, probably with concomitant conformational change and denaturation. Apart from mercury, other metals used as working electrodes, such as gold (Heineman et al., 1975), platinum (Kono and Nakamura, 1958), nickel (Kuznetsov et al., 1979) and silver (Cotton et al., 1980), gave similarly disappointing results.

This marked incompatibility between bare metal electrodes and proteins led to two different approaches in the quest for fast, reversible electron transfer between an electrode and a redox protein: (i) the use of different materials as working electrodes; (ii) the establishment of experimental methods aiming at modifying in some way the surface of existing electrodes. In the mid-seventies, these strategies, independently put into practice by two different groups, allowed a major step forward in this field of research when the reversible behavior of cyt *c* at a tin-doped indium oxide electrode (Yeh and Kuwana, 1977) and at a 4,4'-bipyridyl surface-modified gold electrode was reported (Eddowes and Hill, 1977) (Fig. 1). The latter was later confirmed to be a one-electron quasi-reversible process with no loss of biological activity, implying that the protein kept its biological integrity (Eddowes and Hill, 1979a).

The mechanism of the redox reaction of cyt *c* at a 4,4'-bipyridyl-modified gold electrode encompasses 5 different steps that occur in sequence (Fig. 2) (Hill and Higgins, 1981; Albery et al., 1981):

(i) diffusion of the protein towards the electrode surface;
(ii) adsorption at the electrode surface in an electron-transfer prone orientation;
(iii) electron transfer;
(iv) desorption of the resultant protein from the electrode surface;
(v) diffusion away from the electrode surface.

It is crucial that the association taking place in step (ii) is such that it does not alter the integrity of the protein; instead, the electron transfer ensues and the product is detached from the electrode surface. This process resembles the interaction between proteins and their biological partners in the sense that an effective interaction occurs when several features of the two surfaces, in this case, that of the electrode and the protein, have been matched.

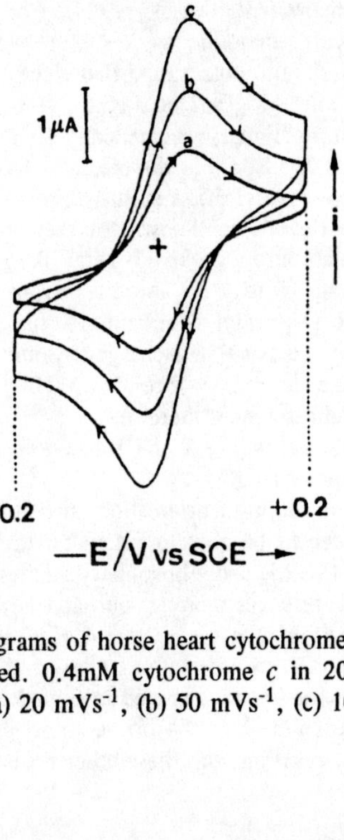

Figure 1.
Direct-current cyclic voltammograms of horse heart cytochrome *c* at a gold electrode upon which 4,4'-bipyridyl is adsorbed. 0.4mM cytochrome *c* in 20 mM phosphate, 100 mM perchlorate, pH 7. Scan rates (a) 20 mVs^{-1}, (b) 50 mVs^{-1}, (c) 100 mVs^{-1} (reproduced from Frew and Hill, 1988).

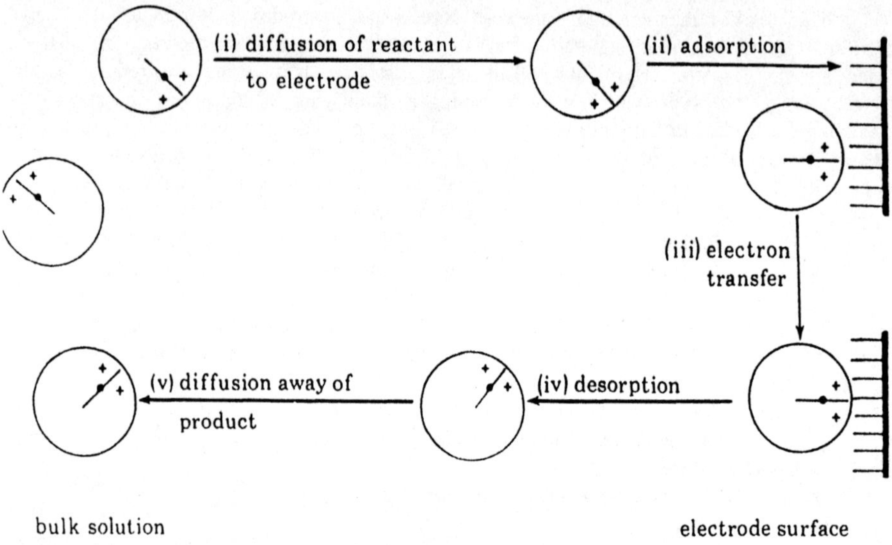

Figure 2.
A scheme showing the proposed mechanism for the reduction of horse heart cytochrome *c* at a gold electrode upon which 4,4'-bipyridyl is adsorbed (reproduced and adapted from Hill and Higgins, 1981).

II. ELECTROCHEMISTRY OF REDOX PROTEINS

A. Gold electrodes

Since the use of 4,4'-bipyridyl as a modifier, several other compounds have been found to promote cyt *c* electrochemistry (Allen et al., 1984). For example, bis(4-pyridyl)disulfide, SS-Bipy, deserves a particular mention, as it was proved to adsorb so strongly on the surface of gold that it is possible to obtain modified electrodes simply by dipping them in an SS-Bipy solution for a few minutes (Taniguchi et al., 1984).

Most compounds are bifunctional, schematically X~Y, having the following properties:

(i) the functional group X adsorbs to the metal surface, ideally in an irreversible way, and usually via a nitrogen (4-pyridyl N), a sulfur (thiol SH or disulfide S) or a phosphorus (phosphine P) atom. These ligands are known for their σ-donor, π-acceptor synergic-type behavior, suited for coordination to a group 11 metal like gold;

(ii) the functional group Y may be anionic (carboxylate, sulfonate, phosphonate or phosphate) or weakly basic (4-pyridyl N or aniline-type amine, Ar-NH_2). These can salt bridge or hydrogen-bond to the positively charged lysine (Lys) residues that surround the solvent exposed heme edge in cyt *c* (Frew and Hill, 1988, and references therein; Armstrong et al., 1988a, and references therein);

(iii) the molecular structure linking X and Y has no special requirements concerning its flexibility, aromatic character or length;

(iv) the modifier has to adopt an orientation such that the Y functionality is directed towards the solution in order to be able to interact with the protein molecules. Surface-enhanced Raman scattering (SERS) and ellipsometry studies suggested that this was, indeed, the case (Hill, 1987a, and references therein; Guo and Hill, 1991, and references therein), and a model for the interaction of cyt *c* with a 4,4'-bipyridyl-modified gold electrode is shown in Figure 3A. The surface conformations of several modifiers known to be effective in promoting cyt *c* electrochemistry are depicted in Figure 3B. It should be noted, however, that sulphide was shown (Allen et al., 1984) to be an effective promoter so that elaborately modified surfaces formed by reaction with these other promoters are not essential.

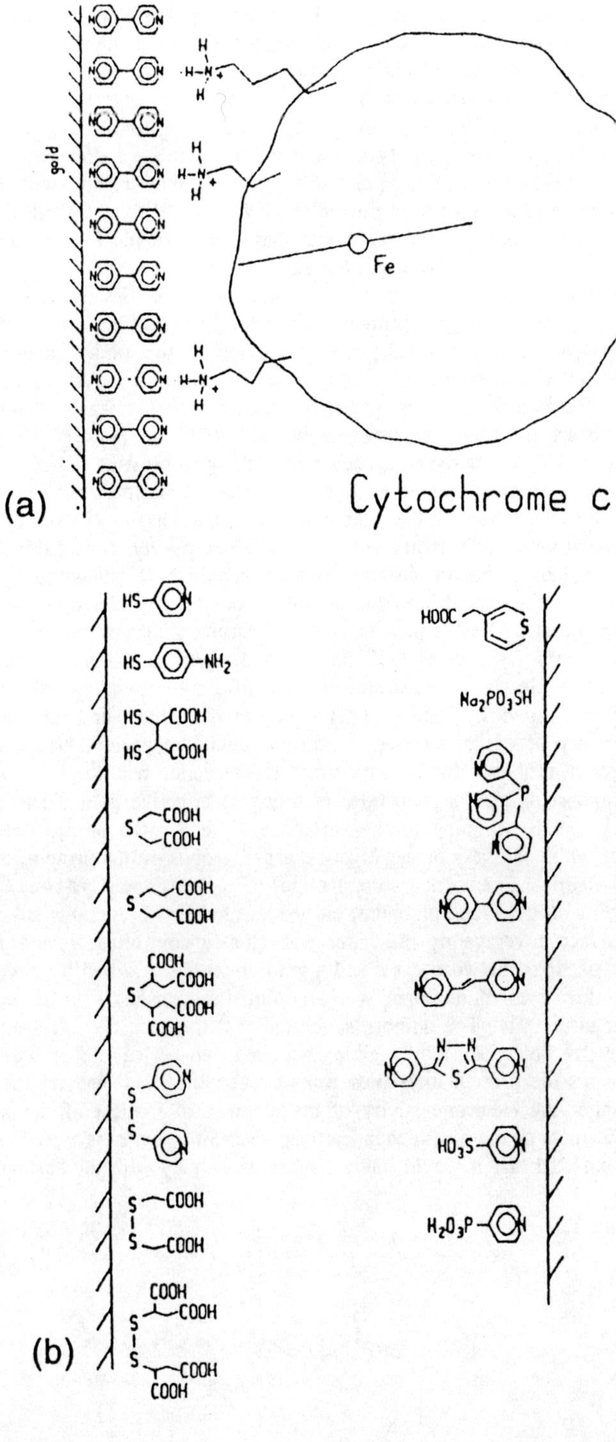

Figure 3.
A. Illustration of the interaction between a 4,4'-bipyridyl-modified gold electrode and a molecule of cyt *c* envisaged for the idealized limiting case of a planar surface (reproduced from Armstrong et al., 1988a).
B. Structures and proposed surface conformations of surface modifiers for the promotion of the direct electrochemistry of cytochrome *c* (reproduced from Allen et al., 1984).

It came as no surprise when it was discovered that the electrochemical process is inhibited by poly-L-lysine. Initially it was thought that this compound binds to the modified surface via similar interactions, blocking the access of cyt *c* molecules to it and therefore preventing electron transfer from occurring. Recent studies suggested that poly-L-lysine displaces the modifier from the electrode surface, destroying its binding domains (Hill et al., 1987b). Interestingly enough, poly-L-lysine is known to be a competitive inhibitor of the cytochrome *c* oxidase reaction. In the interaction of this enzyme with its biological partner, cyt *c*, Lys residues have been proved to play a crucial role. Chemical modification of these residues leads both to the biological and electrochemical inactivation of cyt *c* (Eddowes and Hill, 1979b, and references therein).

It soon became evident that the electrode surface functionalities suitable for interaction with cyt *c*, a positively charged protein, were not at all effective in promoting the electrochemistry of negatively charged proteins. Obviously, these have surface groups that will hydrogen-bond or salt bridge to weakly acidic or positively charged functionalities. However, if the electrode surface were to be modified in such a way as to exhibit positively and negatively charged groups simultaneously, it should be possible to promote the electrochemistry of either positively or negatively charged proteins. This was confirmed employing chemical species known as pyridinealdehydethiosemicarbazides, PATS, that adsorb to the gold surface via both the S atom and the semicarbazide N atom (Fig. 4).

The interaction with proteins in solution is based on pyridyl N (capable of hydrogen-bonding to Lys residues) and on weakly acidic thioamide N-H (suited to hydrogen-bond carboxylate groups). These modifiers successfully promoted the electrochemistry of cyt *c* (positively charged), as well as of plastocyanin and carboxydinitrophenyl derivatives of cyt *c* (negatively charged) (Hill et al., 1985a). The importance of the orientation of the interacting groups is clearly demonstrated by the different promoting capabilities of the three PATS isomers shown in Figure 5 (Armstrong et al., 1988a, and references therein). The electrochemistry of cyt *c* improves markedly when the pyridyl N atom is directed towards the solution, probably due to more effective interaction with the Lys residues.

Several amino acids and oligopeptides emerged as effective gold electrode modifiers (Di Gleria et al., 1986), binding to the surface via the S atom of cysteine (Cys), and interacting either with positively or negatively charged groups on the protein, depending on their own functionalities. Examples are the peptides $(Cys\text{-}Lys)_2$ and $(Cys\text{-}Glu)_2$.

A recent study on gold (and platinum) electrodes modified by various ad-atoms, where the electrode surface coverage by the latter was closely controlled, revealed that direct electron transfer occurred between cyt *c* and a gold electrode modified by silver, mercury, tin, antimony, sulfur or selenium atoms, with an optimal response at a surface coverage of ~ 45% (Shibata et al., 1991). The authors suggested that the modified electrode surface is composed of several holes, each hole giving access to an active site of bare gold. This topography prevents the protein molecules from irreversible adsorbing on the surface. In addition, the higher the electronegativity of the element component of the ad-layer, the higher the enhancement of the electrochemical response. Since cyt *c* is a positively charged protein, it was expected that it would interact more effectively with the surface having the

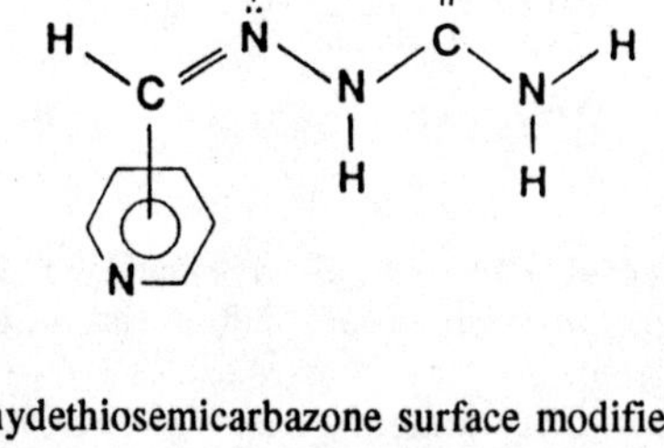

Figure 4.
Structure of pyridine aldehydethiosemicarbazone surface modifiers (reproduced from Frew and Hill, 1988).

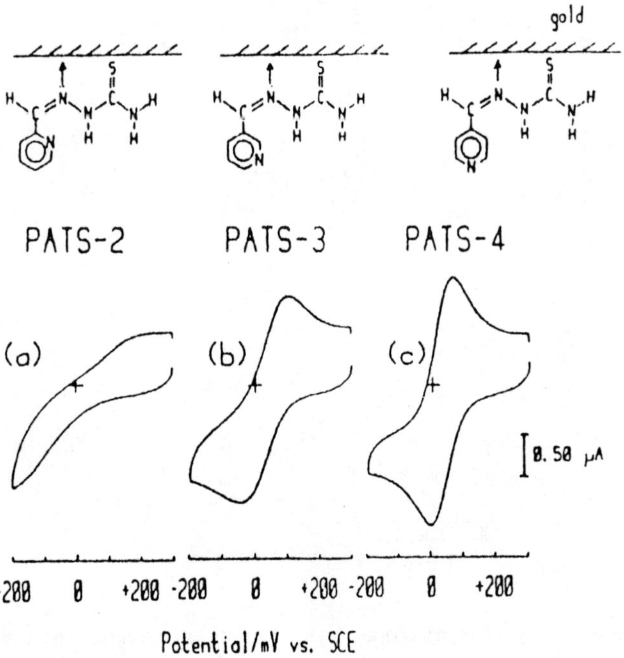

Figure 5.
Systematic improvement in cytochrome *c* electrochemistry observed as the isomer of the surface promoter pyridine-*n*-carboxaldehyde thiosemicarbazone is changed from (a) $n = 2$ to (b) $n = 3$ to (c) $n = 4$ (reproduced from Armstrong et al., 1988a).

highest electron density, imparted by the most electronegative elements: this, indeed, proved to be the case.

B. Pyrolytic graphite electrodes

Pyrolytic graphite is formed by the deposition of carbon from the vapor phase, and has a crystalline structure close to that of an ideal graphite crystal (Walker, 1962). This allotrope of carbon is anisotropic, and two distinctive types of surface can be obtained, depending on the plane along which it is cleaved (Panzer and Elving, 1975): a basal plane surface, hydrophobic in nature and ideally with all the carbon valences satisfied; an edge plane surface, hydrophilic in nature and with several C-O functionalities.

The first studies employed a polished basal plane pyrolytic graphite electrode (BPG) and reported the direct electrochemistry of four negatively charged proteins in the presence of multivalent cations (either Mg^{2+} or $[Cr(NH_3)_6]^{3+}$) (Armstrong et al., 1984a). Later, it was realized that cyt *c*, a positively charged protein, gave rise to markedly different responses, depending on the type of surface employed: sigmoid-shaped or peak-shaped voltammograms, when a BPG or an edge plane pyrolytic graphite electrode (EPG) was used, respectively. It was also noted that no multivalent cations were needed, and that the peak current was higher when an EPG was employed (Armstrong et al., 1984b). Figure 6 shows the cyclic voltammetric response of cyt *c* at various pyrolytic graphite surfaces and its relation to the amount of oxygen on the latter, as revealed by X-ray photoelectron spectroscopy (XPS). It is obvious that the higher the O/C ratio, i.e. the higher the amount of C-O functionalities on the surface, the higher the peak-shaped character of the voltammograms.

These C-O functionalities can be phenolate, carbonyl, ethoxy or carboxylate groups that impart an apparent pK of about 5.6 to the electrode surface. In other words, the electrode surface is negatively charged at pH>5.6, and therefore positively charged

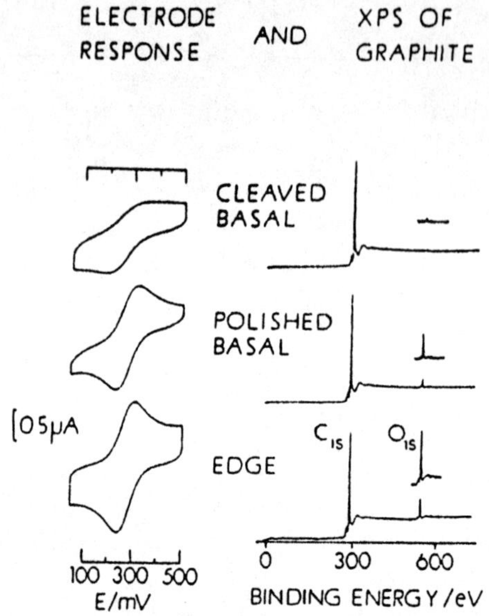

Figure 6.
DC cyclic voltammetry of cytochrome *c* at various surfaces of pyrolytic graphite characterized by X-ray photoelectron spectroscopy (XPS). For the voltammetry, measured on the fourth scan at 20 mVs^{-1}, cytochrome *c* is 0.15 mM in 5 mM Tricine, 0.10 M NaCl, pH 8.0, temperature 20°C. The potential scale is referenced against the normal hydrogen electrode. For XPS, a scale enlargement (x3) is shown for the O_{1s} peak (reproduced and adapted from Armstrong et al., 1988a).

proteins, e.g. cyt *c*, can effectively interact with it, while negatively charged ones, e.g. ferredoxin or plastocyanin, require the presence of multivalent cations to screen the domains of negative charges from each other, allowing protein and electrode surface to come close enough for electron transfer to ensue.

The importance of the surface functionalities was again evident in a study where EPG electrodes were modified by covalently linking chromium complexes to their surface (Armstrong et al., 1985). The modified electrodes, now having a positively charged surface, showed well-behaved electrochemistry with the negatively charged protein plastocyanin, whereas with cyt *c* (positively charged) a marked inhibition of heterogeneous electron transfer was observed.

The effect of multivalent cations on the electrochemistry of negatively charged proteins led to the investigation of the possibility of promotion by more complex molecules. These should bear positively charged groups at definite positions that could better match the protein domains of negative charge accumulation, termed cavities (Armstrong et al., 1988a, and references therein). This role has been successfully fulfilled by several aminoglycosides, which have several $-NH_3^+$ groups on a quasi-rigid framework (Fig. 7), and have been widely employed in the electrochemical study of proteins containing iron-sulfur clusters, ([Fe-S] proteins) (Armstrong et al., 1988b; 1989a; 1989b; George et al., 1989; Thomson et al., 1990; Butt et al., 1993), as well as several enzymes (Armstrong and Lannon, 1987; Hill and Hunt, 1993, and references therein). In many cases, the optimization of the electrochemical response requires that the aminoglycoside with the best 3-dimensional structure and charge distribution is found.

These [Fe-S] proteins are widespread among metabolic pathways involving redox processes. Their electrochemical characterization has largely contributed to the elucidation of both their mechanism of action and their role in vivo. Undoubtedly, the major breakthroughs have been achieved by using EPG electrodes and aminoglycosides as

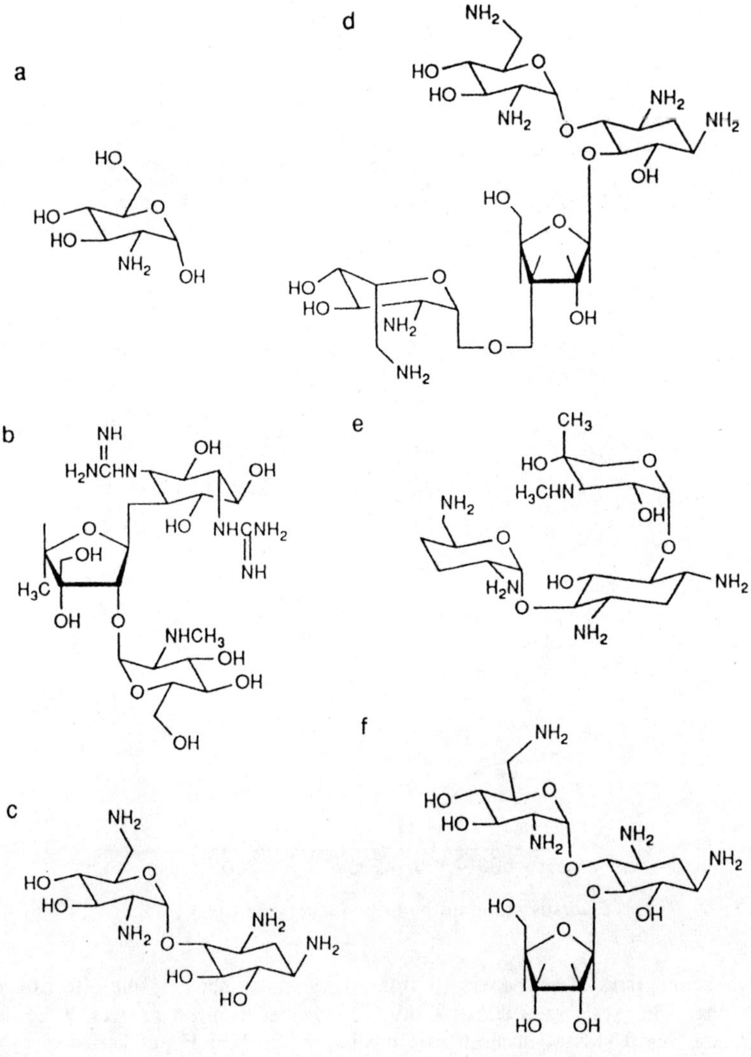

Figure 7.
Members of the aminoglycoside family: (a) glucosamine, (b) dihydrostreptomycin, (c) neamine, (d) neomycin, (e) gentamycin, (f) ribostamycin (reproduced from Hill and Hunt, 1993).

promoters. It is worth mentioning here the contribution of electrochemistry in the study of ferredoxins from different bacteria (Armstrong et al., 1988b; 1989a; 1989b; George et al., 1989; Thomson et al., 1990), that yielded information on the uptake of bivalent cations with concomitant conversion of the [3Fe-4S] cluster into the [M3Fe-4S] (M = Fe, Zn, Cd, Tl) derivative (Fig. 8). A recent paper gives a good account on the functional diversity of [Fe-S] clusters, emphasizing the possibility of modulating their redox potential by exogenous ligands (Butt et al., 1993, and references therein).

Until the end of the last decade, the model describing protein-electrode interaction assumed linear diffusion mass transport of the former to the latter. This means that the electrode was supposed to have a large electroactive surface area relative to the size of the protein, resulting in a peak-shaped voltammogram. Several heterogeneous rates of electron transfer relative to many proteins were calculated according to this model and a wide range of figures was obtained, in contrast to the usually fast values for homogeneous rates of

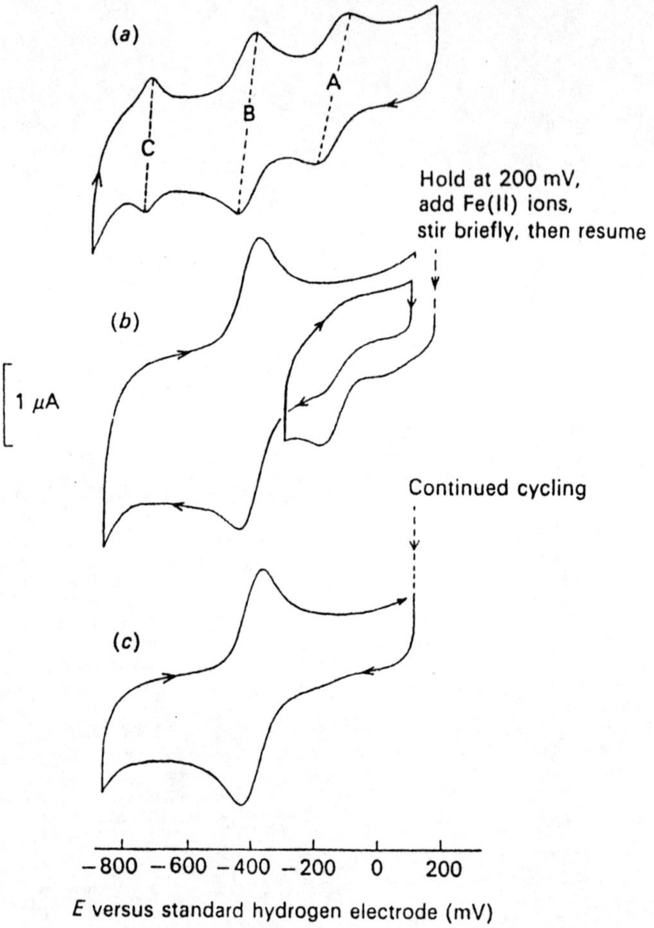

Figure 8.
Cyclic voltammograms of ferredoxin III from *D. africanus* showing the effect of adding iron(II) ions. The scan rate was 16 mVs^{-1} and the temperature was 2°C. Protein concentration was 0.11 mM in electrolyte medium of 20 mM Hepes buffer, pH = 7.4, containing 0.10 M $NaClO_4$, 0.10 mM EGTA and 1.1 mM neomycin. (a) Before the addition of iron(II) ions; (b) after hold at +200 mV for addition of iron(II) ions (to a total of 0.21 mM) and brief stirring by 'microflea'; (c) continued scan from (b) with no re-stirring, shows that the waves associated with the [3Fe-4S] cluster (A and C) have disappeared completely (reproduced from George et al., 1989).

electron transfer. However, in 1989 another model, based on radial diffusion of the proteins to microscopic sites scattered about the electrode surface (Armstrong et al., 1989c), was considered. With this model, sigmoidal current-voltage curves were expected, and the heterogeneous rates of electron transfer calculated were comparable in magnitude to the ones obtained in a detailed study (Albery et al., 1981), consistent with reversible electrochemistry in both situations (Fig. 9). Several experiments supporting the proposed model were conducted with EPG and BPG electrodes (Armstrong et al., 1989c). It was possible to rationalize many of the results obtained in the past if one accepted rapid rates of electron transfer at specific sites on the electrode surface, again stressing the selectivity exhibited by proteins in contrast to l.m.m. chemical species. The observation of what had previously been called either well-behaved or irreversible electrochemistry could be

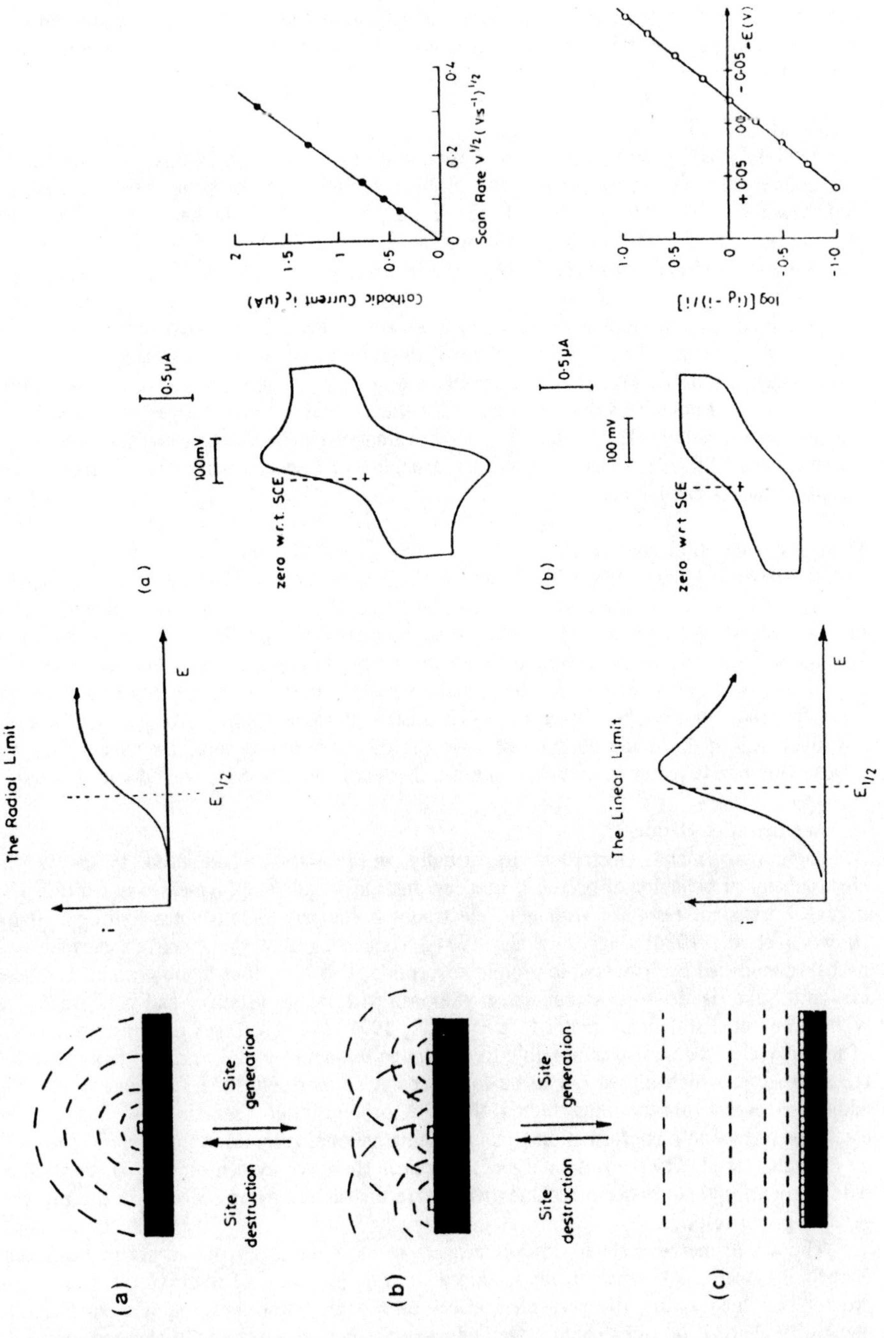

Figure 9 (*turned*).
Left: schematic representation of the conversion of radial to linear diffusion, as the density of specific electroactive surface sites increases; Right: cyclic voltammograms for reduction of 150 μM cyt *c* in a 5 mM Tricine buffer (pH = 8.0) containing 100 mM at (a) polished EPG and (b) freshly cleaved BPG electrodes. Scan rate, 20 mVs^{-1}; temperature, 20°C. Also shown are a plot of peak current for the reduction process at the edge-plane electrode vs the square-root of the scan rate, $v^{1/2}$, verifying the predominance of linear diffusion for (a) and a plot of log $[(i_d - i)/i]$ vs the potential E at the basal-plane electrode verifying the predominance of radial diffusion for (b) (reproduced from Armstrong et al., 1989c).

explained on the basis of the "concentration" of microscopic active sites on the electrode surface, having as limiting situations either the linear or the radial diffusion mass transport, respectively (Fig. 9).

C. Metal oxide electrodes

The well-behaved cyclic voltammetry of cyt *c* at a tin-doped indium oxide electrode was the first successful report concerning protein electrochemistry using metal oxide electrodes (Yeh and Kuwana, 1977). In subsequent years, other materials have been tried: ruthenium dioxide, iridium dioxide and sodium tungstate electrodes in the study of cyt *c*, ferredoxin and rubredoxin (Harmer and Hill, 1984; 1985); tin-doped indium oxide and fluorine-doped tin oxide electrodes in the study of cyt *c* (Bowden et al., 1984). Both positively and negatively charged proteins respond when these electrodes are employed, depending on the conditions of the experiment, namely the pH, the ionic strength and the presence or absence of multivalent cations. This means that there are groups on the electrode surface exhibiting non-neutral acid-base behavior. In fact, it is believed that surface charges arise due to the unequal adsorption of H^+ and HO^- ions by amphoteric hydroxyl groups present on it (Harmer and Hill, 1985). Again, a proper matching of surface charges is required so that electron transfer can take place.

D. Glassy carbon electrodes

The structure of glassy carbon has been described as thin, tangled ribbons of graphite-like sheets, therefore being isotropic (Yamada and Sato, 1962). The surface characteristics of electrodes made of this material are similar to the ones found on EPG electrodes. Hence, it is not surprising that quasi-reversible electrochemistry of cyt *c* has been observed when they have been used (Frew and Hill, 1988, and references therein). Strangely enough, it has recently been reported the direct electrochemistry of three copper proteins, all of them negatively charged, in the absence of any sort of promoter or modifier (Sakurai et al., 1990). This fact is rather remarkable, and a consistent explanation is awaited with interest.

E. Bare metal electrodes

Although bare metal electrodes are usually unsuccessful when used to probe the electrochemical behavior of proteins, in a few instances their employment was fruitful. Cyt *c* gave a transient response with gold electrodes previously cleaned in a hydrogen flame (Bowden et al., 1984; Sagara et al., 1991), suggesting that the electrode surface was probably modified by irreversible protein adsorption. Cyt c_3, a four-heme protein, has been known to give rise to an electrochemical response with mercury, silver and gold electrodes without any mediator or promoter (Sagara et al., 1991, and references therein). In the case of mercury electrodes, it was possible to determine the number of electrons involved in the redox process, which turned out to be four per cyt c_3 molecule (one per heme group). In addition, several observations hinted that the protein might remain adsorbed on the electrode surface in a conformation close to the native one, responding in a quasi-reversible way (Sagara et al., 1991, and references therein). However cytochrome c_3 is an atypical redox protein with four heme residues which are distributed symmetrically in the protein, extending into solution.

The use of mercury electrodes has been reported in another work where two different ferredoxins were polarographically reduced in the presence of multivalent cations as promoters. Once again, the screening effect of the promoter was required so that the negatively charged proteins could effectively interact with the negatively charged electrode surface (Bianco and Haladjian, 1986).

III. ELECTROCHEMISTRY OF METALLOENZYMES

Metalloenzymes are usually more complex entities than redox proteins. They are generally bigger in size and their active centers are commonly located within the 3-dimensional protein structure.

Metalloenzymes often have redox metalloproteins as biological partners. This implies that there must be a route leading the electron(s) from the enzyme surface to the buried active site (or vice versa, depending on the biological process). In principle, these sorts of metalloenzymes will respond electrochemically at an electrode surface and are designated extrinsic enzymes. On the other hand, enzymes that have their active loci well buried in the protein framework and that carry out the whole redox process in that location without the intervention of any electron shuttle are called intrinsic enzymes. These are not expected to give rise to any sort of useful electrochemical response (Frew and Hill, 1988, and references therein; Hill, 1990) unless the active site is close to the surface of the enzyme.

The achievement of direct electrochemistry of metalloenzymes has been more difficult than that of redox proteins. However, it has been claimed to have been observed by several groups in the last few years. These studies involve peroxidases, laccase, flavocytochrome b_2, amino acid oxidase, xanthine oxidase and glucose oxidase (Guo and Hill, 1991, and references therein). Nevertheless, it is not uncommon that the response is due to other electroactive centers in the enzyme, different from the active site, but involved in the redox process, as is the case of the flavin moieties in flavoenzymes.

A. Mediation

In most cases, it is necessary to resort to electrochemically active species that will convey electrons from the electrode surface to the metalloenzyme (or vice versa). The former is called mediator, and thence the process is generically termed mediated electrochemistry. Several species have been employed as mediators (Fultz and Durst, 1982; Kwee, 1986): inorganic complexes, organometallic compounds, organic molecules and redox proteins.

Ferrocene [bis(η^5-cyclopentadienyl)iron], an organometallic compound, proved to be a versatile mediator, not only due to its electrochemical behavior (iron (II) and iron (III) as the accessible oxidation states, slow autoxidation of the low oxidation state, and fast intermolecular electron transfer), but also owing to the fact that substituents can be introduced on either or both of the cyclopentadienyl rings, resulting in a shift of the formal redox potential (Frew and Hill, 1988, and references therein; Hill, 1987a, and references therein). As a consequence, as the number of choices for the most suitable mediator increases, the chances of improving the electrochemical response of the enzyme become higher.

In some cases, it has been possible to attach chemically ferrocene derivatives to enzymes, obviating the use of mediators both free in solution or immobilized on the electrode surface (Frew and Hill, 1988, and references therein; Bartlett et al., 1987).

B. Applications

The electrochemistry of metalloproteins and metalloenzymes has a wide range of applications. Perhaps the most obvious use concerns analytical chemistry and biochemistry. In fact, the high degree of specificity shown by enzymes towards their substrates has led to the development of some methods suitable for clinical analysis. Glucose (Cass et al., 1984) and cholesterol (Ball et al., 1986) sensors, based on ferrocene derivatives and on the enzymes glucose oxidase and cholesterol oxidase, respectively, have been successfully employed in the last years. The field of immunology received also a contribution from enzyme electrochemistry with what can be called immunoelectrodes (Frew and Hill, 1988, and references therein). These systems are again dependent on glucose oxidase and on ferrocene derivatives, specifically on the competition of the latter with free antigen molecules for binding sites on antibodies (Fig. 10). Clearly, the higher the concentration of free antigen, the higher the amount of non-bound ferrocene-antigen derivative, and the more intense the electrochemical response.

Bioelectrochemical synthesis is another realm having enormous potential, since ideally it allows the clean production of a certain compound with high efficiency. The conversion of *p*-cresol into *p*-hydroxybenzaldehyde by *p*-cresol methylhydroxylase, having azurin as mediator, is a classic example (Hill et al., 1985b). Many electrochemical studies have given invaluable contributions for the elucidation of several biochemical processes. Apart from the examples already mentioned above, one should mention:

(i) the study of intermolecular interaction between redox proteins and enzymes (Hill and Walton, 1982; Bagby et al., 1990; Draoui et al., 1991);

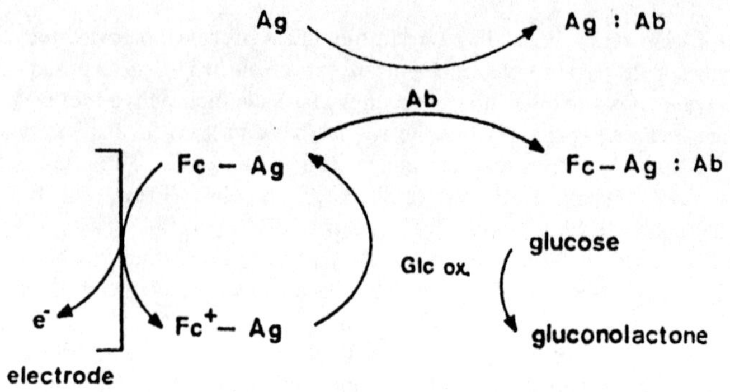

Figure 10.
Schematic representation of a homogeneous amperometric immunoassay. Ag refers to the antigen (i.e. the drug); Fc-Ag is the ferrocene-drug conjugate, Ab is the antibody to the drug, Glc ox. is glucose oxidase (reproduced from Frew and Hill, 1988).

(ii) the use of electrodes modified with amphiphiles aiming at mimicking biological membranes (Guerrieri et al., 1991; Salamon and Tollin, 1991a; 1991b);

(iii) the investigation of respiration using mitochondria through the coupling between cyt *c* and cytochrome oxidase in these organelles (Frew and Hill, 1988, and references therein; Coleman et al., 1983);

(iv) the electrochemical study of genetic hybrids obtained by site-directed mutagenesis (Burrows et al., 1991);

(v) the recent findings concerning metalloenzymes involved in the nitrogen cycle (Kohzuma et al., 1993) have created great expectations towards the achievement of full understanding and control of its steps in the coming years.

IV. CONCLUDING REMARKS

No doubt, the electrochemistry of metalloproteins and metalloenzymes has had an invaluable contribution in understanding several biological processes. Knowing their modus operandi, it might be expected that it will become easier to control them and to make fruitful use of their properties, as was examplified by the development of electrochemical sensors. However, even if it is true that a lot has been achieved since the late seventies, it must be acknowledged that there is still a long way to go before the interaction between the electrode surface and the metalloprotein is understood at the atomic level. This would certainly allow the maximum control over the processes taking place. This breakthrough seems more probable since the introduction of the scanning probe microscopes, especially the scanning tunneling microscope (STM) (Binnig et al., 1982) and the atomic force microscope (AFM) (Binnig et al., 1986) for the study of surfaces, and the results of their application to these systems is awaited with interest in a not too distant future.

ACKNOWLEDGEMENT

P. de Oliveira thanks Fundação Calouste Gulbenkian for its continuing support.

REFERENCES

Albery, W.J., Eddows, M.J., Hill, H.A.O., and Hillman, A.R. (1981). *J. Amer. Chem. Soc.*, *103*: 3905.

Allen, P.M., Hill, H.A.O., and Walton, N.J. (1984). *J. Electroanal. Chem.*, *178*: 69.

Armstrong, F.A., Hill, H.A.O., Oliver, B.N., and Walton, N.J. (1984a). *J. Amer. Chem. Soc.*, *106*: 921.

Armstrong, F.A., Hill, H.A.O., and Oliver, B.N. (1984b). *J. Chem. Soc. Chem. Commun.*, 976.

Armstrong, F.A., Cox, P.A., Hill, H.A.O., Oliver, B.N., and Williams, A.A. (1985). *J. Chem. Soc. Chem. Commun.*, 1236.

Armstrong, F.A., and Lannon, A.M. (1987). *J. Amer. Chem. Soc.*, *109*: 7211.

Armstrong, F.A., Hill, H.A.O., and Walton, N.J. (1988a). *Accounts Chem. Res.*, *21*: 407.

Armstrong, F.A., George, S.J., Thomson, A.J., and Yates, M.G. (1988b). *FEBS Lett.*, *234*: 107.

Armstrong, F.A., George, S.J., Cammack, R., and Hatchikian, E.C. (1989a). *Biochem. J.*, *264*: 265.

Armstrong, F.A., Butt, J.N., George, S.J., Hatchikian, E.C., and Thomson, A.J. (1989b). *FEBS Lett.*, *259*: 15.

Armstrong, F.A., Bond, A.M., Hill, H.A.O., Oliver, B.N., and Psalti, I.S.M. (1989c). *J. Amer. Chem. Soc.*, *111*: 9185.

Bagby, S., Barker, P.D., Guo, L.-H., and Hill, H.A.O. (1990). *Biochemistry*, *29*: 3213.

Ball, M.R., Frew, J.E., Green, M.J., and Hill, H.A.O. (1986). *Electrochemical Sensors for Biomedical Applications*, Proc. Vol. 86 (C.K.N. Li, ed.), Electrochemical Society, New York, p. 16.

Bartlett, P.N., Whitaker, R.G., Green, M.J., and Frew, J. (1987). *J. Chem. Soc. Chem. Commun.*, 1603.

Betso, S.R., Klapper, M.H., and Anderson, L.B. (1972). *J. Amer. Chem. Soc.*, *15*: 8197.

Bianco, P., and Haladjian, J. (1986). *J. Electroanal. Chem.*, *299*: 365.

Binnig, G., Rohrer, H., Gerber, Ch., and Weibel, E. (1982). *Phys. Rev. Lett.*, *49*: 57.

Binnig, G., Quate, F.C., and Gerber, Ch. (1986). *Phys. Rev. Lett.*, *56*: 930.

Bowden, E.F., Hawkridge, F.M., and Blount, H.N. (1984). *J. Electroanal. Chem.*, *161*: 355.

Burrows, A.L., Guo, L.-H., Hill, H.A.O., McLendon, G., and Sherman, F. (1991). *Eur. J. Biochem.*, *202*: 543.

Butt, J.N., Sucheta, A., Armstrong, F.A., Breton, J., Thomson, A.J., and Hatchikian, E.C. (1993). *J. Amer. Chem. Soc.*, *115*: 1413.

Cass, A.E.G., Davis, G., Francis, G.D., Hill, H.A.O., Aston, W.J., Higgins, I.J., Plotkin, E.V., Scott, L.D., and Turner, A.P.F. (1984). *Anal. Chem.*, *56*: 667.

Coleman, J.O.D., Hill, H.A.O., Walton, N.J., and Whatley, F.R. (1983). *FEBS Lett.*, *154*: 319.

Cotton, T.M., Schultz, S.G., and Van Duyne, R.P. (1980). *J. Amer. Chem. Soc.*, *102*: 7960.

Di Gleria, K., Hill, H.A.O., Lowe, V.J., and Page, D.J. (1986). *J. Electroanal. Chem.*, *213*: 333.

Draoui, K., Bianco, P., Haladjian, J., Guerlesquin, F., and Bruschi, M. (1991). *J. Electroanal. Chem.*, *313*: 201.

Eddowes, M.J., and Hill, H.A.O. (1977). *J. Chem. Soc. Chem. Commun.*, 771.

Eddowes, M.J., and Hill, H.A.O. (1979a). *J. Amer. Chem. Soc.*, *101*: 4461.

Eddowes, M.J., and Hill, H.A.O. (1979b). *J. Amer. Chem. Soc.*, *101*: 7113.

Frew, J.E., and Hill, H.A.O. (1988). *Eur. J. Biochem.*, *172*: 261.

Fultz, M.L., and Durst, R.A. (1982). *Anal. Chim. Acta*, *140*: 1.

George, S.J., Armstrong, F.A., Hatchikian, E.C., and Thomson, A.J. (1989). *Biochem. J.*, *264*: 275.

Guerrieri, A., Cataldi, T.R.I., and Hill, H.A.O. (1991). *J. Electroanal. Chem.*, *297*: 541.

Guo, L.-H., and Hill, H.A.O. (1991). *Adv. Inorg. Chem.*, *36*: 341.

Harmer, M.A., and Hill, H.A.O. (1984). *J. Electroanal. Chem.*, *170*: 369.

Harmer, M.A., and Hill, H.A.O. (1985). *J. Electroanal. Chem.*, *189*: 229.

Hawkridge, F.M., and Kuwana, T. (1973). *Anal. Chem.*, *45*: 1021.

Heineman, W.R., Norris, B.J., and Goelz, J.F. (1975). *Anal. Chem.*, *47*: 79.

Hill, H.A.O. (1987a). *Pure Appl. Chem. 59*: 743.

Hill, H.A.O. (1990). *Pure Appl. Chem.*, *62*: 1047.

Hill, H.A.O., and Higgins, I.J. (1981). *Phil. Trans. Roy. Soc. London A*, *302*: 267.

Hill, H.A.O., and Hunt, N.I. (1993). *Meth. Enzymol.*, *227*: 501.

Hill, H.A.O., and Walton, N.J. (1982). *J. Amer. Chem. Soc.*, *104*: 6515.

Hill, H.A.O., Page, D.J., Walton, N.J., and Whitford, D. (1985a). *J. Electroanal. Chem.*, *187*: 315.

Hill, H.A.O., Oliver, B.N., Page, D.J., and Hopper, D.J. (1985b). *J. Chem. Soc. Chem. Commun.*, 1469.

Hill, H.A.O., Page, D.J., and Walton, N.J. (1987b). *J. Electroanal. Chem.*, *217*: 141.

Kohzuma, T., Takase, S., Shidara, S., and Suzuki, S. (1993). *Chem. Lett.*, 149.

Kono, T., and Nakamura, S. (1958). *Bull. Agric. Chem. Soc. Jpn*, *22*: 399.

Kuznetsov, B.A., Mestechkina, N.M., Karuzina, I.I., Karyakin, A.V., and Archakov, A.I. (1979). *Biochemistry Engl. Transl. Biokhimiya*, *44*: 1234.

Kwee, S. (1986). *Bioelectrochem. Bioenerg.*, *16*: 99.

Panzer, R.E., and Elving, P.J. (1975). *Electrochim. Acta*, *20*: 635.

Sagara, T., Nakajima, S., Akutsu, H., and Niki, K. (1991). *J. Electroanal. Chem.*, *297*: 271.

Sakurai, T., Ikeda, O., and Suzuki, S. (1990). *Inorg. Chem.*, *29*: 4715.

Salamon, Z., and Tollin, G. (1991a). *Bioelectrochem. Bioenerg.*, *25*: 447.

Salamon, Z., and Tollin, G. (1991b). *Bioelectrochem. Bioenerg.*, *26*: 321.

Shibata, M., Nakamura, K., Furuya, N., and Watanabe, M. (1991). *J. Electroanal. Chem. 314*: 179.

Taniguchi, I., Yamaguchi, H., and Yasukouchi, K. (1984). *J. Electroanal. Chem.*, *140*: 187.

Thomson, A.J., George, S.J., Armstrong, F.A., Hatchikian, E.C., and Yates, M.G. (1990). *Pure Appl. Chem.*, *62*: 1071.

Walker, P.L., Jr. (1962). *Amer. Sci.*, *50*: 259.

Yamada, S., and Sato, H. (1962). *Nature*, *193*: 261.

Yeh, P., and Kuwana, T. (1977). *Chem. Lett.*, 1145.

ACTIVE SITES OF HEMOPROTEINS

by

Lily Yun Xie and David Dolphin

Department of Chemistry,
University of British Columbia,
Vancouver, BC V6T 1Z1
Canada

I. INTRODUCTION

The essentially identical iron porphyrin core (heme), attached to a protein through a fifth ligand, usually a residue group of an amino acid, carries out a wide variety of chemical and biological processes by combination with different proteins. Hemoproteins function as carriers of both oxygen and electrons. Biological oxidation of substrates catalyzed by enzymes containing the heme prosthetic group is a second major function performed by hemoproteins. For instance, the heme in myoglobin and hemoglobin reversibly bind dioxygen, while in horseradish peroxidase the heme is oxidized by two electrons producing water and an oxidized heme protein which then brings about two single one-electron oxidations of substrates. Cytochromes P-450 which are monooxygenases, on the other hand, perform all three functions in that reversible binding of dioxygen is followed by electron transport and the concomitant reduction of O_2 to peroxide. Oxidation of the heme then occurs upon cleavage of the O-O bond to generate water and an intermediate which transfer oxygen to substrate. It is thus important to understand the structure of the active sites, not only the oxidation state of the metal ion and the geometry of the complexes, but also the topology of the heme pocket. Although high resolution X-ray crystallography provides a direct visualization of the equilibrium structure of the active site in the solid state, spectroscopic methods including NMR, resonance Raman and EXAFS are still extremely valuable for elucidating the dynamics and conformation in solution. The role played by protein in the catalysis is clearly important; the knowledge of the active site structure has tremendous impact on the rational design of synthetic models which mimic the activities of these biological catalysts outside the protein. In this chapter, discussion will be limited to only a few major hemoproteins.

II. ACTIVE SITES OF HEMOPROTEINS

A. Active site of cytochrome P-450

Cytochrome P-450 enzymes are remarkably versatile monooxygenases and have been isolated from a variety of sources, including mammalian tissues (e.g. liver), plants, yeast and bacteria. Substrates that can be oxidized by P-450 include endogenous compounds such as steroids, lipids etc. and exogenous drugs, pesticides and chemical carcinogens (Gunter and Turner, 1991; White, 1990). The products of P-450 oxidation may be excretable metabolites, and initial attention focused on its role in detoxification processes and as an initiator of chemical carcinogenesis (Black and Coon, 1987; Groves, 1985). The overall reaction is summarized in equation 1.

$$RH + O_2 + 2e^- + 2H^+ \rightarrow ROH + H_2O \qquad (1)$$

The high resolution X-ray crystal structure for the eukaryotic membrane bound cytochrome P-450 enzymes is currently unavailable. However, these active sites may bear resemblance to the camphor-metabolizing enzyme P-450$_{cam}$ isolated from *Pseudomonas putida* whose X-ray crystal structure is known (Poulos et al., 1985). The prosthetic group of P-450$_{cam}$, protoporphyrin IX Fe(III) (heme), is attached to the protein through the thiolate group of a cysteine residue (Cys357). The heme pocket where the dioxygen and substrate molecules are bound is composed of entirely lipophilic residue groups of amino acid such as Leu, Val, Phe (Fig. 1). The architecture of the O_2 binding site is closer to that of hemoglobin and myoglobin than to the more polar peroxidase active sites (Poulos, 1986). Hydrophobicity of the active site stabilizes the binding of the non-polar hydrocarbon substrate to protein as well as binding of O_2 to iron. Another feature revealed by X-ray crystallography is that the heme group is buried deep inside the protein pocket; any access to the active site by the substrate or O_2 is only possibly by transient broadening of protein channels (Poulos, 1986).

The generally accepted mechanism of monooxygenation catalyzed by P-450 (McMurry and Groves, 1986) is depicted below (Scheme I). While compounds **1** to **4** of the P-450 cycle are well characterized, evidence for the existence of **5** and **6** is still indirect. The resting state of cytochrome P-450$_{cam}$ is a hexa-coordinate low-spin Fe(III) complex with H_2O as the sixth ligand. Upon substrate binding, water is displaced to give a high-spin penta-coordinate square-pyramidal complex. A one-electron reduction of Fe(III) to Fe(II) is facilitated by the change of spin state and precedes the binding of dioxygen to give a low-spin hexa-coordinate complex. The second one-electron reduction yields an iron peroxo species; subsequent cleavage of the O-O bond liberates H_2O to give a highly reactive iron-oxo species. Substrate is oxidized via oxygen atom incorporation, allowing the enzyme to

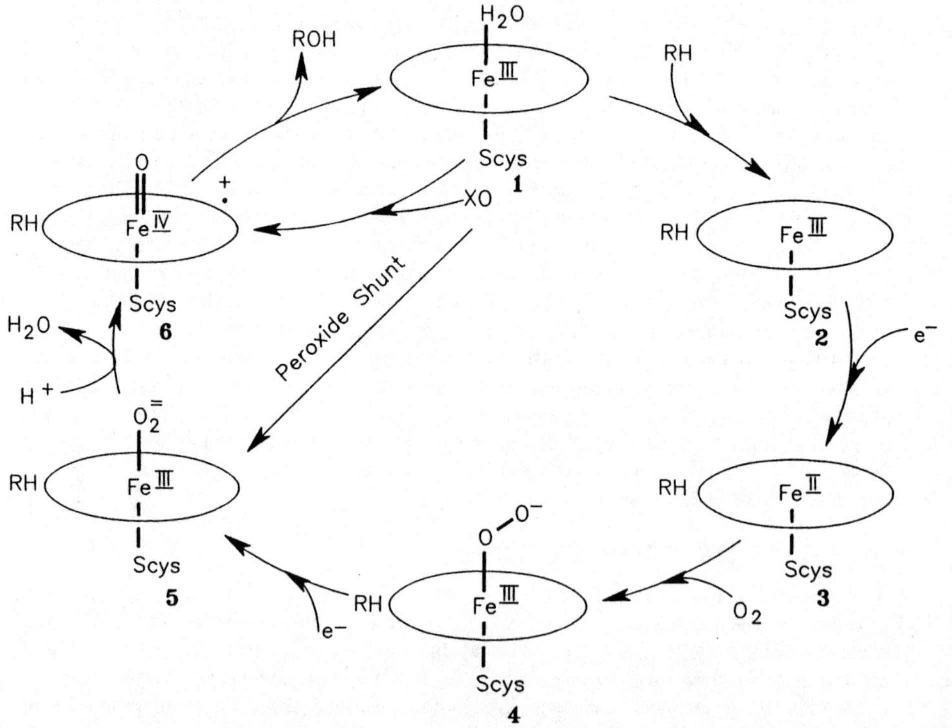

Scheme I

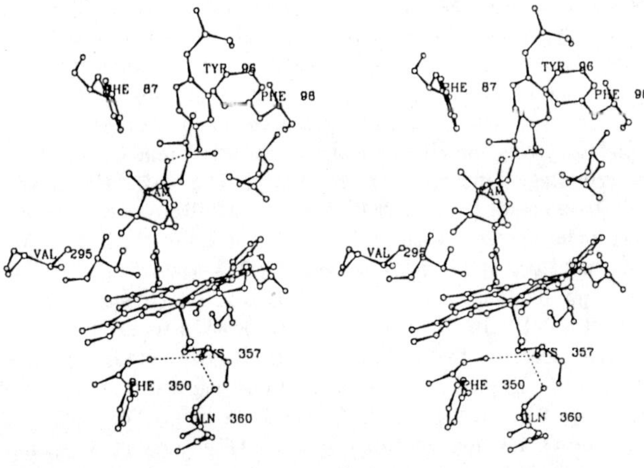

Figure 1.
Stereotopic view of the active site of cytochrome P-450_{cam} with bound camphor. Coordinates (entry 2CPP, version of Aug. 1987) for this molecule (Poulos et al., 1987) were obtained from the Protein Data Bank (Bernstein et al., 1977; Abola et al., 1987) at Brookhaven National Laboratory.

return to the resting state. The details of O-O bond cleavage, the structure of the resulting oxygen complex and the mechanism of oxygen atom transfer are less well understood because the steps that follow the second electron reduction occur too rapidly for any reaction intermediates to have so far been detected (Andersson and Dawson, 1990). The description of the assumed hypervalent intermediate has largely been based on analogy of P-450 with normal peroxidases (Dolphin, 1985). Furthermore, the observation of P-450 chemistry with exogenous two electron oxidizing agents such as iodosoarenes and alkyl hydroperoxides implies that the iron-oxo species is perhaps the viable intermediate in the catalytic cycle, i.e. dioxygen is indeed reduced to a peroxide in the natural sequence (Groves, 1985). Careful studies of various substrates of P-450 which are of diagnostic significance suggested that oxygen atom transfer is in fact the result of hydrogen atom abstraction by iron-oxo species and radical recombination of iron-hydroxo with the organic radicals. It was also suggested that the organic radical formed by hydrogen abstraction could not diffuse out of the heme pocket before it collapsed (Groves et al., 1978; Groves and Nemo, 1983; Traylor et al., 1986).

Designing model compounds that mimic the monooxygenase activity of P-450 without the benefit of protein has been an active area of research for many years (Meunier, 1992; Gunter and Turner, 1991). While there still has not been a system that successfully utilizes molecular oxygen the way the enzyme does, many model compounds are able to hydroxylate and epoxidize substrates efficiently via the peroxide shunt route. Metal ions including Fe (Groves and Nemo, 1983) and Mn (Mansuy et al., 1984), and oxidants including iodosoarenes RIO (Traylor et al., 1984) and peroxides ROOH (Mansuy et al., 1982) have been most widely investigated. The ligand in these model systems are primarily tetraphenylporphyrin derivatives due largely to the fact that the aryl substituents at the meso position protect the porphyrin from destruction upon the formation of the π-cation radical intermediate. Halogenations at ortho positions of the phenyl rings (Traylor et al., 1984; Traylor et al., 1986) and β positions of the pyrroles provide steric effects which avoid the formation of μ-oxo dimers and enhance the electrophilicity (oxidizing power) of the metalloporphyrin entity (Wijesekera et al., 1990).

B. Active site of cytochrome *c* peroxidase
Cytochrome *c* peroxidase (CCP), a non-glycoprotein isolated from baker's yeast (Yonetani, 1970), catalyzes primarily the oxidation of ferrocytochrome to ferricytochrome *c* by H_2O_2 (eq. 2) (Hewson and Hager, 1979a).

$$H_2O_2 + 2\text{cyt-}c\,(Fe^{2+}) + 2H^+ \rightarrow 2H_2O + 2\text{cyt-}c(Fe^{3+}) \qquad (2)$$

The crystal structures of both yeast CCP (Poulos and Finzel, 1984) and its semistable oxidized intermediate, compound **I** (Edwards et al., 1987), are known. In the resting enzyme the heme is attached to the protein through an imidazole ligand of the proximal histidine (His175), and in a highly polar environment consisting of Arg, Trp, Asp, His and H_2O (Poulos and Finzel, 1984) (Fig. 2). Hydrogen bonding occurs between the proximal His175 and Asp235, giving the His175 residue substantial imidazolate character, making it more electron releasing (Smulevich et al., 1990). The distal His52 is believed to be involved in acid-base catalysis and Arg48 provides stabilization of negative charge on the Fe-OOH moiety. The crystal structure of compound **I** reveals little difference from the resting enzyme except that the iron moves about 0.2 Å toward the distal side (Edwards et al., 1987). This supports an unusually short Fe=O bond which is measured to be 1.67 Å by EXAFS (Chance et al., 1986). The heme group is located between two protein helices, and completely covered except for the edges of pyrrole rings A and D; it can be reached by the oxidant through a well-designed channel leading to the edge of ring D (Finzel et al., 1984). The size of the protein channel is 6 Å wide and 11 Å long. (Dasgupta et al., 1989). This unique channel allows for the approach and oxidation of small organic or inorganic molecules such as guaiacol and ferrocyanide to the δ-meso heme edge, but prohibits macromolecule cytochrome *c* from approaching the heme; as a result, the oxidation occurs at a surface site facing the γ- meso edge (DePillis et al., 1991).

CCP and P-450 have very different active sites as shown by their X-ray crystal structures. The heme cavity in CCP is occupied by polar residue groups and the fifth ligand in CCP is an imidazole which is strongly H-bonded to Asp235. The polar residues on the distal side and the partial imidazolate on the proximal side may act collectively to cleave the peroxide bond. On the other hand, P-450 has a thiolate on the proximal side and, despite the lack of polar residues on the distal side, the strongly electron-donating ability of S^- appears to be sufficient to cleave the O-O bond heterolytically.

In a freshly prepared sample of the resting wild type CCP, the Fe(III) atom is penta-coordinate high spin at room temperature over a pH range of 4.3 -7.0 (Dasgupta et al., 1989; Smulevich et al., 1990); aged CCP contains a mixture of penta- and hexa-coordinate species. The sixth ligand in this case is possibly a H_2O molecule in the high-spin species

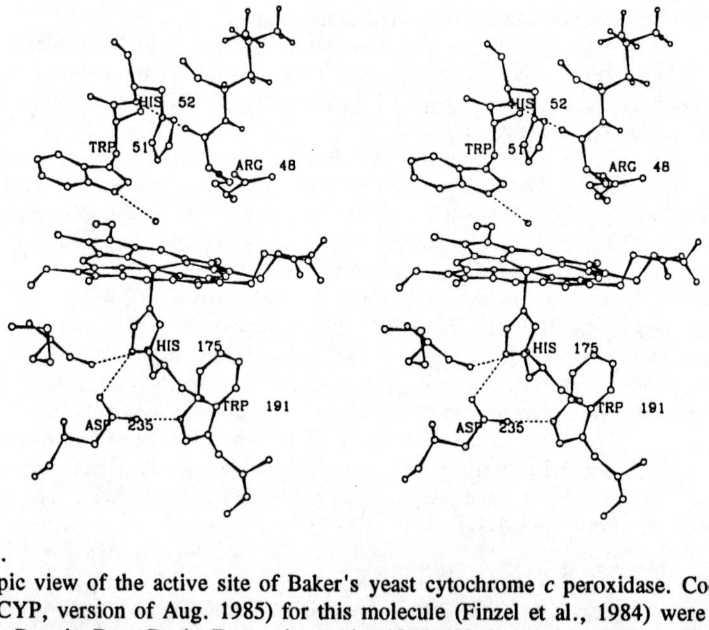

Figure 2.
Stereotopic view of the active site of Baker's yeast cytochrome *c* peroxidase. Coordinates (entry 2CYP, version of Aug. 1985) for this molecule (Finzel et al., 1984) were obtained from the Protein Data Bank (Bernstein et al., 1977; Abola et al., 1987) at Brookhaven National Laboratory.

and is OH^- or distal histidine in the low-spin hexa-coordinate species. A transition takes place from a penta-coordinate to a hexa-coordinate high-spin ferric form when pH inceases (pK_a 5.5) (Dasgupta et al., 1989). The hexa-coordinate ferric CCP is less reactive towards H_2O_2 than the penta-coordinate ferric CCP, due possibly to the rate-limiting dissociation of the sixth ligand (Dasgupta et al., 1989).

The catalytic cycle of CCP involves two single one-electron oxidations of substrate, generally a protein. It is activated by H_2O_2 oxidation to a compound with two oxidation equivalents above the resting state. However, compound **I** of CCP (known as ES) is clearly different from that of compound **I** in horseradish peroxidase (Hewson and Hager, 1979b; Mauro et al., 1989; Smulevich et al., 1988; Erman et al., 1989; Sivaraja et al., 1989). The high valent oxyferryl complexes in the catalytic cycle may well have similar electronic structures, but differ only in the presence of protein radical in the former which likely resides on the indole ring of the proximal Trp191 (Edwards et al., 1987). The UV-vis spectrum of CCP compound **I**, characteristic of the oxyferryl O=Fe(IV)P, is similar to that of HRP compound **II** which is oxidized one equivalent above the ferric state (Coulson et al., 1971). In contrast to HRP compound **I**, the Mössbauer and ESR studies of CCP compound **I** indicate that the protein-based free radical and Fe(IV) are not magnetically coupled (Hoffman et al., 1981).

C. Active site of lignin peroxidase

Lignin peroxidase (LiP) and manganese-dependent peroxidase (MnP), both isolated from *Phanerochaete chrysosporium*, are the two heme peroxidases which are involved in lignin biodegradation (Tien and Kirk, 1983; Glenn et al, 1983; Glenn and Gold, 1985). LiP exists as a series of isozymes with molar masses ranging from 38 to 42 kg mol^{-1}. Each isozyme of LiP consists of a single polypeptide chain and one protoheme IX. Lignin peroxidases, once oxidized by hydrogen peroxide, function by abstracting one electron from the electron-rich oxygenated aromatic units of lignin to give aryl cation radicals. Further non-enzymatic fragmentation of the cation radical leads to the degradation of lignin (Kersten et al., 1985; Harvey et al., 1985). On the other hand, MnP catalyzes the H_2O_2- and Mn^{2+}-dependent oxidation of phenols and other substrates (Glenn and Gold, 1985).

The X-ray crystal structure of lignin peroxidase has recently been solved by Poulos (Poulos, 1992a) (Fig. 3). The active site of LiP is buried in the protein and the iron center is not accessible by substrates. A 10 Å long open channel in LiP allows H_2O_2 to approach the heme, but even veratryl alcohol can best be docked only 6 Å from the heme center (Poulos, 1992a). In the high-spin penta-coordinate form, the iron is approximately 0.2 Å out of the heme plane and the iron-nitrogen (pyrrole) distance is 2.05 Å (Poulos, 1992a). Two calcium-binding sites, one on the proximal side, primarily bound to Ser177, and another on the distal side, primarily bound to Asp48, are also located by X-ray crystallography (Poulos, 1992a). The heme pocket is much like that of CCP, consisting of polar residue groups such as Asp, Arg and His on both proximal and distal sides (Poulos, 1992a; de Ropp et al., 1991). Instead of having Trp51 and Trp191 as in CCP, LiP has Phe46 and Phe193 close to the heme plane. Heterolytic cleavage of the O-O bond of hydrogen peroxide is facilitated by H-bonding on both sides of heme plane (de Ropp et al., 1991). The vinyl groups of the heme are demonstrated by ^1H NMR experiments to take an in plane orientation which would provide more extensive delocalization for the cation radical in compound **I** (de Ropp et al., 1991).

The catalytic cycle of lignin peroxidase involving compound **I** and compound **II** is similar to that of horseradish peroxidase which is discussed below. Compound **I** and compound **II** of LiP have been detected by various spectroscopic techniques and are also similar to those of horseradish peroxidase. At room temperature, the native ferric enzyme is high-spin penta-coordinate with a vacant coordination site opposite the proximal histidine (Andersson et al., 1987). At ≤2°C, a weakly associated ligand, possibly H_2O, was observed (Andersson et al., 1985). However, LiP is unique in that it operates at a low pH optimum (≤3) and exhibits a higher redox potential (Banci and Bertini, 1991). The presence of an ionizable group in the active site with a pK_a <2 has been recently demonstrated by chloride binding studies (Cai and Tien, 1991). Chloride binds only to the protonated form of LiP.

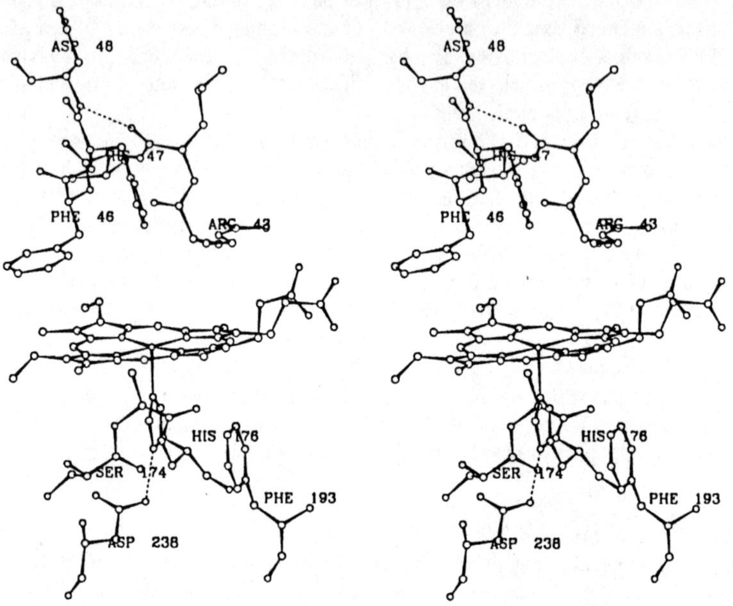

Figure 3.
Stereotopic view of the active site of lignin peroxidase from *P. chrysosporium* (Poulos, 1992b).

Formation of compound **I**, however, requires the deprotonated form of LiP. Binding of chloride shifts the equilibrium toward the protonated form, thereby inhibiting formation of compound **I** at low pH (Cai and Tien, 1991).

The hydrogen bonding of the proximal histidine was recently found to be significantly less in LiP than in HRP by ^1H NMR (Banci and Bertini, 1991). The reduced imidazolate character observed in LiP destabilizes the oxyferryl species which provide a rationale for the higher redox potential and lower stability of the compound **I** of LiP relative to compound **I** of the HRP. A weaker axial histidine bond in LiP relative to HRP has also been shown by resonance Raman spectroscopy (Andersson et al., 1985).

D. Active site of horseradish peroxidase
Horseradish peroxidase (HRP) is a glycohemoprotein with a molar mass of about 40 kg mol^{-1}. It is isolated from the root of the horseradish plant and catalyzes one electron oxidations of a wide range of organic substrates, particularly phenols and aromatic amines (Hewson and Hager, 1979a), and is characteristic of many plant peroxidases.

The catalytic cycle of horseradish peroxidase (Andersson and Dawson, 1990; Dawson, 1988) is depicted in Scheme II.

Efforts to obtain an X-ray crystal structure of HRP have not been successful (Brathwaite, 1976; Aibara et al., 1981). Considerable emphasis has been placed on the use of spectroscopic methods (Thanabal et al., 1987; Smulevich et al., 1990; Terner et al. 1989) and sequence homology to CCP (Sakurada et al., 1986) in attempting to define an active site structure. ^1H NMR experiments have been particularly useful in identifying the catalytically significant amino acid residues which are involved in the H-bonding network on both the proximal and distal sides of the heme plane (Thanabal et al., 1987). The heme prosthetic group in horseradish peroxidase is attached to protein through the imidazole of His170 which remains coordinated throughout the catalytic cycle. The heme edges are buried in the protein as indicated by the formation of the δ-meso alkylated product by phenylhydrazine derivatives instead of the usual N- alkylated product as in the case of cytochrome P-450. The highly oxidizing iron center is completely inaccessible to substrates (Ator and Ortiz de

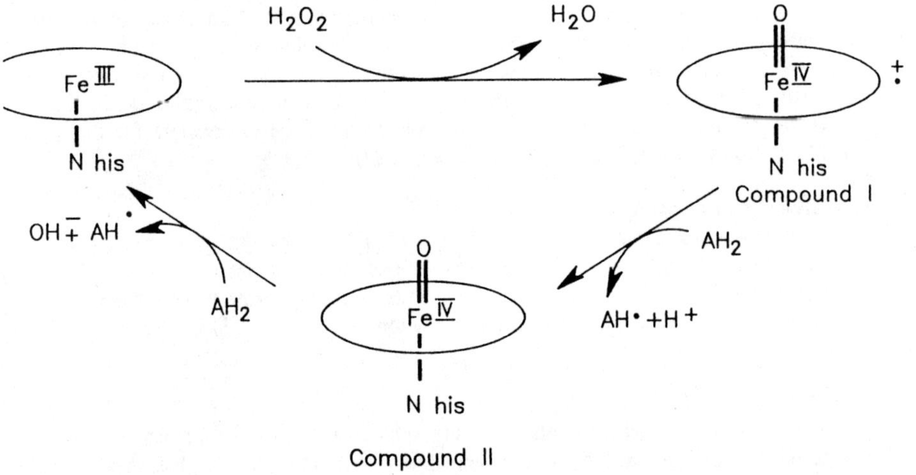

Scheme II

Montellano, 1987; Ortiz de Montellano, 1987). The heme pocket is composed of polar residue groups of amino acid such as Arg38 and His42 capable of forming hydrogen bonds (Sakurada et al., 1986). Arg38 and His42 are believed to play important roles in cleavage of the O-O bond of peroxide for the generation of compound **I** and for its stabilization. The imidazole ring of the distal histidine is protonated upon H_2O_2 oxidation, the imidazolium proton forming a H-bond with and polarizing the O-O bond (Thanabal et al., 1988). Studies show that imidazole of the proximal histidine exhibits two coordination modes. Both ^1H NMR and resonance Raman spectroscopy show hydrogen bonding interactions between the proximal His170 and nearby groups (Thanabal et al., 1987; Smulevich et al., 1990; Terner et al., 1989). The proximal imidazole ligand is neutral when the heme is penta-coordinate, but deprotonates to become a negatively charged imidazolate ligand when the heme becomes hexa-coordinate in compound **I** and compound **II** (Thanabal, 1988). The negatively charged imidazolate donates electron density to iron, and from iron to the O-O bond. This 'push-pull' mechanism involving hydrogen-bonding networks on both the proximal and distal side of the heme explains the ease of the heterolytic cleavage and stability of the resulting $O{=}Fe(IV)P^{+\bullet}$ and $O{=}Fe(IV)$ complexes. The imidazolate character in horseradish peroxidase is also seen in other peroxidases in varying degrees (Finzel et al., 1984; Smulevich et al., 1990; Banci and Bertini, 1991). The degree of deprotonation or imidazolate character is, however, seemingly related to the stability of the ferryl oxygen complexes (Morrison and Schonbaum, 1976; Schonbaum, 1982) and thus to the redox potential of the corresponding peroxidases (Banci and Bertini, 1991).

The Fe(III) ligation of the resting state of horseradish peroxidase is characterized as penta-coordinate high spin (Smulevich et al., 1990). Upon oxidation by hydrogen peroxide, a very reactive intermediate called compound **I** which is two oxidation equivalents above the native ferric state is formed. Compound **I** is a low-spin hexa-coordinated Fe(IV) porphyrin π-cation radical formulated as $O{=}Fe(IV)P^{+\bullet}$. The presence of a π-cation porphyrin radical in compound **I** has been validated by several experimental methods including UV-vis (Dolphin et al., 1971), ESR (Roberts et al., 1981; Rutter et al., 1983) and NMR spectroscopies (La Mar et al., 1981). Magnetic susceptibility data (Theorell and Ehrenberg, 1952) have been interpreted as a ferromagnetically coupled combination of a low-spin Fe(IV) ($S=1$) and a porphyrin π-cation radical ($S=1$). Compound **I** readily undergoes a one electron reduction by organic substrates to compound **II** which is characterized as an $O{=}Fe(IV)P$, supported by compound **II** model studies (Chin et al., 1980). A second substrate molecule can be oxidized by compound **II** which allows the enzyme to return to the resting state. Unlike the organic radical formed in P-450 monooxygenase, substrate free radicals formed during the HRP cycle readily diffuse into solution. Oxygen atoms when incorporated into the oxidized product are not from the H_2O_2 oxidant, but from the solvent or O_2 reacting in a later stage of the reaction steps (Ortiz de Montellano et al., 1987). Although horseradish peroxidase catalyzes some oxygen-incorporating reactions including the expoxidation of styrene, it requires the presence of a cosubstrate or co-oxidant which is capable of forming a radical by a one-electron oxidation of horseradish peroxidase; the peroxy radical formed from the cosubstrate was found to be responsible for the styrene oxide formation (Ortiz de Montellano and Grab, 1987).

E. Active site of catalase

Most mammalian catalases except HPII from *E. Coli* and *Neurosporo crassa* contain iron protoporphyrin IX as the prosthetic group. Beef liver catalase is shown by X-ray crystallography to contain four identical subunits. Catalase protects cell from the toxic effects of H_2O_2 by catalyzing a disproportionation reaction (eq. 3) (Hewson and Hager, 1979a).

$$2H_2O_2 \rightarrow 2H_2O + O_2 \quad (3)$$

The heme site of beef liver catalase in one subunit is accessible by a channel 30 Å long as determined by X-ray analysis. The channel starts with the polar residues Arg, Asp, Gln and Lys at the surface and ends with the hydrophobic residues Val, Pro, Ile, Leu and Phe on the heme distal side (Murphy et al., 1981). The aromatic rings of Phe160 and His74 on the distal side stack parallel to the porphyrin plane. Distal Asn147 and His74 play important roles in substrate binding both in the formation and reduction of compound **I**. Specificity is largely determined by the ability of a substrate capable of forming H-bonding interactions with these two residues (Fita and Rossmann, 1985a). Both propionyl groups are involved in

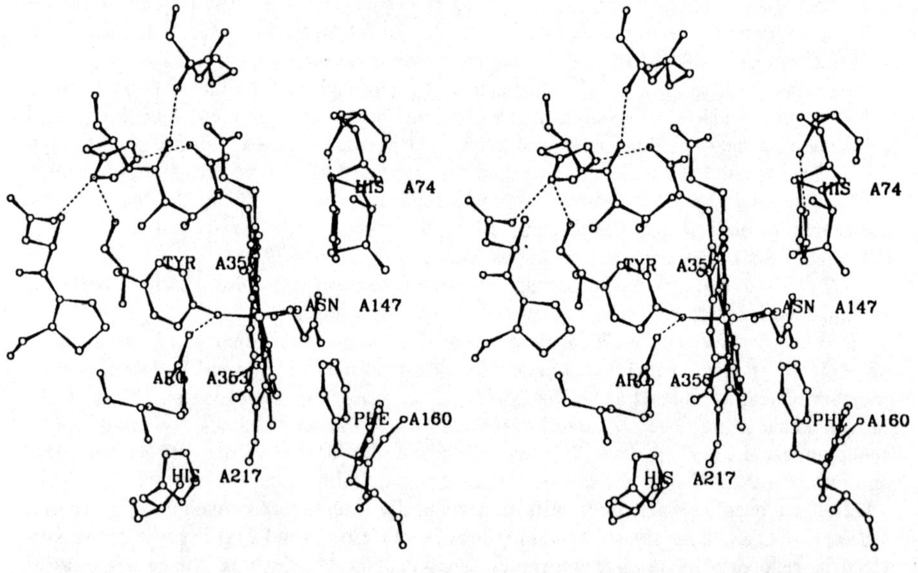

Figure 4.
Stereotopic view of the active site of beef liver catalase. Coordinates (entry 7CAT, version of Nov. 1984) for this molecule (Fita and Rossmann, 1985b) were obtained from the Protein Data Bank (Bernstein et al., 1977; Abola et al., 1987) at Brookhaven National Laboratory.

a complex network of electrostatic and hydrogen bonding interactions. Unlike peroxidases, the fifth ligand of catalase is the phenol ring of Tyr357. As determined by X-ray crystallography, in beef liver catalase, the phenyl plane is tilted 42° inwards to the heme plane with an iron - oxygen distance of 2.2 Å (Vainshtein et al., 1986) (Fig. 4). This suggests that the phenolic side chain is deprotonated and possess a localized charge (Fita and Rossmann, 1985a). Arg353, between Tyr357 and His217, is involved in H-bonding interaction and is suggested to play an important role in the catalytic function (Fita and Rossmann, 1985a).

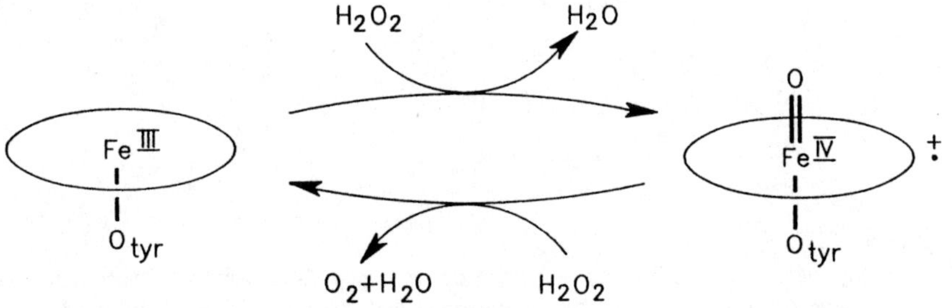

Scheme III

The reaction cycle of catalases (Scheme III) begins with the high-spin ferric state which reacts with one molecule of hydrogen peroxide to form compound **I**. A second hydrogen peroxide reduces compound **I** in one step to the resting state with the concomitant generation of dioxygen (Andersson and Dawson, 1990). The second reaction has been less well studied.

Formation of compound **I** is similar to horseradish peroxidase and the radical is also located at the heme ring (Dolphin et al., 1971). Catalase prefers two electron donors in the reduction of compound **I** to the resting enzyme, whereas peroxidase prefer one electron donors, with the reduction of compound **I** occurring in two distinct reaction steps via compound **II**. In the presence of two-electron donors, e.g. H_2O_2, the formation of compound **II** is never seen. Catalase also has a high degree of specificity and shows a strong preference for H_2O_2 in the reduction. Alkyl or acyl hydroperoxides do not reduce compound **I** (Fita and Rossmann, 1985a). However, N-demethylation analogous to reactions catalyzed by P-450 has been observed in some cases (Kadlubar et al., 1974).

F. Active site of globins

Hemoglobin is the oxygen carrier in vertebrate blood. It consists of two α and two β subunits. The active sites of the α and β subunits are both hydrophobic but not identical and both bind heme. Myoglobin is a similar protein consisting of a single chain and one heme group. It stores O_2 in muscles and releases it to the mitochrondrion for the production of ATP (Ten Eyck, 1979).

The resting state of Hb and Mb is an Fe(II) penta-coordinate high-spin complex which binds O_2 reversibly to form dioxygen complexes which are hexa-coordinate Fe(II) low spin. X-Ray crystallography of human dexoyhemoglobin confirmed that the proximal ligand is His87 and dioxygen binds to the iron in a terminal bent fashion. Binding of dioxygen is stabilized by an H-bonding interaction between the terminal oxygen and the distal His58 (Phillips and Schoenborn, 1981). The H-bond distance in the α-subunit is similar to that in MbO_2 while that in the β-subunit is longer, indicative of a weaker H-bond. The Fe atom is displaced significantly towards the proximal His87 in the deoxy form, but moves into the porphyrin plane upon oxygenation (Perutz et al., 1987) (Fig. 5).

The interaction of O_2 with Mb is a simple bimolecular reaction and the oxygen affinity of Mb is independent of environment. However, the reaction of O_2 with Hb is cooperative and influenced by both heterotropic and homotropic ligands. The cooperativity observed in various hemoglobins has been extensively studied (Perutz et al., 1987).

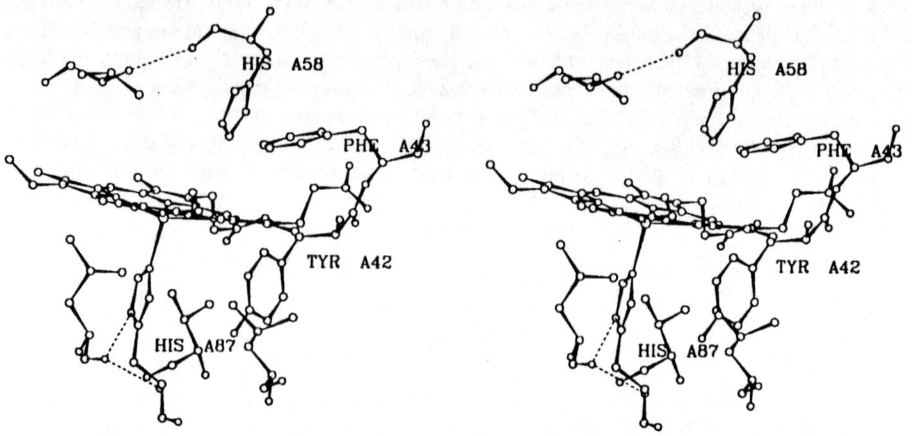

Figure 5.
Stereotopic view of the active site of human deoxyhemoglobin: α-chain. Coordinates (entry 3HHB, version of Mar. 1984) for this molecule (Fermi et al., 1984) were obtained from the Protein Data Bank (Bernstein et al., 1977; Abola et al., 1987) at Brookhaven National Laboratory.

Despite the fact that both myoglobin and hemoglobin have been optimized to store and transport oxygen, they are also capable of catalyzing redox reactions under appropriate conditions (Mieyal, 1985). The fifth ligand in these proteins cannot form hydrogen bonds with nearby residues which makes it a poor electron-donating ligand unlike the proximal histidine in peroxidases. The lack of a strongly polarizing arginine residue on the distal side also renders dioxygen bond cleavage much less facile compared to the peroxidases. The reaction of Hb and Mb with H_2O_2 produces an oxyferryl species and a protein radical similar to that found in CCP but in this case the radical resides on a tyrosine residue (Yonetani and Sihleyer, 1967). Although the active center is accessible to substrate, 70% of the oxygenated product in styrene epoxidation was found to contain oxygen from O_2 rather than H_2O_2. A peroxy radical on Tyr42 formed by O_2 addition to the phenyl radical is perhaps responsible for the formation of styrene oxide (Ortiz de Montellano, 1987).

III. SUMMARY

All the hemoproteins discussed above are involved in interaction with O_2 and/or H_2O_2 of some type. They are closely related in their structures but each performs a different function. The key feature of heme protein catalysis is their ability to form an oxyferryl species by cleavage of the O-O bond formed either by dioxygen or H_2O_2 interaction. The mechanism for the formation of this oxyferryl or compound **I** varies from one system to another, and is governed largely by the nature of the proximal ligand, and the polarity of the proximal and distal heme environment. In peroxidases, the proximal ligands are histidine. The imidazole ring is not a very effective electron donor in the absence of hydrogen bonding to other groups; therefore, subtle difference in the amino acid sequence and polypeptide folding can dramatically alter the hydrogen bonding ability to imidazole. The strength of the imidazole H-bonding to neighboring residues has a significant influence on the imidazolate charcter of the proximal histidine. The heme environment in peroxidases are invariably polar. In addition, the imidazole character is also reflected in the oxidizing power of the enzymes. The greater the negative charge on the imidazole, the more stable the oxyferryl species. The imidazole remains a neutral ligand in hemoglobin and myoglobin because the hydrogen bonding effect to imidazole is absent. The lack of electron-donating effects from nitrogen to iron makes Hb and Mb poor peroxidases. The fifth ligand effect is dominant in P-450 and catalases, with thiolate from cysteine and phenolate from tyrosine respectively. Thiolate and phenolate are much better electron-donating ligands than imidazole, which may account for the readiness of heterolytic cleavage of the O-O bond in these two proteins even though the residue groups on the distal sides are entirely hydrophobic. Partial imidazolate character alone is perhaps not sufficient to cause the cleavage of the O-O bond. Therefore, nature places polarizing residues on the distal side to assist the heterolytic cleavage.

Compound **I**, two oxidation equivalents above resting enzyme, is believed to be the common reactive intermediate formed in hemoprotein catalysis, having two basic types of structures O=Fe(IV) porphyrin cation radical (P-450, HRP, LiP and Cat) or protein radical (CCP on Trp, Hb and Mb on Tyr). The structures of these compounds **I** with different proximal ligands (S, O, N coordination) have been studied by various spectroscopic techniques including X-ray structure for CCP (Edwards et al., 1987). The 1.67 Å bond length of Fe-O measured by EXAFS is consistent with an Fe=O double bond and an iron oxidation state of IV. EPR and Mössbauer spectroscopies are very useful for identifying location of the radical.

It is not surprising that model systems free of protein lack of the substrate binding specificity imposed by the protein structures; therefore, it is still not possible to achieve the catalytic specificity obtained by heme proteins, even though greater turn-over-rates may be observed (Traylor et al., 1984; Traylor et al., 1985).

ACKNOWLEDGEMENTS

This work was supported by the Canadian National Science and Engineering Research Council. Figures were generated on a Silicon Graphics workstation using Biosym InsightII software.

REFERENCES

Abola, E.E., Bernstein, F.C., Bryant, S.H., Koetzle, T.F., and Weng, J. (1987). *Protein Data Bank in Crystallographic Databases - Information Content, Software Systems, Scientific Applications* (F.H. Allen, G. Bergerhoff and R. Sievers, eds.), Data Commission of the International Union of Crystallography, Cambridge, p. 107.

Aibara, S., Kobayashi, T., and Morita, Y. (1981). *J. Biochem. (Tokyo)*, *90*: 489.

Andersson, L.A., and Dawson, J.H. (1990). *Struct. Bonding*, *74*: 1.

Andersson, L.A., Renganathan, V., Chiu, A.A., Loehr, T.M., and Gold, M.H. (1985). *J. Biol. Chem.*, *260*: 6080.

Andersson, L.A., Renganathan, V., Loehr, T.M., and Gold, M.H. (1987). *Biochemistry*, *26*: 2258.

Ator, M.A., and Ortiz de Montellano, P.R. (1987). *J. Biol. Chem.*, *262*: 1542.

Banci, L., and Bertini, I. (1991). *Proc. Nat. Acad. Sci. USA*, *88*: 6952.

Bernstein, F.C., Koetzle, T.F., Williams, G.J.B., Meyer, E.F., Jr., Brice, M.D., Rodgers, J.R., Kennard, O., Shimanouchi, T., and Tasumi, M. (1977). *J. Mol. Biol.*, *112*: 535.

Black, S.D., and Coon, M.J. (1987). *Advances in Enzymology* (A. Mester, ed.), Wiley, New York.

Brathwaite, A. (1976). *J. Biol. Chem.*, *106*: 229.

Cai, D., and Tien, M. (1991). *J. Biol. Chem.*, *266*: 14464.

Chance, M., Powers, L., Poulos, T.L., and Chance, B. (1986). *Biochemistry*, *25*: 1266.

Chin, D.H., Balch, A.L., and La Mar, G.N. (1980). *J. Amer. Chem. Soc.*, *102*: 1446.

Coulson, A.F.W., Erman, J.E., and Yonetani, T. (1971). *J. Biol. Chem.*, *246*: 917.

Dasgupta, S., Rousseau, D.L., Anni, H., and Yonetani, T. (1989). *J. Biol. Chem.*, *264*: 654.

Dawson, J.H. (1988). *Science*, *240*: 433.

de Ropp, J.S., La Mar, G.N., Wariishi, H., and Gold, M.H. (1991). *J. Biol. Chem.*, *266*: 15001.

DePillis, G.D., Sishta, B.P., Mauk, A.G., and Ortiz de Montellano, P.R. (1991). *J. Biol. Chem.*, *266*: 19334.

Dolphin, D., Forman, A., Borg, D.C., Fajer, J., and Felton, R.H. (1971). *Proc. Nat. Acad. Sci. USA*, *68*: 614.

Dolphin, D. (1985). *Phil. Trans. R. Soc. Lond.*, *331B*: 579.

Edwards, S.L., Xuong, N.H., Hamlin, R.C., and Kraut, J. (1987). *Biochemistry*, *26*: 1503.

Erman, J.E., Vitello, L.B., Mauro, J.M., and Kraut, J. (1989). *Biochemistry*, *28*: 7992.

Fermi, G., Perutz, M.F., Shaanan, B., and Fourme, R. (1984). *J. Mol. Biol.*, *175*: 159.

Finzel, B.C., Poulos, T.L., and Kraut, J. (1984). *J. Biol. Chem.*, *259*: 13027.

Fita, I., and Rossmann, M.G. (1985a). *J. Mol. Biol.*, *185*: 21.

Fita, I., and Rossmann, M.G. (1985b). *Proc. Nat. Acad. Sci. USA*, *82*: 1604.

Glenn, J.K., and Gold, M.H. (1985). *Arch. Biochem. Biophys.*, *242*: 329.

Glenn, J.K., Morgan, M.A., Mayfield, M.B., Kuwahara, M., and Gold, M.H. (1983). *Biochem. Biophys. Res. Commun.*, *114*: 1077.

Groves, J.T. (1985). *J. Chem. Edu.*, *62*: 928.

Groves, J.T., and Nemo, T.E. (1983). *J. Amer. Chem. Soc.*, *105*: 5786.

Groves, J.T., McClusky, G.A., White, R.E., and Coon, M.J. (1978). *Biochem. Biophys. Res. Commun.*, *81*: 154.

Gunter, M.J., and Turner, P. (1991). *Coord. Chem. Rev.*, *108*: 115.

Harvey, P.J., Schoemaker, R.M., Bowen, R.M., and Palmer, J.M. (1985). *FEBS Lett.*, *183*: 13.

Hewson, W.D., and Hager, L.P. (1979a). *The Porphyrins*, Vol. 7 (D. Dolphin, ed.), Academic Press, New York, p.295.

Hewson, W.D., and Hager, L.P. (1979b). *J. Biol. Chem.*, *254*: 3182.

Hoffman, B.M., Roberts, J.E., Kang, C.H., and Margoliash, E. (1981). *J. Biol. Chem.*, *256*: 6556.

Kadlubar, F.F., Morton, K.C., and Ziegler, D.M. (1974). *Biochem. Biophys. Res. Commun.*, *54*: 1255.

Kersten, P.J., Tien, M., Kalyanaraman, B., and Kirk, T.K. (1985). *J. Biol. Chem.*, *260*: 2609.

La Mar, G.N., de Ropp, J.S., Smith, K.M., and Langry, K.C. (1981). *J. Biol. Chem.*, *256*: 237.

Mansuy, D., Bartoli, J.F., and Momenteau, M. (1982). *Tetrahedron Lett.*, *23*: 2781.

Mansuy, D., Battioni, P., and Renand, J.P. (1984). *J. Chem. Soc. Chem. Commun.*, 1255.

Mauro, J.M., Miller, M.A., Edwards, S.L., Wang, J., Fishel, L.A., and Kraut, J. (1989). *Metal Ions in Biological Systems*, Vol. 25 (H. Sigel and A. Sigel, eds.), Marcel Dekker, New York, p. 477.

McMurry, T.J., and Groves, J.T. (1986). *Cytochrome P-450: Structure, Mechanism, and Biochemistry* (P.R. Ortiz de Montellano, ed.), Plenum Press, New York, p. 1.

Meunier, B. (1992). *Chem. Rev.*, *92*: 1411.

Mieyal, J.J. (1985). *Reviews in Biochemical Toxicology*, Vol. 7 (E. Hodgson, J.R. Bend and R.M. Philpot, eds.), Elsevier, New York, p. 1.

Morrison, M., and Schonbaum, G.R. (1976). *Annu. Rev. Biochem.*, *45*: 861.

Murphy, M.R., Reid III, T.J., Sicignano, A., Tanaka, N., and Rossmann, M.G. (1981). *J. Mol. Biol.*, *152*: 465.

Ortiz de Montellano, P.R. (1987). *Accounts Chem. Res.*, *20*: 289.

Ortiz de Montellano, P.R., and Grab, L.A. (1987). *Biochemistry, 26*: 5310.

Ortiz de Montellano, P.R., Choe, Y.K., DePillis, G., and Catalano, C.E. (1987). *J. Biol. Chem., 262*: 11641.

Perutz, M.F., Ferml, G., Luisi, B., Shaanan, B., and Liddington, R.C. (1987). *Accounts Chem. Res., 20*: 309.

Phillips, S.E., and Schoenborn, B.P. (1981). *Nature, 292*: 81.

Poulos, T.L. (1986). *Cytochrome P-450: Structure, Mechanism and Biochemistry* (P.R. Ortiz de Montellano, ed.), Plenum Press, New York, p. 505.

Poulos, T.L. (1992a). Symposium on Pulp and Enzymes: New Catalysts for the Environment, Vancouver, B.C., Canada.

Poulos, T.L. (1992b). Private communication.

Poulos, T.L., and Finzel, B.C. (1984). *Pept. Protein Rev., 4*: 115.

Poulos, T.L., Finzel, B.C., Gunsalus, I.C., Wagner, G.C., and Kraut, J. (1985). *J. Biol. Chem., 260*: 16122.

Poulos, T.L., Finzel, B.C., and Howard, A.J. (1987). *J. Mol. Biol., 195*: 687.

Roberts, J.E., Hoffman, B.M., Rutter, R., and Hager, L.P. (1981). *J. Biol. Chem., 256*: 2118.

Rutter, R., Valentine, M., Hendrich, M.P., Hager, L.P., and Debrunner, P.G. (1983). *Biochemistry, 22*: 4769.

Sakurada, J., Takahashi, S., and Hosoya, T. (1986). *J. Biol. Chem., 261*: 9657.

Schonbaum, G.R. (1982). *Oxidases and Related Redox Systems* (E. King, H.S. Mason and M. Morrison, eds.), Pergamon Press, Oxford, England, p. 671.

Sivaraja, M., Goodin, D.B., Smith, M., and Hoffman, B.M. (1989). *Science, 245*: 738.

Smulevich, G., Mauro, J.M., Fishel, L.A., English, A.M., Kraut, J., and Spiro, T.G. (1988). *Biochemistry, 27*: 5477.

Smulevich, G., English, A. M., and Spiro, T.G. (1990). "Laser Applications in Life Sciences", Proceedings of the SPIE-International Society for Optical Engineering '90, Moscow, USSR, pp. 440-447.

Ten Eyck, L.F. (1979). *The Porphyrins*, Vol. 7 (D. Dolphin, ed.), Academic Press, New York, p. 445.

Terner, J.A., Sitter, J., and Shifflett, J.R. (1989). *Charge Field Effect in Biosystems* (M.J. Allen, S.F. Cleary and I.M. Hawkridge, eds.), Plenum Press, New York, p. 31.

Thanabal, V., de Ropp, J.S., and La Mar, G.N. (1987). *J. Amer. Chem. Soc., 109*: 7516.

Thanabal, V., de Ropp, J.S., and La Mar, G.N. (1988). *J. Amer. Chem. Soc., 110*: 3027.

Theorell, H., and Ehrenberg, A. (1952). *Arch. Biochem. Biophys., 41*: 442.

Tien, M., and Kirk, T.K. (1983). *Science, 221*: 661.

Traylor, P.S., Dolphin, D., and Traylor, T.G. (1984). *J. Chem. Soc. Chem. Commun.*, 279.

Traylor, T.G., Marsters, J.C., Nakano, T., and Dunlap, B.E. (1985). *J. Amer. Chem. Soc., 107*: 5537.

Traylor, T.G., Nakano, T., Dunlap, B.E., Traylor, P.S., and Dolphin, D. (1986). *J. Amer. Chem. Soc., 108*: 2782.

Vainshtein, B.K., Melik-Adamyan, W.R., Barynin, V.V., Grebenko, A.I., Borisov, V.V. Bartels, K.S., Fita, I., and Rossmann, M.G. (1986). *J. Mol. Biol., 188*: 49.

White, P.W. (1990). *Bioorg. Chem., 18*:, 440.

Wijesekera, T., Matsumoto, A., Dolphin, D., and Lexa, D. (1990). *Angew. Chem. Int. Ed. Engl., 29*: 1028.

Yonetani, T. (1970). *Advances in Enzymology*, Vol. 33 (F.F. Nord, ed.), Interscience, New York, p. 309.

Yonetani, T., and Schleyer, H. (1967). *J. Biol. Chem., 242*: 1974.

CATALYTIC MECHANISMS IN ZINC ENZYMES

by

Michael F. Dunn

Department of Biochemistry
University of California, Riverside
Riverside, California 92521
U.S.A.

I. INTRODUCTION

The interplay of structural and electronic interactions between the protein and the active site metal ion define the set of structure-function relationships that convey catalytic activity to a metalloprotein. The bioinorganic literature, for the most part, focuses on the metal ion and its inner-sphere coordinated ligands to provide mechanistic rationalizations for catalysis. There are many indications that this approach is too narrow. For example, the Co(II)-R_6 insulin hexamer appears to be an excellent spectroscopic and structural model for Co(II)-substituted carbonic anhydrase (Brader et al., 1991), yet Co(II)-R_6 appears to be devoid of carbonic anhydrase activity. The inability of Co(II)-R_6 to catalyze CO_2 hydration very likely is due to the absence of a CO_2-binding site and/or the absence of a proton relay system for shuttling protons between the site and solvent. In both alcohol dehydrogenase and carbonic anhydrase, site-directed mutations at loci distant from the active site metal ion can exert large effects on individual steps in catalysis (Fierke et al., 1991; Stone et al., 1993). The crystal structures and solution studies on the alcohol dehydrogenases from horse and human liver establish that the interconversion of the enzyme site between conformationally open and closed forms is part of the catalytic cycle (Stone et al., 1993; Hurley et al., 1991; Eklund and Brändén, 1979). Hence, just as is true for other cofactor-requiring enzymes, the metal centers of zinc-metalloproteins are only part of the story. In this brief review, some recent developments in the chemistry and catalytic mechanisms of zinc enzymes are discussed. This treatment is focused on the two most extensively characterized zinc enzyme systems, human carbonic anhydrase II (CAII) and liver alcohol dehydrogenase (LADH). Recent developments in the understanding of the bioinorganic chemistry of Zn(II) are also reviewed. This discourse attempts to place the roles played by the metal ion into a larger structure-function context that includes the contributions of other cofactors and the protein matrix. Due to space limitations, literature citations are limited to a few selected papers from the recent literature and to recently published reviews.

II. THE BIOINORGANIC CHEMISTRY OF Zn(II)

Zn(II) is a Lewis acid that is classified as on the borderline between "hard" and "soft". Hard Lewis acids exhibit low polarizability, soft Lewis acids exhibit high polarizability. The higher the polarizability of a metal ion, the greater the covalency of the bonding interactions with ligands, and hard acids prefer hard bases while soft acids prefer soft bases (see Part One, Chapter 3). As a borderline Lewis acid, Zn(II) shows similar affinities for oxygen, nitrogen and sulfur ligands, and these ligands dominate the bonding of Zn(II) in biological systems (see Glusker, 1991).

Bonding distances in both small molecule and protein complexes are well predicted by summing the values of the ionic radii of Zn(II) and the liganding atom. The appropriate ionic radius for Zn(II) must be used when estimating bond lengths (see Vedani and Huhta, 1990). Bond lengths to N, O and S ligands increase by approximately 0.2 Å as the coordination number is expanded from 4 to 6.

Because Zn(II) has a d^{10} configuration inside a vacant s shell, the stereochemistry and coordination number of Zn(II) complexes are determined only by ligand steric bulk, electrostatic interactions between ligands and covalent constraints on ligand structure. Because Zn(II) complexes are not subject to Jahn-Teller effects, the coordination number and geometry are imposed by the structure of the ligand field. All X-ray structures comparing the apoprotein with the native Zn(II)-protein published to date show that the ligands in the metal-free structure are positioned essentially the same as in the metal-bound structure (Hill et al., 1991; Schneider et al., 1983; Hakansson et al., 1992). Thus, the apoprotein provides a preformed site poised for Zn(II) coordination.

Small molecule Zn(II) complexes generally form structures with coordination number 4, 5 or 6. The zinc enzymes for which X-ray structures have been determined all have an active site zinc ion that is coordinated by the side chains of three amino acid residues and one or two water molecules. These complexes are either 4- or 5-coordinate. For example, carbonic anhydrase is four coordinate at neutral pH (3 His and one water or hydroxide ion), while carboxypeptidase is five coordinate (2 His, both oxygens of one Glu and one water) (see Vallee and Auld, 1990; Glusker, 1991).

Where Zn(II) plays a strictly structural role (e.g. the structural zinc site in alcohol dehydrogenase, zinc finger proteins, and aspartate transcarbamylase), the coordination is usually tetrahedral with all four ligands derived from the protein. The insulin hexamer is an exception; the zinc site of the T_6 hexamer is octahedral and consists of three His residues from the protein and three water molecules (Baker et al., 1988; Hill et al., 1991). The R_6 insulin hexamer has a tetrahedral zinc site consisting of three His residues and a fourth ligand that is usually a monovalent anion derived from solution (Brader et al., 1991; Smith and Dodson, 1992).

By definition, in stable organometallic complexes, the formal charge on the metal ion must be nearly equal to the sum of the strengths of the electrostatic bonds contributed by its ligands. Thus, for an isolated molecule, the ligands to a metal ion must provide bond valence contributions which equal the charge on the metal ion (but are opposite in sign) (Glusker, 1991). In metalloprotein complexes, this need not be true. More distant structural elements (e.g. cofactors, amide linkages, charged residues and helix dipoles) provided by the system can contribute electrostatic fields that partially neutralize the charge on the metal ion. These longer range effects can be modulated by conformational transitions linked to binding and covalent transformations during the catalytic cycle and, therefore, may play a significant role in manipulating the electrostatic microenvironment of a reacting substrate to lower activation energies (Dunn et al., 1986). The Zn(II)-requiring alcohol dehydrogenases appear to require such a conformational transition as an integral component of the catalytic cycle (Stone et al., 1993; Dunn et al., 1986; Eklund and Brändén, 1979).

The three most important ligands derived from protein side chain groups in zinc metalloenzymes are the carboxylate group (usually a Glu), the imidazolyl group of His and the thiolate of Cys. The inner-sphere coordination of other side chain groups is rare.

The Cambridge Structural Database (Allen et al., 1979) contains an enormous amount of high resolution structural data for small molecule-metal ion complexes. Glusker (1991) has reviewed this data base to determine the preferred bonding interactions between metal ions and biologically relevant ligands. Scatterplots of the locations of metal ions about the liganding atom provide a measure of the preferred location of the metal ion with respect to the ligand. These scatterplots give a contoured probability map (Glusker, 1991; Rosenfield et al., 1984) which defines the most probable arrangements of metal ions for a given ligand. The bonding preferences exhibited by metalloproteins are similar to those exhibited by small molecule complexes. Scatterplots for the carboxylate anion show that Zn(II) has a strong preference for in plane *syn* coordination to one oxygen of the carboxylate group, a result that is consistent with the greater basicity of the *syn* lone pair in comparison to the *anti* lone pair (Gandour, 1981; Rebek et al., 1986). Nevertheless, in plane *anti* coordination occurs with a fairly high frequency, but complexes which involve coordination of the same Zn(II) ion to both carboxylate oxygens (with equal bond lengths) are rare. These observations indicate that the carboxylate lone pair electrons are important for Zn(II)-carboxylate bonding. The scatterplot for the imidazolyl group shows there is a high preference for Zn(II) to be aligned with the nitrogen electron lone pair and, therefore, the Zn(II)-imidazole ring complex usually forms a plane. The high degree of polarizability of sulfur and its low electronegativity give the sulfhydryl group bonding characteristics that are somewhat different from hard bases such as O- and N-derived ligands.

III. ELECTROPHILIC STRAIN-DISTORTION EFFECTS IN METALLOENZYME CATALYSIS

Metalloenzymes where the active site metal itself is not directly involved in a redox change are catalysts that employ electrostatic strain-distortion (ESD) effects as the primary means for lowering the activation energy of covalent steps in the catalytic cycle (Dunn et al., 1986). The ESD effect likely is tailored (via evolution) to achieve a compromise between the electrostatic force field needed to facilitate a given chemical transformation and the rates of exchange of inner-sphere coordinated ligands so that a suitable catalytic efficiency is realized. To understand this compromise, consider the following generalized reaction path:

$$\text{E-Me(II)-OH}_2 + \text{S} \rightleftharpoons \underset{\text{Michaelis complex}}{\text{E-Me(II)-OH}_2\text{(S)}} \overset{H_2O}{\rightleftharpoons} \underset{\text{Substrate inner-sphere complex}}{\text{E-Me(II)-S}} \rightleftharpoons$$

$$\underset{\text{Product inner-sphere complex}}{\text{E-Me(II)-P}} \overset{H_2O}{\rightleftharpoons} \underset{\text{Michaelis complex}}{\text{E-Me(II)-OH}_2\text{(P)}} \rightleftharpoons \text{E-Me(II)-OH}_2 + \text{P}$$

Scheme I

In Scheme I, the resting holo-enzyme preexists as a four- or five-coordinate complex wherein at least one water molecule is ligated to the metal ion (Me). Substrate reacts with the holoenzyme via a sequence of steps that involves (i) formation of a Michaelis Complex, (ii) replacement of inner-sphere coordinated water by substrate, (iii) chemical conversion of coordinated substrate to coordinated product, (iv) displacement of coordinated product by water, and (v) dissociation of product from the product Michaelis Complex. As is implied by Scheme I, the chemical activation of substrate/product is a consequence of the ESD effect imposed on the substrate/product by inner-sphere coordination. As is indicated in the accompanying free-energy reaction-coordinate diagram (Fig. 1), if the electrophilicity of the metal center is increased, then the polarizing effect on coordinated substrate also is increased, and the energy barrier for the interconversion of coordinated substrate and product is lowered (compare a and b). However, increasing the electrophilic character of the metal ion also stabilizes the ground states of bound species, and if the electrostatic attraction between the active site metal ion and any of the coordinated species (water, substrate or product) becomes strong enough, then either inner-sphere substitution or product dissociation would become rate determining (viz b and c). Once the catalyst has evolved to a point where inner-sphere substitution or product dissociation (or some other step) has become rate determining (as in b), any further increase in electrophilicity of the metal center (viz c) will result in a reduced catalytic efficiency. As will be seen, in the LADH system, coenzyme product dissociation is generally the rate-determining step in catalysis. However, in the direction of aldehyde reduction, the release of alcohol product for some substrates is slower than, or comparable in rate to, the rate of NAD^+ dissociation. In the CAII system, the turnover cycle in the direction of CO_2 hydration is limited in rate by the transfer of a proton from the inner-sphere coordinated water molecule to His 64. This transfer is the first

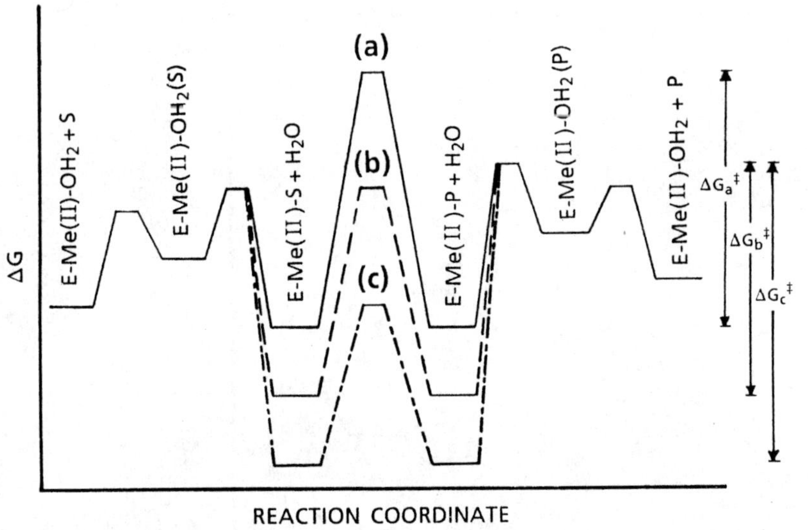

Figure 1.
Free-energy diagram for Scheme I. The profiles (a-c) correspond to enzymes where the active site metal ion exerts different degrees of electrostatic strain-distortion (ESD) catalysis on the interconversion of coordinated substrate and coordinated product. (a) ESD catalysis ($\Delta G_a‡$) is rate determining (-). (b) The ESD effect is so large that product dissociation ($\Delta G_b‡$) is rate determining and catalysis is essentially optimal (- -). (c) The ESD effect is further increased with the result that the activation energy for the chemical step is very low (---). However, the ground states of coordinated substrate and product are so stabilized that this enzyme has reduced catalytic efficiency due to the slow exchange of water for coordinated product ($\Delta G_c‡$).

step in the proton shuttle that regenerates the active form of the enzyme (Tu et al., 1989). Consequently, catalysis has evolved to a stage where further enhancement of the electrophilic character of the active site zinc ions of LADH and CAII would stabilize ground states without lowering the activation energy of the rate determining step.

IV. THE CARBONIC ANHYDRASE CATALYTIC MECHANISM

The zinc metalloenzyme, human carbonic anhydrase isozyme II (CAII), catalyzes the reversible hydration of CO_2 to HCO_3^- and H^+ with unusually high catalytic efficiency. The turnover number, $k_{cat}^{CO_2} = 1.4 \times 10^6\ s^{-1}$ is two to three orders of magnitude larger than most other enzymes, and the ratio $k_{cat}/K_M^{CO_2} = 1.5 \times 10^8\ M^{-1}s^{-1}$ implies a diffusion-limited second order rate constant for CO_2 binding (see Silverman and Lindskog, 1988 and Christianson, 1991). The X-ray crystallographic analysis of the refined CAII structure has recently been reported at 1.54 Å resolution (pH 7.8) and 1.67 Å resolution (pH 6.0) by Hakansson et al. (1992) together with the structures of several inhibitor complexes. These structures show that the active site Zn(II) is coordinated in a distorted tetrahedral ligand field comprised of three His imidazolyl groups (His 94, His 96 and His 119) and one water molecule (Fig. 2).

The prevailing mechanistic consensus (Equations 1-3, see Silverman and Lindskog, 1988) is that the electrostatic field of Zn(II) perturbs the pK_a of the coordinated water molecule to a value of ~7.0. This ESD effect gives a metal-coordinated hydroxide ion species, E-Zn(II)-OH$^-$, that undergoes nucleophilic attack on CO_2 to give a metal-coordinated bicarbonate species (Equation 1), which then is replaced by H_2O in an associative mechanism (Equation 2) to yield bicarbonate and a metal-coordinated water species, E-Zn(II)-OH_2. Regeneration of E-Zn(II)-OH$^-$ involves a proton shuttle (Equation 3)

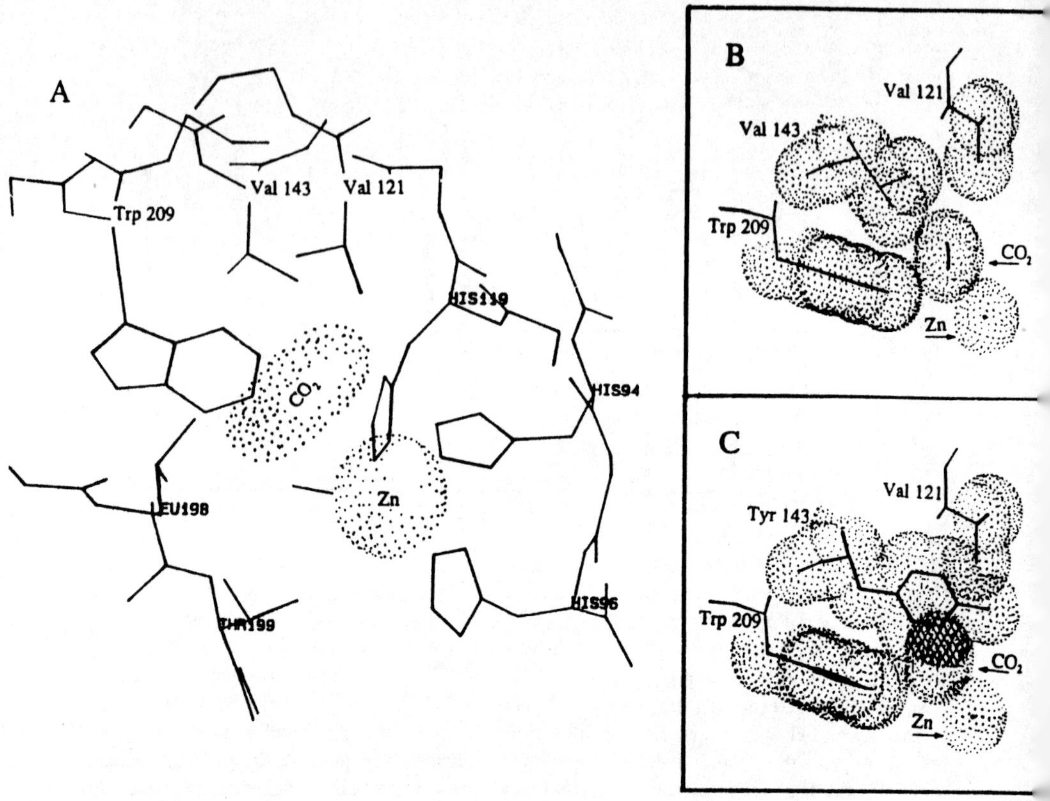

Figure 2.
(A) CAII active site and the hydrophobic pocket proposed to be the CO_2-binding site. The active site Zn(II) and the proposed location of bound CO_2 are shown as stippled space-filling surfaces. The hydrophobic pocket is bounded by Val 121, Val 143, Leu 193, and Trp 209. (B) and (C) are depictions of the active sites of native (B) and V143Y human CAII using the van der Waals surfaces of Zn(II), CO_2 and the residues which define the hydrophobic pocket. The position of CO_2 is modeled into the hydrophobic pocket in (B). In (C), there is a strong overlap between the Tyr 143 side chain and the proposed CO_2-binding site (crosshatched surfaces). Figures were redrawn from Fierke et al. (1991) with permission.

wherein the proton from the coordinated water species is transferred first to His 64 (Tu et al., 1989), and then to solvent buffer ions (Silverman and Tu, 1975).

$$\text{E-Zn(II)-OH}^- \overset{CO_2}{\rightleftharpoons} \text{E-Zn(II)-OH}^-(CO_2) \rightleftharpoons \text{E-Zn(II)-}\overset{\text{H}}{\underset{|}{\text{O}}}\text{-}CO_2^- \quad (1)$$

$$\text{E-Zn(II)-}\overset{\text{H}}{\underset{|}{\text{O}}}\text{-}CO_2^- \overset{H_2O}{\rightleftharpoons} \text{E-Zn(II)-OH}_2 + HCO_3^- \quad (2)$$

$$\text{E-Zn(II)-OH}_2 \rightleftharpoons {}_{\text{His64}}\text{H}^+\text{E-Zn(II)-OH}^- \rightleftharpoons \text{E-Zn(II)-OH}^- + \text{H}^+ \quad (3)$$

Solvent kinetic isotope studies comparing k_{cat} values in H_2O with D_2O establish that the proton transfer to His 64 (Equation 3) is the rate-limiting step in $k_{cat}^{CO_2}$ (Silverman and Lindskog, 1988). Tu et al. (1989) have shown by site-directed mutation that the shuttle occurs via His 64; the mutation H64A causes proton transfer to buffer ion to become the rate-limiting step for $k_{cat}^{CO_2}$ at low buffer ion concentrations. The site-directed mutation studies of Fierke et al. (1991) and Nair et al. (1991) together with the three-dimensional X-ray structures of some of these mutants (Alexander et al., 1991) provide new insight into the nature of the reactive enzyme-CO_2 complex. The 15 Å-deep active site cavity is amphiphilic. The polar residues, Thr 199 and Glu 106, form a hydrogen-bonding lattice on one wall that may stabilize the Zn(II)-coordinated hydroxide ion. The opposing wall contains a hydrophobic pocket (Fig. 2) comprised of the highly conserved residues Val 121, Val 143, Leu 198 and Trp 209 located adjacent to the coordinated hydroxide ion. It has been proposed that this pocket is the CO_2-binding site and that the reactive E-Zn(II)-OH^- (CO_2) complex does not involve inner-sphere coordination of CO_2 (Silverman and Lindskog, 1988). Through site-directed mutation of Val 121 and Val 143, Fierke et al. (1991) and Nair et al. (1991) show that the integrity of this hydrophobic pocket is essential to CAII catalysis of CO_2 hydration. For example, the V143Y mutation fills this pocket (compare Fig. 2B with 2C) and reduces the turnover rate by a factor of $1/(3 \times 10^5)$. These studies are plausibly interpreted as evidence that this hydrophobic pocket is the CO_2-binding site.

V. THE LIVER ALCOHOL DEHYDROGENASE (LADH) CATALYTIC MECHANISM

Liver NAD/NADH requiring alcohol dehydrogenases catalyze the interconversion of aldehydes and alcohols (see Eklund and Brändén, 1979; Cedergren-Zepperzauer et al., 1982; Dunn et al., 1986; Pettersson, 1987). The active site zinc ions of the liver enzymes are coordinated in distorted tetrahedral ligand fields consisting of two Cys sulfhydryl groups, one His imidazolyl group and a water molecule (Fig. 3). The catalytic metal resides at the point of convergence of two deep (20-25 Å-long) clefts, an amphiphilic coenzyme-binding cleft and a hydrophobic substrate-binding cleft. Crystallographic, kinetic and spectroscopic studies (see Dunn et al., 1986; Pettersson, 1987) have established that the active site metal ion plays Lewis acid catalytic roles wherein aldehyde and alcohol ternary complexes are activated for hydride transfer by inner-sphere coordination. The ESD interaction via inner-sphere coordination of the aldehyde carbonyl conveys a carbocation-like character to the carbonyl carbon (Fig. 3), making it an effective electrophile for nucleophilic attack by the hydride ion (or its equivalent) donated from the 1,4-dihydronicotinamide ring of NADH (Dunn et al., 1986). In the reverse reaction (Fig. 3), the inner-sphere coordination of alcohol lowers the pK_a of the hydroxyl group thereby increasing the concentration of the reactive species, the inner-sphere coordinated alcoholate ion (Petterson, 1987). While the magnitude of this pK_a perturbation has been a source of disagreement, there is no doubt that formation of the alcoholate ion ternary complex is obligatory for substrate activation, and this step precedes the transfer of hydride to NAD^+. Because this proton transfer is fast, the rate-determining step for interconversion of the ternary complexes is hydride transfer. However, for most substrates coenzyme product dissociation is rate determining for turnover. The site-directed mutation studies of Ehrig et al. (1991) and Plapp et al. (1991) establish that the proton transfer process occurs by way of

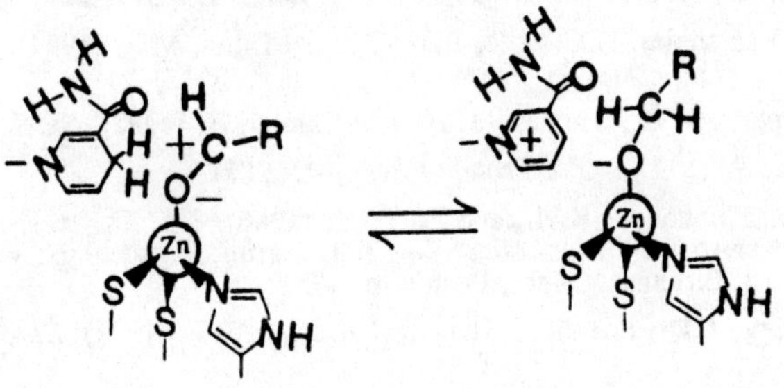

Figure 3.
Structures of the carbocation-like and alcoholate ion-like reactive species in the LADH catalytic mechanism (Dunn et al., 1986).

a proton relay along the hydrogen-bonding network extending from the proton of the coordinated alcohol to the hydroxyl of Ser 48 and the imidazolyl ring of His 51. Disruption of this network in the human liver enzyme (Ehrig et al., 1991) by the H51Q mutation renders the rate of turnover in the direction of aldehyde reduction dependent on the buffer ion concentration.

Crystallographic studies have shown that ternary complex formation is accompanied by the conversion of the liver enzymes from "open" to "closed" conformations and that this conformational transition expels the entire lattice of water molecules detected in the open structure of the coenzyme-free enzyme from the vicinity of the catalytic site (Eklund and Brändén, 1979; Cedergren-Zeppezauer et al., 1982). Dunn et al. (1986) have argued that this conformational transition to a closed structure creates a highly polar site. The effect of the water lattice upon the site in the open structure is to dissipate the electrostatic field extending from the metal ion. Exclusion of water in the closed structures of the ternary complexes increases the effective strength of the electrostatic field. In the NAD^+ complex the positive charge on the nicotinamide ring makes a considerable contribution to the electropositive character of the site. The exclusion of water and inner-sphere coordination of substrate/product gives an electrostatic environment at the site that promotes formation of the carbocation-like aldehyde and alcoholate ion-like species that undergo chemical interconversion. This ESD effect so reduces the activation energy for hydride transfer that coenzyme dissociation becomes rate determining.

VI. CONCLUSIONS

The LADH and CA(II) zinc enzyme paradigms provide two well understood examples of how the ESD effect is used to decrease the activation energies of quite different chemical processes. In CA(II) catalysis of CO_2 hydration, the ESD effect lowers the pK_a of Zn(II)-coordinated water to a value ~7.0, thereby optimizing the concentration of the reactive species, coordinated hydroxide ion. In LADH catalysis of alcohol oxidation, the ESD effect plays a similar role by enhancing the concentration of the reactive hydride donor species, the coordinated alcoholate ion, via a greatly perturbed pK_a. The ESD effect in both enzymes is so perfected that another process limits the rate of turnover. Finally, recent work on these two enzymes to investigate the role of the protein in catalysis has shown that a complete description of their catalytic mechanisms must include the roles played by specific side-chain residues, and, in the case of LADH, the role played by a conformational transition which modulates the ESD effect.

REFERENCES

Alexander, R.S., Nair, S.K., and Christianson, D.W. (1991). *Biochemistry, 30*: 11064.

Allen, F.H., Bellard, S., Brice, M.D., Cartwright, B.A., Doubleday, A., Higgs, H., Hummelink, T., Hummelink-Peters, B.G., Kennard, O., Motherwell, W.D.S., Rodgers, J.R., and Watson, D.G. (1979). *Acta Crystallogr. Sect. B., 35*: 2331.

Baker, E.N., Blundell, T.L., Cutfield, J.F., Cutfield, S.M., Dodson, E.J., Dodson, G.G., Hodgkin, D.C., Hubbard, R.E., Isaacs, N.W., Reynolds, C.D., Sakabe, K., Sakabe, N., and Vijayan, N.M. (1988). *Phil. Trans. Roy. Soc. London, B 319:* 369.

Brader, M.L., Kaarsholm, N.C., Lee, R.W.-K., and Dunn, M.F. (1991). *Biochemistry, 30*: 6636.

Cedergren-Zeppezauer, E., Samama, J.-P., and Eklund, H. (1982). *Biochemistry, 21*: 4895.

Christianson, D.W. (1991). *Adv. Protein Chem., 42*: 281.

Dunn, M.F., MacGibbon, A.K.H., and Pease, K. (1986). *Zinc Enzymes: Progress in Inorganic Biochemistry and Biophysics*, Vol. 1 (I. Bertini, C. Luchinat, W. Maret and M. Zeppezauer, eds.), Birkhäuser, Inc., Boston, p. 485.

Ehrig, T., Hurley, T.D., Edenberg, H.J., and Bosron, W.F. (1991). *Biochemistry, 30*: 1062.

Eklund, H., and Brändén, C.-I. (1979). *J. Biol. Chem., 254*: 3458.

Fierke, C.A., Calderone, T.L., and Krebs, J.F. (1991). *Biochemistry, 30*: 11054.

Gandour, R.D. (1981). *Bioorg. Chem., 10*: 169.

Glusker, J.P. (1991). *Adv. Protein Chem.*, *42*: 1.

Hakansson, K., Carlsson, M., Svensson, L.A., and Liljas, A. (1992). *J. Mol. Biol.*, *227*: 1192.

Hill, C.P., Dauter, Z., Dodson, E.J., Dodson, G.G., and Dunn, M.F. (1991). *Biochemistry*, *30*: 917.

Hurley, T.F., Bosron, W.F., Hamilton, J.A., and Amzel, L.M. (1991). *Proc. Nat. Acad. Sci. USA*, *88*: 8149.

Nair, S.K., Calderone, T.L., Christianson, D.W., and Fierke, C.A. (1991). *J. Biol. Chem.*, *266*: 17320.

Pettersson, G. (1987). *CRC Crit. Rev. Biochem.*, *21*: 349.

Plapp, B.V., Ganzhorn, A.J., Gould, R.M., Green, D.W., Jacobi, T., Warth, E., and Kratzer, D.A. (1991). *Adv. Exp. Med. Biol.*, *284*: 241.

Rebek, J., Jr., Duff, R.J., Gordon, W.E., and Parris, K. (1986). *J. Amer. Chem. Soc.*, *108*: 6068.

Rosenfield, R.E., Jr., Swanson, S.M., Meyer, E.F., Carrell, H.L., and Murray-Rust, P. (1984). *J. Mol. Graphics*, *2*: 43.

Schneider, G., Eklund, H., Cedergren-Zeppezauer, E., and Zeppezauer, M. (1983). *Proc. Nat. Acad. Sci. USA*, *80*: 5289.

Silverman, D.N., and Lindskog, S. (1988). *Accounts Chem. Res.*, *21*: 30.

Silverman, D.N., and Tu, C.-K. (1975). *J. Amer. Chem. Soc.*, *97*: 2263.

Smith, G.D., and Dodson, G.G. (1992). *Biopolymers*, *32*: 441.

Stone, C.L., Bosron, W.F., and Dunn, M.F. (1993). *J. Biol. Chem.*, *268*: 892.

Tu, C.K., Silverman, D.N., Forsman, C., Jonsson, B.-H., and Lindskog, S. (1989). *Biochemistry*, *28*: 7913.

Vallee, B.L., and Auld, D.S. (1990). *Biochemistry*, *29*: 5647.

Vedani, A., and Huhta, D.W. (1990). *J. Amer. Chem. Soc.*, *112*: 4759.

COPPER ACTIVE SITES: PROTEINS AND ENZYMES CONTAINING COUPLED BINUCLEAR COPPER SITES

by

James A. Larrabee

Department of Chemistry and Biochemistry,
Middlebury College,
Middlebury, VT 05753
U.S.A.

and

Edward I. Solomon

Department of Chemistry,
Stanford University,
Stanford, CA 94305
U.S.A.

I. INTRODUCTION

Coupled binuclear copper active sites are found in a variety of proteins and enzymes which either bind or utilize dioxygen in their functional role (Table 1). Hemocyanins are the oxygen transport protein in mullusca and arthropoda, tyrosinases are found in mammals, fungi and plants and are responsible for the "browning" reaction that takes place for example when mushrooms are sliced open and exposed to air. The multi-copper oxidases are found in fungi, plants and animals and reduce dioxygen to water, while playing functionally important roles in substrate oxidation and iron metabolism.

Deoxyhemocyanin and deoxytyrosinase contain reduced binuclear copper sites ([Cu(I)Cu(I)]) which react with dioxygen to form stable oxygenated species with similar spectral properties. While hemocyanins reversibly bind dioxygen, tyrosinase oxygenates monophenols to *o*-diphenols and then further oxidizes this product to the *o*-quinone. The multi-copper oxidases, of which laccase is the simplest example, contain three spectrally distinct copper sites: a type-1 single "blue" copper center, a type-2 "normal" copper, and a type-3 "coupled binuclear" copper center. It is now known that the type-2 and type-3 copper sites are spatially close and can interact so that the type-2 and type-3 sites actually form a trinuclear cluster. The multi-copper oxidases catalyze the irreversible four-electron reduction of dioxygen to water at this trinuclear copper cluster site.

Table 1

Summary of Proteins Containing Coupled Binuclear Copper Active Sites and Their Functions*

Protein	Source/Species	Functions
Hemocyanin	Arthropodal	deoxy + O_2 = oxy
	Mulluscan	deoxy + O_2 = oxy $2H_2O_2 = O_2 + H_2O$
Tyrosinase	mushroom and *Neurospora crassa*	deoxy + O_2 = oxy $2H_2O_2 = O_2 + H_2O$ phenol + $2e^-$ + O_2 + $2H^+$ = *o*-diphenol + H_2O 2*o*-diphenol + O_2 = 2*o*-quinone + $2H_2O$
Multi-Copper Oxidases		$4AH + O_2 \rightarrow 4A + 2H_2O$
Laccase	tree: *Rhus vernicifera* fungal: *Polyporus versicolor*	
Ascorbate oxidase	zuccini	
Ceruloplasmin	mammalian blood plasma	

* Taken from Solomon (1988).

Systematic spectral studies of these proteins, geometrical and electronic structural studies of model compounds and the recently solved X-ray structures of deoxyhemocyanin, oxyhemocyanin and resting ascorbate oxidase have provided a reasonable picture of the active sites in this series of proteins and enzymes. Efforts at correlation of active site geometric and electronic structure with biological function are now underway (Solomon et al., 1992 and Solomon and Lowery, 1992).

II. HEMOCYANIN AND TYROSINASE

Resonance Raman spectroscopy showed that the bound dioxygen in oxyhemocyanin and oxytyrosine is peroxide ($\nu_{O\text{-}O}$ at 750 cm^{-1}) and that the binuclear oxy site is formally [Cu(II)Cu(II)] (Freedman et al., 1976; Eickman et al., 1978). EPR and SQUID magnetic susceptibility studies showed that, unlike monomeric cupric complexes which are paramagnetic with one unpaired electron, the coppers in oxyhemocyanin and oxytyrosinase are diamagnetic through strong antiferromagnetic coupling with an exchange constant, -2J, greater than 600 cm^{-1} (Solomon et al., 1976; Dooley et al., 1978). In addition to the diamagnetism, the optical absorption spectral features of the oxy sites were markedly different from those of normal tetragonal cupric complexes. Instead of weak d-d transitions in the 600-700 nm range, the absorption spectra of the oxy proteins are dominated by two intense bands at approximately 570 nm (ε = 1000 $M^{-1}cm^{-1}$) and approximately 350 nm (ε = 20,000 $M^{-1}cm^{-1}$).

Spectral studies of the unique features of the oxyproteins as well as detailed spectral studies of a series of chemically-generated derivatives (Fig. 1), which allowed the active sites to be systematically varied, led to initial structural insight into the active site of hemocyanin and tyrosinase in the early 1980's. This work has been reviewed (Solomon, 1981; 1988; Solomon et al., 1983; 1992) and will not be discussed in detail here. Only key points will be given to put the most recent results in perspective. Absorption data on all Cu(II) derivatives indicated that the coppers had a tetragonal geometry and were approximately equivalent (Eickman et al., 1979) whereas Raman (Larrabee and Spiro, 1980) and EXAFS studies (Brown et al., 1980; Woolery et al., 1984a; 1984b; 1984c; Co et al.,

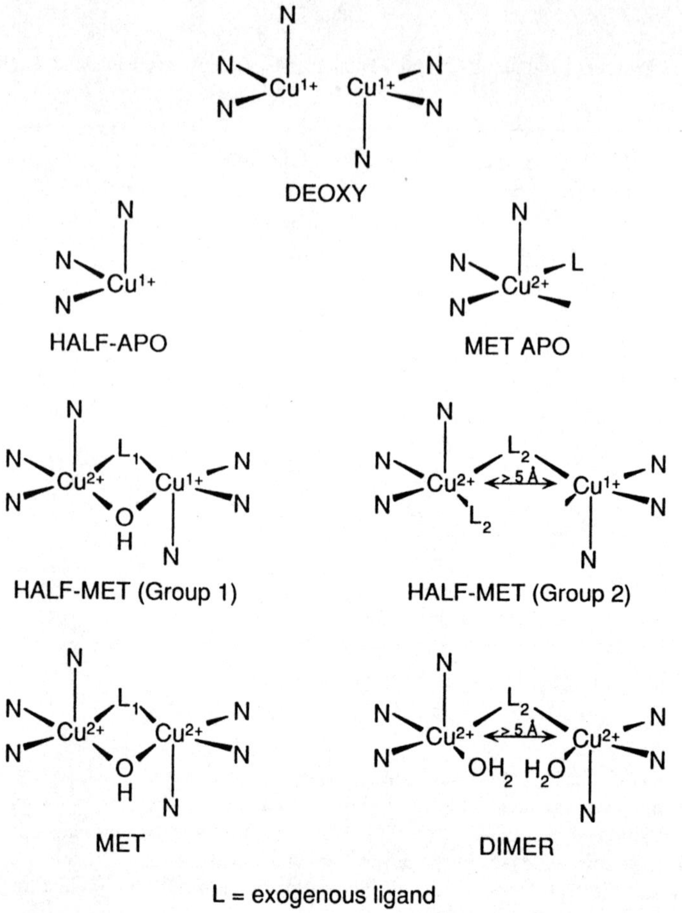

Figure 1.
Models for active site derivatives of hemocyanin and tyrosinase.

1981; Co and Hodgson, 1981; Blackburn, 1990) indicated two to three imidazole (from histidine) nitrogen ligands per copper atom and a Cu-Cu spacing in the range of 345-360 pm in oxy, deoxy and met derivatives. The coppers were strongly exchange-coupled in the oxyproteins as well as the met derivative in which both coppers are divalent but no peroxide is bound. Since the divalent coppers are still strongly coupled in the met derivative, an endogenous bridging ligand, OR^- (phenolate, alkoxide or hydroxide), is present in this derivative. A 1985 crystal structure of deoxyhemocyanin (Gaykema et al.., 1985; Volbeda and Hol, 1989) confirmed three imidazole nitrogen ligands per copper, but eliminated amino acid residues as providing a possible second bridging ligand without a major protein conformational change upon oxidation. Therefore, hydroxide was considered most likely to be the endogenous bridging ligand. The oxyproteins have unique spectral features, two bands in the absorption spectrum and a third in the CD spectrum, which are not present in the met derivative and thus attributable to $O_2^{2-} \rightarrow$ Cu(II) charge transfer. There could only be two charge transfer transitions if the peroxide were bound to only one Cu(II); therefore, the peroxide must bridge the two Cu(II)'s. A transition dipole vector coupling model was used to assign the $O_2^{2-} \rightarrow$ Cu(II) charge transfer spectrum and rule out μ-1,1-O_2^{2-} and *trans*-μ-1,2-O_2^{2-} binding geometries (Solomon et al., 1992). Of the end-on peroxide bridging geometries, only the *cis*-μ-1,2 could be consistent with the peroxide charge transfer spectrum of oxyhemocyanin.

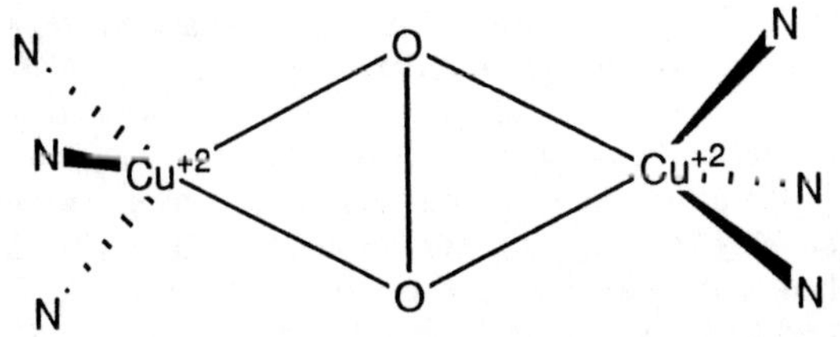

Figure 2.
The side-on bridging, μ-η^2:η^2, peroxide structure of the oxyhemocyanin active site.

There were some problems with this model with respect to the known end-on bridged peroxo copper dimer complex. In the model complex, the $O_2^{2-} \rightarrow$ Cu(II) LMCT transition was at much lower energy and with a factor of two less intensity relative to this transition in oxyhemocyanin and oxytyrosinase. Furthermore, $\nu_{O\text{-}O}$ was higher by 50 to 100 cm^{-1} in the *trans*-μ-1,2-peroxo copper and end-on bridged cobalt models than the 750 cm^{-1} value observed in the oxyproteins (Baldwin et al., 1991; Pate et al., 1987). In 1988, Kitajima et al., (1988; 1989; 1990; 1991; 1992) reported a series of closely related dimeric copper complexes, $[Cu(HB(3,5\text{-}R_2pz)_3)]_2(O_2)$, in which the dioxygen was bound as peroxide in a side-on μ-η^2:η^2-peroxo geometry (Fig. 2). These are the first reported side-on bridging peroxide transition metal complexes. The coppers in the complex with R=*i*-Pr are strongly antiferromagnetically coupled ($-2J > 800$ cm^{-1}), show two intense absorption bands at 350 nm ($\varepsilon = 26{,}000$ $M^{-1}cm^{-1}$) and 538 nm ($\varepsilon = 2000$ $M^{-1}cm^{-1}$), and a $\nu_{O\text{-}O}$ of 749 cm^{-1} (Baldwin et al., 1992). Ross and Solomon (1990; 1991) showed that a novel electronic structure is associated with the side-on μ-η^2:η^2-peroxo geometry which directly explained the origin of the unique spectral, vibrational and magnetic properties of oxyhemocyanin (*vide infra*). It should be noted that the μ-η^2:η^2 bridged model complex is converted to a μ-hydroxy, μ-1,3-azide dimer upon reaction with azide (Kitajima et al., 1992). This is analogous to the structure proposed for the met protein derivative in Figure 1 (Pate et al., 1986; 1989), so there would appear to be a rearrangement of the active site structure upon peroxide displacement from the oxy site to form the met derivative. These structural studies were culminated by Magnus and Ton-That's (1992) solution of the crystal structure of oxyhemocyanin from *Limulus polyphemus* (horseshoe crab) at 240 pm resolution which confirmed the μ-η^2:η^2-peroxo bridging geometry with symmetrically bound dioxygen, coplanar with the tetragonal copper atoms.

The unique spectral features of oxyhemocyanin and oxytyrosinase are now understood in terms of the unusual electronic structure associated with the side-on peroxide binding geometry (Ross and Solomon, 1990; 1991; Baldwin et al., 1992). A combination of electronic structural calculations and spectral analysis shows that peroxide in the μ-η^2:η^2 binding geometry acts as a very strong σ-donor. This explains the high intensity and high energy of the peroxide to copper charge transfer band at 350 nm because peroxide has two σ-donor transitions to each of the two coppers in a side on bridging structure. Also, peroxide was found for the first time to act as a π-acceptor ligand. This involves the σ^* orbital on the peroxide, which is strongly antibonding, and weakens the O-O bond; hence, explaining the unusally low $\nu_{O\text{-}O}$ at 750 cm^{-1}.

Hemocyanin will bind one carbon monoxide, carbon-end-on, to one of the coppers (Fager and Alben, 1972; Van der Deen and Hoving, 1979). The issue of carbon monoxide binding to hemocyanin is important since carbon monoxide binding is not cooperative in hemocyanins while oxygen binding is cooperative. This is different from hemoglobin in which both oxygen and carbon monoxide binding are cooperative (Reed, 1986). Thus, the bridging versus terminal structural difference between carbon monoxide hemocyanin and oxyhemocyanin may be the key to understanding the mechanism of cooperativity in hemocyanin.

From the virtually identical spectral properties of oxyhemocyanin and oxytyrosinase, the active sites must be nearly the same; however, hemocyanin has no (or very little) phenolase activity and the rate of peroxide displacement by azide is roughly 20 times higher in oxytyrosinase than in oxyhemocyanin (Himmelwright et al., 1980), and therefore the active site in tyrosinase is highly accessible. The question arises whether these reactivity differences are solely related to differences in the protein "ligand" or whether there are also small differences in the active site structures. Amino acid sequences for hemocyanins and tyrosinases share a highly homologous region for the three histidines which bind one of the copper atoms, whereas the sequences for the polypeptides which contain the histidine ligands for the second copper vary considerably (Lerch and Germann, 1988).

A structural model for the reaction mechanism for *ortho*-hydroxylation of monophenols and oxidation of the resulting diphenol was first proposed by Winkler et al., (1981). This associative ligand displacement mechanism, shown in Figure 3 and modified (Ross and Solomon, 1991) from the original to incorporate the μ-η^2:η^2 peroxo binding geometry, was based on spectral studies of half-met and met-tyrosinase when reacted with the substrate analogue, mimosine. A study by Wilcox et al., (1985) included carboxylate inhibitor binding to oxy, met and half-met tyrosinase. The carboxylate competitive inhibitors divided into two groups: weak inhibitors in which the carboxylate is not conjugated and strong inhibitors in which the carboxylate is conjugated with an aromatic ring. The spectra of the tyrosinase derivatives with the strong inhibitors bound to the copper site indicated a highly distorted trigonal bipyramidal geometry, leading to the conclusion that the protein pocket must contribute to the interaction of the substrate with the active site. The unique electronic structure of the side-on bridge described above should also make a major contribution to catalysis. The strong σ-donor interaction of peroxide with copper leads to a less negative charge on the peroxide. The π-acceptor nature of the bonding of the peroxide in the μ-η^2:η^2 geometry weakens the O-O bond and activates the dioxygen. Substrate binding causes the

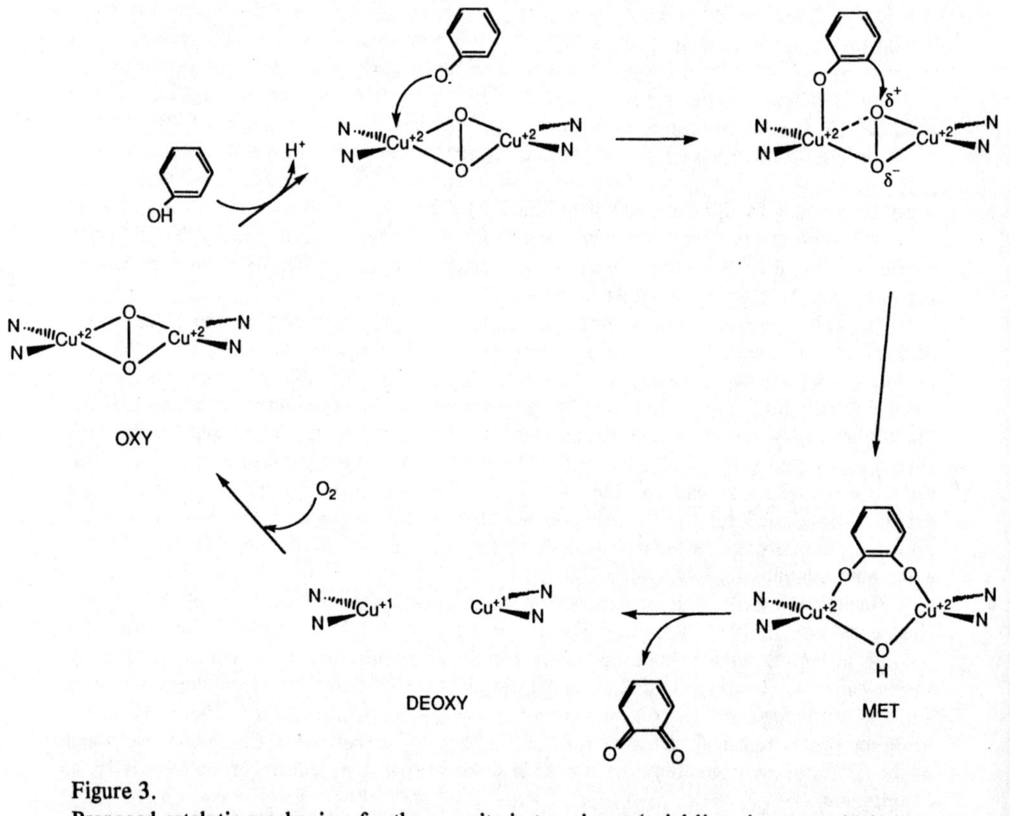

Figure 3.
Proposed catalytic mechanism for the oxy site in tyrosinase (axial ligands are not included for clarity).

peroxide to be polarized, and shifts electron density into the LUMO (Ross and Solomon, 1991), which is antibonding with respect to the O-O and Cu-O bonds, initiating oxygen transfer at the *ortho* position of the substrate.

III. LACCASE AND ASCORBATE OXIDASE

Since 1985 there has been a tremendous increase in our knowledge of the active site structures in the multi-copper oxidases, the interactions among the three different sites and the active site structure/function relationships. Laccase is the simplest of the multi-copper oxidases, having only one of each type of copper center (four coppers total, *vide infra*). It is found in fairly high concentration in tree lacquer from *Rhus vernicifera*, and because it has several spectroscopically accessible derivatives it has been the most studied. Spectral studies of laccase first led to the proposal of a new trinuclear copper cluster active site in the multi-copper oxidases (Allendorf et al., 1985). The recent solution of the X-ray structure of ascorbate oxidase from zuccini at 250 pm resolution (Messerschmidt et al., 1989; 1992; Messerschmidt and Huber, 1990) has focused more research on this enzyme. The strong correlation between the spectral properties and the active site structures in laccase and ascorbate oxidase give confidence that some of the findings can be extended to the most complex multi-copper oxidase, ceruloplasmin, found in mammalian blood plasma. Ceruloplasmin, however, has a copper stoichiometry different from either laccase or ascorbate oxidase. Laccase has one each of type-1, type-2 and type-3 copper sites; ascorbate oxidase has two each of the three different sites (2 x laccase), whereas ceruloplasmin has two type-1, one type-2, one type-3 site, and perhaps additional EPR non-detectable copper.

The copper atoms found in multi-copper oxidases have traditionally been classified (Malkin and Malmstrom, 1970; Malmstrom et al., 1975; Fee, 1975; Solomon, 1981; Solomon et al., 1983) according to their EPR features: type-1 or blue copper ($A_{\parallel} \leq 95 \times 10^{-4}$ cm^{-1}), type-2 or normal ($A_{\parallel} > 140 \times 10^{-4}$ cm^{-1}), and type-3 or coupled binuclear (EPR nondetectable). The type-3 coppers are strongly antiferromagnetically coupled so that the type-3 site is diamagnetic even at room temperature (Dooley et al., 1978; Cole et al., 1990). The superexchange interaction responsible for this antiferromagnetism is mediated by an endogenous bridging ligand which appears to be OH^- on the basis of the ascorbate oxidase crystal structure (Messerschmidt et al., 1989; 1992; Messerschmidt and Huber, 1990). The various copper centers also give rise to electronic spectral features that have been well characterized in laccase (Cole et al., 1991a; Cole et al., 1990; LuBien et al., 1981; Spira-Solomon and Solomon, 1987; Kau et al., 1987; Allendorf et al., 1985; Spira-Solomon et al., 1986) and ascorbate oxidase (Cole et al., 1991b). The type-1 centers in multi-copper oxidases as well as the single "blue" copper proteins exhibit an intense cysteine $S\pi \rightarrow Cu(II)d_{x^2-y^2}$ charge transfer absorption band near 600 nm ($\varepsilon \sim 5000$ M^{-1}cm^{-1}) and weaker features from the UV to the near-IR region (Gewirth and Solomon, 1988). The oxidized type-3 site displays a broad near-UV absorption band at 330 nm, ($\varepsilon = 2800$ M^{-1}cm^{-1}) and several d-d absorption and circular dichroism (CD) features from 585 to 900 nm; the type-2 site has a magnetic circular dichroism (MCD) spectrum which consists of several weak charge transfer bands from 300 to 400 nm and two d-d bands at 720 and 860 nm (Cole et al., 1991a; Cole et al., 1990; LuBien et al., 1981; Spira-Solomon and Solomon, 1987; Kau et al., 1987; Allendorf et al., 1985; Spira-Solomon et al., 1986; Reinhammar, 1972).

The type-1 sites in the multi-copper proteins are nearly identical with those in the small, single blue copper proteins, exemplified by plastocyanin, azurin and stellacyanin because of the close spectral similarities (Gewirth and Solomon, 1988 and reference therein). High resolution structures for several plastocyanins and azurins (see Sykes, 1990, for a recent review) have been solved. The plastocyanin structure consists of a copper with two imidazole ring (histidine) nitrogen ligands at normal Cu-N bond distances (average of 204 pm), a cysteine sulfur ligand with an unusually short Cu-S bond of 213 pm, and a fourth methionine sulfur ligand at a longer (290 pm) bond distance forming a trigonally distorted tetrahedron. The azurin structure has two imidazole nitrogens and a cysteine sulfur at similar distances and a methionine sulfur and a glycine oxygen at long (310 pm) axial distances. The 310 pm distance should be considered a weak bond between the methionine sulfur and copper, but this distance makes a covalent bond between the copper and the glycine oxygen unreasonable, so both type-1 sites can be thought of as distorted four-coordinate with an elongation along the Cu-S(methionine) axis (Lowery and Solomon, 1992). The stellacyanin structure undoubtedly has the same three principal ligands of two imidazole nitrogens and a cysteine sulfur, but it must differ from both plastocyanin and azurin in the axial ligation since stellacyanin has no methionine residue. Recent pulsed ENDOR and model studies show that the fourth ligand of stellacyanin at high pH is an amide nitrogen (Thomann et al., 1991). The type-1 copper spectral features in laccase and ascorbate oxidase are closest to those of plastocyanin, and the type-1 copper in ascorbate oxidase has the plastocyanin structure. The ascorbate oxidase structure places the type-1 site near the protein surface and 1250 pm from the type-2/type-3 cluster (Messerschmidt et al., 1989; Messerschmidt and Huber, 1990), leading to the conclusion that the function of the type-1 site in the multi-copper oxidases is to shuttle electrons into the type-2/type-3 trinuclear cluster where the actual binding and reduction of oxygen take place. This involves long-range electron transfer activated by the thiolate (Lowery et al., 1993). Much of the current research activity deals with the development of the model of the type-2/type-3 trinuclear cluster and how it is involved with the four-electron reduction of oxygen to water.

There is some similarity between the spectral properties of the type-3 sites with those of oxyhemocyanin and oxytyrosinase: namely an EPR silent, coupled binuclear copper site showing near-UV absorbance at 338-350 nm in oxyhemocyanins and oxytyrosinases ($\Delta\varepsilon$ = 20,000 $M^{-1}cm^{-1}$) and at 330 nm in laccase, ascorbate oxidase and ceruloplasmin with a significantly lower intensity ($\Delta\varepsilon \sim$ 3000 $M^{-1}cm^{-1}$). Some authors have referred to the binuclear sites in hemocyanin and tyrosinase as "type-3" copper which is both misleading and, as it turns out, incorrect (Spira-Solomn and Solomon, 1987). The type-3 coppers in the multi-copper oxidases interact with the closeby type-2 copper which gives them unique spectral and chemical properties (Cole et al., 1990; 1991a; 1991b).

Through low-temperature MCD studies it was shown that the anionic inhibitor N_3^- binds to laccase as a bridging ligand between the type-2 and type-3 sites, defining a trinuclear cluster (Allendorf et al., 1985; Spira-Solomon et al., 1986). It was suggested that the type-2/type-3 cluster was the site for binding and reduction of dioxygen. Ascorbate oxidase contains two subunits with two of each type of copper. The 250 pm X-ray structure has confirmed this trinuclear cluster in ascorbate oxidase with three coppers within 390 pm of each other and the type-1 copper approximately 1250 pm away from the trinuclear cluster (Messerschmidt et al., 1989; Messerschmidt and Huber, 1990).

A useful derivative of tree laccase can be prepared by a selective removal and substitution procedure. The type-1 copper may be selectively replaced by Hg(II), leaving the type-2 and type-3 sites intact (Morie-Bebel et al., 1984). This form is referred to as T1Hg. A systematic spectral study of T1Hg laccase by absorption, CD and low-temperature MCD has led to the assignment of the ligand-field and charge transfer transitions associated with the type-2/type-3 cluster (Cole et al., 1990). The d-d transitions in the type-2/type-3 cluster probe geometric and electronic structure which is crucial to understanding the function of the site. T1Hg laccase eliminates the intense Cys S $\rightarrow$ Cu(II) charge tranfer band at ca 600 nm and the weak d-d and charge transfer bands associated with the type-1 copper that would otherwise obscure the even weaker transitions arising from the type-2 and type-3 coppers. While absorption and CD could potentially probe transitions arising from type-2 or type-3 Cu(II)'s, low-temperature MCD discriminates between transitions originating from type-2 and type-3 Cu(II)'s. Since the type-3 site is diamagnetic ($S=0$) it can only give rise to relatively weak, temperature-independent MCD signals (A and B terms), but the paramagnetic ($S=1/2$) type-2 copper can give rise to temperature-dependent MCD bands whose intensities are proportional to $1/T$ for $kT >> g\beta H$ (C terms). Thus, near 4 K the MCD bands from the type-2 copper will be 100-1000 times more intense than features arising from the type-3 coppers. With addition of azide, the endogenous bridging ligand in the type-3 in approximately 10% of the sites is uncoupled, giving rise to additional features in the low-temperture MCD spectrum which are due to uncoupled type-3 copper and are dependent on $[N_3^-]$ (Cole et al., 1990).

In T1Hg laccase, d-d transitions from the type-2 copper were found at 860 and 720 nm, while weak imidazole-to-copper charge transfer transitions were observed at 320, 345 and 370 nm in the MCD spectrum. Fluoride (fluoride strongly binds to the type-2 copper, causing a shift to lower energy of type-3 copper transitions and a lowering of intensity of type-2 copper transitions) dependence of the absorption spectrum of T1Hg laccase revealed a third d-d transition due to the type-2 copper at 620 nm. The energy of these d-d bands indicates that the type-2 copper is tetragonal, and the bands at 620, 720 and 860 nm have been assigned as $d_{xz,yz} \rightarrow d_{x^2-y^2}$, $d_{xy} \rightarrow d_{x^2-y^2}$, and $d_{z^2} \rightarrow d_{x^2-y^2}$, respectively. In the CD spectrum, d-d transitions arising from the type-3 coppers were found at 585, 675 and 900 nm. A fourth d-d band reported by both LuBien et al., (1981) and Tamilarasan and McMillin (1989) appears in the absorption spectrum at 740-745 nm. Ligand field arguments and anion perturbations lead to the conclusion that all four of these d-d transitions cannot be from two equivalent coppers. The type-3 coppers are non-equivalent and the 675 and 585 nm transitions are due to $d_{xy} \rightarrow d_{x^2-y^2}$ and $d_{xz,yz} \rightarrow d_{x^2-y^2}$ transitions, respectively, from type-3 Cu_α and the 900 and 740-745 nm transitions are due to $d_{xy} \rightarrow d_{x^2-y^2}$ and $d_{xz,yz} \rightarrow d_{x^2-y^2}$ transitions, respectively, from type-3 Cu_β.

When azide is reacted with T1Hg laccase, two azides are found to bind with equilibrium constants of 780 M^{-1} for the first azide and 270 M^{-1} for the second azide (Cole et al., 1990). Binding of the first azide uncouples approximately 10% of the type-3 sites, resulting in the appearance of the d-d transitions (slightly shifted to higher energy) associated with the type-3 Cu_α and type-3 Cu_β in the low- temperature MCD spectrum. Upon binding of the second azide, slight upward shifts occur for the d-d transitions associated with the type-2 Cu. The resulting spectroscopically effective model for the type-2/type-3 cluster in laccase is shown in Figure 4. The protonatively uncoupled sites (< 10% in laccase) in Figure 4 may correlate with those observed in the crystal structure of azide-bound ascorbate oxidase (Messerschmidt et al., 1993).

A set of four His-X-His sequences, which are believed to provide the histidine ligation to the type-2 and type-3 coppers, are conserved in *Neurospora crassa* laccase, human ceruloplasmim and zuccini ascorbate oxidase (Germann and Lerch, 1986; Ohkawa et al., 1989; Messerschmidt and Huber, 1990; Messerschmidt et al., 1992), and it is likely that this same set exists in *Rhus vernicifera* laccase. In the latest crystallographic model of ascorbate oxidase, resolved to 190 pm, there are three histidine ligands per type-3 copper

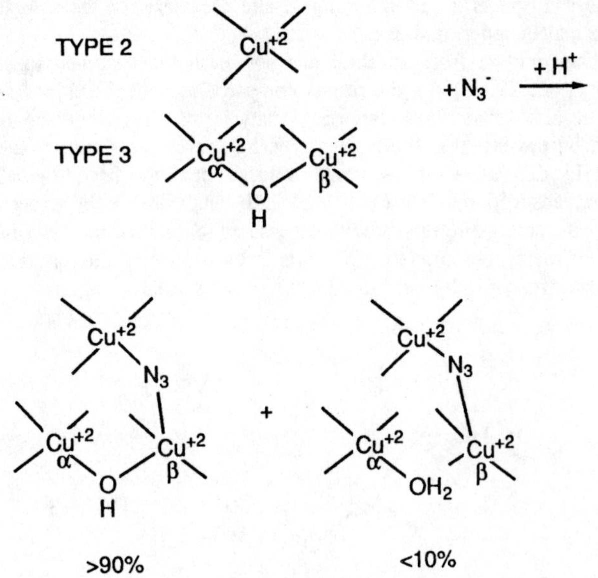

Figure 4.
Spectroscopic model for the type-2/type-3 trinuclear copper cluster site in laccase and its interaction with exogenous azide.

arranged in a trigonal prismatic array and two histidines that ligate the type-2 copper. One hydroxide bridges the type-3 coppers and another hydroxide is bound to the type-2 copper. The spectral inequivalence of the type-3 coppers appears to relate to the fact that of the six imidazole rings bound to the type-3 pair of coppers, one of the imidazole rings is flipped over such that a different ring nitrogen is bound.

The azide binding experiments with T1Hg laccase infer that the type-2/type-3 trinuclear cluster is the site for binding and reduction of dioxygen (Cole et al., 1990). This was confirmed with XAS studies on metal-substituted derivatives. Type-2 depleted (T2D), {T1(oxidized), T3(reduced)} does not react with dioxygen; fully reduced T2D, {T1(reduced), T3(reduced)} also does not react with dioxygen, but fully reduced T1Hg, {T2(reduced), T3(reduced)} does react with dioxygen. Stopped-flow kinetic measurements, CD, MCD and absorption studies have now identified a bound dioxygen intermediate in T1Hg laccase (Cole et al., 1991b). The d-d transitions, which can only arise from Cu(II), reappear upon reaction with oxygen for the type-3 coppers, but not for the type-2 copper. Also, XAS does show that only the two type-3 Cu's have been reoxidized. Thus, the formation of the intermediate involves a two-electron transfer from the type-3 coppers to the dioxygen, forming a peroxide species. The absorption spectrum of the intermediate has peaks at 340 nm (ε = 3700 $M^{-1}cm^{-1}$), 470 nm (ε = 1300 $M^{-1}cm^{-1}$) and 670 nm ($\varepsilon \sim 400$ $M^{-1}cm^{-1}$). The higher energy bands are associated with peroxide to Cu(II) charge transfer while the 670 nm band is d-d in nature. The positions of the peroxide-to-copper charge transfer bands are sufficiently different and the intensities dramatically reduced relative to those in oxyhemocyanin and oxytyrosinase to indicate that peroxide is not binding in the μ-η^2:η^2-peroxo mode. It is likely that the peroxide is bound as hydroperoxide to only one of the oxidized type-3 coppers and does not bridge the coupled binuclear site in contrast to oxyhemocyanin and oxytyrosinase. Figure 5 shows a peroxide binding scheme which has the peroxide as an end-on bound hydroperoxide coordinated to one of the oxidized type-3 coppers, with an additional strong interaction with the reduced type-2 copper as the type-2 copper is needed for the dioxygen reaction (*vida supra*). The second-order rate constant for the formation of the peroxide intermediate is 2.2 x 10^6 $M^{-1}s^{-1}$ while that for the reoxidation of reduced native laccase is nearly the same at 1.7 x 10^6 $M^{-1}s^{-1}$. This indicates that the type-1 copper has little effect on the rate of reoxidation of the trinuclear cluster, and that its role is limited to the very rapid (k > 1000 s^{-1}) transfer of an electron to the trinuclear site. The trinuclear cluster is in fact the minimal site necessary for the reduction of dioxygen to water in the multi-copper oxidases.

Ascorbate oxidase from zucchini has now also been studied in some spectroscopic detail (Cole et al., 1991b) and the results compared to both the crystal structure and to the anologous results from *Rhus* laccase. Analysis of the ascorbate oxidase spectra is complicated by the fact that the intense type-1 features contribute to the spectrum (a well defined T1Hg derivative of ascorbate oxidase has not been available). Despite this complication, absorption, CD, MCD and EPR analysis has been accomplished and the results indicate some differences with laccase with respect to the binding of azide and uncoupling of the type-3 coppers. Otherwise, the trinuclear sites in ascorbate oxidase and laccase are spectroscopically (and structurally) quite similar.

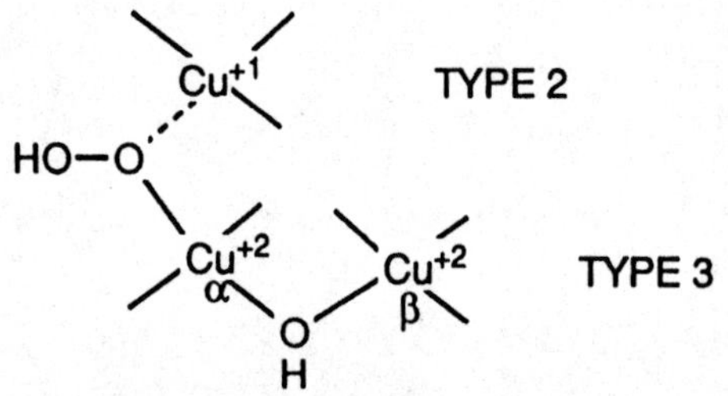

Figure 5.
Spectroscopically effective model for the peroxy intermediate at the trinuclear copper cluster site of laccase.

IV. CONCLUSIONS

The structure/function relationships are now coming into focus for copper proteins in general and copper active sites containing coupled binucler copper specifically. The type-1 blue copper, whose function is to rapidly transfer electrons, has ligands consisting of two imidazole nitrogens and one cysteine sulfur and a weaker fourth ligand in a distorted tetrahedral arrangement with the thiolate activating long-range electron transfer (Lowery et al., 1993). Copper active sites which bind dioxygen as an intermediate or as a stable oxy species have two or three imidazole nitrogen ligands, the bound dioxygen and water-derived ligands in a tetragonal binding arrangement. Hemocyanin and tyrosinase have very similar binuclear copper active site structures, but differences in accessibility and in the protein pocket may account for the phenolase activity in tyrosinase and lack of enzymatic activity in hemocyanin. The peroxide side on μ-η^2:η^2 bridging mode results in a novel electronic structure (very strong peroxide σ donor and π acceptor) which leads to its unique spectral features and plays a key role in the reversible binding and activation of dioxygen. The coupled binuclear type-3 sites in the multi-copper oxidases, once thought to be structurally similar to the active sites in hemocyanin and tyrosinase, are now known to be quite different. The type-3 binuclear coppers actually form a trinuclear cluster with the type-2 copper, and it is the trinuclear cluster which is the minimum structural unit required for O_2 reduction. This structure produces a fundamentally different peroxide intermediate relative to hemocyanin and tyrosinase, an end-on bound hydroperoxide, which correlates to the different biological function of irreversible reduction of dioxygen to water.

ACKNOWLEDGEMENTS

EIS thanks NIH (Grant #DK31450) for support of this research. Funds for JAL's sabbatical at Stanford were provided by the NSF (Grant #DMB-9019752).

REFERENCES

Allendorf, M.D., Spira, D.J., and Solomon, E.I. (1985). *Proc. Nat. Acad. Sci. USA, 82*: 3063.

Baldwin, M.J., Ross, P.K., Pate, J.E., Tyeklar, Z., Karlin, K.D., and Solomon, E.I. (1991). *J. Amer. Chem. Soc., 113*: 8671.

Baldwin, M.J., Root, D.E., Pate, J.E., Fujisawa, K., Kitajima, N., and Solomon, E.I. (1992). *J. Amer. Chem. Soc., 114*: 10421.

Blackburn, N.J. (1990). *Synchrotron Radiation and Biophysics* (S.S. Hasnain, ed.), Halsted Press, New York, p. 63.

Brown, J.M., Powers, L., Kincaid, B., Larrabee, J.A., and Spiro, T.G. (1980). *J. Amer. Chem. Soc., 102*: 4210.

Co, M.S., and Hodgson, K.O. (1981). *J. Amer. Chem. Soc., 103*: 984.

Co, M.S., and Hodgson, K.O. (1981). *J. Amer. Chem. Soc., 103*: 3200.

Co, M.S., Scott, R.A., and Hodgson, K.O. (1981). *J. Amer. Chem. Soc., 103*: 986.

Cole, J.L., Clark, P.A., and Solomon, E.I. (1990). *J. Amer. Chem. Soc., 112*: 9534.

Cole, J.L., Ballou, D.P., and Solomon, E.I. (1991a). *J. Amer. Chem. Soc., 113*: 8544.

Cole, J.L, Avigliano, L., Morpurgo, L., and Solomon, E.I. (1991b). *J. Amer. Chem. Soc., 113*: 9080.

Dooley, D.M., Scott, R.A., Ellinghaus, J., Solomon, E.I., and Gray, H.B. (1978). *Proc. Nat. Acad. Sci. USA, 75*: 3019.

Eickman, N.C., Solomon, E.I., Larrabee, J.A., Spiro, T.G., and Lerch, K. (1978). *J. Amer. Chem. Soc., 100*: 6529.

Eickman, N.C., Himmelwright, R.S., and Solomon, E.I. (1979). *Proc. Nat. Acad. Sci. USA, 76*: 2094.

Fager, L.Y., and Alben, J.O. (1972). *Biochemistry, 11*: 4786.

Fee, J.A. (1975). *Struct. Bonding (Berlin), 23*: 1.

Freedman, T.B, Loehr, J.S., and Loehr, T.M. (1976). *J. Amer. Chem. Soc., 98*: 2809.

Gaykema, W.P.J., Volbeda, A., and Hol, W.G.J. (1985). *J. Mol. Biol.*, *187*: 255.

Gewirth, A.A., and Solomon, E.I. (1988). *J. Amer. Chem. Soc.*, *110*: 3811.

Germann, U.A., and Lerch, K. (1986). *Proc. Nat. Acad. Sci. USA*, *83*: 8854.

Himmelwright, R.S., Eickman, N.C., Lubien, C.D., Lerch, K., and Solomon, E.I. (1980). *J. Amer. Chem. Soc.*, *102*: 7339.

Kau, L-S., Spira-Solomon, D.J., Penner-Hahn, J.E., Hodgson, K.O., and Solomon, E.I. (1987). *J. Amer. Chem. Soc.*, *109*: 6433.

Kitajima, N., Koda, T., Hashimoto, S., Kitagawa, T., and Morooka, Y. (1988). *J. Chem. Soc. Chem. Commun.*, 151.

Kitajima, N., Fujisawa, K., and Morooka, Y. (1989). *J. Amer. Chem. Soc.*, *111*: 8975.

Kitajima, N., Koda, T., Iwata, Y., and Morooka, Y. (1990). *J. Amer. Chem. Soc.*, *112*: 8833.

Kitajima, N., Koda, T., Hashimoto, S., Kitagawa, T., and Morooka, Y. (1991). *J. Amer. Chem. Soc.*, *113*: 5664.

Kitajima, N., Fujisawa, K., Fujimoto, C., Morooka, Y., Hashimoto, S., Kitagawa, T., Toriumi, K., and Nakamura, A. (1992). *J. Amer. Chem. Soc.*, *114*: 1277.

Larrabee, J.A., and Spiro, T.G. (1980). *J. Amer.Chem.Soc.*, *102*: 4217.

Lerch, K., and Germann, U.A. (1988). *Oxidases and Related Redox Systems*, Alan R. Liss, Inc., p. 331.

Lowery, M.D., and Solomon, E.I. (1992). *Inorg. Chim. Acta*, *198-200*: 233.

Lowery, M.D., Guckert, J.A., Gebhard, M.S., and Solomon, E.I. (1993). *J. Amer. Chem. Soc.*, *115*: 2968.

LuBien, C.D., Winkler, M.E., Thamann, T.J., Scott, R.A., Co, M.S., Hodgson, K.O., and Solomon, E.I. (1981). *J. Amer. Chem. Soc.*, *103*: 7014.

Magnus, K.A., and Ton-That, H. (1992). *J. Inorg. Biochem.*, *47*: 20.

Malkin, R., and Malmstrom, B.G. (1970). *Adv. Enzymol. Relat. Areas Mol. Biol.*, *33*: 177.

Malmstrom, B.G., Andreasson, L-E., and Reinhammar, B. (1975). *The Enzymes*, Vol. XII (P.D. Boyer, ed.), Academic, New York.

Messerschmidt, A., and Huber, R. (1990). *Eur. L. Biochem.*, *187*: 341.

Messerschmidt, A., Rossi, A., Ladenstein, R., Huber, R., Bolognesi, M., Gatti, G., Marchesini, A., Petruzzelli, R., and Finazzi-Agro, A. (1989). *J. Mol. Biol.*, *206*: 513.

Messerschmidt, A., Ladenstein, R., Huber, R., Bolognesi, M., Avigliano, L., Petruzzelli, R., Rossi, A., and Finazzi-Agro, A. (1992). *J. Mol. Biol.*, *224*: 179.

Morie-Bebel, M.M., Morris, M.C., Menzie, J.L., and McMillin, D.R. (1984). *J. Amer. Chem. Soc.*, *106*: 3677.

Ohkawa, J., Okada, N., Shinmyo, A., and Takano, M. (1989). *Proc. Nat. Acad. Sci. USA*, *86*: 1239.

Pate, J.E., Thamann, T.J., and Solomon, E.I. (1986). *Spectrochim. Acta*, *42A*: 313.

Pate, J.E., Cruse, R.W., Karlin, K.D., and Solomon, E.I. (1987). *J. Amer. Chem. Soc.*, *109*: 2624.

Pate, J.E., Ross, P.K., Thamann, T.J., Reed, C.A., Karlin, K.D., Sorrell, T.N., and Solomon, E.I. (1989). *J. Amer. Chem. Soc.*, *111*: 5198.

Reed, C.A. (1986). *Biological and Inorganic Copper Chemistry*, Vol. I (K.D. Karlin and J. Zubieta, eds.), Adenine Press, New York, p. 61.

Reinhammar, B.R.M. (1972). *Biochim. Biophys. Acta*, *275*: 245.

Ross, P.K., and Solomon, E.I. (1990). *J. Amer. Chem. Soc.*, *112*: 5871.

Ross, P.K., and Solomon, E.I. (1991). *J. Amer. Chem. Soc.*, *113*: 3246.

Solomon, E.I. (1981). *Copper Proteins* (T.G. Spiro, ed.), Wiley Interscience, New York, p. 41.

Solomon, E.I. (1988). *Oxidases and Related Redox Systems*, Alan R. Liss, Inc., p. 309.

Solomon, E.I., and Lowery, M.D. (1992). *Science*, *259*: 1575.

Solomon, E.I., Dooley, D.M., Wang, R.H., Gray, H.B., Cerdonio, M., Mogno, F., and Romani, G.L. (1976). *J. Amer. Chem. Soc.*, *98*: 1029.

Solomon, E.I., Penfield, K.W., and Wilcox, D.E. (1983). *Struct. Bonding (Berlin)*, *53*: 1.

Solomon, E.I., Baldwin, M.J., and Lowery, M.D. (1992). *Chem. Rev.*, *92*: 521.

Spira-Solomon, D.J., Allendorf, M.D., and Solomon, E.I. (1986). *J. Amer. Chem. Soc.*, *108*: 5318.

Spira-Solomon, D.J., and Solomon, E.I. (1987). *J. Amer. Chem. Soc.*, *109*: 6421.

Sykes, A.G. (1990). *Struct. Bonding (Berlin)*, *75*: 175.

Tamilarasan, R., and McMillin, D.R. (1989). *Biochem. J.*, *263*: 425.

Thomann, H., Bernardo, M., Baldwin, M., Lowery, M.D., and Solomon, E.I. (1991). *J. Amer. Chem. Soc.*, *113*: 5911.

Van der Deen, H., and Hoving, H. (1979). *Biophys. Chem.*, *9*: 169.

Volbeda, A., and Hol, W.G.J. (1989). *J. Mol. Biol.*, *209*: 249.

Wilcox, D.E., Porras, A.G., Hwang, Y.T., Lerch, K., Winkler, M.E., and Solomon, E.I. (1985). *J. Amer. Chem. Soc.*, *107*: 4015.

Winkler, M.E., Lerch, K., and Solomon, E.I. (1981). *J. Amer. Chem. Soc.*, *103*: 7001.

Woolery, G.L., Powers, L., Peisach, J., and Spiro, T.G. (1984a). *Biochemistry*, *23*: 3428.

Woolery, G.L., Powers, L., Winkler, M., Solomon, E I., and Spiro, T.G. (1984b). *J. Amer. Chem. Soc.*, *106*: 86.

Woolery, G.L., Powers, L., Winkler, M., Solomon, E I., Lerch, K., and Spiro, T.G. (1984c). *Biochim. Biophys. Acta*, *788*: 155.

SUPEROXIDE DISMUTASES

by

A.E.G. Cass

Department of Biochemistry,
Imperial College of Science Technology & Medicine,
South Kensington
London SW7 2AY
U.K.

I. INTRODUCTION

Any organism that opts for an anaerobic lifestyle is presented with a tremendous opportunity and a serious challenge; the opportunity comes from the free energy gain in completely oxidizing carbohydrates such as glucose where the overall free energy gain is -2870 kJ mol^{-1} and this can be used for growth, maintenance and other energy-demanding processes. The source of the challenge is an intracellular environment which is reducing and whose constituents are capable, both thermodynamically and kinetically, of reacting with oxygen:

$$O_2 + e^- \rightarrow O_2^- \quad (1)$$

$$O_2 + 2e^- + 2H^+ \rightarrow H_2O_2 \quad (2)$$

Although the great part of the oxygen reduction flux is directly to water, a small fraction partakes of the reactions shown in equations (1) and (2). As both the superoxide ion and hydrogen peroxide are intermediates in the reduction of oxygen, they are by definition oxidizing agents and can potentially take part in uncontrolled redox reactions with cellular components and disrupt their normal function. To counter this threat of random oxidative damage, several strategies have developed; spatial confinement of the enzymes generating O_2^- and H_2O_2 to specific locations, high levels of sacrificial reductants ('anti-oxidants'), enzymes that degrade the products of cellular oxidation (glutathione peroxidase, semidehydroascorbate reductase) and enzymes that disproportionate O_2^- and H_2O_2:

$$2O_2^- + 2H^+ \rightarrow O_2 + H_2O_2 \quad (3)$$

$$2H_2O_2 \rightarrow O_2 + 2H_2O \quad (4)$$

Enzymes that catalyze reaction (4) are catalases whilst it is the superoxide dismutases (SODs) that are catalysts for reaction (3).

The arguments of the previous paragraphs still lack a good quantitative foundation; there are few studies of either the O_2^- flux or its steady state concentrations in cells and tissues under different conditions; making it difficult to assess the in vivo relevance of many in vitro rate constants measured for O_2^- reactions. In contrast to H_2O_2, O_2^- undergoes rapid spontaneous disproportionation (dismutation) with a second order rate constant of 4×10^5 $M^{-1}s^{-1}$ at pH 7 (Bielski and Allen, 1979).

The physiological role(s) of O_2^- and SOD in higher organisms are still not definitively established although many intriguing hints and suggestions abound. A tetrameric extracellular SOD (EC-SOD) has been found with a heparin-binding domain suggesting that it is targetted to cell surfaces and the connective tissue matrix (Sandström et al., 1992). Superoxide appears to be a major factor in reperfusion damage (Murata et al., 1992) and it has been found to react rapidly with nitric oxide, the endothelium-derived relaxing factor, (rate constant = $3.7 \times 10^7 M^{-1} s^{-1}$ (Saran et al., 1990)) and may limit the latter's effective lifetime. Very recently it has been demonstrated that a familial form of the neurodegenerative disease amyotrophic lateral scelerosis is tightly linked to missense mutations in the intracellular Cu,Zn-SOD gene *SOD1* (Rosen et al., 1993).

This review will not be concerned with the physiological roles of O_2^- and by implication of SOD but will concentrate on the structural and mechanistic aspects of the SODs with particular emphasis on the metal coordination sphere and adjacent active site regions.

II. OVERVIEW OF THE ENZYMES

The SODs all contain redox active first row transition metal ions, either copper/zinc, manganese or iron, and the various enzymes have fairly well-defined distributions; the copper/zinc enzymes are found in the cytoplasm of eukaryotes and also in extracelluar fluids whilst the manganese enzymes are found both in mitochondria and in certain bacteria. Iron-containing SODs are present in plants and bacteria.

Comparison of both DNA and protein sequences reveal that the copper enzymes are quite different from the iron and manganese ones whilst the latter show some homology to each other. This similarity also extends to the tertiary structures where the overall fold and coordination geometry of the Mn and Fe enzymes are similar and are quite distinct from the Cu,Zn enzymes.

Structural homologies amongst the Cu,Zn enzymes were also suggested on the basis of similarities in the high resolution ^1H NMR spectra of the reduced forms of bovine and human (Stokes et al., 1973; Hill et al., 1980) and bovine and yeast (Cass et al., 1978a) enzymes.

The basic mechanism of superoxide dismutation involves the metal ions in the proteins cycling between oxidized and reduced states:

$$M^{n+} + O_2^- \rightarrow M^{(n-1)+} + O_2 \quad (5)$$

$$M^{(n-1)+} + O_2^- + 2H^+ \rightarrow M^{n+} + H_2O_2 \quad (6)$$

Thus the redox potentials for SODs must lie between 0.16V (superoxide oxidation) and 0.89V (superoxide reduction). Additional kinetic considerations in the reduction of O_2^-, including the protonation of the developing peroxide dianion and the role of the enzyme's amino acid side chains, will be discussed towards the end of this review.

III. MOLECULAR STRUCTURES

A. Gene cloning

A large number of SODs have been cloned and sequenced, including Cu,Zn enzymes from plants, parasites, fungi, fish and mammals; Mn enzymes from plants, bacteria, fungi and mammals and Fe enzymes from plants and bacteria. The isolation of SOD genes can be readily achieved by complementation of a SOD^- strain of *E. coli*, which is extremely sensitive towards oxygen, and those compounds which increase intracellular concentrations of O_2^- (Carlioz and Touati, 1986).

B. X-ray crystallography

Crystal structures of representatives of all three types of SOD (Cu,Zn, Mn and Fe) have been published (Table 1). The earliest and best resolved structure is that of the bovine Cu,Zn form and, in addition to the native enzyme, there have also been structures of the human, yeast and spinach enzymes as well as cobalt substituted (Cu,Co) and peroxynitrate modified forms of the bovine enzyme.

Table 1

A Summary of the X-ray Crystallographic Results for SODs

Enzyme source	Active center	Resolution (nm)	Reference
Bovine	Cu,Zn	0.2	Tanier et al. (1982)
Bovine	Cu,Co	0.2	Djinovic et al. (1992b)
Bovine	Cu,Zn (Peroxynitrate modified)		Smith et al. (1992)
Yeast	Cu,Zn	0.25	Djinovic et al. (1992a)
Human	Cu,Zn		Parge et al. (1992)
Human	Mn		Borgstahl et al. (1992)
Spinach	Cu,Zn	0.2	Kitagawa et al. (1991)
E. coli	Fe	0.31	Stallings et al. (1983)
B. stearothermophilus	Mn	0.24	Parker and Blake (1988)
T. thermophlus	Mn	0.18	Ludwig et al. (1991)
Ps. ovalis	Fe	0.21	Stoddard et al. (1990a)
Ps. ovalis	Fe/N_3^-	0.29	Stoddard et al. (1990b)

1. *Cu,Zn-SOD*

The structure of this enzyme from bovine erythrocytes has been refined over many years and its detailed analysis has contributed greatly to our understanding of the mechanism of superoxide dismutation (Tanier et al., 1982; 1983). The more recent determination of the structures of the yeast (Djinovic et al., 1992), spinach (Kitagawa et al., 1991) and human (Parge et al., 1992) enzymes confirms the homology deduced from sequence comparisons. The essential features of the molecule; a dimeric structure, an eight-stranded β barrel with stabilizing disulphide and salt bridges, a metal-binding site comprising a distorted square planar copper ion coordinated by 4 histidine residues and an axial water molecule with a bridging histidine to the tetrahedral and solvent inaccessible zinc ion coordinated by 3 histidine and 1 aspartate residue are common to all structures. Also conserved are the charged residues lining the active site cleft, which are known to be important in substrate approach and protonation reactions. Figure 1 shows the backbone structure of one subunit of the bovine enzyme viewed from the entrance to the active site. The copper ion is in the foreground with the zinc ion behind, the histidine and aspartate ligands are shown as is the catalytically important arginine at the entrance to the active site. Figure 2 shows a closer view of the active site, again with the copper ion in the foreground and the metal ligands and arginine 141 picked out.

2. *Fe-SOD*

Two different iron-containing SODs have had their 3-dimensional structures determined, *Ps. ovalis* (Stoddard et al., 1990a) and *E. coli.* (Stallings et al., 1983). The enzymes also show the structural homology expected from sequence comparisons but are quite distinct from the Cu,Zn enzymes although both forms are dimeric. The iron site is best described as a distorted trigonal structure with the iron atom about 0.5 Å above the plane of the equatorial ligands (2 histidine and 1 glutamate) towards an axial histidine ligand. There is no evidence for a second axial ligand such as water, another clear distinction between these enzymes and the Cu,Zn ones. In addition to the coordinating residues, there is a glutamine suitably placed to act as a proton donor to the developing peroxide anion.

3. *Mn-SOD*

Structures of manganese-containing enzymes from *Thermus thermophilus* (Ludwig et al. (1991)) and *Bacillus stearothermophilus* (Parker and Blake, 1988) have been determined and the metal coordination is very similar to that of the Fe-SODs. Other similarities between Fe-

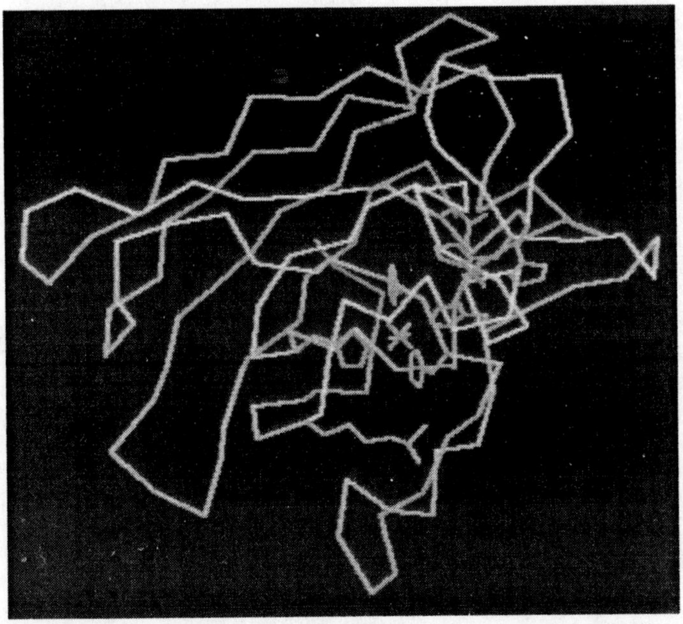

Figure 1.
The structure of one subunit of bovine Cu,Zn-SOD showing the protein backbone and metal ion binding sites as viewed from the entrance to the active site. Also shown is the catalytically important arginine 141.

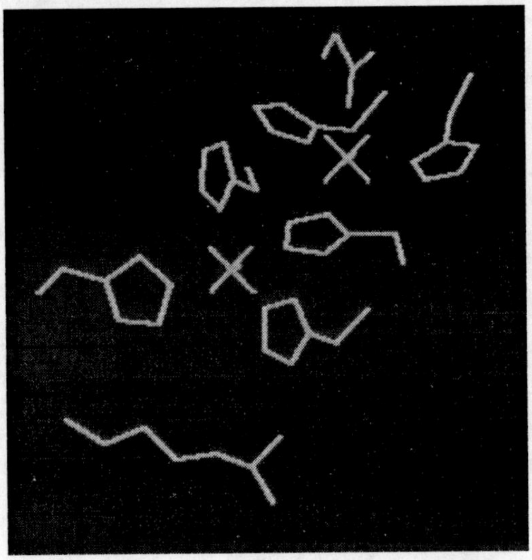

Figure 2.
The ligand arrangement in the metal ion binding site of bovine Cu,Zn-SOD and the relationship of arginine 141 to the copper ion (foreground).

and Mn-SODs include a predominantly hydrophobic region around the substrate-binding domain, and a structurally conserved glutamine residue that is found at a different position in the sequence of the two types of enzyme. This latter observation has led Stoddard et al. (1990a) to postulate that the Fe- and Mn-SODs are second site revertants of one another.

C. Metal-binding properties

1. *Cu,Zn-SOD*

The rich coordination chemistry of the Cu,Zn enzymes has led to many studies of their metal-binding properties. Early experiments were aimed at simply reconstituting enzyme activity, but it soon became apparent that substitution of alternative metals for zinc introduced spectroscopic probes of structure in this region of the protein. In particular, the replacement of zinc(II) by cobalt(II) has been a particularly well studied example (Calabrese et al., 1972; Fee, 1973), taking advantage of the similar coordination geometry of these two metal ions. The close proximity of the copper and zinc sites means that when a paramagnetic ion such as Co(II) is substituted for Zn(II) there is a magnetic interaction between the cobalt and copper. The strength of this magnetic interaction can be readily quantified by susceptibility measurements, and an anti-ferromagnetic coupling is seen at low temperatures (<100K). This appears to be through the bridging imidazolate ion (Moss and Fee, 1975; Morgenstern-Badarau et al., 1986). X-ray crystallography shows that cobalt in the zinc site has a tetrahedral geometry, an observation also suggested by X-ray absortion near edge spectroscopy (XANES). The XANES measurements also revealed that the coordination environment around the cobalt was different in the copper(II) derivative as compared to the copper(I) or apo forms and this was suggested to be a consequence of breaking the imidazolate bridge in the latter 2 derivatives (Desideri et al., 1981).

There are considerable dynamics in the metal-binding sites depending on pH and the presence of anions in the active site, as well as on the type of metal ions present; thus, as the pH is lowered to less than 3.5 in the bovine enzyme (with either the native form or with cobalt or copper ions in the zinc site), the metal in the zinc site is selectively lost (Pantoliano et al., 1979). Studies of both metal depletion and reconstitution of the yeast enzyme seemed to show a more complex behavior than expected from independent metal sites on each monomer, viz a 'half-of-the-sites' reactivity. It appeared that the removal of metal ions from one of the monomers was easier than from the other, resulting in a derivative comprising one intact and one empty monomer per dimer (Dunbar et al., 1982). Reconstitution with cobalt(II) was interpreted as only a single occupancy per dimer, whilst copper(II) titration followed by visible and CD spectroscopy also appeared to be consistent with 2 distinct types of copper occupancy per dimer (Dunbar et al., 1984a; 1984b). More recent studies of the reconstitution of yeast SOD with cobalt(II) by both visible and NMR spectroscopy concluded that at pH 5.5 the cobalt occupied the two zinc sites per dimer whilst at pH 7 the cobalt could also occupy the copper site. No evidence of assymetry in metal binding could be detected (Ming et al., 1988a; Ming and Valentine, 1990a).

Copper(II) can also be substituted into the zinc site with the formation of a 4 copper derivative, with again an anti-ferromagnetic coupling between the two metal ions as judged from EPR spectroscopy (Fee and Briggs, 1975; Pantoliano et al., 1982). A detailed examination of the visible and near IR spectra of the 4 copper form in addition to CD and EPR measurements have been interpreted in terms of LMCT and ligand-field transitions, and appropriate assignments made. Nickel(II) is another ion that favors tetrahedral coordination and can be substituted into the zinc site. Like cobalt, it greatly affects the copper(II) electron spin relaxation time (reducing it by 3 orders of magnitude), which has the consequence that copper no longer efficiently relaxes nuclear spins and a ^{1}H NMR spectrum with isotropically shifted resonances can be observed. These can be assigned on the basis of their relaxation times and the effect of azide ions, and a comparison of the derivative with silver substituted at the copper site as well as nickel at the zinc site has been made (Ming and Valentine, 1987; 1990b; Ming et al., 1988b).

Paramagnetic metals are not the only useful replacements for zinc; cadmium (either as ^{111}Cd or ^{113}Cd) substitution results in full retention of activity. ^{113}Cd is a useful NMR nucleus with a spin of 1/2, large chemical shift dispersion and narrow linewidth. Two early

^{113}Cd NMR studies (Armitage et al., 1978; Bailey et al., 1980) of the bovine enzyme gave contradictory results with very different chemical shifts being assigned to the protein-bound Cd. More recently (Kofod et al., 1991) with the yeast enzyme it has been revealed that when bound to the zinc site the ^{113}Cd nucleus has a chemical shift of 315ppm when the copper site is empty and 328 ppm when it is occupied by copper(I); these values are close to those reported by Bailey et al. (1980). When the conditions are changed to allow cadmium to also occupy the copper site, then its chemical shift is 185ppm more comparable with the results of Armitage et al. (1978). These results again show the mobility of the metal ions.

^{111}Cd is a γ emitter and can be used to study ligand charge and dynamics on the ms timescale by perturbed angular correlation (PAC) spectroscopy. These measurements have shown that the cadmium in the zinc site undergoes a substantial structural change on reduction of the copper, which is interpreted in terms of breaking the bridging histidine linkage (Bauer et al., 1991, Bjerrum et al., 1991).

The other metal in the same group as zinc and cadmium is mercury and this too binds to apo Cu,Zn-SOD although it prefers a linear rather than tetrahedral geometry. Mercury binding was measured indirectly by observing the spin-spin coupling between ^{1}H nuclei on coordinating histidine residues and, ^{199}Hg bound to the protein (Cass et al., 1978b).

The copper site shows as much versatility as the zinc site in binding different metal ions; indeed, given the different coordination geometries of copper (II) and copper(I), appropriate choice of the substituting metal ion can shed light on both these states.

Silver(I) ions proved to be an interesting analog for copper(I) and Roe and Valentine (1990) and Roe et al. (1990) have used complexometric titration to evaluate the binding constants for this metal at both copper and zinc sites. Silver binding at the copper sites is tighter when zinc is present and, at pH values greater than 5.5, silver ions displace copper(II) from the copper site to the zinc site by binding at the copper site Both nickel and cobalt can also occupy the copper site under appropriate conditions and cobalt has proved to be a sensitive indicator of phosphate binding, particularly under alkaline conditions (Banci et al., 1987b). When both copper and zinc sites are available, the location of the cobalt depends on the reconstitution conditions. Salvato et al. (1989) added cobalt in the form of its isothiocyanate complex and characterized the products by fluorescence, absorbance and CD methods.

Quantitative analysis of metal binding has been used to characterize the thermodynamics of metal-protein interactions. Hirose et al. (1981; 1982) used equilibrium dialysis to measure the binding constants for copper at both metal sites as a function of pH, whilst Roe and Valentine (1990) used a complexometric titration with an ion-selective electrode to measure both copper(II) and silver(I) association with the copper site. Although these two approaches gave qualitative agreement, the values for the dissociation constants differed considerably. Hirose et al. (1988) have also separated different reconstituted forms by HPLC.

Zinc will also bind to the coppper site once the zinc site is occupied. Titration of the apo protein with upto 4 equivalents of zinc shows considerable changes in the ^{1}H NMR spectra of the histidine residues. It appears that occupation of the zinc site preforms the copper site (Cass et al., 1979).

In addition to their catalytic roles, the metals in SOD are important in maintaining the compactness of the structure and stabilizing the enzyme against unfolding. This is largely mediated through a network of hydrogen bonds; one of the most sensitive probes of molecular flexibility is backbone amide proton/deuterium (H/D) exchange, and the apo protein shows a much larger rate of exchange than the holo form. The H/D exchange is also sensitive to a structural transition which increases the flexibility of the protein at low pH (<4.5 for the holo and <6.2 for the apo protein) (Dryer et al., 1986).

Differential scanning calorimetry (Lepock et al., 1985) yields direct thermodynamic values for the thermal denaturation mid-point temperature and the enthalpy change, and as expected the copper(II) holo protein shows a greater stability towards thermally induced unfolding than the apo protein (T_m 104° and 60° respectively) with the reduced holo protein being close to the oxidized form in stability (T_m 96°). Very similar results were reported for the human form, although differences were noted in the thermal denaturation of the apo human enzyme (Biliaderis et al., 1987). The presence of metal ions has also been shown to stabilize the protein towards dissociation by detergents (SDS); whereas the holo protein is resistant to dissociation, the apo protein forms monomers (Marmochi et al., 1975). Further evidence for the close interaction of the metal site with the rest of the protein comes from site-directed mutagenesis where an invariant aspartate at position 124 in the human sequence is replaced by either asparagine or glycine. This aspartate residue hydrogen bonds to a copper histidine ligand and a zinc histidine ligand, and when it is replaced the resultant protein is free of zinc although it does contain copper (Banci et al., 1991a). The mutant protein was characterized by various spectroscopic techniques. At high pH, where the chemically prepared copper-only protein shows copper migration to the zinc site, the mutant retains copper in its native site again consistent with impared coordination at the zinc site.

2. *Mn- and Fe-SODs*

Metal substitution in these 2 enzymes has been rather less studied than is the case for the Cu,Zn-SOD; although many metals bind to the coordination sites, only the native ones reconstitute activity (Yamakura and Suzuki, 1976; Brock et al., 1976; Ose and Fridovich, 1976). An intriguing observation is the lack of interchangability of iron and manganese despite their similarity in coordination sites (an exception to this general rule is the report that *Bacillus fragilis* Fe-SOD can be reconstituted with manganese to yield an active enzyme (Gregory and Dapper, 1983). Amongst the metals reported to bind to Fe-SODs are Cd, Cr, Mn (Yamakura and Suzuki, 1980) and copper (Bannister et al., 1985).

IV. SPECTROSCOPIC STUDIES

A. Metal-centered spectroscopy

The metal centers in all the SODs provide a view of the protein which can be seen by many different spectroscopies and some of these have been mentioned in the previous section. Typical metal-centered spectroscopies carry information about ligand type, geometry and local environment with a characteristic timescale and energy range. In this section, we briefly outline the major results of these metal centered spectroscopies.

1. *Cu,Zn-SOD*

EPR, ESE and ENDOR- The unpaired electron density on the copper(II) ion gives rise to signals in electron resonance techniques. EPR is sensitive to the geometry around the copper ion and has been applied to the native enzyme (Lieberman et al., 1982) and also to various mutants and anion-bound states (Bertini et al., 1989a; Banci et al., 1991a).

ESE detects the modulation of electron spin by weakly coupled nuclei and has proven to be a useful probe of whether copper is bound to the bridging histidine (Fee et al., 1981a). ENDOR also detects electron-nuclear spin interactions but in this case strongly coupled ones. It has been used to monitor the different modes of anion coordination in bovine and human enzymes (van Camp et al., 1982; Hutterman et al., 1988).

Water Relaxation Measurements- The paramagnetic copper(II) ion affects the nuclear relaxation times of water via rapid exchange between coordinated molecules and bulk solvent. Measurements are usually made over a range of frequencies; most studies have involved the ^{1}H nuclei of water but ^{17}O can also be employed (Bertini et al., 1981). Several studies of water relaxation in SOD concur in suggesting an axially coordinated water molecule with a dissociation rate constant of 10^6 s^{-1} (Gaber et al., 1972; Terenzi et al., 1974; Boden et al., 1979).

EXAFS- of all the methods for studying the structure of metalloproteins, only EXAFS approaches X-ray crystallography in its precision and resolution in the metal coordination sphere. Several EXAFS studies of Cu,Zn-SOD have been reported. Freeze dried bovine SOD gave an EXAFS pattern for copper(II) that could be best fit by 4 nitrogen imidazole groups at 1.98 Å whilst the zinc(II) was best fit by 3 nitrogen imidazoles at 2.01 Å and an oxygen at 2.03 Å (Blackburn et al., 1983). When copper and zinc edge EXAFS of the reduced enzyme were measured, they showed a minimal perturbation of the zinc site whilst the copper site changed from one of 4 nitrogen ligands at an average distance of 1.99 Å and an oxygen at 2.24 Å to a 3 coordinate form with an average distance between the nitrogens and copper of 1.94 Å (Blackburn et al., 1984). EXAFS of the yeast enzyme was interpreted differently with different metal-ligand distances in each of the 2 sites, an interpretation also advanced for the cadmium-substituted form (Phillips et al., 1984).

2. *Mn- and Fe-SODs*

The Mn is EPR silent but MCD has proven to be a powerful tool to characterize its magnetic properties, and the studies of Whitaker and Whitaker (1991) showed that both the Mn(III) and Mn(II) forms bound fluoride and azide. Whittaker and Solomon (1986,1988) analyzed the absorbance, CD and MCD of Fe-SOD in terms of a 5-coordinate square pyramidal structure for the Fe(III), consistent with EPR data. MCD also suggests no change in coordination geometry upon reduction.

B. Protein-centered spectroscopy

Protein-centered spectroscopic methods may give details at the level of individual amino acid residues (NMR and fluorescence) or at the level of secondary structure (CD). The majority of spectroscopic studies of protein structure have been performed on the Cu,Zn enzyme, although the Mn and Fe-SODs have not been entirely neglected.

Circular dichroism is of value in assessing regular secondary structure elements such as α helix or β sheet and gives an overall view of the protein fold. CD measurements on four Mn-SODs (*S. cerevisiae, B. stearothermophilus, E. coli* and human liver) suggested a common secondary structure content of approximately 60% helix, 13% sheet 11% turn and 8% unstructured (Bjerrum, 1987), which agrees with the content determined by X-ray crystallography. Another type of electronic spectroscopy of use in elucidating protein structures is UV absorption and fluorescence. The human Cu,Zn-SOD, unlike the bovine enzyme, has a tryptophan residue and its fluorescence has been used to probe the internal dynamics of the protein through time resolved measurements (Rosato et al., 1990). First derivative UV absorption spectroscopy was used to distinguish the tryptophan and phenylalanine chromophores in the human enzyme and to investigate their environments. The results were interpreted as an increased solvent exposure of the tryptophan following removal of the copper ions (Mailer et al., 1989). Second derivative UV absorption spectra were employed to investigate the conformational stability of apo, zinc-only and native enzymes towards denaturation by guanadinium chloride. The phenylalanine residues are close to the active site and do not become solvent exposed in any of the three forms so long as the protein remains folded. The kinetics of unfolding as measured by this absorption spectroscopy suggest the existence of metal-stabilized intermediates (Mach et al., 1991).

High resolution NMR studies of the Cu,Zn-SODs have been elements extensively to characterize their structure, especially in the region of the metal-binding site. Early NMR investigations of the oxidized form showed considerable evidence of paramagnetic broadening and loss of resonances close to the copper (Stokes et al., 1973; Lippard et al., 1977; Cass et al., 1977). To overcome this loss of information, the replacement of the zinc with cobalt(II) increases the electron spin relaxation time of the copper(II) and reduces its effectiveness as a relaxation agent, allowing the observation of hyperfine shifted resonances in the oxidized form of the enzyme (Bertini et al., 1985; Ming and Valentine, 1990a; Banci et al., 1990b). Banci et al. (1989b) employed NOE measurements and relaxation time determinations to assign several of the copper ligands. A more recent study of the reduced enzyme at 600MHz using NOESY and truncated driven NOE difference spectra has allowed a more detailed assignment of the metal histidine protons (Bertini et al., 1991a).

V. MECHANISM OF REACTION

A. Cu,Zn-SOD

The reaction catalyzed by the SODs is deceptively simple, but the design of the enzyme shows how the limitations of wrapping an already excellent catalyst (the copper(II) ion) in a protein coat are overcome. The inherent instability of O_2^- in aqueous solution poses some problems for non-steady state measurements of enzyme turnover which have been approached by pulse radiolysis (Klug et al., 1972; Rotilio et al., 1972; Klug-Roth et al., 1973; Fielden et al., 1974), electrochemical (Rigo et al., 1975a; 1975b, Orsega et al., 1982) or stopped-flow techniques (McClune and Fee, 1976). All these methods concur in yielding an essentially diffusion-controlled reaction of the enzyme with its substrate (k_{cat} 10^6 s^{-1}) and a K_m so high (mM) that the enzyme is probably always working under second order kinetics with a rate constant of 10^9 M^{-1} s^{-1}.

Pulse radiolysis experiments have also shown that over the physiological pH range the rate of enzyme reaction, unlike the spontaneous dismutation, is constant (Bielski and Allen, 1977). At high pH, however, there are 2 ionizing groups that control the enzyme's activity with pK_a values of 10.1 and 10.8 (Argese et al., 1987), which probably correspond to the 2 lysine residues which are present in the active site cleft (lysines 120 and 124). As the proportion of oxidized copper during steady state turnover is approximately the same, both above and below these pK_a values the pH dependence is probably related to the rate-limiting diffusion of superoxide into the cleft. A similar pH behavior was also seen in fluoride binding measured by ^{19}F NMR, again consistent with anion accessibility being controlled by these two ionizations.

It has become clear that, although the enzyme is diffusion controlled, the encounter of substrate with the catalytic center is by no means random. A key feature of the enzyme reaction is the electrostatic steering of superoxide ion towards the entrance to the active site and down into the cleft at the bottom of which sits the copper ion. Although the net charge on the protein at pH 7 is negative, a patch surrounding the entry to the active site contains a preponderance of positively charged amino acids, and the lining of the cleft is also positively charged (Tanier et al., 1983). Electrostatic potential calculations based on the crystal structure suggest that the enzyme's electric field enhances the rate of association of the enzyme and substrate at least 30-fold and the calculated effects of ionic strength agree well with those observed (Sharp et al., 1987).

It was the observation of the effect of ionic strength on the enzyme activity (it decreases with increasing ionic strength) (Cudd and Fridovich, 1982) along with chemical modification studies of Arg-141 (in the bovine sequence) (Malinowski and Fridovich, 1979) that showed that the enzyme was more than a copper(II) ion in a protein coat.

Two parallel lines of investigation have enabled a detailed picture of the mechanism to be built up, one is the investigation of anion binding, and the other is site-specific mutagenesis. Sometimes the two are used together.

The interaction of a large number of anions with the enzyme has been investigated by many different techniques and for several metal-substituted forms. Often the anions act as competitive inhibitors (Rigo et al., 1977) and in the copper(II) state various changes in the spectroscopic parameters of the metal ion occur, although interpretation of these is not always straightforward. Originally, the changes brought about by ligand binding were interpreted in terms of their displacing a histidine ligand to the copper although there was some EXAFS evidence that the histidine could still be weakly coordinated in the case of azide binding (Blackburn et al., 1987). More recent ^1H NMR studies of anion binding to the Cu,Co form, which have determined inter-proton distances from NOE measurements, are consistent with all the histidine residues remaining ligated, and the effect of the anion is to reorient the ligand field, leading to a relatively small overlap of the unpaired electron with histidine 48 (Banci et al., 1990a). This structural rearrangement also results in the loss of the water molecule from the coordination sphere as assessed from water relaxation measurements (Banci et al., 1988a).

Fluoride binding is readily followed from the ^{19}F relaxation times which are decreased by the paramagnetic copper(II) center (Viglino et al., 1979) and, because Cu,Zn-SOD is the only paramagnetic protein that also binds F^-, the measurement of the latter's relaxation time has been proposed as a method of assaying the enzyme (Rigo et al., 1979). As the relaxivity of Cu,Zn-SOD depends on the presence of the copper(II) center, then it follows that the reduction to copper(I) will reduce the relaxivity of the enzyme. For a fixed concentration of protein during catalysis, the relaxivity will depend on the steady state concentration of the copper(II) form and hence provides a way of measuring this during turnover (Viglino et al., 1981). Following this argument a little further leads to the idea that as the steady state level of copper(II) during turnover depends on the superoxide concentration, ^{19}F relaxation can also offer a way of measuring this (Rigo et al., 1981). More recently the relaxation of $^{19}F^-$ by yeast SOD has been investigated and this has been complemented by ^1H NMR studies of the Cu,Co protein in the presence of fluoride (Banci et al., 1989a) where the results are consistent with the copper retaining all its protein ligands.

Phosphate acts rather differently, binding to a key arginine residue (141 in the bovine and 143 in the human sequence) that chemical modification studies had previously identified as being critical for activity (Malinowski and Fridovich, 1979) and whose modification perturbs the copper coordination sphere (Paci et al., 1991). Further evidence for phosphate binding to this arginine comes from ^{31}P NMR measurements of the Co,Cu form of the enzyme. The ^{31}P NMR was affected by modification of the arginine residue and phosphate binding perturbed the chemical shift of the NH proton of His-46 (Desideri et al., 1988).

Vanadate is similar to phosphate and its binding to Cu,Zn-SOD has been studied by ^{51}V NMR (Wittenkeller et al., 1991). In solution, vanadate is a complex mixture of monomeric and polymeric forms and it is the tetrameric anion which has the highest affinity, with an association constant of more than 10^6 M. Its interaction appears to be through bridging between 2 of 3 positively charged residues close to the copper (Arg-141, Lys-120 and 134) and forming a 'cap' over the active site.

Although phosphate is the anion which most closely associates with the arginine residue, other anions' affinity for the copper is controlled by this residue as shown by a series of mutants in which the arginine is substituted and the affinity of azide for the protein directly relates to the charge on the substituent (Banci et al., 1988b). Similar conclusions were reached from a study of the interaction of CN^-, NCO^- and phosphate with the mutant of human Cu,Zn-SOD where Arg-143 was replaced by an isoleucine (Bertini et al., 1991b).

Anion binding to the reduced Cu,Zn-SOD has also been investigated, notably in the case of chloride using ^{35}Cl NMR. This nucleus is quadrupolar and, in the assymetric electric field of the protein, its relaxation time is much reduced. Fee and Ward (1976) first described this and recently the effect of chemically modifying the arginine residue on chloride binding has been investigated. As described above, the effect of modifying Arg-141 on the copper(II) form of the enzyme is to reduce the activity and alter the anion binding and spectroscopic properties of the copper site (Paci et al., 1991). The same modification also decreases the enhanced relaxation of $^{35}Cl^-$ in the reduced enzyme, consistent with a weaker binding of this anion (Mota de Freitas et al., 1990). Complementary measurements of the ^1H NMR spectra of ligand histidine resonances showed that these too were less shifted by the addition of chloride or phosphate in the arginine-modified protein.

That anion binding to the reduced form of the enzyme is much weaker than to the oxidized form is suggested from redox potential measurements (Ozaki et al., 1988). Fee and DiCorleto (1973) measured the copper(II)/copper(I) potential by titration with hexacyanoferrate(III)/(II) and obtained a value of 0.42V. However, it was later shown (Morpurgo et al., 1976) that hexacyanoferrate(II) forms a charge transfer complex with copper(II) SOD and this may have distorted the measured potential. Cyanide brings about a marked change in the redox potential of Cu,Zn-SOD (Lawrence and Sawyer, 1979) and Ozaki et al. (1988) measured both the kinetics and thermodynamics of the redox reactions as a function of both pH and coordinating anions. Rates of reduction of the copper(II) form were independent of pH, whilst the reoxidation rates were pH dependent, accounting for the overall pH dependence of the redox potential.

These mechanistic studies along with the spectroscopic work described earlier have built up a picture of a well designed active site where the anionic substrate (O_2^-) is delivered to the catalytic center by electrostatic steering through a hydrogen-bonded network, and the mutagenesis studies of Getzoff et al. (1992) elegantly confirm this by replacing the hydrogen-bonded glutamates (132,133) of the human enzyme by charge-neutralizing glutamines. The resulting mutants have up to a 4-fold faster rate constant than the wild type.

A further mechanistic consideration is the source of protons for the O_2^- reduction step (equation 6). One possibility is the bridging imidazolate ion which is cleaved from the copper and protonated during the reduction step of equation 5. This is illustrated schematically in Figure 3.

Alternatively the water molecules lining the active site cleft could act as proton donors. Recent mutagenesis studies have tended to tell against the importance of the water molecules as even in those enzymes in which water was not present at the catalytic center (as measured

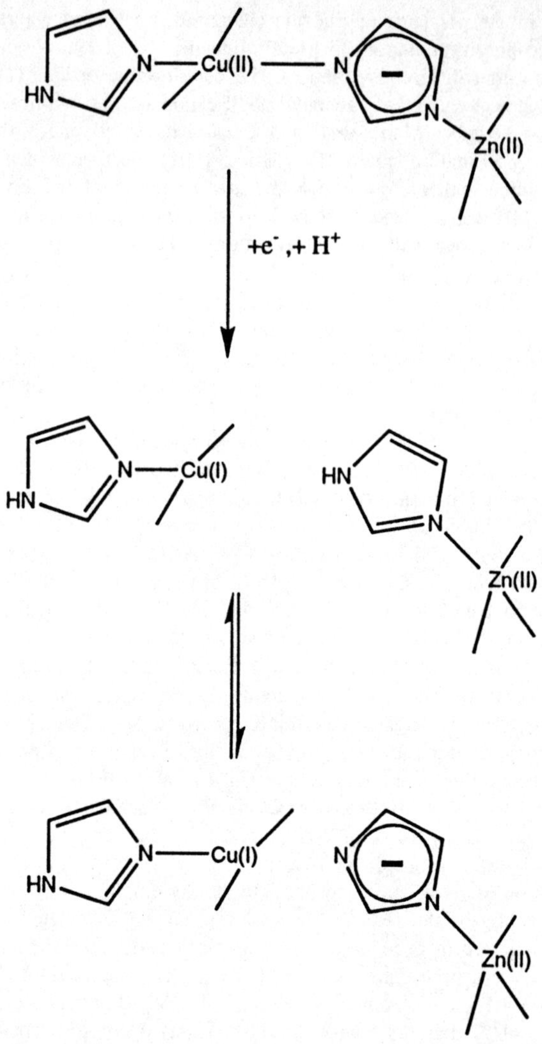

Figure 3.
Proton linked redox reactions of Cu,Zn-SOD showing cleavage of the bridging histidine residue upon reduction and its subsequent deprotonation at high pH.

by water relaxation times) the enzyme activity was retained. The mutants employed were those where Thr-137 (human sequence) was replaced by isoleucine (Bertini et al., 1989a), and serine or isoleucine (Banci et al., 1989c; 1990c). When the spectroscopic properties of the Ile-137 mutant were investigated, considerable distortion of the active site was found to have occurred (Bertini et al., 1989a) but the enzyme's activity at physiological pH was unaffected. Interestingly, there was a 90% loss of water from the copper coordination sphere (Banci et al., 1989c). Although not affecting its physiological activity, the mutation did result in a lowering of the pK_a for the alkaline transition to an inactive form to 9.6 (cf 10.6 in the wild type) (Banci et al., 1991b).

Hydrogen peroxide, a product of the reaction, is also a potent inactivator of the enzyme (Bray et al., 1974; Hodgson and Fridovich, 1975a; 1975b). This inactivation results from the modification of a single histidine residue per subunit and the inactivating species seems to be HO_2^-. The kinetics of this reaction are broadly similar for both yeast (Blech et al., 1983) and bovine (Fuchs et al., 1983) enzymes. Site-specific mutagenesis of the human enzyme has implicated Arg-143 in this reaction also; in this case, it appears to be involved in the approach of HO_2^- to the active site (Horton et al., 1989). When replaced by a non-charged residue, the enzyme is still inactivated by H_2O_2 but through a different mechanism. In contrast to these kinetic studies, the ^1H NMR spectra of hydrogen-peroxide treated Cu,Co-SOD does not show any major change in structure (Bertini et al., 1989b).

B. Mn- and Fe-SODs

Mechanistic studies of these enzymes have been rather less detailed although the results are broadly similar. Pulse radiolysis measurements have yielded rate constants for the enzymes from; *E. coli* Mn-SOD (Pick et al., 1974), *B. stearothermophlus* Mn-SOD (McAdam et al., 1977a; 1977b), *E. coli* Fe-SOD (Fee et al., 1981b) and *Photobacterium leognathi* Fe-SOD (Lavelle et al., 1977).

Fewer anions have been shown to bind to Mn or Fe-SODs, but the interaction of azide with Fe-SOD is particularly well understood thanks to the recent X-ray crystal structure of the *Ps. ovalis* enzyme-azide complex (Stoddardet al., *1*990b) where the ligand binds to the free coordination position and also interacts with the nearby glutamine 70 which in turn may play a role in the O_2^- protonation reaction.

VI. CONCLUSIONS

During the past 10 years, the role of molecular genetics has illuminated much of the biology and chemistry of the superoxide dimutases. Cloning of SOD genes through complementation has shed light on the enzyme's physiological function although there is still much that remains to be done to give a quantitative understanding to the role of O_2^- in vivo.

X-ray crystallography and site-specific mutagenesis have helped to map out the detailed mechanism of the Cu,Zn-SODs and will doubtless also be applied to the less intensively studied Mn and Fe enzymes.

Finally, as the physiological and pathological roles of this enzyme and its substrates become unravelled, then the ability to engineer and manipulate its gene and proteins promises new approaches to therapy.

REFERENCES

Argese, E., Viglino, P., Rotilio, G., Scarpa, M., and Rigo, A. (1987). *Biochemistry, 26*: 3224.

Armitage, I.M., Uiterkamp, A.J.M., Cheblowski, J.F., and Coleman, J.E. (1978). *J. Magn. Res., 29*: 375.

Bailey, D.B., Ellis, P.D., and Fee, J.A. (1980). *Biochemistry, 19*: 591.

Banci, L., Bertini, I., Luchinat, C., Monnanni, R., and Scozzafava, A. (1987). *Inorg. Chem., 26*: 153.

Banci, L., Bertini, I., Luchinat, C., Monnanni, R., and Scozzafava, A. (1988a). *Inorg. Chem., 27*: 107.

Banci, L., Bertini, I., Luchinat, C., and Hallewell, R.A. (1988b). *J. Amer. Chem. Soc., 110*: 3629.

Banci, L., Bertini, I., Luchinat, C., Scozzafava, A., and Turano, P. (1989a). *Inorg. Chem.,* 28: 2377.

Banci, L., Bertini, I., Luchinat, C., Piccioli, M., Scozzafava, A., and Turano, P. (1989b). *Inorg. Chem., 28*: 4650.

Banci, L., Bertini, I., Hallewell, R.A., Luchinat, C., and Viezzoli, M.S. (1989c). *Eur. J. Biochem., 184*: 125.

Banci, L., Bencini, A., Bertini, I., Luchinat, C., and Piccioli, M. (1990a). *Inorg. Chem., 29*: 4867.

Banci, L., Bertini, I., Luchinat, C., and Piccioli, M.S. (1990b). *FEBS Lett., 272*: 175.

Banci, L., Bertini, I., Cabelli, D., Hallewell, R.A., Luchinat, C., and Viezzoli, M.S. (1990c). *Inorg. Chem., 29*: 2398.

Banci, L., Bertini, I., Cabelli, D.E., Hallewell, R.A., Tung, J.W., and Viezzoli, M.S. (1991a). *Eur. J. Biochem., 196*: 123.

Banci, L., Bertini, I., and Turano, P. (1991b). *Eur. Biophys. J.*, *19*: 141.

Bannister, J.V., Desideri, A., and Rotilio, G. (1985). *FEBS Lett.*, *188*: 91.

Bauer, R., Bjerrum, M., Danielsen, E., and Kofod, P. (1990). *Hyperfine Interact.*, *61*: 1201.

Bauer, R., Bjerrum, M.J., Danielsen, E., and Kofod, P. (1991). *Acta Chem. Scand.*, *45*: 593.

Bertini, I., Luchinat, C., and Messori, L. (1981). *Biochem. Biophys. Res. Commun.*, *101*: 577.

Bertini, I., Lanini, G., Luchinat, C., Messori, L., Monnanni, R., and Scozzafava, A. (1985). *J. Amer. Chem. Soc.*, *107*: 4391.

Bertini, I., Banci, L., Luchinat, C., Bielski, B.H.J., Cabelli, D.E., Mullenbach, G.T., and Hallewell, R.A. (1989a). *J. Amer. Chem. Soc.*, *111*: 714.

Bertini, I., Luchinat, C., Viezzoli, M.S., and Wang, Y. (1989b). *Arch. Biochem. Biophys.*, *269*: 586.

Bertini, I., Capozzi, F., Luchinat, C., Piccioli, M., and Viezzoli, M.S. (1991a). *Eur. J. Biochem.*, *197*: 691.

Bertini, I., Lepori, A., Luchinat, C., and Turano, P. (1991b). *Inorg. Chem.*, *30*: 3363.

Bielski, B.H., and Allen, A.O. (1979). *J. Phys. Chem.*, *81*: 1048.

Biliaderis, C.G., Weselake, R.J., Petkau, A., Friesen, A.D. (1987). *Biochem. J.*, *248*: 981.

Bjerrum, M.J. (1987). *Biochim. Biophys. Acta*, *915*: 225.

Bjerrum, M.J., Bauer, R., Danielsen, E., and Kofod, P. (1991). *Free Rad. Res. Commun.*, *12-13*: 297.

Blackburn, N.J., Hasnain, S.S., Diakun, G.P., Knowles, P.F., Binsted, N., and Garner, C.D. (1983). *Biochem. J.*, *213*: 765.

Blackburn, N.J., Hasnain, S.S., Binsted, N., Diakun, G.P., Garner, C.D., and Knowles, P.F. (1984). *Biochem. J.*, *219*: 985.

Blackburn, N.J., Strange, R.W., McFadden, L.M., and Hasnain, S.S. (1987). *J. Amer. Chem. Soc.*, *109*: 7162.

Blech, D.M., and Borders, C.L. (1983). *Arch. Biochem. Biophys.*, *224*: 579.

Boden, N., Holmes, M.C., and Knowles, P.F. (1979). *Biochem. J.*, *177*: 303.

Borgstahl, G.E.O., Parge, H.E., Hickey, M.J., Beyer, W.F., Hallewell, R.A., and Tainer, J.A. (1992). *Cell*, *71*: 107.

Bray, R.C., Cockle, S.A., Fielden, E.M., Roberts, P.B., Rotilio, G., and Calabrese, L. (1974). *Biochem. J.*, *139*: 43.

Brock, C.J., Harris, I.J., and Sato, S. (1976). *J. Mol. Biol.*, *107*: 175.

Calabrese, L., Rotilio, G., and Mondovi,B. (1972). *Biochim. Biophys. Acta*, *263*: 827.

Carlioz, A., and Toutai, D. (1986). *EMBO J.*, *5*: 623.

Cass, A.E.G., Hill, H.A.O., Smith, B.E., Bannister, J.V., and Bannister, W.H. (1977). *Biochemistry*, *16*: 3061.

Cass, A.E.G., Hill, H.A.O., Haseman, V., and Johansen, J.T. (1978a). *Carlsberg Res. Commun.*, *43*: 439.

Cass, A.E.G., Galdes, A., Hill, H.A.O., McCleland, C.E., and Storm, C.B. (1978b). *FEBS Lett.*, *94*: 311.

Cass, A.E.G., Hill, H.A.O., Bannister, J.V., and Bannister, W.H. (1979). *Biochem. J.*, *177*: 477.

Cudd, A., and Fridovich, I. (1982). *J. Biol. Chem.*, *257*: 11443.

Desideri, A., Comin, F., Morpurgo, L., Cocco, D., Calabrese, L., Mondovi, B., Maret, W., and Rotilio, G. (1981). *Biochim. Biophys. Acta*, *670*: 312.

Desideri, A., Paci, M., and Rotilio, G. (1988). *J. Inorg. Biochem.*, *33*: 91.

Djinovic, K., Gatti, G., Coda, A., Antolini, L., Pelosi, G., Desideri, A., Falconi, M., Marmocchi, F., Rotilio, G., and Bolognesi, M. (1992). *J. Mol. Biol.*, *225*: 791.

Dreyer, T., Ottesen, M., and Hvidt, A. (1986). *Carlsberg Res. Commun.*, *51*: 219.

Dunbar, J.C., Johansen, J.T., and Uchida, T. (1982). *Carlsberg Res. Commun.*, *47*: 163.

Dunbar, J.C., Holmquist, B., and Johansen, J.T. (1984a). *Biochemistry*, *23*: 4324.

Dunbar, J.C., Holmquist, B., and Johansen, J.T. (1984b). *Biochemistry*, *23*: 4330.

Fee, J.A. (1973). *J. Biol. Chem.*, *248*: 4229.

Fee, J.A., and Briggs, R.G. (1975). *Biochim. Biophys. Acta*, *400*: 439.

Fee, J.A., and DiCorleto, P.E. (1973). *Biochemistry*, *12*: 4893.

Fee, J.A., and Ward, R.L. (1976). *Biochem. Biophys. Res. Commun.*, *71*: 427.

Fee, J.A., Peisach, J., and Mims, W.B. (1981a). *J. Biol. Chem.*, *256*:1910.

Fee, J.A., McClune, G.J., O'Neill, P.M., and Fielden, E.M. (1981b). *Biochem. Biophys. Res. Commun.*, *100*: 377.

Fielden, E.M., Roberts, P.B., Bray, R.C., Lowe, D.J., Mautner, G.N., Rotilio, G., and Calabrese, L. (1974). *Biochem. J.*, *139*: 49.

Fuchs, H.J.R., and Borders, C.L. (1983). *Biochem. Biophys. Res. Commun.*, *116*: 1107; *Biochim. Biophys. Acta*, *271*: 1.

Getzoff, E.D., Cabelli, D.E., Fisher, C.L., Parge, H.E., Viezzoli, M.S., Banci, L., and Hallewell, R.A. (1992). *Nature*, *358*: 347.

Gregory, E.M., and Dapper, C.H. (1983). *Arch. Biochem. Biophys.*, *220*:293.

Hill, H.A.O., Lee, W.K., Bannister, J.V., and Bannister, W.H. (1980). *Biochem. J.*, *185*: 245.

Hirose, J., Iwazuta, K., and Kidani, Y. (1981). *Biochem. Biophys. Res. Commun.*, *98*: 58.

Hirose, J., Ohhira, T., Hirata, H., and Kidani, Y. (1982). *Arch. Biochem. Biophys.*, *218*: 179.

Hirose, J., Toida, S., Kuno, H., Ozaki, S., and Kidani, Y. (1988). *Chem. Pharm. Bull.*, *36*: 2103.

Hodgson, E.K., and Fridovich, I. (1975a). *Biochemistry*, *14*: 5294.

Hodgson, E.K., and Fridovich, I. (1975b). *Biochemistry*, *14*: 5299.

Horton, P.J., Borders, C.L., and Beyer, W.F. (1989). *Arch. Biochem. Biophys.*, *269*: 114.

Huettermann, J., Kappl, R., Banci, L., and Bertini, I. (1988). *Biochim. Biophys. Acta*, *956*: 173.

Kitagawa, Y., Tanaka, N., Hata, Y., Kusunoki, M., Lee, G. P., Katsube, Y., Asada, K., Aibara, S., and Morita, Y. (1991). *J. Biochem. (Tokyo)*, *109*: 477.

Klug, D., Rabani, J., and Fridovich, I. (1972). *J. Biol. Chem.*, *247*: 4839.

Klug-Roth, D., Fridovich, I., and Rabani, J. (1973). *J. Amer. Chem. Soc.*, *95*: 2786.

Kofod, P., Bauer, R., Danielsen, E., Larsen, E., Bjerrum, M.J. (1991). *Eur. J. Biochem.*, *198*: 607.

Lavelle, F., McAdam, M.E., Fielden, E.M., Roberts, P.B., Puget, K., and Michelson, A.M. (1977). *Biochem. J.*, *161*: 3.

Lawrence, G.D., and Sawyer, D.T. (1979). *Biochemistry*, *18*: 3045.

Lieberman, R.A., Sands, R.H., and Fee, J.A. (1982). *J. Biol. Chem.*, *257*: 336.

Lepock, J.R., Arnold, L.D., Torrie, B.H., Andrews, B., and Kruuv, J. (1985). *Arch. Biochem. Biophys.*, *241*: 243 .

Lippard, S.J., Burger, A.R., Ugurbil, K., Pantoliano, M.W., and Valentine, J.S. (1977). *Biochemistry*, *16*: 1136.

Ludwig, M.L., Metzger, A.L., Pattridge, K.A., and Stallings, W.C. (1991). *J. Mol. Biol.*, *219*: 335.

Mach, H., Dong, Z., Middaugh, C.R., and Lewis, R.V. (1991). *Arch. Biochem. Biophys.*, *287*: 41.

Mailer, K., Addetia, R., and Livesey, D. L. (1989). *J. Inorg. Biochem.*, *37*: 151 .

Malinowski, D.P., and Fridovich, I. (1979). *Biochemistry*, *18*: 5909.

Marmocchi, F., Caulini, G., Venardi, G., Cocco, D., Calabrese, L., and Rotilio, G. (1975). *Physiol. Chem. Phys.*, *7*: 465.

McAdam, M.E., Fox, R.A., Lavelle, F., and Fielden, E.M. (1977a). *Biochem. J.*, *165*: 71.

McAdam, M.E., Fox, R.A., Lavelle, F., and Fielden, E.M. (1977b). *Biochem. J.*, *165*: 81.

McClune, G.J., and Fee, J.A. (1976). *FEBS Lett.*, *67*: 294.

Ming, L.J., and Valentine, J.S. (1987). *J. Amer. Chem. Soc.*, *109*: 4426.

Ming, L.J., and Valentine, J.S. (1990a). *J. Amer. Chem. Soc.*, *112*: 4256.

Ming, L.J., and Valentine, J.S. (1990b). *J. Amer. Chem. Soc.*, *112*: 6374.

Ming, L.J., Banci, L., Luchinat, C., Bertini, I., and Valentine, J.S. (1988a). *Inorg. Chem.*, *27*: 728.

Ming, L.J., Banci, L., Luchinat, C., Bertini, I., and Valentine, J.S. (1988b). *Inorg. Chem.*, *27*: 4458.

Morgenstern-Badarau, I., Cocco, D., Desideri, A., Rotilio, G., Jordanov, J., and Dupre, N. (1986). *J. Amer. Chem. Soc.*, *108*: 300.

Morpurgo, L., Mavelli, I., Calabrese, L., Finazzi Agro, A., and Rotilio, G. (1976). *Biochem. Biophys. Res. Commun.*, *70*: 607.

Moss, T.H., and Fee, J.A. (1975). *Biochem. Biophys. Res. Commun.*, *66*: 799 .

Mota de Freitas, D., Ming, L.J., Ramasamy, R., and Valentine, J.S. (1990). *Inorg. Chem.*, *29*: 3512.

Murata, T., Nakazawa, H., Mori, T., Ohta, Y., and Yamabayashi, A. (1992). *Amer. Rev. Resp. Dis.*, *146* : 1048.

Orsega, E.F., Argese, E.,Viglino, P., Tomat, R., and Rigo, A. (1982). *J. Electroanal. Chem. Interfacial Electrochem.*, *131*: 257.

Ose, D.E., and Fridovich, I. (1976). *J. Biol. Chem.*, *251*: 1217.

Paci, M., Desideri, A., Sette, M., and Rotilio, G. (1991). *Arch. Biochem. Biophys.*, *286*: 222.

Pantoliano, M.W., McDonnell, P.J., and Valentine, J.S. (1979). *J. Amer. Chem. Soc.*, *101*: 6454.

Pantoliano, M.W., Valentine, J.S., and Nafie, L.A. (1982). *J. Amer. Chem. Soc.*, *104*: 6310.

Parge, H.E., Hallewell, R.A., and Tanier, J.A. (1992). *Proc. Nat. Acad. Sci. USA*, *89*: 6109.

Parker, M.W., and Blake, C.C.F. (1988). *J. Mol. Biol.*, *199* :649.

Phillips, J.C., Bauer, R., Dunbar, J., and Johansen, J.T. (1984). *J. Inorg. Biochem.*, *22*: 179.

Pick, M., Rabani, J., Yost, F., and Fridovich, I. (1974). *J. Amer. Chem. Soc.*, *96*: 7329 .

Rigo, A., Viglino, P., and Rotilio, G. (1975a). *Anal. Biochem.*, *68*: 1.

Rigo, A., Viglino, P., and Rotilio, G. (1975b). *Biochim. Biophys. Acta*, *63*: 1013.

Rigo, A., Stevanato, R., Viglino, P., and Rotilio, G. (1977). *Biochim. Biophys. Acta*, *79*: 776.

Roe, J.A., and Valentine, J.S. (1990). *Anal. Biochem.*, *186*: 31.

Roe, J.A., Peoples, R., Scholler, D.M., and Valentine, J.S. (1990). *J. Amer. Chem. Soc.*, *112*: 1538.

Rosato, N., Gratton, E., Mei, G., Finazzi-Agro, A. (1990). *Biophys. J.*, *58*: 817.

Rosen, D.R., Siddique, T., Patterson, D., Figlewicz, D.A., Sapp, P., Hentati, A., Donaldson, D., Goto, J., O'Regan, J.P., Deng, H.-X., Rahmani, Z., Krizus, A., McKenna-Yasek, D., Cayabyab, A., Gaston, S.M., Berger, R., Tanzi, R.E., Halperin, J.J., Herzfeldt, B., van den Bergh, R., Hung, W.-Y., Bird, T., Deng, G., Mulder, D.W., Smyth, C., Laing, N.G., Sriano, E., Pericak-Vance, M.A., Haines, J., Rouleau, G.A., Gusella, J.S., Horvitz, H.R., and Brown, R.H. (1993). *Nature*, *362*: 59.

Rigo, A., Viglino, P., Argese, E., Terenzi, M., and Rotilio, G. (1979). *J. Biol. Chem.*, *254*: 1759.

Rigo, A., Ugo, P., Viglino, P., and Rotilio, G. (1981). *FEBS Lett.*, *132*: 78.

Rotilio, G., Bray, R.C., and Fielden, E.M. (1972). *Biochim. Biophys. Acta*, *268*: 605 .

Salvato, B., Beltramini, M., Ricchelli, F., and Tallandini, L. (1989). *Biochim. Biophys. Acta*, *998*: 14.

Sandström, J., Carlsson, L., Marklund, S.L.and Edlund, T. (1992). *J. Biol. Chem.*, *267*: 18205.

Saran, M., Michel, C., and Bors, W. (1990). *Free Rad. Res. Commun.*, *10*: 221.

Sharp, K., Fine, R., and Honig, B. (1987). *Science (Washington, DC)*, *236*: 1460 .

Smith, C.D., Carson, M., van der Woerd, M., Chen J., Ischiropoulos, H., and Beckman, J.S. (1992). *Arch. Biochem. Biophys.*, *299*: 350.

Stallings, W.C., Powers, T.B., Pattridge, K.A., Fee, J.A., and Ludwig, M.L. (1983). *Proc. Nat. Acad. Sci. USA*, *80*: 3884.

Stallings, W.C., Pattridge, K.A., Strong, R.K., and Ludwig, M.L. (1985). *J. Biol. Chem.*, *260*: 6424.

Stoddard, B.L., Howell, P.L., Ringe, D., and Petsko, G.A. (1990a). *Biochemistry*, *29*: 8885.

Stoddard, B.L., Ringe, D., and Petsko, G. A. (1990b). *Protein Eng.*, *4*: 113.

Stokes, A.M., Hill, H.A.O., Bannister, J.V., and Bannister, W.H. (1973). *FEBS Lett.*, *32*: 119.

Tanier, J.A., Getzoff, E.D., Beem, K.M., Richardson, J.S., and Richardson, D.C. (1982). *J. Mol. Biol.*, *160*: 181.

Tanier, J.A., Getzoff, E.D., Richardson, J.S., and Richardson, D.C. (1983). *Nature*, *306*: 284.

Terenzi, M., Rigo, A., Franconi, C., Mondovi, B., Calabrese, L., and Rotilio, G. (1974). *Biochim. Biophys. Acta*, *351*: 230.

van Camp, H.L., Sands, R.H., and Fee, J.A. (1982). *Biochim. Biophys. Acta*, *704*: 75.

Viglino, P., Rigo, A., Stevanato, R., Ranien, G.A., Rotilio, G., and Calabrese, L. (1979). *J. Magn. Res.*, *34*: 265.

Viglino, P., Rigo, A., Argese, E., Calabrese, L., Cocco, D., and Rotilio, G. (1981). *Biochem. Biophys. Res. Commun.*, *100*: 125.

Whittaker, J.W., and Solomon, E.I. (1986). *J. Amer. Chem. Soc.*, *108*: 835.

Whittaker, J.W., and Solomon, E.I. (1988). *J. Amer. Chem. Soc.*, *110*: 5329.

Whittaker, J.W., and Whittaker, M. M. (1991). *J. Amer. Chem. Soc.*, *113*: 5528.

Whittenkeller, L., Aida, A., Ravichandran, R., and De Freitas, D.M. (1991). *J. Amer. Chem. Soc.*, *113*: 7872.

Yamakura, F., and Suzuki, K. (1976). *Biochem. Biophys. Res. Commun.*, *72*: 1108.

Yamakura, F., and Suzuki, K. (1980). *J. Biochem. (Tokyo)*, *88*: 191.

FUNCTIONAL MODELS FOR METALLOPROTEINS IN DIOXYGEN METABOLISM

by

Clayton R. Randall and Lawrence Que, Jr.

Department of Chemistry,
University of Minnesota,
Minneapolis, MN 55455
U.S.A.

I. INTRODUCTION

Metalloproteins figure prominently in dioxygen metabolism, because it is usually by coordination to a metal center that the oxidative potential of O_2 can be tapped (see Chapter 1-B14). Significant progress towards modeling such proteins has been made in the last twenty years, such that a number of functions carried out by these proteins can be investigated in the absence of the protein constraints. A recurring theme in many of these efforts is the incorporation of sterically hindered groups that control species nuclearity, preventing formation of unwanted oligomeric products. From these studies, valuable insights into the chemistry of metals and dioxygen have been gained. We summarize these results in this section and focus on the efforts that have produced mechanistic information.

II. DIOXYGEN BINDING AND TRANSPORT

A. Hemoglobin and myoglobin

In vertebrates, hemoglobin (Hb) and myoglobin (Mb) function as oxygen transport and storage proteins, respectively. They both contain ferrous heme prosthetic groups and bind dioxygen reversibly (Suslick and Reinert, 1985). These two proteins were among the first metalloproteins to be structurally characterized by X-ray crystallography (Perutz et al., 1987), and were also among the first for which functional models were developed. The synthetic difficulty that had to be overcome was the thermodynamically favorable formation of (μ-oxo)diiron(III) porphyrin dimers from Fe(II) porphyrin-dioxygen adducts.

Several approaches have been taken to prevent oxo-bridged dimer formation and prepare stable O_2 adducts: Wang (1958) embedded an iron(II) porphyrin in a thin polymer film containing 2-phenylethylimidazole under an atmosphere of CO. Purging with N_2 afforded isolated heme-imidazole adducts that were able to bind O_2 reversibly. Subsequently, several groups (Wagner and Kassner, 1974; Brinigar and Chang, 1974; Brinigar et al., 1974; Anderson et al., 1974; Almog et al., 1974) discovered almost simultaneously that low temperatures (-45 to -80 °C) favored reversible dioxygen binding to ferrous hemes. By designing a "molecular cap" on one face of the porphyrin to prevent dimer formation, Hashimoto et al. (1982) demonstrated reversible dioxygen

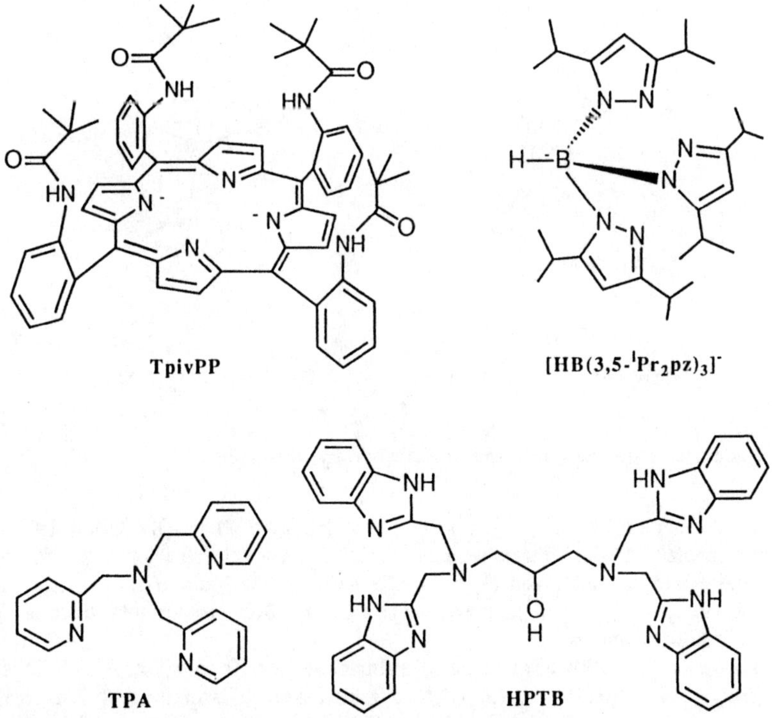

Figure 1.
Examples of ligands used for functional models of metalloproteins.

binding at room temperature. The crowning achievement was the crystallization of an O_2 adduct by Collman and co-workers using sterically hindered "picket fence" porphyrins (Fig. 1). Crystal structures of dioxygen adducts, [Fe(TpivPP)(1-MeIm)(O_2)] (**1**) and [Fe(TpivPP)(2-MeIm)(O_2)·EtOH] (**2**) (TpivPP = 5,10,15,20-tetrakis(2-pivalamidophenyl)-porphyrinate(2-)) provided important insights (Jameson et al., 1978; 1980). In **1**, the Fe-O bond length is 1.75 Å, and the iron atom is displaced 0.03 Å from the pyrrole ring plane toward the O_2 ligand. In **2**, the steric bulk of the 2-methyl-imidazole ligand prevents the iron atom from entering the porphyrin cavity. Thus, the Fe-O bond length is 1.90 Å, and the iron atom is displaced 0.09 Å in the opposite direction, towards the 2-methylimidazole ligand. In both cases, the Fe-O_2 moiety is bent, with Fe-O-O angles of ~130° (Fig. 2).

As a consequence of the structural differences discussed above, **1** has a greater oxygen affinity than **2**, and provides a molecular rationale for the mechanism of cooperativity. In this model, the lower affinity sites in Hb constrain the movement of iron into the porphyrin cavity upon O_2 binding, while the high affinity sites of Hb as well as the Mb site do not impose any constraints on the iron atom (Perutz, 1976). Interestingly, the unsolvated [Fe(TpivPP)(2-MeIm)(O_2)] shows cooperativity in the solid state, which is attributed to crystal packing effects (Jameson et al., 1980). Therefore, **1** can be thought of as a functional model of myoglobin, and **2** as a functional model of hemoglobin.

B. Hemerythrin

Hemerythrin (Hr), the oxygen-carrier protein of some marine invertebrates, possesses a diiron site bridged by a hydroxo or oxo group and two carboxylates and terminally coordinated to five His residues (Holmes et al., 1991; see scheme 2 in Chapter 1-B14). Although a number of synthetic complexes have been prepared that mimic the structural features of Hr (Que and True, 1990), these models do not bind dioxygen.

Nonheme-iron containing complexes related to Hr that bind O_2 have been prepared by the groups of Kitajima, Que, and Suzuki. Kitajima et al. (1990), using sterically hindered

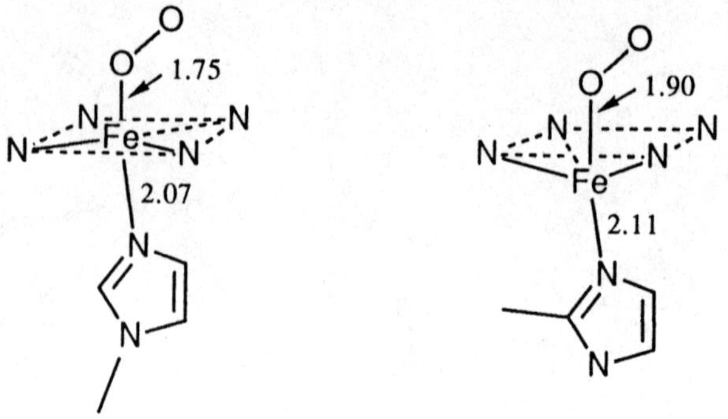

[Fe(TpivPP)(O_2)(1-MeIm)]   [Fe(TpivPP)(O_2)(2-MeIm)]

Figure 2.
The O_2 adducts of sterically hindered iron(II) porphyrin complexes.

tris(pyrazolyl)borate ligands (Fig. 1), prepared [(HB(3,5-^{i}Pr$_2$pz)$_3$)Fe(OBz)(MeCN)] (**3**). This mononuclear complex reacts reversibly with O_2 in a 2:1 ratio below -20 °C, turning from yellow to deep blue-green (λ_{max} = 679 nm). On the basis of its resonance Raman spectrum ($\nu_{O\text{-}O}$ = 876 cm^{-1}), the dioxygen adduct **3**·O_2 is formulated as a (μ-1,2-peroxo)diiron(III) complex.

Ménage et al. (1990) have prepared a dinuclear iron(II) complex, [Fe$_2$(N-Et-HPTB)-(OBz)](BF$_4$)$_2$ (**4**) (HPTB = N,N,N',N'-tetrakis(2-benzimidazolylmethyl)-2-hydroxy-1,3-diaminopropane, Fig. 1) that binds dioxygen irreversibly with a 1:1 stoichiometry at -60 °C, turning from light yellow to deep blue (λ_{max} = 588 nm, ε = 1500 M^{-1}cm^{-1}). The spectroscopic data for **4**·O_2 ($\nu_{O\text{-}O}$ = 900 cm^{-1}) also indicate a (μ-1,2-peroxo)diferric structure. The corresponding HPTP complex (HPTP is the tetrakis(pyridine) analogue of HPTB) also forms an irreversible O_2 adduct (Dong et al., 1993), but the 6-Me-HPTP complex, [Fe$_2$(6-Me-HPTP)(RCO$_2$)(H$_2$O)](BF$_4$)$_2$ (**5**) (R = CF$_3$, Ph) binds O_2 reversibly in acetonitrile or dichloromethane below -20 °C (Hayashi et al., 1992). Since **3** and **5** react reversibly with dioxygen while **4** does not, it appears that steric hindrance around the oxygen atoms may be important for reversible O_2 binding. Unfortunately, none of these diferrous complexes binds O_2 in a manner similar to that found for oxyHr and none of the O_2 adducts has been crystallized.

C. Hemocyanin

In contrast to Hb and Hr, hemocyanin (Hc), the oxygen transport protein of arthropods and molluscs, contains a dinuclear copper site with each copper atom bound by three histidines (Volbeda and Hol, 1989). A preliminary account of the crystal structure of oxyHc indicates that the dioxygen moiety bridges the two copper atoms in a μ-η^2:η^2 fashion (Magnus and Ton-That, 1992).

Using a number of dinucleating ligands, Karlin and co-workers prepared several dicopper(I) complexes capable of reversible O_2 binding at low temperatures (Tyeklár and Karlin, 1989). They succeeded in crystallizing the copper-dioxygen complex, [{Cu(TPA)}$_2$(O_2)] (**6**) (TPA = tris(2-pyridylmethyl)amine, Fig. 1). The two copper(II) atoms are bridged by a μ-1,2-peroxo group, with Cu-O-O angles of 108° (Fig. 3). The complex is unstable above -80 °C (Jacobson et al., 1988).

By using the same approach that proved successful in synthesizing a reversible non-heme iron-O_2 complex, Kitajima et al. (1992) were able to prepare two μ-η^2:η^2 peroxo-dicopper(II) complexes, [{(HB(3,5-R$_2$pz)$_3$)Cu}$_2$(O_2)] (R = ^{i}Pr, **7**; R = Ph, **8**). The metrical details of **7** are shown in Figure 4. The Cu-Cu separation in **7** (3.56 Å) is similar to that in oxyHc (3.4 ± 0.2 Å). Both **7** and **8** display the diamagnetism and low $\nu_{O\text{-}O}$ (759 and 741 cm^{-1}, respectively) characteristic of oxyHc ($\nu_{O\text{-}O}$ ~ 750 cm^{-1}), indicating that the peroxo group in oxyHc is bound in the same manner, which was recently confirmed by the crystal

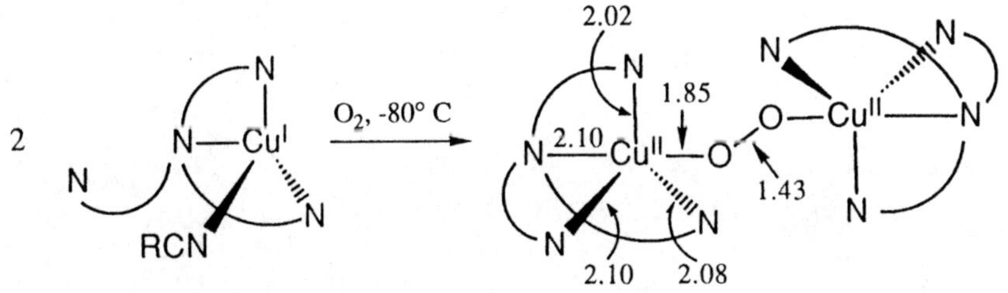

Figure 3.
The reaction of dioxygen with $[(TPA)Cu^I(RCN)]^+$ to form **6.**

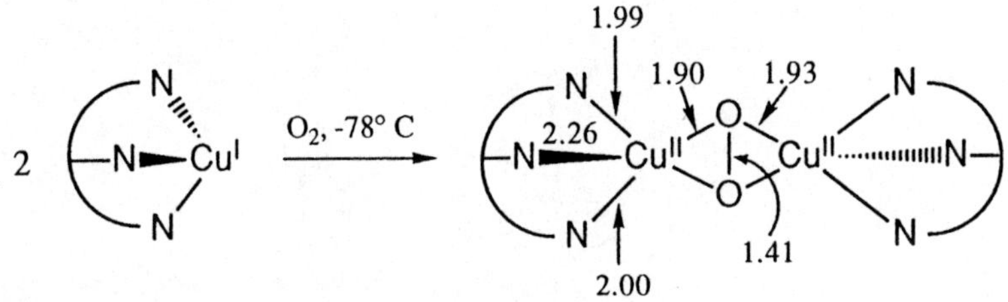

Figure 4.
The reaction of dioxygen with $[(HB(3,5\text{-}^iPr_2pz)_3)Cu]$ to form **7.**

structure (Magnus and Ton-That, 1992). The bound O_2 can be released by treatment with CO or PPh_3, forming $[(HB(3,5\text{-}^iPr_2pz)_3)CuL]$ (L = CO, PPh_3) (Kitajima et al., 1992).

III. DIOXYGEN ACTIVATION

A. Heme peroxidases and cytochrome P450

Cytochrome P450 catalyzes the O_2-dependent oxygenation of hydrocarbons in a mechanism that is postulated to involve a high valent iron-oxo species derived from the breakdown of a ferric peroxide intermediate (Ortiz de Montellano, 1986). While the high valent iron-oxo species has not been directly observed in the cytochrome P450 cycle, an analogous species referred to as Compound I has been observed in the reactions of heme peroxidases with peroxides.

A significant development in the biomimetic chemistry of cytochrome P450 and the heme peroxidases was the generation of a high valent iron-oxo species by the reaction of Fe(III)porphyrin complexes with peroxy acids in CH_2Cl_2 at -80 °C (Groves et al., 1981; Groves and Watanabe, 1988) (Fig. 5). The intermediate species formulated as $[(P^{\cdot})Fe(IV){=}O]^+$ has spectroscopic properties very similar to that of horseradish peroxidase Compound I. These properties include a UV-vis and an NMR spectrum characteristic of a porphyrin cation radical (Groves et al., 1981), a Mössbauer isomer shift (~ 0.1 mm/s) indicative of Fe(IV) (Boso et al., 1983), and an Fe-O bond distance of 1.6 Å from EXAFS analysis (Penner-Hahn et al., 1986). Raman data could not be obtained for $[(P)FeO]^+$ due to its photoinstability in the laser beam, but the one-electron reduced derivative [(P)FeO] exhibits a Raman feature assigned to $\nu_{Fe=O}$ in the 800-850 cm^{-1} region (Su et al., 1988).

$[(P)FeO]^+$ is capable of epoxidizing olefins (Groves et al., 1981). Furthermore, iron(III)porphyrin complexes have been found to be good catalysts for alkane hydroxylation and olefin epoxidation with iodosylbenzene as the oxidant. Subsequently, a number of catalytic systems were developed using different porphyrin complexes and a range of possible oxidants. These studies were recently summarized in a thorough review by Meunier (1992).

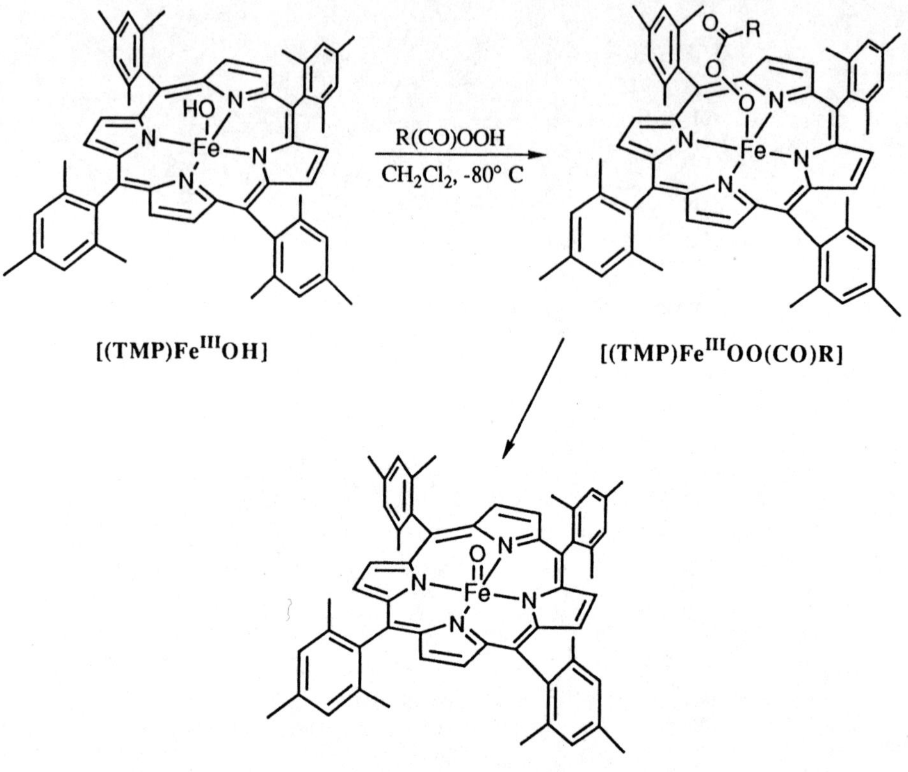

$[(TMP)Fe^{III}OH]$ $[(TMP)Fe^{III}OO(CO)R]$

$[(TMP^{\cdot})Fe^{IV}{=}O]^+$

Figure 5.
The reaction of $[(TMP)Fe^{III}OH]$ with peroxyacids to form $[(TMP^{\bullet})Fe^{IV}{=}O]^+$.

B. Tyrosinase

As noted in an earlier section, a number of dicopper-dioxygen adducts have been obtained by suitable ligand design. Karlin has compared several of these complexes with respect to their reactivity towards a number of reagents; in general, the μ-1,2-peroxo moiety is described as nucleophilic, while the μ-η^2:η^2-peroxo moiety is described as electrophilic (Paul et al., 1991). The best reactivity model for copper oxygenases developed thus far is in fact the first one reported by Karlin, and this model system has evolved into an intricate combination of supporting observations (Karlin et al., 1993). These are illustrated in Figure 6.

$[Cu(I)_2(XYL\text{-}H)]^{2+}$ reacts with O_2 to form $[Cu(II)_2(XYL\text{-}O)(OH)]^{2+}$ in nearly quantitative yield. The use of $^{18}O_2$ results in the incorporation of label into both the bridging phenolate and hydroxide. Low temperature rapid kinetic experiments indicate the reversible formation of a transient O_2 species which is proposed to be a μ-η^2:η^2-peroxo complex on the basis of UV-vis spectral similarities with other μ-η^2:η^2-peroxo complexes. Decomposition of this species affords the hydroxylated product. It is postulated that the bound dioxygen is in close proximity to the aromatic ring and that O-O bond cleavage is synchronous with attack on the arene ring. When the xylyl C2-H is substituted by fluorine, the reaction with O_2 to form a μ-η^2:η^2-peroxo complex still occurs, but hydroxylation of the xylyl ring is blocked and the starting XYL-F ligand can be recovered upon decomposition of the oxygenated intermediate. When the carbon is methylated, C2 hydroxylation occurs, the methyl group migrates to the adjacent carbon on the ring, and the dialkylaminomethyl group is expelled in a retro-Mannich reaction. This latter reaction is an

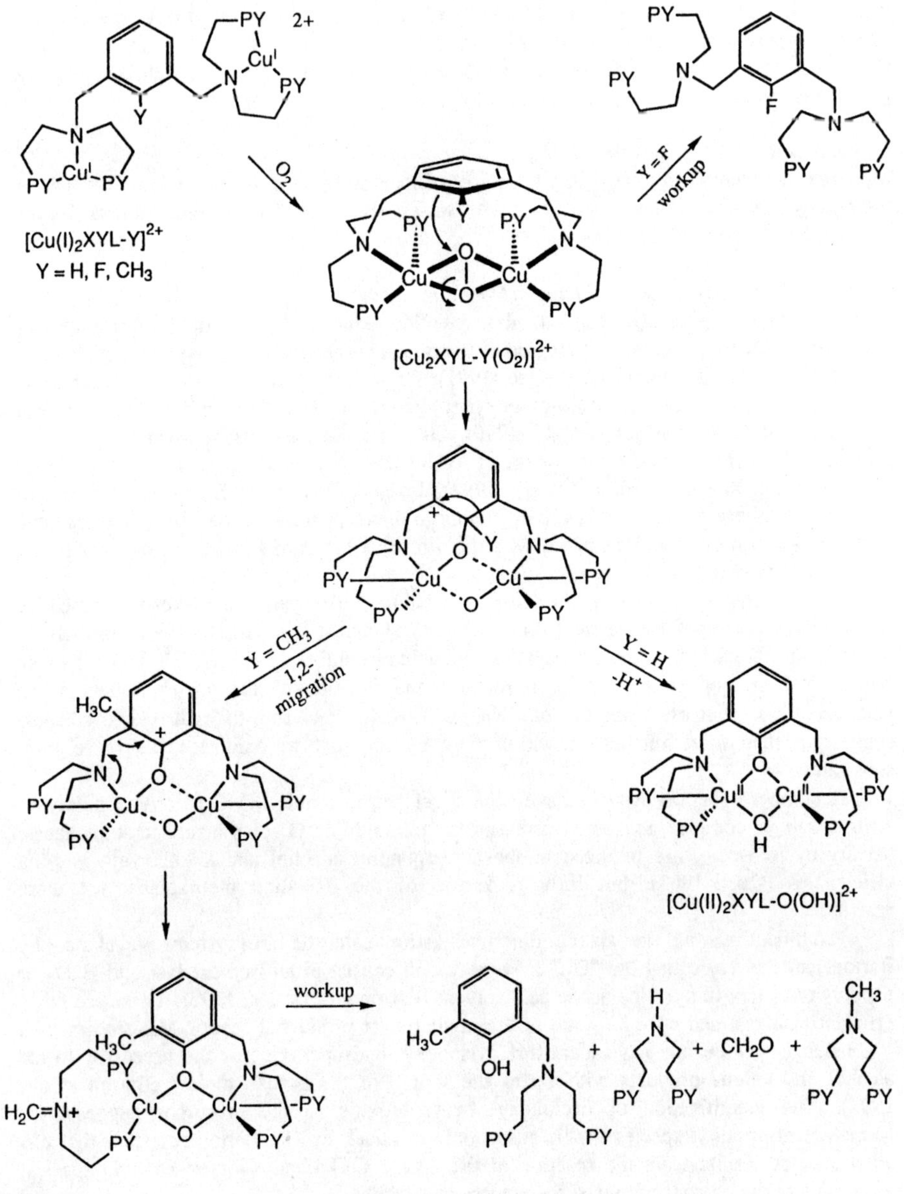

Figure 6.
Reactions of $[Cu_2(XYL\text{-}Y)]^{2+}$ with O_2. Adapted from Karlin et al. (1993).

example of an "NIH shift" characteristic of an electrophilic metal-oxy species and previously observed for iron-containing hydroxylases such as cytochrome P450 and phenylalanine hydroxylase. This model study demonstrates that copper as well is capable of generating similarly electrophilic species.

C. Nonheme-iron oxygenases

Modeling nonheme-iron oxygenases is an area of increasing effort in recent years because of the emergence of a number of enzymes with such active sites. This section summarizes the important results in this field.

1. *Methane monooxygenase and ribonucleotide reductase*

Methane monooxygenase (MMO) and ribonucleotide reductase (RNR) contain diiron centers that activate O_2 to generate oxidizing equivalents that respectively hydroxylate methane and convert Tyr122 to its radical form (Que, 1991). Crystallographic studies of the latter in its oxidized state show a (μ-oxo)(μ-carboxylato)diferric unit with terminal histidine and carboxylate ligands (Nordlund et al., 1990). Sequence comparisons between the RNR R2 protein and the hydroxylase component of MMO have led to the suggestion of a similar diiron site for MMO (Nordlund et al., 1992). Insights into the mechanisms of dioxygen activation by these diiron centers are beginning to develop; transient oxidizing species have been observed in the reactions of O_2 with the diferrous forms of RNR R2 (Bollinger et al., 1991) and MMO (Lee et al., 1993).

Few reactivity studies of the dioxygen binding diferrous complexes discussed in Section IIB have thus far been carried out. The O_2 adduct of $[Fe_2(N\text{-}Et\text{-}HPTB)(OBz)](BF_4)_2$ (**4**) is quite unreactive, while its HPTP analogue can oxidize 2,4-di-tert-butylphenol to its radical form (Dong et al., 1993), mimicking the function of the diiron center in the ribonucleotide reductase from *E. coli.* Neither O_2 adduct is able to hydroxylate alkanes, suggesting that some further activation may be required to oxidize these more inert substrates.

A number of model systems have been developed to mimic MMO reactivity involving an iron salt or complex and an oxidant such as peroxide or O_2 plus a reductant. Alkanes, usually cyclohexane, are oxidized to the corresponding alcohol and ketone with varying efficiencies (Que, 1993), but little is known of the oxidation mechanisms for these reactions.

Prominent among the alkane functionalization catalysts are systems developed by Barton, collectively called the "Gif" systems which consist of an iron catalyst and H_2O_2 or O_2 plus reductant in pyridine/acetic acid solvent (Barton and Doller, 1992). These are fairly efficient catalysts and afford ketone as the predominant product. Recent NMR studies with ^{13}C-labeled cyclohexane have identified cyclohexyl hydroperoxide as the precursor to the alcohol and ketone products with O_2 as the source of the peroxo moiety (Barton et al.. 1992). The identification of cyclohexyl hydroperoxide as the precursor suggests the formation of an alkyl species that is susceptible to attack by O_2. Barton proposes that this alkyl species results from the reaction of the alkane C-H bond with an $Fe^V{=}O$ species, analogous to the oxidant proposed for cytochrome P450.

Mechanistic information is also available for oxidation systems consisting of an iron complex and alkyl hydroperoxide. With both mononuclear and dinuclear Fe^{III}(TPA) complexes as catalysts and t-BuOOH as oxidant, cyclohexane is efficiently functionalized to the corresponding alcohol and ketone (Leising et al., 1993; Que, 1993). These products derive from a two-electron oxidation process because Me_2S inhibits their formation. A high valent iron-oxo species derived from the heterolysis of the alkyl hydroperoxide O-O bond is favored.

The proposed participation of high valent nonheme iron species in these alkane oxidation reactions raises the interesting question of whether a high valent state can be attained in a nonheme environment. The recent work of Leising et al. (1991) provides an example of such a high valent species. They have observed the formation of a transient green species (λ_{max} = 614 nm) in CH_3CN at -40 °C upon treatment of $Fe_2O(TPA)_2(H_2O)_2(ClO_4)_4$ with an equivalent of base and H_2O_2. Based on the initial spectroscopic study, this species was minimally formulated to be $[Fe(TPA)O]^{3+}$, making it analogous in oxidation state to heme peroxidase Compound I. Subsequently, the introduction of a methyl group on the 5-position of each pyridine has allowed the isolation of the intermediate at low temperature (Dong et al., 1994). It is characterized as $[Fe_2O_2(L)_2]^{3+}$, a species formally in the Fe(III)Fe(IV) oxidation state.

2. *Catechol dioxygenases*

The intradiol-cleaving catechol dioxygenases are part of nature's mechanisms for degrading aromatic molecules (Que, 1989; Lipscomb and Orville, 1992). They catalyze the intradiol cleavage of catechols into muconic acids and incorporate the elements of dioxygen into the carboxylate functions of the product. Best characterized of this class are catechol 1,2-dioxygenase and protocatechuate 3,4-dioxygenase. The crystal structure of the latter shows an iron center in a trigonal bipyramidal environment coordinated to two tyrosines, two histidines, and a solvent molecule (Ohlendorf et al., 1988). As a result of a number of spectroscopic and mechanistic studies, a substrate activation mechanism has been proposed to account for the persistence of the Fe(III) oxidation state in the reaction cycle (Fig. 7).

Key points of the substrate activation mechanism have been derived from biomimetic studies (Cox and Que, 1988; Jang et al., 1991). A series of [Fe(L)DBC] complexes (L = tetradentate tripodal ligand based on trimethylamine, DBC = 3,5-di-tert-butylcatecholate) was synthesized and structurally characterized as having six-coordinate environments with chelated catecholates. All of these complexes react with O_2 to yield the desired intradiol cleavage product; for example, $[Fe(TPA)DBC]^+$ is converted within minutes of exposure to O_2 to 3,5-di-tert-butylmuconic anhydride in quantitative yield.

The cleavage yields and their rates of reaction were found to correlate with the Lewis acidity of the ferric center, as indicated by the energies of the catecholate ligand-to-metal charge transfer bands in their visible spectra. As the iron becomes more Lewis acidic, the yield of intradiol cleavage product increases and approaches quantitative conversion. Furthermore, the rate of reaction accelerates by three orders of magnitude from L = nitrilotriacetate to L = TPA. The increased Lewis acidity of the iron center is proposed to enhance the covalency of the iron-catecholate bond and increase the semiquinone character of the bound DBC, the latter being indicated by the progressive upfield shift of the DBC ring protons in the ^1H NMR spectra of the complexes. This semiquinone character on the

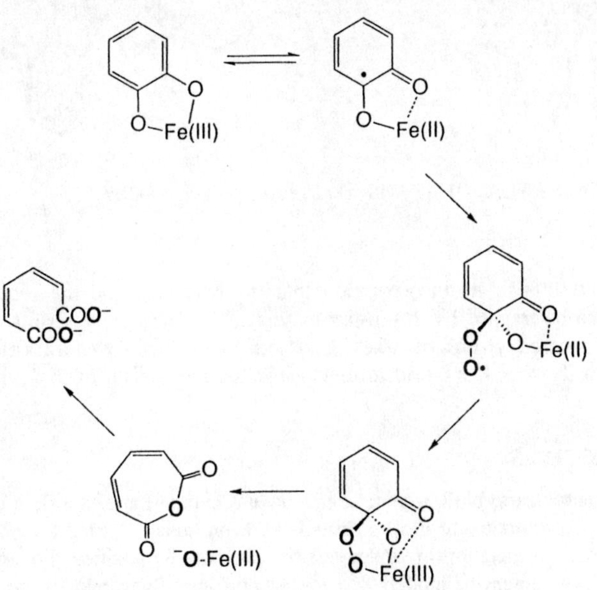

Figure 7.
Proposed reaction cycle for intradiol cleaving catechol dioxygenases.

substrate renders it susceptible to O_2 attack. The viability of the proposed peroxy intermediate was recently strengthened by the crystal structure of the O_2 adduct of [Ir(III)(triphos)(9,10-phenanthrenediolate)]$^+$ (Barbaro et al., 1992). The crystal structure of the Ir adduct reveals that the bidentate catecholate is transformed into a tridentate peroxy ligand as indicated in the proposed enzyme mechanism. The subsequent decomposition of the peroxy species to anhydride is supported by the isolation of a muconic anhydride from the reaction of [Fe(TPA)DBC)]$^+$ with O_2 (Jang et al., 1991).

3. *α-Keto acid-dependent enzymes*
The α-keto acid-dependent enzymes (Que, 1993) catalyze a diverse array of reactions that couple the functionalization of an unactivated C-H bond with the oxidative decarboxylation of a keto acid. For example, in the hydroxylation of prolyl, lysyl and aspartyl residues, one atom of dioxygen is incorporated into the product, while the other ends up on the carboxylate derived from the keto acid, i.e.

$$\text{R-H} + \text{R'COOOH} + \boldsymbol{O_2} \rightarrow \text{R-}\boldsymbol{O}\text{H} + \text{R'CO}\boldsymbol{O}\text{H} + CO_2$$

In all of these reactions, Fe(II) appears to be required, but little is known of its coordination environment or its role in catalysis.

Biomimetic studies have recently been initiated with benzoylformate (BF) as the model for an α-keto acid (Chiou and Que, 1992). The crystal structure of [Fe(TLA)BF]$^+$ exhibits a keto acid chelated to an iron(II) center via one carboxylate oxygen and the carbonyl oxygen atom of its benzoyl moiety (Fig. 8). Exposure of this complex to O_2 results in uptake of O_2 and the formation of [Fe(TLA)(benzoate)]$^+$ (Zang et al., 1993) as product, i.e.

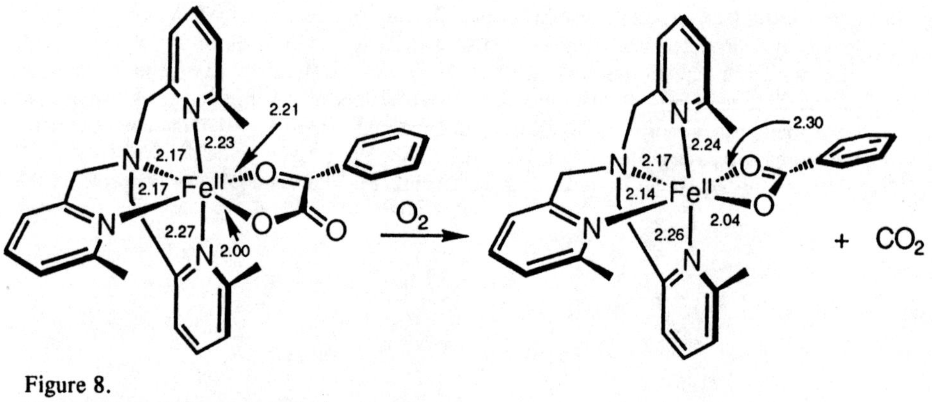

Figure 8.
The reaction of [Fe(TLA)(BF)]$^+$ with O_2 to form [Fe(TLA)(OBz)]$^+$.

Use of $^{18}O_2$ results in its incorporation into the benzoate, and the remaining oxidizing equivalents can be trapped by 2,4-di-tert-butylphenol. These early results demonstrate that the oxidative decarboxylation of α-keto acids can be effected by coordination to an iron(II) center, and much work lies ahead to understand the mechanism of this interesting class of enzymes.

IV. PERSPECTIVES

In the past twenty years, bioinorganic chemistry has come of age as a field of study. During this period, our synthetic and design skills have been honed to such a degree that accurate structural models of metalloprotein active sites can now be attained. Focus is now shifting towards the development of models that mimic both metalloprotein structure and function. The functional models discussed in this chapter represent many of the initial successes in this effort.

ACKNOWLEDGEMENTS

This work was supported by grants from the National Institutes of Health (GM-33162 and GM-38767). C.R.R. is grateful for a predoctoral traineeship from the National Institutes of Health (GM-07323).

REFERENCES

Almog, J., Baldwin, J.E., Dyer, R.L., Huff, J., and Wilkerson, C.J. (1974). *J. Amer. Chem. Soc.*, *96*: 5600.

Anderson, D.L., Weschler, C.J., and Basolo, F. (1974). *J. Amer. Chem. Soc.*, *96*: 5599.

Barbaro, P., Bianchini, C., Linn, K., Mealli, C., Meli, A., Vizza, F., Laschi, F., and Zanello, P. (1992). *Inorg. Chim. Acta*, *198*: 31.

Barton, D.H.R., and Doller, D. (1992). *Accounts Chem. Res.*, *25*: 504.

Barton, D.H.R., Bévière, S., Chavasiri, W., Csuhai, E., Doller, D., and Liu, W.-G. (1992). *J. Amer. Chem. Soc.*, *114*: 2147.

Bollinger, J.M., Jr., Edmondson, D.E., Huynh, B.H., Filley, J., Norton, J.R., and Stubbe, J. (1991). *Science*, *253*: 292.

Boso, B., Lang, G., McMurry, T.J., and Groves, J.T. (1983). *J. Chem. Phys.*, *79*: 1122.

Brinigar, W.S., and Chang, C.K. (1974). *J. Amer. Chem. Soc.*, *96*: 5595.

Brinigar, W.S., Chang, C.K., Geibel, J., and Traylor, T.G. (1974). *J. Amer. Chem. Soc.*, *96*: 5597.

Chiou, Y.-M., and Que, L., Jr. (1992). *J. Amer. Chem. Soc.*, *114*: 7567.

Cox, D.D., and Que, L., Jr. (1988). *J. Amer. Chem. Soc.*, *110*: 8085.

Dong, Y., Ménage, S., Brennan, B.A., Elgren, T.E., Jang, H.G., Pearce, L.L., and Que, L., Jr. (1993). *J. Amer. Chem. Soc.*, *115*: 1851.

Dong, Y., Fujii, H., Pan, G., Randall, C.R., Wilkinson, E.C., Que, L., Jr., Fox, B.G., and Münck, E. (1994). *J. Amer. Chem. Soc.*, submitted.

Groves, J.T., and Watanabe, Y. (1988). *J. Amer. Chem. Soc.*, *110*: 8443.

Groves, J.T., Haushalter, R.C., Nakamura, M., Nemo, T.E., and Evans, B.J. (1981). *J. Amer. Chem. Soc.*, *103*: 2884.

Hashimoto, T., Dyer, R.L., Crossley, M.J., Baldwin, J.E., and Basolo, F. (1982). *J. Amer. Chem. Soc.*, *104*: 2101.

Hayashi, Y., Suzuki, M., Uehara, A., Mizutani, Y., and Kitagawa, T. (1992). *Chem. Lett.*, 91.

Holmes, M.A., Le Trong, I., Turley, S., Sieker, L.C., and Stenkamp, R.E. (1991). *J. Mol. Biol.*, *218*: 583.

Jacobson, R.R., Tyeklár, Z., Farooq, A., Karlin, K.D., Liu, S., and Zubieta, J. (1988). *J. Amer. Chem. Soc.*, *110*: 3690.

Jameson, G.B., Rodley, G.A., Robinson, W.T., Gagne, R.R., Reed, C.A., and Collman, J.P. (1978). *Inorg. Chem.*, *17*: 850.

Jameson, G.B., Molinaro, F.S., Ibers, J.A., Collman, J.P., Brauman, J.I., Rose, E., and Suslick, K.S. (1980). *J. Amer. Chem. Soc.*, *102*: 3224.

Jang, H.G., Cox, D.D., and Que, L., Jr. (1991). *J. Amer. Chem. Soc.*, *113*: 9200.

Karlin, K.D., Tyeklár, Z., and Zuberbühler, A.D. (1993). *Bioinorganic Catalysis* (J. Reedijk, ed.), Marcel Dekker, New York, p. 261.

Kitajima, N., Fukui, H., Moro-oka, Y., Mizutani, Y., and Kitagawa, T. (1990). *J. Amer. Chem. Soc.*, *112*: 6402.

Kitajima, N., Fujisawa, K., Fujimoto, C., Moro-oka, Y., Hashimoto, S., Kitagawa, T., Toriumi, K., Tatsumi, K., and Nakamura, A. (1992). *J. Amer. Chem. Soc.*, *114*: 1277.

Lee, S.-K., Nesheim, J.C., and Lipscomb, J.D. (1993). *J. Biol. Chem.*, *268*: 21569.

Leising, R.A., Brennan, B.A., Que, L., Jr., Fox, B.G., and Munck, E. (1991). *J. Amer. Chem. Soc.*, *113*: 3988.

Leising, R.A., Kim, J., Pérez, M.A., and Que, L., Jr. (1993). *J. Amer. Chem. Soc.*, *115*: 9524.

Lipscomb, J.D., and Orville, A.M. (1992). *Metals Ions in Biological Systems*, Vol. 28 (H. Sigel and A. Sigel, eds.), Marcel Dekker, New York, p. 242.

Magnus, K., and Ton-That, H. (1992). *J. Inorg. Biochem.*, *47*: 20.

Ménage, S., Brennan, B.A., Juarez-Garcia, C., Münck, E., and Que, L., Jr. (1990). *J. Amer. Chem. Soc.*, *112*: 6423.

Meunier, B. (1992). *Chem. Rev.*, *92*, 1411.

Nordlund, P., Sjöberg, B.-M., and Eklund, H. (1990). *Nature (London)*, *345*: 593.

Nordlund, P., Dalton, H., and Eklund, H. (1992). *FEBS Lett.*, *307*: 257.

Ohlendorf, D.H., Lipscomb, J.D., and Weber, C. (1988). *Nature*, *336*: 403.

Ortiz de Montellano, P.R., ed. (1986). *Cytochrome P-450: Structure, Mechanism, and Biochemistry*, Plenum, New York.

Paul, P.P., Tyeklár, Z., Jacobson, R.R., and Karlin, K.D. (1991). *J. Amer. Chem. Soc.*, *113*: 5322.

Penner-Hahn, J.E., Smith Eble, K., McMurry, T.J., Renner, M., Balch, A.L., Groves, J.T., Dawson, J.H., and Hodgson, K.O. (1986). *J. Amer. Chem. Soc.*, *108*: 7819.

Perutz, M.F. (1976). *Brit. Med. Bull.*, *32*: 195.

Perutz, M.F., Fermi, G., Luisi, B., Shaanan, B., and Liddington, R.C. (1987). *Accounts Chem. Res.*, *20*: 309.

Que, L., Jr. (1989). *Iron Carriers and Iron Proteins* (T.M. Loehr, ed.), VCH, New York, Chapter 6.

Que, L., Jr. (1991). *Science*, *253*: 273.

Que, L., Jr. (1993). *Bioinorganic Catalysis* (J. Reedijk, ed.), Marcel Dekker, New York, p. 347.

Que, L., Jr., and True, A.E. (1990). *Prog. Inorg. Chem.*, *38*: 97.

Su, Y.O., Czernuszewicz, R.S., Miller, L.A., and Spiro, T.G. (1988). *J. Amer. Chem. Soc.*, *110*: 4150.

Suslick, K.S., and Reinert, T.J. (1985). *J. Chem. Ed.*, *62*: 974.

Tyeklár, Z., and Karlin, K.D. (1989). *Accounts Chem. Res.*, *22*: 241.

Volbeda, A., and Hol, W.G.J. (1989). *J. Mol. Biol.*, *209*: 249.

Wagner, G.C., and Kassner, R.J. (1974). *J. Amer. Chem. Soc.*, *96*: 5593.

Wang, J.H. (1958). *J. Amer. Chem. Soc.*, *80*: 3168.

Zang, Y., Elgren, T.E., Dong, Y., and Que, L., Jr. (1993). *J. Amer. Chem. Soc.*, *115*: 811.

METAL-ION PROTEIN LABILE COMPLEX EQUILIBRIA IN SOLUTION

by

W. Goumakos and B. Sarkar

Department of Biochemistry Research,
The Hospital for Sick Children and
Department of Biochemistry,
University of Toronto,
Toronto, Ontario
Canada

I. INTRODUCTION

Throughout the biological system we find metals closely associated with proteins and enzymes. Metals are therefore of great interest, not only from a biochemical standpoint but also from a pathological perspective. An understanding of the physiological and indeed pathological interactions of metals with proteins requires an understanding of the complex-forming properties of metals with proteins, especially with respect to their stability and stoichiometry. While all proteins in general can bind metals, the number and the strength of binding vary with the metal and the nature of the protein. In this chapter, we will concern ourselves with a general description of the formation of labile complexes between metal ions and proteins and their investigation primarily by equilibrium dialysis, with special reference to specific examples.

II. GENERAL PROPERTIES OF METALS AND THEIR PROTEIN LIGANDS

While all proteins can bind metals, the exact nature of the binding in terms of number of metal ions bound and nature of the binding interaction depends on the nature of the metal and the ligand. Following is a brief and general description of some of the properties of metals and ligands important to metal-protein complex formation.

A. Metals

One of the earliest classifications of metals was offered by Ahrland et al. (1958) in which metals were classified according to the principle of hard and soft acids and bases, (HSAB), which essentially assumes that in a bond between two atoms, one atom will act as the acid and the other as the base, with electrons holding the bond together. The acid is considered to accept electrons and the base has the tendency to give up electrons. A typical metal ligand reaction based on this principle can be simply described by:

$$P + M \rightleftharpoons PM \quad (1)$$

where metal ions are essentially Lewis acids and protein ligands Lewis bases. The basic underlying principle of HSAB is that strong bonds are formed between strong acids and bases and soft acids and bases, and that generally hard-soft bonds are weak. The terms hard and soft arise from the degree of electron polarizability, that is if the electrons are firmly held to a species then the species is termed hard and vice versa. A hard metal for example is typically of smaller size with a large positive charge and forms relatively strong complexes via predominantly electrostatic interactions, while a soft metal tends towards covalent interactions (see Part One, Chapter 3). Independent of ligand, the order of softness of metal ions is described by the Irving-Williams series (Sigel and McCormick, 1970), which states the binding order for divalent metal ions as follows:

$$Ca < Mg < Mn < Fe < Co < Ni < Cu > Zn$$

which ranges from very hard (Ca) to intermediate (Cu). Furthermore, the complexing ability of a metal ion varies with the charge-to-radius ratio of the metal ion. In that respect, the metal may be ordered as follows:

$$Mn < Fe < Co < Zn < Ni < Cu$$

Metal ions of the (a) or hard type, such as Ca, Mg and Mn, prefer ligands in the following order $R\text{-}COO^- > RNH_2 > RS^- > H_2O$ and have a distinct preference for oxygen to nitrogen. The intermediate metals such as Cu, Ni and Co can be found associated with carboxylate and amino groups (Sillén and Martell, 1964; 1971), while the (b) or soft type, for example Cd, Pb and Hg, prefer $RS^- > RNH_2 > OH^- >> H_2O$ and form very stable complexes with thiols and amines. Due to their preference for oxygen and nitrogen, it is the intermediate metals that are best suited to interact with proteins which provide an abundance of such ligand atoms, most importantly through the oxygen of the carboxylate and imidazole nitrogen of a histidine residue.

B. Protein ligands

Metal-protein interactions tend to follow the HSAB principle. A description of metal-protein labile complexes is essentially a description of metal-amino acid complexes. Metal binding to protein is affected by the nature of the protein side-chain residues, net charge and conformation of the protein. The amino acids in the protein provide the functional groups required for metal binding such as an α-NH_2, which is the most common metal-binding site. This is essentially due to the strong basic character and strong ligand field effect of N-(amino) atoms as compared to other donor atoms. The other very important group is the histidine side chain, the N(imidazole) found to bind metals in many metalloproteins. The side-chain carboxyl group has also been demonstrated in some metal-protein complexes, as have been O(peptide), N(peptide), O(hydroxyl), S(sulphydryl), S(thioether), and bridging S(disulphide) although they are much less common (Sarkar, 1987).

The equilibria of labile metal-ion protein complexes are obtained relatively quickly as opposed to inert complexes where the rate of dissociation is slow. Complexes in general are either kinetically inert or labile. However, it is important to note that the thermodynamic stability of a metal-protein complex as defined by the large formation or stability constant is not necessarily related to inertness so that while a complex may be considered labile it may still have a large stability constant and therefore be considered a strong complex.

III. METHODS OF INVESTIGATION AND DATA ANALYSIS

A. Equilibrium dialysis

Equilibrium dialysis is one of the earliest and most extensively used techniques in the study of metal-protein labile complex equilibria. Equilibrium dialysis is used to determine the binding strength of a metal-protein interaction as well as the number of metal ions bound to the protein. If the experimental design is carefully considered, equilibrium dialysis can yield accurate and useful results despite the somewhat long equilibration periods which may potentially lead to denaturation problems. One must take into account the nonspecific binding of metal to protein and indeed to the dialysis membrane. The effect of metal binding to buffer is also important, since determination of the "free" metal concentration is of paramount importance. So the buffer must be chosen with care. The Gibbs-Donnan effect also has to be considered and preferably eliminated (Baker, 1988).

While there are many variations in the experimental execution of equilibrium dialysis experiments, what follows is typical methodology. Experiments are usually performed in specially made double chamber blocks, typically made of some relatively inert material such as lucite, which are separated by a semi-permeable dialysis membrane which will allow the smaller ions to cross but not the larger proteins. Typically in this procedure, a solution of protein P is placed on one side of this block with an equal volume of a solution containing a metal salt M placed on the opposite chamber; both solutions use a common buffer to maintain a constant pH selected on the basis of its negligible affinity for the metal under study. A neutral sodium or potassium salt is added to both sides to give a final concentration of 0.1 or greater so as to reduce the potential complications due to Gibbs-Donnan effects. At equilibrium, an analysis is carried out to determine the total metal concentration on both sides, usually via determination of "free" metal on the non-protein side by some methods such as atomic absorption spectroscopy or the use of a radiolabelled isotope of the metal. Determination of "free" metal on the non-protein side is convenient from an analytical perspective, since the potential interference of the protein in the determination of the metal concentration is essentially eliminated. If we denote the metal in the protein and non-protein compartment as M_p and M_{np}, respectively, we then have the following relation

$$M_{np} = M_p - M_{bound} = M_p - r[P] \tag{2}$$

where P is the molar concentration of protein and r is the average number of metal ions bound per protein molecule under the experimental conditions. From these studies, the association constant can be calculated by a variety of graphical methods which will be described in the following section. A recent and novel adaptation of the equilibrium dialysis experimental design accommodates the use of several metals and their radioisotopes simultaneously, so as to increase the overall sensitivity of the method (Goumakos et al., 1991).

B. Scatchard analysis

The most common treatment of equilibrium dialysis data has been through graphical treatment by the method first developed by Scatchard (1949). While strict mathematical interpretation of experimentally generated data requires the analysis to be tailored to the specific condition at hand so as to take into account the specific features of the complexation process, we will deal here in general terms in order to present metal-protein complex equilibria in a more simple manner. Typically we need to set up the equation that describes the dissociation constant of a complexation reaction. Through simple algebraic manipulations, we can generate an equation so as to yield a straight line plot from which we can extract the number of metal ions bound and the dissociation constant by measuring the x-intercept and slope, respectively.

1. *Binding of one metal ion to a protein*

If we consider one metal ion M binding to one protein ligand P, the interaction can be expressed by the equilibrium

$$P + M^{n+} \rightleftharpoons PM^{n+} \tag{3}$$

and

$$K = \frac{[PM^{n+}]}{[P][M^{n+}]} \tag{4}$$

where K is the formation constant of the complex, and the accepted practice is to refer to K_d or dissociation constant which is defined as

$$K_d = \frac{[P][M^{n+}]}{[PM^{n+}]} \tag{5}$$

From equation 5, we note that the fraction of protein ligand bound is determined by the concentration of $[M^{n+}]$ and [P]. Of prime importance in Scatchard analysis is the parameter r which is defined as the number of moles (molecules) of metal bound per mole (molecule) of protein as follows:

$$r = \frac{[M]_{bound}}{[P]_{total}} = \frac{[PM]}{[P] + [PM]} \tag{6}$$

if we now combine equations (5) and (6) we have

$$r = \frac{[M]}{K_d + [M]} \tag{7}$$

rearranging this equation to give

$$\frac{r}{[M]} = \frac{1}{K_d} - \frac{r}{K_d} \tag{8}$$

from which a plot of r/[M] vs r will yield a straight line having a slope of $-1/K_d$ as shown in Figure 1.

2. *Multiple metal ions bound to a protein*

It is more common in biological systems for a protein to have more than one binding site for a metal and indeed for several different types of metals. In those situations, in which the metal complexation can be defined as labile, Scatchard analysis has been proven to be very useful for identical but independent sites and for non-identical sites as well.

a. Identical and independent labile metal-protein complexes

In a fashion similar to equation (8) it can be shown that the required equation is

$$\frac{r}{[M]} = \frac{n}{K_d} - \frac{r}{K_d} \tag{9}$$

where in this case K_d represents an average dissociation constant. If we now plot r/[M] vs r, we have a straight line with slope $-1/K_d$ and an x-intercept of n/K_d, where n=number of metal ions bound to one protein ligand. Observe that in the case of equation (8) n is simply equal to 1. This is the plot that is traditionally referred to as the Scatchard plot as seen in Figure 1. An important feature of the Scatchard plot allows identification of situations

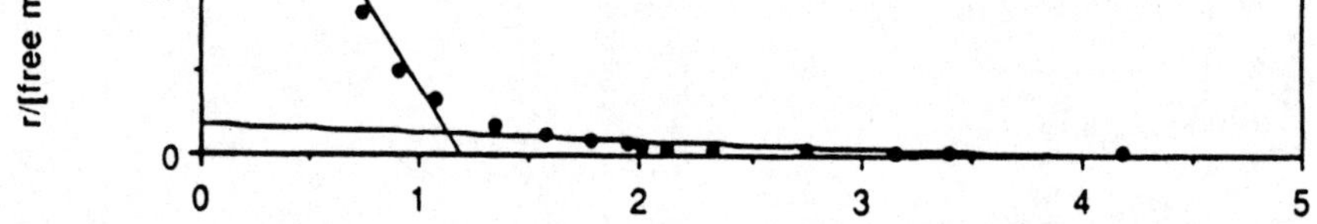

Figure 1.

Scatchard plot of binding of Zn to human serum albumin. All points represent data from one typical experiment. All experiments were carried out in 20mM 2,4,6-trimethylpyridine (collidine) buffer, 0.15 NaCl, pH 7.4, 4°C. Free metal concentration is that concentration of metal on the nonprotein side of the equilibrium dialysis chamber at equilibrium. All metal-to-protein ratios fall between 0.01:1 and 15:1; r, moles of metal bound per mole of protein. (Reproduced with permission from Goumakos et al., 1991).

where the metal binding is neither identical nor independent, in which case the plot will not be a straight line but rather it will be curved (Klotz and Hunston, 1984; Faguet, 1986). This curvature is actually the description of several different types of sites or classes of binding sites and can be mathematically resolved to yield information on each class of sites. Note that if the metal-binding sites are identical but dependent, that is cooperative in binding, one still observes a curved plot.

b. Non-identical and independent labile metal-protein complexes

Several different metal complex sites on one protein ligand, each with its own binding constant, is not unusual or uncommon. If we consider a protein P with a number of distinct types of binding sites, upon addition of metal each class of sites described by r will be filled according to the binding strength of the metal for that class of sites as defined by K_d for that class of binding. In practice, the experimental value of r represents an average for all the classes of metal-binding sites such that

$$r = \frac{n_1[M]}{K_{d1}+[M]} + \frac{n_2[M]}{K_{d2}+[M]} + \cdots\cdots \tag{10}$$

Scatchard analysis remains widely used and yet controversial. If one is careful in the interpretation of the Scatchard plots, much useful information can be obtained through this type of analysis. The major criticism has been that it can easily lead to conclusions largely based on assumptions rather than reality. If Scatchard interpretation can be verified by another technique as has recently been done by Goumakos et al. (1991), this type of analysis can provide reliable results.

C. Multiple and dependent metal sites and the Hill plot

Cooperativity is a term used to describe what is essentially a whole range of interactions with a single common feature, developed by Hill early in the century (Hill, 1910). It can be described as the effect of the binding of one metal ion on the binding of the succeeding metal ion(s). The mechanism by which it is accomplished may vary the overall effect by either increasing (positive cooperativity) or decreasing (negative cooperativity) the relative affinity of succeeding metal ions to the protein.

Let us consider the simplest case of one metal binding with high cooperativity to one protein molecule. In this case, the other metal sites become more readily available to the next metal and so on. With n metal binding sites(M) per protein ligand (P), we have the following

$$P + nM \rightleftharpoons PM_n \tag{11}$$

and

$$K_d = \frac{[P][M]^n}{[PM_n]} \tag{12}$$

where

$$r = \frac{[M]_{bound}}{[P]_{total}} = \frac{n[PM_n]}{[P]+[PM_n]} = \frac{n[M]^n}{K_d + [P]^n} \tag{13}$$

this equation can be rewritten as

$$\frac{r}{n-r} = \frac{[M]^n}{K_d} \tag{14}$$

which is known as the Hill equation. Plotting the log (r/n-r) versus log [M]=log (free metal) yields what is commonly referred to as Hill plot (Fig. 2). From this plot, cooperativity is generally interpreted from the slope n (also known as the Hill coefficient n_H). While not generally as useful as a Scatchard plot, Hill plots are especially important in indicating the absence of cooperativity as in the case of independent sites, where we would expect the slope of the Hill plot to be equal to 1.

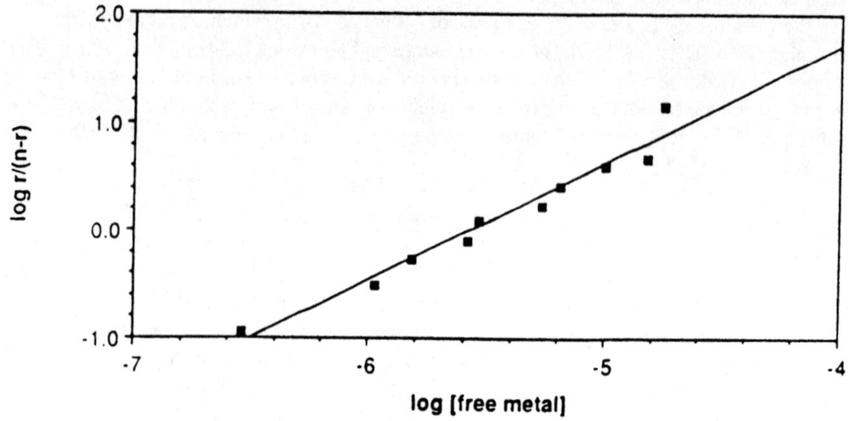

Figure 2.
Hill plot of binding of Zn to human serum albumin. Conditions are the same as in Figure 1. r, moles of metal bound per mole of protein, n, number of binding sites as determined from Scatchard analysis. (Reproduced with permission from Goumakos et al., 1991).

IV. CONCLUSION

Equilibrium dialysis technique still remains a widely used method to study metal-protein labile complex equilibria in solution. The most common shortcoming of this method in practice results from failure to achieve equilibrium. One should also use caution in choosing the background electrolyte. The added salt used by coordination chemists, such as sodium perchlorate, may interact with the protein. Also the use of high ionic strength may result in unexpected binding by the protein, or changes in conformation and solubility of the protein. One should also not overlook the possibility of forming soluble hydroxide complexes or any types of polynuclear complexes that may equilibrate very slowly. The recent adaptation of the traditional equilibrium dialysis method to accommodate the simultaneous use of several metal radioisotopes as described in Goumakos et al. (1991) has been shown to be a very useful method. It has examined the interaction of several different metals with one protein, at the same time providing reliable values for the number of metal-binding sites and their corresponding dissociation constants.

REFERENCES

Ahrland, S., Chatt, J., and Davies, N.R. (1958). *Quart. Rev.*, *12*: 265.

Baker, J.O. (1988). *Methods in Enzymology*, Vol. 158 (J.F. Riordan and B.L. Vallee, eds.), Academic Press, New York, p. 33.

Faguet, G.B. (1986). *J. Cell. Biochem.*, *31*: 243.

Goumakos, W., Laussac, J.P., and Sarkar, B. (1991). *Biochem. Cell Biol.*, *69*: 809.

Hill, A.V. (1910). *J. Physiol.*, *39*: 361.

Klotz, I.M., and Hunston, D.L. (1984). *J. Biol. Chem.*, *259*: 10060.

Sarkar, B. (1987). *Prog. Food Nutr. Sci.*, *11*: 363.

Scatchard, G. (1949). *Ann. N.Y. Acad. Sci.*, *51*: 660.

Sigel, H., and McCormick, D.B. (1970). *Accounts Chem. Res.*, *3*: 201.

Sillén, L.G., and Martell, A.E. (1964). *Stability Constants*, 2nd ed., The Chemical Society, Spec. Publ. No. 17.

Sillén, L.G., and Martell, A.E. (1971). *Stability Constants*, *Suppl. No. 1*, The Chemical Society, Spec. Publ. No. 25.

METAL-ALBUMIN INTERACTIONS

by

C. Harford and B. Sarkar

Department of Biochemistry Research,
The Hospital for Sick Children and
the Department of Biochemistry,
University of Toronto,
Toronto, Ontario
Canada

I. INTRODUCTION

Albumin (molar mass 69 kg mol^{-1}) is the most abundant protein in serum (4.5 g/100 ml) (Sarkar, 1983). It has a wide variety of functions. One of its most prominent functions is transport. Albumin is the main transport plasma protein for several small molecules and ions including metals (Kragh-Hansen, 1990). The binding sites of metals such as Cu(II), Ni(II), Cd(II) and Zn(II) to albumin have been well characterized. Other metals such as Co(II), Hg(II), Mn(II) and cisplatin have also been shown to bind to albumin. There are a variety of metal-binding sites. These range from the highly specific amino terminal Cu(II),Ni(II)-binding motif to non-specific interactions. Individual metals may bind at more than one site. Individual sites may bind more than one type of metal. Smaller molecules such as amino acids may form ternary complexes with metals and albumin.

II. COPPER(II)

A. Primary interactions of Cu(II) and albumin

The most well characterized metal-albumin interactions involve Cu(II). Copper(II)-albumin is the major transport form of Cu(II) in blood. There is one high-affinity binding site for Cu(II) on human albumin which is formed by the amino terminal tripeptide Asp-Ala-His (Laussac and Sarkar, 1984). The ligands of this peptide which are involved in Cu(II) binding are the α-amino nitrogen, the two intervening peptide nitrogens, the N(I) atom of the imidazole group of the histidine residue and the carboxylate group of the aspartate residue (Fig. 1). The four nitrogen ligands are arranged in a planar structure which may be either square planar or rhombic (Rakhit and Sarkar, 1981; Asaturian, 1986). The carboxylate group coordinates Cu(II) from outside the plane of the four other groups.

Experiments with albumins of other mammalian species have demonstrated that the histidine residue in the third position is necessary for the specific binding of Cu(II) to the amino terminal Cu(II)-binding site of albumin. When tyrosine is substituted for histidine, as in dog serum albumin, the ability of albumin to bind Cu(II) specifically is lost (Appleton and Sarkar, 1971; Dixon and Sarkar, 1974). Recently, it has also been demonstrated that when the histidine residue is moved from the third position to the fourth position, the ability

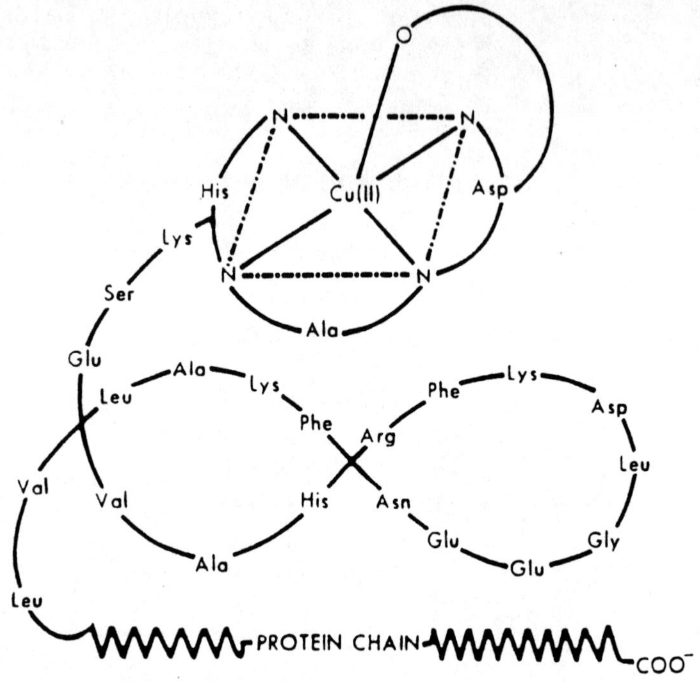

Figure 1.
The amino terminal Cu(II)-binding site of human serum albumin. Nickel(II) binds in the same manner.

to bind Cu(II) specifically is also lost (Predki et al., 1992). It has been shown, however, that the first two residues of the motif can be altered to neutral residues without loss of Cu(II)-binding potential (Shearer et al., 1967; Bradshaw and Peters, 1969; Lau et al., 1974; Camerman et al., 1976). Therefore, it may be predicted that if the amino terminal sequence of the protein is X-Y-His where X and Y are any amino acids and there is also present the α -amino nitrogen and the two intervening peptide nitrogens along with the imidazole nitrogen of the histidine in the third position, a specific Cu(II)-binding motif will be formed. This motif will be further discussed in Section IIC.

The Cu(II)-binding motif of human albumin has also been shown to bind Ni(II) specifically (see Section IIIA). Therefore, this motif has been named the amino terminal Cu(II),Ni(II)-binding motif (Predki et al., 1992).

A small percentage of albumin in normal human sera does not possess the amino terminal aspartyl residue or both the aspartyl and alanyl residues (Brennan et al., 1988). Other variants are known which have the histidyl residue mutated (Takahashi et al., 1987). These albumin variants are unable to bind Cu(II) or Ni(II) with high specificity.

The X-ray crystal structure of albumin demonstrates no defined structure at the amino terminal region of the protein (He and Carter, 1992). This is consistent with the NMR results of the 24-residue amino terminal polypeptide of human albumin (Laussac and Sarkar, 1984). These results have been interpreted to indicate that this region has a relatively flexible conformation. Thus, the region is able to fold around a metal to form the specific binding site for Cu(II).

Studies have also shown that albumin can bind Cu(II) non-specifically after the first equivalent of Cu(II) has been bound (Appleton and Sarkar, 1971; Zgirski and Frieden, 1990). However, the site(s) have not been well characterized.

B. Secondary interactions of Cu(II) and albumin

Cu(II) forms ternary complexes with albumin and single amino acids (Sarkar and Wigfield, 1968). The amino acids histidine and cysteine have been studied with regard to their interactions with Cu(II)-albumin (Tabata and Sarkar, 1985a; 1985b; Suzuki et al., 1989). Histidine is the main Cu(II)-binding amino acid in blood. Under physiological conditions, there is an equilibrium between Cu(II)-albumin and Cu(II)-histidine through the ternary complex albumin-Cu(II)-histidine (Lau and Sarkar, 1971). Studies of the histidine-Cu(II)-Asp-Ala-His system indicate that at physiological pH, 16% of Cu(II) is bound in different forms of the ternary complex (Tabata and Sarkar, 1985a; 1985b). The albumin-Cu(II)-histidine ternary complex acts as an intermediate in the transfer of Cu(II) between Cu(II)-histidine and Cu(II)-albumin.

The growth modulating peptide Gly-His-Lys has also been shown to form ternary complexes with Cu(II)-albumin (Lau and Sarkar, 1981; Laussac et al., 1983). However, the binding strength of this peptide for Cu(II) is not nearly as strong as that of peptides with histidine in the third position.

C. Applications of the amino terminal Cu(II),Ni(II)-binding motif

As it has been shown that the one major requirement for a specific amino terminal Cu(II)-binding site is the histidine residue in the third position, this motif has been used in novel methods other than for the study of metal transport (Predki et al., 1992). In particular, the peptide Gly-Gly-L-His which was originally designed to mimic the Cu(II)-transport site of human albumin has been used in the study of DNA cleavage (Lau et al., 1974; Camerman et al., 1976). In the presence of metal, both the tripeptide alone and the tripeptide linked at the amino terminal of a DNA-binding protein have been used to cleave DNA (Chiou, 1983; Mack et al., 1988). The mechanism of DNA cleavage by the motif is as follows: Cu(II) bound to this motif is able to generate free radicals in the presence of a reducing agent. When the motif is brought in close proximity to DNA, the free radicals are able to cleave the DNA.

III. NICKEL(II)

A. Primary interactions of Ni(II) and albumin

Albumin is a major transport protein for Ni(II) in blood (Sarkar, 1983). Over 95% of Ni(II) in blood has been shown to be bound to albumin. There are two binding sites for Ni(II) on human serum albumin. However, under physiological conditions, only one site demonstrates specificity.

Nickel(II) binds with high affinity to the same amino terminal site as Cu(II). This site includes the α-amino nitrogen, the two intervening peptide nitrogens, the imidazole nitrogen and the aspartate carboxyl group (Fig. 1) (Laussac and Sarkar, 1984). The site forms a square planar or square pyramidal structure. The kinetics of binding of Ni(II) to albumin is very slow (Tabata and Sarkar, 1992).

The second Ni(II) site on albumin is octahedral in structure. This site is less thermodynamically stable than the first site, although it has been suggested that under physiological conditions this site contains approximately 30% of the Ni(II) which is bound to albumin (Laurie and Pratt, 1986).

B. Secondary interactions of Ni(II) and albumin

Nickel(II) forms a significant amount of albumin-Ni(II)-histidine ternary complex in the same manner as Cu(II) (Sarkar, 1983) (see section IIB). However, despite the similarities of Ni(II) and Cu(II) in the metal-binding sites of albumin and the reaction mechanism, the Ni(II)-L-histidine complex is thermodynamically very stable and kinetically inert compared to the Cu(II)-L-histidine complex in the presence of albumin. As a result, very little Ni(II) is transferred from Ni(II)-L-histidine to albumin (Tabata and Sarkar, 1992).

IV. CADMIUM(II)

Few studies have been directed to the understanding of Cd(II) binding to albumin. Therefore, much less is known about the binding of Cd(II) to albumin and its physiologic importance than of other metals.

However, studies have demonstrated that there are three binding sites for Cd(II) on human serum albumin (Goumakos et al., 1991). Two of these sites have high affinity for

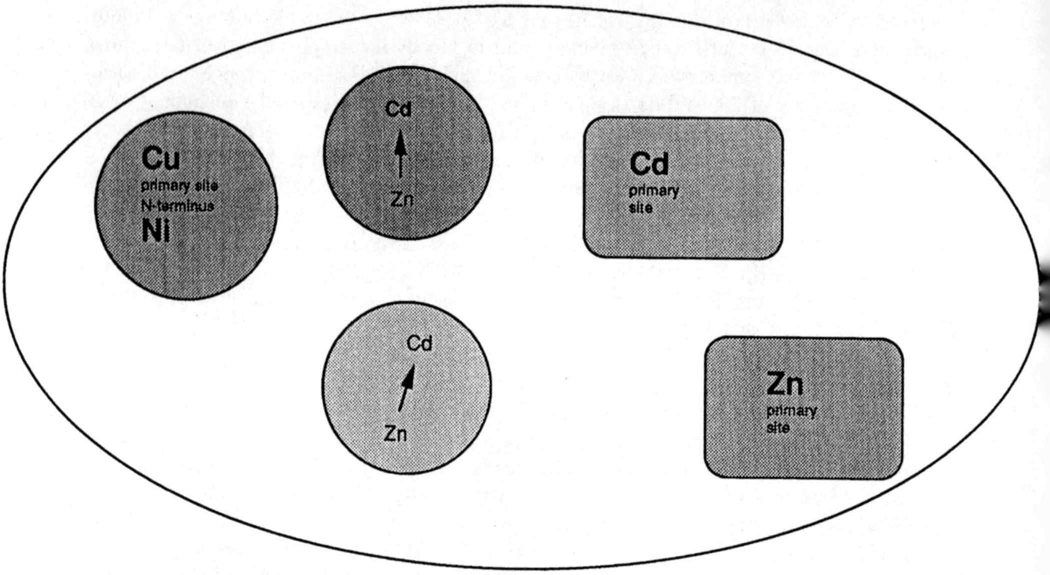

Figure 2.
Sites of human serum albumin-metal interactions.

Cd(II). The first binding site, a high-affinity Cd(II) site, likely involves two or three histidyl residues. The second binding site, a lower-affinity Cd(II) site, likely involves a histidyl residue and three carboxylate oxygens, possibly from glutamyl and/or aspartyl residues. The third site, a high-affinity Cd(II) site, involves four oxygen ligands. Competition studies with Zn(II) demonstrated that Zn(II) has a higher affinity for the first and second sites than Cd(II) does and is thus able to displace Cd(II) from these sites. The third site is a specific, high-affinity Cd(II) binding site.

It was also demonstrated that the third binding site, which along with the first site has high affinity for Cd(II), was not present in dog serum albumin. Therefore, dog serum albumin is missing two high-affinity sites of metal binding, one for Cu(II)/Ni(II) and one for Cd(II).

V. ZINC(II)

A. Primary interactions of Zn(II) and albumin

Albumin is a major transport protein for Zn(II) (Sarkar, 1983). There are approximately three binding sites for Zn(II) in human serum albumin (Guthans and Morgan, 1982; Din and Frazier, 1983; Goumakos et al., 1991). There are two types of Zn(II)-binding sites. One site has a log K of approximately 6.4. The other sites include at least one with a log K of approximately 5. None of these sites are displaceable by Cd(II) (Goumakos et al., 1991). However, it cannot be ruled out that two of the Zn(II) sites may involve some or all of the residues involved in the low-affinity Cd(II) sites of albumin.

The two low-affinity Zn(II) sites prefer Cu(II). Therefore, one of the two low-affinity Zn(II) sites may involve part of the amino terminal Cu(II),Ni(II)-binding motif. However, alteration of the amino terminal Cu(II),Ni(II)-binding motif to a low-affinity Cu(II)-binding site such as that found in dog serum albumin does not change the affinity of albumin for Zn(II). Therefore, either Zn(II) binds to other Cu(II)-binding sites on albumin or histidine is not involved in binding of Zn(II) at the amino terminus. At least one of the Zn(II)-binding sites appears to involve cysteinyl sulfurs (Zhou et al., 1992).

B. Secondary interactions of Zn(II) and albumin

As Zn(II) is an essential metal and an important structural feature in many proteins, the understanding of the transport of this nutrient from the gastrointestinal tract to the bloodstream is of interest. It has been found that an albumin-Zn(II)-gastrin ternary complex likely plays a role in the uptake of Zn(II) from the stomach. Zn(II) and gastrin appear to exhibit synergy in their binding to albumin (Mann et al., 1991). The site of gastrin-Zn(II) binding on albumin was not identified, but it cannot be ruled out that one of the Zn(II) sites which is observed upon Zn(II) binding alone to albumin is responsible.

Similar to albumin-Cu(II)-amino acid complexes (Sarkar and Wigfield, 1968; Lau and Sarkar, 1971), Zn(II)-albumin also forms ternary complexes with amino acids (Giroux and Henkin, 1972). Furthermore, it forms ternary complexes with urate and other purine compounds as well (Fleckenstein and Simkin, 1986). The sites of the binding of these molecules have not been determined. Given that Zn(II) binds many small molecules and ions, it is possible that many other Zn(II)-albumin ternary complexes are also formed.

VI. MERCURY(II)

Albumin is the primary binding site for Hg(II) in serum (Lau and Sarkar, 1979). The primary site for Hg(II) binding involves an albumin dimer. Mercury(II) binds to the free sulfhydryl groups at Cys-34 of two albumin molecules, forming Hg-human serum albumin_2 (Sarkar, 1983). Subsequent metal atoms bind at the amino terminal region.

VII. COBALT(II)

Albumin is the main transport protein of Co(II) (Sarkar, 1983). The metal ions are bound through the carboxylate groups of the protein. The first two Co(II) atoms are bound more firmly than the subsequent atoms.

VII. MANGANESE(II)

Very little is known of Mn(II) binding to albumin. There is controversy over whether transferrin or albumin is the major plasma binding component (Sarkar, 1983). Under in vitro conditions, it has been found that 50% of Mn(II) is bound to transferrin while only 5% is bound to albumin (Scheuhammer and Cherian, 1985).

IX. CISPLATIN

Cisplatin (cis-dichlorodiammineplatinum(II)) is the only non-naturally occurring metal compound whose binding to albumin has been studied in detail. Cisplatin is able to bind to several potential ligands including oxygen and sulfur atoms.

Several studies have demonstrated cisplatin binding to albumin (Pizzo et al., 1988; Yotsuyanagi et al., 1991). However, there has been disagreement over the ratio of moles of cisplatin bound per mole of albumin. Cisplatin has been shown to bind at the single free sulfhydryl group of cysteine (Cys-34) (Pizzo et al., 1988). Others have argued for the binding of up to 20 moles of cisplatin per mole of albumin at unknown sites (Yotsuyanagi et al., 1991).

X. CONCLUSIONS

Albumin is unique in its ability to bind a wide variety of metals at a number of different sites. The binding of metals to albumin may be physiological as in the case of Cu(II) transport or for no other reason than the fact that albumin contains available ligands such as in the case of cisplatin. For many metals, albumin plays an important role in transport.

REFERENCES

Appleton, D.W., and Sarkar, B. (1971). *J. Biol. Chem.*, *246*: 5040.

Asaturian, R.A. (1986). *Gen. Physiol. Biophys.*, *5*: 105.

Bradshaw, R.A., and Peters, T., Jr. (1969). *J. Biol. Chem.*, *244*: 5582.

Brennan, S.O., George, P.M., and Peach, R.J. (1988). *Clin. Chim. Acta*, *176*: 179.

Camerman, N., Camerman, A., and Sarkar, B. (1976). *Can. J. Chem.*, *54*: 1309.

Chiou, S.H. (1983). *J. Biochem. Tokyo*, *94*: 1259.

Din, W.W., and Frazier, J.M. (1983). *FASEB J.*, *42*: 1896.

Dixon, J.W., and Sarkar, B. (1974). *J. Biol. Chem.*, *23*: 2832.

Fleckenstein, J.L., and Simkin, P.A. (1986). *Adv. Exp. Med. Biol.*, *195*: 345.

Giroux, E.L,, and Henkin, R.I. (1972). *Biochim. Biophys. Acta*, *273*: 64.

Goumakos, W., Laussac, J.-P., and Sarkar, B. (1991). *Biochem. Cell Biol.*, *69*: 809.

Guthans, S.L., and Morgan, W.T. (1982). *Arch. Biochem. Biophys.*, *218*: 320.

He, X.M., and Carter, D.C. (1992). *Nature*, *358*: 209.

Kragh-Hansen, U. (1990). *Dan. Med. Bull.*, *37*: 57.

Lau, S.-J., and Sarkar, B. (1971). *J. Biol. Chem.*, *246*: 5938.

Lau, S.-J., and Sarkar, B. (1979). *J. Toxicol. Environ. Health*, *5*: 907.

Lau, S.-J., and Sarkar, B. (1981). *Biochem. J.*, *199*: 649.

Lau, S.-J., Kruck, T.P.A., and Sarkar, B. (1974). *J. Biol. Chem.*, *249*: 5878.

Laurie, S.H., and Pratt, D.E. (1986). *J. Inorg. Biochem.*, *28*: 431.

Laussac, J.-P., and Sarkar, B. (1984). *Biochemistry*, *23*: 2832.

Laussac, J.-P., Haran, R., and Sarkar, B. (1983). *Biochem. J.*, *209*: 533.

Mack, D.P., Iverson, B.L., and Dervan, P.B. (1988). *J. Amer. Chem. Soc.*, *110*: 7572.

Mann, R.J., Howlett, G.J., and Baldwin, G.S. (1991). *Arch. Biochem. Biophys.*, *291*: 311.

Pizzo, S.V., Swaim, M.W., Roche, P.A., and Gonias, S.L. (1988). *J. Inorg. Biochem.*, *33*: 67.

Predki, P., Harford, C., Brar, P., and Sarkar, B. (1992). *Biochem. J.*, *287*: 211.

Rakhit, G., and Sarkar, B. (1981). *J. Inorg. Biochem.*, *15*: 233.

Sarkar, B. (1983). *Life Chem. Rep.*, *1*: 165.

Sarkar, B., and Wigfield, Y. (1968). *Can. J. Biochem.*, *46*: 601.

Scheuhammer, A.M., and Cherian, M.G. (1985). *Biochim. Biophys. Acta*, *840*: 163.

Shearer, W.T., Bradshaw, R.A., Gurd, F., and Peters, T., Jr. (1967). *J. Biol. Chem.*, *242*: 5451.

Suzuki, K.T., Karasawa, A., and Yamanaka, K. (1989). *Arch. Biochem. Biophys.*, *273*: 572.

Tabata, M., and Sarkar, B. (1985a). *Can. J. Chem.*, *63*: 3111.

Tabata, M., and Sarkar, B. (1985b). *Can. J. Chem.*, *63*: 3117.

Tabata, M., and Sarkar, B. (1992). *J. Inorg. Biochem.*, *45*: 93.

Takahashi, N., Takahashi, Y., and Putnam, F.W. (1987). *Proc. Nat. Acad. Sci. USA*, *84*: 7403.

Yotsuyanagi, T., Ohta, N., Futo, T., Ito, S., Chen, D., and Ikeda, K. (1991). *Chem. Pharm. Bull.*, *39*: 3003.

Zgirski, A., and Frieden, E. (1990). *J. Inorg. Biochem.*, *39*: 137.

Zhou, Y., Hu, X., Dou, C., Liu, H., Wang, S., and Shen, P. (1992). *Biophys. Chem.*, *42*: 201.

METAL-TRANSFERRIN INTERACTIONS

by

Timothy J. Egan and Philip Aisen[1]

Department of Physiology and Biophysics,
Albert Einstein College of Medicine,
1300 Morris Park Avenue
Bronx, New York 10461
U.S.A.

I. INTRODUCTION

The transferrins (Chasteen, 1983; Baldwin and Egan, 1987; Harris and Aisen, 1989) are a class of iron-binding glycoproteins composed of serum transferrin (the iron transport protein of the circulation, hereafter referred to as transferrin), lactoferrin (found in milk, tears and other body fluids), ovotransferrin (in avian egg white) and melanotransferrin (Brown et al., 1982; Rose et al., 1986) (found in human melanomas, also known as p97). They have molar masses near 80 kg mol^{-1}, with considerable sequence homology among them (human transferrin/human lactoferrin 59%, human transferrin/chicken ovotransferrin 51%, human transferrin / melanotransferrin 39% (Metz-Boutigue et al., 1984; Rose et al., 1986)). With the exception of melanotransferrin (Baker et al., 1992) they each bind two Fe(III) ions.

Vertebrate transferrins are single polypeptide chains arranged in two approximately equal lobes. There is a large degree of sequence homology between the (human transferrin 41% (MacGillivray et al., 1983; Yang et al., 1984), human lactoferrin 37% (Metz-Boutigue et al., 1984), chicken ovotransferrin 33% (Jeltch and Chambon, 1982)). This internal homology and the organization of the human transferrin genes provide compelling evidence that the modern transferrins arose by duplication and fusion of an ancient gene specifying a single sited precursor protein (MacGillivray et al., 1983).

The major role of transferrin is to solubilize Fe(III) in a non-toxic form. Transferrin may also serve as a cellular growth promoter (independent of its iron transport function) (Eby et al., 1992). Lactoferrin appears to act as a bacteriostatic agent (tightly binding iron, thereby denying it to microorganisms) (Sawatzki, 1987). It may also have a transport role in iron absorption via a lactoferrin-receptor (Kawakami and Lönnerdal, 1991). Ovotransferrin appears to function primarily to protect the egg from microbial growth (Valentini et al., 1981).

II. STRUCTURE

A. From X-ray diffraction analysis

Human lactoferrin (Anderson et al., 1987; 1989) (Fig. 1) and rabbit transferrin (Bailey et al., 1988) (for which high resolution X-ray structures are available) are organized in two

[1] To whom correspondence should be addressed.

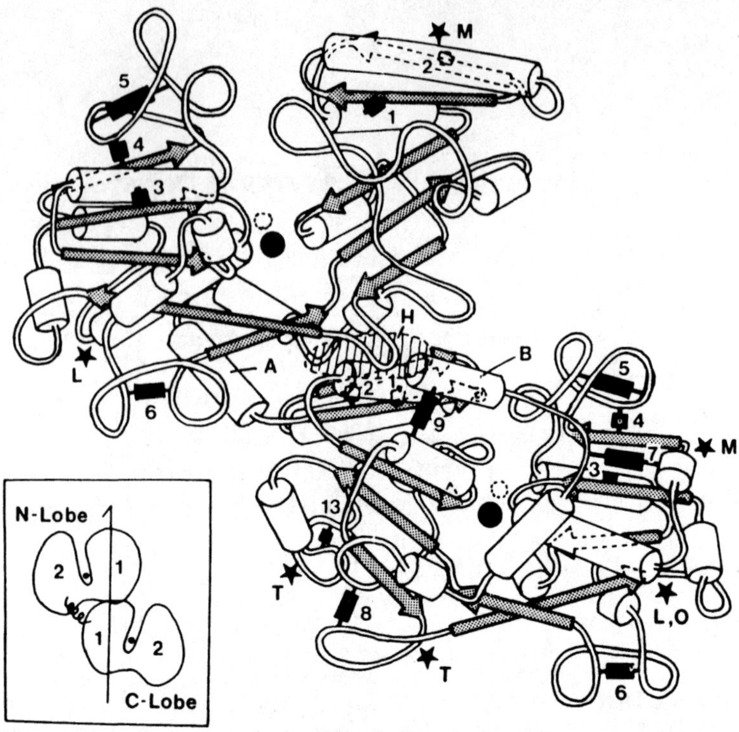

Figure 1.
A schematic diagram of the structure of lactoferrin (reproduced from Baker et al., 1987). Cylinders represent α-helices, arrows represent β-strands, ● = Fe(III), ☼ = carbonate anion, ▮ = disulfide bridge, ★ = carbohydrate attachment site in lactoferrin (L), transferrin (T), ovotransferrin (O) and melanotransferrin (M), hatched area labelled H is a hydrophobic region between the lobes, the connecting helix between the lobes is labelled A and the C-terminal helix is B. In the inset 1 and 2 refer to domains I and II respectively.

very similar lobes with each lobe corresponding to the homologous halves (one containing the amino-terminus of the chain and the other the carboxy-terminus and termed the N- and C-terminal lobes respectively) and each containing an iron-binding site.

Each lobe of transferrin and lactoferrin is folded into two domains. In the N-terminal lobe of lactoferrin, amino acids 1-90 lie in domain I, 91-251 lie in domain II and 252-320 lie in domain I. The portions of the chains linking the two domains are β-strands. The C-terminal lobe is very similar except that it has an α-helical "tail" at its C-terminus which folds across to domain II joining the edges of the two domains with two disulfide bridges. The N-terminal lobe has an α-helix at its C-terminal end which joins the lobes together.

The iron-binding ligands at each site (Fig. 2) are two tyrosines (95,188;426,517) a histidine (249;585), an aspartate (63;392) and a carbonate anion bound in a bidentate manner (N-; C-terminal sites, human transferrin numbering) to complete the coordination shell. The carbonate anion is known as the synergistic anion since specific iron binding does not occur unless carbonate or a suitable substitute anion is present (Bates and Schlabach, 1975). Under CO_2-free conditions, a large number of anions can substitute for the carbonate, but all contain a carboxylate group together with a Lewis base group separated from the carboxylate group by no more than one carbon atom which has no more than one bulky side chain (Schlabach and Bates, 1975). Carbonate binding is very weak in the absence of iron (Harris, 1985). The carbonate anion interacts with the protein via the positive dipole at the N-terminal end of α-helix 5 and the side chain of an arginine (124;455). By providing a negative charge to the binding cleft, it may facilitate entry of iron

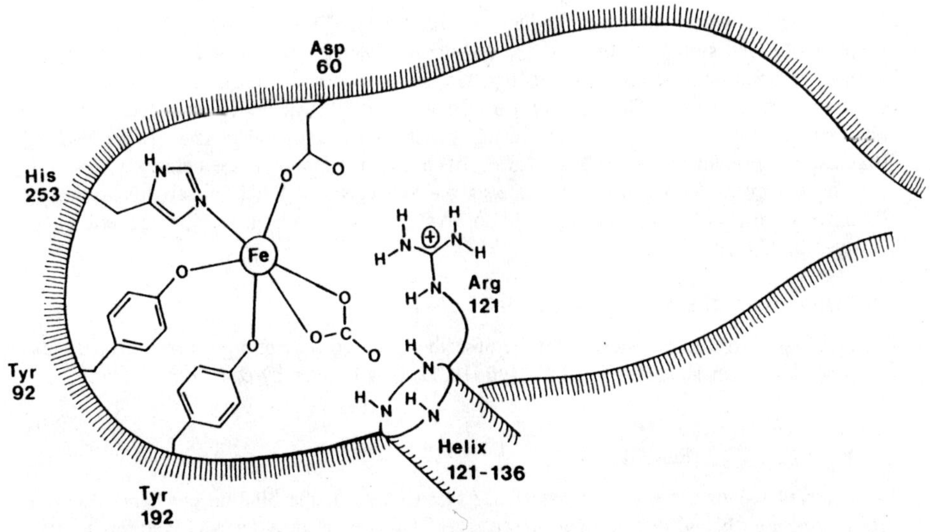

Figure 2.
The iron-binding site of human lactoferrin (reproduced from Anderson et al. 1989). The iron, carbonate, two tyrosines, histidine and aspartate ligands are shown as well as the arginine and N-terminus of α-helix 5 which interacts with the carbonate anion (residue numbering on the figure refers to human lactoferrin).

by neutralizing electrostatic repulsion from the positively charged arginine. The iron of the iron-transferrin-carbonate complex is high spin Fe(III), although at the C-terminal site it can be converted to low spin by addition of 0.5 M NaCN at pH 10.4 apparently with displacement of the carbonate anion and binding of one to three CN^- anions to the Fe(III) (Swope et al., 1988; Snetsinger et al., 1990).

The structure of apolactoferrin has been determined by X-ray diffraction analysis (Anderson et al., 1990), and shows that the N-terminal lobe is in an "open" conformation. The two domains are rotated by about 54° about their adjoining β-strands (with respect to the diferric form) exposing the metal-binding site. By contrast, the C-terminal lobe was found to be closed as in the diferric protein. However, a small angle X-ray scattering study indicates that both lobes of apotransferrin are in the open conformation in solution (Grossmann et al., 1992). There may be an equilibrium between the open and closed conformations in solution with crystal packing forces favoring the closed conformation in the C-terminal lobe.

B. Transferrin spectra
The characteristic salmon pink color of the ferritransferrins apparently arises from $n \rightarrow \pi^*$ charge transfer between the tyrosine ligands and the Fe(III) (Patch and Carrano, 1981). The charge-transfer band is centered at 465 nm (λ_{max}).

The electron paramagnetic resonance (EPR) spectra of Fe(III) transferrins are not completely understood (Harris and Aisen, 1989), but they are sensitive to the binding of non-synergistic anions to transferrin (Foljatar and Chasteen, 1982) and have provided evidence for specific anion binding sites (see Harris and Aisen, 1989 for a detailed discussion of transferrin EPR spectroscopy).

C. Transferrin as an iron-binding ligand
The suitability of transferrin as an Fe(III)-binding ligand can be understood on the basis of the hard and soft acid base (HSAB) theory (Pearson, 1963). Fe(III) is a small cation ($r_+ =$ 0.65 Å) (Shannon and Prewitt, 1970) with a high charge and is thus a hard Lewis acid. It will form strong complexes with hard Lewis bases, typically those containing oxygen donor atoms. The iron in transferrin is coordinated to five oxygen donors. Only the nitrogen atom

of histidine is softer (an intermediate Lewis base). The chelating atoms are thus exceptionally well suited for binding and stabilizing Fe(III); only metal ions of comparable hardness come close in strength of binding (Ga(III) (Harris and Pecoraro, 1983) and Al(III) (Cochran et al., 1984)). The ligands are also apparently structurally arranged to bind a metal ion the size of Fe(III), since Ga(III) which is of comparable size ($r_+ = 0.62$ Å) (Shannon and Prewitt, 1970) is closest to Fe(III) in binding strength (see below).

The strong preference for carbonate as the synergistic anion is apparently structural (Shongwe et al., 1992), as that anion is so arranged as to form a stabilizing network of hydrogen bonds with the protein.

III. THERMODYNAMICS OF IRON BINDING

The binding of iron to transferrin can be described by two macroscopic binding constants. The conditional constants (for fixed pH and $[HCO_3^-]$) are given by equations 1 and 2:

$$K_1 = [FeTf] / [Fe(III)] [Tf] \quad (1)$$
$$K_2 = [Fe_2Tf] / [Fe(III)] [FeTf] \quad (2)$$

The reported values are given in Table 1. As can be seen, the binding constants decrease with decreasing pH and increase with increasing HCO_3^- concentration, and lactoferrin binds iron more strongly than transferrin. Site-specific binding constants for transferrin are given in Figure 3. The C-terminal site binds iron more strongly than the N-terminal site and there is slight cooperativity between the sites.

The Fe(III) complex of transferrin becomes unstable as the pH is decreased (Princiotto and Zapolski, 1975) and iron is lost from the N-terminal site below about pH 6 and from the C-terminal site below about pH 5.5. The critical event in destabilization by acid may be protonation of protein ligands or, more likely, carbonate. By contrast, ferric lactoferrin remains stable down to about pH 3 (Legrand et al., 1990). The reason for this greater stability is not understood.

The E° of transferrin in 2M KCl at pH 7.4 has been estimated by spectroelectrochemistry (Kretchmar et al., 1988) using binding constants for Fe(II) derived from a linear free energy relationship for Ni(II) and Fe(II) complexes (Harris, 1986a). The resulting value, -0.52 V, suggests that physiological reduction at pH 7.4 of transferrin-bound Fe(III) is unlikely.

Recent studies have revealed the effects of non-coordinated amino acids on iron-binding constants using site-directed mutagenesis (Woodworth et al., 1991). Replacement of positively charged groups by negative ones in the vicinity of the iron-binding site increases K, while the inverse has the opposite effect. This result seems to suggest that K is carefully balanced to allow strong binding of iron, but at the same time permitting iron release to occur under the correct conditions of pH (see below).

Table 1

Conditional Iron-Binding Constants for Transferrin and Lactoferrin

Protein	log K_1*	log K_2*	Conditions
Transferrin	20.7	19.4	7.4, atm. CO_2 [a,b]
	21.66	c	7.4, 5 mM HCO_3^- [d,e]
	19.5	17.4	6.7, atm. CO_2 [b]
Lactoferrin	22.36	c	7.4, 5 mM HCO_3^- [e]
	20.1	c	6.4, atm. CO_2 [f]

[a] $pCO_2 = 3.6 \times 10^{-4}$ atm [b] Aisen et al., 1978 [c] not reported
[d] Harris and Pecoraro, 1983 [e] Harris, 1986b [f] Aisen and Leibman, 1972

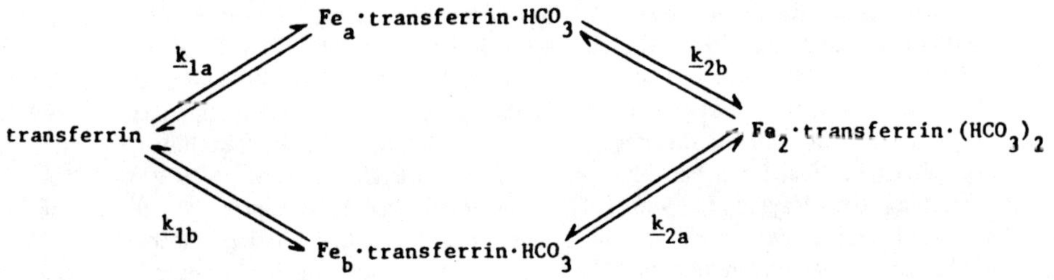

Figure 3.
Site-specific binding constants for human transferrin (reproduced from Aisen et al., 1978), $\underline{k}_{1a}$ = 4 x 10^{20} M^{-1}, $\underline{k}_{2a}$ = 1.6 x 10^{20} M^{-1}, $\underline{k}_{1b}$ = 6.8 x 10^{19} M^{-1}, $\underline{k}_{2b}$ = 2.8 x 10^{19} M^{-1} at pH 7.4 and atm. CO_2 (a refers to the C-terminal site and b to the N-terminal site).

IV. KINETICS OF IRON BINDING AND RELEASE

A. Iron uptake

Iron uptake can occur by direct chelation of Fe(III) from a ferric complex or uptake and oxidation of Fe(II) (only one study: Kojima and Bates, 1981) appears to have been reported on the latter route and first-order plots were presented under second-order conditions, while no cognizance was taken of the possible influence of the metal-catalyzed Haber-Weiss reaction on the kinetics). A number of studies have been performed on the former route using several Fe(III) complexes including ferric pyrophosphate (PP_i) (Cowart et al., 1986), ferric acetohydroxamate (Cowart et al., 1982) and ferric nitrilotriacetate (NTA) (Honda et al., 1980). The following mechanism has been proposed to account for the observed kinetics (Cowart et al., 1982; Kretchmar and Raymond, 1986):

$$\text{Tf} + \text{HCO}_3^- \rightleftharpoons \text{Tf-HCO}_3$$
$$\text{Fe(III)-Chel} \rightleftharpoons \text{Fe(III)-Chel}^\dagger$$
$$\text{Tf-HCO}_3 + \text{Fe(III)-Chel}^\dagger \rightleftharpoons \text{Chel-Fe(III)-Tf-HCO}_3$$
$$\text{Chel-Fe(III)-Tf-HCO}_3 \rightleftharpoons \text{Fe(III)-Tf-HCO}_3^* + \text{Chel}$$
$$\text{Fe(III)-Tf-HCO}_3^* \rightleftharpoons \text{Fe(III)-Tf-HCO}_3$$

Fe(III)-Chel† represents a configurationally excited state of the chelate complex. The quaternary intermediate Chel-Fe(III)-Tf-HCO_3 appears to have been observed spectroscopically (Cowart et al., 1982; 1986) (although this has been disputed (Weaver and Pollock, 1990)). The final step in the mechanism is a conformational change, where Fe(III)-Tf-HCO_3* represents an open conformation of the ferrictransferrin. This step is included to account for the apparent saturation kinetics (with respect to chelator concentration) observed during iron release (where no spectroscopic intermediate is observed) (Cowart et al., 1982; Kretchmar and Raymond, 1986).

An interesting phenomenon of the iron uptake reaction is the effect of various chelators in steering the iron to one or other of the binding sites. For example, NTA directs the Fe(III) to the C-terminal site and citrate and oxalate direct it to the N-terminal site of human transferrin (Aisen et al., 1978). Calorimetric studies on ovotransferrin (Lin et al., 1991) have shown that uptake of Fe(III) from Fe(III)NTA is a two-step process consisting of rapid contact binding of Fe(III)NTA to the protein followed by a much slower process of carbonate insertion. The preference for the N-terminal site of ovotransferrin is due to stronger contact binding and faster carbonate insertion.

In vivo, iron uptake occurs at iron-releasing cells such as the intestinal mucosal cell (Halliday, 1992) and the erythrophagocytosing macrophage (Kondo et al., 1988), but details of the process are unknown.

B. Iron release

Three pathways for iron release from transferrin in vitro can be envisaged:

(i) direct chelation of Fe(III), (ii) reduction of Fe(III) to Fe(II) and complexation of the Fe(II) by a ferrous ion chelator, (iii) release of Fe(III) at low pH.

Iron release by direct chelation has been the subject of extensive study (Baldwin, 1980; Baldwin and de Sousa, 1981; Kretchmar and Raymond, 1986; 1988; Cowart et al., 1986; 1982; Harris, 1984; Harris et al., 1987; Bali and Harris, 1989; 1990; Bali et al., 1991a; Harris and Bali, 1988; Bertini et al., 1988; Marques et al., 1991; Egan et al., 1992). Originally it was believed that the pseudo first-order rate constants exhibited a saturation dependence on ligand concentration (acetohydroxamate (Cowart et al., 1982), 3,4-LICAMS (Kretchmar and Raymond, 1986)). This was interpreted as evidence for the mechanism presented above (with release being the reverse of the binding mechanism). This mechanism, however, predicts that the maximal rate constant for iron release is determined by the rate constant for conversion of the closed conformation to the open conformation. This is an intrinsic property of the protein and should lead to the same limiting rate constants for all chelating agents which is not observed experimentally (Harris, 1984). Furthermore, a number of recent studies (Harris et al., 1987; Bali and Harris, 1989; Marques et al., 1990; Bertini et al., 1988; Marques et al., 1991; Egan et al., 1992) (mostly using PP_i, but also citrate and phosphonates) have demonstrated an apparent saturation-linear dependence of observed rate constant on chelator concentration (k_{obs} saturating to a straight line dependence on chelator concentration).

Anions are known to greatly affect the rate constants of iron release (Baldwin, 1980; Baldwin and de Sousa, 1981; Kretchmar and Raymond, 1988) and it has been shown that anions are an absolute prerequisite for iron release (Kretchmar and Raymond, 1988) (anion binding to specific sites on the protein is apparently required in order to achieve an open conformation), since as ionic strength approaches zero the release rate also approaches zero. A recently proposed mechanism (for the rate determining step) (Marques et al., 1990; 1991)

$$\mathrm{FeTfX} \xrightarrow{k_1[L]} \text{products}$$
$$K_{eq} \quad [L] \updownarrow [X]$$
$$\mathrm{FeTfL} \xrightarrow[k_2[L]]{} \text{products}$$

$$k_{obs} = (k_1\,[L] + k_2\,K_{eq}'\,[L]^2)\,/\,(1 + K_{eq}'\,[L]) \qquad K_{eq}' = K_{eq}\,/\,[X]$$

which can quantitatively account for the dependence of k_{obs} on anion concentration as well as the saturation-linear dependence on chelator concentration is shown in the scheme above.

L is a chelating agent such as PP_i or citrate and X is an anion such as Cl^- or ClO_4^-. The mechanism presupposes binding of an anion (X) to the protein (as shown by Kretchmar and Raymond, 1988). There is then a competition between X and L (the chelator is itself an anion) for the anion-binding site with $k_2 < k_1$. The mechanism is also able to account for previously published data on iron release to EDTA (Baldwin, 1980). Furthermore, if the value of K_{eq} is small, the dependence of k_{obs} on [L] will be experimentally indistinguishable from a saturation-linear dependence (Egan et al., 1992). The location of the (non-synergistic) kinetically significant anion binding site(s) (which appear to be different from those which affect the EPR spectrum) is unknown, but fluorescence energy transfer studies on $Ru(NH_3)_5$-labelled transferrin with Tb(III) in the iron-binding sites (Martin et al., 1991) has suggested histidines 207(535) or 242(577) as possible sites. The first is located on one of the interdomain β-strands and the other in the cleft between the domains (both likely sites for modifications to the protein conformation and hence iron release kinetics). The relevance of these anion-binding effects in vivo is not yet known.

There are striking differences in the effects of anions on the kinetics of iron release from the N- and C-terminal sites (the former becoming more inert with increasing ClO_4^- or Cl-concentration, while the latter becomes more labile). Studies on diferrictransferrin using urea polyacrylamide gel electrophoresis and on monoferric-Co(III)-transferrin (the cobalt being kinetically inert) have demonstrated weak kinetic cooperativity between the sites (Bali and Harris, 1989; 1990).

Far less work has been done on the reductive release route (Kojima and Bates, 1979; Baldwin et al., 1990) and it will not be discussed in depth. Strong anion effects are observed using this route as well (Baldwin et al., 1990). At pH 7.4 release does not occur in the absence of a powerful Fe(II) chelating agent (Kojima and Bates, 1979) which provides the thermodynamic driving force for reduction. Studies in vivo have been performed in the presence of such chelators which are probably responsible for driving the otherwise unfavorable reduction (Thorstensen and Aisen, 1990).

The rate constants for iron release are strongly dependent on pH (Baldwin et al., 1982). Release from the N-terminal site appears to be independent of chelator concentration below about pH 5.6 (at least in the case of EDTA) while the C-terminal site is only weakly dependent (Baldwin et al., 1982). Anions have been shown to accelerate iron release at low pH (Foley and Bates, 1988), but the mechanism has not been studied in detail.

In vivo iron release occurs in a process of receptor-mediated endocytosis (Dautry-Varsat, 1986). Ferric transferrin binds to the transferrin receptor on the cell surface and is endocytosed into a vesicle which is acidified to a pH of about 5.4 (van Renswoude et al., 1982; Dautry-Varsat et al., 1983) in an energy dependent process. Iron is released from the transferrin at the low pH, the apotransferrin remaining bound to the receptor as it is returned to the cell surface. At the pH of the cell surface, 7.4, apotransferrin is released from receptor to the circulation for another cycle of iron transport. The mechanism by which iron is transported across the vesicle membrane to sites of utilization or storage is unknown.

The transferrin receptor modulates the rates of iron release (Bali et al., 1991b; Bali and Aisen, 1991). At pH 7.4 iron release is retarded, while at pH 5.6 it is accelerated by the receptor. A related observation shows that iron release occurs at a higher pH from the transferrin-receptor complex than from free transferrin (Sipe and Murphy, 1991).

V. BINDING OF METAL IONS OTHER THAN IRON

Iron is the most strongly bound metal and no other metal ion is capable of displacing it from the binding site. However, a large number of metals are capable of binding to apotransferrin (Harris and Aisen, 1989) and since transferrin is only about 30% iron saturated in the serum (Leibman and Aisen, 1979) some of these may be significant in vivo. Of these, four will be discussed briefly, Al, Ga, In and Zn:

(i) Al(III) has attracted some attention because of its possible role in Alzheimer's disease (Good et al., 1992). It binds quite strongly to transferrin (log $K_1^* = 15.4$) (Cochran et al., 1984) and its release to citrate is considerably more facile than that of iron (Marques, 1991). It is reportedly capable of being transported inside cells by transferrin (Cannata et al., 1991).

(ii) Ga(III) has the highest binding constant of any metal other than iron (log $K_1^* = 18.1$, atm. CO_2, pH 7.4) (Harris and Pecoraro, 1983), presumably reflecting its similar size and identical charge. Both radioactive Ga(III) (Vallabhajosula et al., 1980) and In(III) (Goodwin et al., 1971) have drawn attention as tumor imaging agents, apparently being transported in the serum by transferrin.

(iii) Transferrin has been suggested as a possible Zn-transporting protein (Evans and Winter, 1975), but it has been shown (Harris and Keen, 1989) that under the conditions of CO_2 partial pressure prevailing in the serum, Zn(II) binding would be greatly decreased by inorganic anions, especially phosphate and carbonate which bind to the iron-binding site. Zinc is likely to be bound predominantly to albumin or other plasma proteins.

VI. CONCLUSIONS

The structure of transferrin and the nature of the ligating amino acid residues are very well suited for selectively binding Fe(III). This powerful binding both solubilizes Fe(III) and results in a large negative reduction potential which probably prevents reduction to Fe(II) in the presence of reducing agents found in vivo (at least at pH 7.4) and hence blocks the metal-catalyzed Haber-Weiss reaction cycle, thwarting $^{\bullet}OH$ production (Baldwin et al., 1984) and rendering the Fe(III) center non-toxic.

To be an efficient iron-transport protein, transferrin must be able to release iron with considerable facility under the desired conditions. Thus its binding strength and lability of iron release are strongly pH-dependent. It is possible that the synergistic anion plays a role in promoting iron release at low pH since the carbonate is likely to convert to a monodentate configuration and then hydrolyze by C-O bond scission under acidic conditions as is observed for low-molar-mass complexes (Palmer and van Eldik, 1983), and this would result in unfavorable interactions between the Fe(III) and the arginine responsible for synergistic anion binding. In addition, the rate of iron release is strongly influenced by the receptor. At the cell surface (pH 7.4) where iron release and resulting Fenton chemistry would be most dangerous release is inhibited, while at the pH of the endosome (pH 5.6) it is considerably enhanced so that iron release can occur within the timescale of endocytosis as long as a suitable concentration of chelating agent is present (about 1 mM in the case of PP_i (Bali et al., 1991b) and probably a similar concentration for other physiologically more relevant polyphosphates such as ATP and/or ADP).

ACKNOWLEDGEMENTS

The preparation of this review was funded in part by Grant DK 37927 from the National Institutes of Health, U.S. Public Health Service.

REFERENCES

Aisen, P., and Leibman, A. (1972). *Biochim. Biophys. Acta*, *257*: 314.

Aisen, P., Leibman, A., and Zweier, J. (1978). *J. Biol. Chem.*, *253*: 1930.

Anderson, B.F., Baker, H.M., Dodson, E.J., Norris, G.E., Rumball, S.V., Waters, J.M., and Baker, E.N. (1987). *Proc. Nat. Acad. Sci. USA*, *84*: 1769.

Anderson, B.F., Baker, H.M., Norris, G.E., Rice, D.W., and Baker, E.N. (1989). *J. Mol. Biol.*, *209*: 711.

Anderson, B.F., Baker, H.M., Norris, G.E., Rumball, S.V., and Baker, E.N. (1990). *Nature*, *344*: 784.

Bailey, S., Evans, R.W., Garratt, R.C., Gorinsky, B., Hasnain, S., Horsburgh, C., Jhoti, H., Lindley, P.F., Mydin, A., Sarra, R., and Watson, J.L. (1988). *Biochemistry*, *27*: 5804.

Baker, E.N., Rumball, S.V., and Anderson, B.F. (1987). *Trends Biochem. Sci.*, *12*: 350.

Baker, E.N., Baker, H.M., Smith, C.A., Stebbins, M.R., Kahn, M., Hellström, K.E., and Hellström, I. (1992). *FEBS Lett.*, *298*: 215.

Baldwin, D.A. (1980). *Biochim. Biophys. Acta*, *623*: 183.

Baldwin, D.A., and de Sousa, D.M.R. (1981). *Biochem. Biophys. Res. Commun.*, *99*: 1101.

Baldwin, D.A., and Egan, T.J. (1987). *S. Afr. J. Sci.*, *83*: 22.

Baldwin, D.A., de Sousa, D.M.R., and von Wandruszka, R. (1982). *Biochim. Biophys. Acta*, *719*: 140.

Baldwin, D.A., Jenny, E.R., and Aisen, P. (1984). *J. Biol. Chem.*, *259*: 13391.

Baldwin, D.A., Egan, T.J., and Marques, H.M. (1990). *Biochim. Biophys. Acta*, *1038*: 1.

Bali, P.K., and Aisen, P. (1991). *Biochemistry*, *30*: 9947.

Bali, P.K., and Harris, W.R. (1989). *J. Amer. Chem. Soc.*, *111*: 4457.

Bali, P.K., and Harris, W.R. (1990). *Arch. Biochem. Biophys.*, *281*: 251.

Bali, P.K., Harris, W.R., and Nesset-Tollefson, D. (1991a). *Inorg. Chem.*, *30*: 502.

Bali, P.K., Zak, O., and Aisen, P. (1991b). *Biochemistry*, *30*: 324.

Bates, G.W., and Schlabach, M.R. (1975). *J. Biol. Chem.*, *250*: 2177.

Bertini, I., Hirose, J., Luchinat, C., Messori, L., and Piccioli, M. (1988). *Inorg. Chem.*, *27*: 2405.

Brown, J.P., Hewick, R.M., Hellström, I., Hellström, K.E., Doolittle, R.F., and Dreyer, W.J. (1982). *Nature*, *296*: 171.

Cannata, J.B., Fernandez-Soto, I., Fernandez-Menendez, M.J., Fernandez-Martin, J.L., McGregor, S.J., Brock, J.H., and Halls, D. (1991). *Kidney Int.*, *39*: 799.

Chasteen, N.D. (1983). *Adv. Inorg. Biochem.*, *5*: 201.

Cochran, M., Coates, J., and Neoh, S. (1984). *FEBS Lett.*, *176*: 129.

Cowart, R.E., Kojima, N., and Bates, G.W. (1982). *J. Biol. Chem.*, *257*: 7560.

Cowart, R.E., Swope, S., Loh, T.T., Chasteen, N.D., and Bates, G.W. (1986). *J. Biol. Chem.*, *261*: 4607.

Dautry-Varsat, A. (1986). *Biochimie*, *68*: 375.

Dautry-Varsat, A., Ciechanover, A., and Lodish, H.F. (1983). *Proc. Nat. Acad. Sci. USA*, *80*: 2258.

Eby, J.E., Sato, H., and Sirbasku, D.A. (1992). *Anal. Biochem.*, *203*: 317.

Egan, T.J., Ross, D.C., Purves, L.R., and Adams, P.A. (1992). *Inorg. Chem.*, *31*: 1994.

Evans, G.W., and Winter, T.W. (1975). *Biochem. Biophys. Res. Commun.*, *66*: 1218.

Foley, A.A., and Bates, G.W. (1988). *Biochim. Biophys. Acta*, *965*: 154.

Folajtar, D.A., and Chasteen, N.D. (1982). *J. Amer. Chem. Soc.*, *104*: 5775.

Good, P.F., Perl, D.P., Bierer, L.M., and Schmeidler, J. (1992). *Ann. Neurol.*, *31*: 286.

Goodwin, D.A., Goode, R., Brown, L., and Imbornone, C.J. (1971). *Radiology*, *100*: 175.

Grossmann, J.G., Neu, M., Pantos, E., Schwab, F.J., Evans, R.W., Townes-Andrews, E., Lindley, P.F., Appel, H., Thies, W.-G., and Hasnain, S.S. (1992). *J. Mol. Biol.*, *225*: 811.

Halliday, J.W. (1992). *Gastroenterology*, *102*: 1071.

Harris, W.R. (1984). *J. Inorg. Biochem.*, *21*: 263.

Harris, W.R. (1985). *Biochemistry*, *24*: 7412.

Harris, W.R. (1986a). *J. Inorg. Biochem.*, *27*: 41.

Harris, W.R. (1986b). *Biochemistry*, *25*: 803.

Harris, D.C., and Aisen, P. (1989). *Iron Carriers and Iron Proteins* (T.M. Loehr, H.B. Gray and A.B.P. Lever, eds.), VCH Publishers, Weinheim.

Harris, W.R., and Bali, P.K. (1988). *Inorg. Chem.*, *27*: 2687.

Harris, W.R., and Keen, C. (1989). *J. Nutr.*, *119*: 1677.

Harris, W.R., and Pecoraro, V.L. (1983). *Biochemistry*, *22*: 292.

Harris, W.R., Rezvani, A.B., and Bali, P.K. (1987). *Inorg. Chem.*, *26*: 2711.

Honda, K., Nishikata, I., and Sasakawa, S. (1980). *Chem. Lett.*, 21.

Jeltch, J.M., and Chambon, P. (1982). *Eur. J. Biochem.*, *122*: 291.

Kawakami, H., and Lönnerdal, B. (1991). *Amer. J. Physiol. Gastrointest. Liver Physiol.*, *261*: G841.

Kojima, N., and Bates, G.W. (1979). *J. Biol. Chem.*, *254*: 8847.

Kojima, N., and Bates, G.W. (1981). *J. Biol. Chem.*, *256*: 12034.

Kondo, H., Saito, K., Grasso, J.P., and Aisen, P. (1988). *Hepatology*, *8*: 32.

Kretchmar, S.A., and Raymond, K.N. (1986). *J. Amer. Chem. Soc.*, *108*: 6212.

Kretchmar, S.A., and Raymond, K.N. (1988). *Inorg. Chem.*, *27*: 1436.

Kretchmar, S.A., Reyes, Z.E., and Raymond, K.N. (1988). *Biochim. Biophys. Acta*, *956*: 85.

Legrand, D., Mazurier, J., Colavizza, D., Montreuil, J., and Spik, G. (1990). *Biochem. J.*, *266*: 575.

Leibman, A., and Aisen, P. (1979). *Blood*, *53*: 1058.

Lin, L.-N., Mason, A.B., Woodworth, R.C., and Brandts, J.F. (1991). *Biochemistry*, *30*: 11660.

MacGillivray, R.T.A., Mendez, E., Shewale, J.G., Sinha, S.K., Lineback-Zins, J., and Brew, K. (1983). *J. Biol. Chem.*, *258*: 3543.

Marques, H.M. (1991). *J. Inorg. Biochem.*, *41*: 187.

Marques, H.M., Egan, T.J., and Pattrick, G. (1990). *S. Afr. J. Sci.*, *86*: 21.

Marques, H.M., Watson, D., and Egan, T.J. (1991). *Inorg. Chem.*, *30*: 3758.

Martin, D.M., Chasteen, N.D., and Grady, J.K. (1991). *Biochim. Biophys. Acta*, *1076*: 252.

Metz-Boutigue, M.-H., Jolles, J., Mazurier, J., Schoentgen, F., Legrand, D., Spik, G., Montreuil, J., and Jolles, P. (1984). *Eur. J. Biochem.*, *145*: 659.

Palmer, D.A., and van Eldik, R. (1983). *Chem. Rev.*, *83*: 651.

Patch, M.G., and Carrano, C.J. (1981). *Inorg. Chim. Acta*, *56*: L71.

Pearson, R.G. (1963). *J. Amer. Chem. Soc.*, *85*: 3533.

Princiotto, J.V., and Zapolski, E.J. (1975). *Nature*, *255*: 87.

Rose, T.M., Plowman, G.D., Teplow, D.B., Dreyer, W.J., Hellstrom, K.E., and Brown, J.P. (1986). *Proc. Nat. Acad. Sci. USA*, *83*: 1261.

Sawatzki, G. (1987). *Iron Transport in Microbes, Plants and Animals* (G. Winkelmann, D. van der Helm and J.B. Neilands, eds.), VCH, Weinheim, p. 477.

Schlabach, M.R., and Bates, G.W. (1975). *J. Biol. Chem.*, *250*: 2182.

Shannon, R.D., and Prewitt, C.T. (1970). *Acta Cryst.*, *B26*: 1046.

Shongwe, M.S., Smith, C.A., Ainscough, E.W., Baker, H.M., Brodie, A.M., and Baker, E.N. (1992). *Biochemistry*, *31*: 4451.

Sipe, D.M., and Murphy, R.F. (1991). *J. Biol. Chem.*, *266*: 8002.

Snetsinger, P.A., Chasteen, N.D., and van Willigen, H. (1990). *J. Amer. Chem. Soc.*, *112*: 8155.

Swope, S.K., Chasteen, N.D., Weber, K.E., and Harris, D.C. (1988). *J. Amer. Chem. Soc.*, *110*: 3835.

Thorstensen, K., and Aisen, P. (1990). *Biochim. Biophys. Acta*, *1052*: 29.

Valentini, P., Orsi, N., and Antonini, E. (1981). *Chemioterapia Antimicr.*, *4*: 12.

Vallabhajosula, S.R., Harwig, J.F., Siemsen, J.K., and Wolf, W. (1980). *J. Nucl. Med.*, *21*: 650.

van Renswoude, J., Bridges, K.R., Harford, J.B., and Klausner, R.D. (1982). *Proc. Nat. Acad. Sci. USA*, *79*: 6186.

Weaver, J., and Pollock, A.S. (1990). *Acta Haematol.*, *84*: 68.

Woodworth, R.C., Mason, A.B., Funk, W.D., and MacGillivray, R.T.A. (1991). *Biochemistry*, *30*: 10824.

Yang, F., Lum, J.B., McGill, J.R., Moore, C.M., Naylor, S.L., van Bragt, P.H., Baldwin, W.D., and Bowman, B.H. (1984). *Proc. Nat. Acad. Sci. USA*, *81*: 2752.

METAL-FERRITIN INTERACTIONS

by

R.R. Crichton

Unité de Biochimie,
Université Catholique de Louvain,
Place Louis Pasteur 1
B-1348 Louvain-la-Neuve
Belgium

I. INTRODUCTION

The ubiquity of ferritin in the biosphere no doubt reflects its role as the principal protein involved in storing iron in a soluble, non-toxic, yet bio available form (reviewed in Crichton, 1991: Harrison et al., 1991: Crichton and Ward, 1992). It is found in bacteria, plants and animals, where it maintains up to 4,500 atoms of hydrolyzed and polymerized Fe(III) in a soluble form within a hollow protein shell (apoferritin). Apoferritin, made up of 24 polypeptide subunits, is synthesized and assembles into the tetracosameric oligomer in the absence of iron. Thereafter the incorporation of iron involves the protein itself both in the nucleation and the growth of the iron core, either as the crystalline mineral ferrihydrite or as an amorphous hydrous oxyphosphate (Theil, 1987; Crichton, 1991; Harrison et al., 1989; 1991). In mammals and plants, iron is also stored in hemosiderins (see Crichton and Ward, 1992 for a recent review).

II. MAMMALIAN FERRITINS

A. Structure of apoferritins

Ferritin is present in all mammalian cells in which its presence has been examined, with a predominance in liver, spleen, bone marrow and muscle (Bothwell et al., 1979). In normal human subjects some 25% of total body iron, around 800-1000 mg, is present in the storage protein ferritin, and to a lesser extent in hemosiderin, the latter essentially in reticuloendothelial cells (Bothwell et al., 1979).

Mammalian ferritins are 24-subunit proteins which are disposed in a highly symmetric protein shell, crystallizing in the space group F 432, of external diameter 12 nm enclosing a cavity of diameter 7-8 nm (Fig. 1). The protein part, apoferritin, is constituted by two subunit types H and L of molar mass respectively 21 kg mol^{-1} and 20 kg mol^{-1}, predominant in heart (H) and liver (L). The 24-subunit oligomer is a heteropolymer containing variable amounts of the two protein subunits. The three-dimensional structure of recombinant human H apoferritin subunits (Lawson et al., 1991) is essentially the same as that of horse spleen apoferritin, rat liver apoferritin and recombinant rat L-chain apoferritin, with backbone atoms being superposable to within 0.1 nm (reviewed in Harrison et al, 1991). This is illustrated in Figure 2 by a stereo α-carbon trace of the

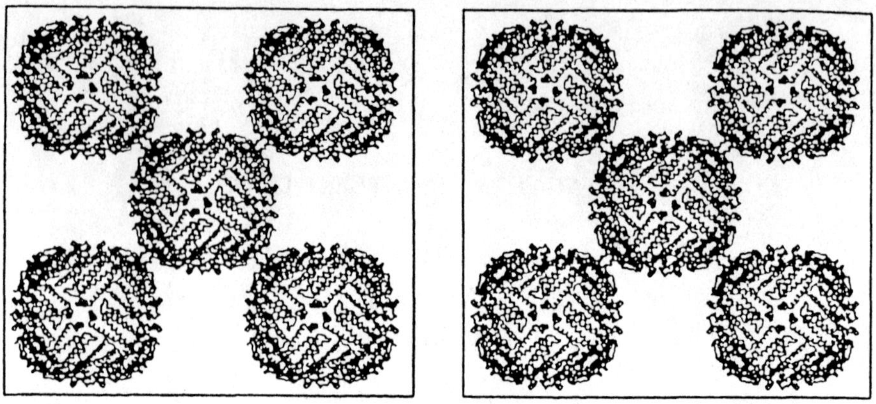

Figure 1.
Stereo diagram showing packing of molecules on a face of the face-centered cubic crystal of human recombinant H-chain apoferritin. Only 16 of the 24 subunits are shown; the 8 outer subunits have been removed to show the hollow protein shell viewed down a 4-fold symmetry axis looking into the interior. It is within this compact protein shell that iron is deposited as a hydrolyzed polymerized ferrihydrite core in mammalian ferritins (from Harrison et al., 1991).

subunit of human recombinant H-chain apoferritin superimposed on that of rat recombinant L-chain apoferritin.

The apoferritin subunit (Fig. 3) is cylindrical, 5 nm long by 2.5 nm wide. The subunit is a typical four α-helical bundle containing two antiparallel helix pairs (A,B and C,D), with an additional helix E situated at one end of the bundle, at about 60° to its axis. Each of the helical pairs is linked by a short turn: the connectivity between the B and C helices is assured by a long non-helical stretch of 18 residues (the BC-loop) which spans the entire length of the bundle. In the H-subunit, the N-terminal 13 residues lie on the outside of the protein shell, and are followed by helix A (14-43), the AB turn (44-49), helix B (50-76) and then the BC-loop (77-95) which stretches along the length of the helical bundle to connect with helix C (96-123). From there the CD turn (124-128) leads into helix D (129-136 and 137-158 with a break at 136 due to a wide turn) which runs along the internal surface of the helix bundle and protrudes from it to terminate after a short non-helical turn (159-162) in the E helix (163-172), which is disposed at about 60° to the major helical bundle. Residues 177-182 at the C-terminus are not visible in the electron density map. Extensive side-chain interactions are present within the subunit, forming a tightly packed hydrophobic core. The subunits can readily form antiparallel dimers removing hydrophobic residues from both helix A and the BC-loop from contact with water. This generates a short region of antiparallel β-sheet within the dimer. In the packed protein shell there are extensive subunit interactions; exceptions are the 4-fold and 3-fold symmetry axes where narrow channels of about 0.15 nm pass through the protein shell. The 4-fold channels, flanked by four roughly parallel E-helices, are about 1.3 nm long and are predominantly hydrophobic in mammalian ferritins, whereas the 3-fold channels are rather polar. The 24-subunit oligomer is very stable, as befits its role in maintaining the insoluble mineral phase in a soluble state within the confines of the protein shell.

H subunits have much more sequence homology with H subunits of different species than they do with L subunits of the same species (Fig. 4 presents the amino acid sequences of H and L subunits of a number of ferritins: the numbering uses L-chain positions). Of particular interest is the central hydrophilic region of the subunit fold (see Section C below) where the four helices A, B, C and D interact. In all H-subunits there are 7 conserved residues, namely Glu-23, Tyr-30, Glu-57, Glu-58, His-61, Glu-103 and Gln-137. The equivalent hydrophilic regions of L-chains only have three of the conserved residues

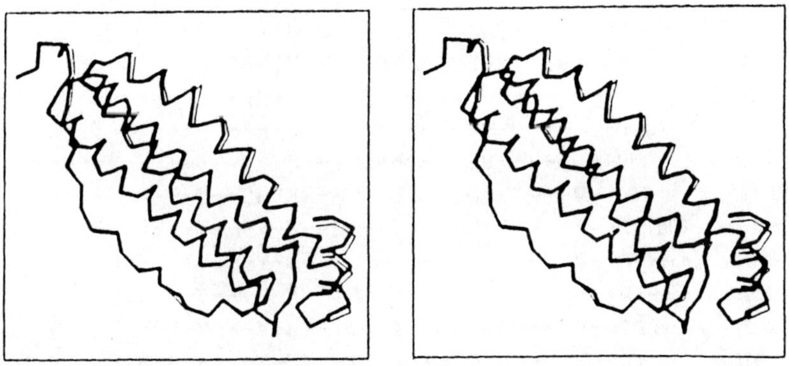

Figure 2.
Stereo α-carbon trace of the subunit of recombinant H-chain apoferritin (heavy line) superimposed on that of rat recombinant L-chain apoferritin. The eight-residue insertion (see Fig. 4) of the rat protein gives no clear electron density (from Harrison et al., 1991).

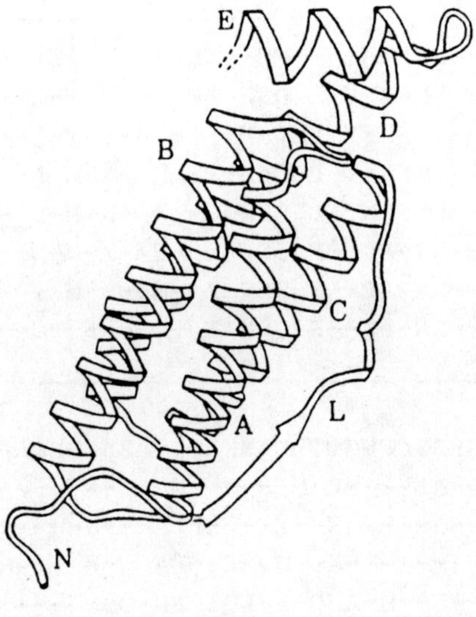

Figure 3.
Ribbon diagram of α-carbon backbone of the apoferritin subunit showing the four helical bundle, the short helix E and the BC-loop (marked L). The N-terminus (N) lies at the other end of the helical bundle to the E helix (from Ford et al., 1984).

```
            1   5    10        20        30        40
HoL         SSQIRQNYSTEVEAAVNRLVNLYLRASYTYLSLGFYFDRD
HuL         ------------D----S------Q---------------
RaL         T---------------------H------------F----
RbL         T--------P-------H----H-----------------
TaL        ME--V---FHQDC--GL--T---KFHS--V---MAS--N--
HuH     TTAST--V----HQDS---I--QI--E-Y---V---MSY-----
RaH     ----P--V----HQDS---I--QI--E-Y---V---MSC-----
MoH     ----P--V----HQDA---I--QI--E-Y---V---MSC-----
ChH     MATPP--V----HQDC---I--QI--E-Y---V---MSY-----
TaH        MD--V---FHRDC---I--M--ME-Y-------MA------
TaM        MV--V----HSDC------ML--E-Y-----S-MYAF----
XeH        MQ--VL--FNSDC-I-I--M---EMY---V---MSY-----

            41       50        60        70        80
Hol         DVALEGVCHFFRELAEEKREGAERLLKMQNQRGGRALFQD
Hul         -------S-------------Y------------------
RaL         -------G-------------------L--E---------
RbL         -------S------------A-------------------
TaL         ----SNFAK----RS--EK-H--K-IEY-------VFL-S
HuH         ----KNFAKY-LHQSH-E--H--K-M-L-------IFL--
RaH         ----KNFAKY-LHQSH-E--H--K-M-L-------IFL--
MoH         ----KNFAKY-LHQSH-E--H--K-M-L-------IFL--
ChH         ----KNFAKY-LHQSH-E--H--K-M-L-------IFL--
TaH         -I--HN-AK--K-QSH-E--H--K-M-D--K----IVL--
TaM         ----HN-AE--K-HSN-E--H--KFM-Y--K----VVL--
XeH         ----NN-AK--K-QSN-E--H--KF--Q--K----VVL--

            81       90        100       110       120
HoL         LQKPSQDEWGTTLDAMKAAIVLEKSLNQALLDLHALGSAQ
HuL         IK--AE----K-P------MA---K--------------R
RaL         V---------K--E-----LA---N---------------
RbL         V---------K--N--E--LA---N--------------H
TaL         VE--ER-D-ANG-E-LQT-LK-Q--V---------VAADK
HuH         IK--DC-D-ESG-N--EC-LH---NV--S--E--K-ATDK
RaH         IK--DR-D-ESG-N--RC-LH----V--S--E--K-ATDK
MoH         IK--DR-D-ESG-N--EC-LH---SV--S--E--K-ATDK
ChH         IK--DR-D-ENG-T--EC-LH---NV--S--E--K-ATEK
TaH         VK--ER----N--E--Q--LQ---TV--------KV--DK
TaM         IK--ER----N--E--Q--LQ---TV--------K-ATDK
XeH         IK--ER---SN--E--Q--LQ---TV------Q-K-A-DK
```

Figure 4.
Amino acid sequences of H- and L-subunits of a number of ferritins.The numbering used is that of the L-subunit.

	121 130 140 150 160
HoL	ADPHLCDFLESHFLDEEVKLIKKMGDHLTNIQRLVGSQAG
HuL	T---------T-------------------LH--G-PE--
RaL	---------------K---------N----LR-WQ-P--S
RbL	T---------N---------L----------R--S-P--S
TaL	S---MT------PY-S-S-ET---L---I-SLKK-WS-HP-
HuH	N-------I-T-Y-N-Q--A--EL---V--LRKMGAPES-
RaH	N-------I-T-Y-N-Q--S--EL---V--LRKMGAPES-
MoH	N-------I-TYY-S-Q--S--EL---V--LRKMGAPEA-
ChH	N-------I-T-Y---Q--A--QL---V--LRKMGAPKY-
TaH	V---------TEY-E-Q--S--QL--YI--LK--GLP-N-
TaM	V----------EY-E-Q--D--RI--FI--LK--GLPEN-
XeH	V--Q-------EY-E-Q--AM-AL--YI--LK--GA*-N-

	161 170 174 178
HoL	LGEYLFERLTLKHD
HuL	--------------
RaL	--------------
RbL	--------------
TaL	MA----NKH--G
HuH	-A----DKH--GDSDNES
RaH	MA----DKH--G-GDES
MoH	M-----------HGD*ES
ChH	MA----DKH--GESDS
TaH	M-----DKH-MGESS
TaM	M-----DKHSV-ESS
XeH	M-----DKHT--GESS

*** 8 residue insertion in rat liver L chain between 157 and 158: QPAQTGVA**

(namely Tyr-30, Glu-57 and Glu-103, Glu-103, however, is replaced by Gln in tadpole L chains).

B. Structure of the iron core

Within the confines of the hollow protein shell, a variable amount of iron is deposited essentially as a ferric oxyhydroxide mineral phase. The cores of individual ferritin molecules are readily visualized by electron microscopy. For mammalian ferritins, the amount of iron in the core is variable and can range from zero in apoferritin up to a maximum of 4,500 iron atoms per molecule of protein (Fischbach and Anderegg, 1965). Both human and horse ferritin molecules contain relatively well ordered iron cores composed of single-domain crystals of a mineral known as ferrihydrite (Fig. 5). This is a well established mineral form of ferric oxyhydroxide of composition $5Fe_2O_3$. $9H_2O$ (Towe and Bradley, 1967). The iron core also contains some phosphate, which seems to be localized to a large extent on the surface of the ferrihydrite core where it interacts with the internal surface of the protein shell. The crystallinity of ferritin iron cores differs from molecule to molecule, some having well-ordered single crystals which occupy much of the cavity whereas others consist of mixtures of small crystals together with poorly crystalline or amorphous regions.

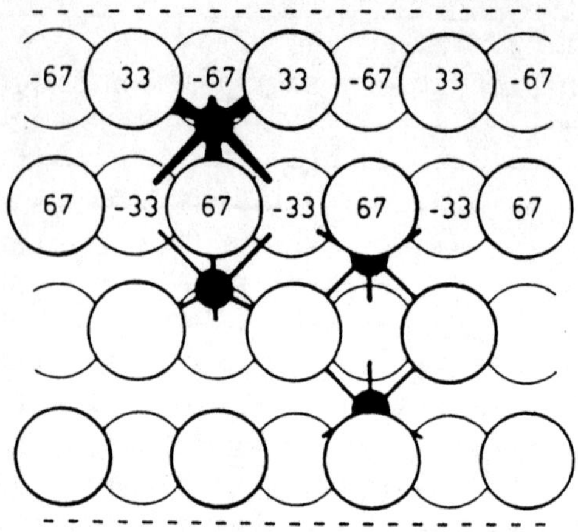

Figure 5.
The structure of ferrihydrite (from Ford et al., 1984).

C. Iron uptake and release
Iron uptake by apoferritin appears to involve oxidation of Fe(II), catalyzed in the initial phase by the protein itself; in the latter stages, iron oxidation and hydrolysis continue on the surface of the growing ferrihydrite iron core. It is suggested that in the initial stages of iron incorporation, Fe(II) binds to specific sites on or close to the inner surface of the protein shell where a ferroxidase activity catalyzes the oxidation of Fe(II) to create localized zones of Fe(III) species. H-subunits appear to have a much greater ferroxidase activity than L-subunits. It has been proposed that the central hydrophilic region of the subunit fold with its seven conserved residues is associated with both metal binding and ferroxidase activity (Lawson et al., 1989). Figure 6 illustrates the ferroxidase center of human recombinant H ferritin as well as the corresponding region of horse spleen apoferritin in which the ferroxidase site is blocked by a salt bridge between Lys-62 and Glu-107, respectively Lys-58 and Glu-103 in L-chain numbering (Harrison et al., 1991). The Fe(III), and perhaps also Fe(II)/ Fe(III) species, present as relatively early intermediates in ferritin core formation may include both peroxo-bridged and μ-oxo-bridged iron dimers (reviewed in Crichton and Ward, 1992). Iron penetration into the interior of the protein shell seems most likely to occur through the 3-fold symmetry axes of the protein shell (reviewed in Crichton and Ward, 1992); although mutation of the Glu and Asp residues of the three-fold channels to His does not prevent core formation, it does reduce the rate of core formation (Treffry et al., 1989). At some as yet undetermined step along the pathway of formation of the iron core of ferritin, the nucleus of Fe(III), perhaps already in a hydrolyzed form, takes over the role of iron oxidation from the protein, and catalyzes the oxidation, hydrolysis and polymerisation which result in the ferrihydrite micellar iron core (Fig. 6) which is characteristic of mammalian ferritins (for more detailed reviews see Crichton and Ward, 1992; St. Pierre et al., 1989; Harrison et al., 1991). The more amorphous forms of the iron core found in plant and microbial ferritins may reflect the higher concentrations of anions such as phosphate in the local environment, which provoke a more rapid precipitation of the micellar iron core (see below).

Iron mobilization from ferritin might proceed by various pathways (although none has been reasonably established in vivo). These include direct chelation of Fe(III), reduction of Fe(III) to Fe(II) and mobilization of the ferrous iron by appropriate chelators (in the absence of chelation the fully electrochemically reduced core of ferritin iron seems to remain within the protein shell - Watt et al., 1985), or perhaps, only after ferritin molecules have been taken up by lysosomes and transformed into hemosiderin, by a still to be discovered mechanism of iron mobilization.

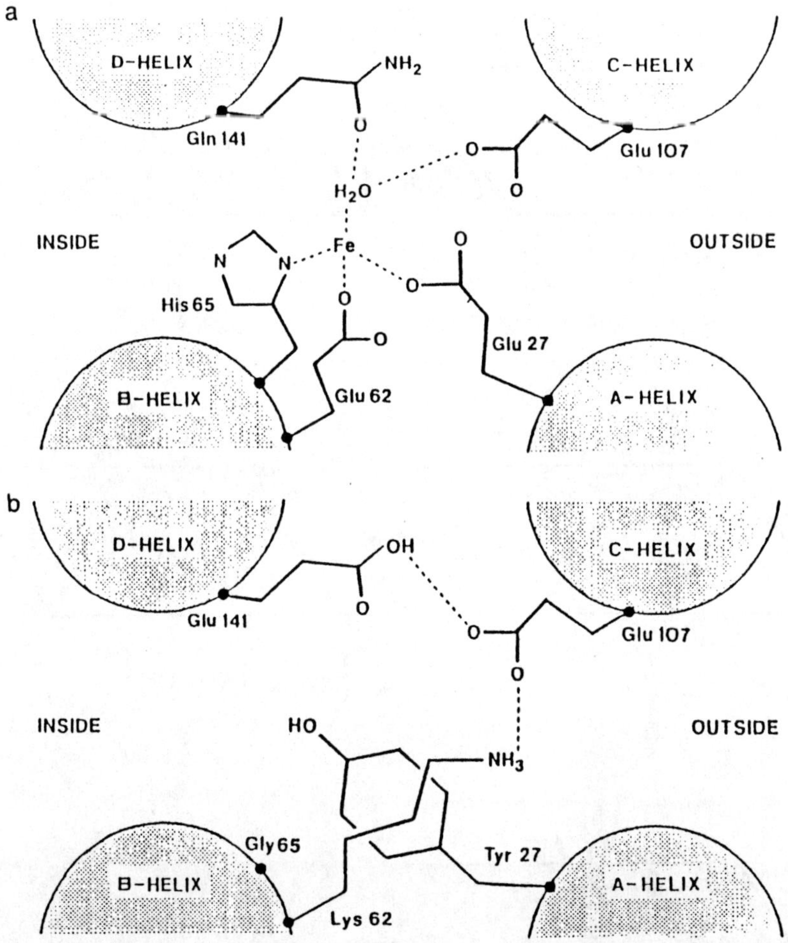

Figure 6.
Schematic diagram of the ferroxidase center of human recombinant L-chain apoferritin (a), and the corresponding region of horse L-chain apoferritin (b) showing the ferroxidase site replaced by a salt bridge (from Harrison et al., 1991). The numbering is that of H-chain ferritins; L-chain numbering as in Figure 4 corresponds to H numbering -4 residues.

D. Biosynthesis

It is well established (Drysdale and Munro, 1966) that cellular ferritin expression is regulated at the level of translation of mRNA, and that there is a pool of untranslated ferritin mRNA within mammalian cells (Zähringer et al., 1976). Ferritin mRNA's are processed from genes which contain four coding exons and three spacer introns (see Fig. 3 for the place of introns in mammalian ferritins). The mature mRNA's for all mammalian ferritins so far examined have a well conserved stem loop sequence (Fig.7) in their 5'-untranslated regions (UTR's), which has been designated as an Iron Regulatory Element (IRE). The IRE's can bind a specific cytoplasmic Iron Regulatory Element Binding Protein (designated IRE-BP), either in a low affinity form (when cells are iron-replete) which allows the translation of ferritin mRNA's, or in a high affinity form (when cells are iron-depleted) which blocks the synthesis of ferritin. It has been proposed that cellular iron homeostasis might be regulated by the oscillation of the IRE-BP between its high and low affinity forms as a function of intracellular iron levels. It is of interest to point out that expression of the mRNA of the transferrin receptor (TfR) also contains IRE's in its 3'-

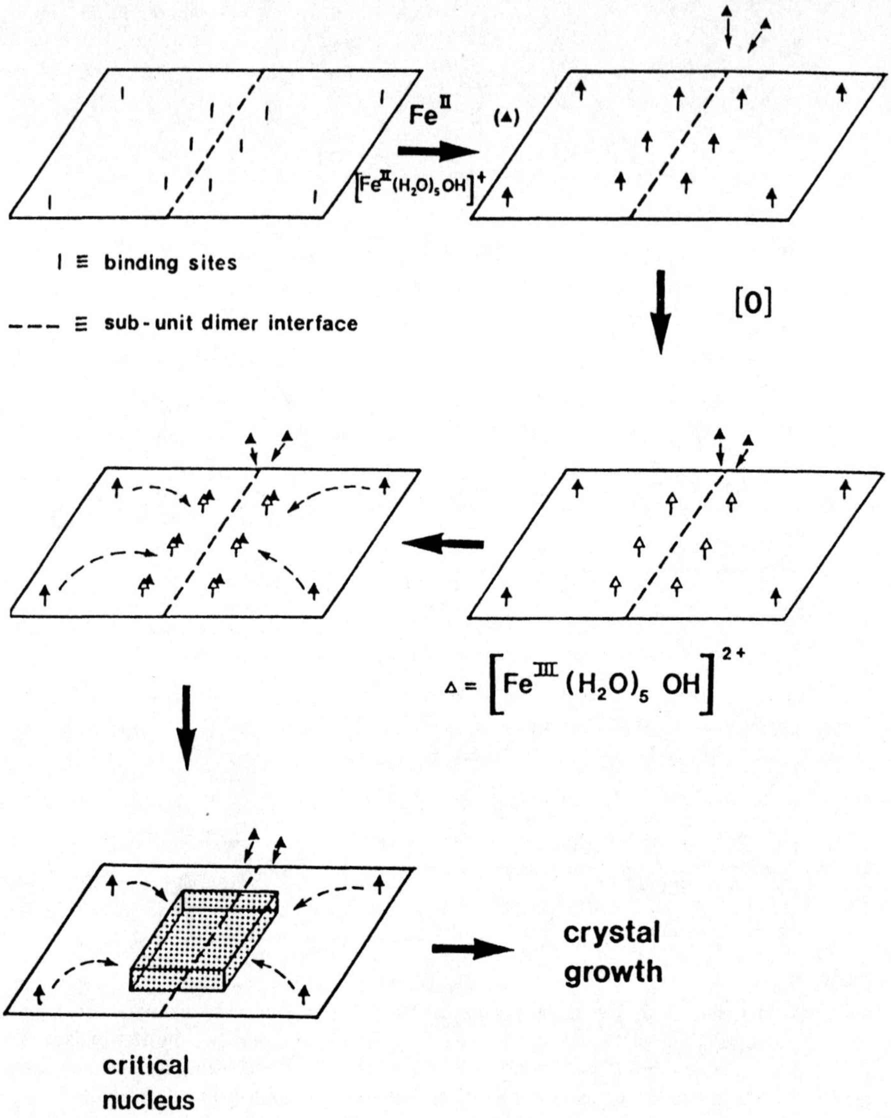

Figure 7.
Diagrammatic representation of a possible mechanism of crystal growth of the mineral core in ferritin. The subunit dimer interface contains a cluster of possible binding sites at which Fe^{2+} is located. Oxidation of Fe^{2+} at specific sites at the interface creates a localized zone of Fe^{3+} species (nucleation zone), which can become a focus for further Fe^{2+} diffusion and oxidation. Once the number of localized Fe^{3+} species attains a critical size, crystal growth continues by further addition and oxidation of Fe^{2+} to the nucleation zone. (From St. Pierre et al., 1989).

UTR. TfR is involved in cellular iron uptake from transferrin (Tf) circulating in the serum and certain extracellular fluids. The expression of TfR mRNA appears to be regulated in the opposite sense to ferritin - when cells are replete in iron TfR, mRNA is down-regulated, and when they are iron-depleted, its expression is up-regulated (for a more detailed review see Rouault et al., 1992). It has very recently been established that this reciprocal regulation of TfR and ferritin mRNA's is operational in physiological conditions of iron overload and iron deficiency in the rat (Ward, R.J., Kühn, L. and Crichton, R.R., in preparation), as distinct from previous studies which used cellular models with pharmacological stimulants of cellular iron status.

We can therefore conclude that ferritin biosynthesis is likely to be an essential element of celullar iron homeostasis in mammalian cells through the regulation of the expression of its mRNA via the IRE-BP.

III. PLANT FERRITINS

The complete primary structures of three plant ferritins have been recently determined (from soybean, French beans and pea - reviewed in Lobreaux et al., 1992). They show extensive sequence homology to animal ferritins, but seem to be constructed from only one type of protein subunit. They have a number of important differences when compared to mammalian ferritins, notably an N-terminal extension of some 70 residues containing a transit peptide, responsible for targetting the plant ferritin precursors to the chloroplasts where it is cleaved before assembly of apoferritin within plastids, and a second part of the N-terminal extension which is cleaved in vivo during germination (Lobreaux and Briat, 1991). From prediction of the three-dimensional structure of pea seed ferritin (Lobreaux et al., 1992), it is proposed that the N-terminal extension is exposed on the outside surface of the protein shell, that the pea ferritin subunit has a ferroxidase center similar to mammalian H-chains and that the 4-fold channel is hydrophilic rather than hydrophobic in mammalian ferritins.

The iron cores of native pea seed ferritin (Wade et al., 1991) are found to be amorphous, in correlation with their high phosphate content (Fe:P = 2.83). In contrast, reconstituted cores obtained by adding iron in the absence of phosphate were crystalline ferrihydrite. This underscores the point made above that the biomineralization of the ferritin core reflects the in vivo milieu in which it is formed.

IV. BACTERIAL FERRITINS

Ferritin is also present in many micro-organisms (for a recent review see Harrison et al., 1991). Bacterioferritins seem to share the same 24-subunit structure of mammalian and plant ferritins, without having any apparent sequence homology (Harrison et al., 1991). However, more recent analysis has indicated that they do have about 20% homology with mammalian ferritins (Andrews et al., 1992), and like the plant ferritins are more similar to the H chains of mammalian ferritins both in conservation of their amino acid sequence and in the conservation of the residues involved in the ferroxidase center. They have one outstanding feature which distinguishes them from the other ferritins, namely that they contain the protoporphyrin IX heme group.

The stoichiometry of heme seems to be about one heme per two subunits, although it has been proposed that a one to one relation might be possible (Kadir and Moore, 1990). It is suggested that the axial ligands to bacterioferritin are the thioether side chains of two methionine residues (Cheesman et al., 1990). The role of heme in bacterioferritins remains unresolved.

The iron cores of bacterioferritins are typically less ordered than their mammalian homologues, reflecting the higher anion (mostly phosphate) concentrations prevalent in the in vivo situations in which they are formed.

V. CONCLUSIONS

We may conclude by affirming the ferritin is the major iron storage protein in all living organisms, that it serves as a soluble, bioavailable and non-toxic form of iron within cells, and that at least in eukaryotes it also participates actively in cellular iron homeostasis. What more could we ask of an iron storage molecule?

REFERENCES

Andrews, S.C., Arosio, P., Bottke, W., Briat, J.-F., von Darl, M., Harrison, P.M., Laulhere, J.-P., Levi, S., Lobreaux, S., and Yewdall, S.J. (1992). *J. Inorg. Biochem.*, *47*: 161.

Bothwell, T.H., Charlton, R.W., Cook, J.D., and Finch, C.A. (1979). *Iron Metabolism in Man*, Blackwell, Oxford.

Cheesman, M.R., Thomson, A.J., Greenwood, C., Moore, G.R., and Kadir, F. (1990). *Nature*, *346*: 771.

Crichton, R.R. (1991). *Inorganic Biochemistry of Iron Metabolism*, Ellis Horwood, Chichester.

Crichton, R.R., and Ward, R.J. (1992). *Biochemistry*, *31*: 11255.

Drysdale, J.W., and Munro, H.N. (1966). *J. Biol. Chem.*, *241*: 3630.

Fischbach, F.A., and Anderegg, J.W. (1965). *J. Mol. Biol.*, *14*: 458.

Harrison, P.M., Artymiuk, P.J., Ford, G.C., Lawson, D.M., Smith, J.M.A., Treffry, A., and White, J.L. (1989). *Biomineralization -Chemical and Biomedical Perspectives* (S. Mann, J. Webb and R.J.P. Williams, eds.), VCH Verlag, Weinheim, p. 257.

Harrison, P.M., Andrews, S.C., Artymiuk, P.J., Ford, G.C., Guest, J.R., Hirzmann, J., Lawson, D.M., Livingstone, J.C., Smith, J.M.A., Treffry, A., and Yewdall, S.J. (1991). *Adv. Inorg. Chem.*, *36*: 449.

Kadir, F.H.A., and Moore, G.R. (1990). *FEBS Lett.*, *271*: 141.

Lawson, D.M., Treffry, A., Artymiuk, P.J., Harrison, P.M., Yewdall, S.J., Luzzago, A., Cesarini, G., Levi, S., and Arosio, P. (1989). *FEBS Lett.*, *254*: 207.

Lawson, D.M., Artymiuk, P.J., Yewdall, S.J., Smith, J.M.A., Livingstone, J.C., Treffry, A., Luzzago, A., Levi, S., Arosio, P., Cesarini, G., Thomas, C.D., Shaw, W.V., and Harrison, P.M. (1991). *Nature*, *349*: 541.

Lobreaux, S., and Briat, J.-F. (1991). *Biochem. J.*, *274*: 601.

Lobreaux, S., Yewdall, S.J., Briat, J.-F., and Harrison, P.M. (1992). *Biochem. J.*, *288*: 931.

Rouault, T.A., Haile, D.J., Downey, W.E., Philpott, C.C., Tang, C., Samaniego, F., Chin, J., Paul, I., Orloff, D., Harford, J.B., and Klausner, R.D. (1992). *Biometals*, *5*: 131.

St. Pierre, T.G., Webb, J., and Mann, S. (1989). *Biomineralization -Chemical and Biomedical Perspectives* (S. Mann, J. Webb and R.J.P. Williams, eds.), VCH Verlag, Weinheim, p. 295.

Theil, E.C. (1987). *Annu. Rev. Biochem.*, *56*: 289.

Towe, K.M., and Bradley, W.F. (1967). *J. Colloid Interface Sci.*, *24*: 384.

Treffry, A., Harrison, P.M., Luzzago, A., and Cesarini, G. (1989). *FEBS Lett.*, *247*: 268.

Wade, V.J., Levi, S., Arosio, P., Treffry, A., Harrison, P.M., and Mann, S. (1991). *J. Mol. Biol.*, *221*: 1443.

Watt, G.D., Frankel, R.B., and Papaefthymiou, G.C. (1985). *Proc. Nat. Acad. Sci. USA*, *82*: 3640.

Zähringer, J., Baliga, B.S., and Munro, H.N. (1976). *Proc. Nat. Acad. Sci. USA*, *73*: 857.

METALLOTHIONEINS

by

Dean L. Pountney, Jeremias H.R. Kägi and Milan Vašák

Biochemistry Institute,
University of Zürich,
Winterthurerstrasse 190
8057 Zürich
Switzerland

I. INTRODUCTION

Metallothionein (MT) is a generic name for a superfamily of ubiquitous proteins or polypeptides possessing a novel type of sulfur-based metal clusters. In these compounds clusters are usually formed through the preferential coordination of d^{10} metal ions by an array of closely spaced cysteine thiolate ligands. MT was discovered in 1957 by Margoshes and Vallee in their search for a macromolecular tissue component responsible for the natural accumulation of cadmium in equine kidney. This protein was subsequently named MT to reflect the extremely high sulfur and metal content that, depending on metal ions present, may together contribute to over 20% of the total molecule by mass (Kägi and Vallee, 1960). Although MTs are still the only biological compounds in which cadmium accumulates naturally, this metal is only one of several optional metallic components, the others being usually zinc and copper. In fact, in subsequent studies, zinc and/or copper were found to be the principal metal constituents under normal physiological conditions, indicating that the major role of MTs might be found in the metabolism and handling of these biologically essential elements (Kägi and Vallee, 1961; Pulido et al., 1966). The in vivo binding of nonessential metals such as Cd, Hg, Pt, Bi, Ag and Au to MT supports its role in cellular protection against the harmful influence of these detrimental metals (cited in Kägi, 1991).

II. OCCURRENCE, DEFINITION AND PRIMARY STRUCTURE

MTs have been isolated from vertebrates, invertebrates, plants, eukaryotic microorganisms and from some prokaryotes (Nordberg and Kojima, 1979; Hamer, 1986; Robinson and Jackson, 1986). In animals, the protein is most abundant in parenchymatous tissues, i.e. liver, kidney, pancreas, and intestines. Wide variations in the levels of MTs have been observed in different species and tissues, relating to effects of age, stage of development, diet and other as yet unidentified factors (see below). Although MT is a cytoplasmic protein, it can also accumulate in lysosomes (Johnson et al., 1981; Freedman et al., 1986), and during development it has been also detected in the nucleus (Nartey et al., 1987).

As early studies dealt exclusively with MT from mammalian sources, the classification of MTs is based on attributes of these proteins. In all mammals, MTs are composed of a single polypeptide chain of about 60 amino acids, 20 of which are cysteine and none of which are aromatic amino acids or histidine and have a total molar mass of just over 6 kg mol^{-1}. All cysteines occur in the reduced form and are usually coordinated to 7 equiv of divalent metal ions through mercaptide bonds, giving rise to spectroscopic features characteristic of metal-thiolate complexes. In view of this quite distinct physico-chemical signature, the phenomenological definition was adopted that any polypeptide resembling mammalian MT in several of these properties can be called an MT. In the absence of a more meaningful system of nomenclature, the metallothioneins have been categorized further according to three arbitrary classes (Fowler et al., 1987). Class I MTs constitute all proteins obviously related to the mammalian MTs. Class II encompasses a growing number of structurally analogous proteins which have little resemblance either to each other or to the mammalian forms. All class I and II MTs so far isolated are single chain proteins. Full or partial primary structures of MTs, derived either by direct protein chemical sequencing or from nucleotide sequencing of cDNA or genomic DNA, are now available for over 80 MTs. A representative selection of class I and II gene product sequences is given in Table 1. The fingerprint of all class I MTs is the close correspondence in the alignment of cysteine residues along the chain. Mammalian forms generally contain 61 or 62 amino acid residues. The *Neurospora crassa* protein is more divergent, having only 26 amino acids, but retaining exactly the same spacing of the seven cysteine residues as found in the 25-residue sequence of the N-terminal domain of mammalian MTs (Lerch, 1979). Class II MTs from pea and sea urchin contain 74 and 64 residues, respectively. Shorter chain proteins are found in invertebrates and in certain fungi, with the shortest so far in *A. bisporus* which has just 25 residues (Münger and Lerch, 1985). Class III MTs, which are not direct gene products and hence not included in Table 1, comprise a novel type of metal-binding polypeptides which are built up of repetitive $(\gamma\text{Glu-Cys})_n$X units, where n = 2-8 and X is most often glycine, and as such are homologues of glutathione. They have been designated as phytochelatins (PC_n). These polypeptides are of variable size, often composed of two or more chains and sometimes also containing inorganic sulfur. As yet, class III MTs have only been characterized in microorganisms and plants exposed to Cd (Kondo et al., 1984; Robinson and Jackson, 1986; Grill et al., 1986a; 1986b).

The characteristic feature of all MTs is the occurrence of Cys-X-Cys tripeptide sequences, where X stands for an amino acid other than Cys. Among mammalian species, the cysteine sequences are the most highly conserved followed by Lys or Arg (Kojima et al., 1976; Kägi and Kojima, 1987) (see Table 1). In contrast, no obvious sequence relationships are discernable among members of class II MTs. For example, the class II MTs from cyanobacterium and yeast (*S. cerevisiae*) contain 10 and 12 cysteines, respectively, but virtually no homology either to mammalian proteins or to each other. Their metal binding capacity is also less, i.e. the protein from cyanobacterium binds 4-5 Zn(II) or Cd(II) ions (Shi et al., 1992).

III. MOLECULAR BIOLOGY

The genes for human and mouse MTs have been most extensively studied. Both species appear to have three functionally distinct genetic cistrons, expressing three major forms of the protein (MT-1, MT-2 and MT-3) that differ by a few amino acids (Searle et al., 1984; West et al., 1990; Hamer, 1986; Palmiter et al, 1992) and usually in their behavior in anion-exchange chromatography (Fowler et al., 1987; Kojima, 1991; Uchida et al., 1991). Additional subforms within these fractions can be separated by reversed-phase HPLC (Klauser et al., 1983) consistent with extensive genetic polymorphism (Karin and Richards, 1982). In human, one functional gene has been cloned for MT-2 and MT-3 and several for MT-1, all or most are on chromosome 16. The mouse probably has only one functional gene each for MT-1, MT-2 and MT-3 all residing on chromosome 8 and are within 85 kilobases of each other (Palmiter et al., 1992).

The three exons composing the mammalian MT genes cover sequentially the highly invariant N-terminal region of the beta domain (exon 1), the rest of the beta domain (from residue 11 or 12) (exon 2) and all of the alpha domain (from residue 31 or 32) (exon 3). Exons 2 and 3 are spliced at the junction of codons for the two lysine residues in the interdomain region. At the nucleotide level, the high degree of homology between the three

Table 1

Representative Sequences of Class-I and -II MT Gene Products

		1 . . 10 . . 20 . . 30 . . 40 . . 50 . . 60 . . 70 . .	
Class I			
human	MT-2	MDP NCSCAAGDSCTCAGSCKCKECKCTSCKKSCCSCCPVGCAKCAQGCICKGA SDKCSCCA	[1]
	MT-3	MDPETCPCPSGGSCTCADSCKCEGCKCTSCKKSCCSCCPAECEKCAKDCVCKGGEAAEAEAEKCSCCQ	[2]
N. crassa		MGD CGCSGASSCNCGSGCSCSNCGSK	[3]
rat	MT-1	MDPN CSCSTGGSCTCSSSCGCKNCKCTSCKKSCCSCCPVGCSKCAQGCVCKGA SDKCTCCA	[4]
chicken		MDPQDCTCAAGDSCSCAGSCKCKNCRCRSCRKSCCSCCPAGCNNCAKGCVCKEPA SSKCSCCH	[5]
oyster		MSDP CNCIETGTCACSDSCPATGCKCGPGCKCGDDCKCAGCKVKCSCTSEGGCKCGEKCTGPATCKCGSGCSCKK	[6]
Class II			
sea urchin	MTa	MPDVKCVCCTEGKECACFGQDCCVTGECCKDGTCCGICTNAACKCANGCKCGSGCSCTEGNCAC	[7]
S. cerevisiae		MFSELINFQNEGHECQCQCGSCKNNEQCQKSCSCPTGCNSDDKCPCGNKSEETKKSCCSGK	[8]
C. elegans		MVCKCDCKNQNCSCNTGTKDCDCSDAKCCEQYCCPTASEKKCCKSGCAGGCKCANCECAQAAH	[9]
Pea		MSGCGCGSSCNCGDSCKCNKRSSGLSYSEMETTETVILGVGPAKIQFEGAEMSAASEDGGCKCGDNCTCDPCNC	[10]
cyanobacterium		MTSTTLVKCACEPCLCNVDPSKAIDRNGLYYCSEACAHGHTGGSKGCGHTGCNC	[11]

[1] Karin and Richards (1982)
[2] Palmiter et al. (1992)
[3] Münger et al. (1987)
[4] Andersen et al. (1983)
[5] Wei and Andrews (1988)
[6] Unger et al. (1991)
[7] Nemer et al. (1985)
[8] Butt et al. (1984)
[9] Imagawa et al. (1990)
[10] Evans et al. (1990)
[11] Robinson et al. (1990)

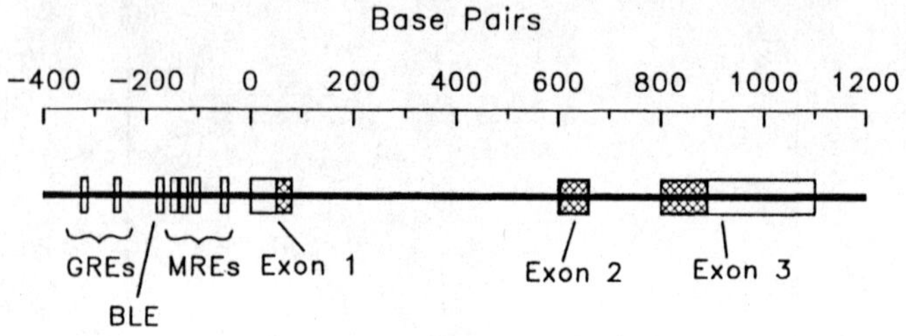

Figure 1.
Schematic representation of the structural Rat MT-1 gene (adapted from Andersen et al., 1987). Shaded areas represent the coding regions inside the three exons.

distinct categories of MT genes inside the coding regions coupled with the comparative lack of sequence conservation in the flanking sequences suggests their common origin in the triplication of a single ancestral MT gene (Palmiter et al., 1992). The induction of MT biosynthesis is regulated at the transcriptional level (Hamer, 1986). The promoter regions of all mammalian MT genes so far isolated have been shown to contain several metal responsive elements (MREs) and glucocorticoid responsive elements (GREs) as well as elements involved in basal level transcription (BLEs) (Palmiter et al., 1983; Palmiter, 1987) (see Fig. 1).

Many of the genes for class II MTs have also been cloned. The biosynthesis of the copper-containing MT from yeast is regulated by the binding of the transcription activating factor ACE1 to a promoter sequence of the MT-gene (CUP1 locus) (Fürst et al., 1988). In order to maintain its DNA binding competence, ACE1 requires the coordination of Cu(I) to a MT-like domain of the protein (Dameron et al., 1991). Consequently, there is an interdependence of the level of MT production on the cellular copper concentration, thus constituting a type of biological feedback loop maintaining physiological metal-ion homeostasis.

IV. BIOLOGICAL FUNCTIONS

Most MTs isolated to date are inducible compounds. In experimental animals and cultured cells, their concentration is raised sharply upon exposure to a diverse range of substances, among them metal ions, hormones and many cytotoxic substances (see Table 2). Such primary effects are also believed to underly the increased synthesis observed under various chemical and physical stress conditions also listed. This involvement of MT in both specific and nonspecific biological response mechanisms makes it difficult to pinpoint a single key biological function. In fact, more than three decades after their discovery, the primary physiological role of MTs remains a topic of discussion (Karin, 1985; Bremner, 1987). For some time, the most popular hypothesis has been that their principal function is to protect organisms by limiting the intracellular concentrations of reactive heavy metal ions thereby shielding cellular structures from the harmful influences of toxic metals such as cadmium, mercury, platinum, bismuth, silver and gold (Cousins, 1983; Bremner, 1987; Webb, 1987). However, the conservation of the structure of MTs in evolution, their ubiquitous occurrence, the redundancy of genes and the programmed synthesis of MTs in development, regeneration and reproduction are strong arguments for suspecting that MT may also serve other and perhaps more specific metal-related cellular roles (Kägi and Schäffer, 1988).

It has been demonstrated in several in vitro studies that MT has the ability to supply zinc and copper to the appropriate apoproteins of a number of metallo-enzymes and metalloproteins, thus supporting the proposal that MT may act as a reservoir for these metals in protein biosynthesis (Brady, 1982; Lerch and Beltramini, 1983). In mammals, where zinc is usually the major and often the sole metallic constituent of MT, it has also been postulated that, by controlling the level of zinc, MT may serve a metalloregulatory role in zinc-dependent processes in replication, transcription and translation (Kägi and Schäffer, 1988). These propositions are substantiated by several recent observations: (i) the

Table 2

Factors that Induce Metallothionein Synthesis in Cultured Cells or In Vivo (cited in Kägi, 1991)

Metal ions	*Inflammatory Agents*	*Cytotoxic agents*
Cd, Zn, Cu, Hg, Au,	Lipopolysaccharide	Hydrocarbons
Ag, Co, Ni, Bi	Carrageenan	Alcohols
Hormones	Dextran	Formaldehyde
Glucocorticoids	Endotoxin	Fatty acids
Progesterone	*Cytokines*	Butyrate
Estrogen	Interleukins	Organo-halides
Catecholamines	Interferons	Mercaptans
Glucagon	Tumor necrosis factor	Metal chelators
Angiotensin II	*Tumor Promoters*	5-azacytidine
Arg-Vasopressin	Phorbol esters	Urethane
Adenosine	*Oncogenes*	*Analgesics*
Second messengers	*ras*	Indomethacin
cAMP	*Vitamins*	Acetaminophen
Diacylglycerol	Ascorbic acid	*Conditions of Stress*
Ca	Retinoate	Inflammation
Growth factors	Dihydroxyvitamin D_3	Laparotomy
Serum factors	*Antiobiotics*	Starvation
Insulin	Streptozotoxin	Infection
IGF-1	Cycloheximide	Radiation exposure
EGF	Mitomycin	Hyperoxymia

halt of cellular replication upon contact inhibition of cultured mammalian cell lines is accompanied by a transitory increase in MT synthesis (Cavigelli, 1992); (ii) apo-MT, or thionein, readily removes zinc from zinc finger transcription factors such as Sp1 and *Xenopus* TFIIIA in vitro, hence eliminating their DNA binding competence and abolishing their ability to activate transcription (Zeng et al., 1991a; 1991b). MT may therefore serve multiple roles as a scavenger of heavy metals, a store of biologically essential metals and a regulator of zinc-dependent processes fundamental to cell activation in proliferation and differentiation.

A new member of the mammalian MT family, classified MT-3, has recently been identified in mouse and human which is specific to brain tissue (Uchida et al., 1991; Palmiter et al., 1992). Unlike the MT-1 or MT-2 groups of isoproteins, the polypeptide chain contains 68 rather than 61 amino acids. This results from two insertions of 1 and 6 additional residues in the N-terminal and C-terminal domains, respectively (see Table 1). A unique feature of this MT-form is its growth inhibitory activity in neuronal cell culture and its absence in the brains of Alzheimer's disease victims (Uchida et al., 1991). This finding lends support to the proposal that MT plays a role in cell signalling mechanisms in the regulation of growth and differentiation, and has spawned the alternative phenomenological name of growth inhibitory factor (GIF).

V. METAL BINDING

In the absence of disulfide bridges, the abundance of cysteine residues in MTs furnishing polarizable thiolate ligands gives this protein a high affinity for d^{10} metal ions. Mammalian MTs have so far been most extensively studied and usually bind a total of 7 equiv of divalent metal ions, e.g. Zn(II), Cd(II), or higher stoichiometries of monovalent ions such as Cu(I) (Vašák and Kägi, 1983; Kägi and Kojima, 1987). In its native form, MT may contain Zn(II), Cd(II), Cu(I) or other heavy metals in varying proportions, depending on the history and physiological state of the organism. The affinity of metal ions for the binding sites of mammalian MTs follows the same order as for thiolate model complexes, i.e. Zn(II) < Pb(II) < Cd(II) < Cu(I), Ag(I), Hg(II), Bi(III). The 10 000-fold higher affinity for Cd(II) compared to Zn(II) is thought to be the reason why Cd tends to be accumulated in MT in cells which have been exposed to Cd (Vašák and Kägi, 1983). Moreover, in the case of Cd-induced hepatic MTs, a Cd,Zn-containing protein is always isolated (Nettesheim et al., 1985).

The mode of metal coordination has been elucidated by a number of spectroscopic techniques. For many of these studies, homogeneously substituted metal derivatives had to be prepared, in which the naturally bound metal ions were replaced by metal ions or isotopes suitable for the chosen spectroscopic method. Zn and Cd are readily released from MT by acidification yielding apoMT and, by subsequent addition of the appropriate metal salts, homogeneously substituted derivatives containing 7 equiv each of Zn(II), Cd(II), Hg(II), Pb(II), Bi(III), Sn(II), Co(II), Fe(II), Ni(II) and TcO(II) ions have been prepared (cited in Kägi and Schäffer, 1988; Vašák, 1991). In particular, replacement studies with Co(II) have proven especially useful in the establishment of the geometry of metal binding. The first unambiguous evidence for the existence of tetrahedral tetrathiolate coordination came from assignment of the spin-allowed $\nu_3[^4A_2 \rightarrow {}^4T^1(P)]$ and $\nu_2[^4A_2 \rightarrow {}^4T_1(F)]$ ligand-field transitions of the Co(II) ion in the absorption and magnetic circular dichroism (MCD) spectra of the Co-MT derivative (Vašák, 1980; Vašák et al., 1981b). In subsequent studies on Zn- and Cd-containing MTs, similar geometries were inferred from the analysis of the CysS-Me(II) charge-transfer transitions in the UV-absorption spectra, sulfur-metal distances obtained from extended X-ray fine structure (EXAFS) measurements and from ^{113}Cd-NMR chemical shifts (Vašák et al., 1981a; Otvos and Armitage, 1980; Abrahams et al., 1986). Furthermore, the propensity of MTs towards the formation of tetrahedral metal sites has been demonstrated in a number of other divalent metal derivatives, such as Ni(II) and Fe(II) (Vašák et al., 1981b; Good and Vašák, 1986; Ding et al., 1988).

In order for the tetrahedral tetrathiolate coordination to be reconciled with the measured stoichiometry of 7 divalent metal ions to 20 cysteines then a clustered arrangement of metal ions must be envisaged. Direct evidence for the existence of discrete metal-thiolate clusters was provided by the elegant homonuclear ^{113}Cd decoupling experiments by Otvos and Armitage (1980) on ^{113}Cd-enriched rabbit liver Cd_7-MT. By analysis of the two-bond ^{113}Cd-^{113}Cd spin coupling connectivities responsible for the multiplet splitting patterns of each resonance, they were able to deduce the presence of two separate clusters containing three and four metal ions in cyclohexane-like $Me(II)_3(Cys)_9$ and bicyclo nonane-like $Me(II)_4(Cys)_{11}$ structures, respectively. Thus, the 20 cysteines are partitioned into eight bridging and twelve terminal thiolate ligands. In addition, magnetic studies on the Co(II)-MT derivative showed the existence of super-exchange of unpaired electrons between neighboring, sulfur-bridged Co(II) ions, thereby confirming the presence of oligonuclear structures in which some thiolate ligands are shared by two adjacent metal ions (Vašák and Kägi, 1981). Moreover, the observation of second-shell back-scattering from Zn(II) ions in EXAFS experiments was also consistent with the existence of such a cluster structure in native Zn_7-MT (Abrahams et al., 1986). That the two clusters were not just separate entities but located in different regions of the protein was demonstrated by the protein-chemical studies of Winge and Miklossy (1982), which revealed that the protein molecule was divided into two domains with the four metal cluster (cluster A) associated with the C-terminus (alpha-domain) and the three metal cluster (cluster B) associated with the N-terminus (beta-domain) (see Fig. 2). Mammalian MTs also appear to be capable of binding around 12 Cu(I) or Ag(I) ions, adding 6 to the beta-domain and about 6 to the alpha-domain by reconstitution of the apoprotein (Nielson et al., 1985; Zelazowski et al., 1989). However, a number of other stoichiometries of copper loading have also been reported in MTs obtained from mammalian tissues under a variety of different physiological conditions (Freedman et al., 1986). Data from EXAFS studies on Cu(I)-containing mammalian MTs are compatible with trigonal coordination of Cu(I) by cysteine thiolate sulfurs (Smith et al., 1986). Similar studies on the isolated copper-containing MT from yeast, which binds 8 equiv of Cu(I) ions through coordination with 12 cysteine residues, are consistent with the presence of a $Cu_8(CysS)_{12}$ core exhibiting analogous geometries of the Cu(I) ions (George et al., 1986; 1988). Subsequent ^{1}H-^{109}Ag heteronuclear multiple-

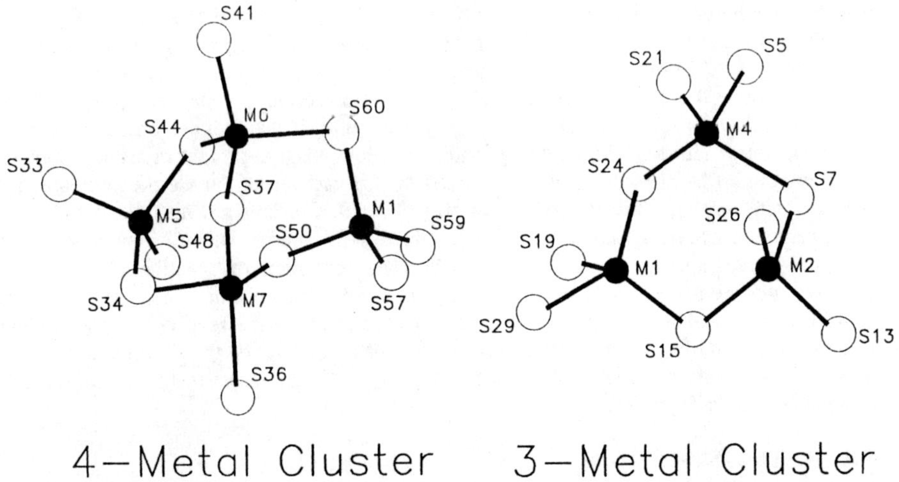

Figure 2.
Ball and stick diagram of the mammalian MT clusters (adapted from Robbins et al., 1991). S = cysteine sulfur, M = divalent metal ion (M1 and M2 in the 3-metal cluster represent Zn1 and Zn2 in Zn_2Cd_5-MT).

quantum coherence transfer experiments on the silver-substituted yeast MT derivative have shown the existence of a cluster structure containing at least 7 bridging thiolates (Narula et al., 1991).

To date, structural and biological studies of mammalian MTs have focused mainly on the monomeric form of this protein containing 7 equiv of divalent metal ions. In general, solely oxidative processes have been considered to account for the presence of dimeric MT, sometimes encountered in the isolation of this protein (Suzuki and Yamamura, 1980). However, recent in vitro studies have shown that exposure of Cd_7-MT from rabbit liver to free Cd(II) ions in phosphate buffer yields non-oxidized dimers which contain additionally 2 less strongly bound Cd(II) ions and 1 inorganic phosphate per monomeric protein unit (Palumaa et al., 1992; Palumaa and Vašák, 1992). Furthermore, it has been found that under certain conditions up to a total of 13 Cd(II) ions can be coordinated together with 2 ions of inorganic phosphate, giving rise to a more tightly packed globular MT form (Palumaa et al., 1993). The physiological relevance of these MT forms is not immediately obvious, although the observation that MT is capable of binding additional divalent metal ions which are not exclusively thiolate coordinated and apparently have a specific affinity for phoshate may perhaps be relevant to some currently unassigned biological activity.

VI. CLUSTER DYNAMICS

In order to function as efficient metalloregulatory agents, MTs should be able to acquire and release metal ions rapidly. Indeed, the metal-sulfur coordinative bonds of MT clusters have been found to be highly labile by a variety of spectroscopic techniques. For example, rapid rates of intramolecular metal exchange have consistently been observed in cluster B (three-metal cluster) with a half-life estimated from ^{113}Cd NMR saturation transfer experiments to be of the order of 0.5 s (Otvos et al., 1989). Comparable processes occurring within cluster A (four-metal cluster) have been inferred from isotope exchange studies which suggest a half-life of about 16 minutes (Kägi, 1991). Such dynamic fluctuations in MT clusters are thought to be a manifestation of the non-rigid nature of the overall protein structure as supported by NMR ^1H-^2H amide exchange studies (Messerle et al., 1990a). Despite the kinetic lability of MT clusters, their structures are thermodynamically rather stable. The process of cluster formation has been studied for Cd(II) and Co(II) MT derivatives. In both instances, the formation of each cluster appears to proceed in a cooperative manner at neutral pH whereas, under alkaline conditions (ca pH 8.6), isolated sites are first formed with up to 4 metal equiv after which cluster formation sets in. This pH-dependent difference in the pathway of cluster formation can be accounted for by the competition between metal ions and protons for the thiolate sulfurs of the cysteine residues. Thus, at neutral pH the binding of metals is accompanied by the displacement of cysteine thiol protons. In this case, the formation of clusters with bridging cysteines at low metal-to-protein ratios will require fewer protons to be displaced per metal bound than if each metal ion were coordinated by four separate terminal thiolates (Good et al., 1988; Bertini et al., 1989). In addition to homometallic clusters, a number of distinct mixed-metal species with divalent metal ions have been characterized (Good et al., 1991). The formation of specific heterometallic clusters, such as the naturally occurring Zn_2Cd_5 species, has not only been shown to be controlled by metal thiolate affinity but is also dependent on the relative volumes of the resulting metal-thiolate clusters. This type of size-match selectivity again suggests that MT is more than simply a sponge which indiscriminantly soaks up excess metal ions in the cell (Pountney and Vašák, 1992).

VII. SPATIAL STRUCTURE

Models for the spatial structure of mammalian MT have been derived from two-dimensional NMR studies on Cd_7-MT from rabbit, rat and human liver in aqueous solution, and from X-ray diffraction data obtained on crystals of Zn_2Cd_5-MT from rat liver (Arseniev et al., 1988; Schultze et al., 1988; Messerle et al., 1990b; Robbins et al., 1991). Both NMR and X-ray structures are in excellent agreement (see Fig. 3). The discontinuity in the NMR structure between Lys-30 and Lys-31 is due to the absence of information as to the mutual orientation of the two domains from the NMR data.

An essential step in the determination of the solution structure by two-dimensional NMR was the assignment of sequence specific cysteine-metal coordinative bonds. This was accomplished using heteronuclear multiple quantum chemical shift correlation experiments which exploit the existence of 3-bond scalar coupling between the cysteine ligand protons and the $^{113}Cd(II)$ ions ($I = 1/2$) (Frey et al., 1985). This information, in conjunction with the sequence-specific proton resonance assignments obtained by ^1H COSY and ^1H NOESY experiments, unequivocally established sequence specific cysteine-metal ligation. With these results and the additional proton distance constraints provided by ^1H NOESY spectra, distance geometry calculations were carried out to determine the three-dimensional solution structures of the three mammalian proteins mentioned above.

Mammalian MTs have a dumb-bell-like shape. The uniformly sized and almost spherical alpha- and beta-domains have a diameter of 15-20 Å and contain at their centres the respective metal thiolate clusters as "mineral cores" around which the polypeptide backbone is wrapped, forming two large helical turns. In the C-terminal alpha-domain the spiral of the peptide fold is left-handed and in the beta-domain it is right-handed. The two domains are connected by a flexible hinge region formed by the conserved Lys-Lys segment in the middle of the peptide chain. Exactly the same Cys-Cd connectivities and structural organization of clusters have been found in mammalian MTs from three different species, which are substantially dissimilar in their amino acid sequences, indicating that the protein fold is largely determined by the conserved arrangement of the 20 Cys residues in the chain (Vašák et al., 1987; Schultze et al., 1988). Hence, the polypetide chain conformation between the cysteines is severely restricted by the metal coordination bonds within the interior of each domain. This feature of mammalian MTs gives rise to a number of unusual secondary structural elements, including two 3_{10} helical segments and several "half turns", a hydrogen-bonded motif so far unique to MTs (Wagner et al., 1986). The preservation of such a novel spatial arrangement of mammalian MTs may be expected to be functionally significant and will perhaps contain elements common to other MTs with more divergent primary structures.

Overall, the MTs, by virtue of their large metal content, their unusual bioinorganic structure and their remarkable kinetic lability constitute a unique class of metalloproteins. These properties are thought to enable them to act as rapid metal transfer and efficient metalloregulatory agents. In addition, the recent findings linking MT with cell signalling

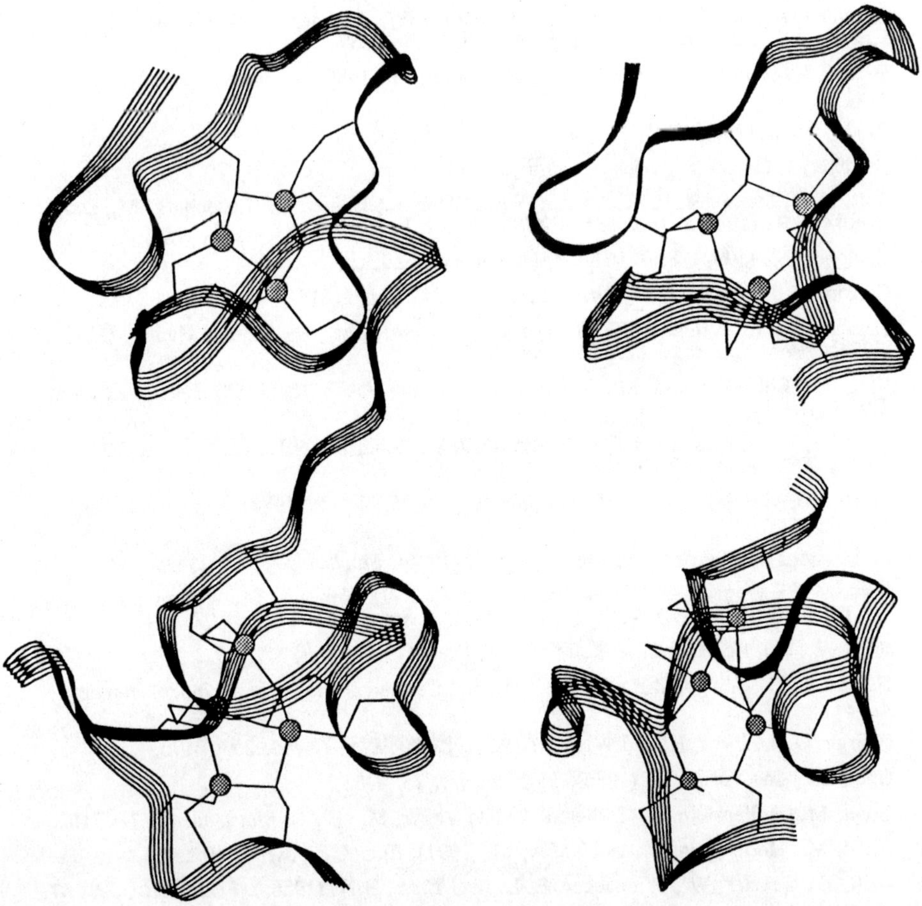

Figure 3.
Ribbon drawing of the rat MT three-dimensional structure as deduced by X-ray crystallography (left; Robbins et al., 1991) and by NMR in solution (right; Schultze et al., 1988). Metals are shown as shaded spheres connected to the protein backbone by cysteine thiolate ligands.

mechanisms opens new avenues of investigation in the search for MT function. Indeed, interest in the study of MTs has grown enormously over the last years, with researchers from a wide spectrum of disciplines including inorganic chemistry, biochemistry, physiology, molecular biology, toxicology and medicine working together to uncover its mysteries.

ACKNOWLEDGEMENT

The authors gratefully acknowledge the financial support of the Swiss National Science Foundation.

REFERENCES

Abrahams, I.L., Bremner, I., Diakun, G.P., Garner, C.D., Hasnain, S.S., Ross, I., and Vašák, M. (1986). *Biochem. J.*, *236*: 585.

Andersen, R.D., Birren, B.W., Ganz, T., Piletz, J.E., and Herschman, H.R. (1983). *DNA*, *2*: 15.

Andersen, R.D., Taplitz, S.J., Birren, B.W., Bristol, G., and Herschman, H.R. (1987). *Experientia Suppl.*, *52*: 373.

Arseniev, A., Schulze, P., Wörgötter, E., Braun, W., Wagner, G., Vašák, M., Kägi, J.H.R., and Wüthrich, K. (1988). *J. Mol. Biol.*, *201*: 637.

Bertini, I., Luchinat, C., Messori L., and Vašák, M. (1989). *J. Amer. Chem. Soc.*, *111*: 7296.

Brady, F.O. (1982). *TIBS*, *7*: 143.

Bremner, I. (1987). *Experientia Suppl.*, *52*: 81.

Butt, T.R., Sternberg, E.J., Gorman, J.A., Clark, P., Hamer, D., Rosenberg, M., and Crooke, S.T. (1984). *Proc. Nat. Acad. Sci. USA*, *81*: 3332.

Cavigelli, M. (1992). Ph.D. Thesis, University of Zürich.

Cousins, R.J. (1983). *J. Inherit. Metab. Dis.*, *6 (Suppl. 1)*: 15.

Dameron, C.T., Winge, D.R., George, G.N., Sansone, M., Hu, S., and Hamer, D.H. (1991). *Proc. Nat. Acad. Sci. USA*, *88*: 6127.

Ding, X., Bill, E., Good, M., Trautwein, A.X., and Vašák, M. (1988). *Eur. J. Biochem.*, *171*: 711.

Evans, K.M., Gatehouse, L.N., Gatehouse, J.A., Robinson, N.J., and Croy, R.R.D. (1990). *FEBS Lett.*, *260*: 29.

Fowler, B.A., Hildebrand, C.E., Kojima, Y., and Webb, M. (1987). *Experientia Suppl.*, *52*: 19.

Freedman, J.H., Powers, L., and Peisach, J. (1986). *Biochemistry*, *25*: 2342.

Frey, M.H., Wagner, G., Vašák, M., Sorensen, O.W., Neuhaus, D., Wörgötter, E., Kägi, J.H.R., Ernst, R.E., and Wüthrich, K. (1985). *J. Amer. Chem. Soc.*, *107*: 6847.

Fürst, R., Hu, S., Hackett, R., and Hamer, D.H. (1988). *Cell*, *55*: 705.

George, G.N., Winge, D.R., Stout, C.D., and Cramer, S.P. (1986). *J. Inorg. Biochem.*, *27*: 213.

George, G.N., Byrd, J., and Winge, D.R. (1988). *J. Biol. Chem.*, *263*: 8190.

Good, M., and Vašák, M. (1986). *Biochemistry*, *25*: 8353.

Good, M., Hollenstein. R., Sadler, P.J., and Vašák, M. (1988). *Biochemistry*, *27*: 7163.

Good, M., Hollenstein, R., and Vašák, M. (1991). *Eur. J. Biochem.*, *197*: 655.

Grill, E., Gekeler, W., Winnacker, E.-L., and Zenk, H.H. (1986a). *FEBS Lett.*, *205*: 47.

Grill, E., Winnacker, E.-L., and Zenk, H.H. (1986b). *FEBS Lett.*, *197*: 115.

Hamer, D.H. (1986). *Annu. Rev. Biochem.*, *55*: 913.

Imagawa, M., Onozawa, T., Okumura, K., Osada, S., Nishihara, T., and Kondo, M. (1990). *Biochem. J.*, *268*: 237.

Johnson, G.F., Morell, A.G., Stockert, R.J., and Sternlieb, I. (1981). *Hepatology (Baltimore)*, *1*: 243.

Kägi, J.H.R. (1991). *Methods Enzymol.*, *205*: 613.

Kägi, J.H.R., and Kojima, Y. (1987). *Experientia Suppl.*, *52*: 25.

Kägi, J.H.R., and Schäffer, A. (1988). *Biochemistry*, *27*: 8509.

Kägi, J.H.R., and Vallee, B.L. (1960). *J. Biol. Chem.*, *235*: 3460.

Kägi, J.H.R., and Vallee, B.L. (1961). *J. Biol. Chem.*, *236*: 2435.

Karin, M. (1985). *Cell*, *41*: 9.

Karin, M., and Richards, R.I. (1982). *Nature*, *299*: 797.

Klauser, S., Kägi, J.H.R., and Wilson, K.J (1983). *Biochem. J.*, *209*: 71.

Kojima, Y. (1991). *Methods Enzymol.*, *205*: 8.

Kojima, Y., Berger, C., Vallee, B.L., and Kägi, J.H.R. (1976). *Proc. Nat. Acad. Sci. USA*, *73*: 3413.

Kondo, N., Imai, M., Isobe, T., Goto, A., Murasugi, C., Wada-Nakagawa, C., and Hayashi, Y. (1984). *Tetrahedron Lett.*, *25*: 3869.

Lerch, K. (1979). *Experientia Suppl.*, *34*: 173.

Lerch, K., and Beltramini, M. (1983). *Chem. Scr.*, *21*: 109.

Margoshes, M., and Vallee, B.L. (1957). *J. Amer. Chem. Soc.*, *79*: 4813.

Messerle, B.A., Schäffer, A., Vašák, M., Kägi, J.H.R., and Wüthrich, K. (1990a). *J. Mol. Biol.*, *214*: 781.

Messerle, B.A., Schäffer, A., Vašák, M., Kägi, J.H.R., and Wüthrich, K. (1990b). *J. Mol. Biol.*, *214*: 765.

Münger, K., and Lerch, K. (1985). *Biochemistry*, *24*: 6751.

Münger, K., Germann, U.A., and Lerch, K. (1987). *Experientia Suppl.*, *52*: 393.

Nartey, N.O., Banerjee, D., and Cherian, M. G. (1987). *Pathology*, *19*: 233.

Narula, S.S., Mehra, R.K., Winge, D.R., and Armitage, I.M. (1991). *J. Amer. Chem. Soc.*, *113*: 9354.

Nielson, K.B., Atkin, C.L., and Winge, D.R. (1985). *J. Biol. Chem.*, *260*: 5342.

Nemer, M., Wilkinson, D.G., Travaglini, E.C., Sternberg, E.J., and Butt, T.R. (1985). *Proc. Nat. Acad. Sci. USA*, *82*: 4992.

Nettesheim, D.G., Engeseth, H.R., and Otvos, J.D. (1985). *Biochemistry*, *24*: 6744.

Nordberg, M., and Kojima, Y. (1979). *Experientia Suppl.*, *34*: 41.

Otvos, J.D., and Armitage, I.M. (1980). *Proc. Nat. Acad. Sci. USA*, *77*: 7094.

Otvos, J., Chen S.-M., and Liu, X. (1989). *Metal Ion Homeostasis: Molecular Biology and Chemistry*, Vol. 98 (D. Winge and D. Hamer, eds.), UCLA Symposia, New Series, A.R. Liss, N.Y., p. 197.

Palmiter, R.D. (1987). *Experientia Suppl.*, *52*: 63.

Palmiter, R.D., Norstedt, G., Gelinas, R.E., Hamer, R.E., and Brinster, R.L. (1983). *Science*, *22*: 809.

Palmiter, R.D., Findley, S.D., Whitmore, T.E., and Durnam, D.M. (1992). *Proc. Nat. Acad. Sci. USA*, *89*: 6333.

Palumaa, P., and Vašák, M. (1992). *Eur. J. Biochem.*, *205*: 1131.

Palumaa, P., MacKay, E., and Vašák, M. (1992). *Biochemistry*, *31*: 2181.

Palumaa, P., Zerbe, O., and Vašák, M. (1993). *Biochemistry*, *32*: 2874.

Pountney, D.L., and Vašák, M. (1992). *Eur. J. Biochem.*, *209*: 335.

Pulido, P., Kägi, J.H.R., and Vallee, B.L. (1966). *Biochemistry*, *5*: 1768.

Robbins, A.H., McRee, D.E., Williamson, M., Collett, S.A., Xuong, N.H., Furey, W.F., Wang, B.C., and Stout, C.D. (1991). *J. Mol. Biol.*, *221*: 1269.

Robinson, N.J., and Jackson, P.J. (1986). *Physiol. Plan.*, *67*: 499.

Robinson, N.J., Gupta, A., Fordham-Skelton, A.P., Croy, R.R., Whitton, B.A., and Huckle, J.W. (1990). *Proc. Roy. Soc. London B*, *242*: 241.

Schultze, P., Wörgötter, E., Braun, W., Wagner, G., Vašák, M., Kägi, J.H.R., and Wüthrich, K. (1988). *J. Mol. Biol.*, *203*: 251.

Searle, P.F., Davison, B.L., Stuart, G.W., Wilkie, T.M., Norstedt, G., and Palmiter, R.D. (1984). *Mol. Cell Biol.*, *4*: 1221.

Shi, J., Linsay, W.P., Huckle, J.W., Morby, A.D., and Robinson, N.J. (1992). *FEBS Lett.*, *303*: 159.

Smith, T.A., Lerch, K., and Hodgson, K.O. (1986). *Inorg. Chem.*, *25*: 4677.

Suzuki, K.T., and Yamamura, M. (1980). *Biochem. Pharmacol.*, *29*: 698.

Uchida, Y., Takio, K., Tatani, K., and Ihara, Y. (1991). *Neuron*, *7*: 337.

Unger, M.E., Chen, T.T., Murphy, C.M., Vestling, M.M., Fenselau, C., and Roesijadi, G. (1991). *Biochim. Biophys. Acta*, *1074*: 371.

Vašák, M. (1980). *J. Amer. Chem. Soc.*, *102*: 3953.

Vašák, M. (1991). *Methods Enzymol.*, *205*: 452.

Vašák, M., and Kägi, J.H.R. (1981). *Proc. Nat. Acad. Sci. USA*, *78*: 6709.

Vašák, M., and Kägi, J.H.R. (1983). *Metal Ions in Biological Systems*, Vol. 15 (H. Sigel, ed.), Marcel Dekker Inc., N.Y., p. 213.

Vašák, M., Kägi, J.H.R., and Hill, H.A.O. (1981a). *Biochemistry*, *20*: 2852.

Vašák, M., Kägi, J.H.R., Holmquist, B., and Vallee, B.L. (1981b). *Biochemistry*, *20*: 6659.

Vašák, M., Wörgötter, E., Wagner, G., Kägi, J.H.R., and Wüthrich, K. (1987). *J. Mol. Biol.*, *196*: 711.

Wagner, G., Neuhaus, D., Wörgotter, E., Vašák, M., Kägi, J.H.R., and Wüthrich, K. (1986). *J. Mol. Biol.*, *187*: 131.

Webb, M. (1987). *Experientia Suppl.*, *52*: 109.

Wei, D., and Andrews, G.K. (1988). *Nucleic Acids Res.*, *16*: 537.

West, A.K., Hildebrand, C.E., Karin, M., and Richards, R.I. (1990). *Genomics*, *8*: 513.

Winge, D.R., and Miklossy, K.A. (1982). *J. Biol. Chem.*, *257*: 3471.

Zelazowski, A.J., Gasyna, Z., and Stillman, M.J. (1989). *J. Biol. Chem.*, *264*: 17091.

Zeng, J., Heuchel, R., Schaffner, W., and Kägi, J.H.R. (1991a). *FEBS Lett.*, *279*: 310.

Zeng, J., Vallee, B.L., and Kägi, J.H.R. (1991b). *Proc. Nat. Acad. Sci. USA*, *88*: 9984.

PROTEIN INTERACTIONS WITH DETRIMENTAL METAL IONS

by

Katherine S. Squibb and Bruce A. Fowler

Program in Toxicology,
University of Maryland at Baltimore,
660 West Redwood Street
Baltimore, MD 21201
U.S.A.

I. INTRODUCTION

The binding of metals to protein functional groups plays a major role in determining the toxicity of metal ions in cells. The reactivity of metals with critical ligands depends upon the chemistry of the individual metal ion and the nature and number of competing ligands present within the cell. Studies over the past 25 years have found that while some metal-protein interactions can lead to adverse effects on cells, others serve as detoxification pathways by preventing undesirable metal binding at sensitive cellular sites. These "detoxification" proteins have evolved as storage proteins which allow cells to maintain pools of available essential metals in "non-toxic" forms. Metal ions such as cadmium and lead, which have no known essential role in mammalian cell processes, are also "detoxified" by these specialized metal-binding proteins, thereby protecting cells against low level exposures to these metals.

This chapter will briefly discuss mechanisms by which protein-metal interactions can cause cell injury and the role played by high-affinity metal-binding proteins in mediating effects of toxic metals.

II. ENZYMES

Many of the detrimental effects of toxic metals occur through interactions with enzymes, causing either a change in substrate specificity or an enhancement or inhibition of enzyme activity. Metalloenzymes, in particular, are prime targets for toxic metals due to competitive binding at essential metal-binding sites. There are presently over seventy known enzymes that require metals as cofactors for activity (Vallee and Auld, 1990). These metalloenzymes participate in a wide range of metabolic pathways within cells, from gluconeogenesis and oxidative phosphorylation to DNA, RNA and protein synthesis. The ability of toxic metal ions to alter the activity of many of these enzymes has been studied in detail in in vitro systems (Vallee and Ulmer, 1972) (Table 1).

Carboxypeptidase A, for example, is a zinc metalloenzyme with esterase and peptidase activity. When cadmium is substituted for zinc, the esterase activity is enhanced and the peptidase activity is inhibited, due to alterations in the tertiary structure of the enzyme molecule (Coleman and Vallee, 1961; Auld and Holmquist, 1974; Yamashita et al., 1990).

Table 1

Enzyme Activities Altered by Toxic Metal Ions[1]

Enzyme	Metals
Acid phosphatase	Cd
Aldehyde oxidase	As*
ATPases	Cd
Alkaline phosphatase	Cd, Pb
δ-Aminolevulinic acid dehydratase	Al, Ag, Cd, Hg, Pb
δ-Aminolevulinic acid synthetase	Cd
Ca^{2+}-dependent endonuclease	Cd
Carbonic anhydrase	Cd
Carboxypeptidase A	Cd
Carboxypeptidase B	Cd
Catalase	Cd
Cholinesterase	Cd
Coproporphyrinogen oxidase	Pb
Cytochrome oxidase	Cd, Pb
Glucose oxidase	Cd
Glucose-6-phosphate dehydrogenase	Cd, Pb
β-Glucuronidase	Cr
Ferrochelatase	Cd, Hg, Pb
Lactic dehydrogenase	Pb
Leucine aminopeptidase	Cd
Malic dehydrogenase	Cd
Phosphopyruvate carboxylase	Cd
Protein kinase C	Pb
Pyruvate dehydrogenase	Cd
Succinic dehydrogenase	As*, Cd, Cr,
Steroid 3 β-ol-dehydrogenase	Pb
Urease	Cd, Cr
Uroporphyrinogen I synthase	Cd, Pb, Hg
Xanthine dehydrogenase	As*, Cd

[1]Compiled from Johnson and Rajagopalan (1978), EPA (1986), Silbergeld and Fowler (1987) and Vallee and Ulmer (1972).
*Metalloid.

Cadmium has also been shown to bind to protein kinase C, potentiating the effect of tumor promotors on the nuclear binding and activation of this enzyme, and increasing its binding to a 105 kg mol^{-1} nuclear protein (Beyersmann et al., 1993). Data suggest that cadmium substitutes for zinc in the regulatory domain of protein kinase C, leading to a more active protein-protein binding site.

The effects of metal ions on enzymes of the heme biosynthetic pathway form the basis for the characteristic porphyrinurias observed following exposure of animals to toxic metals such as Pb, Cd, Hg and the metalloid As (Fowler et al., 1987; Bernard and Lauwerys, 1987). δ-Aminolevulinic acid dehydratase (ALAD), which catalyzes the condensation of two molecules of ALA to form porphobilinogen, is a zinc metalloenzyme composed of eight protein subunits and eight Zn atoms (Wu et al., 1974; Gurne et al., 1977; Anderson and Desnick, 1979; Bevan et al., 1980; Tsukamoto et al., 1979; 1980; Jaffe et al., 1984). Studies have shown that lead inhibits ALAD activity by substituting for one or more of the four Zn atoms present in the active site of the enzyme (Tsukamoto et al., 1979; Finelli et al., 1975; Gibbs et al., 1985; Gibbs and Jordan, 1986). This inhibition is reversible in vitro by incubation with zinc and dithiothreitol (Jaffe et al., 1984; Jaffe and Hanes, 1986; Finelli et al., 1975; Gibbs et al., 1985; Barnard et al., 1977).

Ferrochelatase catalyzes the insertion of Fe into protoporphyrin IX to form protoheme in the final step of the heme biosynthetic pathway. Like ALAD, the activity of this enzyme is sensitive to toxic metals (Dailey, 1987). The binding of Fe^{2+} to ferrochelatase, which occurs via vicinyl sulfhydryl groups, is competitively inhibited by Hg, Pb, Co, and Cd (Dailey, 1987).

III. MEMBRANE PROTEINS

Membrane receptors and transporter proteins are also being studied in detail as target sites of toxic metal action (Chen and Smith, 1992; Eimerl and Schramm, 1993; Shankar et al., 1992). Smith and coworkers (Chen and Smith, 1992; Smith et al., 1993) have reported that low concentrations of a number of toxic metals are capable of triggering the release of stored Ca^{2+} in human fibroblast cells through interactions with an "orphan" Ca-mobilizing membrane receptor protein. Results indicate that Cd, Ni and Co are capable of activating this cell surface sialoprotein receptor at concentrations of 0.04, 0.22 and 0.25 μM, respectively. Based on zinc competition studies, these toxic metals appear to activate the receptor by binding to a Zn site in the exoplasmic domain of the protein (Smith et al., 1993).

IV. NUCLEAR PROTEINS

A. Zinc-finger proteins

Many non-enzymatic proteins are also target sites for metal ion toxicity. Recent evidence indicates that non-essential metals bound to zinc fingers in DNA-binding proteins may alter the binding of these transcription factors to DNA and thus interfere with gene regulation. Sarkar (1993) has observed that while cadmium and cobalt are able to replace zinc in the zinc fingers of the estrogen response element (ERE) binding protein and not alter the binding of this protein to DNA, copper and nickel inhibit DNA binding of this transcription factor when bound to the zinc fingers of the polypeptide.

B. DNA-protein cross-links

Carcinogenic chromium compounds produce a variety of lesions in cellular DNA that may be responsible for their carcinogenicity, but one of the most persistent alterations is the induction of DNA-protein cross-links that are not readily repaired. Based on results obtained with chelating agents and sulfhydryl reducing agents, Costa and coworkers (Costa, 1991; Miller and Costa, 1989) suggest that chromium itself is involved in the DNA-protein binding. Treatment of EDTA substantially decreased the recovery of protein in the DNA-protein complexes, as did treatment with thiourea. On the basis of the binding constants of Cr^{3+} for various ligands, Miller and Costa (1989) have proposed that chromium forms a complex between the N7 position of guanine in DNA and the sulfur group of the amino acid cysteine in nuclear proteins. The identity of the large number of proteins present in these complexes remains to be determined, but they are similar for Cr and cis-platinum, and one of the proteins has been identified as actin (Costa, 1991).

V. CYTOSKELETAL PROTEINS

The cytoskeleton is also a prime target of metal ion toxicity in cells and evidence suggests that protein-metal interactions may play an important role in these effects. The cytoskeletal system is composed of three main types of protein fibrils: microtubules which consist primarily of the protein tubulin, microfilaments which are composed primarily of actin, and intermediate filaments which consist of different tissue specific proteins (Shay, 1986). This system of fibrillar structures crisscrosses the cytoplasm of eukaryotic cells to form an internal network that changes dynamically in response to external and internal stimuli. It functions to maintain the internal structure and shape of individual cells and is also involved in cell processes that involve movement such as exo- and endocytosis, protoplasmic streaming, locomotion, cellular polarity, and cell division (Alberts et al., 1983).

Metal ions such as Cd, Zn, Hg, and the metalloid As (Li and Chou, 1992; Prozialeck and Niewenhuis, 1991; Mills et al., 1992; Mills and Ferm, 1989; Elliget et al., 1991) have been shown to affect microtubule and microfilament organization in cells. These effects could be due to direct interaction of the metal ions with structural proteins within the filaments. Such a mechanism is suggested by Huang (1993) based on in vitro experiments with monomeric G-actin in which concentrations of cadmium as low as 12.5 μM inhibited actin polymerization.

Studies from other laboratories suggest that disruption of cytoskeletal structure could also occur through metal ion activation of the calcium-binding protein, calmodulin (review by Fowler, 1989). Calmodulin is a ubiquitous 16.7 kg mol^{-1} protein with four Ca^{2+}-binding sites formed by carboxyl group ligands. Ca^{2+} binds to calmodulin with a binding affinity of about 10^{-5} to 10^{-6} M (Wong and Cheung, 1979; Cohen et al., 1978; Jamieson et al., 1980; Wang et al., 1980). Upon binding, calcium alters the structure of the protein molecule, allowing it to activate a number of structural and enzymatic proteins, including phosphodiesterases, adenylate cyclase, Ca^{2+},Mg^{2+}-ATPase, phospholipase A2 and phosphorylase kinase (Wallace et al., 1980; Kretsinger, 1992). Through these activation reactions, under normal circumstances, calmodulin appears to play an important role in regulating cell proliferation. Studies also suggest, however, that following cell injury Ca activation of calmodulin may result in cytoskeletal alterations (Trump et al., 1984). Toxic metal ions such as Cd, Pb and Hg have also been shown to activate phosphodiesterase enzymes through binding to calmodulin (Chao et al., 1984; Suzuki et al., 1985; Mills and Johnson, 1985). Thus, direct activation of calmodulin by toxic metal ions may play a role in the cytoskeletal damage caused by these agents.

VI. HIGH-AFFINITY METAL-BINDING PROTEINS

A. Metallothioneins

Transition metals such as zinc and cadmium show high-affinity binding for sulfhydryl groups on proteins. Much of the biological chemistry of these metals is determined by their binding to the low-molar-mass (l.m.m.), cysteine-rich class of proteins called metallothioneins (MT) which are present to varying degrees in most mammalian cells (see preceding Chapter 1-D5). Metallothioneins contain 20 cysteine residues which form two metal-binding sites for 3 and 4 metal ions, respectively. The essential metal ion, Zn, is the metal normally bound to MT; however, due to its higher binding affinity, Cd easily displaces Zn and can occupy up to 5 of the 7 metal-binding sites on the protein molecule following Cd exposure (Vallee and Maret, 1993; Robbins et al., 1991). Binding constants for Cd and Zn have been estimated to be $2x10^{-12}$ and $2x10^{-16}$, respectively (Kägi and Schäffer, 1988). Despite this high thermodynamic stability, however, studies indicate that metals bound to MT are kinetically very labile (Kägi, 1993). This lability, coupled with the flexibility of folding of the MT polypeptide chain, allows the metal ions bound to MT to move easily in and out of the protein (Vašak et al., 1985; Vašak, 1986; Otvos et al., 1989).

Numerous studies have provided evidence that cellular MT can decrease the toxicity of cadmium. In vivo and in vitro studies have shown that increased MT levels increase the tolerance of animals and cells to Cd exposure (Cherian and Nordberg, 1983; Goering and Klaassen, 1984; Webb, 1979; 1987; Templeton and Cherian, 1991; Durnam and Palmiter, 1987; Huang et al., 1987). This protection appears to be due to a combination of the strong metal chelating ability of MT and the high kinetic lability of the bound metals. Acting as a metal-ion buffering system, the concentrations of free, reactive Cd^{2+} ions in the cell are decreased by binding to MT. This serves to protect enzymes active site ligands and other vulnerable cellular proteins (Kägi and Schäffer, 1988). In addition, studies suggest MT can also function to "resurrect" poisoned enzymes and proteins, by metal-ion exchange reactions. Huang and coworkers (Huang, 1993) have reported that the inhibition of actin polymerization by cadmium can be reversed in vitro by the addition of 1-4 μM Zn-MT. Restoration of normal peptidase activity in the carboxypeptidase A enzyme inhibited by cadmium was also achieved by adding Zn-MT to the isolated enzyme (Huang, 1993). Protection against metal toxicity by MT may also include metals other than the Group II metals (Cd and Hg). Studies by Goering and Fowler (1987a; 1987b; 1987c) indicate that Zn-MT is able to decrease the inhibition of δ-aminolevulinic acid (ALAD) by lead by removing lead and restoring zinc at the active site of the enzyme. Thus, MT can serve to prevent and reverse detrimental metal-protein interactions.

B. Low-molar-mass lead-binding proteins

High-affinity Pb-binding proteins (PbBPs) are a second class of l.m.m., metal-binding proteins that are currently under investigation as proteins that may play a role in regulating the effects of detrimental metal ions in cells. Studies by Fowler and coworkers (Goering and Fowler, 1984; 1985; Mistry et al., 1985; 1986; Goering et al., 1986; Duval and Fowler, 1989) have demonstrated that brain and kidney tissues contain PbBPs with dissociation constants for Pb of about 10^{-8} M. In rats, the kidney protein is a specific cleavage product of α_2-microglobulin with a molar mass of approximately 17.6 kg mol^{-1}; it is rich in aspartic and glutamic acids but low in cysteine (Fowler and Duval, 1991). The low cysteine content and its low binding capacity for Cd and Zn, as well as Pb, serve to set it apart from the MT family of l.m.m., metal-binding proteins. The brain PbBP is similar in size and Pb-binding affinity to the kidney protein, but differs antigenically (Fowler and Duval, 1991).

The true function(s) of PbBPs have not been fully elucidated. Some data suggest that, like MT, PbBPs may protect cells against toxic interactions of lead, by chelating free Pb^{2+} ions. In vitro studies with semi-purified PbBPs from kidney and brain have demonstrated that these proteins can confer a dose-related resistance to lead inhibition of the heme enzyme ALAD (Goering and Fowler, 1984; 1985; Mistry et al., 1986). Other studies have also demonstrated that brain and kidney ALAD activities are relatively resistant to lead treatment, which is consistent with the presence of these proteins in these tissues (Moore et al., 1980; Fowler et al., 1980; Oskarsson and Fowler, 1985). Like MT, cellular PbBPs may chelate lead from enzymes such as ALAD and, perhaps, simultaneously donate the essential metal ion (Zn) to the metalloenzyme (Goering and Fowler, 1985; 1987a).

Fowler (1989) has also suggested that these high-affinity PbBPs may serve as receptor proteins for lead in specific cells and play a role in mediating effects of lead on gene expression. Mistry et al. (1985; 1986) have reported that renal PbBP acts as a major binding site for lead in kidney proximal tubule cells and facilitates the movement of lead into the nucleus. It is possible that the PbBPs are metal regulatory proteins that bind to and alter the expression of specific genes. Further studies are needed to verify this hypothesis and to determine whether high-affinity PbBPs are important in mediating effects of other metals as well as lead in mammalian cells.

C. Inclusion body proteins

A second apparent detoxification reaction for some metals such as lead and bismuth in renal tissue is the formation of inclusion bodies, which are electron dense formations composed of metal ions precipitated on a protein matrix (Goyer et al., 1970; Fowler and Goyer, 1975). The protein matrix that accumulates in the presence of Pb is an acidic protein of approximately 30.0 kg mol^{-1} (Shelton and Egle, 1982; Shelton, 1986; Moore et al., 1973). Studies have shown that lead retards the degradation of this nuclear protein, which gives rise to its accumulation in the cell (Shelton et al., 1993).

VII. SUMMARY

Detrimental effects of metals in biological systems occur primarily through metal-protein interactions. In vitro studies have shown that toxic metals can bind to enzymes, membrane receptor proteins, nuclear regulatory proteins, and structural proteins, causing significant alterations in their normal function. These effects come about by chemical and structural changes caused by the binding of the metals to critical ligands in the protein molecules (Table 2). Due to the sensitivity of cell protein systems to metal ions, high-affinity metal-binding proteins, such as MT and l.m.m. PbBPs, have evolved to "buffer" the toxic effects of metals in cells. Thus, the concentration and nature of metal-binding ligands in cells as well as the concentration of the metals themselves are important determinants of metal ion toxicity in vivo. Continuing research in this area is needed to better understand the competitive binding and interactions that occur between proteins and toxic metals in vivo.

Table 2

Protein-Metal Interactions of Detrimental Metals

Mechanisms of metal ion toxicity through protein binding:
Binding to $-S^-$ or NH_3 or COO^- groups in the active sites of enzymes and other proteins.
Displacing essential metal ion cofactors in enzyme active sites.
Altering protein conformation by binding to amino acid protein side groups required for proper structural arrangements.
Mechanisms of metal ion detoxification through protein binding:
Binding of metal ions to soluble metal-storage proteins.
Formation of inclusion bodies by precipitation of metals on protein matrices.

REFERENCES

Alberts, B., Bray, D., Lewis, J., Raff, M., Roberts, K., and Watson, J.D., eds. (1983). *Molecular Biology of the Cell*, Garland Press, New York, p. 549.

Anderson, P.M., and Desnick, R.J. (1979). *J. Biol. Chem.*, *254*: 6924.

Auld, D.S., and Holmquist, B. (1974). *Biochemistry*, *13*: 4355.

Barnard, G.F., Itoh, R., Hohberger, L.H., and Shemin, D. (1977). *J. Biol. Chem.*, *252*: 8965.

Bernard, A., and Lauwerys, R. (1987). *Mechanisms of Chemical-Induced Porphyrinopathies* (E.K. Silbergeld and B.A. Fowler, eds.), *Ann. N.Y. Acad. Sci.*, *514*: 41.

Bevan, D.R., Bodlaender, P., and Shemin, D. (1980). *J. Biol. Chem.*, *255*: 2030.

Beyersmann, D., Block, C., Freyermuth, S., and Malviya, A.N. (1993). *2nd International Meeting on Molecular Mechanisms of Metal Toxicity and Carcinogenesis*, Madonna di Campiglio, Italy, January 10-17, 1992, p. 50.

Chao, S.-H., Suzuki, Y., Zysk, J.R., and Cheung, W.Y. (1984). *Mol. Pharmacol.*, *26*: 75.

Chen, Y.-C., and Smith, J.B. (1992). *Toxicol. Appl. Pharmacol.*, *117*: 249.

Cherian, M.G., and Nordberg, M. (1983). *Toxicology*, *28*: 1.

Cohen, P., Burchell, A., Foulkes, J.G., Cohen, P.T.W., Vanaman, T.C., and Nairn, A.C. (1978). *FEBS Lett.*, *92*: 287.

Coleman, J.E., and Vallee, B.L. (1961). *J. Biol. Chem.*, *236*: 2244.

Costa, M. (1991). *Environ. Health Perspect.*, *92*: 45.

Dailey, H.A. (1987). *Mechanisms of Chemical-Induced Porphyrinopathies* (E.K. Silbergeld and B.A. Fowler, eds.), *Ann. N.Y. Acad. Sci.*, *514*: 81.

Durnam, D.M., and Palmiter, R.D. (1987). *Experientia, Suppl.* *52*: 457.

Duval, G.E., and Fowler, B.A. (1989). *Biochem. Biophys. Res. Commun.*, *159*: 177.

Eimerl, S., and Schramm, M. (1993). *J. Neurochem.*, *61*: 518.

Elliget, K.A., Phelps, P.C., and Trump, B.F. (1991). *Cell Biol. Toxicol.*, *7*: 263.

EPA (U.S. Environmental Protection Agency) (1986). *Air Quality Criteria for Lead*, EPA-600/08-83/028aF-dF, Research Triangle Park, N.C., U.S. Environmental Protection Agency, Vol. 4.

Finelli, V.N., Klauder, D.S., Karaffa, M.A., and Petering, H.G. (1975). *Biochem. Biophys. Res. Commun.*, *65*: 303.

Fowler, B.A. (1989). *Comments Toxicol.*, *3*: 27.

Fowler, B.A., and Duval, G. (1991). *Environ. Health Perspect.*, *91*: 77.

Fowler, B.A., and Goyer, R.A. (1975). *J. Histochem. Cytochem.*, *23*: 722.

Fowler, B.A., Kimmel, C.A., Woods, J.S., McConnell, E.E., and Grant, L.D. (1980). *Toxicol. Appl. Pharmacol.*, *56*: 59.

Fowler, B.A., Oskarsson, A., and Woods, J.S. (1987). *Mechanisms of Chemical-Induced Porphyrinopathies* (E.K. Silbergerd and B.A. Fowler, eds.), *Ann. N.Y. Acad. Sci.*, *514*: 172.

Gibbs, P.N.B., and Jordan, P.M. (1986). *Biochem. J.*, *236*: 447.

Gibbs, P.N.B., Gore, M.G., and Jordan, P.M. (1985). *Biochem. J.*, *225*: 573.

Goering, P.L., and Fowler, B.A. (1984). *J. Pharmacol. Exp. Ther.*, *231*: 66.

Goering, P.L., and Fowler, B.A. (1985). *J. Pharmacol. Exp. Ther.*, *234*: 365.

Goering, P.L., and Fowler, B.A. (1987a). *Arch. Biochem. Biophys.*, *253*: 48.

Goering, P.L., and Fowler, B.A. (1987b). *Biochem. J.*, *245*: 339.

Goering, P.L., and Fowler, B.A. (1987c). *Metallothionein II* (J.H.R. Kägi and Y. Kojima, eds.), Birkhäuser Verlag, Basel, Switzerland, p. 613.

Goering, P.L., and Klaassen, C.D. (1984). *Toxicol. Appl. Pharmacol.*, *74*: 299.

Goering, P.L., Mistry, P. and Fowler, B.A. (1986). *J. Pharmacol. Exp. Ther.*, *237*: 220.

Goyer, R.A., Leonard, D.L., Moore, J.F., Rhyne, B., and Krigman, M.R. (1970). *Arch. Environ. Health*, *20*: 705.

Gurne, D., Chen, J., and Shemin, D. (1977). *Proc. Nat. Acad. Sci. USA*, *74*: 1383.

Huang, P.C. (1993). *Metallothionein III* (K.T. Suzuki, N. Imura and M. Kimura, eds.), Birkhäuser Verlag, Basel, Switzerland, p. 407.

Huang, P.C., Morris, S., Dinman, J., Pine, R., and Smith, B. (1987). *Experientia, Suppl. 52*: 439.

Jaffe, E.K., and Hanes, D. (1986). *J. Biol. Chem.*, *261*: 9348.

Jaffe, E.K., Salowe, S.P., Chen, N.T. and DeHaven, P.A. (1984). *J. Biol. Chem.*, *259*: 5032.

Jamieson, G.A., Jr., Bhonson, D.D., Schachat, F.H., and Vanaman, T.C. (1980). *Ann. N.Y. Acad. Sci.*, *356*: 1.

Johnson, J.L., and Rajagopalan, K.V. (1978). *Bioinorg. Chem.*, *8*: 439.

Kägi, J.H.R. (1993). *Metallothionein III* (K.T. Suzuki, N. Imura and M. Kimura, eds.), Birkhäuser Verlag, Basel, Switzerland, p. 29.

Kägi, J.H.R., and Schäffer, A. (1988). *Biochemistry*, *27*: 8508.

Kretsinger, R. (1992). *Science*, *258*: 50.

Li, W., and Chou, I.-N. (1992). *Toxicol. Appl. Pharmacol.*, *114*: 132.

Miller, C.A., III, and Costa, M. (1989). *Mol. Toxicol.*, 2: 11.

Mills, J.S., and Johnson, J.D. (1985). *J. Biol. Chem.*, *260*: 15100.

Mills, J.W., and Ferm, V.H. (1989). *Toxicol. Appl. Pharmacol.*, *101*: 245.

Mills, J.W., Zhou, J.-H., Cardoza, L., and Ferm, V.H. (1992). *Toxicol. Appl. Pharmacol.*, *116*: 92.

Mistry, P., Lucier, G.W., and Fowler, B.A. (1985). *J. Pharmacol. Exp. Ther.*, *232*: 462.

Mistry, P., Mastri, C., and Fowler, B.A. (1986). *Biochem. Pharmacol.*, *35*: 711.

Moore, J.F., Goyer, R.A., and Wilson, M. (1973). *Lab. Invest.*, *29*: 488.

Moore, M.R., Meredith, P.A., and Goldberg, A. (1980). *Lead Toxicity* (R.L. Singhal and J.A. Thomas, eds.), Urban and Schwarzenberg, Baltimore/Munich, p. 79.

NRC (National Research Council) (1993). *Measuring Lead Exposure in Infants, Children, and Other Sensitive Populations*, National Academy Press, Washington, DC.

Oskarsson, A., and Fowler, B.A. (1985). *Exp. Mol. Pathol.*, *43*: 387.

Otvos, J., Chen, S.-M., and Liu, X. (1989). *Metal Ion Homeostasis: Molecular Biology and Chemistry* (D.H. Hammer and D.R. Winge, eds.), Alan R. Liss, Inc., New York, p. 197.

Prozialeck, W.C., and Niewenhuis, R.J. (1991). *Toxicol. Appl. Pharmacol.*, *107*: 81.

Robbins, A.H., McRee, D.E., Williamson, M., Collett, S.A., Xuong, N.H., Furey, W.F., Wang, B.C., and Stout, C.D. (1991). *J. Mol. Biol.*, *222*: 1269.

Sarkar, V.S. (1993). *Effects of Metal Ion Interaction with DNA Binding Proteins.* Abstract presented at the *2nd International Meeting on Molecular Mechanisms of Metal Toxicity and Carcinogenicity*, Madonna di Campiglio, Italy, Jan. 10-17, 1993.

Shankar, V.S., Bax, C.M.R., Alam, A.S.M.T., Bax, B.E, Huang, C.L.H., Zaidi, M. (1992). *Biochem. Biophys. Res. Commun.*, *187*: 913.

Shay, J.W., ed. (1986). *Cell and Molecular Biology of the Cytoskeleton*, Plenum Press, New York.

Shelton, K.R. (1986). *J. Biol. Chem.*, *261*: 1935.

Shelton, K.R., and Egle, P.M. (1982). *J. Biol. Chem.*, *257*: 11802.

Shelton, K.R., Bigbee, J.W., and Egle, P.M. (1993). *2nd International Meeting on Molecular Mechanisms of Metal Toxicity and Carcinogenicity*, Madonna di Campiglio, Italy, Jan. 10-17, 1992, p. 46.

Silbergeld, E.K., and Fowler, B.A., eds. (1987). *Mechanisms of Chemical-Induced Porphyrinopathies, Ann. N.Y. Acad. Sci.*, *514*: 352.

Smith, J.B., Pijuan, V., Zhuang, Y., Chen, Y.-C., and Smith, L. (1993). *2nd International Meeting on Molecular Mechanisms of Metal Toxicity and Carcinogenicity*, Madonna di Compiglio, Italy, January 10-17, 1992, p. 51.

Suzuki, Y., Chao, S.-H., Zysk, J.R., and Cheung, W.Y. (1985). *Arch. Toxicol.*, *57*: 205.

Templeton, D.M., and Cherian, M.G. (1991). *Meth. Enzymol.*, *205*: 11.

Trump, B.F., Berezesky, I.K., Sato, T., Laiho, U., Phelps, P.C., and DeClaris, N. (1984). *Environ. Health Perspect.*, *57*: 281.

Tsukamoto, I., Yoshinhaga, T., and Sano, S. (1979). *Biochim. Biophys. Acta*, *570*: 167.

Tsukamoto, I., Yoshinhaga, T., and Sano, S. (1980). *Int. J. Biochem.*, *12*: 751.

Vallee, B.L., and Auld, D.S. (1990). *Biochemistry*, *29*: 5647.

Vallee, B.L., and Maret, W. (1993). *Metallothionein III* (K.T. Suzuki, N. Imura and M. Kimura, eds.), Birkhäuser Verlag, Basel, Switzerland, p. 1.

Vallee, B.L., and Ulmer, D.D. (1972). *Annu. Rev. Biochem.*, *41*: 91.

Vašak, M. (1986). *Environ. Health Perspect.*, *65*: 193.

Vašak, M., Hawkes, G.D., Nicholson, J.K., and Sadler, P.J. (1985). *Biochemistry*, *24*: 740.

Wallace, R.W., Tallant, E.A., and Cheung, W.Y. (1980). *Calcium and Cell Function*, Vol. 1 (W.Y. Cheung, ed.), Academic Press, New York, p. 13.

Wang, J.H., Sharma, R.K., Huang, C.Y., Chau, V., and Chock, P.B. (1980). *Ann. N.Y. Acad. Sci.*, *356*: 190.

Webb, M., ed. (1979). *The Chemistry, Biochemistry and Biology of Cadmium*, Elsevier/North Holland, Amsterdam.

Webb, M. (1987). *Metallothionein II* (J.H.R. Kägi and Y. Kojima, eds.), Birkhäuser Verlag, Basel, Switzerland, p. 109.

Wong, P.Y.-K., and Cheung, W.Y. (1979). *Biochem. Biophys. Res. Commun.*, *90*: 473.

Wu, W.H., Shemin, D., Richards, K.E., and Williams, R.C. (1974). *Proc. Nat. Acad. Sci. USA*, *71*: 1767.

Yamashita, M.M., Wesson, L., Eisenman, G., and Eisenberg, D. (1990). *Proc. Nat. Acad. Sci. USA*, *87*: 5648.

COMPLEXES OF METAL IONS WITH VARIOUS NUCLEIC ACID COMPONENTS

by

Helmut Sigel

University of Basel,
Institute of Inorganic Chemistry,
Spitalstrasse 51
CH-4056 Basel
Switzerland

I. INTRODUCTION

Nature has composed nucleic acids, i.e. RNA and DNA, from phosphate, ribose or 2'-deoxyribose, and heteroaromatic ring residues, so-called nucleic (acid) bases (see Fig. 1). The combination of the sugar with a nucleic base is designated as a nucleoside and addition of a phosphate group leads to a so-called nucleoside monophosphate which is also the simplest nucleotide. Other nucleotides are nucleoside diphosphates and nucleoside triphosphates (Fig. 1) (Sigel, 1979). The repeated combination of nucleoside monophosphates via 3',5' linkages at the ribose units leads to the high molecular nucleic acids (Stryer, 1988).

All the mentioned species are ligands, i.e. they offer binding sites for metal ions (Sigel, 1979; 1993). In the remainder of this overview the affinities of various of these components for metal ions are summarized. In nature these kinds of interactions are important (Sigel and Sigel, 1989). Practically all enzymes utilizing nucleotides are metal ion dependent (Mildvan, 1987). This includes adenosine 5'-triphosphate (5'-ATP^{4-}; Fig. 1), an important intermediate in the energy metabolism (Boyer, 1987) which is a substrate usually in the form of its Mg^{2+} complex. Similarly, alkali and alkaline earth ions, i.e. Na^+, K^+, Mg^{2+}, Ca^{2+}, are important regulators of nucleic acid biochemistry; especially Mg^{2+} is prevalent and also required as a cofactor for ribozymes (Cowan, 1992). Moreover, many enzymes involved in the metabolism of nucleotides and nucleic acids contain a further intrinsic metal ion, very often Zn^{2+} (see e.g. Coleman and Giedroc, 1989); in fact, even Cu^{2+} might have a role (Tallineau et al., 1984).

The metals of life, which are of a *hard* or *borderline* type (Fraústo da Silva and Williams, 1991; Sigel and McCormick, 1970), have a rather pronounced affinity to phosphate groups (Sigel, 1989a). As such metal ions, namely the alkaline earth ions and the second half of the divalent 3d series including Zn^{2+} (and in addition also Cd^{2+}), are in the focus of this overview, phosphate-metal ion interactions are first dealt with below. It may be added that deleterious metal ions, such as Hg^{2+}, Pd^{2+} or Pt^{2+}, show a more pronounced affinity toward nitrogen donors and therefore consequently also for the nucleic base residues (regarding Pt^{2+} see e.g. Reedijk, 1992).

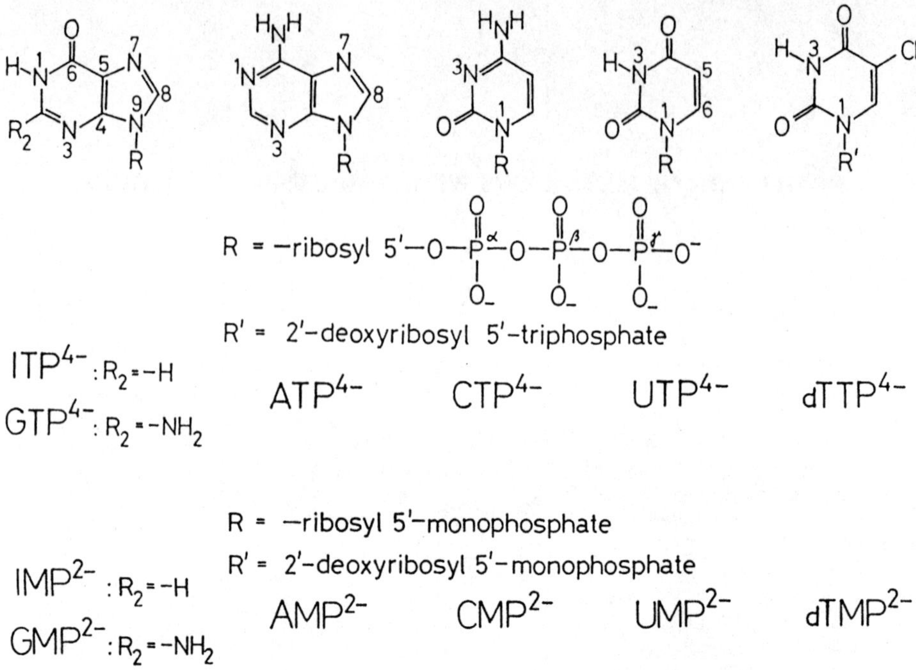

Figure 1.
Chemical structures of various nucleoside 5'-mono- and 5'-triphosphates (5'-NMP^{2-} and 5'-NTP^{4-}) together with the labeling system for the triphosphate chain; note, the phosphate groups in the NTPs are labeled α, β and γ, where γ refers to the terminal phosphate group. Obviously, the analogous nucleoside 5'-diphosphates (5'-NDP^{3-}) have the corresponding structures with two phosphate groups which are labeled α and β. In the above abbreviations the 5' designation is deleted throughout for simplicity; however, in the text this specification is always given when necessary, e.g. 5'-AMP^{2-}, 5'-ATP^{4-}, etc. In addition it may be mentioned that the names of the above shown nucleic (acid) bases, i.e. R = R' = H, are (from left to right) hypoxanthine (R_2 = H), guanine (R_2 = NH_2), adenine, cytosine, uracil, and thymine; the corresponding nucleosides (Ns) are called (R = ribose; from left to right) inosine (Ino; R_2 = H), guanosine (Guo; R_2 = NH_2), adenosine (Ado), cytidine (Cyd), uridine (Urd), and ribosylthymine; combination of thymine with 2'-deoxyribose gives the nucleoside called thymidine (dThd; R' = 2'-deoxyribose in the above structure in the upper right hand corner).

II. METAL ION - PHOSPHATE INTERACTIONS

Phosphate monoesters are an important naturally occurring group of ligands. As in these cases the organic residue is close to the phosphate group, its basicity is affected. For example, the basicity increases in the following series: 4-nitrophenyl phosphate ($NPheP^{2-}$), phenyl phosphate ($PheP^{2-}$), *D*-ribose 5'-monophosphate ($RibMP^{2-}$), and *n*-butyl phosphate (BuP^{2-}). Consequently, the metal ion affinity of these monophosphate esters is expected to rise correspondingly. Indeed, this has been proven for the complexes formed between the mentioned four ligands and Mg^{2+}, Ca^{2+}, Sr^{2+}, Ba^{2+}, Mn^{2+}, Co^{2+}, Ni^{2+}, Cu^{2+}, Zn^{2+}, or Cd^{2+} (Massoud and Sigel, 1988a); for three of these metal ions the results are plotted in Figure 2. At this point it should be mentioned that in this overview only few acidity constants (of protonated ligands; equation 1) and stability constants (of the corresponding complexes; equation 2) are given. However, the literature, where these equilibrium constants have been published, is always cited in the appropriate sections.

$$LH^+ \rightleftharpoons L + H^+ \quad (1a)$$

$$K^{H}_{LH} = [L][H^+]/[LH^+] \quad (1b)$$

$$M^{2+} + L \rightleftharpoons ML^{2+} \quad (2a)$$

$$K^{M}_{ML} = [ML^{2+}]/([M^{2+}][L]) \quad (2b)$$

With regard to Figure 2 the following, in part remarkable, points have to be emphasized:
(i) for a given metal ion the complexes with $NPheP^{2-}$, $PheP^{2-}$ and BuP^{2-} fit within experimental error on a straight line; the value for *D*-ribose 5'-monophosphate ($RibMP^{2-}$) also fits on this line, indicating that $RibMP^{2-}$ behaves like a simple phosphate monoester ($R\text{-}MP^{2-}$) with a non-coordinating organic residue (Massoud and Sigel, 1988a);
(ii) the data pairs for the simplest phosphate ligands one may think of, i.e. hydrogen phosphate (HPO_4^{2-}) and methyl phosphate ($CH_3OPO_3^{2-}$), fit on the same straight-baseline plots, proving that these baselines are free from any steric effects (Massoud and Sigel, 1988a);
(iii) simple phosphonate ligands, like methyl phosphonate and ethyl phosphonate, form complexes for which the data pairs also fit on the mentioned straight lines. By taking into

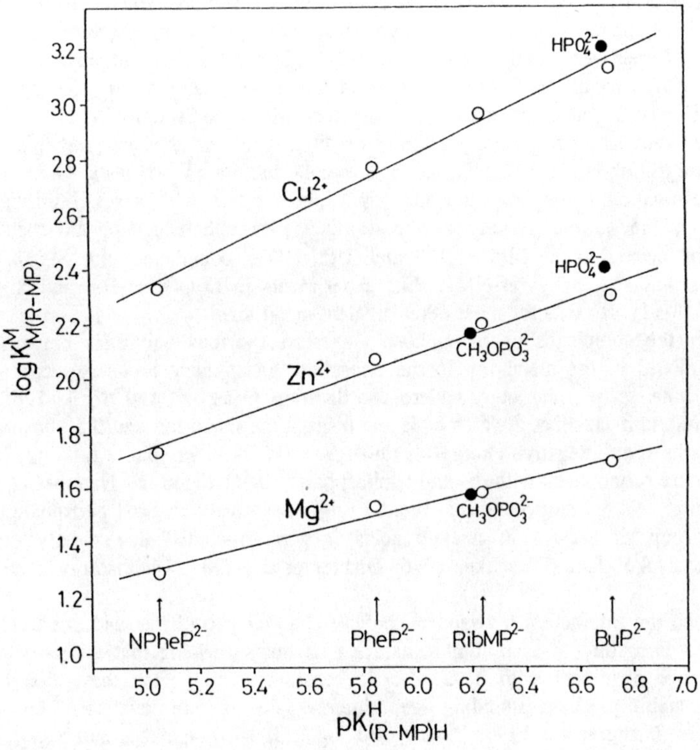

Figure 2.
Relationship between log $K^{M}_{M(R-MP)}$ and $pK^{H}_{(R-MP)H}$ for the Mg^{2+}, Zn^{2+}, and Cu^{2+} 1:1 complexes of 4-nitrophenyl phosphate ($NPheP^{2-}$), phenyl phosphate ($PheP^{2-}$), *D*-ribose 5'-monophosphate ($RibMP^{2-}$), and *n*-butyl phosphate (BuP^{2-}) (from left to right, O). The least-squares lines are drawn through these data sets. The points due to the complexes formed with methyl phosphate ($CH_3OPO_3^{2-}$) and hydrogen phosphate (HPO_4^{2-}) (●) are inserted for comparison. All data refer to aqueous solutions at 25°C and I = 0.1 M ($NaNO_3$). This figure is reproduced by permission of the American Chemical Society from *Inorganic Chemistry* (Massoud and Sigel, 1988a).

account also these results, baseline correlations for M^{2+}-phosphate or -phosphonate complex stabilities and phosphate or phosphonate group basicities (i.e. log $K^{M}_{M(R-MP)}$ versus $pK^{H}_{(R-MP)H}$ plots) are obtained for the pK_a range 5 through 8, allowing calculation of the stability constant of the complex formed between any of the ten metal ions mentioned and a phosphate monoester or a phosphonate ligand with a known pK_a value (for details see Sigel et al., 1992b);

(iv) the data pairs for the complexes of the ten metal ions and cytidine 5'-monophosphate (5'-CMP^{2-}), uridine 5'-monophosphate (5'-UMP^{2-}), or thymidine 5'-monophosphate (5'-TMP^{2-}) also fit on these correlation lines proving that the nucleic base residues of these three nucleoside monophosphates do not participate in complex formation (Massoud and Sigel, 1988a). In other words, the stability of these nucleotide complexes is solely determined by the basicity of the phosphate group (see also Sections IV and V.A).

The indicated results for phosphate monoesters contrast with those obtained for triphosphate monoesters. In the latter case the organic residue is so distant from the terminal γ-phosphate group that it does hardly affect the basicity of this group. Consequently $pK^{H}_{(NTP)H} = 6.50 \pm 0.05$ in aqueous solution (I = 0.1 M, $NaNO_3$ or $NaClO_4$; 25°C) for all the monoprotonated nucleoside 5'-triphosphates (NTP^{4-}) given in Figure 1 (Sigel et al., 1987; Sigel, 1977) as well as for methyl triphosphate (Schneider et al., 1964). It may be added that it was shown by various techniques that in M(5'-UTP)$^{2-}$, M(5'-dTTP)$^{2-}$ and M(5'-CTP)$^{2-}$ complexes the base residues do not participate in complex formation (with the single exception of Cu(5'-CTP)$^{2-}$; Sigel et al., 1987) (Frey and Stuehr, 1978; Mariam and Martin, 1979; Scheller et al., 1981; Sigel et al., 1987). This property of the pyrimidine-NTPs is different from that of the purine-NTPs (Sigel, 1987b); complexes of these latter ligands with various metal ions show an increased stability in accordance with a participation of the purine residue in complex formation (see Section V.C).

Our knowledge about the properties of diphosphate monoesters is at this time rather incomplete (Smith et al., 1991). It may be assumed that the pK_a values are between those of the corresponding monophosphate and triphosphate esters with a small influence of the organic residue on the basicity of the terminal β-phosphate group. No metal ion-base interaction occurs in the M(5'-CDP)$^-$ and M(5'-UDP)$^-$ complexes with Mg^{2+}, Zn^{2+}, or Cd^{2+}, as was shown by ^{1}H-NMR shift experiments in D_2O (Scheller and Sigel, 1983); probably this is also true for most other divalent metal ions.

For the complexes of monophosphate esters, various equilibria between different binding modes of the metal ion to the phosphate group have to be expected, especially between inner-sphere and outer-sphere coordination (Massoud and Sigel, 1988a). Outer-sphere binding means that a water molecule bridges the metal ion and the phosphate group. An increase of the negative charge on the ligand favors inner-sphere binding; hence, this will be more pronounced with di- and triphosphates (Brintzinger and Hammes, 1966). With this in mind it is not surprising that for the single negatively charged phosphate unit of the diester group as present in nucleic acids it was concluded that mostly outer-sphere interactions take place (Pörschke, 1976; Diebler et al., 1987; Dhathathreyan and Diebler, 1991).

Since the interaction between negatively charged phosphates and positively charged metal ions is mainly of a coulombic nature it is not surprising that a decreasing solvent polarity, i.e. a solvent with a dielelectric constant smaller than that of water, favors complex stability. Corresponding experimental details are available for phosphate monoesters (Liang et al., 1989; Bastian and Sigel, 1991; Magalhães and Sigel, 1992) and triphosphate monoesters (Tribolet et al., 1985). Considering that the 'effective' or 'equivalent solution' dielectric constant is reduced in a protein (Rees, 1980; Rogers et al., 1985) or in an active site cavity of an enzyme (Sigel et al., 1985; Bastian and Sigel, 1990), these effects are meaningful because the shift of a phosphate group along the surface of a protein may considerably alter its metal ion affinity.

III. INTERACTIONS OF METAL IONS WITH SUGAR RESIDUES

The ligating properties of carbohydrates or sugar-type ligands have recently been reviewed (Burger and Nagy, 1990; see also Part Three, Chapter 2-D in this volume). The interactions between metal ions and hydroxy groups are weak; in fact, complexes are only formed if certain geometrical arrangements fitting the coordination sphere of the metal ion are provided by the ligand. Indeed, in the preceding Section II (see also Fig. 2) we have seen that *D*-ribose 5'-monophosphate behaves toward Mg^{2+}, Ca^{2+}, Sr^{2+}, Ba^{2+}, Mn^{2+}, Co^{2+}, Ni^{2+}, Cu^{2+}, Zn^{2+}, or Cd^{2+} in aqueous solution like a simple phosphate monoester; in other words, there is no indication for an interaction between any of the hydroxy groups of the ribose residue and one of these metal ions (Massoud and Sigel, 1988a).

The corresponding observation has very recently also been made with dihydroxyacetone phosphate ($DHAP^{2-}$) and glycerol 1-phosphate ($G1P^{2-}$) and the above-mentioned ten metal ions (Liang et al., 1992). The stability of all these M(R-MP) complexes is solely governed in aqueous solution by the basicity of the phosphate group of the ligands. This means, the following isomeric equilibrium is far on its left side:

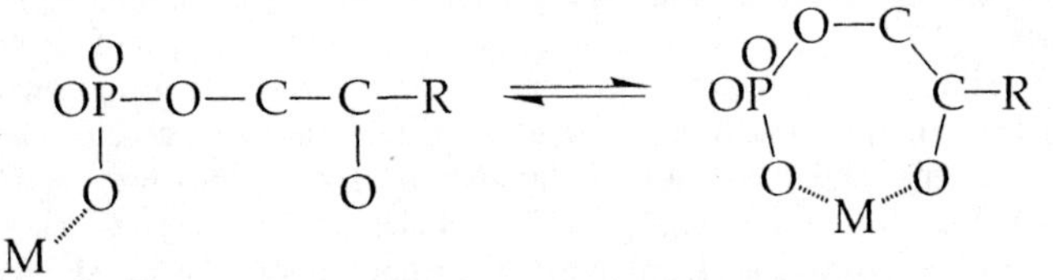

However, under conditions where the activity of water is decreased, i.e. in aqueous organic solvent mixtures, this is different. Corresponding experiments have been carried out with Cu(DHAP) and Cu(G1P) in water containing 30 or 50% (v/v) 1,4-dioxane (Liang et al., 1992). Now the formation degree of the seven-membered chelate shown in the isomeric equilibrium 3 reaches about 40% for both complex systems (Table 1). The formation of this ring-closed isomer, designated as $M(R\text{-}MP)_{cl}$, is reflected (as expected; see Martin and Sigel, 1988) in a somewhat increased stability. Consequently, positive values are observed (Liang et al., 1992) for $\log \Delta_{M(R\text{-}MP)}$, which is defined in equation 4:

$$\log \Delta_{M(R\text{-}MP)} = \log K^{M}_{M(R\text{-}MP)} - \log K^{M}_{M(R\text{-}MP)op} \tag{4}$$

In this equation $K^{M}_{M(R\text{-}MP)}$ represents the experimentally measured (overall) stability constant, while $K^{M}_{M(R\text{-}MP)op}$ represents the stability of the open isomer, $M(R\text{-}MP)_{op}$, in the intramolecular equilibrium 3. This latter value may be calculated via the straight line

Table 1

Extent of chelate formation (equilibrium 3) in the Cu(DHAP) and Cu(G1P) complexes as quantified by the percentage of $Cu(R\text{-}MP)_{cl}$ in dependence on the amount of 1,4-dioxane added to water (I = 0.1 M, $NaNO_3$; 25°C)[a]

$R\text{-}MP^{2-}$	% (v/v) dioxane	mol. fract. dioxane	ε[b]	$\log \Delta_{Cu(R\text{-}MP)}$[c]	% $Cu(R\text{-}MP)_{cl}$
$DHAP^{2-}$	0	0	78.5	0.04 ± 0.08	9 ± 17
	30	0.083	52.7	0.08 ± 0.03	17 ± 5
	50	0.175	35.2	0.26 ± 0.03	45 ± 4
$G1P^{2-}$	0	0	78.5	-0.05 ± 0.09	0(<10)
	30	0.083	52.7	0.06 ± 0.03	13 ± 6
	50	0.175	35.2	0.24 ± 0.04	42 ± 5

[a] Abstracted of Tables III and IV of Liang et al. (1992).

[b] Dielectric constants.

[c] See equation 4. The error limits are *three times* the standard errors (3σ); the errors for % $Cu(R\text{-}MP)_{cl}$ were calculated according to the error propagation after Gauss.

equations (Magalhães and Sigel, 1992; Bastian and Sigel, 1991) of log $K^{M}_{M(R-MP)}$ versus $pK^{H}_{(R-MP)H}$ plots valid for the various solvent mixtures and the acidity constants, $pK^{H}_{(R-MP)H}$, for $H(DHAP)^-$ and $H(G1P)^-$ which were measured in the corresponding solvents (Liang et al., 1992).

These examples (see Table 1) prove that under certain conditions, especially in a low polarity environment (see also the final paragraph in Section II), carbonyl and hydroxy groups of sugars may participate in complex formation and create distinct structures, which in turn may then be responsible for certain reactivities.

IV. INTERACTIONS OF METAL IONS WITH NUCLEOSIDES

It was indicated already in Section II that the complexes of Mg^{2+}, Ca^{2+}, Sr^{2+}, Ba^{2+}, Mn^{2+}, Co^{2+}, Ni^{2+}, Cu^{2+}, Zn^{2+}, or Cd^{2+} and 5'-UMP^{2-} or 5'-$dTMP^{2-}$ showed no increased stability compared to that expected for an exclusive phosphate binding (see also Fig. 4 in Section V.B); i.e. the nucleic base residues do not participate in metal ion binding. The corresponding structures shown in Figure 1 reveal that indeed no remarkable binding sites are present in the uracyl or thymine residues; carbonyl groups have in aqueous solution only a very weak metal ion affinity and the H(N-3) units turn into strong binding sites only after deprotonation (Sigel, 1975; Scheller et al., 1981) which occurs for the nucleosides uridine (Urd) and thymidine (dThd) only above the physiological pH range ($pK^{H}_{(Urd)H}$ = 9.19; $pK^{H}_{(dThd)H}$ = 9.69; I = 0.1 M, $NaNO_3$; 25°C; H. Sigel, results to be published).

The stability of the M(5'-CMP) complexes with the above mentioned metal ions is also completely governed by the basicity of the phosphate group of 5'-CMP^{2-} (cf Section II and Fig. 4 in Section V.B). This result may seem surprising because the cytosine residue (see Fig. 1) offers with N-3 a pyridine-type nitrogen. Certainly, this N-3 is somewhat screened by the neighboring *o*-amino and *o*-carbonyl groups; indeed, the detailed effects of these groups have very recently been carefully quantified (Kinjo et al., 1992) and the fact remains that N-3 of cytidine (Cyd) shows an affinity at least toward the mentioned 3d elements including Zn^{2+} and Cd^{2+}. This apparent discrepancy between the stabilities of the M(5'-CMP) and the corresponding $M(Cyd)^{2+}$ complexes arises from the conformation adopted by 5'-CMP^{2-} in aqueous solution. This species exists predominantly in the *anti* conformation (Martin and Mariam, 1979; Davies et al., 1985), i.e. the carbonyl oxygen at C-2 is directed away from the 'plane' defined by the ribose ring, and consequently N-3 is also pointing away from the phosphate group, preventing such a simultaneous coordination of a given metal ion to these two binding sites.

The adenine residue (see Fig. 1) offers metal ions three aromatic N sites: i.e. N-1, N-3, and N-7; yet binding to N-3 is only exceptionally observed (cf Section V.B): consequently, we are left for adenosine (Ado) with a dichotomy in metal ion binding of N-1 versus N-7 (Martin, 1985); i.e. of a pyridine-like versus an imidazole-like site, respectively. By including the complex-forming properties of tubercidin (Tu) [in Tu the N-7 of Ado is replaced by a CH unit] (Ji et al., 1991) and by constructing baseline plots for the mentioned sites, this dichotomy could largely be resolved (Sigel et al., 1992a): it was concluded that Ni^{2+} and Cu^{2+}, and most probably also Co^{2+} and Cd^{2+}, coordinate preferably via N-7; for Zn^{2+} a more even distribution between the N-1 and N-7 sites of adenosine seems to occur, while Mn^{2+} possibly prefers the N-1 site which definitely is strongly dominating for the binding of H^+.

Regarding inosine (Ino) and guanosine (Guo), the situation is similar to that outlined above for uridine and thymidine. This means, the H(N-1) sites are in general not available for metal ion binding at a pH of about 7 ($pK^{H}_{(Ino)H}$ = 8.76; $pK^{H}_{(Guo)H}$ = 9.22, I = 0.1 M, $NaNO_3$; 25°C; Sigel, H., results to be published). Consequently, for these two purine-nucleosides N-3 and N-7 (see Fig. 1) are left as binding sites in the neutral pH range. Among these N-7 is the dominating site (Sigel, 1989a) while N-3 is only exceptionally involved (see Section V.B).

Based on studies of the indicated kind some tentative conclusions regarding the metal-ion binding properties of the nucleic base residues in single-stranded DNA or RNA have been given (Sigel, 1989a); for the neutral pH range in aqueous solution the following decreasing order for the affinity is suggested: N-7; Guo ≥ N-3; Cyd ≥ N-7; Ado ≥ N-1;Ado > N-3; Ado, Guo. Of course, the type of metal ion has also an influence (Sigel and McCormick, 1970); it determines especially to which extent the oxygen sites of the phosphate groups are also involved in metal ion binding.

V. METAL ION - NUCLEOTIDE COMPLEXES

A. Some summary remarks regarding pyrimidine-nucleotides

As noted already in Sections II and IV (compare also Fig. 4 in Section V.B) there is no metal ion-nucleic base interaction in the complexes formed between Mg^{2+}, Ca^{2+}, Sr^{2+}, Ba^{2+}, Mn^{2+}, Co^{2+}, Ni^{2+}, Cu^{2+}, Zn^{2+}, or Cd^{2+} and 5'-CMP^{2-}, 5'-UMP^{2-}, 5'-$dTMP^{2-}$ (Massoud and Sigel, 1988a), 5'-CTP^{4-}, 5'-UTP^{4-}, or 5'-$dTTP^{4-}$ (Sigel et al., 1987; Sigel, 1987b). Of course, this changes upon ionization of the H(N-3) site of the uracyl or thymine moieties (Sigel, 1975; Scheller et al., 1981). The only known exception is Cu(5'-$CTP)^{2-}$ which exists to approximately 30% as a macrochelated isomer involving N-3 (see also equilibrium 5 below) (Sigel et al., 1987) and in which 5'-CTP^{4-} must have been turned into the *syn* conformation.

B. Macrochelate formation in metal ion complexes of purine-nucleoside monophosphates

At this point it should be emphasized that any study aiming to reveal structural aspects of complexes containing purine-nucleotides in aqueous solution has to be carried out in concentrations below 10^{-3} M due to the rather pronounced tendency for self-association which occurs via stacking of the purine rings (Scheller et al., 1981; Scheller and Sigel, 1983; regarding the effect of protonation see Tribolet and Sigel, 1988; Corfù and Sigel, 1991; for a short review see Sigel, 1989b). This aspect was considered for all the results presented below which were mostly obtained via potentiometric pH titrations (I = 0.1 M, $NaNO_3$; 25°C).

The biologically most important compound among the adenosine monophosphates is probably 5'-AMP^{2-} which exists in aqueous solution predominantly in the *anti* conformation (Fig. 3) allowing therefore a metal ion to coordinate simultaneously to the phosphate group and to N-7 of the purine-base moiety, giving thus rise to the intramolecular equilibrium 5:

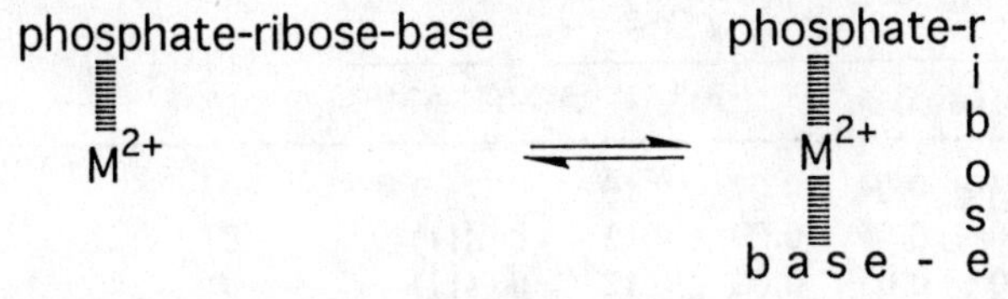

The corresponding dimensionless intramolecular equilibrium constant K_I is defined by equation 6, $M(NMP)_{cl}$ being the macrochelated or 'closed' species and $M(NMP)_{op}$ the 'open' isomer.

$$K_I = [M(NMP)_{cl}]/[M(NMP)_{op}] \quad (6)$$

The formation of any macrochelated species has to be reflected by an increased complex stability (Martin and Sigel, 1988). This means, the values for $\log \Delta_{M(R\text{-}MP)}$, as defined by equation 4, have to be positive. Indeed, the vertical distance between the point due to Cu(5'-AMP) and the corresponding reference line shown in Figure 4 corresponds to the value for $\log \Delta_{Cu(5'\text{-}AMP)}$. Based on such data the formation degree of the macrochelated species may be calculated (for details see Sigel et al., 1988); the results are summarized in Table 2.

From column 2 in Table 2 it is evident that the stability increase $\log \Delta_{M(5'\text{-}AMP)}$ is zero within the error limits for various of the M(5'-AMP) complexes indicating that no or only traces of macrochelates are formed. However, especially for various complexes formed with divalent 3d metal ions, the stability increase is significant and consequently also the formation degree of the macrochelated species (column 4 of Table 2). That macrochelate

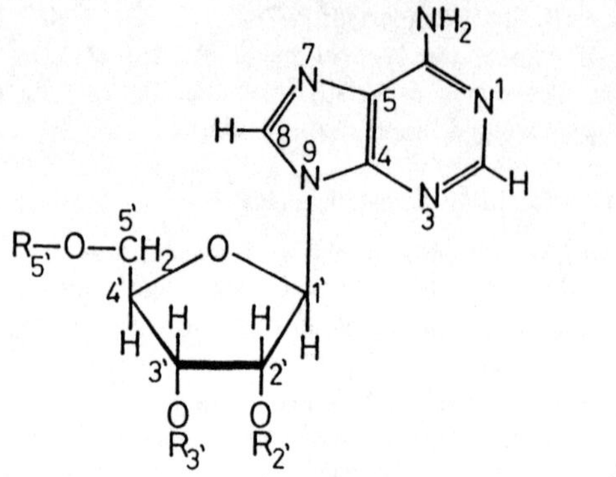

$2'\text{-}AMP^{2-}$: $R_{2'} = -PO_3^{2-}$, $R_{3'} = R_{5'} = -H$

$3'\text{-}AMP^{2-}$: $R_{3'} = -PO_3^{2-}$, $R_{2'} = R_{5'} = -H$

$5'\text{-}AMP^{2-}$: $R_{5'} = -PO_3^{2-}$, $R_{2'} = R_{3'} = -H$

Figure 3.
Chemical structures of the adenosine monophosphates (AMPs) in their dominating *anti* conformation (Tribolet and Sigel, 1987; Martin and Mariam, 1979).

Table 2

Extent of intramolecular macrochelate formation according to equilibrium 5 for various M(AMP) complex systems expressed as the percentage of $M(AMP)_{cl}$ for aqueous solutions at 25°C and I = 0.1 M ($NaNO_3$)[a]

M^{2+}	5'-AMP^{2-}			% $M(3'\text{-}AMP)_{cl}$	% $M(2'\text{-}AMP)_{cl}$
	log $\Delta_{M(5'\text{-}AMP)}$[b]	K_I[c]	% $M(5'\text{-}AMP)_{cl}$		
Mg^{2+}	0.03 ± 0.04	0.07 ± 0.10	0 (7±9/≤15)	0	0
Ca^{2+}	0.00 ± 0.05	0.00 ± 0.12	0 (≤11)	0	0
Sr^{2+}	0.00 ± 0.05	0.00 ± 0.12	0 (≤11)	0	0
Ba^{2+}	0.01 ± 0.05	0.02 ± 0.13	0 (≤13)	0	0
Mn^{2+}	0.07 ± 0.07[d]	0.17 ± 0.19[d]	15 ± 14[d]	0	11 ± 15(<24)
Co^{2+}	0.29 ± 0.07	0.95 ± 0.33	49 ± 9	0	11 ± 16(<24)
Ni^{2+}	0.54 ± 0.06	2.47 ± 0.50	71 ± 4	13 ± 13(<25)	13 ± 13(<25)
Cu^{2+}	0.27 ± 0.08	0.86 ± 0.35	46 ± 10	17 ± 16(<30)	45 ± 10
Zn^{2+}	0.26 ± 0.11	0.82 ± 0.45	45 ± 13	0	15 ± 17(<29)
Cd^{2+}	0.24 ± 0.06	0.74 ± 0.25	43 ± 8	0	13 ± 13(<25)

[a] Columns 2-4 provide additional information on the 5'-AMP^{2-} systems; these data are abstracted from Table VI of Sigel et al., 1988. Columns 5 and 6 are taken from Table 4 of Massoud and Sigel (1989). All error limits correspond to *three times* the standard errors (3σ).

[b] See equation 4. The errors given here and in the other columns to the right were calculated according to the error propagation after Gauss by beginning with the errors obtained for log $K^{M}_{M(AMP)}$ and log $K^{M}_{M(AMP)op}$ (see e.g. Sigel et al., 1988).

[c] See equation 6. From the values for K_I follows: % $M(AMP)_{cl} = 100K_I/(1+K_I)$.

[d] This result is most probably significant; with 2σ as error limits the data are log $\Delta_{Mn(5'\text{-}AMP)}$ = 0.07 ± 0.04, K_I = 0.17 ± 0.11, and % $Mn(5'\text{-}AMP)_{cl}$ = 15 ± 8 (see also Sigel et al., 1988).

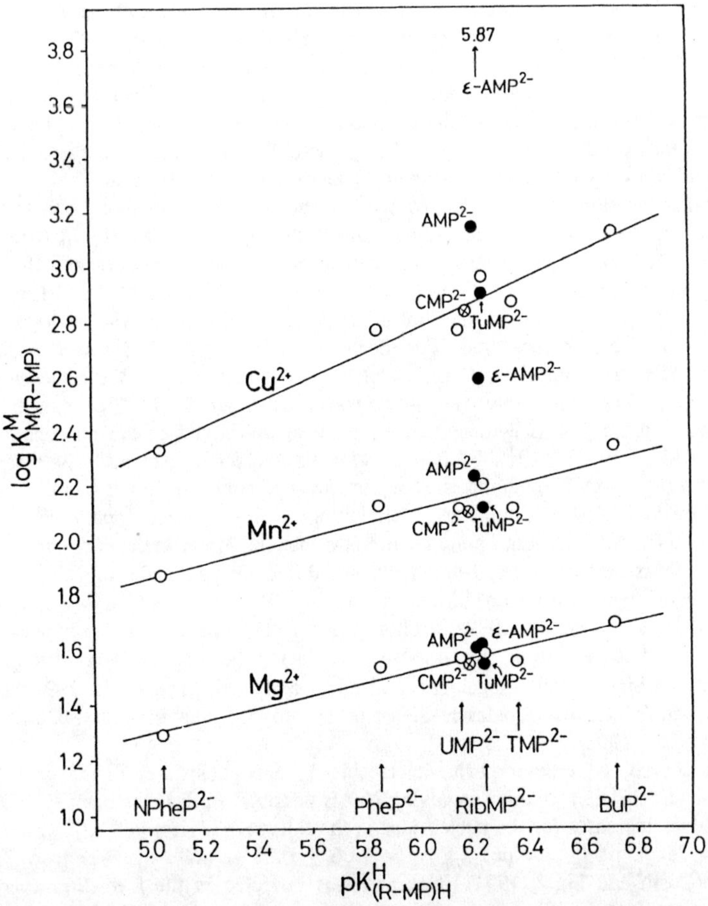

Figure 4.
Relationship between log $K^{M}_{M(R-MP)}$ and $pK^{H}_{(R-MP)H}$ for the 1:1 complexes of Mg^{2+}, Mn^{2+}, and Cu^{2+} with some simple phosphate monoester ligands (R-MP^{2-}): 4-nitrophenyl phosphate (NPheP^{2-}), phenyl phosphate (PheP^{2-}), uridine 5'-monophosphate (UMP^{2-}), *D*-ribose 5'-monophosphate (RibMP^{2-}), thymidine 5'-monophosphate (TMP^{2-}) and *n*-butyl phosphate (BuP^{2-}) (from left to right) (O). The least-squares lines are drawn through the corresponding data sets; these and the equations for the base lines are listed in Massoud and Sigel, 1988a. The points due to the complexes formed with TuMP^{2-}, AMP^{2-} (= 5'-AMP^{2-}), and ε-AMP^{2-} (●; the data are from Sigel et al., 1988) as well as with 5'-CMP^{2-} (⊗; for these data see Massoud and Sigel, 1988a) are inserted for comparison; the stability constants for the M(TuMP) complexes are plotted versus the microconstant $pk^{TuMP}_{TuMP.H}$ = 6.24 (see Figure 2 of Sigel et al., 1988). All plotted equilibrium constant values refer to aqueous solutions at 25°C and I = 0.1 M ($NaNO_3$). This figure is an altered version of Figure 3 from Sigel et al., 1988; it is redrawn by permission of the American Chemical Society.

formation, i.e. base-backbinding of a phosphate-coordinated metal ion, does indeed occur to N-7 could be proved with tubercidin 5'-monophosphate ($TuMP^{2-}$). This nucleotide is identical with 5'-AMP^{2-} except that N-7 is replaced by a CH unit; consequently, M(TuMP) complexes show no increased stability (Sigel et al., 1988) as is also demonstrated by the few examples shown in Figure 4: the data points due to the M(TuMP) complexes fit within the error limits on the reference lines, thus proving that the stability of these complexes is solely determined by the basicity of the phosphate group of $TuMP^{2-}$.

Transformation of 5'-AMP^{2-} into 1,N^6-ethenoadenosine 5'-monophosphate (ε-AMP^{2-}) leads to a phenanthroline-like binding site (Sigel, 1987a). Consequently, the extent of base-backbinding increases quite dramatically for metal ions with a significant affinity toward nitrogen donors (Sigel et al., 1988). This may be seen for Mn(ε-AMP) and Cu(ε-AMP) in Figure 4; Mg^{2+}, with a low affinity toward N donors, forms Mg(ε-AMP), the data pairs of which fit on the reference line. For further details regarding the metal ion binding properties of ε-adenosine derivatives, which are often employed as fluorescence probes in biochemical studies, the literature should be consulted (Sigel, 1987a; Sigel et al., 1988).

From Figure 3 it is evident that for a metal ion bound to the phosphate group of either 2'-AMP^{2-} or 3'-AMP^{2-} N-7 is not accessible. Only N-3 may easily be reached in the case of the M(2'-AMP) complexes as again the *anti* conformation is dominating for 2'-AMP^{2-}. In M(3'-AMP) complexes backbinding to N-3 becomes possible only if the nucleotide turns into the less favored *syn* conformation. These indications are reflected by the results (Massoud and Sigel, 1989) summarized in columns 5 and 6 of Table 2, which are also paralleled by studies with GMPs: % $M(5'\text{-GMP})_{cl}$ > % $M(2'\text{-GMP})_{cl}$ > % $M(3'\text{-GMP})_{cl}$ (Magalhães et al., 1989a; 1989b). That metal ions may interact with N-3 of a purine moiety has become more and more clear during the past few years through solution (Massoud and Sigel, 1989; Magalhães et al., 1989a; 1989b; Sigel et al., 1992b) as well as X-ray structure studies (Raudaschl-Sieber et al., 1985; Sheldrick and Günther, 1988; Marzotto et al., 1992).

Of course, with other purine-nucleoside 5'-monophosphates like 5'-IMP^{2-} and 5'-GMP^{2-}, macrochelate formation involving N-7 is possible. In fact, the basicity of N-7 in the base moieties increases for the nucleosides in the following series: $pk^H_{(N-7/Ado)H} \sim -0.2$ (Sigel et al., 1992a) < $pK^H_{(Ino)H} = pk^H_{(N-7/Ino)H} = 1.06 \pm 0.04$ < $pK^H_{(Guo)H} = pk^H_{(N-7/Guo)H} = 2.11 \pm 0.04$ (Corfù and Sigel, 1991). This series is reflected in the formation degree of the macrochelates (see equilibrium 5) which follow the order: % $M(5'\text{-AMP})_{cl}$ < % $M(5'\text{-IMP})_{cl}$ < % $M(5'\text{-GMP})_{cl}$ (see columns 2 through 4 of Table 3 which is given below in Section V.C).

In this connection one has also to mention a recent IUPAC publication entitled "Stability Constants for Nucleotide Complexes with Protons and Metal Ions" (Smith et al., 1991), which is very helpful to find access to the literature. However, great care should be exercised with regard to the equilibrium constants *recommended* and *not recommended* in this publication; to give a single example: ... "the values of ... (references) ... are tentatively recommended for Cd(5'-CMP), for Cd(5'-UMP), and for Cd(5'-dTMP). The value of ... (reference) ... for Cd(5'-GMP) *is much larger than the above values* and is not recommended". - It is evident that the apparent discrepancy arises because the intramolecular equilibrium 5 is very important for Cd(5'-GMP) while it does not play a role for the Cd^{2+} complexes with the pyrimidine-nucleoside 5'-monophosphates; the macrochelate $Cd(5'\text{-GMP})_{cl}$ occurs with a formation degree of 70 ± 4% (Table 3, column 2) which corresponds to a stability increase (Massoud and Sigel, 1988c) of log $\Delta_{Cd(5'\text{-GMP})} = 0.53 \pm 0.06$!! - This example clearly demonstrates the most unfortunate fact that any user of the mentioned compilation (Smith et al., 1991) has to make her/his own judgement in selecting stability constants to prevent being misguided.

C. Isomeric equilibria in complexes of adenosine 5'-triphosphate and comparison with complexes of some related purine-nucleotides

The equality of the stability constants measured for the complexes of a given metal ion with various pyrimidine-nucleoside 5'-triphosphates and the fact that here no base-backbinding occurs (see Section V.A) allowed to evaluate the increased stability of certain $M(5'\text{-ATP})^{2-}$

complexes; i.e. to define log $\Delta_{M(ATP)}$ (analogous to equation 4) from the stability constants measured via potentiometric pH titrations (Sigel et al., 1987). With these data the percentages of the 'closed' form, $M(5'\text{-}ATP)_{cl}^{2-}$, according to equilibrium 5 could be calculated; the corresponding results are listed in column 6 of Table 3.

It is important to note that these percentages for $M(5'\text{-}ATP)_{cl}^{2-}$ (Table 3, column 6) encompass inner-sphere and outer-sphere coordination to N-7. For example, about 45% of $Ni(5'\text{-}ATP)^{2-}$ exist in the open form, $Ni(5'\text{-}ATP)_{op}^{2-}$, about 30% in a macrochelated form with the metal ion coordinated to the phosphate chain and also inner-sphere to N-7 of the adenine residue, as well as about 25% in another macrochelated form but now with a water molecule between the metal ion and N-7. The quantification of the extent of the inner-sphere coordination is based on UV-difference spectrophotometry and ^{1}H-NMR shift experiments, as these techniques, which detect perturbations on the adenine ring, are sensitive mainly to inner-sphere coordination of the ring by a metal ion (for details see Sigel et al., 1987; Sigel 1987b).

For the M(5'-AMP) species it appears that macrochelate formation to N-7 mainly occurs in an inner-sphere fashion (Sigel et al., 1988); for $M(5'\text{-}ATP)^{2-}$ species both inner-sphere and outer-sphere backbindings to N-7 occur as indicated above, but for nearly all other nucleotide complexes such information is missing. The evaluation of stability data from potentiometric pH titrations furnishes of course only overall stability increases (equation 4) and consequently percentages calculated for macrochelated species according to equilibrium 5 encompass all sorts of backbinding be it inner-sphere or outer-sphere. Hence, all the results listed in Table 3 refer to the total formation degree of macrochelates.

The contents of columns 2 and 3 of Table 3 have already been discussed in Section V.B. Comparison of the data in the other three columns suggests that the total extent of macrochelate formation for the metal ions studied depends on the number of phosphate groups and varies in the series: % $M(5'\text{-}AMP)_{cl}$ < % $M(5'\text{-}ADP)_{cl}^{-}$ > % $M(5'\text{-}ATP)_{cl}^{2-}$. Despite the incompleteness of the data for $5'\text{-}ADP^{-}$ and the fact that the listed values represent rather lower limits, the given order seems to indicate that the macrochelates of $M(5'\text{-}ADP)^{-}$ are less strained than those of M(5'-AMP) and that they also form more easily than those of $M(5'\text{-}ATP)^{2-}$ (Sigel, 1987b).

Table 3

Comparison of the extent of macrochelate formation according to equilibrium 5 for various metal ion complexes of purine 5'-nucleotides (PN) expressed as the percentage of $M(PN)_{cl}$ for aqueous solutions at 25°C and I = 0.1 M ($NaNO_3$)[a]

M^{2+}	$\%M(5'\text{-}GMP)_{cl}$	$\%M(5'\text{-}IMP)_{cl}$	$\%M(5'\text{-}AMP)_{cl}$	$\%M(5'\text{-}ADP)_{cl}^{-}$	$\%M(5\text{-}ATP)_{cl}^{2-}$
Mg^{2+}	26 ± 8	21 ± 8	7 ± 9	0	11 ± 6
Ca^{2+}	15 ± 10	9 ± 11	0		2 ± 6
Mn^{2+}	40 ± 10	29 ± 12	15 ± 14[b]	55	17 ± 10
Co^{2+}	83 ± 3	78 ± 4	49 ± 9	60	38 ± 9
Ni^{2+}	93 ± 1	89 ± 2	71 ± 4	80	56 ± 4
Cu^{2+}	81 ± 4	69 ± 6	46 ± 10	94	67 ± 2
Zn^{2+}	72 ± 5	62 ± 7	45 ± 13	67	28 ± 7
Cd^{2+}	70 ± 4	64 ± 5	43 ± 8		46 ± 4

[a] All error limits refer to *three times* the standard errors (3σ). The values are taken from the following references: column 2 (Massoud and Sigel, 1988b; Corfù et al., 1990), 3 (Kinjo et al., 1991), 4 (Sigel et al., 1988), 5 (Scheller and Sigel, 1983), and 6 (Sigel et al., 1987; Sigel, 1987b).

[b] See the comment in footnote 'd' of Table 2.

For $M(5'\text{-ITP})_{cl}^{2-}$ and $M(5'\text{-GTP})_{cl}^{2-}$ so far only preliminary results are available (Kinjo et al., 1991; Corfù et al., 1990). The percentages of the macrochelated form for a given metal ion are comparable in a first approximation to that determined for $M(5'\text{-ATP})_{cl}^{2-}$.

For some few metal ions, mainly Ni^{2+} and Pt^{2+}, the macrochelate formations as discussed above and in Section V.A have been confirmed for various NMPs and NTPs by specialized techniques (Taylor and Diebler, 1976; Peguy and Diebler, 1977; Thomas et al., 1980; Evans et al., 1987; Green and Miller, 1987; Kuo et al., 1987; Reily et al., 1988).

VI. CONCLUSIONS AND OUTLOOK

It has already been indicated that a decreasing solvent polarity favors phosphate-complex stability (see Section II). Though at this time only a limited amount of results is available, it is evident that the extent of base-backbinding (equilibrium 5) is affected in a more complicated way. For example, the formation degree of $Cu(5'\text{-AMP})_{cl}$ passes through a minimum with increasing concentrations of 1,4-dioxane in water: in water and in water containing 50% (v/v) dioxane, about 50% of Cu(5'-AMP) exist as a macrochelate while the formation degree of this species in water containing 30% dioxane reaches only about 10% (Liang and Sigel, 1990; cf also Chen et al., 1993). Similar results have also to be expected for other nucleotide-metal ion systems, but before generalizations are possible considerably more experimental data are needed. However, nature evidently has here a subtle tool to alter the structure of complexes simply by shifting them from a polar into a less polar region and vice versa.

From the preceding discussions follows that the stability of nucleotide-metal ion complexes is largely determined by the phosphate-metal ion interactions. This means, e.g. for purine-nucleotides, that the N-7 backbinding is the 'weak' point in macrochelate formation. Indeed, the release of N-7 in $M(5'\text{-ATP})^{2-}$ complexes upon mixed-ligand complex formation in solution has been demonstrated for ligands as different as OH^- (Sigel et al., 1984a; Tribolet et al., 1987), NH_3 (Tribolet et al., 1987), imidazole (Tribolet et al., 1987; Sigel et al., 1990), 2,2'-bipyridyl (Mitchell et al., 1979; Sigel et al., 1984b), 1,10-phenanthroline (Mitchell et al., 1979), and tryptophanate (Sigel et al., 1983). Maybe this indicates that the binary $M(5'\text{-ATP})^{2-}$ complexes may exist as closed macrochelates only if no enzyme groups coordinate directly to the metal ion. In any case, the changes in free energy ($\Delta G°$) connected with the intramolecular equilibrum 5 are often small; e.g. a formation degree of 20% for the closed isomer corresponds only to $\Delta G° = -0.6$ kJ/M (Martin and Sigel, 1988; Sigel, 1989a).

However, despite the fact that these nucleic base-metal ion interactions are weak, they are crucial for the reactivity. For example, the oxidation of nucleic base residues by H_2O_2 in the presence of Cu^{2+} depends on the coordination of these residues to the metal ion; i.e. the peroxidase-like reaction proceeds within the coordination sphere (Sigel, 1969). In this way evidence for nucleic base-backbinding in monomeric Cu^{2+} complexes of nucleoside 5'-triphosphates was provided already many years ago (Sigel, 1967). Similarly, native DNA in the presence of H_2O_2 and Cu^{2+} gives rise to a catalase-like activity as well as to the formation of ternary peroxo complexes without an evident degradation of the nucleic bases (Sigel, 1969). However, if Cu^{2+} is added to DNA and the solution is kept for 1.5 days at room temperature, Cu^{2+} penetrates into the double helix and coordinates to the base residues; addition of H_2O_2 to such a solution leads to a peroxidase-like degradation of the nucleic base moieties (Sigel et al., 1967). Native and denatured DNA can thus be distinguished (cf also Erlenmeyer et al., 1968; Sigel, 1969). Similar reactions are also possible with RNA (Sigel et al., 1967).

Another example, and certainly many more will be discovered, demonstrating the importance of metal ion-nucleic base interactions is found in the metal ion-promoted dephosphorylation (Sigel, 1990; 1992). The metal ion-(N-7) interaction is crucial for the reactivity of $5'\text{-ATP}^{4-}$ and other purine-nucleoside 5'-triphosphates even though the reaction proceeds here via a dimeric species. It is evident that a complex with certain structural features, despite a relatively low formation degree, can well be responsible for the high reactivity of a system.

ABBREVIATIONS

Note, only those abbreviations are defined below which are not designated in Figure 1 and its legend or in the legends of Figures 2 and 4.

2'-AMP^{2-}	adenosine 2'-monophosphate (cf Figure 3)
3'-AMP^{2-}	adenosine 3'-monophosphate (cf Figure 3)
$DHAP^{2-}$	dihydroxyacetone phosphate: HO-CH_2-C(O)-CH_2-O-PO_3^{2-}
DNA	deoxyribonucleic acid
2'-GMP^{2-}	guanosine 2'-monophosphate
3'-GMP^{2-}	guanosine 3'-monophosphate
$G1P^{2-}$	glycerol 1-phosphate: HO-CH_2-CH(OH)-CH_2-O-PO_3^{2-}
I	ionic strength of a solution
K_a	acidity constant (see also equation 1)
L	general ligand with undefined charge
M^{2+}	general divalent metal ion
NMR	nuclear magnetic resonance
R-MP^{2-}	phosphate monoester (R may be any organic residue, e.g. phenyl or nucleosidyl; in some instances also phosphonate derivatives, R-PO_3^{2-}, are included into this abbreviation)
RNA	ribonucleic acid
$TuMP^{2-}$	tubercidin 5'-monophosphate (= 7-deaza-AMP^{2-}; i.e. N-7 of the adenine residue is replaced by a CH unit)

ACKNOWLEDGEMENTS

The competent technical assistance of Ms. Rita Baumbusch in the preparation of the manuscript and the financial support of the research of my group on the complexes of phosphates, nucleosides, and nucleotides by the Swiss National Science Foundation is gratefully acknowledged.

REFERENCES

Bastian, M., and Sigel, H. (1990). *Inorg. Chim. Acta, 178*: 249; see p. 258.

Bastian, M., and Sigel, H. (1991). *Inorg. Chim. Acta, 187*: 227.

Boyer, P.D. (1987). *Biochemistry, 26*: 8503.

Brintzinger, H., and Hammes, G.G. (1966). *Inorg. Chem., 5*: 1286.

Burger, K., and L. Nagy (1990). *Biocoordination Chemistry* (K. Burger, ed.), Ellis Horwood, London, p. 236.

Chen, D., Gregán, F., Holý, A., and Sigel, H. (1993). *Inorg. Chem., 32*: 5377.

Coleman, J.E., and Giedroc, D.P. (1989). *Metal Ions in Biological Systems*, Vol. 25 (H. Sigel and A. Sigel, eds.), Marcel Dekker, New York, p. 171.

Corfù, N.A., and Sigel, H. (1991). *Eur. J. Biochem., 199*: 659.

Corfù, N.A., Kinjo, Y., Massoud, S.S., and Sigel, H. (1990). Abstracts of the XXVIIIth I.C.C.C., Gera, GDR, August 13-18, Vol. 2, pp. 1-90.

Cowan, J.A. (1992). *Comments Inorg. Chem., 13*: 293.

Davies, D.B., Rajani, P., and Sadikot, H. (1985). *J. Chem. Soc., Perkin Trans. II*: 279.

Dhathathreyan, K.S., and Diebler, H. (1991). *Ber. Bunsenges. Phys. Chem., 95*: 880.

Diebler, H., Secco, F., and Venturini, M. (1987). *Biophys. Chem., 26*: 193.

Erlenmeyer, H., Griesser, R., Prijs, B., and Sigel, H. (1968). *Biochim. Biophys. Acta, 157*: 637.

Evans, D.J., Green, M., and van Eldik, R. (1987). *Inorg. Chim. Acta, 128*: 27.

Fraústo da Silva, J.J.R., and Williams, R.J.P. (1991). *The Biological Chemistry of the Elements*, Clarendon Press, Oxford.

Frey, C.M., and Stuehr, J.E. (1978). *J. Amer. Chem. Soc., 100*: 134 and 139.

Green, M., and Miller, J.M. (1987). *J. Chem. Soc., Chem. Commun.*: 1864; for a correction see *ibid.*: 404 (1988).

Ji, L.-n., Corfù, N.A., and Sigel, H. (1991). *J. Chem. Soc., Dalton Trans.*: 1367.

Kinjo, Y., Corfù, N.A., Massoud, S.S., and Sigel, H. (1991). *J. Inorg. Biochem.*, *43*: 463.

Kinjo, Y., Ji, L.-n., Corfù, N.A., and Sigel, H. (1992). *Inorg. Chem.*, *31*: 5588.

Kuo, L.Y., Kanatzidis, M.G., and Marks, T.J. (1987). *J. Amer. Chem. Soc.*, *109*: 7207.

Liang, G., and Sigel, H. (1990). *Inorg. Chem.*, *29*: 3631.

Liang, G., Corfù, N.A., and Sigel, H. (1989). *Z. Naturforsch.*, *44b*: 538.

Liang, G., Chen, D., Bastian, M., and Sigel, H. (1992). *J. Amer. Chem. Soc.*, *114*: 7780.

Magalhães, M.C.F., and Sigel, H. (1992). *J. Indian Chem. Soc.*, *69*: 437.

Magalhães, M.C.F., Massoud, S.S., Corfù, N.A., and Sigel, H. (1989a). *J. Inorg. Biochem.*, *36*: 295, J045.

Magalhães, M.C.F., Massoud, S.S., Corfù, N.A., and Sigel, H. (1989b). Abstracts of the Fall-Assembly of the Swiss Chemical Society, Berne, Switzerland, October 20, 1989, p. 99.

Mariam, Y.H., and Martin, R.B. (1979). *Inorg. Chim. Acta*, *35*: 23.

Martin, R.B. (1985). *Accounts Chem. Res.*, *18*: 32.

Martin, R.B., and Mariam, Y.H. (1979). *Metal Ions in Biological Systems*, Vol. 8 (H. Sigel, ed.), Marcel Dekker, New York, p. 57.

Martin, R.B., and Sigel, H. (1988). *Comments Inorg. Chem.*, *6*: 285.

Marzotto, A., Valle, G., and Clemente, D.A. (1992). Abstracts of the XXIXth I.C.C.C., Lausanne, Switzerland, July 19-24, p. 72 (No. P182).

Massoud, S.S., and Sigel, H. (1988a). *Inorg. Chem.*, *27*: 1447.

Massoud, S.S., and Sigel, H. (1988b). Abstracts of the XXVIth I.C.C.C., Porto, Portugal, Aug. 28 - Sept. 2, C49.

Massoud, S.S., and Sigel, H. (1988c). *Bull. Chem. Soc. Ethiop.*, *2*: 9.

Massoud, S.S., and Sigel, H. (1989). *Eur. J. Biochem.*, *179*: 451.

Mildvan, A.S. (1987). *Magnesium*, *6*: *28*.

Mitchell, P.R., Prijs, B., and Sigel, H. (1979). *Helv. Chim. Acta*, *62*: 1723.

Peguy, A., and Diebler, H. (1977). J. Phys. Chem., 81: 1355.

Pörschke, D. (1976). *Biophys. Chem.*, *4*: 383.

Raudaschl-Sieber, G., Schöllhorn, H., Thewalt, U., and Lippert, B. (1985). *J. Amer. Chem. Soc.*, *107*: 3591.

Reedijk, J. (1992). *Inorg. Chim. Acta*, *198-200*: 873.

Rees, D.C. (1980). *J. Mol. Biol.*, *141*: 323.

Reily, M.D., Hambley, T.W., and Marzilli, L.G. (1988). *J. Amer. Chem. Soc.*, *110*: 2999.

Rogers, N.K., Moore, G.R., and Sternberg, M.J.E. (1985). *J. Mol. Biol.*, *182*: 613.

Scheller, K.H., and Sigel, H. (1983). *J. Amer. Chem. Soc.*, *105*: 5891.

Scheller, K.H., Hofstetter, F., Mitchell, P.R., Prijs, B., and Sigel, H. (1981). *J. Amer. Chem. Soc.*, *103*: 247.

Schneider, P.W., Brintzinger, H., and Erlenmeyer, H. (1964). *Helv. Chim. Acta*, *47*: 992.

Sheldrick, W.S., and Günther, B. (1988). *Inorg. Chim. Acta*, *152*: 223.

Sigel, H. (1967). *Helv. Chim. Acta*, *50*: 582.

Sigel, H. (1969). *Angew. Chem.*, *81*: 161; *Angew. Chem., Int. Ed. Engl.*, *8*: 167.

Sigel, H. (1975). *J. Amer. Chem. Soc.*, *97*: 3209.

Sigel, H. (1977). *J. Inorg. Nucl. Chem.*, *39*: 1903.

Sigel, H., ed. (1979). *Metal Ions in Biological Systems,* Vol. 8, Marcel Dekker, New York.

Sigel, H. (1987a). *Chimia*, *41*: 11.

Sigel, H. (1987b). *Eur. J. Biochem.*, *165*: 65.

Sigel, H. (1989a). *Metal-DNA Chemistry* (T.D. Tullius, ed.), ACS Symp. Series, 402, Amer. Chem. Soc., Washington, D.C., p. 159.

Sigel, H. (1989b). *Biol. Tr. Elem. Res.*, *21*: 49.

Sigel, H. (1990). *Coord. Chem. Rev.*, *100*: 453.

Sigel, H. (1992). *Inorg. Chim. Acta*, *198-200*: 1.

Sigel, H. (1993). *Chem. Soc. Rev.*, *22*: 255.

Sigel, H., and McCormick, D.B. (1970). *Accounts Chem. Res.*, *3*: 201.

Sigel, H., and Sigel, A., eds. (1989). *Metal Ions in Biological Systems*, Vol. 25, Marcel Dekker, New York and Basel.

Sigel, H., Prijs, B., and Erlenmeyer, H. (1967). *Experientia*, *23*: 170.

Sigel, H., Fischer, B.E., and Farkas, E. (1983). *Inorg. Chem.*, *22*: 925.

Sigel, H., Scheller, K.H., and Milburn, R.M. (1984a). *Inorg. Chem.*, *23*: 1933.

Sigel, H., Hofstetter, F., Martin, R.B., Milburn, R.M., Scheller-Krattiger, V., and Scheller, K.H. (1984b). *J. Amer. Chem. Soc.*, *106*: 7935.

Sigel, H., Martin, R.B., Tribolet, R., Häring, U.K., and Malini-Balakrishnan, R. (1985). *Eur. J. Biochem.*, *152*: 187.

Sigel, H., Tribolet, R., Malini-Balakrishnan, R., and Martin, R.B. (1987). *Inorg. Chem.*, *26*: 2149.

Sigel, H., Massoud, S.S., and Tribolet, R. (1988). *J. Amer. Chem. Soc.*, *110*: 6857.

Sigel, H., Tribolet, R., and Yamauchi, O. (1990). *Comments Inorg. Chem.*, *9*: 305.

Sigel, H., Corfù, N.A., Ji, L.-n., and Martin, R.B. (1992a). *Comments Inorg. Chem.*, *13*: 35.

Sigel, H., Chen, D., Corfù, N.A., Gregán, F., Holý , A., Strašák, M. (1992b). *Helv. Chim. Acta*, *75*: 2634.

Smith, R.M., Martell, A.E., and Chen, Y. (1991). *Pure Appl. Chem.*, *63*: 1015.

Stryer, L. (1988). *Biochemistry*, W.H. Freeman & Co., New York.

Tallineau, C., Barrière, M., Boulard, M., Boulard-Heitzmann, P., Pontcharraud, R., Reiss, D., and Guillard, O. (1984). *Biochim. Biophys. Acta*, *775*: 51.

Taylor, R.S., and Diebler, H. (1976). *Bioinorg. Chem.*, *6*: 247.

Thomas, J.C., Frey, C.M., and Stuehr, J.E. (1980). *Inorg. Chem.*, *19*: 505.

Tribolet, R., and Sigel, H. (1987). *Eur. J. Biochem.*, *163*: 353.

Tribolet, R., and Sigel, H. (1988). *Eur. J. Biochem.*, *170*: 617.

Tribolet, R., Malini-Balakrishnan, R., and Sigel, H. (1985). *J. Chem. Soc., Dalton Trans.*, 2291.

Tribolet, R., Martin, R.B., and Sigel, H. (1987). *Inorg. Chem.*, *26*: 638.

RECOGNITION OF NUCLEIC ACIDS BY TRANSITION METAL COMPLEXES

by

Ayesha Sitlani and Jacqueline K. Barton*

Division of Chemistry and Chemical Engineering,
California Institute of Technology,
Pasadena, CA 91125
U.S.A

I. INTRODUCTION

The recognition of a nucleic acid site through a myriad of specific non-covalent interactions is a process that is central to a living cell. Proteins recognize DNA and RNA in a sequence-specific fashion to regulate selectively the expression of genetic material. During the last decade several structural studies have been directed towards understanding protein-nucleic acid recognition. X-ray and NMR structural studies have shown us that the conformations of nucleic acids are varied and complex, and that a simple code comprising specific interactions between individual amino acid residues and nucleic acid bases cannot generally account for the sequence specificity observed. We have also learned that nature has evolved a range of structural motifs for nucleic-acid binding proteins (reviewed by Pabo, 1992 and Steitz, 1990). These include helix turn helix proteins, homeodomains, zinc finger proteins, steroid receptors, leucine-zipper proteins, helix loop helix proteins, and β-sheet proteins. In comparison, less rapid progress has been made on determining the structures of protein complexes with folded RNA molecules and on the significance of RNA-binding motifs (Steitz, 1990).

In addition to providing a distinct DNA-binding motif, zinc fingers have recently also been implicated in the recognition of RNA by retroviral proteins and synthetases (Miller et al., 1991; Fitzgerald and Coleman, 1991; Delahunty et al., 1992). However, unlike other nucleic-acid binding motifs, zinc fingers are unique in that they contain a central metal ion which, due to predisposed coordination geometry requirements, serves to organize structurally the folding of protein around it. Several types of zinc-containing motifs have been discovered which have different shapes depending on the length of folded peptide around the metal ion, the stoichiometry of zinc ions, and the nature of the amino acids involved in ligation to the zinc ion. Zinc finger proteins therefore exemplify how nature has very judiciously used the properties of metal ion-ligand interactions to design structures which recognize specific nucleic acid sites. Perhaps analogously, transition metal complexes, which like zinc fingers have well-defined shapes owing to their distinct and rigid coordination geometries, have been designed in our laboratory to probe the basic principles of nucleic acid recognition.

* to whom correspondence should be addressed

Unlike most small molecules and natural products which bind nucleic acids, the rigid coordination geometries of transition metal complexes provide us with a well-defined scaffold on which to append moieties that may contribute to site-selective recognition. In this way, we may build complexes to match the shape and symmetry of specific nucleic acid sites. This strategy lends itself potentially to the general design of transition metal complexes as site-specific chemotherapeutic agents that would discriminate among specific nucleic acid sites in a predictable manner. The design of metal complexes with shapes that match the local structure of nucleic acid sites has taught us some of the features which are important to nucleic-acid recognition.

In the first section of this article, we illustrate how various characteristic features of transition metal complexes may be exploited to probe nucleic acids. For example, the inherent reactivities and electronic properties of transition metal complexes provide sensitive spectroscopic assays to detect their binding to nucleic acids. Importantly, the well-defined shapes of transition metal complexes, as determined by their distinct and rigid coordination geometries, make them sensitive probes of nucleic acid conformation. These features have allowed us to study the nucleic-acid binding and recognition properties of a wide range of transition metal complexes, as described in the second section of this article. Interestingly, a correlation between site-specific discrimination by metal complexes and by proteins has been observed. This correlation supports the notion that transition metal-nucleic acid interactions may be viewed as a model to delineate some basic principles of protein-nucleic acid recognition. We have learned that the overall shape-complementarity between metal complex and its cognate nucleic acid site is a crucial element of site-specificity, and it is likely that similar steric factors are significant in protein-nucleic acid interactions as well. Recognition of nucleic acids, however, is far from being completely understood, and important issues involving the predictability of nucleic acid recognition and the rational design of drugs still remain. We are currently addressing these issues by exploring the use of transition metal complexes as recognition building blocks in the development of site-specific chemotherapeutic agents.

II. METHODS TO PROBE NUCLEIC ACID-METAL COMPLEX INTERACTIONS

Here we describe the methods used to study nucleic-acid binding and recognition by transition metal complexes. The rich spectroscopic properties and chemical reactivities of transition metals offer several methods to probe metal complex-nucleic acid interactions. Spectroscopic perturbations in the metal complex on binding to its target provide sensitive assays to study nucleic acid binding. For example, simple absorption and luminescence measurements have been useful in elucidating the overall nucleic-acid binding modes of metal complexes, while NMR spectroscopy has provided more detailed structural information on site-specific binding characteristics. Furthermore, the redox and photochemical reactivities of transition metal complexes have also been exploited to examine nucleic-acid binding interactions. Chemical reactions of the metal complexes bound to nucleic acids have provided sensitive chemical reporters that mark the specific nucleic-acid binding site.

A. Spectroscopic assays used to measure binding

As illustrated by Pyle et al. (1989a), UV-visible and luminescence measurements on a family of Ru(II) and Rh(III) mixed-ligand complexes (Fig. 1) provide a description of the general characteristics of how these metal complexes interact with double helical DNA. Except for the complex $[Ru(bpy)_3]^{2+}$, binding of simple polypyridyl complexes of Ru(II) complexes to DNA is associated with hypochromism and red shifting of the metal-to-ligand charge transfer bands in the visible absorption spectra of the bound forms. In the case of Ru(II) and Rh(III) complexes containing the 9,10-phenanthrenequinone diimine (phi) ligand, substantial hypochromism and shifts to longer wavelengths of the phi centered π-π^* transition are observed upon binding to DNA. The magnitude of red shifts is greatest for phi complexes (> 16 nm), followed by DIP complexes (> 6 nm), phen complexes (> 2 nm), and bpy complexes (no red shift). If red shifts upon binding are taken as a measure of the stacking interaction, a trend is observed in which the optimal shape for metal complex intercalation into DNA is phi > DIP > phen > bpy.

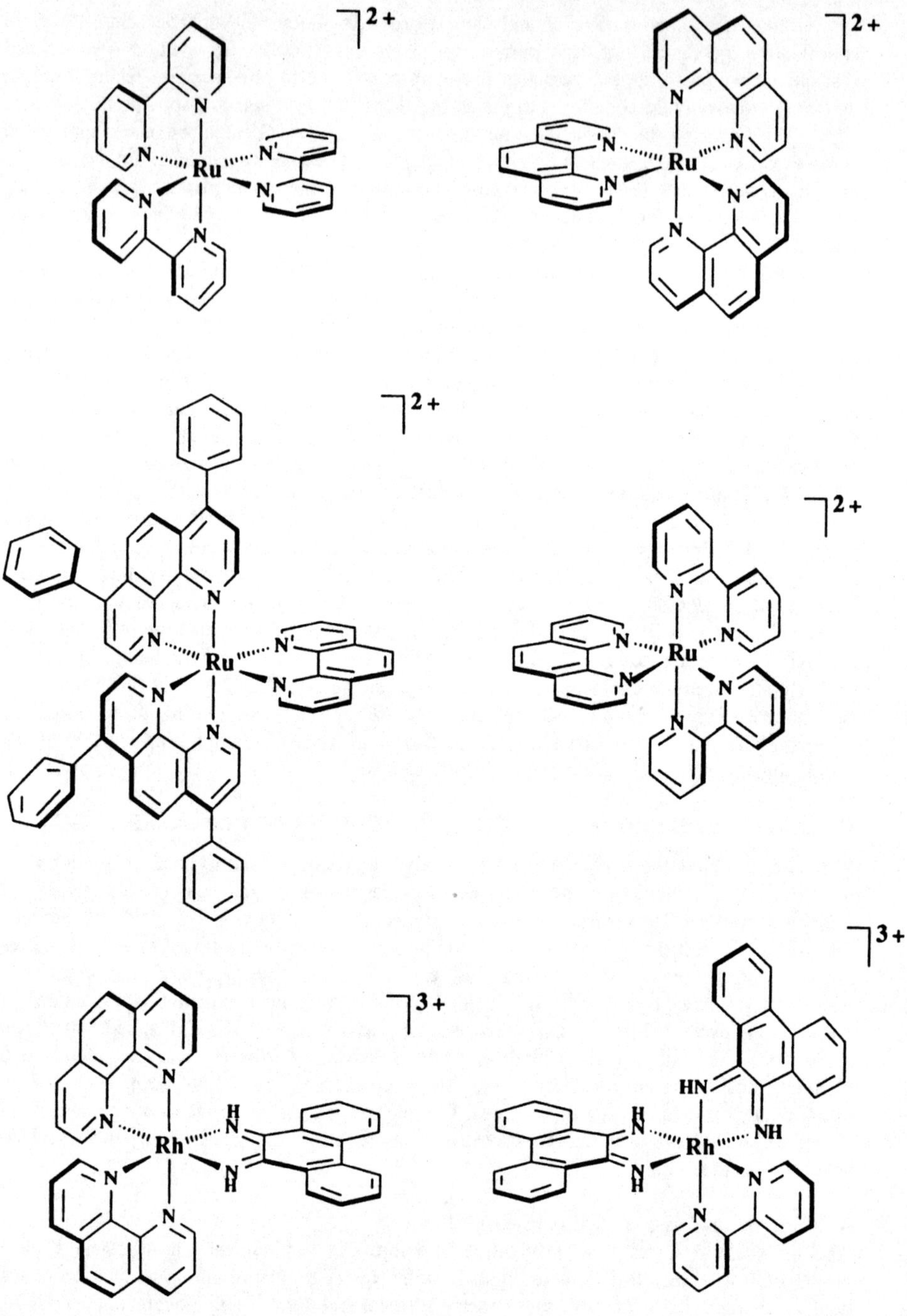

Figure 1.
Illustration of several mixed-ligand Ru(II) and Rh(III) complexes: $[Ru(bpy)_3]^{2+}$ (top left); $[Ru(phen)_3]^{2+}$ (top right); $[Ru(DIP)_2(phen)]^{2+}$ (middle left); $[Ru(bpy)_2(phen)]^{2+}$ (middle right); $[Rh(phen)_2(phi)]^{3+}$ (bottom left); $[Rh(phi)_2(bpy)]^{3+}$ (bottom right). Bpy = 2,2'-bipyridine, phen = 1,10-phenanthroline, DIP = 4,7-diphenylphenanthroline, phi = 9,10-phenanthrenequinone diimine.

A

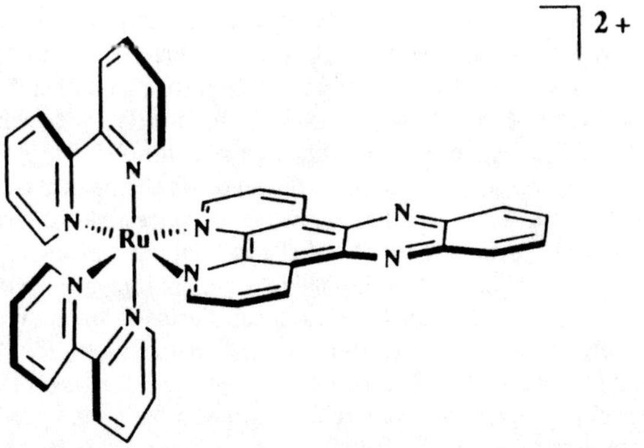

bis(bipyridyl)(dipyrido[3,2-a:2',3':c]phenazine)ruthenium(II)

B

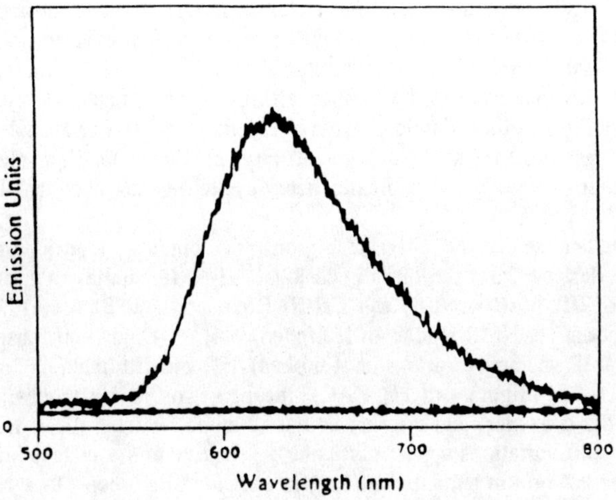

Figure 2.
Luminescence enhancement of $[Ru(bpy)_2(dppz)]^{2+}$ upon binding to DNA.

A. Structure of $[Ru(bpy)_2(dppz)]^{2+}$, (bpy = 2,2'-bipyridine, dppz = dipyrido [3,2-a: 2',3'-c] phenazine).

B. Steady-state emission spectrum of $[Ru(bpy)_2(dppz)]^{2+}$ in the presence and absence of B-form poly d(GC).d(GC). In the absence of polynucleotide, there is no detectable luminescence; however, intense emission centered at 628 nm is observed in the presence of polynucleotide.

(From Friedman et al., 1990).

Paralleling the measurements of absorbance hypochromism, for the Ru(II) complexes that luminesce, emission enhancements are apparent with DNA binding (Pyle et al., 1989a). Depending on the mixed-ligand complex examined (Fig. 1), red shifts or blue shifts in the emission spectra are observed. Complexes such as $[Ru(bpy)_3]^{2+}$ that bind through weak electrostatic modes do not exhibit appreciable changes in luminescence or absorption intensity. The most dramatic change in luminescence upon binding to nucleic acids, as reported by Friedman et al. (1990), is observed for Ru(II) polypyridyl complexes containing a dipyrido[3,2-a:2',3'-c]phenazine (dppz) ligand (Fig. 2). The dppz complexes show no photoluminescence in aqueous solutions; however, in the presence of double-stranded DNA they luminesce intensely. The luminescence enhancement observed upon nucleic acid binding is attributed to the sensitivity of the excited state of dppz complexes to quenching by water. Upon intercalation into the helix, the dppz ligand is protected from the aqueous solvent, thereby preserving the luminescence. Furthermore, the work of Jenkins et al. (1992) has illustrated that the luminescent parameters of $[Ru(phen)_2dppz]^{2+}$ and $[Ru(bpy)_2dppz]^{2+}$ are remarkably sensitive to the DNA structure at the intercalation site and, more specifically, to how the structure protects the dppz ligand from interactions with the solvent. In particular, the highest level of emission is observed upon binding to triple-helical DNA, where substantial overlap and stacking between the dppz ligand and base triple may occur.

The extent of luminescence enhancement, absorption hypochromism, and red shifts associated with spectra of the bound forms all seem to reflect the extent of overlap of the ligand with the base pairs of the helix. The phi ligand is particularly well-suited for intercalation, likely because the ligand most closely mimics the spacial arrangement of the aromatic regions of a base pair, thereby allowing greater helical overlap; in comparison, the phenanthroline ligand permits a substantially smaller surface area for intercalative stacking. Consistent with this observation, the binding affinity of phi complexes are three orders of magnitude greater than other simple polypyridyl complexes of Ru(II) and Rh(III) (Pyle et al., 1990b). Therefore, to achieve high affinity in DNA binding and to favor the intercalative mode, large planar aromatic heterocyclic ligands such as dppz or phi are preferred.

More detailed structural characterizations of binding modes and site-selective recognition are derived from the one dimensional ^{1}H-NMR studies of tris(phenanthroline) complexes of Rh(III), Ru(II), Ni(II) and Cr(III) (Rehmann and Barton, 1990a; 1990b) and the two-dimensional ^{1}H-NMR studies of $[Rh(phen)_2phi]^{3+}$ (David and Barton, 1993) bound to DNA. The NMR spectral titrations of $[Ru(phen)_3]^{2+}$ and $[Rh(phen)_3]^{3+}$ indicate that the complexes bind to the oligomer $d(GTGCAC)_2$ through two binding modes: the intercalative mode favoring the Δ isomer, and the surface bound mode favoring the Λ isomer. However, NMR chemical shift variations appear particularly sensitive to the surface-bound interaction, with the adenine H2 proton showing the largest upfield shifts upon DNA-binding by the Λ isomers of the metal complexes. The major groove thymine methyl protons (TMe) shift downfield to a lesser extent, and more so for the Δ isomers. Selective paramagnetic relaxation of oligonucleotide protons of $d(GTGCAC)_2$ by Δ- and Λ-$[Ni(phen)_3]^{3+}$ and by Δ- and Λ-$[Cr(phen)_3]^{3+}$ support these observations. Two binding modes are necessary to account for the binding of each isomer to the oligonucleotide. Direct evidence for major groove intercalation by metal complexes is apparent in the two-dimensional NOESY experiments performed on $[Rh(phen)_2phi]^{3+}$-$[d(GTCGAC)_2]$. Dramatic upfield shifts are observed only for the phi ligand, while the ancillary phenanthroline protons are now only slightly shifted, which points to preferential intercalation of the phi ligand. Furthermore, evidence for specific intercalation comes from an observed disruption in the sequential NOE connectivities at the 5'-CG-3' step (Fig. 3a). Parallel to this observation, site-specific cleavage at the 5'-CG-3' step is observed upon irradiation of the oligonucleotide substrate in the presence of Δ-$[Rh(phen)_2phi]^{3+}$ (Fig. 3b). Clearly, these studies indicate that replacing a phenanthroline with a phi ligand not only favors the intercalative mode of binding but also enhances the overall site-specific recognition of the metal complexes.

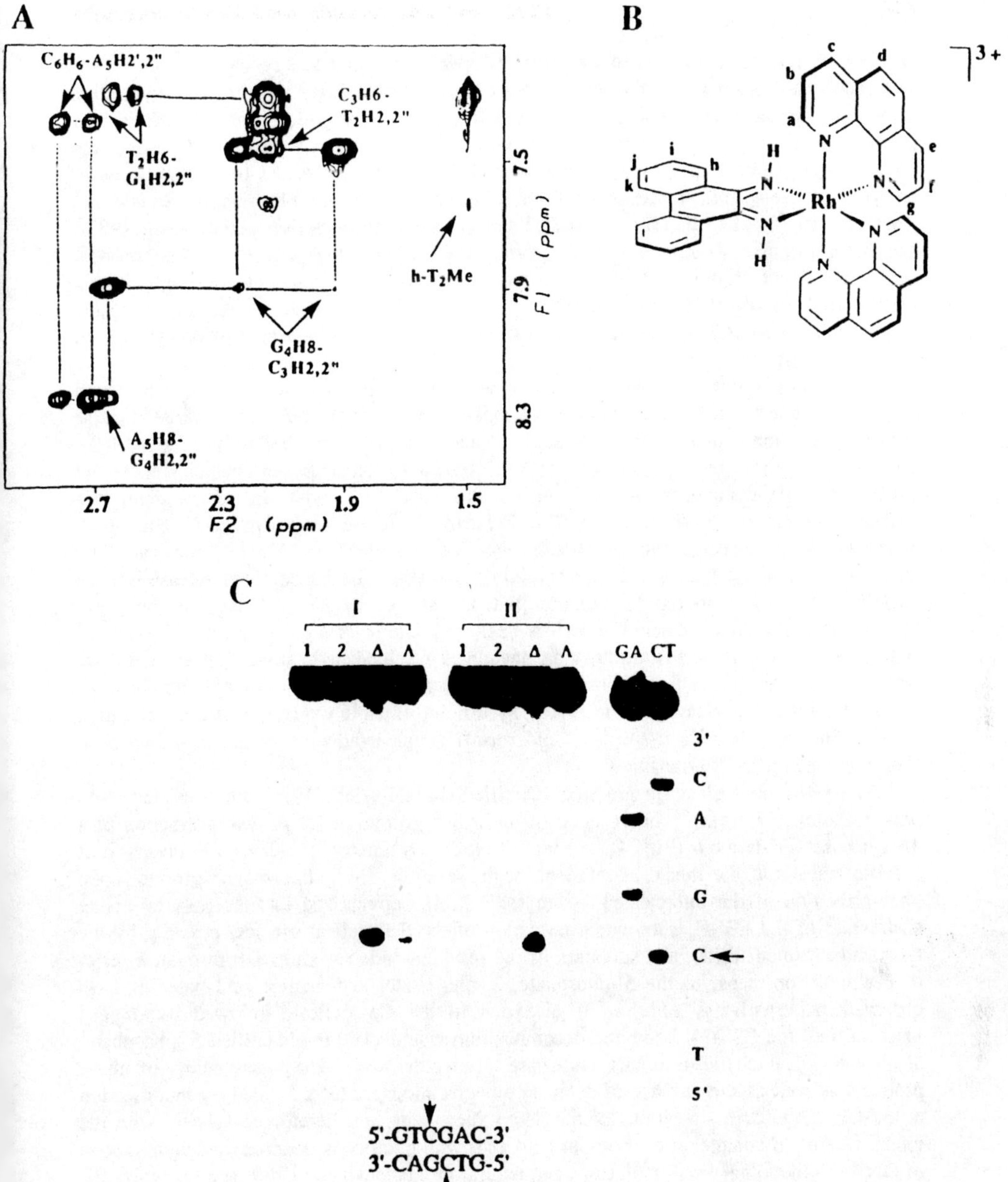

Figure 3.
Site-specific intercalation of Δ $[Rh(phen)_2phi]^{3+}$ into $d(GTGCAC)_2$.

A. Expanded NOESY contour plot of Δ $[Rh(phen)_2phi]^{3+}$-$[d(GTCGAC)_2]$. The labeled cross peaks correspond to the internucleotide NOEs between the base protons and the H2'2" protons of the sugar residue of the 5' nucleotide. The numbering scheme for the bases is given from the 5' end of the nucleotide. A loss in intensity at the 5'-CG-3' step indicates site-specific binding of the complex.

B. Structure of Δ-$[Rh(phen)_2phi]^{3+}$, (phen = 1,10-phenanthroline, phi = 9,10-phenanthrenequinone diimine), with each unique proton labeled.

C. Site-specific photocleavage by Δ-$[Rh(phen)_2phi]^{3+}$ on a 5'-^{32}P-end labeled hexamer, 5'-GTCGAC-3'. Cleavage is seen at the 5'-CG-3' step, which is consistent with the site of intercalation shown by the NOESY plot.

(From David and Barton, 1993).

B. DNA cleavage assays used to measure site-specific binding and recognition

The redox and chemical reactivities of transition metal complexes also provide a sensitive means to examine nucleic-acid binding interactions. Reactions which promote DNA strand scission are powerful in marking sites of binding along the DNA polymer. Coupling the variety of reactivities of transition metal complexes to differing modes of binding leads to an array of site-specific reactions on DNA. Agents such as Fe(II)-bleomycin (Stubbe and Kozarich, 1987), MPE-Fe(II) and $Fe(EDTA)^{2-}$ (Dervan, 1986; Moser and Dervan, 1987), $Cu(phen)_2$ (Sigman, 1986), and metalloporphyrins (Pasternack et al., 1983) bind to most sites in the minor groove of DNA and react primarily with the DNA sugars through either a metal-bound or diffusing radical. Our laboratory has taken advantage of inorganic photochemistry to couple reactions which promote DNA strand scission to recognition in the major groove of DNA.

Rhodium polypyridyl complexes are known to be powerful photooxidants; it has been proposed that irradiation into ligand transitions leads to a ligand-to-metal charge transfer producing a redox active excited state species (Indelli et al., 1984; Ballardini et al., 1980). This species may be responsible for the oxidative cleavage of nucleic acids by tris(polypyridyl) complexes of rhodium(III) and cobalt(III), with the target being the backbone sugars (Fleisher et al., 1988). Cleavage by these tris(polypyridyl) complexes produces sharp cleavage bands which have been assigned as 5'-phosphates and 3'-phosphates based on their gel electrophoretic mobilities. The concomitant release of free nucleic acid bases supports the proposal that the site of oxidative damage is the sugar moiety and not the nucleic acid bases. It is likely that photocleavage of nucleic acids by phi complexes of Rh(III) occurs through an analogous mechanism involving an excited state redox active species. The production of sharp cleavage bands (Fig. 3b) by phi complexes of Rh(III) suggests that cleavage is mediated by a non-diffusible oxidant, likely a phi cation radical. The reactive radical species could form through irradiation of the ligand-to-metal charge transfer (LMCT) transition.

The DNA photocleavage products identified (Sitlani et al., 1992) are consistent with photoreaction of $Rh(phi)^{3+}$ intercalated in the major groove of DNA, via abstraction of a deoxyribose C3'-H atom (Fig. 4). A single base 5'-asymmetry in cleavage is observed at specific sites on the helix, consistent with reaction from the major groove upon photoactivation of the intercalated ligand radical. As summarized in Figure 4, following abstraction of the C3'-H, subsequent oxidation of the C3'-radical species, possibly by the transiently formed Rh(II), and solvation of the resultant cation would lead, through a series of β-elimination steps, to the 5'-phosphate, 3'-phosphate, and nucleic acid base products observed. Alternatively, addition of dioxygen to the C3'-radical, followed by oxygen insertion into the C3'-C4' bond and decomposition would yield the identified 5'-phosphate, 3'-phosphoglycaldehyde terminus and base propenoic acid. The quantitation of these products as well as the absence of detectable degradation products formed by modification at the C1'-, 4'-, or 5'- positions of the deoxyribose ring are therefore consistent with the model where phi complexes of rhodium bind and, with photolysis, react in the major groove of DNA. Unlike other small molecules and natural products which bind to and recognize the DNA minor groove, these metal complexes therefore appear to be unique in that they bind in the DNA major groove. Furthermore, since most proteins bind in the major groove of DNA, the study of site- and sequence-specific recognition by these metal complexes may be viewed as a model to understand some aspects of protein-DNA interactions.

While the mechanistic studies have been carried out on phi complexes containing polypyridyl ancillary ligands, the cleavage products obtained by $Rh(polyamine)phi^{3+}$ complexes are consistent with a parallel mechanism of DNA strand scission. In general, then, the exploration of the mechanism of DNA photocleavage by these complexes has allowed us to use photocleavage as an assay to probe site-specific recognition. Since cleavage is mediated by a non-diffusible species, the point of strand scission represents the site of bound metal complex.

This mechanism contrasts with the photocleavage reactions found with $[Ru(phen)_3]^{2+}$. Here photolysis in the MLCT band sensitizes the formation of a diffusible singlet oxygen species that oxidatively adds to the DNA bases; addition of piperidine to the modified DNA

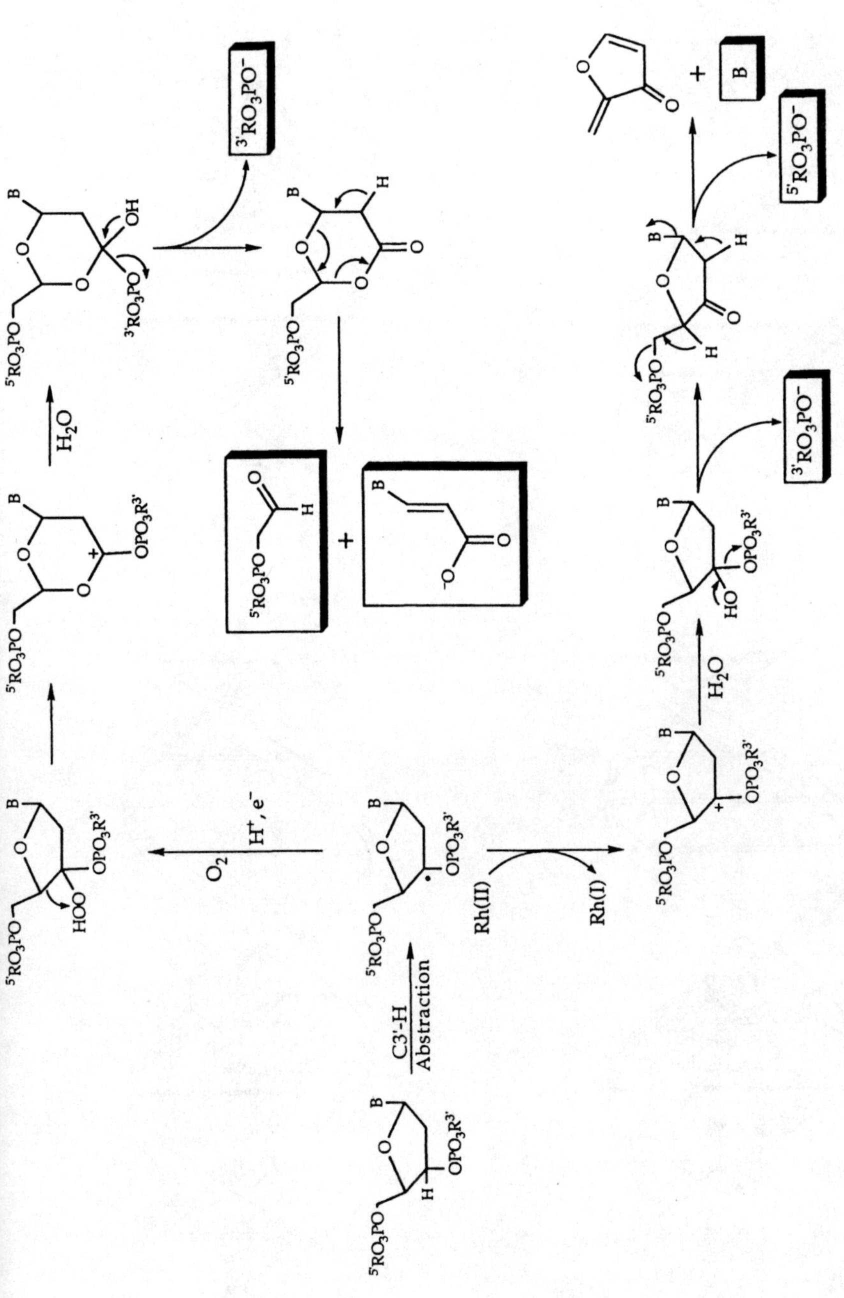

Figure 4.
Proposed mechanism of DNA cleavage by phi complexes of rhodium(III). Photolysis of the metal complexes intercalated from the major groove of DNA leads to selective abstraction of the C3'-H atom, with formation of the C3'-radical. This reaction can promote degradation through oxygen-independent and oxygen-dependent pathways to form the identified products (boxed). (From Sitlani et al., 1992).

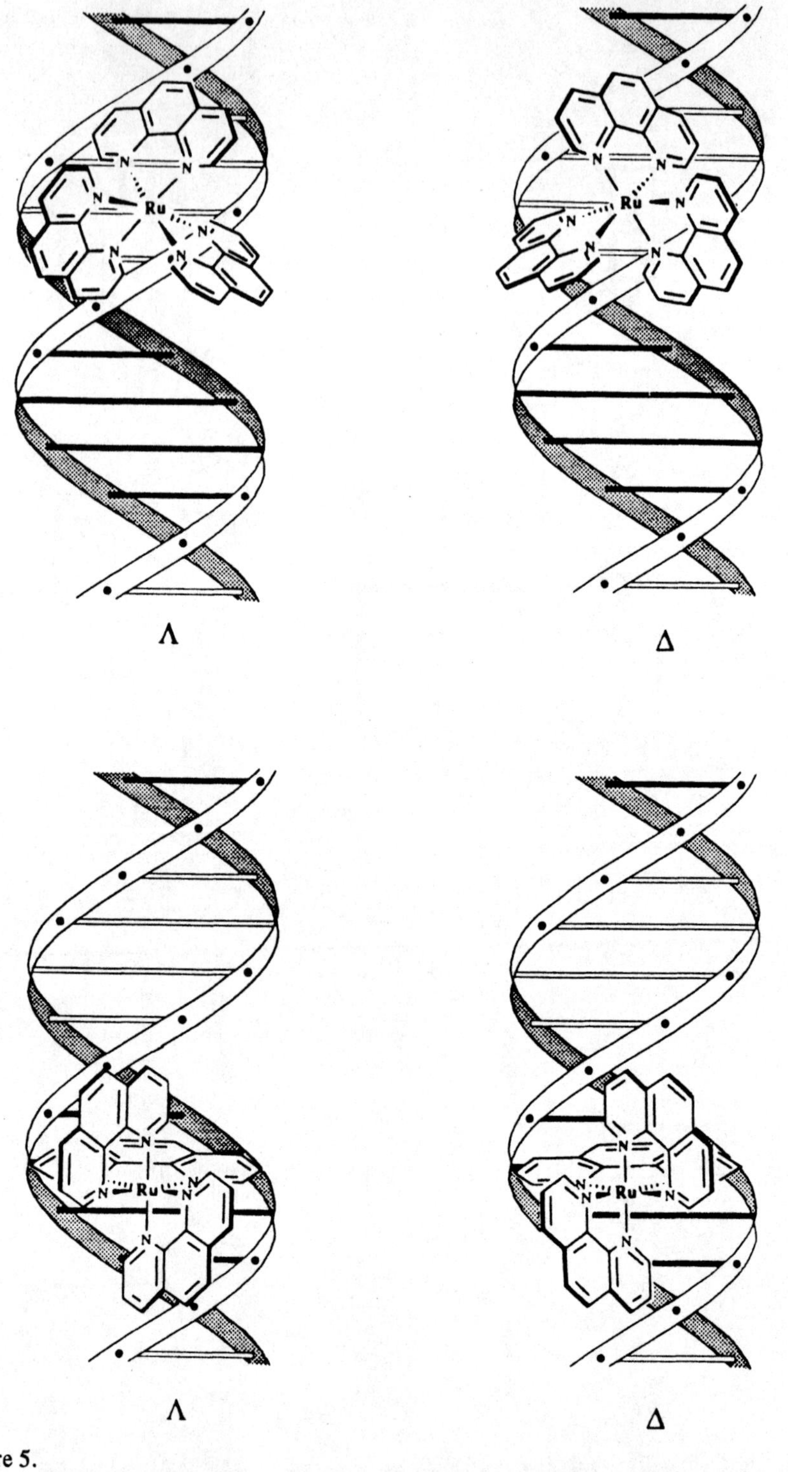

Figure 5.
Binding modes of the Δ and Λ isomers of $Ru(phen)_3^{2+}$ to B-form DNA. Bottom panel shows that intercalation in the major groove of DNA is sterically favored by the Δ isomer since the disposition of the non-intercalated ligands follows the right-handed groove. The top panel illustrates the basis for preferential surface binding by the Λ isomer in the minor groove of DNA, since here the disposition of its ligands follows the right-handed groove.

strand leads to DNA strand scission. However, singlet oxygen reacts with different efficiencies with different bases (G > > T > A,C). Therefore in this case, normalization of the differing cleavage efficiencies must first be made before photocleavage is directly correlated with site-specific binding (Fleisher et al., 1986). In the case of the rhodium(III) and cobalt(III) complexes, since the site of oxidative damage is the sugar, to a first approximation, the sugars at each site should be equally reactive and thus photocleavage can provide a direct, unweighted marker for sites of binding.

III. FACTORS IMPORTANT IN METAL COMPLEX-NUCLEIC ACID RECOGNITION

A. Steric factors

A simple understanding of the surface and intercalative binding modes of metal complexes to nucleic acids has allowed us to design metal complexes that can preferentially bind to and recognize specific nucleic acid structures. The nucleic acid photocleavage reactions of metal complexes have served as an important assay to examine nucleic acid recognition. In this section we illustrate, with examples, that nucleic acid recognition by metal complexes is dominated by steric and symmetry factors. Shape-selection, where the three-dimensional structure of the metal complex is matched in shape and symmetry to a nucleic acid site, may serve as a primary recognition element in distinguishing not only global features of nucleic acid tertiary structure but also subtle variations in the local structure associated with single base pair steps.

1. *Recognition of global features of nucleic acids*

The studies of $[Ru(phen)_3]^{2+}$ binding to B-DNA illustrates that the overall shape and chirality of octahedral, coordinatively saturated metal complexes determine the nature of nucleic acid-metal complex binding interactions. The complementarity between B-form DNA and $[Ru(phen)_3]^{2+}$ is such that surface binding in the minor groove is favored by the Λ isomer and intercalative binding in the major groove is favored by the Δ isomer (Fig. 4). Mei and Barton (1986) have shown that replacing the phenanthroline ligands with 3,4,7,8-tetramethylphenanthroline ligands (TMP) precludes both intercalation and surface binding to B-form DNA; instead, $[Ru(TMP)_3]^{2+}$ binds preferentially to A-form helices. Figure 6 illustrates that the bulky, sterically hindered shape of $[Ru(TMP)_3]^{2+}$ prevents it from binding in the deep and narrow minor groove of B-form DNA, but allows a hydrophobic association with the shallow and wide minor groove of A-form helices. Furthermore, owing to the right-handedness of A-form helices, preferential surface binding by the Λ isomer of $[Ru(TMP)_3]^{2+}$, with complementary symmetry, is observed.

2. *Recognition of sequence-specific secondary structures and tertiary interactions*

On a more detailed level, shape complementarity also appears to govern sequence-specific structural recognition by metal complexes. The complexes $[Rh(phen)_2phi]^{3+}$ and $[Rh(phi)_2bpy]^{3+}$ are highly shape-selective in their DNA sequence recognition characteristics (Pyle et al., 1989b). Both complexes contain the phenanthrenequinone diimine (phi) ligand, which promotes tight intercalation and which allows them to associate intimately with the helix and to sense the local structure at the site of intercalation. Consistent with their DNA photocleavage chemistry (Sitlani et al., 1992), both complexes bind in the major groove of DNA. However, $[Rh(phen)_2phi]^{3+}$ and $[Rh(phi)_2bpy]^{3+}$ display remarkably different DNA recognition characteristics. With $[Rh(phen)_2phi]^{3+}$, upon intercalation of the phi ligand, the steric constraints imposed by the ancillary phenanthroline ligands lead to selective binding at sites which are more open in the major groove (Fig. 7a). For $[Rh(phi)_2bpy]^{3+}$, no similar steric constraints arise, because the steric bulk associated with the ancillary ligands is pulled back from the axial direction; therefore, this complex binds to most sites with equal affinity (Fig. 7b). Thus, despite the possibility of hydrogen bonding between the imine protons of the ancillary phi ligand in $[Rh(phi)_2bpy]^{3+}$ and the DNA bases, the complex is largely sequence-neutral in its recognition characteristics. In this instance, the simple comparison between $[Rh(phen)_2phi]^{3+}$ and $[Rh(phi)_2bpy]^{3+}$ underscores the importance of shape rather than hydrogen bonding in site-specific recognition of DNA.

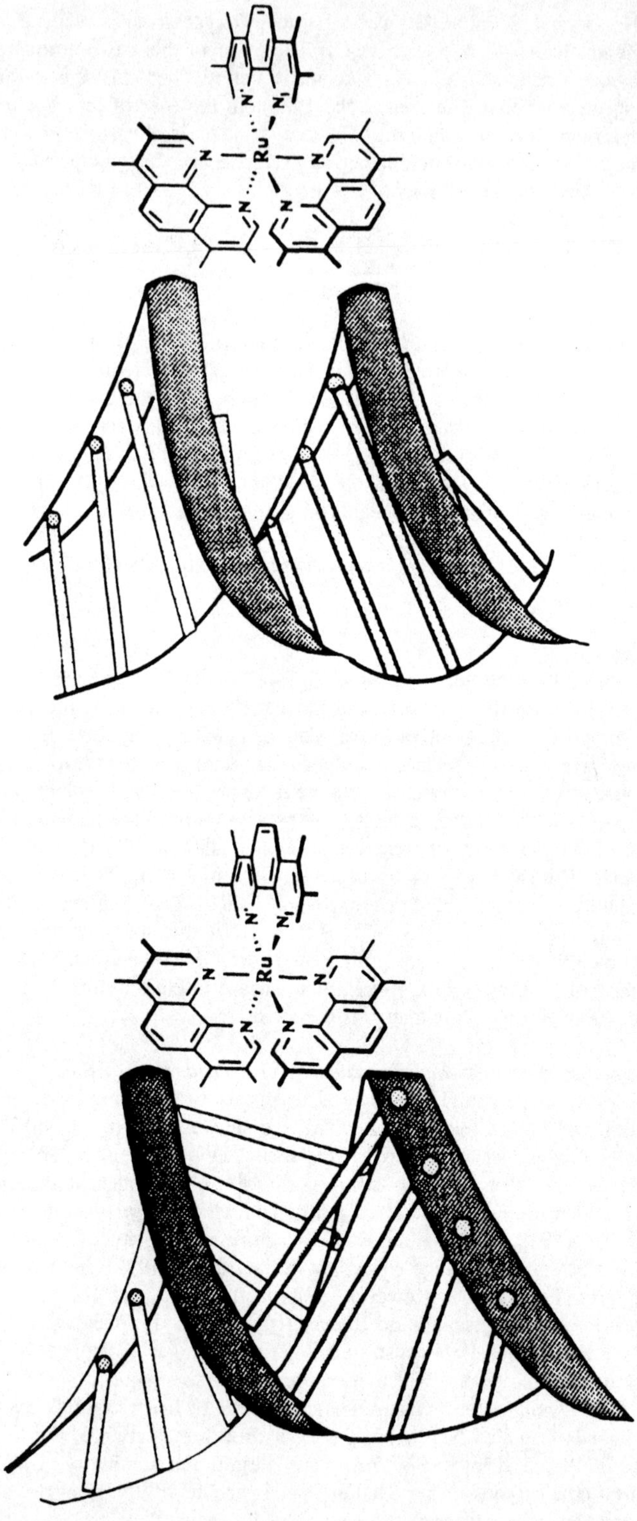

Figure 6.
An illustration of preferential binding by $Ru(TMP)_3^{2+}$ in A-form over B-form helices. The shape of the complex is well-matched to bind against the shallow and wide minor groove surface of A-form helices (left) but is too large to surface bind in the deep and narrow minor groove of B-form DNA (right).

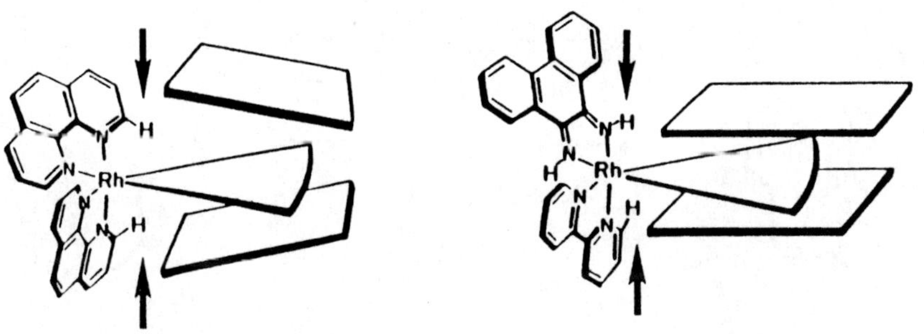

$[Rh(phen)_2(phi)]^{3+}$ $[Rh(phi)_2(bpy)]^{3+}$

Figure 7.
Shape-selective recognition of local DNA structure by $Rh(phen)_2phi^{3+}$ and $Rh(phi)_2bpy^{3+}$. The schematic illustrates intercalation of $Rh(phen)_2phi^{3+}$ and $Rh(phi)_2bpy^{3+}$ into either an opened or canonical major groove DNA site, respectively. For $Rh(phen)_2phi^{3+}$, the steric constraints imposed by its ancillary phenanthroline ligands promote selective intercalation of this complex at such opened sites. In $Rh(phi)_2bpy^{3+}$, the lack of similar steric requirements by its ancillary phi ligand, which is pulled back from the metal center, allows the complex to bind to most sites without selectivity. (From Pyle et al., 1989b).

$[Rh(phen)_2phi]^{3+}$ also displays chiral recognition within 5'-pyrimidine-purine-3' base-pair steps on the basis of its shape and symmetry (Pyle et al., 1990a; Campisi et al., 1994). Strong cleavage by the Δ isomer is observed at 5'-py-py-pu-3' steps where the shape of Δ -$[Rh(phen)_2phi]^{3+}$ matches the chiral pocket created by differential propeller twisting at these steps. The Λ isomer, with the opposite disposition of ancillary ligands upon intercalation, clashes with the pyrimidines at these steps; hence little or no cleavage is observed for this isomer. $[Rh(phen)_2phi]^{3+}$ also recognizes 5'-py-py-py-3' steps, but since this step has no C_2 axis along the intercalative dyad axis, little or no enantiomeric discrimination is observed (Fig. 8); likely these sites are recognized owing to an opening in the major groove which results from base tilting. Thus, based on steric and symmetry considerations, $[Rh(phen)_2phi]^{3+}$ shows selectivity in the recognition of sites within B-DNA.

$[Rh(phen)_2phi]^{3+}$ also displays shape selectivity in its recognition of crystallographically well-characterized tRNA molecules (Chow and Barton, 1990). While double helical regions of RNA adopt primarily the A-conformation, RNA molecules also contain complex tertiary interactions which establish the three-dimensional folding of the nucleic acid. Consistent with its DNA recognition characteristics, $[Rh(phen)_2phi]^{3+}$ does not intercalate into the deepened major groove of A-like conformations. Instead the complex, which binds avidly by intercalation from the major groove, specifically targets sites in tRNA in which three bases are hydrogen bonded to another. As is evident in crystal structures of tRNAs, the third base in triply hydrogen-bonded array interacts in the major groove through Hoogsteen base hydrogen-bonding with a purine of a normal Watson Crick base pair. This stabilized base position then creates a filled major groove site which may serve as a platform for intercalative stacking by the metal complex. Mutations which preserve the structures of the triply bonded region of the tRNA are still cleaved by $[Rh(phen)_2phi]^{3+}$ (Chow et al., 1992a) indicating that the molecule recognizes the structure generated by the triply bonded array rather than individual nucleotides (Fig. 9). Selective depurination of this third base position furthermore decreases selective targeting by $[Rh(phen)_2phi]^{3+}$. Interestingly, competition experiments have shown that the complex $[Ru(bpy)_2dppz]^{2+}$, owing to its avidly intercalating dppz ligand, also binds to some of the tRNA triple sites targeted by $[Rh(phen)_2phi]^{3+}$ (Jenkins et al., 1992).

3. *Recognition of folded RNA and DNA molecules of unknown structure*

Based upon the cleavage characteristics observed on the crystallographically well-characterized tRNAs, $[Rh(phen)_2phi]^{3+}$ appears to target sites of tertiary interactions and

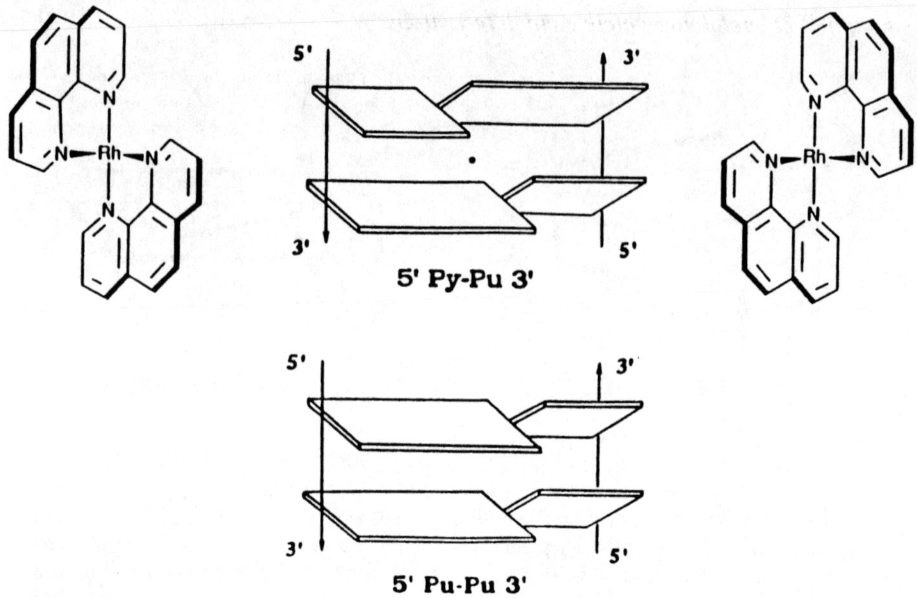

Figure 8.
Chiral discrimination in binding to propeller-twisted sites on DNA by the Δ and Λ isomers of $Rh(phen)_2phi^{3+}$. As shown in the top panel, a highly differentially propeller-twisted 5'-py-pu-3' step creates a chiral pocket which matches the shape of the Δ isomer, but clashes with the orientation of the ancillary ligands of the Λ isomer. Thus this C_2 symmetric step (with an intercalative dyad axis marked by the ·), promotes enantiospecific recognition by the isomers of $Rh(phen)_2phi^{3+}$. As shown in the bottom panel, a 5'-py-py-3' step lacks a C_2 dyad axis; enantiomeric discrimination in binding to this step is not observed. (From Pyle et al., 1990a).

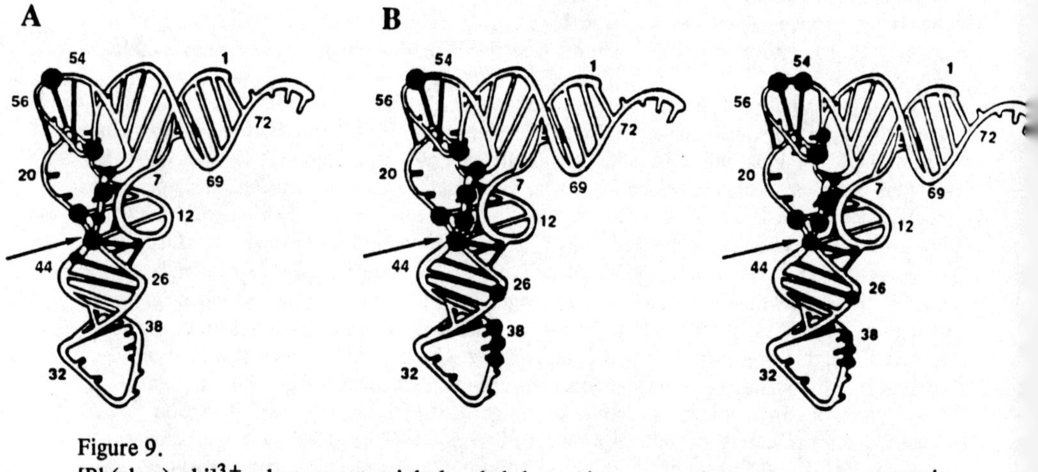

Figure 9.
$[Rh(phen)_2phi]^{3+}$ cleavage at triply-bonded base sites on native and mutant tRNA^{phe} substrates.

A. Schematic showing cleavage sites on native yeast tRNA^{phe}. The arrow points to cleavage at the G45-[G10-C25] tertiary interaction.

B. Cleavage on tRNA^{phe} substrates containing a single G45U mutation (left) and a double G10C, C25G mutation (right). The arrows point to cleavage at the mutated triple base sites. Both these mutations should preserve the triple-base interaction at this site by forming the base triplets U45-[G10-C25] and G45-[C10-G25], respectively. Cleavage at this site in both mutants is comparable to cleavage on the native tRNA^{phe} substrate.

(From Chow et al., 1992a).

hence may be a valuable chemical probe in structurally characterizing the folding of RNA molecules. In that context, $[Rh(phen)_2phi]^{3+}$ has been useful in probing the tertiary structure of *Xenopus* oocyte 5S rRNA (Chow et al., 1992b). A map of the photocleavage patterns of $[Rh(phen)_2phi]^{3+}$ on the proposed secondary structure of 5S rRNA (Fig. 10) shows that the rhodium complex recognizes several families of localized tertiary interactions, including stem-loops, structured loops, and sites with mismatched bases. Consistent with its recognition of crystallographically characterized tRNAs, $[Rh(phen)_2phi]^{3+}$ does not bind to double-helical or single stranded regions of 5S RNA. Instead, the complex targets regions containing complex tertiary structure in which the major groove has become more accessible. Stem-loop junctions, multiple bulged nucleotides and base pair mismatches in helical regions all appear to cause a widening of the major groove. These sites of localized tertiary interactions are thus likely to facilitate intercalation via the phi ligand and to accommodate the steric bulk offered by the ancillary phenanthroline ligands. Furthermore, as shown in Figure 10, strong cleavage by $[Rh(phen)_2phi]^{3+}$ is seen in the loop regions of C and E, indicating that the two loops are intimately and intrinsically structured by noncanonical base pairing or base stacking interactions. This conclusion is consistent with the results that both loops are not only resistant to single- and double- strand-specific nucleases but also inaccessible to chemical probing. Cleavage on a truncated *Xenopus* oocyte 5S rRNA, comprising of stems IV and V and loops E and D, is almost identical to that obtained for wild-type 5S rRNA. This observation is consistent with a model for the RNA which adopts an opened T-structure and where the rhodium complex recognizes independent tertiary structures in which the major groove has become accessible to intercalative stacking, as in helix IV and loop E.

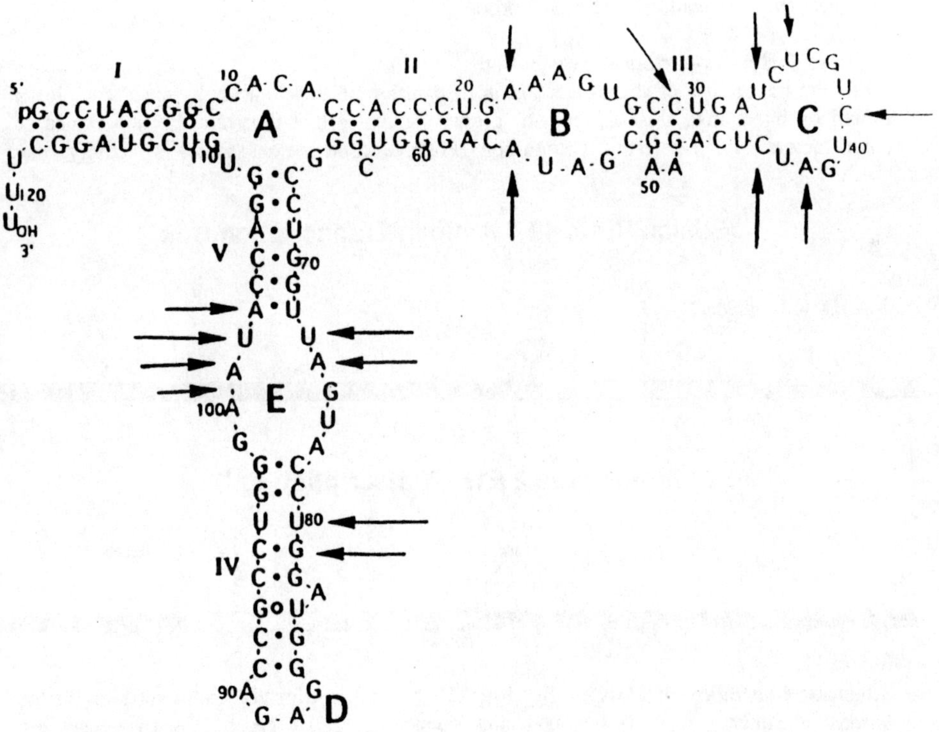

Figure 10.
Recognition of stem-loop junctions and base-pair mismatches in *Xenopus* 5S rRNA by $[Rh(phen)_2phi]^{3+}$. The arrows indicate the positions of $[Rh(phen)_2phi]^{3+}$ promoted strand scission with length corresponding to relative cleavage intensity. (From Chow et al., 1992b).

Interestingly, in support of these observations, recent NMR studies by Wimberly et al. (1993) have suggested that the two strands of loop E are involved in base-pairing or stacking interactions.

Shape-selective recognition has also been useful in probing folded DNA structures which are held together by tertiary interactions. The complex $[Rh(DIP)_3]^{3+}$ yields specific double-stranded cleavage at the base of cruciforms extruded by supercoiled plasmids (Kirshenbaum et al., 1988). No specific cleavage is observed in the corresponding linearized plasmids, which illustrates that the complex is recognizing a folded structure and not local structures associated with particular sequences. Clearly, the complex has no hydrogen bond donors or acceptors; instead the well-defined hydrophobic structure may recognize a small pocket created by cruciform helices that interact with each other. $[Rh(DIP)_3]^{3+}$ also cleaves specifically at the junction point of Holliday junctions (Waldron and Voulgaris), which are believed to be structural intermediates in genetic recombination and bear some resemblance to the folding of cruciforms.

The complex has also been a powerful shape-selective tool in probing completely unknown tertiary DNA structures. Site-specific targeting by $[Rh(DIP)_3]^{3+}$ has provided evidence for a distinct structure delineating introns on the DNA level (Lee and Barton, 1993). The complex cleaves specifically adjacent to the branch points of introns of Simian Virus 40 T-antigen and Adenovirus 2 E1A genes inserted in supercoiled double-stranded plasmids (Fig. 11). Also cleavage at the donor and acceptor sites of the intron of the SV40 - T antigen gene is observed. Importantly, while the cleavage sites in the coding strand of the two mammalian viral genes lack sequence homology, they are similar in that they demarcate functionally important sites. Targeting by the metal complex may suggest a similar targeting by proteins within the cell. In conclusion, shape-selective metal complexes have the potential to recognize novel and unknown sites of tertiary structures, which are prone to structural characterization by chemical probes.

B. Other factors important in recognition

While the importance of shape-selection in recognition of nucleic acids by metal complexes cannot be overstated, this section will present examples that illustrate the importance of other factors in site-selective recognition. Positive contributions to site-selectivity may

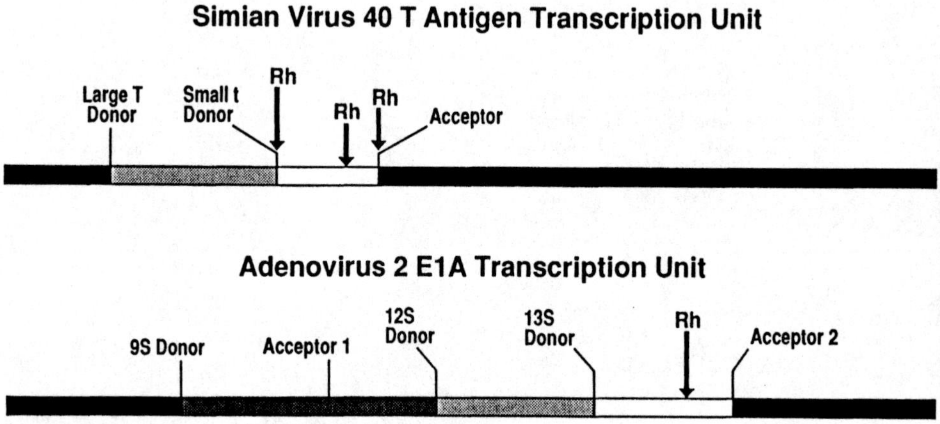

Figure 11.
Schematic illustration of cleavage by $[Rh(DIP)_3]^{3+}$ at functionally important sites in the introns of Simian Virus 40 T-antigen and Adenovirus 2E1A genes. The two genes are represented, roughly to scale, by bars. Solid shaded elements represent exons and the unshaded elements, the introns of interest. The arrows point to the sites of $[Rh(DIP)_3]^{3+}$ cleavage. On the E1A gene, substantial cleavage is observed near the branch point of the intron; however, on the SV40 gene cleavage is most strongly at the donor site, and less strongly at the acceptor site and near the branch site. (From Lee and Barton, 1993).

emanate not only through specific hydrogen bonding interactions to base positions but also as the result of site-specific stacking interactions or even through the cooperative non-covalent associations of two metal complexes bound at adjacent sites along the polymer.

The study of phi complexes of rhodium(III) containing saturated amines and macrocyclic thioethers as ancillary ligands (Fig. 12) has pointed to the importance of hydrogen bonding and van der Waals interactions to DNA site specificity (Krotz et al., 1993). Phi complexes of rhodium containing saturated amine ligands also bind avidly to DNA by intercalation, and with photoactivation promote efficient strand cleavage. A 5'-GC-3' selectivity in cleavage is observed with the complexes $[Rh(en)_2phi]^{3+}$, $[Rh(NH_3)_4phi]^{3+}$, and $[Rh([12]aneN_4)phi]^{3+}$ and this site-selectivity is likely to arise from hydrogen bonding between the axial amines in their ancillary ligands and the O6 of guanines (Fig. 13a). The cyclic thioether complex $[Rh([12]aneS_4)phi]^{3+}$, which lacks hydrogen bonding functionalities, shows no preference for 5'-GC-3' steps. Instead, the complex has a remarkably high level of specificity for 5'-ATG, 5'-ACG and 5'-ACT steps, with cleavage occurring at the highlighted residues, presumably again based upon shape-selection. It is interesting that besides targeting 5'-GC-3' steps, Λ-$[Rh(en)_2phi]^{3+}$ also enantioselectively targets 5'-TX-3' sites. The preference of the Λ isomer of $[Rh(en)_2phi]^{3+}$ for 5'-TA-3' steps

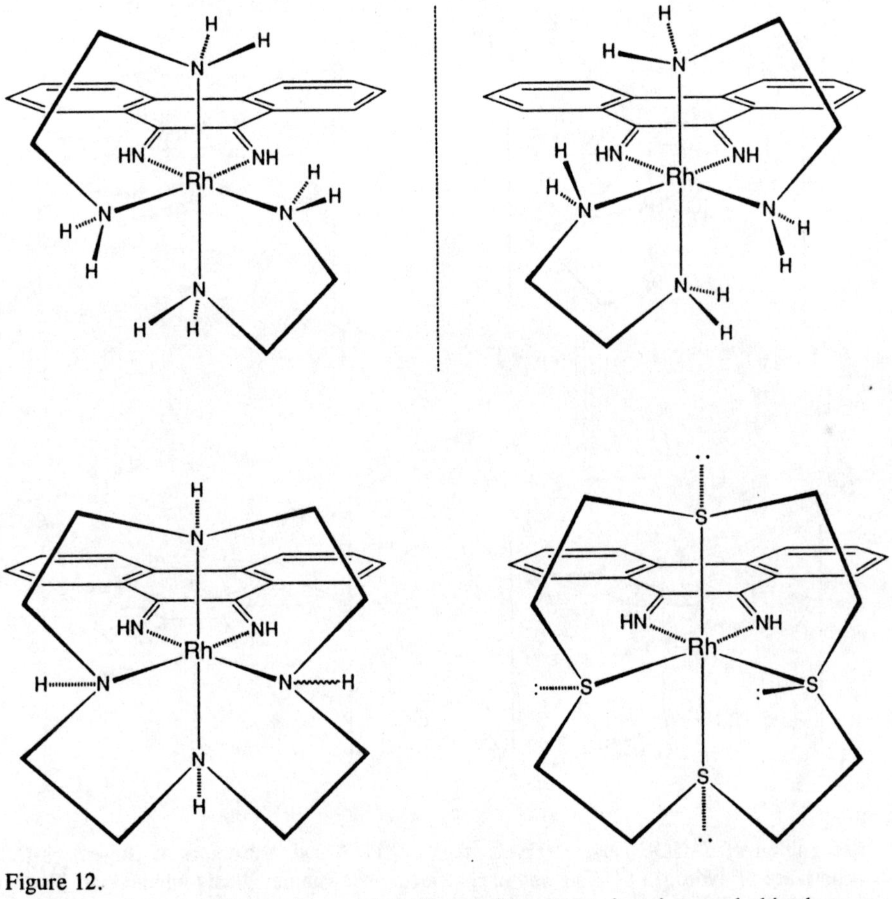

Figure 12.
Structures of some phi complexes of rhodium with saturated amines and thioethers as ancillary ligands. These complexes all bind DNA avidly by intercalation and promote strand cleavage upon photoactivation.

The top panel shows the Δ and Λ isomers of $[Rh(en)_2phi]^{3+}$ (en = ethylenediamine, phi = 9,10-phenanthrenequinone diimine). The bottom panel shows the structures of $[Rh([12]andN_4)phi]^{3+}$ (left) and $[Rh([12]aneS_4)phi]^{3+}$ (right).

suggests the possibility of stabilizing van der Waals contacts between the methylene groups of the ancillary ethylene diamine ligand and the thymine methyl group, located in the major groove (Fig. 13b). Moreover, replacement of a thymine by uracil, which lacks the methyl group, abolishes the enantioselectivity in cleavage at the 5'-UA-3' site on an oligonucleotide. Modeling studies have shown that analogous interactions between the Δ isomer and a 5'-AT-3' step are not available owing to the right-handed helicity of DNA.

It is noteworthy in this discussion that $[Rh([12]aneS_4)phi]^{3+}$ shows a strong preference for 5'-AXY sites with the complex intercalated between the XY base step, even though molecular modeling has shown that the complex is small enough to span only a single base step in an intercalated DNA site. This observation points to the role that sequences flanking the binding site may play in stabilizing the post-intercalative binding state. Intercalation and unwinding of the helix via the phi ligand may promote significant structural changes in the binding site which must be accommodated through stacking interactions with the neighboring base pairs. It is likely that the structural change induced by this complex is

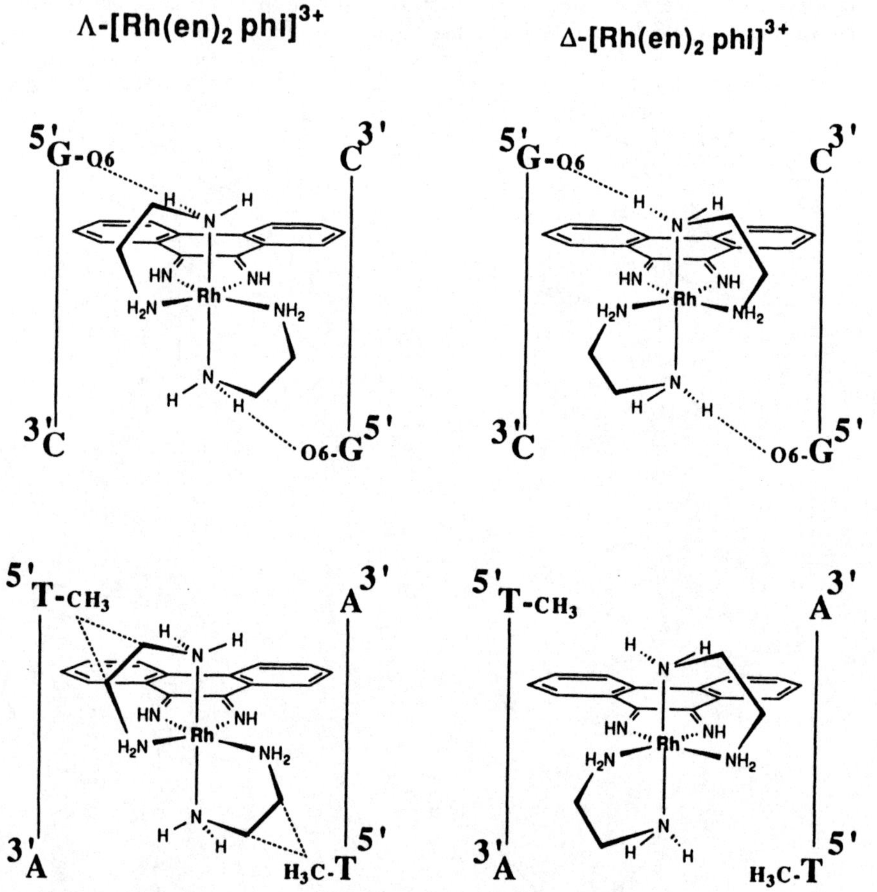

Figure 13.
Recognition of 5'-GC-3' and 5'-TA-3' steps by the Δ and Λ isomers of $[Rh(en)_2phi]^{3+}$: importance of hydrogen bonding and methyl-methylene van der Waals interactions. The top panel illustrates the basis for GC specificity by both isomers. The O6 of guanines in the GC base step is well-positioned to make H-bonding contacts with the axial amines of the ethylenediamine ligands. The bottom panel illustrates the basis for enantioselective recognition of the TA step. While the Λ isomer is well-positioned to afford positive van der Waals interactions between its methylene groups and the thymine methyl groups, the symmetry of the intercalated Δ isomer does not afford similar interactions.

stabilized by the neighboring 5'-AT base pair. However, the importance of flanking sequences in recognition by metallointercalators is likely to be a more general phenomenon. As is typical for classical intercalating agents, intercalation by the octahedral metal complexes causes the DNA helix to unwind by 15-25 degrees; the contribution of neighboring base pairs is likely to depend on the structural changes induced at the binding site. The ability of flanking base pairs to accommodate structural perturbations at the binding site may be closely related to their level of deformability in the context of a particular sequence. While recognition by metallointercalators of the pre-intercalation state may be more directly related to the local structural variations, stabilization of the post-intercalation state may be more directly related to the extent of sequence-deformability.

The energy gained through base-pair stacking interactions is the driving force for intercalation by metal complexes. This intercalative stacking may also play an important role in site-specific recognition. Preferential intercalative binding at unstacked or weakly stacked regions of nucleic acids may occur; conversely, destacked regions may form a distinct structure that is preferentially stabilized by the metal complex. For example, the complex $[Rh(DIP)_3]^{3+}$ shows a remarkable level of specificity for guanine-uracil (G-U) mismatches in double helical regions of folded RNAs (Chow and Barton, 1992). In particular, specific cleavage is observed by the complex at the residue which lies to the 3'-side of the wobble-paired U. Changing the orientation of the G-U wobble pair changes the

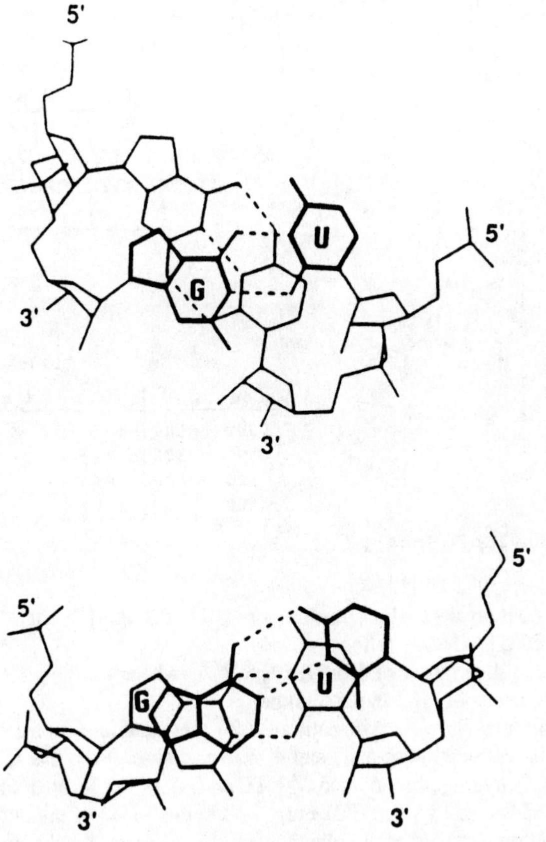

Figure 14.
The stacking configuration at the 3'-end (top) and 5'-end (bottom) of the G4-U69 wobble pair in the middle of the acceptor stem of yeast $tRNA^{phe}$. Compared to the 5'-end of the mismatch, the 3'-side of the U exhibits a highly destacked structure which may be recognized by the shape-selective $[Rh(DIP)_3]^{3+}$ metal complex. (From Mizuno and Sudaralingam, 1978).

position of cleavage. Interestingly, crystal structures (Mizuno and Sudaralingam, 1978) have revealed that the geometry of the G-U wobble base pair results in an unusual stacking of its bases with the neighboring bases. The stacking interactions on either side of the G-U mismatch differ greatly, with the base-pair on the 3'-side of the U exhibiting highly destacked structures with the U extended into the major groove of the RNA (Fig. 14). It appears that the highly hydrophobic complex $[Rh(DIP)_3]^{3+}$ is able to detect these differences in base-stacking with the flanking base pairs without specific hydrogen bonding interactions. Shape-selective targeting of the 3'-side of the G-U mismatch is favored.

Finally, there is reason to believe that like DNA-binding proteins, metal complexes have the potential to associate or non-covalently dimerize in order to afford high levels of specificity. The complex $[Rh(4,4'\text{-diphenylbipyridyl})_2phi]^{3+}$ has a level of specificity for the eight base-pair palindromic site 5'-CTCTAGAG-3' that rivals restriction enzymes (Sitlani and Barton, 1993). This complex, which is highly hydrophobic and lacks hydrogen-bonding functionalities, is able to shape-selectively recognize the specific sequence-dependent structure associated with the eight base pair 5'-CTCTAGAG-3' site. Competitive thermodynamic experiments have suggested that two molecules of the complex $[Rh(4,4'\text{-diphenylbipyridyl})_2phi]^{3+}$ bind simultaneously and cooperatively through non-covalent associations one with another on an eight base-pair site on linear DNA. As shown in Figure 15, at low concentrations of the metal complex, cleavage is observed at the 8 base-pair dimeric 5'-CTCTAGAG-3' site preferentially over the monomeric 6 base-pair 5'-CTCTAG-

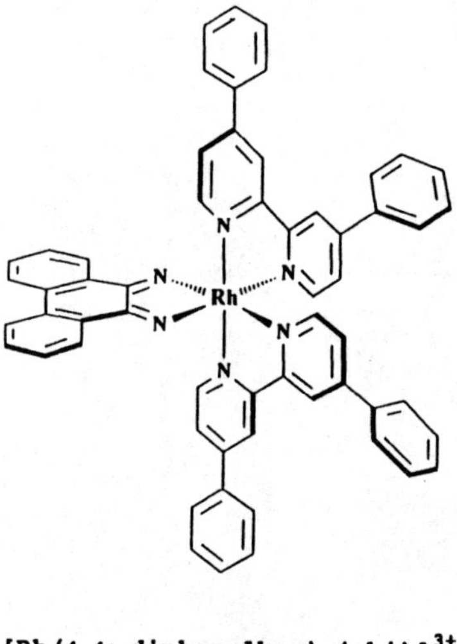

$[Rh(4,4'\text{-diphenylbpy})_2(phi)]^{3+}$

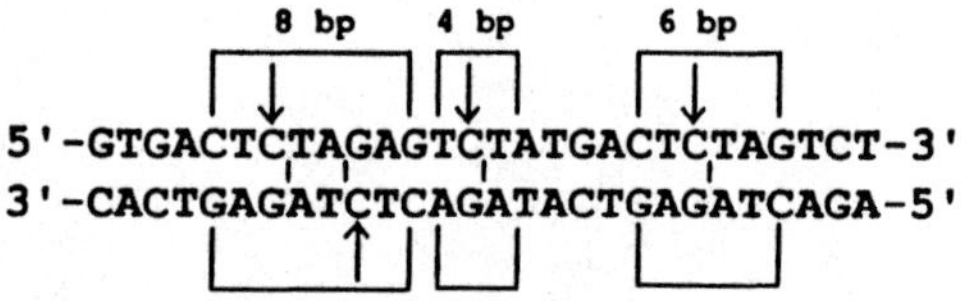

Conc. (nM)	Cleavage 8 bp	4 bp	6 bp
1000 - 500	+++++	+	+++++
500 - 300	++++	-	+++
300 - 100	+++	-	++
100 - 50	+	-	-

Figure 15.
Evidence for non-covalent dimerization of $[Rh(DPB)_2phi]^{3+}$ at the eight-base pair palindromic 5'-CTCTAGAG-3' site.

A. Schematic illustration of $[Rh(DPB)_2phi]^{3+}$, where DPB = 4,4'-diphenylbipyridyl and phi = 9,10-phenanthrenequinone diimine.

B. The top panel shows a schematic of a 28 mer double-stranded oligomer containing the indicated 8 base-pair, 6 base-pair, and 4 base-pair sites. Note that the 8 base-pair site is a composite of 2 overlapping 6 mer 5'-CTCTAG-3' sites located on opposite strands. Arrows refer to cleavage by $[Rh(DPB)_2phi]^{3+}$ on both strands of the oligomer and the short lines between the two strands of the oligonucleotide represent the site of intercalation by the metal complex. A table summarizing the cleavage at each site as a function of DNA/Rh concentration (where the DNA/Rh ratio is kept constant at 2 duplexes/ 1 Rh-complex) is shown in the bottom panel. At low concentrations, the cleavage per strand at the 8 mer site is greater than that observed for the 6 mer and 4 mer sites. This result suggests that the 2 complexes bind *simultaneously and cooperatively* at the 2 overlapping 6 mer 5'-CTCTAG-3' sites contained in the composite 8 mer site.

3' site. This added stabilization, derived through dimerization on the DNA polymer, therefore provides a valuable strategy to increase sequence specificity of the small transition metal complexes on DNA.

IV. PARALLELS BETWEEN METAL COMPLEX- AND PROTEIN-NUCLEIC ACID RECOGNITION: BIOLOGICAL IMPLICATIONS

Correlations between the sites specifically targeted by metal complexes and important protein-binding sites on DNA were initially observed in studies of cleavage by the chiral complex Λ-$Co(DIP)_3{}^{3+}$ on the Simian Virus-40 genome (Muller et al., 1987). The sites of Λ-$Co(DIP)_3{}^{3+}$ cleavage include the origin of replication, transcriptional promoter and enhancer regions, and the ends of coding regions for early and late genes. Interestingly, all these sites flank protein-coding regions and perhaps the specific structures associated with these cleavage sites may play a role in protein recognition and gene expression. Similarly, several rhodium intercalators appear to target specific sites in RNA and DNA that correspond to important protein-binding sites. Perhaps, then, these metal complexes may serve as models to understand basic principles governing the recognition of nucleic acids by proteins.

In particular, the complex $[Rh(4,4'\text{-diphenylbpy})_2phi]^{3+}$ is specific for the palindromic 8 base-pair 5'-C<u>TCTAGA</u>G-3' site, with the central 6 base-pair sequence 5'-TCTAGA-3' forming the central core for the overlapping recognition boxes of the *E. Coli met* repressor protein (Somers and Philips, 1992). While there is evidence to suggest that two molecules of this complex dimerize across the central CTAG sequence, the *E. Coli met*- DNA crystal structure shows that the *met* dimerization domain contains β-sheets which lie across a CTAG site, central to the operator sequence. The CTAG sequence appears to be a recognition motif for other DNA-binding proteins such as the *E. Coli trp* repressor and S1 and micrococcal nucleases. Crystal structures have shown that this sequence is characterized by an unusually distorted phosphate backbone, which may be an important recognition signal for both the metal complex $[Rh(4,4'\text{-diphenylbpy})_2phi]^{3+}$ and DNA-binding proteins. Importantly, the co-crystal structures of the *met*-DNA and *trp*-DNA complexes (Somers and Philips, 1992; Otwinowski et al., 1988) show no direct amino acid-base specific contacts in this region, thereby implicating the importance of indirect-readout or shape-selection in DNA recognition at these sites.

Similarly, the study of the interactions between rhodium complexes and RNA has shown that shape selection may also be an important criterion in determining protein-RNA recognition. Interestingly, the sites generally recognized by $[Rh(phen)_2phi]^{3+}$ and $[Rh(DIP)_3]^{3+}$ appear to correspond to critical elements for RNA recognition by proteins: triple-base sites, stem-loop junctions, base-pair mismatches, and bulged residues. The rhodium complexes we have studied appear to bind at sites where the normally deep and narrow major groove of RNA has become more accessible and open, due to tertiary interactions. Similarly, recent studies have suggested a model for RNA-protein recognition which involves the distortion of the A-form double helix in RNA by bulged nucleotides, thereby permitting protein binding in the open major groove (Weeks and Crothers, 1991). Additionally, the importance of stem-loop junctions and G-U base mismatches have been implicated as recognition motifs for tRNA binding yeast aspartyl-tRNA synthetase and *E. Coli* alanyl-tRNA synthetase respectively (Ruff et al., 1991). Hence sites of unique structural distortion may not only punctuate sites for the specific binding by transition metal complexes but also mark structural elements that are recognized by proteins.

V. PROSPECTS FOR THE FUTURE

The examples we have described illustrate how novel metal complexes may be constructed to achieve site-specific targeting. The primary lesson we have learned thus far has been the critical importance of a shape-selective matching of the structure of the metal complex to its nucleic acid binding site. Indeed the correlations between nucleic acid sites recognized by rhodium complexes and sites specifically bound by proteins highlight the value of shape-selective recognition.

Could this shape-selective design be useful in the construction of novel therapeutics? Inorganic pharmaceuticals such as cis-platin and metallobleomycin have proven to be remarkably successful as anticancer agents, but for a truly new generation of inorganic compounds the goal is clearly a new level of selective targeting and specificity. Certainly the first step in this rational design depends upon a chemical understanding of what hierarchy of factors governs the ensemble of non-covalent interactions which together lead to site-specific recognition. Transition metal chemistry has already been proven to be valuable in elucidating these factors. Shape-selection is surely one important element. It has now been demonstrated that shape-selection may be exploited in the construction of metal complexes which bind DNA with both affinities and specificities that rival DNA-binding proteins. New designs based upon shape-selection, however, require a better three-dimensional understanding of the local sequence-dependent structure of DNA. Here the transition metal complexes already prepared may be useful in probing local variations in structure, but much work needs still to be done.

What strategies provide solutions with a greater level of predictability? Considerations of hydrogen bonding and positive stabilizing van der Waals interactions surely offer a route to more predictive approaches to site-specific design, with the metal center providing a coordinate system upon which to append site-selecting functionalities in three-dimensional space. Furthermore the strategy of DNA-promoted dimerization, again mimicking a tool well-used in site-specific DNA recognition by proteins, provides a general approach to enhancing specificity in binding by small metal complexes. Therefore, schemes to exploit general strategies for metal-based dimerization need to be developed.

What is apparent thus far is the utility of transition metal chemistry to this endeavor, not only in the construction of new molecules of defined structure and stereochemistry but in the sensitive detection and characterization of sites and modes of binding. Transition metal chemistry therefore provides the arena not only in which to learn subtle lessons regarding site-selective recognition, but also in which, with versatility, to create new, rationally designed molecules which may serve as the next generation of bioinorganic pharmaceutical agents. Metal-nucleic acid chemistry offers promise and challenge for new site-selective recognition.

ACKNOWLEDGEMENT

We thank the National Institutes of Health (GM33309) for their financial support. We also thank our co-workers and collaborators, as cited, for their efforts in this research.

REFERENCES

Ballardini, R., Varani, G., and Balzani, V. (1980). *J. Amer. Chem. Soc.*; *102*: 1719.

Campisi, D.M., Morii, T., and Barton, J.K. (1994). *Biochemistry*, *33*: 4130.

Chow, C.S., and Barton, J.K. (1990). *J. Amer. Chem. Soc.*, *112*: 2839.

Chow, C.S., and Barton, J.K. (1992c). *Biochemistry*, *31*: 5423.

Chow, C.S., Behlen, L.S., Uhlenbeck, O.C., and Barton, J.K. (1992a). *Biochemistry*, *31*: 972.

Chow, C.S., Hartmann, K.M., Rawlings, S.L., Huber, P.W., and Barton, J.K. (1992b). *Biochemistry*, *31*: 3534.

David, S.S., and Barton, J.K. (1993). *J. Amer. Chem. Soc.*, *115*: 2984.

Delahunty, M.D., South, T.L., Summers, M.F., and Karpel, R.L. (1992). *Biochemistry*, *31*: 6461.

Dervan, P.B. (1986). *Science*, *232:* 464.

Fitzgerald, D.W., and Coleman, J.E. (1991). *Biochemistry*, *30:* 5195.

Fleisher, M.B., Waterman, K.C., Turro, N.J., Barton, J.K. (1986). *Inorg. Chem.*, *25*: 3549.

Fleisher, M.B., Mei, H.Y., and Barton, J.K. (1988). *Nucl. Acids Mol. Biol.*, *2*: 65.

Friedman, A.E., Chambron, J.-C., Sauvage, J.-P., Turro, N.J., and Barton, J.K. (1990). *J. Amer. Chem. Soc.*, *112*: 4960.

Indelli, M.T., Carioli, A., and Scandola, F. (1984). *J. Phys. Chem.*, *88*: 2685.

Jenkins, Y., Friedman, A.E., Turro, N.J., and Barton, J.K. (1992). *Biochemistry*, *31*: 10809.

Kirshenbaum, M.R., Tribolet, R., and Barton, J.K. (1988). *Nucl. Acids Res.*, *16*: 7943.

Krotz, A., Kuo, L., Sheilds, T.P., and Barton, J.K. (1993). *J. Amer. Chem. Soc.*, *115*: 3877.

Lee, I., and Barton, J.K. (1993). *Biochemistry*, *32:* 6121.

Mei, H.-Y., and Barton, J.K. (1986). *J. Amer. Chem. Soc.*, *108*: 7414.

Miller, W.T., Hill, K.A.W., and Schimmel, P. (1991). *Biochemistry*, *30:* 6970.

Mizuno, H, and Sudaralingam, M. (1978). *Nucl. Acids Res.*, *5:* 4451.

Moser, H.E., and Dervan, P.B. (1987). *Science*, *238:* 645.

Muller, B., Raphael, A.L., and Barton, J.K. (1987). *Proc. Nat. Acad. Sci. USA*, *84*: 1764.

Otwinowski, Z., Schevitz, R.W., Zhang, R.-G., Lawson, C.L., Joachimiak, A, Marmorstein, R.Q., Luisi, B.F., and Siglar, P.B. (1988). *Nature*, *355*: 321.

Pabo, C.O. (1992). *Annu. Rev. Biochem.*, *61*: 1053.

Pasternack, R.F., Gibbs, E.J., and Villafranca, J.J. (1983). *Biochemistry*, *22*: 2406.

Pyle, A.M., Rehmann, J.P., Meshoyrer, R., Kumar, C.V., Turro, N.J., and Barton, J.K. (1989a). *J. Amer. Chem. Soc.*, *111*: 3051.

Pyle, A.M., Long, E.C., and Barton, J.K. (1989b). *J. Amer. Chem. Soc.*, *111*: 4520.

Pyle, A.M., Morii, T., and Barton, J.K. (1990a). *J. Amer. Chem. Soc.*, *112*: 9432.

Pyle, A.M., Chiang, M.Y., and Barton, J.K. (1990b). *Inorg. Chem.*, *29*: 4487.

Rehmann, J.P., and Barton, J.K. (1990a). *Biochemistry*, *29*: 1701.

Rehmann, J.P., and Barton, J.K. (1990b). *Biochemistry*, *29*: 1710.

Ruff, M., Krishnaswamy, S., Boeglin, M., Poterszman, A., Mitschler, A., Podjarny, A., Rees, B., Thierry, J.C., and Moras, D. (1991). *Science*, *252*: 1682.

Sigman, D.S. (1986). *Accounts Chem. Res.*, *19*: 180.

Sitlani, A., Long, E.C., Pyle, A.M., and Barton, J.K. (1992). *J. Amer. Chem. Soc.*, *114*: 2303.

Sitlani, A., Dupureur, C.M., and Barton, J.K. (1993). *J. Amer. Chem. Soc.*, *115*: 12589.

Somers, W.E., and Philips, S.E.V. (1992). *Nature*, *359*: 387.

Steitz, T. (1990). *Quart. Rev. Biophys.*, *23*: 205.

Stubbe, J., and Kozarich, J.W. (1987). *Chem. Rev.*, *87*: 1107.

Waldron, K., and Voulgaris, T., unpublished results.

Weeks, K., and Crothers, D.M. (1991). *Cell*, *66*: 577.

Wimberly, B., Varani, G., and Tinoco, I. (1993). *Biochemistry*, *32*: 1078.

ZINC PROTEINS INVOLVED IN DNA REPLICATION AND TRANSCRIPTION

by

Joseph E. Coleman

Department of Molecular Biophysics and Biochemistry,
Yale University,
New Haven, CT 06510
U.S.A.

I. INTRODUCTION

Until the late 1980's DNA-binding proteins were sparsely represented among the nearly 300 zinc-containing metalloproteins discovered. The multisubunit bacterial RNA polymerases were discovered to contain zinc in the early 1970's (Scruton et al., 1971; Halling et al., 1977; Speckhard et al., 1977). Following these findings, the eukaryotic RNA polymerases, containing many more subunits than the bacterial enzymes, but whose largest subunits conserve much of the amino acid sequence found in the β and β' subunits of the *E. coli* enzyme, were also found to be zinc enzymes (Lewis and Burgess, 1982; Coleman, 1983 for summary). The RNA polymerases remained the only examples of zinc-containing DNA-binding proteins until the 1980's. In 1983 the transcription factor TFIIIA[1] required for activating the transcription of the 5S RNA gene by RNA polymerase III in *Xenopus* oocytes, was shown to contain zinc (Hanas et al., 1983). The presence of zinc was necessary for the binding of TFIIIA to the DNA sequence it specifically recognized within the 5S RNA gene (Wu, 1986). TFIIIA was the first of several hundred transcription factors now recognized to be zinc proteins. In 1985 the DNA sequence of TFIIIA became available, and from the translated amino acid sequence Miller et al., (1985) pointed out that TFIIIA contained 9 sequential segments of amino acid sequence, each of which maintained almost the same spacing for two Cys and two His residues which could act as ligands to a zinc ion. Each of these sequences, $\text{-C-X}_{4\text{-}5}\text{-C-X}_{12\text{-}13}\text{-H-X}_{3\text{-}4}\text{-H-}$, was postulated to coordinate zinc in

---

[1] Abbreviations used in this chapter are: Transcription Factors; TFIIIA = the factor IIIA activating the expression of the 5S RNA gene in Xenopus; GAL4 = the factor activating the expression of a number of genes coding enzymes in the galactose metabolic pathway in yeast, the numbers as in GAL4(62) refer to the number of amino acids in the fragment under study; UAS_G = the upstream activation DNA sequence recognized by GAL4; the hormone receptors are: GR = glucocorticoid, ER = estrogen, TR = thyroid, MR = mineralocorticoid, PR = progesterone, AR = androgen and RAR = retenoic acid receptors; HRE = hormone receptor element, the DNA sequence recognized by the various hormone receptors with the initial letter, e. g. GRE, designating the particular hormone, as in the first letter of the receptor abbreviations above; EXAFS = extended X-ray absorption fine structure; PMPS = paramercuriphenylsulfonate.

a S_2N_2 tetrahedral complex and the motif was called a "zinc finger" (Miller et al., 1985; see Rhodes and Klug, 1993 for a recent review). The 7S complex from *Xenopus* oocytes, which contains both TFIIIA and 5S RNA, was shown to contain 7-11 zinc atoms and EXAFS confirmed the postulated ligands and coordination geometry (Diakun et al., 1986).

A. Zinc-containing transcription factors

The field of Zn-containing transcription factors containing the C_2H_2 ligand arrangement, as found in TFIIIA, has recently taken a "quantum jump" in terms of numbers. One of the first human zinc-finger genes to be sequenced was that for Zfy, a transcription factor whose gene is located on the X chromosome (Page et al., 1987). It is made up of 13 TFIIIA-type fingers, whose precise spacing alternates between the even and the odd fingers (see Table 2). The spacing of the odd fingers is C-X_2-C-X_{12}-H-X_3-H, a spacing also known as the Kruppel-type, since this spacing was first found in the Drosophila transcription factor Kruppel (Rosenberg et al., 1986). An additional feature recognized after the sequencing of a number of zinc-finger proteins is that the spacers, i.e. the amino acid residues connecting adjacent fingers, known as the H/C link (TGEKPYE/K in the Kruppel-type), were more highly conserved than the non-ligand residues within the fingers. It has been suggested that this spacer may be important in adjusting the spacing of multiple fingers along the DNA (Pavletich and Pabo, 1991). A DNA oligonucleotide made up of the coding sequence for this spacer attached to the coding sequence encompassing only the two ligand H residues has been used as a hybridization probe to isolate genes for C_2H_2 zinc-finger proteins from cDNA libraries (Bellefroid et al., 1989; Thiesen, 1990) and from human (Bray et al., 1991; Hoovers et al., 1992) and mouse (Chowdhury et al., 1987) genomic libraries.

From the number of non-overlapping hybridizations observed with this probe and the human DNA libraries, it is now estimated that there may be 300-700 different C_2H_2 zinc-finger proteins of the Kruppel-type alone (Bellefroid et al., 1989; Crossley and Little, 1991). While the genes are located on several different chromosomes, an unusually large number map to chromosome 19 (Bellefroid et al., 1993). These zinc-finger genes frequently occur in clusters on any one chromosome. A number of the genes for the zinc-finger proteins isolated by the above hybridization methods have been sequenced and the zinc finger sequence almost uniformly shows the C-X_2-C-X_{12}-H-X_3-H spacing of the Kruppel-type. The genes isolated by these hybridization experiments uniformly code for multiple fingers, which suggests that positives require that multiple hybridizations occur within a gene. This and the conserved connector sequence may explain why this probe has identified a subgroup of zinc-finger proteins which contain exclusively one particular spacing. Therefore, the total number of proteins containing all variants of the zinc-finger motif may be much larger (see below for other ligand arrangements). The above work has led to suggestions that as much as 1% of human genomic DNA may code for zinc fingers and possibly 8% of the DNA of chromosome 19 (Rhodes and Klug, 1993).

B. Zinc-containing single-stranded nucleotide binding proteins

In 1986 the classical single-stranded DNA-binding protein, gene 32 protein, coded for by the bacteriophage genome and participating in phage DNA replication was discovered to contain a single Zn ion coordinated to 3 Cys and 1 His side chains in the central DNA-binding domain of this three domain protein (Giedroc et al., 1986). This class of zinc-containing single-stranded nucleic-acid binding proteins now includes the nucleocapsid proteins from mammalian retroviruses (Roberts et al., 1989; South et al., 1989; 1990; Fitzgerald and Coleman, 1991; Summers et al., 1992; South and Summers, 1993).

C. Structural variations among the zinc binding motifs found in eukaryotic transcription factors

While many of the transcription factors contain the zinc-finger motif as originally described for TFIIIA, two other zinc-binding motifs occurring in eukaryotic transcription factors have been well characterized. The second zinc-binding motif to emerge is that contained in the hormone receptors, transcription factors which activate (sometimes repress) a variety of eukaryotic genes in response to the binding of steroid hormones, thyroid hormones, retinoic acid and vitamin D (Hollenberg et al., 1987; Arriza et al., 1987). These proteins contain two independent $ZnCys_4$ complexes separated within the polypeptide chain by ~16 residues.

The third zinc-binding motif to emerge is represented by a series of fungal transcription factors each of which activate a set of genes which code for enzymes required for specific metabolic pathways (Balzi et al., 1987; Marczak and Brandriss, 1991; Beri et al., 1987; Baum et al., 1987; Friden and Schimmel, 1987; Zhou et al., 1987; Andrianopoulos and Hynes, 1990; André, 1990; Messenguy et al., 1986; Pfeifer et al., 1989; Kim and Michels, 1988; Sollitti and Marmur, 1988). The DNA-binding domains of these proteins are centered on a Zn_2Cys_6 binuclear cluster in which two of the $-S^-$ ligands act as bridging ligands (Pan and Coleman, 1989; 1990a; 1990b; 1991; Gardner et al., 1991; Marmorstein et al., 1992; Baleja et al., 1992; Kraulis et al., 1992). The transcription factor GAL4 is the prototype.

Two additional ligand arrangements, C_2HC and C_3HCC_4, have recently been described as zinc-binding motifs occurring in transcription factors as well as a diverse group of proteins (containing the C_3HC_4 ligand arrangement), whose influences on transcription are not clear as yet. While these are not classed here as separate subgroups of transcription factors, they are discussed in some detail below.

Enough evidence has now accumulated to suggest that a reasonably large number of the proteins involved in gene expression and its control have evolved to incorporate one or more zinc ions. While the zinc confers structural stability, the metal ion also induces substantial folding of the polypeptide chain, a folding required for sequence-specific binding to DNA. Significant metal ion-induced folding has not been a prominent feature of zinc sites in metalloenzymes, but appears to be a frequent feature of the zinc-containing DNA-binding proteins. Zn(II) is the ideal biologically active metal ion to place in locations close to the genome, since it is not redox active and therefore not likely to generate electron-rich species prone to damage the polynucleotide structure. There are examples of specific nucleotide-binding proteins which carry redox active metal ions, a Cu(I) cluster in the case of the transcription factor for the yeast metallothionein gene (Furst et al., 1988; Furst and Hamer, 1989) and a Fe(II)-sulfur cluster in the case of the protein (cytoplasmic aconitase) recognizing the specific RNA secondary structure known as the iron-responsive elements (IRE's) found in the mRNA's for ferritin and transferrin (Rouault et al., 1992; Haile et al., 1992). These can, however, be considered as special cases where the gene product under control is directly involved in the metabolism of these metal ions.

II. ZINC-CONTAINING DNA-BINDING DOMAINS OF EUKARYOTIC TRANSCRIPTION FACTORS COMPARED TO THEIR TRANSACTIVATING DOMAINS

Investigations of the zinc-containing transcription factors have naturally focused on the structure of the zinc domains, since these domains coincide with their DNA-binding and nucleotide sequence recognition functions. The structural studies have made rapid progress using relatively small subdomains of these transcription factors which contain little more than the intact DNA-binding domain. For example, all of the 2D NMR determinations of the "zinc-finger" structure in solution are of single "zinc-fingers" induced to fold by relatively large Zn:peptide ratios (Parraga et al., 1988; Lee et al., 1989; Omichinski et al., 1990). Generally these single fingers do not recognize DNA sequences specifically (Parraga et al., 1988). In contrast, the native transcription factors are large proteins, 400-2800 amino acids, in which the polypeptide chain is involved in interactions with other proteins, most frequently some component of a large transcription complex consisting of multiple polypeptides. A given transcription factor often interacts with other control proteins as well. In order to place the structural work on the DNA-binding and sequence-recognition domains in its appropriate context, a summary of the functional domains, for appropriate examples of complete zinc-containing transcription factors, diagrammed in terms of the positions of these domains in the linear polypeptide sequence, is given in Figure 1.

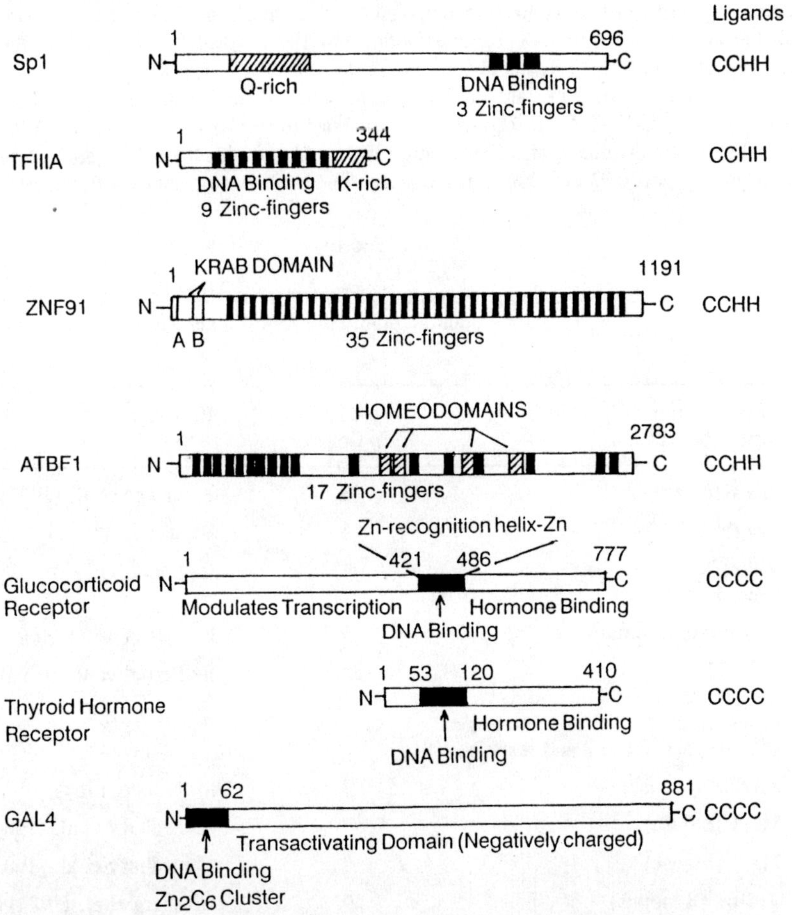

Figure 1.
Diagrams of the functional domains of zinc-containing eukaryotic transcription factors expressed as positions along the polypeptide chains.

III. TRANSCRIPTION FACTORS CONTAINING THE ZINC-FINGER MOTIF AS FOUND IN TFIIIA

The interactions of the first zinc-finger protein to be discovered, TFIIIA, with nucleic acid are somewhat unusual in that the protein not only recognizes the internal control region of ~ 35[2] bp in the center of the 5S RNA gene, but TFIIIA itself is bound to the product, 5S RNA, to form the 7S RNA-protein particle. Both the 7S particle and the RNA-free TFIIIA protein were initially reported to contain 2 Zn ions per molecule (Hanas et al., 1983; Wu, 1986). The zinc was resistant to removal by chelating agents in the 7S particle, but could be readily removed from the free protein (Wu, 1986). It was necessary to add zinc to the apoTFIIIA in order to reconstitute the 7S particle. It was also demonstrated by DNAase footprinting that only the Zn-containing TFIIIA bound specifically to the 5S RNA gene; removal of zinc from the protein abolished specific binding to the gene.

When the amino acid sequence of TFIIIA became available as translated from the nucleotide sequence of the TFIIIA gene, Miller et al. (1985) recognized the 9 approximate repeat amino acid sequences within the protein which contained the two C and two H

---

2 The site size for the binding of TFIIIA was originally reported as 35 bp, but recent data and models suggest it may be up to 45 bp.

residues now proved to contribute the ligands to a zinc ion. The 9 approximate repeat polypeptide sequences of ~ 30 residues within the TFIIIA protein are based on the amino acid sequence: $C\text{-}X_{2\text{-}5}\text{-}C\text{-}X_{12\text{-}13}\text{-}H\text{-}X_{3\text{-}4}\text{-}H$. Within the $X_{12\text{-}13}$ spacers there is an additional pair of conserved residues, an aromatic amino acid, F or Y, and a branched aliphatic, usually L; i.e. $\text{-}C\text{-}X_3\text{-}F\text{-}X_5\text{-}L\text{-}X_{2\text{-}3}\text{-}H\text{-}$. Zinc analysis of the 7S particle showed it to contain 7-11 zinc atoms (Miller et al., 1985), more in agreement with the 9 "fingers" present than the original report of 2 zinc ions per molecule. The precise stoichiometry or the reasons for

Table 1

Zinc-containing Transcription Factors of the TFIIIA type
A. Proteins Containing Only the Zinc-Finger Motif

Transcription factor (Source)	No. of Zn fingers	Reference
Sp1 (Human)	3	Kadonaga et al. (1987)
Zfy (Human)	13	Page et al. (1987)
Wilms' tumor gene (Human)	4	Call et al. (1990)
GLI (glioblastoma)[2]	5	Kinzler et al. (1988)
ZNF 91 (one of 300-700 related human genes which code C_2H_2 zinc-finger proteins, see Table 2 and text)	35	Bellefroid et al. (1993)
Zif268[1] (Mouse)	3	Milbrandt (1987)
Mkr1 (Mouse)	7	Chowdhury et al. (1987)
Mkr2 (Mouse)	9	Chowdhury et al. (1987)
TFIIIA (Xenopus)	9	Ginsberg et al. (1984)
Xfin (Xenopus)	37	Ruis et al. (1987)
p43 (5S RNA Binding Protein) (Xenopus)	9	Joho et al. (1990)
Serendipity β (Drosophila)	6	Vincent et al. (1985)
Serendipity δ	7	Vincent et al. (1985)
Kruppel (Drosophila)	5	Rosenberg et al. (1986)
Snail (Drosophila)	5	Boulay et al. (1987)
Krh (Drosophila)	6	Schuh et al. (1986)
Hunchback (Drosophila)	6	Tautz et al. (1987)
Terminus (Drosophila)	1	Baldarelli et al. (1988)
CF2 (Drosophila) (Three forms, reflecting alternate RNA splicing events)	3 6 7	Hsu et al. (1992)
TRS-1 (Trypanosome)	5	Pays and Murphy (1987)
ADR1 (Yeast)	2	Hartshorne et al. (1986)
SW15 (Yeast)	3	Stillman et al. (1988)

(cont'd)

Table 1 (continued)

B. Proteins Containing Both Zinc Fingers and Homeodomain DNA-binding Motifs

Transcription factor (Source)	No. of Zn fingers	No. of homeodomains	Reference
Zfh-1 (Drosophila)	9	1	Fortini et al. (1991); Lai et al. (1991)
Zfh-2 (Drosophila)	16	3	Fortini et al. (1991); Lai et al. (1991); Lundell and Hirsh (1992)
ATBF-1 (Human)	17	4	Morinaga et al. (1991)

[1]This factor is also known as Krox-20, NGFI-A and Egr 1. Egr 1 has recently been shown to be required for differentiation of macrophages (Nguyen et al., 1993). The human Egr 1 gene is located on chromosome 5 in a region often deleted in humans with therapy-induced acute myeloid leukemia (Sukhatme et al., 1988; Nagarajan et al., 1990). The three fingers in these proteins also bear close resemblance in structure and DNA recognition properties to another nerve growth factor-induced early-response gene NGFI-C (Crosby et al., 1991) and to Egr 2 (Joseph et al., 1988).

The protein coded for by the DNA sequence from the Wilms' tumor locus on human chromosome 11 shows significant homology to the Egr 1 and Egr 2, but contains 4 rather than 3 zinc fingers (Call et al., 1990).

[2]The GLI protein is a zinc-finger protein expressed in embryonyl carcinoma cells but not in most adult tissues. Its expression has been found to be 75-fold amplified in a cell line derived from a glioblastoma (Kinzler et al., 1988).

variable metal ion stoichiometry in different TFI!IA preparations remains a matter of debate, but more recent analytical data still show significant differences in zinc:protein stoichiometry from different laboratories (Shang et al., 1989). Extended X-ray absorption edge spectroscopy (EXAFS) shows the Zn absorption edge in TFIIIA to be fit by a tetrahedral model with 2S and 2N atoms as ligands (Diakun et al., 1986).

A. Solution structures of zinc fingers

There are a number of solution structures of single C_2H_2 zinc-fingers determined by 2D NMR methods (Parraga et al., 1988; Lee et al., 1989; Omichinski et al., 1990; Neuhaus et al., 1990; Kochoyan et al., 1991). These include single fingers from ADR1 (Parraga et al., 1988), Xfin (Lee et al., 1989), Zfy (Kochoyan et al., 1991), human enhancer binding protein (Omichinski et al., 1990), and several "mutant" fingers produced by substituting residues in the native sequence. Those transcription factors containing one or more copies of this motif are listed in Table 1. The finger consists of an N-terminal short antiparallel β-sheet of 9-11 residues including the β-turn encompassing the $C\text{-}X_{2\text{-}4}\text{-}C$ group, a second turn at the top of the finger followed by ~ 3 turns of α-helix (9 residues), although some of this helix is a 3_{10} helix in some zinc fingers (see X-ray structure below). The Zn is held in a tetrahedral complex by the two $\text{-}S^{-}$ ligands from the C residues and the N3 nitrogens of the imidazole side chains of the $H\text{-}X_{2\text{-}5}\text{-}H$ sequence toward the C-terminal end of the helix. The two conserved aromatic residues, Y (or F) and L, were discovered to be in contact forming a hydrophobic interaction or sort of "strut", important for stabilizing the distal portion of the finger. This fold of the zinc-finger was predicted on theoretical grounds by Berg (1988) prior to completion of any of the solution structures. The amino acid sequences of several zinc fingers found in native proteins, chosen to illustrate the variation in spacing found among the C_2H_2-type of ligand structure, are collected in Table 2.

Table 2

Variation in Amino Acid Sequences of Zinc Fingers Found in Proteins Generally Containing the C_2H_2-type of Motif as Found in TFIIIA

TFIIIA[1]	Finger 1	CSFAD-C GAAYNKNWKLQ-A HLC-KH
	Finger 6	CKKDDSC SFVGKTWTLYLK- HVAECH
	Finger 7	C---DVC NRKFRHKDYLR-D HQK-TH
Zfy	Finger 1 (odd)	CMIC GKKFKSRGFLKR HMKNH
	Finger 2 (even)	CKFC EYETAEQGLLNR HLLAVH
SW15 (Yeast)	Finger 1	CLFPGC TKTFKRRYNIRS HIQTH
	Finger 3	CP---C GKKFNREDALVV HRSRMIC
Kruppel (Drosophila)	Finger 1	CKIC SRSFGYKHVLQN HERTH
	Finger 2	CPEC DKRFTRDHHLKT HMRLH
ZNF91[2] (Human)	Finger 1	CKKC VKSFCIRL---- HKTQH
	Finger 3	CEEC GKAFKQLSTLTT HKIIC
	Finger 6	CEEC GKAFSHSSALAK HKR1H
	Finger 35	CEKC CKAFNQSSILTN HKKIH

[1]The dashes are deletions. These fingers illustrate the spectrum of length of spacers between ligands as well as a possible change in ligand type (Finger 2, SWI5; Finger 3, ZNF91) found in "classical" zinc fingers of the TFIIIA-type.

[2]The genes for several hundred human zinc-finger proteins have been identified by hybridization of cCNA or genomic DNA libraries with a probe specific for the "Kruppel-like" zinc-finger spacing (see text). ZNF91 is one of these containing 35 fingers as determined by the DNA sequence (Bellefroid et al., 1993). While most of its "fingers" are the Kruppel-type, there are interesting deviations in loop size in finger 1 and putative ligands in finger 3. The $C\text{-}X_2\text{-}C\text{-}X_{12}\text{-}H\text{-}X_3\text{-}H$ arrangement is preserved in 33 of the 35 fingers of ZNF91, a spacing also found in the odd fingers of Zfy.

B. Crystal structure of three zinc fingers, the Zif 268-DNA complex

The three C_2H_2-type zinc fingers of the mouse transcription factor Zif 268 (73 amino acids corresponding to residues 349 to 421 of the native protein) have been crystallized in a complex with a 12 bp DNA fragment carrying the 9 bp sequence recognized by Zif 268 (Pavletich and Pabo, 1991). The protein fragment was cloned, overproduced, extracted with 6.4 M guanidinium-HCl and refolded in the presence of Zn or Co. The structure of the complex was solved at 2.1 Å resolution (Fig. 2A). The three "zinc fingers" are inserted sequentially into the major groove of the DNA such that if the complex is viewed down the axis of the DNA double helix, the protein appears as a C-shaped clamp surrounding the DNA. The N-terminal end of the α-helix projects into the major groove, but the axis of the α-helix is only approximately aligned with the major groove and is tipped at ~ 45° with respect to the plane of the base pairs. The result is that the zinc ions and their ligands are located outside the 20 Å perimeter of the DNA. The N-terminal end of each α-helix contributes the side chains which make the specific contacts with the base pairs. The helix of each zinc finger interacts with a 3 bp subset of the 9 bp recognition sequence. The short β-sheet of each zinc finger is folded on the back of the helix away from the base pairs. The N-terminal strand of each β-sheet makes no contact with the DNA, while the more C-terminal strand contacts the sugar phosphate backbone along one DNA strand.

The consensus nucleotide-binding site for Zif 268 is 5'-GCGGGTGCG-3'. The three fingers of the protein make a total of 11 hydrogen bonds between the three zinc fingers and

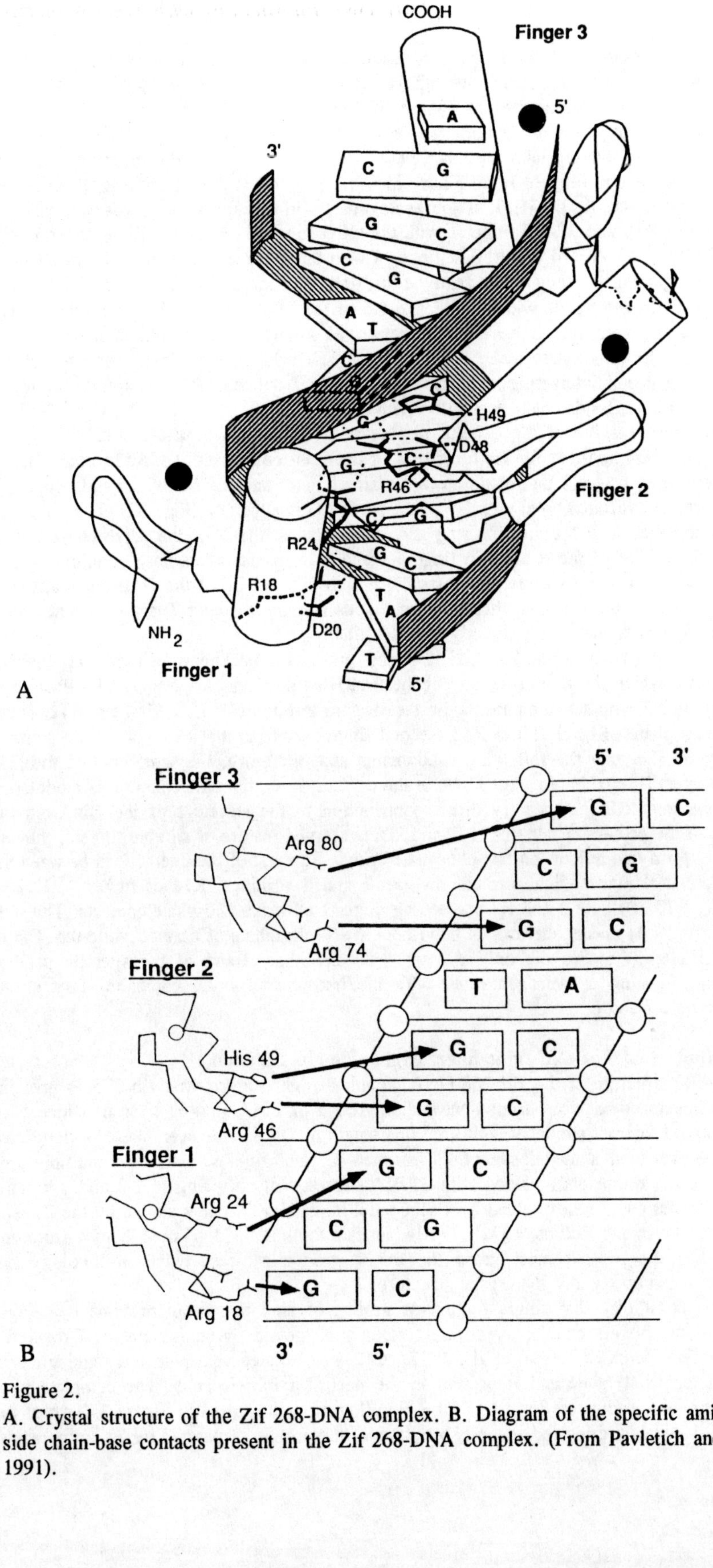

Figure 2.
A. Crystal structure of the Zif 268-DNA complex. B. Diagram of the specific amino acid side chain-base contacts present in the Zif 268-DNA complex. (From Pavletich and Pabo, 1991).

the bases of the DNA, all within the major groove. All of these are made with one strand of the DNA, the G-rich sequence shown above. The contacts between the protein side chains and the bases are diagrammed in Figure 2B. On each finger, the residue immediately preceding the start of the α-helix and the second, third and sixth residues of the α-helix are critical to the interaction with the DNA bases, although details are not precisely the same for all three fingers. Fingers 1 and 3 have identical residues at these positions, R, D, E, R, while finger 2 has R, D, H, T at the analogous positions. Each finger has two primary interactions, R18 and R24 of finger 1 with the GCG subsite, R46 and H49 of finger 2 with the GGT subsite and R74 and R80 of finger 3 with the GCG subsite (Fig. 2B). In all three fingers the guanidino group of the arginine just preceding the start of the α-helix, R18, R46 and R74, hydrogen bonds with the N7 and O6 of the G at the 3' end of the subsite. The second residue within each α-helix, the conserved aspartic acid, is placed such that the oxygens of the carboxylate make hydrogen bond-salt bridges with the N_ε and N_η of the guanidinium group interacting with the first G of each subsite. This apparently serves to stabilize the long R-side chain interaction with the base.

The third residue of the α-helix in fingers 1 and 3 is a glutamic acid. While this residue may be important for conformation, it does not contact the DNA. In contrast, the third residue in finger 2 is a histidine (not a zinc ligand) and the N3 of the imidazole side chain forms a hydrogen bond with N7 of the second G in subsite 2 (Fig. 2B). The imidazole ring is coplanar with the guanine ring and stacks against the T of the adjacent base pair. The sixth residue of the α-helix in fingers 1 and 3 is an arginine whose guanidino group again makes a typical hydrogen-bond structure with N7 and O6 of the G residues at the 5' ends of the first and third subsites. The analogous residue in finger 2 is a threonine which does not appear to participate directly in DNA binding.

The contacts between Zif 268 and the phosphate backbone include one unusual structure in which the imidazole ring of the first His-ligand in each finger, coordinating to Zn through N3, contacts the phosphodiester oxygens through N1. H25 (first finger) contacts the 5'-phosphate of bp 7, while H53 (second finger) contacts that of bp 4. These contacts actually overlap into the following subdomains and, therefore, the same contact for H81 (third finger) is outside the short DNA in the crystal. Thus, the zinc complex is modulating the DNA interaction by a very direct contribution to the structure of the DNA-binding surface of the protein. Since this H-ligand is a conserved feature of all zinc fingers, they all probably form this unique Zn-His-phosphate bridge. The rest of the interactions between the phosphate backbone and the protein involve R and S residues. R14 of finger 1, R42 of finger 2, R70 of finger 3 and R3 preceeding finger 1 all make phosphate contacts. The -OH of S45 (finger 1) makes a hydrogen bond with the 5'-phosphate of base 6, while the -OH of Ser75 (finger 3) makes the only contact with the other strand of the specific dsDNA sequence, forming a hydrogen bond with the phosphate on the complementary strand between bp's 7 and 8.

C. Multiple modes of C_2H_2 zinc finger expression and function

The crystal structure of the Zif 268-DNA complex might suggest that zinc-finger proteins with n fingers could wrap around the major groove of DNA in a manner in which each finger would interact similarly with the same length of DNA. However, there is significant evidence that such a model cannot be assumed for all proteins containing multiple zinc fingers. DNA footprints obtained with TFIIIA suggest that only fingers 1-3 and 7-9 show footprints that could be explained by a sequential interaction like that shown by the Zif 268 complex (Hayes and Tullius, 1992). TFIIIA appears to bind the DNA double helix from one side. Thus the protein must cross the minor groove at least twice, a topology not accommodated by the Zif 268 crystal structure.

The possibility that single fingers or groups of zinc fingers in proteins containing multiple zinc fingers could play different roles is suggested by studies on the Drosophila transcription factor CF2 (Hsu et al., 1992). CF2 was isolated as a potential regulator of a chorion (eggshell) gene and is present in the nuclei of follicle cells. The complete gene contains coding sequences for a total of 7 zinc fingers as illustrated in Figure 3. Two of the fingers, 1 and 2, are located at the N-terminus of the protein, separated by 29 amino acids.

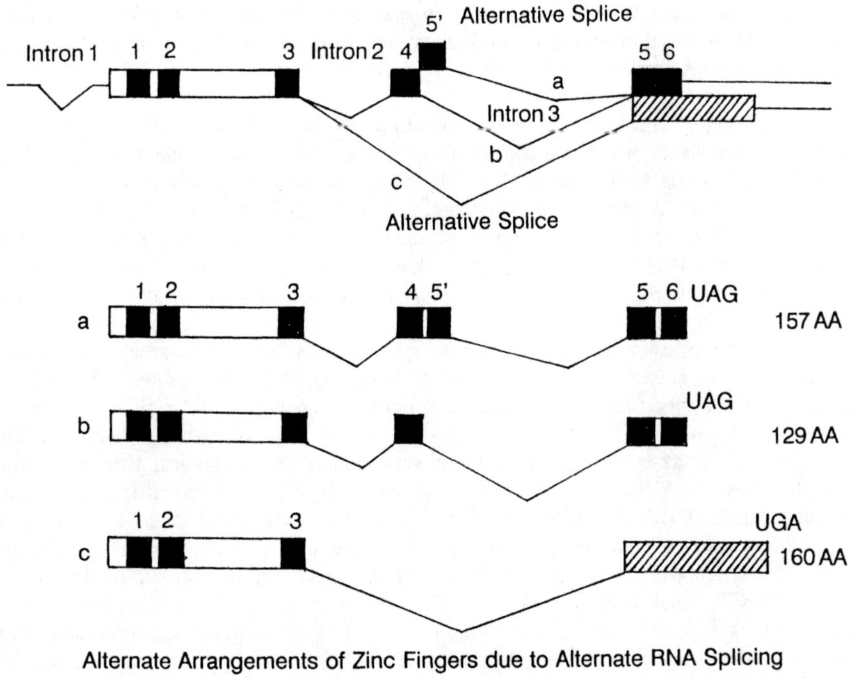

Figure 3.
Genomic structure of the Drosophila chorionic transcription factor, CF2 (top), indicating the three introns and the two alternative RNA splicing patterns, a and c, which exist. The three possible RNA splices give proteins with the length and zinc finger domain distribution shown in a, b and c (below). The numbered black boxes indicate the zinc finger domains (see text for details). The data is diagrammed from Hsu et al., 1992.

Surprisingly, deletion studies show that these N-terminal fingers appear to have no effect on the binding of CF2 to the chorion gene promoters it regulates (Hsu et al., 1992). Similarly, no evidence for recognition of a specific DNA sequence by finger 3 has been found. DNA recognition appears to reside exclusively in the C-terminal zinc fingers.

The story is even more complex, however, in that the coding region of the gene is interrupted by two introns (Fig. 3), one splicing out most of the RNA between the codons for fingers 3 and 4 (intron 2) and a second splicing out most of the RNA between the codons for fingers 4 and 5 (intron 3). These two introns introduce a potential for an alternative splice of the hnRNA from the donor of intron 2 to the acceptor of intron 3. Thus, in addition to the protein containing fingers 1, 2, 3, 4, 5, and 6, there could be a form containing fingers 1, 2, 3, 5 and 6. However, the exclusion of the exon coding for finger 4 by the alternate donor-acceptor pair also results in a frame shift downstream of the splice. Hence, the coding sequence for fingers 5 and 6 is not coded in the mRNA, and this alternate form of the protein will contain only fingers 1, 2 and 3. The sequence of the hnRNA introduces an additional complexity, since the 5'-end of intron 3 contains a sequence encoding a 7th finger, 5', which is followed by an alternative splice donor consensus sequence. Thus a third form of the protein is possible if this alternate splice donor site were used instead of that upstream of the coding sequence for 5'. The resultant protein would contain fingers 1, 2, 3, 4, 5', 5 and 6. Isolation of the polyadenylated RNAs from the appropriate Drosophila cells suggests that all three of the proposed mRNA variants are produced, and the relative quantities of the three mRNA's appear to be both cell specific and developmentally regulated (Hsu et al., 1992). It is possible that fingers 1, 2 and 3, rather than participating in recognizing a specific DNA sequence, help stabilize the interaction by relatively non-specific interactions with DNA which facilitate the specific recognition provided by the C-terminal fingers. It seems clear that a single model of protein-DNA interaction will not accommodate all zinc finger-DNA recognitions, even if limited to proteins containing the classical C_2H_2-ligand motif.

D. A "zinc-finger" transcription factor containing C_2HC as the ligands

A transcription factor whose Zn-containing structural motif appears similar to the TFIIIA-type but with a C_2HC instead of C_2H_2 ligand arrangement has recently been cloned by virtue of the ability of the protein to recognize a specific DNA sequence from the upstream region of a glia-specific gene, the myelin proteolipid protein (PLP) gene (Kim and Hudson, 1992). A human fetal brain-derived λgt11 cDNA library was used to derive a library of fusion proteins with β-galactosidase which were then tested for their ability to gel-retard the specific DNA sequence.

A cDNA corresponding to a protein designated as MyTI (myelin transcription factor I) was isolated and cloned into *E. coli* (Kim and Hudson, 1992). The protein of 725 amino acids contains 6 zinc-finger-like sequences grouped as an N-terminal pair and a more C-terminal group of four fingers. The conserved sequence is $-C-X_4-C-X_{12}-H-X_5-C-$. Thus MyTI appears to have a zinc-binding motif very similar to the classical zinc-finger motif first described in TFIIIA, but in which the fourth ligand is C rather than H. Ligand arrangements of CHC_2 and C_2HC for Zn-binding motifs are found in the single-stranded nucleotide-binding protein gene 32 and the retroviral nucleocapsid proteins (see Section VIII). The spacing of ligands in the latter proteins, however, is quite different from that found in MyTI which is similar to the TFIIIA-like zinc fingers. The binding of the MyTI protein to its specific DNA (gel retardation assays) is inhibited by 1,10-phenanthroline, suggesting that a metal ion is required for binding (Kim and Hudson, 1992). The MyTI gene is expressed predominantly in the nervous system at embryonic day 15 in the mouse and 8 months after birth in humans.

E. Transcription factors containing both zinc fingers and homeodomains

The genes for three proteins have now been cloned and sequenced which show coding regions for both the classical C_2H_2 zinc-finger motif and one or more homeodomains (the helix-turn-helix motif first described for transcription factors from Drosophila and similar to the helix-turn-helix motif in prokaryotic proteins such as CAP). Two of these, zfh-1 and zfh-2, were isolated from a Drosophila λgt11 retinal cDNA expression library using radiolabeled oligonucleotides containing the sequences for cis-acting elements of the opsin gene promoter (Fortini et al., 1991; Lai et al., 1991). Zfh-1 contains 9 zinc fingers and one homeodomain, while zfh-2 contains 16 zinc fingers and 3 homeodomains. Zfh-2 was isolated independently as a factor binding to the upstream region of the gene for DOPA decarboxylase (Lundell and Hirsh, 1992). The zfh-1 gene is expressed in the mesoderm of early embryos and the mesodermally derived structures of late embryos including the dorsal vessel, support cells of the gonads, the segment-specific arrays of adult muscle precursors, and most motor neurons of the developing CNS (Lai et al., 1991). Zfh-2 has been shown to be widely expressed in the Drosophila larval CNS (Lundell and Hirsh; 1992). While it is possible that there are alternative hnRNA processing events that result in protein products shorter than predicted from the complete expression of the gene (3005 amino acids for zhf-2), there is reasonable evidence that for zfh-1 the expected 145 kD complete gene product is expressed (Fortini et al., 1991).

The arrangement of zinc fingers and homeodomains in zhf-2 was discovered to be quite similar to that of the human transcription factor, ATBF1, isolated from a hepatoma cell line (Morinaga et al., 1991). The latter was identified because it binds to an enhancer element of the α-fetoprotein gene. ATBF1 contains 17 zinc fingers and 4 homeodomains. The arrangement of these DNA-binding elements is illustrated in Figure 1.

IV. THE HORMONE RECEPTOR ZINC-CONTAINING TRANSCRIPTION FACTORS

It has been known since the early 1970s that non-peptide hormones like adrenal steroids, estrogens, androgens and the thyroid hormones bind to intracellular proteins (receptors) which ultimately are found associated with chromosomal proteins. The glucocorticoid hormone receptor became a prototype and tracer studies showed that the receptor was initially found in the cytoplasm, but that upon binding of the hormone, the complex moved into the nucleus and became associated with some part of chromosome structure, perhaps involved in direct binding to the DNA. It is now known that the initial cytoplasmic location of the glucocorticoid receptor derives from a complex formed between heat shock protein HSP90 and the unliganded receptor, a complex which dissociates on hormone binding to the receptor. In contrast, the thyroid hormone receptors are found initially in the nucleus. Similar receptor molecules have now been discovered for thyroid hormones, vitamin D, retinoic acid, estrogen, progesterone and androgenic steroids, in addition to the adrenal cortical steroids (Table 2).

The hormone receptors are relatively large proteins, 400-800 amino acids, and recognize specific DNA sequences located upstream from the promoters for the genes which they regulate. As illustrated by two examples in Figure 1, the DNA-binding domain consists of a stretch of ~ 90 amino acids located within the central portion of the polypeptide chain. The amino acid sequence within the DNA-binding domain contains two sequences, $C-X_2-C-X_{13}-C-X_2-C$ and $C-X_5-C-X_{12}-C-X_4-C$, (this example from the vitamin-D receptor) which were postulated initially to form zinc fingers analogous to those of TFIIIA, except with C_2C_2 rather than C_2H_2 contributing the ligands. Both zinc analyses and ^{113}Cd NMR of ^{113}Cd-substituted derivatives of the glucocorticoid receptor showed these regions to bind two Zn or Cd ions (Freedman et al, 1988; Pan et al., 1990). Both solution structures determined by 2D NMR methods (Hard et al., 1990; Schwabe et al., 1990) and the crystal structure of the glucocorticoid receptor-specific DNA complex (Luisi et al., 1991) show the protein fold and the DNA-binding modes of the hormone receptors to be significantly different from those of the "classical" TFIIIA-type zinc finger. In addition, the specific DNA sequences to which the hormone receptors bind, known as hormone responsive elements or HRE's, are found to be palindromic and these transcription factors bind to these DNA sequences as dimers. Thus the spacing determined by the central 2-fold axis of the palindromic DNA becomes an important element for enhancing the specificity of DNA recognition.

A. Solution structures of the glucocorticoid receptor (GR) and the estrogen receptor (ER)

The 2D NMR structures of the DNA-binding domains of the glucocorticoid receptor (Hard et al., 1990) and the estrogen receptor (Schwabe et al., 1990; Schwabe and Rhodes, 1991) are similar to each other and both are reasonably well represented by the ribbon diagram in Figure 4. They consist of two separate, but similar zinc-binding subdomains folded together to form a compact domain. The two helices found in the structure, one from the N-terminal zinc subdomain and one from the C-terminal zinc subdomain are packed against each other at right angles. The helices are amphipathic with the side chains of their hydrophobic faces packing the core of the domain, while the surface of the vertical helix facing the reader carries two lysines and an arginine (estrogen receptor) that appear to be involved in DNA binding.

The 4 C residues of the N-terminal half of the estrogen receptor domain and 4 of the 5 C residues in the N-terminal half of the glucocorticoid receptor domain form one tetrahedral Zn complex. Four of the 5 C residues in the C-terminal half of both the ER and GR form the second tetrahedral Zn complex. The presence of C residues within these sequences which are not ligands suggests caution in assigning the ligand arrangement from sequence data alone. Even in the NMR structures there can be potentially some doubt, since the NMR data do not directly supply information on the Zn coordination. The possible alternate structures must be entered in the structure determination and examined to see if one converges better than the other as was done for the GR fragment (Hard et al., 1990). ^{113}Cd-substitution can give direct information on the coordination arrangement by the detecting $^{113}Cd-^{1}H$ coupling (see GAL4 studies below).

In both zinc subdomains of a hormone receptor, the α-helix begins with the residue following the third C ligand and continues for 11 to 13 residues. The loops of peptide between the pairs of C ligands, originally thought to be fingers analogous to those of the

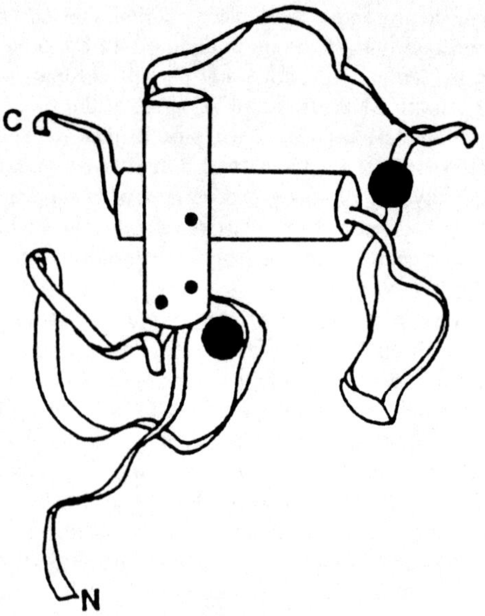

Figure 4.
Sketch of the solution structure of the 84 residue DNA-binding domain of the estrogen receptor as determined by 2D NMR methods. Sketch of the structure as presented by Schwabe et al., 1990.

TFIIIA type, form irregular folded structures on either side of the central core formed by the packed helices (Fig. 4). In the domain from ER, the N-terminal loop is 13 residues in length and well structured (Hard et al., 1990; Schwabe and Rhodes, 1991), while the analogous C-terminal loop consists of only 9 residues and is less well structured. As the crystal structure of the DNA complex of the GR reveals, these loops, in contrast to those of the C_2H_2-type zinc fingers, do not form the major DNA-binding surface. This conclusion was indicated earlier by site-directed mutagenesis. Site mutation showed that residues E25, G26 and A29 of ER (see Table 2) and an analogous triple in GR (G458, S459 and V462) determine whether the receptor will recognize the palindromic ERE or the GRE DNA sequence (Mader et al., 1989; Umesono and Evans, 1989; Danielsen et al., 1989)). These residues are in the so-called "knuckle" of the first zinc complex. These three residues begin the N-terminal helix. The crystal structure of the GR-DNA complex shows, however, that with the exception of V462 these are not the three residues that interact directly with the bases in GR. Rather the side chains of K461, V462 and R466 outside the "knuckle" make the interactions of GR with three specific bases of the GRE (see crystal structure below). However, this apparent discrepancy between the site mutagenesis data and the crystal structure of the GR complex appears to be explained by the finding that the carboxyl of E25 (a first "knuckle" residue in ER) does indeed interact with and modify the nature of the hydrogen bonding between the ε-NH_2 of K28 (corresponding to K461 in the GR) and the G and C bases of the ERE. The replacement of T's 5 and 6 in the GRE by A and C in the ERE also accommodates the change in the water-hydrogen bonding interaction present with K461 in the GR to that involving E25 without the water in the ER (Xu, W.X., Freedman, L.P., and Sigler, P.B., in preparation).

The "knuckle" region of the second zinc complex, residues P44 to Q48 in the ER, were found to be important in enabling the protein to discriminate between the different half-site spacings in the ERE and TRE DNA (Umesono and Evans, 1989). This must result from the fact that this region forms the dimerization domain which also makes a contribution to the recognition of the correct spacing of the palindrome in the DNA. Swapping of this "second finger" sequence forming this dimerization domain between the

Table 3

Amino-acid Sequences Between the Zinc Ligands in the Hormone-Receptor Proteins

Hormone receptor[1]	Amino-acid sequence	Reference
GR	C LV C SDEASGCHYGVLT C GS C **KV**FFK**R**A C AGRND C IIDKIRRKN C PA C	Hollenberg et al. (1987)
MR	C -- C G----------V- C -- C KVFFKRA C ----- C -V-------- C -- C	Arriza et al. (1987)
PR	C -I C G------------ C -- C KVFFKRA C ----- C --------- C -- C	Misrahi et al. (1987)
		Huckaby et al. (1987)
AR	C -I C G---------A-- C -- C KVFFKRA C -S--- C T---F---- C -S C	Chang et al. (1988)
		Lubahn et al. (1988)
ER	C A- C N-Y---Y----WS C EG C KAFFKRS C PAT-Q C T---N---- C Q- C	Walter et al. (1985)
TR	C V- C G-K-T-Y--RCI- C EG C KSFFRRT C TYDGC C V----T-NQ C QL C	Sap et al. (1986)
		Weinberger et al. (1986)
erbA[2]	C V- C G-K-T-Y--R C IT C EGCKSFFRRT C TYDGC C V----T-NQ C QL C	Debuire et al. (1984)
VitD	C G- C G-R-T-F-FNAM- C EG C KGFFRRS C PFDG- C K-T-DN-RH C E- C	McDonnell et al. (1987)
RAR	C FV C Q-KS--Y----SA C EG C KGFFRRS C HRDKN C --N-VT-NR C QY C	Giguere et al. (1987)
		Petkovich et al. (1987)

[1]The dashes indicate residues that are identical to those found in GR and spaces indicate deletions compared to the GR sequence. The sequences do not include the amino acids outside the zinc domains except for the five residues just distal to the first finger. The latter include the residues known to interact specifically with the bases in the GR (**K, V and R**). The two residues between the second pair of Cys ligands in the first domain (GS in the GR and EG in the ER), begining the recognition helix are also thought to be involved in specific sequence recognition, since the GS to EG mutation is associated with a switch from GRE to HRE recognition (Danielsen et al., 1989). While the crystal structure of the GR complexed with the GRE DNA does not suggest a very direct interaction of the GS side chains with the DNA, this is probably not true of the EG in the estrogen receptor complex, thus accounting for the switch (see text). GR=glucocorticoid receptor, MR=mineralocorticoid receptor, PR=progesterone receptor AR=androgen receptor, ER=estrogen receptor, TR=thyroid hormone receptor, erbA=oncogene from the avian erythroblastosis retrovirus with much amino acid sequence shared by TR, Vit-D=vitamin D receptor, RAR=rentenoic acid receptor.

[2]The erbA oncogene appears to have an insertion of three residues, EGC, before the KSFFRRT sequence. The latter appears to correspond to the --FFRR- or --FFKR- sequence that follows immediately after the 4th C residue in all the hormone receptors.

GR and ER, however, does not result in loss of the original specificity, in contrast to the swap of the three residues in the "knuckle" of the N-terminal helix (Green et al., 1988). DNA footprinting studies of the bound ER and GR, either as the entire receptors or their isolated DNA-binding domains, show that dimers of the ER and GR contacted two adjacent major grooves on one face of the DNA helix and do so with exact rotational symmetry (Tsai et al., 1988; Klein-Hitpass, 1989). This conclusion was modeled by Hard et al. (1990) using their NMR-determined solution structure of the GR with the base contacts being made with side chains from the N-terminal helix. The major features of the model were confirmed by the crystal structure of the GR-DNA complex (see below). The model building suggested that the C-terminal zinc-domain and its less well ordered 9 residue loop must be involved in dimerization of the two monomers bound to the palindromic sequence, since placement of the N-terminal helices of two monomers symmetrically in adjacent major grooves resulted in close contact of the two C-terminal zinc domains.

The overall structure of the DNA-binding domains of the steroid receptors can be described as a stretch of polypeptide forming a 3-4 turn α-helix (the DNA recognition helix) followed by a 5-7 residue strand (shown to be a β-strand in the crystal structure). This structure is anchored at both ends by the Zn complexes whose associated loops form supporting structures rather than direct DNA-binding surfaces. The C-terminal Zn subdomain also constitutes the dimerization domain. Parts of these Zn loops, however, are important to the DNA interactions as shown by the crystal structure of the GR complex (see below). The C-terminal helix then projects back to support the DNA recognition helix and helps form the hydrophobic core. This motif has been termed a "zinc twist" (Vallee et al., 1991) to distinguish it from the original zinc finger. The amino acid sequences of the immediate N- and C-terminal Zn domains found in the family of hormone receptors are collected in Table 3.

B. Crystal structure of the DNA-binding domain of the glucocorticoid receptor complexed with the DNA of the glucocorticoid responsive element (GRE)

Crystal structures have been determined for a 86 amino-acid residue fragment of the GR (residues 440 through 525 of the rat GR) complexed to two different oligonucleotides containing a consensus palindromic sequence corresponding to the GRE (Luisi et al., 1991). The two oligonucleotides differ in that one (GRE_{S3}) contains a 3 bp spacer between the half-sites of the palindrome, while the other oligo (GRE_{S4}) contains a 4 bp spacer between the half-sites. The latter spacing, one bp larger than a natural GRE, has exact 2-fold symmetry. This was expected not to influence the binding, since the 86 amino-acid subdomain is a monomer in solution up to mM concentrations as proven by the 2D NMR studies (Hard et al., 1990). Thus, two monomers were expected to bind noncooperatively but slightly farther apart to the oligonucleotide containing the 4 vs the 3 bp spacer. The GRE_{S4} complex has been solved to 2.9 Å resolution, while the GRE_{S3} complex is currently solved to 4.0 Å resolution (Luisi et al., 1991). Comparison of the two structures gives a surprising result. While each of the oligonucleotides complexes two molecules of the 86 amino acid GR fragment, only the GRE_{S3} complex shows a strictly symmetrical interaction of the protein dimer with the palindrome. In the case of the complex of GRE_{S4}, for which the resolution of 2.9 Å provides significant detail, only one monomer of the protein dimer interacts with the half-site in a highly specific manner. Because of the 4 rather than 3 bp spacing, the other protein monomer is "out of register" and is forced to dock with the other half of the palindrome as if there were a 3 bp spacer rather than a 4 bp spacer. The strong monomer-monomer protein interaction occurs on the DNA despite the failure of the protein to dimerize in solution. When at least one monomer interacts in the correct way with a specific DNA half-site, a "tight" dimer is induced to form which then apparently dictates a 3 bp spacing between half-sites. Since the complex with the GRE_{S3} is arranged symmetrically, it shows that the protein has assumed at both halves of the palindrome a binding mode characteristic of only one of the two halves of the GRE_{S4} complex. The half-site interaction common to both S3 and S4 complexes clearly shows the specific base-protein interactions.

The general structure of the complex between the dimer of the 86 residue GR subdomain and the GRE_{S4} is shown in Figure 5. The N-terminal helices of the two

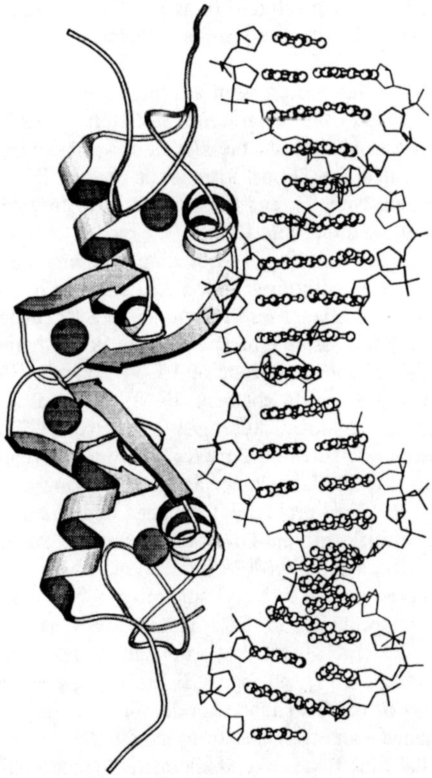

Figure 5.
Representation of the crystal structure of the dimer of the 86 residue DNA-binding domain of the glucocorticoid receptor bound to the GRE DNA sequence. This figure is drawn from the coordinates deposited in the Brookhaven data bank representing the structure published by Luisi et al., 1991. The drawing was produced by Molscript: a program to produce both detailed and schematic plots of protein structures (Kraulis, 1991).

monomers dock from one side of the DNA double helix and lie in adjacent major grooves. In the specific binding mode (the lower part of Fig. 5), the N-terminal helix lies deeper in the major groove which has widened by ~ 2 Å. The N-terminal rather irregular zinc-finger-like structure can be seen at the lower or upper central part of Figure 5 and is similar to that found in the NMR structure. The zinc is tetrahedrally coordinated by the C residues 440, 442, 457 and 460. This amino terminal finger has a short segment of anti-parallel β-sheet which helps to orient some of its residues such that they make contacts with the phosphate backbone. The N-terminal helix begins with the second pair of C ligands to the zinc. This helix contributes two major functions in agreement with the general conclusions of the NMR studies. The hydrophobic face opposite the DNA contributes the side chains of F163 and F464 to the hydrophobic core of the molecule which also receives hydrophobic side chains from the C-terminal helix. The hydrophilic face of this N-terminal helix makes the specific base contacts with the DNA.

The second or C-terminal zinc-finger-like domain has a much better defined structure in the DNA complex than found in the NMR solution structure of the monomers of either the GR or ER DNA-binding subdomains. In the crystal structure it consists of a short β-strand forming the proximal half and a distorted α-helix forming the distal half (seen at the dimer interface in the middle of Fig. 5). The ligands are C476, 482, 492 and 495, not C500. The C-terminal zinc-finger-like structure forms the entire interface responsible for the dimerization of this GR subdomain, largely through contacts of the residues located between C476 and C482, the loop referred to as the "D box" (Umesono and Evans, 1989). Like the first zinc subdomain, it also provides contacts to the phosphates of the DNA backbone.

The N-terminal zinc-sulfur coordination has the S-configuration like the zinc fingers of the TFIIIA type. In contrast, the C-terminal zinc-sulfur coordination complex has the opposite chirality of the ligands around the zinc, i.e. an R-configuration. This chirality is associated with the fact that the D-box assumes a reverse β-turn configuration. The R-stereochemistry of the zinc complex and the reverse turn provide the important structural elements for formation of the dimer interface.

The specific contacts between the DNA-binding subdomain of the GR receptor and the bases of the GRE are all made with three amino acid side chains K461, V462 and R466, on the face of the N-terminal α-helix. These residues interact principally with bases G4, T5, T6 and G7 of the GRE. R466 forms typical hydrogen bonds between the guanidino group and the N7 and O6 of G4. The methyl group of T5 (invariant in GRE's) makes a "glove fit" van der Waals contact with the side chain of V462. G7 is also a constant base in most HRE's and K461 makes one direct hydrogen bond with G7 and one water-mediated hydrogen bond. The same water molecule makes a second hydrogen bond to the O4 of T6. The latter is the second thymine base important in specifying a GR-specific element. It is the absence of T6 in the ERE and probably the change of V462 to the corresponding A29 in the ER that alters the specific protein-DNA interaction of the ER such that E25 in the "knuckle" of ER contributes a carboxyl-K28 interaction, which modifies the base binding of K28 so that it is a more typical carboxyl-supported ε-NH_2 group interaction than one mediated by a water molecule as in GR. This appears to account for the important specificity function of E25 (ER) in shifting from the recognition of GRE to that of ERE DNA (Xu, W.X., Freedman, L.P., and Sigler, P.B., in preparation).

Since the monomers of the GR subdomain do not dimerize in solution, it is interesting to find that the monomer-monomer interface interactions consequent to the additional organization of the distal zinc-finger-like subdomains induced by the DNA result in such strong dimerization forces that this dimer interaction controls the base pair "register" of the whole structure. This finding would correlate well with the observation that the spacer length in the center of the palindromic HRE recognition sequences seems as important as specific base recognition in controlling the discrimination required for recognition of specific upstream activation DNA sequences. In addition to the specific base contacts, the GR subdomain makes an array of phosphate backbone interactions, several of which come from the C-terminal zinc subdomain or dimerization domain.

V. TRANSCRIPTION FACTORS CONTAINING ZINC BINUCLEAR CLUSTERS

The third distinct type of structural motif for a zinc complex found in eukaryotic transcription factors is represented by a set (now 20 examples) of fungal transcription factors. The amino acid sequences for their DNA-binding domains are listed in Table 4. These proteins bind as dimers to palindromic DNA sequences located in the 5'-untranslated region of the genes whose transcription they control. The amino acid sequences of the DNA-binding domains of these transcription factors (determined by deletion mutagenesis) showed two pairs of C-X_2-C sequences separated by ~ 13 residues (Table 4). This observation initially led to the assumption that these were zinc-finger proteins of the TFIIIA class containing a single zinc-finger with C_2C_2 ligands (Johnston, 1987).

Metal binding studies of the cloned fragment consisting of the N-terminal 149 residues of the prototype of this group, GAL4, showed the binding of this fragment to the 17 bp palindromic DNA sequence (UAS_G) recognized by GAL4 to be zinc dependent (Pan and Coleman, 1989). Either Zn or Cd restored specific DNA recognition to the apoprotein. Analysis of homogeneous samples of the GAL4(149) dialyzed against metal-free buffers showed zinc contents from 1.2 to 1.6 zinc ions per mole of protein (Pan and Coleman, 1989; 1990). In contrast, the CdGAL4(149) consistently bound two Cd ions per mole when all Zn was removed (Pan and Coleman, 1989; 1990). The presence of two metal-binding sites on GAL4(149) was confirmed by ^{113}Cd NMR of CdGAL4(149) which showed two ^{113}Cd NMR signals, 669 and 707 ppm downfield of the signal for $Cd(ClO_4)_2$ (Pan and

Table 4

Amino-acid Sequence in the DNA-binding Domains of Fungal Transcription Factors Containing A Cluster of Six Cysteine Residues

Protein	Initial residue #	Sequence of DNA-binding Domain	Reference
GAL4	10	A C DI C RLKKLK C SK--EK-P-K C AK C LKNNWE- C	Laughton and Gesteland (1984)
LAC9	94	A C DA C RKKKWK C SK--TV-P-T C TN C LKYNLD- C	Salmeron and Johston (1986)
PDR1	45	S C DN C RKRKIK C NG--KF-P-- C AS C EIYSCE- C	Balzi et al. (1987)
PUT3	33	A C LS C RKRHIK C P---GGNP-- C QK C VTSNAI- C	Marczak and Brandriss (1991)
QUTA	48	A C DS C RSKKCK C DG--AQ-P-I C ST C ASLSRP- C	Beri et al. (1987)
qa-1F	75	A C DQ C RAAREK C DG--IQ-P-A C FP C VSQGRS- C	Baum et al. (1987)
LEU3	36	A C VE C RQQKSK C DAH-ERAPEP C TK C AKKNVP- C	Friden and Schimmel (1987) Zhou et al. (1987)
amdR	19	A C VH C HRRKKR C DARLVGLP-- C SN C RSAGKTD C	Andrianopoulos and Hynes (1990)
UGA3	16	G C IT C KIRKKR C SE--DK-P-V C RD C RRLSFP- C	André (1990)
ARGRII	20	G C WT C RGRKVK C DL--RH-P-H C QR C EKSNLP- C	Messenguy et al. (1986)
HAP1	63	S C TI C RKRKVK C DK--LR-P-H C QQ C TKTGVA- HLCH	Pfeifer et al. (1989)
MAL63	7	S C DC C RVRRVK C DR--NK-P-- C NR C IQRNLN- C	Kim and Michels (1988)
MAL6R	7	S C DC C RVRRVK C DR--NK-P-- C NR C IQRNLN- C	Sollitti and Marmur (1988)
nirA	42	A C IA C RRRKSK C DG--NL-P-S C AA C SSVYHTT C	Burger et al. (1991)

Coleman, 1989; 1990). Each signal integrated to one Cd nucleus. The magnitude of these downfield chemical shifts implied coordination of each ^{113}Cd(II) to 4 sulfur donor atoms based on ^{113}Cd chemical shifts in proteins of known ligand composition. Since the six conserved C residues are the only sulfur donors within the DNA-binding domains, the presence of a binuclear cluster, Cd_2Cys_6, involving all six cysteine residues with two bridging -S^- ligands was postulated to account for the cadmium data (Pan and Coleman, 1989; 1990). Cloning of the first 62 residues of the protein, GAL4(62), resulted in a soluble monomer showing the same ^{113}Cd NMR spectrum as the $^{113}Cd_2$GAL4(149). A number of ^{1}H-^{113}Cd heteronuclear 2D NMR experiments on $^{113}Cd_2$GAL4(62) have now clearly established the presence in GAL4(62) of the Cd_2Cys_6 binuclear cluster shown in Figure 6. A crystal structure of Cd_2GAL4(65) and the 17 bp UAS_G DNA sequence (contained on a 19 bp fragment) has been determined (data at 3.2-3.8 Å resolution) which confirms the presence of the binuclear cluster (Fig. 7) (Marmorstein et al., 1992).

As expected, GAL4(65) binds to the 17 bp UAS_G DNA sequence as a dimer. Residues 50-64 form a short dimerization helix, each of which is connected by a relatively long linker (residues 41-49) to the compact metal cluster structure. The latter contributes the only specific base interactions with a 5'-CCG-3' triplet, nucleotide residues 6, 7 and 8 outward from the AT at the center of the palindrome. Although GAL4(65) is a monomer in solution, it dimerizes when bound to the DNA. The protein motif forming the metal cluster contains two one and one half turn helical segments, an N-terminal one straddling C11 and C14 and a C-terminal one straddling C28 and 31. The N-terminal helical segment is placed such that its C-terminal end points into the major groove and contributes the specific contacts to the CCG sequence. The main chain carbonyl groups of K17 and K18 are exposed at the end of the helix and that of K17 accepts a hydrogen bond from the N4 of cytosine 8, while that of K18 accepts hydrogen bonds from the N4 nitrogen of both cytosine 8 and cytosine 7. The side chain of K18 donates hydrogen bonds to the O6 of guanine 6 and to the N7 of guanine 7. These are the only specific DNA contacts, although the structure is anchored to the sugar-phosphate backbone by hydrogen bonds made by Q9 and R15 with phosphate 5 and

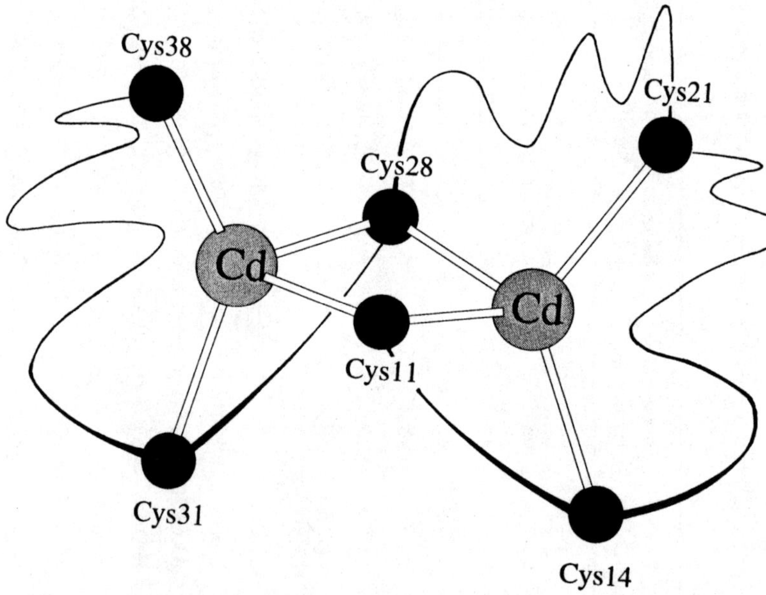

Figure 6.
Cysteine-metal connectivity in the binuclear (Zn or Cd) cluster found in the DNA-binding domain of GAL4. (From Gardner et al., 1991).

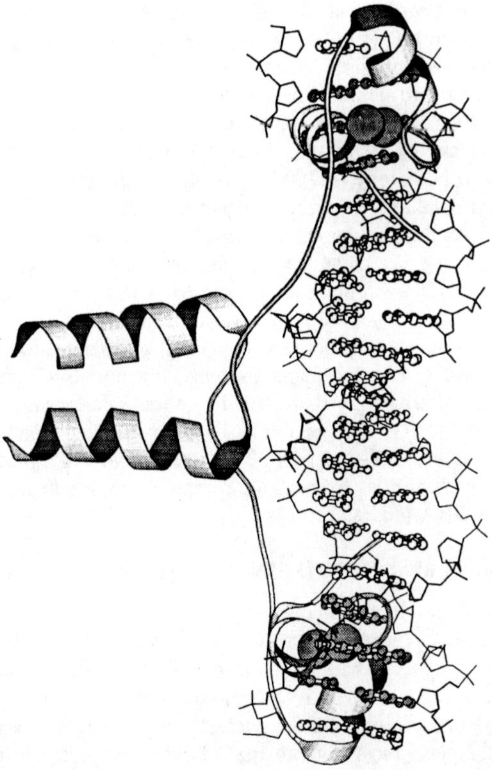

Figure 7.
Representation of the crystal structure of the 65 residue N-terminal DNA-binding domain of GAL4 complexed with the UAS_G DNA, a 17 bp palindromic sequence contained in a 19 bp oligonucleotide. This figure is drawn from the coordinates deposited in the Brookhaven data bank representing the structure published by Marmorstein et al., 1992. The drawing was produced by the program Molscript as described in the caption to Figure 5.

by K20 and K23 with phosphate 6. In addition, a contact is made between phosphate 6 and the main chain NH of C21 just outside the helix (Marmorstein et al., 1992).

While the above constitute the major interactions with the DNA, there are interactions between phosphates 2 and 3 on each side of the AT center and the side chains of the K51's extending to the backbone from the N-terminal end of the dimerization helix. The dimerization helix, directed perpendicular to the axis of the DNA, and the long connecting linker leave almost 11 bp of DNA exposed between the two metal clusters. This suggests the possibility that part of another protein might interact with the DNA or another subdomain of the large GAL4 molecule (881 residues) could loop back and make additional contacts. Solution structures determined by 2D NMR are also completed for Cd_2GAL4(65) (Baleja et al., 1992), Zn_2GAL4(7-49) (Kraulis et al., 1992) and Cd_2LAC9(94-125) (Gardner et al., 1991). All three confirm the presence of a binuclear metal ion cluster.

VI. PUTATIVE C_3HC_4 ZINC-BINDING MOTIFS IN DNA BINDING PROTEINS-TRANSCRIPTION FACTORS

There is a rather diverse group of 13 proteins which have recently been found to share a highly conserved spacing of 7C and 1H residues in the sequence $\text{-C-X}_2\text{-C-X}_{12\text{-}27}\text{-C-X-H-X}_2\text{-C-X}_2\text{-C-X}_{10\text{-}16}\text{-C-X}_2\text{-C}$ which has been proposed to form Zn complexes (Everett, 1988; Brunk et al., 1991; van Lohuizen et al., 1991b). Among these proteins are the murine oncogene, bmi-1 (van Lohuizen et al., 1991a; Haupt et al., 1991), the herpes simplex transcription factor Vmw110 (Everett, 1988) and two drosophila proteins, posterior sex combs and supressor two of zeste (Brunk et al., 1991; van Lohuizen et al., 1991b). Expression of the latter two genes is needed to maintain the expression pattern of certain homeotic selector genes in drosophila and thus is required for the maintenance of segmental determination. These are large proteins from ~ 700 to over 1300 amino acids. They all share a region of high homology, ~ 200 amino acids in length, which includes the above-conserved 7C and 1H residue. The specific sequences containing these potential zinc ligands vary from 41 to 57 residues in length among the 13 proteins and are listed in Table 5. Residues of the loop regions must have undergone deletions for some of them if this arrangement of ligands is evolutionarily connected within this group of proteins. Little structural work is available as yet on members of this group and no zinc analyses have been reported. The possible ligand arrangements suggested by this sequence could be analogous to some of the classical C_2H_2 zinc-finger proteins, but possibly with C_3H and C_4 ligand arrangements as well. Alternatively there could be some relationship to the structure found in the hormone receptors. C-X-H is not a ligand sequence found in Zn-DNA-binding proteins thus far; however, C-X-C is part of the metallothionein ligand arrangement. If this $C_3H_1C_4$ ligand arrangement does form a zinc complex, this group will establish a 5th zinc-binding motif among DNA-binding proteins.

VII. ZINC PROTEINS BINDING TO SINGLE-STRANDED NUCLEIC ACIDS

A. Gene 32 protein from bacteriophage T4

Gene 32 protein, the classic accessory protein required for DNA replication, recombination and repair in bacteriophage T4, was shown to contain one atom of Zn in 1986 (Giedroc et al., 1986). The protein binds cooperatively and non-specifically to the single-stranded DNA at the replication fork and protects it from nuclease attack and helps assemble other proteins of the replication complex, most notably the T4 DNA polymerase (Chase and Williams, 1986, see this review for extensive references). The protein contains three domains. The positively charged N-terminal domain, residues 1-21, is required for cooperative binding to DNA, while the C-terminal, residues 253-275, is negatively charged and interacts with the DNA polymerase. The core region, residues 22-252, contains the single-stranded DNA-binding site and the zinc ion coordinated to C77, H81, C87 and C90 in the sequence shown below (Giedroc et al., 1986).

Gene 32		-C-S-S-T-H-G-D-Y-D-S-C-P-V-C-
HIV NC Protein	1	-C-F-N-C-G-K-E-G-H-T-A-R-N-C-
	2	-C-W-K-C-G-K-E-G-H-Q-M-K-D-C-

Scheme 1

Spectral, and NMR data support this assignment of the zinc ligands (Giedroc et al., 1987; 1989; Pan et al., 1989). One study of site mutations in gene 32 protein suggests that the His ligand is probably H64 rather than H81 (Giedroc et al., 1992). A crystal structure has not yet been obtained on the protein, although crystals have been obtained (Shamoo and Königsberg, personal communication). Studies comparing the binding of the zinc and zinc-free proteins to DNA show that the apo protein binds less tightly to DNA, protects the DNA less well from nuclease digestion, and the cooperativity of binding is significantly reduced (Giedroc et al., 1987; Nadler et al., 1990). Gene 32 protein also functions at the translational level in that it represses the translation of its own mRNA. It does so by binding to a secondary structure present near the 5'-end of the message (McPheeters et al., 1988). It has recently been shown that this binding and translational repression depend critically on the presence of the Zn (Shamoo et al., 1991).

Table 5

Amino-acid Sequences of the Putative Zinc-binding Domains for Proteins Containing A Postulated $C_3H_1C_4$ Zinc-Ligand Arrangement

Su(z)2[1]	IT CRLC RGYMIDPTTVD---------------Y CYH TY CRSC ILKHLLRAV------Y CPEC K
Psc	II CHLC QGYLINATTIV---------------E CLH SF CHSC ILKHLLRAV------F CPRC E
BMI-1	LM CVLC GGYFIDATTII---------------E CLH SF CKTC IVRYLETSK------Y CPIC D
RAD18	LR CHIC KDFLKVPVLT----------------P CGH TF CSLC IRTHINNOP------N CPLC L
RAG1	IS CQIC EHILADPVET----------------S CKH LF CRIC ILRCLKVMGS-----Y CPSC R
HUMRFP	TT CPVC LQYFAEPMML----------------D CGH NF CCAC LARCWGTAETNV---S CPQC R
HUMROS	VT CPIC LDPFVEPVSI----------------E CGH SF CQEC ISQVGKGGGS-----V CPVC R
MUSRPT	VT CPIC LELLKEPVSA----------------D CNH SF CRAC ITLNYESNRNTDGKGN CPVC R
WZBE61	NT CTIC MSTVSDLGKTM---------------P CLH DF CFVC IRAWTSTSV------Q CPLC R
Vmw110	DV CAVC TDEIAPHLRCDTF-------------P CMH RF CIPC MKTWMQLRN------T CPLC N
TLR	GM CAVC REPWAEGAVELL--------------P CMH VF CTAC VVQRW----------R CPSC Q
RING1	LM CPIC LDMLKNTMTTK---------------E CLH RF CSDC IVALRSGNK------E CPTC R
CG30	LQ CNIC FSVAEIKNYLFQPIDRLTIIPVLELDT CKH QF CSMC IRKIRKRKKV-----P CPLC R

1. The dashes indicate deletions compared to CG30 for the first long intervening amino acid sequence and compared to MUSRPT for the second intervening sequence. The underlined and bold letters indicate the postulated residues contributing the zinc ligands as well as two hydrophobic residues, an aromatic residue (Y or F) and a P residue conserved in all these proteins.

2.Su(z) (supressor two of zeste) and Psc (posterior sex combs) are drosophila gene products required to maintain expression patterns of a group of homeotic selector genes (Brunk et al., 1991; van Lohuizen et al., 1991a). BMI-1 is the product of a proto-oncogene that can be activated in mice by the insertion of the Moloney murine leukemia virus (Haupt et al., 1991; van Lohuizen et al., 1991b). RAD18 is the product of a gene, mutations in which cause increased sensitivity to agents which damage DNA (Jones et al., 1988). RAG1 is a gene from humans and mice whose product activates V(D)J recombination (Schatz et al., 1989). HUMRFP is a translated sequence found in an oncogene which results from a protein fusion (Takahashi et al., 1988). HUMROS is a protein component of the Ro antigen (Itoh et al., 1991). MUSRPT is a gene whose product negatively regulates the interleukin-2 receptor gene (Patarca et al., 1988). WZBEG1 is translated from an open reading frame found in the varicella zoster virus genome (Davison and Scott, 1987). Vmw110 is a protein coded for by the herpes simplex virus type I genome which activates the expression of certain other herpes simplex genes (Everett, 1988). TLR is the translation of a trypanosome gene (Smiley et al., 1990). Ring I is the translation of a gene located near the DP region of the human major histocompatibility complex (Freemont et al., 1991). CG30 is the translation of a gene isolated from baculovirus (Theim and Miller, 1989).

B. Retroviral nucleocapsid proteins

The genome of retroviruses code for single-stranded RNA-binding proteins known as nucleocapsid (NC) proteins which are found in the mature virus particle in 2000-3000 copies (Meric et al., 1984). These NC proteins contain one or two copies of an amino acid sequence containing 3C and 1H residue whose spacing suggests that the proteins could form Zn complexes similar to that found in gene 32 protein (Giedroc et al., 1986). These repeats are shown above for the NC protein from the HIV virus. The NC protein of the human immunodeficiency virus (HIV), also known as p15, is a protein of 149 amino acids and is coded for by the 3'-end of the GAG gene of the AIDS virus (Ratner et al., 1985; Guyader et al., 1987).

Several functions have been reported for the retroviral NC proteins. The protein nonspecifically coats the single-stranded RNA genome (Fleissner and Tress, 1973; Meric et al., 1984) and it may aid reverse transcription when the virus particle enters the cell (Sykora and Moelling, 1981; Prats et al., 1988). Mutations in the C residues of the putative metal-binding sequence show these residues to be essential for the specific packaging of viral RNA. Mutations of a single one of these C residues can lead to viral particles which either do not contain the viral RNA or contain cellular RNA (Gorelick et al., 1988; 1990; Meric and Goff, 1989; Aldovini and Young, 1990). The NC proteins also appear to bind specifically to a region near the 5'-end of the retroviral genome designated the Psi sequence. This sequence, located between a 5'-splice donor site and the ATG translational start signal for the GAG gene, is also necessary for the correct packaging of viral RNA (Mann et al., 1983; Lever et al., 1989), and this may explain the requirement for viral RNA packaging.

When the severe oxidation problems accompanying isolation of a native cloned NC protein of HIV were solved, it became possible to isolate the protein containing 2 Zn, 2 Cd or 2 Co ions per molecule (Fitzgerald and Coleman, 1991). The NC protein from bovine leukemia virus contains only one copy of the putative metal-binding sequence and is only 59 amino acids. The latter protein had been synthesized earlier and shown to bind one Zn, Co or Cd ion (Roberts et al., 1989). The 18 residue peptide containing the putative CCHC ligands from the HIV virus had also been synthesized and shown to form a zinc complex (South et al., 1989). Simple titrations of RNA with the NC protein do not reveal large differences in binding affinity between the metallo and apo forms of the protein. However, the effect of the metal ions on the more specific functions postulated for the NC protein remains to be tested.

A solution structure of the N-terminal half of the zinc-containing form of the NC protein from HIV has been published (Summers et al., 1992). The only structured amino acid residues are the 8 or so residues immediately surrounding each Zn-binding site. While this protein does not have much organized structure except for the Zn-sites, it has recently been demonstrated that a 18 residue peptide with the amino acid sequence corresponding to one of the zinc domains does form a complex with the oligonucleotide dACGDD, the DNA analogue of part of the Psi sequence. While more work must be done to understand the structure-function relationships of the NC proteins, EXAFS of whole virus particles from two animal retroviruses have shown there is enough Zn in the virus particles to completely complement all the NC protein Zn-binding sites (Summers et al., 1992). In addition, the EXAFS spectra are very similar to that generated by the Zn-complex of the peptide derived from the HIV NC protein.

VIII. MISCELLANEOUS ZINC PROTEINS BINDING TO DNA

The multisubunit RNA polymerases from bacteria and eukaryotes are zinc metalloenzymes which bind to DNA and are capable of specifically recognizing promoter sequences. Thus they are properly included among zinc-containing proteins which bind to DNA. The data concerning the zinc sites in the bacterial RNA polymerases which have the subunit structure $\alpha_2\beta\beta'$, are more complete than for the eukaryotic polymerases with 12 or more subunits. Comparisons of the amino acid sequences, however, show the largest and next largest subunits of yeast RNA polymerases II and III to contain many regions of homology to the β ' and β subunits of *E. coli* RNA polymerase (Allison et al., 1985). Thus the two zinc sites of the *E. coli* (Scruton et al., 1971; Wu et al., 1977; Chatterji and Wu, 1982; Giedroc and

Coleman, 1986) and *B. subtilis* (Halling et al., 1977; Speckhard et al., 1977) enzymes are probably present in the eukaryotic RNA polymerases as well.

One of the zinc ions of the *E. coli* RNA polymerase appears to be associated with β' (Wu et al., 1977). This zinc can only be replaced with Co or Cd by growing *E. coli* on media containing these metals (Giedroc and Coleman, 1986). The other zinc-binding site appears either to be on the β subunit or associated with the ββ' interface (Wu et al., 1977). The second zinc can be dissociated by titration of the enzyme with PMPS (Giedroc and Coleman, 1986). The one zinc enzyme remains fully active after removal of the PMPS. Spectroscopic data, including near ultraviolet charge transfer bands from S → Co in the Co(II) enzyme and the PMPS displacement of one of the zinc ions suggest that both sites contain sulfur atoms as donors (Giedroc and Coleman, 1986). Recent EXAFS data confirm the presence of multiple sulfur ligands at both sites (Wu et al., 1992). The β' subunit of an *E. coli* enzyme has the C-rich amino sequence between residues 58 and 88 as shown below (Ovchinnikov et al., 1982). This is the only C-rich sequence in β' and a very similar sequence is conserved in the N-terminal end of the polypeptide chains for the largest subunits of eukaryotic polymerases (Allison et al., 1985). This sequence probably contributes the ligands to the zinc located in the β' subunit. While sulfur-liganded zincs tend to be structural, there is some evidence that one site may be near nucleotides bound at the active center (Chatterji and Wu, 1982).

RNA Pol, β'	$-\underline{C}-X_{11}-\underline{C}-X_2-\underline{C}-X_{12}-\underline{C}-X_2-\underline{C}-$
UVrA Zn-Domain #1	-<u>C</u>-P-I-<u>C</u>-G-Y-S-M-R-E-L-E-P-R-L-F-S-F- N-N-P-A-G-A-<u>C</u>-P-T-<u>C</u>-
UvrA Zn-Domain #2	-<u>C</u>-E-A-<u>C</u>-Q-G-D-G-V-I-K-V-E-M-H-F-L-P- D-I-Y-V-P-<u>C</u>-D-Q-<u>C</u>-
XPAC Zn-Domain	-<u>C</u>-E-E-<u>C</u>-G-K-E-F-M-D-S-Y-L-M-N-H-F-D- L-P-T-<u>C</u>-D-N-<u>C</u>-

Scheme 2

UvrA is one of three subunits of the *E. coli* (A)BC excinuclease (Sancar and Sancar, 1988; Selby and Sancar, 1990) required for the repair of UV damage to DNA. The UvrA, UvrB and UvrC proteins have all been purified (Thomas et al., 1985). UvrA has a higher affinity for damaged than undamaged DNA (Seeberg and Steinum, 1982), particularly DNA containing pyrimidine dimers produced by UV irradiation. UvrA also possesses an ATPase activity. UvrA contains 2 zinc-finger-like amino acid sequences of 21-23 amino acids each including two $C-X_2-C$ sequences which could form a C_2C_2 ligand arrangement to zinc (Doolittle et al., 1986). The molecule was found to contain 2 zinc atoms which appear to be bound to two finger motifs (Navaratnam et al., 1989). The amino acid sequences of these motifs are shown above along with that from the RNA polymerase β' subunit. EXAFS confirm that each zinc ion is coordinated to 4 C residues, with sulfur-zinc bond lengths of 2.32 ± 0.2 Å (Navaratnam et al., 1989). While there is another $C-X_2-C$ sequence in the molecule, it does not appear to be part of a zinc-binding domain. The first zinc domain is in the N-terminal half of the molecule and the second zinc domain is in the C-terminal one third. However, neither the 70 kg mol^{-1} N-terminal fragment nor the 35 kg mol^{-1} C-terminal fragment demonstrate the specific DNA recognition of damaged DNA displayed by the whole protein (Myles and Sancar, 1991). While the zinc-binding amino acid sequences of UvrA bear some relationship to the zinc finger sequences, they are sufficiently different to suggest that the three dimensional structure of the zinc domains in UvrA can not be easily predicted.

Repair of UV damage (pyrimidine dimers) in mammalian DNA is a more complex process than that in bacteria and involves at least eight polypeptides. The latter conclusion is based on the number of separate complementation groups observed using cultured cells from humans with the inherited disease, Xerodema pigmentosum (XP). The latter is due to a defective DNA repair system which results in sunlight-induced skin cancer. One of the DNA repair genes, the XPAC gene, whose protein product complements group A of XP cells, one of the defects which gives rise to XP, has been sequenced (Tanaka et al., 1990). The gene codes for a protein which contains one zinc finger rather similar to the C-terminal zinc domain of UvrA as shown above.

X. SUMMARY

Zinc has been known as an important component of enzymatic protein molecules since the late 1930's. Over 300 enzymes containing one or more zinc ions are now known (Vallee and Auld, 1990). The discovery in the last several years of the roles of multiple zinc ions in the structure and function of a large number of DNA-binding proteins, including a great variety of eukaryotic transcription factors may, by virtue of the shear number of different macromolecules involved, make the role of zinc in gene expression the predominant one for zinc in the eukaryotic organism. The number of individual zinc proteins playing roles in gene expression appears to be approaching 1000 and it is likely that many more remain to be discovered. The relative predominance of zinc-sulfur ligands in the DNA-binding proteins is worth noting. While zinc-sulfur ligation does occur in zinc metalloenzymes, mixtures of nitrogen and oxygen ligands predominate at the active sites of enzymes. Zinc appears to be involved in the folding of the DNA-binding surface of many DNA-binding zinc proteins.

REFERENCES

Aldovini, A., and Young, R.A. (1990). *J. Virol.*, *64*: 1920.

Allison, L.A., Mayle, M., Shales, M., and Engles, C.J. (1985). *Cell*, *42*: 599.

André, B. (1990). *Mol. Gen. Genet.*, *220*: 269.

Andrianopoulos, A., and Hynes, M.J. (1990). *Mol. Cell Biol.*, *10*: 3194.

Arriza, J.L., Weinberger, C., Cerelli, G., Glaser, T.M., Handelin, B.L., Housman, D.E., and Evans, R.M. (1987). *Science*, *237*: 268.

Baldarelli, R.M., Mahoney, P.A., Salas, F., Gustavson, E., Boyer, P.D., Chang M.-F., Roark, M., and Lengyel, J.A. (1988). *Develop. Biol.*, *125*: 85.

Baleja, J.D., Marmorstein, R., Harrison, S.C., and Wagner, G. (1992). *Nature*, *356:* 450.

Balzi, E., Chen, W., Ulaszewski, S., Capieaus, E., and Goffeau, A. (1987). *J. Biol. Chem.*, *262*: 16871.

Baum, J.A., Greever, R., and Giles, N.H. (1987). *Mol. Cell Biol.*, *7*: 1256.

Bellefroid, E.J., Lecocq, P., Benhida, A., Poncelet, D.A., Belayew, A., and Martial, J.A. (1989). *DNA-J. Molec. Cell Biol.*, *8*: 377.

Bellefroid, E.J., Marine, J.-C., Ried, T., Lecocq, P.J., Riviere, M., Amemiya, C., Poncelet, D.A., Coulie, P.G., de Jong, P., Szpirer, C., Ward, D.C., and Martial, J.A. (1993). *EMBO J.*, *12*: 1363.

Berg, J.M. (1988). *Proc. Nat. Acad. Sci. USA*, *85*: 99.

Beri, R.K., Whittington, H., Roberts, C.R., and Hawkins, A.R. (1987). *Nucl. Acids Res.*, *15*: 7991.

Boulay, J.L., Dennefeld, C., and Alberga, A. (1987). *Nature*, *330*: 395.

Bray, P., Lichter, P., Thiesen, H.-J., Ward, D.C., and Dawid, I.B. (1991). *Proc. Nat. Acad. Sci. USA*, *88*: 9563.

Brunk, B.P., Martin, E.C., and Adler, P.N. (1991). *Nature*, *353*: 351.

Burger, G., Strauss, J., Scazzocchio, C., and Lang, B.F. (1991). *Mol. Cell. Biol.*, *11*: 5746.

Call, K.M., Glaser, T., Ito, C.Y., Buckler, A.J., Pelletier, J., Haber, D.A., Rose, E.A., Kral, A., Yeger, H., Lewis, W.H., Jones, C., and Housman, D.E. (1990). *Cell*, *60*: 509.

Chang, C., Kokonitis, J., and Liao, S. (1988). *Science*, *240*: 324.

Chase, J.W., and Williams, K.R. (1986). *Annu. Rev. Biochem.*, *55*: 103.

Chatterji, and Wu, F.Y.-H. (1982). *Biochemistry*, *21*: 4651.

Chowdhury, K., Deutsch, U. and Gruss, P. (1987). *Cell, 48*: 771.

Coleman, J.E. (1983). *Zinc Enzymes* (T.G. Spiro, ed.), Wiley Interscience, New York, NY, p. 220.

Crosby, S.D., Puetz, J.J., Simburger, K.S., Fahrner, T.J., and Milbrandt, J. (1991). *Mol. Cell. Biol., 11*: 3835.

Crossley, P.H., and Little, P.F.R. (1991). *Proc. Nat. Acad. Sci. USA, 88*: 7923.

Danielsen, M., Hinck, L., and Ringold, G.M. (1989). *Cell, 57*: 1131.

Davison, A.J., and Scott, J.E. (1987). *J. Gen. Virol., 67*: 1759.

Debuire, B, Henry, C., Benaissa, M., Biserte, G., Claverie, J.M., Saule, S., Martin, P., and Stehelin, D. (1984). *Science, 224:* 1456.

Diakun, G.P., Fairall, L., and Klug, A. (1986). *Nature, 324*: 698.

Doolittle, R.F., Johnson, M.S., Husain, I., Van Houten, B., Thomas, D.C., and Sancar, A. (1986). *Nature, 323*: 451.

Everett, R.D. (1988). *J. Mol. Biol., 202*: 87.

Fitzgerald, D.W., and Coleman, J.E. (1991). *Biochemistry, 30*: 5196.

Fleissner, E., and Tress, E. (1973). *J. Virol., 12*: 1612.

Fortini, M.E., Lai, Z., and Rubin, G.M. (1991). *Mech. Develop., 34*: 113.

Freedman, L.P., Luisi, B.F., Korszun, Z.R., Basavappa, R., Sigler, P.B., and Yamamoto, K. (1988). *Nature, 334*: 543.

Freemont, P.S., Hanson, I.M., and Trowsdale, J. (1991). *Cell, 64*: 483.

Friden, P., and Schimmel, P. (1987). *Mol. Cell Biol., 7*: 2708.

Furst, P., and Hamer, D.H. (1989). *Proc. Nat. Acad. Sci. USA, 86*: 5267.

Furst, P., Hu, S., Hackett, R., and Hamer, D.H. (1988). *Cell, 55*: 705.

Gardner, K.H., Pan, T., Narula, S., Rivera, E., and Coleman, J.E. (1991). *Biochemistry, 30*: 11292.

Giedroc, D.P., and Coleman, J.E. (1986). *Biochemistry, 25*: 4969.

Giedroc, D.P., Keating, K.M., Williams, K.R., Konigsberg, W.H., and Coleman, J.E. (1986). *Proc. Nat. Acad. Sci. USA, 83*: 8452.

Giedroc, D.P., Keating, K.M., Williams, K.R., and Coleman, J.E. (1987). *Biochemistry, 26*: 5251.

Giedroc, D.P., Johnson, B.A., Armitage, I.M., and Coleman, J.E. (1989). *Biochemistry, 28*: 2410.

Giedroc, D.P., Qiu, H., Khan, R., King, G.C., and Chen, K. (1992). *Biochemistry, 31*: 765.

Giguere, V., Ong, E.S., Segui, P., and Evans, R.M. (1987). *Nature, 330*: 624.

Ginsberg, A.M., King, B.O., and Roeder, R.G. (1984). *Cell, 39:* 479.

Gorelick, R.J., Henderson, L.E., Hanser, J.P., and Rein, A. (1988). *Proc. Nat. Acad. Sci. USA, 85*: 8420.

Gorelick, R.J., Nigida, S.M., Jr., Bess, J.W., Jr., Arthur, L.O., Henderson, L.E., and Rein, A. (1990). *J. Virol., 64*: 3207.

Green, S., Kumar, V., Theulaz, I., Wahli, W., and Chambon, P. (1988). *EMBO J., 7*: 3037.

Guyader, M., Emerman, M., Sonigo, P., Clavel, F., Montagnier, L., and Alizon, M. (1987). *Nature, 326*: 662.

Haile, D.J., Rouault, T.A., Tang, C.K., Chin, J., Harford, J.B., and Klausner, R.D. (1992). *Proc. Nat. Acad. Sci. USA, 89:* 11735.

Halling, S.M., Sanchez-Anzaldo, F.J., Fukuda, R., Doi, R.H., and Meares, C.F. (1977). *Biochemistry, 16*: 2880.

Hanas, J.S., Hazuda, D.J., Bogenhagen, D.F., Wu, F.Y.-H., and Wu, C.-W. (1983). *J. Biol. Chem., 258:* 14120.

Hard, T., Kellenbach, E., Boelens, R., Maler, B.A., Dahlman, K., Freedman, L.B., Carlstedt-Duke, J., Yamamoto, K.R., Gustafsson, J.-A., and Kaptein, R. (1990). *Science, 249*: 157.

Hartshorne, T.A., Blumberg, H., and Young, E.T. (1986). *Nature, 320*: 283.

Haupt, Y., Alexander, W.S., Barri, G., Klinken, S.P., and Adams, J.M. (1991). *Cell, 65*: 753.

Hayes, J.J., and Tullius, T.D. (1992). *J. Mol. Biol., 227*: 407.

Hollenberg, S.M., Giguere, V., Segui, P., and Evans, R.M. (1987). *Cell, 49*: 39.

Hoovers, J.M.N., Mannens, M., John, R., Bliek, J., Van Heyningen, V., Porteous, D.J., Lesehot, N.J., Westerveld, A., and Little, P.F.R. (1992). *Genomics, 12*: 254.

Hsu, T., Gogos, J.A., Kirsh, S.A., and Kafatos, F.C. (1992). *Science, 257*: 1946.

Huckaby, C.S., Connelly, O.M., Beattie, W.G., Dobson, A.D.W., Tsai, M.J., and O'Malley, B.W. (1987). *Proc. Nat. Acad. Sci. USA, 84*: 8380.

Itoh, K., Itoh, Y., and Frank, M.B. (1991). *J. Clin. Invest., 87*: 177.

Johnston, M. (1987). *Nature, 328*: 353.

Joho, K.E., Darby, M.K., Crawford, E.T., and Brown, D.D. (1990). *Cell, 61*: 293.

Jones, J.S., Weber, S., and Prkash, L. (1988). *Nucl. Acids Res., 16*: 7119.

Joseph, L.R., Le Beau, M.M., Jamieson, G.M., Jr., Acharya, S., Shows, T.B., Rowley, J.D., and Sukhatme, V.P. (1988). *Proc. Nat. Acad. Sci. USA, 85*: 7164.

Kadonaga, J.T., Camer, K.C., Masiarz, F.R., and Tjian, R. (1987). *Cell, 51*: 1079.

Kim, J., and Michels, C.A. (1988). *Curr. Genetics, 14*: 319.

Kim, J.G., and Hudson, L.D. (1992). *Mol. Cell. Biol., 12*: 5632.

Kinzler, K.W., Ruppert, J.M., Bigner, S.H., and Vogelstein, B. (1988). *Nature, 332*: 371.

Klein-Hitpass, L. (1989). *Mol. Cell. Biol., 9*: 43.

Kochoyan, M., Havel, T.F., Nguyen, D.T., Dahl, C.E., Keutman, H.T., and Weiss, M.A. (1991). *Biochemistry, 30*: 3371.

Kraulis, J. (1991). *J. Applied Cryst., 24*: 946.

Kraulis, P.J., Raine, A.R.C., Gadhavi, P.L., and Laue, E.D. (1992). *Nature, 356*: 448.

Lai, Z., Fortini, M.E., and Rubin, G.M. (1991). *Mech. Dev., 34*: 123.

Laughton, A., and Gesteland, R.F. (1984). *Mol. Cell. Biol., 4*: 260.

Lee, M.S., Gippert, G.P., Soman, K.V., Case, D.A., and Wright, P.E. (1989). *Science, 245*: 635.

Lever, A., Gottlinger, H., Haseltine, W., and Sodroski, J. (1989). *J. Virol., 63*: 4085.

Lewis, M.K., and Burgess, R.R. (1982). *The Enzymes*, 3rd ed., Vol. 14, Part B, Academic Press, New York, p. 109.

Lubahn, D.B., Joseph, D.R., Sullivan, P.M., Willard, H.F., French, F.S., Wilson, E.M. (1988). *Science, 340*: 327.

Luisi, B.F., Xu, W.X., Otwinowski, Z., Freedman, L.P., Yamamoto, K.R., and Sigler, P.B. (1991). *Nature, 352*: 497.

Lundell, M.J., and Hirsh, J. (1992). *Develop. Biol., 154*: 84.

Mader, S., Kumar, V., de Verneuil, H., and Chambon, P. (1989). *Nature, 338*: 271.

Mann, R., Mulligan, R.C., and Baltimore, D. (1983). *Cell, 33*: 401.

Marczak, J.E., and Brandriss, M.C. (1991). *Mol. Cell. Biol., 11*: 2609.

Marmorstein, R., Carey, M., Ptashne, M., and Harrison, S.C. (1992). *Nature, 356:* 408.

McDonnell, D.P., Mangelsdorf, D.J., Pike, J.W., Haussler, M.R., and O'Malley, B.W. (1987). *Science, 235*: 1214.

McPheeters, D.S., Stormo, G.D., and Gold, L. (1988). *J. Mol. Biol., 201*: 517.

Messenguy, R.F., Dubois, E., and Descamps, F. (1986). *Eur. J. Biochem., 157*: 77.

Meric, C., and Goff, S. (1989). *J. Virol., 63*: 1558.

Meric, C., Darlix, J.L., and Spahr, P.F. (1984). *J. Mol. Biol., 174*: 531.

Milbrandt, J. (1987). *Science, 238*: 797.

Miller, J., MacLachlan, A.D., and Klug, A. (1985). *EMBO J., 4*: 1609.

Misrahi, M., Atger, M., D'Auriol, L., Loosfelt, H., Meriel, C., Friedlansky, F., Guichon-Mantel, A., Galibert, F., and Milgrom, E. (1987). *Biochem. Biophys. Res. Comm.*, *143*: 740.

Morinaga, T., Yasuda, H., Hashimoto, T., Higashio, K., and Tamaoki, T. (1991). *Mol. Cell. Biol.*, *11*: 6041.

Myles, G.M., and Sancar, A. (1991). *Biochemistry*, *30*: 3834.

Nadler, S.G., Roberts, W.J., Shamoo, Y., and Williams, K.R. (1990). *J. Biol. Chem.*, *265*: 10389.

Nagarajan, L., Lange, B., Cannizzaro, L., Finan, J., Novell, P.C., and Huebner, K. (1990). *Blood*, *75*: 82.

Navaratnam, S., Myles, G.M., Strange, R.W., and Sancar, A. (1989). *J. Biol. Chem.*, *264:* 16067.

Neuhaus, D., Nakaseko, Y., Nagai, K., and Klug, A. (1990). *FEBS Lett.*, *262*: 179.

Nguyen, H.Q., Hoffman-Liebermann, B., and Liebermann, D.A. (1993). *Cell*, *72*: 197.

Omichinski, J.G., Clore, G.M., Appella, E., Sakaguchi, K., and Gronenborn, A.M. (1990). *Biochemistry*, *29*: 9324.

Ovchinnikov, Y.A., Monastyrskaya, G.S., Gubanov, V.V., Guryev, S.O., Salomatina, I.S., Shuvaeva, T.M., Lipkin, V.M., and Sverdlov, E.D. (1982). *Nucl. Acids Res.*, *10*: 4035.

Page, D.C., Mosher, R., Simpson, E.M., Fisher, E.M.C., Mardon, G., Pollack, J., McGillivray, B., de la Chapelle, A., and Brown, L.G. (1987). *Cell*, *51*: 1091.

Pan, T., and Coleman, J.E. (1989). *Proc. Nat. Acad. Sci. USA*, *86*: 3145.

Pan, T., and Coleman, J.E. (1990a). *Biochemistry*, *29*: 3023.

Pan, T., and Coleman, J.E. (1990b). *Proc. Nat. Acad. Sci. USA*, *87*: 2077.

Pan, T., and Coleman, J.E. (1991). *Biochemistry*, *30*: 4212.

Pan, T., Giedroc, D.P., and Coleman, J.E. (1989). *Biochemistry*, *28*: 8828.

Pan, T., Freedman, L.P., and Coleman, J.E. (1990). *Biochemistry*, *29*: 9218.

Parraga, G., Horvath, S.J., Eisen, H., Taylor, W.E., Hood, L., Young, E.T., and Klevit, R. (1988). *Science*, *241*: 1489.

Patarca, R., Schwartz, J., Singh, R.P., Kong, Q.-T., Murphy, E., Anderson, Y., Wei Sheng, F.-Y., Singh, P., Johnson, K.A., Guarnagia, S.M., Durfee, T., Blattner, F., and Cantor, H. (1988). *Proc. Nat. Acad. Sci. USA*, *85*: 2733.

Pavletich, N.P., and Pabo, C.O. (1991). *Science*, *252*: 809.

Pays, E., and Murphy, N.B. (1987). *J. Mol. Biol.*, *197*: 147.

Petkovich, M., Brand, N.J., Krust, A., and Chambon, P. (1987). *Nature*, *330*: 444.

Pfeifer, K., Kim, K.-S., Kogan, S., and Guarente, L. (1989). *Cell*, *56*: 291.

Prats, A.C., Sarik, L., Gabus, C., Litvak, S., Keith, G., and Darlix, J.L. (1988). *EMBO J.*, *7*: 1777.

Ratner, L., Haseltine, W., Patarca, R, Livak, K.J., Staricich, B., Josephs, S.F., Doran, E.R., Rafalski, J.A., Whitehorn, E.A., Baumeister, K., Ivanoff, L., Petteway, S.R., Jr., Pearson, M.L., Lautenberger, J.A., Papas, T.S., Ghrayeb, J., Chang, N.T., Gallo, R.C., and Wong-Staal, F. (1985). *Nature*, *313*: 277.

Rhodes, D., and Klug, A. (1993). *Sci. Amer.*, *268*: 56.

Roberts, W.J., Pan, T., Elliott, J.I., Coleman, J.E., and Williams, K.R. (1989). *Biochemistry*, *28*: 10043.

Rosenberg, U.B., Schröder, C., Preiss, A., Kienlin, A., Cote, S., Riede, I., and Jäckle, H. (1986). *Nature*, *319*: 336.

Rouault, T.A., Haile, D.J., Downey, W.E., Philpott, C.C., Tang, C., Samaniego, F., Chin, J., Paul, I., Orloff, D., Harford, J.B., and Klausner, R.D. (1992). *Biometals*, *5:* 131.

Ruis, I., Altaba, A., Perry-O'Keefe, H., and Melton, D.A. (1987). *EMBO J.*, *6*: 3065.

Salmeron, J.M., and Johnston, S.A. (1986). *Nucl. Acids Res.*, *14*: 7767.

Sancar, A., and Sancar, G.B. (1988). *Annu. Rev. Biochem.*, *57*: 29.

Sap, J., Murioz, A., Damm, K., Goldberg, Y., Ghysdael, J., Leutz, A., Beug, H., and Vennström, B. (1986). *Nature, 324*: 635.

Schatz, D.G., Oettinger, M.A., and Baltimore, D. (1989). *Cell, 59*: 1035.

Schuh, R., Aicher, W., Gaul, U., Cote, S., Preiss, A., Maier, D., Seifert, U., Nauber, C., Schröder, C., Kemler, R., and Jäckle, H. (1986). *Cell, 47*: 1025.

Schwabe, J.W.R., and Rhodes, D. (1991). *Trends Biochem. Sci., 16*: 291.

Schwabe, J.W.R., Neuhaus, D., and Rhodes, D. (1990). *Nature, 348*: 458.

Scruton, M.C., Wu, C.W., and Goldthwait, D.A. (1971). *Proc. Nat. Acad. Sci. USA, 68*: 2497.

Seeberg, E., and Steinum, A.L. (1982). *Proc. Nat. Acad. Sci. USA, 79*: 988.

Selby, C.P., and Sancar, A. (1990). *Mutat. Res., 236*: 203.

Shamoo, Y., Webster, K.R., Williams, K.R., and Konigsberg, W.H. (1991). *J. Biol. Chem., 266*: 7967.

Shang, Z., Liao, Y.-D., Wu, F.Y.-H., and Wu, C.-W. (1989). *Biochemistry, 28*: 9790.

Smiley, B.L., Stadnyk, A.W., Myler, P.J., and Stuart, K. (1990). *Mol. Cell Biol., 10*: 6436.

Sollitti, P., and Marmur, J. (1988). *Mol. Gen. Genet., 213*: 56.

South, T.L., and Summers, M.F. (1993). *Protein Sci., 2:* 3.

South, T.L., Kim, B., and Summers, M.F. (1989). *J. Amer. Chem. Soc., 111*: 395.

South, T.L., Blake, P.R., Sowder, R.C., Arthur, L.E., Henderson, L.E., and Summers, M.F. (1990). *Biochemistry, 29*: 7786.

Speckhard, D.C., Wu, F.Y.-H., and Wu, C.-W. (1977). *Biochemistry, 16*: 5228.

Stillman, D.J., Banktier, A.T., Seddon, A., Groenhout, E.G., and Nasmyth, K.A. (1988). *EMBO J., 7*: 485.

Sukhatme, V.P., Cao, X., Chang, L.C., Tsai-Morris, C.-H., Stamenkovich, D., Ferreira, P.C.P., Cohen, D.R., Edwards, S.A., Shows, T.B., Curran, T., Le Beau, M.M., and Adamson, E.D. (1988). *Cell, 53*: 37.

Summers, M.F., Henderson, L.E., Chance, M.R., Bess, J.W., South, T.L., Blake, P.R., Sagi, I., Perez-Alvarado, Sowder, R.C., Hare, D.R., and Arthur, L.O. (1992). *Protein Sci., 1:* 563.

Sykora, K.W., and Moelling, K. (1981). *J. Gen. Virol., 55*: 379.

Takahashi, M., Inaguma, Y., Hiai, H., and Hirose, F. (1988). *Mol. Cell. Biol., 8*: 1853.

Tanaka, K., Miura, N., Satokata, I., Miyamoto, I., Yoshida, M.C., Satoh, Y., Kondo, S., Yasui, A., Okayama, H., and Okada, Y. (1990). *Nature, 348:* 73.

Tautz, D., Lehmann, R., Schnürch, H., Schuh, R., Seifert, E., Kienlin, A., Jones, K., and Jäckle, H. (1987). *Nature, 327*: 383.

Theim, S., and Miller, L.K. (1989). *J. Virol. 63*: 4489.

Thiesen, H.-J. (1990). *New Biol., 2*: 363.

Thomas, D.C., Levy, M., and Sancar, A. (1985). *J. Biol. Chem., 260*: 9875.

Tsai, S.Y., Carlstedt-Duke, J., Weigel, N.L., Dahlman, K., Gustafsson, J.-A., Tsai, M.-J., and O'Malley, B.W. (1988). *Cell, 55*: 361.

Umesono, K., and Evans, B.M. (1989). *Cell, 57*: 1139.

Vallee, B.L., and Auld, D.S. (1990). *Biochemistry, 29*: 5647.

Vallee, B.L., Coleman, J.E., and Auld, D.S. (1991). *Proc. Nat. Acad. Sci. USA, 88*: 999.

van Lohuizen, M., Frasch, M., Wientjens, E., and Berns, A. (1991a). *Nature, 353*: 353.

van Lohuizen, M., Verbeek, S., Scheijen, B., Wientjens, E., van der Gulden, H., and Berns, A. (1991b). *Cell, 65*: 737.

Vincent, A., Colot, H.V., and Rosbach, M. (1985). *J. Mol. Biol., 186*: 149.

Walter, P., Green, S., Greene, G., Krust, A., Bornert, J.-M., Jeitsch, J.-M., Staub, A., Jensen, E., Scrace, G., Waterfield, M., and Chambon, P. (1985). *Proc. Nat. Acad. Sci. USA, 82*: 7889.

Weinberger, C., Thompson, C.C., Ong, E.S., Lebo, R., Gruol, D.J., and Evans, R.M. (1986). *Nature*, *324*: 641.

Wu, C.-W. (1986). *Zinc Enzymes* (I. Bertini, C. Luchinat, W. Maret, and M. Zeppezauer, eds.), Birkhäuser, Boston, p. 563.

Wu, C.-W., Wu, F.Y.-H., and Speckhard, D.C. (1977). *Biochemistry*, *16*: 5449.

Wu, F.Y.-H., Huang, W.J., Sinclair, R.B. and Powers, L. (1992). *J. Biol. Chem.*, *267*: 25560.

Zhou, K., Brisco, P.R.G., Hinkkanen, A.E., and Kohlhaw, G.B. (1987). *Gene*, *15*: 5261.

COMPLEX FORMATION CONSTANTS

by

Peter William Linder

Department of Chemistry,
University of Cape Town,
Rondebosch
7700 South Africa

I. INTRODUCTION

Formation constants have important applications in the area of metal ion interactions with low-molar-mass ligands, as discussed in various other sections of this handbook, especially Part Five. The term, formation constant, which is synonymous with stability constant, is a special name for the equilibrium constant of a metal complex formation reaction in solution (usually, but not necessarily, aqueous). The latter may be written generally as

$$pM + qL + rH \rightleftharpoons M_pL_qH_r \quad (1)$$

where M = a metal ion species;
L = a ligand;
H = hydrogen ions;
electrical charges are omitted for clarity of presentation;
p,q,r = the stoichiometric coefficients of the reactants; r may be positive or negative, if negative the implication is that protons are removed or that hydroxide ligands are added.

It is understood that M and H in reaction (1) are hydrated (e.g. $M(H_2O)_6$, H_3O^+) and that L may be hydrated.

Thus when $p = 1$ and $r = 0$, reaction (1) refers to the formation of mononuclear binary complexes, ML_1, ML_2,... These reactions are actually to be regarded as ligand substitution reactions, in which a coordinated water molecule is substituted by a coordinated ligand molecule. When $p = 1$ and $r > 0$, the complexes concerned are the protonated species, MLH, MLH_2, ML_2H... Hydroxo complexes are catered for when $r < 1$ yielding MLOH, $ML(OH)_2$... and the hydroxo metal species (corresponding to $q = 0$), M(OH), $M(OH)_2$... Provision is made for oligonuclear and polynuclear complexes by setting $p > 1$.

When a given metal species, M, interacts with a given ligand, L, in general, several complex forming reactions (1) take place concomitantly, corresponding to several sets of values of the stoichiometric coefficients, p, q and r. Thus, in general, several complexes, each with its own distinctive stoichiometry, are formed and co-exist in equilibrium.

II. THERMODYNAMIC FORMATION CONSTANTS

The thermodynamic equilibrium constant, $\beta_{T(pqr)}$, for reaction (1) is defined in equation (2) and is expressed in terms of the activities, a_s, of the reacting species, S. (S = M, L, H or $M_pL_qH_r$)

$$\beta_{T(pqr)} = a_{M_pL_qH_r}/a_M^p\, a_L^q\, a_H^r \qquad (2)$$

III. STOICHIOMETRIC FORMATION CONSTANTS

It is not practicable to measure the activities of all the reacting species present in a given solution. Thus, the custom is to measure and report stoichiometric formation constants, which are quotients involving the molar concentrations, c_s, of the reacting species, s, as defined in equation (3).

$$\beta_{pqr} = c_{M_pL_qH_r}/c_M^p\, c_L^q\, c_H^r \qquad (3)$$

The relationship between a thermodynamic formation constant and the corresponding stoichiometric formation constant is shown in equation (4)

$$\beta_{T(pqr)} = \beta_{pqr}(\gamma_{M_pL_qH_r}/\gamma_M^p\, \gamma_L^q\, \gamma_H^r) \qquad (4)$$

where γ_s is the activity coefficient of the reacting species, S. The activity, a_s, molar concentration, c_s, and activity coefficient, γ_s, are related according to equation (5)

$$a_s = \gamma_s c_s \qquad (5)$$

or strictly,

$$a_s = \gamma_s c_s/c^\circ$$

where c° is a standard molar concentration commonly taken as 1 mol dm^{-3}.

The activity coefficients in equation (4) vary with the ionic strength, I, of the solution which is defined by equation (6)

$$I = 0.5\sum_i c_i z_i^2 \qquad (6)$$

where c_i is the molar concentration and z_i is the charge of the i th ionic species. The summation is carried out over all the ionic species present in the solution. Essentially, the ionic strength provides a measure of the concentration of ionic charges in the solution.

When the set of complex-forming reactions (1) proceeds, there is a concomitant change in the concentrations of ionic charges and, hence, also of the ionic strength of the solution. This change in ionic strength brings about changes in the activity coefficient quotients of equation (4). Since, at a fixed temperature of and a fixed pressure applied to the solution, the thermodynamic formation constants, $\beta_{T(pqr)}$, are true constants which are independent of the ionic strength, it follows that the corresponding stoichiometric formation constants, β_{pqr}, vary with the degree of advancement of the complex-forming reactions. To overcome this problem in practice, the custom is to include as a solute in the experimental solutions an inert background electrolyte with a concentration about 100 to 1000 times greater than the initial concentrations of metal and ligand. By this means, the variation in ionic strength due to the complex-forming reactions becomes swamped and the overall ionic strength of the solution is maintained at an essentially constant level. The activity coefficients of equation (4) and hence the stoichiometric formation constants are accordingly maintained at constant values in accordance with the principle of Brønsted (Beck and Nagypál, 1990a). The latter states that the activity coefficient of all solutes present as small fractions of the total electrolyte concentration is constant at total electrolyte concentration.

IV. CHOICE OF IONIC MEDIUM

The inert background electrolyte must be such that the cation and anion do not interact with M, L or H of reaction (1). Examples are sodium perchlorate and potassium nitrate. These have been used commonly with concentrations ranging from 0.1 to 3 mol dm^{-3}. There are many stoichiometric formation constants which have been measured in 0.15 mol dm^{-3} sodium chloride, however. The intention here has been to simulate human blood plasma. In these measurements, the complexing tendency of chloride to the transition metal ions, manganese(II), iron(II), iron(III), cobalt(II), copper(II), zinc(II) etc. tends to be ignored. In consequence, the formation constants concerned are actually conditional constants because they incorporate, implicitly, the corresponding chloride complex formation constants.

V. DIFFERENT TYPES OF FORMATION CONSTANTS

From this point, it is assumed that all formation constants considered are stoichiometric constants, that is, that they have been determined as concentration quotients in specific ionic media. The constant defined in equation (3) corresponds to reaction (1) and is termed an overall, a cumulative or a gross formation constant. In the literature on metal-ligand equilibria, results are often reported as overall constants. Nevertheless other, alternative, ways of expressing the equilibria are also used. One example is as *stepwise formation constants*, K_n, corresponding to writing the reactions as given below (n = 1, 2,...)

$$ML_{n-1} + L \rightleftharpoons ML_n \,, \quad K_n = [ML_n]/[ML_{n-1}][L] \tag{7}$$

The overall constants are related to the stepwise constants through the expression,

$$\beta_{1n0} = \prod_{i=1}^{n} K_i \tag{8}$$

When p = 0 in reaction (1), the equilibria involve protonation of the ligand which may be expressed in terms of both overall and stepwise constants, as in equations (9) and (10), respectively (j = 1, 2,...)

$$L + jH \rightleftharpoons LH_j \,, \quad \beta_{01j} = [LH_j]/[L][H]^j \tag{9}$$

$$LH_{j-1} + H \rightleftharpoons LH_j \,, \quad K_{H,j} = [LH_j]/[LH_{j-1}][H] \tag{10}$$

Again, the overall and stepwise protonation constants are related through the expression,

$$\beta_{01j} = \prod_{i=1}^{j} K_{H,i} \tag{11}$$

The protonation equilibria, above, are written as associations. In some texts, however, especially in organic chemistry, this type of equilibrium is written as the reverse, that is, as acid dissociation. The acid dissociation constants, $K_{a,j}$ are inversely related to the stepwise protonation constants: e.g. for an acid LH_J

$$K_{a,1} = [LH_{J-1}][H]/[LH_J] = 1/K_{H,J};$$
$$K_{a,2} = [LH_{J-2}][H]/[LH_{J-1}] = 1/K_{H,J-1};$$
$$\cdots\cdots$$
$$K_{a,J} = [L][H]/[LH] = 1/K_{H,1}. \tag{12}$$

Hydroxo complexes arise when r < 1 in reaction (1) and may be regarded as products of the removal of hydrogen ions from water molecules coordinated to the metal. Thus the species, MLH_{-1} and MLOH, are compositionally equivalent with overall formation constants,

$$\beta_{11-1} = [MLH_{-1}][H]/[M][L] \tag{13}$$

and

$$\beta_{MLOH} = [MLOH]/[M][L][OH], \tag{14}$$

respectively. The two constants are related by the expression

$$\beta_{11-1} = \beta_{MLOH}K_w \tag{15}$$

or $\log \beta_{11-1} = \log \beta_{MLOH} - pK_w$

where K_w = [H][OH]

= the ionic product of water,

and $pK_w = -\log K_w$.

VI. EQUILIBRIUM APPROXIMATION IN MODELING OF BIOLOGICAL FLUIDS

In so far as formation constants are used in constructing speciation models of biological fluids, such as is described in Part Five, a question arises regarding the validity of assuming biological fluids to be at equilibrium. Perrin and Agarwal (1973) point out that a biological system is never truly in a state of equilibrium but they justify the assumption of equilibrium as a first approximation because to achieve high efficiencies of energy conversion, most biological systems operate near to reversible equilibria. In the present context, moreover, rates of complex formation and dissociation are usually very fast. Jones and May (1987) consider the assumption of equilibrium in treating chelation therapy by means of computer simulation. These authors enumerate the difficulties in developing kinetic models of human blood plasma and contiguous body compartments, in the present state of the art, and indicate a way forward. They imply, however, that even in the development of appropriate kinetic models, the preliminary creation of an equilibrium model is a sine qua non.

VII. APPLICABILITY OF FORMATION CONSTANTS

In applying formation constants to speciation calculations as described in Part Five, close attention needs to be given to certain points. First, as shown in section III, the value of a stoichiometric formation constant generally depends on the ionic strength of the solutions in which the constant is determined. It so happens that in some instances the value of a stoichiometric formation constant depends also on the identity of the ionic medium in which the reacting system is investigated. For example, sodium perchlorate and potassium nitrate when used as background electrolytes of the same concentration may yield stoichiometric formation constants with significantly different values, for a given M-L system.

The second point is that in characterizing a given metal-ligand system, different researchers may propose differing chemical models as best fitting their respective experimental data. By a chemical model is meant the specific set of species or complexes (each characterized by a given set of p, q and r values in reaction (1)) assumed to be formed. The differences, in general, tend to affect the identities of the minor complex species; usually there is agreement on the identities of the major species although the values of the formation constants of the latter may vary somewhat from one chemical model or set of experimental conditions, to another.

Thus in applying a formation constant to a given speciation modeling problem, it is of paramount importance to ensure that the constant has been measured under conditions which correspond as closely as possible to the ionic composition, reactant concentration ranges, temperature etc of the biological fluid under consideration.

VIII. SPECIES DISTRIBUTION DIAGRAMS

The importance of selecting appropriate concentration ranges for the reactants may be illustrated by considering species distribution diagrams (Baxter and Williams, 1974; Perrin and Sayce, 1967; Beck and Nagypál, 1990b). The distribution of concentrations of the individual species co-existing in equilibrium in a metal-ligand-hydrogen ion system (three components) can be plotted, in principle, only in four dimensions (Beck and Nagypál, 1990b). In practice, two-dimensional plots tend to be presented showing the distribution with two of the total concentrations of the components fixed. The plot then shows the concentrations of all the species concerned as a function of the equilibrium concentration of the third component. The distribution curves in the resulting two-dimensional diagram then reflect the equilibrium relations which apply only to the values of the two fixed total concentrations. The variation of the distribution with the values of the two fixed total concentrations is illustrated by the example of the copper(II)-alanine system (Linder and Torrington, 1980) as shown in Figures 1 and 2. Here it can be seen that two different sets of values of total copper and total alanine concentrations yield diagrams showing quite different distributions of the complexes formed. It is clear, therefore, that two-dimensional distribution diagrams representing three-component systems are not to be considered unique in that they are applicable only under rather narrow ranges of conditions.

Another example which illustrates the caution that needs to be exercised when considering speciation distribution diagrams, is presented in the book by Beck and Nagypál

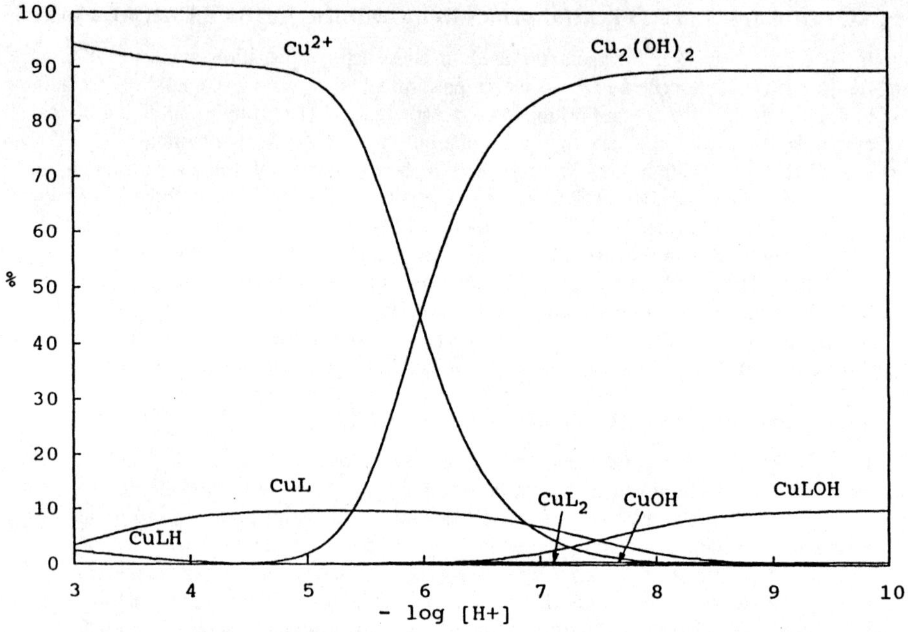

Figure 1.
Distribution of copper(II)-alanine complexes as a function of pH. Total concentrations/ mol dm^{-3}: copper(II), 0.05; alanine, 0.005.

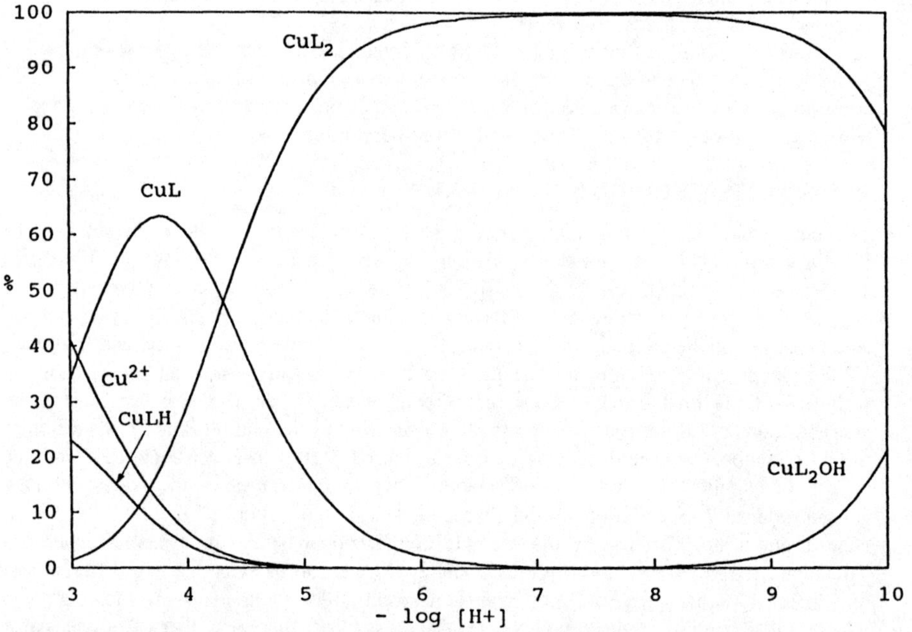

Figure 2.
Distribution of copper(II)-alanine complexes as a function of pH. Total concentrations/ mol dm^{-3}: copper(II), 0.005; alanine, 0.05.

(1990c). These authors indicate the distribution of complexes in the copper(II)-glycine system using constants determined photometrically on the one hand and potentiometrically on the other hand. In the diagram constructed from the potentiometric data, the total concentration of glycine greatly exceeds that actually used in the experimental measurements. Indeed, the potentiometric diagram has a distinctly different character from the photometric diagram, illustrating how misleading distribution curves can be if the curves are calculated for concentrations which differ appreciably from those used in the determination of the constants.

IX. CONCLUSION

It is justifiable as a good approximation to assume chemical equilibrium when modeling the distributions of low-molar-mass metal complexes in biological fluids. In calculating such distributions, use is made of a database consisting of stoichiometric formation constants determined for individual metal-ligand systems. The stoichiometric formation constants are concentration quotients determined in ionic media containing specific inert background electrolytes with concentrations 100 - 1000 times greater than the concentrations of the complex-forming reactants. Caution should be exercised to ensure that the constants comprising a database applicable to a given biological fluid have been determined in an ionic medium which resembles as closely as possible the biological fluid being modeled. When using two-dimensional distribution diagrams to ascertain the significant complexes formed in a given metal-ligand system, caution should be exercised to ensure that the concentration ranges for the metal and the ligand, that have been used to calculate the distribution diagram, are comparable with the respective concentration ranges in the system being modeled.

REFERENCES

Baxter, A.C., and Williams, D.R. (1974). *J. Chem. Soc., Dalton. Trans.*, 1117.

Beck, M.T., and Nagypál, I. (1990a). *Chemistry of Complex Equilibria*, Ellis Horwood Limited, Chichester, p. 17.

Beck, M.T., and Nagypál, I. (1990b). *Chemistry of Complex Equilibria*, Ellis Horwood Limited, Chichester, p. 74.

Beck, M.T., and Nagypál, I. (1990c). *Chemistry of Complex Equilibria*, Ellis Horwood Limited, Chichester, p. 75.

Jones, M.M., and May, P.M. (1987). *Inorg. Chim. Acta, 138*: 67.

Linder, P.W., and Torrington, R.G. (1980). *S. Afr. J. Chem., 33*: 55.

Perrin, D.D., and Agarwal, R.P. (1973). *Metal Ions in Biological Systems*, Vol. 2 (H. Sigel, ed.), Marcel Dekker, Inc., New York, p. 200.

Perrin, D.D., and Sayce, I.G. (1967). *Talanta, 14*: 833.

EXPERIMENTAL TECHNIQUES FOR DETERMINING COMPLEX STABILITY CONSTANTS

by

Peter William Linder

Department of Chemistry,
University of Cape Town,
Rondebosch
7700 South Africa

I. INTRODUCTION

In the investigation of a given metal-ligand system in solution, the aim is two-fold, namely to identify the various complexes formed (as implied by reaction (1) of Section A1, above) and to determine the formation constant of each. The approach, generally, is to prepare a series of solutions each with accurately known total concentrations of the three components, M, L and H (charges omitted for clarity) and to measure in each solution the equilibrium concentration of at least one of the set: free (i.e. uncomplexed) M, free L, free H or the equilibrium concentration of one or more of the complexes. These measurements usually provide sufficient information to fulfil the two aims of the investigation. A variety of techniques has been developed for carrying out the relevant measurements and these are described, in detail, in several publications (Martell and Motekaitis, 1988; Meloun et al., 1988; Beck and Nagypál, 1990; Nancollas and Tomson, 1982; Linder et al., 1984; Rossotti and Rossotti, 1961; Hartley et al., 1980; Rossotti, 1978; Braibanti et al., 1987). Amongst the techniques in current use are potentiometry, spectrophotometry, nuclear magnetic resonance (NMR), polarography and associated methods, ion exchange, conductance measurements, solvent extraction methods, solubility measurements. Of these, potentiometry is the most widely used and spectrophotometry is probably the second most popular. Potentiometry may be regarded as comprising two divisions, namely glass electrode potentiometry for determining hydrogen ion concentrations (pH-metry) and potentiometry involving ion-selective electrodes other than hydrogen ion-sensitive electrodes. Here, emphasis is given to pH-metry with some attention being devoted to ion-selective electrode potentiometry, spectrophotometry, nuclear magnetic resonance spectrometry and polarography. The reader is referred to the references given above for descriptions of other techniques.

II. EXPERIMENTAL PRACTICALITIES

Whatever method is to be used, all the chemicals required should be of the best reagent grade available. In general, further purification should be carried out, e.g. by recrystallization. Special attention should be given to the material to be used as the background electrolyte, especially when the higher concentrations are adopted since trace impurities may then become all the more significant. This applies, in particular, to commercially available sodium perchlorate which is not sufficiently pure to be used directly as a background electrolyte. Purification of sodium perchlorate should be carried out as described by Jones and Williams (1970).

All water used to prepare the solutions should be distilled (or doubly distilled) de-ionized (or doubly de-ionized) water. The purified water should be boiled for five minutes or longer to expel carbon dioxide. The cooled product should have a conductivity not greater than 10^{-6} siemens cm^{-1}.

The laboratory in which the equilibrium measurements are made should be temperature-controlled as finely as possible in order to ensure reliability of volumetric measurements. Glassware should preferably be grade A and should be calibrated at the temperature of the laboratory. A circulating water thermostat should be used to control the temperature of the experimental solutions while the equilibrium measurements are being made.

In the case of pH-metry, the glass electrodes should be stored in sealed tubes containing solutions of the background electrolyte at the same ionic strength as the experimental solutions and at a pH close to the lowest pH of the experimental measurements. The reference electrode should be stored in a sealed tube containing a solution identical to the salt bridge solution of the electrode. Two or three of each type of electrode should be stored in the manner described and should be instantly available. All should be stored in the water thermostat used in the experimental procedure.

III. pH-METRY

The crux of this method is measurement of the e.m.f. between a hydrogen ion-sensitive electrode and a suitable reference electrode immersed in an experimental solution under investigation. Through application of calibration data obtained for the aforementioned pair of electrodes, the measured e.m.f. is used to obtain the concentration of free (unbound) hydrogen ions in the experimental solution. The latter, obtained for a series of experimental solutions, are combined with the known total analytical concentrations of metal ion, ligand and hydrogen ions in the solutions to obtain values of formation constants. (The algebra and computer programs concerned are described in Section A3 of this Chapter 1).

Although different types of hydrogen ion-sensitive electrodes, such as the hydrogen gas electrode and the quinhydrone electrode, have been used in the past, nowadays virtually all pH-metrically determined formation constants are obtained with the use of the glass electrode (Linder et al., 1984). This can be attributed not only to convenience and robustness but also to the high quality and reliability of modern glass electrodes. Numerous commercial firms supply glass electrodes and, presumably, the competition involved tends to ensure the maintenance of good standards. In the investigation of a given metal-ligand system, the custom is first to carry out a series of experiments aimed at determining the protonation constants of the ligand (p = 0 in reaction (1), section 1). In a second series of experiments, the metal complexes are investigated. These experiments are carried out in a cell of the type,

Glass electrode	Experimental solution of constant I	Bridge solution at the same I as experimental solution but with partial replacement of the anion by Cl^-	Reference electrode

The experimental solution contains the ligand and mineral acid in the first series of experiments or ligand, mineral acid and metal ions in the second series of experiments. The background eletrolyte should preferably be a salt such as sodium perchlorate with an anion that does not form significant complexes. There exist, however, situations in which there are reasons to use a background electrolyte such as sodium chloride which does possess a complex-forming anion. The experimental solution is contained in a jacketed, sealed titration vessel. Thermostatted water (± 0.05°C) is circulated through the jacket and the solution is protected from atmospheric oxygen and carbon dioxide by the use of purified nitrogen. The latter is either bubbled through or passed over the experimental solution. Titration vessels meeting these requirements are obtainable commercially from several manufacturers. The nitrogen should be presaturated by passing it through background electrolyte at the same concentration and the same temperature as the experimental solution. The experimental solution should be stirred with a magnetic stirrer.

The composition of the experimental solution is adjusted in a series of steps by adding an appropriate titrant, from a piston burette or a microsyringe. After making each addition of titrant, the solution is allowed to reach equilibrium and the steady e.m.f. of the electrode pair is recorded. The titrant solution must contain the same inert background electrolyte at the same concentration as the experimental solution. Usually, the titrant consists of standardized alkali, so that upon making each addition, the total hydrogen ion concentration of the experimental solution is adjusted. In some applications the titrant solution may contain the metal ion and/or the ligand under investigation. Thus, it is possible, upon each addition of titrant, to adjust not only the total hydrogen ion concentration but also the total metal and/or the total ligand concentration. Alternatively, the total metal and total ligand concentrations of the experimental solution may be maintained at a predetermined constant level while adjustments are made to the total hydrogen ion concentration. This latter approach has the advantage of minimizing the number of parameters which are varied thereby providing appropriate conditions for revealing more sharply the effects of the parameter being varied, that is, the total hydrogen ion concentration. In every type of titration, the pH range should be limited to 2 - 12.

In order to ensure detection of all the complexes which can potentially be formed in the system under investigation, each series of titrations should cover wide ranges of total metal and total ligand concentrations. After having identified all possible complexes, in this way, further definitive titrations can be carried out with optimum total component concentrations. An approach to estimating these optima is discussed by Beck and Nagypál (1990).

The reference eletrode must have a stable potential and is usually either a calomel electrode or a silver/silver chloride electrode. The reference electrode may be either a commercially available one or home made (Willard et al., 1988; Biedermann, 1956). When sodium perchlorate is used as the background electrolyte and as one of the components of the bridge solution, contact with potassium ions must be avoided owing to problems which arise from the low solubility of potassium perchlorate. In such an instance, the potassium chloride electrolyte of the reference electrode or its internal bridge solution should be replaced by sodium chloride.

IV. CALIBRATION OF THE ELECTRODES IN PH-METRY

The e.m.f. of the electrodes discussed in section III is affected by:
(i) the hydrogen ion concentration in the experimental solution;
(ii) the activity coefficient of the hydrogen ions;
(iii) the liquid junction potentials at all the interfaces between pairs of adjacent solutions; and
(iv) interference from alkali metal ions (May et al., 1985).

In the determination of formation constants, the object is to make (i) predominant and either to eliminate the effects of (ii), (iii) and (iv) or to ensure that these factors are rigorously reproducible so that they do not bias inferred hydrogen ion concentrations. Factor (ii), the hydrogen ion activity coefficient, may be kept effectively constant through the use of a suitable background electrolyte for controlling the ionic strength of the experimental solution. Factor (iii) is somewhat complicated in that the liquid junction potentials arise from diffusion of all the types of ions concerned, across the solution interfaces. Thus it is important to make certain that the liquid junctions are reproducible in order to ensure constancy of the liquid junction potentials. Over and above this, the contributions of hydrogen and hydroxyl ions to the liquid junction potential at the interface between the bridge and experimental solutions are pH-dependent. This dependency is significant, however, only at lower pH's (in the case of hydrogen ions) of at higher pH's (in the case of hydroxyl ions). Taking all the foregoing into account, the hydrogen ion concentration-dependence of the e.m.f. may be represented by the Nernst-type equation,

$$E = E_{const} + s\log[H^+] \quad (1)$$

provided the ionic strength is maintained at a constant level and provided measurements are limited to the range,

$$2 < -\log[H^+] < 11.$$

With these restrictions in mind, it follows that calibration procedures are aimed at determining the values of E_{const} and s for a particular glass electrode-reference electrode pair. The value obtained for E_{const} applies strictly to the medium of the experiment and has to be re-determined in the event of changing to a new ionic strength. Indeed, E_{const} incorporates not only the intrinsic electrode parameter but also hydrogen ion activity coefficients and liquid junction potentials. The value obtained for s should approximate closely to the Nernstian value, namely 2.303RT/F where R is the gas constant, T the temperature in kelvin and F the Faraday constant. The experimental value of s is invariably smaller than the Nernstian value, but caution should be exercised in the event of the value turning out to be smaller than 95% of 2.303RT/F, since this is indicative of a deteriorating glass electrode.

The aim should be to determine the electrode parameters, E_{const} and s, in the pH range to be used in the subsequent equilibrium investigation of the metal-ligand system of interest. A sound approach is to carry out titrations of the ligand in question, at the required ionic strength, over the pH-range concerned. The titration data are processed by an appropriate program, such as ESTA (May et al., 1985; 1988) May and Murray (1988a; 1988b) to yield values of E_{const}, s, β_{011}, β_{012}, ...β_{01j}, where j is the maximum number of dissociable protons bindable by the ligand molecule. It is wise, of course, to carry out a series of titrations covering a reasonable range of concentrations of the ligand and replicates of each concentration. In this way, two birds are killed with one stone, in that the electrode parameters and the protonation constants of the ligand are determined simultaneously. If the ligand happens to have a pK_a less than 2 or greater than 11, the simple equation, (1), is not applicable and a more elaborate approach needs to be followed. The papers on ESTA (May et al., 1985; 1988; May and Murray, 1988a; 1988b) provide information which is useful for designing procedures to deal with such special cases.

A reality that has to be taken into account is that glass electrodes have a variable life-span and once deterioration sets in, the e.m.f. readings can no longer be relied upon to give accurate hydrogen ion concentrations. Hence, it is necessary to check the performance of the electrode system at reasonably frequent intervals, say once a fortnight during continuous use. Certainly, if the titration apparatus has been dormant for any length of time, the performance of the electodes needs to be checked immediately prior to the resumption of any programme of measurements. Three performance-checking procedures are worth carrying out. The first is to measure the potential of the glass electrode against that of another identical electrode which is known to be reliable; upon changing the pH of the solution into which the two electrodes are immersed, there should be no change in the e.m.f. reading. The second performance-checking procedure to be recommended is to determine the buffer line for the glass electrode-reference electrode pair. Here e.m.f. readings obtained after immersing the electrodes in a series of standard buffers are plotted against the buffer pH. A straight line should be obtained with a slope approximating closely to 2.303RT/F. When the slope falls to approximately 95% of the theoretical value, the electrode should be discarded or attempts should be made to regenerate it (Linder et al., 1984; Linnet, 1970). The third useful method of monitoring electrode performance is to examine the residuals (i.e. differences between observed and calculated e.m.f., total concentrations, titre or other suitable parameter) obtained from an optimization program, such as ESTA (May et al., 1988) when the calibration data are processed. A tendency for residuals to increase may be indicative of a deteriorating electrode.

V. ION-SELECTIVE ELECTRODE POTENTIOMETRY

Several different types of electrode have been developed, the potential of which is dependent upon the activity of a given metal aqua ion in solution. These comprise:
(i) metal or metal-amalgam electrodes (Biedermann and Sillén, 1953; Clayton and Vosburgh, 1937);
(ii) redox electrodes (Dellien, 1977; Rossotti and Whewell, 1977a; 1977b);
(iii) special types of glass electrode which respond selectively to alkali metal ions (Eisenman, 1967);
(iv) ion-selective electrodes (Covington, 1980; Bailey, 1976; Guilbault et al., 1976; Buck, 1974; Pungor and Toth, 1972; 1973; Moody and Thomas, 1972).

All of these have been applied to the determination of formation constants, the basic principle of application being the same as in pH-metry. Much attention has been given in recent years to the development of the ion-selective electrodes in class (iv) and, nowadays, these seem to be the most widely used in determining formation constants. Covington (1980) specifies precisely the term "ion-selective electrodes" as being applied to a range of membrane electrodes which respond selectively towards one (or several) ion species in the presence of others. The latter definition applies, also, to a range of electrodes the potential of which depends on the activity of given anions (Beck and Nagypál, 1990; Covington, 1980). Anion-selective electrodes may also be applied to the determination of formation constants for appropriate systems. Thus, in principle, an ion-selective electrode may be used instead of a glass electrode in a cell of the type described in section III, for investigating metal complex equilibria. In practice, however, it is not usual to do so because ion-selective electrodes tend to have a considerably smaller dynamic range than a glass electrode. This restricts the metal and ligand concentration ranges that can be covered in equilibrium titrations. Nevertheless, in instances where ion-selective electrodes have been applied it has been customary to couple their application with a glass electrode thereby yielding measurements of either free hydrogen and metal aqua ions or free hydrogen ions and free ligand concentrations (Beck and Nagypál, 1990). The advantage of this double electrode approach is an increase in the reliability of the information obtained from the e.m.f. measurements.

When ion-selective electrodes are used for the purpose of determining formation constants, it is vital to calibrate them appropriately. In this regard, it is essential to make use of metal ion buffers (Perrin and Dempsey, 1974) or free ligand buffers (Hartley et al., 1980). The detailed procedure is given lucidly by Hartley et al. (1980).

VI. SPECTROPHOTOMETRY

There are many familiar examples of the color changes which occur, as a result of complex formation. The deepening and intensifying of the blue color upon adding ammonia to copper sulfate solution is one such example. Others are the red color of hemoglobin (an iron(II) complex) and the green color of chlorophyll (a magnesium(II) complex). These color changes suggest the possibility of applying spectrophotometry to the determination of complex formation constants. Indeed spectrophotometry can be so applied to systems whose products, $M_pL_qH_r$ (refer to reaction (1) of section 1), have absorption spectra which differ characteristically and significantly from the absorption spectra of the reactants, M and L. The underlying principle involved in spectrophotometric determinations of formation constants is based on the Beer-Lambert-Bouguer law (eq. 2) which relates the absorbance ("optical density"), A, at a given wavelength of light to the molar concentration, c, of a light-absorbing species.

$$A = \varepsilon c l \qquad (2)$$

ε is a constant characteristic of the light-absorbing species and is termed its molar decadic absorption coefficient ("extinction coefficient");
l is the optical path length through the solution containing the light-absorbing species.

The absorbance is defined by equation (3)

$$A = \log_e(I_o/I) \qquad (3)$$

where I_o = the intensity of the incident monochromatic beam of light which passes through the solution under investigation, and I = the intensity of the emergent beam.

In the case of there being several, say n, species present in the solution, which absorb light of a given wavelength, the absorbance is given by equation (4)

$$A = \varepsilon_1 c_1 l + \varepsilon_2 c_2 l + \ldots \varepsilon_n c_n l \tag{4}$$

Some of the species may have ε values which are so small that the species makes no significant contribution to the absorbance of the solution. The molar decadic absorption coefficients of the components, M and L, are readily determined using solutions containing the pure component. In general, however, it is not possible to determine, directly, the ε values of the complexes. The concentration terms in equation (4) are related to the respective formation constants of the complexes concerned through mass action expressions such as equation (3) of section 1. The recommended experimental approach is to measure the absorbance, A, as a function of wavelength for a series of solutions involving wide ranges of total metal concentrations, total ligand concentrations and (metal concentration):(ligand concentration) ratios. The recently introduced photodiode-array spectrophotometric detectors (Lobinski and Marczenko, 1992; Grossman, 1989) should afford the opportunity of making a very large number of such measurements very efficiently. The data so obtained may then be processed as described in section 3 to yield formation constants for the system under investigation together with, incidentally, the values of the molar decadic absorption coefficients of the complexes.

In cases for which the investigator feels sure that only a single mononuclear binary complex is formed, the data may be processed by means of Job's method of continuous variations or the molar ratio method (Beck and Nagypál, 1990). Examples of such systems are evidently quite rare, however. For systems in which relatively few (say two or three) mononuclear binary complexes are formed, the method of corresponding solutions may be applied (Beck and Nagypál, 1990). On the other hand, since fast digital computers are readily available, it would appear to be preferable to use the more sophisticated non-linear least squares approach as described in Section A3 of this Chapter 1. There is, however, merit in carrying out a manual analysis as a preliminary to computer evaluation.

In most applications of the spectrophotometric method to the determination of formation constants, measurements have been made within the spectral range, 200 nm to 500 nm, that is, in the ultraviolet/visible region of the electromagnetic spectrum. In certain circumstances, measurements may be usefully made in the infra-red range. Further, there are instances where Raman spectral measurements may be usefully applied. Reference should be made to the book by Hartley et al. (1980) for discussions of the applicability of infra-red spectrophotometry and Raman spectroscopy.

VII. POLAROGRAPHY

Numerous formation constants have been determined by means of polarography and various elaborations of this technique, such as differential pulse polarography (Hogfeldt, 1982; Perrin, 1979; Anderegg, 1982; Killa, 1992; Rodriguez Placeres et al., 1992; Urbanska, 1992; Urbanska et al., 1992; Kurzak et al., 1992). Details of this method are described by Heyrovsky and Kuta (1966), Nancollas (1966), Hartley et al. (1980), Bard and Faulkner (1980), Nancollas and Thomson (1982), Beck and Nagypál (1990). The principle involved is to measure the shift in half-wave potential of the metal ion under investigation upon stepwise addition of the ligand. Under certain conditions and subject to some assumptions, equation (5) is valid, in many instances (Heyrovsky and Kuta, 1966; Kolthoff and Lingane, 1952; Nancollas and Thomson, 1982; Marin and Mendicuti, 1988).

$$\Delta E_{1/2} = (E_{1/2})_M - (E_{1/2})_{MLn} = (RT/nF)(\log_e \beta_{1n0} - n\log_e[L]) \tag{5}$$

Since the free ligand concentration is not readily determinable, a large excess of ligand over metal is used so that approximation (6) becomes applicable.

$$\text{Free ligand concentration, [L],} \sim \text{total ligand concentration, } T_L \tag{6}$$

Thus a linear plot of $\Delta E_{1/2}$ vs T_L yields the composition and the overall formation constant of the highest mononuclear binary complex formed in the system. The conditions under which equation (5) may legitimately be applied are discussed by Beck and Nagypál (1990) and Nancollas (1966). This approach illustrates one of the powers of polarography, in the present context, namely the availability of direct evidence for the maximum number of ligand molecules which can be bound by a given metal. The procedures can be modified to yield consecutive formation constants (Ringbom and Erikson, 1953; De Ford and Hume, 1951). Even when conditions deviate from those under which equation (5) is applicable, it is still possible to extract formation constants from the $\Delta E_{1/2}$ vs T_L data. The relevant approaches are described by Hartley et al. (1980) and Beck and Nagypál (1990).

One of the factors which has been recognized as contributing to errors in polarographically determined formation constants is adsorption of the ligand and/or the complexes at the surface of the mercury drop (Heyrovsky and Kuta, 1966; Bond and Hefter, 1971; Lovric, 1989). Suitable corrections for the adsorption effect can be applied, however. A major advantage of polarography is its usefulness as a complement to potentiometry for determining formation constants (Kurzak et al., 1992). Polarography can be used to obtain additional evidence for the model derived from potentiometric data collections. A general disadvantage is that experimental $\Delta E_{1/2}$ values are often small and difficult to measure accurately. Another frequently encountered disadvantage is that the polarograms obtained may be irreversible. Both of these factors lead to significant errors in the resulting formation constants.

VIII. NUCLEAR MAGNETIC RESONANCE SPECTROSCOPY

Applications of nuclear magnetic resonance spectroscopy to metal-ligand equilibrium studies are increasing rapidly. This technique can be used not only to determine formation constants but also to provide information on structures of the complexes formed as well as binding sites on the ligand for hydrogen and metal ions (Beck and Nagypál, 1990; Hartley et al., 1980; Crisponi et al., 1984; Rizkalla et al., 1988; Dhansay et al., 1990; Farkas and Kiss, 1990; Uguagliati and Canovese, 1991; Kurzak et al., 1992).

IX. CONCLUSION

pH-metry is the most powerful method for determining formation constants. This is testified to by the information given by Martell and Motekaitis (1990) that 90% or more of formation constant data have been obtained by pH-metry. The method is sensitive, precise and reliable. The accuracy of the formation constants obtained depends on the control of systematic errors. Indeed, a disadvantage of the method sometimes arises in cases when a postulated complex species turns out to be an error-absorbing parameter, there actually being no chemical justification for having postulated the species concerned. Such false species tend to appear when the experimental conditions have not been properly controlled or when the ligand has not been purified sufficiently.

The other methods discussed in this section should usually be regarded as useful complements to the pH-metric method. Thus increased confidence in a postulated chemical model can sometimes be obtained by coupling glass-electrode e.m.f. measurements with e.m.f. measurements obtained from an appropriate ion-selective electrode which is sensitive to either the metal ion or the ligand of the system being investigated. Spectrophotometry can, likewise, be used for confirming a chemical model postulated on the basis of pH-metric measurements. Spectrophotometry is also useful, in some instances, for investigating solutions with properties outside of the range of applicability of pH-metry, for example when the pH is smaller than about 2 or exceeds 12. There are also instances in which spectrophotometry can be used when concentrations of metal ion or ligand are limited to very small values, e.g. by solubility.

Polarography, too, more often than not provides useful data which are complementary to pH-metric data. Thus, by means of polarography, the maximum coordination number of a given metal towards a given ligand can be ascertained. Further, polarographic measurements can be applied to confirming a chemical model postulated on the basis of pH-metric measurements. The actual formation constant values obtained from polarography should approximate to the pH-metric formation constants, but the errors tend to be appreciably greater.

Finally, attention must be drawn to NMR techniques which are uniquely useful for identifying the proton- and metal-binding sites on the ligand molecule.

REFERENCES

Anderegg, G. (1982). *Pure Appl. Chem.*, *54*: 2693.

Bailey, P.L. (1976). *Analysis with Ion-Selective Electrodes*, Heyden, London, New York, Rheine.

Bard, A.J., and Faulkner, L.R. (1980). *Electrochemical Methods*, Wiley, New York.

Beck, M.T., and Nagypál, I. (1990). *Chemistry of Complex Equilibria*, Ellis Horwood Limited, Chichester.

Biedermann, G. (1956). *Rec. Trav. Chim. Pays-Bas*, *75*: 716.

Biedermann, G., and Sillén, L.G. (1953). *Arkiv Kemi*, *5*: 425.

Bond, A.M., and Hefter, G.T. (1971). *J. Electroanal. Chem. Interfacial Electrochem.*, *31*: 477.

Braibanti, A., Ostacoli, G., Paoletti, P., Pettit, L.D., and Sammartano, S. (1987). *Pure Appl. Chem.*, *59*: 1721.

Buck, R.P. (1974). *Anal. Chem.*, *46*: 28R.

Clayton, W.J., and Vosburgh, W.C. (1937). *J. Amer. Chem. Soc.*, *59*: 2414.

Covington, A.K. (1980). *Ion-Selective Electrode Methodology*, CRC Press, Inc., Boca Raton, Florida.

Crisponi, G., Caminiti, R., Biagini, S., Casu, M., and Lai, A. (1984). *Polyhedron*, *3*: 1105.

De Ford, D.D., and Hume, D.N. (1951). *J. Amer. Chem. Soc.*, *73*: 5321.

Dellien, I. (1977). *Acta Chem. Scand.*, *A31*: 473.

Dhansay, M.A., Linder, P.W., Torrington, R.G., and Modro, T.A. (1990). *J. Phys. Org. Chem.*, *3*: 248.

Eisenman, G. (1967). *Glass Electrodes for Hydrogen and Other Cations*, Edward Arnold, London; Marcel Dekker Inc., New York.

Farkas, E., and Kiss, T. (1990). *J. Chem. Soc., Perkin Trans.* 2: 1255.

Grossman, W.E.L. (1989). *J. Chem. Educ.*, *66*: 697.

Guilbault, G.G., Durst, R.A., Frant, M.S., Freiser, H., Hansen, E.H., Light, T.S., Pungor, E., Rechnitz, G., Rice, N.M., Rohm, T.J., Simon, W., and Thomas, J.D.R. (1976). *Pure Appl. Chem.*, *48*: 127.

Hartley, F.R., Burgess, C., and Alcock, R.M. (1980). *Solution Equilibria*, Ellis Horwood Limited, Chichester.

Heyrovsky, J., and Kuta, J. (1966). *Principles of Polarography*, Academic Press, New York.

Hogfeldt, E. (1982). *Stability Constants of Metal-Ion Complexes*, IUPAC, Pergamon Press, Oxford, New York, Toronto.

Jones, A.D., and Williams, D.R. (1970). *J. Chem. Soc. A*, 3138.

Killa, H.M. (1992). *Transition Met. Chem.*, *17*: 55.

Kolthoff, I.M., and Lingane, J.J. (1952). *Polarography*, Wiley Interscience, New York.

Kurzak, B., Spychala, J., and Swiatek, J. (1992). *J. Coord. Chem.*, *25*: 95.

Linder, P.W., Torrington, R.G., and Williams, D.R. (1984). *Analysis Using Glass Electrodes*, Open University Educational Enterprises Ltd.

Linnet, N. (1970). *pH Measurements in Theory and Practice*, Radiometer A/S, Copenhagen.

Lobinski, R., and Marczenko, Z. (1992). *Crit. Rev. Anal. Chem.*, *23*: 55.

Lovric, M. (1989). *Anal. Chim. Acta*, *218*: 7.

Marin, D., and Mendicuti, F. (1988). *J. Chem. Educ.*, *65*: 916.

Martell, A.E., and Motekaitis, R.J. (1988). *Determination and Use of Stability Constants*, VCH.

Martell, A.E., and Motekaitis, R.J. (1990). *Coord. Chem. Rev.*, *100*: 323.

May, P.M., and Murray, K. (1988a). *Talanta*, *35*: 927.

May, P.M., and Murray, K. (1988b). *Talanta*, *35*: 933.

May, P.M., Murray, K., and Williams, D.R. (1985). *Talanta*, *32*: 483.

May, P.M., Murray, K., and Williams, D.R. (1988). *Talanta*, *35*: 825.

Meloun, M., Havel, J., and Hogfeldt, E. (1988). *Computation of Solution Equilibria*, Ellis Horwood Ltd, Chichester.

Moody, G.J., and Thomas, J.D.R. (1972). *Talanta*, *19*: 623.

Nancollas, G.H., and Tomson, M.B. (1982). *Pure Appl. Chem.*, *54*: 2675.

Nancollas, G.H. (1966). *Interactions in Electrolyte Solutions*, Elsevier Publishing Company, Amsterdam, London, New York.

Perrin, D.D. (1979). *Stability Constants of Metal-Ion Complexes*, IUPAC, Pergamon Press, Oxford.

Perrin, D.D., and Dempsey, B. (1974). *Buffers for pH and Metal Ion Control*, Chapman and Hall, London.

Pungor, E., and Toth, K. (1972). *Pure Appl. Chem.*, *31*: 521.

Pungor, E., and Toth, K. (1973). *Pure Appl. Chem.*, *34*: 105.

Ringbom, A., and Eriksson, L. (1953). *Acta Chem. Scand.*, *7*: 1105.

Rizkalla, E.N., Niu, C., and Choppin, G.R. (1988). *Inorg. Chim. Acta*, *146*: 135.

Rodriguez Placeres, J.C., Miquel, T.B., Cabrera, G.R., and Sanz Alaejos, M.T. (1992). *Talanta*, *39*: 613.

Rossotti, F.J.C., and Whewell, R.J. (1977a). *J. Chem. Soc., Dalton Trans.*, 1223.

Rossotti, F.J.C., and Whewell, R.J. (1977b). *J. Chem. Soc., Dalton Trans.*, 1229.

Rossotti, H. (1978). *The Study of Ionic Equilibria*, Longman, London and New York.

Rossotti, J.C., and Rossotti, H. (1961). *The Determination of Stability Constants*, McGraw-Hill Book Company, Inc.

Uguagliati, P., and Canovese, L. (1991). *Talanta*, *38*: 697.

Urbanska, J. (1992). *Anal. Chim. Acta*, *259*: 311.

Urbanska, J., Kozlowski, H., and Kurzak, B. (1992). *J. Coord. Chem.*, *25*: 149.

Willard, H.H., Merritt, L.L., Dean, J.A., and Settle, F.A. (1988). *Instrumental Methods of Analysis*, Wadsworth Publishing Company, California.

COMPUTATIONAL DETERMINATION OF FORMATION CONSTANTS

by

David J. Leggett and Gene G. Wu

Dow USA, Texas Operations
Analytical & Engineering Sciences,
Freeport, Texas 77541
U.S.A.

I. BACKGROUND

This chapter will concentrate on potentiometry, spectrophotometry, calorimetry and nuclear magnetic resonance spectroscopy as measurement processes that provide data from which formation constants may be obtained. The authors feel that these four techniques can generate data of sufficient precision to permit reliable estimation of the overall formation constants for the system(s) of interest. Two of the techniques also provide direct information regarding the environment and/or the thermodynamics of the binding site(s).

We will use the following notation to refer to overall formation constants, β. The general and cumulative reaction:

$$aA + bB + cC + \ldots \rightleftharpoons A_a B_b C_c \ldots$$

involves the uncomplexed components A, B, C, ...; the general product species $A_aB_bC_c$ ... ; and the stoichiometric indices a,b,c,....Therefore the overall formation constant, β_{abc}, is defined as:

$$\beta_{abc\ldots} = \frac{[A_aB_bC_c\ldots]}{[A]^a[B]^b[C]^c\ldots}$$

where $[A_aB_bC_c\ldots]$ is the concentration of the complex species and [A], [B], [C], are the concentrations of the uncomplexed components. Each discrete datum point is described by the following set of mass balance equations:

$$C_A = [A] + \Sigma\, a\, \beta_{abc\ldots}\, [A]^a\, [B]^b\, [C]^c$$

$$C_B = [B] + \Sigma\, b\, \beta_{abc\ldots}\, [A]^a\, [B]^b\, [C]^c$$

$$C_C = [C] + \Sigma\, c\, \beta_{abc\ldots}\, [A]^a\, [B]^b\, [C]^c$$

II. NON-LINEAR LEAST SQUARES

The majority of programs used to estimate formation constants employ some variant of the non-linear squares algorithm. Using initial estimates of the formation constants, a set of simultaneous normal equations is built up then solved in order to obtain the corrections to be applied to the overall formation constants. Their new values are a better approximation for the final values and are employed in the next refinement cycle. Such an iterative

Table 1

Observable Parameters Used in the Objective Function for Minimization of Sum of Squares

Technique	Observables minimized
Potentiometry	pH, E, C_H, C_M, C_L, V_B, all $C_{component}$
Calorimetry	q_r, [all $C_{component}$]
Spectrophotometry	A, pH for aqueous systems
Polarography	$E_{1/2}$, [+ pH for aqueous systems]
Nuclear Magnetic Resonance	Average chemical shift of nuclei, δ_i

procedure ultimately provides values for β_{abc} that give the best agreement between calculated and experimental data, based on the equilibrium model described by the stoichiometric indices, a,b,c, ... The least squares criterion is used to drive the iterative process:

$$U(x) = \sum_{k=1}^{NPTS} w_k (Y_k^{obs} - Y_k^{calc})^2$$

where Y_k^{obs} is the observed data; Y_k^{calc} is the value of the objective function, defined in general terms as $f(x, t)$; w_k is the weighting factor for the objective function. The weighting question is an ongoing debate (May and Murray, 1988).

The parameter vector (or dependent variable vector), **x**, will always contain β_{abc} and may also include additional parameters relating to the equilibrium model (molar absorptivity or heats of formation of the complexes, etc) and/or parameters relating to the experimental conditions (analytical concentrations, electrode calibration constants, etc). The vector **t** (independent variable vector) contains the data determined during the experiment. Thus we seek to locate the dependent variable vector that will minimize U; in other words, provide the best set of variables that will reliably predict the observations.

During the last twenty years, the number of non-linear least squares algorithms used to estimate formation constants has increased. The major ones are, Newton-Raphson, Gauss-Newton, Levenberg-Marquardt, Fletcher-Powell, Variable Metric Method, and Pit-Mapping (Leggett, 1985).

A review of the literature reveals that a number of observable parameters are used in the objective function for minimization of sum of squares (see Table 1).

III. TECHNIQUES TO GENERATE DATA FOR REFINEMENT

A. Potentiometric methods

Of the several electrode processes available for use in determining formation constants, potentiometric techniques are the most widely used and are generally agreed to be accurate and precise. Measurements of $[H^+]$ are made using an appropriately calibrated glass and reference electrode system (May and Williams, 1985).

Other electrode processes include electrodes sensitive to the metal ion or the ligand, for example the many varieties of ion-selective electrodes. Such information, if available, is invaluable in providing direct access to p[M] and/or p[L]. The reader is directed to excellent monographs on the subject of experimental techniques by Rossotti (1978), Martell and Motekaitis (1990), and Linder et al. (1984). Polarography, although a limited technique due to the requirement of electrode reversibility, provides another way of obtaining p[M] data.

Potentiometric methods rely on the familiar Nernst equation and an electrochemical cell that contains at least one salt bridge:

reference with salt bridge $|_J$ Test Solution $|_I$ probe with salt bridge

$$E_{Cell} = E_{Probe} - E_{Ref} + E_J + E_I$$

The presence of liquid junction potentials becomes, in high precision potentiometric work, an operational difficulty that must be taken into account. Linder et al. (1984) provide an admirable account of the practical aspects of using the pH electrode.

Various objective functions have been defined for the non-linear least squares search for the best set of formation constants from potentiometric data.

SUPERQUAD (Gans et al., 1985) employs the weighted electrode potential in the objective function to set up the sum of squares of residuals:

$$U(\beta) = \sum_{i=1}^{NPTS} \left[(E^{obs} - E^{calc})^2 / \sigma^2(E)\right]_i$$

$$\sigma^2(E)_i = \sigma_c^2 + \left(\sigma_V \cdot \frac{dE_i}{dV_i}\right)^2$$

Motekaitis and Martell (1982) use hydrogen ion concentrations:

$$U(\beta) = \sum_{i=1}^{NPTS} w_i (p[H]^{obs} - p[H]^{calc})_i^2$$

$$w_i = 1/\left[p[H]_{i+1} - p[H]_{i-1}\right]^2$$

MINIQUAD82 (Vacca and Sabatini, 1985) minimizes the full component concentration array:

$$U(\beta) = \sum_{i=1}^{NPTS} \sum_{j=1}^{NCOMP} (C_{i,j}^{obs} - C_{i,j}^{calc})^2$$

SCOGS (Sayce, 1968) and SCOGS2 (Perrin and Stünzi, 1985) minimize the hydrogen ion concentration special case, the volume of titrant:

$$U(\beta) = \sum_{i=1}^{NPTS} (V_{titrant}^{obs} - V_{titrant}^{calc})_i^2$$

As the convergence reliability of least squares algorithms and experience with their use has increased, the number of parameters that have been refined has risen. For potentiometry, in addition to formation constants, several other parameters have been refined:

$$E^\circ ;\ s_k ;\ pK_w ;\ C_{init} ;\ C_{titr} ;\ V_{init} ;\ j_H ;\ j_{OH}.$$

Meloun and Havel (1985) present a comparison of twelve programs for potentiometric data and provide the results of data analysis of ten sets of real data using eight of the twelve programs. These authors provide a valuable comparison of the significant programs for processing potentiometric data. Casassas et al. (1986) have compared the principle programs for potentiometric data. Their observations on model selection and weighting schemes are well taken.

B. Spectrophotometric methods

The Beer-Lambert-Bouguer law is used to relate the concentration of complexes, and hence formation constants:

$$A_{i,k} = \text{path} \cdot \sum_{j=1}^{NSP} [\text{species}]_{i,j} \cdot \varepsilon_{j,k}$$

for k wavelengths, and i solutions (spectra). Typically the UV and visible spectral regions are studied. The absorbance range of the solutions should be limited to 0.01 - 0.8 in order to minimize measurement error. Typical accuracy of absorbance values is ± 0.0005 to ± 0.002, a precision comparable to potentiometry. The error sum square to be minimized is, for absolute error estimation:

$$U(\beta,\varepsilon) = \sum_{k=1}^{NWVL} \sum_{i=1}^{NSOLN} (A_{i,k}^{obs} - A_{i,k}^{calc})^2$$

and for relative error estimation, U is defined as:

$$U(\beta,\varepsilon) = \sum_{k=1}^{NWVL} \sum_{i=1}^{NSOLN} \left(\frac{A_{i,k}^{obs} - A_{i,k}^{calc}}{\sqrt{A_{i,k}^{calc} \cdot A_{i,k}^{obs}}} \right)^2$$

The concentration range for complexes in solution is given by the general relationship concentration range $(0.01 \rightarrow 0.8) / \varepsilon_{j,k}$. Since, from quantum mechanical considerations, molar absorptivities are not greater than 1.5×10^5, concentrations of less than 10^{-5}M can be measured - an advantage where limited solubility makes it difficult to obtain reliable potentiometric data. The technique also can provide valuable information derived from the molar absorptivity of each complex.

C. Calorimetric methods

Since biological reactions occur with an appreciable change in the heat content, calorimetry is a universal tool for the determination of formation constants and the corresponding enthalpy changes. Over the last fifteen years, advances have occurred in the development of sensitive microcalorimeters and computational procedures to extract information, such as ΔH and ΔS. The majority of biological complexing reactions occur in an essentially constant temperature environment. Therefore, isothermal heat conduction calorimetry (HCC), isoperibolic titration calorimetry (ITC), and isothermal power compensation calorimetry (IPC) have become the methods of choice for these studies. (Hansen et al., 1985; Martin and Marini, 1979).

Since there are many ways for heat to be gained or lost in equipment designed to measure thermodynamic parameters, considerable care must be taken when developing data reduction methods for the calorimetric measurements. Corrections equations for heat effects, not due to the reaction under study, may be found elsewhere (Eatough et al., 1982). Once all heat sources are known, the heat of reaction, q_r , is given by:

$$q_r = q_{tot} - [\, q_{dil} + q_{hl} + q_{td} \,]$$

where q_{tot} is the total heat measured; q_{dil} is the heat of dilution of reactants; q_{hl} is the heat leakage during the course of the reaction; q_{td} is the heat resulting from the temperature differences between various reactants. This equation can be simplified by dropping the last two terms if HCC or IPC techniques are used. The objective function for the array of reaction heats, q_r, is:

$$q_r = \sum_{i=1}^{NPTS} (\Delta H_j \cdot [\text{species}]_{j,i})$$

and the sum of squares to be minimized is:

$$U(\beta_j,\Delta H_j) = \sum_{i=1}^{NPTS} \left((q_r^{obs})_i - \sum_{j=1}^{NSP} \Delta H_j \cdot [\text{species}]_{j,i} \right)^2$$

Arena et al. (1989) have provided a review of the pros and cons of using calorimetry to estimate formation constants. Their observations, in particular on the ability of the technique to distinguish multiple equilibria, are well taken.

The best values for β_j and ΔH_j are those that minimize $U(\beta_j, \Delta H_j)$ or satisfy both

$$\frac{\partial U(\beta_j,\Delta H_j)}{\partial \Delta H}=0=\sum_{i=1}^{NPTS}\left[(q_r)_i\cdot\Delta[\text{species}]_{j,i}-\Delta[\text{species}]_{j,i}\cdot\sum_{i=1}^{NSP}\Delta H_j\cdot\Delta[\text{species}]_{j,i}\right]$$

and

$$\frac{\partial U(\beta_j,\Delta H_j)}{\partial \Delta H}=0=\sum_{i=1}^{NPTS}\left[(q_r)_i-\sum_{i=1}^{NSP}\Delta H_j\cdot\Delta[\text{species}]_{j,i}\right]\cdot\sum_{j=1}^{NSP}\left(\frac{\partial \Delta H_j\cdot\Delta[\text{species}]_{j,i}}{\partial \beta}\right)$$

The complete solution for β_j and ΔH_j involving the non-linear least squares approach have been discussed in detail (Eatough et al., 1972).

The particular problem of reaching a minimum value of U for calorimetric data has been of great interest. Three methods have been used frequently:

(i) mapping of U using a non-least squares minimization. This method has been used extensively for systems involving one or two equilibria; programs for this purpose are by Fenby and Hepler (1974) and Eatough et al. (1974); EQDH-REACTIONS (Nakatsuji et al., 1985) and NET (Mullens et al., 1985) have been used for many complexation systems;

(ii) pit-mapping to find U_{min} by direct differentiation proposed by Sillén (1962);

(iii) simultaneous solution of equations. ES5CM (Alessandro et al., 1989) can deal with any complex coordination system. Program MULTIK (Oscarson et al., 1988) has been used for acid gas treatment and geothermal systems where both K_j and ΔH_j were sought.

Although the versatile and non-destructive nature of calorimetry is a significant advantage for biological systems, there are two inherent limitations for the simultaneous calculation of β_j and ΔH_j when compared with potentiometry. First, $\beta_{overall}$ can be determined simultaneously with $\Delta H_{overall}$ only if the magnitude of $\beta_{overall}$ is within certain limits (Christensen et al., 1972). For equilibrium systems with log $\beta_{overall} < 0.5$ the calorimetric signals become indistinguishable from the background. On the other hand, for the systems with log $\beta_{overall} > 4.5$, the reaction(s) have approached completion and the differences in heats due to minor changes in equilibrium position also become indistinguishable.

Second, the number of parameters (β_j, ΔH_j) to be determined is twice as many as in potentiometry. The inherently small differences between calorimetric titration curves makes it more difficult to distinguish the different reaction systems. The problem is alleviated by ensuring that the system of interest is significantly overdetermined. This is usually accomplished by generating several independent data sets, with different titration ratios and concentrations, combined together for data refinement. Also, the magnitude of $\beta_{overall}$ to be determined needs to be within the limits stated above. It is also advantageous to perform multi-technique studies: calorimetry to determine $\Delta H_{overall}$ for the model that has been determined separately by a potentiometric study.

D. Nuclear magnetic resonance methods

The majority of formation constant measurements have been carried out in aqueous solution, where electrochemical techniques are particularly effective. However, electrochemical methods are often less effective in non-aqueous or mixed solvents. Although spectrophotometry or calorimetry provides part of the solution to this problem, these techniques cannot provide formation constants in conjunction with microscopic binding constants for specific sites of the molecule. NMR fills this gap and is well-suited to the task.

Spin-lattice relaxation times, T_1, and chemical shift methods have been used to determine constants (James 1975; Connors 1987). However, since standard errors for T_1 are normally about 5 - 10%, this method is often used for semi-quantitative binding strength studies (Wu et al., 1991).

Under conditions of fast exchange on the NMR time scale, the average chemical shift of a nucleus is the weighted average chemical shift of the nucleus on the free and complexed ligands

$$\delta_{avg}=X_i\delta_i+\sum_{j=1}^{NSP}X_{c,j}\,\delta_{c,j}$$

where δ_{avg} is the theoretical average chemical shift of the nucleus of interest; δ_l is the chemical shift of the specific nucleus (site) of the free ligand; δ_c is the chemical shift of that site for the complexed ligand; X_l is the corresponding mole fraction.
For nuclear magnetic resonance the SSR is defined by:

$$U(\beta,\delta_{avg}(\delta_c)) = \sum_{i=1}^{NPTS}(\delta_i^{obs} - \delta_{avg,i}^{calc})^2$$

For the simplest case, the formation of ML, δ_{avg} is defined by:

$$\delta_{avg} = \delta_l + (\delta_c - \delta_l)X_c$$

For multiple equilibria, more complicated forms of δ_{avg} are derived (Popov and Boss, 1985).

The accuracy of the method depends heavily on the chemical shift range of the nucleus and the broadness of the resonance lines. Similar to calorimetry, the NMR method has an upper restriction of $\log \beta \leq 5$; competitive methods provide access to $\log \beta > 5$. Good examples are given using the program KITFIT (Dye and Nicely, 1971) and LETAGROP-SPEFO (Glaser et al., 1986).

Among many NMR chemical shift methods including ^1H, ^{13}C, and metal ion isotope NMR, ^1H NMR chemical shift method provides a promising means to estimate β_{abc} for a biochemical system in which only a few milligrams of material are available. However, the proton signal used as a probe for the complexation reaction is not necessarily directly involved in bond breaking/forming and may also be far from the coordination site(s). Therefore, the observed proton shift may not be directly linked to the thermodynamic stability of the complex.

By assuming appropriate β's for the complexing system (single or multiple equilibria), an array of X_c can be calculated using the known initial concentrations and solving the polynomial equation in X_c derived from the mass balance equations. For the systems in which the value of δ_c cannot be obtained experimentally, an iterative procedure, similar to the calorimetric approach, must be used. The best values for β and δ_c are those that minimize $U(\beta_j, \delta_{c,j})$ or satisfy

$$\partial U(\beta_j, \delta_{c,j}) / \partial \delta_c = 0 \quad \text{and} \quad \partial U(\beta_j, \delta_{c,j}) / \partial \beta = 0$$

Using calculated δ_c and an array of the observed δ_i, U_{min} can be obtained.

E. Programs to estimate formation constants from data

A variety of established, well tested programs are available to extract constants from the primary data. A reasonably comprehensive list is shown in Table 2.

Despite the use of sophisticated data processing, with or without the use of graphical pre-processing, questions still remain which are fundamental to any model-fitting process. Briefly the questions are:

- When is the fitted model "correct"?
- What is a "good" fit to the data?
- How does one more species improve the fit?

Clearly, answers to these basic questions are dependent upon quality of the data, overall goodness-of-fit, and the statistics for the best model. However, relative to the advances of data generation and processing over the last 30 years, model selection still remains a relatively subjective procedure and a rich area for research.

F. Simultaneous estimation of model and formation constants

These least squares algorithms require that the equilibrium model be defined prior to executing the program. An approach pioneered by Hedström and Österberg, recently reviewed (Leggett, 1985), has the unique capability of simultaneously estimating both the formation constants and the model. The elegance of the approach, and the fact that "no a

Table 2

Programs to Estimate Formation Constants from Experimental Data

Program	Author(s)	Primary data
LETAGROP (VRID)[1]	Sillén et al.	Pot, Spectro, Calori
SCOGS(2)[2]	Perrin / Sayce / Stünzi	pH/pM-metric
TITFIT[3]	Kaden / Zuberbühler	pH-metric
MINIQUAD82,[4] SUPERQUAD[5]	Gans / Vacca / Sabatini	Potentiometric
STBLTY[6]	Avdeef	pH/pM-metric
ESTA[7]	May / Williams	Potentiometric
ACBA[8]	Arena et al	Potentiometric
POLET[9]	Havel / Meloun	Potentiometric
NONLIN15[10]	Taylor	Potentiometric
DALSFEK[11]	Hartley / Alcock	Spectro or Potentiometric
SQUAD[12]	Leggett	Spectrophotometric
SPECFIT[13]	Gampp / Zuberbühler	Spectrophotometric
CFTSP[14]	Rotunno / Meites	Spectrophotometric
LESSDAD[15]	Leporati	Spectrophotometric
SQUAD-MINUIT[16]	Havel / Korinek	Spectrophotometric
ES5CM[17]	Alessandro	Calorimetric
MULTIK[18]	Oscarson	Calorimetric
EQDH-REACTIONS[19]	Izatt / Eatough	Calorimetric
KITFIT[20]	Dye / Nicely	NMR
LETAGROP-SPEFO[21]	Glaser	NMR
POLAG[22]	Leggett	Polarographic

1. Ingri and Sillén (1964)
2. Perrin and Stünzi (1985)
3. Zuberbühler and Kaden (1980)
4. Vacca and Sabatini (1985)
5. Gans et al. (1985)
6. Avdeef et al. (1992)
7. May et al., (1985)
8. Arena et al., (1979)
9. Havel and Meloun (1986)
10. Taylor et al., (1988)
11. Hartley et al., (1980)
12. Leggett and McBryde (1975)
13. Gampp et al., (1985)
14. Lampugnani et al., (1987)
15. Leporati (1990)
16. Havel and Korinek (1987)
17. Alessandro et al. (1989)
18. Oscarson et al. (1988)
19. Nakatsuji et al. (1985)
20. Dye and Nicely (1971)
21. Glaser et al. (1986)
22. Leggett (1980)

priori assumption, whatsoever, is made as to the composition, or abundance, of the complexes formed" (Hedstrom, 1955) is well demonstrated by Avdeef et al. (1992).

G. In conclusion

Since more than just formation constants are unknowns in the parameter equations for spectrophotometry, calorimetry, and NMR, these techniques are less favored than potentiometry from a mathematical viewpoint due to the greater number of unknowns that must be extracted from the data.

However, spectrophotometry, calorimetry and NMR provide experimental routes to formation constants that are not possible if only e.m.f. measurements are considered. Also, it is well worthwhile obtaining estimates of the formation constants by potentiometry followed by the latter three techniques to substantiate the estimates and provide reliable information on other properties of the system.

IV. ALTERNATIVE STRATEGIES TO ESTIMATE FORMATION CONSTANTS

Generally, computational methods for the estimation of formation constants implies the use of computer programs to process the data generated from experimental procedures, such as those described in the preceding section. Therefore, it could be concluded that without this data no equilibrium information about the system is available. The final part of this chapter will review an alternative approach.

A. Previewing the equilibrium system using graphical methods

In 1993 the question could be asked 'Given the ease of use, power, and availability of computer programs is there any value in graphical methods these days?' Maybe not as the primary route to the estimation of formation constants. However, there is value in the use of graphical techniques, aided by spreadsheet PC programs, to assess the equilibrium picture, *prior to full computer analysis*.

The graphical methods are derived from the familiar formation function for the average number of ligands and protons bound to the metal and ligand respectively:

$$\bar{n} = \frac{C_L - \sum_{j=0}^{J}[LH_j]}{C_M} = \frac{C_L - \left\{\frac{C_H - [H] + [OH]}{\bar{n}_H}\right\}}{C_M} = \frac{\sum_{n=0}^{N} n\beta_n [L]^n}{\sum_{n=0}^{N} \beta_n [L]^n}$$

where

$$\bar{n}_H = \frac{C_H - [H] + [OH]}{C_L} = \frac{\sum_{j=0}^{J} j\beta_j^H [H]^j}{\sum_{j=0}^{J} \beta_j^H [H]^j}$$

Originally, these equations were accompanied by several restrictions. However, these "deviations from ideality" provide diagnostic information about the prevailing equilibria. Formation function curves that do not overlay, irrespective of the metal to ligand concentration ratio, have the following characteristics: (i) curl backs signify the presence of hydroxy species; (ii) humps indicate the presence of protonated species; (iii) curve crossovers and spreading are caused by polynuclear species.

A recent publication by Findlow et al. (1990) illustrates the point. Additional formation functions are defined to provide extra information from an initial data analysis, prior to a full non-linear least squares analysis of the data.

B. Other methods to estimate formation constants

The overall premise is that by the appropriate selection of unique sets of descriptors, basic thermodynamic information may be estimated. This approach is, of course, not new. Linear free energy relationships, and more recently structure activity relationships relating $\log K_1$ or $\log K_2$ for a single complex with a property related usually to the ligand have been available for at least 50 years.

In the area of solution equilibria other than linear free energy studies, a number of focused studies have been published; three are presented briefly below:

Christensen et al. (1969) devised an additive approach to the entropy of protonation of amine; Barbucci et al. (1970) extended the additive approach to the enthalpy of diamines. Their equation is shown below:

$$\Delta H = \Delta H_{ref} + \Sigma\{0.5^{n-1}\,\delta(C) + 0.5^{n-1}\,\delta(NH_2) + 0.5^{n-1}\,\delta(NH_3^+)$$

Harris (1977) has cast the additivity equation in terms of stability terms for functional groups, and 5- and 6-membered ring contributions. The equation is shown below along with Table 3 that provides information concerning the values used.

Table 3

Parameters Used in Harris' Equation

Parameter X_i	Metal ion			
	Ni	Zn	Cd	Fe(III)
Amine	3.0	2.5	2.6	4.7
Carboxylate	2.1	1.6	1.7	3.4
Phenol	-	-	-	9.4
r_5	0.8	0.9	0.09	0.4
f_5	1.0	1.1	2.4	1.0
r_6	-0.3	-0.2	-0.9	-
f_6	0.9	1.7	1.4	-

$$\log K_{ML} = \sum_{i=1}^{n} n_i X_i + \left[r_5 + \sum_{i=2}^{n_5} (r_5 f_5^{i-1}) \right]_{n_5 \neq 0} + \left[r_6 + \sum_{i=2}^{n_6} (r_6 f_6^{i-1}) \right]_{n_6 \neq 0}$$

Tanaka (1974) has approached the question of predicting formation constants from basic information for ML_2 once K_{ML} is known. He has incorporated interaction terms between like atoms of the first and second ligand as well as unlike atoms.

$$\log K_{ML_2} = \log K_{ML} + \Delta + \log\left[\frac{\#\ H_2O\ ML}{\#\ H_2O\ M}\right] + \delta_{NN}N^2 + \delta_{NO}NO + \delta_{ON}ON + \delta_{OO}O^2 - \log 2$$

where Δ = outer sphere complexation difference between ML_2 and ML; δ_{NN} = interaction between N amine in 1st ligand and N amine in 2nd ligand; δ_{NO} = interaction between N in 1st ligand and O in 2nd ligand; N = number of coordinating amine nitrogens in ligand; O = number of coordinating carboxylates in ligand.

This brief overview has shown that this approach has much to recommend it. Although the unequivocal estimation of the formation constant still comes from direct measurements, this process is relatively slow. Optimistically, only 10-20 systems can be fully characterized by one researcher per year. On the other hand this "first-principles" approach, once fully developed, could generate the same formation constants in a matter of hours. For the present, however, experimentally determined formation constants remain *de rigueur*.

REFERENCES

Alessandro, D.R., Concetta, D.S., and Carmelo, R. (1989). *Thermochim. Acta, 138*: 141.

Arena, G., Rizzarelli, E., and Sammartano, S. (1979). *Talanta, 26*: 1.

Arena, G., Cali, R., Maccarronne, G., and Purrello, R. (1989). *Thermochim. Acta, 155*: 353.

Avdeef, A., Hartenstein, F., Chemotti, A.R., and Brown, J.A. (1992). *Inorg. Chem., 31*: 3701, and references therein.

Barbucci, P., Paoletti, P., and Vacca, A. (1970). *J. Chem. Soc.*, 2202.

Casassas, E., Tauler, R., and Filella, M. (1986). *Anal. Chim. Acta, 191*: 399.

Christensen, J.J., Izatt, R.M., Wrathal, D.P., and Hansen, L.D. (1969). *J. Chem. Soc.*, 1212.

Christensen, J.J., Ruckman, J., Eatough, D.J., and Izatt, R.M. (1972). *Thermochim. Acta, 3*: 203.

Connors, K.A. (1987). *Binding Constants*, Wiley, New York, p. 189.

Dye, J.L., and Nicely, V.A. (1971). *J. Chem. Educ.*, *48*: 443.

Eatough, D.J., Christensen, J.J., and Izatt, R.M. (1972). *Thermochim. Acta*, *3*: 233.

Eatough, D.J., Christensen, J.J., and Izatt, R.M. (1974). *Experiments in Thermometric Titration Calorimetry*, Brigham Young University Press, p. 109.

Eatough, D.J., Izatt, R.M., and Christensen, J.J. (1982). *Comprehensive Analytical Chemistry* (G. Svehla, ed.), Elsevier, New York, p. 112.

Fenby, D.V., and Hepler, L.G. (1974). *Chem. Soc. Rev.*, *3*: 193.

Findlow, J.A., Duffield, J.R., and Williams, D.R. (1990). *Bioavailability*, 2: 3.

Gampp, H., Maeder, M., Meyer, C.J., and Zuberbühler, A.D. (1985). *Talanta*, *32*: 257.

Gans, P., Sabatini, A., and Vacca, A. (1985). *J. Chem. Soc., Dalton Trans.*, *1985*: 1195.

Glaser, J., Henriksson, U., and Klason, T. (1986). *Acta Chem. Scand., Ser. A*, *A40*: 344.

Hansen, L.D., Lewis, E.A., and Eatough, D.J. (1985). *Analytical Solution Calorimetry*, (J.K. Grime, ed.), J. Wiley, New York, p. 57.

Harris, W.R. (1977). *J. Coord. Chem.*, *13*: 163.

Hartley, F.R., Burgess, C., and Alcock, R. (1980). *Solution Equilibria*, Ellis Horwood, Chicester.

Havel, J., and Korinek, P. (1987). *Scripta Fac. Sci. Nat. Univ. Purk. Brun.*, *17*: 327.

Havel, J., and Meloun, M. (1986). *Talanta*, *33*: 525.

Hedstrom, B. (1955). *Acta Chem. Scand.*, *9*: 613.

Ingri, N., and Sillén, L.G. (1964). *Arkiv Kemi*, *23*: 97.

James, T.L. (1975). *Nuclear Magnetic Resonance in Biochemistry*, Academic Press, New York, p. 119.

Lampugnani, L., Meites, L., Papoff, P., and Rotunno, T. (1987). *Anal. Chim. Acta*, *194*: 77.

Leggett, D.J. (1980). *Talanta*, *27*: 787.

Leggett, D.J. (1985). *Computational Methods for the Determination of Formation Constants* (D.J. Leggett, ed.), Plenum Press, New York, p. 1.

Leggett, D.J., and McBryde, W.A.E. (1975). *Anal. Chem.*, *47*: 1075.

Leporati, E. (1990). *Anal. Chim. Acta*, *234*: 381.

Linder, P.W., May, P.M., Torrington, R.G., and Williams, D.R. (1984). *Analysis Using Glass Electrodes*, Open University, Milton Keynes.

Martell, A.E., and Motekaitis, R.J. (1990). *Coord. Chem. Rev.*, *100*: 323.

Martin, C.J., and Marini, M.A. (1979). *CRC Critical Reviews in Analytical Chemistry*, *8*: 221.

May, P.M., and Murray, K. (1988). *Talanta*, *35*: 927.

May, P.M., and Williams, D.R. (1985). *Computational Methods for the Determination of Formation Constants* (D.J. Leggett, ed.), Plenum Press, New York, p. 37.

May, P.M., Murray, K., and Williams, D.R. (1985). *Talanta*, *32*: 483.

Meloun, M., and Havel, J. (1985). *Computation of Solution Equilibria*, Univ. J.E. Purkyne V Brne Press, p. 45.

Motekaitis, R.J., and Martell, A.E. (1982). *Can. J. Chem.*, *60*: 168.

Mullens, J., Yperman, J., Francois, J.P., and Van Poucke, L.C. (1985). *J. Phys. Chem.*, *89*: 2937.

Nakatsuji, Y., Bradshaw, J.S., Tse, P.K., Arena, G., Wilson, B.E., Dalley, N.K., and Izatt, R.M. (1985). *J. Chem. Soc. Chem. Commun.*, *1985*: 749.

Oscarson, J.L., Izatt, R.M., Brown, P.R., Pawlak, Z., Gillespie, S.E., and Christensen, J.J. (1988). *J. Sol. Chem.*, *17*: 841.

Perrin, D.D., and Stünzi, H. (1985). *Computational Methods for the Determination of Formation Constants* (D.J. Leggett, ed.), Plenum Press, New York, p. 71.

Popov, A.I., and Boss, R.D. (1985). *Inorg. Chem.*, *24*: 3660.

Rossotti, H. (1978). *The Study of Ionic Equilibria*, Longman Press, London.

Sayce, I.G. (1968). *Talanta*, *22*: 1397.

Sillén, L.G. (1962). *Acta Chem. Scand.*, *16*: 159.

Tanaka, M. (1974). *J. Inorg. Chem. Nucl.*, *36*: 151.

Taylor, P.D., Morrison, I.E.G., and Hider, R.C. (1988). *Talanta*, *35*: 507.

Vacca, A., and Sabatini, A. (1985). *Computational Methods for the Determination of Formation Constants* (D.J. Leggett, ed.), Plenum Press, New York, p. 99.

Wu, G.G., Jiang, W., Lamb, J.D., Bradshaw, J.S., and Izatt, R.M. (1991). *J. Amer. Chem. Soc.*, *113*: 6539.

Zuberbühler, A.D., and Kaden, Th.A. (1980). *Talanta*, *27*: 513.

DETERMINATION OF ENTHALPY CHANGES

by

Piero Paoletti and Antonio Bianchi

Università degli Studi,
Dipartimento di Chimica,
Via Maragliano 75/77
50144 Firenze
Italy

I. INTRODUCTION

The word enthalpy is derived from Greek words meaning "warmth within", and accordingly may be described as heat content. It is represented by the symbol H and can be defined by the equation (1), where U is the internal energy, P is the pressure and V is the volume.

$$H = U + PV \tag{1}$$

The enthalpy change that occurs during a chemical reaction at constant pressure is given in equation (2), where Q_p is the heat change involved in the process.

$$\Delta H = \Delta E + P\Delta V = Q_p \tag{2}$$

ΔH is negative for exothermic processes and positive for endothermic processes. There are two ways in which the enthalpy change accompanying chemical reactions such as those between metal ions and ligands in solution can be obtained. One way is to measure, by means of a calorimeter, the amount of heat absorbed or evolved in the reaction. In the indirect method, equilibrium constants are measured at various temperatures, and the Van't Hoff isochore is used to derive the value of ΔH.

In an equilibrium system the standard free energy change is given by equation (3). At 25°C ΔG^o is equal to $-1.3637 \cdot \log K$ when it is expressed in kcal mol^{-1},

$$\Delta G^o = - RT\log K \tag{3}$$

and $-5.706 \cdot \log K$ when it is expressed in kJ mol^{-1}. From the relationship $\Delta G^o = \Delta H^o - T\Delta S^o$ it can be seen that there are two contributions to the free energy change, an enthalpic contribution, ΔH^o, and an entropic term $T\Delta S^o$. Thus, knowledge of both ΔH^o and K leads to a knowledge of the standard entropy change. As we shall see in a following chapter, the combined examination of both the enthalpy and entropy terms can lead to useful information concerning the structures of complexes formed in solution equilibria.

II. THE VAN'T HOFF METHOD

Direct substitution of equation (3) into the Gibbs-Helmholtz equation results immediately in the Van't Hoff equation, or isochore, (4).

$$d(R\log_e K_p(T))/d(1/T) = -\Delta H^o(T) \quad (4)$$

To be used, this equation must be integrated; if ΔH^o is assumed to be independent of temperature, the integration is straightforward, and leads to equation (5).

$$\log K(T_1) = -\Delta H^o/2.303R(1/T) + \text{const} \quad (5)$$

However, $d(\Delta H^o)/dT = \Delta C_p$, so ΔH^o it is not exactly constant if the change in heat capacity, ΔC_p, is not constant. If the heat capacity change is expressed as a polynomial (6), the integration can be performed without the assumption of constant ΔH^o.

$$\Delta C_p = a + bT + cT^2 \quad (6)$$

The Van't Hoff method, though widely employed, must be used with a certain amount of caution. To illustrate the problem, let us consider a simple example in which the equilibrium constant has been measured at two temperatures, 283 and 323K. From equation (5) we can obtain the following expression

$$-\Delta H^o = 2.303R(\log K(T_1) - \log K(T_2))/(1/T_1 - 1/T_2) \quad (7)$$

Now suppose that the equilibrium constants were found to be $\log K(283) = 9.36 \pm 0.02$ and $\log K(323) = 8.75 \pm 0.02$. It follows that

$-\Delta H^o = 10462(9.36 - 8.75) = 6.38$ kcal mol^{-1},

$\sigma = 10462(0.02^2 + 0.02^2)^{1/2} = 0.28$ kcal mol^{-1}, i.e. $\Delta H^o = 6.1 \pm 0.3$ kcal mol^{-1}.

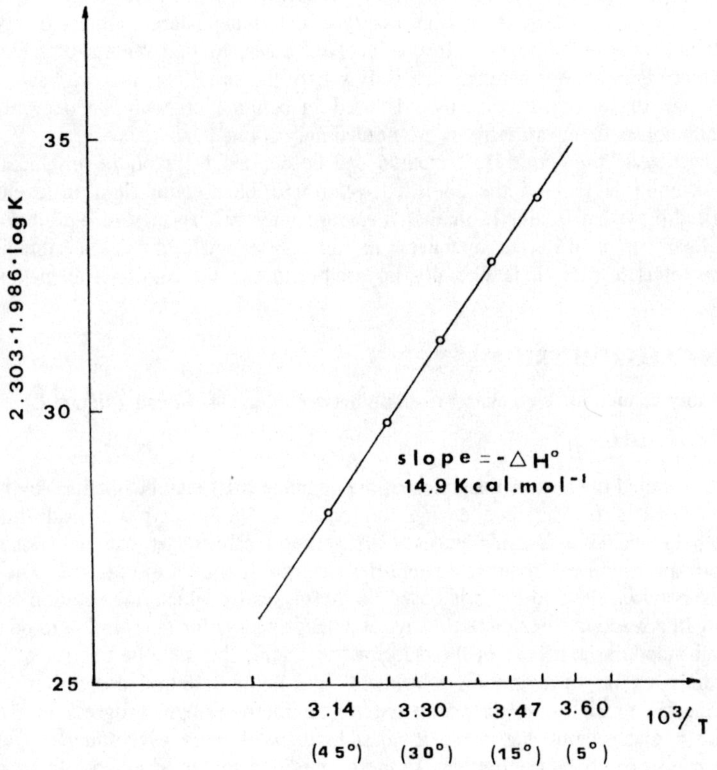

Figure 1.
Van't Hoff plot.

This very simple example shows that the error on ΔH^o is magnified by an enormous factor (10462 in this instance) when the Van't Hoff isochore is used directly.

In order to improve the precision of the determination of ΔH^o, two things are necessary. Firstly, the equilibrium constants should be determined as precisely as possible and secondly, the equilibrium constants should be measured at a number of different temperatures in order to apply the method of least-squares to equation (5) (see Fig. 1). The error on the slope is somewhat reduced by the least-squares process.

For the best results, the temperature range of the equilibrium constant determinations should be as large as possible. However, the wider the temperature range the less likely it is that the enthalpy change can be approximated by a constant value. In this case it becomes necessary to know the coefficients of the expansion of ΔC_p as expressed in equation (6).

It should also be remembered that the value of log K depends on the medium and its ionic strength. The formation constants of many complexes have been determined in a variety of ionic media, and in some cases the data were extrapolated to zero ionic strength. By contrast, little data are available concerning the effect of the ionic medium on the enthalpy changes accompanying complex formation. A rare example is the work of Everett and Pinsent (1952) who measured the two protonation constants of ethylenediamine by means of a hydrogen electrode without liquid junction in the temperature range 0-60°C: the two enthalpy changes were calculated by the Van't Hoff method, and the results were extrapolated to zero ionic strength. The values obtained for ΔH^o at 25°C are in excellent agreement with the best data obtained calorimetrically.

Unfortunately, the determination of the stability constants of metal complexes is often a long operation, and also often not an easy one. Commonly the ligand is both polybasic and potentially polydentate, so that the variety of species that may be formed in solution is rather extensive: the free ligand and its protonated and deprotonated forms, the free metal ion and metal ion complexes which may be mononuclear or polynuclear, protonated or deprotonated, or hydrolyzed. In addition, the ligand:metal ratio in the complexes may be 1, 2, ... As a consequence it is often necessary to determine a large number of equilibrium constants and it is therefore rare, in the involved cases, to find reports of measurements made at more than one temperature that fully satisfy the conditions that we have illustrated above. When the calorimetric method is used in conjunction with the determination of equilibrium constants, measurements are needed only at one temperature.

Nevertheless, the Van't Hoff method can be applied in favorable instances. A well known example is that of the so-called yellow to blue equilibrium in complexes of nickel(II); the system is simple in that there are only two species in equilibrium which absorb light in a different manner. In this case, ΔH has been obtained from spectrophotometric data; it is virtually impossible to use the calorimetric method in this example.

III. THE CALORIMETRIC METHOD

The enthalpy change for the general reaction between a metal, M and a ligand L:

$$M + L \rightleftharpoons ML \qquad (8)$$

(improperly called the heat of formation of the complex ML) may be obtained by measuring the heat liberated or absorbed during the reaction. Various types of calorimeters are commercially available for this purpose. A solution calorimeter can be classified as a reaction or a titration calorimeter; a simplified diagram is shown in Figure 2. The principal difference between the two types is based on the means by which one solution is added to the other. In a reaction calorimeter, the usual arrangement is for the solution to be contained in a small ampoule immersed in the calorimetric liquid, that may be opened or broken to initiate the reaction. In a titration calorimeter, the liquid is introduced through a narrow tube connected to an automatic piston burette or micro-syringe. Progress in the field of electronic miniaturization has made it possible to work with ever smaller quantities of solution at ever lower concentrations. In these cases one speaks of micro-calorimetry and it is not rare to use 2-3 cm^3 of final volume of solutions at concentrations of 10^{-2}-10^{-3}M or less.

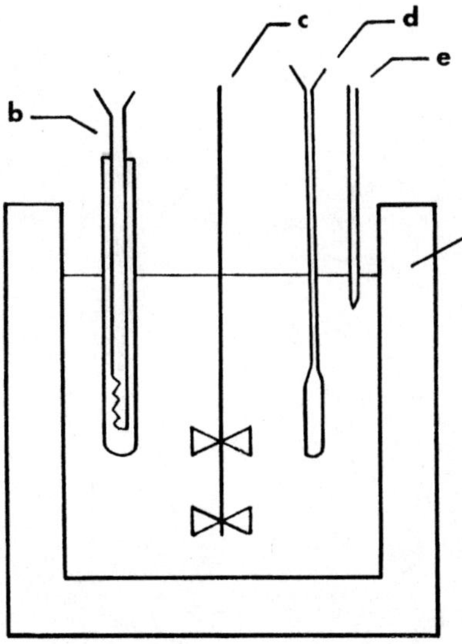

Figure 2.
A simplified diagram of a calorimeter.

A calorimeter is an instrument that allows an unknown quantity of heat to be determined by comparison with a known amount. The quantity of heat is measured by monitoring something like temperature which depends on it.

$$Q'_{(unknown)} : Q''_{(known)} = \Delta T' : \Delta T'' \quad (9)$$

$$Q'_{(unknown)} = (Q''_{(known)}/\Delta T'')\Delta T' \quad (10)$$

The quantity in parenthesis in equation (10) is the thermal capacity of the calorimetric system. The known quantity of heat is usually produced electrically ($Q'' = RI^2t$) and is precise to about 10 ppm. The temperature can also be measured with high precision, and calorimetric sensors such as thermistors or thermocouples exist which are sensitive to temperature changes as small as 10^{-6} degrees centigrade. However, the precision with which heats of formation of complexes can be measured is not very high. Thus, a heat of 10 kcal (41.84 kJ) with an error between ± 0.01 and ± 0.05 is considered to be good, whilst with an error between ± 0.1 and ± 0.2 it can still be useful.

There are many sources of error. The first are the so-called "heat losses". Although the reaction may be rapid, the temperature change cannot be instantaneous (see Fig. 3) but takes place over a period of time during which the calorimeter (assumed to be thermostatted) takes in or gives out heat to the surroundings. There are various graphical and numerical methods, some more sophisticated than others, which can be used to correct for this "heat loss" on the basis of the instrument characteristics. Furthermore, while some reactions are so fast as to be complete during the time taken for the mixing of the solutions, others are so slow as to proceed for some time after mixing is complete. In the latter cases, integration methods have to be adopted such as that illustrated in Figure 4, which refers to a batch calorimeter in which the heat flow is measured by a series of thermocouples which surround the calorimetric cell.

For a single stage reaction which is instantaneous and complete, ΔH^o can be obtained with great precision. For example, the standard enthalpy change for the reaction $H^+ + OH^- \rightleftharpoons H_2O$ has been found to be -13,34$\pm$0.01 kcal mol^{-1} at infinite dilution. By an instantaneous reaction, we mean one that is complete within the time taken to mix the solutions, assuming that the mixing in the calorimeter is both rapid and efficient.

If, however, the reaction is fast but only reaches a point of equilibrium, the equilibrium concentrations of the species involved in the reaction must be known, and these

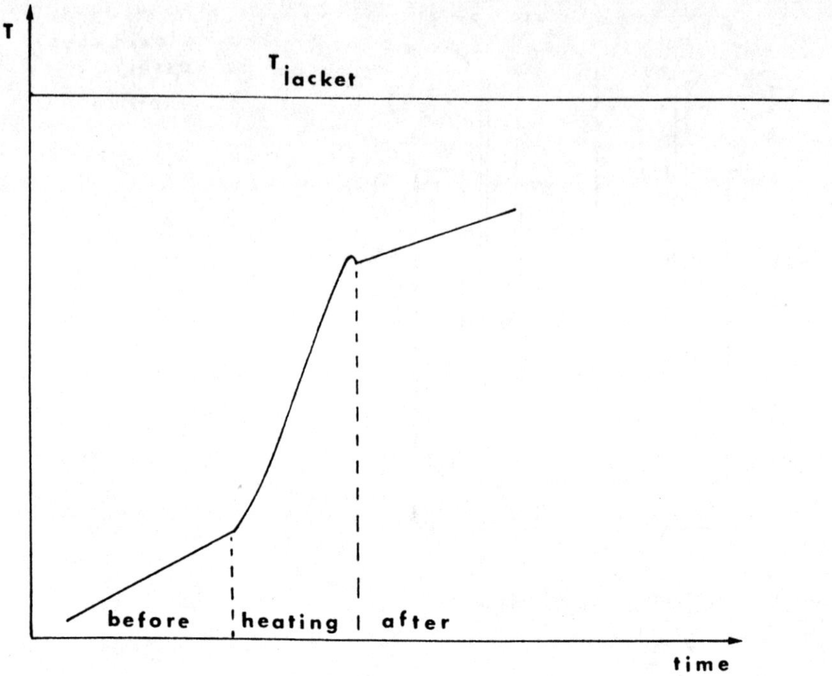

Figure 3.
A calorimetric curve of an isoperibolic calorimeter.

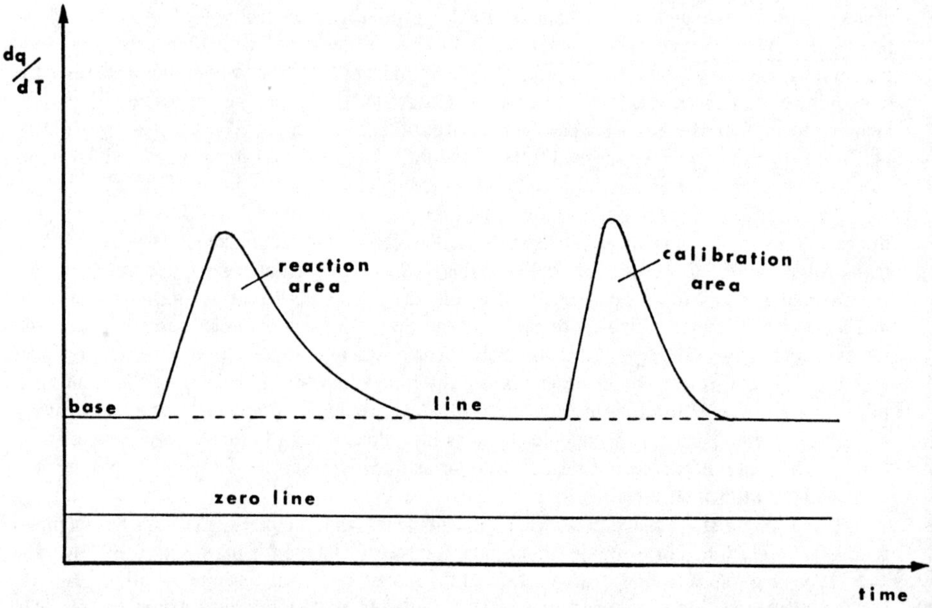

Figure 4.
A calorimetric curve for a "batch" calorimeter. dq/dt is the heat flow. The heat effects are obtained by integrating the areas under the peaks.

are usually calculated from the equilibrium constant which has been determined in a separate experiment. In such a case, the conditions (ionic medium, concentrations, temperature) used for calorimetry should ideally be as near as possible to the conditions used in the equilibrium constant determination.

Most complexation reactions are equilibria, and furthermore there are often many of these occurring simultaneously. When there are two or more competing reactions to consider, calorimetric measurements have to be made under a variety of conditions in which the position of the equilibria is different., i.e. conditions of varying pH and/or ligand:metal ratio. Let us consider a common example: the determination of the heats of formation of complexes of copper(II) with a basic ligand such as ethylenediamine (en) in 1M KNO_3 at 25°C. The simplest, though somewhat inconvenient, system to use is the following. A concentrated solution of the ligand (en) in 1M KNO_3 is added by means of a suitable device (see Fig. 3) to a dilute solution of the other reagent (HCl or $Cu(NO_3)_2$) also in 1M KNO_3. The determination proceeds in stages:

(i) Determine the heat of formation of water

$$H^+ + OH^- \rightleftharpoons H_2O \qquad (11)$$

under the chosen reaction conditions (1M KNO_3, 25°C) by adding a solution of HCl (1M KNO_3) to a slight excess of a solution of carbonate-free NaOH (also 1M KNO_3). The heats of reaction must be corrected for the heat of dilution of HCl.

(ii) determine the heats of dilution by adding concentrated solutions (e.g. en in 1M KNO_3) to the ionic medium, 1M KNO_3 .

(iii) Determine the heats of protonation of the ligand (en)

$$en + H^+ \rightleftharpoons enH^+; \qquad en + 2H^+ \rightleftharpoons enH_2^{2+} \qquad (12)$$

by adding various amounts of the ligand (in 1M KNO_3) to a solution of HCl in 1M KNO_3

(iv) Determine the heats of formation of the complexes

$$Cu^{2+} + en \rightleftharpoons (Cu(en))^{2+}; \qquad Cu^{2+} + 2en \rightleftharpoons (Cu(en)_2)^{2+} \qquad (13)$$

by adding various amounts of the ligand (en) in 1M KNO_3 to solutions of $Cu(NO_3)_2$ in 1M KNO_3 .

All the heats have to be corrected for the effects of dilution, although these corrections are often ignored as regards the most dilute solutions (NaOH and $Cu(NO_3)_2$ in the example). At each stage it is necessary to know the quantities (generally in mmoles) of the relevant species present both before and after the reaction, which are, in the example, Cu^{2+} (hydrolysis is suppressed by the addition of a known amount of HCl), en, enH^+, enH_2^{2+}, $Cu(en)^{2+}$, $Cu(en)_2^{2+}$, and finally H^+ and OH^-. Obviously, to calculate these quantities, the volumes of the reagents, the concentrations of metal ion, of ligand, and of added acid, the ionic product of water, the two protonation constants of the ligand, and the two complex formation constants must be known. The heats measured in (iv), after correction for dilution effects (ii), are adjusted for the protonation and deprotonation reactions (11) and (12) whose heats are determined in stages (i) and (iii). Finally, the adjusted and corrected heats in *n* measurements are used to set up a system of *n* equations in the two unknowns $\Delta H^o_{(Cu+en)}$ and $\Delta H^o_{(Cu+2en)}$

$$\begin{aligned}
&(\text{mmoles Cu en})_1 \Delta H^o_{(Cu+en)} + (\text{mmoles Cu en}_2)_1 \Delta H^o_{(Cu+2en)} = q_1 \\
&(\text{mmoles Cu en})_2 \Delta H^o_{(Cu+en)} + (\text{mmoles Cu en}_2)_2 \Delta H^o_{(Cu+2en)} = q_2 \\
&\cdots\cdots\cdots\cdots\cdots\cdots\cdots\cdots\cdots\cdots\cdots\cdots \\
&(\text{mmoles Cu en})_n \Delta H^o_{(Cu+en)} + (\text{mmoles Cu en}_2)_n \Delta H^o_{(Cu+2en)} = q_n \qquad (14)
\end{aligned}$$

The solution of these equations using the method of least-squares provides the values and standard deviations of the enthalpy changes. Naturally these calculations have been programmed into minicomputers.

It is clear that the errors on the values of ΔH^o depend not only on the errors of the calorimetric measurements, but also on the errors of all the quantities used in the calculation: volumes, concentrations and equilibrium constants. The error propagation calculation is long and complex, and is beyond the scope of this article.

The type of measurement described above is inconvenient in respect of the fact that basic solutions must be protected against atmospheric carbon dioxide, and also that their concentrations must be obtained by titration. These problems can be avoided if hydrochloride or another soluble salt is used in place of the pure amine. Such salts can be purified by recrystallization and can be accurately weighed, whereas the liquid amine must be purified by distillation, and is hygroscopic. When an acid salt is used in place of the basic ligand, the calorimetric measurements are made by adding an alkali (NaOH, KOH, Me_4NOH ...) to solutions containing it with or without the metal ion being also present. This procedure is rather similar to the one used in potentiometric measurements. The calculation of ΔH^o is in general terms the same as that just described, and differs only in respect of the heats of dilution. With a computer-controlled calorimeter and an automated addition system, large amounts of calorimetric data can be obtained in a period of 24 hours.

When the reaction of complex formation is relatively slow, the methods described above cannot be used without modification. If the equilibrium is reached within the space of an hour or two, it is still possible to use the "batch" calorimetric method: the curve dq/dt (Fig. 4) is obtained over the span of the reaction time and the heat change is derived by integration of the area under this curve.

The inconveniences caused by slow reaction of formation of complexes can be avoided by studying the reverse reaction, that is, by measuring the heat change accompanying decomposition rather than formation. To study a decomposition reaction, the complex is first prepared and placed in the calorimeter; to the solution containing the complex is added a reagent which destroys it.

For example, the ion Ni^{2+} reacts in two stages with the ligand triethylene-tetramine (trien). In the first stage, three nitrogen atoms of the ligand become rapidly attached to the nickel ion and in the second stage the fourth nitrogen becomes attached slowly. If one follows the reaction calorimetrically, the first -exothermic- phase is followed by the slow second phase -also exothermic-, in such a way that, unless the data are treated with particular attention, the second phase may be entirely missed and the measured heat change will be much less than the total. By contrast, the complex $Ni(trien)^{2+}$ may be decomposed rapidly by concentrated hydrochloric acid. The heat of formation of the complex may be obtained from the decomposition reaction in the same way as before, except that a different dilution correction will be needed.

Various reagents can be used to decompose complexes or, in other words, demetallate them. The cyanide ion is particularly well adapted for this purpose as the resulting cyanide complexes are soluble. Naturally when another complex is formed as a result of a reaction with a second ligand, this is not, strictly speaking, a decomposition, and the heat of formation has to be calculated in a different manner. In particular, the heat of formation of the resulting (e.g. cyanide) complex will need to be determined and taken into account. This method has been used to determine the heats of formation of macrocyclic complexes.

Sometimes the calorimetric measurements and calculations can be simplified by imposing chemical conditions in which certain reactions can be ignored. For example, if the first (log) protonation constant of a polybasic ligand is less than ca 10.2, the measurements can be made in solutions of sufficiently high pH for the concentration of protonated ligand to be negligible. Then the situation resembles that in which the ligand exists in a single form and the corrections given in stage (iii) above will not have to be applied. If the metal complex is formed under these conditions of pH, its heat of formation can be found without having to make allowance for ligand protonation reactions. It should be noted that when a metal ion solution is added to a strongly alkaline solution, the latter is significantly diluted and the heat of this dilution may be comparable to the heat of the principal reaction.

There are other methods which have been or may be used to obtain the heat of formation of a complex. The competition method, well known in potentiometry, may be applied to weak complexes. For example, the heat of formation of a zinc complex with an amine ligand may be measured by displacing the zinc by copper, which forms a stronger complex. Another possibility is to replace a weak ligand by another which forms a stronger complex. In both cases, the reactions which are occurring in the calorimeter must be correctly established in order that thermal contributions can be correctly apportioned to the appropriate reaction. It would be helpful to be able to follow such reactions visually, but unfortunately none of the commercially available calorimeters are transparent.

REFERENCE

Everett, D.H., and Pinsent, B.R.W. (1952). *Proc. Roy. Soc. (London)*, *215A*: 416.

KINETIC ASPECTS OF COMPLEX FORMATION

by

R. Bruce Martin

Chemistry Department,
University of Virginia,
Charlottesville, Virginia 22903
U.S.A.

I. INTRODUCTION

Belying the title word *formation* that might imply an addition reaction, complex formation in solution is almost always a *substitution* (occasionally with a reduction in coordination number). In aqueous systems, bound water undergoes substitution either in the reactant complex or as an intermediate in the route from reactants to products.

II. SUBSTITUTION REACTIONS

The greater complexity of kinetics over thermodynamics is shown in the simple appearing ligand replacement reaction with *n* fixed ligand donor atoms, L, bound to metal ion M (charges not specified).

$$L_nMX + Y \rightarrow L_nMY + X$$

While thermodynamics deals with the species above, kinetics also includes additional species involved in the mechanism of the reaction. Substitution of unidentate ligand X by unidentate ligand Y may proceed by a reduction of coordination number (dissociation) to give the intermediate L_nM, by an increase in coordination number (association) to yield the intermediate $L_nM(X)Y$, or by a process in which the incoming and outgoing ligands interchange in a more or less concerted fashion. From the wholly dissociative mechanism through the varying degrees of the interchange route to the wholly associative mechanism, there is a continuous spectrum of possibilities. Moreover, in aqueous systems, virtually all such substitution reactions occur via an aquo intermediate $L_nM(OH_2)$. Thus for the most general case where X and Y are not water, the question just posed as to the coordination number of the intermediates needs to be answered twice: for both formation and reaction of the aquo intermediate.

Detailed answers about the mechanism of substitution reactions in metal ion complexes have been the object of numerous investigations for many years (Wilkins, 1991; Jordan, 1991; Margerum et al., 1978). Some general conclusions now seem reliable, but universal agreement on all details remains to be reached. *Two statements sum up many of the results:* (i) Since room is available for approach of an axial ligand, most substitutions in linear, trigonal, and planar complexes proceed by the associative mechanism with an increase in coordination number; (ii) As the environment about the metal ion is more crowded, most substitutions in octahedral complexes proceed in the interchange-dissociative region of the continuum.

III. LIGAND EXCHANGE

Even if aquo complexes do not appear as reactants or products, their intermediacy in most substitution reactions whatever the detailed mechanism of substitution in any particular case makes their properties of primary importance.

The characteristic rate constant for exchange of inner-sphere water (both X and Y are H_2O in the above reaction) gives insight as to the differences among metal ions. Increasing water exchange rates on wholly aquo metal ions follow the order (Ducommun and Merbach, 1986)

Cr^{3+}, Ru^{3+} << Pt^{2+} << Ru^{2+} < Co^{3+} < Al^{3+}
<< V^{2+}, Fe^{3+} < VO^{2+}, Ga^{3+}, V^{3+}, Pd^{2+}, Be^{2+}
< Ni^{2+} < Mg^{2+} < Co^{2+}, Fe^{2+} < Mn^{2+}, Zn^{2+}, Sc^{3+}
< Ca^{2+}, Cd^{2+}, Gd^{3+} < alkali metal ions, Cr^{2+}, Cu^{2+}, Hg^{2+}, Pb^{2+}

with each inequality sign indicating an approximate ten-fold increase in rate.

The first row list cations with slowly exchanging water, the second and third rows rapidly exchanging cations, and the last row virtually instantaneously exchanging cations. The series spans a phenomenal 15 powers of ten from a lifetime at 25°C of about 5 days for the mean time of a water molecule on aqueous Cr^{3+} to 10^{-9} seconds for the metal ions at the end of the list. Indeed, exchange lifetimes are so short for the cations in the last line that some values are not known with certainty: some of the lifetimes may be 10^{-10} seconds or even shorter (Friedman, 1985).

The above series is useful in several ways. The contrast between the presence of Zn^{2+} in numerous mammalian enzymes and of Ni^{2+} in only a few plant enzymes has proved puzzling. Ni^{2+} differs by neither size nor most complex stabilities from Zn^{2+}. The above exchange rate series explains the difference since Zn^{2+} undergoes ligand exchange about 10^3 times faster than Ni^{2+}, a significant factor for a metal ion bound at enzyme active sites (Martin, 1986).

In four instances reduction of the metal ion substantially increases the rate. An approximate 10^4 rate increase occurs for the $d^5 \rightarrow d^6$ reduction in both Ru^{3+} and Fe^{3+}. An even greater 10^7-fold rate increase takes place upon the reduction $Co^{3+} \rightarrow Co^{2+}$. Finally, the 10^{15} span of the entire series is gained upon reduction of $Cr^{3+} \rightarrow Cr^{2+}$.

Though these lifetimes refer to water exchange in the wholly aquated metal ions, they also reflect the relative rates of exchange of other ligands. When bound to a metal ion, other ligands, including hydroxide ion, increase modestly the release rate of bound water, typically by about 10^2. As a poorer leaving group, hydroxide ion itself exchanges much more slowly than water. Chelated ligands exchange more slowly, but again the relative order prevails. In the case of chelates, binding occurs first by one donor group followed by chelate ring closing. In some cases, such as five-membered rings, the chelate ring closes rapidly, and in others, especially with larger rings, ring closure may become the slow step. Selective broadening of lines in NMR spectroscopy that have been used to indicate the binding site of a paramagnetic metal ion at stoichiometric concentrations often fail because chelate ring closure is slow compared to rapid passage of the metal ion to other ligands present in large excess which bind only in a unidentate mode (Espersen and Martin, 1976).

Contrary to common expectation, the above series demonstrates that some of the strongest binding metal ions undergo the most rapid ligand exchange. Examples from the high rate end of the series include Pb^{2+}, Hg^{2+}, Cu^{2+}, and Cd^{2+}, all strong ligand binders. A significant feature of the toxicology of Hg^{2+} and CH_3Hg^+ is the rapid exchange of ligands in and out of coordination to the metal ion. For CH_3Hg^+ binding to the sulfhydryl group of glutathione (present in blood plasma at 4 mM) the stability constant logarithm is an extraordinary log K = 15.9 (Rabenstein, 1978). For such strong binding the half-life for exchange by a dissociative mechanism is about 10 days. Yet the average lifetime for the CH_3HgSR complex in the presence of excess glutathione is less than 0.01 sec (Rabenstein et al., 1982). This comparision shows that exchange does not occur by a dissociative mechanism. The ability of linear two-coordinated Hg^{2+} and CH_3Hg^+ to associate weakly with additional donor atoms accounts for rapid metal ion exchange among donor atoms (Cheesman et al., 1988). Additional ligands are bound much more weakly than the first two and exhibit longer Hg to donor atom bond lengths. Excess ligand participates in nucleophilic attack at an uncoordinated site on the metal ion with rearrangement in the coordination geometry and release of a formerly bound ligand. This addition-elimination mechanism accounts for the very rapid exchange in Hg^{2+} and CH_3Hg^+ complexes. This topic appears again in the discussion of toxicology (see Part Four, Chapter 2-B).

IV. NUCLEOPHILICITY OF METAL ION BOUND HYDROXIDE

Complexes commonly bear at least one water molecule, and only a pH increase separates the water molecule from becoming a bound hydroxide ion. On one hand, many reactions, such as ester and amide hydrolysis, involve nucleophilic attack by hydroxide ion, and on the other, many metal ions, either aquo or with some bound ligands, bear bound hydroxide. How effective a nucleophile is metal-ion bound hydroxide compared to unbound hydroxide?

The relative effectiveness of bound and unbound hydroxide as nucleophiles has been probed by studies comparing the ability of metal-ion bound and free hydroxide to effect various hydrolysis reactions (Martin, 1976; Buckingham and Engelhardt, 1975). The pK_a for deprotonation of the associated water molecule measures the basicity of the metal-ion bound hydroxide. If the nucleophilicity corresponded to the basicity of the bound hydroxide, a plot of log k_2, where k_2 is the second order rate constant for hydrolysis, versus pK_a for a series of metal ion complexes would yield a slope of unity. Observed slopes in such linear free energy plots for the hydration of carbon dioxide and the hydrolysis of esters are much less than unity, about 0.25. This low value implies that metal-ion bound hydroxide is a far more effective nucleophile than expected from its basicity. Thus even though metal-ion bound hydroxide is a weaker nucleophile than unbound hydroxide, in neutral solutions a complex may provide a more than offsetting higher concentration of metal-ion bound hydroxide. For example, a metal-ion bound OH^- with $pK_a = 7.0$ for the conjugate acid form with bound water is 8.7 log units less basic than free OH^- with $pK_a = 15.7$. With a slope of 0.25 in the linear free energy plot, the metal-ion bound hydroxide is 1/150 as nucleophilic as unbound OH^-. Thus, for any concentration of the metal ion complex greater than 40 μM, it surpasses unbound OH^- at pH 7.4 as the nucleophilic agent.

The nucleophilic effectiveness of metal-ion bound hydroxide becomes reduced upon its protonation, when the pH becomes less than the pK_a for deprotonation of the bound water. A higher pK_a offers greater nucleophilicity, but this increase may be offset by protonation for a pK_a greater than the operating pH. It may be shown that for a given pH the optimum value of $pK_a = \mathrm{pH} - \log[(1-\mathrm{slope})/\mathrm{slope}]$, where the slope is that of a log k_2 versus pK_a plot. For a slope of 0.25 we find $pK_a = \mathrm{pH} - 0.5$. For a system operating at the blood pH = 7.4, the optimum pK_a is about 6.9 for deprotonation of metal-ion bound water. This argument supports the recent reassignment of the lower of the two observed pK_a values of 7 and 9 for the blood metalloenzyme carbonic anhydrase to the zinc-bound water. If the pK_a were 9, the observed catalysis of carbon dioxide hydration would be only 10% as effective as that for a pK_a of 7. Of course, this discussion accounts for only one factor in the activity enhancement of carbonic anhydrase.

V. CONSECUTIVE VERSUS COMPETITIVE REACTIONS

If the concentration of a substance first rises and then falls during a reaction, it is often assumed that the substance serves as an intermediate on the pathway from reactants to products. Indeed, even stronger statements appear in the literature: since a substance concentration passes through a maximum, the compound *must* be an intermediate. However, it is difficult to prove that a substance whose concentration passes through a maximum is necessarily an intermediate on the pathway from initial reactants to final products. In many cases the result may fit equally well a scenario in which the "intermediate" forms in a reversible side reaction and the final products are produced from the reactants by another course. In this scenario, the "intermediate" is not an intermediate on the main reaction pathway from reactants to final products, but a metastable side product whose concentration builds initially and then reverts to reactants as they convert to final products. Thus assumed intermediates used to delineate mechanisms of reactions may not be intermediates at all, but the results of unproductive side reactions.

Compare the scheme of consecutive reactions A $\rightleftharpoons$ B → C with that for parallel or competitive reactions A $\rightleftharpoons$ B and A → C. To strengthen the comparison, the first step in the consecutive scheme is here made reversible, though not required. In both schemes, assignment of appropriate rate constants to each of the three arrows yields an increase and then a decrease in the concentration of the species B, which exhibits the same form in both schemes. Thus, even though species B is not on the pathway from reactant A to final product C in the competitive scheme, its concentration rises and falls like that expected for consecutive reactions. The disappearance of starting material A also follows a similar pattern in the two schemes. Only by careful monitoring of final product C may the two schemes be distinguished. In the consecutive mechanism, the time course for appearance of the product C displays an induction period with an inflection point just as the maximum appears in the concentration of true intermediate B. No inflection point is possible for product C in the competitive scheme (Martin, 1985).

The importance of the differentiation between the two mechanisms is shown by the claim that a metal-ion complexed α,α'-carbanion is an intermediate in transamination between pyridoxal and pyridoxamine Schiff bases. With a starting solution containing pyridoxamine, pyruvate, and Al(III) in a 1:1:1 mole ratio, Figure 1 shows in the open circles buildup and decline of a species B, and in the closed circles, appearance of final complexed aldimine product C. Clearly, there is no induction period nor inflection point in the solid points for aldimine product C. Rate constants determined in the fit of the open circles for species B completely determine the product C curve in the consecutive scheme. The resulting dashed curve for product C in Figure 1 does not fit the solid points, ruling out the consecutive scheme (Martin, 1987).

On the other hand, rate constants determined from the behavior of species B in the competitive scheme fit the solid points for production of aldimine product C as shown by

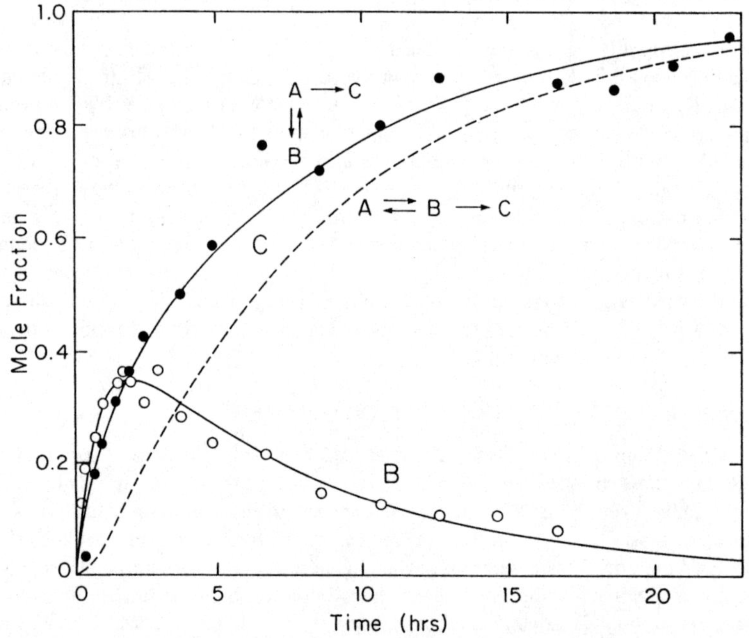

Figure 1.
Mole fraction versus time in hours for species B (open circles) and final aldimine product C. The upper, solid C curve is for the competitive scheme and the lower, dashed C curve for the consecutive scheme. From Martin (1987).

the solid line that passes through the solid points. Thus species B is not an intermediate but a metastable side product not on the pathway from reactants to aldimine. Species B is probably complexed carbanion. Thus the carbanion complex is a metastable byproduct of a reversible side reaction. Though the metal-ion complexed carbanion does not lie on the reaction pathway, it is likely that uncomplexed carbanion does. Metal ions serve as catalysts in transamination reactions by coordinating carbinolamine intermediates and changing the rate-limiting step (McDonnell et al., 1970).

This example shows that care must be taken in assessing whether a species that rises and falls during the course of a reaction is an intermediate or a metastable side product not on the pathway from reactants to final products. Probably other unknown examples exist in the literature, but only infrequently is the requisite data presented so that the induction period and inflection in the product C required in the consecutive mechanism may be tested.

Twenty years ago, it was suggested that oxygenated hemoglobin is an unreactive by-product not on the main pathway for oxidation of iron(II) hemoglobin to iron(III) methemoglobin (Stadtherr et al., 1973). This suggestion appears borne out by recent site-directed mutagenesis and binding studies (Springer et al., 1989).

REFERENCES

Buckingham, D.A., and Engelhardt, L.M. (1975). *J. Amer. Chem. Soc.*, *97*: 5915.

Cheesman, B.V., Arnold, A.P., and Rabenstein, D.L. (1988). *J. Amer. Chem. Soc.*, *110*: 6359.

Ducommun, Y., and Merbach, A.E. (1986). *Inorganic High Pressure Chemistry* (R. vonEldik, ed.), Elsevier, Amsterdam.

Espersen, W.G., and Martin, R.B. (1976). *J. Amer. Chem. Soc.*, *98*: 40.

Friedman, H.L. (1985). *Chem. Scr.*, *25*: 42.

Jordan, R.B. (1991). *Reaction Mechanisms of Inorganic and Organometallic Systems*, Oxford University Press, New York.

Margerum, D.W., Cayley, G.R., Weatherburn, D.C., and Pagenkopf, G.K. (1978). *Coordination Chemistry* (A.E. Martell, ed.), American Chemical Society, Washington, DC, p. 1.

Martin, R.B. (1976). *J. Inorg. Nucl. Chem.*, *38*: 511.

Martin, R.B. (1985). *J. Chem. Educ.*, *62*: 789.

Martin, R.B. (1986). *Metal Ions in Biological Systems*, Vol. 20 (H. Sigel, ed.), Marcel Dekker, New York, p. 21.

Martin, R.B. (1987). *Inorg. Chem.*, *26*: 2197.

McDonnell, C.V., Michailidis, M.S., and Martin, R.B. (1970). *J. Phys. Chem.*, *74*: 26.

Rabenstein, D.L. (1978). *Accounts Chem. Res.*, *11*: 100.

Rabenstein, D.L., Isab, A.A., and Reid, R.S. (1982). *Biochim. Biophys. Acta*, *696*: 53.

Springer, B.A., Egeberg, K.D., Sligar, S.G., Rohlfs, R.J., Mathews, A.J., and Olson, J.S. (1989). *J. Biol. Chem.*, *264*: 3057.

Stadtherr, L.G., Prados, R, and Martin, R.B. (1973). *Inorg. Chem.*, *12*: 1814.

Wilkins, R.G. (1991). *Kinetics and Mechanism of Reactions of Transition Metal Complexes*, VCH, Weinham, Germany.

THERMODYNAMIC QUANTITIES AND STRUCTURAL DETERMINATIONS OF LOW-MOLAR-MASS METAL COMPLEXES IN SOLUTION

by

Antonio Bianchi and Piero Paoletti

Department of Chemistry,
University of Florence,
Via Maragliano 75/77
50144 Florence
Italy

I. INTRODUCTION

Chemical thermodynamics only relates the macroscopic observable properties of bulk matter with its behavior in physico-chemical processes. It leads to an understanding of chemical reactivity based on the properties of individual reagents and products, but it does not require any assumption about their atomic or molecular structure.

However, the analysis of thermodynamic properties of wide classes of compounds presenting similar structural features define some general behaviors and, therefore, correlations between thermodynamic quantities and structural properties can be outlined. Compounds which are tentatively assigned to such classes, but deviate from the expected trends, usually present structural characteristics disagreeing with the general features of those classes.

The first general information characterizing a chemical species is given by stoichiometry, which contains also the primary structural data about that species. For metal complexation in solution, the determination of complex stoichiometry is currently part of the selection process leading to the determination of stability constants. Stability constants, or $\Delta G°$ values, account for the extent of complex formation and correlate with $\Delta H°$ and $\Delta S°$ values. On the other hand, heats of reaction in aqueous solution relate with the strength of chemical bonding between reacting species and can be used to investigate the nature of the metal-ligand interaction.

The study of the chemical equilibrium in solution is complicated by the presence of the solvent. Solvent molecules can crowd or disperse around the species consumed or formed during a chemical reaction, affecting very much the thermodynamic parameters for the equilibrium attainment. In particular, in the case of complexation reactions, solvent molecules can be directly involved in both first and second coordination spheres of metal ions and pass from one to the other during the course of complexation. The availability of $\Delta S°$ values should provide insight into the interaction between the equilibrium species and the solvent.

We can distinguish some favorable and unfavorable contribution to $\Delta H°$ and $\Delta S°$. Are favorable contributions to the enthalpy change the formation of metal-donor bonds, the crystal field stabilization energy (when it is present), the hydration of the complex. Are unfavorable the dehydration of both ligand and metal ion, the steric repulsion and the molecular strain imposed in the ligand upon coordination. The increase in the total number of free particles and the loss of degrees of freedom of the ligand which can accompany the complexation reaction are favorable and unfavorable contributions to the entropy change, respectively.

The main connections between the thermodynamic data for a complexation process and the structural properties of the species involved deal with the type, the number, and the geometrical arrangement of donor atoms around the coordinated metal ion.

II. TYPE OF DONOR ATOMS

Complexation of metal ions in aqueous solution is a process involving the replacement of coordinated water molecules in the solvated cation. Aquo-ions present very different affinities for different ligands. Depending on the ligands a metal ion prefers (Table 1), we can distinguish two types of characters (Ahrland et al., 1958). Class (*a*) metal ions form more stable complexes with F^- than with other halogenide anions; they prefer oxygen-containing ligands to ligands with sulfur, selenium or tellurium donor atoms, and similarly they prefer nitrogen donors to phosphorus, arsenic and antimony donors. They have high positive charges and small radii and, generally, they present no unpaired electron in the valence shell. Due to their low polarizability they have been called "hard" (Pearson, 1968). On the other hand, class (*b*) metal ions present an almost reverse trend of affinity for these ligand donor atoms (Table 1). They form most stable complexes with low electronegative donor atoms (I, S, P). They have lower positive charges and larger radii. Due to their greater polarizability they have been called "soft" (see Part One of this volume). Table 2 reports some typical hard and soft metal ions together with some borderline cations having mixed hard and soft character. Based on the observation that hard metal ions prefer to coordinate to hard donors and soft metal ions to soft donors, ligands can also be classified (Table 2). According to the so-called hard-and-soft-acids-and-bases (HSAB) classification, H_2O is a hard ligand and therefore the formation of aquo-ions competes more strongly in the complexation of hard than soft metal ions.

Association of hard metal ions with hard ligands principally involves electrostatic interactions; their reaction is generally accompanied by charge neutralization and release of water molecules. On the other hand, the interaction between soft metal ions and soft ligands gives rise to the formation of essentially covalent bonds. From a thermodynamic point of view, the hard-character association is driven by a large positive entropy change; enthalpy contributes little or nothing and the reactions may even be endothermic. In contrast, the soft-character association is due to a dominant enthalpy decrease, the entropy contribution being of lower importance.

The HSAB principle may be of help in distinguishing the donor atoms involved in the coordination to metal ions. An interesting example is given by the nucleotide adenosine triphosphate (H_4ATP). Table 3 reports selected thermodynamic parameters (Smith et al.,

Table 1

Sequences of Affinities of *a*- and *b*-type Metal Ions for the Various Donor Atoms.

a-type metal ions	*b*-type metal ions
F > Cl > Br > I	F < Cl < Br > I
O >> S > Se > Te	O << S ~ Se ~ Te
N >> P > As > Sb	N << P > As > Sb

Table 2

Hard-Soft Classification of Metal ions and Ligands

Hard metal ions	Borderline	Soft metal ions
H^+, Li^+, Na^+, K^+, Be^{2+}, Mg^{2+}, Ca^{2+}, Sr^{2+}, Mn^{2+}, Al^{3+}, Sc^{3+}, Ga^{3+}, In^{3+}, La^{3+}, Gd^{3+}, Lu^{3+}, Cr^{3+}, Co^{3+}, Fe^{3+}, Si^{4+}, Ti^{4+}, Sn^{4+}, WO^{4+}, VO^{2+}	Fe^{2+}, Co^{2+}, Ni^{2+}, Cu^{2+}, Zn^{2+}, Pb^{2+}, Sn^{2+}, Sb^{3+}, Bi^{3+}	Cu^+, Ag^+, Au^+, Tl^+, Hg_2^{2+}, Hg^{2+}, Pd^{2+}, Pt^{2+},Pt^{4+}, Tl^{3+}
Hard ligands	**Borderline**	**Soft ligands**
H_2O, OH^-, F^-, Cl^-, $CH_3CO_2^-$, PO_4^{3-}, SO_4^{2-}, CO_3^{2-}, ClO_4^-, NO_3^-, ROH, R_2O, NH_3,	$C_6H_5NH_2$, C_5H_5N, N_3^-, Br^-, NO_2^-, SO_3^{2-}, N_2	R_2S, RS^-, I^-, SCN^-, $S_2O_3^{2-}$, R_3P, $(RO)_3P$, R_3As, CN^-, RNC, CO, H^-, R^-, C_2H_4, C_6H_6

1991) for the coordination of ATP^{4-} to some hard metal cations (Mg^{2+}, Ca^{2+}, Mn^{2+}) in comparison with those obtained for the softer Cu^{2+}. The enthalpic and entropic contribution to the formation of Mg^{2+}, Ca^{2+} and Mn^{2+} ATP^{4-} complexes are as expected for a hard-character association, indicating that these cations interact mainly with the hard phosphate groups of ATP^{4-}. In contrast, the complexation reaction of the nucleotide with the borderline Cu^{2+} cation is almost athermal suggesting an additional covalent interaction between the metal ion and the adenine nitrogens. A more useful comparison is made when the thermodynamic parameters (Smith and Martell, 1974-1989) for the complexation of these metal ions by triphosphate anion (Table 3) are considered. In fact, in the absence of the adenine group Cu^{2+} gives rise to an highly endothermic reaction. A number of experimental techniques (Weser, 1968) indicates that while Mg^{2+}, Ca^{2+} and Mn^{2+} bind

Table 3

Thermodynamic Parameters for the Formation of Some Complexes with Adenosine Triphosphate and Triphosphate Ligands in Aqueous Solution at 25 °C. Values in kJ mol^{-1}

Reactants	ΔG°	ΔH°	TΔS°
Mg^{2+} + ATP^{4-}	-26.0	18.8	44.8
Ca^{2+} + ATP^{4-}	-23.4	14.2	37.5
Mn^{2+} + ATP^{4-}	-29.3	19.2	48.5
Cu^{2+} + ATP^{4-}	-36.8	3.3	40.1
Mg^{2+} + $P_3O_{10}^{5-}$	-33.0	18.0	51.0
Ca^{2+} + $P_3O_{10}^{5-}$	-31.0	13.8	44.8
Mn^{2+} + $P_3O_{10}^{5-}$	-40.6	11.7	52.3
Cu^{2+} + $P_3O_{10}^{5-}$	-47.3	20.5	67.8

Table 4

Logarithms of the Equilibrium Constants for the Formation of some Complexes of Cysteine, 2-Mercaptoethylamine and Glycine in Aqueous Solution at 25°C

Metal ions	cysteine	2-mercaptoethylamine	glycine
Mn^{2+}	4.7		2.80
Ni^{2+}	9.82	10.05	5.78
Zn^{2+}	9.9	9.9	4.96
Hg^{2+}	14.4		10.3
Pb^{2+}	11.6	11.1	5.47

only to the triphosphate portion of ATP^{4-}, Cu^{2+} also bind to the N7 nitrogen of the adenine moiety. The involvement of different binding sites in metal coordination by nucleotides has been extensively argumented (Eichhorn, 1973; see also Part Two, Chapter 2-A in this volume).

In some cases, the results obtained or predicted by means of the HSAB principle about the types of donor atoms of a ligand involved in complex formation can be confirmed or strengthened by comparison with the ligational properties of related ligands. For instance, cysteine ($HSCH_2CH(NH_2)COOH$) contains, in its dianionic form, three groups ($-NH_2$, $-COO^-$ and $-S^-$) which can potentially act as binding sites toward metal ions. On the basis of the HSAB principle, one could expect the $-S^-$ and probably the $-NH_2$ groups to be the effective binding sites to softer metal ions. The equilibrium constants for the formation of some complexes with this aminoacid (Smith and Martell, 1974-1989) are reported in Table 4. In the same Table, the equilibrium constants for the formation of the analogous complexes of 2-mercaptoethylamine ($HSCH_2CH_2NH_2$) and glycine (H_2NCH_2COOH) have been also reported for the sake of comparison. The structural features of the last two ligands are closely related to those of cysteine. As can be seen, the coordinating ability of cysteine toward the reported metal ions is very similar to those of 2-mercaptoethylamine, whilst it differs significantly from those presented by glycine. This suggests that cysteine does not coordinate like many other aminoacids through $-NH_2$ and $-COO^-$ but, instead, through $-NH_2$ and the softer $-S^-$ group. Nevertheless, it seems (Tomita et al., 1968) that the carboxylate group also can participate in the coordination to hard metal ions like Fe^{3+}. A similar comparison allows one to conclude that also penicillamine ($HSC(CH_3)_2CH(NH_2)COOH$) coordinates to Ni^{2+}, Zn^{2+}, Cd^{2+}, Hg^{2+} and Pb^{2+} through its $-NH_2$ and $-S^-$ groups.

III. NUMBER OF DONOR ATOMS

One of the most intriguing points encountered when dealing with the complexation of metal ions by polydentate ligands in solution is the determination of the number of donor atoms actually involved in the coordination. Polydentate ligands have largely contributed to characterize the structural and thermodynamic properties of metal complexation in solution; striking correlations between these properties led to the definition of "chelate" and "macrocyclic" effects (Frausto da Silva, 1983). Chelating and macrocyclic ligands have been widely used as models for biological metal-containing systems whose activity is often largely determined by the number of donor atoms coordinated to the metal center. Correlations between logK values for the formation of polydentate ligand complexes and the total number of donor atoms in these ligands have been used to deduce the number of coordinated donors. For example, it has been shown (Bencini et al., 1991) that, in the series of polyazacycloalkanes containing secondary nitrogen atoms connected by ethylenic chains, the equilibrium constant for the formation of Co^{2+}, Ni^{2+}, Cu^{2+} and Zn^{2+} complexes increases as the number of ligand nitrogens increases up to five for Cu^{2+} and Zn^{2+} and six

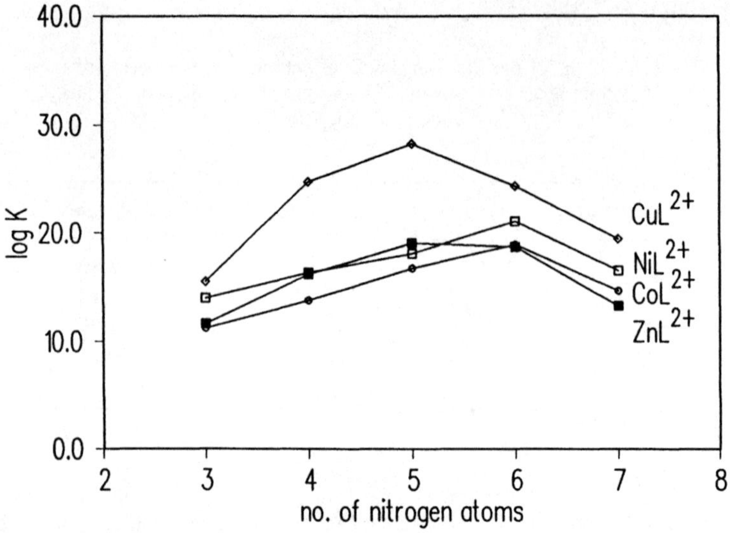

Figure 1.
Variation in log K for the reactions of Co^{2+}, Ni^{2+}, Cu^{2+} and Zn^{2+} ions with polyazacycloalkanes, as a function of the number of nitrogen atoms, in aqueous solution at 25 °C.

for Co^{2+} and Ni^{2+} (Fig. 1). This role has been ascribed to a larger and larger stabilizing contribution due to the involvement of an increasing number of nitrogen atoms in the coordination. The observed stability trends for Ni^{2+} and Cu^{2+} are paralleled by the enthalpy contributions (Fig. 2) which largely predominate over the entropic ones, confirming that these trends are due to an increasing number of coordinative bonds. Therefore no more than five nitrogen atoms bind to the Cu^{2+} and Zn^{2+} ions, while six nitrogen atoms can be involved in the coordination to Ni^{2+} and most likely also to Co^{2+}. A more complicated situation, which can be solved by analyzing the entropy contributions, is presented by the Mn^{2+} complexes (Bencini et al., 1990). In this case a steady logK value was observed as a maximum of stability for the complexes with the macrocycles containing from five to seven donor atoms. Calorimetric measurements revealed that the stability of the complex of the hard Mn^{2+} ion with the heptaazaligand is mainly entropic in nature ($\Delta H° =$ -21 kJ mol^{-1}, $T\Delta S° = 34$ kJ mol^{-1}), which is in accordance with the large increase in translational entropy expected for the heptacoordination of this ligand. A crystal structure determination demonstrated that the heptaazaligand involves all its seven donor atoms in the coordination to Mn^{2+}, at least in the solid state.

It is worth noting that metal complexes of polyamine ligands in which one or more nitrogen atoms are not involved in the binding of the metal ion generally form important protonated species in solution. As a matter of fact, the previous heptaazamacrocyclic ligand form protonated species with all the metal ions studied, Mn^{2+} excluded. In general, the analysis of thermodynamic parameters for the protonation of metal complexes is of great help to establish the number and sometimes the type of donor atoms involved in the coordination.

Correlations of equilibrium constants of metal complexes with basicity or acidity constants of free ligands have been considered for analyzing the binding mode of large classes of ligands. As an example, the protonation constants of the $-NH_2$ group of some aminoacids have been correlated with the equilibrium constants for the formation of their Zn^{2+}, Fe^{2+} and Fe^{3+} complexes (Sillén and Martell, 1964; 1971; Perrin, 1958; 1959) as reported in Figure 3. With the exception of aspartic acid, histidine and cysteine with the three metal ions above and glutamic acid with Fe^{3+}, a good correlation was observed. On the basis of the noticeable deviation of aspartic acid, it was suggested that this aminoacid

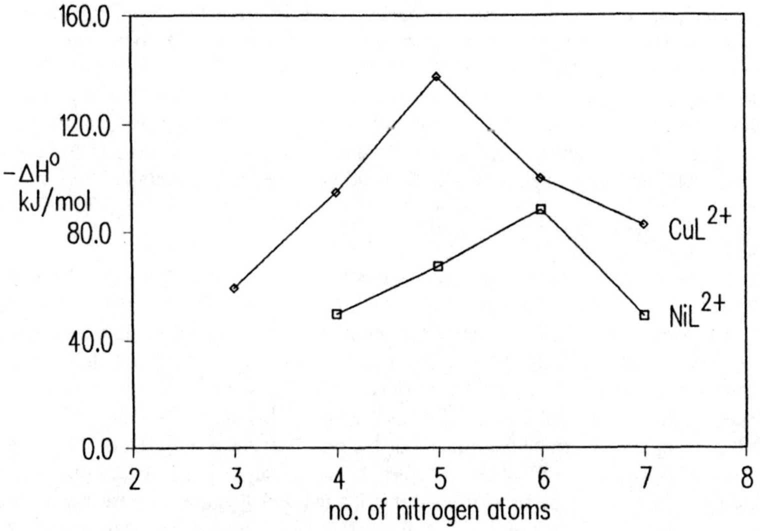

Figure 2.
Variation in -ΔH° for the reactions of Ni^{2+} and Cu^{2+} ions with polyazacycloalkanes, as a function of the number of nitrogen atoms, in aqueous solution at 25 °C.

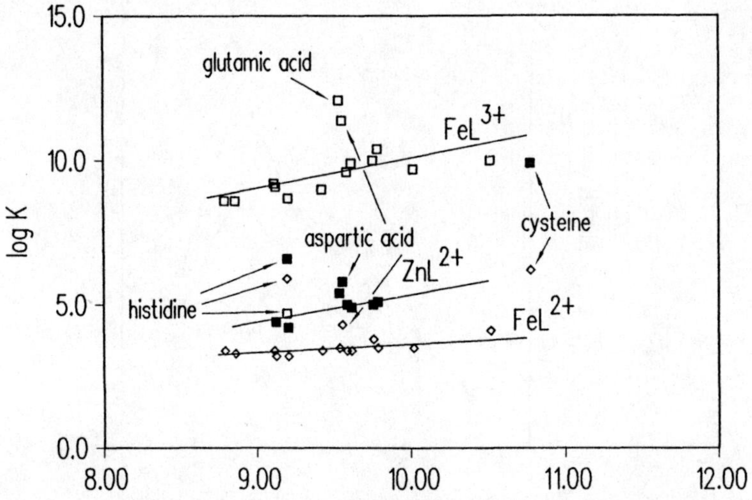

Figure 3.
Correlation between stability constants for the reaction of some aminoacids with Zn^{2+}, Fe^{2+} and Fe^{3+} and pK_a values of the $-NH_2$ group in the aminoacids. The aminoacids considered are: arginine, methionine, threonine, asparagine, tryptophan, serine, valine, leucine, alanine, sarcosine, glycine, proline, glutamin acid, aspartic acid, histidine, cysteine.

behaved as a tridentate ligand. A similar behavior was inferred for glutamic acid in its Fe^{3+} complex. The tridentate nature of aspartic acid was also suggested (Wellman et al., 1968; Ho et al., 1970) for its Cu^{2+} and Ni^{2+} complexes, and confirmed (Freeman, 1967) in the solid state by the crystal structure determination of Co^{2+}, Ni^{2+} and Zn^{2+} complexes. On the other hand, the deviation of histidine and cysteine strongly suggests, as already noted, that alternative donor sites, i.e. imidazole and thiol groups, dominate the competition among the potential binding sites in histidine and cysteine, respectively.

IV. GEOMETRICAL ARRANGEMENT

It has so far been shown that the analysis of the thermodynamic parameters for complexation reactions in aqueous solution can provide some information on the types and number of donor atoms for the ligand(s) involved in the coordination process. In some cases, the main characteristics of the geometrical arrangement of donor atoms around the metal center can also be deduced. The properties of metal complexes are often characterized by means of stability trends. One of these is the "natural order" of stability of the high-spin octahedral complexes formed by the d^5-d^{10} dipositive metal ions from Mn^{2+} to Zn^{2+} ($Mn^{2+} < Fe^{2+} < Co^{2+} < Ni^{2+} < Cu^{2+} > Zn^{2+}$). An example is illustrated in Figure 4 for the binding of the first molecule of ethylenediamine and glycine to these metal ions. This sequence of stability, better known as the Irving-Williams order (Irving and Williams, 1948,1953), is to be expected also for the d^0-d^5 metal ions of the first transition row ($Ca^{2+} < Sc^{2+} < Ti^{2+} < V^{2+} < Cr^{2+} > Mn^{2+}$); however, the coordination properties of some of these cations in aqueous solution cannot be studied because of their reducing character. Along this series, especially in the case of ligands which contain amino donors, the entropy contribution to complex formation appears to be small and roughly constant, so that the stability order is largely determined by the enthalpic term (Fig. 5). The enthalpy profile can be interpreted in terms of crystal-field (CF), ligand-field (LF) or molecular-orbital (MO)

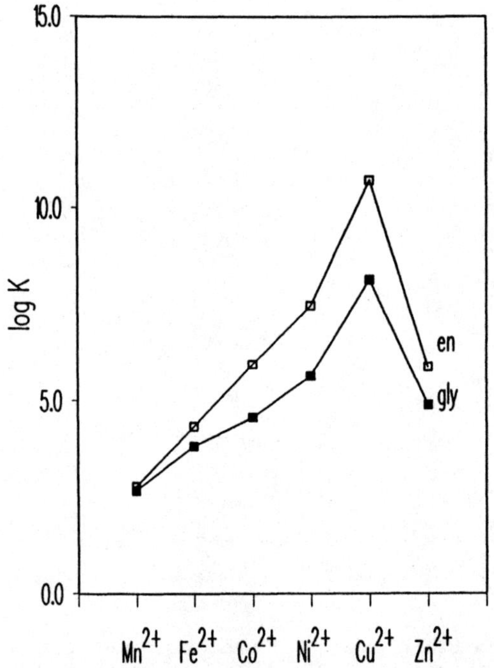

Figure 4.
Stability trends for the complexation reactions of M^{2+} first row transition metal ions with one molecule of ethylenediamine (en) or glycine (gly) in aqueous solution at 25°C.

theories. In the sense of the CF theory, ligand coordination to transition metal ions causes a splitting in the energy levels of d electrons of the metal, the characteristics of such a splitting depending on the charge of the metal ion, the position of the ligand in the spectrochemical series and the geometry around the metal ion. Therefore, if the stability constants or the enthalpy changes for the formation of Mn^{2+}-Zn^{2+} complexes with the same ligand, having the same stoichiometry and presenting almost constant complexation entropies, follow the Irving-Williams order, one can assume that the complexes are high-spin octahedral provided the water molecules in the first coordination sphere are taken into account.

The Irving-Williams sequence of stabilities is also observed for carboxylate ligands complexes, although, in that case, the entropy contribution generally predominates over the enthalpic one.

An interesting deviation from the Irving-Williams order is presented by the complexes of the hexadentate ligand ethylenediaminetetraacetate (EDTA), $(^{-}OOCCH_2)_2$-NCH_2CH_2N-$(CH_2COO^{-})_2$ (Fig. 6). As shown in this figure, the marked increase in stability expected for the Cu^{2+}-EDTA complex is not observed, the Ni^{2+} complex having a very similar stability. Along the series, the enthalpy and entropy contributions alternate in determining the stability profile. However, from Co^{2+} to Zn^{2+} the entropic term keeps constant allowing the attribution of the deviation from the Irving-Williams order to the enthalpic one. In this sense, Cu^{2+} is not able to interact with the six binding sites of EDTA as strongly as expected on the base of the observed trend for the other metal ions.

More deviations from this stability order are encountered among complexes of ligands with strong fields, namely when low-spin instead of high-spin complexes are formed.

Other stability and enthalpy change sequences for Mn^{2+}-Zn^{2+} complexes have been proposed (Bianchi and Paoletti, 1985) for tetrahedral and trigonal bipyramidal geometries on the bases of thermodynamic data for the formation of $(MBr_4)^{2-}$ and $M(Me_5dien)Br_2$ (Me_5dien = $(CH_3)_2NCH_2CH_2N(CH_3)CH_2CH_2N(CH_3)_2$) complexes. $\Delta G°$ and $\Delta H°$ values

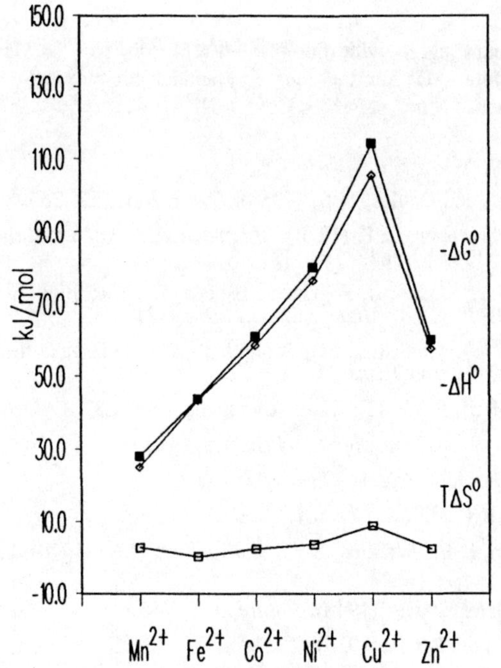

Figure 5.
-$\Delta G°$, -$\Delta H°$ and $T\Delta S°$ trends for the complexation reactions of M^{2+} first row transition metal ions with two molecule of ethylenediamine in aqueous solution at 25°C.

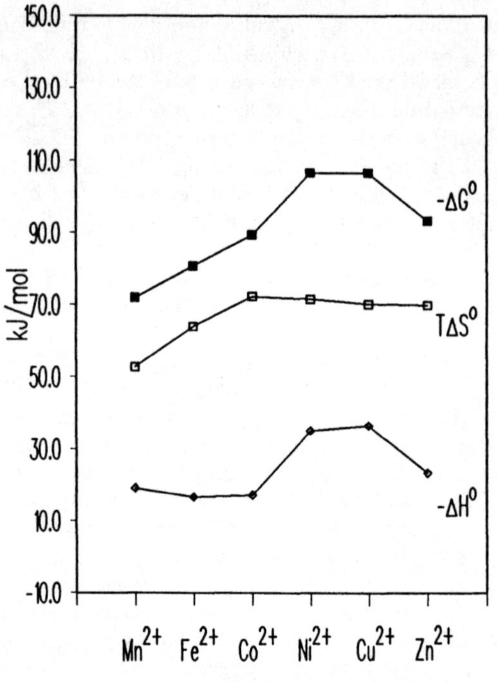

Figure 6.
-ΔG°, -ΔH° and TΔS° trends for the complexation reactions of M^{2+} first row transition metal ions with ethylenediaminetetraacetate (EDTA) ligand in aqueous solution at 25°C. "Best values" from Ashcroft and Mortimer (1970).

for $(MBr_4)^{2-}$ complexes agree with the $Ni^{2+} << Mn^{2+} < Co^{2+} < Mn^{2+} < Cu^{2+} < Zn^{2+}$ order of stability, while -ΔH° for the square-pyramidal complexes $M(Me_5dien)Br_2$ varies according to the sequence $Mn^{2+} < Fe^{2+} < Co^{2+} > Ni^{2+} << Cu^{2+} > Zn^{2+}$.

REFERENCES

Ahrland, S., Chatt, J., and Davies, N.R. (1958). *Quart. Rev.*, *12*: 265.

Ashcroft, S.F., and Mortimer, C.T. (1970). *Thermochemistry of Transition Metal Complexes*, Academic Press, London, p. 117.

Bencini, A. Bianchi, A., Dapporto, P., Garcia-España, E., Marcelino, V., Micheloni, M., Paoletti, P., and Paoli, P. (1990). *Inorg. Chem.*, *29*: 1717.

Bencini, A., Bianchi, A., Micheloni, M., Paoletti, P., Garcia-España, E., and Niño, M.A. (1991). *J. Chem. Soc., Dalton Trans.*, 1171.

Bianchi, A., and Paoletti, P. (1985). *Inorg. Chim. Acta*, *96*: L37.

Eichhorn, G.L., ed. (1973). *Inorganic Biochemistry*, Elsevier.

Freeman, H.C. (1967). *Adv. Protein Chem.*, *22*: 258.

Fraústo da Silva, J.J.R.(1983). *J. Chem. Ed.*, *60*: 390.

Ho, F.F.L., Erickson, L.E., Watkins, S.R., and Reilley, C.N. (1970). *Inorg. Chem.*, *9*: 1139.

Irving, H., and Williams, R.J.P. (1948). *Nature*, *162*: 746.

Irving, H., and Williams, R.J.P. (1953). *J. Chem. Soc.*, 3192.

Pearson, R.M. (1968). *J. Chem. Ed.*, *45*: 581, 643.

Perrin, D.D. (1958). *J. Chem. Soc.*, 3125.

Perrin, D.D. (1959). *J. Chem. Soc.*, 290.

Sillén, L.G., and Martell, A.E. (1964). *Stability Constants of Metal-Ion Complexes*, The Chemical Society, London.

Sillén, L.G., and Martell, A.E. (1971). *Stability Constants of Metal-Ion Complexes*, The Chemical Society, London.

Smith, R.M., and Martell, A.E. (1974-1989). *Critical Stability Constants* (R.M. Smith and A.E. Martell, eds.), Plenum Press, New York.

Smith, R.M., Martell, A.E., and Chen, Y. (1991). *Pure Appl. Chem.*, *63*: 1015.

Tomita, A., Hirai, H., and Makishima, S. (1968). *Inorg. Chem.*, *7*: 760.

Wellman, K.M., Mecca, T.G., Mungal, W., and Hare, C.R. (1968). *J. Amer. Chem. Soc.*, *90*: 805.

Weser, U. (1968). *Struct. Bonding*, *5*: 41.

SPECTROSCOPIC TECHNIQUES AND STRUCTURAL DETERMINATIONS OF LOW-MOLAR-MASS METAL COMPLEXES IN SOLUTION

by

Henryk Kozlowski[1] and Giovanni Micera[2]

[1]Instytut Chemii,
Uniwersytet Wroclawski,
ul. F. Joliot-Curie 14
50383 Wroclaw
Poland

[2] Dipartimento di Chimica,
Università di Sassari,
Via Vienna 2
07100 Sassari
Italy

I. INTRODUCTION

The investigation of metal-ligand interactions in solution, even for low-molar-mass (l.m.m.) compounds, represents a rather difficult challenge from both the structural and thermodynamic points of view. The identification of metal-binding sites in usually multi-donor organic molecules, or the structure assignment to a metal site or to a bound ligand (conformation, configuration), may often be incomplete and tentative, even when very sophisticated and expensive techniques are used. This is because all methods employed, among which spectroscopic ones, tend to measure the resulting effects of the formation of a complex rather than to directly investigate its structure.

The coordination environment in l.m.m. bioinorganic complexes is usually very complicated because of the irregular symmetries of the metal sites and of the various possible conformations for the ligands. This may additionally complicate the study. Also, the equilibria taking place in solutions containing l.m.m. bioinorganic complexes usually involve several species with different structures and stabilities. The concentrations of these are often pH-dependent and, in most cases, it is impossible to crystallize a particular complex to investigate its structure through X-ray analysis techniques.

Thus, spectroscopic analysis represents an attractive approach for studying metal-binding sites if the metal ion is "colored" (electronic spectra may help deduce the environment of such metals, e.g. copper, nickel, iron, molybdenum, vanadium, etc) or contains "unpaired electrons" (EPR and ENDOR spectra may help in assigning the electronic structure, bond characters, geometry and donor atoms around a central metal ion, etc). In addition, recent advances in nuclear magnetic resonance (NMR) spectroscopy have made it one of the most fertile physicochemical tools for the structural investigation of various biological objects including l.m.m. molecules. The possibility of directly monitoring different nuclei of atoms (^{1}H, ^{13}C, ^{31}P, ^{15}N, etc as well as different metal ions) located in different positions of a molecule allows the molecular structure to be evaluated quite precisely in diamagnetic and paramagnetic systems. A careful application of the above spectroscopic techniques (preferably more than one) to the investigation of metal-l.m.m. ligand interactions often results in the reliable determination of the structure of the resulting complexes, especially when spectroscopy is combined with other approaches, e.g. potentiometry or voltamperometry.

Given the limited space available, only the basic principles of selected spectroscopic techniques are introduced in this short review, along with a critical discussion of structure determinations in l.m.m. complexes.

II. ABSORPTION SPECTRA

The application of the simple crystal field theory to various stereochemistries yields a *d*-orbital splitting diagram which can be used to predict the electronic spectrum of a metal ion in a particular environment. Although this point charge approximation does not assume any metal-ligand bond covalency, it introduces the basic observable in electronic spectroscopy, the *d-d* transition. More or less developed ligand field theories as well as the more complete molecular orbital (MO) theory help in understanding metal-ligand bonds not only qualitatively but also quantitatively. The MO approach allows also to consider ligand (atomic or molecular) orbitals which are critical in the evaluation of electronic transitions (e.g. charge transfer or intra-ligand transitions) or electron delocalization within a complex molecule.

Energy levels of the *d* orbitals deduced from crystal field approximation for octahedral (O_h), tetrahedral (T_d), and tetragonal (D_{4h}) symmetries are given in Figure 1. These one-electron energy levels can be used to predict the transition assignment and energies for multi-electron systems. The construction of the term diagram (many-electron energy levels) is a proper approach, but space limits force us to use rather simple experimental descriptions and to send readers to the excellent monograph of Lever (1984) for details.

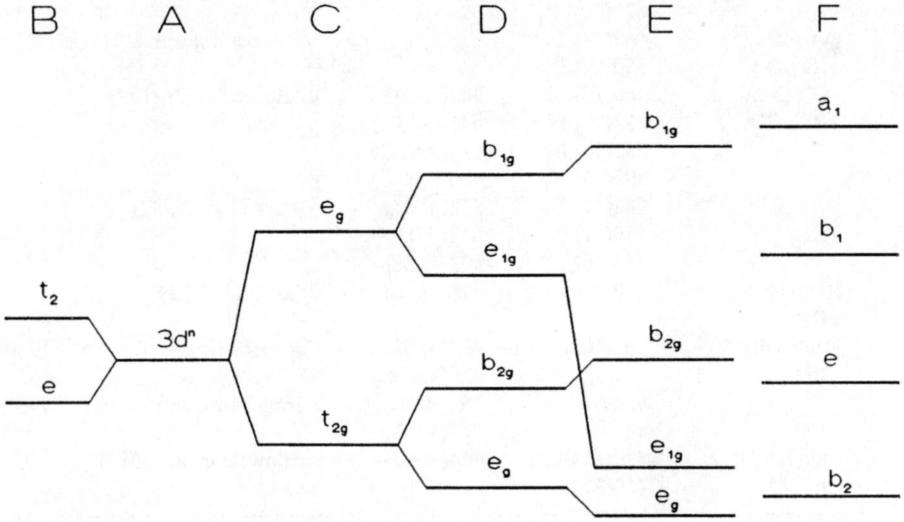

Figure 1.
Energy levels in the point charge approximation; **A-** free metal ion, **B-** tetrahedral geometry (T_d), **C-** octahedral geometry (O_h), **D-** weak tetragonal distortion of O_h (D_{4h}), **E-** strong tetragonal distortion of O_h or square planar geometry (D_{4h}), **F-** energy diagram for VO(IV) ion (C_{4v}).

A. d-d spectra

It is clear from *d-d* orbital schemes shown in Figure 1 that electronic transitions within metal *d* orbitals depend on (i) the number of *d* electrons (*d* configuration), (ii) the splitting pattern (stereochemistry), and (iii) the energy differences between different *d* orbitals (energy of transition). The following discussion is devoted to three selected metal ions V(IV) (d^1), Ni(II) (d^8) and Cu(II) (d^9). All these metal ions are of biological importance, and display a specific spectroscopic behavior in their l.m.m. complexes with bioligands.

1. *Vanadium(IV), d^1*

Vanadyl species containing the V=O unit exhibit very specific spectra which are quite distinct from those of other vanadium(IV) complexes. The highest possible symmetry around the metal ion would be C_{4v} (Fig. 1) with three potential intra *d*-shell transitions observed below 330 nm. Their precise assignment is sometimes controversial (Lever, 1984) because the relative positions of the b_1 ($d_{x^2-y^2}$) and e (d_{xz}, d_{yz}) levels are mainly determined by a balance between σ donor strength of equatorial ligands and V-O π bonding, theoretical approaches being necessary to establish the correct energy-level scheme (Deeth, 1991). Distortion from C_{4v} symmetry to C_{2v} usually occurs for chelate complexes and an additional *d-d* transition may be observed. Considering the data presented in Table 1, the three transitions (C_{4v}) may correspond to b_2 (d_{xy}) ⇒ e (d_{xz}, d_{yz}) (*E*), b_2 ⇒ b_1 ($d_{x^2-y^2}$) (B_1), and b_2 ⇒ a_1 (d_{z^2}) (A_1). The presence of a fourth *d-d* transition indicates splitting of *e* orbitals (C_{2v}). A band observed near 400 nm could correspond to the b_2 ⇒ a_1 transition,

Table 1

Absorption and CD Spectra for Selected Complexes of VO(IV) and Planar Ni(II)

Species[a] {binding mode}	Absorptions[b] λ [nm], (ε)	CD λ [nm], (Δε)	Reference
$[VO(H_2O)_5]^{2+}$	763(16) E		Bridgland et al. (1965)
	625(7.5) B_1		
$[VO(cat)_2]^{2-}$	662(77) E		Jezowska-Bojczuk et al. (1990)
2{O⁻, O⁻}	550(50) B_1		
$[VO(LacA)_2]^{2-}$	760(16) E'	780(-0.01)	Kozlowski et al. (1991)
2{O⁻, COO⁻}	622(19) E''	640(+0.06)	
	532(20) B_1	520(+0.025)	
	392(26) A_1	400(+0.03)	
$[VO(GalA)_2]^{4-}$	680(27) E	618(-0.85)	Micera et al. (1989)
2{O⁻, O⁻}	520(12) B_1	518(+0.28)	
	415(26) A_1	403(-0.38)	
[Ni(Ang)]	412(320)	476(-3.10)	Pettit et al. (1989)
{4N}			
[Ni(TRF)]	458(105)	494(-0.32)	Formicka-Kozlowska et al. (1980)
{3N}		426(+1.10)	
[Ni(MIF)]	470sh	455(-1.50)	Formicka-Kozlowska et al. (1981)
{3N}	400	379(+0.15)	
[Ni(GGC)]	520(124)sh	474(-0.03)	Kozlowski et al. (1987)
{3N, S}	410(336)sh		

[a]Ligands: cat = catechol, LacA = *D*-lactobionic acid, GalA = *D*-galacturonic acid, Ang = angiotensin II, TRF = pGlu-His-Pro-NH_2, MIF = Pro-Leu-Gly-NH_2, GGC = Gly-Gly-Cys; {3N} and {4N} indicate the number of bound nitrogens.

[b]Transitions: E, b_2 ⇒ e, B_1, b_2 ⇒ b_1, and A_1, b_2 ⇒ a_1; E' and E'' correspond to split E transition (C_{2v}).

but it is often very intense and is sometimes assigned to a forbidden charge transfer band (Selbin, 1965; Lever, 1984).

2. *Nickel(II), d^8*

Ni(II) complexes exist in all coordination numbers from 2 to 7. The most popular structures in l.m.m. biological systems are, however, six- and four-coordinate complexes.

Six-coordinate octahedral species exhibit three spin-allowed transitions from $^3A_{2g}$ (in term diagram) to $^3T_{2g}$, $^3T_{1g}$ and $^3T_{1g}(P)$. All of these transitions are relatively weak (ε < 30 mol^{-1} dm^3 cm^{-1}) and their energies fall within the ranges 1430-770, 910-500, and 520-370 nm. Some examples are given in Table 1. The splitting of the middle octahedral band into two transitions indicates tetragonal distortion (D_{4h}) in complexes.

The spectra of four-coordinate tetrahedral complexes are much more intense than those of octahedral complexes. The visible band observed usually over the 900-500 nm region has intensity about 10^2-10^3 mol^{-1} dm^3 cm^{-1} comparable with that of similar complexes of Co(II).

The d^8 configuration favors formation of four-coordinate diamagnetic square-planar complexes, especially with stronger field ligands (amide nitrogens of peptides). Such complexes exhibit usually a single band at 550-400 nm with intensity around 50-500 mol^{-1} dm^3 cm^{-1} (Table 1). They are yellow, red, or orange, and are easily distinguished from complexes with other geometries. A second more intense band may be observed below 430 nm. However, it often has a charge transfer origin. Diamagnetism is a consequence of eight electrons being paired in the four lower-lying *d* orbitals (Fig. 1, D_{4h}) and may be easily established by magnetic moment measurements (NMR).

3. *Copper(II), d^9*

There is a huge amount of spectroscopic information on copper in the literature, although it is rarely possible to make definitive statements. Most complexes exhibit a single broad poorly resolved absorption band in the visible region arising from the three *d-d* transitions characteristic for tetragonal structures (Fig. 1). Since a number of copper(II) complexes exist in heavily distorted stereochemistries, the number of *d-d* transitions can increase to four but the observed band may remain single. Even so, the average energy of this band strongly depends on the donor set involved in coordination. The large amount of experimental data obtained for Cu(II) complexes with peptides provides a simple method for identifying the number of nitrogens (amine and amide) coordinated to the metal ion (Sigel and Martin, 1982; Pettit et al., 1991). Typical energy values for *d-d* bands observed in Cu(II) complexes of peptides with different nitrogen atoms bound to cupric ion are collected in Table 2. The approximate values listed in this table cannot be treated as rigid limits for a particular number of bound nitrogens, as the observed energy depends strongly on stereochemistry and nature of nitrogens bound to the metal ion. Substitution of the water oxygen by nitrogen in the Cu(II) coordination sphere shifts considerably the *d-d* band towards higher energies while the intensity of these bands increases from about 10 for the aqua ion to about 150 mol^{-1} dm^3 cm^{-1} for 4N species.

Table 2

Typical Values of Transition Energies for *d-d* Bands in Cu(II) Complexes with 1, 2, 3 and 4-Coordinated Nitrogens[a]

Complex	λ_{max} [nm], (ε)
1N	680-730 (30-60)
2N	620-670 (40-120)
3N	540-590 (100-150)
4N	500-530 (120-200)

[a]See, e.g., Pettit et al. (1991).

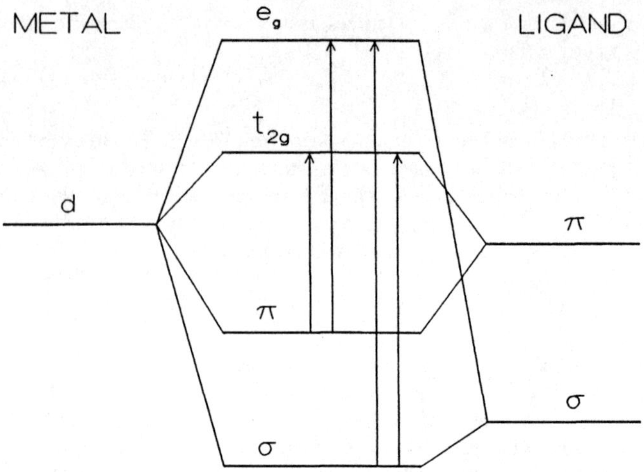

Figure 2.
A simplified molecular orbital diagram showing ligand to metal charge-transfer transitions for O_h symmetry.

B. Charge transfer transitions

Besides *d-d* transitions discussed above, the transfer of an electron from an orbital centered primarily on the ligand to one centered primarily on the metal, or vice versa, can take place. Charge transfer transitions lie often at the blue end of the visible spectra or in the ultraviolet region. The energy of the charge transfer (CT) transition depends on the particular donor involved in coordination as shown in a schematic molecular orbital diagram illustrating the potential ligand-to-metal CT transitions (Fig. 2). The energy of ligand σ and π orbitals strongly depends on the ligand nature, and the same donor atom, if involved in different groups, can even lead to different CT energy values. For instance, nitrogen of amine, amide or imidazole bound to the Cu(II) ion will give CT bands at 260-290, 300-320, and about 330 nm, respectively. In the latter case, two other bands around 220 and 250 nm should be also observed, but only that at 220 nm, corresponding to a σ(im) ⇒ $\sigma^*(d_{x^2-y^2})$ transition, is sharply defined (Fawcett et al., 1980). A red shift for the imidazole ⇒ Cu(II) CT is expected in type 1 chromophores. All these transitions are poorly resolved in absorption spectra while circular dichroism spectra show most of these transitions as well separated bands (vide infra).

Sulphur and phenolic oxygen donors bound to cupric ions exhibit their CT bands at lower energies than those of nitrogen donors and can be easily seen in absorption spectra, even with complicated ligands (Schugar, 1982; Pettit et al., 1985; Varnagy et al., 1988). The CT band of the phenolate oxygen to Cu(II) transition lies around 400 nm yielding usually yellow solutions, while S ⇒ Cu(II) CT bands, depending on the nature of sulphur donor and complex formed, may be centered between 350 and 450 nm, and are found even at wavelengths as high as 600-700 nm in type 1 chromophores (Brader et al., 1992).

Non-oxo vanadium(IV) complexes with phenolic ligands are also characterized by intense O ⇒ V(IV) charge-transfers in the visible region (Hambley et al., 1987).

III. CIRCULAR DICHROISM SPECTRA

Many biological molecules like amino acids, peptides or sugars are optically active species. An optically active substance interacts differently with left and right circularly polarized light. This interaction can be measured as a differential absorption of the left and right circularly polarized light beams. This effect is known as circular dichroism (CD) spectroscopy and may provide data which are complementary to those obtained from electronic absorption spectroscopy. CD spectra are especially useful in the ultraviolet

Table 3

Characteristic Absorption and CD Spectra for Cu(II) Complexes with l.m.m. Bioligands

Ligand (Species)	Absorptions λ [nm], (ε)	CD λ [nm], (Δε)	Reference
Pro-Leu-GlyNH_2			
{2N}	660(40)	695(-0.12)[a]	Formicka-Kozlowska et al. (1981)
		565(-0.10)[b]	
		310(+0.44)[c]	
		260(-1.20)[d]	
{4N}	503(130)	503(-0.36)[b]	
		293(+0.58)[c]	
		260(-1.28)[d]	
Ala-Asp-Ser-Gly			
{2N}	630(83)	615(+0.07)[b]	Decock-Le Révérend et al. (1986)
		300(-0.75)[c]	
		262(+0.60)[d]	
{4N}	506(150)	522(-0.90)[b]	
		298(+0.75)[c]	
		270(-0.50)[d]	
Phe-Gly-Pro-Tyr			Livera et al. (1988)
{2N, O^-_{Tyr}}	650(126)	620(+0.27)[e]	
	407(821)[g]	528(-0.25)[f]	
		430(+0.10)[g]	
		330(-0.15)[c]	
		303(+0.10)[d]	
D-Lactobionic acid			
(CuL_2H_{-2})		700(+0.08)[e]	Bouhsina et al. (1991)
2{COO^-, O^-}		560(-0.02)[f]	
		284(-1.15)[h]	
(CuL_2H_{-4})		765(+0.06)[a]	
2{O^-, O^-}		600(+0.11)[b]	
		280(-1.80)[h]	

[a,b,e,f]A, B+E, B, and E *d-d* transitions, respectively (see refs. for details); [c]N^- ⇒ Cu(II) CT transition; [d]NH_2 ⇒ Cu(II) CT transition; [g]phenolate O^- ⇒ Cu(II) CT transition; [h]sugar O^- ⇒ Cu(II) CT transition.

region, giving much more resolved sets of CT or intra-ligand transition bands than those seen in absorption spectra. Much better resolution is also observed for *d-d* transitions.

The characteristic sets of CD bands for Cu(II)-oligopeptide complexes are collected in Table 3. According to the *d*-orbital energy scheme for a tetragonal Cu(II) complex (Fig. 1, D_{4h}), three transitions are possible within metal orbitals. For a CD band to be observed, transition should be both electronically and magnetic dipole allowed. Only two of the three Cu(II) transitions are usually observed in CD spectra as the $d_{z^2} \Rightarrow d_{x^2-y^2}$ transition is magnetically forbidden in D_{4h} symmetry (Tsangaris and Martin, 1970). A symmetry lower than D_{4h} may split e_g orbitals (Fig. 1) into two singlets and the number of *d-d* transitions increases even to 4.

Besides improving resolution of the *d-d* absorptions, CD spectroscopy gives also much better resolved charge transfer bands in the UV region. A representative example is provided by Cu(II) complexes with peptides. As shown in Table 3, CD spectra make it possible to easily distinguish a NH_2 ⇒ Cu(II) CT band from a N^- ⇒ Cu(II) CT transition.

It is interesting to note that CT transitions occurring at around 280 nm, which are usually assigned to $NH_2 \Rightarrow Cu(II)$, may also correspond to the hydroxyl $O^- \Rightarrow Cu(II)$ charge transfer found in the Cu(II) complexes of sugar acids (Table 3; see Bouhsina et al., 1991). Thus, although resolution of CD spectra can improve the information on the system studied, the proper assignment of a particular transition could provide major problems especially in mixed systems, e.g. for ligands containing both NH_2 and hydroxyl donors (e.g. amino sugars, see Chapter 2-G).

CD spectra of oxovanadium(IV) complexes with peptides or amino acids are scarce as peptides are not particularly powerful ligands for this metal ion. However, the complexes with sugar acids give good examples for the application of CD (Table 3).

IV. FLUORESCENCE SPECTRA

After an electronic transition has taken place, the excited molecule must relax, i.e. return to the ground state with a release of energy. One of the mechanisms by which the excited state can lose energy is the emission of light (luminescence). If the excited state is of the same spin as the ground state, the emission is generally an allowed and rapid process (lifetime of excited states: 10^{-6}-10^{-9} s) and is called fluorescence (Fig. 3).

In bioinorganic complexes, the fluorophores can be aromatic amino acids (e.g. tyrosine or tryptophan) or metal ions like lanthanides which are often used as luminescence structural probes of biological macromolecules. Many other biologically active molecules, like anticancer anthracycline drugs (e.g. adriamycin), are also very effective fluorophores exploited in structural studies.

The interaction of metal ions with ligands containing fluorophores may change drastically the fluorescence spectra, for example causing its quenching. Antracyclines give a characteristic fluorescence spectrum (Fig. 4) which is usually quenched through metal ion coordination (Fiallo and Garnier-Suillerot, 1986). It is, however, also possible that metal coordination enhances the fluorescence of a bound fluorophore. As an example, coordination of Al^{3+} to adriamycin (LH_2) leads to the formation of a minor octahedral complex in which the ligand binds the metal ion through two anthracycline ring oxygens. This complex, [Al(LH)], gives very specific and strong fluorescence (Fig. 4). The fluorescence, however, is almost completely quenched when two water molecules bound to

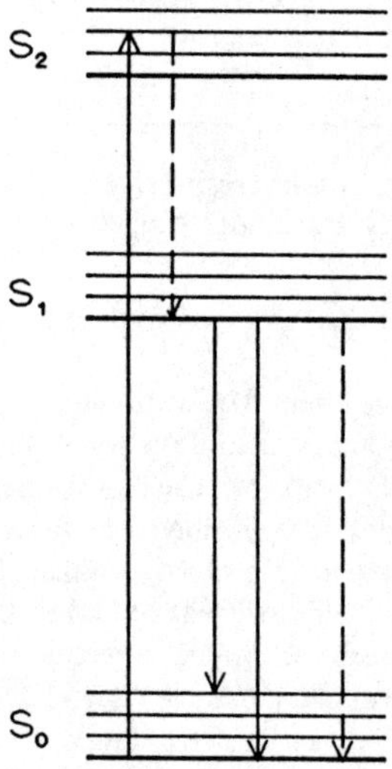

Figure 3.
Schematic representation of transitions occurring during fluorescence. Excitation ($S_0 \Rightarrow S_2$), radiationless transitions (dotted lines, $S_2 \Rightarrow S_1$ and $S_1 \Rightarrow S_0$), fluorescence (solid lines, $S_1 \Rightarrow S_0$).

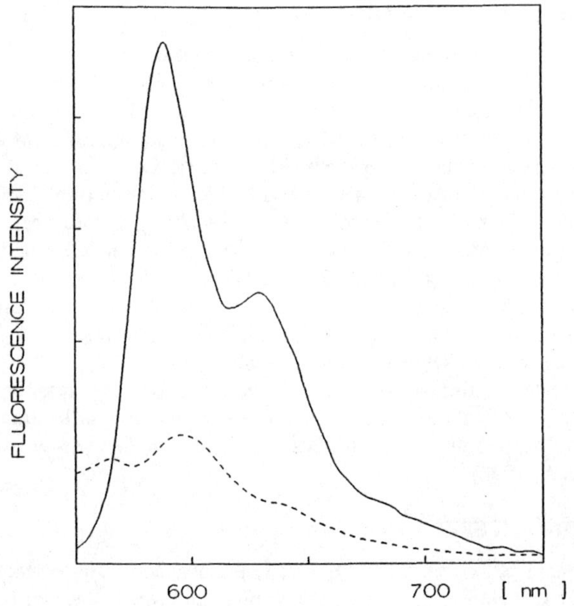

Figure 4.
Uncorrected fluorescence spectrum of Al(LH) complex (solid line, LH^- adriamycin protonated on amino group), and metal-free adriamycin (dotted line) at pH 5.

aluminum dissociate protons and the complex $[Al(LH_{-1})]$ changes its geometry to tetrahedral. It is noteworthy that the donor system of adriamycin involved in Al^{3+} coordination is exactly the same in both octahedral and tetrahedral complexes (Pereira et al., 1993).

When both a luminescent energy donor and an absorbing energy acceptor exist in the same complex and there is a significant overlap between the donor emission spectrum and acceptor absorption spectrum, energy transfer of a non-radiative kind can take place between the two moieties. It is possible to measure the efficiency of energy transfer between donor and acceptor and this can be related to the donor acceptor distance, *r*, by Förster equation (Schwyzer and Caviezel, 1971):

$$r = (T^{-1} - 1)^{1/6} R_0 \quad (1)$$

where T is the efficiency of energy transfer and R_0 the donor-acceptor distance for which T = 0.5.

This approach can be very useful, for example in metal-peptide systems in which the peptide molecule contains two fluorophores (aromatic residues) (Siemion et al., 1985). Complexation of metal ions may change the ligand conformation and thus the distance between fluorophores. Measurement of distances in such systems may help in evaluating complex conformation and stoichiometry. The energy transfer measurements can also be used to calculate the distance between "inorganic" fluorophores (e.g. lanthanides, see Horrocks and Sudnick, 1981) or between inorganic and organic fluorophores. The evaluation of distances between fluorophores is especially useful for large biomolecules like proteins (Horrocks and Sudnick, 1981).

Several processes contribute to the decay of an excited state, and an exponential decrease of luminescence intensity following excitation is usually observed. Luminescence decay lifetimes may also supply very interesting information; for example, they can provide a direct measure of the number of water molecules bound to lanthanides (Horrocks and Sudnick, 1981).

V. FT-INFRARED SPECTROSCOPY

The introduction of Fourier transformation into infrared spectroscopy has brought enormous advances in infrared instruments. The tremendous increase of sensitivity (more than three orders of magnitude) coupled with the development of very fine methods of data treatment have made the Fourier transform infrared (FT-IR) technique one of most universally used methods for chemical and structural analysis (Mantsch, 1991).

One of the advantages of the FT-IR method is the possibility of obtaining good spectra in aqueous solution. Of particular interest here is the ability to measure conformations of peptides and proteins in aqueous solutions by analysis of the amide I vibration bands. This predominantly C=O stretching mode is both sensitive to the backbone conformation and well removed from other absorption bands. These amide I bands are, however, not well resolved from one another, and Fourier self-deconvolution technique was found extremely useful in separating these closely spaced vibrations.

Besides peptide conformation, amide I vibration is very sensitive to metal ion coordination. The coordination of Cu(II) ions to dipeptides results in dramatic changes in the 1550-1700 cm^{-1} range due to the displacement of the amide hydrogen by the metal ion (Shaw and Mantsch, 1992).

VI. EPR AND ENDOR SPECTRA

The combined application of EPR (Electron Paramagnetic Resonance) and ENDOR (Electron Nuclear Double Resonance) spectroscopies can provide detailed information about metal ions having unpaired electrons in *d* orbitals. Usually, simple EPR measurements on paramagnetic complexes can probe only the metal ion and its closest environment. Most of the information is extracted from the analysis of the spin Hamiltonian parameters, *g*-values and hyperfine (hf) coupling constants. The deviation of the *g*-values from the free-electron value (2.0023) as well as their orientation-dependence are distinctive of the orbital distribution of the unpaired electron(s), as affected by the ligand donor properties and the coordination geometry. Hyperfine structure is associated with the interaction of the unpaired electron with those magnetic nuclei having net magnetic moment and yields the coupling constants (*A*) of the EPR spectrum. When resolved, hf structure is a rich source of structural information on the metal ion and its closest vicinity. For instance, metal hf coupling is related to the electronic structure of the complex, its symmetry and bonding parameters, whose detailed trends have been established for several metal ions. Ligand hf structure (e.g. due to ^1H or ^{14}N atoms) permits to obtain valuable information on the ligand, for example to distinguish directly the presence of some nuclei, either bound to metal or not, as well as their number, and, to a more advanced level, to estimate spin population and hybridization of the interacting orbitals.

In principle, any of the hf interactions with surrounding magnetic nuclei should contribute to the EPR spectrum through a set of $2nI+1$ spectral components (I = nuclear spin magnetic number and n = number of equivalent nuclei), but most of them are not resolved. ENDOR spectroscopy can resolve even weak magnetic interactions within a distance of 5-6 Å, making it possible to regain the missing hf details of the EPR spectrum. According to theory, each class of equivalent nuclei contributes a set of ENDOR resonances. For a system with a single unpaired electron, this set consists of a pair of transitions at $\nu_\pm = |\nu_0 \pm A/2|$, where ν_0 is the free nuclear Larmor frequency and A is the same hf constant as that expected in the EPR spectra. For $\nu_0 > |A/2|$ these pairs are equally spaced around ν_0, whereas for $\nu_0 < |A/2|$ they are centered about $|A/2|$ and separated by $2\nu_0$. Therefore, advantages are obtained by use of ENDOR because the interacting nuclides may be distinguished by the proper frequency ranges and each group of equivalent nuclei contributes only two ENDOR lines (for nuclei with $I \geq 1$ quadrupole splitting is also expected).

Hf coupling constants A are made up of two main contributions. The isotropic contribution A_i, mainly due to Fermi-contact and pseudo-contact interaction, does not depend on from the molecular orientation with respect to the magnetic field. The anisotropic contribution A_a depends on the molecular orientation and in most cases, for example with d^1

ions and distant interacting nuclei, can be approximated by the point-dipole equation:

$$A_a = gg_n\beta\beta_n(3\cos^2\alpha - 1)/hr^3 \quad (2)$$

where g and g_n are the electron and nuclear g-factors, β and β_n are the Bohr and nuclear magnetons, h is the Planck constant, r is the nucleus-electron distance and α the angle between the nucleus-electron axis and the magnetic field direction. Hence, an analysis of the hf interaction requires the knowledge of the molecular orientation and this is possible in single-crystal measurements. However, one can obtain crystal-like information for randomly oriented molecules (powders or frozen solutions) provided g or A anisotropy is large. In such a case, the EPR spectrum identifies well-defined molecular directions (the principal axes of g- and A-tensors) along which the hf interaction can be measured by ENDOR.

Because of the complexity of the field, only selected applications of these techniques are presented for simple paramagnetic probes of biological interest.

A. Vanadium(IV)

Vanadyl ion exhibits an orbitally non-degenerate ground state for which sharp and simple EPR spectra are expected. Typical eight-line patterns are observed due to ^{51}V hf interaction (I= 7/2 and natural abundance greater than 99%). In C_{4v} symmetry, the ground state is predominantly d_{xy} in character and the following spectral features are predicted: $g_{\parallel} < g_{\perp} < 2.0023$ and $A_{\parallel} > A_{\perp}$, where the subscripts $\parallel$ and $\perp$ refer to directions that are normally parallel and perpendicular to the V=O bond, respectively. In complexes of chelated ligands, lowering of symmetry to C_2 or C_{2v} and mixing of d_{z^2} and $d_{x^2-y^2}$ in the ground state do not significantly affect this spectral trend (Hitchman et al., 1969). Correlations between EPR parameters and ligand properties have been derived for a large variety of square pyramidal and six-coordinate complexes. With increasing the in-plane donor strength, there is a trend to decrease the isotropic metal coupling constant, $A_o = (2A_{\perp}+A_{\parallel})/3$ and to increase the average g-factor ($g_o = 2g_{\perp}+g_{\parallel})/3$, which are measured in fluid solution. The same trend is valid for $g_{\perp}$, $g_{\parallel}$, $A_{\perp}$, and $A_{\parallel}$ recorded in frozen solution. These parameters cluster in domains according to the identity of equatorial donor atoms, e.g. O_4, O_2N_2, O_2S_2, N_4, S_4, and they may aid in assigning metal coordination. Coupling constants, mainly A_o and $A_{\parallel}$, may be predicted to a reasonable approximation by use of additive contributions for equatorial donors (L_i), i.e. $A = (A_{L1}+A_{L2}+A_{L3}+A_{L4})/4$ (Chasteen, 1981).

EPR spectroscopy can contribute to the study of l.m.m. complexes in biological systems. For instance, mono- and bis-chelated VO(IV) species of l.m.m. catecholic ligands formed in plants upon adsorption of vanadium have been identified and distinguished from the VO^{2+} ions bound to macromolecules (Micera and Dessì, 1989). The intracellular reduction of vanadium(V) leading to formation of VO(IV) glutathione complexes was also shown in adipocytes (Degani et al., 1981).

Hexa-coordinate non-oxo vanadium(IV) complexes exhibit substantial distortion from the octahedral geometry. In five-membered chelated species the geometry is close to the trigonal-prismatic one and there is a switch of the ground state to d_{z^2}. In this case, theory requires $g_x \sim g_y < g_z \sim 2.0023$ and $A_z < A_y \sim A_x$ (Desideri et al., 1978). Therefore, EPR can be used as a diagnostic tool to distinguish oxo- from non-oxo vanadium(IV) chelated complexes formed in solution, for example by catecholic and hydroxamic ligands.

EPR spectroscopy detects polynuclear complexes and provides structural information, for example on the metal-metal distance. Hydroxo-bridged VO(IV) dinuclear species are EPR-silent because of the strong magnetic coupling between the metal ions. Weakly coupled dimeric complexes, with a metal-metal distance around 4 Å, are formed by optically active or racemic tartrates (e.g. Belford et al., 1969). In this case, distinctive fifteen-line patterns, with a magnitude of the hf splitting half of that expected for a monomeric species, may be observed in the EPR spectra. The dinuclear structure yields also a "zero-field" splitting, which is measured in anisotropic spectra and permits an estimate of the metal-metal distance. A similar species, with the ligand in the open-chain form, has been found with D-galacturonic acid (Branca et al., 1990c).

VO(IV) is a suitable probe for ENDOR spectroscopy. ^1H, ^{14}N and ^{15}N ENDOR studies are useful in assigning binding sites in VO(IV) complexes of l.m.m. ligands in solution, for instance imidazoles and sugars (Mulks et al., 1982; Branca et al., 1989). As an example, the quite straightforward ^1H ENDOR analysis for VO(IV) catecholates and salicylates is described here (Branca et al., 1990b). In these complexes, which are nearly axial, the ligands are almost coplanar with the metal and the proton-vanadium distance is large enough to make the interaction purely dipolar. Therefore, the ^1H couplings measured along the principal directions of the A(V)-tensor can allow to locate the protons and evaluate their distance from the metal in terms of the point-dipole model (eq. 2). For instance, in ENDOR spectra of bis(catecholato)oxovanadate(IV), aromatic protons adjacent to the donor groups exhibit hf couplings with absolute magnitudes and orientation dependence well consistent with X-ray crystallographic data. In the bis-chelated complex of salicylic acid, the above protons, being non-equivalent, yield distinct couplings that permit accurate estimates of the distances. Since the spectra are diagnostic of substituents near the donor groups, the technique is sensitive to changes in the structure of ligands even when they adopt the same coordination mode. Deuteration removes the signals of exchangeable protons and reveals, for example, inner- and outer-sphere water molecules as well as undissociated phenolic groups.

B. Chromium(V)

Cr(V) is another d^1 ion that, in the form of oxochromium(V), is well suited for EPR and ENDOR studies. Attention has been paid to this ion because it is formed in all known oxidations by Cr(VI) and seems involved in the mutagenic and carcinogenic properties of the element. Cr(V) was found by EPR spectroscopy as a stable intermediate in the interaction of Cr(VI) with rat liver microsomes in the presence of NADPH (Jennette, 1982). EPR shows that Cr(V) formed upon reduction of Cr(VI) in living systems may be associated with l.m.m. ligands, for example sugars or polyols in plants (Micera and Dessì, 1988) and glutathione in blood cells (Branca et al., 1990a). A range of long-lived Cr(V) species derived from Cr(VI) have been studied in aqueous media, mainly those with oxygen and sulphur donors like sugars, polyols, hydroxycarboxylic acids, ribonucleotides, thiols, etc. In spite of this large amount of data, only limited information is available on the spectral properties of the ion. Anisotropic EPR spectra of Cr(V) complexes have been less studied compared with those of V(IV) and correlations which could be useful for a direct assignment of the structure are generally lacking, also due to the small anisotropy of the g factors which may have significant contribution from charge-transfer transitions. EPR spectra exhibit metal hf structure due to the only isotope with a non-zero nuclear spin (^{53}Cr, I = 3/2; 9.5%). In fluid solutions, Cr(V) species may be discriminated by g_o and $A_o(^{53}Cr)$ values. For instance, the complexes of sugars and polyols, compared with those of α-hydroxycarboxylic acids, exhibit significantly different values for g_o and $A_o(^{53}Cr)$. The latter ones are greater for mono- than for bis-chelated complexes and are particularly small when sulphur is involved in coordination. The ligand hf structure of the EPR spectra is even more distinctive; in the case of ethylene glycolate(2-) for example, the number of ^1H components corresponds well to the number of ligand protons. Unfortunately, ^1H hyperfine interaction is often unresolved, for instance because of the presence of overlapping couplings. With this ion too, hydrolysis of the mono-chelated complexes may lead to EPR-silent hydroxo-bridged species, as in the case of α-hydroxycarboxylic ligands.

The Cr(V) probe is almost unique among transition metal ions in that it is suitable for ENDOR studies at temperatures as high as 180-200 K (Möhl et al., 1982). This makes it possible to measure isotropic couplings with ligand nuclei in fluid (e.g. CH_3OH) solution even when hf structure is lacking in EPR spectra. For instance, in the complexes of ethylene glycol derivatives, isotropic ^1H ENDOR clearly indicates nonequivalent axial, equatorial and methyl protons in the ligand (Branca et al., 1992). The magnitude of the single ^1H coupling observed for the ethylene glycolate complex is nearly the average of those expected for equatorial and axial protons, suggesting that these are magnetically equivalent at 200 K because of a fast interconversion process. In spectra recorded in frozen solution, "complete" orientation selection for the proton hf interaction has been obtained along the principal axes of the ^{52}Cr and ^{53}Cr g/A tensors. By taking into account the isotropic contributions measured in fluid solution, all the metal-proton distances have been extracted from the anisotropic couplings and a rather complete picture of the molecular structure has been obtained.

VII. MULTINUCLEAR NMR

Multinuclear NMR has rapidly become a powerful tool in the study of model biosystems containing metal ions. Since signal acquisition capabilities have been extraordinarily improved by the developments in instrumentation, good spectra can be obtained in acceptably short times for many nuclei. A whole new range of information can be gleaned by use of metal NMR which probes directly the metal center (e.g. Lambert and Riddel, 1983; Benn and Rufinska, 1986; Mason, 1987; 1989). NMR active nuclei can be divided into dipolar (I = 1/2) and quadrupolar (I > 1/2) isotopes. The parameters to which one has access through NMR experiments are chemical shifts, spin-spin coupling constants, and relaxation times.

Many metal nuclei exhibit a much larger range of chemical shifts than is observed for ^{1}H, ^{13}C, or ^{31}P, e.g. 13000 ppm for ^{195}Pt and 18000 ppm for ^{59}Co. An obvious benefit is the excellent spectral resolution and the possibility of monitoring distinct resonances even when the metal ion takes part in multiple coordination sites of very similar structure. However, in some cases, the number of structurally defined compounds is insufficient to provide a data base for the interpretation of chemical shift scales, and this is a real bar particularly to the in situ characterization of complexes formed in biological systems. However, at least qualitative structure-shift correlations are available for the most commonly used nuclei. The oxidation state exerts the largest influence on the chemical shift of a metal ion. Many elements have more than one diamagnetic oxidation state and thus exhibit NMR resonances spanning quite distinct spectral ranges. Normally, lower oxidation states exhibit upfield (or negative) shifts compared with higher oxidation states. The scalar chemical shift, representing the isotropic part of the chemical shift tensor, is often discussed, in terms of Ramsey's shielding equation, as the sum of "diamagnetic" and "paramagnetic" contributions. The diamagnetic term (σ_d) involves only the ground state electronic distribution and reflects the electron density in the immediate vicinity of the nucleus. The paramagnetic contribution (σ_p) is a second-order effect which also depends on the excited states that are mixed into ground state by the external magnetic field. This term vanishes for *s* electrons and may become very large when there is an asymmetric distribution of *p* and *d* electrons. Both the diamagnetic and paramagnetic terms contribute to the variation of chemical shift in atoms such as C, N, and F. For nuclei heavier than F, variations in σ_d are small compared with those in σ_p and the major contribution to the chemical shift is due to the paramagnetic term which is a function of the average radius $<r>$ of the valence orbitals and the average energy ΔE of low-lying excited states:

$$\sigma_p \propto - [<r^{-3}>_p P_i + <r^{-3}>_d D_i] \Delta E^{-1} \qquad (3)$$

Here P_i and D_i measure the degree of charge imbalance in the valence *p*- and *d*-shells which yields the orbital angular moment. Since the shift δ and nuclear shielding σ are defined with opposite sign, an increase in the absolute magnitude of σ_p corresponds to an increase in δ.

In main group elements, low-energy $n \Rightarrow \pi^*$ transitions yield large negative values of σ^p which cause nuclear deshielding, whereas the involvement of the lone pairs in bonding removes the paramagnetic contribution and increases the nuclear shielding. In transition metal ions, nuclear deshielding is mainly due to $d\pi \Rightarrow d\sigma^*$ or $d\sigma \Rightarrow d\pi^*$ transitions. Even with d^0 and d^{10} configurations, metal-ligand interactions populate outer *p* or *d* orbitals (donation) or produce holes in *d* orbitals (back-donation) leading to paramagnetic contributions (Becker and Berlage, 1983; Nakatsuji et al., 1984). Since the chemical shift is dependent on ΔE values, shielding tends to increase from weak to strong ligands and can be correlated to ligand-field parameters. In low-spin Co(III) complexes, a linear relation exists between $\delta(^{59}Co)$ and the wavelength of the $^{1}A_{1g} \Rightarrow {}^{1}T_{1g}$ transition multiplied by the nephelauxetic ratio which takes into account the delocalization of *d* electrons (Bramley et al., 1985).

Line width measurements can provide direct information concerning the degree of symmetry of the electric field at the nucleus (electron distribution about the metal). In nuclei with I > 1/2 the quadrupole moment Q couples with local electric field gradients and this interaction provides an efficient process for nuclear relaxation. In the limit of fast motion, longitudinal and transverse relaxation rates $1/T_1$ and $1/T_2$ are equal and related to the line width $\Delta\nu_{1/2}$ (at half-height) by the equation:

$$\frac{1}{T_1}=\frac{1}{T_2}=\frac{3\pi^2(2I+3)}{10I^2(2I-1)}\left(1+\frac{\eta^2}{3}\right)\left(\frac{e^2Qq_{zz}}{h}\right)^2\tau_c=\pi\Delta\nu_{1/2} \qquad (4)$$

Therefore, line widths depend on the nuclear properties I (spin quantum number) and Q (electric quadrupole moment), but also on molecular properties such as q_{zz} (the largest component of the electric field gradient at the nucleus, EFG), η (an asymmetry parameter for EFG), the correlation time for the molecular tumbling τ_c (which, on the other hand, is affected by molecular size, medium viscosity, and temperature). The EFG depends upon the arrangement and nature of bound ligands. Relaxation is least effective where the EFG is small and so sharp lines are observed for tetrahedral or octahedral environments. As the metal environment is distorted by differences in ligand properties, the line width increases accordingly. However, as shown by Akitt and McDonald (1984) who described rules for predicting point charges arrangements which give a zero EFG, many non-cubic molecules, even with irregular symmetry, can give narrow lines (low-EFG situations), the effective factor being the electronic density at the nucleus. Ligand-exchange processes contribute to the line widths. Nevertheless, in many cases, $\Delta\nu_{1/2}$ provides useful information.

In addition to the chemical shift and line width information, isotropic "scalar" couplings are often available. Basic interactions contributing to the spin-spin couplings are of orbital, dipole-dipole, and contact origin, perturbation approximations being necessary for a detailed description (Lambert and Riddel, 1983). However, a large data base of information is available, for example correlations of couplings with the nature of ligating atoms, bond connectivity, size of chelated rings, oxidation state, coordination number, geometry, substituent effects, etc (Mason, 1987; 1989). Usually, spin-spin couplings between I = 1/2 nuclei can be measured; for example ^{15}N couplings to ^{195}Pt or ^{113}Cd are well documented in complexes of bioinorganic relevance. Couplings are not resolved if one of the nuclei has a relaxation rate which greatly exceeds the spin coupling constant (in frequency units).

In the following sections, brief overviews of the NMR properties of ^{195}Pt and ^{27}Al, two representative nuclides with I = 1/2 and I > 1/2, respectively, are given. Other metal nuclei of biological relevance are easily detectable by NMR; for example, ^{51}V and ^{113}Cd. ^{51}V NMR spectroscopy has been recently reviewed by Rehder (1990) and many of the biological applications of ^{113}Cd were exhaustively described by Summers (1988). However, the fundamental contribution that is provided by studies on ligand nuclei, ^{17}O (I = 5/2), ^{15}N (I = 1/2), and ^{31}P (I = 1/2) besides ^{1}H and ^{13}C routine nuclides, must not be overlooked. In most cases, the multinuclear approach yields quite complete information on the stoichiometry, donor sites and geometry of complexes, the structure of ligands itself, numbers and lifetime of coordinated solvent molecules, etc.

A. ^{195}Pt NMR

Chemical shift for ^{195}Pt (I = 1/2) is sensitive primarily to the donor set and secondarily to the metal geometry and ligand structure or deprotonation. Regular trends for δ_{Pt} are often observed with replacement of ligands, e.g. $[Pt(H_2O)_4]^{2+}$ +31 ppm, *cis*-$[Pt(NH_3)_2(H_2O)_2]^{2+}$ -1593 ppm, and $[Pt(NH_3)_4]^{2+}$ -2579 ppm, relative to $[PtCl_6]^{2-}$ (Boreham et al., 1981; Gröning et al., 1982). Apart from the range, the overall order of ligand effects on δ_{Pt} is similar for Pt(II) and Pt(IV). According to theory (Dean and Green, 1968) the chemical shift of Pt(II) complexes is determined mainly by the contribution of *d* orbitals to σ bonding and π back-donation as well as by excitation energies and $<r^{-3}>$ values. Ammine-platinum(II) complexes have been extensively studied by ^{195}Pt NMR, often with use of ^{15}N-enriched ammonia. Detailed studies on the interaction between *cis*-$[Pt(NH_3)_2(H_2O)_2]^{2+}$ and α-amino acids or simple peptides were presented by Appleton et al. (1985; 1988; 1989a; 1989b). Distinctive δ_{Pt} values for PtN_2O_2 species are at ca -1580 ppm. PtN_3O complexes are distinguished by δ_{Pt} from -2000 to -2130 ppm. Sulphur coordination exerts a pronounced deshielding effect, approximate δ_{Pt} ranges being -2650 to -2800, -2800 to -3200, and -3200 to -3700 ppm for PtN_2OS, PtN_3S, and PtN_2S_2, respectively. Shielding decreases in dinuclear species where double hydroxo or thiolate bridges yield four-membered rings (Appleton et al., 1989b; Boreham et al., 1981).

Like their counterparts in ^1H, ^{15}N, and ^{31}P NMR spectra, Pt couplings are also very useful in structure assignment. ^{15}N-coordination splits ^{195}Pt resonances, while the interaction with the quadrupolar ^{14}N nucleus ($I = 1$) broadens the peaks, permitting a direct observation of metal binding to isotopically normal nitrogen donors. One-bond coupling constants are related to the Pt *6s* orbital contribution to bonding (Pregosin et al., 1973). Therefore, Pt couplings with ^{15}N nuclei are sensitive to the nature of ligands completing the coordination sphere (*trans* and *cis* influence). Pt-Pt couplings may be also detected in dimeric species with nonequivalent environments at the metal nuclei.

^{195}Pt NMR was used to obtain conclusive evidence about the binding modes of Pt(II) to nucleotides and nucleosides (Clore and Gronenborn, 1982; Reily and Marzilli, 1986; Bancroft et al., 1990). Signals for *cis*-$[Pt(II)(NH_3)_2LCl]$ and *cis*-$[Pt(II)(NH_3)_2L_2]$ complexes where L is a purine N7-donor are at -2280 to -2350 and -2430 to -2480 ppm, respectively. The difference between a N7 and a N1 donor is ca -40 ppm upfield; for example, two different metal environments are detected in a dimer where N1 and N7 donors bridge two *cis*-$[Pt(NH_3)_2Cl]$ units (Miller and Marzilli, 1985). This set of data made it possible to use ^{195}Pt NMR for monitoring the interaction of *cis*- and *trans*-$[Pt(NH_3)_2Cl_2]$ with DNA (Bancroft et al., 1990). Mono- and bi-functional adducts, predominantly those formed upon Pt(II) binding to the N7 donor of guanosine, were identified. Furthermore, shifts indicative of PtN_3S coordination were observed after reaction of glutathione with the mono adducts.

B. ^{27}Al NMR

The high natural abundance and sensitivity make the quadrupolar ^{27}Al ($I = 5/2$) probe favorable for NMR studies. Al(III) octahedral and tetrahedral coordination geometries cover different ranges of chemical shift, which depends also on the nephelauxetic effect of the donor atoms and on the solvent. In aqueous solutions, octahedral complexes with O- or N-donors extend approximately from -10 to 40 ppm, while tetrahedral species span the range from about 60 to 80 ppm, though some overlap exists (Akitt, 1989). Typical δ_{Al} values for $[Al(H_2O)_6]^{3+}$, $[Al_2(OH)_2(H_2O)_8]^{4+}$, $[Al_{13}(OH)_{24}O_4]^{7+}$, and $[Al(OH)_4]^-$, are 0, 3.5, 62.5, and 80 ppm, respectively. The multinuclear ^1H, ^{13}C, ^{17}O, and ^{27}Al NMR approach is specific for studying the Al(III) coordination to carboxylic, hydroxycarboxylic and polyaminocarboxylic acids. Poor spectral resolution resulting from quadrupolar broadening can be improved by measuring spectra in a high magnetic field. High-field ^{27}Al NMR spectra presented by Venema et al. (1990) showed a very regular trend for δ_{Al} in octahedral species formed by glycolic acid (LH). Namely, each ligand chelating via the carboxylate and unionized α-hydroxyl groups yields a downfield shift increment of about 9 ppm, while the ionization of the Al-bonded hydroxyl group affords a further contribution of about 5 ppm, cf δ_{Al} (ppm) for AlL (9.0), AlL_2 (17.5), AlL_3 (27.0), $AlLH_{-1}$ (14.0), AlL_2H_{-1} (22.0), and AlL_3H_{-3} (34.0). Tetrahedral species (AlL_2H_{-2} and $AlLH_{-3}$) exhibit pronounced deshielding (60.5 and 55.5 ppm).

Line width is an important parameter in ^{27}Al NMR and its dependence on temperature may be very indicative. High ligand-exchange rates make broader the signals as temperature is increased; if these processes are slow, for example in polynuclear species, the line width decreases due to a decrease in τ_c (eqn. 4). This effect was useful in assigning nuclearity to complexes (Venema et al., 1992). From shift and EFG measurements on Al^{3+} complexes of hydroxamic cyclopeptides, information was derived on the metal coordination and the nonuniform charge distribution in the hydroxamate chelating moiety (Llinás and De Marco, 1980). On the other hand, because of different EFGs, geometrical isomers can be distinguished. For example, $\Delta\nu_{1/2}$ is 900 Hz for *mer*-tris(maltolato(1-))Al(III) and 580 Hz for the more symmetrical *fac*-tris(3-hydroxy-1,2-dimethyl-4-pyridonato(1-))Al(III) species (Nelson et al., 1987).

REFERENCES

Akitt, J.W., and McDonald, W.S. (1984). *J. Magn. Res.*, *58*: 401.

Akitt, J.W. (1989). *Prog. Nucl. Magn. Reson. Spectr.*, *21*: 1.

Appleton, T.G., Hall, J.R., and Ralph, S.F. (1985). *Inorg. Chem.*, *24*: 4685.

Appleton, T.G., Connor, J.W., and Hall, J.R. (1988). *Inorg. Chem.*, *27*: 130.

Appleton, T.G., Hall, J.R., and Prenzler, P.D. (1989a). *Inorg. Chem.*, *28*: 815.

Appleton, T.G., Connor, J.W., Hall, J.R., and Prenzler, P.D. (1989b). *Inorg. Chem.*, *28*: 2030.

Bancroft, D.P., Lepre, C.A., and Lippard, S.J. (1990). *J. Amer. Chem. Soc.*, *112*: 6860.

Becker, K.D., and Berlage, U. (1983). *J. Magn. Res.*, *54*: 272.

Belford, R.L., Chasteen, N.D., So, H., and Tapscott, R.E. (1969). *J. Amer. Chem. Soc.*, *91*: 4675.

Benn, R., and Rufinska, A. (1986). *Angew Chem. Int. Ed.*, *25*: 861.

Boreham, C.J., Broomhead, J.A., and Fairlie, D.P. (1981). *Aust. J. Chem.*, *34*: 659.

Bouhsina, S., Decock, P., Kozlowski, H., Micera, G., and Swiatek, J. (1991). *J. Inorg. Biochem.*, *42*: 57.

Brader, M.L., Borchardt, D., and Dumm, M.F. (1992). *J. Amer. Chem. Soc.*, *114*: 4480.

Bramley, R., Brorson, M., Sargeson, A.M., and Schäffer, C.E. (1985). *J. Amer. Chem. Soc.*, *107*: 2780.

Branca, M., Micera, G., Dessì, A., and Kozlowski, H. (1989). *J. Chem. Soc., Dalton Trans.*, 1283.

Branca, M., Dessì, A., Kozlowski, H., Micera, G., and Serra, M.V. (1990a). *FEBS Lett.*, *257*: 52.

Branca, M., Micera, G., and Dessì, A. (1990b). *J. Chem. Soc., Dalton Trans.*, 457.

Branca, M., Micera, G., Sanna, D., Dessì, A., and Kozlowski, H. (1990c). *J. Chem. Soc., Dalton Trans.*, 1997.

Branca, M., Micera, G., Segre, U., and Dessì, A. (1992). *Inorg. Chem.*, *31*: 2404.

Bridgland, B.E., Fowles, G.W.A., and Walton, R.A. (1965). *J. Inorg. Nucl. Chem.*, *27*: 383.

Chasteen, N.D. (1981). *Biological Magnetic Resonance*, Vol. 3 (L.J. Berliner and J. Reuben, eds.), Plenum Press, New York, p. 53.

Clore, G.M., and Gronenborn, A.M. (1982). *J. Amer. Chem. Soc.*, *104*: 1369.

Dean, R.R., and Green, J.C. (1968). *J. Chem. Soc. A*, 3047.

Decock-Le Révérend, B., Andrianarijaona, L., Livera, C., Pettit, L.D., Steel, I., and Kozlowski, H. (1986). *J. Chem. Soc., Dalton Trans.*, 2221.

Deeth, R.J. (1991). *J. Chem. Soc., Dalton Trans.*, 1467.

Degani, H., Gochin, M., Karlish, S.J.D., and Shechter, Y. (1981). *Biochemistry*, *20*: 5795.

Desideri, A., Raynor, J.B., and Diamantis, A.A. (1978). *J. Chem. Soc., Dalton Trans.*, 423.

Fawcett, T.G., Bernarducci, E.E., Krogh-Jespersen, K., and Schugar, H.J. (1980). *J. Amer. Chem. Soc.*, *102*: 2598.

Fiallo, M.M.L., and Garnier-Suillerot, A. (1986). *Biochemistry*, *25*: 924.

Formicka-Kozlowska, G., Kozlowski, H., and Kupryszewski, G. (1980). *Inorg. Chim. Acta*, *46*: 29.

Formicka-Kozlowska, G., Kozlowski, H., Bezer, M., Pettit, L.D., Kupryszewski, G., and Przybylski, J. (1981). *Inorg. Chim. Acta*, *56*: 79.

Gröning, Ö., Drakenberg, T., and Elding, L.I. (1982). *Inorg. Chem.*, *21*: 1820.

Hambley, T.W., Hawkins, C.J., and Kabanos, T.A. (1987). *Inorg. Chem.*, *26*: 3740.

Hitchman, M.A., Olson, C.D., and Belford, R.L. (1969). *J. Chem. Phys.*, *50*: 1195.

Horrocks, W.D., Jr., and Sudnick, D.R. (1981). *Accounts Chem. Res.*, *14*: 384.

Jennette, K.W. (1982). *J. Amer. Chem. Soc.*, *104*: 874.

Jezowska-Bojczuk, M., Kozlowski, H., Zubor, A., Kiss, T., Branca, M., Micera, G., and Dessì, A. (1990). *J. Chem. Soc., Dalton Trans.*, 2903.

Kozlowski, H., Decock-Le Révérend, B., Ficheux, D., Loucheux, C., and Sóvágó, I. (1987). *J. Inorg. Biochem.*, *29*: 187.

Kozlowski, H., Bouhsina, S., Decock, P., Micera, G., and Swiatek, J. (1991). *J. Coord. Chem.*, *24*: 319.

Lambert, J.B., and Riddel, F.G., eds. (1983). *The Multinuclear Approach to NMR Spectroscopy*, Reidel, Dordrecht.

Lever, A.B.P. (1984). *Inorganic Electronic Spectroscopy*, Elsevier, Amsterdam.

Livera, C., Pettit, L.D., Bataille, M., Krembel, J., Bal, W., and Kozlowski, H. (1988). *J. Chem. Soc., Dalton Trans.*, 1357.

Llinás, M., and De Marco, A. (1980). *J. Amer. Chem. Soc.*, *102*: 2226.

Mantsch, H.H. (1991). *Kemia-Kemi*, *18*: 183.

Mason, J., ed. (1987). *Multinuclear NMR*, Plenum Press, New York.

Mason, J. (1989). *Polyhedron*, *8*: 1657.

Micera, G., and Dessì, A. (1988). *J. Inorg. Biochem.*, *34*: 157.

Micera, G., and Dessì, A. (1989). *J. Inorg. Biochem.*, *35*: 71.

Micera, G., Dessì, A., Kozlowski, H., Radomska, B., Urbanska, J., Decock, P., Dubois, B., and Olivier, I. (1989). *Carbohyd. Res.*, *188*: 25.

Miller, S.K., and Marzilli, L.G. (1985). *Inorg. Chem.*, *24*: 2421.

Möhl, W.C., Winscom, C.J., Plato, M., Möbius, K., and Lubitz, W. (1982). *J. Phys. Chem.*, *86*: 149.

Mulks, C.F., Kirste, B., and van Willigen, H. (1982). *J. Amer. Chem. Soc.*, *104*: 5906.

Nakatsuji, H., Kanda, K., Endo, K., and Yonezawa, T. (1984). *J. Amer. Chem. Soc.*, *106*: 4653.

Nelson, W.O., Rettig, S.J., and Orvig, G. (1987). *J. Amer. Chem. Soc.*, *109*: 4121.

Pereira, E., Fiallo, M.M.L., Garnier-Suillerot, A., Kiss, T., and Kozlowski, H. (1993). *J. Chem. Soc., Dalton Trans.*, 455.

Pettit, L.D., Steel, I., Kowalik, T., Kozlowski, H., and Bataille, M. (1985). *J. Chem. Soc., Dalton Trans.*, 1201.

Pettit, L.D., Pyburn, S., Kozlowski, H., Decock-Le Révérend, B., and Liman, F. (1989). *J. Chem. Soc., Dalton Trans.*, 1471.

Pettit, L.D., Gregor, J.E., and Kozlowski, H. (1991). *Perspectives on Bioinorganic Chemistry*,Vol. 1 (R.W. Hay, J.R. Dilworth and K.B. Nolan, eds.), JAI Press, London, p. 1.

Pregosin, P.S., Omura, H., and Venanzi, L.M. (1973). *J. Amer. Chem. Soc.*, *95*: 2047.

Rehder, D. (1990). *Vanadium in Biological Systems* (N.D. Chasteen, ed.), Kluwer, Dordrecht, p. 173.

Reily, M.D., and Marzilli, L.G. (1986). *J. Amer. Chem. Soc.*, *108*: 6785.

Schugar, H.J. (1982). *Copper Coordination Chemistry. Biochemical and Inorganic Perspectives* (K.D. Karlin and J. Zubieta, eds.), Academic Press, New York, p. 43 and refs. therein.

Schwyzer, R., and Caviezel, M. (1971). *Helv. Chim. Acta*, *54*: 1395.

Selbin, J. (1965). *Chem. Rev.*, *65*:153.

Shaw, R.A., and Mantsch, H.H. (1992). *Metal Ions in Biology and Medicine*, Vol. 2 (J. Anastassopoulou, Ph. Collery, J.C. Etienne and Th. Theophanides, eds.), J. Libbey Eurotext, p. 10.

Siemion, I.Z., Jankowski, A., Sobczyk, K., and Szewczuk, Z. (1985). *Int. J. Peptide Prot. Res.*, *25*: 280.

Sigel, H., and Martin, R.B. (1982). *Chem. Rev.*, *82*: 385.

Summers, M.F. (1988). *Coord. Chem. Rev.*, *86*: 43.

Tsangaris, J.M., and Martin, R.B. (1970). *J. Amer. Chem. Soc.*, *92*: 4255.

Varnagy, K., Kozlowski, H., Sóvágó, I., Kowalik-Jankowska, T., Kruszynski, M., and Zboinska, J. (1988). *J. Inorg. Biochem.*, *34*: 83.

Venema, F.R., Peters, J.A., and van Bekkum, H. (1990). *J. Chem. Soc., Dalton Trans.*, 2137.

Venema, F.R., Peters, J.A., and van Bekkum, H. (1992). *Inorg. Chim. Acta*, *197*: 1.

MIXED LIGAND COMPLEX FORMATION

by

Osamu Yamauchi

Department of Chemistry,
Faculty of Science,
Nagoya University,
Chikusa-ku, Nagoya 464-01
Japan

I. MIXED-LIGAND COMPLEXES IN BIOLOGICAL SYSTEMS

Formation of low-molar-mass (l.m.m.) complexes in biofluids may be described as multimetal-multiligand equilibria involving essential exchangeable metal ions and amino acids, peptides, nucleotides, or other biomolecules. Because potential bioligands greatly exceed biological transition metal ions in number and quantity, mixed-ligand complexes must be more common than binary and multinuclear complexes. They play important roles in metal ion transport, enzyme reactions, and other biological functions.

Histidine and other amino acids are involved in copper(II) transport in blood and exchange interactions with serum albumin (Sarkar and Kruck, 1966; Sarkar and Wigfield, 1968; Neumann and Sass-Kortzak, 1967; Sarkar, 1981). A mixed amino acid complex containing histidine (His) and threonine (Thr), Cu(L-His)(L-Thr), was isolated from normal human blood serum (Sarkar and Kruck, 1966). Enzyme-metal-substrate (EMS) complexes as intermediates in enzymatic reactions may be regarded as mixed-ligand complexes. A classical example of EMS complexes is the complex of carboxypeptidase A with a model substrate, glycyl-L-tyrosine, whose X-ray crystal structure analysis revealed the zinc center structure involving the coordinated substrate and the enzyme amino-acid residues interacting with each other (Quiocho and Lipscomb, 1971; Christianson and Lipscomb, 1989). Possible EMS complex formation has been suggested for Cu-containing galactose oxidase, which may bind galactose at an equatorial position of the Cu coordination plane specifically through hydrogen bonds to the arginine residue (Arg 330) of the enzyme (Ito et al., 1991). Similar hydrogen bonds have been characterized for superoxide binding to Cu,Zn-superoxide dismutase (Cudd and Fridovich, 1982; Getzoff et al., 1983; Mota de Freitas et al., 1987; Banci et al., 1988). Such enzyme-substrate interactions around the metal center are essential for the efficiency and specificity of metalloenzymes (Jencks, 1975; Frieden, 1975; Alberts et al., 1989).

Mixed-ligand complex formation and ligand-ligand interactions may thus be implied in a variety of intermolecular reactions ranging from molecular recognition and catalysis by metalloenzymes to information transfer by neurotransmitters and hormones whenever a metal ion is required.

Amino acids are by far the most effective and important l.m.m. biological ligands. Considering the available coordination sites of a metal ion and the multidentate nature of amino acids, peptides, and other potential bioligands, mixed-ligand complexes with two different ligands, i.e. ternary complexes, are most important in biological systems. In the following sections some interesting aspects of ternary complexes will be briefly described, with emphasis on ligand-ligand interactions and their biological implications. General aspects of mixed-ligand complex formation by amino acids and related biomolecules (Sigel, 1973) and the stability of mixed-ligand complexes (Marcus and Eliezer, 1969; Martin et al., 1973; Sigel, 1975) have been previously reviewed. In addition, there are excellent reviews on binary and ternary amino acid and peptide complexes (Freeman, 1973; Sigel, 1979; Kiss, 1990).

II. EVALUATION OF STABILITY INCREMENT IN TERNARY COMPLEX FORMATION

Ternary complexes are stabilized relative to the parent binary complexes by various factors such as statistical factors, steric effects, electronic or through-metal ligand-ligand interactions, and through-space ligand-ligand interactions (Sigel, 1975).

Formation of a mixed-ligand complex $M_pA_qB_rH_s$, where M refers to a metal ion, A and B to ligands and H to proton, with p, q, r, and s representing the stoichiometric numbers of the respective components, is expressed by equation 1 (charges being omitted for clarity):

$$pM + qA + rB + sH \overset{\beta_{pqrs}}{\rightleftharpoons} M_pA_qB_rH_s$$

$$\beta_{pqrs} = \frac{[M_pA_qB_rH_s]}{[M]^p[A]^q[B]^r[H]^s} \tag{1}$$

The β_{pqrs} value reflects the effects of the above-mentioned factors. There are several criteria for evaluating the stability enhancement due to mixed-ligand complex formation, and some points of the criteria related with statistical factors (Marcus and Eliezer, 1969; Sigel, 1975) and ligand-ligand interactions are described below.

A. Statistical factors

A commonly used method of evaluation is based on the equilibrium for the formation of MAB from MA and MB according to equation 2:

$$MA + MB \overset{10^{\Delta \log K_M}}{\rightleftharpoons} MAB + M$$

$$\Delta \log K_M = \log \beta_{1110} - (\log \beta_{1100} + \log \beta_{1010}) \tag{2}$$

where the statistical $\Delta \log K_M$ values for bidentate ligands are -0.4 and -0.6 for octahedral and square-planar complexes, respectively, and -0.9 for a distorted octahedral complex (Sigel, 1975).

B. Ligand-ligand interactions

Through-metal ligand-ligand interactions stabilize ternary complexes involving a π-donating (π-base) and a π-accepting (π-acid) ligand. Thus, 2,2'-bipyridine (bpy) as a π-acid accepts the π-back donation by M, which in turn acquires a high affinity for a π-base such as pyrocatecholate (pyr). Combination of a π-acid and a π-base around M enhances the $\Delta \log K_M$ values significantly, e.g. $\Delta \log K = 0.43$ for Cu(bpy)(pyr) (Sigel, 1975), and this may be regarded as a ligand recognition or substrate discrimination by enzyme metal centers with imidazole ligands (Banerjea et al., 1981; Saha and Sigel, 1982).

In addition to electronic effects, through-space ligand-ligand interactions are also an important complex-stabilizing factor (Sigel, 1975; 1980; 1981; 1989a; Yamauchi, 1984). The extra stabilization due to the interactions may be given by equation 2 and also by the equilibrium constant K for the following hypothetical equilibrium (Yamauchi and Odani, 1985a; 1985b; Odani and Yamauchi, 1987; Sigel et al., 1990):

$$MAB' + MA'B \overset{K}{\rightleftharpoons} MAB + MA'B'$$

$$\log K = \log \beta_{MAB} + \log \beta_{MA'B'} - \log \beta_{MAB'} - \log \beta_{MA'B} \tag{3}$$

where β_{MAB} etc are equal to the β_{1110} values of MAB etc. In this formulation, ligand-ligand interactions are possible in MAB complexes where both A and B have interacting groups but not in MAB', MA'B, and MA'B' ones where A' and B' are devoid of such groups. The log K value is comparable to the $\Delta\Delta\log K_M$ defined by Sigel et al. (Fischer and Sigel, 1980; Sigel et al., 1990). On the assumption that log K is equal to 0 in the absence of interactions, it is possible to evaluate to what extent these stabilize MAB relative to MA'B', or shift the equilibrium to the right. When A and B have coordinating groups similar to those of A' and B', respectively, and differ from each other only by the side chain groups, the coordination structures and thus the ligand fields of the relevant complexes are nearly the same. A positive log K value then indicates that there is stabilization of MAB, which is mainly due to the attractive interaction between A and B.

C. Intramolecular equilibrium for ligand-ligand interactions

As a consequence of ligand-ligand interactions, a macrocyclic ring is formed in addition to stable chelate rings and, because of the weakness of these interactions, the complex species with and without this ring, i.e. the closed and open species, are in equilibrium as schematically shown by equation 4 for MAB (Mariam and Martin, 1979; Sigel, 1980; 1981; 1982; Martin and Sigel, 1988; see also Part Two, Chapter 2-A):

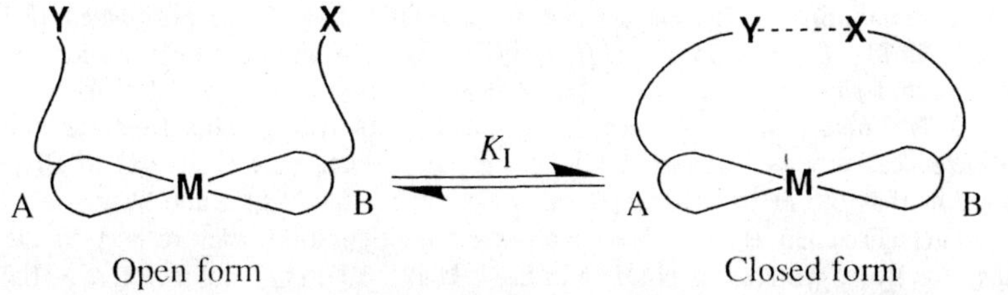

(4)

where K_I is the constant for the intramolecular equilibrium. The interactions may be electrostatic or hydrophobic, depending on the nature of X and Y groups. When one of the ligands is an aromatic or an aliphatic amino acid and the other is a heteroaromatic compound such as bpy and 1,10-phenanthroline (phen), aromatic-ring stacking or hydrophobic interactions may occur under favorable conditions. Aromatic-ring stacking and hydrophobic interactions within complex molecules have been extensively studied for nucleotides and aromatic carboxylate (Sigel, 1981; 1982; 1987) and amino acids and peptides (Fischer and Sigel, 1980; Yamauchi and Odani., 1985a; 1985b; Odani et al., 1986; Masuda et al., 1991). The K_I value for MAB is related to $\Delta\log K_M$ (equation 5) (Sigel, 1981) and log K (equation 6) (Yamauchi and Odani, 1988):

$$K_I = \frac{10^{\Delta\log K_M}}{10^{\Delta\log K_{M/op}}} - 1 \qquad (5)$$

$$K_I = 10^{\log K} - 1 \qquad (6)$$

where $\Delta\log K_{M/op}$ refers to the $\Delta\log K_M$ for a ternary complex without ligand-ligand interactions, such as MAB'. These equations are used for estimating the population of the closed species and thus the effectiveness of ligand-ligand interactions.

III. STRUCTURES AND STABILITIES OF TERNARY COMPLEXES WITH LIGAND-LIGAND INTERACTIONS

Through-space ligand-ligand interactions in dilute solution occur most probably within complex molecules and lead to selective formation of ternary complexes. The interactions are noncovalent in nature and are classified as follows (Frieden, 1975): (i) steric hindrance arising from bulky groups, (ii) electrostatic interactions or hydrogen bonds between charged and/or polar groups, and (iii) hydrophobic interactions between aliphatic and/or aromatic groups. While steric hindrance (i) favors ternary complex formation leading to strain-free complexes and thus gives rise to ligand selectivity, the interactions (ii) and (iii) contribute to stabilization of ternary complexes relative to those without such interactions. Presence of ligand-ligand interactions may be inferred from (i) facile isolation of ternary complexes with side chains or coordinated groups capable of interaction, (ii) stability enhancement as evaluated by the methods described in II, (iii) spectral changes due to complex formation, and (iv) X-ray crystal structure analysis. For the complexes containing amino acids or peptides, deviation of circular dichroism (CD) spectral magnitude from magnitude additivity (Tsangaris and Martin, 1970; Martin, 1974) is a clear indication of ligand-ligand interactions causing conformational changes and increasing asymmetry (Yamauchi et al., 1975; Sakurai et al., 1976a; 1976b; Yamauchi and Odani, 1981; 1985a; 1985b). Rotamer population as calculated from ^{1}H NMR coupling constants (Pachler, 1964; Feeney, 1976) gives the information of the side-chain conformation as previously reported for Pd(II)-peptide complexes (Kozlowski and Jezowska-Trzebiatowska, 1976; Kozlowski, 1978; Kozlowski et al., 1983; Vestues and Martin, 1980; Scheller-Krattiger et al., 1982).

Intramolecular noncovalent interactions have been reviewed as a complex-stabilizing factor and as a source of ligand and chiral selectivity in ternary complexes and in binary peptide complexes (Pettit and Hefford, 1979; Sigel, 1980; 1981; Martin and Sigel, 1988; Okawa, 1988; Cucinotta et al., 1990). The following examples illustrate the effects of ligand-ligand interactions on the structure and stability of mixed-ligand complexes.

A. Electrostatic interactions

Hydrogen bonds or electrostatic interactions were found to stabilize the ternary Cu(II) complexes containing L-His and a basic amino acid L-BH with a protonated side chain, Cu(L-His)(L-BH) (B = arginine (Arg), lysine (Lys), or ornithine (Orn)), as compared with the diastereomeric complexes, Cu(D-His)(L-BH), by ca 0.1 log unit (Brookes and Pettit, 1975; 1977). The stability difference was ascribed to the hydrogen bond between the axially coordinated carboxylate oxygen of His and the protonated side chain (X) of BH. On the assumption that the His-containing ternary complexes have the same structure as Cu(L-His)(L-Thr) (Freeman et al., 1969) with a *cis* configuration with respect to the amino groups, this hydrogen bond is possible in Cu(L-His)(L-BH) (**1**) but not in Cu(D-His)(L-BH) (**2**):

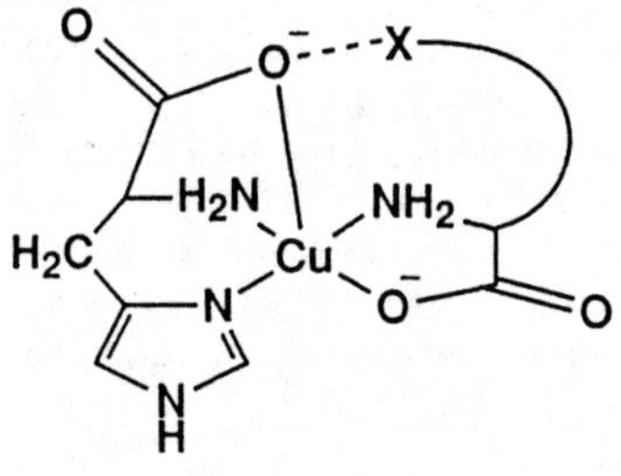

1

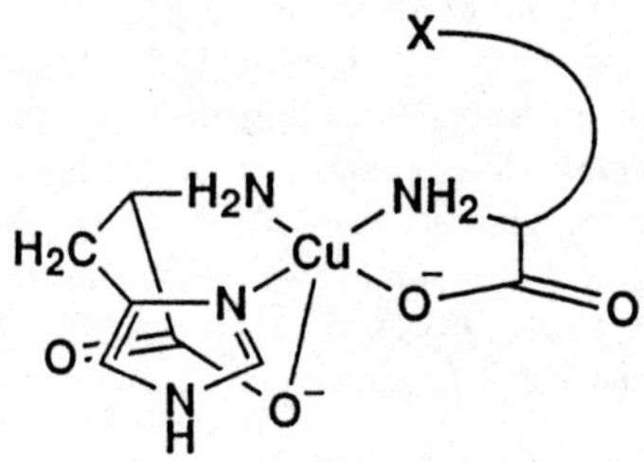

2

Although the His carboxylate group is not always axially coordinated, the ternary Cu(II) complexes containing His and an amino acid reported to date have the same *cis* configuration in the solid state (Ono et al., 1979; Ono and Sasada, 1981; Sasada et al., 1983; Yamauchi and Odani, 1988) as Cu(L-His)(L-Thr), which supports the above conclusion.

From a comparative ^{64}Cu tracer study it was concluded that, in predialyzed human blood serum, His forms mixed-ligand Cu(II) complexes with Thr, asparagine (Asn), and glutamine (Gln) more favorably than with the other amino acids (Neumann and Sass-Kortzak, 1967; Sarkar, 1981). In contrast, sophisticated computer simulation of the equilibria in blood plasma (Hallman et al., 1971; May et al., 1977; Berthon et al., 1978) first suggested the most predominant species to be Cu(His)(cystinate) and $Cu(His)_2$. This discrepancy, however, was attributable to the poor reliability of the stability constants, since the above complexing tendency and higher stability, as expressed e.g. by $\Delta \log K_M$, has recently been confirmed using improved stability constants for all the relevant complex species (Berthon et al., 1984; 1986; Brumas et al., 1993). Interestingly, Thr, Asn, and Gln, which were classified as the amino acids of significant cooperativity in the above tracer study, have a polar side chain group, i.e. an OH or a $CONH_2$ group, and may form an intramolecular hydrogen bond with the COO^- group of His axially coordinated. This was supported by isolation of a number of ternary complexes, Cu(L-His)(L-AA), where AA = Asn, Gln, serine, homoserine, and citrulline, all having a polar side group (Yamauchi et al., 1979) and optical resolution of DL-His by formation of a ternary complex with L-Asn, which incorporated 98% opically pure L-His (Sakurai et al., 1977). X-ray analysis of Cu(L-His)(L-Asn) indicated flexibility in the amide side chain, which corroborated the hydrogen bond proposed in solution (Ono et al., 1979; Yamauchi et al., 1979). Complexes with intramolecular hydrogen bonds or electrostatic interactions may be preferred in Cu(II) transport through membrane.

Electrostatic interactions were shown to take place in ternary complexes containing an acidic (A) and a protonated basic (B) amino acid, M(L-A)(L-BH) (M = Cu(II), Pd(II); A = aspartate (Asp), glutamate (Glu), ethylenediamine-*N*-monoacetate (EDMA); B = Arg, Lys, Orn), from the anomaly characterizing the CD spectral magnitude, the isolation of M(L-A)(L-BH) as crystals (Yamauchi et al., 1975; Sakurai et al., 1976a; 1976b), the optical resolution of DL-A via isolation of Cu(II) complexes with L-BH (Yamauchi et al., 1977), and the rotamer populations of L-AA in the complex, as determined by NMR spectra and CD spectral magnitude additivity (Yamauchi and Odani, 1981). Similar electrostatic interactions and ring formation have been reported for Cu(DL-8-*N*-trimethylornithinate)(DL-homocysteinate) (Tribolet et al., 1983). X-ray analysis of Cu(EDMA)(L-Arg) (Fig. 1) revealed intermolecular hydrogen bonds between the axially bound carboxylate group of EDMA and the guanidinium group of L-Arg, which suggests that the same type of bonding occurs within a single complex molecule in solution (Masuda et al., 1993). From thermodynamic considerations, the stereoselectivity in the formation of Cu(II)-dipeptide-L- or D-Asp (or Orn) was attributed to electrostatic interactions (Gergely and Farkas, 1982). Preferential complex formation has been observed e.g. between Rh(III)-porphyrin complexes and amino acid esters (Aoyama et al., 1990) and between Zn(II)-porphyrin complexes and aromatic or acidic amino acids (Verchère-Béaur et al., 1990) through a hydrogen bond and an electrostatic or a stacking interaction, respectively, in addition to the coordinate bond at an axial position.

A stability enhancement has been observed for complexes with electrostatic ligand-ligand interactions involving a phosphoester group and the side chain of BH (Yamauchi et al., 1992): *O*-Phosphotyrosine (Ptyr) and *O*-phosphoserine (Pser) can be involved in electrostatic interactions with the guanidinium and the ammonium group of Arg and Lys, respectively, in Cu(L-Ptyr or Pser)(L-BH). The log *K* values were calculated to be 0.36-0.73 at I = 0.1 and 0.03-0.40 at I = 1.0 (25 °C) according to equation (4) with A = Ptyr

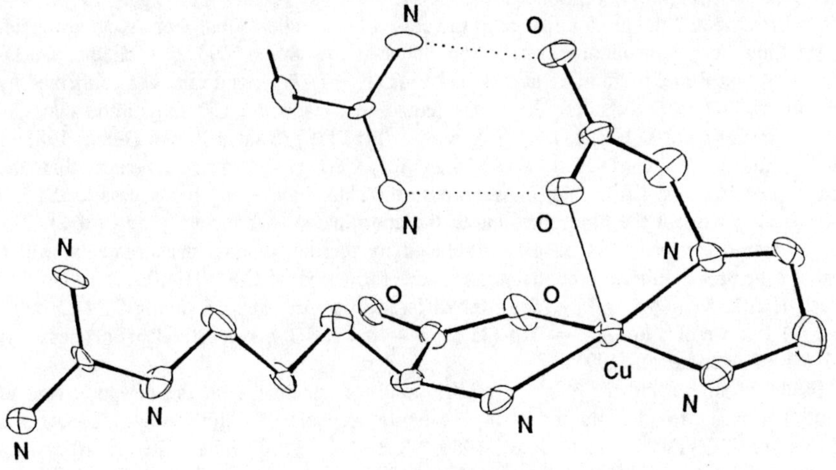

Figure 1.
Molecular structure of [Cu(EDMA)(L-Arg)] showing intermolecular hydrogen bonds (Masuda et al., 1993).

or Pser, B = Arg or Lys, A' = tyrosine, and B' = alanine. The ionic strength dependence clearly shows that the increase in stability is due to attractive electrostatic interactions between oppositely charged side chains.

The Arg guanidinium group has recently attracted much attention owing to its importance in the sequence-specific DNA-binding by zinc finger proteins (Miller et al., 1985; Berg, 1989). Recent X-ray crystal structure analysis of a peptide containing the DNA-binding domain from Zif268 revealed that the zinc finger-DNA binding occurs through guanidinium-guanine and other weak interactions which are regarded as metal-ion-assisted noncovalent interactions and add to the significance of metal ions in biological mechanisms (Pavletich and Pabo, 1991). The carboxylate-guanidinium hydrogen bonds are established in the solid state for the carboxypeptidase A-substrate binding at Arg 145 (Quiocho and Lipscomb, 1971). Similar bonding has been revealed for the Asp residue near the active site of *Pseudomonas putida* cytochrome P-450 (Poulos et al., 1986).

B. Hydrophobic and aromatic ring stacking interactions

The presence of aliphatic and/or aromatic groups in ternary complexes leads to hydrophobic ligand-ligand interactions. In aqueous solution, these groups have an intrinsic tendency to form an aggregate with other aliphatic or aromatic groups, which may be explained by the solvophobic force (Sinanoglu, 1982). For example, 1-alkanesulfonate $C_nH_{2n+1}SO_3^-$ interacts with the phenanthroline ring of $Fe(phen)_3^{2+}$ in aqueous solution (Masuda et al., 1982; Tachiyashiki and Yamatera, 1984), and $Pt(phen)(en)^{2+}$ (en = ethylenediamine) forms an adduct with nucleotides such as AMP and GMP in dilute aqueous solution (Odani et al., 1991). The interaction between an aliphatic group and an aromatic ring, which is referred to as a CH-π interaction (Nishio and Hirota, 1989), has been shown by Okawa et al. (Okawa, 1988) to give rise to stereoselectivity for Co(III) and other metal complexes of ligands containing a chiral group such as 1-*l*-methyloxy-3-benzoylacetone, which favors the *cis*-Δ isomer of the tris complex. An interesting enantioselectivity has been reported for cyclodextrin functionalized with histamine, whose Cu(II) complex preferentially forms the ternary complex with D-Trp as complared with L-Trp (Impellizzeri et al., 1991). This has been ascribed to the indole-cyclodextrin interaction, which is possible for the D-Trp but not for L-Trp on the assumption that the ternary complex assumes the *cis* configuration with respect to the amino groups as in Cu(L-His)(L-Thr) etc (vide supra).

Ternary complexes involving aromatic and/or aliphatic amino acids M(A)(B), such as M(Phe)(Tyr) (M = Co(II), Ni(II), Cu(II)) and Cu(L- or D-His)(L-AA) (AA = Phe, Trp, valine (Val)), exhibit an increase in stability as evaluated by $\Delta\log K_M$ (equation 2), and the percentages of the M(A)(B) closed form have been calculated from the K_I values (equation 5) for over 70 complexes, e.g. Cu(D-His)(L-Trp) (90%) (Fischer and Sigel, 1980). The stability increase and the predominance of the closed form indicate that there is an aromatic-ring stacking or a hydrophobic interaction between the amino acid side chains. Cu(D-His)(L-Trp) was found to be more stable than Cu(L-His)(L-Trp), and this was confirmed by the relationship between the log K values (equation 3) and the CD magnitude anomaly observed for Cu(L- or D-His)(L-AA) (AA = Phe, Tyr, Trp) (Yamauchi and Odani, 1985b). Higher stability for Cu(D-His)(L-AA) than for Cu(L-His)(L-AA) indicates that the interaction occurs with Cu(D-His)(L-AA) where the side chain group has access to the His imidazole ring without the hindrance due to the coordinated carboxylate group (cf 2). The observed stereoselectivity has been substantiated by thermodynamic measurements, which revealed a greater exothermic contribution to the formation of Cu(D-His)(L-AA) than that of Cu(L-His)(L-AA) (AA = Phe, Trp), the difference in $-\Delta H^\circ$ and ΔS° being 5.94 kJ mol^{-1} and -11.3 J K^{-1} mol^{-1} for AA = Trp (25°C; I = 0.1 KNO_3), respectively (Borghesani et al., 1990; cf Arena et al., 1991).

Evidence supporting the stacking of the imidazole nucleus with the aromatic ring of Phe or Tyr was provided by the X-ray structure analysis of Cu(histamine)(L-Phe) and Cu(histamine)(L-Tyr) (Yamauchi et al., 1989; Sigel et al., 1990), where the aromatic rings are located above the Cu(II) coordination plane to be in close contact (3.45-3.49 Å) with the imidazole ring. The interactions between coordinated heteroaromatic rings (DA = bpy, phen, histamine, etc) and the side chain of amino acids (AA = Val, Phe, Tyr, Trp, 5-

hydroxytryptophan (Htrp)) have been established by the log K values (equation 3) for Cu(DA)(AA) (Yamauchi and Odani, 1985a; 1988) and the ^{1}H NMR studies of dipeptide complexes, Pd(DA)(L) (DA = bpy, etc; L = TyrGly, TrpGly, PheGly) (Odani et al., 1986), the stability of Cu(bpy)(L-AA) being in the following sequence of AA:

AA =	Val	Phe	Tyr	Trp	Htrp

le chain group

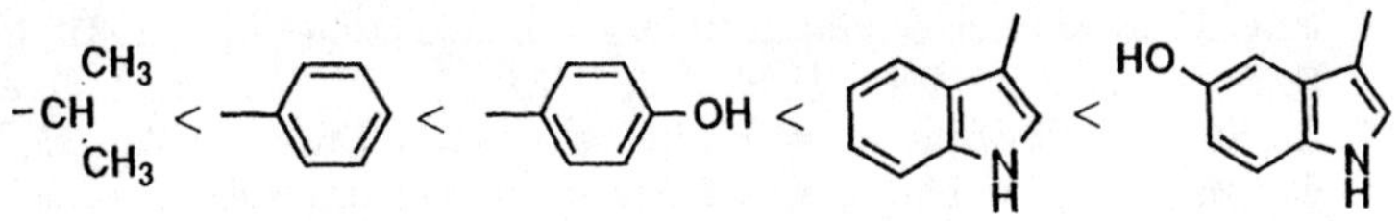

log K =	0.02	0.60	0.90	1.19	1.80

(25°C; I = 0.1(KNO_3))

Δlog K_M values have been reported for several other systems (e.g. Prasad and Mohan, 1987; Ullah and Bhattacharya, 1991). A similar stabilization was characterized from the Δ log K_M values for DOPA-containing ternary complexes (Patel and Bhattacharya, 1984; Kiss and Gergely, 1985), although a different explanation has been proposed for the stabilization. A stability increase was revealed for binary metal-dipeptide complexes, which is due to hydrophobic interactions between peptide side chains (Kiss and Szücs, 1986; Xiao et al., 1990). Fujii et al. reported the chiral recognition of Phe and Trp by a Co(III) complex (Fujii et al., 1990), and analogous stereoselectivity has also been reported (Strasak et al., 1990).

Figure 2 illustrates the molecular structure of Cu(bpy)(L-Trp) with the bpy-indole stacking (Masuda et al., 1991), which together with the X-ray crystal structure

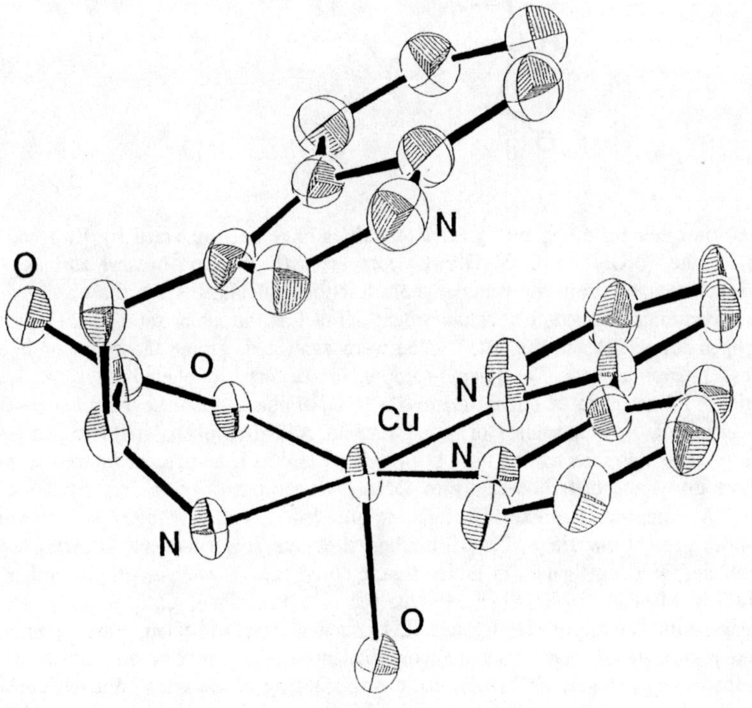

Figure 2.
Molecular structure of [Cu(bpy)(L-Trp)] with the bpy-indole stacking (Masuda et al., 1991).

determinations of Cu(phen)(L-Trp) (Aoki and Yamazaki, 1987; Masuda et al., 1988), and Cu(bpy)(L-Tyr) (Yamauchi et al., 1992) establishes the aromatic-ring stacking in Cu(DA)(AA) and related complexes in the solid state, and strongly supports it in aqueous solution.

Sigel et al. have shown by potentiometric and NMR spectroscopic measurements that ternary complexes M(DA)(L) (M = Cu(II), Zn(II)) with L = alkyl and aryl carboxylate such as valerate (Va) and phenylacetate (PAc) form the closed species due to intramolecular interactions between coordinated DA and L side chain (Sigel, 1981; 1982; 1989a; Malini-Balakrishnan et al., 1985; Sigel et al., 1985a; 1985b; Liang et al., 1988). Thermodynamic studies on Zn(bpy)(L) (L = *n*-butylmalonate, etc) revealed the $-\Delta H°$ increase and $\Delta S°$ decrease which are typical for solvophobic interactions, substantiating the presence of such hydrophobic interactions (Amico et al., 1989). The closed species for M(phen)(Va), of probable structure **3**, amounts to ca 25% for both M = Cu(II) and Zn(II) (Sigel, 1982), whereas for M(phen)(PAc) it reaches 48±5% with M = Cu(II) and 21±8% with Zn(II) (Sigel et al., 1985a). It is interesting to note that in less polar solvents such as 50% aqueous ethanol and 50% aqueous dioxane the percentage of closed species of some of these complexes increases with respect to water, which may be due to the cooperative effect of the hydrophobic environment on the polar metal-carboxylate bond. Highly stable and inhibitory binding of phenylpropionate to the Zn(II) active site of carboxypeptidase A can be explained by a "cooperative mechanism" due to hydrophobic interactions and reduction of the effective dielectric constant by the phenylalkyl moiety (Sigel et al., 1985a; Sigel, 1989a).

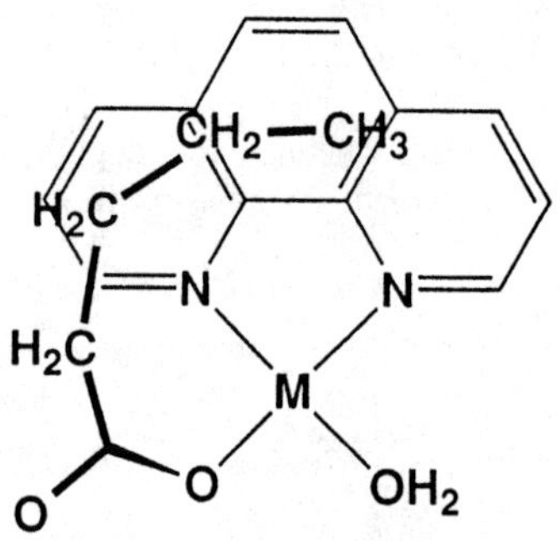

3

On the other hand, ligand-ligand interactions have been detected by Kim and Martin (1984) for the Pd(GlyPhe or GlyGly)(L) complexes (L = alkyl or aryl amines) through NMR spectroscopy. From the rotamer mole fractions calculated from the observed vicinal coupling constants, the stability enhancement defined as the molar ratio of the closed form to the open form (K_I) and the $-\Delta G°$ value were evaluated, giving the order of interaction energies: phenyl-aromatic > phenyl-(propyl or larger) > phenyl-ethyl > isoleucyl-aromatic > Pd-aromatic > phenyl-methyl >> Pd-aliphatic ~ isoleucyl-aliphatic ~ 0.

Because of the importance of metal ions in ATP hydrolysis in biological systems, metal ion-nucleotide interactions have been studied in detail, and recent interest in platinum anticancer drugs and metal-ion-catalyzed DNA cleavage reactions has greatly advanced the metal-DNA complex chemistry (Reedijk et al., 1987; Tullius, 1989). Nucleosides and nucleotides present a variety of metal-binding sites that are dependent on metal ions, base moieties, and existing ligands in mixed-ligand complexes as well as pH and other factors (Martin and Mariam, 1979; Sigel, 1989b; see also Part Two, Chapter 2-A). In ternary complexes with heteroaromatic ligands (DA) such as bpy and phen, binding through the terminal phosphate oxygens is most favored owing to electronic or through-metal ligand-ligand interactions (Sigel, 1975). Aromatic ring stacking of the uncoordinated nucleic base moiety of ATP with DA (= phen) as shown by **4** or with an aromatic amino acid has been established from stability enhancement as evaluated by $\Delta\log K_M$, the charge transfer band at

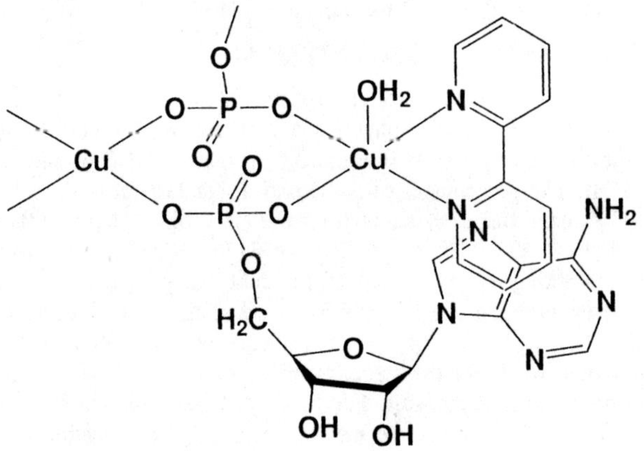

Figure 3.
Molecular structure of dimeric [Cu(AMP)(bpy)(H_2O)] showing the adenine-bpy stacking (Aoki, 1978).

~300 nm, and the upfield shifts of the ^1H NMR signals due to the ring current effect. Aoki and Yamazaki (1978) reported the structure of [Cu(AMP)(bpy)(H_2O)] (Fig. 3), which serves as the first structural evidence for the intramolecular AMP-bpy stacking deduced from solution studies. Other complexes such as Zn(ATP)(bpy) (Orioli et al., 1981) have also been reported (reviews cited below).

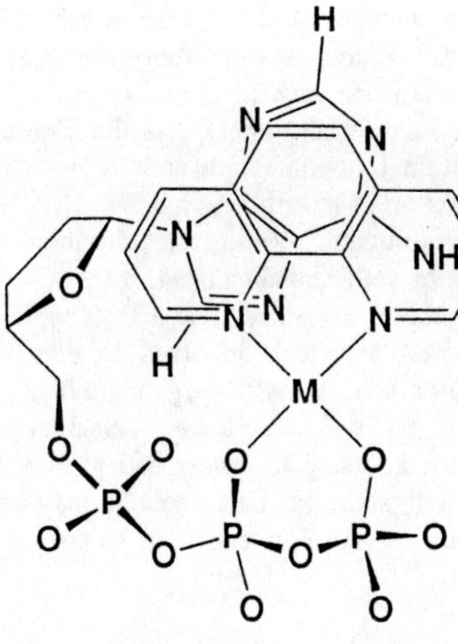

4

The closed form of the ternary species is in equilibrium with the open form as shown below (Sigel, 1981; 1989a; 1989b):

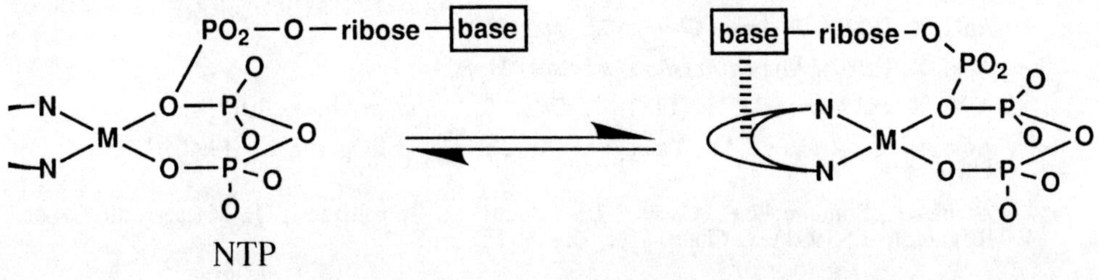

Positive $\Delta \log K_M$ values (= 0.05-0.6) for the equilibrium

$$M(NTP) + M(phen) \rightleftharpoons M(phen)(NTP) + M \qquad (2')$$

indicate that stacking is present in M(NTP)(phen) (M = Ca(II), Mg(II), Mn(II), Zn(II), Cu(II); NTP = nucleotide 5'-triphosphate) and that the Mg(II) and Ca(II) complexes of ATP are most stabilized ($\Delta \log K_M$ = 0.4 and 0.6, respectively) (Mitchell and Sigel, 1978; Fukuda et al., 1978). The percentages of the closed or stacked form of Zn(DA)(NTP) (NTP = ATP, ITP, UTP) have been evaluated from the ^{1}H NMR upfield shifts to be ≥95, 48, and ~40% for Zn(phen)(ATP), Zn(bpy)(ITP), and Zn(bpy)(UTP), respectively (Mitchell et al., 1979). Similar stacking has been detected for the ternary complexes of NMP (= AMP, GMP, IMP, etc) (Massoud and Sigel, 1990; Massoud et al., 1990; Bastian and Sigel, 1991) and for M(AA)(ATP) (AA = Trp, leucine) (Sigel et al., 1983). For Cu(DA)(NMP) systems, the following trends for the stacking ability of nucleic bases and nucleotides were obtained: for the bases uracil ≤ cytosine ≤ thymine < < adenine, and for purine nucleotides IMP < AMP ~ GMP. Thermodynamic measurements of the formation of Cu(Trp)(ATP), for example, revealed a larger positive $-\Delta H°$ value than that for Cu(Ala)(ATP), supporting the stabilization due to stacking (Arena et al., 1983; 1984).

Structural studies on the coordination modes and stacking interactions reported for various binary and ternary complexes involving nucleic bases have been summarized and reviewed (Aoki, 1981; Cini, 1992; Terrón, 1993; see also Part Two, Chapter 2-A). A review on Pd(II)-peptide-nucleotide complexes as nucleic acid-protein interaction models has also been published (Kasselouri et al., 1990).

IV. CONCLUDING REMARKS AND BIOLOGICAL RELEVANCE

In a number of cases where small bioligands play a crucial role, metal ions present may form mixed-ligand complexes ranging from an enzyme-metal-cofactor-substrate complex to a neurotransmitter-metal-receptor complex to mediate or activate the interaction between bound molecules. Present knowledge of the functions of metal ions is by no means adequate, but recent investigations have revealed more important roles than expected. The function of zinc in neural mechanisms such as opioid-receptor binding has been supported by various lines of evidence (Gulya et al., 1991), and the high copper content in brain has been reported to affect the brain function (Kolb and Sadee, 1983) and is believed to be associated with the analgesic action of morphine (Iwata et al., 1970a; 1970b). Specific DNA binding by zinc finger proteins dramatically illustrates the importance and possibility of the functions of biological metal ions yet to be elucidated.

Studies on mixed-ligand complexes have shown that, while the Tyr phenol moiety is stacked with bpy in Cu(bpy)(L-Tyr), it is involved in electrostatic interactions with a positively charged group such as in Cu(Ptyr)(L-Arg) upon Tyr phosphorylation (Yamauchi et al., 1992). This may give an explanation to the structural changes of proteins due to Tyr phosphorylation considered as a key step in cancer cell growth (Lau et al., 1989). Simple model ternary complexes with ligand-ligand interactions may serve as a prototype of the mode of reactions at the active sites and thus enable us to get a deeper insight into complex biological reaction mechanisms.

REFERENCES

Alberts, B., Bray, D., Lewis, J., Raff, M., Roberts, K., and Watson, J.D. (1989). *Molecular Biology of the Cell*, 2nd ed., Garland Publishing, New York.

Amico, P., Bonomo, R.P., Cali, R., Cucinotta, V., Daniele, P.G., Ostacoli, G., and Rizzarelli, E. (1989). *Inorg. Chem.*, *28*: 3555.

Aoki, K. (1978). *J. Amer. Chem. Soc.*, *100*: 7106.

Aoki, K. (1981). *Nippon Kessho Gakkaishi*, *23*: 309.

Aoki, K, and Yamazaki, H. (1987). *J. Chem. Soc., Dalton Trans.*, 2017.

Aoyama, Y., Asakawa, M., Yamagishi, A., Toi, H., and Ogoshi, H. (1990). *J. Amer. Chem. Soc.*, *112*: 3145.

Arena, G., Bonomo, R.P., Casella, L., Gullotti, M., Impellizzeri, G., Maccarrone, G., and Rizzarelli, E. (1991). *J. Chem. Soc., Dalton Trans.*, 3203.

Arena, G., Cali, R., Cucinotta, V., Musumeci, S., Rizzarelli, E., and Sammartano, S. (1983). *J. Chem. Soc., Dalton Trans.*, 1271.

Arena, G., Cali, R., Cucinotta, V., Musumeci, S., Rizzarelli, E., and Sammartano, S. (1984). *J. Chem. Soc., Dalton Trans.*, 1651.

Banerjea, D., Kaden, T.A., and Sigel, H. (1981). *Inorg. Chem.*, *20*: 2586.

Banci, L., Bertini, I., Luchinat, C., and Hallewell, R.A. (1988). *J. Amer. Chem. Soc.*, *110*: 3629.

Bastian, M., and Sigel, H. (1991). *J. Coord. Chem.*, *23*: 137.

Berg, J.M. (1989). *Metal Ions in Biological Systems*, Vol. 25 (H. Sigel and A. Sigel, eds.), Marcel Dekker, New York, p. 235.

Berthon, G., May, P.M., and Williams, D.R. (1978). *J. Chem. Soc., Dalton Trans.*, 1433.

Berthon, G., Blais, M.-J., Piktas, M., and Houngbossa, K. (1984). *J. Inorg. Biochem.*, *20*: 113.

Berthon, G., Hacht, B., Blais, M.-J., and May, P.M. (1986). *Inorg. Chim. Acta*, *125*: 219.

Borghesani, G., Pulidori, F., Remelli, M., Purrello, R., and Rizzarelli, R. (1990). *J. Chem. Soc., Dalton Trans.*, 2095.

Brookes, G., and Pettit, L.D. (1975). *J. Chem. Soc. Chem. Commun.*, 385.

Brookes, G., and Pettit, L.D. (1977). *J. Chem. Soc., Dalton Trans.*, 1918.

Brumas, V., Alliey, N., and Berthon, G. (1993). *J. Inorg. Biochem.*, *52*: 287.

Christianson, D.W., and Lipscomb, W.N. (1989). *Accounts Chem. Res.*, 22: 62.

Cini, R. (1992). *Comments Inorg. Chem.*, *13*: 1.

Cucinotta, V., Purrello, R., and Rizzarelli, E. (1990). *Comments Inorg. Chem.*, *11*: 85.

Cudd, A., and Fridovich, I. (1982). *J. Biol. Chem.*, *257*: 11443.

Feeney, J. (1976). *J. Magn. Reson.*, *21*: 473.

Fischer, B.E., and Sigel, H. (1980). *J. Amer. Chem. Soc.*, *102*: 2998.

Freeman, H.C. (1973). *Inorganic Biochemistry*, Vol. 1 (G.I. Eichhorn, ed.), Elsevier, Amsterdam, p. 121.

Freeman, H.C., Guss, J.M., Healy, M.J., Martin, R.-P., Nockolds, C.E., and Sarkar, B. (1969). *Chem. Commun.*, 225.

Frieden, E. (1975). *J. Chem. Educ.*, *52*; 754.

Fujii, Y., Yoshikawa, Y., Syoji, M., and Shinohara, H. (1990). *Bull. Chem. Soc. Jpn.*, *63*: 138.

Fukuda, Y., Mitchell, P.R., and Sigel, H. (1978). *Helv. Chim. Acta*, *61*: 638.

Gergely, A., and Farkas, E.(1982). *J. Chem. Soc., Dalton Trans.*, 381.

Getzoff, E.D., Tainer, J.A., Weiner, P.K., Kollmann, P.A., Richardson, J.S., and Richardson, D.C. (1983). *Nature*, *306*: 287.

Gulya, K., Kovács, G.L., and Kása, P. (1991). *Life Sci.*, *48*: PL-57.

Hallman, P.S., Perrin, D.D., and Watt, A.E. (1971). *Biochem. J.*, *121*: 549.

Impellizzeri, G., Maccarrone, G., Rizzarelli, E., Vecchio, G., Corradini, R., and Marchelli, R. (1991). *Angew. Chem., Int. Ed. Engl.*, *30*: 1348.

Ito, N., Phillips, S.E.V., Stevens, C., Ogel, Z.B., McPherson, M.J., Keen, J.N., Yadav, K.D.S., and Knowles, P.F. (1991). *Nature*, *350*: 87.

Iwata, H., Watanabe, K., and Matsui, Y. (1970a). *Eur. J. Pharmacol.*, *11*: 298.

Iwata, H., Watanabe, K., Miichi, H., and Matsui, Y. (1970b). *Pharmacol. Res. Commun.*, *2*: 213.

Jencks, W.P. (1975). *Adv. Enzymol. Relat. Areas Mol. Biol.*, *42*: 219.

Kasselouri, S., Garoufis, A., Lamera-Hadjiliadis, M., and Hadjiliadis, N. (1990). *Coord. Chem. Rev.*, *104*: 1.

Kim, S.-H., and Martin, R.B. (1984). *J. Amer. Chem Soc.*, *106*: 1707.

Kiss, T. (1990). *Biocoordination Chemistry* (K. Burger, ed.), Ellis Horwood, London, p. 56.

Kiss, T., and Gergely, A. (1985). *J. Inorg. Biochem.*, *25*: 247.

Kiss, T., and Szücs, Z. (1986). *J. Chem. Soc., Dalton Trans.*, 2443.

Kolb, V.M., and Sadeé, W. (1983). *Res. Commun. Psychol. Psychiatry Behav.*, *8*: 343.

Kozlowski, H. (1978). *Inorg. Chim. Acta*, *31*: 135.

Kozlowski, H., and Jezowska-Trzebiatowska, B. (1976). *Chem. Phys. Lett.*, *42*: 246.

Kozlowski, H., Decock-Le-Révérend, B., Delaruelle, J.-L., and Loucheux, C. (1983). *Inorg. Chim. Acta*, *78*: 31.

Lau, K.-H., Farley, J.R., and Baylink, D.J. (1989). *Biochem. J.*, *257*: 23.

Liang, G., Tribolet, R., and Sigel, H. (1988). *Inorg. Chem.*, *27*: 2877.

Malini-Balakrishnan, R., Scheller, K.H., Häring, U.K., Tribolet, R., and Sigel, H. (1985). *Inorg. Chem.*, *24*: 2067.

Marcus, Y., and Eliezer, I. (1969). *Coord. Chem. Rev.*, *4*: 273.

Mariam, Y.H., and Martin, R.B. (1979). *Inorg. Chim. Acta*, *35*: 23.

Martin, R.B. (1974). *Metal Ions in Biological Systems*, Vol. 1 (H. Sigel, ed.), Marcel Dekker, New York, p. 129.

Martin, R.B., and Mariam, Y.H. (1979). *Metal Ions in Biological Systems*, Vol. 8 (H. Sigel, ed.), Marcel Dekker, New York, p. 57.

Martin, R.B., and Sigel, H. (1988). *Comments Inorg. Chem.*, *6*: 285.

Martin, R.P., Petit-Ramel, M.M., and Scharff, J.P. (1973). *Metal Ions in Biological Systems*, Vol. 2 (H. Sigel, ed.), Marcel Dekker, New York, p. 1.

Massoud, S.S., and Sigel, H. (1990). *Chimia*, *44*: 55.

Massoud, S.S., Tribolet, R., and Sigel, H. (1990). *Eur. J. Biochem.*, *187*: 387.

Masuda, Y., Tachiyashiki, S., and Yamatera, H. (1982). *Chem. Lett.*, 1065.

Masuda, H., Matsumoto, O., Odani, A., and Yamauchi, O. (1988). *Nippon Kagaku Kaishi*, 783.

Masuda, H., Sugimori, T., Odani, A., and Yamauchi, O. (1991). *Inorg. Chim. Acta*, *180*: 73.

Masuda, H., Odani, A., Yamazaki, T., Yajima, T., and Yamauchi, O. (1993). *Inorg. Chem.*, *32*: 1111.

May, P.M., Linder, P.W., and Williams, D.R. (1977). *J. Chem. Soc., Dalton Trans.*, 588.

Miller, J., McLachlan, A.D., Klug, A. (1985). *EMBO J.*, *4*: 1609.

Mitchell, P.R., and Sigel, H. (1978). *J. Amer. Chem. Soc.*, *100*: 1564.

Mitchell, P.R., Prijs, B., and Sigel, H. (1979). *Helv. Chim. Acta*, *62*: 1723.

Mota de Freitas, D., Luchinat, C., Banci, L., Bertini, I., and Valentine, J.S. (1987). *Inorg. Chem.*, *26*: 2788.

Neumann, P.Z., and Sass-Kortsak, A. (1967). *J. Clin. Invest.*, *46*: 646.

Nishio, M., and Hirota, M. (1989). *Tetrahedron*, *45*: 7201.

Odani, A., and Yamauchi, O. (1987). *Nippon Kagaku Kaishi*, *1987*: 336.

Odani, A., Deguchi, S., and Yamauchi, O. (1986). *Inorg. Chem.*, *25*: 62.

Odani, A., Masuda, H., Yamauchi, O., and Ishiguro, S. (1991). *Inorg. Chem.*, *32* : 4484.

Okawa, H. (1988). *Coord. Chem Rev.*, *92*: 1.

Ono, T., and Sasada, Y. (1981). *Bull. Chem. Soc. Jpn.*, *54*: 90.

Ono, T., Shimanouchi, H., Sasada, Y., Sakurai, T., Yamauchi, O., and Nakahara, A. (1979). *Bull. Chem. Soc. Jpn.*, *52*: 2229.

Orioli, P., Cini, R., Donati, D., and Mangani, S. (1981). *J. Amer. Chem. Soc.*, *103*: 4446.

Pachler, K.G.R. (1964). *Spectrochim. Acta*, *20*: 581.

Patel, V.K., and Bhattacharya, P.K. (1984). *J. Inorg. Biochem.*, *21*: 169.

Pavletich, N.P., and Pabo, C.O. (1991). *Science*, *252*: 809.

Pettit, L.D., and Hefford, R.J.W. (1979). *Metal Ions in Biological Systems*, Vol. 9 (H. Sigel, ed.), Marcel Dekker, New York, p. 173.

Poulos, T.L., Finzel, B.C., and Howard, A.J. (1986). *Biochemistry*, *25*: 5314.

Prasad, K., and Mohan, M.S. (1987). *J. Coord. Chem.*, *16*: 1.

Quiocho, F.A., and Lipscomb, W.N. (1971). *Adv. Protein Chem.*, *25*: 1.

Reedijk, J., Fichtinger-Schepman, A.M.J., Van Oosterom, A.T., and Van de Putte, P. (1987). *Struct. Bonding*, *67*: 53.

Saha, N., and Sigel, H. (1982). *J. Amer. Chem. Soc.*, *104*: 4100.

Sakurai, T., Yamauchi, O., and Nakahara, A. (1976a). *Bull. Chem. Soc. Jpn.*, *49*: 169.

Sakurai, T., Yamauchi, O., and Nakahara, A. (1976b). *Bull. Chem. Soc. Jpn.*, *49*: 1579.

Sakurai, T., Yamauchi, O., and Nakahara, A. (1977). *J. Chem. Soc. Chem. Commun.*, 718.

Sarkar, B. (1981). *Metal Ions in Biological Systems*, Vol. 12 (H. Sigel, ed.), Marcel Dekker, New York, p. 233.

Sarkar, B., and Kruck, T.P.A. (1966). *The Biochemistry of Copper* (J. Peisach, P. Aisen and W.E. Blumberg, eds.), Academic Press, New York, p. 183.

Sarkar, B., and Wigfield, Y. (1968). *Can J. Biochem.*, *46*: 601.

Sasada, Y., Takenaka, A., and Furuya, T. (1983). *Bull. Chem. Soc. Jpn.*, *56*: 1745.

Scheller-Krattiger, V., Scheller, K.H., and Martin, R.B. (1982). *Inorg. Chim. Acta*, *59*: 281.

Sigel, H., ed. (1973). *Metal Ions in Biological Systems*, Vol. 2, Marcel Dekker, New York.

Sigel, H. (1975). *Angew. Chem., Int. Ed. Engl.*, *14*: 394.

Sigel, H., ed. (1979). *Metal Ions in Biological Systems*, Vol. 9, Marcel Dekker, New York.

Sigel, H. (1980). *Coordination Chemistry-20* (D. Banerjea, ed.), Pergamon Press, Oxford, p. 27.

Sigel, H. (1981). *Advances in Solution Chemistry* (I. Bertini, L. Lunazzi and A. Dei, eds.), Plenum Publishing, New York, p. 149.

Sigel, H. (1982). *Angew. Chem., Int. Ed. Engl.*, *21*: 389.

Sigel, H. (1987). *Chimia*, *41*: 11.

Sigel, H. (1989a). *Pure Appl. Chem.*, *61*: 923.

Sigel, H. (1989b). *Metal-DNA Chemistry* (T.D. Tullius, ed.), ACS Symposium Series 402, ACS, Washington.

Sigel, H., Fischer, B.E., and Farkas, E. (1983). *Inorg. Chem.*, *22*, 925.

Sigel, H., Malini-Balakrishnan, R., and Häring, U.K. (1985a). *J. Amer. Chem. Soc.*, *107*: 5137.

Sigel, H., Tribolet, R., and Scheller, K.H. (1985b). *Inorg. Chim. Acta*, *100*: 151.

Sigel, H., Tribolet, R., and Yamauchi, O. (1990). *Comments Inorg. Chem.*, *9*: 305.

Sinanoglu, O. (1982). *Molecular Interactions*, Vol. 3 (H. Ratajczak and W.J. Orville-Thomas, eds.), John Wiley, New York, Chapter 6.

Strasak, M., Novomesky, P., and Butvin, P. (1990). *Collect. Czech. Chem. Commun.*, *55*: 1518.

Tachiyashiki, S., and Yamatera, H. (1984). *Bull. Chem. Soc. Jpn.*, *57*: 1061.

Terrón, A. (1993). *Comments Inorg. Chem.*, *14*: 63.

Tribolet, R., Sigel, H., and Trefzer, K. (1983). *Inorg. Chim. Acta*, *79*: 278.

Tsangaris, J.M., and Martin, R.B. (1970). *J. Amer. Chem. Soc.*, *92*, 4255.

Tullius, T.D., ed. (1989). *Metal-DNA Chemistry*, ACS Symposium Series 402, ACS, Washington.

Ullah, M.R., and Bhattacharya, P.K. (1991). *Bull. Chem. Soc. Jpn.*, *64*: 3659.

Verchère-Béaur, C., Mikros, E., Perrée-Fauvet, M., and Gaudemer, A. (1990). *J. Inorg. Biochem.*, *40*: 127.

Vestues, P.I., and Martin, R.B. (1980). *J. Amer. Chem. Soc.*, *102*: 7906.

Xiao, L., Jouini, M., Fan, B.T., Lapluye, G., and Huet, J. (1990). *J. Chem. Soc., Dalton Trans.*, 1137.

Yamauchi, O. (1984). *J. Mol. Catal.*, *23*: 255.

Yamauchi, O., and Odani, A. (1981). *J. Amer. Chem. Soc.*, *103*: 391.

Yamauchi, O., and Odani, A. (1985a). *J. Amer. Chem. Soc.*, *107*: 5938.

Yamauchi, O., and Odani, A. (1985b). *Inorg. Chim. Acta*, *100*: 165.

Yamauchi, O., and Odani, A. (1988). *Nippon Kagaku Kaishi*, *1988*: 369.

Yamauchi, O., Nakao, Y., and Nakahara, A. (1975). *Bull. Chem. Soc. Jpn.*, *48*: 2572.

Yamauchi, O., Odani, A., Kohzuma, T., Masuda, H., Toriumi, K., and Saito, K. (1989). *Inorg. Chem.*, *28*: 4066.

Yamauchi, O., Odani, A., and Masuda, H. (1992). *Inorg. Chim. Acta*, *198-200*: 749.

Yamauchi, O., Sakurai, T., and Nakahara, A. (1977). *Bull. Chem. Soc. Jpn.*, *50*: 1776.

Yamauchi, O., Sakurai, T., and Nakahara, A. (1979). *J. Amer. Chem. Soc.*, *101*: 4164.

STRUCTURAL ASPECTS OF METAL ION INTERACTIONS WITH SYNTHETIC CARRIER LIGANDS

by

Vincent L. Pecoraro and Brian R. Gibney

Department of Chemistry,
University of Michigan,
Ann Arbor, MI 48109-1055
U.S.A.

I. INTRODUCTION

Complexation of metal ions is greatly influenced by the molecular architecture of a chelating agent. In this chapter, we discuss the practical factors that lead to thermodynamically robust metal/chelating agent complexes. One may organize these factors into five main categories: ligand donor type, denticity, allowed coordination geometry, preformed cavities and ion stereochemistry. Optimizing specificity and selectivity requires that attention be paid to each of these considerations.

As we pointed out in an earlier chapter in this volume (see Part One, Chapter 3-C1), one of the most important considerations in the design of a metal chelator is the choice of heteroatoms used to coordinate the ion of interest. The general rule is that small, highly charged, non-polarizable electron pair acceptors (Hard acids) (Pearson, 1963) such as Fe(III), Mg(II) and Al(III) form the most stable compounds with small, highly charged, non-polarizable electron pair donors (Hard bases) such as carboxylates, phenolates, catecholates, hydroxamates and phosphonates. Conversely, large, weakly charged and highly polarizable electron pair acceptors (Soft acids) such as Cu(I), Hg(II), Cd(II) and Pd(II) form the most stable compounds with large, weakly charged, highly polarizable electron pair donors (Soft bases) such as thioethers, phosphines and the larger halides. Metals of intermediate acid strength (e.g. Zn(II)) will prefer bases of intermediate strength which often include nitrogen donors. Enterobactin, desferrioxamine (DFO or Desferal®) and EDTA are prototypical chelating agents for hard metal ions since these ligands include extremely hard donor groups (catecholates, hydroxamates and carboxylates, respectively). In contrast, crown thiaethers show a marked selectivity for the softer metals.

Although the affinity of a ligand for a chosen metal ion is critically dependent on the type of heteroatom donor, there are considerations beyond simple acid/base chemistry that markedly impact the efficacy of a chelating agent. In the examples of enterobactin, DFO and EDTA, each ligand satisfies the preferred coordination number of the target metal, Fe(III) (usually six coordinate). Metals such as Gd(III) and Pu(IV) tend to form complexes of higher coordination number than six and would not be expected to form maximally stable complexes with enterobactin or DFO. To circumvent this problem, ligands that have eight donor groups such as DTPA or 3,4,3-LICAM have been prepared. The optimal coordination number for some metals is often or always less than six and in these cases, with the possible exception of Hg(II) that has a propensity to form only two bonds and linear coordination, usually is four coordinate. Metals that fall in this class include Cu(II), Pt(II), Zn(II) and Li(I). It is not a simple matter to design a chelating agent based solely on coordination number for these metal atoms since coordination *geometry* becomes as important as coordination number in these compounds.

While nearly all six-coordinate complexes are octahedral, or a rhombic or axial distortion from an octahedral basis, four-coordinate compounds may be either tetrahedral or square planar. One would predict all four-coordinate compounds would be tetrahedral if electrostatic repulsion was the only factor defining the geometry; however, this geometric distinction derives from electronic considerations of the metal (usually due to a severe Jahn-Teller distortion as in d^9 Cu(II) or to a splitting of energy levels when 10Dq is great for d^8 ions such as Pt(II)). A thermodynamic price will be paid if a metal with a tetrahedral preference is restricted by a ligand that positions donors in a square planar geometry (e.g. porphyrin). The energy deficit will be small, however, since most tetrahedral ions are d^0, high spin d^5 or d^{10} and will not be destabilized as a result of electronic changes in the metal. In contrast, a ligand that presents donors in a tetrahedral orientation will be ineffective towards metals with a strong preference for a square planar polyhedron. Of course, one may extract opportunity out of adversity in this case since one should be able to design an agent that chelates tetrahedral ions preferentially even in the presence of other metals that are four coordinate (e.g. Zn(II) vs Pt(II)).

Metals with coordination numbers greater than six do not have strong geometrical preferences. Metals with seven coordination are found in "piano stool", face capped octahedral, pentagonal bipyramidal or site-deficient dodecahedral environments while eight-coordinate geometries include square prism and antiprisms and dodecahedral polyhedra. Because the energy difference between each of these geometries is small, basing a ligand design on coordination geometry for higher coordination number metals is not essential, as long as there are a sufficient number of donors available that can encapsulate the ion.

To this point, we have considered explicitly ligand donor type, number and arrangement and implicitly suggested that selectivity and specificity is influenced by the denticity of the ligand. This latter point deserves more attention since only in multidentate ligands is true specificity manifested in a useable way. The literature on chelating agents is daunting and cannot be covered rigorously in so short a space; however, we will present the primary guiding principles using important classes of metal binding agents. Specific attention will be given to crown ethers (Pedersen, 1988), cyclams, cryptands (Lehn, 1973), supramolecules (Lehn, 1988), polyaminocarboxylates (Hancock and Martell, 1989) and siderophores. The interested reader should consult the indicated articles for leading references that more completely describe each system.

II. CROWN ETHERS

Crown ethers are the paradigm of low-molar-mass (l.m.m.) chelating agents that show specificity for hard, low valent metal ions such as Li(I), Na(I), K(I), Mg(II) and Ca(II). Molecules such as 1,4,7,10,13,16-hexaoxacyclooctadecane (18-crown-6) provide an appropriate heteroatom donor set for coordinating to the hard alkali and alkaline ions by grouping 6 ether oxygen atoms into a cyclic structure, generating a preformed cavity that is flexible enough to satisfy the coordination preferences of these metals (Pedersen, 1967). Smaller crowns (e.g. 15-crown-5 and 12-crown-4) also will bind these metals; however, because the radius of the molecular cavity that is formed is smaller than the ionic radius of the metals and the denticity of these ligands is too low to completely satisfy the coordination preference of the larger ions, one routinely obtains 2:1 crown ether to metal complexes (Pedersen, 1970). In an effort to impart higher stability to the metal ligand complex, lariat ethers that incorporate additional "pendant arms" have been prepared to circumvent both coordination number and cavity size deficiencies (Schultz et al., 1981). As expected, these molecules have been shown to bind ions more efficiently than the parent crowns. Crown ethers and derivatives have become important tools in analytical chemistry since they can be used to measure cation concentrations in complex mixtures of ions (Hiraoka, 1982; Izatt et al., 1985). They also can greatly enhance the lipid/organic solubility of metal ions.

III. CYCLAMS

Substitution of nitrogen for oxygen in the crown core leads to molecules (CYCLAMS) with dramatically different metal specificities. For example, triazacyclononane (TACN) has a very low affinity for the alkali metal or alkaline earth cations; however, this ligand forms very stable metal complexes with high valent metals such as Mn(III), Mn(IV) and V(V). It is also an excellent ligand for Fe(III). As was the case for the crowns, TACN cannot satisfy the coordination number nor does it have an appropriate cavity to completely encapsulate the metal. Therefore, clusters of metals, rather than discrete mononuclear complexes, are often generated when this ligand is used (Wieghardt et al., 1983). A new molecular class called "metallacrowns" show specificities that are intermediate between those of conventional crowns and cyclams (Lah and Pecoraro, 1990). Metallacrowns have preformed cavities that are structurally analogous to the crown but substitute metal and nitrogen atoms into positions that are occupied by methylene groups in the crowns or cyclams (Lah et al., 1989). This modification leads to metal clusters that have relatively basic oxygen donors that can bind metals ranging from Li(I) to Cu(II) to Fe(III) (Lah and Pecoraro, 1989). The crown thiaethers are structurally related but include sulfur in the heteroatom position. These compounds show a greater preference for softer metal ions (Bradshaw et al., 1973). Summarizing this class of molecule, crowns, cyclams and crown thiaethers, all use a prearranged cavity to provide a docking site for a metal to the ligand. Hard, relatively low valent ions are best bound by ether oxygens of crowns, higher valent hard acids are best bound by amine nitrogens of cyclams, and soft metals are best bound by the soft, thioether linkages offered by the crown thiaethers.

IV. CRYPTANDS

Cryptands (Lehn, 1978) are molecules that are related to crown ethers and cyclams in having preformed cavities composed of oxygen and nitrogen donors; however, they differ significantly in the level of preorganization (Lehn, 1988). The rigidly preorganized spheroidal cavity of the cryptands results in more thermodynamic and kinetically favored ion encapsulation. Whereas crown compounds possess rotational freedom in solution and X-ray structures of vacant crowns have demonstrated the lack of a preformed ion binding cavity (Hiraoka, 1982), the cryptands possess a rigid binding cavity. The rigidity in the binding cavity effects the kinetics and thermodynamics of ion binding. Kinetically the formation of the [2.2.2], NaCl cryptate complex is about 10^2 times slower than the analogous crown, but the dissociation rate of the complex is 10^6 slower. The stability constants of cryptands are 10^3 to 10^5 higher than analogous crown compounds (Izatt et al., 1991). Cryptands follow the hole size-ion size relationship of ion binding much more strictly than the crowns, which leads to increased ion selectivity. Cryptate cavities of tetrahedral, octahedral and linear binding geometries offer ion selectivity based on coordination geometry. The association of receptor and substrate results in a supramolecule which includes entities of higher organization resulting from the association of coordinately saturated species (Lehn, 1988).

V. POLYAMINOCARBOXYLATES

The polyaminocarboxylate ligands such as IDA (iminodiacetic acid), EDTA (ethylenediamine tetraacetic acid) and DTPA (diethylenetriamine pentaacetic acid) represent another important class of metal chelating agents. These molecules differ markedly from crown ethers, cryptands and supramolecules. Incorporation of carboxylates into polyamine ligands allows for a reduction in complex charge and stabilization of higher oxidation states (Evers et al., 1989). The tridentate IDA has a log $K_1=5.5$ for Fe^{2+} and log $K_1=10.7$ for Fe^{3+} and nitrilotriacetic acid, NTA, shows 12 orders of magnitude greater stability for ferric ion than ferrous ion (Martell and Smith, 1982). The carboxylates and polyaminocarboxylates show moderate stabilities with these metals, but in most cases offer little in the selectivity for a specific metal ion. For example IDA has less than a log unit higher stability constant for the hard acid Fe^{3+} over the borderline hard acid Cu^{2+} (Martell and Smith, 1981). This poor selectivity has led to some clinical concerns since important divalent cations can be depleted by the administration of Na_2H_2EDTA or Na_3H_2DTPA. To circumvent this problem the chelate ligands are formulated as zinc or other trace metal salts. Another disadvantage of chelation agents such as EDTA and DTPA is that they have several ionized carboxylate groups at physiological pH and do not pass through cell membranes to any extent (Jones, 1992). While effective at removing metal ions from serum, they are excreted before reaching equilibrium with intracellular metal ions.

In an effort to design In^{3+} complexes for radiolabelling monoclonal antibodies directed against tumor-associated antigens, Mäcke has fused the polyaminocarboxylate and cyclam architectures to form seven- and eight-coordinate polyaminocarboxylates based on cyclen and cyclam (Riesen et al., 1989). After N-functionalizing the ring system with three or four acetates and complexation with In^{3+}, their metal exchange rates with transferrin were evaluated. The log of the stability constants were evaluated to be >25 and the ligands based on cyclen showed 0.4% and 0.35% metal exchanged for the tri- and tetraacetate ligands respectively in the first 24 hours. The tri- and tetraacetates of cyclam showed much higher exchange rates in vitro, 12 and 9% respectively. These are compared to 1.5% for DTPA which is most often used in ^{111}In labeling studies.

VI. SIDEROPHORES

Nature has provided clues for the selective encapsulation of ferric ion with the siderophores, a class of l.m.m. ferric chelation agents that possess exceptional stability and selectivity for this ion. Siderophores are categorized into two broad classes based on the presence of catechol or hydroxamic acid moieties. The affinity and specificity of siderophores is related to the high affinity of catechols and hydroxamic acids (Kurzak et al., 1992) for hard acids, e.g. Fe^{3+}. Acetohydroxamic acid forms a tris complex with Fe^{3+} with a stability constant of $10^{28.3}$ and catechol, another siderophore moiety, forms a tris complex with a stability constant of $10^{43.8}$ (Martell and Smith, 1982). Not only do these donor groups provide high stability but also good selectivity for Fe^{3+} as the log β's for other metals are considerably lower. For example, the log β_2 of bis(ligand) complexes with the softer Ni^{2+} are 9.3 and 14.4 for acetohydroxamic acid and catechol (Martell and Smith, 1982).

Of the naturally occurring siderophores, enterobactin is the most powerful Fe^{3+} chelator known with a log K = 49 (Loomis and Raymond, 1991). This prototype catechol-based siderophore is a cyclic triester of 2,3-dihydroxybenzoyl-L-serine. The high binding constant of enterobactin with Fe^{3+} is a result of the high affinity of catechol moieties for Fe^{3+} and the conformational rigidity enforced by the hydrogen bond belt through the adjacent amide linkages. Recently a X-ray crystal structure determination of the V(IV) enterobactin complex has been reported and we present it here to provide insight into the biomimetic approach to ligand design (Karpishin and Raymond, 1992). Although V(IV) is slightly smaller than Fe(III), the major features of the enterobactin complexes are the same. The twelve-membered triester platform is ideal for the preorganization of the catechol oxygens for complexation. The V-O1 and V-O2, the vanadium - catecholate oxygen, bond lengths are 1.964Å and 1.939Å, respectively, demonstrating the cavity is not strained. The hydrogen bonds from the amide hydrogens of the platform to the catechol oxygens form six-membered rings which enhance the metal complex stability. The chirality of the platform (λ) dictates the chirality of the metal center (Λ) which determines the bioactivity of the complex (Nielands et al., 1981). When designing synthetic ligands with high selectivity using a biomimetic approach, these features of enterobactin need to be kept in mind. A key requirement is the design of a structure which will more effectively form a set of donor groups that completely envelops the metal of interest.

Rather than using the hydrolytically unstable triester platform of enterobactin, Raymond had prepared a series of Fe^{3+} chelation agents which utilize 1,3,5-tris(methylamino)benzene as a platform. The catechol groups are coupled to the platform with amide bonds from 2,3-dihydroxybenzoic acid. The Fe^{3+} complex of the ligand, MECAM, shows a high pM value of 29.1 and has been shown to remove Fe^{3+} from transferrin in vitro (Harris et al., 1979). Sulfonation on the catechol moieties yields the ligand MECAMS, and allows for greater aqueous solubility and a reduction in the basicity of the catechol protons while not significantly altering the pM value (Harris et al., 1981). In an attempt to block microbial protease activity on MECAMS, the nitrogens were methylated and the resulting ligand, Me_3MECAMS, forms a complex with Fe^{3+} which is thermodynamically (pM=26.9) and kinetically able to remove Fe^{3+} from transferrin (Pecoraro et al., 1981). The use of non-cyclic platforms for these catechol moieties, such as in spermidine (3,4-LICAMS), also yields Fe^{3+}-sequestering agents of high stability (pM=28.5). Extension of this biomimetic approach to rigid ion binding cavities has improved the pM values. The bicapped TRENCAM has a pM of 30.7 which is almost 3 log units better than its monocapped analogue TRENCAM (Garrett et al., 1991). The increase in pM is a result of preformation of the binding cavity, although the catecholate oxygens are exo to the cavity in the free ligand.

Due to the similar charge-to-radius ratios of Fe^{3+} (CRR=4.6) and Pu^{4+} (CRR=4.2), the lessons learned from the siderophores can be applied to the selective complexation of actinides. The catechol-based 3,4,3 LICAMS has been shown to bind Pu^{4+} and Am^{3+} in vitro (Kappel et al., 1985). Hydroxypyridinones (HOPO) which are isoelectronic with catechols and have a lower pKa are effective Fe^{3+} and Pu^{4+} chelators (Sheppard and Kontoghiorghes, 1991). The ferric complex of 3,3,4-LIHOPO has been shown to effectively and selectively remove Pu^{4+} in vivo (White et al., 1988) .

VII. THE CONCEPT OF STEREOGNICITY

A recent wrinkle in the design of selective metal sequestration agents is the concept of stereognocity (Franczyk et al., 1992) whereby the three dimensional properties of an ion can be exploited to confer an additional level of recognition. Candidates for this approach include complex ions such as VO^{2+} (vanadyl), VO_2^+ (pervanadyl) and UO_2^+ (uranyl) or metals with stereochemically active lone pairs such as Pb^{2+}. Practically, one attempts to design a ligand that upon complexation of the metal places a hydrogen-bond donor in a position to interact with the oxygen of the complex ion or the lone pair. This approach enhances the selectivity of the ligand in two ways. First, one gains an additional 1-3 kcal per hydrogen bond in complex stability that would be unavailable to an ion lacking an oxygen or lone pair in the appropriate position. This could amount to as much as 4-12 kcal for a tetrahedral MO_4^{n-}. Second, the absence of a hydrogen-bond acceptor position on the metal would place the proton's positive charge in close proximity to the positively charged sequestered metal. As an example, one could prepare a ligand that would form hydrogen bonds only with complex ions that have linear oxo groups. Favorable interactions would be seen with the uranyl ion that would form two hydrogen bonds. In contrast, the pervanadyl ion (with a O-V-O angle ~110°) could form one hydrogen bond but would gain little from the second proton donor since the second oxo group is directed nearly orthogonal to the hydrogen-bond donor. Similarly, the vanadyl ion would form one hydrogen bond but would also suffer from electrostatic repulsion. Thus, stereognostic recognition demonstrates that one can begin to incorporate more subtle, but quite effective principles to modify metal selectivity.

The future may hold even greater advances in selective metal-chelating agent design due to advances in molecular biology and peptide chemistry. Very recent studies have shown that peptides can be designed that will bind cations such as Cu(II) or Hg(II) on either the exterior or interior of synthetic four or three helix bundle motifs. Advances such as these promise truly biocompatible reagents that can target specific organs and have higher biological efficacy. However, for these next generation of sequestration agent to truly impact medicine in a meaningful way will require that fundamental questions in de novo metallopeptide design be addressed. We anticipate that the next decade should provide these answers.

VIII. SUMMARY

Optimization of molecular architecture for the purposes of selective ion encapsulation requires that ligand donor type, denticity, coordination geometry, cavity preformation and ion stereochemistry be optimized. Optimization conditions can be inferred from synthetic and natural ionophores. The crown compounds provide insight into the areas of ligand donor type, denticity and cavity preformation in a systematic manner. Nature provides a siderophores and ionophores which aid in ligand donor type, coordination geometry and cavity preformation. Design of the most selective and specific ion sequestering agents requires the molecular architect heed these considerations.

REFERENCES

Bradshaw, J.S., Hui, J.Y., Chan, Y., Haymore, B.L., Christensen, J.J., and Izatt, R.M. (1973). *J. Heterocyclic Chem., 10:* 1.

Evers, A., Hancock, R.D., Martell, A.E., and Motekaitis, R.J. (1989). *Inorg. Chem., 28:* 2189.

Franczyk, T.S., Czerwinski, K.R., and Raymond, K.N. (1992). *J. Amer. Chem. Soc., 114:* 8138.

Garrett, T.M., McMurry, T.J., Hosseini, M.W., Reyes, Z.E., Hahn, F.E., and Raymond, K.N. (1991). *J. Amer. Chem. Soc., 113:* 2965.

Hancock, R.D., and Martell, A.E. (1989). *Chem. Rev., 89:* 1875.

Harris, W.R., Carrano, C.J., Cooper, S.R., Sofen, S.R., Avdeef, A.E., McArdle, J.V., and Raymond, K.N. (1979). *J. Amer. Chem. Soc., 101:* 6097.

Harris, W.R., Raymond, K.N., and Weitl, F.L. (1981). *J. Amer. Chem. Soc., 103:* 2667.

Hiraoka, M. (1982). *Crown Compounds: Their Characteristics and Applications*, Elsevier Scientific, New York.

Izatt, R.M., Bradshaw, J.S., Nielsen, S.A., Lamb, J.D., and Christensen, J.J. (1985). *Chem. Rev., 85:* 271.

Izatt, R.M., Pawlak, K., Bradshaw, J.S., and Bruening, R.L. (1991). *Chem. Rev., 91:* 1721.

Jones, M.M. (1992). *Comments Inorg.Chem., 13:* 91.

Kappel, M.J., Nitsche, H., and Raymond, K.N. (1985). *Inorg. Chem., 24:* 605.

Karpishin, T.B., and Raymond, K.N. (1992). *Angew. Chem., Int. Ed. Engl., 31:* 466.

Kurzak, B., Kozlowski, H., and Farkas, E. (1992). *Coord. Chem. Rev., 114:* 169.

Lah, M.S., Kirk, M.L., Hatfield, W., and Pecoraro, V.L. (1989). *J. Chem. Soc. Chem. Commun., 21:* 1606.

Lah, M.S., and Pecoraro, V.L. (1989). *J. Amer. Chem. Soc., 111:* 7258.

Lah, M.S., and Pecoraro, V.L. (1990). *Comments Inorg. Chem., 11:* 59.

Lehn, J.-M. (1973). *Structure and Bonding*, Vol. 16, Springer-Verlag, New York.

Lehn, J.-M. (1978). *Accounts Chem. Res., 11:* 49.

Lehn, J.-M. (1988). *Angew. Chem., Int. Ed. Engl., 27:* 89.

Loomis, L.D., and Raymond, K.N. (1991). *Inorg. Chem., 30:* 906.

Martell, A.E., and Smith, R.M. (1982). *Critical Stability Constants*, Vols. I-VI, Plenum, New York.

Nielands, J.B., Erickson, T.J., and Rastetter, W.H. (1981). *J. Biol. Chem., 256:* 3831.

Pearson, R.G. (1963). *J. Amer. Chem. Soc., 85:* 3533.

Pecoraro, V.L., Weitl, F.L., and Raymond, K.N. (1981). *J. Amer. Chem. Soc., 103:* 5133.

Pedersen, C.J. (1967). *J. Amer. Chem. Soc., 89:* 7017.

Pedersen, C.J. (1970). *J. Amer. Chem. Soc., 92:* 386.

Pedersen, C.J. (1988). *Angew. Chem., Int. Ed. Engl., 27:* 1021.

Riesen, A., Kaden, T.A., Ritter, W., and Mäcke, H.R. (1989). *J. Chem. Soc. Chem. Commun.*, 460.

Schultz, R.A., Dishong, D.M., and Gokel, G.W. (1981). *Tetrahedron Lett., 104:* 2623.

Sheppard, L.N., and Kontoghiorghes, G.J. (1991). *Inorg. Chim. Acta, 188:* 177.

White, D.L., Durbin, P.W., Jeung, N., and Raymond, K.N. (1988). *J. Med. Chem., 31:* 11.

Wieghardt, K., Pohl, J., and Gebrt, W. (1983). *Angew. Chem., Int. Ed. Engl., 22:* 727.

METAL-AMINO ACID COMPLEXES

by

Stuart H. Laurie

School of Applied Sciences,
De Montfort University,
The Gateway,
Leicester LE1 9BH
U.K.

I. INTRODUCTION

Biological systems have developed a number of ways of keeping the concentrations of 'free' (aquated) metal ions to very low levels. Too high a concentration of these free metal ions, even for the essential elements, is clinically damaging and even fatal. Proven examples are the effects of excess Cu ions in Wilson's disease and Fe in thalassaemia. The primary function of the amino acids is, of course, as building blocks for peptides and proteins. However, within blood serum is a low-molar-mass (ultra-filtrable) fraction composed mainly of amino acids and complexed metal ions. Although only a small amount, often less than 1%, of the essential metals occurs in this fraction, it serves an important role in chelating any non-protein bound metal ions and, because of the kinetically labile nature of such complexes, it provides a readily available source of the essential metal ions (that is, the amino acid pool serves as a 'metal-buffer' system). Furthermore, because of their small size and neutrality, these chelates have the potential to cross membranes. In fact, the amino acids themselves are readily taken up by tissues because their intracellular concentrations are much higher than that in blood plasma.

Not surprisingly, there has been a large amount of research into all aspects of metal-amino acid complexing. Excellent reviews of earlier work can be found in the series 'Metal Ions in Biological Systems', edited by Sigel. There is also an annual review in the Royal Society of Chemistry's Specialist Periodicals Report on 'Amino Acids and Peptides', Jones (1991). In this chapter, the coverage will be restricted to the more commonly occurring amino acids and their complexation in mainly aqueous media, and the solids formed from such media. Not covered, for example, are the large number of studies on derivatives of amino acids.

The amino acids have the general formula $^{+}H_3NCH(R)CO_2^{-}$. There is a large number of possible R groups. The most important of these considered in this chapter are given in Table 1. The trivial naming rather than the systematic organic nomenclature is used since these and their associated abbreviations are still in general use, e.g. in peptide sequences and genetic codes. The central carbon atom in the above formula, referred to as the α-carbon, is a center of chirality (optical activity) and hence each amino acid can exist as a pair of enantiomeric forms; the exception to this is glycine (R=H). Apart from some bacteria and snake venoms, only one enantiomer is found in nature. This is shown in Figure

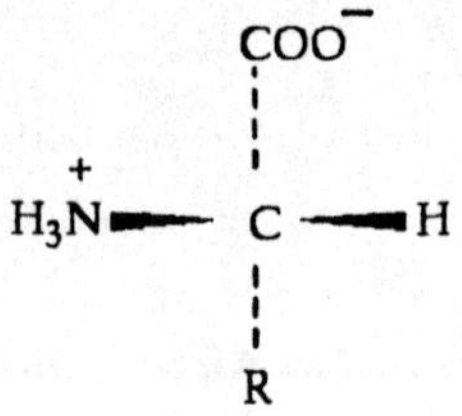

Figure 1.
Fischer projection of the naturally occuring S (or L) configuration of an amino acid.

1, and referred to as the S (or L) enantiomer, the mirror image form being the R (or D) enantiomer. Both R - S and D - L systems will be used - according to the original author's usage.

The configuration at the α-C center can influence the stereochemistry of coordination compounds (see next chapter). Hence, in studies of metal-amino acids complexation it is important to state the chiral nature of the amino acid used; this is often omitted in the literature. To illustrate this point with regard to the common $M(aa)_2$ complex*, use of the S enantiomer (or the R enantiomer) would give just $M(S\text{-}aa)_2$, (or $M(R\text{-}aa)_2$). However, use of the racemic form commonly available from commercial suppliers would give a mixture of $M(S\text{-}aa)_2$, $M(R\text{-}aa)_2$ and M(S-aa)(R-aa). In the absence of other factors (vide infra) the statistical ratio of these forms would be 1:1:2 (Laurie and James, 1983).

II. SOLUTION EQUILIBRIUM STUDIES

Ignoring any possible ionizing contributions from the side-chain R groups, then the amino acids at neutral pH exist in bipolar (zwitterionic) form. Protonation of the amino and carboxyl groups means the nature of the amino acid in solution is pH dependent.

$$\underset{pH<3\ (H_2aa^+)}{^+H_3NCH(R)COOH} \rightleftharpoons \underset{pH \sim 7\ (Haa)}{^+H_3NCH(R)COO^-} \rightleftharpoons \underset{pH>9\ (aa^-)}{H_2NCH_2COO^-}$$

As can be seen from Table 1, the pKa values of the amino and carboxyl groups are approximately 9 and 2, respectively. Consequently, depending upon the solution pH, the amino acids can coordinate to metal ions through either or both of the deprotonated amino or carboxyl groups. In addition, those with polar R groups, e.g. Asp, Cys, His, Glu, Pen, offer additional pH-dependent coordinating sites.

The equilibria in solution and associated stability constants (β_{pqr}) are generally expressed in the forms

$$pM + q(aa) + rH \rightleftharpoons M_p(aa)_qH_r \quad (1)$$

$$\beta_{pqr} = [M_p(aa)_qH_r] / [M]^p[aa]^q[H]^r \quad (2)$$

where p, q and r are the stoichiometric coefficients and [ ] denotes equilibrium concentrations. The experimental techniques and computational methods to obtain β_{pqr} values are described in the previous chapter. From these values is obtained the speciation, i.e. the number of complexes which can coexist and their concentrations at any specified conditions, e.g. pH, ionic strength, temperature and metal and amino acid concentrations. Aspects regarding speciation are covered in Part Five. There are several compilations of relevant equilibrium data, viz Martell and Smith (1974; 1982), Sillén and Martell (1964; 1971), Laurie (1987). The International Union of Pure and Applied Chemistry (IUPAC) is currently sponsoring a computer compilation of all published stability constants for the period 1974-89 and this should be available soon. Of particular value are the IUPAC commissioned series on the critical evaluation of stability constants of metal complexes in

* M = a metal ion, aa = amino acidate anion form from the parent aaH_n zwitterion

Table 1

The More Common Amino Acids, $H_3{}^+NCH(R)CO_2H$, and their Ionization Constants

Name	Symbol	R	pK_a*
Alanine (α)	Ala	Me	2.31, 9.70
Arginine	Arg	$CH_2(CH_2)_2NHC(^+NH_3)=NH$	2.03, 9.02 (12.1)
Asparagine	Asn	CH_2CONH_2	2.15, 8.72
Aspartic acid	Asp	CH_2CO_2H	1.94, 3.70, 9.62
Cysteine	Cys	CH_2SH	1.91, 8.16, 10.29
Dihydroxyphenyalanine	Dopa	CH_2 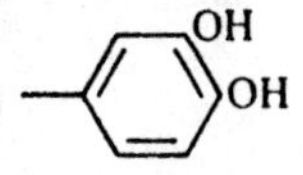	2.20, 8.72, 9.78, (13.4)
Glutamine	Gln	$CH_2CH_2CONH_2$	2.16, 9.96
Glutamic acid	Glu	$CH_2CH_2CO_2H$	2.18, 4.20, 9.59
Glycine	Gly	H	2.36, 9.56
Histidine	His	CH_2 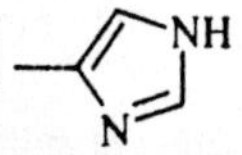	1.7, 6.02, 9.09
Isoleucine	Ile	CHMeEt	2.21, 9.56
Leucine	Leu	CH_2CHMe_2	2.27, 9.28
Lysine	Lys	$CH_2(CH_2)_3{}^+NH_3$	2.19, 9.12, 10.68
Methionine	Met	CH_2CH_2SMe	2.10, 9.06
Penicillamine	Pen	CMe_2SH	1.90, 7.92, 10.6
Phenylalanine	Phe	CH_2 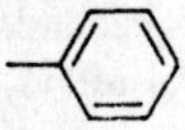	2.17, 9.09
Serine	Ser	CH_2OH	2.13, 9.05
Threonine	Thr	CH(OH)Me	2.20, 8.96
Tryptophan	Trp	CH_2 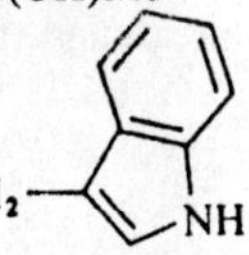	2.34, 9.32
Tyrosine	Tyr	CH_2 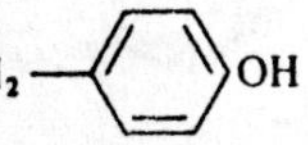	2.17, 9.04, 10.11
Valine	Val	$CHMe_2$	2.26, 9.49
β-Alanine	β-Ala	$H_3N^+CH_2CH_2CO_2H$	3.53, 10.10
Proline	Pro	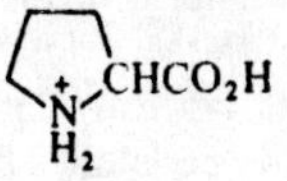	1.9, 10.41

* at 25°C, 0.1 mol dm^{-3} (Martell and Smith, 1974; 1982).

solution. Published so far are the evaluation of DOPA, His, Phe, Trp and Tyr data (Pettit, 1984) and Gly (Kiss et al., 1991), others covering the amino acids are currently in press.

Where stability constants are not available, then a method for estimating such values has been described (Smith et al., 1985). This is particularly valuable for metal ions which are difficult to handle experimentally, e.g. the actinides. The factors which influence stability constants have been discussed by Yatsimirskii (1986).

A. Amino acids with non-coordinating side-chains

For the kinetically *labile* metal ions (main group metals and the divalent cations of the first row transition elements) where equilibria are obtained within the time of mixing, the determination of monodentate coordination is difficult to establish. It is generally seen as a minor species at low pH (4 -5). Monodentate coordination is more easily found with the kinetically *inert* metal ions such as Cr(III), Co(III), Rh(III), Pd(II), Pt(II); in these cases, coordination is usually via the amino N atom rather than the carboxyl O atom.

The much more common mode of coordination in all cases is bidentate chelation giving rise to the more thermodynamically stable five-membered chelate ring **(1)**, six membered for β-alanine.

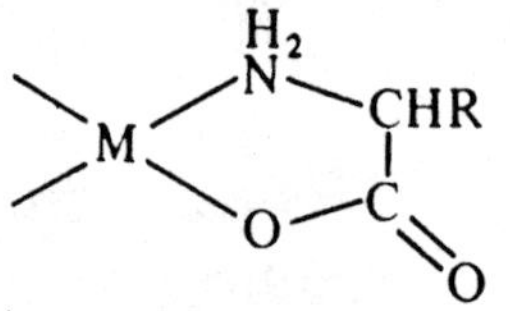

Scheme **(1)**

A selection of thermodynamic stability constants for the major chelate species M(aa), $M(aa)_2$ and $M(aa)_3$ is given in Table 2. Conclusions made on differences in β_{pqr} values should be treated with caution for several reasons: (i) conditions of temperature, ionic strength and methods differ between workers, (ii) differences in basicities between ligands (vide infra), (iii) two or more species may coexist under the same conditions. A large β_{pqr} value does not necessarily imply a predominant species because of the concentration and the pH dependence of its formation.

Taking the above factors into consideration, the data in Table 2 reveal two important properties: (i) at neutral pH the predominant species for the important divalent transition metal ions and Zn(II) is the neutral $M(aa)_2$ complex, (ii) for the same amino acids the order of the stability constants follows the Irving-Williams order, i.e. Mn(II) < Fe(II) < Co(II) < Ni(II) < Cu(II) > Zn(II). The largest stability constants in this series are thus obtained with Cu(II). What is not found from the β values is the nature of the geometry of the donor atoms around the Cu(II) ions, which can be either *cis* or *trans* **(2)**.

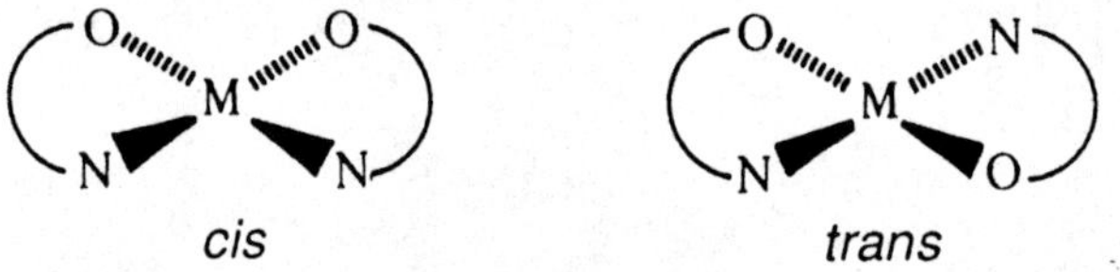

Scheme **(2)**

Here, N⌒O represents the chelate form of the amino acidate as in **(1)**. Undoubtedly, with small R groups there is little difference in stabilities, i.e. (trans) ~ (cis) and both will therefore coexist with a rapid cis ⇌ trans interconversion. However, this may not be the same with larger R groups and is certainly not true in the solid state. Sigel et al. (1985) have produced evidence for interligand interactions in aqueous solutions in the $M(aa)_2$ complexes of Co(II) and Ni(II) with alkyl and aryl side-groups, these hydrophobic interactions between R groups depending on the nature of the geometry around the metal center.

With the kinetically inert metal ions, such geometric isomerism is more easily established as their interconversion rates are sufficiently slow as to allow their separation (e.g. by chromatographic means). For the octahedral $M(aa)_3$ complexes, the situation becomes even more complicated as the metal ions are now chiral centers; thus four geometric-chiral isomers **(3)** can coexist.

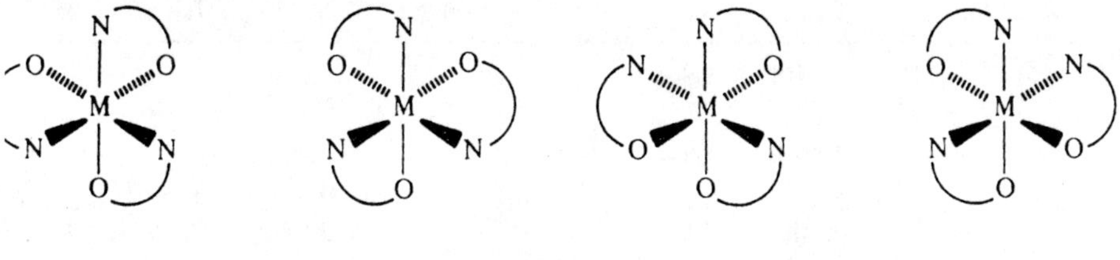

Δ-*fac* Λ-*fac* Δ-*mer* Λ-*mer*

Scheme **(3)**

Here, Δ and Λ are enantiomeric forms and fac(facial) and mer(meridional) are geometric isomers. All four isomers have been identified for the $Co(aa)_3$ complexes of Gly, L-Ala, L-Leu, L-Val and L-Asp. There appears to be no significant differences in their relative stabilities in solution.

B. Amino acids with potential coordinating side-chains

As can be seen from Table 1, a number of the amino acids contain polar side chains which have the potential to form a further bond to the metal ions, thus giving tridentate coordination. These include Asp and Glu (carboxylate O), Asn and Gln (amide N or O), Cys and Met (thiol S and thioether S respectively), Ser or Thr (hydroxyl O) and His (imidazole N). Penicillamine is included in this discussion, even though it is not a naturally occurring amino acid, because of its therapeutic use in treating Wilson's disease. Its chelating ability is strongly linked to its side-chain thiol group.

The enhanced thermodynamic stability of chelated complexes as compared to the monodentate complexes (the so-called chelate effect) is linked to the size of the cyclic structure so formed. In general, 5- and 6-membered rings are the more common and more stable. This factor, together with donor strength considerations (e.g. the poor donicity of the thioether S and amide N and O atoms), reduces the potentially tridentate amino acids to Asp, Glu, Cys, Pen, Ser, Thr and His.

Coordination of these side-groups also depends on the nature of the metal ion. A convenient classification is that known as the theory of hard and soft acids and bases (HSAB). In essence, a 'hard' metal ion (high charge:radius ratio) prefers to bind to a 'hard' base, i.e. prefers O-donors to S-donors and N-donors to P-donors. Conversely, a 'soft' metal ion (low charge:radius ratio) forms stronger bonds to the 'soft' or larger donor atoms, especially S (thiol) atoms. Some metal ions are borderline. The classification into these groups is given in Table 3. Although suffering limitations (see Part One, Chapter 3-A), this classification explains a number of important observations such as the affinity of Fe(III) and Al(III) for the O-donor atoms in Tyr and DOPA, and the very strong preferential binding of the toxic heavy metal ions for the Cys thiol group.

The more obvious indication of tridentate coordination is an increase in stability constants relative to those of the truly bidentate amino acids. Examination of Table 2 shows, for example, that for bidentate $Ni(aa)_2$ complexes $\log\beta \sim 10$ as compared to values of approximately 15, 20 and 23 for $Ni(His)_2$, $Ni(Cys)_2$ and $Ni(Pen)_2$ respectively. As mentioned earlier, simple comparisons of these values need to be treated with caution, especially with ligands of different basicities. In order to take this latter aspect into consideration, Martin (1979) has advocated comparisons of the equilibrium constant (K) for the competition reaction (3)

$$M^{n+} + LH \rightleftharpoons ML^{(n-1)+} + H^+ \tag{3}$$

Table 2

Log Formation Constants (β_{pqr})[a] of some Metal Amino Acid Complexes in Aqueous Solution at 25°C.[b]

Metal Ion	pqr	Gly	Ala	β-Ala	Dopa	Ser	His	Met
Mg^{2+}	110	1.34(b)	1.96(c)	-	-	1.03(e)	-	-
	111	-	-	-	4.71(e)	-	-	-
Ca^{2+}	110	1.4(c)	1.24(c)	-	-	1.43(c)	-	-
Mn^{2+}	110	2.60(b)	3.24(c)	-	-	2.50(a)	3.58(a)	2.77(a)
	120	4.5(b)	6.05(c)	-	-	3.98(a)	7.74(c)	4.57(a)
	130	5.3(b)	-	-	-	-	-	-
Fe^{2+}	110	4.13(a)	3.54(d)	~4(c)	-	4.30(e)	5.88(e)	-
	120	7.65(a)			-	7.38(c)	7.39(e)	-
	130	-	-	-	-	10.30(e)	-	-
Co^{2+}	110	4.67(a)	4.31(a)	~7(c)	-	4.58(e)	6.87(a)	4.14(a)
	112				3.75(a)	-	-	-
	120	8.46(a)	7.8(a)	-	-	8.57(e)	12.38(a)	7.28(a)
	124	-	-	-	7.25(a)	-	-	-
	130	-	-	-	-	11.55(e)	-	-
Ni^{2+}	110	5.78(a)	5.40(a)	4.54(a)	-	5.32(a)	8.66(a)	5.33(a)
	112	-	-	-	4.88(a)	-	15.52(a)	9.90(a)
	120	10.58(a)	9.9(a)	7.87(a)		9.74(a)	-	-
	124	-	-	-	14.1(a)	-	-	-
	130	14.0(a)	12.9(a)	9.7(a)	-	12.7(a)		
Cu^{2+}	110	8.13(a)	8.15(a)	7.04(a)	-	7.09(a)	10.16(a)	7.85(a)
	112	-	-	-	7.55(a)	-	-	-
	120	15.13(a)	14.9(a)	12.54(a)	-	14.49(a)	18.10(a)	14.52(a)
	124	-		-	14.1(a)	-	-	-
Zn^{2+}	110	4.96(a)	4.56(a)	3.90(b)	-	4.94(a)	6.51(e)	4.38(a)
	112	-	-	-	3.77(a)	-	-	-
	120	9.19(a)	8.55(a)	7.20(b)	-	9.22(a)	12.05(a)	8.35(a)
	122				18.06(c)	-	-	-
	130	-	-	-	-	-	-	-
Hg^{2+}	110	10.3(c)	-	-	-	11.7(a)	-	6.52(a)
	120	19.2(c)	-	-	-	19.1(a)	-	11.45(a)

[a] For the equilibrium pM + qL + rH $\rightleftharpoons$ MpLqHr (see text)
[b] Taken from Martell and Smith (1974; 1982); ionic strengths/ mol dm^{-3} (a) 0.1, (b) 0.5, (c) 0.0, (d) 1.0, (e) 3.0, (f) 0.2.

Table 2 (Continued)

Log Formation Constants (β_{pqr})[a] of some Metal Amino Acid Complexes in Aqueous Solution at 25°C.[b]

Metal Ion	pqr	Thr	Asp	Asn	Glu	Gln	Cys	Pen
Mg^{2+}	110	-	2.43(a)	-	1.9(a)	-	-	-
	111	-	-	-	-	-	-	-
Ca^{2+}	110	-	1.60(a)	-	1.43(a)	-	-	-
Mn^{2+}	110	2.58(a)	3.7(a)	3.10(e)	-	2.86(e)	4.6(a)	-
	120	3.98(a)	-	5.22(e)	-	4.6(e)	-	-
	130	-	-	-	-	-	-	-
Fe^{2+}	110	7.7(c)	-	4.37(e)	-	-	-	-
	120	-	-	7.57(e)	-	-	11.77(c)	-
	130	-	-	10.26(e)	-	-	-	-
Co^{2+}	110	4.25(e)	5.95(a)	4.51(a)	4.56(a)	4.04(a)	8.00(f)	-
	112	-	-	-	-	-	-	-
	120	8.18(a)	10.23(a)	8.01(a)	7.86(a)	7.32(a)	14.20(f)	-
	124	-	-	-	-	-	-	-
	130	-	-	9.96(a)	-	11.41(e)	-	-
Ni^{2+}	110	5.47(a)	7.15(a)	6.15(e)	5.60(a)	5.56(e)	8.7(f)	10.63(a)
	112	-	-	-	-	-	20.16(f)	22.97(a)
	120	9.99(a)	12.39(a)	11.16(e)	9.76(a)	10.28(e)	-	-
	124	-	-	-	-	-	-	-
	130	-	-	14.55(e)	-	13.82(e)	-	-
Cu^{2+}	110	7.98(a)	8.89(a)	7.83(a)	8.33(a)	7.76(a)	-	-
	112	-	-	-	-	-	-	-
	120	14.66(a)	15.93(a)	14.36(a)	14.84(a)	14.23(a)	-	-
	124	-	-	-	-	-	-	-
Zn^{2+}	110	4.53(f)	5.58(b)	5.07(e)	4.49(b)	4.83(e)	9.19(a)	-
	112	-	-	-	-	-	-	-
	120	8.38(f)	10.1(b)	9.43(e)	8.25(b)	9.17(e)	18.18(a)	19.40(a)
	122	-	-	-	-	-	-	-
	130	-	-	12.30(e)	9.8(b)	11.84(e)	-	-
Hg^{2+}	110	11.7(a)	-	-	-	-	37.8(a)	16.1(a)
	120	18.7(a)	-	-	-	-	44.0(a)	-

Table 3

Hard and Soft Acids and Bases Classification of Metal Ions and some Ligands[a]

	Hard	Borderline	Soft
Metals	Li^+, Na^+, K^+ Be^{2+}, Mg^{2+}, Ca^{2+} Cr^{3+}, Fe^{3+}, Al^{3+}	Mn^{2+}, Fe^{2+}, Co^{2+} Ni^{2+}, Cu^{2+}, Zn^{2+}	Pd^{2+}, Pt^{2+}, Pb^{2+} Hg^{2+}
Ligands	F^-, OH^-, H_2O NH_3, CO_3^{2-} O^{2-}, PO_4^{3-}	N_2, Cl^-, Br^-	CN^-, I^-, CO R_2S, C_6H_6

[a] taken from Pearson, R.G. (1973). *Hard and Soft Acids and Bases*, Dowden, Hutchinson and Ross, Stroudsburg.

from which

$$\log K = \log \beta_{ML} - pK_{a(LH)} \qquad (4)$$

Examination of the equilibrium data is this way, for example with Co(II) and Ni(II), then reveals that Arg, Lys and Trp are bidentate (i.e. glycine-like), Glu^{2-} and Gln^- display weak tridentate behavior and, that His^-, Cys^{2-} and Pen^{2-} are most obviously strongly tridentate.

For the kinetically inert metal ions where stability constants are difficult to obtain, the evidence for tridentate behavior comes from spectroscopic measurements, in particular NMR spectroscopy. A classic example of this is the early work by McDonald and Phillips (1963) who used NMR line-broadening measurements to demonstrate tridentate coordination of His to Co(II) ions. Circular dichroism (CD) is of particular value with chiral ligands because of its sensitivity to both the ligand conformation and the configuration at the metal center. A method which might more accurately reflect conditions at all membrane surfaces is the microemulsion technique which allows the study of specific interactions between amino acid side-chains and metal ions (Smith et al., 1978).

Since these potentially tridentate amino acids are of such importance, they will now be considered individually, except for Cys and Pen which are covered in more detail by Sovago in Section 3.

1. *Histidine*

Earlier reviews of the complexation of histidine were given by Martin (1979) and Sundberg and Martin (1974). The imidazole ring **(4)**

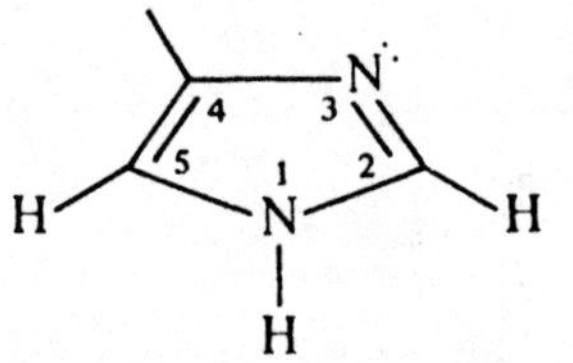

Scheme **(4)**

is an important ligand for metal ions, particularly in proteins. The pyrrole N(N-1) has a pK_a of ca 14 whilst the pyridine N(N-3) has a much lower pK_a near 7 and hence is mainly deprotonated at physiological pH. Chelation to give tridentate behavior occurs therefore via the N-3 atom as depicted in **(5)**.

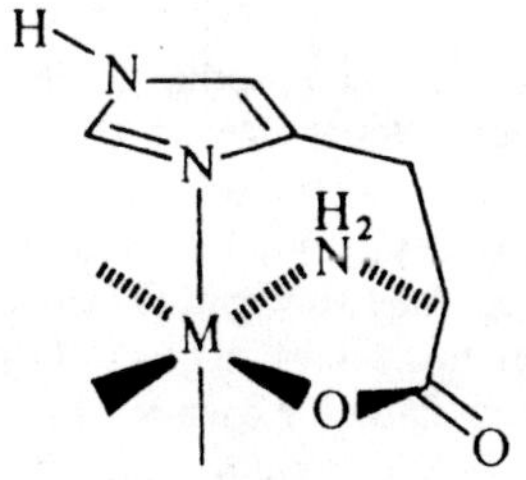

Scheme **(5)**

Tridentate behavior has been established in both solution and the solid state for a number of metal ions, e.g. Cr(III), Co(II), Co(III), Ni(II), Zn(II), Cd(II) and Mo(V). The neutral $Cu(His)_2$ complex has been identified from computer simulation studies as the second most important low-molar-mass Cu species in blood plasma (Berthon et al., 1986). It is also used in the treatment of Menkes' disease. The enhanced stability (see Table 2) of this species in solution is consistent with tridentate coordination, as in **(5)**. The stereochemical requirement for Cu(II), however, is for a tetragonally distorted geometry with weak axial coordination and a predominant square-planar bonding of four ligand atoms. Explaining this dichotomy and the nature of the $Cu(His)_2$ species has taxed the minds of a number of research workers! An extensive array of spectroscopic techniques have been used: UV/VIS, CD, ESR, ^1H and ^{13}C NMR. The consensus of opinion from these studies is that at neutral pH $Cu(His)_2$ is a mixture of linkage isomers which are in rapid equilibrium with each other (see, e.g. Pasenkiewicz-Gierula et al., 1987). The likely isomers are shown in Figure 2. None of the techniques can distinguish between *cis* and *trans* isomers in solution, nor can they establish the nature of the axial coordinating ligands.

With heavier metal ions, other modes of coordination have been observed, e.g. metal-carbon (imidazole) bonding for Hg(II) and Ru(II), and imidazole bridging of Ru centers through the two N atoms. This latter observation may be of relevance to spectroscopic changes observed for many metal histidine solution at the high pH region in which the N-1 proton **(4)** dissociates (Gulka and Isied, 1980).

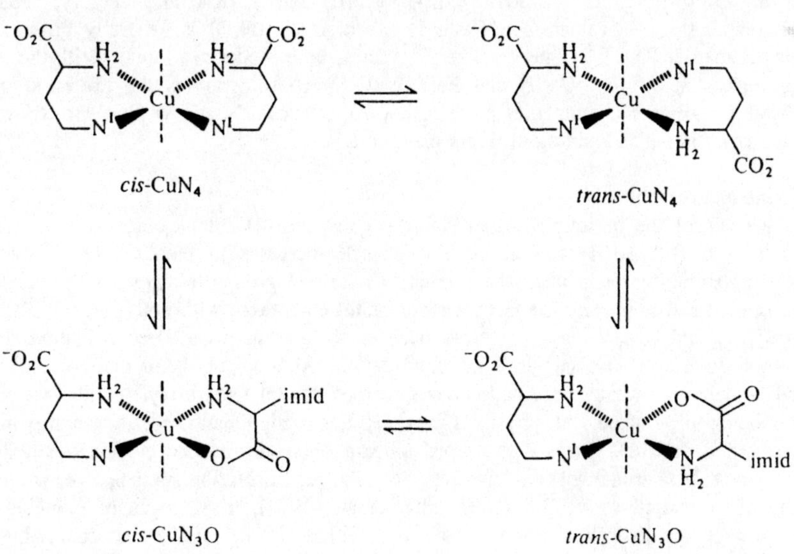

Figure 2.
Proposed forms (linkage isomers) of $Cu(L\text{-}His)_2$ in neutral aqueous solution; N^I = imidazole N-3 atom.

2. *Aspartic and glutamic acids*

Martin (1979) has earlier reviewed the complexing ability of these two acidic amino acids. Not surprisingly, there is evidence for tridentate coordination of Asp^{2-} with transition metals in solution, but not for Glu^{2-} which has an extra methylene group in its side-chains making chelation through its β-carboxyl group less favorable. The group can, however, lead to polynuclear complex formation. The two negative carboxyl groups make these two amino acids attractive ligands for other metal ions, e.g. the lanthanides and the group II metal ions. In fact with Mg(II) there is a claim for chemotherapeutic properties, the Asp and Glu 1:1 complexes being thought to be able to cross membrane barriers including the blood-brain barrier.

3. *Asparagine and glutamine*

From a comparison of the basicity-adjusted stability constants (4), Martin (1979) has concluded that at neutral pH the γ-amide oxygen of Gln^- only weakly interacts with Co(II) and Ni(II), but not at all with Cu(II). With Asn^- the interactions are much stronger, particularly with Cu(II) and Ni(II). However, NMR measurements indicate no coordination of the Asn^- amide group with Cu(II) at neutral pH (Lomozik and Wojeiechowska, 1989). No such coordination has been found either in solid $M(Asn)_2$ compounds, where M=Cd(II), Cu(II) or Ni(II). At pH>9.5, solutions of Cu(II) and Asn undergo further deprotonation with corresponding spectroscopic changes suggesting tridentate behavior. No such changes are observed with Gln. With $Pd(en)^{2+}$ as the coordinating center, bidentate coordination of both Asn (at pH ca 6.4) and Gln (at pH ca 9.0) occurs via the N(amino) and N^- (amide) atoms.

4. *Dihydroxyphenylalanine (DOPA)*

L-Dopa is included because of its biological and therapeutic importance, and this may well be linked to its versatile coordination behavior. The coordination properties were earlier reviewed by Gergely and Kiss (1979). The various potentiometric and spectroscopic studies show that DOPA chelates either as a normal bidentate amino acid (amino N, carboxyl O atoms) or through the two phenolate oxygen atoms (O,O); the ligand is therefore classed as being ambidentate. Either (N,O) or (O,O) coordination can occur or a mixture of species of the two types - depending upon the metal ion and the solution pH. Rearrangement (linkage isomerism) from (N,O) to (O,O) has been observed, the tendency for rearrangement following the order Cu(II) ~ Zn(II)>Co(II)>Mn(II), Ni(II), (Kiss and Gergely, 1983). In keeping with the pK_a values (see Table 1) chelation via (N,O) is generally favored at a lower pH than is (O,O); at intermediate pH values, both modes are found. With the 'hard' metal ions, e.g. Mg(II), Ca(II) and Fe(III), (O,O) coordination is the preferred mode. Predictably, polymeric complexes are also known, particularly at low DOPA:metal ratios, the two chelating ends bonding different metal ions.

5. *Serine and threonine*

On the basis of the basicity-adjusted stability constants (4) complexation of these amino acids with Co(II), Ni(II) and Zn(II) show small increases in thermodynamic stability, suggesting some involvement of the side-chain hydroxyl groups (Martin, 1979). However, these same metal ions can also form tris(bidentate) complexes with both Ser and Thr with normal β_{120} and β_{130} values. There has been much interest in Cu(II) coordination since a report of Cu(His)(Thr) occurrence in blood serum. At low pH, Pettit and Swash (1976) found evidence of weak coordination via O(carboxyl) and O(hydroxyl) atoms, the amino group being protonated. At physiological pH, the predominant species are the neutral $Cu(Ser)_2$ and $Cu(Thr)_2$ forms with normal bidentate behavior and no evidence of hydroxyl involvement. Numerous workers have reported changes in solution spectroscopic properties as the pH is raised above 9.5 (Grenouillet et al., 1973). These changes coincide with ionization of the hydroxyl groups. However, Gillard et al. (1980) also found high-pH spectroscopic changes for $Cu(aa)_2$ complexes with the normal bidentate ligands Ala, Pro and Val.

C. Ternary complex formation

Ternary complex formation involving chelation of two different amino acids to the same metal ions is to be expected in biological fluids. Statistically such complexes are favored over related binary complexes. Frequently their formation is greater than that expected on statistical consideration. Computer simulation studies point to complexes of the type M-His-aa, M=Cu(II),Zn(II) being major species of these metal ions in solutions such as blood serum and synthetic nutrition fluids (Berthon et al., 1984a; 1984b). Evidence from chromatograghic measurements points to ternary complexes M(II)-aa-nucleotide as also being of importance in biological fluids. Formation constants have been determined for Cu(II)-aa-EDTA** ternary systems and point to the formation of such species during EDTA administration (e.g. in chelation therapy). The stability of these species shows the importance of considering such ternary complexes in assessing the behavior of any chelating drugs (Arena et al., 1985). The enhanced stabilities in solution of some ternary complexes, mentioned above, frequently occur when the ligands possess aromatic side-chains, e.g. imidazole (His) or indole (Trp). Surprisingly perhaps for aqueous media, the enhanced stabilities can be attributed to favorable hydrophobic interactions between the π-aromatic systems, often referred to as ring-stacking (Sigel, 1985). These interactions can also be detected by CD and NMR techniques. There have been many investigations of such interactions involving amino acids as ligands, principally M-$(aa)_1$-$(aa)_2$; M-(aa)-1,10 Phenanthroline (or bipyridyl); and M-(aa)-nucleotides systems.

All the work referred to in this section involves the divalent transition metal ions and Zn(II). Because of the labile nature of their equilibria, such species need to be studied in situ without disturbance of the equilibria. With the kinetically inert metal ions, particularly Co(III) and Pt(II), ternary complexes are more familiar and can be easily obtained as solids. For example, numerous ions of the types $[Co(aa)(amine)]^{2+}$ and $[Pt(aa)(amine)]^{+}$ have been well characterized where (amine) represents a number of neutral amine ligands.

III. REACTIONS OF COORDINATED AMINO ACIDS

By coordination to a metal ion, the properties of amino acids are altered and can be exploited, e.g. for peptide bond formation, for carrying out reactions under less drastic conditions and therefore avoiding problems such as racemization, and for stereochemical control. This latter aspect will be dealt with by Pettit in the following chapter. In this section, the reactions covered are those which occur in solution. For earlier reviews in this area, see Laurie (1987).

A. Kinetics and mechanisms of formation reactions

Determination of the rate constants for the formation of divalent metal ion-amino acid complex formation requires fast reaction techniques, such as relaxation measurements. These measurements show that, as found in general for bidentate complex formation, a two-step mechanism occurs in which the initial monodentate coordination (5) is followed by ring closure (6)

$$(H_2O)_n\,M + A^{\frown}B \rightleftharpoons (H_2O)_{n-1}\,M{:}A^{\frown}B + H_2O \qquad (5)$$

$$(H_2O)_{n-1}\,M{:}A^{\frown}B \rightleftharpoons (H_2O)_{n-2}\,M\left\langle {}^{A}_{B} \right) + H_2O \qquad (6)$$

With Cu(II) and Ni(II), it has been demonstrated that the zwitterionic forms of the amino acids (protonated amino group) are extremely unreactive, indicating that the initial coordination in (5) must involve the amino N atom. This may not be generally true and certainly with the thiol-containing amino acids the evidence points to initial coordination via O or S donor atoms (Letter and Jordan, 1975).

With the more kinetically inert metal ions, more conventional and informative techniques can be used to follow reactions. Using ^{195}Pt NMR, for example, Appleton et al.

** EDTA = ethylenediaminetetra-acetate ion

(1986) have revealed interesting differences between glycine and other amino acids (even alanine) on chelation to Pt(II) centers. Thus, Hgly reacts with $[Pt(NH_3)_2(H_2O)_2]^{2+}$ to initially give the O-bonded complex followed by rapid ring closure and H^+ elimination (7)

$$[Pt(NH_3)_2(H_2O)_2]^{2+} + HGly \rightarrow [Pt(NH_3)_2(H_2O)(O\text{-}GlyH)]^{2+} + H_2O$$
$$\downarrow$$
$$[Pt(NH_3)_2(Gly)]^{+} + H^{+}_{aq} \qquad (7)$$

In contrast, other amino acids gave mixtures of the O-bonded ions $[Pt(NH_3)_2(H_2O)(O\text{-}aaH)]^+$ and $[Pt(NH_3)_2(O\text{-}aaH)_2]^{2+}$.

Co(III) centers can be used to promote the synthesis of amino acids via imine formation between a coordinated aminate ion and a 2-keto acid followed by reduction (Lawson et al., 1982). The steric and electronic effects of the amino acids side-chains on the formation of complexes with Co(III) have been reviewed by Jursik and Hajek (1987). Co(III) complexation of amino acid derivatives can also be used in the synthesis of peptide bonds, a problem in this erstwhile important reaction is the appreciable degree of racemization in the peptide products (Mensi and Isied, 1986).

B. Activation of αC-H bonds

The electron-withdrawing effect of the metal ion causes enhancement of the acidity of the α C-H bond of coordinated amino acids. The more important consequence of this is that *most* transition metals, except Ni(II), increase the rate of racemization of amino acids under basic conditions. With Co(III) complexes, ^1H NMR shows that the rate of H - D exchange at the αC is similar to that of the rate of racemization, a result in keeping with the formation of a symmetrical carbanion intermediate, as shown in Figure 3. With $Cu(L\text{-}Ala)_2$, however, the rates of racemization and H - D exchange differ, and a different mechanism is advocated involving a Schiff base intermediate (Gillard et al., 1977). Surprisingly, while metal ions such as Cr(III), Co(III), Cu(II), Pd(II) and Pt(II) have been shown to accelerate the racemization of amino acids under basic conditions, Ni(II) retards this reaction; a result confirmed by two different groups (Reddy et al., 1989; Khatib and Engel, 1989).

One of the earliest demonstrations of the activation of the αC-H was the aldol condensation reaction of glycine, coordinated to Cu(II), with acetaldehyde to yield threonine. The reaction which is base-catalyzed proceeds under far milder conditions than for free glycine. It has also been observed with other metal ions and aldehydes; in all cases, the intermediate appears to be a carbanion.

C. Hydrolytic reactions

Through coordination to a metal ion, the catalytic hydrolysis of peptide bonds and amino acid esters are accelerated; enhancement of reaction rates of up to 10^7 over the metal-free systems have been reported. As Hay and Morris (1976) pointed out in their earlier review, the problem with the kinetically labile systems is establishing the nature of the reaction intermediates in solution. With the kinetically inert systems, e.g. Co(III), Pd(II), Pt(II), such problems do not generally arise and, consequently, studies of these metal ions have been the more popular.

Amino acid esters may coordinate as monodentate (via the amino N) or as bidentate (N,O) ligands. In the latter case, polarization of the ester bond contributes to the accelerated hydrolysis. A likely mechanism is shown in Figure 4. Earlier work had established (by ^{18}O labelling) that the reaction proceeded without opening of the ester chelate ring. When the

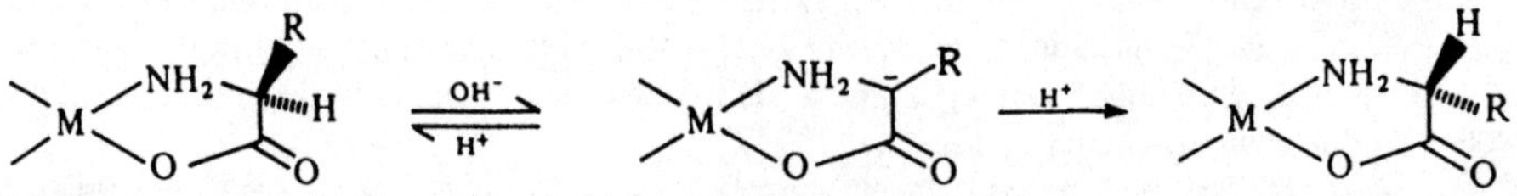

Figure 3.
Mechanism for the racemization of a coordinated amino acid via a symmetrical carbanion intermediate.

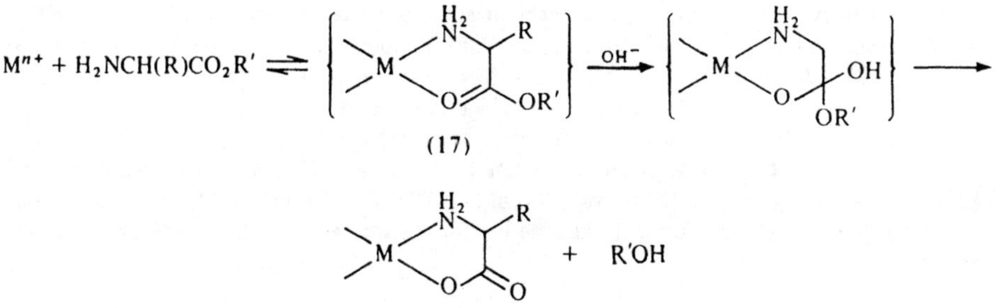

(17)

Figure 4.
Mechanism for the hydrolysis of a coordinated amino acid ester by an internal nucleophilic attack.

hydroxy nucleophile is also coordinated to the metal ion, then intramolecular nucleophilic attack (8) occurs to give rate enhancements of the order 10^7 - 10^{11}. These systems are good models for metalloenzyme-catalyzed reactions.

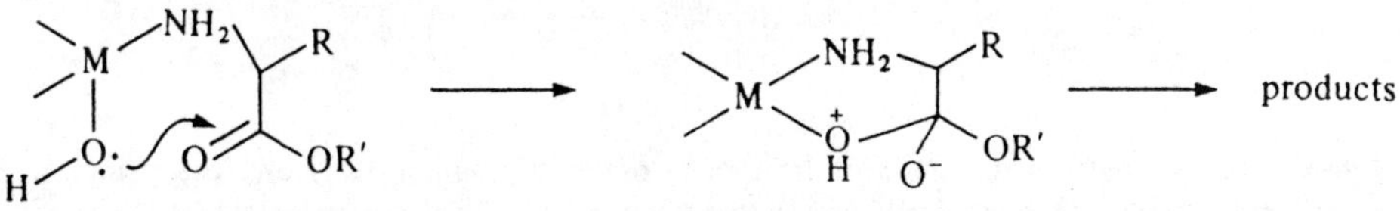

Coordination of O-acetyl or O-sulphonyl serine to Co(III) accelerates their hydrolysis by a factor of 10^7 (Chong et al., 1985). In this, elimination of ROH occurs (R= CH_3CO or SO_3) to give a chelated 2-iminopropanoate (6).

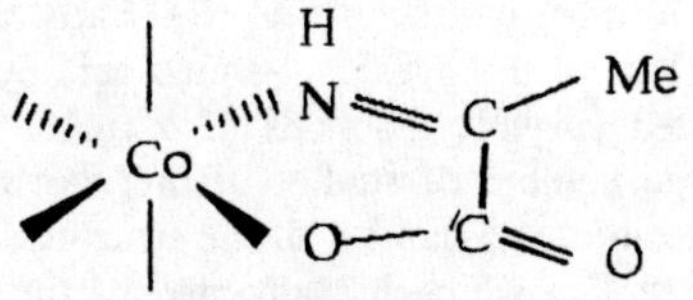

Scheme (6)

D. Oxidation-reduction reactions
There is a wide range of such reactions involving metal amino acid complexes either as reactants or as reaction intermediates. The more familiar of these and biologically important is the rapid oxidation of cysteine to the disulphide cystine. This oxidation is carried out by a number of high oxidation-state metal ions, e.g. Mo(VI), Fe(III), Cu(II). The stoichiometry of the reaction with Cu(II) is shown in (9)

$$2\ Cu^{2+}{}_{aq} + 4\ CysSH \rightarrow 2\ Cu^{I}CysS + CysS\text{-}SCys + 4\ H^{+}{}_{aq} \qquad (9)$$

where CysSH represents the zwitterionic form of cysteine. More information on this and related reactions is given by Sovago in the following chapter. The autocatalytic oxygenation of several amino acids in the presence of metal ions is well known. Of particular interest is the oxygenation catalysis by Fe(III) salts in solution of L-DOPA. On the other hand, the reaction is inhibited by Cu(II) ions. This difference is related to the different modes of bonding of these metals to the L-DOPA ion (see II.B.4). Many metal salts and their aqueous amino acid complexes catalyze the disproportionation of hydrogen peroxide and superoxide ion, reminiscent of the metalloenzymes catalase and superoxide dismutase. Studies on the catalyzes by the complex ions $[Cu(aa)L]^+$, L = bipyridyl or 1,10-phenanthroline, suggest that a critical step in the reactions is the exchange of the amino acidate ligand with H_2O_2 or O_2^{-}.

The ability of many Co(II) complexes in solution to reversibly absorb molecular O_2 has interested many groups for some years. One requirement is that the ligands contain three or more donor N atoms. This is satisfied by Co(L-His)$_2$, whose solutions rapidly absorb a half mole of O_2 to give a peroxo-bridge intermediate $(His)_2Co^{III}\text{-}O_2^{2-}\text{-}Co^{III}(His)_2$. Oxidation of amino acids under more forcing conditions can be accomplished, such as the oxidation of L-ornithine, side-chain R = $-(CH_2)_3{}^+NH_3$, with CuO to the corresponding α-keto acid with cyclization (Yorkovsky et al., 1989) and the oxidation of amino acids coordinated to Co(III) with thionyl chloride in dimethylformamide solutions (Hammershi et al., 1988).

E. Schiff base formation

In this context, Schiff-base formation results from the condensation of a reactive carbonyl group with the amino group of an amino acid. The formation of these bases proceeds readily in neutral aqueous solution in the presence of a coordinating metal ion, as represented in (10),

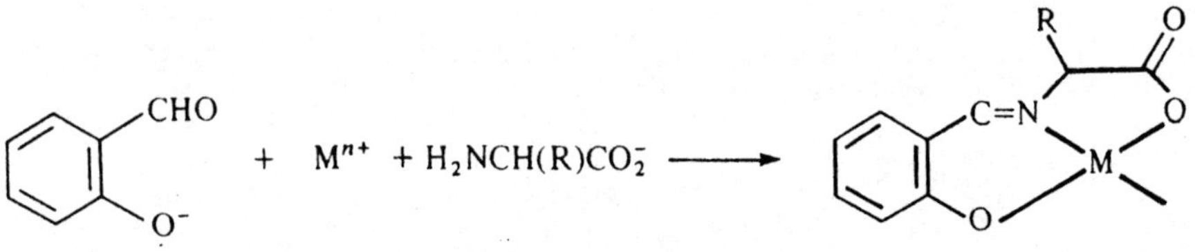

where M can be a wide range of transition- or main-group metal ions. The principal carbonyl reagents of biological interest that have been studied are salicylaldehyde, as in reaction (10), pyridoxal and pyruvic acid.

The interest in these compounds stems from the observations, made some 50 years ago by Pfeiffer, that some reactions of amino acid esters (notably ester exchange, racemization and oxygenation) are effected very readily when their Schiff bases with salicylaldehyde are complexed to transition metal ions, particularly Cu(II). Early work also established a close similarity between the reactions of the pyridoxal-amino acid Schiff base complexes and the pyridoxal phosphate-promoted enzymic reactions of α amino acid metabolism. Since that time there have been a large number of studies of the above reactions involving a wide range of Schiff bases and metal ions, a number of the structures of such compounds are also known (Laurie, 1987). A reaction which has attracted particular attention is non-enzymic transamination, as outlined in (11).

$$R^1CH(NH_2)COO^- + R^2COCOO^- \xrightarrow{M^{n+}} R^1COCOO^- + R^2\,CH(NH_2)COO^- \qquad (11)$$

The mechanisms of such reactions have been delineated by the work of Holm and his group (e.g. Weinstein and Holm, 1972). Particular use was made of NMR spectroscopy. Their work shows the key factors to be the formation of a Schiff base intermediate which undergoes tautomerism between ketoimine (7) and aldimine (8) forms.

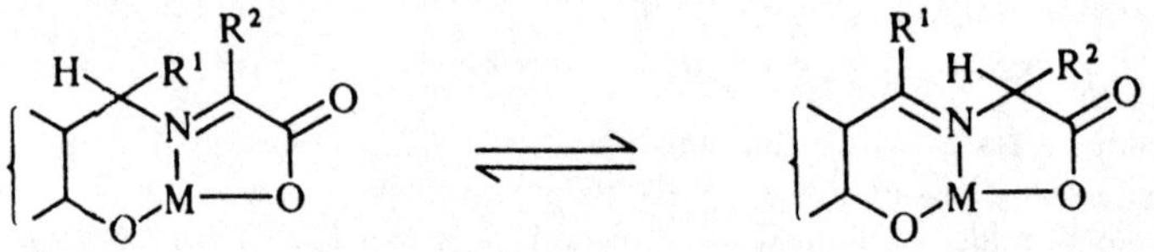

Schemes (7) ⇌ (8)

Synthetic applications, e.g. synthesis of other amino acids by alkylation of glycine, are well known. More recently this has been used to prepare the pure S-chiral isomers of Ala, Val and Trp by means of a chiral Schiff base (Belokon et al., 1986). In this and all above reactions, a key feature is the activation of the α-CHR group of the amino acid.

Schiff base complexes formed from the tridentate amino acids are ambidentate. With histidine, for example, only with pyruvate of those mentioned does the His bond through N(amino) and N-3 (imidazole); with the others, glycine-like coordination occurs. These different forms can readily be distinguished by CD spectroscopy. Histidine appears to be more effective than other amino acids in promoting transamination.

IV. STRUCTURAL ASPECTS

Much of the definitive structural information, particularly as regards stereochemistry, bond lengths and bond angles, comes from the analysis of X-ray diffraction of single crystals. A large number of structures of metal-amino acid complexes are now known; encompassing compounds of main group, lanthanides, actinides, as well as the transition-metal group. Structure reports prior to 1967 were reviewed by Freeman (1967) who also presented a detailed analysis of the geometries adopted. His comments are still of relevance today. A later compilation is given by Laurie (1987). For recent reports, the Royal Society of Chemistry's Specialist Report, mentioned earlier, is a very useful source. A technique now available for obtaining information on a metal ion's chromophoric group, i.e. the nature and number of coordinating atoms and the metal-atoms bond lengths, is that of extended X-ray absorption fine structure spectroscopy (EXAFS). This technique has the advantage that it does not require single crystalline material and may also be used with solutions.

Correlation between the solid-state structures and those in the corresponding solution phases is generally good for the kinetically inert metal ion complexes. However, care is needed with the more labile systems because the solid complex isolated from solution may only be a minor species in equilibrium in that solution. This is merely an example of Le Chatelier's principle in practice. Furthermore, the species present in the concentrated solutions required for crystallization may not necessarily be the same as those present in the more dilute solutions of physiological relevance. A recent example of the disparity between such solution and solid phases is the report of the crystal structure of $[Cu(Gly)(NO_3)(H_2O)]$. No such species has been reported in solution (Davies et al., 1992). In other cases, the agreement between the two phases is good. For example, the ternary complex [Cu(L-His)(L-Thr)] was initially reported as being an important species in human blood serum. It was then established as being a major species in aqueous solutions containing Cu(II), L-His and L-Thr at neutral pH. From the aqueous mixture, the crystalline complex [Cu(L-His)(L-Thr)] was isolated and found to have the same (3N,O) donor atoms as predicted in solution from spectroscopic measurements. A more unusual example comes from the complex formation between Cu(II) and D-Pen. The species causing the intense violet color in aqueous solution was unknown until the crystal structure of a violet compound isolated from aqueous solution was reported. The structural analysis showed the violet-colored species to be the, then unusual, mixed-valence ion $[Cu^{I}_{8}Cu^{II}_{6}(D\text{-}pen)_{12}Cl]^{5-}$. Subsequent studies showed the same species to be present in neutral aqueous solutions (Laurie and Prime, 1979).

Structural techniques which can be applied to aqueous solutions and have been used in attempts to define metal-amino acid structures are ESR with isotopic enrichment (Goodman and McPhail, 1985), ^{13}C NMR relaxation (Henry et al., 1985), as well as more general techniques such as CD.

Another driving force for many structural investigations is the possibility of physiological activity or diagnostic use. There have been numerous mixed ligand complexes of Pd(II) and Pt(II) reported and characterized following the discovery of the antitumor activity of cis-platin, cis-$[Pt(NH_3)_2Cl_2]$. However, there is no evidence that amino acid ligands have any beneficial effects. Ruthenium compounds are of interest as metallopharmaceuticals because of the availability of the isotopes ^{97}Ru, ^{103}Ru and ^{106}Ru. Compounds of the type $[CpRuL]PF_6$ have been isolated and characterized, Cp = cyclopentadienyl, L = amino acid or derivative; the L-Ala containing compound has been evaluated as a pancreatic imaging agent. Likewise, amino acid chelates of Tc have been studied for their ability to target ^{99m}Tc to specific organs (Chattopadhyay and Banerjee, 1988).

REFERENCES

Appleton, T.G., Hall, J.R., and Ralph, S.F. (1986). *Aust. J. Chem.*, *39*: 1347.

Arena, G., Musumeci, S., Rizzarelli, E., and Sammartano, S. (1985). *Trans. Met. Chem.*, *10*: 399.

Belokon, Y.N., Chernoglazova, N.I., Garbalinskaya, N.S., Saporovskaya, M.B., Kochetkov, K.A., and Belikov, V.M. (1986). *Izv. Akad. Nauk. SSSR, Ser. Khim.*, 2340.

Berthon, G., Piktas, M., and Blais, M.-J. (1984a). *Inorg. Chim. Acta*, *93*: 117.

Berthon, G., Blais, M.-J., Piktas, M., and Houngbossa, K. (1984b). *J. Inorg. Biochem.*, *20*: 113.

Berthon, G., Hacht, B., Blais, M.-J., and May, P.M. (1986). *Inorg. Chim. Acta*, *125*: 219.

Chattopadhyay, M., and Banerjee, S. (1988). *J. Inorg. Biochem.*, *34*: 25.

Chong, E.K., Harrowfield, J.M., Jackson, W.G., Sargeson, A.M., and Springborg, J. (1985). *J. Amer. Chem. Soc.*, *107*: 2015.

Davies, H.O., Gillard, R.D., Hursthouse, M.B., Mazi, M.A., and Williams, P.A. (1992). *J. Chem. Soc. Chem. Commun.*, 226.

Freeman, H.C. (1967). *Adv. Protein Chem.*, *22*: 257.

Gergely, A., and Kiss, T. (1979). *Metal Ions in Biological Systems*, Vol. 9 (H. Sigel, ed.), Marcel Dekker, New York, p. 143.

Gillard, R.D., O'Brien, P., Norman, P.R., and Phipps, D.A. (1977). *J. Chem. Soc., Dalton Trans.*, 1988.

Gillard, R.D., Lancashire, R.J., and O'Brien, P. (1980). *Trans. Met. Chem.*, *5*: 340.

Goodman, B.A., and McPhail, D.B. (1985). *J. Chem. Soc., Dalton Trans.*, 1717.

Grenouillet, R., Martin, R.-P., Rossi, A., and Ptak, M. (1973). *Biochim. Biophys. Acta*, *322*: 185.

Gulka, R., and Isied, S.S. (1980). *Inorg. Chem.*, *19*: 2842.

Hammershi, A., Hartshorn, R.M., and Sargeson, A.M. (1988). *J. Chem. Soc. Chem. Commun.*, 1226.

Hay, R.W., and Morris, P.J. (1976). *Metal Ions in Biological Systems*, Vol. 5 (H. Sigel, ed.), Marcel Dekker, New York, p. 174.

Henry, B., Boubel, J.C., and Delpuech, J.J. (1985). *Polyhedron*, *4*: 1069.

Jones, J.H., ed. (1991). *Amino Acids and Peptides*, Royal Society of Chemistry Specialist Periodical Reports, London, Vol. 23.

Jursik, F., and Hajek, B. (1987). *Chem. Listy*, *81*: 941.

Khatib, A.A., and Engel, M.H. (1989). *Inorg. Chim. Acta*, *166*: 273.

Kiss, T., and Gergely, A. (1983). *Inorg. Chim. Acta*, *78*: 247.

Kiss, T., Sóvágó, I., and Gergely, A. (1991). *Pure Appl. Chem.*, *63*: 597.

Laurie, S.H. (1987). *Comprehensive Coordination Chemistry*, Vol. 2 (G. Wilkinson, R.D. Gillard and J.A. McCleverty, eds.), Pergamon Press, p. 739.

Laurie, S.H., and James, C. (1983). *Inorg. Chim. Acta*, *78*: 225.

Laurie, S.H., and Prime, D.M. (1979). *J. Inorg. Biochem.*, *11*: 229.

Lawson, P.J., McCarthy, M.G., and Sargeson, A.M. (1982). *J. Amer. Chem. Soc.*, *104*: 6710.

Letter, J.E., and Jordan, R.B. (1975). *J. Amer. Chem. Soc.*, *97*: 2381.

Lomozik, L., and Wojeiechowska, A. (1989). *Polyhedron*, *8*: 1.

Martell, A.E., and Smith, R.M. (1974). *Critical Stability Constants*, Plenum, New York, Vol. 1.

Martell, A.E., and Smith, R.M. (1982). *Critical Stability Constants*, Plenum, New York Vol. 5.

Martin, R.B. (1979). *Metal Ions in Biological Systems*, Vol. 9 (H. Sigel, ed.), Marcel Dekker, New York, p. 1.

McDonald, C.C., and Phillips, W.D. (1963). *J. Amer. Chem. Soc.*, *92*: 5365.

Mensi, N., and Isied, S.S. (1986). *Inorg. Chem.*, *25*: 147.

Pasenkiewicz-Gierula, M., Froncisz, W., Basosi, R., Antholine, W.E., and Hyde, J.S. (1987). *Inorg. Chem.*, *26*: 801.

Pettit, L.D. (1984). *Pure Appl. Chem.*, *56*: 247.

Pettit, L.D., and Swash, J.L.M. (1976). *J. Chem. Soc., Dalton Trans.*, 2416.

Reddy, G.S., Reddy, G.V., and Smith, G.G. (1989). *Inorg. Chim. Acta*, *166*: 55.

Sigel, H., ed. (1973-1992). *Metal Ions in Biological Systems*, Vols. 1-27, Marcel Dekker, New York.

Sigel, H. (1985). *Proc. Sch. Symp. Inorg. Biochem. Mol. Biophys.*, 57.

Sigel, H., Tribolet, R., and Scheller, K.H. (1985). *Inorg. Chim. Acta*, *100*: 151.

Sillén, L.G., and Martell, A.E. (1964). *Stability Constants*, Special Publication No. 17, The Chemical Society, London.

Sillén, L.G., and Martell, A.E. (1971). *Stability Constants*, Supplement No. 1, Special Publication No. 25.

Smith, G.D., Barden, R.E., and Holt, S.L. (1978). *J. Coord. Chem.*, *8*: 157.

Smith, R.M., Motekaitis, R.J., and Martell, A.E. (1985). *Inorg. Chim. Acta*, *103*: 73.

Sundberg, R.J., and Martin, R.B. (1974). *Chem. Rev.*, *74*: 471.

Weinstein, G.N., and Holm, R.H. (1972). *Inorg. Chem.*, *11*: 2553.

Yatsimirskii, K.B. (1986). *Russ. J. Inorg. Chem.*, *31*: 1.

Yorkovsky, Y., Kapon, M., and Dori, Z. (1989). *Inorg. Chim. Acta*, *158*: 115.

STEREOSELECTIVITY IN METAL-AMINO ACID AND OLIGOPEPTIDE COMPLEX FORMATION

by

Leslie D. Pettit

School of Chemistry,
The University of Leeds,
Leeds, LS2 9JT
U.K.

I. INTRODUCTION

Stereoselectivity in metal complex formation results from differences in the affinities of metal ions for chiral isomers of the ligand. Many biological systems involving metal ions approach stereospecificity in their interaction with chiral ligands whereas model systems, using smaller and simpler molecules, show much smaller differences. Amino acids and peptides are one of the most important groups of biologically active small chiral molecules which form stable complexes with a wide range of metal ions at biological pH; hence most work on model systems has been based on these ligands in either binary or ternary complexes.

By convention, chirality in amino acids (AA) and peptides is always defined in terms of L- and D-forms, rather than R- and S-forms. When a ligand has only one chiral center, there cannot be stereoselectivity in formation of either the proton complexes or the mono-complex with a metal ion. However, the bis-complexes, $[M(L\text{-}AA)_2]$ (or $[M(D\text{-}AA)_2]$) and [M(L-AA)(D-AA)] will be diastereoisomeric and differences in stability and reactivity are to be expected. Peptides in which the chirality of the component amino acid residues differ will themselves be diastereoisomeric (e.g. L-Ala-L-Ala and L-Ala-D-Ala). Hence differences are to be expected between both their protonation constants and their mono-complexes with metal ions.

Reviews of stereoselectivity, including explanations of its origin, have been published by Bernauer (1976) and Pettit (1979); therefore, this account will concentrate on more recent thermodynamic studies.

II. DEFINITIONS

A. Amino acid complexes

With a chiral amino acid, AA, a metal ion can form three possible bis-complexes: $[M(L\text{-}AA)_2]$, $[M(D\text{-}AA)_2]$ and [M(L-AA)(D-AA)] in addition to any geometrical or conformational isomers. The first two complexes will have identical stabilities while the third (a meso or mixed complex) may well have a different stability. If it is assumed that there is no intrinsic stabilization or destabilization in the meso complex, the bis-complexes would exist in the statistically expected ratios of 1:1:2 since the complex [M(L-AA)(D-AA)] can be formed from a mono-complex by two different routes, namely: M(L-AA) + D-AA $\rightleftharpoons$ M(L-AA)(D-AA) and: M(D-AA) + L-AA $\rightleftharpoons$ M(L-AA)(D-AA), while the optically pure species can only be formed from one mono-complex each (Advani et al., 1970). As a result, the equilibrium constant for the ligand exchange reaction:

$$0.5\ [M(L\text{-}AA)_2] + 0.5\ [M(D\text{-}AA)_2] \rightleftharpoons [M(L\text{-}AA)(D\text{-}AA)]$$

would have a value of K=2. Hence:

$$\beta[M(L\text{-}AA)(D\text{-}AA)] = 2\ \beta[M(L\text{-}AA)_2] \text{ or}$$
$$\log \beta[M(L\text{-}AA)(D\text{-}AA)] = \log \beta[M(L\text{-}AA)_2] + 0.30.$$

Thermodynamic stereoselectivity, expressed as $\Delta\log\beta$, may therefore be defined quantitatively as the deviation from this relationship. i.e.

$$\Delta\log\beta = \log \beta[M(L\text{-}AA)(D\text{-}AA)] - \log \beta[M(L\text{-}AA)_2] + 0.30.$$

Stability constants calculated from pH-metric titrations of solutions of racemic amino acids (equal concentrations of L-AA and D-AA) are macro-constants and cannot distinguish between the chiral bis-complexes. However it can be shown that:

$$\beta_2(\text{macro}) = 0.25\ \beta[M(L\text{-}AA)(D\text{-}AA)] + 0.5\ \beta[M(L\text{-}AA)_2].$$

This means that, in the absence of stereoselectivity, titration curves for a metal ion in the presence of a racemic amino acid would be expected to superimpose those of the optically pure amino acid while, in the case of total stereoselectivity (e.g. $\beta[M(L\text{-}AA)(D\text{-}AA)] = 0$), the value for log β(macro) calculated from the racemic mixture would be only 0.30 log units lower than that from the optically pure ligand.

B. Peptide complexes

Peptides which contain amino acids of differing chirality will be diastereoisomeric with the result that stereoselectivity in both protonation and metal complex formation constants is to be expected. With metal ions, the major species are mono-complexes with differing numbers of nitrogens coordinated; hence statistical factors do not present the complications found with bis-complexes of amino acids.

The above arguments only apply to labile complexes. Kinetic stereoselectivity, resulting from different activation energies, can mean that one particular chiral complex may form in preference to another. In some cases (e.g. many Co(III) complexes) this can result in almost complete stereospecificity. While these will not be covered in this review, stereoselectivity in the syntheses of several ternary complexes which include amino acids or peptides have been reported recently.

III. ORIGINS OF STEREOSELECTIVITY

With amino acids, it has been shown above that stereoselectivity is likely only in stepwise reactions: ML + L $\rightleftharpoons$ ML_2 and ML_2 + L $\rightleftharpoons$ ML_3, and that statistical factors will cause the stability constants of optically pure and *meso* complexes to differ as a result of different values for the entropy of formation. However, other factors could influence either the entropy or the enthalpy changes, or both. The two most significant factors are:

(i) Changes in the apparent coordination number as a result of altering the chirality of a ligand or, in the case of a peptide, of an amino acid residue within a ligand. This will tend to have most influence on the enthalpy of complex formation.

(ii) Changes in the complex-solvent interface (hydrophobic interactions). This will have most influence on entropy changes although the introduction of 'ring-stacking' or other inter-ligand interactions can also cause enthalpy changes. In practice, it is difficult to separate the enthalpy and entropy effects, and in most cases where stereoselectivity occurs, both contribute towards the overall effect.

Many amino acids are potentially tridentate but, initially, bind as bidentate ligands, especially with a metal ion such as Cu(II) which forms stable planar complexes. In a planar bis-complex of an amino acid, either *cis-* and *trans-*isomers are possible. If the two amino acids are of the same chirality, both side chains will be on the same side of the coordination

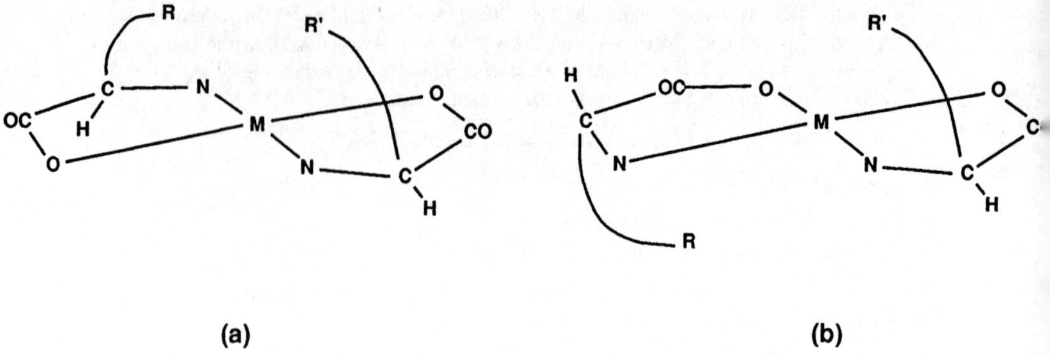

Figure 1.
Geometrical isomers of Cu(L-AA)(L-AA'); (a) *trans*-isomer, (b) *cis*-isomer.

plane in the *trans*-isomer and on opposite sides in the *cis*-isomer, while the reverse will be true in the *meso* complex. If the side chains can interact with the metal ion along the z-axis, then stereoselectivity is to be expected if the *cis*- and *trans*-isomers differ in stability. This is shown in Figure 1. The major complex formed will depend on the relative energy differences between the *cis*- and *trans*-isomers on one hand and the five and six coordinated complexes on the other, and steric control of the reaction will be governed by whichever is the larger.

In some ternary complexes, the side chains may be oppositely charged. While the negatively charged side chain may interact with the metal ion along the z-axis, the complex in which the positively charged side chain of the other amino acid can interact electrostatically with the negatively charged center will be stabilized. In this case, the effect will be the opposite to that described above, i.e. the *trans*-bis-isomer of the *meso* complex will be stabilized relative to the *cis*-isomer as shown in Figure 2.

It is not necessary for the side chains to be charged for side chain interactions to occur. It is found that side chains with hydrophobic groups tend to concentrate on the same side of the coordination plane although no classical bonding can be present. Hydrophobic interactions may result from both favorable changes in solvation energy and also the minimization of the energy of the hydrophobic region-water interface, and will tend to be largely entropy stabilized although some enthalpy contribution will always be present. Aromatic rings, in particular, frequently 'stack' one above the other, so encouraging stereoselectivity in favor of the chiral isomers which optimize these stacking interactions and contributing towards a favorable enthalpy change. Alternatively the π-orbitals of an aromatic ring can interact along the z-axis with the d-orbitals of a metal ion such as Cu(II). This has been demonstrated in a number of crystal structures and the resulting stabilization (caused by the favorable enthalpy change of this interaction) has been measured quantitatively to give a value of -9.5 kJ mol^{-1} (Xiao et al., 1990). These interactions will give the ligand a tridentate character and so lead to stereoselectivity.

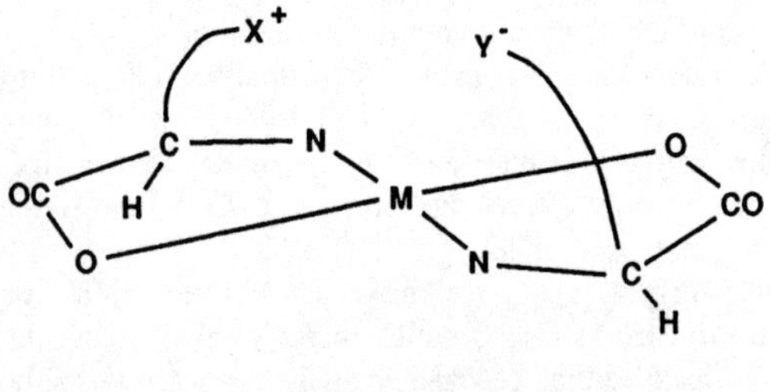

Figure 2.
The trans-isomer of Cu(L-AA)(L-AA') when the side chains have opposite charges.

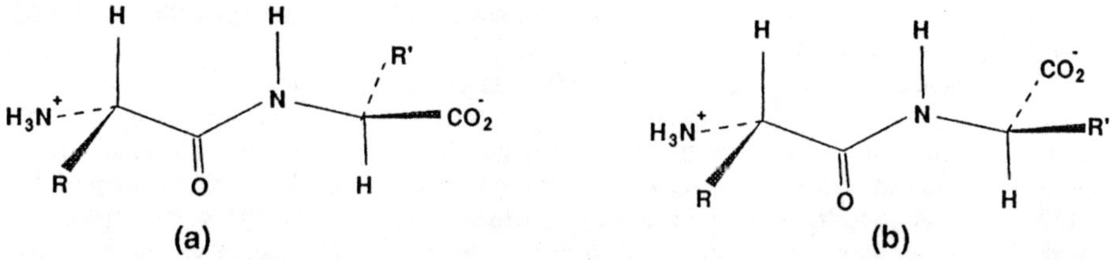

Figure 3.
Preferred β-conformations of neutral dipeptides; amino acids of (a) opposite chirality, (b) same chirality.

With peptides, stereoselectivity is to be expected in both proton and metal complex formation. As a simple example, consider protonation of the dipeptides L-Ala-L-Ala and L-Ala-D-Ala which, around neutral pH, exist as neutral zwitterions with the preferred β-conformations shown in Figures 3(a) and (b). The conformation (b) (the *meso* isomer) will be stabilized since the charged functional groups are both on the same side of the molecule, while they are on opposite sides in (a), the optically pure isomer. As a result, deprotonation of the $-NH_3^+$ and protonation of the $-CO_2^-$ groups are made more difficult in the *meso* isomer, since this would lead to the removal of the electrostatic stabilization. With Ala-Ala the stereoselectivity is about -0.8 kJ mol^{-1} in each case and the magnitude of the effect increases as the non-coordinating side chains increase in size (Bonomo et al., 1986).

With simple peptides, the initial coordination centers for metal ions are the terminal amino-N and the neighboring carbonyl-O, and no stereoselectivity would be expected. However, the peptide-N atoms are able to deprotonate and bind to metal ions such as Cu(II), thus involving more than one amino acid residue in the coordination scheme. A dipeptide can then behave as a tridentate ligand forming the complex $[CuLH_{-1}]$ bonded through both nitrogen donors and the carboxyl-O^-, with the peptide being held in a near planar conformation as shown in Figure 4. In a similar way, tetrapeptides and higher will give $[CuLH_{-3}]^{2-}$ complexes. In all these complexes, the side chains of optically homogeneous peptides will all be located on the same side of the coordination plane, and stereoselectivity would be expected when an amino acid residue is replaced by one of opposite chirality. It is generally greater than that found with amino acids.

IV. RECENT RESULTS

A. Thermodynamic studies

1. *Amino acids and peptides with non-coordinating side chains*

Stereoselectivity in the formation of bis-complexes of amino acids with non-coordinating side chains is generally too small to detect, although it has been reported in the formation of ternary Cu(II) complexes with aminophosphonic acids, where the formation of mixed complexes containing ligands of the same optical chirality is favored (Nabirkina et al.,

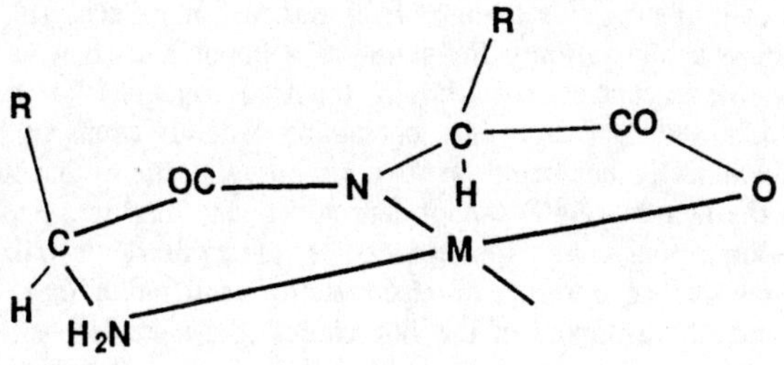

Figure 4.
The $[CuLH_{-1}]$ complex of L-Leu-L-Leu.

1990). With dipeptides, on the other hand, stereoselectivity is found in both proton and metal complex formation.

In protonation reactions, stereoselectivity can be up to 0.4 log units (Lyons et al., 1984; Bonomo et al., 1987). Calorimetric studies have shown that stereoselectivity in deprotonation of the $-NH_3^+$ group of simple *meso* dipeptides with small side chains (e.g. Ala-Ala) is almost entirely a result of enthalpy stabilization, while that of protonation of the $-CO_2^-$ group is a result of entropy stabilization (Bonomo et al., 1987). This is in agreement with the electrostatic stabilization expected to arise from the conformation shown in Figure 3(b). When larger side chains were introduced, other contributory factors were detected, particularly in enthalpy changes. These were interpreted as resulting from the algebraic sum of the contribution due to solvophobic (hydrophobic) interactions between the side chains and the contribution due to electrostatic interactions between the terminal functional groups.

With metal ions (e.g. Cu(II)) there is significant stereo-selectivity in the formation of the amide bonded complex $[CuLH_{-1}]$. The complex with a chiral ligand (e.g. L-Ala-L-Ala) is generally more stable than that formed with a *meso* ligand, and differences ranging up to 0.7 log units have been found with Leu-Phe. Calorimetric studies have shown this to be due to a more favorable enthalpy of formation of the LL complex as a result of hydrophobic interactions (Bonomo et al., 1986). Similar hydrophobic stabilization has been found in complexes with Ni(II) and Co(II) as well as with Cu(II), and the effects of aliphatic and aromatic side chains quantified (Xiao et al., 1990). It was found that the aromatic ring of a Phe residue in a dipeptide is only significant (contributing -9.5 kJ mol^{-1} to the enthalpy change) when it is the C-terminal amino acid.

The tripeptides Leu-Leu-Gly and Gly-Leu-Leu also show stereoselective stabilization, favoring the optically pure isomers as a result of hydrophobic interactions. The stabilization is lost, however, when the aliphatic side chains are separated as in Leu-Gly-Leu (Hanaki et al., 1988).

Stereoselectivity between Arg^8-vasopressin (AVP),

Cys-Tyr-Phe-Gln-Asn-Cys-Pro-Arg-Gly-NH_2
└── S ── S ──┘

and D-Val^4-Arg^8-vasopressin is particularly large. The D-Val residue replaces the L-Gln^4 residue in AVP and, since neither participate in the bonding, the change would not be expected to have a significant effect on coordination reactions. However, the 4N bonded complex of AVP with Cu(II) is particularly stable because the disulfur-bridged ring presents the four N-donors in a suitable conformation for coordination without significant conformational reorganization. Insertion of a D-amino acid residue destroys this favorable conformation with the result that the stability constant of the $[CuLH_{-2}]$ complex is 1.8 log units less than with AVP. With Arg^8-vasotocin (similar to AVP but with an Ile^3 residue) the stereoselectivity is 1.84 log units (a factor of 80) (Kozlowski et al., 1989).

2. *Peptides with S-containing side chains*

The S-donors of amino acid residues such as methionyl or S-Me-cysteinyl do not bind to Cu(II) to any significant extent, and formation of the $[CuLH_{-1}]$ complex of L-Met-L-Met has been shown to be favored enthalpically over that of the *meso* isomer as a result of hydrophobic interactions (Bonomo et al., 1987). It therefore behaves similarly to other peptides with non-coordinating side chains. Metal ions with more class (b) character such as Ag(I), however, have a high affinity for sulfur as a donor and so would be expected to behave differently. Significant stereoselectivity (up to 1 log unit) has been found in the Ag(I) complexes of a series of dipeptides containing S-donor atoms in both side chains, complexes with the *meso* ligands being favored; i.e. the opposite to that found with Cu(II). This is a result of the fact that Ag(I) cannot deprotonate and bind to a peptide-N, but binds strongly to the S-donors only. As explained earlier, the zwitterionic [HL] complex of the DL dipeptide is favored as a result of electrostatic stabilization (see Fig. 3) and this conformation presents the S-donors of the side chains in a suitable position for bidentate, near-linear coordination to give the complex $[AgLH]^+$. The preferred conformation of the LL isomer would have to be changed to allow chelation.

The $[CuLH_{-1}]$ complex of L-Met-L-Met has both $CH_3.S.CH_2.CH_2$- side chains on the same side of the coordination plane, without Cu-S binding, while they are on opposite sides with L-Met-D-Met. In the LL complex, both S-donors can coordinate to Ag(I) to form $[CuAgLH_{-1}]^+$ while in the DL complex this is impossible. As a result, stereoselectivity would be expected to be large and a value of over 1.2 log units has been found (Pettit et al., 1986).

3. *Histidine and peptides containing the His residue*

Ternary complexes of Cu(II) with histidine and other amino acids have been shown to form stereoselectively as a result of the electrostatic effects outlined above (Brookes et al., 1977). A recent calorimetric and potentiometric study of ternary complexes with amino acids which have non-coordinating side chains has shown that, when these side chains are aromatic, the ternary complexes are more stable when the ligands have opposite chirality, while the opposite is true when the side chains are aliphatic. This was shown to result from the tridentate nature of the histidinate ligand, and from hydrophobic interactions with the side chains. This stereoselectivity was used as a model to present a possible explanation of molecular recognition assisted by Cu(II) (Borghesani et al., 1990).

The His residue plays an even more important role in stereoselectivity in peptide complexes. The conformations of D-His-L-His and L-His-L-His in aqueous solution differ considerably, and stereoselectivity in the protonation constants calculated from NMR studies (0.48 log units, Tanokura, 1983) is almost identical to that calculated potentiometrically (0.44 log units, Livera et al., 1987). Stereoselectivity in the formation of Cu(II) complexes is large, as expected, with the formation of both monomeric and dimeric complexes, $[Cu_2L_2H_{-2}]$ complexes being the major species above pH 6.5, and the DL complex being more stable than the LL isomer. This stereoselectivity would be expected as a result of the preferred trans conformation of the peptide chain.

Ternary complexes of Cu(II) with amino acids and cyclo-(L-His-L-His) have also been studied using a combination of potentiometry, calorimetry and ESR spectroscopy (Arena et al., 1991). With L-amino acids, ternary complex formation was found to be more entropy

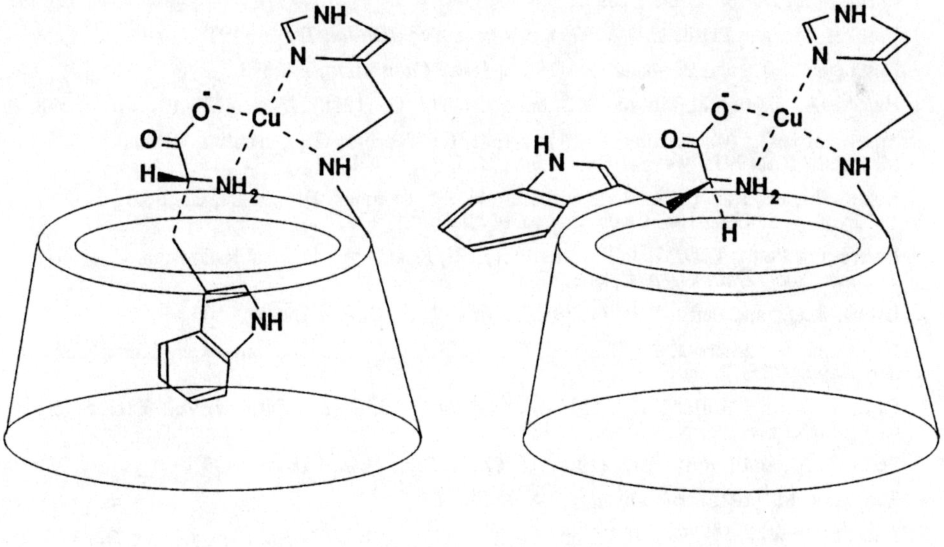

(a) (b)

Figure 5.
Model for the ternary complex of Cu(II)-(His-cyclodextrin)-tryptophan; (a) D-Trp, (b) L-Trp. After Impellizzeri et al. (1991).

and less enthalpy favored than with the analogous D-amino acids, the effect being greater with tryptophan than with phenylalanine. Enthalpy changes were found to be the driving force for the stereoselectivity found.

B. Optical resolution and stereoselective synthesis

Stereoselective interactions in metal complexes of amino acids may be used (i) for selective binding of one chiral isomer to achieve the resolution of a racemic mixture, or (ii) to perform an asymmetric synthesis by means of a stereoselective reaction in the coordination sphere of a mixed-ligand complex, and the principles behind these methods have been reviewed by Bernauer (1976). Since this review, several stereoselective syntheses have been reported using, for example, mixed-ligand Cu(II)-Schiff base complexes (Deschenaux et al., 1984).

Recent work using the Cu(II) complex of a histidyl-cyclodextrin has introduced a new class of media for the chromatographic resolution of chiral amino acids, particularly those containing aromatic side chains, with the discrimination process occurring in the aqueous mobile phase. Origin of the stereoselectivity is demonstrated in Figure 5 which represents the cyclodextrin as an annular hydrophobic cavity. The relative stability of the L- and D-tryptophan complexes is governed by the position of the aromatic side chain relative to the cavity, that with the D-isomer being favored (Impellizzeri et al., 1991).

REFERENCES

Advani, A.T., Irving, H.M., and Pettit, L.D. (1970). *J. Chem. Soc. (A)*, 2649.

Arena, G., Bonomo, R., Casella, L., Gullotti, M., Impellizzeri, G., Maccarrone, G., and Rizzarelli, E. (1991). *J. Chem. Soc., Dalton Trans.*, 3205.

Bernauer, K. (1976). *Topics Curr. Chem.*, *65*: 1.

Bonomo, R.P., Cali, R., Cucinotta, V., Impellizzeri, G., and Rizzarrelli, E. (1986). *Inorg. Chem.*, *25*: 1641.

Bonomo, R.P., Maccarrone, G., Rizzarrelli, E., and Vidali, M. (1987). *Inorg. Chem.*, *26*: 2893.

Borghesani, G., Pulidori, F., Remelli, M., Purrello, R., and Rizzarelli, E. (1990). *J. Chem. Soc., Dalton Trans.*, 2095.

Brookes, G., and Pettit, L.D. (1977). *J. Chem. Soc., Dalton Trans.*, 1918.

Deschenaux, R., and Bernauer, K. (1984). *Helv. Chim. Acta*, *67*: 373.

Hanaki, A., Ikota, N., Motono, K., and Yamauchi, O. (1988). *Nippon Kagaku Kaishi*, 578.

Impellizzeri, G., Maccarrone, G., Rizzarrelli, E., Vecchio, G., Corradini, R., and Marchelli, R. (1991). *Angew. Chem. Int. Ed. Engl.*, *30*: 1347.

Kozlowski, H., Radomska, B., Kupryszewski, G., Lammek, B., Livera, C., Pettit, L.D., and Pyburn, S. (1989). *J. Chem. Soc., Dalton Trans.*, 173.

Livera, C., Pettit, L.D., Bataille, M., Perly, B., Kozlowski, H., and Radomska, B. (1987). *J. Chem. Soc., Dalton Trans.*, 661.

Lyons, A.Q., and Pettit, L.D. (1984). *J. Chem. Soc., Dalton Trans.*, 2305.

Nabirkina, E., Ignateva, T., Raevskii, O., and Belov, Yu. (1990). *Izv. Akad. Nauk. SSSR, Ser. Khim.*, 755.

Pettit, L.D., and Hefford, R.J. (1979). *Metal Ions in Biological Systems*, Vol. 9 (H. Sigel, ed.), Marcel Dekker, New York, p. 173.

Pettit, L.D., and Lyons, A.Q. (1986). *J. Chem. Soc., Dalton Trans.*, 499.

Tanokura, M. (1983). *Biopolymers*, *22*: 2563.

Xaio, L., Jouini, M., Fan, B.T., and Lapluye, G. (1990). *J. Chem. Soc., Dalton Trans.*, 1137.

METAL COMPLEXES OF SULFUR-CONTAINING AMINO ACIDS AND DERIVATIVES

by

Imre Sóvágó

Kossuth Lajos Tudományegyetem,
Szervetlen és Analitikai Kémia Tanszék,
4010 Debrecen
Hungary

I. INTRODUCTION

Sulfur-containing amino acids are the most common metal binding sites in the various metalloproteins. This is demonstrated by the large number of iron-sulfur proteins, blue copper proteins and various zinc(II)-containing enzymes (e.g. horse liver alcohol dehydrogenases or zinc fingers). The outstanding role of the cysteinyl residue in metalloproteins is not surprising because transition metal ions and sulfur donor atoms have a high affinity for one another, as illustrated by the ready formation and high stability of metal sulfides. Consequently, a wide spectrum of biological phenomena is believed to be dependent on thiols and thiol derivatives. Two factors probably account for the outstanding role of thiol compounds. The first is the relative ease with which the reactions of this group can occur at neutral pH and moderate temperature. The second factor is the great variety of the possible chemical modifications of these compounds.

There are four different structural possibilities for the appearance of sulfur donor atoms in natural amino acids and derivatives:

- thioethers (methionine and S-methyl-L-cysteine),
- thiols (L-cysteine, D-penicillamine and derivatives),
- disulfides (L-cystine and D-penicillamine disulfide),
- thioamides (amino acids with thiocarbonyl groups).

II. METAL COMPLEXES OF AMINO ACIDS CONTAINING A THIOETHER GROUP

Metal complexes of thioether-containing ligands have already been extensively studied and the results are discussed in several review papers (McCormick et al., 1974; Martin, 1979; Decock and Kozlowski, 1985; Kiss, 1990). A literature survey clearly demonstrates that the thioether group is an extremely weak base (pK ~ -6.8), which does not provide a general basis for its metal ion coordination (Bonvicini et al., 1972). On the other hand, sulfur atoms are soft bases and thioether groups can readily coordinate to soft metal ions (e.g. silver(I), platinum(II), palladium(II), mercury(II), etc), whereas the amino acid locus is the main binding site for the divalent 3d-transition metal ions.

The naturally occurring amino acid L-methionine (Met) and a derivative of cysteine (S-methyl-L-cysteine, SMC) are the best studied examples for the coordination of thioether-containing amino acids. A comparison of the stability constants of aliphatic amino acids and Met or SMC clearly indicates that both SMC and Met coordinate primarily as substituted glycinates to the borderline transition metal ions. The presence of the thioether side chain, however, results in a slightly enhanced stability of the metal complexes. The stability increase is more pronounced in the case of SMC complexes, while it is almost negligible for the metal complexes of Met. This is due to the differences in the sizes of the chelate rings formed with the participation of the thiother sulfur and amino nitrogen donor atoms: fused 5-membered chelate rings can be formed with SMC, while the less favoured 6-membered ring is characteristic of Met. The weak apical coordination of the thioether group of SMC was first suggested by Espersen and Martin (1976), and it was supported by the fact that the thioether group does not prevent the formation of tris complexes with octahedral transition metal ions (Swash and Pettit, 1976; Sóvágó et al., 1986; 1987). On the basis of spectroscopic studies, direct coordination of the thioether group of Met was suggested to occur under certain experimental conditions (low pH, low temperature and non-aqueous solution) (Kozlowski and Kowalik, 1978; 1979). On the other hand, crystal structure studies gave no evidence of the interaction of the thioether group of SMC or Met with borderline metal ions (Dubler et al., 1986).

A large number of publications support the finding that soft metal ions readily form stable complexes with SMC or Met (see the rewiev by Decock and Kozlowski, 1985). Absorption spectra and crystal structure studies clearly indicate the formation of bis complexes with palladium(II) and platinum(II), in which sulfur and nitrogen atoms are coordinated to form 5- and 6-membered chelate rings for SMC and Met, respectively. Silver(I) is also a soft metal ion and capable of linear coordination, which results in the formation of only sulfur coordinated species (monodentate binding) in acidic solution (Pettit et al., 1981). After deprotonation of the amino group, various binuclear or heterobinuclear complexes can be present in solution (Pettit and Lyons, 1986).

III. METAL COMPLEXES OF L-CYSTEINE AND D-PENICILLAMINE

The coordination chemistry of L-cysteine (Cys) and its derivative D-penicillamine (β,β-dimethyl-D-cysteine or β-mercapto-D-valine, Pen) is determined by the presence of the mercapto group, which is very soft in character. Via the mercapto group, these compounds may participate in both acid-base and redox reactions and these processes frequently occur in biological systems, too. The importance of Pen is enhanced by its independent therapeutic activity. Its most common use has been the treatment of Wilson's disease caused by the accumulation of copper, but due to the outstanding metal binding capability of the ligand it has also been employed for the elimination of other heavy metals (May and Bulman, 1983).

Cys and Pen contain three dissociable protons, and deprotonation of the ammonium and mercapto groups takes place in overlapping processes. Details of the deprotonation scheme will be discussed in a separate chapter, together with the cysteinyl peptides (see Part Three, Chapter 2-B3).

As a result of the presence of the three functional groups in the molecules, Cys and Pen are real ambidentate ligands, and the coordination chemistry of the two compounds is rather complicated and largely depends on the nature of the metal ion. Consequently, investigation of the metal complexes of Cys and Pen has a long past and most of the results have already been published in several reviews (McAuliffe and Murray, 1972; Gergely and Sóvágó, 1979; Kiss, 1990). In this chapter we summarize the most important features of the ligands as concerns their interactions with metal ions.

Because of the ambidentate nature of Cys and Pen, various types of coordination can occur in their metal ion complexes:

(i) Since the -SH and $-NH_2$ groups have relatively high pK values, hard metal ions can coordinate to the carboxylate group alone. This type of coordination is characteristic of only alkaline and alkaline earth metal ions and results in low stability in the same way as with other amino acids.

(ii) Monodentate binding of Cys and Pen is possible via the mercapto group, too. Copper(I), silver(I) and mercury(II) prefer this type of coordination and they readily form protonated complexes with Cys and Pen in acidic medium, where the metal ions are bonded monodentately via the sulfur atom, and the amino (or even carboxylate) groups remain protonated. At higher pH, however, there is a possibility for the formation of polynuclear species via sulfur bridges and the amino group also can take part in the coordination (Van der Linden and Beers, 1973).

(iii) Formation of chelate rings is the most characteristic metal binding mode of Cys and Pen, especially with d-block elements. There are three different types of bidentate chelates: the 5-membered (N,O) and (N,S) and the 6-membered (S,O) chelates. Finally, most of the transition elements can form complexes where Cys and Pen are coordinated tridentately: (N,S,O) fused chelates.

This classification of the binding mode of Cys and Pen clearly shows that the complex equilibria of these ligands are different from those of other amino acids, because in addition to the simple parent complexes, various protonated and polynuclear species can be present in solution. The complexity of the metal ion thiol interaction is enhanced by the possibility of redox reactions. It is a common feature of all thiol compounds that they readily oxidize to disulfides. Several metal ions (e.g. copper(II) and iron (III)) catalyze this process, which results in the occurrence of parallel acid-base and redox reactions. This means that in most cases the metal ion-Cys or -Pen systems are multicomponent systems, because both the oxidized and reduced forms of the metal ions and ligand can form stable complexes and the stoichiometric composition of the end-products will largely depend on the metal-to-ligand ratio, as will be demonstrated in connection with the copper(II)-D-penicillamine interaction.

A. Complexes of copper

As a consequence of the remarkable therapeutic application of Pen, its interaction with copper(II) is the best studied as regards the metal ions. Solution studies have clearly demonstrated that the stoichiometric composition of the final product formed in the parallel redox and complexation processes of copper(II) and Pen depends on the metal-to-ligand ratio and on the presence of various inorganic anions. In the presence of excess of the ligand, the reduction of copper(II) to copper(I) is complete and copper(I) forms a stable polynuclear complex with the unoxidized Pen (Gergely and Sóvágó, 1978). An equilibrium study on the direct copper(I)-D-penicillamine interaction supported the existence of Cu-S bonded chain polymers, among them $[Cu_5L_4]^{3-}$ being the most favored in solution. The existence of sulfur-coordinated mononuclear protonated complexes in dilute aqueous solution has also been suggested (Österberg et al., 1979).

Most of the studies on the copper(II)-Pen interaction had the aim of understanding the formation and structure of a deep red-violet complex, first identified by Sugiura and Tanaka (1970) and by Wilson and Martin (1971). Visible spectroscopic and EPR studies in solution revealed that it is a mixed valence complex and its formation largely depends on the metal-to-ligand ratio (Pen/Cu(II) = 1.45 being the most favored) and on the presence of halide ions (0.2 mol dm^{-3} Cl^- being the most favored) (Wright and Freiden, 1975; Musker and Neagley, 1975; Gergely and Sóvágó, 1978). The structure of the complex $Tl_5[Cu(I)_8Cu(II)_6(Pen)_{12}Cl]$ was determined by Birker and Freeman (1976), and its binding mode is shown in Figure 1.

Similar mixed valence complexes were identified in solution for various cysteine derivatives, but their lifetime proved to be very short in the absence of alkyl substituents on the β-carbon atom of L-cysteine (Sóvágó et al., 1980). An analogous mixed valence species of β,β-dimethylcysteamine was also obtained in the solid phase (Schugar et al., 1980).

The mixed valence complexes of Pen and its derivatives are fairly stable both in neutral solution and in the solid phase, but their slow decomposition was suggested to occur in serum (Deuschle and Weser, 1987). Decomposition of the red-violet complex is very fast in acidic media and the formation of a short-lived intense blue species can be observed. Products with the same absorption spectra are transiently formed by mixing copper(II) with various derivatives of Cys in 1:2 ratio. It has been proposed that the intense blue species are the bis complexes of the thiol compounds containing unreduced copper(II) ions and their spectral parameters are reminiscent of those of blue copper proteins.

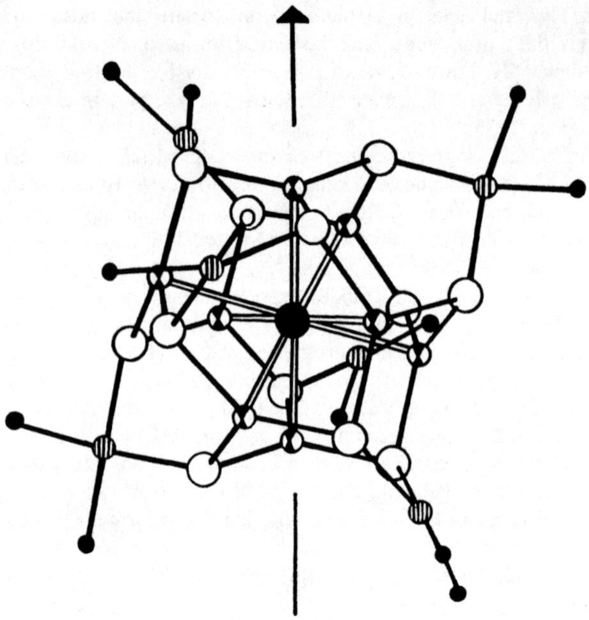

Figure 1.
Structure of the mixed-valence copper complex of D-penicillamine.

The unique structure of the copper-D-penicillamine mixed valence complex has attracted the preparation of derivatives by metal ion substitution. It was shown that copper(I) can be replaced by silver(I), and copper(II) by nickel(II) or palladium(II) (Burnier et al., 1981). A new tetranuclear mixed metal complex of Pen containing gold(I) and nickel(II) has also been obtained with the composition of $Na_2[Au(I)_2Ni(II)_2(Pen)_4]$ (Birker and Verschoor, 1982).

B. Complexes of other metal ions

A large number of publications are available on the complexation with main group elements and among them the complexes of lead(II) are the best studied. Most authors agree that Pen forms very stable 1:1 complex with lead(II), which makes possible the application of D-penicillamine for the treatment of lead intoxication. Tridentate coordination of the ligand was suggested to occur in the mono complex and it sterically hinders the formation of bis complexes (Doornbos and Faber, 1964; Gruenwedel and Hao, 1973; Freeman et al., 1974). Hard metal ions of the main group elements generally form complexes of low stability with amino acids, including Cys or Pen.

As concerns the early transition metal ion complexes of Cys and Pen, the interaction with vanadium, chromium and molybdenum should be mentioned. A detailed equilibrium and spectroscopic study of the vanadyl(IV)-Cys and -Pen systems has recently been published (Pessoa et al., 1990). The formation of various protonated complexes was suggested and the structure of the species involving sulfur coordination was discussed. The similarity between (S,O) and (O,O) coordination including -SH or alcoholic-OH groups was demonstrated in another study (Kiss et al., 1992).

Thiols generally reduce chromium(VI) and details of this process will be discussed in another chapter, on glutathione (see Part Three, Chapter 2-B3). The coordination of amino and carboxylate groups was suggested in the chromium(III) complexes of Cys in acidic solution (Abdullah et al., 1985), while the tridentate coordination of Pen was described in a $[Cr(Pen)_2]^-$ complex (Hojo et al., 1977).

The interaction of molybdenum with Cys and Pen has been extensively studied because the Cys complex of molybdenum(V) with composition $[Mo_2O_4(Cys)_2]^{2-}$ (Knox and Prout, 1968) was regarded as a model for the molybdenum-containing enzyme nitrogenase (Schrauzer et al., 1974). The catalytic activity of molybdenum(V)-complexes was assumed to involve the existence of species containing both molybdenum(IV) and molybdenum(III) (Lamache-Duhameaux, 1977; 1981). The formation of complexes $[Mo_3(Cys)]^{2-}$ containing molybdenum (VI) or tungsten(VI) and tridentately bonded L-cysteine has recently been described (Cavaleiro et al., 1990).

Investigation of the iron and cobalt complexes of thiols is hampered by the occurrence of various redox reactions. Although both iron(II) and iron(III) form stable complexes with Cys and Pen, their compositions and structures have not been clarified unambiguously because of the iron(III)-catalyzed oxidation of the ligands. The results obtained from earlier studies are discussed in previous reviews (Gergely and Sóvágó, 1979; Kiss, 1990).

Potentiometric and spectroscopic studies on the cobalt(II) complexes of Cys and Pen revealed the formation of stable mono and bis complexes containing octahedral cobalt(II) ions and tridentately coordinated ligands. Various polynuclear species were also detected in solution. Spectral and magnetic data clearly indicated that in the predominating trinuclear species $[Co_3(Cys)_4]^{2-}$ two octahedral units ($[CoL_2]^{2-}$) are linked by a tetrahedral Co^{2+} ion via sulfur bridges (Harman and Sóvágó, 1983). The reaction of L-cysteine with cobalt(III) was reported to yield tris complexes containing (S,N) coordinated ligands, while the stereospecificity in the synthesis of $[Co(Cys)_3]^{3-}$ was described recently (Arnold and Jackson, 1990).

The nickel(II) complexes of both Cys and Pen have been extensively studied. The results of equilibrium measurements revealed the formation of very stable bis complexes, with a negative ratio for the logarithms of the stepwise stability constants (Sóvágó et al., 1979). The crystal structures of the square planar diamagnetic bis complexes were published recently (Baidya et al., 1991a; 1991b). The results on the nickel(II) complexes also revealed the differences between the complex-forming abilities of Cys and Pen. Formation of a trinuclear species $[Ni_3L_4]^{2-}$ was characteristic of Cys, while its concentration was negligible in the case of Pen, due to the steric hindrance caused by the methyl substituents on the β-carbon atom.

As regards the complexes of platinum metals, several publications are available on the complexes of platinum and palladium. Similarly as for nickel(II), bidentate (S,N) chelation was suggested and the existence of various polynuclear species was also proposed (Levason et al., 1977). Complexation between ruthenium and Cys or its derivatives has been reported recently. Tridentate coordination of the ligands and the existence of sulfur bridges was suggested in the diene-Ru(II) complexes of Cys and Pen (Sheldrick and Exner, 1990). On the other hand, a ruthenium(VI) complex of Cys has been reported involving bidentate (S,O) chelation in the species $[Ru(O)_2L_2(Cys)]$ (where L = pyridine or dipyridine). The possibility of redox reactions between ruthenium(VI) and thiols was ruled out (Bigham and Shapley, 1991).

Similarly to nickel(II), zinc(II) also forms stable bis complexes with Cys and Pen. In addition to the parent complexes, the formation of various protonated and polynuclear species has also been suggested. The existence of protonated complexes in the medium pH range can be attributed to the formation of 6-membered (S,O) chelates, whereas tridentate binding of the ligands is probable in the bis complexes (Sóvágó et al., 1979; Bammel and Evilia, 1984). The formation of polynuclear complexes is favored in the case of Cys, and this is especially so for complexation with cadmium(II) ions. Cadmium(II) complexes of a series of cysteine derivatives have recently been studied (Kozlowski et al., 1990). The equilibrium data revealed that cadmium(II) readily forms stable complexes with sulfur-containing ligands, with the stability sequence (S,N,O) > (S,N) > (S,O,O) > (S,O). Polynuclear complexes were detected with all ligands and in the case of Cys polynuclear complex formation resulted in precipitation. The tendency of cadmium(II) to form sulfur-bridged complexes has also been observed with other thiol ligands (Avdeef and Kearney, 1982; Gonzalez-Duarte and Vives, 1989).

The therapeutic agents most commonly used for the removal of ingested mercury compounds are thiols. Consequently, a large number of publications are available on the complexation of mercury(II) or alkyl-Hg(I) with Cys and Pen. The data reveal that mercury(II) (similarly to copper(I) or silver(I)) readily reacts with these ligands to form protonated complexes in which the metal ions are bonded monodentately via sulfur atoms. After deprotonation of the amino group, however, there is a chance for the formation of polynuclear complexes with involvement of the amino group in the coordination (Van der Linden and Beers, 1973; Reid and Rabenstein, 1981; Wille et al., 1990).

IV. COMPLEXES OF DISULFIDES

As mentioned in the previous section, thiols are readily oxidized to disulfides and this reaction is generally catalyzed by the higher oxidation states of various transition metal ions. Among them, the iron(III)- or copper(II)-catalyzed reactions are the best studied and the resulting disulfides contain two separated amino acid-like binding sites and a disulfide moiety.

Due to the low solubility of L-cystine in neutral aqueous solution, the complexes of D-penicillamine disulfide have mainly been studied. It has been suggested that the disulfide bond is an even weaker base than a thioether group. Thus, if other metal binding sites are available in disulfides, as in D-penicillamine disulfide or cystine, the amino acid locus will be the primary coordinating center. Solution equilibria and spectroscopic studies on the cobalt(II), nickel(II), copper(II) and zinc(II) complexes evidently support the formation of dimeric species M_2L_2, in which the metal ions are coordinated via the amino and carboxylate groups (Laurie et al., 1981; Várnagy et al., 1988a). From an X-ray study of the dimeric species, however, a weak apical interaction of the disulfide moiety has also been suggested (Tich et al., 1974).

The disulfide group does not have a significant effect on the coordination of small peptide molecules. The amino and carboxylate groups of the glutamyl residue of oxidized glutathione and the terminal amino and amide nitrogens of L-cysteinylglycine have been suggested to be the main binding sites in the complexes with $3d^7$-$3d^{10}$ transition metal ions. It should also be mentioned that disulfides disproportionate in basic solution, this process being again catalyzed by metal ions (Ostern et al., 1980). Details of this reaction are discussed in Part Three, Chapter 2-B3.

V. COMPLEXES OF THIOAMIDES

Thioamides are derived from the amino acid amides (or peptides) by replacement of the carbonyl oxygen with a sulfur atom. It is well known (see thiourea and derivatives) that thiocarbonyl groups are quite strong donors to soft or borderline metal ions, and their presence in the molecule significantly increases the metal binding ability of the ligands.

The copper(II) and nickel(II) complexes of various amino acid and peptide amides and their analogues have recently been studied (Kowalik et al., 1987; Várnagy et al., 1988b). It was found that coordination of the amino acid amides starts with the participation of amino-N and carbonyl-O donors (ML, ML_2), and this is followed by the deprotonation and coordination of amide nitrogens, resulting in the species ML_2H_{-2} with 4N-coordination. The same species are formed with the corresponding thioamides, but the complexes containing (N,S) chelates (ML, ML_2) have much higher stability constants than those of the simple amino acid amides. This is well illustrated by the species distribution of the complexes formed in the copper(II)-leucine-amide and thioamide systems (Fig. 2).

It should also be emphasized that, although the presence of a thioamide group increases the metal-binding ability of amino acid or peptide derivatives, it does not prevent the deprotonation and coordination of amide nitrogens in basic solution.

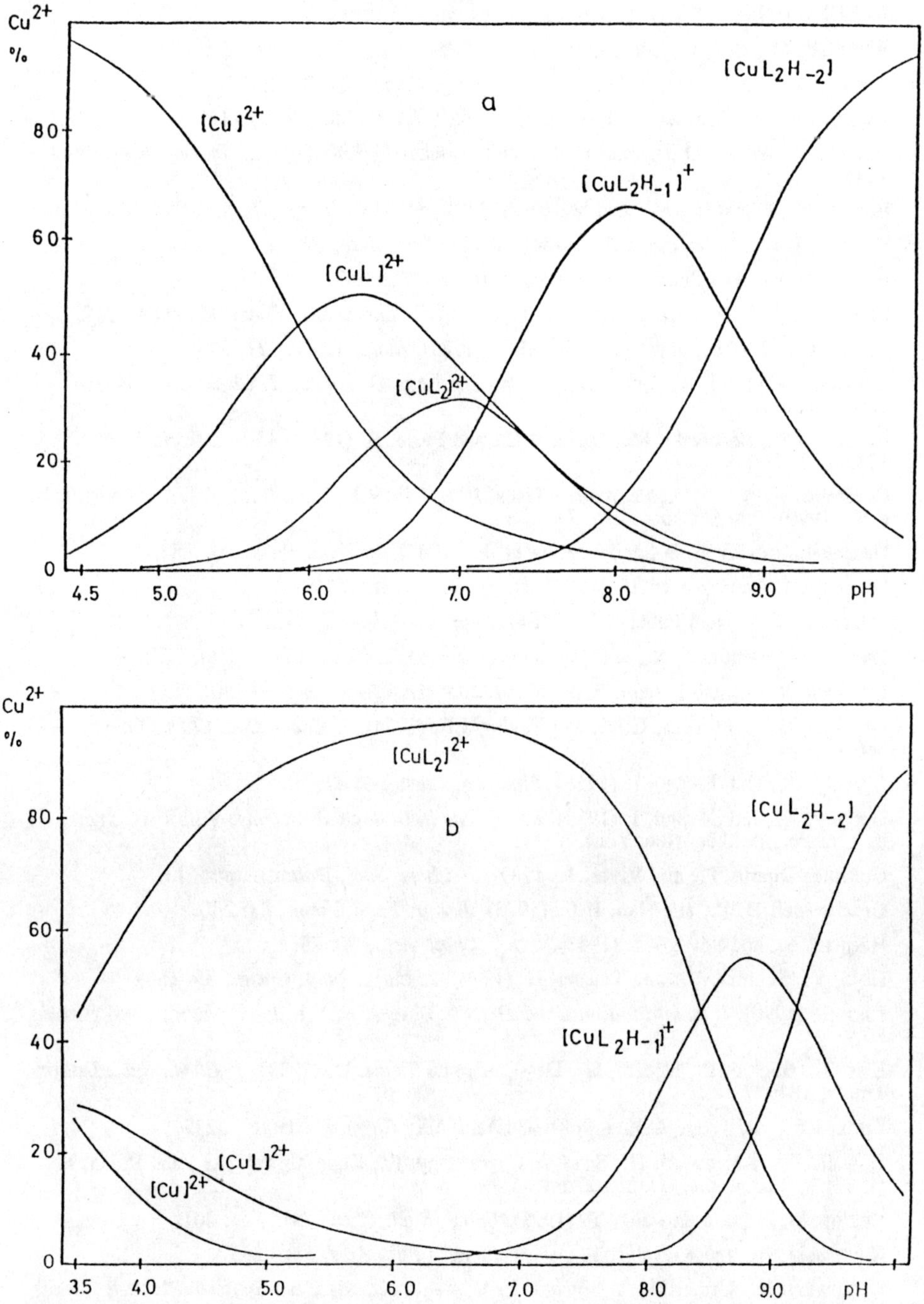

Figure 2.
Species distribution of the complexes formed in the copper(II)-leucine-N-methylamide (a) and copper(II)-leucine-N-methylthioamide (b) systems.

REFERENCES

Abdullah, M., Barret, J., and O'Brien, P. (1985). *Inorg. Chim. Acta, 96*: L35.

Arnold, A.P., and Jackson, W.G. (1990). *Inorg. Chem., 29*: 3618.

Avdeef, A., and Kearney, D.L. (1982). *J. Amer. Chem. Soc., 104*: 7212.

Baidya, N., Ndreu, D., Olmstead, M., and Mascharak, P.K. (1991a). *Inorg. Chem., 30*: 2448.

Baidya, N., Olmstead, M., and Mascharak, P.K. (1991b). *Inorg. Chem., 30*: 3967.

Bammel, B.P., and Evilia, R.F. (1984). *Inorg. Chim. Acta, 81*: L5.

Bigham, W.S., and Shapley, P.A. (1991). *Inorg. Chem., 30*: 4093.

Birker, P.J.M.W.L., and Freeman, H.C. (1976). *J. Chem. Soc. Chem. Commun.*, 312.

Birker, P.J.M.W.L., and Verschoor, G.C. (1982). *Inorg. Chem., 21*: 990.

Bonvicini, P., Levi, A., Lucchini, V., and Scorrano, G. (1972). *J. Chem. Soc., Perkin Trans.*, 2267.

Burnier, J.P., Borowski, M., Furie, B.C., and Furie, B. (1981). *Mol. Cell Biochem., 39*: 191.

Cavaleiro, A.M.V.S.V., Pedrosa de Jesus, J.D., Gil, V.M.S., Gillard, R.D., and Williams, P.A. (1990). *Inorg. Chim. Acta, 172*: 25.

Decock-Le Révérend, B., and Kozlowski, H. (1985). *J. Chim. Phys., 82*: 883.

Deuschle, U., and Weser, U. (1987). *Inorg. Chim. Acta, 135*: 5.

Doornbos, D.A., and Faber, J.S. (1964). *Pharm. Weekblad., 99*: 289.

Dubler, E., Cathomas, N., and Jameson, G.B. (1986). *Inorg. Chim. Acta, 123*: 99.

Espersen, W.G., and Martin, R.B. (1976). *J. Amer. Chem. Soc., 98*: 40.

Freeman, H.C., Stevens, G.N., and Taylor, I.F. (1974). *J. Chem. Soc. Chem. Commun.*, 366.

Gergely, A., and Sóvágó, I. (1978). *Bioinorg. Chem., 9*: 47.

Gergely, A., and Sóvágó, I. (1979). *Metal Ions in Biological Systems*, Vol. 9 (H. Sigel, ed.), Marcel Dekker, New York, p. 77.

Gonzalez-Duarte, P., and Vives, J. (1989). *J. Chem. Soc., Dalton Trans.*, 13.

Gruenwedel, D.W., and Hao, H.C. (1973). *J. Agr. Food Chem., 21*: 246.

Harman, B., and Sóvágó, I. (1983). *Inorg. Chim. Acta, 80*: 75.

Hojo, Y., Sugiura, Y., and Tanaka, H. (1977). *J. Inorg. Nucl. Chem., 39*: 1859.

Kiss, T. (1990). *Biocoordination Chemistry* (K. Burger, ed.), Ellis Horwood, New York, p. 56.

Kiss, T., Buglyó, P., Micera, G., Dessì, A., and Sanna, D. (1993). *J. Chem. Soc., Dalton Trans.*, 1849.

Knox, J.R., and Prout, C.K. (1968). *J. Chem. Soc. Chem. Commun.*, 1227.

Kowalik, T., Kozlowski, H., Sóvágó, I., Várnagy, K., Kupryszewski, G., and Rolka, K. (1987). *J. Chem. Soc., Dalton Trans.*, 1.

Kozlowski, H., and Kowalik, T. (1978). *Inorg. Nucl. Chem. Lett., 14*: 201.

Kozlowski, H., and Kowalik, T. (1979). *Inorg. Chim. Acta, 34*: L231.

Kozlowski, H., Urbanska, J., Sóvágó, I., Várnagy, K., Kiss, A., Spychala, J., and Cherifi, K. (1990). *Polyhedron, 9*: 831.

Lamache-Duhameaux, M. (1977). *J. Inorg. Nucl. Chem., 39*: 2081.

Lamache-Duhameaux, M. (1981). *J. Inorg. Nucl. Chem., 43*: 208.

Laurie, S.H., Mohammed, E.S., and Prime, D.M. (1981). *Inorg. Chim. Acta, 56*: 135.

Levason, W., McAuliffe, C.A., and Johns, D.M. (1977). *Inorg. Nucl. Chem. Lett., 13*: 123.

Martin, R.B. (1979). *Metal Ions in Biological Systems*, Vol. 9 (H. Sigel, ed.), Marcel Dekker, New York, p. 1.

May, P.M., and Bulman, R.A. (1983). *Progress in Medicinal Chemistry*, Vol. 20 (G.P. Ellis and G.B. West, eds.), Elsevier Science Publishers.

McAuliffe, C.A., and Murray, S.G. (1972). *Inorg. Chim. Acta Rev.*, *6*: 103.

McCormick, D.B., Griesser, D.B., and Sigel, H. (1974). *Metal Ions in Biological Systems*, Vol. 1 (H. Sigel, ed.), Marcel Dekker, New York, p. 213.

Musker, W.K., and Neagley, C.H. (1975). *Inorg. Chem.*, *14*: 1728.

Ostern, M., Pelczar, J., Kozlowski, H., and J.-Trzebiatowska, B. (1980). *Inorg. Nucl. Chem. Lett.*, *16*: 251.

Österberg, R., Ligaarden, R., and Persson, D. (1979). *J. Inorg. Biochem.*, *10*: 341.

Pessoa, J.C., Vilas Boas, L.F., and Gillard, R.D. (1990). *Polyhedron*, *9*: 2101.

Pettit, L.D., Siddiqui, K.F., Kozlowski, H., and Kowalik, T. (1981). *Inorg. Chim. Acta*, *55*: 87.

Pettit, L.D., and Lyons, A.Q. (1986). *J. Chem. Soc., Dalton Trans.*, 499.

Reid, R.S., and Rabenstein, D.L. (1981). *Can. J. Chem.*, *59*: 1505.

Schrauzer, G.N., Kiefer, G.W., Tano, K., and Doemeny, P.A. (1974). *J. Amer. Chem. Soc.*, *96*: 641.

Schugar, H.J., Ou, C.C., Tich, J.A., Potenza, J.A., Felthouse, T.R., Haddad, M.S., Hendrickson, D.N., Furey, W., and Lalancette, R.A. (1980). *Inorg. Chem.*, *19*: 543.

Sheldrick, W.S., and Exner, R. (1990). *J. Org. Chem.*, *386*: 375.

Sóvágó, I., and Petöcz, Gy. (1987). *J. Chem. Soc., Dalton Trans.*, 1717.

Sóvágó, I., Gergely, A., Harman, B., and Kiss, T. (1979). *J. Inorg. Nucl. Chem.*, *41*: 1629.

Sóvágó, I., Harman, B., and Gergely, A. (1980). *Inorg. Chim. Acta*, *14*: L107.

Sóvágó, I., Várnagy, K., and Bényei, A. (1986). *Magy. Kem. Foly.*, *92*: 114.

Sugiura, Y., and Tanaka, H. (1970). *Chem. Pharm. Bull.*, *18*: 368.

Swash, J.L., and Pettit, L.D. (1976). *Inorg. Chim. Acta*, *19*: 19.

Tich, J.A., Mastropaolo, D., Potenza, J.A., and Schugar, H.J. (1974). *J. Amer. Chem. Soc.*, *96*: 726.

Van der Linden, W.E., and Beers, C. (1973). *Anal. Chim. Acta*, *68*: 143.

Várnagy, K., Sóvágó, I., and Kozlowski, H. (1988a). *Inorg. Chim. Acta*, *151*: 117.

Várnagy, K., Kozlowski, H., Sóvágó, I., Kovalik-Jankowska, T., Kruszynski, M., and Zboinska, J. (1988b). *J. Inorg. Biochem.*, *34*: 83.

Wille, K., Moran, J.K., and Roundhill, D.M. (1990). *Polyhedron*, *9*: 597.

Wilson, E.W., and Martin, R.B. (1971). *Arch. Biochem. Biophys.*, *142*: 445.

Wright, J.R., and Freiden, E. (1975). *Bioinorg. Chem.*, *4*: 163.

METAL-PEPTIDE COMPLEX FORMATION

by

Leslie D. Pettit and Rebecca A. Robbins

School of Chemistry,
The University of Leeds,
Leeds, LS2 9JT
U.K.

I. INTRODUCTION

Amino acids, H_3N^+-CHR-COO$^-$, are the essential building blocks of life. Condensation between the O- and N-terminals of two amino acids results in the formation of a peptide bond (-CONH-) and this condensation reaction, repeated a number of times, leads to the formation of an oligopeptide and ultimately a protein. The actual sequence in which amino acids bind together to form peptides and then proteins is boundless, hence accounting for the diversity of functions that various peptides and proteins fulfil. Recent reviews of metal complexes of peptides have been presented by Laurie (1987), Sóvágó (1990) and Pettit et al. (1991); complexes of nickel have been reviewed by Martin (1988), of the platinum metals by Kozlowski and Pettit (1991) and of palladium by Kasselouri et al. (1990). These reviews should be consulted for details of much of the earlier work on peptide complexes.

II. METAL COORDINATION

At neutral pH, peptides [LH] exist in solution as zwitterions similar to amino acids, the amino group being protonated (-NH_3^+) while the carboxyl group is ionized (-CO_2^-). The carboxyl group protonates in acid solution (pH 3-4) to give the $[LH_2]^+$ complex whereas the amino group loses its proton in basic solution (pH 7-9) to give L^-. Much interest has been devoted to the study of the coordination of peptides with a variety of metal ions, due to the diversity of the possible bonding centers and to the resultant changes in reactivity of the peptide. Among factors influencing the stability of peptide complexes, one of the most important is the number and spatial arrangement of donor atoms, particularly in side chains. Also important is the actual backbone conformation of the ligand and its ability to accommodate the preferred coordination geometries of the metal ions. If chelate rings are formed, then their number and size can have a marked effect on the stability of the resulting complex. Finally, for transition metal ions, crystal field effects are significant and play an important role in determining the stability of the metal complexes formed.

Peptides contain a number of potential binding sites for metal ions as shown in Figure 1, and the actual donor sites chosen are dictated largely by the type of metal ion involved, generally classified as 'hard' or 'soft' acceptors (see Part One, Chapter 3-A). The former tend to form more stable complexes with ligands in which the donor atom is a first row element (N,O) while the latter (e.g. Ag(I) or platinum group metals) prefer second row donors (e.g. P,S).

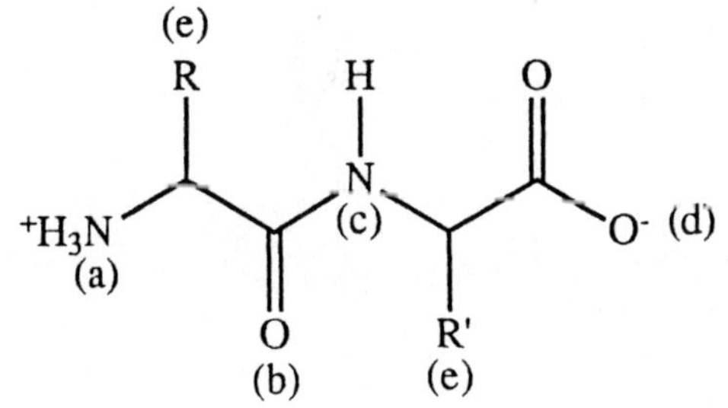

Figure 1.
Possible donor centers in a peptide molecule.

In general, alkali and alkaline earth metal ions bind most effectively to O-donors while transition metal ions bind to both N- and O-donors. The basicity of the donor site is important in determining its donor properties. Within peptides the potential sites, indicated in Figure 1, are: (a) terminal-NH_2; (b) carbonyl=O; (c) amide-N^--; (d) carboxyl-O^-, (e) side chain donor atoms.

The terminal amino nitrogen is generally the most important anchoring site, and most metal ions coordinate to this nitrogen and the oxygen of the neighboring peptide carbonyl group. In the absence of further coordination, the complexes will be weak and usually unimportant since, unlike the situation with amino acids, steric requirements exclude the formation of chelate rings involving the simultaneous bonding of both the terminal amino-N and carboxylate-O^-. However, other donors (viz (b), (c), (e)) are present and are able, at least in principle, to bind to metal ions to form chelated complexes.

Resonance occurs between the carbonyl group and amide nitrogen of the peptide bond (see Fig. 2) and this has the effect of decreasing the basicity of the amide nitrogen; hence dissociation of the amide proton (c) does not occur within the normal pH range - it has an estimated pK value of about 15. However certain metal ions, notably Pd(II), Cu(II) and Ni(II), are able to facilitate the removal of this proton with simultaneous formation of a M-N^- covalent bond. For this to happen, the metal must normally be already bound to the peptide through another donor, and the ease with which it is achieved is dependent on the actual metal ion involved, e.g.:

Pd(II) [pH 2-3] > Cu(II) [pH 4-5] > Ni(II) [pH 8-9] > Co(II) [pH 9-10]

Cr(III) and Co(III) can also bond in this way, forming inert complexes (Ama et al., 1991), but Zn(II) is probably unable to promote peptide ionization in the measurable pH range, apart from in some peptides containing His where NMR evidence suggests deprotonation at high pH (Rabenstein et al., 1985).

Side chain donors can be very important also, particularly an imidazole-N of His and the sulfur atom of Cys, which bond most strongly to metal ions with soft character.

Complexes with Cu(II) and Ni(II) have been studied in most detail. Copper(II) forms very stable complexes with simple oligopeptides such as tetraglycine, LH. Initial binding is through the terminal amino-N and neighboring carbonyl-O, forming a $[CuL]^+$ complex with

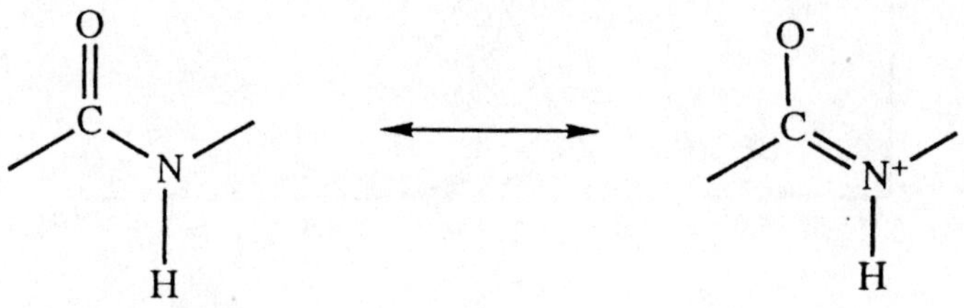

Figure 2.
Resonance in a peptide bond.

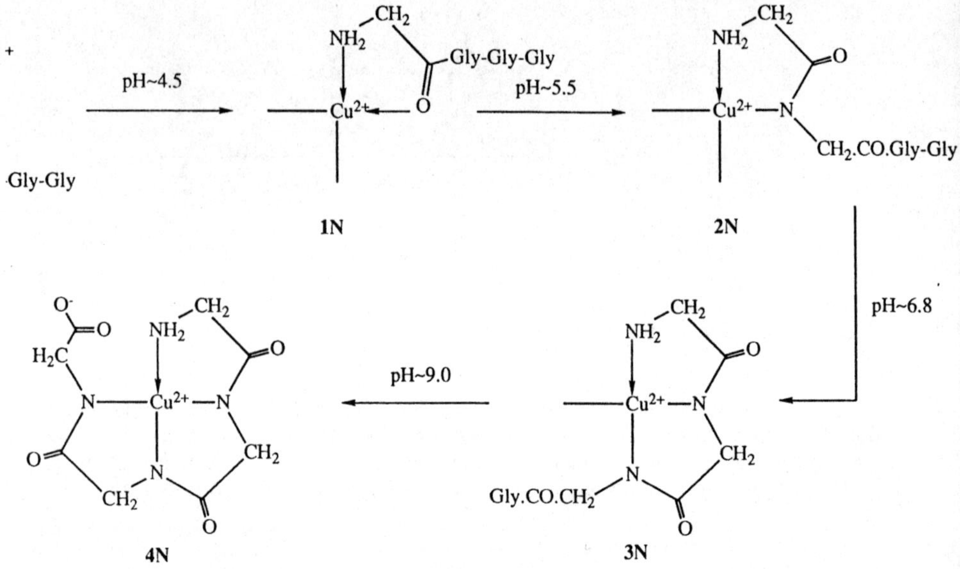

Figure 3.
Coordination scheme for tetraglycine with Cu(II).

a 5-membered chelate ring as shown in Figure 3, and this is followed by the facile deprotonation and coordination of the neighboring peptide-N around pH 5 to give $[CuLH_{-1}]$ which also has a 5-membered chelate ring. This is followed by coordination to successive peptide-N donors to result in a 4N complex, $[CuLH_{-3}]^{2-}$, by pH 9 in which the peptide wraps itself around the metal ion thus adopting a very specific conformation. The speciation of Cu(II) with tetraglycine as a function of pH is shown in Figure 4. With tripeptides (e.g. triglycine) the final complex is a 3N species and the 4th coordination position on Cu(II) is occupied by the carboxylate-O^-. This is displaced around pH 11 by a hydroxyl ion, formed by the hydrolysis of a coordinated water molecule.

Cu(II) complexes may be studied both potentiometrically and spectrophotometrically. Complexes with nitrogen donors generally range from deep blue to red-purple in color,

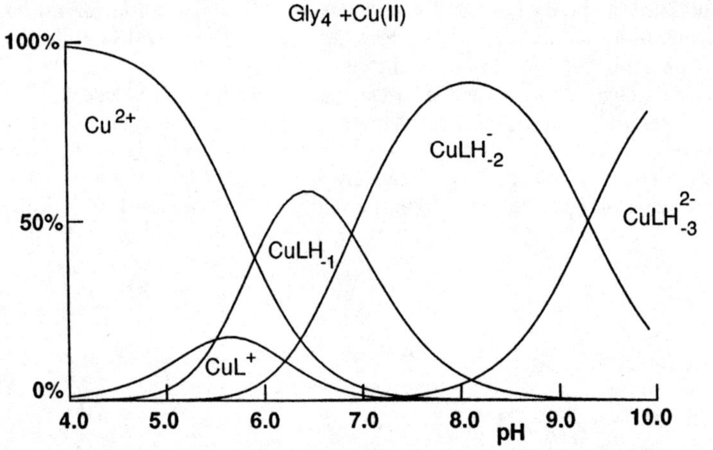

Figure 4.
Species distribution curves for 1:1 mixtures of tetraglycine with Cu(II) (0.001 mol dm^{-3}) as a function of pH.

Table 1

Typical Values for λ_{max} for d-d Transitions in Cu(II) Complexes and Corresponding Values for $A_{||}$ from ESR Spectroscopy.

Nitrogens coordinated	λ_{max} (nm)	$A_{||}$ (x 10^4T)
1N	680-730	140-160
2N	620-670	160-180
3N	540-590	180-205
4N	500-530	205-220

depending on the number of nitrogens coordinated and whether they are amino-N or peptide-N^- donors. Empirical formulae for calculating the values for λ_{max} for the d-d transitions have been proposed by Sigel and Martin (1982), and representative values are given in Table 1. Even more reliable information on the bonding centers is provided by CD and ESR spectroscopy when values for $A_{||}$ are particularly useful.

Ni(II) differs from Cu(II) in that it is capable of forming peptide complexes with two distinct geometries; paramagnetic hexacoordinate octahedral and diamagnetic tetracoordinate planar. Initial coordination is similar to that with Cu(II) but forming 6 coordinate rather than 4 coordinate complexes. Following coordination of the first deprotonated peptide-N^- however, there is a cooperative and virtually simultaneous deprotonation of up to two successive peptide-N donors to give planar $[NiLH_{-3}]^{2-}$, as shown in the species distribution curves for Ni(II)-tetraglycine (Fig. 5). The species $[NiLH_{-1}]$ and $[NiLH_{-2}]^-$ are of almost insignificant concentration because the constants for the successive amide deprotonation reactions (pK values) actually show a decrease in value, being 9.3, 7.4 and 7.8 respectively for Ni(II)-tetraglycine and 8.7 and 8.0 for Ni(II)-triglycine (Martin, 1988).

The change from hexacoordinate octahedral to tetracoordinate square-planar geometry about Ni(II) is accompanied by a color change from blue (λ_{max}=660) to yellow (λ_{max}= 410-450) and a change in magnetic behavior from paramagnetic to diamagnetic. This cannot

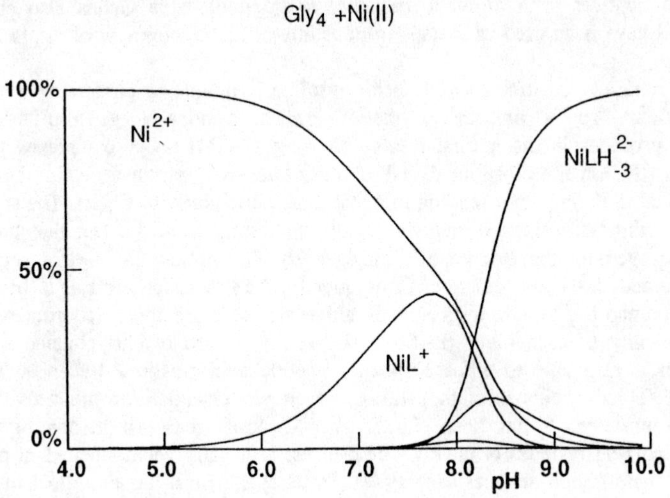

Figure 5.
Species distribution curves for 1:1 mixtures of tetraglycine with Ni(II) (0.001 mol dm^{-3}) as a function of pH.

happen with dipeptides, and these form the paramagnetic octahedral bis-complexes, $[NiL_2H_{-2}]^{2-}$, at high pH. Compared with other divalent metal ions, the kinetics of Ni(II)-peptide coordination are slow, especially the octahedral-planar transition (Kennedy and Margerum, 1985), making a quantitative study of the thermodynamics of complex formation much more difficult than with Cu(II). Pd(II) promotes peptide deprotonation below pH 3 and forms only square-planar diamagnetic complexes.

Because peptides differ considerably in their ability to coordinate to metal ions, they will be divided into seven groups for the purpose of this review. Complexes of peptides containing sulfhydryl groups and of peptide hormones will be treated as separate topics.

A. Peptides with non coordinating side chains

All amino acid residues apart from glycine have side chains and at least one chiral center. With many peptides, these side chains do not contain any donor centers, e.g. α- and β-alanine, leucine, isoleucine, valine, proline, tryptophan and phenylalanine. Generally the mode of bonding of such peptides is similar to that of tetraglycine, although the stabilities of the resulting complexes varies depending on a number of factors.

The position of the non-coordinating side chain residue within a peptide chain can have a significant effect on its stability. Sigel and Martin (1982) have reported that the Cu(II) and Ni(II) complexes of dipeptides containing a non-glycine N-terminal residue (X-Gly) form less stable $[ML]^+$ complexes than those of Gly-Gly, while with a C-terminal Gly residue (Gly-X) there is a slight increase in stability. Metal complexes of dipeptides of the type X-X consequently have comparable stability to those of Gly-Gly. The bis complex $[Cu(LH_{-1})_2]^{2-}$ is only observed in dipeptides of the type X-Gly, and its absence in systems with Gly-X is probably due to steric hindrance.

The possibility of interactions occurring between the aromatic ring of the side chain and the metal ion must be taken into account when considering the formation of metal complexes with peptides containing aromatic amino acid residues. 'Ring stacking' (a hydrophobic interaction) can influence the nature of the complexes formed as well as their stability. The presence of an aromatic ring in the C-terminal residue of a dipeptide results in enhanced stability of the 2N $[MLH_{-1}]$ copper(II) complex. From calorimetric, potentiometric and spectroscopic studies, this has been attributed to an interaction between the d-orbitals of Cu(II) and the π-electron system of the aromatic ring, and the magnitude of the effect evaluated as -9.5 kJ mol^{-1} (Xiao et al., 1990). This effect is not observed for N-terminal aromatic residues since here the rigid Cu-N bond formed is stronger than the weak, easily distorted $Cu\text{-}O^-$ bond in the former. Ternary complexes of Pd(II) with aromatic diamines and peptides with aromatic or other hydrophobic side chains also show ring alignment and have been used as models in the study of the cytotoxicity of cisplatin (Odani et al., 1986).

Among naturally occurring amino acids, proline is unique in possessing a secondary nitrogen. When Pro is present as the N-terminal residue (e.g. Pro-Gly-Gly-Gly) coordination proceeds in the normal fashion forming 4N $[MLH_{-3}]^{2-}$ complexes with both Cu(II) and Ni(II). On the other hand, a Pro residue inserted within a peptide chain has a profound effect on the mode of binding to metal ions, particularly to Cu(II). The same effect is also found with sarcosine (N-methylglycine) and results from the fact that there is no ionizable hydrogen on the Pro peptide nitrogen so interrupting the normal sequence of deprotonation and $Cu\text{-}N^-$ coordination. Consequently, the Pro residue acts as a 'break-point' to coordination and the peptide fragments on either side of it are able to coordinate to metal ions independently of each other (Pettit et al., 1985). A Pro residue situated within the peptide chain is also able to enforce specific peptide conformations by encouraging the formation of folded chains through β-turns, which are essential for the active globular structures of proteins. It has been shown by ^{13}C NMR that Pro is one of the most frequently occurring residues at the sites of β-turns, especially when situated in position 2 of a peptide chain (Chou and Fasman, 1977). In tetrapeptides it has the effect of bringing the donor atoms of the N- and C-terminals closer together and this, in turn, favors the construction of large but surprisingly stable chelate ring bridging these positions (Livera, 1988). A large ring of this type is not possible with Gly-Pro-Gly-Pro-Gly since there is now a secondary-N in the Pro^4 residue, hence the only complexes formed with Cu(II) are $[CuL]^+$ and $[CuL_2]$ (Pettit et al., 1991). The presence of two neighboring Pro residues (e.g. Gly-Pro-Pro-Gly-Gly) has an even more dramatic effect, forming a $[CuLH_{-1}]$ complex with 2N coordination and a large chelate ring (10- or 13-membered) between the amino-N and carbonyl-O of the Gly^1 residue and the peptide-N of Gly^4 or Gly^5 (Pettit et al., 1991). The stability of this large chelate ring is unexpectedly high, being only an order of magnitude less than that for the 2N complex of tetraglycine which has stable 5-membered chelate rings. The influence of Pro residues on peptides containing Tyr or Lys residues will be described later.

B. Serine and threonine residues

Ser and Thr residues contain the side chains: $-CH_2OH$ and $-CH_2CH_2OH$ respectively. The alcohol groups of these side chains are usually classed as non-coordinating but they do enhance the stabilities of many transition metal complexes, probably through axial bonding (Sigel et al., 1977) and possibly via a bridging water molecule. As expected, stabilization in the $[CuL]^+$ species is most significant when Ser is the N-terminal residue. This axial interaction also promotes amide binding as implied by the low value for the pK of amide deprotonation. At high pH, potentiometry and spectrophotometry show that with N-terminal Ser and Thr dipeptides, $[Cu(LH_{-1})_2]^{2-}$ complexes occur with bidentate 4N coordination via amino-N and amide-N donors. The enhanced stability of these complexes may be a result of hydrogen bonding between the non-coordinated carboxylate and alcohol-OH groups (Farkas and Kiss, 1989). No such species were detected with the C-terminal dipeptides; instead, direct interaction between deprotonated alcoholic-O^- groups and Cu(II) above pH 12 has been proposed in complexes of Gly-Ser and Gly-Thr (Farkas and Kiss, 1989).

Acceptors of pronounced hard character would be expected to coordinate best to Ser or Thr side chains; hence it is likely that Ca^{2+} would coordinate effectively with peptide T, Ala-Ser-Thr-Thr-Thr-Asn-Tyr-Thr, an HIV inhibitor (Pert et al., 1986). Mg^{2+} would not be expected to bind as effectively as Ca^{2+} because it has a smaller coordination number and a lower affinity for uncharged oxygen donors.

C. Aspartic acid, glutamic acid, aspartamine and glutamine residues

Asp, Glu, Asn and Gln residues contain the side chains $-CH_2CO_2H$, $-CH_2CH_2CO_2H$, $-CH_2CONH_2$ and $-CH_2CH_2CONH_2$, and these differ considerably in their ability to interact with metal ions. The side chain amido groups of Asn and Gln do not deprotonate and coordinate as readily as the nitrogen of a peptide linkage, but that of Gln can do so above pH 10 (Gergely et al., 1982). When an amido group is part of the C-terminal residue (e.g. $-CHRCONH_2$) deprotonation and coordination to Cu(II) can take place by pH 9, as with Pro-Leu-Gly-NH_2 (Formicka-Kozlowska et al., 1981). When it is part of the N-terminal residue (e.g. pyroglutamyl- in LHRH) it can take place by pH 5.5 (Gerega, 1988).

Studies performed on the Cu(II) complexes of a series of dipeptides provided the following order for stability:

Gly-Ala < Gly-Asn < Gly-Glu < < Gly-Asp

with the β carboxylate of the Asp residue having by far the greatest effect (Gergely and Farkas, 1982). Chelation through the carboxylate-O^- of an Asp residue gives a 6-membered chelate ring whereas with Glu the ring is 7-membered and much less stable.

Studies on the Cu(II) complexes of a series of tetrapeptides containing Asp and Glu residues in various positions in a peptide chain based on tetraalanine have demonstrated the possible influences of the carboxyl side chains (Galey, 1991). The influence of a Glu residue is either small or negligible and can usually be dismissed as a primary contributor to complex stability. When Asp is the N-terminal residue, the [CuL] complex is stabilized by about 0.6 log units through coordination of the β-carboxyl-O^- as demonstrated in Figure 6(a) where the species distribution curves for Asp-Ala-Ala-Ala are compared with those for Ala_4. With Asp^2 the 2N $[CuLH_{-1}]^-$ species is stabilized to a greater extent (2.4 log units) while, with Asp^3, the 3N $[CuLH_{-2}]^{2-}$ species is stabilized even more (3.2 log units). This stabilization is completely lost when the β-carboxylate-Cu(II) bond is broken to allow coordination of a further peptide-N donor. As a result, the 2N $[CuLH_{-1}]^-$ complex for peptides with an Asp^2 residue is the major species over a wide pH range (5-9.8) as shown in Fig. 6(b) and with Ala-Ala-Asp-Ala, a 4N complex could not be detected (see Fig. 6(c)). The influence of an Asp^1 residue is also apparent in the Ni(II) and Cu(II) complexes of Asp-Arg-Val-Tyr, the N-terminal tetrapeptide fragment of angiotensin II (Decock-le-Révérend et al., 1988; Pettit et al., 1989) and in the Cu(II) and Ni(II) complexes of Asp-Phe-NH_2 (Varnagy et al., 1988). The influence of an Asp^2 residue was found in complexes of Ala-

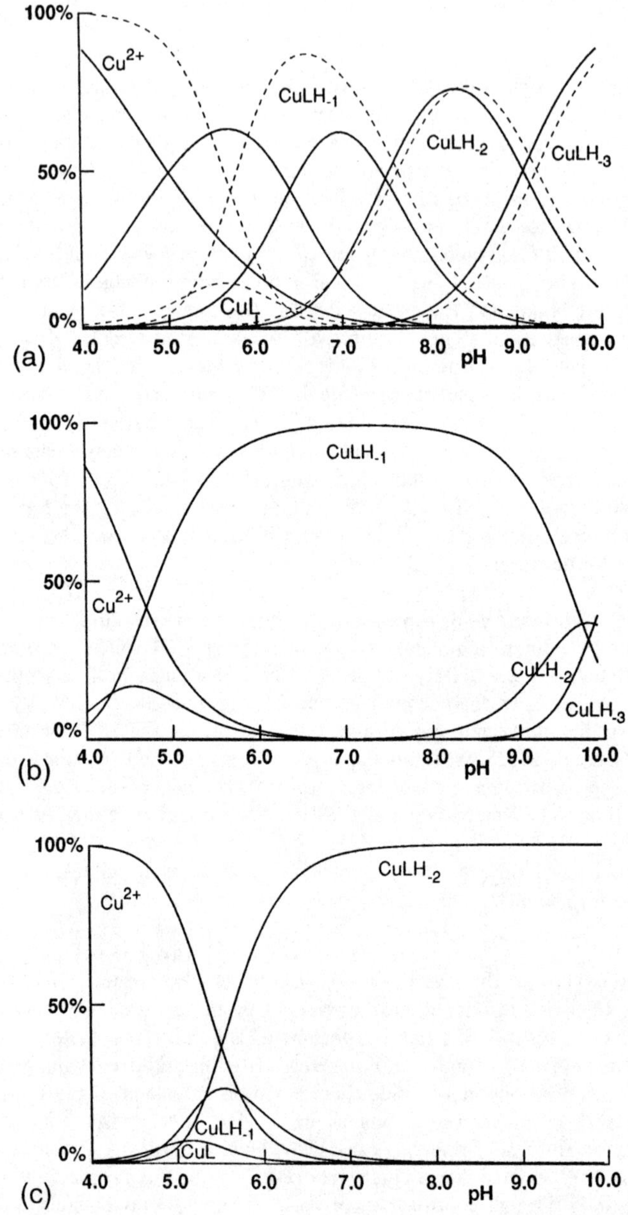

Figure 6.
Species distribution curves for 1:1 mixtures of (a) Asp-Ala-Ala-Ala, (b) Ala-Asp-Ala-Ala and (c) Ala-Ala-Asp-Ala with Cu(II) (0.001 mol dm^{-3}) as a function of pH. Dotted curves are for complexes of Ala_4.

Asp-Ser-Gly, the N-terminal fragment of fibrinopeptide A (Decock-le-Révérend, 1986) and of an Asp^3 residue in the pentapeptide Arg-Lys-Asp-Val-Tyr, a biologically active fragment of the peptide hormone thymopoietin (Sóvagó et al., 1990). Potentiometric and spectroscopic studies have shown that its $tri^{(1-3)}$, $tetra^{(1-4)}$ and $penta^{(1-5)}$ fragments are all able to form 1:1 complexes with Cu(II), Ni(II) and Zn(II) (Sóvágó et al., 1984; Veber et al., 1985). They all exhibit triglycine-like coordination via the terminal amino, two deprotonated peptide-N donors and the β-carboxylate function. It is therefore concluded that the β-carboxylate group of Asp successfully blocks coordination of the peptide-N donors of the fourth or fifth residues.

D. Lysine residues

Lys ($-NHCH(CH_2CH_2CH_2CH_2NH_2)CO-$) is the only amino acid residue with an amino-side chain to be found in naturally occurring peptides. The ε-terminal $-NH_2$ group cannot normally bond to a metal ion as a result of the large chelate ring required for multidentate coordination (cf amino acid complexes), hence there is no evidence of ε-amino-N coordination to Cu(II) in complexes of Gly-His-Lys. However, chelation is possible either if the conformation of the peptide is constrained to bring the ε-NH_2 group sufficiently close to other donor centers to allow chelation or if binuclear complexes are formed. Both possibilities have been found in a number of systems studied recently.

Chelation between N-terminal and side chain ε-amino-N donors can be favored by the presence of a Pro residue in the peptide chain, which encourages the formation of a β-turn. For example, with Gly-Gly-Pro-Lys (Bataille et al., 1985) and with Arg-Pro-Lys-Pro-Gln and Substance P (Pettit et al., 1991) the ε-amino-N is deprotonated and coordinated at about pH 6 to form very large chelate rings. Dimeric complexes with ε-amino coordination from a Lys residue to Cu(II) have been detected in dilute solution with a Tyr-Leu and Leu-Tyr (Radomska et al., 1990).

E. Residues with thioether donors

Methionyl (Met), $-NH-CH(CH_2CH_2.SCH_3)CO-$, is the most important peptide residue with a thioether side chain. Although thioether-S donors of peptides bond effectively to Ag(I) or the platinum metals, there is little evidence for coordination to other metal ions, and a detailed calorimetric study of complexes of Met-Met with Cu(II) gave no evidence of such interaction (Bonomo et al., 1987). In the binuclear complex of Cu(II) and Ag(I) with Met-Met ($[CuAgLH_{-1}]^+$) the Cu(II) is NNO coordinated while the Ag(I) is S coordinated (Pettit and Lyons, 1986). In the binary complexes between Ag(I) and several dipeptides containing thioether side chains, the Ag(I) ion tends to adopt linear coordination to S-donors through the formation of dimeric species (Lyons and Pettit, 1984). In biological systems, coordination from a thioether sulfur can be important, e.g. in cytochrome the iron is coordinated to five nitrogen donors and a thioether sulfur.

F. Histidine residues

Coordination from a nitrogen donor of the imidazole ring of the His residue (N-Im) is probably the most important side chain bonding to metal ions in the bioinorganic chemistry of peptides and proteins. The pyridine-like π-nitrogen of the imidazole ring (N^3-Im) protonates at about pH 6 and, when deprotonated, is able to form a very stable six-membered chelate ring to a metal ion in cooperation with the deprotonated amide-N^- of the His residue. The N^3-Im is able to compete with other amino acid binding sites (usually the terminal amino-N) to become the primary anchoring site for metal ions, and coordination of a metal ion at N^3-Im has the affect of lowering the pK of peptide-nitrogen deprotonation (Sóvágó et al., 1982).

Complex formation with N-terminal His peptides is dependent on the metal to ligand ratio. In equimolar solutions at low pH the N^3-Im and the amino-N are able to form a six-membered chelate ring. Around pH 6, Cu(II) promotes the ionization and coordination of the amide nitrogen with the result that a dimeric complex $[Cu_2L_2H_{-2}]$ is formed with glycylglycine-like coordination (NH_2, N^-, O^-), the fourth binding site being occupied by the N^3-Im of another $[CuLH_{-1}]$ complex. In the presence of excess ligand, deprotonation of the peptide linkage is reduced and the formation of bis complexes (e.g. $[CuL_2]$) is favored.

Complexation of peptides with a C-terminal His residue with Cu(II) leads to the formation of several different complexes, including dimers. Steric requirements preclude histamine-like coordination and initial coordination at N^3-Im is immediately followed by deprotonation of the peptide nitrogen to allow Cu-N^- bond formation by pH 4, much lower than in simple dipeptides.

His-His, (Livera et al., 1987), His-His-Gly and His-Gly-His (Udo et al., 1987), exhibit similar general behavior, with Cu-N^- bond formation occurring below pH 5 and with the formation of binuclear species at intermediate pH. The tripeptide Gly-Gly-His has been widely studied as it is a model for human serum albumin (Farkas et al., 1984). With Cu(II) and Ni(II), the predominant species is $[MLH_{-2}]^-$, a 4N coordinated complex bonded via the terminal amino-N, both peptide-N^-donors and N^3-Im; X-ray evidence reveals the absence of any C-terminal carboxylate bonding (Camerman et al., 1976). The negligible concentration of $[CuLH_{-1}]$ may be explained by the tendency of the amide groups to undergo deprotonation in a cooperative manner. Complexes of Cu(II) with Gly-His-Gly have been studied in particular detail using potentiometry, calorimetry and visible and CD spectroscopy. Above pH 8, a tetrameric complex $[Cu_4L_4H_{-8}]^{4-}$ was identified, probably consisting of four $[CuLH_{-2}]^-$ monomers linked by imidazolate bridges via the pyrrole-like N^1-Im bonded to Cu(II) (Daniele, 1991). Zn(II) also has the ability to promote amide ionization in some His peptides, resulting in polynuclear complex formation.

The preference of Cu(II) for initial coordination to an imidazole-N donor in preference to a terminal amino-N has been investigated by studying Ala-Gly-Gly-His and its analogues with the various donor centers protected (Pettit et al., 1990). It was concluded that an imidazole-N of the His residue is the preferred binding site, while the next nitrogen to bind can be the terminal amino-N forming a 2N complex with a macrocyclic, 15-membered chelate ring. This is then followed by peptide-N deprotonation and coordination to give successively 3N and 4N complexes.

When a His residue which occupies an intermediate position within the peptide chain is coordinated to Cu(II), binding of peptide nitrogens following the His residue is hindered sterically. Thus in the Cu(II) complex of Gly-His-Gly there is no deprotonation of the amide-N of the Gly^3 residue, the predominant species being $[CuLH_{-1}]$. A ligand excess promotes formation of the bis $[CuL_2]$ complex (Farkas et al., 1984). For the same reason, 4N coordinated complexes are not found with LHRH (Gerega et al., 1988). The major complex of the biologically important blood plasma peptide Gly-His-Lys with Cu(II) and Ni(II) is $[MLH_{-1}]$, a 3N coordinated species without any bonding to the ε-amino group of the lysine residue. The high stability of the Cu(II) complex of Gly-His-Lys has been utilized by incorporating the tripeptide into a synthetic model of an anthraquinone anticancer drug, the efficacy of which is increased by the coordination of Cu(II). The Cu(II) catalyzes reactions which promote the formation of free radicals which have a deleterious effect on DNA (Pettit et al., 1992).

Other biologically important peptides in which N^3-Im is the primary anchor site are discussed in the section on peptide hormones and many examples of metalloproteins involving M-N^3-Im bonding (e.g. myoglobin, carboxypeptidase and bleomycin) will be found in other sections.

Substitution of amide hydrogens by metal ions is less favorable in *cyclo*-dipeptides than it is in their linear analogues. A potentiometric-calorimetric study of the Cu(II) complexes of *cyclo*-(His-His) shows that it forms extremely large (11-13 membered) chelate rings, bonded through the N^3-Im donors, as a consequence of the preferred conformation of the molecule (Arena et al., 1987). In *cyclo*-Gly-His-Gly-His-Gly-His-Gly, coordination is initiated at N^3-Im followed by amide deprotonation at higher pH and movement of the Cu(II) ion into the plane of the cyclo-peptide ring as shown in Figure 7 (Laussac et al., 1986).

G. Tyrosine residues

The phenolic group of tyrosine usually ionizes around pH 10 and, for steric reasons, is generally not a binding site for metal ions. There is, however, a possible interaction between the π-electron system of the aromatic ring and the empty d-orbitals of copper. This interaction can promote deprotonation and subsequent coordination of the peptide nitrogen

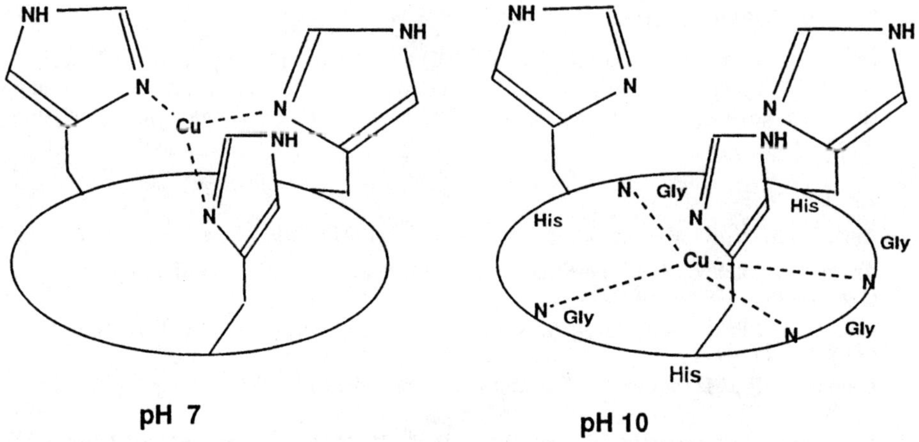

Figure 7.
Bonding centers in the Cu(II) complexes of *cyclo*-(Gly-His-Gly-His-Gly-His-Gly) at pH 7 and pH 10. After Laussac et al. (1986).

and, as a result, increase the stability of the resulting $[CuLH_{-1}]^-$ complex (Radomska et al., 1990). NMR, ESR and crystal structures reveal this is also the case for Pd(II) and Ni(II) (Vestues and Martin, 1980).

The Cu(II) complexes of peptides containing the Tyr residue are generally similar to simple peptides, but when Tyr is the N-terminal residue in dipeptides such as Tyr-Gly, Tyr-Leu and Tyr-Phe, a green dimeric species linked via deprotonated phenolate bridges, $[Cu_2LH_{-2}]^{2-}$, forms between pH 8.5-10.5 (Yamauchi, 1985; Radomska, 1990). The color is a result of an absorption band at about 420 nm caused by the phenolate-O^--Cu charge transfer transition. Dimerization of Tyr-Phe is greater than that of Tyr-Gly, because the folded dimer structure is stabilized by the stacking between the two parallel aromatic rings. When tyrosine is in the C-terminal position, dimeric species are not formed for steric reasons. With Co(II) and Zn(II), Gly-Tyr forms the expected bis complexes with amino-N and carbonyl-O coordination. With Cu(II) tri- and larger oligopeptides containing Tyr residues exhibit simple oligoglycine-like coordination since in $[CuLH_{-2}]$ the coordination sphere of the metal is satisfied and thus the formation of polynuclear complexes is excluded. This is demonstrated by Met- and Leu-enkephalins (Tyr-Gly-Gly-Phe-Met, Tyr-Gly-Gly-Phe-Leu) which exhibit 4N coordination without any significant Cu(II) to phenolate-O^- or Cu(II) to thioether interactions (Formicka-Kozlowska et al., 1985).

The introduction of a Pro residue into the peptide backbone can constrain the conformation of the peptide to allow the formation of very large chelate rings incorporating the phenolate-O^-; e.g. Tyr-Pro-Gly-Tyr (H_3L) forms an unusually stable [CuLH] complex coordinated via the terminal amino-N of Tyr^1, the neighboring peptide C=O and the phenolate-O^- of Tyr^4 resulting in a 17-membered chelate ring (Pettit et al., 1985). Tyr-Gly-Pro-Tyr (LH_3) initially forms a 1N $[CuLH_2]^+$ followed by a 2N [CuLH] complex in the normal fashion. However, Pro^3 acts as a break point to any further coordination, although a dimeric complex, $[(CuL)_2]^{2-}$, is present consisting of two [CuL] species linked by tyrosinate-O^--Cu bonds. Tyr-Gly-Pro-Phe forms a dimer between pH 8-10 while Phe-Gly-Pro-Tyr, unlike Tyr-Gly-Pro-Tyr, forms a monomeric complex with a large chelate ring (Livera et al., 1988).

III. PEPTIDE HYDROXAMATES

Two biologically active peptide hydroxamates, Pro-Leu-NHOH and Pro-Leu-Gly-NHOH, have been studied with Cu(II), Ni(II), Co(II), Zn(II) and Fe(III). The amino-N of the Pro^1 residue is the site of initial coordination followed by bonding to the hydroxamate-N. Both Cu(II) and Ni(II) are able to deprotonate and bond to peptide-N donors with Ni(II) forming diamagnetic complexes (Farkas and Kiss, 1990).

IV. DEHYDROPEPTIDES

Complexes of α,β-dehydropeptides with Cu(II) have been studied by Jezowska-Bojczuk et al. (1991). Complexes which include coordination to a deprotonated peptide-N are more stable than their dipeptide analogues, the effect being greatest with short side chains (e.g. Ala).

REFERENCES

Ama, T., Maki, R., Kawaguchi, H., and Yasui, T. (1991). *Bull. Chem. Soc. Jpn.*, *64*: 459.

Arena, G., Bonomo, R.P., Impellizzeri, G., Izatt, R., Lamb, J.D., and Rizzarelli, E. (1987). *Inorg. Chem.*, *26*: 795.

Bataille, M., Pettit, L.D., Steel, I., Kozlowski, H., and Tatarowski, T. (1985). *J. Inorg. Biochem.*, *24*: 211.

Bonomo, R.P., Maccarone, G., Rizzarelli, E., and Vidali, M. (1987). *Inorg. Chem.*, *26*: 2893.

Camerman, N., Camerman, A., and Sarkar, B. (1976). *Can. J. Chem.*, *54*: 1309.

Chou, P.Y., and Fasman, G.D. (1977). *J. Mol. Biol.*, *115*: 135.

Daniele, P.G. (1991). *J. Chem. Soc., Dalton Trans.*, 2771.

Decock-Le Révérend, B., Andrianarijaona, L., Livera, C., Pettit, L.D., Steel, I., and Kozlowski, H. (1986). *J. Chem. Soc., Dalton Trans.*, 2221.

Decock-Le Révérend, B., Lebkiri, A., Livera, C., and Pettit, L.D. (1986) *Inorg. Chim. Acta*, *124*: L19.

Decock-Le Révérend, B., Liman, F., Livera, C., Pettit, L.D., Pyburn, S., and Kozlowski, H. (1988). *J. Chem. Soc., Dalton Trans.*, 887.

Farkas, E., and Kiss, T. (1989). *Polyhedron*, *8*: 2463.

Farkas, E., and Kiss, T. (1990). *J. Chem. Soc., Dalton Trans.*, 749.

Farkas, E., Sóvágó, I., Kiss, T., and Gergely, A. (1984). *J. Chem. Soc., Dalton Trans.*, 611.

Formicka-Kozlowska, G., Kozlowski, H., Bezer, M., Pettit, L.D., Kupryszewski, G., and Przybylski, J. (1981). *Inorg. Chim. Acta*, *56*: 79.

Formicka-Kozlowska, G., Pettit, L.D., Steel, I., Livera, C.E., Kupryszewski, G., and Rolka, K. (1985). *J. Inorg. Biochem.*, *24*: 299.

Galey, J.-F., Decock-Le Révérend, B., Lebkiri, A., Pettit, L.D., Pyburn, S.I., and Kozlowski, H. (1991). *J. Chem. Soc., Dalton Trans.*, 2281.

Gerega, K., Kozlowski, H., Masiukiewicz, E., Pettit, L.D., Pyburn, S., and Rzeszotarska, B. (1988). *J. Inorg. Biochem.*, *33*: 11.

Gergely, A., and Farkas, E. (1982). *J. Chem. Soc., Dalton Trans.*, 381.

Jezowska-Bojczuk, M., Kozlowski, H., Sóvágó, I., Kubica, Z., Rzeszotarska, B., and Smelka, L. (1991). *Polyhedron*, *10*: 2331.

Kasselouri, S., Garoufis, A., Lamera-Hajiliadis, M., and Hadjiliadis, N. (1990). *Coord. Chem. Rev.*, *104*: 1.

Kennedy, W.R., and Margerum, D.W. (1985). *Inorg. Chem.*, *24*: 2490.

Kozlowski, H., and Pettit, L.D. (1991). *Chemistry of Platinum Group Metals, Recent Developments* (F.R. Hartley, ed.), Elsevier, p. 529.

Laurie, S.H. (1987). *Comprehensive Coordination Chemistry*, Vol. 2 (G. Wilkinson, ed.), Pergamon Press, Oxford, p. 739.

Laussac, J.-P., Robert, A., Haran, R., and Sarkar, B. (1986). *Inorg. Chem.*, *25*: 2760.

Livera, C.E., Pettit, L.D., Bataille, M., Perly, B., Kozlowski, H., and Radomska, B. (1987). *J. Chem. Soc., Dalton Trans.*, 661.

Livera, C., Pettit, L.D., Bataille, M., Krembel, J., Bal, W., and Kozlowski, H. (1988). *J. Chem. Soc., Dalton Trans.*, 1357.

Lyons, A., and Pettit, L.D. (1984). *J. Chem. Soc., Dalton Trans.*, 2305.

Martin, R.B. (1988). *Metals Ions In Biological Systems*, Vol. 23 (H. Sigel, ed.), Marcel Dekker, New York, p. 123.

Odani, A., Deguchi, S., and Yamauchi, O. (1986). *Inorg. Chem.*, *25*: 62.

Pert, C.B., Hill, J.M., Ruff, M.R., Beman, R.M., Robert, W.G., Arthur, L.O., Ruscetti, F.W., and Farrar, W.L. (1986). *Proc. Nat. Acad. Sci. USA*, *83*: 9254.

Pettit, L.D., and Lyons, A.Q. (1986). *J. Chem. Soc., Dalton Trans.*, 499.

Pettit, L.D., Steel, I., Formicka-Kozlowska, G., Tatarowski, T., and Bataille, M. (1985). *J. Chem. Soc., Dalton Trans.*, 535.

Pettit, L.D., Steel, I., Kowalik, T., Kozlowski, H., and Bataille, M. (1985). *J. Chem. Soc., Dalton Trans.*, 1201.

Pettit, L.D., Pyburn, S., Kozlowski, H., Decock-Le Révérend, B., and Liman, F. (1989). *J. Chem. Soc., Dalton Trans.*, 1471.

Pettit, L.D., Pyburn, S., Bal, W., Kozlowski, H., and Bataille, M. (1990). *J. Chem. Soc., Dalton Trans.*, 3565.

Pettit, L.D., Bal, W., Bataille, M., Cardon, C., Kozlowski, H., Leseine-Delstanche, M., Pyburn, S., and Scozzafava, A. (1991). *J. Chem. Soc., Dalton Trans.*, 1651.

Pettit, L.D., Gregor, J.E., and Kozlowski, H. (1991). *Perspectives on Bioinorganic Chemistry*, Vol. I (R.W Hay, J.R. Dilworth and K.B. Nolan, eds.), Jani Press, London, p. 1.

Pettit, L.D., Ueda, J.-I., Morier-Teisser, E., Helbecque, N., Bernier, J.-L., Hénichart, J.-P., and Kozlowski, H. (1992). *J. Inorg. Biochem.*, *45*: 203.

Rabenstein, D.L., Daignault, S.A., Isab, A.A., Arnold, A.P., and Shoukry, M.M. (1985). *J. Amer. Chem. Soc.*, *107*: 6435.

Radomska, B., Sóvágó, I., and Kiss, T. (1990). *J. Chem. Soc., Dalton Trans.*, 289.

Sigel, H., and Martin, R.B. (1982). *Chem. Rev.*, *82*: 385.

Sigel, H., Naumann, C.F., Prijs, B., McCormick, D.B., and Falk, M.C. (1977). *Inorg. Chem.*, *16*: 790.

Sóvágó, I. (1990). *Biocoordination Chemistry* (K. Burger, ed.), Ellis Horwood, London, p. 135.

Sóvagó, I., Farkas, E., and Gergely, A. (1982). *J. Chem. Soc., Dalton Trans.*, 2159.

Sóvágó, I., Kiss, T., and Gergely, A. (1984). *Inorg. Chim. Acta.*, *93*: L53.

Sóvágó, I., Radomska, B., Schon, I., and Nyeki, O. (1990). *Polyhedron*, *9*: 825.

Udo, J., Ikota, N., Hanaki, A., and Koga, K. (1987). *Inorg. Chim. Acta*, *135*: 43.

Varnagy, K., Kozlowski, H., Sóvagó, I., Jankowska, T.K., Kruszynski, M., and Zboinska, J. (1988). *J. Inorg. Biochem.*, *34*: 83.

Veber, M., Horvath, I., and Burger, K. (1985). *Inorg. Chim. Acta*, *108*: L21.

Vestues, P.I., and Martin, R.B. (1980). *J. Amer. Chem. Soc.*, *102*: 7906.

Xiao, L., Jouini, M., Fan, B.T., and Lapluye, G. (1990). *J. Chem. Soc., Dalton Trans.*, 1137.

Yamauchi, O., Tsujide, K., and Odani, A. (1985). *J. Amer. Chem. Soc.*, *107*: 659.

INTERACTION OF METAL IONS WITH PEPTIDE HORMONES

by

Leslie D. Pettit and Rebecca A. Robbins

School of Chemistry,
The University of Leeds,
Leeds, LS2 9JT
U.K.

I. INTRODUCTION

In all multicellular organisms, communication between cells is necessary to coordinate the activities of the cells. Communication can take place by cell-to-cell contact, allowing the transfer of small molecules or ions, but communication over greater distances requires extracellular molecules which can act as signals. These specific molecules, hormones, are synthesized and released by the signalling cell and then travel to the target cell where they interact with specific receptor sites. In a multicellular biological system, the signalling molecule (hormone) interacts with a specific receptor site through a combination of steric, conformational and electrostatic interactions; it therefore carries both the information to be transmitted and the address information of its destination. If the conformation of the hormone is changed by interaction with a metal ion then this address information may also be changed, and with it the biological activity of the hormone. Long distance signalling hormones, many of which are peptides or peptide derivatives, are usually carried to the target sites by the blood. Information transmission from one nerve cell to another across a synaptic gap is an example of short range signalling and is carried by neurotransmitters, many of which are short-chain peptides (neuropeptides). For the purpose of this review the term peptide will be restricted to oligopeptides of less than about 20 amino acid residues and the formulae of some are shown in Table 1. Oligopeptide hormones may be formed as a result of the cleavage of larger peptide hormones such as pro-opiocortin, which is the precursor to several opiate peptide hormones.

II. INTERACTION WITH METAL IONS

A. Aqueous media

Hormone oligopeptides are often flexible and usually have terminal donor groups (in particular an N-terminal amino group) able to act as initial anchoring sites for metal ions. From the point of view of metal coordination, they can therefore be distinguished from the many long chain polypeptide hormones (e.g. insulin) in which terminal donor groups are few and comparatively inaccessible, and which have recognizable 3-dimensional structures. Most peptide hormones contain amino acid residues which have effective side chain donor groups (e.g. His, Lys or Asp) and, as a result, would be expected to coordinate with a range of metals, especially transition metal ions. In vivo, there is virtually no evidence for

Table 1

Some Peptide Hormones for which Interaction with Metal Ions Has Been Studied

Peptide	Sequence
TRH Thyrotropin releasing hormone	pGlu-His-Pro-NH_2
Arg^8-Vasopressin	Cys-Tyr-Phe-Gln-Asn-Cys-Pro-Arg-Gly-NH_2 (S — S bridge Cys1–Cys6)
Arg^8-Vasotocin	Cys-Tyr-Ile-Gln-Asn-Cys-Pro-Arg-Gly-NH_2 (S — S bridge Cys1–Cys6)
Oxytocin	Cys-Tyr-Ile-Gln-Asn-Cys-Pro-Leu-Gly-NH_2 (S — S bridge Cys1–Cys6)
LHRH Luteinizing hormone-releasing hormone	pGlu-His-Trp-Ser-Tyr-Gly-Leu-Arg-Pro-Gly-NH_2
Fibrinopeptide A	Ala-Asp-Ser-Gly-Glu-Gly-Asp-Phe-Leu-Ala-Glu-Gly-Gly-Gly-Val-Arg
$Thymopoietin^{32\text{-}36}$	Arg-Lys-Asp-Val-Tyr
Met-Enkephalin	Tyr-Gly-Gly-Phe-Met
Leu-Enkephalin	Tyr-Gly-Gly-Phe-Leu
Substance P	Arg-Pro-Lys-Pro-Glu-Glu-Phe-Phe-Gly-Leu-Met-NH_2
β-Casomorphin-5	Tyr-Pro-Phe-Pro-Gly
Bradykinin	Arg-Pro-Pro-Gly-Phe-Ser-Pro-Phe-Arg
Angiotensin II	Asp-Arg-Val-Tyr-Ile-His-Pro-Phe
Somatostatin	Ala-Gly-Cys-Lys-Asn-Phe-Phe-Trp-Lys-Thr-Phe-Thr-Ser-Cys

the direct coordination of metal ions as part of the biological activity of peptide hormones, although there is often a relationship between their activity and high concentrations of some biologically important metals.

Coordination of a peptide to a metal ion, particularly one able to deprotonate and bond to peptide-N^- donors, would be expected to have a profound effect on the conformation of the peptide and, as a result, on its ability to interact with receptor sites (Pettit and Formicka-Kozlowska, 1984). With short-range hormones (e.g. neuropeptides) the kinetics of complex formation would have to be fast compared to the time-scale of hormone signalling if coordination is to have any biological significance and, with Cu(II), this is the case. Studies of the thermodynamics of complex formation in aqueous solution have only been performed with Cu(II) and, to a lesser extent, with Ni(II) and Zn(II). The only biologically important metal ion able to deprotonate peptide nitrogens around neutral pH is Cu(II), hence this is the ion most likely to influence the conformation of peptide hormones. With other metal ions, which will tend to form weaker complexes, significant concentrations of chelated complex are unlikely to be present, unless the concentrations of both free metal ion and peptide are in the millimolar range or complex formation is strongly favored by the conformation of the peptide molecule. Transient species at low concentration, however, may have a biological effect and coordination centers have been identified; e.g. Mn(II) has been shown to interact with an imidazole nitrogen in angiotensin II (Basosi et al., 1984).

B. Effects of lipid surfaces

Stability constants calculated in aqueous solution with an inert background electrolyte (e.g. NaCl) may not provide a satisfactory model for metal-peptide interaction in biological fluids and in the presence of lipid surfaces and micelles. Recent work has shown that the conformations of linear oligopeptides are profoundly affected by the presence of sodium dodecyl sulphate (SDS) which forms hydrophobic micelles able to interact with peptides (especially if they contain positively charged side chains) and so act as a model for a lipid surface. Conformational changes will clearly influence metal complex stability constants and hence metal-ligand speciation, and very large changes have been found in a number of systems with Cu(II).

III. COMPLEXES WITH ALKALI AND ALKALINE EARTH METAL IONS

Metal ions from groups 1 and 2 form only very weak complexes with simple peptides; thus direct complex formation would not be expected to be significant in biological systems. However, an excess of Mg^{2+} inhibits the activity of a variety of peptide hormones and these, in turn, can affect the biological transport of Mg^{2+} (Altura and Altura, 1990). The biological functions of angiotensin II have been linked to Li^+, Na^+, Ca^{2+} and Mg^{2+} (Günther et al., 1980) but coordination does not appear to influence the peptide conformation significantly (Leninski and Stephens, 1981). Complexation between Ca^{2+} and Leu-enkephalin and some related ligands has been studied fluorimetrically (Haran et al., 1982) and, in the gas phase, complexes of Mg^{2+}, Ca^{2+} and Ba^{2+} have been investigated using fast atom bombardment and collisional activation mass spectrometry (Teesch and Adams, 1990).

Ca^{2+} is undoubtedly the most important metal ion in signalling mechanisms (Rasmussen, 1989) and binds to the protein calmodulin (148 amino acid units) through Asp, Glu, Asn, Ser and Thr amino acid residues (Cheung, 1982), which are all potential O-donors. On coordination the conformation of the calmodulin is changed, so influencing its interaction with other proteins. However, there are no quantitative results for specific calcium complexes of oligopeptides.

IV. COMPLEXES WITH ZINC IONS

The interaction of Zn(II) with proteins is well established (e.g. carboxypeptidase and zinc fingers) but complexes with oligopeptides are generally weak and there is little evidence for significant binding to peptide hormones. With LHRH, Zn(II) bonds through an imidazole-N and the His-Trp peptide carbonyl-O (Bal et al., 1989) while with human fibrinopeptide A (a hexadecapeptide with an Asp^2 residue) there is little binding below pH 7, the major complex being a partially hydrolysed species (Galey et al., 1991).

V. COMPLEXES WITH COPPER IONS

In the body, copper is concentrated in the brain where the concentration is 10 times that in the blood. It does not appear to have a defined role although it is probably involved in the in vivo synthesis of neuropeptides, particularly the α-amidation step (Eipper et al., 1992). The activity of some peptide hormones is enhanced by the presence of copper, e.g. Met-enkephalin (Sharrock et al., 1982) and LHRH (Hartter and Barnea, 1988) while with others it is decreased, e.g. TRH (Tonoue et al., 1979) and, in other cases, there is a relationship between copper deficiency and peptide hormone activity, e.g. atrial natriuretic factor (Bhathena et al., 1988). However, there is no direct evidence that specific complex formation between the peptide and Cu(II) is the cause of modified hormone activity. Quantitative studies of the complexation of peptide hormones with Cu(II) have been performed on the following systems, which will be considered separately.

A. Thyrotropin Releasing Hormone (TRH)

One of the simplest peptide hormones is the N- and O-protected tripeptide pyro-Glu-His-Gly-NH_2 (TRH, L) and two independent studies have given almost identical results (Formicka-Kozlowska et al., 1983; Farkas et al., 1984). The initial anchoring site is an imidazole nitrogen to give a 1N coordinated complex. This is only a minor species and is followed by the simultaneous deprotonation of two amide protons to give a neutral complex with 3N coordination, $[CuLH_{-2}]$, which is the major species at physiological pH. In this

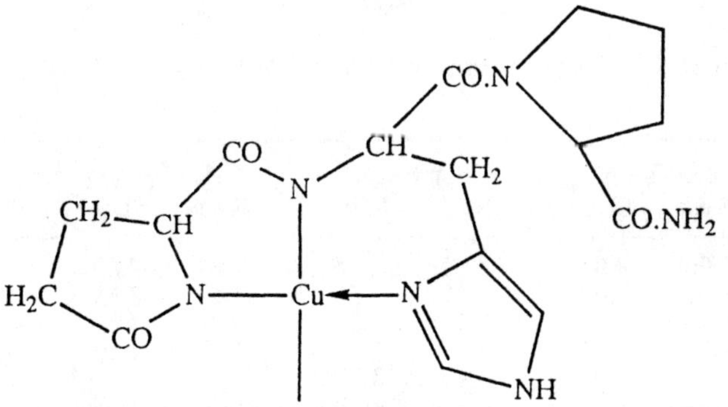

Figure 1.
The 3N complex of TRH with Cu(II), $[CuLH_{-2}]$.

complex, the conformation of the peptide is rigidly constrained as shown in Figure 1, possibly adversely affecting its biological activity (Tonoue et al., 1979).

B. Luteinizing Hormone Releasing Hormone (LHRH)

LHRH (HL) is a decapeptide which, like TRH, is doubly protected and with the same two N-terminal residues. Its activity is also influenced by the presence of Cu(II) which can stimulate the in vivo release of LHRH, particularly when coordinated to histidine or cysteine (Barnea et al., 1984; 1986). Coordination to Cu(II) has been shown, from potentiometric and spectroscopic studies, to proceed in a similar way to that of TRH (Gerega et al., 1988). The 1N complex (bonded through an imidazole-N) readily loses two amide protons simultaneously to form a 3N, $[CuLH_{-1}]$, complex at about pH 5.5 with the Tyr residue protonated as shown in the species distribution curves in Figure 2. There was no evidence for 4N coordination, even at high pH.

In the presence of 0.1 mol dm^{-3} SDS, LHRH (LH) behaves very differently (Bal et al., 1993). The proton complexes of both the tyrosinate-O^- and imidazole-N donors are more stable by about 0.75 log units while the 3N complex, $[CuLH_{-1}]$, is less stable as shown in Table 2. The decrease in the stability of the 3N Cu(II) complex of LHRH is, in all probability, a result of conformational changes forced on the LHRH molecule when its hydrophobic side chains (especially Trp) are drawn into the micelle and the positively charged Arg residue is held by the anionic SDS 'head'.

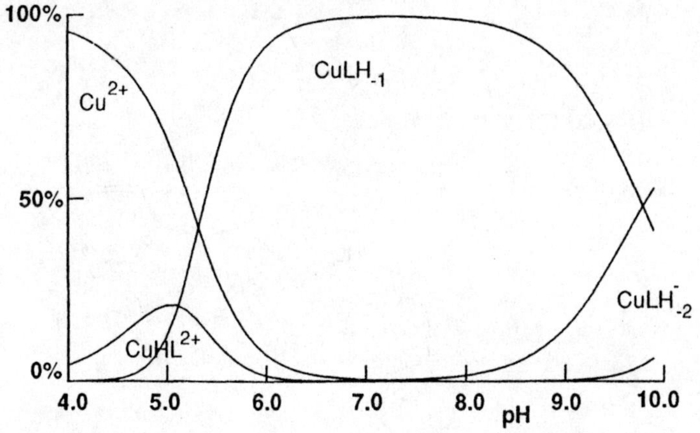

Figure 2.
Species distribution curves for 1:1 mixtures of LHRH and Cu(II) (0.001 mol dm^{-3}) as a function of pH.

Table 2

The Effect of SDS on Complex Formation of some Peptides with H^+ and Cu(II) (25°C)

Peptide	log $K_{(LH)}$		log $K_{(LH_2)}$		log $\beta_{(3N)}$		log $\beta_{(4N)}$	
	NaCl	SDS	NaCl	SDS	NaCl	SDS	NaCl	SDS
Gly^4	7.96	8.01	3.31	3.42	-7.47	-7.69	-16.78	-16.93
Ala^4	8.04	8.09	3.39	3.63	-7.69	-8.63	-16.84	-17.86
AASG	8.12	8.27	3.76	4.02	-6.87	-7.81	-16.06	-17.10
LHRH	9.74	10.49	6.27	7.18	3.38	2.49		
AVP	9.33	10.93	6.40	8.28			-14.76	-20.09

C. Melanostatin

Melanostatin, Pro-Leu-Gly-NH_2 (L), is a hypothalamic hormone which forms a stable 4N complex with Cu(II) by pH 9 (Formicka-Kozlowska et al., 1981). This neutral species, $[CuLH_{-2}]$, is intensely red in color and involves coordination through the deprotonated C-terminal amido-N^- (see Chapter 2-B1).

D. Vasopressin, vasotocin and oxytocin

Arg^8-Vasopressin (AVP), vasotocin and oxytocin are all nonapeptide-amides locked into a ring by a disulphide bridge between Cys^1 and Cys^6 residues (see Table 1). All three ligands form abnormally stable complexes with Cu(II), forming 4N species by pH 6.4 to 7 (Kozlowski et al., 1989; Bal et al., 1992). This must be a result of the favorable conformation in which the four nitrogen donors are held in the uncoordinated molecule, creating a cavity site ideally suited to the Cu(II) ion (see Fig. 3). If the ring conformation is

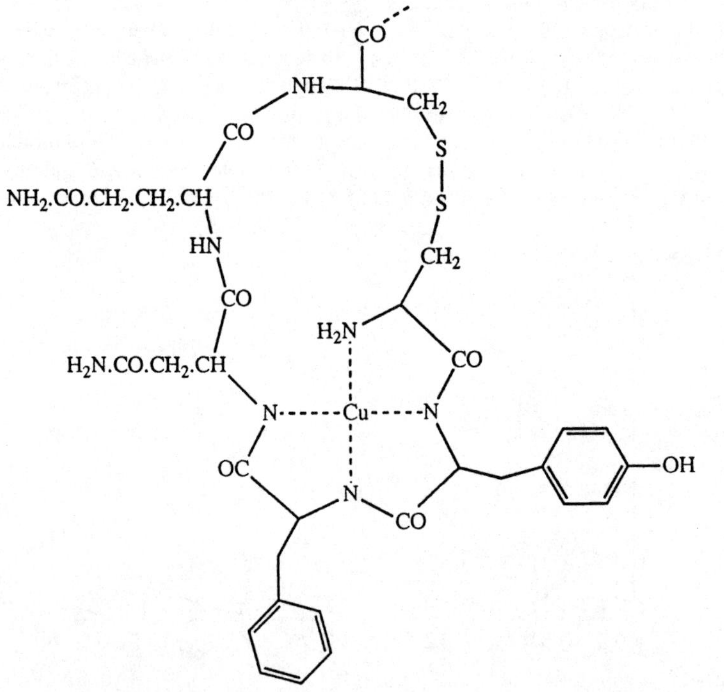

Figure 3.
The proposed 4N complex of Arg^8-vasopressin with Cu(II).

changed by introducing an amino acid residue of opposite chirality (see Chapter 2-A2) much of this stabilization is lost. The increased stability is also mostly removed by the condensation of an Ala residue on the N-terminal Cys residue to form Ala-Arg^8-vasopressin (Bal et al., 1992). Somatostatin and related peptides also contain disulphide bridges and so may also form chelated complexes of unexpected stability.

The effect of 0.1 mol dm^{-3} SDS is even more marked than with LHRH (Pettit et al., 1992) as shown in Table 2. Protonation constants for Tyr-O$^-$ and terminal-NH_2 are increased by over 1.6 log units and the speciation of complex formation is more similar to that for tetraalanine (Ala^4) and Cu(II). For example, in aqueous solution with an NaCl background, the 4N complex of AVP forms by pH 6.4 and of Ala^4 around pH 9 while in SDS the corresponding values are both around 9.1. These effects are illustrated in Figure 4 which demonstrates that while Ala^4 is hardly affected by the presence of SDS micelles, the effect on AVP is profound.

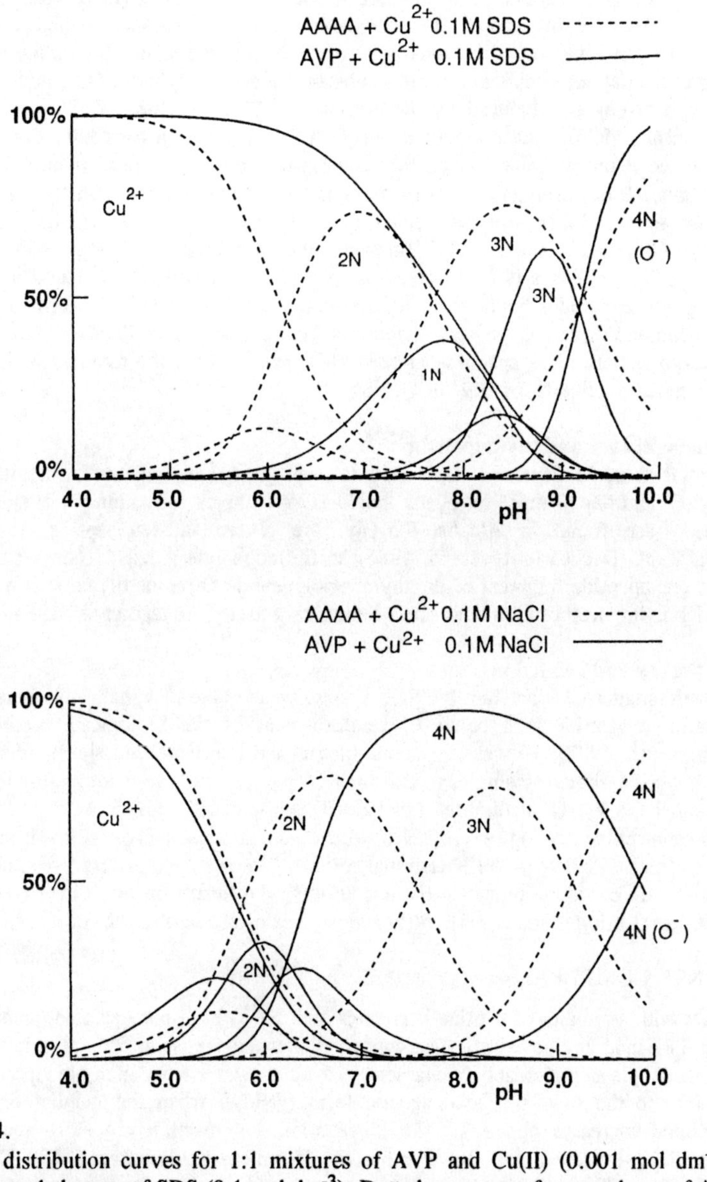

Figure 4.
Species distribution curves for 1:1 mixtures of AVP and Cu(II) (0.001 mol dm^{-3}) in the presence and absence of SDS (0.1 mol dm^{-3}). Dotted curves are for complexes of Ala^4.

E. Angiotensin II

Angiotensin II is an octapeptide with both a terminal amino donor center and a His^6 residue. With Cu(II), initial coordination takes place through the imidazole nitrogen, followed by two peptide deprotonations and coordination to give a 3N complex. Around pH 8.7 the Cu(II) ion appears to change coordination sites completely and move along the peptide backbone to form a 4N complex starting at the N-terminal amino group. This complex is closely similar to that formed by the N-terminal tetrapeptide fragment, Asp-Arg-Val-Tyr, alone (Decock-le-Révérend et al., 1988). A similar reorganization of coordination sites from imidazole-N to peptide-N donors on increase in pH has been reported in the Cu(II) complexes with *cyclo*-(Gly-His-Gly-His-Gly-His-Gly) illustrated in Chapter 2-B1 (Laussac et al., 1986).

F. Substance P and bradykinin

These hormone peptides contain few side chains able to coordinate to metal ions, so would not be expected to form very stable complexes. The Pro residue, which contains a secondary nitrogen atom, acts as a break-point to coordination and also favors the formation of β-turns in the peptide backbone. In both Substance P and bradykinin, the conformation of the peptide is strongly influenced by the presence of SDS micelles (Woolley and Deber, 1987; Lee et al., 1990), with a resultant effect on the protonation constants. For example, with Substance P the protonation constant is increased by 2 log units. In the absence of SDS, Substance P has been shown to form 1N and 2N complexes with Cu(II) which are of virtually the same stability as those formed by the N-terminal pentapeptide fragment, Arg-Pro-Lys-Pro-Gln (Pettit et al., 1991). These involve coordination to amino-N donors only (N-terminal-NH_2 and Lys-ε-NH_2) to give a large chelate ring without any Cu(II) to peptide-N^- bonding. Pentapeptides beginning with the amino acid sequence X-Pro-Pro- are models for bradykinin and these behave very differently from peptides with X-Pro-Y-Pro- residues, forming large chelate rings spanning at least 10 atoms between the terminal amino-N and the deprotonated peptide N^- of Gly^4 or Gly^5.

G. Fibrinopeptide A and thymopoietin$^{32\text{-}36}$

These peptides contain Asp^2 and Asp^3 residues respectively, and chelate formation between a carboxyl-O^- and one or more nitrogens should favor complex formation. Such stabilization has indeed been found in Ala-Asp-Ser-Gly, the N-terminal tetrapeptide fragment of fibrinopeptide A (Decock-le-Révérend, 1986), in fibrinopeptide A itself (Galey et al., 1991) and in the pentapeptide fragment of the thymus polypeptide hormone thymopoietin (residues $^{32\text{-}36}$) which is important to the hormone's biological activity (Sóvágó et al., 1984; 1990).

H. Enkephalins and related peptides

It has been suggested that the biological activity of Met-enkephalin is enhanced by complexation to Cu(II) as a result of reinforcement of the biologically active β-turn (Sharrock et al., 1982). However, potentiometric and spectroscopic studies demonstrate coordination speciation very similar to that for tetraglycine, with no evidence for interaction between either the Tyr-O^- or the Met-S and Cu(II) (Formicka-Kozlowska et al., 1985).

β-Casomorphins are opioid peptides which contain the X-Pro-Y-Pro- fragment (cf Substance P) but with Tyr as the N-terminal residue. The presence of this Tyr residue has a marked effect on coordination to Cu(II), encouraging the formation of a dimeric complexes with Cu-O^-(Tyr) bridges above pH 6.5 (Formicka-Kozlowska et al., 1984).

VI. NICKEL COMPLEXES

Complexes with a number of peptide hormones with Ni(II) have been studied quantitatively, including TRH and angiotensin II. The coordination sequence is, in general, similar to that found with simple peptides and is characterized by a switch from high spin paramagnetic Ni(II) (blue) to the low spin, diamagnetic form (yellow) when the number of nitrogen donors bound increases above 2. The kinetics of this reaction are slow and there is generally little evidence for 3N complexes (Formicka-Kozlowska, 1983; Decock-le-Révérend, 1986). Since Ni(II) cannot deprotonate peptide-N atoms as readily as Cu(II) can, it would not be expected to have as much influence on peptide conformation, and there are few Ni-N$^-$ bonds at physiological pH. This may account for the very different effects which Cu(II) and Ni(II) have on the biological activity of TRH (Tonoue et al., 1979).

Complexes of angiotensin II with Ni(II) have been studied using potentiometry and CD and NMR spectroscopies. The results show that, above pH 8, Ni(II) coordinates to four N-donors, starting at the Asp residue, without any bonding through an imidazole-N (Pettit et al., 1989).

VII. CONCLUSION

Many peptide hormones form stable complexes with transition metal ions in aqueous solution. In biological systems, the concentrations of both ligand and metal ion are generally low and other competing ligands are present, with the result that significant concentrations of metal complexes will be found with only strongly coordinating metal ions such as Cu(II), or where there are locally high concentrations of peptide. On coordination, the conformation of a peptide would, in all probability, be altered significantly and this must affect its biological activity by increasing (or possibly decreasing) its ability to interact with a receptor site. Preliminary results of studies of the coordination of peptide hormones to metal ions in the presence of models for lipid surfaces suggest that complex formation may be less important than studies in aqueous solution might imply.

REFERENCES

Altura, B.M., and Altura, B.T. (1990). *Metal Ions in Biological Systems*, Vol. 26 (H. Sigel and A. Sigel, eds.), Marcel Dekker, New York, p. 359 and refs. therein.

Bal, W., Kozlowski, H., Masiukiewicz, E., Rzeszotarska, B., and Sóvágó, I. (1989). *J. Inorg. Biochem.*, *37*: 135.

Bal, W., Kozlowski, H., Lammek, B., Pettit, L.D., and Rolka, K. (1992). *J. Inorg. Biochem.*, *45*: 193.

Bal, W., Kozlowski, H., Lisowski, M., Pettit, L., Robbins, R., and Safavi, A. (1994). *J. Inorg. Biochem.*, in press.

Barnea, A., and Cho, G. (1984). *Endocrinology*, *115*: 936.

Barnea, A., Cho, G., and Colombani-Vidal, M. (1986). *Brain Res.*, *384*: 101.

Basosi, R., Gaggelli, E., and Valensin, G. (1984). *J. Inorg. Biochem.*, *20*: 263.

Bhathena, S.J., Kennedy, B.W., Smith, P.M., Fields, M., and Zamir, M. (1988). *J. Tr. Elem. Exp. Med.*, *1*: 199.

Cheung, W.Y. (1982). *Sci. Amer.*, *246*(6): 48.

Decock-le-Révérend, B., Andrianarijaona, L., Livera, C., Pettit, L.D., Steel, I., and Kozlowski, H. (1986). *J. Chem. Soc., Dalton Trans.*, 2221.

Decock-le-Révérend, B., Liman, F., Livera, C., Pettit, L.D., Pyburn, S., and Kozlowski, H. (1988). *J. Chem. Soc., Dalton Trans.*, 887.

Eipper, B.E., Stoffers, D.A., and Mains, R.E. (1992). *Annu. Rev. Neurosci.*, *15*: 57.

Farkas, E., Sóvágó, I., Kiss, T., and Gergely, A. (1984). *J. Chem. Soc., Dalton Trans.*, 611.

Formicka-Kozlowska, G., Kozlowski, H., Bezer, M., Pettit, L.D., Kupryszewski, G., and Przybylski, J. (1981) *Inorg. Chim. Acta*, *56*: 79.

Formicka-Kozlowska, G., Bezer, M., and Pettit, L.D. (1983). *J. Inorg. Biochem.*, *18*: 335.

Formicka-Kozlowska, G., Pettit, L.D., Steel, I., Hartrodt, B., Neubert, K., Rekowski, P., and Kupryszewski, G. (1984). *J. Inorg. Biochem.*, *22*: 155.

Formicka-Kozlowska, G., Pettit, L.D., Steel, I., Livera, C., Kupryszewski, G., and Rolka, K. (1985). *J. Inorg. Biochem.*, *24*: 299.

Galey, J.-F., Kozlowski, H., and Pettit, L.D. (1991). *J. Inorg. Biochem.*, *44*: 149.

Gerega, K., Kozlowski, H., Masiukiewicz, E., Pettit, L.D., Pyburn, S., and Rzeszotarska, B. (1988). *J. Inorg. Biochem.*, *33*: 11.

Günther, S., Gimbourne, M.A., and Alexander, R.W. (1980). *Circ. Res.*, *47*: 267.

Haran, R., Sharrock, P., Gairin, J., and Mazarguil, H. (1982). *Int. J. Peptide Prot. Res.*, *20*: 380.

Hartter, D.E., and Barnea, A. (1988). *Synapse*, *2*: 412.

Kozlowski, H., Radomska, B., Kupryszewski, G., Lammek, B., Livera, C., Pyburn, S., and Pettit, L.D. (1989) *J. Chem. Soc., Dalton Trans.*, 173.

Lee, S.C., Russell, A.F., and Laidig, W.D. (1990). *Int. J. Peptide Prot. Res.*, *35*: 367.

Leninski, R.E., and Stephens, R.L. (1981). *J. Inorg. Biochem.*, *15*: 95.

Laussac, J.-P., Robert, A., Haran, R., and Sarkar, B. (1986). *Inorg. Chem.*, *25*: 2760.

Pettit, L.D., and Formicka-Kozlowska, G. (1984). *Neurosci. Lett.*, *50*: 53.

Pettit, L.D., Pyburn, S., Kozlowski, H., Decock-Le-Révérend, B., and Liman, F. (1989). *J. Chem. Soc., Dalton Trans.*, 1471.

Pettit, L.D., Bal, W., Bataille, M., Cardon, C., Kozlowski, H., Leseine-Delstanche, M., Pyburn, S., and Scozzafava, A. (1991). *J. Chem. Soc., Dalton Trans.*, 1651.

Rasmussen, H. (1989). *Sci. Amer.*, *261*(4): 44.

Sharrock, P., Day, R., Lemaire, S., St-Pierre, S., Mazarguil, H., Gairin, J.E., and Haran, R. (1982). *Inorg. Chim. Acta*, *66*: 91.

Sóvágó, I., Kiss, T., and Gergely, A. (1984). *Inorg. Chim. Acta*, *93*: L53.

Sóvágó, I., Radomska, B., Schon, I., and Nyeki, O. (1990). *Polyhedron*, *9*: 825.

Teesch, L.M., and Adams, J. (1990). *J. Amer. Chem. Soc.*, *112*: 4110.

Tonoue, T., Minagawa, S., Kata, N., and Ohki, K. (1979). *Pharmacol. Biochem. Behav.*, *10*: 201.

Woolley, G.A., and Deber, C.M. (1987). *Biopolymers*, *26*: S109.

METAL ION COMPLEXES WITH SULFHYDRYL-CONTAINING PEPTIDES

by

Imre Sóvágó

Kossuth Lajos Tudományegyetem,
Szervetlen és Analitikai Kémia Tanszék,
4010 Debrecen
Hungary

I. INTRODUCTION

Proteins and the various poly- and oligopeptides comprise the major groups of organic substances that bind metal ions in living systems. The coordination chemistry of peptide molecules was discussed in the previous chapters and the results clearly indicate that the binding mode in the transition metal complexes largely depends on the presence and on the location of the side-chain donor groups of peptide molecules. Among them the imidazolyl group of the histidyl residue and the mercapto group of the cysteinyl residue have the most pronounced effect on the metal binding ability of peptide molecules.

The importance of the sulfhydryl group of L-cysteine has already been demonstrated by the amino acid complexes of the ligand as discussed in a previous Chapter (Part Three, Chapter 2-A1). On the other hand, metalloenzymes provide many examples on the role of cysteinyl residues in proteins (e.g. blue copper proteins, iron-sulfur proteins, metallothioneins, zinc fingers and other zinc- or copper-containing metalloproteins; see Part Two, Chapter 1). The outstanding role of the L-cysteine residue in metalloenzymes is due to its unique chemical behavior as compared to other amino acids. This is reflected in the soft character of the thiol compounds and the capability of these for redox reactions. Because of the high frequency of the occurrence of metal-sulfur bond in enzymes, it is out of the scope of this review to describe all important features of thiol coordination in biological substances. The interaction of metal ions with peptides containing L-cysteine and its derivatives will only be discussed, and these results can serve as models for the deeper understanding of the structure and function of metalloenzymes. Some of the results have already been reviewed by several authors (Sigel and Martin, 1982; Sóvágó, 1990; Pettit et al., 1991).

II. SULPHYDRYL GROUP AS AN "ANCHOR"

The coordination chemistry of simple oligopeptide molecules has already been discussed elsewhere in this book (Part Three, Chapter 2-B1). It is obvious from these studies that peptide molecules can easily form stable complexes with various transition metal ions, and generally the terminal amino and deprotonated amide nitrogen and carboxylate oxygen donors take part in metal ion coordination. The formation of the amide-bonded metal

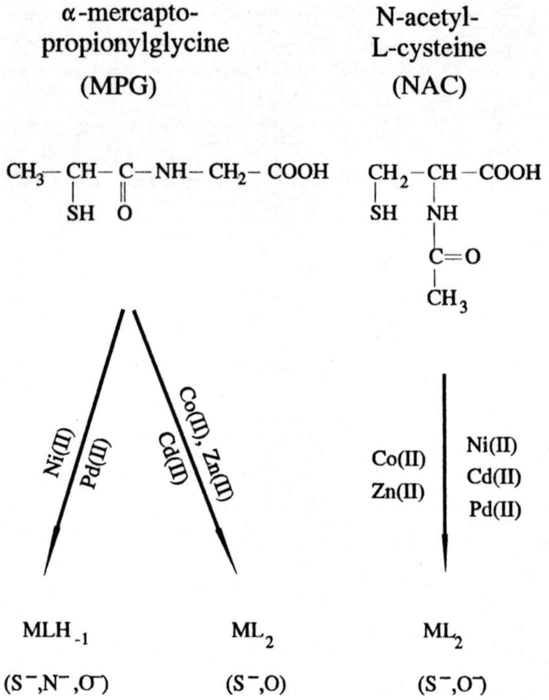

Figure 1.
Complex formation processes of MPG and NAC with transition metal ions.

complexes requires the presence of an anchoring donor group, which should be in chelatable position relative to the amide or peptide bond. The terminal amino group is the most common anchor, but imidazole nitrogen or phosphate oxygen have also been reported to work as anchors (Bryce et al., 1966; Kiss et al., 1989).

Taking into account the high affinity of the deprotonated sulfhydryl groups to soft metal ions, the sulfur donors should be considered as effective candidates of anchors for amide binding. N-acetyl-L-cysteine (NAC) (or N-acetyl-D-penicillamine) and α-mercaptopropionylglycine (MPG) are the simplest ligands which contain both the amide and the thiol groups, and they can be used to demonstrate the anchoring ability of sulfhydryl-containing molecules. The possible complex formation processes and the binding modes of the ligands in the various metal complexes are shown in Figure 1.

It can be seen from Figure 1 that formation of bis complexes is exclusive during the reaction of metal ions with N-acetyl-L-cysteine (Sóvágó and Martin, 1981; Harman and Sóvágó, 1983; Kozlowski et al., 1990). Metal ions are coordinated via sulfur and carboxylate oxygen donors, forming six-membered chelate rings in the bis complexes. Polynuclear species via sulfhydryl bridges were also detected in some cases (e.g. nickel(II), cobalt(II) and especially with cadmium(II)), which support the finding that the sulfur atom is the main binding site in this ligand too, but it does not promote deprotonation and coordination of the amide nitrogens if other donor groups (e.g. carboxylate) are present in chelatable position with the sulfhydryl group.

Further details on the anchoring ability of sulfhydryl group can be obtained from the results obtained for the metal complexes of MPG. Figure 1 clearly demonstrates that complex formation processes of the ligand largely depend on the nature of the metal ions. It has been proved by several authors that nickel(II) and palladium(II) (which are soft in character) are able to induce the deprotonation and coordination of the amide group in MPG (Sugiura et al., 1975a; Sóvágó and Martin, 1981; Filella and Williams, 1985). The (S,N,O)-chelation, however, will not saturate the coordination sphere of the metal ions in the $NiLH_{-1}$ square planar, diamagnetic species; therefore there are possibilities for bis complex formation (NiL_2H_{-1}) in which the second ligand is bonded monodentately, or for polynuclear complex formation via sulfhydryl bridges. The latter is supported by recent structural studies, where a trimeric nickel(II) complex of MPG was isolated and characterized (Baidya et al., 1989). The formation of the amide-coordinated species was, however, ruled out during the complexation with other transition elements, e.g. cobalt(II), zinc(II), cadmium(II) and lead(II) (Sugiura et al., 1975a; Harman and Sóvágó, 1983; Filella and Williams, 1985, Kozlowski et al., 1990). The formation of bis (or tris) complexes via the coordination of sulfur and carbonyl-O donors (five-membered chelate) was detected with these metal ions.

On the basis of the results obtained for the metal complexes of NAC and MPG it can be concluded that the sulfhydryl group is the primary binding site in the ligands, but it can induce amide deprotonation and coordination if other stable chelatable donor groups are not present in the molecule and the trigonal carbon atom of the peptide linkage is located inside the chelate ring formed with the participation of sulfur and amide nitrogen donors.

Further interesting conclusion can be drawn from the studies on the copper(II)-MPG interaction. In contrast with other thiol compounds, the interaction of copper(II) with MPG results in the formation of a green-colored copper(II) species, in which the coordination occurs via the mercapto, deprotonated amide nitrogen and carboxylate oxygen donors, similarly to the nickel(II) or palladium(II) complexes. The same species were obtained with several derivatives of MPG, and these compounds are believed to mimick the active sites of blue copper proteins (Sugiura et al., 1975b; Sugiura and Hirayama, 1976; McLendon, 1978). Especially stable copper(II) and nickel(II) complexes were obtained with the ligands containing both imidazolyl and sulfhydryl donor groups in the same molecules (Sugiura and Hirayama, 1977; Sugiura, 1978a; 1978b).

III. OLIGOPEPTIDES CONTAINING L-CYSTEINE

A. Protonation equilibria

L-cysteine and its di- and tripeptides contain three dissociable protons. The first relates to the dissociation of the carboxylic group, while the deprotonation of the ammonium and sulfhydryl groups takes place in overlapping processes. Therefore, pK_2 and pK_3 can not be ascribed unambiguously to the dissociation of ammonium or sulfhydryl groups. Thus, the real acidity of the individual groups must be characterized by microconstants which are collected in Table 1.

It can be seen from Table 1 that ammonium groups of peptides are more acidic than the -SH groups ($k_2/k_3 < 1$), whereas the acidity sequence is just opposite for L-cysteine ($k_2/k_3 = 1.62$). This difference can be attributed to the greater distances of the strong electron-releasing carboxylate groups from the amino groups in peptides. On the other hand, the data of AlaAlaCys reveal that elongation of the peptide chain does not change the protonation microconstants significantly, although the acidity of carboxylic group is slightly decreased.

B. Metal complexes of oligopeptides of L-cysteine

In spite of the outstanding significance of L-cysteine residues in metalloproteins, only a few publications are available on the complexes of small peptides containing cysteinyl moieties. The results obtained from equilibrium and structural studies in solution for the complexation with some transition elements are summarized in Figure 2 (Kozlowski et al., 1987; Sóvágó et al., 1988; Cherifi et al., 1990; Kozlowski et al., 1990).

Figure 2 reveals that complex formation processes of CysGly are very similar to those of the amino acids L-cysteine or D-penicillamine. Namely, square planar diamagnetic bis complexes were formed with all metal ions studied (cobalt(II), nickel(II), zinc(II), cadmium(II) and palladium(II)), in which bidentate coordination of the N-terminal cysteinyl residue was supposed via the amino and sulfhydryl donor atoms. The results suggest that the N-terminal cysteinyl residue prevents the deprotonation and coordination of the amide groups, but significantly increases the metal binding ability of peptide molecules as compared to that of glycylglycine.

Table 1

Dissociation Macroconstants (pK) and Microconstants (pk) of L-Cysteine and Its Di- and Tripeptides (T = 298 K; I = 0.2 mol dm^{-3} KCl)

$$LH_3^+ \underset{K_1}{\rightleftharpoons} LH_2 \underset{K_2}{\rightleftharpoons} LH^- \underset{K_3}{\rightleftharpoons} L^{2-}$$

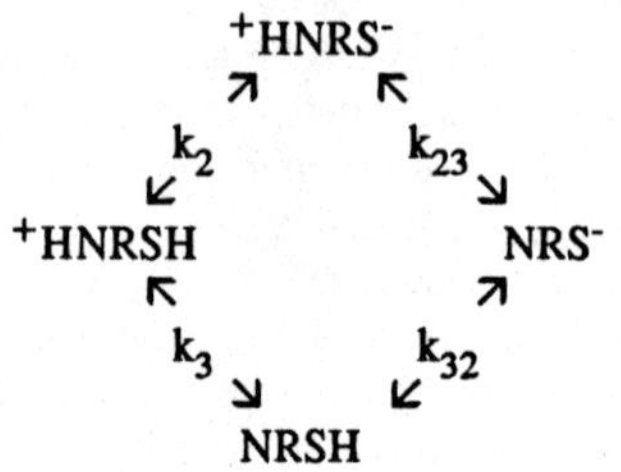

	Cys[a]	CysGly[a]	GlyCys[a]	AlaAlaCys[b]
pK_1	1.86	3.13	2.73	3.20
pK_2	8.10	6.94	8.04	7.96
pK_3	10.16	9.37	9.48	9.73
pk_2(SH)	8.31	7.78	9.14	9.37
$pk_3(NH_3^+)$	8.52	7.01	8.08	7.98
pk_{23}	9.95	8.53	8.38	8.32
pk_{32}	9.74	9.30	9.44	9.72
k_2/k_3	1.62	0.17	0.09	0.04

[a] Sóvágó et al. (1988)
[b] Cherifi et al. (1990)

In case of GlyCys, steric requirements rule out the coordination of terminal amino and sulfhydryl groups to the same metal ion; therefore, complex formation processes of the ligand will depend on the nature of metal ions. Studies on the metal complexes of various thiol compounds revealed that cobalt(II), zinc(II) and cadmium(II) can form rather stable complexes with both (S,N) and (S,O) donor atoms, whereas the (S,N) coordination is preferred for the complexation with nickel(II) or palladium(II) (see the results obtained for MPG in the previous paragraph). Therefore, cobalt(II), zinc(II) and cadmium(II) form bis complexes with GlyCys, in which the metal ions are coordinated via the sulfur and carboxylate-O donor atoms of the C-terminal cysteinyl residue (six-membered chelate). The terminal amino group does not take part in coordination in the bis complexes, which results in the formation of protonated complexes in acidic and medium pH ranges.

Palladium(II) and nickel(II) ions prefer the coordination of nitrogen donors, which results in the formation of sulfur-bridged dimeric complexes ($M_2L_2H_{-2}$) in equimolar solution. In the presence of ligand excess, the species ML_2 and ML_2H_{-1} are formed containing both tridentately (S,N,O) and monodentately (via sulfur atom) coordinated peptide molecules.

Peptides of histidine are discussed elsewhere in this book, and the reader can easily recognize the similarities between the complex forming abilities of His- and Cys-containing dipeptides. Namely, both HisGly and CysGly form stable bis complexes with the coordination of the N-terminal moieties of the peptide molecules, and the amide nitrogens

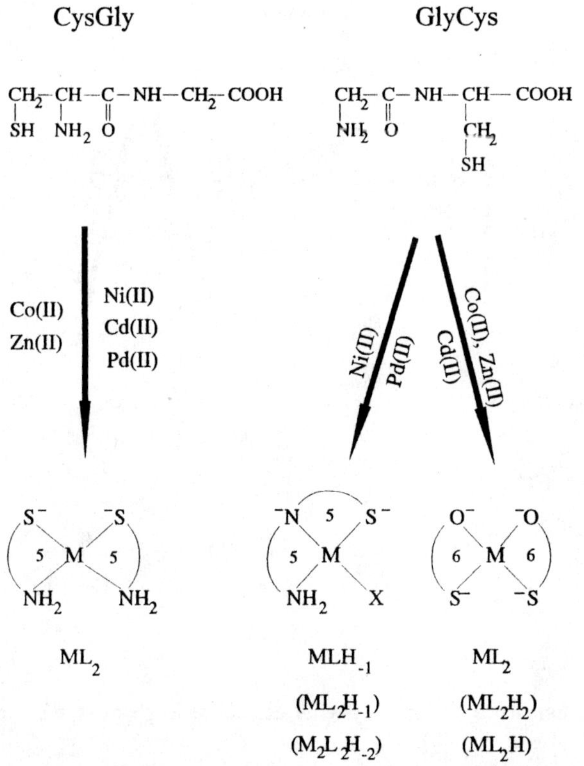

Figure 2.
Complex formation processes of L-cysteine containing dipeptides with transition metal ions.

do not take part in coordination, with the exception of the copper(II)-HisGly system where the dimeric species $Cu_2L_2H_{-2}$ was detected in equimolar solution. It should be noted that a fast redox reaction between copper(II) and cysteine-containing dipeptides has been observed, but detail of the process has not been clarified, yet.

The metal complexes of tripeptides containing C-terminal (GlyGlyCys and AlaAlaCys) or intermediate (AlaCysAla) cysteinyl residues have been studied recently (Cherifi et al., 1990; Várnagy et al., 1992), and their coordination modes are shown in Figure 3.

Comparison of Figures 2 and 3 reveals that complex formation processes of GlyCys and AlaAlaCys are very similar to each other. Namely, palladium(II) and nickel(II) induces again the deprotonation and coordination of amide nitrogens, which results in the formation of the species MLH_{-2} with (N,N,N,S) coordination. In the case of cobalt(II), zinc(II) and cadmium(II), stable bis complexes are formed and only the C-terminal cysteinyl residue takes part in coordination.

The sulfhydryl group of AlaCysAla can form stable chelate rings neither with the terminal amino (eight-membered chelate) nor with the carboxylate groups (nine-membered chelate), due to steric requirements. Consequently, bis complex formation is not characteristic of this ligand. The nickel(II) ion easily promotes deprotonation and coordination of amide nitrogens and the same species are formed as with GlyCys. This type of coordination is characteristic of the zinc(II) and cobalt(II) complexes, too, but amide deprotonation occurs at higher pH values.

Complex formation processes of some other oligopeptides containing L-cysteinyl residue(s) have also been studied with several other metal ions, and the results are

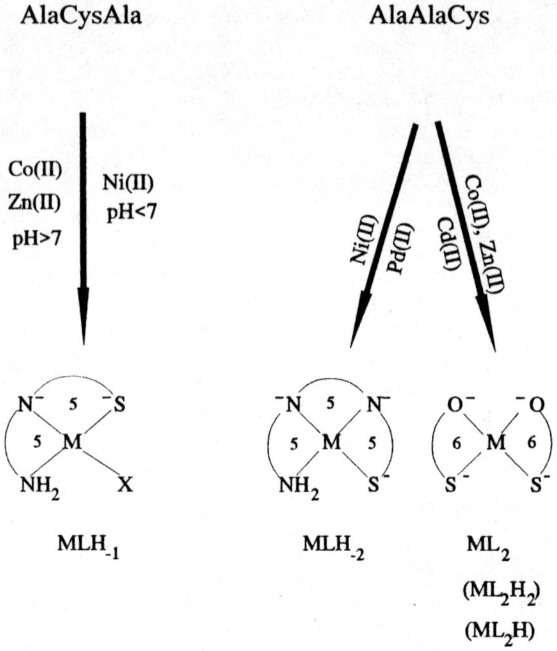

Figure 3.
Complex formation processes of L-cysteine containing tripeptides with transition metal ions.

discussed in a review published recently (Pettit et al., 1991). Complexes of iron are best studied and the results can be used for modeling the iron-sulfur proteins. Ueyama and Nakamura prepared a series of peptides (Z-Cys-X-X-CysOMe and related ligands) and the electrochemical behavior and catalitic activity of the iron(II) complexes were studied (Ueyama et al., 1985a; 1985b; 1987; Nakata et al., 1983; Ohno et al., 1990; Sugawara et al., 1991; Sun et al., 1991). Palladium(II), iron(III) and mercury(II) complexes of the same tetrapeptide sequences have also been studied and coordination of the sulfhydryl groups without amide binding was suggested (Ueyama et al., 1981; 1985c; 1987).

IV. METAL COMPLEXES OF GLUTATHIONE

Glutathione (γ-glutamyl-L-cysteinylglycine) is one of the most common naturally occuring tripeptides. In the fully protonated form it contains four dissociable protons, which indicates the great variety of microscopic dissociation processes. Acidity constants of glutathione were determined by NMR (Rabenstein, 1973). The results indicate that the two carboxylic groups ionize simultaneously in the acidic range, while the sulfhydryl and ammonium groups do so at a slightly basic pH. The α-carboxylic group of the glutamyl residue was found to be 10.8 times more acidic than the C-terminal carboxylic group from the glycyl residue, whereas the sulfhydryl group was 1.57 times more acidic than the ammonium group. This value is quite close to that of L-cysteine (Table 1), suggesting the increased basicity of the terminal amino group as compared to other peptide molecules.

There are several possibilities for the metal ion coordination of glutathione. An amino-acid binding site is available at the α-glutamyl locus of the molecule and it is well separated from the peptide bonds and sulfhydryl group. On the other hand, the sulfur atom is also in chelatable position with the neighboring carbonyl oxygen or deprotonated amide nitrogens. Therefore, the metal complex formation equilibria of glutathione are rather complicated and speciation of metal ion-glutathione systems or the structure of the complexes have not been satisfactorily clarified, yet. Some reviews have already been published on the coordination chemistry of glutathione (Rabenstein et al., 1979; Sóvágó and Gergely, 1981; Sóvágó, 1990). Therefore, only the most important findings and some recent results will be discussed in this survey.

The interaction of glutathione was studied with a series of 3d elements and among them nickel(II), zinc(II) and cobalt(II) are the best candidates for metal ion-promoted amide coordination. All authors agree that the interaction of glutathione with nickel(II) starts at the amino acid locus in slightly acidic pH range. It is followed by the formation of an intensely brown colored species suggesting the formation of square planar complexes (Martin and Edsall, 1959). Some authors concluded that deprotonation and coordination of amide nitrogen take place above pH 7 (Trzebiatowska et al., 1976; Kozlowska et al., 1980), while the formation of sulfur-bridged polynuclear species was suggested in another study (Sóvágó and Martin, 1981). The investigations carried out in our laboratories recently revealed that the nickel(II) ion promotes the oxidation of glutathione to disulfide, which has not been considered in the previous equilibrium and spectroscopic studies. Thus, it seems reliable that both amide coordination and polynuclear complex formation are responsible for the intense color of the nickel(II)-glutathione complexes.

The coordination of both the amino acid locus and the sulfhydryl group was described in the zinc(II)-glutathione system and zinc(II)-promoted amide deprotonation and coordination was suggested to occur above pH 10.5 (Fuhr and Rabenstein, 1973). In the cobalt(II)-glutathione system, the parallel coordination of the amino acid locus and the sulfhydryl group results in the formation of various polynuclear species containing both tetrahedral and octahedral cobalt(II) ions. Deprotonation and coordination of the amide nitrogens of glutathione was, however, ruled out in this case (Harman and Sóvágó, 1983).

The coordination of the sulfur donor atom is the governing factor in the reaction of glutathione with various soft metal ions. The formation of sulfur-coordinated polynuclear species was suggested with these elements, including mercury(II), silver(I), lead(II) and cadmium(II) (Rabenstein et al., 1979). Studies on the mixed-ligand complexes of cadmium(II)-nitrilotriacetate system with various thiols led to the conclusion that glutathione had a strong tendency to form mixed-ligand complexes with sulfur coordination in biological systems (Kadima and Rabenstein, 1990). $CH_3Hg(I)$ complexes of glutathione were studied by NMR spectroscopy. The sulfhydryl group was suggested to be the only binding site in acidic media, whereas both the amino and sulfhydryl groups were coordinated between pH 4 and 8 in a $(CH_3Hg)_2GS$ complex (Rabenstein and Fairhurst, 1975). A large number of papers has been published recently on the interaction of the anticancer drug cis$[Pt(Cl)_2(NH_3)_2]$ and related compounds with thiol ligands including glutathione. The results can help to understand the metabolism and toxic side effects of cisplatin drugs. Formation of various Pt-S bonded polynuclear species was suggested from both in vitro and in vivo experiments (Berners-Price and Kuchel, 1990a; 1990b; Djuran et al, 1991).

The sulfhydryl group of glutathione may reduce several metal ions. Among them, the interactions of glutathione with copper(II), iron(III) and chromium(VI) are the best studied. Glutathione is a more effective reducing agent of copper(II) than the amino acids L-cysteine or D-penicillamine. Therefore, transient copper(II) or mixed valence complexes cannot be unambiguously identified in this interaction. Copper(I) complexes of reduced glutathione were studied by potentiometric methods and the existence of mononuclear species (CuLH and CuL_2H_2) was suggested (Österberg et al., 1979). The reaction of iron(III) with glutathione clearly demonstrated the formation of iron(II) complexes, but the binding sites of reduced glutathione in these complexes seem to be questionable (Hamed and Silver, 1983).

There are now many papers which report the generation of the highly toxic chromium(V) complexes during the reduction of chromate with glutathione (Goodgame and Joy, 1986). The product $Na_4[Cr(GSH)_4].8H_2O$ has been prepared recently, and it was found that chromium(V) is coordinated via the carboxylate and thiolate groups of glutathione (O'Brien et al., 1990).

Oxidation of glutathione results in the formation of the disulfide (oxidized glutathione), which can undergo disproportionation in basic solution, as follows (Ostern et al., 1980):

$$2\ \text{R-S-S-R} + 2\ OH^- \rightleftharpoons 2\ RS^- + RSH + RSO_2H$$

Several metal ions (especially copper(II)) catalyze this disproportionation of disulfides, which enhances the complexity of metal ion-glutathione interactions.

REFERENCES

Baidya, N., Olmstead, M.M., and Mascharak, K.P. (1989). *Inorg Chem.*, *28*: 4326.

Berners-Price, S.J., and Kuchel, P.W. (1990a). *J. Inorg. Biochem.*, *38*: 305.

Berners-Price, S.J., and Kuchel, P.W. (1990b). *J. Inorg. Biochem.*, *38*: 327.

Bryce, G.F., Roeske, R.W., and Gurd, F.R.N. (1966). *J. Biol. Chem.*, *241*: 1072.

Cherifi, K., Decock-Le Révérend, B., Várnagy, K., Kiss, T., Sóvágó, I., and Kozlowski, H. (1990). *J. Inorg. Biochem.*, *38*: 69.

Djuran, M.I., Lempers, E.L.M., and Reedijk, J. (1991). *Inorg. Chem.*, *30*: 2648.

Filella, M., and Williams, D.R. (1985). *Inorg. Chim. Acta*, *106*: 49.

Fuhr, B.J., and Rabenstein, D.L. (1973). *J. Amer. Chem. Soc.*, *95*: 6944.

Goodgame, D.M.L., and Joy, A.M. (1986). *J. Inorg. Biochem.*, *10*: 341.

Hamed, M.Y., and Silver, J. (1983). *Inorg. Chim. Acta*, *80*: 115.

Harman, B., and Sóvágó, I. (1983). *Inorg. Chim. Acta*, *80*: 75.

Kadima, W., and Rabenstein, D.L. (1990). *J. Inorg. Biochem.*, *38*: 277.

Kiss, T., Farkas, E., Kozlowski, H., Siatecki, Z., Kafarski, P., and Lejczak, B. (1989). *J. Chem. Soc., Dalton Trans.*, 1053.

Kozlowska, G., May, P.M., and Williams, D.R. (1980). *Inorg. Chim. Acta*, *46*: L51.

Kozlowski, H., Decock-Le Révérend, B., Ficheux, D., Loucheux, C., and Sóvágó, I. (1987). *J. Inorg. Biochem.*, *29*: 187.

Kozlowski, H., Urbanska, J., Sóvágó, I., Várnagy, K., Spychala, J., and Cherifi, K. (1990). *Polyhedron*, *9*: 831.

Martin, R.B., and Edsall, J.T. (1959). *J. Amer. Chem. Soc.*, *81*: 4044.

McLendon, G. (1978). *J. Inorg. Nucl. Chem.*, *40*: 755.

Nakata, M., Ueyama, N., Terakawa, T., and Nakamura, A. (1983). *Inorg. Chim. Acta*, *56*: 3647.

O'Brien, P., Pratt, J., Swanson, F.J., Thornton, P., and Wang, G. (1990). *Inorg. Chim. Acta*, *169*: 265.

Ohno, R., Ueyama, N., and Nakamura, A. (1990). *Inorg. Chim. Acta*, *169*: 253.

Ostern, M., Pelczar, J., Kozlowski, H., and Trzebiatowska, B. (1980). *Inorg. Nucl. Chem. Lett.*, *16*: 251.

Österberg, R., Ligaarden, R., and Persson, D. (1979). *J. Inorg. Biochem.*, *10*: 341.

Pettit, L.D., Gregor, J.E., and Kozlowski, H. (1991). *Perspectives in Bioinorganic Chemistry* (R.W. Hay, J.R. Dilwortth and K.B. Nolan, eds.), Jai Press Ltd, London, p. 1.

Rabenstein, D.L. (1973). *J. Amer. Chem. Soc.*, *95*: 2797.

Rabenstein, D.L., and Fairhurst, M.T. (1975). *J. Amer. Chem. Soc.*, *97*: 2086.

Rabenstein, D.L., Guevremont, R., and Evans, C.A. (1979). *Metal Ions in Biological Systems*, Vol. 9 (H. Sigel, ed.), Marcel Dekker, New York, p. 173.

Sigel, H., and Martin, R.B. (1982). *Chem. Rev.*, *82*: 385.

Sóvágó, I. (1990). *Biocoordination Chemistry* (K. Burger, ed.), Ellis Horwood, New York, p. 56.

Sóvágó, I., and Gergely, A. (1981). *Agents Actions, Suppl. 8*: 289.

Sóvágó, I., and Martin, R.B. (1981). *J. Inorg. Nucl. Chem.*, *43*: 425.

Sóvágó, I., Farkas, E., and Gergely, A. (1982). *J. Chem. Soc., Dalton Trans.*, 685.

Sóvágó, I., Kiss, T., Várnagy, K., and Decock-Le Révérend, B. (1988). *Polyhedron, 7*: 1089.

Sugawara, T., Ueyama, N., and Nakamura, A. (1991). *J. Chem. Soc., Dalton Trans.*, 249.

Sugiura, Y. (1978a). *Inorg. Chem.*, *17*: 2176.

Sugiura, Y. (1978b). *Bioinorg. Chem.*, *8*: 453.

Sugiura, Y., and Hirayama, Y. (1976). *Inorg. Chem.*, *15*: 679.

Sugiura, Y., and Hirayama, Y. (1977). *J. Amer. Chem. Soc.*, *99*: 1581.

Sugiura, Y., Hirayama, Y., Tanaka, H., and Sakurai, H. (1975a). *J. Inorg. Nucl. Chem.*, *37*: 2367.

Sugiura, Y., Hirayama, Y., Tanaka, H., and Ishizu, K. (1975b). *J. Amer. Chem. Soc.*, *97*: 5577.

Sun, W.Y., Ueyama, N., and Nakamura, A. (1991). *Inorg. Chem.*, *30*: 4026.

Trzebiatowska, B., Kozlowska, G., and Kozlowski, H. (1976). *Chem. Phys. Lett.*, *42*: 242.

Ueyama, N., Nakata, M., and Nakamura, A. (1981). *Inorg. Chim. Acta*, *55*: L61.

Ueyama, N., Nakamura, M., Fuji, M., Terakawa, T., and Nakamura, A. (1985a). *Inorg. Chem.*, *24*: 2190.

Ueyama, N., Nakata, M., and Nakamura, A. (1985b). *Polym. J.*, *17*: 721.

Ueyama, N., Nakata, M., and Nakamura, A. (1985c). *Bull. Chem. Soc. Jpn*, *58*: 464.

Ueyama, N., Ueno, S., and Nakamura, A. (1987). *Bull. Chem. Soc. Jpn*, *60*: 283.

Várnagy, K., Sóvágó, I., and Djemil, S. (1992). "Abstracts", EUROBIC I, Newcastle.

METAL ION INTERACTIONS WITH NEUROTRANSMITTERS

by

Tamás Kiss

Kossuth Lajos Tudományegyetem,
Szervetlen és Analitikai Kémiai Tanszék
H-4010 Debrecen
Hungary

I. INTRODUCTION

Neurotransmitters are the classic neuromessengers that are released rapidly by the presynaptic neuron, diffuse across the synaptic cleft and have either an excitatory or an inhibitory effect on a postsynaptic neuron.

There are three classes of neurotransmitters: biogenic amines, amino acids and peptides, and a few examples are given in Figure 1. The biogenic amines consist of three catecholamines (dopamine, norepinephrine and epinephrine), an indolamine (serotonin), a quaternary amine (acetylcholine) and an ethylamine (histamine). The biogenic amines account for only 5 to 10 percent of the synapses of the human brain, whereas the amino acid neurotransmitters account for up to 60 percent. The best studied amino acid neurotransmitters are γ-aminobutyric acid (an inhibitory neurotransmitter) and glutamic acid (an excitatory amino acid neurotransmitter). Other inhibitory amino acids are glycine, and probably taurine and β-alanine. Other excitatory amino acids include aspartic acid and probably cysteine. Over 100 neuroactive peptides have been identified including endogenous opioids (e.g. enkephalins, endorphins), "gut" peptides (e.g. substance P, neurotensin), hypothalamic hormones etc.

These neurotransmitters are synthesized in the nerve terminals (biogenic amines and amino acids) or in the neuronal cell body (peptides) and stored in the synaptic vesicles within the nerve terminals. They are released by a calcium-dependent mechanism. Due to the relatively high metal ion content of the brain and neurons, Na^{+}, K^{+}, Ca^{2+}, Mg^{2+}, Cu^{2+}, Mn^{2+}, Fe^{2+}, Fe^{3+}, and Zn^{2+} being the most important, the metal ion - neurotransmitter interactions may be involved in or may influence neurotransmitter processes. Similarly, exogenous administration of trace metals or absorption of toxic metals may also affect either advantageously or disadvantageously the neurological processes.

The biological activity of the metal ions can take place either through ionic interactions or metal complex formation. In this chapter we deal with the metal binding ability of neurotransmitters, focusing on that of the biogenic amines. The metal complexation properties of amino acid neurotransmitters is reviewed in detail in Chapters 2-A1 and 2-A3. As there is a close overlap between peptide hormones and peptide neurotransmitters, many neuropeptides also have both hormone and neurotransmitter functions; here we only refer to Chapter 2-B2, where their interactions with metal ions are discussed. Reviews have also

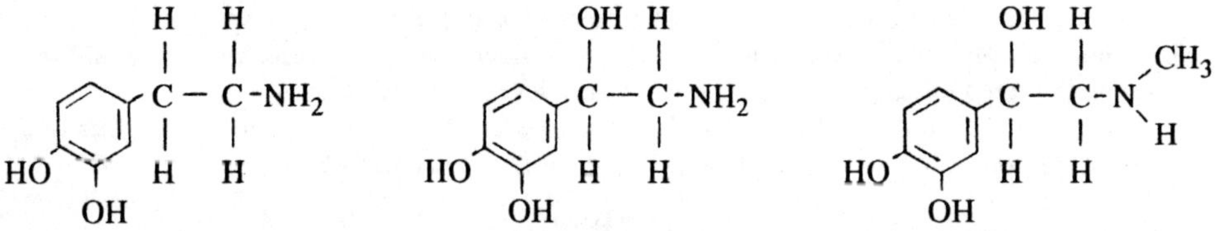

Dopamine Norepinephrine Epinephrine

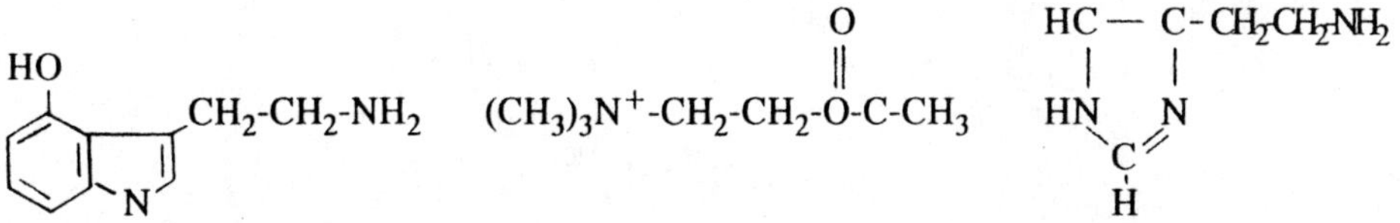

Serotonin Acetylcholine Histamine

AMINO ACID NEUROTRANSMITTERS (examples):

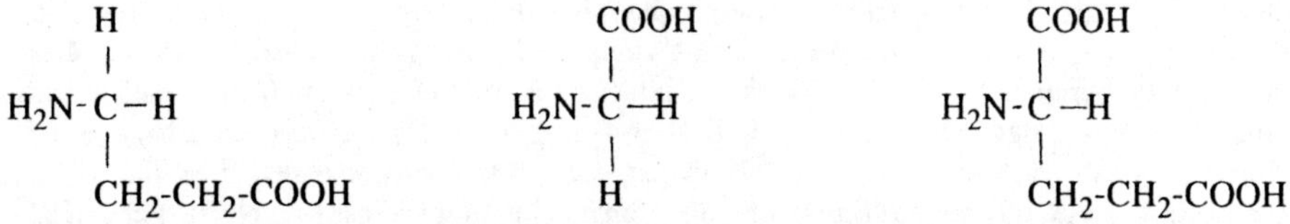

γ-aminobutyric acid Glycine Glutamic acid

PEPTIDE NEUROTRANSMITTERS (examples):

NEUROTENSIN:
Glu-Leu-Tyr-Glu-Asn-Lys-Pro-Arg-Arg-Pro-Tyr-Ile-Leu

THYROTROPIN RELEASING HORMONE (TRH):
Glu-His-Pro

CHOLECYSTOKININ OCTAPEPTIDE (CCK-8):
Asp-Tyr-Met-Gly-Trp-Met-Asp-Phe

Figure 1.
The three classes of neurotransmitters.

appeared on the interactions of biogenic amines with metal ions, for example on the role of metal chelates in the storage and transport of neurotransmitters (Rajan et al., 1976a), or on the biocoordination chemistry of L-dopa and related ligands (Gergely and Kiss, 1979).

II. BIOGENIC AMINES

A. Catecholamines and dopa

In their reactions with metal ions, catecholamines present three or four potential binding sites: two phenolate groups in ortho position (catecholate chelating site), and a side-chain amino group. In addition to these groups, epinephrine and norepinephrine have an alcoholic hydroxyl, and dopa a carboxylate group in a chelatable position with respect to the amino group. The protonation constants of these basic donor groups describing their acid-base properties at both macroscopic and molecular levels are listed in Table 1. A common characteristic of these ligands is that the side-chain amino group and one of the phenolate groups protonate in overlapping parallel processes in the pH range 8-10.

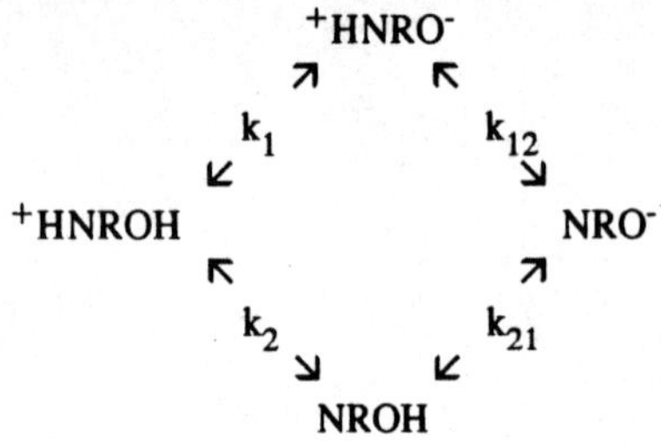

$$LH_2^+ \underset{}{\overset{K_{LH2}}{\rightleftharpoons}} LH \overset{K_{LH}}{\rightleftharpoons} L^-$$

The molar concentration ratio $R=[^{+}HNRO^{-}]/[NROH]=k_1/k_2$ of the two protonation isomeric forms of LH indicates that the phenolic hydroxy group is more acidic (its real acidity constant, $\log k_1$ is around 8.7-8.9 and nearly independent of the side-chain donors) due to hydrogen bond formation between the phenolic groups. Hence, these so-called catecholamines are more aptly described as ammonium catecholates (Martin, 1971). The interactivity parameter $S=k_1/k_{21}=k_2/k_{12}$ characteristic of the separation of these proton binding sites is around 0.4 log units, indicating some interaction between the two sites (if there is no interaction then S=1). Knowledge of the microconstants, i.e. the exact protonation state of the molecules, is of primary importance as the physico-chemical properties (e.g. membrane transport properties) or physiological role (e.g. enzyme effect) of these biomolecules is strictly structure dependent. The type and the extent of metal-ligand interactions are also considerably influenced by the protonation state of the ligands. Incorrect conclusions may result from misassignment of macroscopic protonation constants when they are used in the calculation of metal complex formation constants which characterize the metal binding ability of these bioligands. The use and significance of microscopic acidity constants of polyfunctional biomolecules has recently been reviewed by Noszál (1990).

These ligands are typical ambidentate ligands, i.e. they can bind metal ions at different parts of the molecules. Besides the chelate-forming catecholate site, dopamine contains a separate monodentate amino group, epinephrine and norepinephrine a weakly chelating aminoethanolate site, and dopa a strongly chelating aminocarboxylate site. Which donor groups take part predominantly in the coordination of a metal ion is determined primarily

Table 1

Acid-Base Properties of the Catecholamines*

Ligand	$\log K_{ROH}$	$\log K_{ArOH}$	$\log K_{LH}$	$\log K_{LH2}$	$\log k_1$	$\log k_2$	R	S
dopm	-	13.11	10.41	8.89	8.90	9.98	12	2.5
epi	>14	13.1	9.84	8.64	8.67	9.38	5.1	2.7
nepi	>14	12.9	9.53	8.58	8.69	9.11	2.6	2.1
dopa†	-	13.4	9.83	8.80	8.98	9.19	1.6	2.8

dopm: dopamine; epi: epinephrine; nepi: norepinephrine
* Data taken from Kiss et al. (1989). †$\log K_{COOH}$= 2.22.

by the strength of interaction between metal ion and binding sites, but also by pH, as the extent of proton competition at the separate binding sites is different because of differences in the overall basicity of the sites (i.e. $\log\beta_{LH_2}$ for the catecholate moiety is ~22, while for the side-chain donors it is ~10.4 for dopamin, ~11.5 for dopa, and ~23.5 for epinephrine and norepinephrine). Furthermore, the enhanced formation of a chelate over a monodentate complex is of primary importance when considering the strength of metal ion binding.

1. *Dopamine*

Dopamine contains only one pair of chelating donor groups and a monodentate side-chain NH_2 group. Due to the large differences in metal ion affinity for these two separate sites, formation of the (O,O)-coordinated catecholate-type complexes is expected to be much more favored and in fact the participation of the side chain N donor in the complexation processes seems to be negligible. The stability constants of the (O,O)-coordinated complexes listed in Table 2 indicate that in agreement with the HSAB principle, dopamine (as a catechol derivative containing the hard O donors) preferentially binds hard metal ions such as Fe^{3+}, Al^{3+}, and VO^{2+}.

The formation of mono and bis (O,O)-coordinated complexes is well established with several 3d transition metal ions such as Mn^{2+}, Co^{2+}, Ni^{2+}, Cu^{2+}, and Zn^{2+}, and their stabilities follow the Irving-Williams order (Gergely and Kiss, 1979). As the formation of these complexes and the deprotonation of the non-coordinated side-chain ammonium groups are more or less separated in the pH range, the stoichiometry of the complexes is M(LH), ML, $M(LH)_2$, $M(L_2H)$ and ML_2. Spectral (UV-VIS and EPR) studies unambiguously prove that the metal ion coordinates in these species solely at the catecholate site (Monduzzi et al., 1982; Lai et al., 1982; Burlamacchi et al., 1983). Al^{3+} is found to bind to dopamine via the formation of octahedral 1:1, 1:2 and 1:3 complexes (Kiss et al., 1989). The Al(III)-(O,O) interaction is so strong that in the presence of an excess of ligand not only the precipitation of $Al(OH)_3$ can be avoided in the pH range 6-8, but also the formation of the very stable tetrahedral $Al(OH)_4^-$ complex remains negligible even in very basic solutions (pH > 10-11). In the case of Fe^{3+}, potentiometric, EPR and NMR data demonstrate the existence of the mixed hydroxo complexes FeL(OH) and $FeL_2(OH)$ as well as the normal (O,O)-coordinated mono and bis complexes (Crisponi et al., 1983; Lai et al., 1980;

Table 2

Stability Constants of the Stepwise Formation of (O,O) Coordinated Complexes of Dopamine with Several Metal Ions

Metal	$\log K_{ML}$	$\log K_{ML_2}$	$\log K_{ML_3}$
Ca^{2+}	very weak		
Mg^{2+}	3.59	-	-
Mn^{2+}	7.73	4.85	-
Co^{2+}	8.74	6.73	-
Ni^{2+}	9.01	5.93	-
Cu^{2+}	13.86	11.25	-
Zn^{2+}	9.85	8.36	-
Fe^{2+}	7.5	-	-
Fe^{3+}	18.6	12.5	-
Al^{3+}	15.63	12.98	8.95
VO^{2+}	16.84	15.09	1.8*
H_3BO_3	4.86†	-	-

* $VOL_2 + LH_2 \rightleftharpoons VL_3 + H_2O$

† $H_3BO_3 + LH_2 \rightleftharpoons (HO)_2BL^- + H_3O^+$

Burlamacchi et al., 1983). On the basis of the potentials of the Fe^{3+}/Fe^{2+} and catechol/o-quinone redox couples, a direct redox reaction (even if a reversible one) between Fe^{3+} and catechol derivatives has to be assumed leading to the formation of well defined Fe^{2+}-semiquinone complexes (Hider et al., 1981; 1983) . This is supported by the dramatic change of colour with pH of the solutions from green to blue and finally to red (Crisponi et al., 1983).

The exceptional chelating ability of *ortho*-diphenolates and their great affinity for highly charged metal ions results in an oxygen displacement reaction of oxovanadium(IV) by catecholates, leading to the formation of a very stable tris complex, VL_3, with trigonal prismatic geometry (Cooper et al., 1982; Branca et al., 1990; Buglyó and Kiss, 1991). This kind of bonding mode, with the presence of a bare vanadium(IV) ion in the coordination center, might be involved in various naturally occurring vanadium-containing biomolecules such as the tunichromes, and might be important in their unusual redox properties (Klime-Hunt et al., 1991). Only very weak interaction of dopamine with Ca^{2+} and Mg^{2+} was detected (Rajan et al., 1971) although Mg^{2+} was found to affect the catecholamine uptake, while Ca^{2+} affects the catecholamine release of the amine-storing granules of adrenal medulla (Rajan et al., 1976a).

2. *L-3,4-Dihydroxyphenylalanine*

As L-dopa contains two separate chelating sites, it can coordinate to a metal ion as a typical ambidentate ligand via both the amino acid side chain and the *ortho*-phenolic hydroxy groups. Assuming that the two binding sites are nearly independent from each other (see the interactivity parameter, S, of the ligands in Table 1) the following sequences can be established for their metal binding affinity from the stability data on the amino acid and catechol complexes (Kiss, 1990):

(NH_2, COO^-): $Al^{3+} < VO^{2+} < Mn^{2+} < Co^{2+} < Zn^{2+} < Ni^{2+} < Cu^{2+}$
(O^-, O^-): $Mn^{2+} < Ni^{2+} < Co^{2+} < Zn^{2+} < Cu^{2+} < Al^{3+} < VO^{2+}$

With the hard metal ions, e.g. Mg^{2+} (Gorton and Jameson, 1972a; Rajan et al., 1978), Ca^{2+} (Rajan et al., 1978), Al^{3+} (Kiss et al., 1989), Cr^{3+} (Davidov et al., 1975), Fe^{3+} (Mentasti et al., 1976; Linert et al., 1991), VO^{2+} (Buglyó et al., 1993), VO_2^{+} (Kustin et al., 1974), W(VI) (Kustin and Liu, 1973), and Mo(VI) (Gilbert and Kustin, 1976), (O,O)-coordination is the preferred bonding mode, provided that the ligand is in excess. The amino acid-type (N,O)-coordination is rather weak and practically negligible with these metal ions. Of the L-dopa complexes, those of several 3d transition metal ions, such as Mn^{2+}, Co^{2+}, Ni^{2+}, Cu^{2+}, and Zn^{2+} are the best studied. Besides detailed pH-metric investigations (Gorton and Jameson, 1972a; 1972b; Bogges and Martin, 1975; Gergely and Kiss, 1976; Gergely et al., 1979; Amico et al., 1983; Kiss and Gergely, 1983a) to determine the stoichiometry and stability of the complexes formed, spectral measurements (Pilbrow et al., 1970; Kwik et al., 1974; Daniele et al., 1990) have also been employed to clarify the bonding modes in these complexes. The basic bonding modes of the metal ion-L-dopa complexes, together with what metal ions are able to form a given complex, are depicted in Figure 2.

At low pH these metal ions form amino acid-type (N,O)-coordinated 1:1 complexes, $M(LH_2)$; this bonding mode is so favored with Ni^{2+} and Cu^{2+} that 1:2 complexes, $M(LH_2)_2$ are also formed (in (LH_2) the two protons are on the non-coordinating phenolic hydroxy groups). Catecholate-type (O,O)-coordinated 1:1 complexes M(HL) are formed only with Mn^{2+}, Co^{2+}, and Zn^{2+} (in HL the proton is on the non-coordinating side-chain amino group). The mixed bonding mode, containing both (N,O) and (O,O)-coordinated ligands, is preferred for all metal ions within the physiological pH range. As the pH is raised, with decreasing proton competition on the phenolic hydroxyls, the catecholate-type coordination becomes dominant. A combined pH-spectrophotometric method (Kiss and Gergely, 1983b) monitoring the changes of the protonation/coordination state of the catecholate site separately on the UV band of the phenolate group was used to determine the metal ion dependence of the tendency to rearrangement from the mixed bonding mode to the purely (O,O) coordination; it was found to follow the sequence Cu(II) ~ Zn(II) > Co(II) >

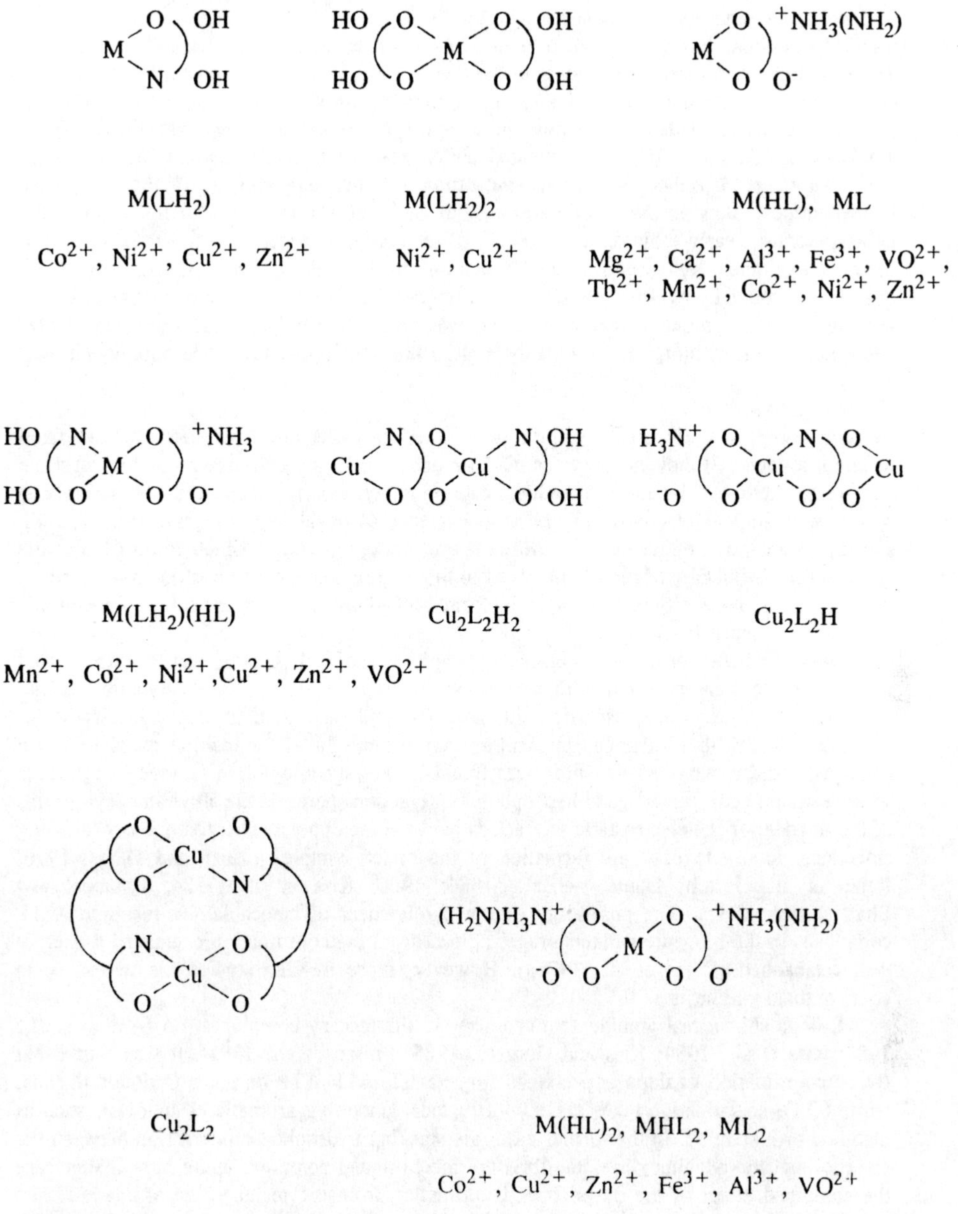

Figure 2.
The basic bonding modes of metal ions in their complexes formed with L-dopa.

Mn(II) ~ Ni(II) (Kiss, 1990). Because of the ambidentate nature of L-dopa, a metal ion can coordinate simultaneously at both separate metal-binding sites, which can result in the formation of dimeric complexes. Formation of chain-like dimers and a cyclic dimer has been assumed (Gorton and Jameson, 1972b) and confirmed by visible and EPR spectral measurements (Bogges and Martin, 1975; Gergely and Kiss, 1976) in the Cu(II)-L-dopa system. Heterobinuclear complexes of L-dopa with Cu(II) and Zn(II) and Cu(II) and Cd(II) couples were also detected (Daniele et al., 1983). In these complexes, different metal ions are coordinated at the different binding sites. Although the distribution of the metal ions at each site has not been studied, it can be assumed that it is not constant but changes with the pH of the solution.

3. *Epinephrine and norepinephrine*
Similarly to dopa, epinephrine and norepinephrine also contain two separate chelating sites. There is, however, a great difference in the metal binding ability of the catecholate and the ethanolamine sites, the former being much more favored, and as a consequence the participation of the side-chain donors in complex formation is more subordinate. Thus, similarly to dopamine, (O,O)-coordinated complexes are formed with most of the metal ions. There is a rather ambiguous indication of the involvement of the side-chain ethanolamine moiety in the coordination. With Cu^{2+} at 1:1 metal ion to ligand ratio the EPR spectrum clearly indicates a Cu(II)-Cu(II) interaction (Gergely et al., 1981) which can be best explained by assuming the formation of a cyclic tetramer Cu_4L_4 containing presumably mixed (N,O)(O,O)-coordinated copper(II) centers. An analogous mode of bonding, albeit in mononuclear complexes, was proposed for Ni^{2+} (Jameson and Neille, 1966) and also for MoO_4^{2-} (El-Hendawy et al., 1988) with epinephrine and norepinephrine.

4. *Ternary complexes*
Ternary complexes of catecholamines with other biomolecules have been studied from different aspects: (i) they can serve as useful models for in vivo enzyme-metal ion-substrate complexes involved in their metalloenzyme (e.g. tyrosinase, dopamine β-hydroxylase) catalyzed biological reactions; (ii) the co-occurrence of metal ions, biogenic amines, ATP and membrane-bound phospholipids in the amine storage granules, which in turn led to the speculation that metal-chelation is involved in the binding and storage of these monoamines; (iii) to find a more efficient therapeutic form of L-dopa in the treatment of Parkinson's disease for which it is widely used.

Only (O,O)-coordinated ternary complexes of dopamine, epinephrine and norepinephrine were reported with various small biomolecules (such as glycine, alanine, arginine, ornithine, lysine, aspartic acid, glutamic acid, histidine, glycylglycine, tyrosine, malonic acid, 2,2'-bypiridyl and phenanthroline) and the $3d^6$-d^{10} transition metal ions. An enhanced ternary complex formation was found in the physiological pH range in the cases where intramolecular interligand hydrophobic interactions between the alkyl and aryl groups of the ligands, or the electrostatic interaction between the opposite charged non-coordinating side-chain donors favored the formation of the mixed complex (Rajan and Davis, 1976; Rajan et al., 1976b; Daniele et al., 1984; 1986; Kiss et al., 1984; Emanual and Bhattacharya, 1987). The possibility of the involvement of catecholamine-metal ion-ATP complexes in the biogenic amine storage, at least from the coordination chemistry aspect, is well established (Rajan et al., 1976a). However, more in vivo studies are necessary to confirm this hypothesis.

L-dopa shows real ambidentate character in the ternary complexes too (Amico et al., 1983; Kiss et al., 1984; Kiss and Gergely, 1985; Prasad et al., 1987). It was found that (N,O) coordination of dopa is preferred for example with ATP or other O donor ligands, while (O,O) coordination takes place with ligands containing aromatic N donor(s), such as histidine or histamine. In the former case, the stacking hydrophobic interaction between the catechol and the adenine rings stabilizes the mixed-ligand complex, while in the latter case the combined effect of the dπ-pπ back-coordination from the metal to the aromatic-N site together with the strong ligand → metal ion electron transfer at the metal-catecholate site gives the same effect. With 2,2-bypiridyl and 1,10-phenanthroline, both the (N,O) and the (O,O) coordination sites of dopa are preferentially involved in the formation of mixed complexes due to the simultaneous occurrence of the two effects.

In order to achieve a therapeutic effect in the treatment of Parkinsonism, L-dopa must be administered in high daily doses, because a considerable proportion of it is decarboxylated in the peripheral regions and accordingly cannot reach the brain where its dopamine level increasing action is to be exerted. It was assumed (Rajan et al., 1986; 1987) that metal ion coordination at the aminocarboxylate moiety of L-dopa reduces the peripheral decarboxylation and a greater proportion of L-dopa is thus transported to the brain. This hypothesis is confirmed on thermodynamic grounds as a strong correlation can be detected between the catecholamine levels in the brain after administration of L-dopa or its binary or ternary chelates and the relative ratio of the (N,O)-bound L-dopa in the same binary or ternary systems at physiological pH. Metal ion coordination, however, can cause an opposite effect when metal ion is coordinated at the catecholate site (e.g. with iron(III), Campbell and Hasinoff (1989)).

B. Serotonin

Serotonin (5-hydroxytryptamine) contains two metal-binding sites, the phenolic hydroxylate at the indole ring and the side-chain amino group. The assignment of the pK values: 9.98 and 11.26 (Weber and Simeon, 1971) is not unambiguous because of the possible overlap between the two protonation processes, although it is very likely that the phenolic hydroxy group has a higher acidity. The indole-NH group is extremely weak both as a base (pK ~ -6.2) (Weber, 1974) and as an acid (pK ~ 16.8) (Yagil, 1967) cannot be regarded as a potential site of metal binding. Metal ion coordination can occur at the two separate monodentate binding sites. As serotonin forms remarkably more stable complexes with several metal ions such as Ni^{2+}, Cu^{2+}, Zn^{2+}, Cd^{2+} and Pb^{2+} than its parent compound tryptamine, the participation of the hydroxy group in complexation can also be assumed. Weber and Simeon (1971) reported the possibility of the formation of both types of complexes with these metal ions. The metal-binding ability of the two basic sites does not differ significantly, although the amino-type coordination may be a little more favored due to the stabilizing effect of the interaction between the empty d orbitals of the metal ion and the 6π electron system of the aromatic ring (Odani and Yamauchi, 1984).

C. Acetylcholine

The interacting capability of acetylcholine with metal ions is rather limited due to the lack of a coordinating electron pair at the quaternary ammonium group and to the extremely weak coordinating strength of the ester group. Very weak binding of metal ions, especially hard metal ions, cannot, however, be fully excluded at this end of the molecule. Electrostatic interactions with negatively charged biomolecules, e.g. with nucleotides and nucleic acids, have been observed (Kopell and Westhead, 1982). Hydrophobic forces between the protons of the quaternary ammonium group and 6π electron systems of the aromatic biomolecules are also assumed to play a role in some of its biological reactions (Wilson, 1952; Minch et al., 1979).

D. Histamine

The histamine molecule presents two potential coordination sites in aqueous solution: the imidazole nitrogen (pK=6.2) and the side-chain amino nitrogen (pK=9.9) become available for metal ion binding as the pH increases. The overlap between the two protonation steps is negligible; the molar concentration of the imidazole-protonated ^{+}HL species is about 2% of that of the ammonium-protonated species LH^{+} at any pH (Sundberg and Martin, 1974). The neutral imidazole ring of the histamine residue may exist in two tautomeric forms, the N1-H tautomer and the N3-H tautomer (see Fig. 3). This was confirmed by a Raman scattering study (Ashikawa and Itoh, 1978) for various imidazole derivatives. ^{13}C NMR (Reynolds et al., 1973) and ^{1}H NMR (Tanokura, 1983) studies provided conclusive evidence that the N1-H tautomer is the predominant form of the imidazole ring of various histidine derivatives including histidine methyl ester both in acidic and basic solutions. It is reasonable to assume that this tautomer preference is maintained in histamine too.

The basic bonding mode of histamine is chelation via the side-chain amino N and the tertiary nitrogen N3 of the aromatic ring (Eilbeck et al., 1970; Walker et al., 1973; Gergely and Sóvágó, 1976; Itabashi et al., 1982). As a purely N-donor ligand, it forms fairly stable complexes with middle and late transition metal ions, but is less able to bind the early transition metal ions, the hard non-transition metal ions, or the lanthanide ions. Due to the π-acceptor properties of aromatic amines (Sigel, 1975) histamine forms more stable six-membered chelates than its aliphatic counterpart 1,3-diaminopropane. This is clearly reflected in the spectral (VIS and EPR) parameters. For instance, the shift of the d-d transition of Cu(II)-histamine complexes to higher energy values and the larger ligand field splitting in the case of the Ni(II)-histamine complexes clearly indicate stronger metal-ligand interactions. The chelation enthalpy and entropy changes also support this conclusion

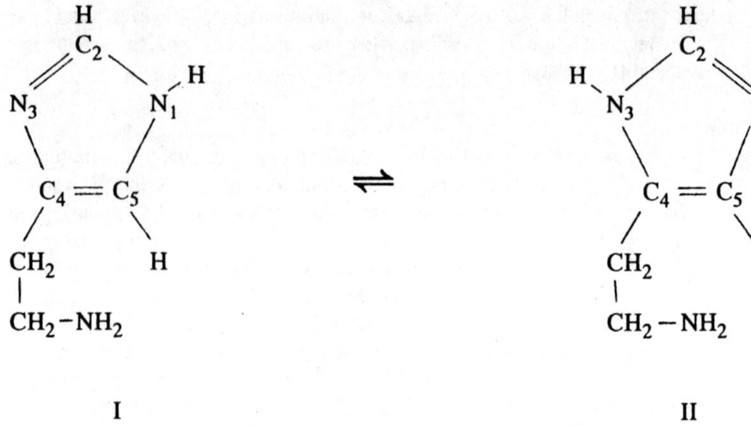

Figure 3.
The two tautomeric structures of the imidazole ring of histamine.

(Gergely and Sóvágó, 1976). Speciation studies revealed that beside the ML_n type chelates, protonated complexes can also be formed with several transition metal ions in the acidic pH range (Eilbeck et al., 1970; Gergely and Sóvágó, 1976; Dasgupta and Jordan, 1985). On the basis of Raman spectral studies, Itabashi et al. (1982) concluded that the protonated ligand in these complexes is monodentately coordinated to the metal ion through the N1 donor atom whereas the side-chain nitrogen is protonated, existing as NH_3^+.

It was observed (Morris and Martin, 1970; Aiba et al., 1974) that in certain transition metal complexes of imidazole derivatives the N1-H hydrogen may undergo ionization at a few units lower pH than for the free ligand ($pK_{N1\text{-}H} > 14$). Bogges and Martin (1974) found that the ionization of 2-(2'-pyridyl)imidazole is promoted by the 3d metal ions according to the sequence, Fe(II) > Cu(II) > Co(II) > Zn(II) > Ni(II). The importance of the presence of other ligands was demonstrated by Sóvágó et al. (1978) in the Cu(II)-histamine-B ligands systems. They found that mixed complex formation with π-acceptor ligands such as 2,2-bypiridyl promoted further the ionization of the N1-H proton, whereas with catechol derivatives, which are capable of a charge transfer process in the opposite direction, this process is significantly suppressed.

Mixed complex formation of histamine was found to be pronounced with various small biomolecules, including amino acids, hydroxycarboxylic acids, catecholamines, nucleotides (Gergely and Sóvágó, 1973; 1976; Sóvágó and Gergely, 1976; 1979; Daniele et al., 1976; Daniele and Ostacoli, 1978; 1987; Kayali and Berthon, 1980; 1982; Blais et al., 1981; Berthon and Kayali, 1982; Berthon and Germonneau, 1982; Berthon et al., 1987; Nair et al., 1980; 1982; Reddy and Rao, 1985; 1986; Reddy and Reddy, 1986; Grasso et al., 1980; Arena et al., 1984; Borghesani et al., 1990; Jackson et al., 1978). On the basis of the main factors influencing mixed complex formation (see Chapter 1-C3), enhanced ternary complex formation can be expected to a different extent with ligands which contains O donor atoms (Sigel, 1980), those which are able to form 5-membered chelate rings (Gergely and Sóvágó, 1976), those with which complexation is less hindered than the coordination of a second histamine molecule (Sóvágó and Gergely, 1979), those which are able to increase the electron density on the metal ions (reduced due to the π-acceptor behavior of histamine) via a strong ligand $\rightarrow$ metal charge transfer process (Daniele et al., 1987). The role of these factors is nicely demonstrated in the formation percentages of the ternary species in different systems containing equimolar amounts of Cu(II) and each of the two ligands: 68% in the Cu(II)-histamine-ethylenediamine, 78% in the Cu(II)-histamine-glycine, 83% in the Cu(II)-histamine-oxalic acid, and 98% in the Cu(II)-histamine-dopamine systems.

Besides mixed-ligand complexes, histamine (hm) can also form mixed metal complexes, e.g. $CuCd(hm)_2$ or $CuZn(hm)_2$, in which histamine behaves as a bridging ligand (Amico et al., 1981).

Table 3

Plasma Distribution of Histamine (Computer Simulation)*

Species	Electrical charge	Concentration (mol dm^{-3})	Percentage of the complexed histamine fraction
Zn(hm)(cys)	0	2.09 10^{-13}	75.75
Zn(hm)(cys)H	+1	2.88 10^{-14}	10.41
Zn(hm)(his)	+1	2.18 10^{-14}	7.89
Cu(hm)(his)	+1	4.16 10^{-15}	1.50
Zn(hm)	+2	4.13 10^{-15}	1.49
Zn(hm(citrate)	-1	2.68 10^{-15}	0.97
Zn(hm)(OH)	+1	2.59 10^{-15}	0.94
Zn(hm)(glu)	+1	1.84 10^{-15}	0.67
Zn(hm)(thr)	+1	1.08 10^{-15}	0.39

* Data taken from Kayali and Berthon (1980).

Mixed complex formation was extensively studied by Berthon et al. (Kayali and Berthon, 1980; 1982; Blais et al., 1981; Berthon and Kayali, 1982; Berthon and Germonneau, 1982; Berthon et al., 1987) in a simulation of histamine distribution under blood plasma conditions and to obtain information in this way on the possible role of important plasma metal ions, such as Cu(II) and Zn(II) on the physiological activity and catabolism of histamine.

Table 3 shows the computer simulation results of plasma distribution of histamine among the ternary complexes formed with the most important low-molar-mass plasma constituents. It has been found (Kayali and Berthon, 1982) that the Zn(II)-histamine-cysteinate complex, which is electrically neutral, represents ~75 % of the metal-complexed fraction of histamine in normal blood plasma. Although its concentration is about hundred times smaller than that of free histamine (5.6×10^{-11} mol dm^{-3}), the latter is highly polar and hence not likely to diffuse through lipidic cell membranes. Therefore it has been proposed that the Zn(II)-histamine-cysteinate complex plays an important role in the above processes via ready diffusion from plasma into the environmental tissues, in which histamine is catabolized.

REFERENCES

Aiba, H., Yokoyama, A., and Tanaka, H. (1974). *Bull. Chem. Soc. Jpn.*, *47*: 1003.

Amico, P., Arena, G., Daniele, P.G., Ostacoli, G., Rizzarelli, E., and Sammartano, S. (1981). *Inorg. Chem.*, *20*: 772.

Amico, P., Daniele, P.G., Ostacoli, G., and Zelano, V. (1983). *Ann. Chim. (Rome)*, *73*: 253.

Arena, G., Cali, R., Cucinotta, V., Musumeci, S., Rizzarelli, E., and Sammartano, S. (1984). *J. Chem. Soc., Dalton Trans.*, 1651.

Ashikawa, I., and Itoh, K. (1978). *Chem. Lett.*, 681.

Berthon, G., and Germonneau, P. (1982). *Agents Actions*, *12*: 619.

Berthon, G., and Kayali, A. (1982). *Agents Actions*, *12*: 398.

Berthon, G., Varsamidis, A., Blaquière, C., and Rigal, D. (1987). *Agents Actions*, *22*: 231.

Bhattacharya, P.K., Emanual, N.A., and Kulkarni, N.D. (1986). *Proc. Indian Acad. Sci.*, *97*: 529.

Blais, M.-J., Kayali, A., and Berthon, G. (1981). *Inorg. Chim. Acta*, *56*: 5.

Bogges, R.K., and Martin, R.B. (1974). *Inorg. Chem.*, *13*: 1525.

Bogges, R.K., and Martin, R.B. (1975). *J. Amer. Chem. Soc.*, *97*: 3076.

Borghesani, G., Pulidori, F., Remelli, M., Purrello, R., and Rizzarelli, E. (1990). *J. Chem. Soc., Dalton Trans.*, 2095.

Branca, M., Micera, G., Dessì, A., Sanna, D., and Raymond, K.N. (1990). *Inorg. Chem.*, *29*: 1586.

Buglyó, P., and Kiss, T. (1991). *J. Coord. Chem.*, *22*: 259.

Buglyó, P., Dessì, A., Kiss, T., Micera, G., and Sanna, D. (1993). *J. Chem. Soc., Dalton Trans.*, 2057.

Burlamacchi, L., Lai, A., Monduzzi, M., and Saba, G. (1983). *J. Magn. Res.*, *55*: 39.

Campbell, N.R.C., and Hasinoff, B. (1989). *Clin. Pharmacol. Ther.*, *45*: 220.

Cooper, S.R., Koh, Y.B., and Raymond, K.N. (1982). *J. Amer. Chem. Soc.*, *104*: 5092.

Crisponi, G., Lai, A., Monduzzi, M., and Saba, G. (1983). *Inorg. Chim. Acta*, *80*: 85.

Daniele, P.G., and Ostacoli, G. (1976a). *Ann. Chim. (Rome)*, *66*: 387.

Daniele, P.G., and Ostacoli, G. (1976b). *Ann. Chim. (Rome)*, *66*: 537.

Daniele, P.G., and Ostacoli, G. (1978). *J. Inorg. Nucl. Chem.*, *40*: 1273.

Daniele, P.G., Amico, P., Ostacoli, G., and Zelano, V. (1983). *Ann. Chim. (Rome)*, *73*: 199.

Daniele, P.G., Amico, P., and Ostacoli, G. (1984). *Ann. Chim. (Rome)*, *74*: 105.

Daniele, P.G., Amico, P., Zerbianti, O., and Ostacoli, G. (1986). *Ann. Chim. (Rome)*, *76*: 393.

Daniele, P.G., Zerbinati, O., Negro, G., and Ostacoli, G. (1987). *Ann. Chim. (Rome)*, *77*: 879.

Daniele, P.G., Ostacoli, G., and Zerbinati, O. (1990). *Ann. Chim. (Rome)*, *80*: 89.

Dasgupta, P., and Jordan, R.B. (1985). *Inorg. Chem.*, *24*: 2721.

Davidov, J.P., Voronik, N.I., and Skripcova, A.V. (1975). *Radiokhim.*, 597.

Eilbeck, W.J., Holmes, F., and Thomas, T.W. (1970). *J. Chem. Soc. (A)*, 2062.

El-Hendawy, A.M., Griffith, W.P., and Pumphrey, C.A. (1988). *J. Chem. Soc., Dalton Trans.*, 1817.

Emanual, N.A., and Bhattacharya, P.K. (1987). *Polyhedron*, *6*: 845.

Gergely, A., and Kiss, T. (1976). *Inorg. Chim. Acta*, *16*: 51.

Gergely, A., and Kiss, T. (1979). *Metal Ions in Biological Systems*, Vol. 9 (H. Sigel, ed.), Marcel Dekker, New York, p. 143.

Gergely, A., and Sóvágó, I. (1973). *J. Inorg. Nucl. Chem.*, *35*: 4355.

Gergely, A., and Sóvágó, I. (1976). *Inorg. Chim. Acta*, *20*: 19.

Gergely, A., Kiss, T., and Deák, Gy. (1979). *Inorg. Chim. Acta*, *36*: 113.

Gergely, A., Kiss, T., Deák, Gy., and Sóvágó, I. (1981). *Inorg. Chim. Acta*, *56*: 35.

Gilbert, K., and Kustin, K. (1976). *J. Amer. Chem. Soc.*, *98*: 5502.

Gorton, J.E., and Jameson, R.F. (1972a). *J. Chem. Soc., Dalton Trans.*, 310.

Gorton, J.E., and Jameson, R.F. (1972b). *J. Chem. Soc., Dalton Trans.*, 304.

Grasso, M., Musumeci, S., Rizzarelli, E., and Sammartano, S. (1980). *Ann. Chim. (Rome)*, *70*: 193.

Hider, R.C., Mohd-Nor, A.R., Silver, J., Morrison, I.E.G., and Rees, L.V.C. (1981). *J. Chem. Soc., Dalton Trans.*, 609.

Hider, R.C., Howlin, B., Miller, J.R., Mohd-Nor, A.R., and Silver, J. (1983). *Inorg. Chim. Acta*, *80*: 51.

Itabashi, M., Shoji, K., and Itoh, K. (1982). *Inorg. Chem.*, *21*: 3484.

Jackson, G.E., May, P.M., and Williams, D.R. (1978). *J. Inorg. Nucl. Chem.*, *40*: 1227.

Jameson, R.F., and Neille, W.F.S. (1966). *J. Inorg. Nucl. Chem.*, *28*: 2667.

Kayali, A., and Berthon, G. (1980). *J. Chem. Soc., Dalton Trans.*, 2374.

Kayali, A., and Berthon, G. (1982). *Polyhedron*, *4*: 371.

Klime-Hunt, E., Spartalian, K., Holmes, S., Mohan, M., and Carrano, C.J. (1991). *J. Inorg. Biochem.*, *41*: 125.

Kiss, T. (1990). *Biocoordination Chemistry* (K. Burger, ed.), Ellis Horwood, Chichester, p. 84.

Kiss, T., and Gergely, A. (1983a). *Acta Chim. Hung.*, *114*: 249.

Kiss, T., and Gergely, A. (1983b). *Inorg. Chim. Acta*, *78*: 247.

Kiss, T., and Gergely, A. (1985). *J. Inorg. Biochem.*, *25*: 247.

Kiss, T., Deák, Gy., and Gergely, A. (1984). *Inorg. Chim. Acta*, *91*: 269.

Kiss, T., Sóvágó, I., and Martin, R.B. (1989). *J. Amer. Chem. Soc.*, *111*: 3611.

Kopell, W.N., and Westhead, E.W. (1982). *J. Biol. Chem.*, *257*: 5707.

Kustin, K., and Liu, S.T. (1973). *Inorg. Chem.*, *12*: 1486.

Kustin, K., Liu, S.T., Nicolini, C., and Toppen, D.L. (1974). *J. Amer. Chem. Soc.*, *96*: 7410.

Kwik, W-L., Purdy, E., and Stiefel, E.J. (1974). *J. Amer. Chem. Soc.*, *96*: 1638.

Lai, A., Monduzzi, M., and Saba, G. (1980). *Adv. Mol. Rel. Interact. Proc.*, *18*: 221.

Lai, A., Monduzzi, M., Saba, G., Casu, M., and Crisponi, G. (1982). *Chem. Phys.*, *71*: 271.

Linert, W., Jameson, R.F., and Herlinger, E. (1991). *Inorg. Chim. Acta*, *187*: 239.

Martin, R.B. (1971). *J. Phys. Chem.*, *75*: 2657.

Mentasti, E., Pellizetti, E., and Saini, G. (1976). *J. Inorg. Nucl. Chem.*, *38*: 785.

Minch, M.J., Sevenair, J.P., and Henling, C. (1979). *J. Org. Chem.*, *44*: 3247.

Monduzzi, M., Lai, A., Saba, G., Casu, M., and Crisponi, G. (1982). *Adv. Mol. Rel. Interact. Proc.*, *24*: 233.

Morris, P.J., and Martin, R.B. (1970). *J. Amer. Chem. Soc.*, *92*: 1543.

Nair, M.S., Santappa, M., and Natarajan, P. (1980a). *J. Chem. Soc., Dalton Trans.*, 1312.

Nair, M.S., Santappa, M., and Natarajan, P. (1980b). *J. Chem. Soc., Dalton Trans.*, 2138.

Nair, M.S., Venkatachalapathi, K., and Santappa, M. (1982). *J. Chem. Soc., Dalton Trans.*, 555.

Noszàl, B. (1990). *Biocoordination Chemistry* (K. Burger, ed.), Ellis Horwood, Chichester, p. 18.

Odani, A., and Yamauchi, O. (1984). *Inorg. Chim. Acta*, *93*: 13.

Pilbrow, J.R., Carr, S.G., and Smith, T.D. (1970). *J. Chem. Soc. (A)*, 723.

Prasad, K., Rao, A.K., and Mohan, M.S. (1987). *J. Coord. Chem.*, *16*: 251.

Rajan, K.S., and Davis, J.M. (1976). *J. Inorg. Nucl. Chem.*, *38*: 897.

Rajan, K.S., Davis, J.M., and Colburn, R.W. (1971). *J. Neurochem.*, *18*: 345.

Rajan, K.S., Davis, J.M., Colburn, R.W., and Jarke, F.H. (1972). *J. Neurochem.*, *19*: 1099.

Rajan, K.S., Colburn, R.W., and Davis, J.M. (1976a). *Metal Ions in Biological Systems*, Vol. 6 (H. Sigel, ed.), Marcel Dekker, New York, p. 292.

Rajan, K.S., Skripkus, A., Marks, G.E., and Davis, J.M. (1976b). *Bioinorg. Chem.*, *6*: 93.

Rajan, K.S., Mainer, S., and Davis, J.M. (1978). *Bioinorg. Chem.*, *9*: 187.

Rajan, K.S., Diamond, B.I., Borison, R.L., and Davis, J.M. (1986). *Biochem. Arch.*, *2*: 107.

Rajan, K.S., Wiehle, R.D., and Davis, J.M. (1987). *Biochem. Arch.*, *3*: 113.

Reddy, P.R., and Rao, V.B.M. (1985). *Polyhedron*, *4*: 1603.

Reddy, P.R., and Rao, V.B.M. (1986). *J. Chem. Soc., Dalton Trans.*, 2331.

Reddy, P.R., and Reddy, B.M. (1986). *Polyhedron*, *5*: 1947.

Reynolds, W.F., Peat, I.R., Freedman, M.H., and Lyerla, J.R. (1973). *J. Amer. Chem. Soc.*, *95*: 328.

Sigel, H. (1975). *Angew. Chem. Int. Ed. Engl.*, *14*: 394.

Sigel, H. (1980). *Coordination Chemistry-20* (D. Banerjea, ed.), Pergamon Press, New York, p. 27.

Sóvágó, I., and Gergely, A. (1976). *Inorg. Chim. Acta*, *20*: 27.

Sóvágó, I., and Gergely, A. (1979). *Inorg. Chim. Acta*, *37*: 233.

Sóvágó, I., Kiss, T., and Gergely, A. (1978). *J. Chem. Soc., Dalton Trans.*, 964.

Sundberg, R.J., and Martin, R.B. (1974). *Chem. Rev.*, *74*: 471.

Tanokura, M. (1983). *Biochim. Biophys. Acta*, *742*: 576.

Walker, W.R., Shaw, Y.L., and Li, N.C. (1973). *J. Coord. Chem.*, *3*: 77.

Weber, O.A. (1974). *J. Inorg. Nucl. Chem.*, *36*: 1341.

Weber, O.A., and Simeon, Vl. (1971). *J. Inorg. Nucl. Chem.*, *33*: 2097.

Wilson, N.B. (1952). *J. Biol. Chem.*, *197*: 215.

Yagil, G. (1967). *Tetrahedron*, *23*: 2855.

METAL COMPLEXES WITH CARBOHYDRATES

by

Henryk Kozlowski and Malgorzata Jezowska-Bojczuk

Instytut Chemii,
Uniwersytet Wroclawski,
ul. F. Joliot-Curie 14
50383 Wroclaw
Poland

I. INTRODUCTION

Complex formation between metal ions and carbohydrates is an old subject although the species formed are weak and monosaccharides are unlikely to have any important biological significance (Angyal, 1989). In aqueous solution, metal ions are coordinated to water molecules and the unionized hydroxyl groups of carbohydrates would be only weak competitors for metal ion coordination. Monodentate coordination is unlikely to be able to displace water and only a suitably arranged combination of two or three hydroxyl groups would be expected to lead to more direct sugar coordination. As a result, the stability constants of complexes formed with monosaccharides are so small that complex formation is difficult to detect.

In biological systems, metal complexation is more likely to occur with polysaccharide ligands and it is widely believed that metal ion-carbohydrate systems may play a role in the transport of cations through membranes. In the crystalline state, the situation may be different as a result of low solvent concentration and the direct involvement of sugar unionized hydroxyl is more likely.

Thus, solution studies on metal ion binding to monosaccharides is a difficult task and the results obtained are often found to be controversial. Up-to-date discussion of low-molar-mass (l.m.m.) metal complexes in solution with carbohydrates has been published very recently in several detailed review articles (Angyal, 1989; Burger and Nagy, 1990; Dill and Carter, 1989).

From these critical reviews it can be concluded that, among biologically essential metal ions, only Ca^{2+} (which occurs in substantial amounts) may be affected by carbohydrates while trace elements will not form complexes to any significant extent. It is, however, possible that metal ions like VO^{2+} or Fe^{3+} may be directly bound to deprotonated carbohydrate hydroxyl groups. There are chemically significant complexes formed with metal containing anions such as molybdates or tungstates, but these formally should be considered as esters. Thus, in this review the general description of Ca^{2+} complexes as well as some characteristics of tungstate and molybdate species formed with simple carbohydrates will be given. Other metal ions complexes will be mentioned only briefly because of space limitations.

Acidic carbohydrates (sugar acids) are much more effective ligands for metal ions. These ligands (e.g. uronic acids) contain the carboxylate anchor function which is able to act as a primary binding site. Coordination to carboxylate allows easier deprotonation of the vicinal hydroxyl group and stable chelates may be formed at relatively low pH. These complexes will be considered in the second part of this chapter.

Cyclodextrins, cyclic oligosaccharides with internal cavities capable of forming inclusion complexes with small organic and metal compounds will be also mentioned.

II. COMPLEXES OF CARBOHYDRATES

The spatial arrangement of hydroxyl oxygen atoms is of critical importance for metal ion interactions with unionized carbohydrates (Angyal et al., 1974). The most commonly found sequence which is favorable for metal ion binding is an axial (**a**), an equatorial (**e**) and an axial hydroxyl group on a six-membered ring in a chair conformation (Fig. 1). The rare 1,3,5-triaxial arrangement is also very suitable for complex formation.

The size of the metal ion has also been found to be a critical factor in binding. The optimum radius for complex formation is around 100 pm as is found for Na^+, Ca^{2+} and La^{3+}. Larger and smaller cations interact with unionized carbohydrates less strongly. Generally, cations with higher charges (+3 or +2) form stronger complexes than univalent ions (Angyal, 1989). It should be mentioned, however, that e.g. Al^{3+} may not form stronger complexes than Na^+ (Hämäläinen and Lönnberg, 1991), which suggests that ionic radius is a more crucial factor than cationic charge.

Thermodynamic data obtained for metal ions interacting with D-ribose could not be related either to the size or to the charge of the complexed ion in a simple way, suggesting rather high complexity of complexation (Alvarez et al., 1987). An additional difficulty in the evaluation of any thermodynamic property in solution results from the fact that the values obtained are "macro" constants and so represent overall values characteristic to all complexing forms (isomers) of carbohydrate present in solution.

A measure of the strength of complex formation between a metal ion and a carbohydrate is the stability constant. However, accurate determination is a difficult task and it is often unreasonable to compare the results of various experimental methods since the observed variables obtained from these different techniques cannot be transformed into stability constants without special thermodynamic assumptions (Hämäläinen and Lönnberg, 1991). Nevertheless, it is agreed that values found for calcium with some carbohydrates (of the order of 10 mol^{-1}) indicate comparatively strong complex formation (Angyal, 1989).

In the crystalline state, metal ions like calcium are coordinated directly to oxygen atoms of sugar molecules as well as to water molecules. Crystal structures show clearly the binding modes of metal ions with carbohydrates and provide a valuable insight into the nature of metal ion coordination to hydroxyl groups. It is believed that geometrical features established from single crystal studies may help to understand especially the structural chemistry of calcium interacting with polysaccharides (Dheu-Andries and Pérez, 1983). Considerations based on known structures allow to conclude that a calcium carbohydrate complex must satisfy the overall geometry of the site rather than merely a conformation favoring electrostatic contacts. Possible calcium-carbohydrate interactions may be, to some extent, specific so that they can participate in the control of different biologically important processes such as calcification, gelation or cell-cell adhesion.

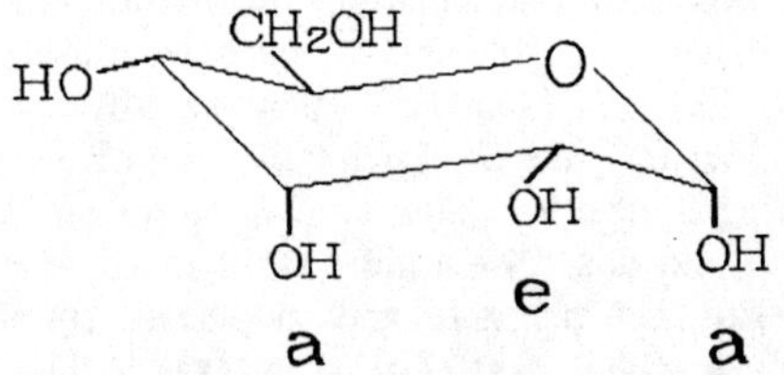

Figure 1

It is also interesting to note that this study (Dheu-Andries and Pérez, 1983) has shown that, in the solid state only, three general types of calcium coordination polyhedra have been found. The most common is close to a square-antiprism arrangement of eight oxygen atoms, followed by seven- and nine-fold coordination with a geometry which can be obtained by slight alterations of the eight-fold coordination.

The crystal structure data, however, do not reflect the situation in solutions precisely because, in the solid state where there are only a few or no water molecules, a cation such as calcium is forced to accept the donor site which may be weakly bound or not coordinated at all in solution. The X-ray structures of Sr^{2+} (Angyal et al., 1989) and Ba^{2+} (Koziol and Lis, 1991) complexes with carbohydrates indicate that both these metal ions may form similar complexes, although their solution behavior may differ from that of Ca^{2+}.

Other divalent metal ions, such as the zinc-group metal ions, do not form any major complexes with carbohydrates in aqueous solution although in the solid state or in non-aqueous solutions they are found to be of great importance (Tajmir-Riahi, 1989). The divalent metal ion Pb^{2+} forms stable complexes with carbohydrates comparable to those of Ca^{2+} (Lönnberg and Vesala, 1980).

The binding of VO^{2+} or Fe^{3+} ions to carbohydrates may differ considerably from those discussed above. Both metal ions induce hydroxyl group deprotonation and bind simple sugar ligands very effectively in neutral and basic solutions (Branca et al., 1992; Nagy et al., 1989). This effective coordination depends, however, on positions of the adjacent hydroxyl groups which bind only when in a cis arrangement (**a,a** or **e,e**) thus allowing chelate complex formation with metal ions. One of the most effective ions to bind two consecutive deprotonated hydroxyl groups of a carbohydrate has been found to be the Cr^{5+} ion (Branca et al., 1990a) where the bound sugars stabilize effectively the +5 oxidation state of the metal ion.

Vanadate, tungstate and molybdate form very specific complexes with carbohydrates due to the possible formation of stable di- and tri-esters (Chapelle and Verchère, 1991; Verchère and Chapelle, 1989; Geraldes and Castro, 1989). The stability of the resulting complexes basically depends on the positions of the hydroxyl groups of the sugar ligand involved in the ester bond formation. Dinuclear complexes are often formed in these systems.

III. COMPLEXES OF SUGAR ACIDS

With sugar acids, the ability to form complexes is usually established more easily than with the sugars discussed above. The carboxylate function, having pK values between 3 and 4, is an effective binding site for many metal ions. Metal ion anchored at this site may induce the deprotonation of an adjacent hydroxyl group, creating stable chelated complexes. Even Ca^{2+} complexes with sugar acids with all hydroxyl groups protonated are much more stable than those obtained with neutral carbohydrates due to the involvement of the carboxylate function in metal ion binding (Angyal, 1989).

The uronic acids D-galacturonic and D-glucuronic acids have carboxyl groups of very similar acidity (pK = 3.28 and 3.06, respectively, Micera et al., 1989). The carboxylate initiates binding of cupric or vanadyl ion at pH around 3 and one or two sugar hydroxyl groups are then deprotonated and coordinated. The position HO4 is critical for the resulting equilibria as is demonstrated by the fact that while D-galacturonic acid forms a series of monomeric complexes (the O4 atom can easily form a chelate together with the adjacent carboxylate, as shown in Table 1; Micera et al., 1989; Branca et al., 1989) D-glucuronic acid forms oligomeric species which are difficult to evaluate precisely from potentiometric or spectroscopic data. Complexes of the vanadyl ion are much more stable than those of Cu^{2+} because of the lower affinity of the latter ion for oxygen donors.

The affinity of the vanadyl ion for deprotonated hydroxyl groups is so high that, in basic media, it is able to stabilize the open structure of D-galacturonic acid by forming a dimeric species (Fig. 2; Branca et al., 1990b).

The coordination equilibria between Cu^{2+} and VO^{2+} ions and D-lactobionic acid are exactly the same as those found for galacturonic acid, i.e. coordination starts at carboxylate site and then continues to one or two deprotonated hydroxyl groups. However, the stability

Table 1

Stability Constants (logβ) for Proton, Cu^{2+} and VO^{2+} Complexes with D-Lactobionic (LacA) and D-Galacturonic (GalA) Acids

Metal Species (M L H)	D-LacA	D-GalA	Binding mode
Cu^{2+}			
0 1 1	3.53	3.28	
1 1 0	3.10[a]	1.80[a]	{COO⁻}
1 2 0	5.90[a]	3.00[a]	2 x {COO⁻}
1 2 -1	-0.43	-3.02	{COO⁻}{COO⁻,O⁻}
1 2 -2	-7.05	-10.06	2 x {COO⁻,O⁻}
1 2 -4	-27.06	-29.69	2 x {O⁻,O⁻}
VO^{2+}			
1 2 0	6.07		2 x {COO⁻}
1 2 -1	2.32	0.47	{COO⁻}{COO⁻,O⁻}
1 2 -2	-1.92	-4.01	2 x {COO⁻,O⁻}
1 2 -3	-10.31	-12.98	{COO⁻,O⁻}{O⁻,O⁻}
1 2 -4	-18.56	-22.56	2 x {O⁻,O⁻}

[a] Stability obtained from polarographic data; all other values were obtained from potentiometric data (Kozlowski et al., 1991; Bouhsina et al., 1991; Micera et al., 1989).

constants found for both linear and cyclic sugars differ considerably (Table 1; Kozlowski et al., 1991; Bouhsina et al., 1991). The results clearly show that, even when strong bonds are formed via deprotonated hydroxyl groups, the most critical factor is again the position of the respective hydroxyl groups involved in metal ion binding. It also seems evident that metal interactions with protonated sugar OH groups may additionally stabilize the coordination to carboxylate or deprotonated hydroxyl functions.

Sugar acids are also effective ligands for other metal ions such as Pb^{2+} (Cesaro et al., 1988; Whitfield and Sarkar, 1991), Co^{3+} (Tajmir-Riahi, 1988), Zn^{2+} (Whitfield and Sarkar, 1991) and Cd^{2+} (Blomquist and Still, 1985; Whitfield and Sarkar, 1991). The coordination pattern is, in most cases, similar to that discussed above, i.e. it starts at the carboxylate group and then proceeds to one or two hydroxyl groups which are deprotonated before binding to the metal ion. In some cases, however, deprotonation of a water molecule coordinated to the metal ion leads to the formation of mixed hydroxo-complexes.

Al^{3+} ions, which are ineffective in binding to simple carbohydrates, may easily react with sugar acids (e.g. gluconic or saccharic acid) to form stable equimolar complexes (Motekaitis and Martel, 1984). Al^{3+} ions have a strong tendency to displace the protons of

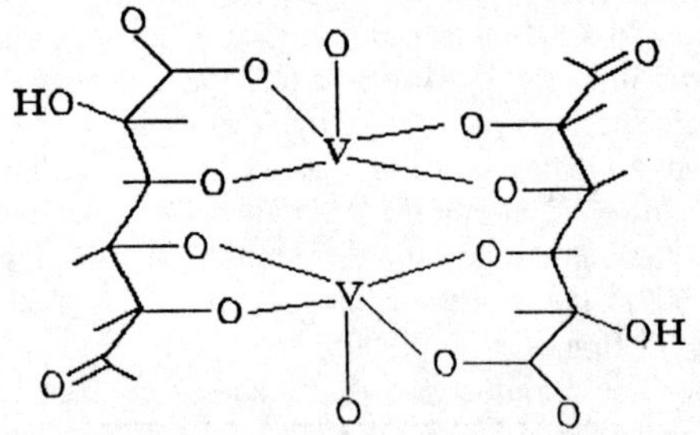

Figure 2

hydroxy-groups of sugar acids, but they need to be anchored to the ligand to avoid total metal ion hydrolysis.

The relatively high stability of Fe^{3+} complexes with sugar acids prevents hydrolysis of these ions in biological fluids and can be useful in iron therapy (Burger and Nagy, 1990). Gluconic acid is found to be the most effective ligand for Fe^{3+}, although evaluation of the concentration distribution of the complexes over the physiological pH range has suggested that galacturonic and malonic acids may be the most promising ligands in oral iron therapy (Cape et al., 1974).

The anchor sites may be introduced to the carbohydrate in chemically independent moieties, e.g. peptide. The recent study on the coordination of metal ions with sugar containing thiazolidine derivatives has shown that the carboxyl group may act as an anchor and, in basic media, some metal ions would also bind to carbohydrate hydroxyl groups (Gajda et al., 1992).

IV. INTERACTION OF METAL IONS WITH CYCLODEXTRINS

The torus-shaped cyclodextrins (CD) are a family of cyclic oligo saccharides consisting of 6 (α-CD), 7 (β-CD), 8 (γ-CD) or 9 (δ-CD) α-D-glucopyranoses connected by α-(1-4) interglucose bonds. The direct coordination of a metal ion to CD rarely happens because of the poor coordinating ability of CD hydroxy groups, and it has been found only for Cu^{2+} (Matsui et al., 1975), Mn^{3+} (Nair and Dismukes, 1983) and Co^{3+} (Yamanari et al., 1989). Having internal cavities of 5-8 Å, however, they can form inclusion complexes with small organic or organometallic compounds which are able to fit into these. In such inclusion complexes, cyclodextrins act as second-sphere ligands influencing, however, the structure of the guest molecules. CD molecules have been proposed to be artificial enzymes as their inclusion complexes show regio- and stereo-specificity with respect to the substrate and the resulting complexes can be useful models for understanding molecular recognition (Kotake and Janzen, 1989 and refs. therein). Many of the inclusion complexes crystallize readily and they are easily studied using X-ray and neutron techniques (Steiner et al., 1991).

Inclusion complexes containing metal complex inside the CD cavity frequently have properties dependent on the cavity size. The bis(2-pyridylcarbinolato)copper(II) complex forms a 1:1 inclusion complex with α-CD while a more stable 2:1 inclusion complex is formed with γ-CD. In the latter case, the copper(II) species exists as a dimer (Yokoi et al., 1991).

The functionalized cyclodextrins are, in many respects, more efficient models for biological systems (enzymes, receptors) than the parent CD molecules (Breslow, 1982; Anslyn and Breslow, 1989). Furthermore, for more sophisticated molecular recognition, multiple recognition systems can be obtained by means of the metal complexes of functionalized CD where the additivity of cavity and metal-ligand interaction has been demonstrated (Tabushi, 1986).

The functionalized 6-deoxy-6-(N-histamino)-β-cyclodextrin (CDhm) shows a protonation-promoted inclusion of the imidazole ring, while Cu^{2+} coordination is markedly affected by the presence of the CD cavity (Bonomo et al., 1991). Thus, the formation of the

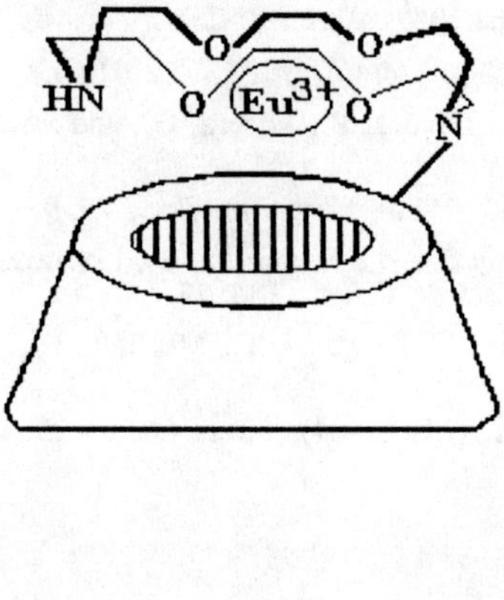

Figure 3

[Cu(CDhm)] complex (logK = 7.26) is much less favored than the [Cu(hm)] species (logK = 9.57) due to a smaller exothermic enthalpy contribution and a less positive entropy change (stiffing effect induced by the cavity).

In the crown-modified cyclodextrin (Fig. 3), Eu^{3+} ions bound at the *"aza"* receptor site could be a useful luminescence probe to follow basic photochemical processes involving the CD cavity. For example, the enhanced europium emission observed upon the introduction of benzene into a solution of the supramolecular assembly permits the study of the photochemical process between a benzene molecule occupying the cavity and a Eu^{3+} ion residing in the appended *"aza"* macrocycle (Pikramenou and Nocera, 1992).

REFERENCES

Alvarez, A.M., Morel-Desrosiers, N., and Morel, J.P. (1987). *Can. J. Chem., 65:* 2656.

Angyal, S.J. (1989). *Adv. Carbohyd. Chem. Biochem., 47:* 1.

Angyal, S.J., Greeves, D., and Pickles, V.A. (1974). *Carbohyd. Res., 35*: 165.

Angyal, S.J., Craig, D.C., Defaye, J., and Gadelle, A. (1989). *Can. J. Chem., 68*: 1140.

Anslyn, E., and Breslow, R. (1989). *J. Amer. Chem. Soc., 111*: 8931.

Blomquist, K., and Still, E.R. (1985). *Anal. Chem., 57*: 749.

Bouhsina, S., Decock, P., Kozlowski, H., Micera, G., and Swiatek, J. (1991). *J. Inorg. Biochem., 42*: 57.

Bonomo, R.P., Cucinotta, V., D'Alessandro, F., Impellizzeri, G., Maccarrone, G., Vecchio, G., and Rizzarelli, E. (1991). *Inorg. Chem., 30*: 2708.

Branca, M., Micera, G., Dessì, A., and Kozlowski, H. (1989). *J. Chem. Soc., Dalton Trans.*, 1283.

Branca, M., Dessì, A., Kozlowski, H., Micera, G., and Swiatek, J. (1990a). *J. Inorg. Biochem., 39*: 217.

Branca, M., Micera, G., Sanna, D., Dessì, A., and Kozlowski, H. (1990b). *J. Chem. Soc., Dalton Trans.*, 1997.

Branca, M., Micera, G., Dessì, A., and Sanna, D. (1992). *J. Inorg. Biochem., 45*: 169.

Breslow, R. (1982). *Science, 218*: 532.

Burger, K., and Nagy, L. (1990). *Biocoordination Chemistry, Coordination Equilibria in Biologically Active Systems* (K. Burger, ed.), Ellis Horwood, New York, p. 236.

Cape, J.N., Cook, D.H., and Williams, D.R. (1974). *J. Chem. Soc., Dalton Trans.*, 1849.

Cesaro, A., Delben, F., Flaibani, A., and Paoletti, S. (1988). *Carbohyd. Res., 181*: 13.

Chapelle, S., and Verchère, J.F. (1991). *Carbohyd. Res., 211*: 279.

Dheu-Andries, M.L., and Pérez, S. (1983). *Carbohyd. Res., 124*: 324.

Dill, K., and Carter, R.D. (1989). *Adv. Carbohyd. Chem. Biochem., 47*: 125.

Gajda, T., Nagy, L., Rozlosnik, N., Korecz, L., and Burger, K. (1992). *J. Chem. Soc., Dalton Trans.*, 475.

Geraldes, C.F.G.C., and Castro, M.M.C.A. (1989). *J. Inorg. Biochem., 35*: 79.

Hämäläinen, M.M., and Lönnberg, H. (1991). *Carbohyd. Res., 215*: 357.

Kotake, Y., and Janzen, E.G. (1989). *J. Amer. Chem. Soc., 111*: 2066.

Koziol, A.E., and Lis, T. (1991). *Acta Cryst., C47*: 2076.

Kozlowski, H., Bouhsina, S., Decock, P., Micera, G., and Swiatek, J. (1991). *J. Coord. Chem., 24*: 319.

Lönnberg, H., and Vesala, A. (1980). *Carbohyd. Res., 78*: 9.

Micera, G., Dessì, A., Kozlowski, H., Radomska, B., Urbanska, J., Decock, P., Dubois, B., and Olivier, I. (1989). *Carbohyd. Res., 188*: 25.

Matsui, Y., Kurita, T., Yagi, M., Okayama, T., Mochida, K., and Date, Y. (1975). *Bull. Chem. Soc. Jpn, 48*: 2187.

Motekaitis, R.J., and Martell, A.E. (1984). *Inorg. Chem., 23*: 18.

Nagy, L., Ohtaki, H., Yamaguchi, T., and Nomura, M. (1989). *Inorg. Chim. Acta, 159*: 201.

Nair, B.U., and Dismukes, G.C. (1983). *J. Amer. Chem. Soc., 105*: 124.

Pikramenou, Z., and Nocera, D.G. (1992). *Inorg. Chem., 31*: 532.

Steiner, Th., Mason, S.A., and Saenger, W. (1991). *J. Amer. Chem. Soc., 113*: 5676.

Tabushi, I. (1986). *Pure Appl. Chem., 58*: 1529.

Tajmir-Riahi, H.A. (1988). *J. Inorg. Biochem., 32*: 79.

Tajmir-Riahi, H.A. (1989). *Carbohyd. Res., 190*: 29 .

Verchère, J.F., and Chapelle, S. (1989). *Polyhedron, 8*: 333.

Whitfield, D.M., and Sarkar, B. (1991). *J. Inorg. Biochem., 41*: 157.

Yamanari, K., Nakamichi, M., and Shimura, Y. (1989). *Inorg. Chem., 28*: 248.

Yokoi, H., Satoh, M., and Iwaizumi, M. (1991). *J. Amer. Chem. Soc., 113*: 1530.

METAL COMPLEXES WITH POLYAMINES

by

Lechoslaw Lomozik

Faculty of Chemistry,
Adam Mickiewicz University,
60-780 Poznan
Poland

I. INTRODUCTION

Polyamines (PA, biogenic amines): putrescine (Put) - $NH_2(CH_2)_4NH_2$, spermidine (Spd) - $NH_2(CH_2)_3NH(CH_2)_4NH_2$ and spermine (Spm) - $NH_2(CH_2)_3NH(CH_2)_4NH(CH_2)_3NH_2$ - found in most living organisms play an essential role in a number of biochemical processes. There are also other naturally occurring polyamines such as $NH_2(CH_2)_3NH_2$ (diaminopropane), $NH_2(CH_2)_5NH_2$ (cadaverine), $NH_2(CH_2)_3NH(CH_2)_3NH_2$ (norspermidine), $NH_2(CH_2)_4NH(CH_2)_4NH_2$ (homospermidine), $NH_2(CH_2)_3NH(CH_2)_3NH(CH_2)_3NH_2$ (norspermine), $NH_2(CH_2)_3NH(CH_2)_3NH(CH_2)_4NH_2$ (thermospermine). However, they are found in lesser amounts. Widespread presence of amines, their considerable concentration in living organisms, and, in particular, a significantly increased concentration in fast growing cells have stimulated intense investigation of this group of compounds (Williams-Ashman and Canellakis, 1979; Tabor and Tabor, 1984; Zappia and Pegg, 1988). The concentration varies significantly depending on the species and the type of tissue, and in some tissues reaches a millimolar level (Raina and Jänne, 1975; Cooper et al., 1976; Rennert and Shukla, 1978; El Baze et al., 1983). Cancer patients show an increase in PA content (Durie et al., 1977; Jänne et al., 1978; Potkonjak and Pasic, 1983). Therefore measurements of PA level may be helpful in making a diagnosis or for monitoring the treatment.

The mechanism of polyamine biosynthesis has been well established (Metcalf et al., 1978; Tabor and Tabor, 1985; Pegg, 1986; Morgan, 1987; Shukla, 1990; Jänne et al., 1991). Regulation of this biosynthesis is an important investigative method in the strategy of cancer treatment (Porter et al., 1987; Celano et al., 1988; Pegg, 1988a; Pegg et al., 1988b).

II. INTERACTION OF POLYAMINES WITH BIOMOLECULES

In a physiological medium, polyamines are found in the protonated cation form. Due to their high charge, biogenic amines react with negatively charged fragments of amino acids, proteins, fatty acids, phospholipids and nucleic acids. They also affect the activity of enzymes and the rate of metabolic reactions (Jänne et al., 1991). PA cations influence many biological processes such as pH regulation and ionic adaptation (Heby, 1981; Flores and Galston, 1982; Young and Galston, 1983).

Particularly important are the high affinity of biogenic amines to DNA, RNA and ribosomes and their influence on biosynthesis of nucleic acids and proteins. (Cohen, 1978;

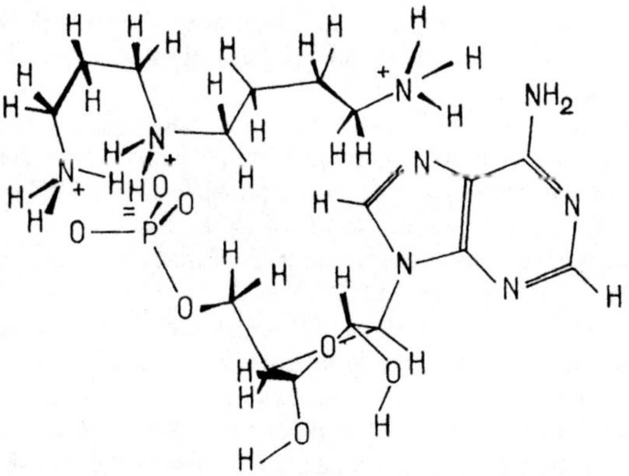

Figure 1.
Model of spermidine with 5'-AMP interaction (from Bunce and Kong, 1978).

Tabor and Tabor, 1984; Marton and Feuerstein, 1986). The interactions of amines with nucleic acids determine the role of PA in genetic processes and most probably lead to structural changes of DNA on a few levels of organization. Both the charge and the structure of the counterion (protonated amines) influence conformational stability of nucleic acids (Wilson and Bloomfield, 1979; Allison et al., 1981; Schellman et al., 1984; Bratek-Wiewiórowska et al., 1987; Srivenugopal et al., 1987; Vertino et al., 1987; Tomita et al., 1989; Feuerstein et al., 1990; Plum et al., 1990; Ohishi et al., 1991; Thomas et al., 1991). Recently, a number of works have been published on the interactions of PA with RNA. (Basu et al., 1988; Kucan et al., 1988; Frydman et al., 1990; Kusama-Eguchi et al., 1991; Wiewiórowski et al., 1991).

In the field of PA investigation, their influence on biological membranes was one of the earliest observed effects (Tabor, 1960). The efficiency of the interaction depends mainly on the charge and structure of the polyamine (Harold, 1964; Powell and Reidenberg, 1982; Souzu, 1986; Schuber, 1989). Spd and Spm play an important role in controlling Ca(II) transportation in mitochondria (Nicchitta and Williamson, 1984; Jensen et al., 1987). PA inhibit the activity of kinase C in vitro (Qi et al., 1983; Thams et al., 1986; Moruzzi et al., 1987; Tadolini and Hakim, 1988).

A convincing model of Spd interactions with 5'-AMP on the basis of NMR investigations was proposed by Bunce and Kong (1978) (Fig. 1). The polyamine reacted with a phosphate group and electron-rich centers of the nucleotide.
Results of equilibrium studies (Nakai and Glinsmann, 1977; Kimura et al., 1982) indicate that unlike the binding of the Mg(II) ion, the binding of polyamines to nucleotides appears to be affected not only by the charge of the species but also by the structural features of the molecules.

Structural information on the polyamine - amino acid complexes has been also obtained (Ramaswamy and Murthy, 1990; 1991).

III. COMPLEXES OF POLYAMINES

One of the most important and the earliest studied chelate ligands is ethylenediamine (en). Free en can adopt one of two configurations: trans or gauche (cis).

In complex compounds, two nitrogen atoms separated by two carbon atoms bind to the central cation to form a five-membered puckered ring. Two enantiomeric conformations are possible: λ (left-handed helicity) and δ (right-handed helicity) with a low energy of inversion barrier (Gollogly et al., 1971; Hambley and Hawkins, 1981a). Although en is a typical chelating ligand, cases are known where this ligand acts in the monodenate or bridging mode (Alexander and Kilcrease, 1973; Ogino, 1980a; 1980b). The number of stereochemical conformations increases in the bis(en) systems. When both en molecules are in trans arrangement, the following isomers are possible: λλ,δδ and δλ. The first two are energetically equivalent and about 4 kJ mol^{-1} more stable than diastereoisomers δλ (House, 1987 and references therein).

1,3-propanediamine (tn) coordinates with the formation of a 6-membered ring. These complexes prefer a chair type coordination, although a twist type conformation has also been reported. The twist conformation is more probable in alkyl substituents (Hambley et al., 1981b).

Linear triamines react with the formation of meridional or facial conformation, whereas the amines of tripodal type: $RC(CH_2NH_2)_3$ (R= H,Me,Et) studied both in solution (Micheloni et al., 1977; Boloni et al., 1983) and in solid state (Flückiger et al., 1980; Ahlgren and Turpeinen, 1982) can only form a facial structure.

A number of reports have been devoted to equilibrium investigations (Barbucci et al., 1973; Anderegg and Blauenstein, 1982; Emanuel and Bhattacharya, 1986; Hague and Moreton, 1989) (in non-aqueous solutions, as well, Cassol et al., 1988) and kinetic studies (Hague and Moodley, 1983; Weatherburn, 1983; Diaz et al., 1987; House and Tobe, 1989; Massoud and Milburn, 1989) of the metal/amine systems. A thorough review of coordination complexes of polyamines was recently given by House (1987). The review quotes about one thousand references on the reactions of this group of ligands with metals. However, only a few of them dealt with biogenic amines: Put, Spd, and Spm.

A. Protonation of polyamines

Table 1 presents stepwise protonation constants of the polyamines.

Protonation of the primary amine group is more exothermic than that of the secondary one (Hedwig and Powell, 1973). The difference between subsequent values of ΔH for diamines $NH_2(CH_2)_nNH_2$ decreases with the increasing chain length. NH_2 groups separated by an appropriately long chain behave as two isolated donor groups. Different from that of 3,2,3-tet* and 3,3,3-tet, protonation of spermine occurs with negligible transmission of inductive effect through a tetramethyl group (Palmer and Powell, 1974a; Palmer and Powell, 1974b). If a carbon chain of polyamine is shorter than 3, at least one of the pK values is lower than 7 and PA is partly deprotonated in a physiological medium. This explains the weak biological activity of this type of amines relative to biogenic ones (Takeda et al., 1983).

Based on analysis of entropy and enthalpy changes, Palmer and Powell (1974a; 1974b) and Anichini et al. (1977) suggested a different thermodynamic nature of protonation of symmetric and asymmetric amines. In the case of symmetric compounds (2,2-tri, 3,3-tri), the first step of protonation can be treated as a process of a tautomeric equilibrium between the two forms:

$$NH_3{}^+\text{-}(CH_2)_n\text{-}NH\text{-}(CH_2)_n\text{-}NH_2 \rightleftharpoons NH_2(CH_2)_n\text{-}NH_2{}^+\text{-}(CH_2)_n\text{-}NH_2$$

As a consequence of the electrostatic repulsion of the charged amine group, in the second step the hydrogen cation is transferred from a secondary to a primary atom. According to Barbucci and Budini (1976) and Anichini et al. (1977), the sequence of protonation enthalpy values for asymmetric amines (2,3-tri, 2,4-tri, Spd: $|\Delta H_1| > |\Delta H_2| > |\Delta H_3|$) excludes the tautomeric mechanism. For symmetric amines, the protonation enthalpy sequence is $|\Delta H_1| < |\Delta H_2| > |\Delta H_3|$. On the basis of NMR studies, however, Kimberly and Goldstein (1981) and Onasch et al. (1984) have concluded that the first step of Spd protonation occurs at the primary and secondary nitrogens atoms. In contrast, Delfini et al. (1980) and Takeda et al. (1983) suggest that the protonation of secondary nitrogens of polyamines begins only after the protonation of primary nitrogen atoms is complete.

B. Metal complexes of putrescine

Until recently it was believed that putrescine cannot form complexes with copper, nickel or zinc in aqueous solution (Bertsch et al., 1958; Paoletti, 1980). However, the results of

---

* in this text, tri stands for triamine, tet for tetramine; 2,3,4,... signifies the number of CH_2 groups between nitrogen atoms

Table 1

Stepwise Protonation Constants of Polyamines (PA) Obtained by Potentiometry

PA	$\log K_1$*	$\log K_2$	$\log K_3$	$\log K_4$	t (°C)	μ (mol dm^{-3})	ref.
Put	10.80	9.35			20	?	a
	10.72	9.24			25	0	b
	10.83	9.68			20	0.1($NaClO_4$)	c
	10.67	9.49			25	0.1(NaCl)	i
	10.80	9.63			25	0.5(KNO_3)	k
Spd	11.16	10.06	8.51		27	0.1(NaCl)	d
	10.95	9.98	8.56		25	0.15(NaCl)	e
	10.97	10.06	8.58		20	0.1($NaClO_4$)	c
	10.83	9.83	8.35		25	0.1(NaCl)	f
	10.89	9.81	8.34		25	0.1(NaCl)	g
Spm	10.97	10.27	9.04	8.03	27	0.1(NaCl)	d
	10.94	10.12	9.04	7.97	25	0.15(NaCl)	e
	10.91	10.37	9.11	8.28	20	0.1($NaClO_4$)	c
	10.86	10.05	8.82	7.95	25	0.1(NaCl)	f
	10.80	10.02	8.85	7.96	25	0.1(NaCl)	h

* for the sake of simplicity, statistical parameters are omitted

a-Bertsch et al. (1958); b-Koskinen and Nikkilä (1972); c-Wojciechowska et al. (1991); d-Takeda et al. (1983); e-Templeton and Sarkar (1985); f-Aikens et al. (1983); g-Palmer and Powell (1974b); h-Palmer and Powell (1974a); i-Ogino (1980a); k-Barbucci et al. (1970)

recent potentiometric and spectral investigation have proven the presence of a few copper complex species (Lomozik and Wojciechowska, 1989; Wojciechowska et al., 1991; Lomozik and Gasowska, 1993). Table 2 presents stability constants of the copper(II) polyamine complexes.

At pH near 7, a protonated $[CuHPut)]^{3+}$ complex of monodentate coordination is formed while at high pH two different hydroxocomplexes of the $[Cu(Put)_2(OH)]^+$ type occur. One of these has two chelate rings, whereas, in the other, two nitrogen atoms of the first ligand and one nitrogen atom of the second ligand are involved in coordination. The reaction of copper(II) with diamines (including Put) is confirmed by the results of ^{13}C NMR studies (Voelter et al., 1975).

The formation of cobalt complexes with Put was noticed by Craighead et al. (1975) and Tešic et al. (1989). Juranic et al. (1986) found that in the diamine series (en, tn, Put) increasing size of the ring in complexes with Co(III) caused a downfield shift of ^{59}Co NMR due to the weakening of the ligand field. They calculate that in larger chelate rings, the overlap of the metal-ligand δ-bond is smaller.

For the series of platinum complexes, Romeo et al. (1978) observed a clear dependence of the ring-closing reaction rate on the ring size. The rate constants, k^f, of the reaction (at 30°C):

$$\text{trans-}[PtCl_2(NH_3)(N\text{-}NH)]^+ \rightleftharpoons [PtCl_2(NH_3)(N\text{-}N)] + H^+ \xrightarrow{k^f} [Pt(N\text{-}N)(NH_3)Cl]^+ + Cl^-$$

(N-NH = the protonated monodentate form of the diamine)

for en, tn and Put are 8.17, 0.62 and 0.0041 s^{-1} (respectively) (Tobe et al., 1982). Differences in reactivity in the series of amine are a consequence of changes in enthalpy of activation rather than entropy of activation. When compared to the 7-membered Put ring, the process of the 8-membered ring closing in trans-$[Pt(Me_2SO)(N\text{-}N)Cl_2]$ (N-N = cadaverine) is 100 times slower (Schwab et al., 1982).

Table 2

Overall Stability Constants** [logβ] of $Cu_p(PA)_qH_r$ Complexes

pqr	110	120	111	121	112	11-1	12-1
Put	8.62[d]	13.40[d]	15.83[d]				0.065[b]
Spd	11.7[a]	16.0[a]	18.9[a]	26.3[a]		1.67[a]	6.18[a]
	11.70[b]	17.13[b]	18.91[b]			2.90[b]	6.72[b]
	11.61[f]		6.83[f*]			4.48[f*]	
Spm	14.7[a]	18.9[a]		29.0[a]	27.8[a]		
	14.66[b]			29.32[b]	27.63[b]		
	14.44[c]	17.2[c]	20.4[c]				
	14.70[e]		9.99[e*]				

* equilibrium constant

** for the sake of simplicity statistical parameters are omitted

a) Templeton and Sarkar (1985): 25°C; μ=0.15 (NaCl)
b) Wojciechowska et al. (1991): 20°C; μ=0.1 ($NaClO_4$)
c) Antonelli et al. (1988): 25°C; μ=0.1 (KNO_3)
d) Lomozik and Gasowska (1993): 20°C; μ=0.1 ($NaClO_4$)
e) Palmer and Powell (1974a): 25°C; μ=0.1 (NaCl)
f) Palmer and Powell (1974b): 25°C; μ=0.1 (NaCl)

In certain platinum complexes, amines react as a bridging reagent (Romeo et al., 1977). Investigation of the reactions of bis(platinum) complexes have proven that dimers are kinetically more active than their monomeric analogs (Farrell and Qu, 1989). Formation enthalpies of ternary complexes [$Cu(hfaa)_2PA$] (hfaa= hexafluoropentane-2,4-dionato) in benzene were found to change with the increasing length of the methylene chain (Graddon and Ong, 1975). The value reflects the differences between the energy of newly formed Cu-N bonds and diminished energy of Cu-O bonds translocated to the axial positions.

Spectral changes in ternary complexes $[Co(NH_3)_5\{NH_2(CH_2)_nNH_2\}]^{3+}$ (n = 2-8,10) when compared to the spectra of $[Co(NH_3)_5\{NH_2(CH_2)_nNH_3\}]^{4+}$ were attributed to the formation of the charge-transfer band. This is a consequence of intramolecular hydrogen bonding between a free amine group and N-H protons of the coordinated ammines or amine groups (Ogino, 1980a). This type of non-covalent interaction plays an important role in biological systems (Frieden, 1975). The analysis of polyamine properties should also take into consideration differences in the number of methylene groups determining their hydrophobic character. The alkyl groups from the coordinating ligands affect the nature of water molecules in their vicinity including those in the solvation domains of an outer-sphere ligand (Craighead et al., 1975).

After many attempts to prepare metal complexes of 1,ω-diamines with a large ring formation, Pfeiffer (1948) succeeded in isolating a bis(Put) complex with copper. As follows from an analysis of molecular models, a 7-membered ring is conformationally labile and transitions among certain conformers are possible (Kojima et al., 1981). In fact the two chelate rings in the solid trans-$[CoCl_2(Put)_2]Cl.H_2O$ have different conformations. The $[Co(Put)_3]Br_3$ complex of the chair conformation of 7-membered rings obtained by Fujita and Ogino (1974) has the absolute Δ (λλλ) configuration with the CoN6 chromophore trigonally elongated (Sato et al., 1974; Sato and Saito, 1975). In crystals of metal complexes, the 7-membered chelate ring is conformationally stabilized. However, due to solvation effects and ion pair formation, the conformation may change in solution (Kojima et al., 1981).

In the reaction of $[Co(H_2O)(NH_3)_5](ClO_4)_3$ or $[CoCl_2(en)_2]Cl$ with diamines $NH_2(CH_2)_nNH_2$ (n=2-14) (in DMSO), a number of monomeric and dimeric complexes were obtained. The yield of this reaction greatly depends on the length of the methylene ring (a medium-size chelate ring is unstable) (Eliel et al., 1965; Ogino and Fujita, 1975; Ogino, 1977). Monomers are not formed in ternary systems containing en or NH_3 when n varies between 5-10. Put still forms a complex, but another naturally occurring one, cadaverine, does not. When the number of methylene groups in the chain is higher than 10, monomers start to form again. With an increasing number of CH_2 groups in the amine, a tendency of the ligand to coordinate in the monodentate mode or to act as a bridging agent increases (Ogino, 1977; Kapanadze et al., 1990).

Bis and tris chelate nickel complexes: $[Ni(Put)_3]Cl_2.H_2O$, $[Ni(Put)_2(H_2O)_2]Br_2$, $[Ni(Put)_2(NCS)_2]$ and several other Ni complexes pyrolytically synthesized from the above ones have been recently described by De and Chaudhuri (1985). All parent complexes are of O_h symmetry; the symmetry of $[Ni(Put)Cl_2]$ is T_d, while that of $[Ni(Put)(NCS)_2]$ is O_h. Diamine complexes: 1,3-propanediamine, putrescine and cadaverine with Zn(II), Cd(II) and Hg(II) halides and pseudohalides were isolated by Ahuja and Singh (1975). These amines act as bidentate bridging or chelating ligands. On the other hand, Bañares et al. (1983) suggest that $NH_2(CH_2)_nNH_2$ diamines with n=2-6,8 act as monodentate ligands with Mo(V). The diamagnetic character of the complexes indicates the formation of dimeric structures.

C. Metal complexes of spermidine and spermine

As follows from the investigation of copper complexes with linear tetramines, the most stable complexes are those with the ligand involving an alternating sequence of five- and six-membered rings (Weatherburn et al., 1970; Hedwig and Powell, 1973; Paoletti et al., 1973; Palmer and Powell, 1974a). For the series of amines, the values of $\log K_{CuL}$ increase in the following sequence: 2,3,2-tet > 3,2,3,-tet > 2,2,2-tet > 3,3,3-tet > 3,4,3-tet (Spm) and depend on a cumulative ring strain and the possibilities of assuming the favored chair or twist-chair conformation. This, however, does not explain the low stability of Spm complexes (Nakayama et al., 1973).

The formation heat of the protonated Spd complex is similar to that of $[Cu(tn)]^{2+}$ but considerably lower than those of the formations of analogous compounds with a 5-membered ring. This suggests the formation of a 6-membered ring in the $[CuHSpd]^{3+}$ complex, with the protonated nitrogen of the amine group attached to the tetramethylene chain (Holmes and Williams, 1967; Anichini et al., 1977). According to thermodynamical and spectral data, the 6-membered ring in a simple $[CuSpd]^{2+}$ complex is fused to the 7-membered one, with 3 nitrogen atoms involved in coordination. The formation enthalpy of the complex is about 12.5 kJ higher than that of the $[CuHSpd]^{3+}$ complex, and also larger that the heat of formation of $[Cu(en)]^{2+}$ with two nitrogen atoms involved in coordination. For the series of homologous linear triamines the low stability of CuL complexes with a seven-membered ring arises from a low enthalpy term (the entropy term is equal) while, for tetramines, the low stability of the complexes with 7-membered rings is due to a very low entropy term - the enthalpy term is favorable and reflects the high basicity of the ligand (Gold and Powell, 1976).

On the basis of computer analysis of potentiometric data, the presence of several complex species in the Cu(II)/Spd and Cu(II)/Spm systems was detected (Templeton and Sarkar, 1985; Antonelli et al., 1988; Wojciechowska et al., 1991) (Table 2). The complex formation begins near pH 5.5, with the formation of a six-membered ring as the anchor side in $[CuHSpd]^{3+}$ (or in $[CuH_2Spm]^{4+}$). The less favored seven-membered ring is found in subsequent forms, stabilized by one or two six-membered rings (in Spd and Spm, respectively). The three-chelate ring effect is reflected in the increase in thermodynamic stability for $[CuSpm]^{2+}$ ($\log\beta=14.7$) with respect to $[CuSpd]^{2+}$ ($\log\beta=11.7$). Absorption energy of the $[CuSpm]^{2+}$ complex, λ = 564nm, suggests coordination with the N4 chromophore. This has been confirmed by crystallographic studies, whose results indicate the formation of a square-planar structure with four nitrogen atoms of the same spermine molecule (Boggs and Donohue, 1975). Decrease in the ligand field strength in $[CuSpd]^{2+}$

(N3 coordination) leads to a decrease in absorption energy (λ = 626nm), and then to λ_{max} = 655 nm in $[CuHSpd]^{3+}$.

The $[Cu(Spd)X_2]$ (with X= Cl,Br,I) complex isolated by Barbucci et al. (1980) has a square pyramidal structure with one ligand X and three nitrogen atoms in one plane, and another atom X in apical position. The six-membered ring assumes the chair conformation, while on the seven-membered ring, the N-C amine bonds are almost coplanar. Recently, Surma et al. (1994) have prepared a Cu(II) complex with norspermine. The coordination polyhedron is close to square pyramidal with four nitrogens on the basal plane and a water molecule in apical position. In the series of synthesized rhodium complexes with linear polyamines, it was found that the smallest of the ligands, 2,2,2-tet, forms only the cis-α isomer: 2,3,2-tet -cis-α and trans isomers. In contrast, the longest ligands studied: 3,2,3-tet, 3,3,3-tet, and 3,4,3-tet (Spm) assumed the trans configuration, which is explained by the differences in steric strain (Halliday and Court, 1974). The size of chelate rings also affects the kinetic character of the complexes. In acidic solution, $[Co(Spd)Cl_3]$ undergoes rapid hydrolysis and increased lability of the coordinated chloride relative to 2,3-tri and 3,3-tri complexes is observed. For a series of triamines, the changes in the reduction rate of $[Co(PA)(OH_2)_3]^{3+}$ to Co(II) are: 3,4-tri (Spd) $<$ 3,3-tri $<$ 2,2-tri, following the changes in the activation energy (House and Moon, 1981).

Due to the interaction of the ClO_4^- and NO_3^- anions with coordination compounds of spermine with copper (II) or zinc (II), an equilibrium shift in the solution was observed. Outer-sphere complex formation is more likely for octahedral complexes of copper than tetrahedral complexes of zinc (Antonelli et al., 1984).

In the model system Cu-AMP-Spm (which might be very useful in understanding some problems of genetic information transfer involving polyamines), the N7 atom of the purine ring is the favored binding site. But a CD investigation testified to the formation of ternary complexes (Weser et al., 1974). A low ability of biogenic amines to form mixed complexes was found for the system of palladium(II)-Spd-2,3-diaminopropionic (or 2,4-diaminobutyric) acid. At a 1:1:1 molar ratio of the reagents, the main complexes are binary compounds: $[Pd(AA)_2]$ (AA=amino acid) and $[Pd(Spd)_2]^{2+}$, while polyamine remains uncoordinated in the system with the excess of acid. CD investigation suggests that the protonated group of Spd interacts electrostatically with a COO^- group of the $[Pd(AA)_2]$ complex (Lomozik et al., 1994). Non-covalent intermolecular interactions play an important role in a number of biological processes (e.g. "zinc finger" structures, Klug and Rhodes (1987), molecular recognition). Formation of a similar type of compounds M-L-L' has been suggested by Antonelli et al. (1988) for the Cu-AMP-Spm system, in which Cu(II) binds to the N7 atom of the purine ring, while the protonated tetramine reacts electrostatically with the phosphate di-anion. In the Cu-ATP-Spm system, however, the same authors conclude that hydrogen bonds are formed between either the N1 or the N7 atom of the base and the protonated nitrogen of Spm (Antonelli et al., 1989).

D. Polyamines in antitumor compounds

A number of PA complexes of Pt(II) were synthesized with a view to finding better antitumor agents than those widely applied in medicine - cis-diamminedichloroplatinum(II), (cis-DDP) (Brunner et al., 1990; Farrell et al., 1990a). Compounds of putrescine (7-membered ring) exhibit higher antitumor activity toward some tumors than en or tn compounds (with 5- or 6-membered rings, respectively) (Nowatari et al., 1989). Complexes of bis(platinum) with bridging diamine are very active against certain tumor cells resistant to cis-DDP (Farrell et al., 1990b). Studies of the interactions of such complexes with DNA indicate that unique interstrand cross-links are formed through the binding of one Pt to each strand of DNA. Dimers have a wider range of clinical activity than the monomeric analogues (Roberts et al. 1989). Antitumor properties are also observed with the palladium compounds. Pd(II) complexes with putrescine and spermidine have ID_{50} values better or similar to cis-DDP while acting against certain cancer cells (Navarro-Ranninger et al., 1992; 1993). This review does not cover, however, the problems of antitumor activity of metal complexes (see Pt. Four, Ch. 5; see also *Medicine*, Vol. 2, Pt. Five, Ch 2).

IV. FINAL REMARKS

(i) It is generally accepted that polyamines play an essential role in many biological processes but the molecular mechanism of their action is still unclear. There is also no reliable estimate of the extent to which metal ions are involved in these processes.
(ii) There is a correlation between the structure of PA and their mode of interaction with biomolecules and metal ions.
(iii) The sites responsible for PA interactions with biomolecules are the same as those involved in complex formation and protonation.
(iv) Formation of metal complexes changes the nature of polyamines (e.g. structure, kinetic behavior, net charge, electron distribution) and, consequently, influences their interactions with biomolecules, particularly their dynamic control of DNA/RNA conformation.
(v) Only a few attempts have been made to unravel the interactions in more complicated (e.g. ternary) systems in which, in addition to metal-ligand bond formation, non-covalent ligand-ligand interactions (very important in biological systems) may also be present.

ACKNOWLEDGEMENT

This work was supported by the KBN Grant Nr 2 P303 074 04.

REFERENCES

Ahlgren, M., and Turpeinen, U. (1982). *Acta Crystallogr.*, *B38*: 276.

Ahuja, I.S., and Singh, R. (1975). *J. Coord. Chem.*, *4*: 181.

Aikens, D., Bunce, S., Onasch, F., Parker, R., III, Hurwitz, C., and Clemans, S. (1983). *Biophys. Chem.*, *17*: 67.

Alexander, M.D., and Kilcrease, H.G. (1973). *J. Inorg. Nucl. Chem.*, *35*: 1583.

Allison, S.A., Herr, J.C., and Schurr, J.M., (1981). *Biopolymers*, *20*: 469.

Anderegg, G., and Blauenstein, P. (1982). *Helv. Chim. Acta*, *65*: 913.

Anichini, A., Fabbrizzi, L., Barbucci, R., and Mastroianni, A. (1977). *J. Chem. Soc., Dalton Trans.*, 2224.

Antonelli, M.L., Balzamo, S., Carunchio, V., and Cernia, E. (1984). *Thermochim. Acta*, *78*: 1.

Antonelli, M.L., Balzamo, S., Carunchio, V., Cernia, E., and Purrello, R. (1988). *J. Inorg. Biochem.*, *32*: 153.

Antonelli, M.L., Carunchio, V., Cernia, E., and Purello, R. (1989). *J. Inorg. Biochem.*, *37*: 201.

Bañares, M.A., Angoso, A., and Criado, J.J. (1983). *Synth. React. Inorg. Met. - Org. Chem.*, *13*: 145.

Barbucci, R., and Budini, M. (1976). *J. Chem. Soc., Dalton Trans.*, 1321.

Barbucci, R., Paoletti, P., and Vacca, A. (1970). *J. Chem. Soc. A*, 2202.

Barbucci, R., Fabbrizzi, L., and Paoletti, P. (1973). *J. Chem. Soc., Dalton Trans.*, 1763.

Barbucci, R., Campbell, M.J.M., Cannas, M., and Marongiu, G. (1980). *Inorg. Chim. Acta*, *46*: 135.

Basu, H.S., Feuerstein, B.G., Zarling, D.A., Shafer, R.H., and Marton, L.J. (1988). *J. Biomol. Struct. Dyn.*, *6*: 299.

Bertsch, C.R., Fernelius, W.C., and Block, B.P. (1958). *J. Phys. Chem.*, *62*: 444.

Boggs, R., and Donohue, J. (1975). *Acta Cryst.*, *B31*: 320.

Boloni, L., Micheloni, M., Sabatini, A., and Vacca, A. (1983). *Inorg. Chim. Acta*, *70*: 117.

Bratek-Wiewiórowska, M.D., Alejska, M., Figlerowicz, M., Barciszewski, J., Wiewiórowski, M., Jaskólski, M., Zielenkiewicz, W., Zielenkiewicz, Z., and Kaminski, M. (1987). *Pure Appl. Chem.*, *59*: 407.

Brunner, H., Hankofer, P., Holzinger, U., Treittinger, B., and Schoenenberger, H. (1990). *Eur. J. Med. Chem.*, *25*: 35.

Bunce, S., and Kong, E.S.W. (1978). *Biophys. Chem.*, *8*: 357.

Cassol, A., Di Bernardo, P., Zanonato, P., Portanova, R., Tolazzi, M., and Tomat, G. (1988). *J. Chem. Soc., Dalton Trans.*, 1781.

Celano, P., Baylin, S.B., Giardiello, F.M., Nelkin, B.D., and Casero, R.A., Jr., (1988). *J. Biol. Chem.*, *263*: 5491.

Cohen, S.S. (1978). *Nature*, *274*: 209.

Copper, K.D., Shukla, J.B., and Rennert, O.M. (1976). *Clin. Chim. Acta*, *73*: 71.

Craighead, K.L., Jones, P., and Bryant, R.G. (1975). *J. Phys. Chem.*, *79*: 1868.

De, G., and Chaudhuri, N.R. (1985). *Bull. Chem. Soc. Jpn.*, *58*: 715.

Delfini, M., Serge, A.L., Conti, F., Barbucci, R., Barone, V., and Ferruti, P. (1980). *J. Chem. Soc., Perkin Trans.*, 900.

Diaz, G., Bustos, C., and Shepherd, R.E. (1987). *Inorg. Chim. Acta*, *133*: 23.

Durie, B.G.M., Salmon, S.E., and Russell, D.H. (1977). *Cancer Res.*, *37*: 214.

El Baze, P., Milano, G., Verrando, P., Renee, N., and Ortonne, J.P. (1983). *Arch. Dermatol. Res.*, *275*: 218.

Eliel, E.L., Allinger, N.L., Angyal, S.L., and Morrison, G.A. (1965). *Conformational Analysis*, John Wiley and Sons, Inc., New York.

Emanuel, N., and Bhattacharya, P.K. (1986). *Indian J. Chem.*, *25A*: 561.

Farrell, N., and Qu, Y. (1989). *Inorg. Chem.*, *28*: 3416.

Farrell, N., Qu, Y., and Hacker, M.P. (1990a). *J. Med. Chem.*, *33*: 2179.

Farrell, N., Qu, Y., Feng, L., and Van Houten, B. (1990b). *Biochemistry*, *29*: 9522.

Feuerstein, B.G., Pattabiraman, N., and Marton, L.J. (1990). *Nucl. Acid. Res.*, *18*: 1271.

Flores, H.E., and Galston, A.W. (1982). *Science*, *217*: 1259.

Flückiger, J.R., Schlapfer, C.W., and Couldwell, C. (1980). *Inorg. Chem.*, *19*: 2493.

Frieden, E. (1975). *J. Chem. Educ.*, *52*: 754.

Frydman, B., De los Santos, C., and Frydman, R.B. (1990). *J. Biol. Chem.*, *265*: 20874.

Fujita, J., and Ogino, H. (1974). *Chem. Lett.*, 57.

Gold, M., and Powell, H.K.J. (1976). *J. Chem. Soc., Dalton Trans.*, 230.

Gollogly, J.R., Hawkins, C.J., and Beattie, J.K. (1971). *Inorg. Chem.*, *10*: 317.

Graddon, D.P., and Ong, W.K. (1975). *J. Inorg. Nucl. Chem.*, *37*: 469.

Halliday, R.W., and Court, R.H. (1974). *Can. J. Chem.*, *52*: 3469.

Hambley, T.W., and Hawkins, C.J. (1981a). *Aust. J. Chem.*, *34*: 2505.

Hambley, T.W., Hawkins, C.J., Palmer, J.A., and Snow, M.R. (1981b). *Aust. J. Chem.*, *34*: 45.

Harold, F.M. (1964). *J. Bacteriol.*, *68*: 1416.

Hague, D.N., and Moodley, K.G. (1983). *S. Afr. J. Chem.*, *36*: 10.

Hague, D.N., and Moreton, A.D. (1989). *J. Chem. Soc., Dalton Trans.*, 1171.

Heby, O. (1981). *Differentation*, *19*: 1.

Hedwig, G.R., and Powell, H.K.J. (1973). *J. Chem. Soc., Dalton Trans.*, 793.

Holmes, F., and Williams, D.R. (1967). *J. Chem. Soc. A*, 1702.

House, D.A. (1987). *Comprehensive Coordination Chemistry*, Pergamon Press, Oxford, Vol. 2, p. 23.

House, D.A., and Moon, R.P., (1981). *J. Inorg. Nucl. Chem.*, *43*: 257.

House, D.A., and Tobe, M.L. (1989). *J. Chem. Soc., Dalton Trans.*, 853.

Jänne, J., Pösö, H., and Raina, A. (1978). *Biochim. Biophys. Acta*, *473*: 241.

Jänne, J., Alhonen, L., and Leinonen, P. (1991). *Ann. Med.*, *23*: 241.

Jensen, J.R., Lynch, G., and Baundry, M. (1987). *J. Neurochem.*, *48*: 765.

Juranic, N., Malinar, M. J., Radivoj a, P. N., Juranic, J., and Celap, M. P. (1986). *J. Serb. Chem. Soc.*, *51*: 417.

Kapanadze, T.Sh., Kokunov, Yu.V., Tsintsadze, G.V., and Buslaev, Yu. A. (1990). *Koord. Khim. (Russ.)*, *16*: 666.

Kimberly, M.M., and Goldstein, J.H. (1981). *Anal. Chem.*, *53*: 789.

Kimura, E., Kodama, M., and Yatsunami, T. (1982). *J. Amer. Chem. Soc.*, *104*: 3182.

Klug, A., and Rhodes, D., (1987). *Trends Biochem. Sci.*, *12*: 464.

Kojima, M., Morita, K., and Fujita, J. (1981). *Bull. Chem. Soc. Jpn.*, *54*: 2947.

Koskinen, M., and Nikkilä, J. (1972). *Suomen Kemi.*, *B45*: 89.

Kucan, Z., Naranda, T., Plohl, M., Nöthig-Laslo, V., and Weygand-Dura evic, J. (1988). *Progress in Polyamine Research* (V. Zappia and A.E. Pegg, eds.), Plenum Press, New York, p. 525.

Kusama-Eguchi, K., Watanabe, S., Irisawa, M., Watanabe, K., and Igarashi, K. (1991). *Biochem. Biophys. Res. Commun.*, *177*: 745.

Lomozik, L., and Gasowska, A. (1993). *Monatsh. Chem.*, *124*: 109.

Lomozik, L., and Wojciechowska, A. (1989). *Polyhedron*, *8*: 2645.

Lomozik, L., Odani, A., and Yamauchi, O. (1994). *Inorg. Chim. Acta*, submitted.

Marton, L.J., and Feuerstein, B.G. (1986). *Pharmaceut. Res.*, *3*: 311.

Massoud, S.S., and Milburn, R.M. (1989). *Polyhedron*, *8*: 415.

Metcalf, B.W., Bey, P., Danzin, C., Jung, M.J., Casara, J., and Vevert, J.P. (1978). *J. Amer. Chem. Soc.*, *100*: 2551.

Micheloni, M., Sabatini, A., and Vacca, A. (1977). *Inorg. Chim. Acta*, *25*: 41.

Morgan, D.M.L. (1987). *Essays Biochem.*, *23*: 82.

Moruzzi, M.S., Barbiroli, B., Monti, M.G., Tadolini, B., Hakim, G., and Mezzetti, G. (1987). *Biochem. J.*, *247*: 175.

Nakai, C., and Glinsmann, W. (1977). *Biochemistry*, *16*: 5636.

Nakayama, Y., Matsumoto, K., Ooi, S., and Kuroya, H. (1973), *J. Chem. Soc. Chem. Commun.*, 170.

Navarro-Ranninger, C., Zamora, F., Pérez, J.M., López-Solera, I., Martinez-Carrera, S., Masaguer, J.R., and Alonso, C. (1992). *J. Inorg. Biochem.*, *46*: 267.

Navarro-Ranninger, C., Pérez, J.M., Zamora, F., González, V.M., Masaguer, J.R., and Alonso, C. (1993). *J. Inorg. Biochem.*, *52*: 37.

Nicchitta, C.V., and Williamson, J.R. (1984). *J. Biol. Chem.*, *259*: 12978.

Nowatari, H., Kuroda, Y., Hayami, H., Okamoto, K., Ekimoto, H., and Takahashi, K. (1989). *Chem. Pharm. Bull.*, *37*: 2406.

Ogino, M., and Fujita, J. (1975). *Bull. Chem. Soc. Jpn.*, *48*: 1836.

Ogino, H. (1977). *Bull. Chem. Soc. Jpn.*, *50*: 2459.

Ogino, H. (1980a). *Inorg. Chem.*, *19*: 1619.

Ogino, H. (1980b). *Inorg. Chem.*, *19*: 3178.

Ohishi, H., Kunisawa, S., van der Marel, G., van Boom, J.H., Rich, A., Wang, A.H.J., Tomita, K., and Hakoshima, T. (1991). *FEBS Lett.*, *284*: 238.

Onasch, F., Aikens, D., Bunce, S., Schwartz, H., Nairn, D., and Hurwitz, C. (1984). *Biophys. Chem.*, *19*: 245.

Palmer, B.N., and Powell, H.K.J. (1974a). *J. Chem. Soc., Dalton Trans.*, 2086.

Palmer, B.N., and Powell, H.K.J. (1974b). *J. Chem. Soc., Dalton Trans.*, 2089.

Paoletti, P., Fabbrizzi, L., and Barbucci, R. (1973). *Inorg. Chem.*, *12*: 1961.

Paoletti, P. (1980). *Adv. Mol. Rel. Interact. Proc.*, *18*: 73.

Pegg, A.E. (1986). *Biochem. J.*, *234*: 249.

Pegg, A.E. (1988a). *Cancer Res.*, *48*: 759.

Pegg, A.E., Jones, D.B., and Secrist, J.A. (1988b). *Biochem. J.*, *263*: 285.

Pfeiffer, P. (1948). *Naturwissenschaften*, *35*: 190.

Plum, G.E., Arscott, P.G., and Bloomfield, V.A. (1990). *Biopolymers*, *30*: 631.

Porter, C.W., Herrera-Ornelas, L., Pera, P., Petrelli, N.F., and Mittelman, A. (1987). *Cancer, 60*: 1275.

Potkonjak, D., and Pasic, J. (1983). *Iugosl. Physiol. Pharmacol. Acta, 19*: 170.

Powell, J.H., and Reidenberg, M.M. (1982). *Biochem. Pharmacol., 31*: 3447.

Qi, D., Schatzman, R.C., Mazzei, G.J., Turner, R.S., Raynor, R.L., Liao, S., and Kuo, J.F. (1983). *Biochem. J., 213*: 281.

Raina, A., and Jänne, J. (1975). *Rev. Art. Med. Biol., 53*: 121.

Ramaswany, S., and Murthy, M.R.N. (1990). *Curr. Sci., 59*: 379.

Ramaswany, S., and Murthy, M.R.N. (1991). *Curr. Sci., 61*: 410.

Rennert, O.M., and Shukla, J.B. (1978). *Adv. Polyamine Res., 2*: 195.

Roberts, J.D., Van Houten, B., Qu, Y., and Farrell, N.P. (1989). *Nucl. Acid. Res., 17*: 9719.

Romeo, R., Minniti, D., Lanza, S., and Tobe, M.L. (1977). *Inorg. Chim. Acta, 22*: 87.

Romeo, R., Lanza, S., Minniti, D., and Tobe, M.L. (1978). *Inorg. Chem., 17*: 2436.

Sato, S., and Saito, Y. (1975). *Acta Crystallogr., B31*: 1378.

Sato, S., Saito, Y., Fujita, J., and Ogino, H. (1974). *Inorg. Nucl. Chem. Lett., 10*: 669.

Schellman, J.A., and Parthasarathy, N. (1984). *J. Mol. Biol., 175*: 313.

Schuber, F. (1989). *Biochem. J., 260*: 1.

Schwab, A.P., Tobe, M.L., and Romeo, R. (1982). *Inorg. Chim. Acta, 58*: 161.

Shukla, O.P. (1990). *J. Sci. Ind. Res., 49*: 263.

Souzu, H. (1986). *Biochim. Biophys. Acta, 861*: 361.

Srivenugopal, K.S., Wemmer, D.E., and Morris, D.R. (1987). *Nucl. Acid. Res., 15*: 2563.

Surma, K., Perkowska, A., and Gawron, M. (1994). *Polish J. Chem., 68*: 481.

Tabor, C.W. (1960). *Biochem. Biophys. Res. Commun., 2*: 117.

Tabor, C.W., and Tabor, H. (1984). *Annu. Rev. Biochem., 53*: 749.

Tabor, C.W., and Tabor, H. (1985). *Microbiol. Rev., 49*: 81.

Tadolini, B., and Hakim, G. (1988). *Progress in Polyamine Research* (V. Zappia and A.E. Pegg, eds.), Plenum Press, New York, p. 481.

Takeda, Y., Samejima, K., Nagano, K., Watanabe, M., Sugeta, H., and Kyogoku, Y. (1983). *Eur. J. Biochem., 130*: 383.

Templeton, D.M., and Sarkar, B. (1985). *Can. J. Chem., 63*: 3122.

Te ic, Z. Lj., Janjic, T.J., Malinar, M.J., and Celap, M.B. (1989). *J. Chromatogr., 481*: 471.

Thams, P., Capito, K., and Hedeskov, C.J. (1986). *Biochem. J., 237*: 131.

Thomas, T.J., Gunnia, U.B., and Thomas, T. (1991). *J. Biol. Chem., 266*: 6137.

Tobe, M.L., Schwab, A.P., and Romeo, R. (1982). *Inorg. Chem., 21*: 1185.

Tomita, K., Hakoshima, T., Inubushi, K., Kunisawa, S., Ohishi, H., van der Marel, G. A., van Boom, J.H., Wang, A.H.J., and Rich, A. (1989). *J. Mol. Graphics, 7*: 71.

Vertino, P.M., Bergeron, R.J., Cavanaugh, P.F., Jr., and Porter, C.W. (1987). *Biopolymers, 26*: 691.

Voelter, W., Strobel, G.J., and Weser, U. (1975). *Bioinorg. Chem., 5*: 183.

Weatherburn, D.C. (1983). *Aust. J. Chem., 36*: 433.

Weatherburn, D.C., Billo, F.J., Jones, J.P., and Margerum,D.W. (1970). *Inorg. Chem., 9*: 1557.

Weser, U., Strobel, G.-J., Rupp, H., and Voelter, W. (1974). *Eur. J. Biochem., 50*: 91.

Wiewiórowski, M., Zielenkiewicz, A., Zielenkiewicz, W., and Zólkiewski, M. (1991). *Thermochim. Acta, 182*: 153.

Wilson, R.W., and Bloomfield, V.A. (1979). *Biochemistry, 18*: 2192.

Williams-Ashman, H.G., and Canellakis, Z.N. (1979) *Prospect. Biol. Med.*, *22*: 421.

Wojciechowska, A., Bolewski, L., and Lomozik, L. (1991). *Monatsh. Chem.*, *122*: 131.

Young, N.D., and Galston, A.W. (1983). *Plant Physiol.*, *71*: 767.

Zappia, V., and Pegg, A.E., eds. (1988). *Progress in Polyamine Research*, Plenum Press, New York.

METAL ION INTERACTIONS WITH LIPIDS

by

Joachim Seelig

Department of Biophysical Chemistry,
Biocenter of the University of Basel,
Klingelbergstrasse 70
CH-4056 Basel
Switzerland

I. LIPIDS IN BIOLOGICAL MEMBRANES

Biological membranes are composed of proteins and lipids. The latter are arranged in a bilayer leaflet with the polar groups in contact with water and the hydrocarbon chains forming a non-polar interior. The phospholipid bilayer acts as a permeability barrier for charged molecules such as metal ions which cannot penetrate into the hydrophobic bilayer interior due to their large Born charging energy. The exchange of ions between the cell exterior and interior is highly regulated and requires different types of proteins such as carriers, channels, exchange proteins and ion pumps.

A second functional role of lipids is to provide a "grease" for the protein machinery. Under physiological conditions the lipid bilayer is in the *liquid-crystalline state*, characterized by a high-rate of lateral diffusion of the lipid molecules. In addition, the hydrocarbon chains of the phospholipid molecules undergo rapid flexing motions. These are caused by trans-gauche isomerizations around carbon-carbon single bond, creating a disordered, fluid-like membrane interior. However, due to the anchoring of the phospholipid headgroups at the water interface, the hydrocarbon chain disordering is different from that of an isotropic fluid. The entropy increase (per CH_2 segment) accompanying the crystal-to-liquid crystal transition of a lipid bilayer is only half the value of a crystal-to-fluid transition (Phillips et al., 1969; Seelig, 1977).

A direct consequence of hydrocarbon chain disordering is the *lateral expansion* of the lipid membrane. A hydrocarbon chain in the all-trans conformation has a cross-section of A ~ 24 Å^2, thus yielding a minimum area of A = 48 Å^2 for a crystalline phospholipid molecule with two hydrocarbon chains. However, in the liquid-crystalline state the average cross-section of this lipid increases to A ~ 60 Å^2 (Büldt et al., 1978; 1979). Moreover, most biological lipids usually have an unsaturated fatty acyl chain at the sn-2 position of the glycerol backbone, containing one or more *cis*-double bonds. *Cis*-double bonds perturb the lateral packing of the hydrocarbon chains and induce a further expansion of the lattice by about 10-15 Å^2. The typical cross sectional area of a simple lecithin such as POPC (1-palmitoyl-2-oleoyl-*sn*-glycero-3-phosphocholine, the main component in egg lecithin) thus is A ~ 68 Å^2 (Altenbach and Seelig, 1984; Evans et al., 1987). It should be realized that it is not the size of the lipid headgroups but the cross section of the hydrocarbon chains which determines the area requirement of the phospholipids in the bilayer. Model building demonstrates that with an area of A = 68 Å^2 it is easily possible to align even the phosphocholine headgroup with its bulky $-N^+(CH_3)_3$ residue well within the plane of the membrane, with sufficient space between the $^-P-N^+$dipoles to allow a free rotational movement.

II. LIPID SURFACE AREA AND METAL ION COMPLEXATION

In the context of metal ion-phospholipid binding, the surface area plays a functional role at different molecular levels: (i) the surface area per lipid molecule together with its electric charge determines the membrane surface potential, ψ_o. The latter regulates the attraction or repulsion of cations and anions at the lipid-water interface; (ii) for sufficiently large and energetically stable surface areas, metal ions can form only *short-lived* coordination complexes with individual phospholipid molecules. This situation is typical for most biological membranes and also for model membranes with *cis*-unsaturated hydrocarbon chains and a low percentage (<25 %) of negatively charged lipids; (iii) the tendency of metal ions to form several coordination bonds can reduce the distance between phospholipid headgroups, stretching the hydrocarbon chains into an all-*trans* conformation. Such "crystallization" phenomena have indeed been reported for pure phosphatidylglycerol and phosphatidylserine membranes in the presence of sufficient amounts of Ca^{2+} (van Dijck et al., 1978; Sacré et al., 1979; Findlay and Barton, 1978; Boggs and Rangaraj, 1983; Jacobson and Papahadjopoulos, 1975; Newton et al., 1978; Harlos and Eibl, 1980; Roux and Bloom, 1991; Mattai et al., 1989). Alternatively, the presence of Ca^{2+} may also induce a bilayer → hexagonal phase II transition (de Kruijff et al., 1982). However, if a negatively charged phosphatidylserine membrane (POPS) is sufficiently diluted out with zwitterionic phosphatidylcholine (POPC), the membrane remains in the liquid-crystalline state up to 1 M $CaCl_2$ (Roux and Bloom, 1991).

Unwarranted generalization has been made because the role of the surface area has been ignored. Many of the early studies on phospholipid-metal ion binding were performed with lipids containing two saturated hydrocarbon chains since these were the only commercially available synthetic lipids. However, the phase behavior of, for example, *saturated* phosphatidylglycerol membranes in the presence of Ca^{2+} is quite different from the phase behavior of phosphatidylglycerol with *unsaturated* hydrocarbon chains (Borle and Seelig, 1985).

Conflicting results on metal ion binding to lipids have been published. In critically evaluating the earlier literature, it should be borne in mind that lipid preparation, lipid purity, and lipid handling have been a problem for many years and, in part, still are. Traces of lyso-lipids and free fatty acids in lipid preparation can change the electric surface potential; exposure to oxygen and sunlight could induce peroxidation of the hydrocarbon chains; purification of lipids from natural sources may still produce a chemically heterogeneous mixture.

III. THERMODYNAMICS OF METAL ION-LIPID INTERACTION. ELECTROSTATIC ATTRACTION AND CHEMICAL BINDING

Biological membranes usually contain a small percentage of negatively charged lipids such as phosphatidylglycerol, phosphatidylserine, or cardiolipin, creating a negative surface potential of about -50 mV (in 0.15 M NaCl). The binding of metal ions to the membrane surface will thus occur in at least two steps. First, the metal ion will be *attracted electrostatically* to the negatively charged membrane surface. The metal ion concentration will increase near the lipid-water interface compared to the equilibrium concentration in bulk solution. Secondly, the metal ion will *penetrate* into the lipid headgroup region and will interact with the lipid headgroups. Binding sites on the membrane are either the phosphate or the carboxylate group and this interaction can be described by a *chemical equilibrium*, characterized by a stoichiometry and a binding constant. However, as mentioned above, metal ions appear to form rather short-lived complexes with phospholipids in biological membranes. Indeed, using deuterium nuclear magnetic resonance it could be demonstrated that the residence time of a metal ion such as Ca^{2+} at an individual phospholipid binding site is less than 10^{-5} s for lipids in the fluid-like state

(Akutsu and Seelig, 1981). Hence it could be argued that for sufficiently large surface areas the metal ions are simply trapped in a trough of the chemical potential comparable, for example, to the phenomenon of counterion condensation observed with polyelectrolytes (Manning, 1978).

To proceed, we consider the two binding steps in more detail. If we denote with ψ_o the membrane surface potential, then the concentration of a particular ion "i" at the membrane-water interface "I" is given by the Boltzmann relation

$$C_{i,I} = C_{i,eq}\exp(-z_i\psi_oF_o/RT) \quad (1)$$

where z_i is the signed charge number of the ion, F_o is the Faraday constant, $C_{i,eq}$ is the ion concentration in bulk solution, and RT has its usual meaning. *All* cations and anions present in the solution must be taken into account. In particular, protons also accumulate near the membrane interface which thus becomes more acidic. For a bulk pH of pH 7.4, the interfacial pH drops to pH 6.6 if the membrane contains 25 % negatively charged lipid (in 0.1 M NaCl).

Next, the metal ion can enter the headgroup region. In the simplest case, which holds true for the case of weak ion binding and low degree of membrane loading, the "chemical" binding can be described by a simple *surface partition equilibrium* according to

$$X_b = K_p.C_I \quad (2)$$

where the degree of binding, X_b, is the molar amount of metal ion bound per mole of total lipid. This binding model does not single out specific binding sites but corresponds to the "trough" of the electrochemical potential mentioned above. More common in the analysis of metal ion binding data is the *Langmuir adsorption model* in which a fixed number of metal ion binding sites are defined on the membrane surface. If each lipid constitutes a lipid binding site, this model may be written as

$$X_b/(1-X_b) = K.C_I \quad (3)$$

If n lipids define the binding site, then

$$X_b/(1-nX_b) = K.C_I \quad (4)$$

More complicated models can be defined. Knowledge of the binding model (for all types of ions participating in the equilibrium!), the binding constant(s), and the surface potential ψ_o allows the calculation of $X_{b,i}$, i.e. the amount of bound metal ion of species "i". The electric surface charge density, σ, can then be calculated as the sum of the (negative) lipid surface charge density plus the (positive) surface charge density contributed by the bound metal ions. In addition, a second relationship exists between the surface charge density, σ, and the membrane surface potential, ψ_o, dating back to Gouy and Chapman who independently solved the Poisson equation for the case of a flat, charged surface immersed into a salt solution. The Gouy-Chapman equation (for detailed discussions see Aveyard and Haydon, 1973; McLaughlin 1977; 1989) is given by

$$\sigma^2 = 2000\,\varepsilon_o\varepsilon_rRT\Sigma C_{i,eq}(e^{-z_iF_o\psi_o/(RT)} \quad (5)$$

where ε_o is the permittivity of free space and ε_r is the dielectric constant of water. In solving the binding problem, one will hence first choose a binding model and then search numerically for a set of parameters K and ψ_o which will simultaneously satisfy the "chemical" binding equations and the Gouy-Chapman theory.

The combination of electrostatic and "chemical" binding is not limited to charged membrane surfaces but must also be applied to neutral membranes composed of zwitterionic lipids such as phosphatidylcholine and phosphatidylethanolamine (which together account for about 80 % of the total lipid content in many biological membranes). Only the ion-free membrane is electrically neutral. Once the first metal ion is bound, a positive electrical surface potential will gradually build up, repelling ions of like charge and making metal ion binding increasingly more difficult.

The effect of electrostatic repulsion or attraction is demonstrated in Figure 1 for the binding of Ca^{2+} to neutral and negatively charged membranes in 0.15 M NaCl salt solution

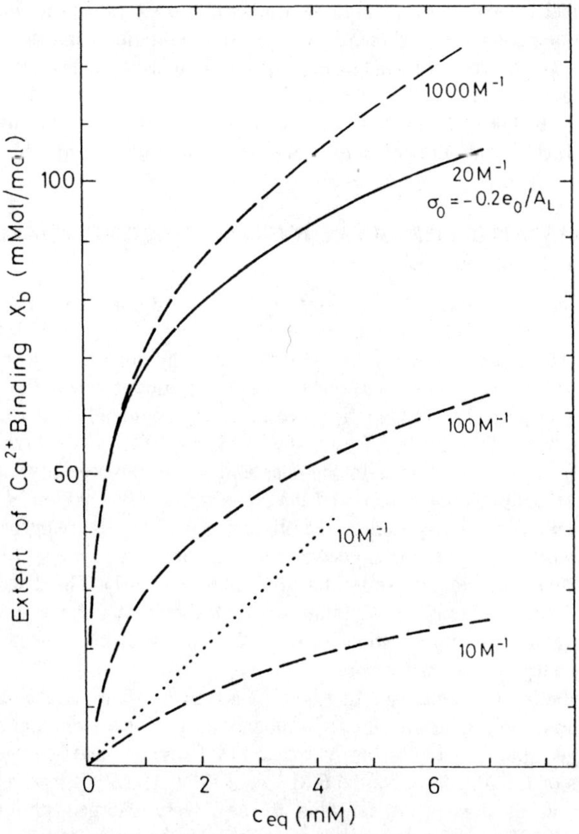

Figure 1.
Ca^{2+} binding to neutral (broken lines) and negatively charged (solid line) phospholipid bilayers. For the "chemical" binding step, a Langmuir adsorption isotherm was assumed and the corresponding binding constants are given in the figure. Electrostatic attraction (negatively charged membranes) or repulsion (zwitterionic membrane) was taken into account by means of the Gouy-Chapman theory. The dotted line corresponds to a Langmuir adsorption isotherm *without* electrostatic interactions. The solid line, calculated for a membrane composed of 75 mol% phosphatidylcholine/25 mol% phosphatidylglycerol (with lipid surface areas of 68 $Å^2$ each), is relevant for Ca^{2+} binding under physiological conditons. All calculations at 25°C and 0.154 M NaCl. Taken from Seelig (1990).

(Seelig, 1990). The "chemical" part of the binding process is described by the Langmuir adsorption isotherm $X_b/(1-X_b) = K.C_I$ and the
intrinsic binding constant was varied between 10 M^{-1} to 1000 M^{-1}. First, if electrostatic interactions are neglected, the surface concentration C_I is identical with the bulk concentration C_{eq}. The actual Ca^{2+} binding isotherm is obtained from equation (3) by replacing C_I with C_{eq}. For Ca^{2+} concentrations in the mM range and a binding constant $K = 10\ M^{-1}$, the initial part of the Langmuir isotherm yields a straight line with slope K as indicated by the dotted line in Figure 1. Next, we investigate Ca^{2+} binding to neutral membranes *including electrostatic repulsion*. The dashed line ($K = 10\ M^{-1}$) demonstrates that the resulting binding isotherm deviates considerably from the simple Langmuir adsorption isotherm and predicts a considerably reduced Ca^{2+} binding. The experimentally determined Ca^{2+} binding data for POPC membranes correspond closely to this lowest adsorption isotherm with $K = 10\ M^{-1}$ (Altenbach and Seelig, 1984).

The biologically most relevant binding isotherm is that calculated for a mixture of phosphatidylcholine/phosphatidyl-glycerol (80/20) using a binding constant of 20 M^{-1} (solid line in figure) which corresponds to experimental data obtained with this system (Macdonald and Seelig, 1987a). Even though the binding constant is small, the electrostatic attraction of Ca^{2+} to the membrane surface makes the initial part of the binding isotherm almost superimposable with that of a neutral membrane with a 100-fold higher binding constant.

IV. BINDING AFFINITIES OF METAL IONS OF CHARGE Z=+1 TO Z=+3 FOR PHOSPHOLIPID MEMBRANES

Pioneering work on the binding of metal ions to charged and non-charged lipid bilayers has been performed by McLaughlin and co-workers (Eisenberg et al., 1979; Lau et al., 1981) using ζ-potential measurements as an experimental tool. The migration of lipid vesicles in a constant electric field was observed under the light microscope. The initially rapid movement of the negatively charged lipid vesicles was gradually slowed down by the addition of metal ions. For divalent ions, when added at sufficiently high concentrations (0.1 M - 0.2 M), a reverse in sign in the ζ-potential was observed. At the point of no movement, the surface potential is zero and the complications of the Gouy-Chapman theory can be ignored. To a first approximation, the binding constant is the reciprocal of the cation concentration at which charge reversal occurs. Binding constants between 5 M^{-1} and 10 M^{-1} for the earth alkaline metal ions were estimated (Lau et al., 1981). The disadvantage of this simple approach is the necessity to use rather high concentrations of metal ions in order to induce charge reversal. At concentrations of 0.1-0.2 M of metal ions, membranes may start to fuse or may undergo phase transitions.

However, binding constants can also be obtained at low ion concentrations by fitting the ζ-potential measurements to the Gouy-Chapman theory. For the binding of alkali ions to negatively charged phosphatidylserine vesicles, McLaughlin and co-workers reported binding constants of Li^+ 0.8 M^{-1}, Na^+ 0.6 M^{-1}, and K^+ 0.15 M^{-1} (Eisenberg et al., 1979). For the adsorption of divalent metal ions to negatively charged phosphatidylglycerol membranes, the intrinsic 1:1 association constants (M^{-1}) were Mn^{2+} 11.5, Ca^{2+} 8.5, Mg^{2+} 6.0, Ba^{2+} 5.5. We have confirmed these results using a centrifugation assay combined with atomic absorption spectroscopy. We found a Ca^{2+}-binding constant for mixed phosphatidylglycerol/phosphatidylcholine membranes of ~ 19 M^{-1} and a weaker sodium-binding constant of 0.6-0.85 M^{-1} (Macdonald and Seelig, 1987a), using again a 1:1 stoichiometry. The Ca^{2+}-binding constant for mixed phosphatidylcholine/cardiolipin membranes was K = 19 M^{-1} (Macdonald and Seelig, 1987b) and for zwitterionic phosphatidylcholine membranes K = 13.5 M^{-1} (Altenbach and Seelig, 1984), all calculated by means of the Gouy-Chapman theory. A slightly higher binding constant of K = 30 M^{-1} was found for the Ca^{2+} binding to phosphatidylserine *monolayer* (Ohki and Kurland, 1981).

The question of metal ion-lipid stoichiometry deserves a special comment. Langmuir adsorption isotherms of the form of eq. (3) or (4) degenerate into a simple partition equilibrium (2) at low degrees of metal ion binding. In order to determine the exact mode of the binding, binding isotherms must be measured up to very high ion concentrations. For Ca^{2+} binding to neutral POPC membranes, we have measured the binding isotherm up to concentrations of 2 M $CaCl_2$ and could describe the binding isotherm (after correcting for electrostatic effects) over the whole concentration range with a 1 Ca^{2+}:2 phospholipid stoichiometry using a binding constant of 13.8 M^{-1} (0.1 M NaCl) (Altenbach and Seelig, 1984). However, re-evaluating the previous data and limiting the analysis to concentrations c_{CaCl_2} < 100 mM yields an equally good fit to the data with a simple partition equilibrium, $X_b = K_p C_l$, with a slightly smaller partition constant of K ~ 9.5 M^{-1}. In hindsight, the analysis of the binding data over the whole concentration range of $0 \le C_{CaCl2} < 2M$ appears to be unwarranted because (i) the simple Gouy-Chapman approach is no longer valid at these high concentrations, and (ii) the structure of the membrane itself can be changed under these conditions.

Using an NMR-method to detect Ca^{2+} binding and evaluating the variation of the phosphorus chemical shift data with the Gouy-Chapman theory, a 1:1 binding stoichiometry and a Ca^{2+} binding constant of K = 2.2 M^{-1} was deduced for a Ca^{2+} concentration up to 0.4 M (in the presence of 0.1-1.0 M NaCl) which is in broad agreement with the above measurements, even though the binding constant appears to be somewhat small (Grasdalen et al., 1977).

The problem of metal ion-phospholipid stoichiometry has been re-investigated recently with ζ-potential measurements (Tatullan, 1987). Binding isotherms were measured up to metal ion concentrations of up to 1 M. The Gouy-Chapman theory was used to fit the data over the whole concentration range. The author concludes that a metal-ion binding site comprises several molecules. However, the physical relevance of this analysis remains doubtful. First, the outcome of the analysis is heavily influenced by the high concentration range. Secondly, the binding of monovalent cations and anions is included in the analysis and requires a multi-parameter fit with a large number of free parameters.

Trivalent metal ions such as La^{3+} exhibit a distinctly larger binding constant for neutral lipids. Binding of La^{3+} and related transition metal ions can be detected by high-resolution NMR due to strong chemical shift effects which make the lanthanides analytically useful as so-called *shift-reagents*. While the initial studies with lanthanides applied the law of mass action to evaluate the data in terms of a binding model (Hauser et al., 1975; Chrzeszczyk et al., 1981), a more refined analysis took advantage of the Gouy-Chapman theory along the lines discussed above (Grasdalen et al., 1977). The latter approach yielded intrinsic binding constants of 500-1000 M^{-1} for a 1:1 stoichiometry. The adsorption of Al^{3+} at pH 3.8 (in 0.12 M KCl) was investigated with equilibrium dialysis and electrophoresis experiments (Akeson et al., 1989). Evaluating these data in terms of the combined surface partition/Gouy Chapman model, we obtain a surface partition coefficient of $K_p \sim$ 5000 M^{-1}. Likewise, a partition coefficient of $K_p \sim$ 4000 M^{-1} was evaluated for the binding of La^{3+} to phosphatidylcholine membranes (Altenbach, 1985).

Little is known about the *binding enthalpy* of metal ion-phospholipid interactions since in many calorimetric experiments the actual binding step is masked by a phase transition of the hydrocarbon chains, i.e. either by the liquid crystal → gel or the lamellar → hexagonal phase II transition (de Kruijff et al., 1982). Calorimetric titrations of phosphatidylserine vesicles with earth alkali ions at low Me^{2+}:phospholipid rations ($<$0.35) showed only small heat effects, consistent with the complexation of either carboxylate or phosphate ligands (Rehfeld et al., 1982). We have measured the binding of Ca^{2+} to mixed phosphatidylglycerol/phos-phatidylcholine membranes (POPG/POPC) with high sensitivity titration calorimetry and observed endothermic heats of reaction of $\Delta H \sim$ + 0-.7 kcal/mol for small sonified vesicles. For La^{3+} similar measurements yielded $\Delta H \sim$ 1.0 kcal/mol (Lehrmann and Seelig, 1994). Since the free energy change, ΔG = RTln (55.5K), is of the order of -3.7 kcal/mol for Ca^{2+} and -7 kcal/mol for La^{3+} and Al^{3+}, the binding of metal ions to the membrane surface is *entropy-driven*. (The factor 55.5 in the evaluation of ΔG takes into account the cratic contribution; cf Cantor and Schimmel, 1980).

V. STRUCTURAL ASPECTS OF METAL ION-PHOSPHOLIPID INTERACTION

The effect of metal ions on the structure of lipid membranes is strongly dependent on the lipid composition. The Ca^{2+}-induced liquid crystal → gel phase transition and the liquid crystal → hexagonal H_{II} phase transition have already been mentioned. In addition, Ca^{2+} binding to phosphatidylserine membranes may trigger a membrane fusion process (Papahadjopoulos, 1977). Ca^{2+} ions have also been shown to mediate the binding of proteins to the membrane surface, forming a ternary lipid/metal ion/protein complex as, for example, in the blood clotting process.

At the molecular level, Ca^{2+} binding is accompanied by a dehydration and an immobilization of the phospholipid headgroup as has been documented in detail for pure phosphatidylserine membranes (Hauser and Shipley, 1984; Casal et al., 1987). Here we are concerned with the effect of metal ions on phospholipid headgroups in the *liquid crystalline state*. A suitable method to study lipid headgroups is to introduce deuterium labels at specific sites and to measure the ^{2}H-NMR spectra (Seelig, 1977). The latter are characterized by so-called *deuterium quadrupole splittings* which provide insight into the lipid headgroup orientation and motion. As an example, Figure 2 shows the variation of the

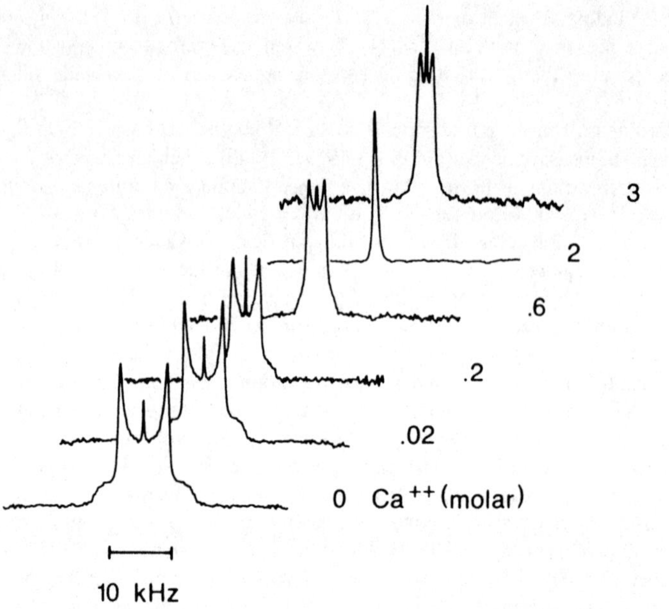

Figure 2.
Deuterium nuclear magnetic resonance spectra of mixed phosphatidylcholine/phosphatidylglycerol bilayers (80/20 mol%/mol%) at various concentrations of $CaCl_2$ (plus 100 mM NaCl and 10 mM Tris-HCl, pH 7.4, 25 °C). Phosphatidylcholine was deuterated at the α-choline segment ($-POCD_2CH_2N-$). The figure shows the variation of the quadrupole splitting due to a conformational change of the $^-P-N^+$ dipole into a more upright position upon adsorption of Ca^{2+} to the membrane surface (taken from Macdonald and Seelig, 1987a).

quadrupole splitting with the $CaCl_2$ concentration in solution (Altenbach and Seelig, 1984). In these measurements, phsophatidylcholine membranes were selectively deuterated at the α -methylene segment of the choline moiety. The figure demonstrates (i) that a single quadrupole splitting is seen at all $CaCl_2$ concentrations, and (ii) that this quadrupole splitting varies continuously as more and more Ca^{2+} ions become bound to the membrane surface. Similar results have also been obtained for other ions (Brown and Seelig, 1977; Akutsu and Seelig, 1981; Macdonald and Seelig, 1987a; 1987b) and the following molecular picture is suggested. In the absence of metal ions, the $^-P-N^+$ dipole of the phosphocholine moiety is almost parallel to the membrane surface. As metal ions bind, this orientation is changed and the N^+ end of the dipole moves towards the water phase. For divalent metal ions these effects are small, for trivalent ions the angular change is somewhat larger. In fact, as detailed elsewhere, this reorientation is not only caused by metal ions but can also be induced by cationic peptides, local anesthetics or charged amphiphiles (cf Beschiaschvili and Seelig, 1991; Marassi and Macdonald, 1992 and references cited therein). Because the $^-P-N^+$ dipole responds to the membrane surface charge, it is described as behaving like a *molecular voltmeter* (Seelig et al., 1987). Metal ion binding to bilayer membranes will hence not only change the electric surface charge but will also reorient the phospholipid headgroup and, in turn, affect the membrane dipole potential. Strong dipolar fields in the vicinity of membrane proteins could induce conformational changes and thus contribute to signal transmission.

REFERENCES

Akeson, M.A., Munns, D.N., and Burau, R.G. (1989). *Biochim. Biophys. Acta*, *986*: 33.

Akutsu, H., and Seelig, J. (1981). *Biochemistry*, *20*: 7366.

Altenbach, Ch. (1985). PhD Thesis, University of Basel.

Altenbach, Ch., and Seelig, J. (1984). *Biochemistry*, *23*: 3913.

Aveyard, R., and Haydon, D.A. (1973). *An Introduction to the Principles of Surface Chemistry*, Cambridge University Press, London.

Beschiaschvili, G. and Seelig, J. (1991). *Biochim. Biophys. Acta*, *1061*: 78.

Boggs, J.M., and Rangaraj, G. (1983). *Biochemistry*, *22*: 5425.

Borle, F., and Seelig, J. (1985). *Chem. Phys. Lipids*, *36*: 263.

Brown, M.F., and Seelig, J. (1977). *Nature*, *269*: 721.

Büldt, G., Gally, H.U., Seelig, A., Seelig, J., and Zaccai, G. (1978). *Nature*, *271*: 182.

Büldt, G., Gally, H.U., Seelig, J., and Zaccai, G. (1979). *J. Mol. Biol.*, *134*: 673.

Cantor, C.R., and Schimmel, P.R. (1980). *Biophysical Chemistry*, Vol. I, Freeman, San Francisco, p. 283.

Casal, H.L., Mantsch, H.H., Paltauf, F., and Hauser, H. (1987). *Biochim. Biophys. Acta*, *919*: 275.

Chrzeszczyk, A., Wishnia, A., and Springer, C.S. (1981). *Biochim. Biophys. Acta*, *648*: 28.

de Kruijff, B., Verkleij, A.J., Leunissen-Bijvelt, J., van Echfeld, C.J.A., Hille, J., and Rijnbout, H. (1982). *Biochim. Biophys. Acta*, *693*: 1.

Eisenberg, M., Gresalfi, T., Riccio, T., and McLaughlin, S. (1979). *Biochemistry*, *18*: 5213.

Evans, R.W., Williams, M.A., and Tinoco, J. (1987). *Biochem. J.*, *245*: 455.

Findlay, E.J., and Barton, P.G. (1978). *Biochemistry*, *17*: 2400.

Grasdalen, H., Eriksson, L.E.G., Westman, J., and Ehrenberg, A. (1977). *Biochim. Biophys. Acta*, *469*: 151.

Harlos, K., and Eibl, H. (1980). *Biochemistry*, *19*: 895.

Hauser, H., Phillips, M.C., Levine, B.A., and Williams, R.J.P. (1975). *Eur. J. Biochem.*, *58*: 133.

Hauser, H., and Shipley, G.G. (1984). *Biochemistry*, *23*: 34.

Jacobson, K., and Papahadjopoulos, D. (1975). *Biochemistry*, *14*: 152.

Lau, A., McLaughlin, A., and McLaughlin, S. (1981). *Biochim. Biophys. Acta*, *645*: 279.

Lehrmann, R., and Seelig, J. (1994). *Biochim. Biophys. Acta*, *1189*: 89.

Macdonald, P.M. and Seelig, J. (1987a). *Biochemistry*, *26*: 1231.

Macdonald, P.M., and Seelig, J. (1987b). *Biochemistry*, *26*: 6292.

Manning, G.S. (1978). *Quart. Rev. Biophys.*, *11*: 179.

Marassi, F.M., and Macdonald, P.M. (1992). *Biochemistry*, *31*: 10031.

Mattai, J., Hauser, H., Demel, R.A., and Shipley, G.G. (1989). *Biochemistry*, *28*: 2322.

McLaughlin, S. (1989). *Annu. Rev. Biophys. Biophys. Chem.*, *18*: 113.

McLaughlin, S.A. (1977). *Curr. Top. Membrane Transp.*, *9*: 71.

Newton, C., Pangborn, W., Nir, S., and Papahadjopoulos, D. (1978). *Biochim. Biophys. Acta*, *506*: 281.

Ohki, S., and Kurland, R. (1981). *Biochim. Biophys. Acta*, *645*: 170.

Papahadjopoulos, D. (1977). *J. Coll. Interf. Sci.*, *58*: 459.

Phillips, M.C., Williams, R.M., and Chapman, D. (1969). *Chem. Phys. Lipids*, *3*: 234.

Rehfeld, S., Hansen, L.D., Lewis, E.A., and Eatough, D.J. (1982). *Biochim. Biophys. Acta*, *691*: 1.

Roux, R., and Bloom, M. (1991). *Biophys. J.*, *60*: 38.

Sacré, M., Hoffmann, W., Turner, M., Tocanne, J.F., and Chapman, D. (1979). *Chem. Phys. Lipids*, *25*: 69.

Seelig, J. (1977). *Quart. Rev. Biophys.*, *10*: 353.

Seelig, J. (1990). *Cell. Biol. Rep.*, *14*: 353.

Seelig, J., Macdonald, P.M., and Scherer, P.G. (1987). *Biochemistry*, *26*: 7535.

Tatulian, S.A. (1987). *Eur. J. Biochem.*, *170*: 413.

van Dijck, P.W.M., de Kruijff, B., Verkleij, A.J., van Deenen, L.L.M., and de Gier, J. (1978). *Biochim. Biophys. Acta*, *512*: 84.

METAL ION-AMINO SUGAR COMPLEXES

by

Giovanni Micera

Dipartimento di Chimica,
Università di Sassari,
Via Vienna 2
07100 Sassari
Italy

and

Henryk Kozlowski

Institute of Chemistry,
University of Wroclaw,
F. Joliot-Curie 14
50-383 Wroclaw
Poland

I. INTRODUCTION

In spite of its relevance in the fields of biochemistry and industry, the interaction of metal ions with amino sugars is still largely unexplored. Amino sugars are the building blocks of many bioactive polysaccharides, e.g. the structural polymers that give form to living organisms, and also play important roles in the interactions between cells and their environment. For instance, chitin, the homopolymer of GlcNAc, is a major component in the walls of most fungi, moulds and yeasts and in the hard exoskeleton of insects and crustaceans (Muzzarelli, 1977). Chitosan, which is mainly composed of poly-(1→ 4)-β-*D*-GlcN, is generally obtained from chitin after partial or full *N*-deacetylation, although it may also be found to occur naturally. Almost pure poly-(1→ 4)-α-*D*-GalN is produced by some microorganisms (Takagi and Kadowaki, 1986). Amino sugars, either *N*-sulfated (heparin) or *N*-acetylated, are present in the acid polysaccharides glycosaminoglycans, which are found in connective tissues.

Due to their non-toxicity and fairly good solubility, amino sugar residues are found in many of the anti-cancer reagents and antibiotics that have been widely used clinically (bleomycin, adriamycin, hygromycin, etc).

Their extensive stereochemistry as well as the fact they contain multiple oxygen and nitrogen donor atoms make simple and polymeric amino sugars ideal ligands for precise interactions with metal ions, which may influence either their structure or function (or both) at the biological level. Their metal-binding ability may be exploited for industrial and environmental purposes, e.g. ion removal from solution and waters, including sea waters (Muzzarelli and Rocchetti, 1986). Proper chemical derivatization can make the ligands selective for the capture of target metal ions.

II. DETERMINATION OF THE ACTIVE BINDING SITES IN L.M.M. AMINO SUGARS

The presence of both amino and alcoholic groups provides amino sugar ligands with a variety of potential coordinating atoms and donor set arrangements. The main binding site is the amino group which anchors the metal ions. Undissociated hydroxyls may assist the nitrogen donor in chelation and their ligand effectiveness is enhanced by metal ion-promoted deprotonation.

In the measurable pH range, the protonated form of simple amino sugars (LH^+) undergoes only an acid dissociation process corresponding to the deprotonation of the ammonium into the amine group. Depending on the ligand structure and ring substituent, pK_a values over the range 7-8 have been generally measured. Alcoholic groups are much less acidic ($pK_a > 12$, see Rendleman, 1973).

Most of the studies on the complexes of simple amino sugars have been concerned with the complexing behaviour of *D*-glucosamine (GlcN, Scheme 1). Ishidate et al. (1960) first described qualitatively the complex formation of this ligand with copper(II) and concluded that a chelated species with a molar ratio of 1:1 was formed in neutral aqueous solution. Tamura and Miyazaki (1965a; 1965b) studied the complexes of several divalent metal ions with GlcN by pH-metric and spectrophotometric methods and found the following stability sequence: Cu > Pb > Zn > Ni, Co, Cd, Mn >> Ca, Mg. Species with 1:1 and 1:2 metal-to-ligand ratios involving NH_2 coordinated amino sugar molecules were proposed. However, the role of hydroxyl groups of the ligand in complex formation was not documented.

Further pH-metric investigations showed that the stability of the amino sugar complexes is dependent on the ligand structure and conformation, substantiating the role of hydroxyl groups (Tamura and Miyazaki, 1965c; Miyazaki et al., 1979). In particular, the rather high stability of the 1:1 complex (the only one reported in these papers) formed by TalN may be explained by the 1,3 diaxial mutual arrangement of OH(4) and NH_2 groups. Such an arrangement permits the formation of a six-membered chelated ring without any significant distortion of the stable ligand conformation. The change of OH(4) from an axial to an equatorial position makes the structure less favorable for chelation and leads to weaker chelated species. Meanwhile, the presence of an amino group in the equatorial position of the pyranose ring is less favorable for complexation. Similarly, selective *O*-methylation leads to a decrease in the complex stability. Also, *N*-methylation, while increasing the basicity of the amino nitrogen atom, makes the formed complexes weaker because of the steric hindrance at the donor group.

The first unambiguous evidence supporting the participation of deprotonated sugar hydroxyl groups in metal coordination was obtained by Inaki et al. (1978). Based on absorption spectra, the copper(II) ion in the 1:2 complex of GlcN was assigned with coordination by two amino groups up to pH 6.5 and by two amino together with two alcoholate groups between pH 6.5 and 9.

Re-examination of copper(II) complexation by GlcN in aqueous solution demonstrated the step-by-step formation of chelated species involving the metal coordination by amino and hydroxyl groups, the latter being undissociated or not (Appelman-Lippens et al., 1985; Micera et al., 1985). The speciation model was further refined by use of a combination of spectroscopic, potentiometric, and polarographic techniques (Lerivrey et al., 1986; Urbanska et al., 1988) and enabled the proposal of the complexation mechanism described in Scheme 2 and Table 1.

According to such a model, complexation starts with the formation of CuL and CuL_2 species where the amino nitrogen is the primary donor bound to the metal ion. Deprotonation of CuL_2 leads to CuL_2H_{-1} in which one of the bound ligands is chelated via

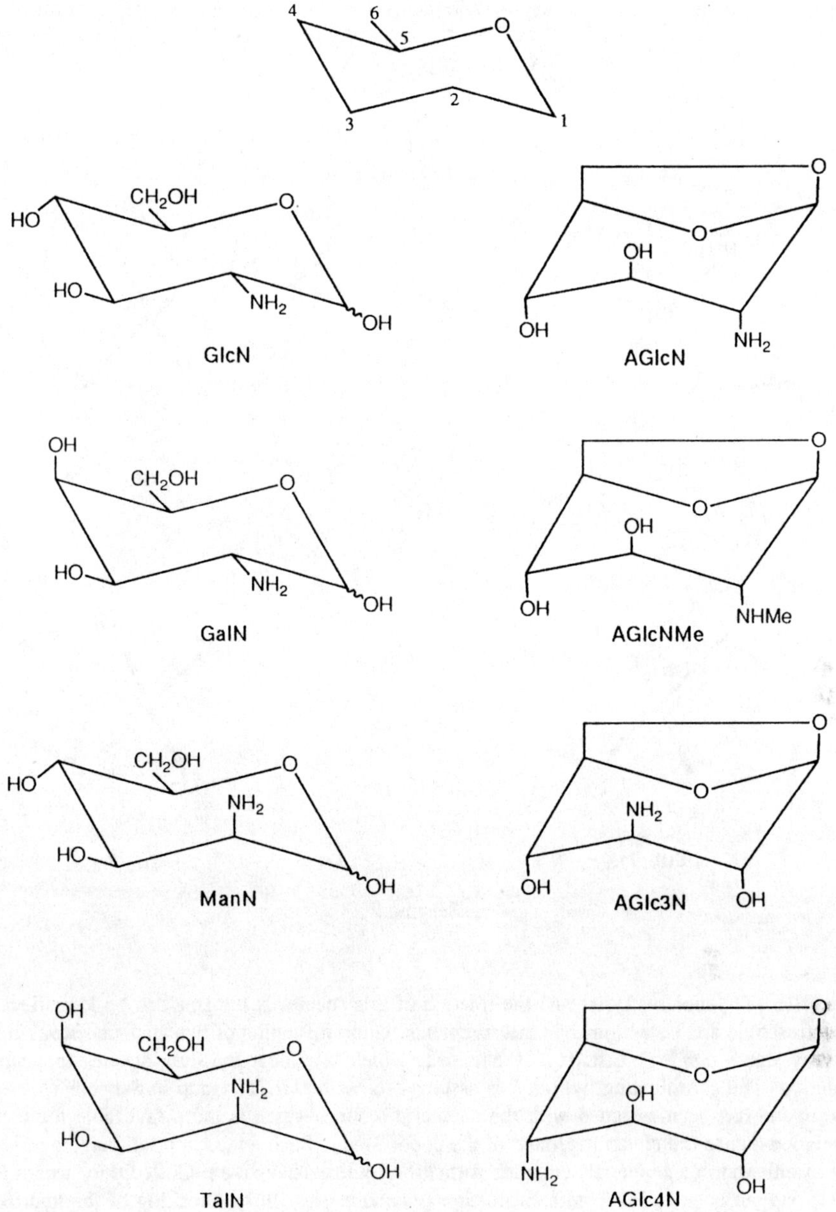

Scheme 1

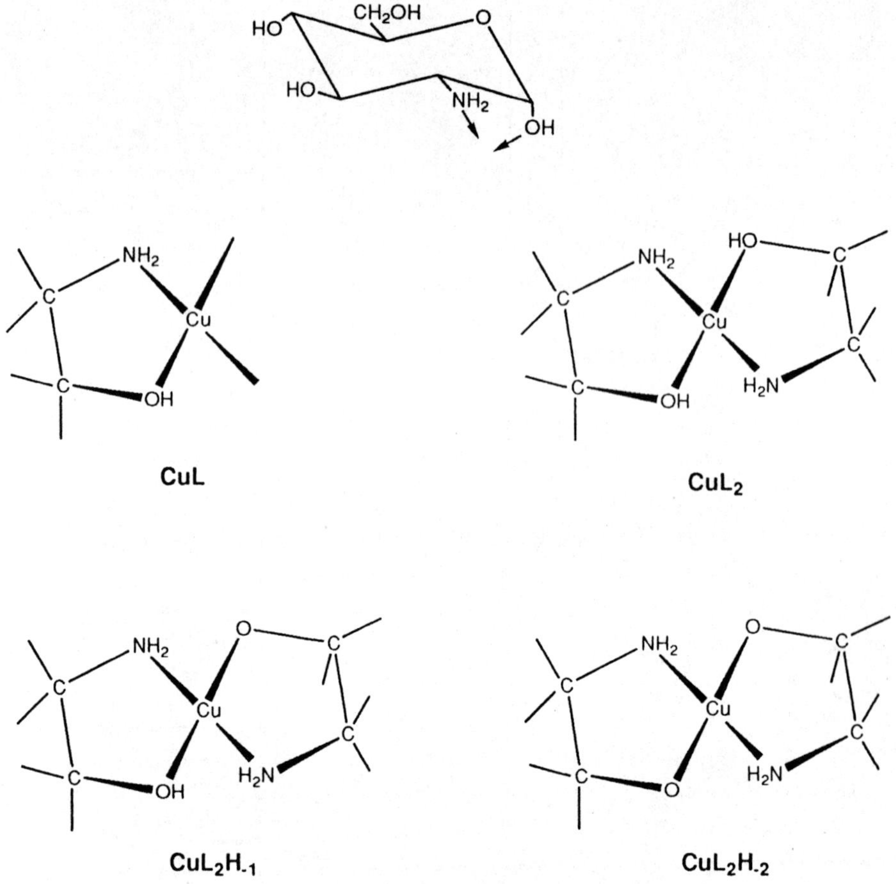

Scheme 2

the (NH_2, O^-) moiety. However, the fraction of this species is too small to be identified by spectroscopic and potentiometric measurements. Upon formation of two such chelated rings, a very stable complex CuL_2H_{-2} is obtained, which is almost the only species in neutral solution. The coordinating hydroxyl is assumed to be the OH(1) group in Scheme 2, where the ligand has been assigned with the anomeric α-form which is more favorable for metal chelation due to the *cis*-arrangement of the donor atoms. Such an assumption may be proved by examination of copper(II) complex formation by the derivative β-GlcN-OMe, where the O(1) oxygen is excluded from coordination (Pusino et al., 1987). Blocking of the hydroxyl group at C(1) significantly affects the species distribution. In fact, the CuL and CuL_2 complexes formed by β-GlcN-OMe have a similar coordination but a considerably lower stability compared with those of GlcN. The obvious explanation for this finding is the participation of the OH(1) group in the metal coordination of GlcN and its exclusion in the glucopyranoside derivative β-GlcN-OMe, where the oxygen donor may be derived from either the OH(3) or OH(4) groups. Since OH(3) is in *trans*-position with respect to NH_2, whereas OH(4) would participate in a six-membered ring, chelated species weaker than those of GlcN are formed. A major difference is observed in neutral and basic solutions where CuL_2H_{-1} becomes the main species because of the low stability of the CuL_2 complex of the glycopyranoside.

A comparison of the coordination abilities of the amino sugars adopting galacto- or manno-configuration confirms that various hydroxyl groups can be involved in the binding of metal ions (Radomska et al., 1988; Urbanska et al., 1988; Kozlowski et al., 1990;

Table 1

Stability Constants (log β) for Copper(II) Complexes of Amino Sugars[a]

Species	GlcN	GalN	ManN	β-GlcN-OMe	α,β-GalN-OMe	α-ManN-OMe	β-AGlcN	β-AGlcNMe	β-AGlc3N	β-AGlc4N
LH	7.70	7.84	7.59	7.69	7.84	7.47	7.07	7.30	7.32	7.48
CuL	5.12[b]	4.20, 5.23[b]		4.13	4.40	4.81				
CuL_2	9.02, 8.85[b]	9.13, 9.02[b]	9.68	7.52	8.40				6.86	
Cu_2L_4							18.63	17.92		20.03
CuL_2H_{-1}		2.37	2.72	1.38	2.27	2.91	0.71	0.28	-0.36	1.49
CuL_2H_{-2}	-5.26	-5.21	-3.66	-6.85	-5.18	-4.29	-8.05	-8.05	-8.20	-6.89
CuL_2H_{-3}	-13.77	-15.44	-13.0			-13.4				-17.25
CuL_2H_{-4}						-23.7				

[a] References are given in the text; [b] From polarographic measurements.

Urbanska and Kozlowski, 1990). These ligands exhibit the same complexation scheme as GlcN. The CuL_2 species of GalN and GlcN have similar stability. The same trend is observed for the corresponding CuL_2H_{-2} complexes suggesting that the same hydroxyl donor OH(1) is active in both ligands. However, blocking this group in α,β-GalN-OMe does not significantly affect the binding ability of the ligand, and the high stabilities of the chelated complexes indicate that the availability of a hydroxyl group for chelation in α,β-GalN-OMe is almost the same as in GalN and GlcN. This behavior suggests that the conformation of GalN and/or α,β-GalN-OMe may allow competition between OH(3) and OH(1) in metal binding, whereas GlcN has a preference for OH(1). This is possible if the formation of chelated rings alters the sugar conformation, for example if it moves the N and O donor atoms away from the neighboring substituents in the α-galacto structure, while bringing them closer in the β-gluco structure.

The most effective chelates are formed by ManN, and the CuL_2H_{-2} complex of this ligand is more stable than those of GalN and GlcN by more than 1.5 orders of magnitude. This major difference in stability indicates that the hydroxyl group of ManN involved in the coordination is not OH(1). Since the ligand exists predominantly in the anomeric α-form and does not coordinate through O(1), complexation at O(3) becomes important. Accordingly, blocking OH(1) in α-ManN-OMe does not change the stability of the complexes formed significantly.

A characteristic feature of these systems is the formation of the deprotonated species CuL_2H_{-3} and CuL_2H_{-4}. These could arise from the dissociation of other hydroxyl groups and their subsequent interaction in the apical positions of the metal ion. It may be tentatively suggested that, in α-ManN-OMe, the complexes involve O(6) atoms of the ligands in a distorted $^1C_4$ chair conformation.

The stabilities of the Ni(II) and Co(II) complexes of ManN, GlcN and GalN are similar but considerably lower than those of the Cu(II) species (Lerivrey et al., 1986; Urbanska et al., 1988; Urbanska and Kozlowski, 1990). With these ions too, ManN is the most effective ligand.

The critical influence of the sugar structure and the donor position on the complex stability becomes evident on examination of the ligand properties of the 1,6-anhydro derivatives of simple amino sugars, β-AGlcN, β-AGlcNMe, β-AGlc3N, and β-AGlc4N (Scheme 1) in which OH(1) is blocked and which also have a limited ring flexibility (Jezowska-Bojczuk et al., 1991). The chair conformation and the position of the donors (1,3-diaxial) in β-AGlcN, β-AGlcNMe, β-AGlc4N force the formation of six-membered [O(2), N(4)]- or [N(2), O(4)]-chelated complexes which are less stable than those of the parent amino sugars by one or more orders of magnitude. The *trans*-arrangement of the donors makes the β-AGlc3N ligand the least favorable for the formation of chelated species, and the interaction of copper(II) with a pair of amino and hydroxyl groups is possible only with the less stable $B_{0,3}$ boat form. Another consequence of this chelating arrangement is that β-AGlc3N does not yield ESR-silent dimeric Cu_2L_4 species. Complexes of this type have been observed with the other 1,6-anhydro amino sugars and seem to be favored by the interaction of the metal ion with the 1,3-dioxolane ring oxygens which mediate the metal-metal interaction.

The acetamide nitrogen in *N*-acetyl amino sugars (e.g. GlcNAc) is not an effective donor toward metal ions (Appelman-Lippens et al., 1985; Micera et al., 1985). This finding is consistent with the NMR observation that with the GlcNAc-1-OAr-6S glucopyranoside the binding to Zn^{2+} and Ni^{2+} is very weak and takes place only with sulfate or lactone groups (Whitfield and Sarkar, 1991). This behavior may be assumed also for polysaccharides containing *N*-sulphated or *N*-acetylated amino sugars. Accordingly, the metal ion interaction with these macromolecules is mainly electrostatic in origin unless carboxylate groups are present (Wright et al., 1981; Balt et al., 1983).

III. SOLID-STATE STUDIES

There are few studies on amino sugar complexes in the solid-state since difficulties in obtaining single crystals hinder the acquisition of structural data by X-ray diffraction.

A solid copper(II) complex of GlcN with assumed CuL(OH)Cl stoichiometry (L being the neutral form of the ligand) was reported by Tamura and Miyazaki (1965b).

The preparation of amino complexes of Co(III) containing GlcN as second ligand was also described (Bunel and Ibarra, 1985).

Cisplatinum-type complexes of amino sugars have been studied. Due to its high solubility in water, *cis*-dichloro[bis(*D*-GlcN)]platinum(II) was proposed as an antitumor drug less toxic than *cis*-dichlorodiaminoplatinum(II) (Haroutounian et al., 1986; Kuduk-Jaworska and Jezowska-Trzebiatowska, 1986). A high anti-tumor activity was demonstrated in vivo for the [$PtLCl_2$] complexes formed by *N,N*-bidentate amino sugars, GlcNN, β-GlcNN-OMe, GalNN, ManNN, α-ManNN-OMe, and β-AraN,N-OBz (Tsubomura et al., 1986a; 1990).

Cu^{2+}, Co^{2+}, and Zn^{2+} complexes with the salicylaldehyde Schiff's base derived from β-GlcN-1-OMe-3,4,6-OAc were reported (Adam and Hall, 1979). Schiff's bases of amino sugars easily undergo hydrolysis. In fact, the reaction between $PtCl_4^{2-}$ and the 4-methoxy-benzylidene Schiff's base of GlcN-1,3,4,6-OAc yielded mainly complexes containing the free sugar (L), namely K[$PtLCl_3$] and a mixture of *cis*- and *trans*-[PtL_2Cl_2], indicating hydrolysis of the primary ligand once bound to the metal (Kuduk-Jaworska and Jezowska-Trzebiatowska, 1986).

The technetium(V) complex of the *N*-salicylidene Schiff's base of GlcN (H_2L) was synthesized in order to obtain a ^{99m}Tc radiopharmaceutical with the active probe linked to a metabolically active molecule (Duatti et al., 1987). In this case a metal-bound ligand also underwent hydrolysis and yielded the neutral species TcOL(sal) where sal is salicylaldehydato(1-). Coordination of the intact Schiff's base molecule occurs through the aldimine nitrogen, the phenolate oxygen and the deprotonated hydroxyl at the C(1) position in the amino sugar moiety.

Extensive information has been obtained concerning the Ni(II) complexes of *N*-glycosides, i.e. amino sugars bearing an aminoalkyl group at C(1) or C(2) in the aldohexose or ketohexose sugar rings. These ligands may be generally derived by reaction between an amine and a sugar molecule. However, complexes of *N*-glycosides are directly obtained by reaction of simple aldose monosaccharides with the amine centers of $[Ni(en)_3]^{2+}$or $[Ni(tn)_3]^{2+}$, e.g. $[NiL_2]X_2$, where L = Glc-en, Man-en, Man-tn or Rha-tn and X = Cl or Br (Takizawa et al., 1980). These sugar derivatives behave as tridentate ligands coordinating through the amino nitrogens of the diamine residue and the oxygen of the undissociated C(2) sugar hydroxyl. The use of amino sugars instead of simple monosaccharides yielded $[NiL_2]X_2$ complexes with 3N-donor *N*-glycoside ligands, namely GlcN-en, GalN-en, and ManN-en (Yano et al., 1985). These species are octahedral and exhibit coordination of the metal ion to six nitrogen atoms. Interestingly, as shown by the crystal structure determination, in the complex derived from GlcN, the ligand is in its stable β-$^4C_1$ conformation and no significant strain is produced upon metal coordination. Two amino groups occupy equatorial positions with respect to the pyranose ring of the sugar residue suggesting that the *eq-eq* combination of donor groups is suitable for five-membered chelation.

Also *N*-glycosides derived from an aldose with an amino acid (e.g. Glc-β-ala) bind Ni(II) and behave as tridentate through the carboxylate oxygen, the nitrogen atom and the OH(2) sugar hydroxyl. This coordination stabilizes the ligand which, due to the Amadori rearrangement, is not stable in itself (Tsubomura et al., 1986b).

[Ni(en)L]Cl_2 amino sugar-like complexes are obtained with *N*-glycoside ligands derived from ketoses, e.g. Fru-en, Sor-en, Sor-tn, Sor-pn, Fru-pn, Sor-ampr, and Fru-ampr (Shioi et al., 1983; Tsubomura et al., 1983; 1985). The ligands, in their stable α-$^2C_5$ conformation, are tetradentate and the ketose residue is bound to the metal through the oxygens of OH(1) and OH(3) groups.

A dinuclear nickel(II) complex contains three mannose residues forming glycosides with the diamine N,N'-Me_2en, N-Man-N,N'-Me_2-en and N,N'-$(Man)_2$-N,N'-Me_2-en (Tanase et al., 1987). Two of these bind Ni(II) through the C(2)-hydroxyl and exhibit the stable β-$^4C_1$ pyranose form. The third mannose residue is an unusual furanose conformation, approximately ^3E, in which all the OH groups are involved in the coordination, including the one on C(3) which acts as a bridge between two metal ions.

Co(III) complexes of these N-glycoside ligands have been also synthesized (Ishida et al., 1986; 1989). The tridentate glycoside ligands Man-en, Rha-en, and Rib-en bind cobalt through the amino groups and the deprotonated C(2) and C(3) hydroxyls. A conformational analysis indicates that Man and Rha units adopt the unusual $^3S_5$ skew-boat structure, while ribose is probably in the ^2E envelope form.

Obviously, because of the key role played by the metal-anchoring diamine residue in determining the observed structures, the above data cannot be directly extended to natural systems. Nevertheless, they are further supporting that sugars and amino sugars are versatile and flexible ligands provided with multiple donor sites which can be easily adapted to fulfil coordination requirements of metal ions.

IV. CONCLUSIONS

Amino sugars are found to be very effective ligands for metal ions having amine nitrogen as an anchor site. The set of hydroxyl groups may participate in metal ion coordination both as OH or more effective deprotonated O^- donors. The variety of sugar ring conformations leads to metal complexes with different structures and stabilities.

List of ligands

AGlcN = 2-amino-1,6-anhydro-2-deoxy-*D*-glucopyranose
AGlc3N = 3-amino-1,6-anhydro-3-deoxy-*D*-glucopyranose
AGlc4N = 4-amino-1,6-anhydro-4-deoxy-*D*-glucopyranose
AGlcNMe = 2-aminomethyl-1,6-anhydro-2-deoxy-*D*-glucopyranose
ampr = 2(*S*)-(aminomethyl)pyrrolidine
AraN,N-OBz = benzyl 3,4-diamino-3,4-dideoxy-*D*-(or *L*)-arabinopyranoside
en = ethylenediamine
Fru-ampr = 2-[(2(*S*)-pyrrolidinylmethyl)amino]-2-deoxy-*D*-fructopyranose
Fru-en = 2-[(2-aminoethyl)amino]-2-deoxy-*D*-fructopyranose
Fru-pn = *N*-*D*-fructopyranosyl-(*S*)-propylenediamine
GalN = 2-amino-2-deoxy-*D*-galactopyranose (*D*-galactosamine)
GalN-en = 1-[(2-aminoethyl)amino]-2-amino-1,2-deoxy-*D*-galactose
GalNN = 2,3-diamino-2,3-dideoxy-*D*-galactopyranose
GalN-OMe = methyl 2-amino-2-deoxy-*D*-galactopyranoside
Glc-ala = *N*-*D*-glucopyranosyl-3-aminopropionate
Glc-en = 1-[(2-aminoethyl)amino]-1-deoxy-*D*-glucopyranose
GlcN = 2-amino-2-deoxy-*D*-glucopyranose (*D*-glucosamine)
GlcNAc = 2-acetamido-2-deoxy-*D*-glucopyranose (*N*-acetyl-*D*-glucosamine)
GlcNAc-1-OAr-6S = 7-(4-methylumbelliferyl) 2-acetamido-2-deoxy-6-*O*-sulpho-*D*-glucopyranoside
GlcN-en = 1-[(2-aminoethyl)amino]-2-amino-1,2-deoxy-*D*-glucopyranose
GlcNN = 2,3-diamino-2,3-dideoxy-*D*-glucopyranose
GlcNN-OMe = methyl 2,3-diamino-2,3-dideoxy-*D*-glucopyranoside
GlcN-1,3,4,6-OAc = 1,3,4,6-tetra-*O*-acetyl-2-amino-2-deoxy-*D*-glucopyranose
GlcN-OMe = methyl 2-amino-2-deoxy-*D*-glucopyranoside
GlcN-1-OMe-3,4,6-OAc = methyl 3,4,6-tri-*O*-acetyl-2-amino-2-deoxy-*D*-glucopyranoside
Man-en = 1-[(2-aminoethyl)amino]-1-deoxy-*D*-mannopyranose
ManN = 2-amino-2-deoxy-*D*-mannopyranose (*D*-mannosamine)
ManN-en = 1-[(2-aminoethyl)amino]-2-amino-1,2-deoxy-*D*-mannopyranose
N-Man-*N,N'*-Me_2-en = 1-(*N*-methyl-mannosylamino)-2-(methylamino)ethane
N,N'-$(Man)_2$-*N,N'*-Me_2-en = 1,2-bis(*N*-methyl-mannosylamino)ethane
ManNN = 2,3-diamino-2,3-dideoxy-*D*-mannopyranose
ManNN-OMe = methyl 2,3-diamino-2,3-dideoxy-*D*-mannopyranoside
ManN-OMe = methyl 2-amino-2-deoxy-*D*-mannopyranoside
Man-tn = 1-[(3-aminopropyl)amino]-1-deoxy-*D*-mannopyranose
N,N'-Me_2en = *N,N'*-dimethylethylenediamine
Rha-en = 1-[(2-aminoethyl)amino]-1,6-dideoxy-*L*-mannopyranose
Rha-tn = 1-[(3-aminopropyl)amino]-1,6-dideoxy-*L*-mannopyranose
Rib-en = 1-[(2-aminoethyl)amino]-1-deoxy-*D*-ribofuranose
Sor-ampr = 2-[(2(*S*)-pyrrolidinylmethyl)amino]-2-deoxy-*L*-sorbopyranose
Sor-en = 2-[(2-aminoethyl)amino]-2-deoxy-*L*-sorbopyranose
Sor-pn = *N*-*L*-sorbopyranosyl-(*S*)-propylenediamine
Sor-tn = 2-[(3-aminopropyl)amino]-2-deoxy-*L*-sorbopyranose
TalN = 2-amino-2-deoxy-*D*-talopyranose (*D*-talosamine)
tn = 1,3-diaminopropane

REFERENCES

Adam, M.J., and Hall, L.D. (1979). *J. Chem. Soc. Chem. Commun.*, 234.

Appelman-Lippens, E.B.V., de Bolster, M.W.G., Tiemersma, D.N., and Visser-Luirink, G. (1985). *Inorg. Chim. Acta, 108*: 209.

Balt, S., de Bolster, M.W.G., Booj, M., van Herk, A.M., and Visser-Luirink, G. (1983). *J. Inorg. Biochem., 19*: 213.

Bunel, S., and Ibarra, C. (1985). *Polyhedron, 4*: 1537.

Duatti, A., Marchi, A., Magon, L., Deutsch, E., Bertolasi, V., and Gilli, G. (1987) *Inorg. Chem., 26*: 2182.

Haroutounian, S.A., Georgiadis, M.P., and Bailar, J.C., Jr. (1986). *Inorg. Chim. Acta, 124*: 137.

Inaki, Y., Otsuru, M., and Takemoto, K. (1978). *J. Macromol. Sci.-Chem., A12*: 953.

Ishida, K., Yano, S., and Yoshikawa, S. (1986). *Inorg. Chem., 25*: 3552.

Ishida, K., Nonoyama, S., Hirano, T., Yano, S., Hidai, M., and Yoshikawa, S. (1989). *J. Amer. Chem. Soc., 111*: 1599.

Ishidate, M., Sakaguchi, T., Taguchi, K., and Kanao, S. (1960). *Anal. Chim. Acta, 22*: 452.

Jezowska-Bojczuk, M., Kozlowski, H., Decock, P., Cerny, M., and Trnka, T. (1991). *Carbohyd. Res., 216*: 453.

Kozlowski, H., Decock, P., Olivier, I., Micera, G., Pusino, A., and Pettit, L.D. (1990). *Carbohyd. Res., 197*: 109.

Kuduk-Jaworska, J., and Jezowska-Trzebiatowska, B. (1986). *Inorg. Chim. Acta, 123*: 209.

Lerivrey, J., Dubois, B., Decock, P., Micera, G., Urbanska, J., and Kozlowski, H. (1986). *Inorg. Chim. Acta, 125*: 187.

Micera, G., Deiana, S., Dessl, A., Decock, P., Dubois, B., and Kozlowski, H. (1985). *Inorg. Chim. Acta, 107*: 45.

Miyazaki, M., Nishimura, S., Yoshida, A., and Okubo, N. (1979). *Chem. Pharm. Bull., 27*: 532.

Muzzarelli, R.A.A. (1977). *Chitin*, Pergamon Press, Oxford.

Muzzarelli, R.A.A., and Rocchetti, R. (1986). *Trace Metal Removal from Aqueous Solution* (R. Thompson, ed.), Special Publication No 61, The Royal Society of Chemistry, London, p. 44 and refs. therein.

Pusino, A., Droma, D., Decock, P., Dubois, B., and Kozlowski, H. (1987). *Inorg. Chim. Acta, 138*: 5.

Radomska, B., Kozlowski, H., Decock, P., Dubois, B., and Micera, G. (1988). *J. Inorg. Biochem., 33*: 153.

Rendleman, J.A., Jr. (1973). *Adv. Chem. Ser., 117*: 51.

Shioi, H., Yano, S., Toriumi, K., Ito, T., and Yoshikawa, S. (1983). *J. Chem. Soc. Chem. Commun.*, 201.

Takagi, H., and Kadowaki, K. (1986). *Chitin in Nature and Technology* (R.A.A. Muzzarelli, C. Jeuniaux and G.W. Gooday, eds.), Plenum Press, New York, p. 121.

Takizawa, S., Sugita, H., Yano, S., and Yoshikawa, S. (1980). *J. Amer. Chem. Soc., 102*: 7969.

Tamura, Z., and Miyazaki, M. (1965a). *Chem. Pharm. Bull., 13*: 333.

Tamura, Z., and Miyazaki, M. (1965b). *Chem. Pharm. Bull., 13*: 345.

Tamura, Z., and Miyazaki, M. (1965c). *Chem. Pharm. Bull., 13*: 387.

Tanase, T., Kurihara, K., Yano, S., Kobayashi, K., Sakurai, T., Yoshikawa, S., and Hidai, M. (1987). *Inorg. Chem., 26*: 3134.

Tsubomura, T., Yano, S., Yoshikawa, S., Toriumi, K., and Ito, T. (1983). *Polyhedron, 2*: 123.

Tsubomura, T., Yano, S., Toriumi, K., Ito, T., and Yoshikawa, S. (1985). *Inorg. Chem., 24*: 3218.

Tsubomura, T., Yano, S., Kobayashi, K., Sakurai, T., and Yoshikawa, S. (1986a). *J. Chem. Soc. Chem. Commun.*, 459.

Tsubomura, T., Yano, S., and Yoshikawa, S. (1986b). *Inorg. Chem.*, *25*: 392.

Tsubomura, T., Ogawa, M., Yano, S., Kobayashi, K., Sakurai, T., and Yoshikawa, S. (1990). *Inorg. Chem.*, *29*: 2622.

Urbanska, J., and Kozlowski, H. (1990). *J. Coord. Chem.*, *21*: 175.

Urbanska, J., Kozlowski, H., Delannoy, A., and Hennion, J. (1988). *Anal. Chim. Acta*, *207*: 85.

Whitfield, D.M., and Sarkar, B. (1991). *J. Inorg. Biochem.*, *41*: 157.

Wright, W.M., Feigan, J., Denny, W., Leupin, W., and Klarus, D.R. (1981). *J. Magn. Res.*, *45*: 514.

Yano, S., Sakai, Y., Toriumi, K., Ito, T., Ito, H., and Yoshikawa, S. (1985). *Inorg. Chem.*, *24*: 498.

METAL COMPLEXES OF AMINOPHOSPHONIC ACIDS

by

Tamás Kiss

Kossuth Lajos Tudományegyetem,
Szervetlen és Analitikai Kémiai Tanszék,
H-4010 Debrecen
Hungary

I. INTRODUCTION

Aminophosphonic acids are broadly defined as analogues of aminocarboxylates (including polyamino and/or polycarboxylates, alkyl, aryl or cyclic derivatives) in which one or more carboxylic groups are replaced by a phosphonic or related function (phosphonous, phosphinic, etc). A common characteristic of these compounds is that all contain a direct carbon-phosphorus (C-P) bond. They were first synthesized in the forties and were found to occur in living organisms in the late fifties, but the study of aminophosphonates has accelerated only in the past fifteen-twenty years, when it was discovered that many of the natural and synthetic aminophosphonates display diverse and useful biological properties. Since then, many aspects of this ligand group have been reviewed, with attention focused on their occurrence, properties, reactions and biological activity; here only some of the most recent publications are referred to (Rizkalla, 1983; Kafarski and Mastalerz, 1984; Kukhar and Solodenko, 1987; Kafarski and Lejczak, 1991). Although a few of them (e.g. 2-aminoethanephosphonic acid (β-Ala-P) and its N-methyl derivatives, or 2-amino-3-phosphonopropionic acid (Asp-β-P)) are found as free compounds in tissues, the most frequent forms of occurrence of aminophosphonic acids are derivatives, such as phosphonopeptides, phosphonolipids and phosphonoglycolipids. The several hundred known natural and mainly synthetic aminophosphonic acids and their derivatives include neuroactive compounds, antibacterial agents, growth factors, anticancer drugs and pesticides; their possible applications therefore range from medicine to agriculture. As structural analogues of aminocarboxylic acids, aminophosphonic acids usually act as their antagonists and compete with their carboxylic counterparts for the active sites of enzymes or other cell receptors. It seems obvious that this enzyme (sometimes metalloenzyme) regulatory ability of aminophosphonates is connected with the differences in size (the phosphonic function is considerably larger), shape (flat carboxylate and tetrahedral phosphonate), basicity (phosphonate is more basic than carboxylate) and charge (phosphonate is binegative, while carboxylate is mononegative) of the carboxylate and phosphonate groups.

Following the original analytical applications (Rizkalla, 1983), multidentate aminopolyphosphonates and polyaminopolyphosphonates are now receiving considerable attention as metal-sequestering drugs, e.g. for the treatment of iron-overload anemias

(Taliaferro and Martell, 1984; 1985), as radiopharmaceuticals involving Ga(III) or In(III) (Hayes and Hubner, 1983) or as NMR imaging agents (Lauffer, 1987).

In this review, special emphasis is given to a comparison of analogous carboxylate and phosphonate systems.

II. PHOSPHONIC DERIVATIVES OF α-AMINO ACIDS

When the protonation constants of the simple α-aminocarboxylates and their various phosphoric derivatives (see Fig. 1) are compared, it can be seen that the pK of the acidic OH increases in the sequence phosphonous (-PO(H)(OH)) ~ phosphinic (-PO(R)(OH)) < carboxylic (-CO(OH)) < phosphonic ($-PO_3H^-$) ~ phosphoric ($-OPO_3H^-$). The first proton in the phosphoric moiety of phosphonic and phosphoric acid derivatives is very acidic; similarly as in the phosphonous and phosphinic acids, pK is ~ 1. The acidity sequence of the ammonium group of these ligands corresponds to the variations in the electron-withdrawing effects of the other group; this gives the following sequence: PO_3^{2-} < OPO_3^{2-} < CO_2^- < PO_2R^- < PO_2H^-. Similarly as for the α-aminocarboxylate analogues, a cyclic conformation via internal hydrogen-bonding between the phosphonate and ammonium groups is strongly supported by the results of NMR studies (Appleton et al., 1984). The acid-base character of the further side-chain donor groups of the phosphonic derivatives of tridentate amino acids such as aspartic acid (Asp), glutamic acid (Glu), ornithine (Orn), lysine (Lys), tyrosine (Tyr) and 3,4-dihydroxyphenylalanine (Dopa) shows the same pattern as their aminocarboxylate analogues (Kiss, 1990).

As concerns the metal-binding ability of these α-phosphono derivatives, bidentate (N,O)-chelation is most likely with most of the biologically important transition metal ions. The stability of the complexes roughly corresponds to the overall basicity sequence of the coordinating donor groups; the higher the basicity of the donor groups the higher the stability of the complexes formed. Owing to the relatively high basicity of the PO_3^{2-} group, there is also a possibility for monodentate coordination of the phosphonate group in acidic solution. The equilibrium constants characteristic for the reaction $M^{n+} + H_2L \rightleftharpoons ML^{(n-2)+} + 2H^+$, which express the competition between proton and metal ion for a ligand chelating site, may be a more realistic measure of the relative metal ion-binding strength. These basicity-adjusted stability constants reveal an opposite sequence for these ligands: aminocarboxylates > aminophosphates > aminophosphonates >

R | $CH-NH_2$ (8.5-9.5) | COOH (2.0-3.0)	R | $CH-NH_2$ (9.5-10.5) | $PO(OH)_2$ (5.0-6.0) (0.5-1.5)	R | $CH-NH_2$ (7.5-8.5) | HPO(OH) (0.5-1.5)
α-amino-carboxylic acid	α-amino-phosphonic acid	α-amino-phosphonous acid
R | $CH-NH_2$ (8.0-9.0) | R'PO(OH) (0.5-1.5)	R | $CH-NH_2$ (10.0-11.0) | CH_2 | $PO(OH)_2$ (5.5-6.5) (0.5-1.5)	R | $CH-NH_2$ (9.5-10.5) | CH_2 | $OPO(OH)_2$ (5.0-6.0) (0.5-1.5)
α-amino-phosphinic acid	β-amino-phosphonic acid	β-amino-phosphoric acid

Figure 1.
Structural formulas of aminocarboxylates and their phosphono derivatives (the pK range of the donor groups are given in parenthesis).

aminophosphinates; i.e. the electrostatic and steric hindrance due to the binegative charge and the larger size of the PO_3^{2-} group overcompensates the stability increase arising from the higher basicity of the phosphonate (Kiss, 1990). In alkaline solution, monodentate N-bonded complexes are formed with *cis*-$Pt(NH_3)_2(H_2O)^{2+}$ and *cis*-$Pt(NH_3)_2(OH)_2$ as well as with Pd(II), as indicated by detailed multinuclear NMR studies (Appleton et al., 1986; Matczak-Jon and Wojciechowski, 1990).

The presence of further coordinating donor groups in the molecules may modify the metal-binding ability of these ligands. The various phosphonic and phosphinic acid derivatives of the potentially tridentate Asp and Glu, some of them potent enzyme inhibitors, display typical ambidentate (aminocarboxylate and/or aminophosphonate) coordinating capability towards transition metal ions. For the two phosphono analogues of Asp, for instance, it was established that both aminocarboxylate and aminophosphonate coordination of the ligands can occur. This results in the co-existence of complexes with the same stoichiometry, but with different bonding modes. This behavior was observed to be more marked when the PO_3^{2-}/CO_2^- substitution was in the α-position (Kiss et al., 1989a).

Microorganisms have been found to utilize the phosphorus of aminoalkylphosphonic acids for growth, this being an indication that the C-P bond can be cleaved. To demonstrate this, the metal ion and pyridoxal-5'-phosphate-catalyzed transamination and dephosphonylation reactions of Asp-β-P (one of the few aminophosphonates occurring freely in living organisms) have been widely studied (Martell et al., 1985; Szpoganicz and Martell, 1986). A basically analogous mechanism was proposed for β-dephosphonylation to that generally accepted for the enzyme-catalyzed β-decarboxylation of Asp. Although no enzyme has been reported so far that catalyzes this transaminative dephosphonylation reaction, these results suggest that enzymes which perform C-P bond fission in biological systems probably do exist (Martell et al., 1985).

As mentioned above, aminophosphonic acids are potent inhibitors of many enzymes, most of them involved in the metabolism of amino acids. A good example is the interaction of the phosphonic analogues of 3,4-dihydroxyphenylalanine (Dopa) with copper-containing tyrosinase. A comparative study (Balla et al., 1990a) of the complex forming properties of dopa, dopa-P (1-amino-1-(3',4'-dihydroxyphenyl)ethylphosphonic acid) and other phosphonic derivatives (1-amino-1-(3',4'-dihydroxyphenyl)-methylphosphonic acid and 1-amino-1-(2',3'-dihydroxyphenyl)methylphosphonic acid) with copper(II) revealed a significant similarity in the metal-binding ability of these aminophosphonates and their aminocarboxylate analogues. Namely, the 3,4-dihydroxyphenyl derivatives which contain two chelating functions separated within the molecule are coordinated via the aminophosphonate side-chain at lower pH, but via the catecholate moiety at higher pH. In the intermediate pH range, various monomeric and dimeric species involving the participation of both chelating centers are produced. For the 2,3-dihydroxy derivative, where the chelate-forming donor group pairs are not fully separated from each other, the tendency to dimer formation is more favoured. Thus, it is not surprising that dopa-P serves as a synthetic substrate for tyrosinase and it does not differentiate between the carboxylate and phosphonic functions. A significant change in activity was achieved, however, by shortening the side chain by one methylene group, and the phosphonic derivative of 3,4-dihydroxyphenylglycine proved to be one of the most powerful known inhibitors of this enzyme. It can be assumed that both substrate and inhibitor molecules coordinate to the active center of tyrosinase via the catecholate binding site with similar strengths, but the differences in charge and steric conditions of the side-chain donors, which are capable of strong electrostatic binding to the appropriate portion of the enzyme, seems to be the decisive factor in the enzyme-regulatory effect.

1-Amino-2-(N-alkylamino)alkylphosphonic acids are a new class of aminopeptidase inhibitors, which prevent zinc-containing exopeptidases from catalyzing the hydrolysis of the N-terminal peptide bonds in polypeptide chains (Lejczak et al., 1989). It was found that the (N,N) chelating site was able to compete with the aminophosphonate site in metal ion binding, although this might happen only at high pH beyond the physiological range (Balla et al., 1990a). Thus, the inhibitor is expected to act through chelation to the metal ion in the active center via the phosphonate moiety, which may be stabilized by an electrostatic interaction between the chain ammonium groups of the inhibitor molecule and the negatively charged peptide moieties of the enzyme.

III. PHOSPHONIC DERIVATIVES OF DIPEPTIDES

The phosphonopeptides hitherto isolated from living organisms or synthesized as bioactive drugs contain the phosphonate group in either the C-terminal or the N-terminal position. Peptides containing phosphonamide linkages could not be find or obtained, probably because of their high sensitivity toward acids and bases (Hariharan et al., 1975). Many synthetic phosphonopeptides exhibit promising biological activity. Only two widely studied representatives of the two types will be mentioned here: Alafosfalin [(N-L-alanyl-L-1-aminoethyl)phosphonic acid, Ala-Ala-P (1)] and PALA [(N-phosphonoacetyl)-L-aspartic acid, (2)].

$CH_3CHCO(NH)CHCH_3$ (with NH_2 on the first CH and PO_3H_2 on the second CH)

(1)

$CH_2CO(NH)CHCOOH$ (with PO_3H_2 on CH_2 and CH_2COOH on CH)

(2)

The metal-binding ability of these phosphonopeptides has been studied so far only with copper(II) (Kiss, 1989b). It may be extremely interesting to establish how phosphonate substitution of the terminal groups affects the coordination properties of these ligands, with special emphasis on the metal ion-induced deprotonation and coordination of the peptide-amide, this bonding mode being particularly important for oligopeptides. On the basis of potentiometric and spectral studies on the coordination of several monophosphono-dipeptides (in which the C-terminal carboxylate group is replaced by a phosphonate moiety) and diphosphono-dipeptides (in which both the C-terminal carboxylate and the N-terminal amino groups are replaced by phosphonate moieties) with copper(II), it was revealed (Kiss, 1989b) that neither PO_3^{2-}/CO_2^- nor PO_3^{2-}/NH_2 substitution significantly changes the basic bonding modes of the simple dipeptides (cf Section B1 in this Chapter 2). The extent of the copper(II)-induced deprotonation and simultaneous coordination of the peptide amide decreases in the following sequence: dipeptides > monophosphono-dipeptides > diphosphono-dipeptides. It is noteworthy, however, that, similarly to the NH_2 group, the less basic PO_3^{2-} can also act as an anchor donor group and can bind copper(II) strongly enough to induce deprotonation of the peptide amide.

IV. AMINOPOLYPHOSPHONATES

Aminopolyphosphonates as multidentate complexing agents with high specificity for various cations are widely used for various purposes. They are applied among others in analytical chemistry, in the paper and textile industries for the removal of trace amounts of metal ions from bleaching baths, and also in medicine for the removal of a metal overload from living organisms (Motekaitis et al., 1971a; 1971b; 1976; Rizkalla and Zaki, 1979; 1980a; 1980b; 1980c; Taliaferro and Martell, 1984; 1985; Motekaitis and Martell, 1985; Sawada et al., 1987; Gillard et al., 1989). A few important representatives of this ligand group can be seen in Figure 2, and their stability constants with several metal ions are listed in Table 1.

The successive protonation constants cannot provide definite information on the protonation sites of the particular species involved in each step. The microscopic constants obtained mainly via NMR techniques can unambiguously prove that the relative basicity sequence of the donor groups (see paragraph 2) is overridden by electrostatic factors. A rather complex cyclic configuration between the protons bridging the phosphonate groups and the nitrilo group(s) favorably reduces the repulsive forces between the negatively charged moieties. In some cases, this results in the full protonation of the nitrogens at a lower pH than with the corresponding amino-carboxylates.

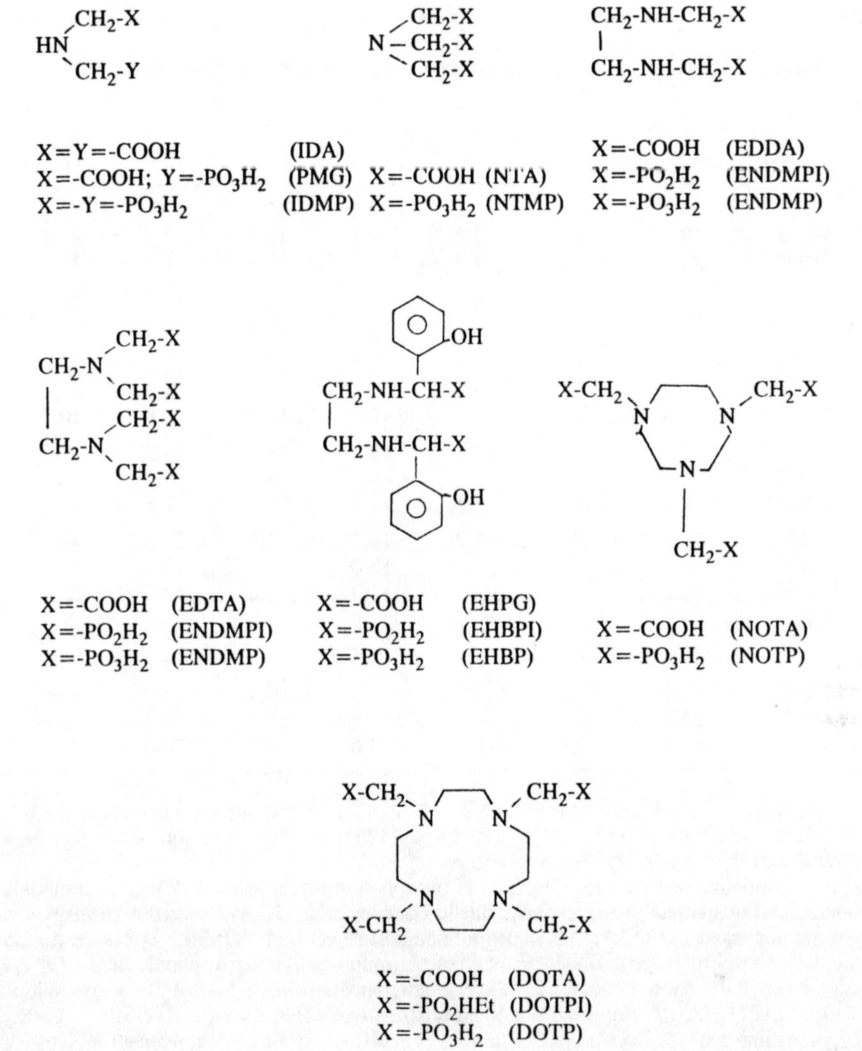

Figure 2.
Structural formulas of several aminopolyphosphonic acids and aminopolyphosphinic acids.

These multidentate ligands typically form ML complexes, although molecules of lower denticity (e.g. imino derivatives) can also form bis complexes. Complexes with ter- and tetravalent metal ions may undergo hydrolysis and then olation at higher pH (Motekaitis and Martell, 1980; 1985; Taliaferro and Martell, 1984; 1985).

A comparison of the complexing properties of mixed phosphonate and carboxylate ligands (see e.g. lines 1-3 in Table 1) indicates that the stability of the complexes increases with the number of phosphonate moieties. However, this stability increase is much lower than would be expected from the increase in basicity of the potentially coordinating groups. Namely, the more carboxylate is replaced by phosphonate, the higher the electrostatic and steric hindrance that compensate or overcompensate the former effect (Motekaitis and Martell, 1985). The mutual repulsion between the binegative phosphonate groups tends to lower the coordinating ability and denticity of the aminophosphonates with respect to their carboxylate analogues, which results in only a slightly higher, or sometimes even lower, stability of the phosphono complexes (see Table 1). The incomplete

Table 1

Log β_{MA}* Values for Some Aminocarboxylate, Phosphinate and Phosphonate Complexes

Ligand †	ΣpK_i	Mg^{2+}	Ca^{2+}	Ni^{2+}	Cu^{2+}	Zn^{2+}	Fe^{3+}
IDA	13.8	3.0	2.6	8.3	10.6	7.2	10.7
PMG	17.8	3.3	3.3	8.1	11.9	8.7	16.1
IDMP	22.8	4.3	3.8	8.3	12.8	9.0	>17
NTA	14.8	5.5	6.4	11.5	13.1	10.7	15.9
NTMP	31.7	7.2	7.5	11.1	17.4	16.4	-
EDDA	20.1	3.9	-	13.5	16.2	11.1	-
ENDMPI	13.1	-	-	7.5	10.7	6.2	-
ENDMP	28.6	2	-	11.7	17.5	12.0	10.3
EDTA	21.0	8.7	10.7	18.6	18.8	16.3	25.0
ENTMPI	13.3	1.9	-	8.4	9.8	7.6	-
ENTMP	46.7	8.4	9.6	16.4	23.2	18.8	22.5
EHPG	36.9	8.0	7.2	19.7	23.9	16.8	39.7
EHBPI	34.3	-	-	15.0	20.1	13.4	31
EHBP	46.3	-	-	12.3	18.6.	14.5	25
NOTA	22.0	8.9	8.8	-	-	18.3	13.7‡
NOTP	35.4	11.0	6.4	19.4	21.3	24.9	29.6
DOTA	30.0	11.0	15.9	-	19.1	18.9	24.6‡
DOTPI	22.9	4.4	9.4	-	16.6	15.8	16.5‡
DOTP	43.0	7.3	10.3	19.0	22.9	24.8	23.7

* Values are obtained from Martell (1974; 1975; 1982; 1989) for the open-chain ligands, and from Kabachnik (1984a; 1984b) and Lazar (1991) for the cyclic ligands. † Structural formulas of the ligands are shown in Figure 2.
IDA, iminodiacetic acid; PMG, N-phosphonomethylglycine; IDMG, iminobis-(methylenephosphonic acid); NTA, nitrilotriacetic acid; NTMP, nitrilotris(methylene-phosphonic acid); EDDA, ethylenediiminodiacetic acid; ENDMPI, etylenebis(imino-metylenephosphinic acid); ENDMP, ethylenebis(iminomethylenephosphonic acid); EDTA, ethylenenitrilotetraacetic acid; ENTMPI, ethylenedinitrilotetrakis(methylene-phosphinic acid; ENTMP, ethylenedinitrilotetrakis(methylenephosphonic acid); EHPG, ethylenediiminobis[(2-hydroxyphenyl)acetic acid]; EHBPI, ethylenediiminobis[(2-hydroxybenzyl)methylenephosphinic acid]; EHBP, ethylenediiminobis[(2-hydroxy-benzyl)methylenephosphonic acid]; NOTA, 1,4,7-triazacyclononanetriacetic acid; NOTP, 1,4,7-triazacyclononanetris(methylenephosphonic acid); DOTA, 1,4,7,10-tetraazacyclododecanetetraacetic acid; DOTPI, 1,4,7,10-tetraazacyclododecane-tetrakis(methylenephosphinic acid); DOTP, 1,4,7,10-tetraazacyclododecanetetrakis-(methylenephosphonic acid). ‡ Gd^{3+} complexes

participation of donor sites in metal binding, however, may make them more selective for cations with an enhanced capacity for coordination. Complexes with aminopolyphosphinates are generally less stable than both carboxylate and phosphonate analogues, although the partial coordination of the phosphinic groups is supported by both the stability and spectral data (Motekaitis et al., 1971a; 1971b; 1976; Dyatlova et al., 1970; Belugin et al., 1972). These polyaminopolyphosphono derivatives readily form protonated complexes with similar stability to that of the parent complexes; this strongly suggests that protonation occurs on the non-coordinated phosphonic moieties. This situation is different from that of the aminocarboxylate analogues, where all donor groups are generally coordinated to the metal ion, and thus protonation takes place through proton displacement of the metal ion from the most basic nitrogen site, resulting in a significant decrease in the stability of the protonated complexes. Because of the high stability of the complexes and fulfillment of the required selectivity at physiological pH, organophosphonates may be promising candidates for the treatment of iron overload anemias (Taliaferro and Martell, 1984; 1985).

A large number of phosphonic derivatives of azamacrocycles and aza-crown ethers have been synthesized and described in the literature (Tazaki et al., 1982; Kabachnik et al., 1984a; 1984b; Geraldes et al., 1989; Broan et al., 1990; Delgado et al., 1990; Lazar et al., 1991; Ramasamy et al., 1991). The cyclic structure may further modify the normal basicity sequence of the donor groups (see paragraph 2), although some contradictions exist in the exact assignment of the protonation constants (Geraldes et al., 1989; Delgado et al., 1990). It is assumed, for instance, that in the triaza- and tetraazaphosphonic acid derivatives (see Fig. 2) the most basic sites are two ring nitrogens, followed by the phosphonate oxygens which are protonated to different degrees depending on the ring structure. In the tetraaza ligand, the protonation of the pendant phosphonate oxygens is more extensive than that in the triaza ligand before further protonation of the ring nitrogens occurs (Geraldes et al., 1989). Besides the basicity of the coordinating donors, as is the case for the open-chain polyaminopolyphosphonates, the size of the ring comprising the metal ion is the main factor governing the metal-binding ability. This is clearly indicated by the fairly high selectivity of Ca^{2+} over Mg^{2+} with the tetraaza derivative DOTP, and the reverse sequence with the triaza derivative NOTP, or the rather similar stability with bivalent transition metal ions (see Table 1). Although structural data are not available, it is reasonable to assume that PO_3^{2-}/CO_2^- substitution decreases the denticity of coordination especially to alkaline earth cations, which is manifested in the lower stability of the phosphonates, although their basicity is higher. Macrocyclic aminophosphonates exhibit a much stronger tendency to form binuclear complexes than the aminocarboxylate analogues, which can be explained by electrostatic and steric reasons. The binuclear structure allows the binegative phosphonate groups to be kept further away and thus to minimize electrostatic repulsion (Delgado et al., 1990). The potential use of the Gd^{3+} complex of DOTA as a contrast agent for NMR imaging in medicine has also drawn attention to its phosphono derivatives (Geraldes et al., 1989; Delgado et al., 1990; Broan et al, 1990; Ramasamy et al., 1991; Lazar et al., 1991). Metal complex-based contrast agents have to meet multiple and exact requirements, such as high thermodynamic stability, kinetic inertness, good water solubility, low osmolality, high relaxivity, specific in vivo distribution, fast excretability and lack of toxicity. The above properties make polyaminopolyphosphono complexes good candidates for further exploration in this field.

REFERENCES

Appleton, T.G., Hall, J.R., Harris, A.D., Kimlin, H.A., and McMahon, I.J. (1984). *Aust. J. Chem.*, *37*: 1833.

Appleton, T.G., Hall, J.R., and McMahon, I.J. (1986). *Inorg. Chem.*, *25*: 720.

Balla, J., Jezowska-Bojczuk, M., Kiss, T., Kozlowski, H., Lejczak, B., and Matczak-Jon, E. (1990a). *J. Inorg. Biochem.*, *40*: 37.

Balla, J., Kiss, T., Jezowska-Bojczuk, M., Kozlowski, H., and Kafarski, P. (1990b). *J. Chem. Soc., Dalton Trans.*, 1861.

Belugin, Yu.F., Dubrov, Yu.N., Marov, I.N., and Dyatlova, N.M. (1972). *Russ. J. Inorg. Chem.*, *17*: 1773.

Broan, C.J., Jankowski, K.J., Kataky, R., and Parker, D. (1990). *J. Chem. Soc. Chem. Commun.*, 1738.

Delgado, R., Siegfried, L.C., and Kaden, T.A. (1990). *Helv. Chim. Acta*, *73*: 140.

Dyatlova, N.M., Medved, T.Ya., Rudomino, M.V., and Kabachnik, M.I. (1970). *Izv. Akad. Nauk SSSR*, *4*: 815.

Geraldes, C.F.G.C., Sherry, A.D., and Cacheris, W.P. (1989). *Inorg. Chem.*, *28*: 3336.

Gillard, R.D., Newman, P.D., and Collins, J.D. (1989). *Polyhedron*, *8*: 2077.

Hariharan, M., Motekaitis, R.J., and Martell, A.E. (1975). *J. Org. Chem.*, *40*: 470.

Hayes, R.L., and Hubner, K.F. (1983). *Metal Ions in Biological Systems*, Vol. 16 (H. Sigel, ed.), Marcel Dekker, New York, p. 279.

Kabachnik, M.I., Medved, T.Ya., Polikarov, Yu.M., Shcherbakov, B.K., Belskii, F.I., Matrosov, E.I., and Pasechnik, M.P. (1984a). *Izv. Akad. Nauk. SSSR.*, 835.

Kabachnik, M.I., Medved, T.Ya., Belskii, F.I., and Pisareva, S.A. (1984b). *Izv. Akad. Nauk. SSSR.*, 844.

Kafarski, P., and Lejczak, B. (1991). *Phosp. Sulf. Silicon*, *63*: 193.

Kafarski, P., and Mastalerz, P. (1984). *Beitr. Wirkst. Forsch.*, *21*: 1.

Kiss, T. (1990). *Biocoordination Chemistry* (K. Burger, ed.), Ellis Horwood, Chichester, p. 104.

Kiss, T., Farkas, E., and Kozlowski, H. (1989a). *Inorg. Chim. Acta*, *155*: 281.

Kiss, T., Farkas, E., Kozlowski, H., Siatecki, Z., Kafarski, P., and Lejczak, B. (1989b). *J. Chem. Soc., Dalton Trans.*, 1053.

Kukhar, V.P., and Solodenko, V.A. (1987). *Usp. Khim.*, *56*: 1504.

Lauffer, R.B. (1987). *Chem. Rev.*, *87*: 901.

Lazar, I., Sherry, A.D., Brucher, E., and Kiraly, R. (1991). *Inorg. Chem.*, *30*: 5016.

Lejczak, B., Kafarski, P., and Zygmut, J. (1989). *Biochemistry*, *28*: 3549.

Martell, A.E., and Smith, R.M. (1974, 1975, 1982, 1989). *Critical Stability Constants*, Vols. 1,2,5,6, Plenum Press, New York.

Martell, A.E., Langohr, M.F., and Tatsuimoto, K. (1985). *Inorg. Chim. Acta*, *108*: 105.

Matczak-Jon, E., and Wojciechowski, W. (1990). *Inorg. Chim. Acta*, *173*: 85.

Motekaitis, R.J., and Martell, A.E. (1980). *Inorg. Chem.*, *19*: 1646.

Motekaitis, R.J., and Martell, A.E. (1985). *J. Coord. Chem.*, *14*: 139.

Motekaitis, R.J., Murase, I., and Martell, A.E. (1971a). *J. Inorg. Nucl. Chem.*, *33*: 3353.

Motekaitis, R.J., Murase, I., and Martell, A.E. (1971b). *Inorg. Nucl. Chem. Lett.*, *7*: 1103.

Motekaitis, R.J., Murase, I., and Martell, A.E. (1976). *Inorg. Chem.*, *15*: 2303.

Ramasamy, R., Lazar, I., Brucher, E., Sherry, A.D., and Malloy, C.R. (1991). *FEBS Lett.*, *280*: 121.

Rizkalla, E.N. (1983). *Rev. Inorg. Chem.*, *5*: 223.

Rizkalla, E.N., and Zaki, M.T.M. (1979). *Talanta*, *26*: 507.

Rizkalla, E.N., and Zaki, M.T.M. (1980a). *Talanta*, *27*: 423.

Rizkalla, E.N., and Zaki, M.T.M. (1980b). *Talanta*, *27*: 709.

Rizkalla, E.N., and Zaki, M.T.M. (1980c). *Talanta*, *27*: 769.

Sawada, K., Araki, T., and Suzuki, T. (1987). *Inorg. Chem.*, *26*: 1199.

Szpoganicz, B., and Martell, A.E. (1986). *Inorg. Chem.*, *25*: 327.

Taliaferro, C.H., and Martell, A.E. (1984). *J. Coord. Chem.*, *13*: 249.

Taliaferro, C.H., and Martell, A.E. (1985). *Inorg. Chem.*, *24*: 2408.

Tazaki, M., Nita, K., Takagi, M., and Ueno, K. (1982). *Chem. Lett.*, 571.